MOLECULAR BIOLOGY OF
Food and Water Borne Mycotoxigenic and Mycotic Fungi

# Food Microbiology Series

Series Editor
Dongyou Liu

Molecular Biology of Food and Water Borne Mycotoxigenic and Mycotic Fungi, *edited by R. Russell M. Paterson, and Nelson Lima* (2015)

Biology of Foodborne Parasites, *edited by Lihua Xiao, Una Ryan, and Yaoyu Feng* (2015)

Food Microbiology Series

MOLECULAR BIOLOGY OF
Food and Water Borne Mycotoxigenic and Mycotic Fungi

R. Russell M. Paterson
Nelson Lima

CRC Press
Taylor & Francis Group
Boca Raton London New York

CRC Press is an imprint of the
Taylor & Francis Group, an **informa** business

CRC Press
Taylor & Francis Group
6000 Broken Sound Parkway NW, Suite 300
Boca Raton, FL 33487-2742

© 2016 by Taylor & Francis Group, LLC
CRC Press is an imprint of Taylor & Francis Group, an Informa business

No claim to original U.S. Government works

Printed on acid-free paper
Version Date: 20150601

International Standard Book Number-13: 978-1-4665-5986-8 (Hardback)

This book contains information obtained from authentic and highly regarded sources. Reasonable efforts have been made to publish reliable data and information, but the author and publisher cannot assume responsibility for the validity of all materials or the consequences of their use. The authors and publishers have attempted to trace the copyright holders of all material reproduced in this publication and apologize to copyright holders if permission to publish in this form has not been obtained. If any copyright material has not been acknowledged please write and let us know so we may rectify in any future reprint.

Except as permitted under U.S. Copyright Law, no part of this book may be reprinted, reproduced, transmitted, or utilized in any form by any electronic, mechanical, or other means, now known or hereafter invented, including photocopying, microfilming, and recording, or in any information storage or retrieval system, without written permission from the publishers.

For permission to photocopy or use material electronically from this work, please access www.copyright.com (http://www.copyright.com/) or contact the Copyright Clearance Center, Inc. (CCC), 222 Rosewood Drive, Danvers, MA 01923, 978-750-8400. CCC is a not-for-profit organization that provides licenses and registration for a variety of users. For organizations that have been granted a photocopy license by the CCC, a separate system of payment has been arranged.

Trademark Notice: Product or corporate names may be trademarks or registered trademarks, and are used only for identification and explanation without intent to infringe.

Library of Congress Cataloging-in-Publication Data

Molecular biology of food and water borne mycotoxigenic and mycotic fungi / Robert Russell Monteith Paterson, Nelson Lima, editors.
    pages cm
"A CRC title."
Includes bibliographical references and index.
ISBN 978-1-4665-5986-8 (hardcover : alk. paper) 1. Mycoses. 2. Mycotoxins. 3. Foodborne diseases. 4. Waterborne infection. I. Paterson, Robert Russell Monteith, editor. II. Lima, Nelson, editor.

RC117.M65 2015
616.9'6--dc23   2015021171

Visit the Taylor & Francis Web site at
http://www.taylorandfrancis.com

and the CRC Press Web site at
http://www.crcpress.com

Contents

Series Preface ix
Preface xi
Editors xiii
Contributors xv

Chapter 1 Introduction 1

                 R. Russell M. Paterson and Nelson Lima

Chapter 2 Mycotoxin-Producing and Clinically Important Fungi: Their Classification and Naming 5

                 David L. Hawksworth

Chapter 3 Phylogenetic Analysis Especially in Relation to Fungi 15

                 Anna Muszewska and Krzysztof Ginalski

Chapter 4 Fungal DNA Barcoding 37

                 V. Robert, G. Cardinali, B. Stielow, T.D. Vu, F. Borges dos Santos, W. Meyer, and C. Schoch

Chapter 5 Metabolomics of Food- and Waterborne Fungal Pathogens 57

                 Danny Alexander, Adam D. Kennedy, Nalini Desai, Elizabeth Kensicki, and Kirk L. Pappan

Chapter 6 Systems Biology in Fungi 69

                 Oscar Dias and Isabel Rocha

Chapter 7 Brief History of Fungal Genomics from Linkage Maps to Sequences 93

                 Kevin McCluskey and Scott E. Baker

Chapter 8 Recommendations for Quantitative PCR *Aspergillus* Assays 103

                 Stéphane Bretagne, Odile Cabaret, and Jean-Marc Costa

Chapter 9 *Acremonium* 115

                 Richard C. Summerbell and James A. Scott

Chapter 10 Alternaria Mycoses 129

                 Giuliana Lo Cascio and Marco Ligozzi

Chapter 11 *Alternaria* spp. and Mycotoxins 139
  Miguel Ángel Pavón, Isabel González, Rosario Martín, and Teresa García

Chapter 12 *Aspergillus* and Aspergillosis 151
  Malcolm D. Richardson and Riina Richardson

Chapter 13 *Aspergillus* Mycotoxins 165
  János Varga, Sándor Kocsubé, Gyöngyi Szigeti, Nikolett Baranyi, and Beáta Tóth

Chapter 14 *Aureobasidium* 187
  Hasima Mustafa Bamadhaj, Giek Far Chan, and Noor Aini Abdul Rashid

Chapter 15 *Candida* as Foodborne Pathogens 197
  Sónia Silva, Cláudia Botelho, and Mariana Henriques

Chapter 16 *Chaetomium* 211
  Vit Hubka

Chapter 17 *Claviceps*: The Ergot Fungus 229
  Janine Hinsch and Paul Tudzynski

Chapter 18 *Curvularia* 251
  Jeannette Guarner

Chapter 19 *Encephalitozoon* 267
  Carmen del Águila de la Puente, Soledad Fenoy Rodríguez, and Nuno Henriques-Gil

Chapter 20 *Enterocytozoon* 293
  Olga Matos and Maria Luisa Lobo

Chapter 21 Mycotoxins of *Fusarium* spp.: Biochemistry and Toxicology 323
  Anthony De Lucca and Thomas J. Walsh

Chapter 22 *Lichtheimia* (ex *Absidia*) 355
  Volker U. Schwartze and Kerstin Kaerger

Chapter 23 *Microascus/Scopulariopsis* 375
  Sean P. Abbott

Contents

Chapter 24 *Mucormycosis* 387

Luis Zaror, Patricio Godoy-Martínez, and Eduardo Álvarez

Chapter 25 *Paecilomyces*: Mycotoxin Production and Human Infection 401

Cintia de Moraes Borba and Marcelly Maria dos Santos Brito

Chapter 26 *Penicillium* Mycotoxins: Physiological and Molecular Aspects 423

Rolf Geisen

Chapter 27 *Phoma* spp. as Opportunistic Fungal Pathogens in Humans 451

Mahendra Rai, Vaibhav V. Tiwari, and Evangelos Balis

Chapter 28 Yeasts Previously Included in the Genus *Pichia* 463

Volkmar Passoth

Chapter 29 Medically Important *Rhodotorula* Species 483

Sahar Yazdani, Audrey N. Schuetz, Ruta Petraitiene, Malcolm D. Richardson, and Thomas J. Walsh

Chapter 30 Saccharomyces and Kluyveromyces Infections 495

Firas A. Aswad, Vibhati V. Kulkarny, Kingsley Asare, and Samuel A. Lee

Chapter 31 *Trichoderma* Mycoses and Mycotoxins 521

Christian P. Kubicek and Irina S. Druzhinina

Chapter 32 *Trichosporon, Magnusiomyces,* and *Geotrichum* 539

Guillermo Quindós, Cristina Marcos-Arias, Elena Eraso, and Josep Guarro

Chapter 33 *Wallemia* 569

Janja Zajc, Sašo Jančič, Polona Zalar, and Nina Gunde-Cimerman

Chapter 34 Vaccine Development against Fungi 583

James I. Ito

Chapter 35 Fungi in Drinking Water 597

Ida Skaar and Gunhild Hageskal

Index 607

Series Preface

Microorganisms (including viruses, bacteria, molds, yeasts, protozoa, and helminthes) represent abundant and diverse forms of life that occupy various ecological niches of Earth. Those utilizing food and food products for growth and maintenance are important to human society due not only to their positive and negative impacts on food supply, but also to their potential pathogenicity to human and animal hosts.

On the one hand, foodborne microorganisms are known to play a critical role in fermentation and modification of foods, leading to a variety of nutritious food products (e.g., bread, beverage, yogurt, and cheese) that have contributed to the sustainment of human civilization from time immemorial. On the other hand, foodborne microorganisms may be responsible for food spoilage, which, albeit a necessary step in keeping up ecological balance, reduces the quality and quantity of foods for human and animal consumption. Furthermore, some foodborne microorganisms are pathogenic to humans and animals, which, besides creating havoc on human health and animal welfare, decrease the availability of meat and other animal-related products.

Food microbiology is a continuously evolving field of biological sciences that addresses issues arising from the interactions between food/waterborne microorganisms and foods. Topics of relevance to food microbiology include, but are not limited to, the adoption of innovative fermentation and other techniques to improve food production; the optimization of effective preservation procedures to reduce food spoilage; the development of rapid, sensitive, and specific methods to identify and monitor foodborne microbes and toxins, helping alleviate food safety concerns among consumers; the use of omic approaches to unravel the pathogenicity of foodborne microbes and toxins; the selection of non-pathogenic foodborne microbes as probiotics to inhibit and eliminate pathogenic viruses, bacteria, fungi, and parasites; and the design and implementation of novel control and prevention strategies against foodborne diseases in human and animal populations.

The Food Microbiology series aims to present a state-of-the-art coverage on topics central to the understanding of the interactions between food/waterborne microorganisms and foods. The series consists of individual volumes, each of which focuses on a particular aspect/group of foodborne microbes and toxins, in relation to their biology, ecology, epidemiology, immunology, clinical features, pathogenesis, diagnosis, antibiotic resistance, stress responses, treatment and prevention, etc. The volume editors/authors are professionals with expertise in the respective fields of food microbiology, and the chapter contributors are scientists directly involved in foodborne microbe and toxin research.

Extending the contents of classical textbooks on food microbiology, this series serves as an indispensable tool for food microbiology researchers, industry food microbiologists, and food regulation authorities wishing to keep abreast with latest developments in food microbiology. In addition, the series offers a reliable reference for undergraduate and graduate students in their pursuit to becoming competent and consummate future food microbiologists. Moreover, the series provides a trustworthy source of information to the general public interested in food safety and other related issues.

Don Liu
Series Editor

Preface

There are vast numbers of fungi, and a conservative estimated is 1.5–3.0 million with 100,000 described. Fungi consist of 36 classes, 140 orders, 560 families, and 8283 genera with 5101 synonyms. A wide range of mycotoxins are produced within certain groups, and some fungi cause mycoses of humans. Fungi (a) reproduce sexually or asexually with the sexual phase generally short-lived and (b) are saprobic, mutualistic, or parasitic. A majority of true fungi are haploid and are capable of potentially unlimited growth provided nutrients are available. This book discusses fungi isolated from food and their involvement in human diseases. It is quite surprising the number that fit both categories and there are obvious connections in some cases. The understanding of fungal population genetics and fungal evolution has increased beyond recognition from that possible 30 years ago: an understanding of population biology, speciation, phylogeny, and evolution has lagged far behind that of other groups of organisms. It is true to state that fungi's unique structure and way of life is so very different that it has been a deterrent to their study.

Some of the advances in studying fungi include the following: improvements in the formal genetics of many genera involved in human diseases, advances in molecular biology techniques as described herein, the ability to sequence DNA at very low cost, and the development of alternative techniques. Previously, phylogenetic speculation concerning the origin and evolution of fungi was entirely based on comparative morphology. There were serious imperfections in fungal taxonomy at almost all levels often due to the continuing dearth of fungal taxonomists. However, the application of the more modern approaches has permitted a more objective interpretation of the data. The naming of fungi to enable "one fungus:one name" is currently causing something of a revolution in taxonomy, although this has been resisted in some quarters, for example, in medical mycology.

Invasive infections caused by a large number of fungal species are a major cause of severe diseases in a variety of critically ill patients. Human pathogenic fungi were well defined toward the end of the twentieth century, some of which were limited to geographical regions and were known to clinicians. New infectious agents are continually appearing, with around 20 species yearly even in 1999. Food mycology is an area of increasing interest because of mycotoxin production, which causes human diseases when mycotoxins enter the food chain, and the organoleptic spoilage of food by fungi. The presence of fungi in water deserves special attention.

The use of molecular biology has increased tremendously in mycology in relation to the detection and aspects of pathogenicity, and it is in these areas that the book has been conceived. However, the use of PCR in the detection of fungal mycoses and fungi from food is not common practice despite PCR's existence for years. The details of the methods are often naive from the points of view of inappropriate selection of genes, the lack of internal amplification controls, and the choice of growth media in culture-dependent PCR.

The molecular biology of food-borne toxigenic and mycotic fungi of humans is devoted to revealing similarities between fungi that are present in food and water and those that cause human fungal diseases. The most obvious association is from the production of mycotoxins in food and the cause of diseases in human from the consumption of the mycotoxins in food. In the series, this book differs from other books, inter alia, by dealing specifically with food-borne mycotoxigenic fungi in details. Most importantly, we explain which fungi can be present in particular food and water and how similar ones cause mycosis in humans. This is a unique aspect of the book, and in some cases, fungi in food are considered to cause human mycoses. However, each chapter stands on its own as an up-to-date reference for the diseases caused by that respective taxon and provides optimal methods for identification using the molecular biology techniques in particular. Some immediate innovations are that fungi in drinking water and chapters on the important mycotoxin producers are considered.

I cannot finish without acknowledging the huge contribution the authors have made to the book. All of them were extremely willing to contribute, and I was gratified that so many world-class scientists gave off their time so readily. The breadth of the scientists involved can be seen from the wide range of countries from which chapters have been obtained. I especially thank the authors for being patient and willing to persevere with the revisions: Those who stepped up to the mark when others "dropped out" are appreciated particularly. The publisher, CRC Press, who provided an extremely high level of professionalism and patience throughout the whole process, is also greatly commended.

R. Russell M. Paterson
Nelson Lima
Centre of Biological Engineering
University of Minho
Braga, Portugal

Editors

R. Russell M. Paterson has a BSc Honours (2:1) in applied microbiology from the University of Strathclyde, United Kingdom. He earned his MSc and PhD in chemistry from the University of Manchester, United Kingdom. He undertook a 2.5 year postdoctoral appointment at the Boyce Thompson Institute at Cornell University, New York, United States. Dr. Paterson worked at the Centre for Industrial Innovation, Strathclyde University, for two years before being employed at the international nongovernmental organization CABI, United Kingdom, where he was the senior scientist researching fungal natural products. He was awarded the IOI Professorial Chair on Plant Pathology at the Universiti Putra Malaysia, Malaysia, in 2008. Currently, he is employed as a researcher at the Centre for Biological Engineering, University of Minho, Portugal. He has worked in many countries, particularly in Southeast Asia, and was central to a fungal bioprospecting project in the Iwokrama rain forest of Guyana. The European Research Council confirmed that he is an external reviewer of projects (2012, 2013). He was coordinating editor of *Mycopathologia* and is an editor of *Current Enzyme Inhibition*, *Journal of Earth Science & Climatic Change* and *Current Opinion in Food Science*. He has written numerous papers, books, and chapters on health aspects of fungi toward humans, particularly in relation to mycotoxin and fungal contamination of food and drinking water. He has a total publication list of c. 200.

Nelson Lima earned his PhD in engineering sciences (biotechnology) from the University of Minho, Portugal, in 1993 and has been a full professor of the University of Minho since March 2004. His primariy research is related to food and environmental mycology with the integration of polyphasic approaches for fungal identification. He has also been involved for more than 25 years in the educational research field, mainly in science and environmental promotion and education. Since 1996, he is head of the fungal culture collection, Micoteca da Universidade do Minho (MUM). He was member of the executive board of the World Federation of Culture Collections (WFCC) from 2007 to 2010 and the collection officer of the European Culture Collections' Organisation (ECCO) from 2003 to 2006. Currently, he is the president of ECCO. He has been the evaluator or consultant of different funding agencies such as in Chile (CONICYT), Brazil (FINEP), Belgium (BELSPO), and the European Commission. He has been partner of several EU and Brazilian research funding projects, as well as coordinator or partner of several Portuguese research projects. In Portugal, Brazil, Mexico, and Chile, he has supervised 21 PhD and 28 master theses.

Contributors

Sean P. Abbott
Natural Link Mold Lab, Inc.
Reno, Nevada

Carmen del Águila de la Puente
Faculty of Pharmacy
San Pablo CEU University
Madrid, Spain

Danny Alexander
Metabolon, Inc.
Research Triangle Park, North Carolina

Eduardo Álvarez
Mycology Unit
Faculty of Medicine
Institute of Biomedical Sciences
University of Chile
Santiago, Chile

Kingsley Asare
New Mexico Veterans Healthcare System
and
Health Science Center
University of New Mexico
Albuquerque, New Mexico

Firas A. Aswad
New Mexico Veterans Healthcare System
and
Health Science Center
University of New Mexico
Albuquerque, New Mexico

Scott E. Baker
Environmental Molecular Sciences Laboratory
Pacific Northwest National Laboratory
Richland, Washington

Evangelos Balis
Pulmonology Department of "Evaggelismos"
   General Hospital
Athens, Greece

Hasima Mustafa Bamadhaj
Faculty of Biosciences and Bioengineering
Universiti Teknologi Malaysia
Johor, Malaysia

Nikolett Baranyi
Faculty of Science and Informatics
Department of Microbiology
University of Szeged
Szeged, Hungary

Cintia de Moraes Borba
Laboratory of Taxonomy, Biochemistry and
   Bioprospecting of Fungi
Oswaldo Cruz Institute
Oswaldo Cruz Foundation
Rio de Janeiro, Brazil

F. Borges dos Santos
Bioinformatics Group
CBS-KNAW Fungal Biodiversity Center
Utrecht, the Netherlands

Cláudia Botelho
Centre of Biological Engineering
University of Minho
Braga, Portugal

Stéphane Bretagne
Parasitology-Mycology Laboratory
Lariboisière-Saint Louis Hospital
Assistance Publique-Hôpitaux de Paris
Paris Diderot
Sorbonne Paris Cité University
and
Molecular Mycology Unit
Pasteur Institute
National Reference Center of Invasive Mycoses
   and Antifungals
Paris, France

Marcelly Maria dos Santos Brito
Laboratory of Taxonomy, Biochemistry and Bioprospecting of Fungi
Oswaldo Cruz Institute
Oswaldo Cruz Foundation
Rio de Janeiro, Brazil

Odile Cabaret
Department of Medical Biology and Pathology
Gustave Roussy
Villejuif, France

G. Cardinali
University of Perugia
Perugia, Italy

Giek Far Chan
School of Applied Science
Temasek Polytechnic
Singapore, Singapore

Jean-Marc Costa
Cerba Laboratory
Cergy-Pontoise, France

Anthony De Lucca
Transplantation-Oncology Infectious Diseases Program
Weill Cornell Medical Center
New York, New York

Nalini Desai
Metabolon, Inc.
Research Triangle Park, North Carolina

Oscar Dias
Centre of Biological Engineering
University of Minho
Braga, Portugal

Irina S. Druzhinina
Research Division Biotechnology and Microbiology
Institute of Chemical Engineering
Vienna University of Technology
Vienna, Austria

Elena Eraso
Facultad de Medicina y Odontología
Departamento de Inmunología
Microbiología y Parasitología
Universidad del País Vasco/Euskal Herriko Unibertsitatea
Bilbao, Spain

Soledad Fenoy Rodríguez
Faculty of Pharmacy
San Pablo CEU University
Madrid, Spain

Teresa García
Department of Food Science and Technology
Faculty of Veterinary Medicine
Universidad Complutense de Madrid
Madrid, Spain

Rolf Geisen
Department of Safety and Quality of Fruit and Vegetables
Max Rubner-Institut
Karlsruhe, Germany

Krzysztof Ginalski
Laboratory of Bioinformatics and Systems Biology
Centre of New Technologies
University of Warsaw
Warsaw, Poland

Patricio Godoy-Martínez
Institute of Clinical Microbiology
Faculty of Medicine
University Austral of Chile
Valdivia, Chile

Isabel González
Department of Food Science and Technology
Faculty of Veterinary Medicine
Universidad Complutense de Madrid
Madrid, Spain

Jeannette Guarner
Department of Pathology and Laboratory Medicine
Emory University
Atlanta, Georgia

Josep Guarro
Facultat de Medicina i Ciències de la Salut
Unitat de Microbiologia
Universitat Rovira i Virgili
Reus, Spain

Contributors

Nina Gunde-Cimerman
Biotechnical Faculty
Department of Biology
University of Ljubljana
and
Centre of Excellence for Integrated Approaches in Chemistry and Biology of Proteins (CIPKeBiP)
Ljubljana, Slovenia

Gunhild Hageskal
Department of Biotechnology
Biotechnology and Nanomedicine Sector
SINTEF Materials and Chemistry
Trondheim, Norway

David L. Hawksworth
Departamento de Biología Vegetal II
Facultad de Farmacia
Universidad Complutense de Madrid
Plaza Ramón y Cajal
Madrid, Spain
and
Department of Life Sciences
Natural History Museum
London, United Kingdom
and
Mycology Section
Royal Botanic Gardens
Surrey, United Kingdom

Mariana Henriques
Centre of Biological Engineering
University of Minho
Braga, Portugal

Nuno Henriques-Gil
Faculty of Medicine
San Pablo CEU University
Madrid, Spain

Janine Hinsch
Institut für Biologie und Biotechnologie der Pflanzen
Westfälische Wilhelms Universität Münster
Münster, Germany

Vit Hubka
Faculty of Science
Department of Botany
Charles University in Prague
and
Laboratory of Fungal Genetics and Metabolism
Institute of Microbiology of the AS CR
Prague, Czech Republic

James I. Ito
Division of Infectious Diseases
City of Hope
Duarte, California

Sašo Jančič
Biotechnical Faculty
Department of Biology
University of Ljubljana
Ljubljana, Slovenia

Kerstin Kaerger
German National Reference Center for Invasive Mycoses
Leibniz Institute for Natural Product Research and Infection Biology
Hans Knoell Institute
Jena, Germany

Adam D. Kennedy
Metabolon, Inc.
Research Triangle Park, North Carolina

Elizabeth Kensicki
Metabolon, Inc.
Research Triangle Park, North Carolina

Sándor Kocsubé
Faculty of Science and Informatics
Department of Microbiology
University of Szeged
Szeged, Hungary

Christian P. Kubicek
Institute of Chemical Engineering
Vienna University of Technology
Vienna, Austria

Vibhati V. Kulkarny
New Mexico Veterans Healthcare System
and
Health Science Center
University of New Mexico
Albuquerque, New Mexico

Samuel A. Lee
New Mexico Veterans Healthcare System
and
Health Science Center
University of New Mexico
Albuquerque, New Mexico

Marco Ligozzi
Dipartimento di Patologia
Sezione di Microbiologia
Università di Verona
Verona, Italy

Nelson Lima
Centre of Biological Engineering
University of Minho
Braga, Portugal

Giuliana Lo Cascio
Servizio di Microbiologia e Virologia
Azienda Ospedaliera Universitaria Integrata di Verona
Verona, Italy

Maria Luisa Lobo
Medical Parasitology Unit
Global Health and Tropical Medicine
Instituto de Higiene e Medicina Tropical
Universidade NOVA de Lisboa
Lisbon, Portugal

Cristina Marcos-Arias
Facultad de Medicina y Odontología
Departamento de Inmunología, Microbiología y Parasitología
Universidad del País Vasco/Euskal Herriko Unibertsitatea
Bilbao, Spain

Rosario Martín
Department of Food Science and Technology
Faculty of Veterinary Medicine
Universidad Complutense de Madrid
Madrid, Spain

Olga Matos
Medical Parasitology Unit
Global Health and Tropical Medicine
Instituto de Higiene e Medicina Tropical
Universidade NOVA de Lisboa
Lisbon, Portugal

Kevin McCluskey
Department of Plant Pathology
Kansas State University
Manhattan, Kansas

W. Meyer
Molecular Mycology Research Laboratory
Centre for Infectious Diseases and Microbiology
Sydney Medical School-Westmead Hospital
Westmead Millennium Institute
The University of Sydney
Westmead, New South Wales, Australia

Anna Muszewska
Institute of Biochemistry and Biophysics
Polish Academy of Sciences
Warsaw, Poland

Kirk L. Pappan
Metabolon, Inc.
Research Triangle Park, North Carolina

Volkmar Passoth
Department of Microbiology
Uppsala Biocenter
Swedish University of Agricultural Sciences
Uppsala, Sweden

R. Russell M. Paterson
Centre of Biological Engineering
University of Minho
Braga, Portugal

Miguel Ángel Pavón
Department of Food Science and Technology
Faculty of Veterinary Medicine
Universidad Complutense de Madrid
Madrid, Spain

Ruta Petraitiene
Division of Infectious Diseases
Department of Medicine
Weill Cornell Medical Center
Cornell University
New York, New York

Guillermo Quindós
Facultad de Medicina y Odontología
Departamento de Inmunología, Microbiología y Parasitología
Universidad del País Vasco/Euskal Herriko Unibertsitatea
Bilbao, Spain

Mahendra Rai
Department of Biotechnology
Sant Gadge Baba Amravati University
Amravati, India

Noor Aini Abdul Rashid
Faculty of Biosciences and Bioengineering
and
Nanoporous Materials for Biological Application Research Group
Sustainability Research Alliance
Universiti Teknologi Malaysia
Johor, Malaysia

Malcolm D. Richardson
Mycology Reference Centre Manchester
Education and Research Centre
University Hospital of South Manchester (Wythenshawe Hospital)
Manchester, United Kingdom

Riina Richardson
Mycology Reference Centre Manchester
Education and Research Centre
University Hospital of South Manchester (Wythenshawe Hospital)
Manchester, United Kingdom

V. Robert
Bioinformatics Group
CBS-KNAW Fungal Biodiversity Center
Utrecht, the Netherlands

Isabel Rocha
Centre of Biological Engineering
University of Minho
Braga, Portugal

C. Schoch
NIH/NLM/NCBI
Bethesda, Maryland

Audrey N. Schuetz
Division of Infectious Diseases
Department of Medicine
Weill Cornell Medical Center
Cornell University
and
Department of Pathology and Laboratory Medicine
Weill Cornell Medical Center
New York-Presbyterian Hospital
New York, New York

Volker U. Schwartze
Department of Microbiology and Molecular Biology
Jena Microbial Resource Collection
Institute of Microbiology
University of Jena
and
Department of Molecular and Applied Microbiology
Leibniz Institute for Natural Product Research and Infection Biology
Hans Knöll Institute
Jena, Germany

James A. Scott
Sporometrics Inc.
and
Dalla Lana School of Public Health
University of Toronto
Toronto, Ontario, Canada

Sónia Silva
Centre of Biological Engineering
University of Minho
Braga, Portugal

Ida Skaar
Section of Mycology
Norwegian Veterinary Institute
Oslo, Norway

B. Stielow
Bioinformatics Group
CBS-KNAW Fungal Biodiversity Center
Utrecht, the Netherlands

Richard C. Summerbell
Sporometrics Inc.
and
Dalla Lana School of Public Health
University of Toronto
Toronto, Ontario, Canada

Gyöngyi Szigeti
Faculty of Science and Informatics
Department of Microbiology
University of Szeged
Szeged, Hungary

Vaibhav V. Tiwari
Department of Biotechnology
Sant Gadge Baba Amravati University
Amravati, India

Beáta Tóth
Cereal Research Nonprofit Ltd.
Szeged, Hungary

Paul Tudzynski
Institut für Biologie und Biotechnologie der Pflanzen
Westfälische Wilhelms Universität Münster
Münster, Germany

János Varga
Faculty of Science and Informatics
Department of Microbiology
University of Szeged
Szeged, Hungary

T.D. Vu
Bioinformatics Group
CBS-KNAW Fungal Biodiversity Center
Utrecht, the Netherlands

Thomas J. Walsh
Division of Infectious Diseases
Department of Medicine
and
Department of Pediatrics and Microbiology
and
Department of Immunology
Weill Cornell Medical Center
Cornell University
New York, New York

Sahar Yazdani
Division of Infectious Diseases
Department of Medicine
Weill Cornell Medical Center
Cornell University
New York, New York

Janja Zajc
Biotechnical Faculty
Department of Biology
University of Ljubljana
Ljubljana, Slovenia

Polona Zalar
Biotechnical Faculty
Department of Biology
University of Ljubljana
Ljubljana, Slovenia

Luis Zaror
Medical Technology
University Mayor
Temuco, Chile

1 Introduction

R. Russell M. Paterson and Nelson Lima

CONTENT

References 3

This book (1) considers fungi isolated from food and (2) indicates the involvement of these fungi in human diseases. It is quite surprising how many fungi fit both categories, and there are obvious connections in some cases. Methods for identifying the fungi are also covered particularly by PCR. Some of the chapters are general and introductory in nature—more so than in some previous books in the series. For example, there are introductions to systems biology, metabolomics, and phylogenetics: advances in the understanding of fungal population genetics and fungal evolution have increased beyond recognition from that possible even 30 years ago [1].

Fungi consist of 36 classes, 140 orders, 560 families, and 8283 genera with 5101 synonyms [2]. However, the true fungi belong to the Kingdom of Eukaryota and are without plastids. Nutrition is absorptive and never phagotrophic, and they lack an amoeboid phase. The cell walls contain chitin and β-glucans, whereas cell membranes universally contain ergosterol, which can be used to identify and quantify fungi, for example, as pathogens of plants [3], and, by extrapolation, could be employed in human diseases as an unspecific method for identifying fungal infections. A wide range of secondary metabolites are produced. Mitochondria have flattened cristae and peroxisomes that are almost always present; Golgi bodies or individual cisternae are present. They are unicellular or filamentous and consist of multicellular coenocytic haploid hyphae (homo- or heterokaryotic). They are mostly nonflagellate. They reproduce sexually or asexually with the sexual phase generally short lived and are saprobic, mutualistic, or parasitic [4]. The majority of plant or animal populations studied exist as groups of diploid individuals of limited growth that reproduce sexually. In contrast, the majority of true fungi are haploid, capable of potentially unlimited growth provided that nutrients are available. There are vast numbers of fungi and a conservative estimate is 1.5–3.0 million with (more than) 100,000 described [5].

There has been little attention paid to fungi by biologists and evolutionists to this major group of haploid Eukaryotes. Consequently, an understanding of population biology, speciation, phylogeny, and evolution has lagged far behind that of other groups of organisms. The unique structure and way of life of fungi is so very different that it has been a deterrent to their study. Increasingly, population biology, genetics, molecular biology, ecology, and speciation have been extrapolated from other organisms to fungi. However, it is unclear how far fungal behaviors can parallel evolutionary behaviors in the eukaryotes [1].

The situation has changed rapidly. Some of the advances include improvements in formal genetics of many genera involved in human disease, advances in molecular biology techniques as described herein, the ability to sequence DNA at very low cost, and development of alternative techniques. Toward the end of the nineteenth century and sporadically in the twentieth century, phylogenetic speculation concerning the origin and evolution of fungi were based almost entirely upon comparative morphology. Also, often due to the continuing dearth of fungal taxonomists, there were serious imperfections in fungal taxonomy at almost all levels. Little attention was given to modern species concepts. Application of the modern approaches has permitted a more objective interpretation of the data [1].

The fungi that are studied by mycologists are placed in the aforementioned group, but others belong to the kingdoms Chromista and Protozoa. Some authors use Eumycota as the kingdom name that has the advantage of avoiding confusion with *fungi*. Others use the name *fungi* as most fungi belong here and is well recognized and in almost common use. The Deuteromycotina are rejected as a formal taxonomic category as they are not a monophyletic unit but are fungi that have lost a sexual phase or that are anamorphs of other phyla. With molecular or ultrastructural methods, such fungi can be assigned to existing taxa. Most of the reported taxa-causing infections may have been misdiagnosed due to inadequate technology, and this situation created a growing interest in fungal systematics. Fungal taxonomy is a dynamic, progressive discipline that consequently requires changes in nomenclature, although these changes are often difficult for clinicians and clinical microbiologists to understand [6]. However, the naming of fungi to enable *one fungus: one name* is currently causing something of a revolution in taxonomy as discussed in Chapter 2 by David Hawksworth. The acceptance of such radical changes has been resisted in some quarters, for example, in medical mycology [7].

Invasive infections caused by a large number of fungal species are a major cause of severe disease in a variety of critically ill patients. The numbers of patients with mycoses are multiplying largely from advances in the treatment of malignant illnesses, with patients undergoing severe immunosuppression as part of intensive chemotherapeutical regimes and stem cell/organ transplantation. AIDS patients are also at high risk. Other risk factors include (1) long exposure to broad-spectrum antibiotic therapy, (2) long-term presence of catheters, and (3) the general condition and geographical location of patients [2]. Medical mycology is the scientific study of fungi that cause diseases in humans and the characteristics and epidemiology of the diseases that they cause. Fungi affect different organs, and according to the disorder syndromes, the diseases are named chromoblastomycoses, dermatophytoses, mycetoma, onychomycoses, otomycoses, phaeohyphomycoses, and rhinosporidioses.

Incorrect identifications of clinical strains may result in poor recovery of patients and additional costs to health-care institutions. It is essential that the wide range of fungi involved in mycoses can be identified accurately to facilitate appropriate and specific treatments. Current identification methods for the fungi are often based on morphological characters that can give unreliable identifications. The development of PCR and mass and IR spectroscopic techniques may contribute to improved diagnoses for clinical strains, although they are not used routinely for identifications.

Human pathogenic fungi were well defined toward the end of the twentieth century, some of which were limited to geographical regions and were known thoroughly by clinicians. New infectious agents are continually appearing, with around 20 species yearly even in 1999. These new opportunistic pathogens have increased the knowledge base of medical mycology, and unexpected changes have been seen in the pattern of fungal infections in humans. Another difficulty for microbiologists inexperienced in mycology is that fungi are mostly classified on the basis of their appearance rather than on the nutritional and biochemical differences that are of such importance in bacterial classification. This implies that different concepts have to be applied in fungal taxonomy. Generally, medical mycologists are familiar with only one aspect of pathogenic fungi, that is, the stage that develops by asexual reproduction. Usually, microbiologists ignore or have sparse information about the sexual stages of these organisms. However, the sexual stages are precisely the baseline of fungal taxonomy and nomenclature. It is evident that modern molecular techniques would allow some pathogenic and opportunistic fungi to be connected to their corresponding sexual stages and integrated into a more natural taxonomic scheme [6].

Food mycology is an area of increasing interests because of mycotoxin production and the organoleptic spoilage of food by fungi. Also, the presence of fungi in water deserves special consideration. Fungi produce secondary metabolites that are low molecular weight compounds often considered as unessential for the growth of the fungi. Many of these compounds have potent biological activities of a beneficial or detrimental nature to humans. Pharmaceuticals are examples of the beneficial metabolites such as antibiotics (e.g., penicillin). The compounds can be toxic to humans and animals and can be produced in foodstuff from growth by particular fungi. These are referred to as mycotoxins [8] and are discussed in the appropriate chapter in this book. Furthermore, mycotoxicoses are being recognized particularly in developing countries. A useful innovation for mycotoxin fungi is using PCR to detect genes involved in the metabolic pathway for particular mycotoxins, rather than simply general genes shared with all fungi. Readers will notice that I have not included mushroom poisoning per se, largely because they are *the food* rather than being *foodborne*. Nevertheless, it is a fascinating field of study.

Molecular biology has of course increased tremendously in mycology in relation to detection and aspects of pathogenicity, and it is in these areas that the book was conceived. However, the use of PCR for detection of fungal mycoses and fungi from food is far from routine despite it being in existence for many years. The details of the methods are often naive from the points of view of inappropriate selection of genes, the lack of internal amplification controls and the choice of growth media in culture-dependent PCR [9,10]. It is not just conventional PCR that is innovative, but other technologies involving bar coding, metabolomics, and systems biology will all play their part in affecting the field of investigations. We trust that you, the reader, will find the book informative and enjoyable.

REFERENCES

1. Burnett J H (2003) *Fungal Population Genetics and Species*. Oxford University Press, Oxford, U.K.
2. Liu D (2011) *Molecular Detection of Human Fungal Pathogens*. CRC Press, New York.
3. Muniroh M S, Sariah M, Zainal Abidin M A, Lima N, Paterson R R (2014) Rapid detection of *Ganoderma*-infected oil palms by microwave ergosterol extraction with HPLC and TLC. *J Microbiol Methods*, 100, 143–147.
4. Kirk P, Cannon P, Minter D, Stalpers J (2008) *Dictionary of the Fungi*. CABI, Wallingford, U.K.
5. Paterson R R M, Lima N (2013) Biochemical mutagens affect the preservation of fungi and biodiversity estimations. *Appl Microbiol Biotechnol*, 97, 77–85.
6. Guarro J, Gené J, Stchigel AM (1999) Developments in fungal taxonomy. *Clin Microbiol Rev*, 12, 454–500.
7. de Hoog G S, Haase G, Chaturvedi V, Walsh T J, Meyer W, Lackner M (2013) Taxonomy of medically important fungi in the molecular era. *Lancet Infect Dis*, 13, 385–386.
8. Paterson R R M, Lima N (2010) Toxicology of mycotoxins. *EXS*, 100, 31–63.
9. Paterson R R M (2007) Internal amplification controls have not been employed in diagnostic fungal PCR hence potential false negative results. *J Appl Microbiol*, 102, 1–10.
10. Paterson R R M, Lima N (2014) Self mutagens affect detrimentally PCR analysis of food fungi by creating potential mutants. *Food Control*, 35, 329–337.

2 Mycotoxin-Producing and Clinically Important Fungi
Their Classification and Naming

David L. Hawksworth

CONTENTS

2.1 Introduction 5
2.2 Classification 5
2.3 Names 8
    2.3.1 Publication of New Scientific Names 8
    2.3.2 Determining the Correct Name 10
    2.3.3 Conserved, Sanctioned, and Protected Lists of Names 10
    2.3.4 Abandonment of the Separate Naming of Morphs of a Species 11
2.4 Future Prospects 12
Acknowledgments 13
Postscript 13
References 13

2.1 INTRODUCTION

Scientists faced with issues relating to mycotoxin-producing and clinically important fungi frequently complain about changing classifications, changes of names, and difficulties of identification. These are all critical when considering fungi that can be harmful to humankind. A sound classification based on evolutionary relationships can be predictive of the kinds of products a fungus may produce, its substrates, and physiological properties. Scientific names are crucial to all communication about, and accessing the accumulated research base on, species. And reliable identifications are essential when considering risks, remedial action, and prophylaxis. The underlying explanation for these understandable irritations is threefold: ignorance of the fungi present on Earth, historical nomenclatural practices, and difficulties of identification. This contribution can be no panacea but may assist in understanding and contending with these frustrations. Attention is also drawn to actions currently being implemented to streamline the internationally agreed naming systems and prospects for the future.

2.2 CLASSIFICATION

Fungal organisms are among the most ancient eukaryotic organisms on Earth, and it has recently been argued that they may well have arisen in the Precambrian prior to the animals and plants [1]. The bulk of them form a well-defined kingdom distinct from plants and part of the same major superkingdom as the animals, termed *Opisthokonta*. Some other organisms traditionally studied by mycologists prove to belong in different kingdoms, the most important of which are the downy mildews and their relatives that belong to the kingdom *Straminipila* (sometimes termed *Chromista* or *Heterokonta*), and the slime molds that are members of *Protista* (sometimes termed *Protoctista*). At the same time, some organisms not traditionally studied by mycologists are now known to belong to the kingdom *Fungi*, notably *Cryptomycota* and *Microsporidia*. Today, the word fungi is generally used for all cellular and filamentous organisms studied by mycologists that have a heterotrophic absorptive nutritional lifestyle. However, the scientific term *Fungi* is confined to all true members of the kingdom.

The major constraint to a stable classification of fungal organisms is the paucity of knowledge about them. Around 100,000 species are currently known and named [2], but it is now recognized that there are at least 1.5 million, and probably 3 million, species on Earth [3], with some estimates suggesting as many as 10 million. This means that we know only 3.3%–6.6% of the species existing today. Such a huge number of still unnamed fungi not only have implications for identification, discussed in the following text, but also mean that current classifications are based on just a fraction of the actual variation present. Frequently, newly discovered fungi cannot be pigeonholed into already named orders, families, or genera. Further, even hitherto unrecognized phyla and classes are coming to light, especially as a result of increased investigations of environmental samples using the latest sequencing facilities.

In recent years, in order to provide a robust system for as many of the known fungi as possible, there has been a concerted international effort to assemble a comprehensive fungal tree of life down to the categories of order and, in some cases, families. This development is based on molecular sequence data from as comprehensive a sample of fungi as could be obtained. This drive, in which numerous mycologists around the world have been involved, has led to several major multiauthored publications, among which can be highlighted those of Hibbett et al. [4], Spatafora et al. [5], Schoch et al. [6], Kurtzman et al. [7], Seifert et al. [8], and Hyde et al. [9]. Other overviews are currently in preparation, some going down the classification hierarchy as far as the rank of genus.

Perhaps the greatest surprise to emerge from molecularly based classifications is the extent to which there has been a convergence of morphological characters in unrelated evolutionary lines. This is especially so in the case of the structures in which spores are formed, the sporophores. Sporophore form, for example, the nature of ascomata in ascomycetes and basidiomata in basidiomycetes, was given a major emphasis in earlier classification systems. For example, among the ascomycete fungi, classes termed *Hemiascomycetes* (no ascomata), *Plectomycetes* (closed ascomata, i.e., cleistothecia), *Pyrenomycetes* (flask-shaped ascomata, i.e., perithecia), and *Discomycetes* (open cuplike ascomata, i.e., apothecia) have had to be abandoned [10]. These terms are, nevertheless, sometimes still used in a colloquial sense to refer to fungi with a particular ascoma form, but are not then italicized or used with an initial capital letter, for example, pyrenomycetes for all ascomycetes with perithecia dispersed through different classes and orders today. Some characters, such as the production of closed sporophores, have arisen independently in numerous separate evolutionary lineages, and in *Basidiomycota*, there are numerous orders that include stalked mushroom sporocarps as well as ones that are resupinate or truffle-like [11].

The fungi with no known sexual stages have been variously termed *Deuteromycetes*, *Deuteromycota*, or *Fungi Imperfecti* and generally subdivided into *Agonomycetes* (no spores known, *Mycelia Sterilia*), *Coelomycetes* (spores formed in disklike or flask-like sporocarps), and *Hyphomycetes* (spores formed exposed with no enclosing sporocarps). These categories have now disappeared from formal classifications as molecular methods enable all fungi to be classified in a single system whether or not their sexual spore-forming structures are known. Again, as in the case of the *Ascomycota*, the terms coelomycetes and hyphomycetes are sometimes used colloquially, especially in identification manuals. It also sometimes still comes as a surprise to mycologists to find that lichens are covered in the fungal system, even though that integration has been in progress since the 1970s; the names given to lichens are ruled as referring only to the fungal partner, and the photosynthetic partners have separate names and classification systems.

Each fungal species may be placed in a series of hierarchical categories, termed *taxonomic ranks*, starting with genus and proceeding through family, order, class, and phylum. In some cases, named subdivisions of those principal ranks are used, such as section, subgenus, subfamily, suborder, and subphylum. The categories at ranks above genus are not always used, and in many cases, the actual relationships of particular genera remain unknown; in order to indicate this uncertainty, the phrase *incertae sedis* is commonly used. The terminations of scientific names are a valuable indicator of the rank of a name. In the fungi, phylum names end in *-mycota*, those of class in *-mycetes*, order in *-ales*, and family in *-aceae*. Recent editions of the internationally agreed rules governing fungal nomenclature (e.g., McNeill et al. [12]) place all scientific names in italics, regardless of rank. This is a useful device that makes clear at a glance if a term being used is a scientific name or not; the practice is commended, but not ruled on or mandatory, and is now well established in many leading mycological journals.

Classifications of the fungi will necessarily remain in flux for the foreseeable future as representatives of more genera are sequenced and fungi new to science are discovered in nature. Fortunately, as a result of concerted international efforts, the situation is becoming more stable at the ranks of order and above, at least in *Ascomycota* and *Basidiomycota*. In order to ascertain the current view on the position of a particular genus, and so relationships, the most valuable single work is the latest edition of *Ainsworth & Bisby's Dictionary of the Fungi* [2]; changes subsequent to that work are incorporated in the *Species Fungorum* and *Index Fungorum* online databases (http://www.speciesfungorum.org) as resources permit.

It is unfeasible to discuss the characters of the classes, orders, and families of fungi here, but I have endeavored to indicate the positions of the genera that are the subject of the chapters in this book, as currently understood, in Table 2.1.

TABLE 2.1
Current Disposition of the Principle Genera of Mycotoxin-Producing and Clinically Important Fungi Treated in This Book by Phylum, Order, and Family

Phylum	Order	Family	Genus
Ascomycota	*Chaetothyriales*	*Herpotrichiellaceae*	*Phialophora*
	Eurotiales	*Trichocomaceae*	*Aspergillus*
			Paecilomyces
			Penicillium
	Dothideales	*Dothioraceae*	*Aureobasidium*
	Hypocreales	*Clavicipitaceae*	*Claviceps*
		Hypocreaceae	*Trichoderma*
		Nectriaceae	*Fusarium*
	Microascales	*Microascaceae*	*Microascus*
	Pleosporales	*Pleosporaceae*	*Alternaria*
			Curvularia
	Saccharomycetales	*Cryptococcaceae*	*Candida*
		Dipodascaceae	*Geotrichum*
		Pichiaceae	*Pichia*
		Saccharomycetaceae	*Debaryomyces*
			Saccharomyces
	Sordariales	*Chaetomiaceae*	*Chaetomium*
Basidiomycota	*Sporidiobolales*	*incertae sedis*	*Rhodotorula*
	Tremellales	*Trichosporonaceae*	*Trichosporon*
Microsporidia	*Incertae sedis*	*Incertae sedis*	*Encephalitozoon*
	Enterocytozoonidae	*Incertae sedis*	*Enterocytozoon*
Zygomycota	*Mucorales*	*Cunninghamellaceae*	*Cuninghamella*
		Mucoraceae	*Lichtheimia*
			Mucor
			Rhizomucor
			Rhizopus

2.3 NAMES

The procedures for allocating scientific names to fungi have been governed by an internationally agreed set of rules since 1866, the current edition of which is the *International Code of Nomenclature for algae, fungi, and plants* (ICN; [12]). These are now revised at a special Nomenclature Section meeting convened immediately prior to each International Botanical Congress (IBC); those congresses are currently held at intervals of 6 years, the latest being in Melbourne (Australia) in 2011 and the next in Shenzhen (China) in 2017. Each congress appoints a series of committees concerned with different groups of organisms treated under the ICN to deal with matters related to its groups between congresses and to advise on proposals that relate to them; that for fungi is the Nomenclature Committee for Fungi (NCF). Subsequent to the initiation of International Mycological Congresses (IMCs) in 1971, debates on issues related to the naming systems for fungi have formed an increasing component of IMCs [13]. This has enabled a greater number of mycologists to make their views known, as few are generally able to be present at IBC Nomenclature Section meetings. In the case of IMC9 in Edinburgh in 2010, this led to the development of a series of fundamental proposals to make several important changes in the naming of fungi, which were subsequently refined and adopted by the 2011 Nomenclature Section meetings in Melbourne.

These far-reaching changes aim to streamline and simplify fungal nomenclature, progressing toward a more stable and efficient system appropriate for the electronic and molecular age. It is intended to construct a system that will be less time consuming and easier for those with limited experience of nomenclatural matters to adopt. The changes include ones that relate to procedures for the formal publication of new scientific names (at all ranks of the taxonomic hierarchy), the registration of all categories of new names, the development of protected lists of names, and the abandonment of separate naming for different morphs of the same species. As many of these modifications, especially the last, impact on the names to be used for toxigenic and clinically important fungi, their key elements and effects are summarized here. For fuller information, summaries of changes pertinent to mycologists made at the Melbourne IBC in 2011 should be consulted [14,15].

Some of the key provisions pertinent to the naming of fungi are summarized in the following text, but the ICN itself should always be consulted and also has helpful examples showing what a, sometimes rather abstrusely worded, rule means in practice [12]. In addition, there is also now a most helpful and authoritative user's guide, *The Code Decoded* [16], which provides guidance and explanations for those using the ICN.

2.3.1 Publication of New Scientific Names

There are two aspects to the publication of new scientific names under the ICN: effective publication and valid publication. Effective publication is concerned with the medium of delivery and its availability to the scientific community, while valid publication relates to the content of the article itself.

1. *Effective publication*: Prior to January 1, 2012, effective publication could only be achieved by publication of printed works, whether books or periodicals, but not in trade catalogues, newspapers, or oral presentations at meetings. From that date, however, electronic publication is also permitted as an alternative medium, provided certain criteria are met. For an electronic publication to be effective, it must be available via the World Wide Web, be in a book or journal with an International Standard Book Number (ISBN) or International Standard Serial Number (ISSN), respectively. It must also be in a final and unalterable PDF. Consequently, preliminary versions made available by publishers are not considered effective, and no corrections can be made retroactively. Where journals are made available in both print and electronic formats, it is the version that is made public first that fixes

the date of publication. Most journals now indicate the date of online publication on the hard copies. Interestingly, if final page numbers of an article are not known, but the text is otherwise treated by the publisher as the archived version of record, the unpaginated article is nevertheless considered effectively published on the date it was released; such works can be referred to by their unique digital object identifier (DOI) number. Further explanatory material on electronic publication is provided by Knapp et al. [17].

2. *Valid publication*: Over the last century, there have been various changes to the requirements for valid publication. Because of this, dates that apply to particular elements are required:

   a. *Description*: A diagnosis (a statement of how a new organism differs from similar ones previously described) or a description (an account of the characters of the organism) are fundamental requirements. From January 1, 1935, to December 31, 2011, one or both had to be in Latin, but from January 1, 2012, English can also be used instead of Latin. In consequence, Latin is disappearing from accounts of new species, though the gender of the generic name will continue to affect the termination of the species epithet. However, mycologists will continue to find that they need to understand pre-2012 Latin descriptions, and the text of Short and George [18] is particularly helpful in this regard. A practice several mycological journals are now adopting, however, is to require authors to supply a separate diagnosis (usually 3–5 lines) as well as a description (often half a page or more); this is done so that the distinguishing features of the new fungus are obvious and readily extracted.

   b. *Typification*: Fundamental to the system of nomenclature, and a requirement since January 1, 1958, is fixing the application of names by name-bearing types. The type can be a specimen, slide preparation, permanently preserved culture (provided it is lyophilized or in liquid nitrogen), or in some cases, an illustration. It is important to remember that the type is merely a nomenclatural device and that it need not be representative of a species but just fall within that species concept (the circumscription). I recall Luella K. Weresub's (1918–1979) dictum: a name has a type but no circumscription, and a species a circumscription but no type.

   Only a single name-bearing type can be cited, and from January 1, 1990, that has to be referred to as the *type* or *holotype* and accompanied by an indication of the collection in which it is permanently preserved. If more than one collection is cited as the type, the name is treated as not validly published. Duplicates of the holotype are termed "isotypes," and living cultures derived from a permanently preserved culture are referred to as *ex-type*. If the holotype is lost or destroyed, a *lectotype* can be selected to fix the name at a later date from other material the original author used; if there is no original material available, a new type (a *neotype*) has to be chosen. In cases where a name-bearing type exists but does not show features needed for conclusive identification with a particular species circumscription, for example, one lacking spores, an interpretative type, an *epitype*, can be designated to supplement information evident in the name-bearing type. Epitypes are increasingly being designated in mycology today for sequenced material where DNA has not or cannot be recovered from the name-bearing type. Later acts of typification have to be stated explicitly from January 1, 2001, and registration of such acts in nomenclatural databases, notably *MycoBank* and *Index Fungorum*, is now actively encouraged and may become mandatory in due course.

   Illustrations are only acceptable as types before December 31, 2006. Exceptions are for nonfossil microfungi where a specimen either cannot be preserved or would not show the diagnostic features. Although some have argued they should be, illustrations are not mandatory when new species of fungi are being described, but it is good practice to provide them.

c. *Changes in position or rank*: When a fungus is being moved to a different genus (a *new combination*) or having the rank changed (a *new status*), the full publication details of the name being moved have to be given for the new name to be validly published.

In addition to all the aforementioned requirements for effective and valid publication, there is also an additional criterion that applies only to fungi. There has to be an act of registration.

d. *Registration:* From January 1, 2013, in order to be validly published, the key information about a new fungus name at any rank, whether a new taxon, a change in generic position, or a change in rank, has to be deposited in a recognized repository. The accession number in the repository cited must be stated in the place of effective publication for the name to be accepted as valid. Recognition of repositories is the responsibility of the NCF (see p. 8 above). At present, there are three: *MycoBank* (http://www.mycobank.org), *Index Fungorum* (http://www.indexfungorum.org), and *Fungal Names* (http://fungalinfo.om.ac.cn/fungalname/fungalname.html). It is recommended that accession numbers for new names are obtained only after papers are accepted for publication as changes may be required while papers are being reviewed prior to acceptance for publication.

2.3.2 Determining the Correct Name

The ICN regulates which name should be used as the correct one for a particular species or other concept (i.e., a circumscription) in a particular position in the classification system, that is, it is something that is applied after the taxonomic decisions on relationships and ranks have been made. Consequently, a single fungus species can have more than one correct name if different classifications are adopted.

In establishing the correct name in the rank of family and below, the fundamental principle is that of priority of publication. The earliest date of publication generally takes precedence, starting from May 1, 1753. Priority is not mandatory for names at ranks above family, such as order, class, and phylum. Prior to the 1981 IBC, some fungal groups had later starting point dates, but that is no longer the case.

As a strict application of priority can lead to name changes in well-known species, fortunately there are mechanisms by which these can be avoided, as discussed in the following text.

2.3.3 Conserved, Sanctioned, and Protected Lists of Names

There are several mechanisms by which familiar and well-established names can be retained when their existence is contrary to the rules. Sadly, these are not always used by taxonomists because of the bureaucracy involved and the time it takes for a decision.

1. *Conserved and rejected names*: Names can be proposed for conservation, rejection, or to have the type species (for generic names) or name-bearing types (for species names) changed and fixed. This involves publishing the case, normally in *Taxon*, allowing a period for response, a discussion, and then a vote by members of the NCF. A report from the NCF is published and, in turn, considered by another committee appointed by the last IBC (the General Nomenclature Committee). These committees report to the next congress, after which updated cumulative lists of conserved and rejected names are issued. Traditionally, these lists have been published as appendices in the back of each edition of the code, but as they have become so bulky, they are not bound in the back of the new ICN but are to be made available separately.
2. *Sanctioned names*: In 1981, the starting point date for the naming of all fungi became May 1, 1753. In order to minimize the disruption that could have occurred in those asco- and

basidiomycetes that had had much later dates, names that had been used in the particular previous starting point works of Persoon or Fries are said to be *sanctioned*. That is, names adopted in the specified works were given preference over earlier names that otherwise would have priority. Further, if necessary, sanctioned names could be typified with material conforming to the concepts of Persoon or Fries, as appropriate, even if that differed from the concept represented by the original name-bearing type.

3. *Protected and suppressed names*: With the abandonment of the separate naming of fungal morphs of the same species (see below), a new provision was introduced into the ICN in 2011. This enables lists of names for protection or suppression to be drawn up and approved en masse. This is a novel, and potentially most valuable, step toward improved nomenclatural stability. Various committees and subcommittees (mainly established under the auspices of the International Commission on the Taxonomy of Fungi [ICTF]; see p. 12) and some individual specialists are currently working on drafts of such lists for generic and specific names for publication and submission to the NCF for approval. This will be a protracted process, but it is anticipated that the first draft lists will be available for discussion at the 10th IMC in Bangkok in August 2014. The aim is to secure a status for these lists, which will mean that listed names are protected against any unlisted names. Further information on this process and the selection of names for inclusion in the lists is provided elsewhere [19].

2.3.4 ABANDONMENT OF THE SEPARATE NAMING OF MORPHS OF A SPECIES

The most fundamental change in the naming of fungi that occurred in 2011 was the ending of the system that allowed separate scientific names to be used for the asexual and sexual morphs of the same species. This practice had been adopted in ascomycetes (except lichen-formers) and basidiomycetes since 1910 but proved difficult to regulate and had undergone several major revisions. In several cases, the revisions led to a large number of name changes that were unpopular and confusing for plant pathologists and industrial or medical mycologists. Prior to July 30, 2011, even if different morphs were known to belong to the same species through single-spore culture or molecular sequence data, the rules required separate names to be applied to the different morphs. The name with a name-bearing type showing the sexual morph (the *teleomorph*), however, was used for the fungus overall, and could be used even if just the asexual morph was under consideration, to further add to the confusion. Now, all names compete equally for the establishment of priority, regardless of whether they are typified by a type with the sexual or the asexual morph (the *anamorph*). Further, there is no obstacle for fungi that are not known to produce any spores, whether sexual or asexual, being placed in a genus based on molecular sequence data alone. Any names newly proposed for separate morphs of the same species after January 1, 2013, are ruled as *illegitimate* and are not to be taken up.

Two examples can illustrate what this means in practice. In the case of *Hypocrea rufa* (Pers.) Fr. 1849, the earliest species epithet is *Trichoderma viride* Pers. 1794 (typified by the asexual morph) and not *Sphaeria rufa* Pers. 1796 (typified by the sexual morph). Further, *Trichoderma* Pers. 1794 is earlier than *Hypocrea* Fr. 1825. Consequently, on priority of date, the single correct name for the fungus is now *Trichoderma viride*, irrespective of the kind of spores formed. When monographing *Penicillium*, Pitt [20] was forced by the rules then in force to create numerous names that could not be used under *Penicillium* as the sexual morph was present in the type. Consequently, for instance, he described *P. dodgei* Pitt 1980 for the asexual morph of *Eupenicillium brefeldianum* (B.O. Dodge) Stolk & D.B. Scott 1967, as the last name was based on *Penicillium brefeldianum* B.O. Dodge 1933 that included the sexual morph in the type. Under the new rules, *Penicillium* Link 1809 predates *Eupenicillium* F. Ludw. 1892, and the earliest species name available is that of Dodge, so the correct name becomes *Penicillium brefeldianum*.

In cases where a name typified by an asexual morph was published at an earlier date than a widely used one typified by a sexual morph, the ICB indicates that the asexually typified name should not be taken up automatically but should be formally proposed for rejection (see p. 10). Some guidance is given by examples in the ICN, and issues to be taken into consideration in deciding which names should be adopted have been flagged [19]. Fortunately, most taxonomic mycologists are proceeding responsibly, and controversial cases are under discussion by those most intimately concerned with different genera or families through committees/commissions or subcommittees/subcommissions linked to the ICTF. These include, for example, ones concerned with *Hypocreaceae* (which includes *Trichoderma*), *Trichocomacaeae* (which includes *Aspergillus* and *Penicillium*), *Fusarium*, and medically important fungi. In addition, a draft list of protected generic names for all fungi is under active development as I write (see the postscript, p. 13).

Some fine-tuning in these provisions is still required. For instance, where in the past an author was forced by the rules to introduce a new name for the morph of what was realized to be the same species, the same species epithet was commonly used, but with a different name-bearing type with the other morph. In some case where there are other asexually typified names, this means that familiar species epithets would have had to be changed. It is proposed that where authors have deliberately used the same species epithet, those names are treated as new combinations and not new species [21]. For example, *Neosartorya fumigata* O'Gorman et al. 2009 would become *Neosartorya fumigata* (Fresen.) O'Gorman et al. 2009, based on *Aspergillus fumigatus* Fresen. 1863. Without this provision, if a name was still required in *Neosartorya*, a taxonomic decision on which there are different views, pre-2009 epithets such as *A. aviaries* Peck 1891 and *A. bronchialis* Blumentritt 1901, could theoretically threaten the continued use of *fumigatus*—something that will be overcome if the envisaged protected lists (see p. 11) can protect against unlisted names.

The abandonment of the dual nomenclature system for ascomycete and basidiomycete fungi with pleomorphic life cycles has been widely welcomed and is now in the process of being adopted by a broad spectrum of applied and systematic mycologists. Further, one species/one name will simplify often currently confusing legislation concerned with plant and public health and facilitate communication and data retrieval among mycologists. It will also simplify fungal teaching and place fungal nomenclature in the molecular era, where monophyletic clades can have a single generic name.

2.4 FUTURE PROSPECTS

The prospects for a more efficient and more stable system for the naming of fungi are currently more encouraging than at any previous time. There will be some uncertainty and anguish while lists of names for protection are being drawn up and agreed by the international mycological community, but that can be compared with the anguish of a new birth and the prospect of subsequent joy. The new nomenclatural procedures that came into effect over the years 2011–2013 mean that at last we have a system more fit for purpose than ever before. With so many fungi still having to be collected and described, a slicker system has been much needed.

It is important to appreciate that the new provisions are concerned with improved nomenclatural practices and will not constrain taxonomists from proposing alternative systems of classification or describing novel fungi. This is not to say that mycologists should proceed headlong into introducing name changes without due consideration and pertinent research, as cautioned in particular by Braun [22]. The ICTF has already given some advice on good practice in the description of fungi new to science [23], and hopefully that body will have an increasing role in improving the standards of fungal taxonomy in the years ahead.

There is currently no comprehensive set of keys to all known genera of fungi. Such a work would greatly facilitate not only identifications but also the recognition of new genera. This has not been attempted since 1931 [24], though a major step toward that goal was provided in 1973 [25]. Keys to all known families were attempted in the eighth edition of *Ainsworth & Bisby's Dictionary of the Fungi* [26], and there is a fine set of descriptions and illustrations of accepted families available [27]. However, it is the genera that are of primary concern to most mycologists, and the prospect of lists of protected generic names may now facilitate the production of a new work to fill this gap.

ACKNOWLEDGMENTS

I am indebted to my wife, Patricia Wiltshire-Hawksworth, for suggestions to render the text somewhat more digestible to nonnomenclaturalists than was my first draft. This contribution was prepared while in receipt of funding from the Spanish Ministerio de Ciencia e Innovación project CGL2011–25003.

POSTSCRIPT

This chapter was submitted on February 21, 2014, before the *10th International Mycological Congress* in Bangkok was held in August. Many aspects of fungal nomenclature were debated in three Nomenclature Sessions, and participants also expressed their views in a questionnaire. The proposals which were overwhelmingly supported are likely to be adopted and included in the next edition of the *International Code of Nomenclature for algae, fungi, and plants*, expected in 2018. A detailed report is now available [28].

REFERENCES

1. Moore D (2013) *Fungal Biology in the Origin and Emergence of Life*. Cambridge, U.K.: Cambridge University Press.
2. Kirk PM, Cannon PF, Minter DW, Stalpers JA (2008) *Ainsworth & Bisby's Dictionary of the Fungi*, 10th edn. Wallingford, U.K.: CAB International.
3. Hawksworth DL (2012) Global species numbers of fungi: Are tropical studies and molecular approaches contributing to a more robust estimate? *Biodiversity and Conservation* **21**: 2425–2433.
4. Hibbett DS, Binder M, Bischoff JF, Blackwell M, Cannon PF, Eriksson OE, Huhndorf S et al. (2007) A higher-level phylogenetic classification of the *Fungi*. *Mycological Research* **111**: 509–547.
5. Spatafora JW, Hughes KW, Blackwell M (eds.) (2006) A phylogeny for kingdom *Fungi*: Deep hypha issue. *Mycologia* **98**: 829–1103.
6. Schoch CL, Sung G-H, Lopez-Giraldez FL, Townsend JP, Miadlikowska J, Hofstetter VR, Roberts B et al. (2009) The *Ascomycota* tree of life: A phylum-wide phylogeny clarifies the origin and evolution of fundamental reproductive and ecological traits. *Systematic Biology* **58**: 224–239.
7. Kurtzman CP, Fell JW, Boekhoet T (eds.) (2011) *The Yeasts: A Taxonomic Study*, 3 vols. Amsterdam, The Netherlands: Elsevier.
8. Seifert K, Morgan-Jones G, Gams W, Kendrick B (2011) *The Genera of Hyphomycetes*. [CBS Biodiversity Series no. 9.] Utrecht, The Netherlands: CBS-KNAW Fungal Biodiversity Centre.
9. Hyde KD, Jones EBG, Liu J-K, Ariyawansha H, Boehm E, Boonmee S, Braun U et al. (2013) Families of *Dothideomycetes*. *Fungal Diversity* **63**: 1–313.
10. Lumbsch HT, Huhndorf SM (2007) Whatever happened to the pyrenomycetes and loculoascomycetes. *Mycological Research* **111**: 1064–1074.
11. Hibbett DS (2007) After the gold rush, or before the flood? Evolutionary morphology of mushroom-forming fungi (*Agaricomycetes*) in the early 21st century. *Mycological Research* **111**: 1001–1018.
12. McNeill J, Barrie FR, Buck WR, Demoulin V, Greuter W, Hawksworth DL, Herendeen PS et al. (eds.) (2012) *International Code of Nomenclature for algae, fungi, and plants (Melbourne Code)*. [Regnum vegetabile no. 154.] Königstein, Germany: Koeltz Scientific Books.
13. Norvell LL, Hawksworth DL, Petersen RH, Redhead SA (2010) IMC9 Edinburgh Nomenclature Sessions. *Mycotaxon* **113**: 503–511; *IMA Fungus* **1**: 143–147; *Taxon* **59**: 1867–1868.
14. Hawksworth DL (2011) A new dawn for the naming of fungi: Impacts of decisions made in Melbourne in July 2011 on the future publication and regulation of fungal names. *MycoKeys* **1**: 7–20; *IMA Fungus* **2**: 155–162.
15. Norvell LL (2011) Melbourne approves a new Code. *Mycotaxon* **116**: 481–490.
16. Turland NJ (2013) *The Code Decoded: A User's Guide to the International Code of Nomenclature for algae, fungi, and plants*. [Regnum vegetabile no. 155.] Königstein, Germany: Koeltz Scientific Books.
17. Knapp S, McNeill J, Turland NJ (2011) Changes to publication requirements made at the XVIII International Botanical Congress in Melbourne—What does e-publication mean for you? *Taxon* **60**: 1498–1501; *Mycotaxon* **117**: 509–515; *MycoKeys* **1**: 21–28.

18. Short E, George A (2013) *A Primer of Botanical Latin with Vocabulary.* Cambridge, U.K.: Cambridge University Press.
19. Hawksworth DL (2012) Managing and coping with names of pleomorphic fungi in a period of transition. *Mycosphere* **3**: 52–64; *IMA Fungus* **3**: 15–24.
20. Pitt JI (1980) ["1979"] *The Genus Penicillium and Its teleomorphic States Eupenicillium and Talaromyces.* London, U.K.: Academic Press.
21. Hawksworth DL, McNeill J, de Beer ZW, Wingfield MJ (2013) Names of fungal species with the same epithet applied to different morphs: How to treat them. *IMA Fungus* **4**: 53–56.
22. Braun U (2012) The impacts of the discontinuation of dual nomenclature of pleomorphic fungi: The trivial facts, problems, and strategies. *IMA Fungus* **3**: 81–86.
23. Seifert KA, Rossman AT (2010) How to describe a new fungal species. *IMA Fungus* **1**: 109–111.
24. Clements FE, Shear CL (1931) *The Genera of Fungi.* New York: HW Wilson.
25. Ainsworth GC, Sparrow FK, Sussman AS (1973) *The Fungi: An Advanced Treatise*, Vol. 4: *A Taxonomic Review with Keys.* Parts A–B. New York: Academic Press.
26. Hawksworth DL, Kirk PM, Sutton BC, Pegler DN (1995) *Ainsworth & Bisby's Dictionary of the Fungi*, 8th edn. Wallingford, U.K.: CAB International.
27. Cannon PF, Kirk PM (2007) *Fungal Families of the World.* Wallingford, U.K.: CAB International.
28. Redhead SA, Demoulin V, Hawksworth DL, Seifert KA, Turland NJ (2014) Fungal nomenclature at IMC10: Report of the Nomenclature Sessions. *IMA Fungus* **5**: 449–462.

3 Phylogenetic Analysis Especially in Relation to Fungi

Anna Muszewska and Krzysztof Ginalski

CONTENTS

- 3.1 Introduction: Why We Want to Make Phylogenetic Trees 15
  - 3.1.1 Homology and Homoplasy 16
  - 3.1.2 Paralogs, Orthologs, and Xenologs 16
  - 3.1.3 Reading Phylogenetic Trees 17
- 3.2 Major Steps of a Phylogenetic Analysis 19
  - 3.2.1 Input Format Issues 19
  - 3.2.2 Dataset Selection 20
  - 3.2.3 Protein versus DNA Sequences 20
  - 3.2.4 Sequence Alignment 20
  - 3.2.5 Positive Selection and Purifying Selection 21
- 3.3 Inferring Phylogeny 22
  - 3.3.1 Distance Methods 23
    - 3.3.1.1 Neighbor Joining 23
  - 3.3.2 Evolutionary Models 23
    - 3.3.2.1 Nucleic Acid Substitution Models 23
    - 3.3.2.2 Amino Acid Substitution Models 24
  - 3.3.3 Maximum Parsimony 25
  - 3.3.4 Bayesian Methods 25
  - 3.3.5 Maximum Likelihood Methods 26
  - 3.3.6 Bootstrap, Jack-Knifing, and aLTR Scores to Test Tree Topologies 26
  - 3.3.7 Super Trees and Supermatrices 27
  - 3.3.8 Tree Formats 27
  - 3.3.9 Tree Visualization 28
- 3.4 Applications of Phylogenetic Analysis 30
  - 3.4.1 Species Classification and Identification 30
    - 3.4.1.1 Molecular Marker Selection for Species Trees 30
  - 3.4.2 Gene Phylogeny versus Species Phylogeny 30
  - 3.4.3 Phylogeography 30
  - 3.4.4 Fungal Tree of Life 31
  - 3.4.5 Horizontal Gene Transfer, Gene Clusters, and Supernumerary Chromosomes 32
- 3.5 Concluding Remarks 32
- References 33

3.1 INTRODUCTION: WHY WE WANT TO MAKE PHYLOGENETIC TREES

Initially, phylogenetic trees were mostly used to provide a robust taxonomical classification based on the inference of phylogenetic relationships among species. Taxonomy is a formalized system of describing and naming objects. This topic remains current in modern mycology[1,2] since the dual nomenclature applied for many fungal lineages still remains unresolved. For example, parallel usage of *Deuteromycota* based on different morphological features and the *Eumycota* based on sexual morphology leads to misunderstanding and confusion to nonspecialists. This issue led to the formulation of the Amsterdam Declaration on Fungal Nomenclature[1] as well as the formation of *One Fungus = One Name* movement.[2] Furthermore, pathogenic and food-borne fungi are often classified into *species complex* and other artificial taxa. Currently, phylogenetic analyses are applied to detect patterns of natural selection, describe orthology/paralogy in gene families, estimate divergence times, identify specimens, determine histories of populations, or compare whole genomes and metagenomes. Phylogenetic methods have become a powerful tool for comparative genomics with the advent of whole-genome sequencing. We will concentrate on species and gene trees in this chapter.

3.1.1 Homology and Homoplasy

Homology is a central term in evolutionary biology and is a key factor in phylogenetic analyses. Homology implies common ancestry whereas homoplasy implies shaping by similar constrains. Similar characteristics are not necessarily indicators of homology but they may be a sign of convergence. Convergence can be found at various levels of organization from body shape suitable for a certain environment to the molecular level. For example, the adaptation to a similar function leads to the formation of S–D–H and D–H–S catalytic triads in serine proteases independently.[3] Similar functions of orthologs are a mere consequence of slow changes rather than a biological law. One should *never* describe homology as a scalable feature. Terms *percentage of homology* and *strong/weak homology* violate the logic of the homology definition stating common ancestry. Common ancestry is a binary characteristic that can either be present or absent. On the other hand, similarity is a scalable feature and in consequence can be described effectively with percentage values and with strong/weak adjectives.

3.1.2 Paralogs, Orthologs, and Xenologs

Homologous sequences are divided into *paralogs* and *orthologs* based on their origin. Orthologous sequences are a result of *speciation* and paralogous sequences are a product of *duplication*.[4] Orthologous sequences are likely to display similar functions due to shared ancestry. However, there might be exceptions as a consequence of the adaptation of the organism to its niche. Paralogs are divided into in-paralogs and out-paralogs. In-paralogs are the result of a duplication after speciation, while out-paralogs are a product of a duplication preceding a speciation event. Two paralogs in one species along with one ortholog in another species make co-orthologs. Figure 3.1 shows a simple visual representation of paralog–ortholog relationships. Let us consider there was an ancestral species with an ancestral gene shown as a square. This gene became duplicated into two descendant genes A and B, marked, respectively, as a triangle and circle. After a speciation event, each of the resulting species, 1 and 2, obtained a copy of both genes. Hence, the A and B genes are out-paralogs because speciation followed duplication. Additionally, in species 1 there was a duplication of gene A leading to in-paralogs A1 and A2. Genes B in species 1 and 2 are orthologous. The same applies to genes A1, A2, and A, which are orthologs.

There is a third major category of genes, namely, the xenologs, which are horizontally transferred genes. The presence of numerous xenologs makes some of the gene phylogenies hard to interpret, although these are predominantly in prokaryotic organisms. See Ref. [5] for a review on ortholog prediction methods and Ref. [6] for a recent tool dedicated to large-scale ortholog annotation. There are several databases with tree datasets of orthologous groups of genes/proteins: TreeFam,[7] EggNOG,[8] PhylomeDB,[9] OrthoMCL,[10] COG (http://www.ncbi.nlm.nih.gov/COG/), and Fungal Orthogroups.[11]

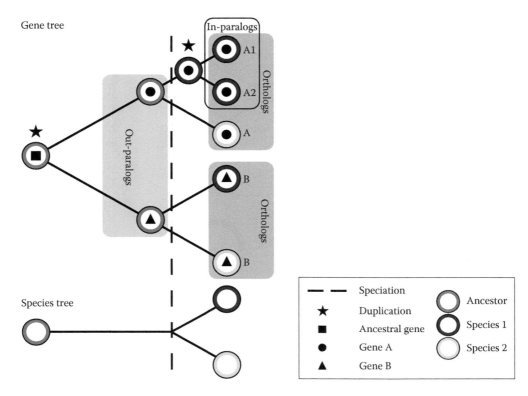

FIGURE 3.1 This schema tells the story of one gene (the ancestral gene on the gene tree) in one species (the ancestral species on the species tree) during evolution at gene and species levels correspondingly. The species tree displays one speciation event. The gene tree shows that paralogs arise by duplication and orthologs by speciation.

In conclusion, one should be very cautious making functional observations from the mere fact of orthology as organisms evolve and protein functions change with time. Single amino acid changes might shift substrate specificity with an overall sequence similarity remaining very high.

3.1.3 Reading Phylogenetic Trees

Phylogenetic trees are graphs that address evolutionary questions regarding orthology and paralogy that are based on the assumption of homology between all taxa in the dataset. Tree descriptions usually encompass specific nomenclature. In order to illustrate some of the most common phylogenetic terms, a schematic tree will be used (Figure 3.2). Trees consist of nodes and branches (vertices in graph nomenclature). Nodes without descendants can be called *leaves* or *terminal nodes* and are associated with the data points subjected to the tree calculations. They denote taxa existing today and are usually labeled. The internal nodes describe predicted intermediate, ancestral states that cannot be observed currently. For specific applications, internal nodes are sometimes labeled. Nodes are connected by branches, which correspond to the amount of change that occurred between the ancestral and descendant node. Most trees assume a *bipartition (bifurcation) model* what means that each node has only two immediate descendants. For example, the node leading to the A, B, and C clade has only two descendants the taxon A and a branch leading to B and C. Bifurcation is a consequence of the assumption that speciation is a process leading to splitting one species into two. However, not all trees are resolved with reliable support and result in nodes with more than two descendants. This is called a *politomy (multifurcation)*. One of the best-studied cases

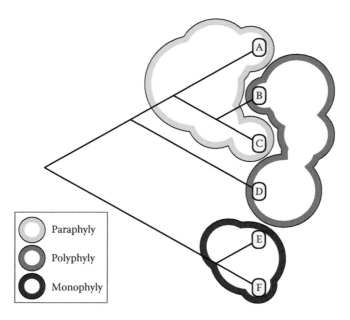

FIGURE 3.2 This image illustrates monophyly, paraphyly, and polyphyly terms. A monophyletic clade harbors the descendants of a node. In a paraphyletic group, some of the node's descendants are not included. A paraphyletic group encompasses some nodes without the common ancestor of those nodes.

of unresolved phylogenies comes from the animal world where the divergence of placental mammals into three lineages—Afrotheria, Xenarthra, and Boreotheria—seems to have occurred nearly simultaneously.[12] Rapid radiation events and lack of sufficient phylogenetic signal are the main causes of politomies.

From a technical point of view, a phylogenetic tree can be either *unrooted* or *rooted*. A rooted tree has a particular node, the *root*, from which each taxon can be reached by moving solely in the direction toward the leaves. The root represents the common ancestor of all taxa in the tree. Each unrooted tree can be rooted in multiple ways, at each branch from any internal or terminal node. The tree itself is unrooted and the positioning of the root is one of the responsibilities of the researcher that require prior knowledge of the analyzed dataset. Each of the taxa in the analysis should be more closely related to each other than to the members of the outgroup. One can imagine analyzing a group of Eumycota and rooting the dataset with Glomeromycota because it is a well-established relation. In other words, the Glomeromycota are an *outgroup* for Eumycota. A *sister group* or *sister clade* is the closest relative of a clade, what comes with the assumption of common ancestry. For example, Basidiomycota are a sister clade of Ascomycota. It is important to find a group that is easily distinguishable but not too distant to the analyzed dataset. Using a very distant group like Microsporidia as an outgroup for Ascomycota would reduce the phylogenetic signal due to multiple ambiguously aligned residues in the sequence alignment used to derive a tree. One of the common errors in phylogenetic analyses is the use of unrelated objects as roots. Trees are collections of nested clades, where a *clade* is any group of nodes that includes the ancestor and all of its descendants. Each terminal node is then a clade because it encompasses all of the descendants of a node. A clade is *monophyletic* when it encompasses all descendants. For example, the clade composed of E and F in Figure 3.2 forms a monophyletic group. The spatial organization of the branches, including the order of the nodes, is the tree topology. A clade without all descendants is *paraphyletic*. For example, A and C with their last common ancestor forms a paraphyletic group because they do not contain B, which is one of the descendants of the last common ancestor of A and C. The clade A, which does not contain the last common ancestor of all of the members, is *polyphyletic*. This kind of grouping is the most counter intuitive and should be avoided. One may bring B and D together and consider them as a clade. However, this grouping is a polyphyletic clade because it omits the last common ancestor of these taxa.

3.2 MAJOR STEPS OF A PHYLOGENETIC ANALYSIS

Figure 3.3 summarizes all the major steps to take into account when designing a phylogenetic analysis,[13] and this section introduces software to perform all the essential steps required. One of the most user-friendly books is Ref.[13], which provides a step-by-step introduction to tree building. A more detailed but intuitive book is Ref.[14] The key steps in a phylogenetic analysis are usually repeated several times with improved dataset selection and tuning of phylogenetic inference. Usually different evolutionary models are tried and more than one phylogeny inference method is used. None of the current methods guaranties a perfect output and the key steps should be repeated using different software.

3.2.1 INPUT FORMAT ISSUES

Some convenient tools accept only sequences in specific formats where most of the technical problems with feeding the data to the program are related to taxon labeling. When naming a sequence, it

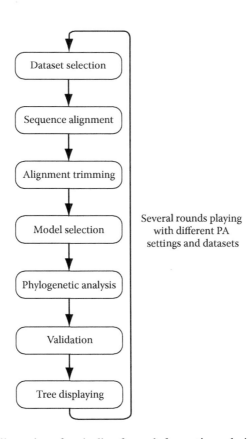

FIGURE 3.3 Schematic illustration of a pipeline for a phylogenetic analysis. When performing a typical phylogenetic analysis, one follows a sequence of steps starting from a dataset selection to tree displaying. These steps are often repeated in order to obtain a robust tree.

is recommended to avoid characters other than A–Z and 0–9 (e.g., whitespaces should be avoided) and some software requires unique sequence labels no longer than 10 characters. Other common problems relate to incorrect gap representation, disallowed characters in sequences, and using DNA instead of protein sequences or *vice versa*.

3.2.2 Dataset Selection

Dataset selection is the most crucial step in all analyses: phylogenetic analysis software makes the following assumptions: (1) data (i.e., the sequence) are correct, (2) sequences are homologous, (3) each position in a sequence alignment is homologous with the corresponding position in other sequences, (4) taxon sampling is adequate to answer the scientific question of interest, and (5) the phylogenetic signal is sufficient to produce a reliable answer. Violation of these assumptions might result in a false, yet statistically well-supported phylogenetic tree. Phylogenetic analyses normally involve usage of biological sequences from different researchers and the publication process requires deposition of sequenced material in a publicly available database. Contemporary databases provide querying and retrieval tools that facilitate resource usage. Database records are often cross-referenced to other databases and the databases store sequence data with additional information about the organism, investigators, sequencing project and function if applicable. However, all annotations should be verified in the corresponding literature due to limited curation of the database resources. Sound strategies to retrieve sequences of interest include (1) selecting sequences described by a certain keyword, (2) searching for a set of homologous sequences by applying the criterion of sequence similarity, or (3) combining these.

3.2.3 Protein versus DNA Sequences

Whether to choose protein or corresponding coding DNA sequence is an open question. Both carry all the necessary information to produce a functional protein. However, they are shaped by different constrains in time. The protein sequence changes over time depending on the selective pressure acting on this protein. DNA reflects both the selective pressure acting on the protein and the genetic code and codon usage bias of the host organism. At large evolutionary distances, amino acid sequences permit phylogeny resolution with more accurate resolution than nucleic acids because of substitution saturation. When analyzing protein coding regions, it is recommended to use the protein sequence because of the uncertainty of the genetic code and codon usage bias. However, for closely related taxa, DNA sequence may carry all of the phylogenetic signal. Additionally, closely related organisms are likely to have similar codon usage and genetic code. DNA sequences permits the splitting of the sequence into categories based on codon position, hence improving the overall performance and strengthen the phylogenetic signal.

Different protein sequences evolve at variable rates, which results in different look back time, that is, the time when two related sequences are separated by a number of mutations that still permits their correct identification from random similarities.[14] Housekeeping proteins, such as histones, cytoskeleton components, enzymes of core metabolism, and replication machinery, change very slowly and their functions are conserved over time, so their look back time is longer. They incorporate few mutations per millions of years making them applicable to interkingdom phylogenetic comparisons. It is recommended to perform phylogeny inference for both datasets when both DNA and protein sequences are available. Another general rule is to choose protein sequences when greater evolutionary distances are expected and DNA sequences for closely related taxa.

3.2.4 Sequence Alignment

An alignment is a representation of two or more biological sequences, which aligns homologous positions and introduces deletions or insertions in the nonhomologous sites. An alignment can be named pairwise or in multiple form depending on the number of aligned sequences. In a pairwise approach, only two sequences are compared and it is frequently used for sequence database searches. Multiple sequence alignments agree upon homologous position within a bunch of sequences. Based on the portion of the sequence aligned, alignments may be local or global. The first aligns sequence regions of the highest similarity, while the second covers the entire sequences. Local alignments, often used as heuristics for a global alignment approach, match only these substrings of the sequences that are the most similar. However, pairwise alignments overestimate the sequence similarity, thus the correct method to obtain sequence identity is multiple sequence alignment.

If no substitutions in a column alignment occur, then this might mean that this position cannot mutate because it is critical for protein function. Different sequence fragments might be informative for phylogeny inference depending on the evolutionary distance of the analyzed taxa. For closely related taxa, only variable regions will harbor the phylogenetic signal, namely, the sites with changed states.

There are many aligning programs but none produce perfect alignments. Alignment is considered as one of the most difficult computational problems, thus the tools used in computational biology use heuristic approaches when building a multiple sequence alignment. The tree derived from a sequence alignment cannot be more reliable than the alignment itself. Consequently, the quality of the alignment is the bottleneck of most phylogenetic analyses and it is essential to use the best-performing aligning tools. More than one aligning tool should be employed and the best resulting alignment selected. Alignment selection should take into account that available methods use residue substitutions models that are built on the assumption of independent evolution of sites. Multiple alignments of the same sequence dataset, calculated with the same program using identical settings, might differ also because of the heuristic approach applied in all aligning tools. Results from different programs can be successfully combined, and this function is provided by M-Coffee.[15] Sometimes, automatically calculated alignments are manually adjusted that should be performed using all knowledge about the analyzed proteins including their secondary, tertiary, and quaternary structures. Conserved features such as polarity/hydrophobicity patterns, position of secondary structure elements (alpha-helices, beta-strands and loops), and key amino acids should be carefully considered. Gaps representing insertion and deletion events are more likely to appear in loop regions than within alpha-helices or beta-strands. Charged residues (i.e., D, E, R, H, K) are replaced usually by similarly charged amino acids and often play a role in active site formation. Furthermore, proline and cysteine play a crucial role in protein folding. Proline often ends an alpha-helix, whereas pairs of cysteines may form disulfide bonds. Hydrophobic amino acids (i.e., A, I, L, M, F, W, Y, V) easily mutate to other hydrophobic amino acids and are involved in protein folding as they tend to locate in the inner part of the protein.

Whole-genome aligners require even more efficient algorithms than protein sequence aligners. Currently, Mafft,[16] ProbCons,[17] Muscle,[18] and PCMA[19] are the most popular alignment tools. Table 3.1 presents some of the multiple alignment calculating programs. In addition, there are web services offering multiple aligning tools online, for example, http://toolkit.tuebingen.mpg.de/sections/alignment. It is possible to remove ambiguously aligned regions of an alignment using Gblocks[20] or TrimAl.[21] This procedure usually enhances the reliability of phylogeny inference as it removes dubious signal. However, for very divergent sequences, it is hard to obtain long blocks of aligned sequences that are required by the trimmers. Consequently, after trimming there may be too few columns remaining to analyze, although it is possible to remove poor quality fragments of the alignment. Finally, inspecting data before feeding them into subsequent programs is recommended.

3.2.5 POSITIVE SELECTION AND PURIFYING SELECTION

A multiple sequence alignment together with a species tree can be used to describe the selective pressure acting on each codon in a protein coding gene. Measurement is based on the ratio of the number of nonsynonymous substitutions (Ka) to the number of synonymous substitutions (Ks)

TABLE 3.1
Some of the Available Multiple Sequence Alignment Tools

Tool	Characteristic	URL	References
Mafft	Multiple variants of iterative or progressive aligning algorithms.	http://mafft.cbrc.jp/alignment/software/index.html	[16]
ProbCons	Progressive protein multiple sequence alignment based on probabilistic consistency, a novel scoring function for multiple sequence comparisons.	http://probcons.stanford.edu/	[17]
Muscle	Fast aligning tool, iterative refinement usually improves accuracy. Sensitive to differences in sequence length.	http://www.drive5.com/muscle/index.htm	[18]
PCMA	A hybrid approach combining fast alignment of similar sequences and profile–profile comparison for distant sequences.	http://prodata.swmed.edu/pcma/pcma.php	[19]
T-Coffee	Many variants, accurate progressive alignments. Can combine multiple sequences alignments obtained previously and integrate structural information.	http://www.tcoffee.org/	[15]
DIALIGN	Greedy and progressive algorithm for segment-based alignment of both proteins and nucleic acid sequences.	http://bibiserv.techfak.uni-bielefeld.de/dialign/	[71]
Probalign	Dynamic algorithm for a global alignment.	http://probalign.njit.edu/standalone.html	[72]
Kalign	A fast and accurate, progressive alignment.	http://msa.sbc.su.se/cgi-bin/msa.cgi	[73]
ClustalW	Tree-based alignments. Currently a massive computing flavor has been developed—Omega.	http://www.clustal.org/	[74]

per site. A Ka/Ks ratio of a neutrally evolving site should be close to 1. A gene under purifying selection will have a Ka/Ks $\ll$ 1, whereas a gene under positive selection may have a Ka/Ks $\gg$ 1. This simplification is limited by the decreasing amount of information encompassed by an alignment with a growing timescale. Multiple substitutions at each position might lead to saturation and information loss in a long evolutionary perspective. GC content bias also leads to problems with accurate Ka/Ks estimation. Transitions are more likely to be synonymous than transversions at the third codon position and codon usage bias results in a lower number of synonymous sites. The R package seqinr makes Ka/Ks calculations straightforward (http://cran.r-project.org/web/packages/seqinr/index.html) and another Ka/Ks tool is the codeml program in PAML.[22]

3.3 INFERRING PHYLOGENY

There are four main algorithmic methods for calculating phylogenetic trees[13]: Distance, maximum parsimony, Bayesian analysis, and maximum likelihood (ML) analyses. Phylogenetic methods can be classified into distance and character based. Distance methods are based on pairwise comparisons between all sequences whereas character-based methods (maximum parsimony, Bayesian inference, and ML) compare one site in an alignment in all analyzed sequences. The latter comparison results in a score for each sampled/possible tree. After all characters have been analyzed the best scoring tree is returned. In theory, this search should compare all possible trees, but this is unfeasible computationally for large datasets and instead heuristic searches limit the space of trees being sampled and compared. A second way of classifying phylogeny reconstructing methods is based on the usage of evolutionary models. Maximum parsimony calculates trees without a defined evolutionary model. Bayesian inference, ML, and neighbor joining (NJ) rely on an evolutionary model when calculating a phylogenetic tree.

3.3.1 DISTANCE METHODS

Distance methods use a distance matrix of sequence similarity, pairwise scores derived from an alignment. The distance matrix is calculated for each pair of sequences as the distance or fraction of differences between them.

3.3.1.1 Neighbor Joining

NJ is a clustering method operating on a distance matrix. NJ starts with the star tree topology and choosing pairs of taxa based on their distance, so that they minimize an estimate of tree length. Based on the distance matrix, a new modified matrix is calculated from the net divergence of each taxon from all other taxa. The connected taxa are then represented as their ancestral node and replaced in the matrix by the ancestral node. The number of taxa deriving directly from the root is decremented, which is performed iteratively. NJ does not assume that all taxa are equally distant from the root and the method is computationally efficient. The method performs reasonably well for big datasets of moderate sequence divergence and selection of a suitable substitution model can enhance performance. However, NJ should be avoided for highly diverged sequences and alignments with missing data.

3.3.2 EVOLUTIONARY MODELS

Evolutionary models involve sets of assumptions about the substitution process occurring in nucleotide or protein sequences. Models are used to calculate branch lengths that illustrate the amount of changes accumulated from the ancestor to its descendant. The most intuitive way of calculating branch lengths would be to count the distance between the ancestor and its descendant as the number of differences in the sequences. Often the difference is expressed as a proportion rather than as a raw count of changes. However, this simplistic approach does not take into account multiple substitutions at a single site or mutations reverting to the same state. For very closely related taxa, this simple intuitive approach may produce reasonable results, but for more diverged sequences increasingly realistic evolutionary models are required. Mutations occurring in ancestral sequences are fixed by selection and by drift resulting in changes in the nucleotide sequence of the descendant taxa. Models are collections of assumptions about the substitution processes and are employed to estimate branch lengths. The method by which the model scores substitutions may also influence the topology of the tree. Heterologous datasets combining, for example, coding and noncoding regions require special treatment. One might decide either to split the data into partitions with separate models or combine them using a mixed model with different models of evolution for specific regions of an alignment.

3.3.2.1 Nucleic Acid Substitution Models

These models differ in the number of estimated parameters. Apart from rates of change from one nucleotide to another, models can assess a proportion of invariable sites (sites that remain unchanged in the dataset) and rate variation among sites with a gamma distribution. These have been developed as mathematical methods incorporating no empirical datasets. The simplest is the Jukes–Cantor model,[23] which assumes that the probability of changing any nucleotide to another is always equal. The unique parameter required to calculate the probability of a particular nucleotide at a site mutating into a particular nucleotide over certain time is the substitution rate. Base frequency differences can be considered, as implemented in the Felsenstein 81 model (F81).[24] The single parameter models are not very realistic because some changes are more common than others. The Kimura 2-parameter model (K2P)[25] applies two different rates of substitution for transversions and transitions: the Tamura 3-parameter model adds a correction for composition bias for sequences with base ratios frequencies far removed from equal. A more detailed scenario distinguishing transitional substitution in purines from transversional substitutions in pyrimidines is proposed by Tamura and Nei.[26] The most complex scenario is described by the general time reversible model (GTR), which estimates nine different parameters for six different substitution rates.[27] The substitution process is symmetrical in time because the probability P(A → B) is identical to P(B → A). The aforementioned model treats all sites equally; however, biological sequences tend to have different substitution rates at particular sites, for example, the third codon is more likely to mutate than the second codon.

Substitution matrices and evolutionary models for nucleic acids and amino acid sequences predict the probability of mutation leading to a change visible at each position in a sequence alignment. One type of nucleotide (or amino acid) might be substituted by another type with different, yet independent probabilities. Sequences that code for (1) internal regions of a protein, (2) the core of the folds, (3) active sites, and (4) ligand binding pockets tend to evolve at slower rates than those coding for the less conserved parts of the protein. These regions are distributed along the nucleotide sequence, and resulting in protein stretches of conserved and variable residues. A model may be considered where each site has its own specific substitution rate. However, this would be computationally inefficient. To address this problem, a gamma distribution of rates across sites is widely exploited, which is feasible here because it is mathematically well studied. Biological phenomena were not observed that make gamma distribution more useful over other distributions. From the user's perspective, introduction of gamma distribution means an additional parameter in the model—alpha, which is often referred to as the shape parameter, because it determines the shape of the gamma distribution. Additionally, some models allow for a fraction of sites that do not change, that is, the invariant sites, which is a special case of site-specific variation equaling zero. However, some of the equal positions appearing in the alignment are not necessarily invariant sites and might be conserved simply because the analyzed sequences are not diverse enough to display observable differences. Some programs estimate the proportion of invariant sites as a part of the model. Eventually, a model might combine multiple approaches and can be specified as GTR + G + I or GTR + InvGamma what means a GTR model with rate variation across sites estimated from gamma distribution and estimated portion of invariant sites.

3.3.2.2 Amino Acid Substitution Models

Amino acid substitution models describe the divergence between sequences based on matrices coding probabilities of changing each amino acid into another. Such models assume an independent evolution at each site. The substitution matrices usually consider differences in amino acid psychochemical properties such as size, charge, and hydrophobicity. Most amino acid substitution models have been built based on empirical data, calculated on datasets of alignments, which leads to replacement matrices with realistic properties. Additionally, the uneven frequency of specific amino acids in biological sequences can be taken into account. The most widely used matrices are based on BLOSSUM[28] and PAM[29] matrices, and their main application includes sequence searches. PAM and BLOSSUM matrices series are scaled for more similar and more distant sequence comparisons (lower BLOSSUM numbers and higher PAM numbers should be used for more divergent sequences). JTT,[30] Dayhoff,[29] WAG,[31] and LG[32] are the most popular substitution models for phylogenetic purposes. All these models are derived from the first Markov model proposed by Dayhoff, which uses 20 states.[29,33] Because the estimation of 20 states is time and resource consuming, the models use empirical data leading to fixed relative rates of amino acid substitution (e.g., JTT[30] or WAG[31]). These models treat sites independently without taking into consideration the variable evolutionary forces acting on specific parts of a given protein. However, model selection should not be performed based on sequence source or method of construction.[34] Models are used to make complex problems computationally tractable and a good model can make accurate predictions despite its simplicity. There is a general rule that parameter-rich models fit data better than simple models only because they retain many parameters. Nevertheless, choosing the most complex model has its pitfalls not least from elevated computation time. Also, estimation of more parameters from the same amount of information can increase the error rate.

The golden rule is to find a model that explains the data without unnecessary complexity. Model selection for protein alignments can be carried out with ProtTest[33] and for nucleotide alignments with jModelTest.[35]

3.3.3 Maximum Parsimony

Parsimony assumes that the most likely evolutionary scenario is the one that requires least changes. There are three major sources of conflicts (i.e., homoplasies) with this assumption: (1) reversal when a character changed but mutated back to its ancestral state, (2) convergence when unrelated objects developed the same state of a character independently, and (3) parallelism when two similar objects have a predisposition to develop similar states of a characteristic.

Parsimony considers the *homoplasies* less probable and less frequent than common ancestry. Figure 3.4 illustrates the homoplasy, apomorphy, and related terms. An ancestral state shared by multiple taxa is referred to as a plesiomorphy and a newly derived state is called an autoapomorphy. This new state, when shared with other related taxa, can be named a synapomorphy or simply apomorphy. A homoplasy is a new state shared by distantly related taxa.

Parsimony chooses the tree (or trees) with the minimal number of evolutionary steps indispensable to explain the data. For a given alignment of sequences, not all sites are informative for constructing a parsimony tree and only those appearing in more than one sample are useful. N.B. There are usually many equally parsimonious trees, that is, trees requiring the same number of changes. PAUP,[36] MEGA,[37] and PHYLIP[38] implement many phylogenetic methods including maximum parsimony. Finally, TNT[39] is a program that specialized in parsimony and performs very well.

3.3.4 Bayesian Methods

Bayesian phylogeny inference is one of the character state methods. Bayesian methods are based on the Bayesian rule where the conditional probability P(A|B) of event A given B is expressed by P(B|A)P(A)/P(B). It derives the posterior probability as a consequence of a prior probability and probability of the data. A Bayesian approach (BA) searches for a set of likely trees verifying

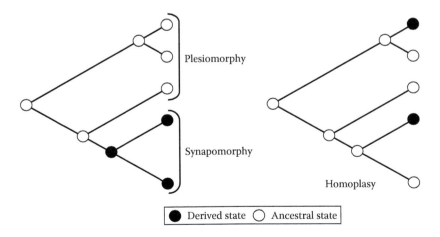

FIGURE 3.4 Two schematic trees are used to show ancestral and derived character states. A character shared by all ancestors and some of the descendants is likely an ancestral state called a plesiomorphy. A synapomorphy is a character state present in some related descendants, it is a derived state. A character shared by distantly related organisms that arose via convergence is called a homoplasy.

the prior distribution of model parameters, that is, branch lengths, tree topology, and substitution model. The model is updated based on the information provided by the dataset. Bayesian inference uses a uniform prior on the topology distribution, which means that initially each tree topology is equally likely. This distribution is then modified based on the provided dataset. Posterior probabilities are inferred from the tree space sampling by using Markov chain Monte Carlo (MCMC). The sampling procedure starts with the generation of a set of random states. These parameters are changed in each step and the new set of parameters is compared to the current state. The better likelihood ratio and prior ratio values mean that the new parameter values are accepted. The worse parameters are rejected with the probability inversely proportional to *how much worse* is the new state. After a phase of exploration, called burn in, it is expected that the sampling process converges and continues sampling from a set of posterior probability distributions. The initial set of probable parameters collected in the burn in phase should be discarded as it will contain some very unlikely values. The frequency of each set of parameters is then sampled and is reflected by its posterior probability value. Based on this value, a maximum *a posteriori* tree can be selected. Bayesian methods are implemented in MrBayes[40] and BEAST,[41] which are the two best-known BA calculating tools.

3.3.5 MAXIMUM LIKELIHOOD METHODS

ML is a method to estimate initially unknown parameters in a model so that the parameter values maximize the likelihood. The likelihood function describes how probable the data are given the parameters. Felselstein was the first to develop ML algorithms for DNA sequences. The search is performed iteratively with two optimizations steps: one for branch lengths and the other for the tree that maximizes the likelihood. The substitution models described for distance methods can be also applied in ML. ML searches for an evolutionary model to describe the data and usually the likelihood is calculated for each alignment position. The cumulative score (likelihood) of a tree under a given substitution model for a dataset is the multiplication of likelihoods at all sites. The likelihood ratio test can be performed to test the fit of evolutionary models.[13] Modern tools with performances outcompeting the all in one packages like PAUP,[36] MEGA,[37] and PHYLIP,[38] are RAxML,[42] PhyML,[43] GARLI,[44] and PAML[31] (Table 3.2). More phylogenetic software sources can be found at http://evolution.gs.washington.edu/phylip/software.html.

3.3.6 BOOTSTRAP, JACK-KNIFING, AND aLTR SCORES TO TEST TREE TOPOLOGIES

Bootstrap and approximate likelihood ratio (aLTR) scores are used to assess the reliability of a tree topology. Bootstrap is the most commonly used method of tree validation and can be applied to trees constructed with all algorithms. Bootstrapping produces scores for each branch present in the tree that are often referred to as branch support. The validation is performed by recalculating a tree using random sampling with replacement of the original data and counting how often each bifurcation from the original tree was reconstructed in the bootstrap replicates. Bootstrap is very time consuming because the original phylogenetic analysis is recalculated several hundred times, which is unfeasible for very large genomic data. However, Jack-knife performs random sampling of the original data without replacement. The aLTR test validates the significance of each branch by computing the log-likelihood value optimized only over the branch of interest and the four adjacent branches.[45] A standard likelihood ratio test calculates twice the difference of the likelihoods of the most likely arrangement and of the branch of length 0. Each branch has only three arrangements if it exists. The approximation is to calculate the statistic using twice the difference of the likelihoods of the two most likely arrangements. This makes the aLTR calculation faster than any bootstrap. The test is implemented in PhyML and is computed during the phylogeny inference. The aLTR score is described as more conservative and robust to certain violations of model assumption than the bootstrap values.

TABLE 3.2
Phylogeny Programs Mentioned in This Chapter

Tool	Characteristic	URL	References
PAUP	Commercial, still a beta version, widely used, implements parsimony, distance matrix, invariants, and maximum likelihood methods and statistical tests.	http://paup.csit.fsu.edu/	[36]
MEGA	Implements parsimony, distance matrix and likelihood methods, consensus tree computing, data editing tasks, sequence alignment and statistical validation.	http://www.megasoftware.net	[37]
PHYLIP	All in one software package. Implements parsimony, distance matrix, and maximum likelihood methods. Offers many data manipulation tools.	http://evolution.gs.washington.edu/phylip.html	[38]
RAxML	Fast reconstruction of phylogenies by maximum likelihood. It has a light version for large DNA datasets.	http://www.exelixis-lab.org/	[42]
PhyML	Fast reconstruction of phylogenies by maximum likelihood, also available as a web service.	http://www.atgc-montpellier.fr/phyml/	[43]
PAML	Implements maximum likelihood phylogeny inference, calculates substitution rates at particular sites, reconstructs ancestral sequences, simulates DNA and protein sequence evolution, and computes distances across a wide range of models.	http://abacus.gene.ucl.ac.uk/software/paml.html	[31]
GARLI	Genetic algorithms for maximum likelihood phylogeny inference.	https://www.nescent.org/wg_garli/	[44]
TNT	Fast parsimony inference.	http://www.zmuc.dk/public/phylogeny/tnt/, http://www.lillo.org.ar/phylogeny/tnt	[39]
MrBayes	Bayesian MCMC analysis of sequences and model choice across a wide range of phylogenetic and evolutionary models.	http://mrbayes.sourceforge.net/	[40]
BEAST	Bayesian MCMC analysis of sequences. Additionally is suitable for evolutionary hypotheses testing without conditioning on a single tree topology.	http://beast.bio.ed.ac.uk/	[41]

Note: More phylogeny programs are described at http://evolution.genetics.washington.edu/phylip/software.html.

3.3.7 Super Trees and Supermatrices

For large datasets, often with missing loci for a subset of genes/proteins in some of the species of interest, additional approaches may be used. Phylogenetic analyses of multiple genes/proteins can be performed either as a *supertree* or a *supermatrix*. The supermatrix stands for a concatenated set of sequences used to generate a common, merged dataset. This results in an implicit assumption of even evolutionary dynamics of the concatenated genes/proteins. When different evolutionary dynamics are taken into account, it is equivalent to a separate analysis. It is an efficient approach to estimate a common phylogeny for a set of genes/proteins. The approach is used to reconstruct a supertree based on separated subtrees for individual genes/proteins. This enables the detection of horizontal gene transfer (HGT) and the differential analyses of groups of genes/proteins.

3.3.8 Tree Formats

For batch processing of annotated trees, it is recommended to use the XML format. Single trees are usually stored in *newick* and nexus files that are plain text files and so can be edited with any text editor. The newick notation consists of node names, commas, and parentheses. A tree (see Figure 3.5) can be stored as a simple string (((A,(B,C)),(D,E)),F); where a semicolon ends the tree. Commas

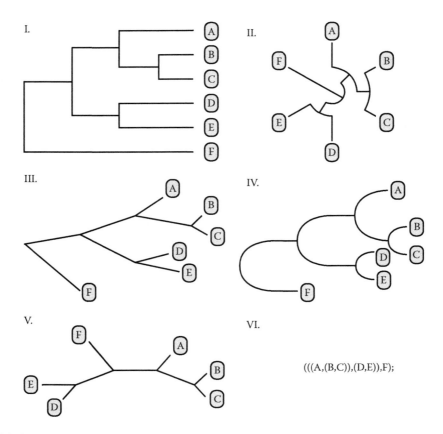

FIGURE 3.5 Multiple graphical representations of the same phylogenetic tree with six taxa. I, rectangular without branch lengths; II, circular; III, triangular; IV, curved; V, unrooted; VI, a newick string encoding the tree.

separate the subtrees linked to a node and paired parentheses represent an internal node. Branch lengths are assigned to nodes using a number preceded by a : (obviously a node has no length), this value refers to the branch linking to the immediately parental node. The length from a terminal node should be calculated as a sum of branch lengths leading from this node *via* all internal nodes to the root. Unfortunately, there are many problems encountered with newick string formatting. A tree string might not be accepted as an input because it contains some unclosed parentheses, special characters (others than letters, numbers, and underscores), white spaces in taxon names, and end-of-line signs within the string. Some programs have problems with polytomies, although these can be overcome by inserting *zero length* in each node.

3.3.9 Tree Visualization

Tree visualization is a crucial step in phylogenetic analysis. The same tree may be displayed in different ways depending on the original scientific question. When deep branches are to be shown, collapsing the leaves can be considered to make the base of the tree more visible. With some minor differences, zooming in on a subset of the tree with deeper branches collapsed or not shown, should make the message clearer. Branch order can be swapped, collapsed clades, the use of arrows and other graphical techniques can be employed to gain better resolution. However, it is important not to (1) combine results from separate analyses, (2) manipulate branch lengths or scores, or (3) remove/add data from/to an existing tree. The choice of tree representation is subjective, and Figure 3.5

shows five different ways of drawing the same simple tree. On a circular tree representation, all branches get the same amount of space making them equally visible. One should consider displaying a tree in this format when many taxa need to be observed. An unrooted tree enables quick perception of major divisions in a dataset. If the aim is to split data into categories and not to obtain too much detail, or to show how data is disjointed, an unrooted tree is a good choice. Tree drawing tools are summarized in Table 3.3.

TABLE 3.3
Tree Drawing Tools

Tool	Characteristic	URL	References
Archaeopteryx and PhyloXML	A number of different shapes of tree display, enables edition, can integrate different data sources.	http://www.phylosoft.org/archaeopteryx/, https://sites.google.com/site/cmzmasek/home/software/archaeopteryx	[75]
Bosque	Client–server to run, manage, and visualize multiple trees.	http://bosque.udec.cl/	[76]
Dendroscope	It provides a number of methods for drawing and comparing rooted phylogenetic networks and to compute rooted phylogenetic networks from trees.	http://www.dendroscope.org, http://ab.inf.uni-tuebingen.de/software/dendroscope/, http://ab.inf.uni-tuebingen.de/software/dendroscope/	[77]
ETE	Automated manipulation, analysis, and visualization of phylogenetic trees.	http://ete.cgenomics.org/node	[78]
EvolView	Online visualizing, annotating, and managing phylogenetic tree.	http://www.evolgenius.info/evolview.html	[79]
iTol	Web-based tool for the display, manipulation, and annotation of phylogenetic trees.	http://itol.embl.de/	[80]
HyperTree	Visualize and navigate large trees in hyperbolic space.	http://kinase.com/tools/HyperTree.html	[81]
PhyloWidget	Web-based tool for the visualization and manipulation of trees.	http://www.phylowidget.org	[82]
TreeDyn	Tree manipulation with integration of meta-information.	http://www.treedyn.org/	[83]
TreeGraph 2	Combining and visualizing annotation from different phylogenetic analyses.	http://treegraph.bioinfweb.info/	[84]
TopiaryExplorer	Display of large trees with associated metadata.	http://topiaryexplorer.sourceforge.net/	[85]
Treevolution	Highly interactive tree visualizations.	http://carpex.usal.es/~visusal/treevolution/	[86]
jsPhyloSVG	An open-source javascript library specifically built for rendering highly extensible, customizable phylogenetic trees.	http://www.jsphylosvg.com/	[87]
NJplot	Plots trees in PDF and PostScript formats.	http://pbil.univ-lyon1.fr/software/njplot.html	[88]
Phyutility	Not only standard tree operation but also consensus trees, measures of leaf stability, varying attachment locations of lineages, tree support.	https://code.google.com/p/phyutility/	[89]
GenGIS	Phylogenetic trees or cluster relationships can be drawn in two or three dimensions.	http://kiwi.cs.dal.ca/GenGIS/Main_Page	[90]

3.4 APPLICATIONS OF PHYLOGENETIC ANALYSIS

Here, we describe some of the applications of phylogenetic inference in different mycological projects. All these attempts may be applied to food fungi and fungal pathogens. The most widely used application is in taxonomy, but other fields make ample use of molecular phylogenetics.

3.4.1 Species Classification and Identification

Operational species concepts provide a means for species identification and classification. However, some operational species concepts are not applicable to fungi due to their morphological changeability, hidden sexual processes, and limited cultivation. Morphological species recognition has often not enough resolution resulting in grouping several species into one. Biological species classification requires meiospores and is inefficient for homothallic species that produce offspring without outcrossing. When applying the phylogenetic species concept, multiple gene trees should be analyzed in order to establish the species borders.[46,47]

3.4.1.1 Molecular Marker Selection for Species Trees

Molecular marker selection for species trees is one of the key steps influencing the resolution of the phylogenetic tree. However, there is no single and universal gene applicable to all analyses. For taxonomical purposes, the internal transcribed spacer (ITS) was proposed for a primary fungal barcode marker to the Consortium for the Barcode, although additional supplementary barcodes might be considered for specific taxonomic groups.[48] The fungal tree of life was built based on "five to seven loci, including the nuclear ribosomal small and large subunits (nrSSU and nrLSU), the elongation factor 1 alpha (tef1), the largest and the second largest subunits of RNA polymerase II (rpb1 and rpb2), beta-tubulin (tub), and mitochondrial ATP6 (atp6)."[49] These are highly conserved loci with very slow substitution rates that were considered optimal for whole kingdom comparisons. However, the nuclear ribosomal subunits are not variable enough to distinguish closely related taxa.[48]

3.4.2 Gene Phylogeny versus Species Phylogeny

Gene and protein trees enable not only paralogy/orthology inferences, but also HGT detection and duplication/loss patterns descriptions. The fate of a gene family does not need to follow exactly the species evolutionary history. However, multiple gene/protein families should corroborate a robust taxonomy. Some protein families get expanded *via* multiple duplication events during host adaptation to the ecological niche. This phenomenon was observed for carbohydrate and protein degrading enzymes. For example, the expansions of laccases and manganese peroxidases are signs of *Ceriporiopsis subvermispora* adaptation to lignin degradation.[50] Abundant protease families are usually an indication of adaptation toward animal hosts no matter whether dead or alive. Onygenales group both pathogenic (Coccidioides, Blastomyces, Arthroderma) and saprotrophic (Uncinocarpus) fungi often associated with warm blooded animals including those involved in human infections. Similar protease expansions have been described for saprotrophic, skin-infecting, and systemic infection–causing fungi.[51] Furthermore, a very rich and diverse group of ABC transporters has expanded frequently and independently in different lineages of the fungal kingdom. These enzymes are often associated in secondary metabolite clusters leading to increased adaptivity of the host. However, ABC transporter classification is problematic due to uncertain ortholog prediction and the evolutionary pattern consisting of multiple convergent events.[52] These examples are the *peak of an iceberg* of interesting gene phylogenies.

3.4.3 Phylogeography

Phylogeography studies, integrating population genetics and phylogenetics with biogeography, commenced approximately 20 years ago with Avise and colleagues advocating the use of mtDNA in the wider biological context.[53] It has become a separate field based on statistical analysis, with the development of the coalescent theory that enabled model-based parameter estimation and hypothesis testing.[54] The discipline apart from integrating historical and contemporary genetic analysis, uses knowledge from geospatial analysis, geology, climatology, and computer science.

In relation to mycology, phylogeography is an emerging way of studying geographic distributions of lineages.[55] Initially fungi were considered as widely distributed in large unstructured populations. An example of such a ubiquitous organism is *Aspergillus flavus* present in different habitats worldwide. Detailed studies corroborated the initial hypothesis that *everything is everywhere*. All isolates are composed of a single population, with no significant differentiation of marine versus terrestrial isolates, clinical versus environmental isolates, or association with substrate or geographic origin.[56] Nevertheless, the mating-type locus results were more interesting. In environmental isolates, the mating type ratio was 1:1 and in clinical sample the MAT-1 variant constituted 85%, suggesting linkage of the pathogenicity gene. Also, plant pathogenic fungi showed unexpected evolutionary potential for expansion and colonization of new hosts. *Lasiodiplodia theobromae* and *L. pseudotheobromae* isolates were analyzed by Begoude Boyogueno et al.[57] These populations display high levels of diversity, outcrossing, and gene flow even between distant regions and different host plants. Phylogeographic studies of African rusts revealed a high rate of endemic species. There are fewer rust species in this region compared to other lands probably due to environmental conditions. Interestingly, the application of molecular phylogenetics enabled Moncalvo and Buchanan[58] to advocate that *Ganoderma* species probably dispersed after the division of Gondwana to reach current wide and disjoint distribution on the Northern and Southern Hemispheres. Phylogeographic studies of symbiotic fungi lead to different conclusions and either host shift or comigration scenarios are favored. A study focused on Pacific boletes,[59] which are obligate symbionts of different plant families and are found in Australia and Southeast Asia. The analysis of plant symbiotic boletes based on the RPB1 and LSU analysis revealed shifts of symbiotic partners during migration.[59] Cortinarius species in section Calochroi are ectomycorrhizal fungi that occur throughout the geographical range of the host tree or have disjoint intercontinental distributions.[60] The population structure suggests that host distribution is a key factor shaping the current distribution of Cortinarius. Most fungi have not been studied in phylogeography terms, but we expect significant expansion of this field with the availability of next generation sequencing techniques.

3.4.4 FUNGAL TREE OF LIFE

The kingdom represented by the highest number of individual sequenced genomes is that of the fungi. However, the sampling is strongly biased toward the Dikarya/Eumycota and there are only a few genomes of basal fungi. Extremely scarce fossil records make estimation of their time of divergence a true challenge. Furthermore, the fungal tree of life was calculated by using 6 genes from 200 specimens and was proposed in 2006 by James et al.[61] One of the main conclusions from this analysis was the independent loss of the flagellum characteristic for the chytrid-like ancestors that occurred at least four times in fungal evolution. Lucking et al. managed to improve previous inconsistent estimations by applying uniform calibration points. Their results suggest that the origin of fungi took place between 760 million and 1.06 billion years ago and the origin of the Ascomycota was at 500–650 million years ago.[62] Furthermore, Microsporidia are the earliest diverging clade of the sequenced fungi according to recent phylogenomic analyses.[63] They are known for their accelerated evolutionary rates and many phylogenetic artifacts had to be taken into account during this 53 concatenated gene phylogeny. Species-rich taxa can also be difficult to analyze in the light of (over) abundant data. Finally, phylogenetic analysis of multiple genes and supporting morphological information for Cordyceps and other Clavicipitaceae leads to the introduction of a revised classification of this very species-rich taxon.[49]

The biggest challenges, however, are related to deep fungal branches within the so-called Zygomycota. They are a polyphyletic basal group of the fungal kingdom, characterized by their sexual reproduction structures, the zoosporangia, where spores are formed. Currently, their taxonomy is strongly debated and comparative genomics will help to solve this taxonomical puzzle.

3.4.5 Horizontal Gene Transfer, Gene Clusters, and Supernumerary Chromosomes

HGT is one of the major causes of gene tree topologies inconsistencies within genome phylogenies. Indeed, HGT is blamed for disproving the tree of life theory! However, this seems to be a large oversimplification.[64] On the other hand, one of the main applications of phylogenetic analysis is HGT detection. The HGT hypothesis can be additionally supported by genomic context analysis and comparisons of GC content with other regions of the donor and recipient genomes. Genomic context scattered with mobile elements is often a sign of HGT origin. A significant deviation of the local GC content and codon usage from the average calculated for the whole genome of interest suggest its recent evolution might have been shaped by different factors including horizontal acquisition. However, a transfer from organisms with a similar GC content would not leave this kind of fingerprint. Ancient transfers usually get domesticated leading to unified GC content and codon usage. This area of research concentrates mostly in Prokaryota. HGT can be efficiently detected by an automated approach when very distant HGT origin is considered. Marcet-Houben and Gabaldon conducted a massive analysis scanning 60 fungal genomes to detect the acquisition of bacterial 713 genes.[65] All these findings were supported by phylogenetic analyses.

Recently, well-documented cases of HGT in pathogenic fungi brought public attention to eukaryotic HGT.[66] Researchers from the BROAD Institute managed to show the transfer of conditionally dispensable chromosomes between different *Fusarium oxysporum* strains providing adaptation to a new niche, and converting a nonpathogenic strain into a pathogen.[66] Furthermore, conditionally dispensable chromosomes controlling the host specificity have been described for *Alternaria alternata*[67] and *Nectia haematoccocca*.[68] This seems to be a more widespread phenomenon in plant pathogenic fungi.[69]

Cases of HGT have been described for the wine making yeasts,[70] from Saccharomyces and non-Saccharomyces sources. It is expected that host shifts, shared niche, and microbial community interactions are circumstances favoring HGT. Quantitative estimates of HGT in fungi are unknown.

3.5 CONCLUDING REMARKS

For phylogenetics, the main message is to formulate a hypothesis prior to analyses because the key point in all projects will be the dataset selection. Secondly, it is essential to carefully examine any input and output before running calculations. A tiny mistake in the initial steps might result in false, and hard to interpret, results. The main bottleneck in most phylogenetic analyses is the quality of the alignment and the availably of sequence data. However, the latter seems to be to decreasing in significance with the advent of next generation sequencing and related techniques. Phylogeny inference cannot be more reliable than the data on which it is based. Repeating the phylogenetic analysis by changing conditions might save a lot of time spent on interpreting dubious results.

Phylogenetic analyses shed light on the evolutionary history of organisms. Comparative genomics, phylogenomic, and phylogeography reveal how populations of fungi evolved, migrated, and adapted to new ecological niches, including host switching. Parallel analyses of hosts and pathogen species elucidate the complex coevolution and coadaptation patterns. In the following years, more genomes will become available to the scientific community. This will enable detailed and complex comparisons of evolutionary traits specific to fungi.

REFERENCES

1. Hawksworth, D.L. et al. The Amsterdam declaration on fungal nomenclature. *IMA Fungus* 2, 105–112 (2011).
2. Taylor, J.W. One fungus = One name: DNA and fungal nomenclature twenty years after PCR. *IMA Fungus* 2, 113–120 (2011).
3. Hartley, B.S. Evolution of enzyme structure. *Proc R Soc Lond B Biol Sci* 205, 443–452 (1979).
4. Koonin, E.V. Orthologs, paralogs, and evolutionary genomics. *Annu Rev Genet* 39, 309–338 (2005).
5. Gabaldon, T. Large-scale assignment of orthology: Back to phylogenetics? *Genome Biol* 9, 235 (2008).
6. Lechner, M. et al. Proteinortho: Detection of (co-)orthologs in large-scale analysis. *BMC Bioinform* 12, 124 (2011).
7. Ruan, J. et al. TreeFam: 2008 Update. *Nucleic Acids Res* 36, D735–D740 (2008).
8. Powell, S. et al. eggNOG v3.0: Orthologous groups covering 1133 organisms at 41 different taxonomic ranges. *Nucleic Acids Res* 40, D284–D289 (2012).
9. Huerta-Cepas, J. et al. PhylomeDB v3.0: An expanding repository of genome-wide collections of trees, alignments and phylogeny-based orthology and paralogy predictions. *Nucleic Acids Res* 39, D556–D560 (2011).
10. Fischer, S. et al. Using OrthoMCL to assign proteins to OrthoMCL-DB groups or to cluster proteomes into new ortholog groups. *Curr Protoc Bioinform* Chapter 6, Unit 6(12), 1–19 (2011).
11. Zhou, X., Lin, Z., and Ma, H. Phylogenetic detection of numerous gene duplications shared by animals, fungi and plants. *Genome Biol* 11, R38 (2010).
12. Nishihara, H., Maruyama, S., and Okada, N. Retroposon analysis and recent geological data suggest near-simultaneous divergence of the three superorders of mammals. *Proc Natl Acad Sci USA* 106, 5235–5240 (2009).
13. Yang, Z. and Rannala, B. Molecular phylogenetics: Principles and practice. *Nat Rev Genet* 13, 303–314 (2012).
14. Lemey, P., Salemi, M., and Vandamme, A.-M. *The Phylogenetic Handbook: A Practical Approach to Phylogenetic Analysis and Hypothesis Testing* (Cambridge University Press, Cambridge, U.K., 2009).
15. Moretti, S. et al. The M-Coffee web server: A meta-method for computing multiple sequence alignments by combining alternative alignment methods. *Nucleic Acids Res* 35, W645–W648 (2007).
16. Katoh, K. and Toh, H. Parallelization of the MAFFT multiple sequence alignment program. *Bioinformatics* 26, 1899–1900 (2010).
17. Do, C.B., Mahabhashyam, M.S., Brudno, M., and Batzoglou, S. ProbCons: Probabilistic consistency-based multiple sequence alignment. *Genome Res* 15, 330–340 (2005).
18. Edgar, R.C. MUSCLE: A multiple sequence alignment method with reduced time and space complexity. *BMC Bioinform* 5, 113 (2004).
19. Pei, J., Sadreyev, R., and Grishin, N.V. PCMA: Fast and accurate multiple sequence alignment based on profile consistency. *Bioinformatics* 19, 427–428 (2003).
20. Castresana, J. Selection of conserved blocks from multiple alignments for their use in phylogenetic analysis. *Mol Biol Evol* 17, 540–552 (2000).
21. Capella-Gutierrez, S., Silla-Martinez, J.M., and Gabaldon, T. trimAl: A tool for automated alignment trimming in large-scale phylogenetic analyses. *Bioinformatics* 25, 1972–1973 (2009).
22. Moretti, S. et al. gcodeml: A Grid-enabled tool for detecting positive selection in biological evolution. *Stud Health Technol Inform* 175, 59–68 (2012).
23. Jukes, T.H. Evolution of protein molecules. In *Manmmalian Protein Metabolism*, Munro, H.N. ed., pp. 21–132 (Academic Press, New York, 1969).
24. Felsenstein, J. Evolutionary trees from DNA sequences: A maximum likelihood approach. *J Mol Evol* 17, 368–376 (1981).
25. Kimura, M. A simple method for estimating evolutionary rates of base substitutions through comparative studies of nucleotide sequences. *J Mol Evol* 16, 111–120 (1980).
26. Tamura, K. and Nei, M. Estimation of the number of nucleotide substitutions in the control region of mitochondrial DNA in humans and chimpanzees. *Mol Biol Evol* 10, 512–526 (1993).
27. Tavaré, S. Some probabilistic and statistical problems in the analysis of DNA sequences. In *American Mathematical Society: Lectures on Mathematics in the Life Sciences*, Miura, R.M. ed., Vol. 17, pp. 57–86 (American Mathematical Society, Providence, RI, 1986).
28. Henikoff, S. and Henikoff, J.G. Amino acid substitution matrices from protein blocks. *Proc Natl Acad Sci USA* 89, 10915–10919 (1992).

29. Dayhoff, M.O., Schwartz, R.M., and Orcutt, B.C. A model of evolutionary change in proteins. In *Atlas of Protein Sequence and Structure*, Dayhoff, M.O. ed., pp. 345–352 (National Biomedical Research Foundation, Washington, DC, 1978).
30. Jones, D.T., Taylor, W.R., and Thornton, J.M. The rapid generation of mutation data matrices from protein sequences. *Comput Appl Biosci* 8, 275–282 (1992).
31. Yang, Z. PAML 4: Phylogenetic analysis by maximum likelihood. *Mol Biol Evol* 24, 1586–1591 (2007).
32. Le, S.Q., Lartillot, N., and Gascuel, O. Phylogenetic mixture models for proteins. *Philos Trans R Soc Lond B Biol Sci* 363, 3965–3976 (2008).
33. Darriba, D., Taboada, G.L., Doallo, R., and Posada, D. ProtTest 3: Fast selection of best-fit models of protein evolution. *Bioinformatics* 27, 1164–1165 (2011).
34. Keane, T.M., Creevey, C.J., Pentony, M.M., Naughton, T.J., and McLnerney, J.O. Assessment of methods for amino acid matrix selection and their use on empirical data shows that ad hoc assumptions for choice of matrix are not justified. *BMC Evol Biol* 6, 29 (2006).
35. Posada, D. Selection of models of DNA evolution with jModelTest. *Methods Mol Biol* 537, 93–112 (2009).
36. Wilgenbusch, J.C. and Swofford, D. Inferring evolutionary trees with PAUP*. *Curr Protoc Bioinform* Chapter 6, Unit 6, 4 (2003).
37. Kumar, S., Stecher, G., Peterson, D., and Tamura, K. MEGA-CC: Computing core of molecular evolutionary genetics analysis program for automated and iterative data analysis. *Bioinformatics* 28, 2685–2686 (2012).
38. Retief, J.D. Phylogenetic analysis using PHYLIP. *Methods Mol Biol* 132, 243–258 (2000).
39. Goloboff, P.A., Farris, J.S., and Nixon, K.C. TNT, a free program for phylogenetic analysis. *Cladistics* 24, 774–786 (2008).
40. Ronquist, F. et al. MrBayes 3.2: Efficient Bayesian phylogenetic inference and model choice across a large model space. *Syst Biol* 61, 539–542 (2012).
41. Drummond, A.J., Suchard, M.A., Xie, D., and Rambaut, A. Bayesian phylogenetics with BEAUti and the BEAST 1.7. *Mol Biol Evol* 29, 1969–1973 (2012).
42. Stamatakis, A. et al. RAxML-Light: A tool for computing terabyte phylogenies. *Bioinformatics* 28, 2064–2066 (2012).
43. Guindon, S., Delsuc, F., Dufayard, J.F., and Gascuel, O. Estimating maximum likelihood phylogenies with PhyML. *Methods Mol Biol* 537, 113–137 (2009).
44. Zwickl, D.J. Genetic algorithm approaches for the phylogenetic analysis of large biological sequence datasets under the maximum likelihood criterion. PhD dissertation, The University of Texas at Austin (2006).
45. Anisimova, M. and Gascuel, O. Approximate likelihood-ratio test for branches: A fast, accurate, and powerful alternative. *Syst Biol* 55, 539–552 (2006).
46. Taylor, J.W. et al. Phylogenetic species recognition and species concepts in fungi. *Fungal Genet Biol* 31, 21–32 (2000).
47. Walker, D.M., Castlebury, L.A., Rossman, A.Y., and White, J.F., Jr. New molecular markers for fungal phylogenetics: Two genes for species-level systematics in the Sordariomycetes (Ascomycota). *Mol Phylogenet Evol* 64, 500–512 (2012).
48. Schoch, C.L. et al. Nuclear ribosomal internal transcribed spacer (ITS) region as a universal DNA barcode marker for Fungi. *Proc Natl Acad Sci USA* 109, 6241–6246 (2012).
49. Sung, G.H. et al. Phylogenetic classification of *Cordyceps* and the clavicipitaceous fungi. *Stud Mycol* 57, 5–59 (2007).
50. Fernandez-Fueyo, E. et al. Comparative genomics of *Ceriporiopsis subvermispora* and *Phanerochaete chrysosporium* provide insight into selective ligninolysis. *Proc Natl Acad Sci USA* 109, 5458–5463 (2012).
51. Muszewska, A., Taylor, J.W., Szczesny, P., and Grynberg, M. Independent subtilases expansions in fungi associated with animals. *Mol Biol Evol* 28, 3395–3404 (2011).
52. Kovalchuk, A. and Driessen, A.J. Phylogenetic analysis of fungal ABC transporters. *BMC Genomics* 11, 177 (2010).
53. Avise, J.C. et al. Intraspecific phylogeography: The mitochondrial DNA bridge between population genetics and systematics. *Annu Rev Ecol System* 18, 489–522 (1987).
54. Hickerson, M.J. et al. Phylogeography's past, present, and future: 10 Years after Avise, 2000. *Mol Phylogenet Evol* 54, 291–301 (2010).
55. Thorsten Lumbsch, H., Buchanan, P.K., May, T.W., and Mueller, G.M. Phylogeography and biogeography of fungi. *Mycol Res* 112, 423–424 (2008).

56. Ramirez-Camejo, L.A., Zuluaga-Montero, A., Lazaro-Escudero, M., Hernandez-Kendall, V., and Bayman, P. Phylogeography of the cosmopolitan fungus *Aspergillus flavus*: Is everything everywhere? *Fungal Biol* 116, 452–463 (2012).
57. Begoude Boyogueno, A.D., Slippers, B., Perez, G., Wingfield, M.J., and Roux, J. High gene flow and outcrossing within populations of two cryptic fungal pathogens on a native and non-native host in Cameroon. *Fungal Biol* 116, 343–353 (2012).
58. Moncalvo, J.M. and Buchanan, P.K. Molecular evidence for long distance dispersal across the Southern Hemisphere in the Ganoderma applanatum-australe species complex (Basidiomycota). *Mycol Res* 112, 425–436 (2008).
59. Halling, R.E., Osmundson, T.W., and Neves, M.A. Pacific boletes: Implications for biogeographic relationships. *Mycol Res* 112, 437–447 (2008).
60. Garnica, S., Spahn, P., Oertel, B., Ammirati, J., and Oberwinkler, F. Tracking the evolutionary history of *Cortinarius* species in section Calochroi, with transoceanic disjunct distributions. *BMC Evol Biol* 11, 213 (2011).
61. James, T.Y. et al. Reconstructing the early evolution of Fungi using a six-gene phylogeny. *Nature* 443, 818–822 (2006).
62. Lucking, R., Huhndorf, S., Pfister, D.H., Plata, E.R., and Lumbsch, H.T. Fungi evolved right on track. *Mycologia* 101, 810–822 (2009).
63. Capella-Gutierrez, S., Marcet-Houben, M., and Gabaldon, T. Phylogenomics supports microsporidia as the earliest diverging clade of sequenced fungi. *BMC Biol* 10, 47 (2012).
64. Doolittle, W.F. The practice of classification and the theory of evolution, and what the demise of Charles Darwin's tree of life hypothesis means for both of them. *Philos Trans R Soc Lond B Biol Sci* 364, 2221–2228 (2009).
65. Marcet-Houben, M. and Gabaldon, T. Acquisition of prokaryotic genes by fungal genomes. *Trends Genet* 26, 5–8 (2010).
66. Ma, L.J. et al. Comparative genomics reveals mobile pathogenicity chromosomes in Fusarium. *Nature* 464, 367–373 (2010).
67. Hatta, R. et al. A conditionally dispensable chromosome controls host-specific pathogenicity in the fungal plant pathogen *Alternaria alternata*. *Genetics* 161, 59–70 (2002).
68. Rodriguez-Carres, M., White, G., Tsuchiya, D., Taga, M., and VanEtten, H.D. The supernumerary chromosome of Nectria haematococca that carries pea-pathogenicity-related genes also carries a trait for pea rhizosphere competitiveness. *Appl Environ Microbiol* 74, 3849–3856 (2008).
69. Mehrabi, R. et al. Horizontal gene and chromosome transfer in plant pathogenic fungi affecting host range. *FEMS Microbiol Rev* 35, 542–554 (2011).
70. Novo, M. et al. Eukaryote-to-eukaryote gene transfer events revealed by the genome sequence of the wine yeast *Saccharomyces cerevisiae* EC1118. *Proc Natl Acad Sci USA* 106, 16333–16338 (2009).
71. Subramanian, A.R. et al. DIALIGN-TX and multiple protein alignment using secondary structure information at GOBICS. *Nucleic Acids Res* 38, W19–W22 (2010).
72. Chikkagoudar, S., Roshan, U., and Livesay, D. eProbalign: Generation and manipulation of multiple sequence alignments using partition function posterior probabilities. *Nucleic Acids Res* 35, W675–W677 (2007).
73. Becker, E., Cotillard, A., Meyer, V., Madaoui, H., and Guerois, R. HMM-Kalign: A tool for generating sub-optimal HMM alignments. *Bioinformatics* 23, 3095–3097 (2007).
74. Li, K.B. ClustalW-MPI: ClustalW analysis using distributed and parallel computing. *Bioinformatics* 19, 1585–1586 (2003).
75. Han, M.V. and Zmasek, C.M. phyloXML: XML for evolutionary biology and comparative genomics. *BMC Bioinform* 10, 356 (2009).
76. Ramirez-Flandes, S. and Ulloa, O. Bosque: Integrated phylogenetic analysis software. *Bioinformatics* 24, 2539–2541 (2008).
77. Huson, D.H. and Scornavacca, C. Dendroscope 3: An interactive tool for rooted phylogenetic trees and networks. *Syst Biol* 61, 1061–1067 (2012).
78. Huerta-Cepas, J., Dopazo, J., and Gabaldon, T. ETE: A python Environment for Tree Exploration. *BMC Bioinform* 11, 24 (2010).
79. Zhang, H., Gao, S., Lercher, M.J., Hu, S., and Chen, W.H. EvolView, an online tool for visualizing, annotating and managing phylogenetic trees. *Nucleic Acids Res* 40, W569–W572 (2012).
80. Letunic, I. and Bork, P. Interactive Tree Of Life v2: Online annotation and display of phylogenetic trees made easy. *Nucleic Acids Res* 39, W475–W478 (2011).

81. Bingham, J. and Sudarsanam, S. Visualizing large hierarchical clusters in hyperbolic space. *Bioinformatics* 16, 660–661 (2000).
82. Jordan, G.E. and Piel, W.H. PhyloWidget: Web-based visualizations for the tree of life. *Bioinformatics* 24, 1641–1642 (2008).
83. Chevenet, F., Brun, C., Banuls, A.L., Jacq, B., and Christen, R. TreeDyn: Towards dynamic graphics and annotations for analyses of trees. *BMC Bioinform* 7, 439 (2006).
84. Stover, B.C. and Muller, K.F. TreeGraph 2: Combining and visualizing evidence from different phylogenetic analyses. *BMC Bioinform* 11, 7 (2010).
85. Pirrung, M. et al. TopiaryExplorer: Visualizing large phylogenetic trees with environmental metadata. *Bioinformatics* 27, 3067–3069 (2011).
86. Santamaria, R. and Theron, R. Treevolution: Visual analysis of phylogenetic trees. *Bioinformatics* 25, 1970–1971 (2009).
87. Smits, S.A. and Ouverney, C.C. jsPhyloSVG: A javascript library for visualizing interactive and vector-based phylogenetic trees on the web. *PLoS One* 5, e12267 (2010).
88. Perriere, G. and Gouy, M. WWW-query: An on-line retrieval system for biological sequence banks. *Biochimie* 78, 364–369 (1996).
89. Smith, S.A. and Dunn, C.W. Phyutility: A phyloinformatics tool for trees, alignments and molecular data. *Bioinformatics* 24, 715–716 (2008).
90. Parks, D.H. et al. GenGIS: A geospatial information system for genomic data. *Genome Res* 19, 1896–1904 (2009).

4 Fungal DNA Barcoding

*V. Robert, G. Cardinali, B. Stielow, T.D. Vu,
F. Borges dos Santos, W. Meyer, and C. Schoch*

CONTENTS

4.1	Introduction	37
4.2	Barcoding for Identification	39
	4.2.1 ITS Chosen as the Fungal Barcode	40
	4.2.2 Alternative Barcoding Markers	43
	4.2.3 Availability of Reference Sequences	45
4.3	Databases	47
4.4	Future	51
Acknowledgment		52
References		52

4.1 INTRODUCTION

Fungal systematics and the delimitation of species trace their origins back to the time when Linneus and then Persoon and Fries started compiling data on visible fungi. They were using morphological characteristics to describe what they perceived to be species, but the list of observable characters was always very limited, however, these descriptions now fit many modern species. Over the past two centuries, mycologists continued to improve methods for enhanced visual observation, utilizing tools such as more advanced light microscopes (and eventually electron microscopes). Descriptions became more detailed and included both macroscopic and microscopic criteria. However, morphology, mainly focusing on sexual characteristics, remained the ultimate criteria to classify and identify fungi. Like bacteria, microfungi have a limited set of distinctive morphological features. This is especially notable in the single cellular yeasts. Given their importance for the production of bread, wine, and beer, these fungi were among the groups whose biology was intensively studied. Since most yeast are superficially very similar, most distinctions rely on determining differences in growing habit and metabolism. Therefore, a number of yeast taxonomists employed criteria such as the presence of fermentation or assimilation features [1–5]. This was subsequently expanded to an array of various tests employed in standard testing panels used for identification and classification [6–11]. Chemical and biochemical properties were also used to distinguish closely related organisms. Even if such biochemical tests are still commonly used in medical diagnostics with API or Vitek (bioMérieux, France), they do not determine phylogenetic relatedness [12].

The advent of molecular biology allowed the utilization of methods based on genetic variation. Once again yeast taxonomists were among the early adaptors. Electrophoresis was used to predict relationships between strains or closely related species [13]. Random amplification of polymorphic DNA (RAPD) [14,15], restriction fragment length polymorphism (RFLP) [16], and pulse field methods [17], among many others, are still in limited use to distinguish closely related strains on a population level [18,19]. Amplified fragment length polymorphism (AFLP) has been used to enable analysis with larger genome coverage and is quite effective to get a broad view on the global molecular similarities between strains [20,21]. One-dimensional electrophoresis methods are rather inexpensive, easy to implement, and fast. However, such methods do not provide any phylogenetic information that can be used for reliable classifications and subsidiary identifications.

Early attempts to employ molecular techniques to determine relatedness included the determination of the percentages of guanine and cytosine (G + C) content of DNA, described by Price et al. [22], Kurtzman [23], and Tamaoka and Komagata [24]. These were not very useful since the method was only able to exclude that two strains are conspecific if the G + C content differs by more than a given threshold. It was noted that ascomycetous yeasts have a lower G + C content (28%–50%) than basidiomycetes (50%–70%) [22,25] and differences of 1%–2% in G + C content between strains were considered sufficient to distinguish species. An even earlier method was DNA–DNA re-association. This was used to assess the relationship between closely related species. It was developed and applied in the 1970s and used by a number of bacteriologists and yeast taxonomists [23,26–29]. Measurements of DNA complementarity were expressed as a percentage of relatedness, which provided an approximation of overall genome similarity between two organisms [30].

Sanger DNA sequencing and the polymerase chain reaction (PCR) opened up new possibilities. Early pioneers in mycology such as White et al. [31], Kurtzman [32], Fell [33], and Fell et al. [34], as well as many others, employed DNA sequencing to classify and later identify fungal specimens or strains. In fungal taxonomy, the early adopted *loci* were all part of the nuclear rRNA cistron. The internal transcribed spacer region (ITS) consists of two spacer regions flanking the 5.8S nuclear rRNA gene, which get removed in post-translational processes. The second region, commonly used in yeast systematics, is the variable D1/D2 region of the 26S or 28S nuclear rRNA gene, often referred to as the large subunit (LSU). Another region of the nuclear rRNA cistron, the 18S rRNA gene or small subunit (SSU) is also used in fungal systematics. Several other *loci* or protein coding genes are popular with specific groups of taxonomists. These include beta tubulin (*TUB2*), actin (*ACT1*), the largest and second largest subunits of RNA polymerase II (*RPB1, RPB2*), elongation factor 1-alpha (*TEF1*), and a minichromosome maintenance protein (*MCM7*). Mitochondrial genes include cytochrome oxidase subunit 1 (CO1), and 3 (CO3), NADH dehydrogenase subunit 6, and ATPase subunit 6–9 [35]. It should be noted that many of the genes and loci are referred to with different labels depending on the organism they reside in.

Yeast taxonomists such as Cletus Kurtzman and Jack Fell created the first exhaustive set of LSU reference sequences for accepted type strains of all known yeast species [34,36]. This allowed researchers to use one of the three International Nucleotide Sequence Database Collaboration (INSDC; NCBI-GenBank in the United States, EMBL in the United Kingdom, and DDBJ in Japan), or more specialized databases such as the one of the CBS-KNAW collection [37–39], to assign unknown sequences to type strains of yeast species. While the INSDC are generalist databases, also containing archives with any sequences from any origin and quality, other databases such as the one of the UNITE/PlutoF group [40–42] are based on ITS sequences only and they hold a strong selection of specimens associated with ecological studies. At present, the UNITE database contains 6775 ITS sequences of 1977 species from 418 genera (January 2013). Another important specialized database is the Ribosomal Database Project (RDP), which houses the naïve Bayesian rRNA Classifier [43]. This database includes bacterial data, but also hosts a reference set of fungal LSU sequences, which can be queried in order to identify unknown isolates to the species level [44]. The Assembling the Fungal Tree of Life (AFTOL 1) project started in 2002 and was prolonged in 2007 until 2011 (AFTOL 2). The original idea was to significantly enhance the understanding of the evolution of the Fungal Kingdom by improving the backbone phylogeny of all fungi by comparing eight *loci*. A number of collaborative papers using multiple *loci* resulted in broad phylogenetic analyses (e.g., [45–47]). A database and the associated software called WASABI (Web Accessible Sequence Analysis for Biological Inference) [48] and Hal (an automated pipeline for phylogenomic analyses) [49] were released. At the same time, several groups started to use multiple *loci* to produce more reliable phylogenies and this trend continues [50–54].

With the emergence of next-generation sequencing (NGS), the costs associated with sequencing have been drastically reduced, while the speed to retrieve data has been multiplied by several orders of magnitude. The trend is now to sequence genomes, either completely or partially, in order to produce genome-wide phylogenies and gain insights into the potential of sequenced organisms [55]. The 1000 fungal genomes project (Spatafora, http://1000.fungalgenomes.org) and other initiatives [56,57] clearly indicate the tendency for the future. This being said, the large amount of data generated by these projects continues to pose a challenge for nonspecialists, and it seems clear that smaller sequence comparisons will continue to be used in the near future. Similarly, currently available analytical pipelines do not allow to easily and correctly handle the huge amount of sequence data obtained from (e.g., ecological) NGS analyses with known taxa.

Altogether these considerations suggest that identifications, or species assignments, will still be based on a small number of *loci* for a number of applications. Hence, the careful selection of the most effective fungal DNA barcoding region(s) remain a valid objective for the forthcoming years (e.g., [58–60]). In 2011, a group of mycologists created a consortium to compare various loci and concluded that the ITS *locus* was the most appropriate for fungal identification [61]. We will expand on this proposition in the next section.

4.2 BARCODING FOR IDENTIFICATION

The use of molecular data for specimen identification is well established starting before the first published paper mentioning the term *barcode* in association with DNA sequencing [62]. The barcoding analogy was certainly a good one, since it is an easy concept to explain and it initiated a number of major initiatives leading to the elaboration of major national and international projects as well as the rejuvenation of taxonomy to some extent. The original idea behind DNA barcoding was to use a single mitochondrial *locus* (*CO1*) to identify all eukaryotes with a standardized protocol, allowing the identification of any organism using a single method and a limited number of primer sets. This ensured that sequence identity could be compared across all species, as the use of several markers only tied to specific groups and lineages have made this impossible before. Beyond the first experience with *CO1*, a number of standards were proposed in order to increase the ability to correlate sequences to specimens identity. The following information is required [63]:

- Species name (although this can be *ad interim*)
- Voucher data (catalogue number and institution storing)
- Collection record (collector, collection date, and location with GPS coordinates)
- Identifier of the specimen
- *CO1* sequence of at least 500 bp
- PCR primers used to generate the amplicon
- Raw trace files obtained from the sequencer (ABI files usually)

Most of the criteria mentioned earlier are usually part of the information available in taxonomic and classifications revisions. The idea of using vouchered specimens and indicating typification (using a type specimen as representative of a newly described taxon) is central to taxonomy and relies on a complicated set of rules [64]. The stability brought about by this set of standards allows researchers to refer to previously studied material on which to base their revisions and include newly discovered species. Depositing such vouchered material in reference herbaria or culture collections allows subsequent studies to access the original material and use it for further complementary analyses.

However, since the introduction of DNA barcoding, a number of misconceptions have arisen on what it should and should not do. The most important distinction is in understanding the two ways DNA barcoding has been applied: specimen identification and species discovery [65]. The first application relies on an adequate database of reference sequences and well-validated multilocus phylogenies to define species that remains an elusive ideal except for a small number of fungal species. This implies that the second function, species discovery, is the only way to survey the majority of fungal lineages. At best these DNA barcode surveys can be treated as a first approximation of diversity until species can be tested with multiple loci comparisons. Arguments about the validity of DNA barcoding as the tool in either role (but especially in the second) continues.

The functionality of the marker gene plays an important role. In some groups, ITS can identify specimens quite well, while in several groups it can only identify species complexes or genera. This is discussed in more detail later. A number of taxonomists are strongly against ITS since the chosen barcode (*CO1*) is a mitochondrial and not a nuclear gene: it is not able to distinguish some closely related species in many clades and some feel that DNA barcoding is an oversimplification of taxonomy. Some of the criticisms are certainly grounded and one cannot defend the notion that a single gene could do everything from (1) providing the ultimate classification, (2) to identifying any living organism accurately under any circumstances, and (3) providing additional information about the specific properties of the identified specimen or strain. To do all this with minimal effort, at minimal costs, and rapidly for numerous samples.

4.2.1 ITS Chosen as the Fungal Barcode

ITS has been chosen as the primary fungal barcode marker. However, it is interesting to consider how this arose. The first designated barcode locus, *CO1*, was selected on the basis of its efficiency to separate animal species, with a special focus on insects [62]. Ideally, a barcode locus should be universal to all forms of life, including fungi. It should also be noted that DNA barcodes have until now only focused on eukaryotic species, although there is a long tradition in bacteriology of using the 16S ribosomal subunit as a reference sequence for species identification. The use of *CO1* has proven to be problematic in several groups and in 2009 plant taxonomists published a paper where they compared the performance of seven leading candidate plastid DNA regions (*atpF–atpH* spacer, *matK* gene, *rbcL* gene, *rpoB* gene, *rpoC1* gene, *psbK–psbI* spacer, and *trnH–psbA* spacer). On the basis of recoverability, sequence quality, and levels of species discrimination, a two-marker system based on *rbcL* and *matK* has been chosen as an official barcode for plants [66]. However, the use of ITS as a barcode marker has since seen a resurgence [67,68]. Several studies have shown the potential of *CO1*s for fungi and fungal-like organisms [69]. In fact, ITS and *CO1* were selected as barcodes for the stramenopile group, Oomycota, which is unrelated to fungi but often studied by mycologists [70]. In many other groups of fungi, *CO1* has been shown to be of limited use, especially from high numbers of introns complicating PCR amplification [60].

Several papers suggested that ITS could be the best choice for a fungal barcode [58–61] but no complete comparison of ITS with other loci has been attempted. In April 2011, a consortium of approximately 75 experienced mycologists gathered in Amsterdam and decided to create a dataset of 4366 strains/specimens belonging to 1983 fungal species using 12,904 sequences [61] for ITS, LSU, SSU, *RPB1*, *RPB2*, and *MCM7*. Sequence fragments from the latter two genes were included as optional comparisons in this study where the main focus was a universal comparison of the first four. Pairwise alignments of all specimens and strains not belonging to the same species indicated that the number of base pair differences between species was much lower for SSU than for LSU, ITS, and *RPB1*, in increasing order (see Figure 4.1). As expected, many closely related species shared the same sequence with SSU, while the resolution was much higher with ITS and higher still with *RPB1* (Figure 4.1).

When comparing the ratios of within—versus among—species base pair differences for the four *loci*, it was obvious that *RPB1* was the best with about 12 times more chance to have no difference between strains belonging to the same species than for species belonging to the different species (Figure 4.2). ITS was the second best overall *locus* with a ratio just above 5. LSU and SSU do not perform well, with ratios below 1, indicating that identical sequences are extremely frequent for strains belonging to different species. When combining *RPB1* with ITS, ratios improved slightly when no differences were present. The four *loci* together improved the overall differentiation power since the ratio rose to nearly 20 (Figure 4.2).

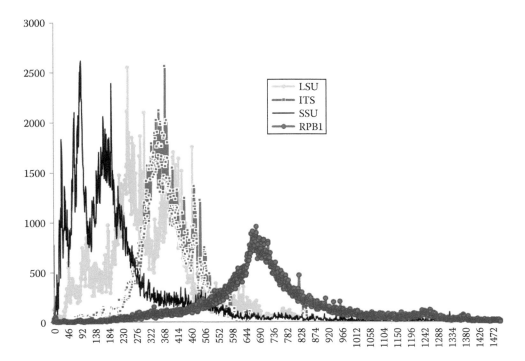

FIGURE 4.1 Number of base pair differences between specimens or strains belonging to different species for four loci. The x axis shows numbers of base pair changes in pairwise comparisons, and the y axis shows numbers of sequence pairs.

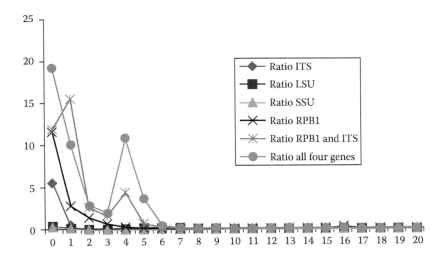

FIGURE 4.2 Ratios of within- versus among-species (y axis) base pair differences for the four loci. The x axis shows the number of base pair differences.

Although *RPB2* performed better than ITS in terms of correct species identification percentages, the latter was selected on the basis of the (1) obvious presence of a barcoding gap (Figure 4.3), (2) higher PCR amplification and sequencing success rates, and (3) ITS sequences being available for a large number of fungal species. Although ITS has been chosen as the primary fungal barcode marker, significant problems remain with high intraspecific sequence variation in some

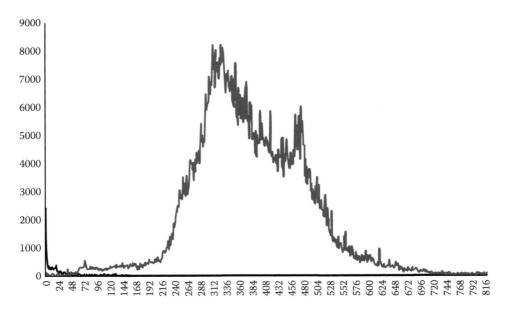

FIGURE 4.3 Barcoding gap of ITS locus. The x axis show numbers of base pair changes in pairwise comparisons, and the y axis shows numbers of sequence pairs. Within-species differences are in black. Among-species differences are in gray.

species [71–78]. Intragenomic ITS variation among the repeated copies of the ribosomal array has been recognized in various fungal species and in a wide range of other organisms, from animals [79–81] to bacteria and flagellates [82,83]. This intraspecific variation is believed to be due to a relaxation of concerted evolution [84–87], an evolutionary effect that homogenizes variation among the rDNA arrays. It has been suggested that concerted evolution interferes through unequal crossing between repeating units, although the exact mechanisms that governs this process remain largely unknown. In fungi, rDNA tandem arrays usually contain ca. 40–200, or even more, copies of the ribosomal cluster, thus allowing for an extended variation within a single specimen [78]. The overall variation in the ITS region at the population level presents significant challenges for phylogenetic analyses and species identification but strongly depends on the overall threshold applied to delimitate species using rDNA sequence similarity, and to a larger extent whether cloned sequence data render an OTU (therefore, a species in the common sense) para- or polyphyletic.

Another issue is the protocol being used to produce ITS sequences. Due to the presence of multiple copies of the same locus within the same genome as discussed earlier, primers and PCR conditions might have important effects on the amplified region. This is especially true when sequencing after cloning or when using direct or shotgun sequencing. The latter can become a serious issue with the emergence of NGS environmental sequencing. Multiple copies of the ITS locus can be amplified and could lead to overestimations of species diversity or to the erroneous presence of new species.

The following protocols are commonly used and can be recommended for PCR amplification and sequencing reactions. In general, the sticky and blunt end cloning systems from Life Technologies (TOPO series), Promega (pGEM series), and Thermo Scientific (Fermentas-Pjet series) have been successfully applied and used recently for cloning of PCR products derived from fungal genomic DNA. Direct sequencing of PCR products (derived via the standard ITS rDNA primers) displaying a high quantity of intraspecific variation usually results in clear overlaid double peaks (ambiguities) that cannot be solved by any molecular biology technique other than cloning, or high coverage using NGS.

Data obtained from 1983 macromycetous species present in the fungal barcoding reference database (http://www.fungalbarcoding.org) show an obvious barcode gap for the ITS locus (Figure 4.3).

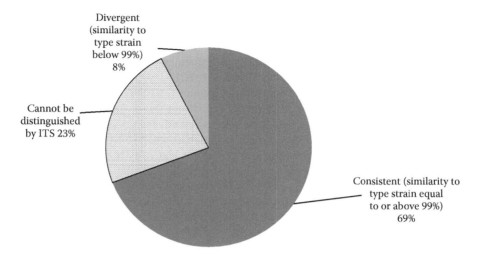

FIGURE 4.4 Statistics based on ITS pairwise distances within 1283 yeasts species.

From another study, not yet published, we analyzed the intraspecific ITS pairwise distances of 3456 strains of the CBS yeast collection (http://www.cbs.knaw.nl) belonging to 1283 species. We found that 888 species (69.2%) had consistent ITS sequences (similarity to type strain equal to or above ca. 99%), and 298 (23.2%) species shared the same ITS sequence with at least one separate species. Finally, 97 species (7.6%) had highly divergent ITS sequences suggesting that splitting of the species might be needed (Figure 4.4).

In some species such as *Candida parapsilosis* or *C. tropicalis*, all strains had exactly the same ITS sequences, while divergence in *C. albicans* increased to almost 30%, as also stated by Schoch et al. [61]. The within-species average similarity is approximately 99.3% for the ITS locus based on 3456 yeast strains from the CBS collection.

4.2.2 Alternative Barcoding Markers

As mentioned before, alternative regions to the ITS have been investigated by a number of researchers and, again, some regions are certainly performing as well as, or better in some clades than ITS. However, the selection of such alternative regions was usually based on rather subjective decisions. When the first large molecular phylogenetic studies were completed, it was obvious that many clades were poorly supported statistically when only one or two genes were used. Recently, many authors explored possibilities for analyzing several genes to obtain the *true phylogeny* [47,50,88–94 and many more]. Some [89] suggested that a few genes (5–20) randomly selected were sufficient to obtain the *true phylogeny*. Others [93] found that excellent classifications and identifications could be obtained using single loci present in most known fungal clades.

In a first approach, Robert et al. [93] tried to find the *Ideal Locus*. Such a *locus* would provide a phylogeny as close as possible to the one putatively expected from the whole genome phylogeny and would distinguish distantly and closely related species. With the *Ideal Locus* method, the authors used all available fungal genomes and classified all protein coding segments in euKaryote Orthologous Groups (KOGs) of proteins. As many of these genes were not found in all species, it was necessary to be selective and keep only those found in all species of interest. At this point, one or several copies of the same KOG and for each species/genome were obtained. Another selection step was introduced to reduce the number of copies to one by keeping the copy that maximized the similarity with the other copies of the same KOG in the other analyzed genomes. Multiple alignments for each KOG were produced using one KOG copy per genome. The resulting multiple alignments were transformed into distance matrices. A reference matrix (i.e., the *true matrix*) was created by concatenating all multiple alignments of the different KOGs to build a large multiple alignment representing the KOGs present in all of the selected genomes. All single-gene distance matrices were compared to the reference matrix using the Pearson's correlation algorithm (Mantel test) and ranked according to how well they fitted with the ideal or reference phylogeny. From there, the best possible genes for phylogenetic analyses on the group of interest were selected, from which potential barcoding candidates were then selected. Results based on 25 genomes and 531 KOGs in common showed that one-third of the genes (29.8%) produced a phylogeny that was highly correlated with the reference matrix. Seventy percent of the gene's matrices had a correlation higher than 0.70 with the reference matrix. Only a few genes (25) showed no or a very low correlation with the reference matrix. Neither the length (Pearson's coefficient of correlation = 0.28) nor the evolutionary rates (Pearson's coefficient of correlation = 0.018) of the KOGs were related to, or could explain, the level of correlation with the reference matrix.

The loci showing the highest level of correlation with the reference matrix are presented in Table 4.1. Their predicted functions were obtained from the National Center for Biotechnology Information (NCBI) website. The best possible KOG (1234) had a correlation of 0.986 with the reference matrix (i.e., the *true phylogeny*). A large number of *loci* showed such high correlation levels with the reference matrix and could therefore be used to reconstruct a robust phylogeny from very diverse taxa. However, for routine identification it was not possible to find *universal primers* that could be used for PCR amplification of phylogenetically diverse groups.

In a second approach, called best pair of primers (BPP) method, Robert et al. [93] mined 77 Fungal genomes to find short conserved regions (coding or not) that could be used as forward and reverse primers. The variable regions between these two primers were analyzed and subsequently

TABLE 4.1
List of the First 10 KOGs Showing the Highest Correlation with the Super Matrix and Therefore Being the Best Candidate Genes for Phylogeny Reconstruction and Potentially for Barcoding

Ranking	KOG Number	Correlation Level	Length	Predicted Function
1	KOG 1234	0.986674	435	ABC (ATP binding cassette) 1 protein
2	KOG 0724	0.981528	435	Zuotin and related molecular chaperones (DnaJ superfamily), contains DNA-binding domains
3	KOG 2472	0.98067	587	Phenylalanyl-tRNA synthetase beta subunit
4	KOG 0714	0.98045	309	Molecular chaperone (DnaJ superfamily)
5	KOG 2002	0.978686	513	TPR-containing nuclear phosphoprotein that regulates K(+) uptake
6	KOG 3844	0.977597	366	Predicted component of NuA3 histone acetyltransferase complex
7	KOG 2369	0.976687	449	Lecithin:cholesterol acyltransferase (LCAT)/Acyl-ceramide synthase
8	KOG 0363	0.976442	523	Chaperonin complex component, TCP-1 beta subunit (CCT2)
9	KOG 0362	0.976349	522	Chaperonin complex component, TCP-1 theta subunit (CCT8)
10	KOG 1450	0.976006	382	Predicted Rho GTPase-activating protein

Source: ftp://ftp.ncbi.nih.gov/pub/COG/KOG/kog.

TABLE 4.2
Some of the Optimal Primer Pairs and Their Location/Function

Primer Pair	Location/Function
acaagcgtttct <> catcaagttcca	Unknown function
acatggagaaga <> catcaaggagaa	Actin gene
accttcttgatg <> catgttcttgat	Elongation factor 1-alpha
agtacttgtagg <> cttggccttgta	60S ribosomal protein L15
ggaacttgatgg <> agaaacgcttgt	Unknown function
ggtatcaccatc <> caacaagatgga	Elongation factor 1-alpha
gtccatcttgtt <> gatggtgatacc	Elongation factor 1-alpha
gtccatcttgtt <> tacttgaaggaa	Elongation factor 1-alpha
gttcttggagtc <> gtccatcttgtt	Elongation factor 1-alpha
aacaagatggac <> ctccaagaacga	Elongation factor 1-alpha

the ability of the amplified regions to be used as reliable phylogenetic representatives and/or as potential barcode candidates for identification was assessed. A number of interesting loci were found and are listed in Table 4.2. More candidates are currently under investigations and seem to be extremely promising as secondary or even primary barcodes for fungi.

4.2.3 Availability of Reference Sequences

As explained previously, while ITS is the currently accepted DNA barcode locus for fungi and many reference databases have been created to perform online identifications on the basis of this locus, a large number of described species are still not characterized molecularly. Between 2006 and 2010, only 28.6% of newly described fungal species were characterized by at least one sequence (Figure 4.5). So, almost three quarters of the known species were not included in sequence reference databases used for identifications. From Figure 4.5, it can be seen that ITS

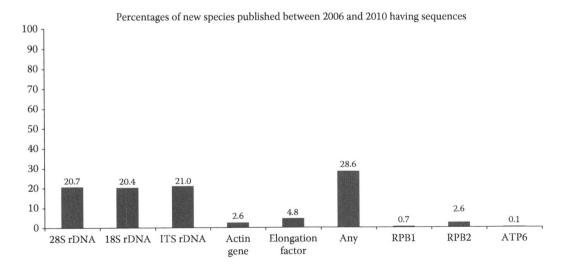

FIGURE 4.5 Percentages of new species of fungi published between 2006 and 2010 having associated sequences for the following loci: 28S rDNA, 18S rDNA, ITS rDNA, *Actin, Elongation factor 1-alpha, RPB1, RPB2, ATP6* and others.

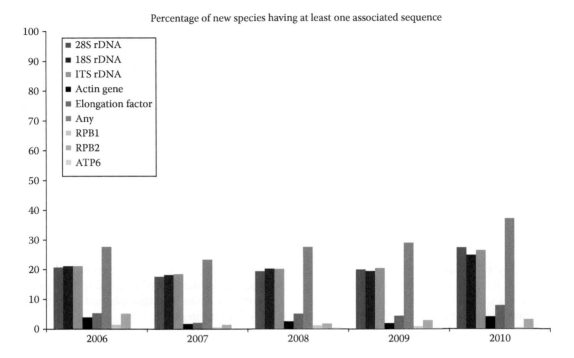

FIGURE 4.6 Evolution of the number of sequences deposited in INSDC for new species of fungi.

is the most commonly sequenced region with 21% of the newly described species, followed by 28S (20.7%), 18S (20.4%), *Elongation factor 1-alpha* (4.8%), *Actin* (2.65%), *RPB2* (2.64%), *RPB1* (0.74%), and *ATP6* (0.09%).

Even worse, only ca. 20%–25% of the newly described fungal species between 2006 and 2010 had associated ITS sequences (Figure 4.6). During this period, there was a slight increase in the total number of sequences deposited in INSD from 27.5% in 2006 to 36.9% in 2010. At the same time, fungal ITS sequences deposited rose from 20.9% to 26.2%.

These numbers clearly indicate the magnitude of the problem and the efforts remaining to be done in order to provide reliable identifications. Of course, some fungal groups like medical fungi or industrially important species and associated clades have been more studied and there are several good and complete databases. The problem remains nevertheless serious for a large number of applications, especially those related to ecological studies.

One of the possible solutions to this issue is to enforce deposition of sequence data in INSD when publishing new fungal species. In 2011, the International Code of Nomenclature for Algae, Fungi, and Plants has been changed and it is now mandatory to deposit every new species in one of the three recognized online repositories such as MycoBank (http://www.mycobank.org). Such tools can be used to easily track new species that do not have any associated sequences. Furthermore, The International Commission on the Taxonomy of Fungi (ICTF, http://www.fungaltaxonomy.org/) could suggest that the authors not having sequenced their type material do so when possible. Some microscopic and noncultivable material might be difficult to sequence and serious hurdles can be encountered in some groups, but everything should be done to provide reference sequences. When financial or technical resources are an issue, mechanisms should be implemented to send type material to culture collections, herbaria or institutions such as the BOLD (Barcode of Life Database) system (http://www.barcodinglife.com/) that are technically and financially able to sequence large numbers of specimens.

4.3 DATABASES

A number of databases have been created over the years to store and publicize DNA based data. The largest and most commonly used are the sets of databases for nucleotide and protein data created by the NCBI at the National Institutes of Health (NIH) funded by the U.S. government and called GenBank (http://www.ncbi.nlm.nih.gov/genbank). This is the default central repository for all DNA sequences and most journals require authors preparing manuscripts to submit their sequences to GenBank. The three INSD repositories (EMBL, GenBank, and DDBJ) exchange all of their new data releases, sequence, and annotation updates on a daily basis. The flat file formats are different: in particular, EMBL has a very different display format than GenBank and the DDBJ—but the underlying sequence data and feature table annotation should be identical between all of the three of the INSD collaboration partners. So, alignments done on one of the three INSD repositories should deliver the same results, in principle at least. This being said, while BLAST implementations are present on all repositories, the associated default parameters used in the different web portals can sometimes be quite different. Hence, the outcome of a pairwise sequence alignment can differ quite significantly between the three websites. Also, even if sequences can be part of the repository, some sequences are not part of the reference databases used for pairwise sequence alignments, complicating the interpretations of the end-users.

The major advantages of INSD reference databases are

- Largest reference database
- All types of organisms included (microbes, plants, animals, etc.)
- Any *locus* can be deposited
- Genome data present
- Meta genomic data can be deposited

Despite attempts by GenBank and its sister databases to verify submitted data, a number of issues are however present:

- Data are archived and only partially curated. It means that INSD curators do not change the original deposited data. They can only contact the original submitters and ask them to update their records. However, they will exclude records from the blast interface if no reply is forthcoming in cases where they are aware of erroneous names. All records will still be archived.
- Taxonomic names are not always updated to reflect the latest changes in literature due to the large number of names submitted. Taxonomists do verify most new names at the time of submission.
- Many records may contain errors due to the incorrect taxonomy submitted by the user.
- Type strain information may not be clearly indicated.
- Many records have no associated, or only vaguely defined, taxon names.

When trying to identify unknown sequences using one of the INSD databases, one should be cautious regarding the proposed taxon names. However, some common sense precautions can help. One should consider the depositor's profile and always try to trace back the origin of the strain or specimen from which the sequence was obtained. Also, see if multiple deposits from different authors agree. If original material is stored in reference collections or museums, there might be a link to a specific website where the current name is maintained. Where possible, type specimens or strain sequences should be used as references since these are the only acceptable species anchors. All other sequences might be incorrect or slightly different from the central and type sequence. This remark is of course valid for any of the reference databases and ideally, only type material, or very closely related curated sequences should be used for identification. Nilsson et al. [95] attempted to simplify the submission process especially focusing on ITS sequences by indicating common errors committed by submitters at public databases. INSD contain more than 170,000 ITS sequences, with more than half being associated with Latin binomials, representing more than 15,500 species and 2,500 genera [61]. Multiple or batch submissions are possible.

The RefSeq database (http://www.ncbi.nlm.nih.gov/refseq/) is a sister database at NCBI that consists of curated sequences from GenBank containing integrated, nonredundant, and well-annotated sets of reference sequences. A smaller set of LSU (28S) and ITS sequences from type strains and AFTOL were selected from GenBank and re-annotated and verified with additional information where possible. It currently contains more than 200 ex-type ITS sequences and is being expanded as new ex-type sequences are released in GenBank. This database will eventually have a customized BLAST interface.

The BOLD is focused on barcode data (http://www.barcodinglife.com). However, the identification of fungi based on ITS sequences is not working adequately since an extremely limited number of fungal ITS sequences is present in the database. Most of the data are related to the animal or plant kingdoms. This is likely to evolve in the future since a number of fungal institutions such as CBS-KNAW are dedicated to produce large numbers of fungal ITS sequences in the near future and the latter will be submitted to BOLD and GenBank.

The AFTOL database (http://aftol.org and http://wasabi.lutzonilab.net/pub/blast/publicblastSequence) contains approximately 3800 fungal species and more than 2000 ITS sequences from verified species. Identifications are rapid and the database has been well curated. However, it remains uncertain whether this database will be updated as the project is finished.

The naive Bayesian classifier hosted at RDP (http://rdp.cme.msu.edu/classifier/classifier.jsp) uses a training set of well-identified sequences. It currently only uses LSU but an expansion is planned for ITS sequences.

The UNITE database for molecular identification of fungi (http://unite.ut.ee) was conceived by a North European team of researchers (working in Sweden, Denmark, Estonia, Norway, Finland, Germany, etc.) with a common interest in mycorrhizal mycology. It was built as a response to the difficulties encountered when identifying environmental samples of fungi to species level using molecular data with INSD [42]. UNITE provides a curated ITS-only database that is targeted toward the identification of environmental samples. It has a strong representation of basidiomycetes, but is expanding to ascomycetes and other fungi. In total, 1077 species and 6816 ITS sequences are available in this reference database (January 2013). The system provides an efficient user interface allowing to Blast sequences against UNITE only or UNITE + INSD sequences. Sequences present in UNITE only are restricted to the ecological arena, allowing for a faster identification of OTUs for focused studies. Medically and industrially important species are poorly represented in this online database. In principle, the system allows batch submission of sequences under the Fasta format.

The Fungal Barcoding database (http://www.fungalbarcoding.org) was created for the community paper written by Schoch et al. [61]. It has been further updated and now contains 4,366 strains of 1,999 species and 12,904 sequences (including 2,136 ITS sequences). The taxonomic diversity is high and includes representatives of most fungal lineages. It is a well-curated database, but it remains uncertain whether it will be further maintained or developed in the future. This database will likely be merged with the upcoming EUBOLD database (European mirror of the BOLD system) and data are likely to be uploaded to BOLD central. Batch sequence submissions are not yet possible. Polyphasic or multilocus identifications are possible and constitute a major advantage over previously cited systems when identifying closely related organisms requiring more than one locus to be distinguished.

MycoBank (http://www.mycobank.org), another major source of information in fungal taxonomy, contains only a few sequences, but it can be used as central hub for fungal species identifications since it allows aligning sequences remotely and centrally against a large number of reference databases. In fact, most of the reference databases cited in Table 4.3 can be used for remote pairwise

TABLE 4.3
List of Websites and Databases Where Pairwise Alignments or Polyphasic Identifications Are Possible

Organisms	Address	Type of Data	Number of ITS Sequences	Number of Fungal Species	Curated	Metagenomics Ready	Pairwise Alignments	Polyphasic IDs	Multiple Submissions
AFTOL/Wasabi	http://aftol.org and http://wasabi.lutzonilab.net/pub/blast/publicblastSequence	D	ca 5200	ca 2,100	Yes	No	Yes	Yes	Yes/100 max
Aspergillus and Penicillium	http://www.cbs.knaw.nl/aspergillus	DMS	3239	341	Partly	No	Yes	Yes	No
BOLD Systems	http://www.barcodinglife.com	DEMS	5302 mainly COI	650	Partly	No	Yes	No	Yes with registration
CBS collection	http://www.cbs.knaw.nl/collections	DEMS	36682	5,180	Yes	No	Yes	Yes	No
Ceratocystis	http://www.q-bank.eu/fungi	DEMS	94	30	Yes	No	Yes	Yes	No
Colletotrichum	http://www.q-bank.eu/fungi	DEMS	680	107	Yes	No	Yes	Yes	No
Dermatophytes	http://www.cbs.knaw.nl/dermatophytes	DEMS	638	49	Yes	No	Yes	Yes	No
Fungal barcoding	http://www.fungalbarcoding.org	DMS	2136	1,999	Yes	No	Yes	Yes	No
Fusarium ID	http://isolate.fusariumdb.org/	DS	5559	76	Yes	No	Yes	No	Yes
Fusarium MLST	http://www.cbs.knaw.nl/fusarium	DS	5559	76	Yes	No	Yes	Yes	No
GenBank NCBI	http://www.ncbi.nlm.nih.gov/genbank	DS	170000	15,500	No	Yes	No	No	Yes
Indoor fungi	http://www.cbs.knaw.nl/indoor	DS	1646	364	Yes	No	Yes	Yes	No
Medical fungi	http://www.cbs.knaw.nl/medical	DS	2231	315	Yes	No	Yes	Yes	No
Medical fungi ITS ID	http://www.mycologylab.org/	DS	2235	380	Yes	No	Yes	Yes	No
Medical fungi MLST	http://MLST.mycologylab.org/	DS	624	4	Yes	No	Yes	Yes	No
MLST imperial college	http://www.mlst.net	D	0	4	Yes	No	Yes	No	No
Monilinia	http://www.q-bank.eu/fungi	DEMS	52	7	Yes	No	Yes	Yes	No
Morchella	http://www.cbs.knaw.nl/morchella	DS	279	21	Yes	No	Yes	Yes	No
Mycosphaerella	http://www.cbs.knaw.nl/mycosphaerella	DMS	2051	4,373	Yes	No	Yes	Yes	No

(Continued)

TABLE 4.3 (Continued)
List of Websites and Databases Where Pairwise Alignments or Polyphasic Identifications Are Possible

Organisms	Address	Type of Data	Number of ITS Sequences	Number of Fungal Species	Curated	Metagenomics Ready	Pairwise Alignments	Polyphasic IDs	Multiple Submissions
MycoBank	http://www.mycobank.org	S	Hub	466,063 names	Yes	No	Yes	No	No
Penicillium subgenus *Penicillium*	http://www.cbs.knaw.nl/penicillium	MS	156	58	Yes	No	Yes	Yes	No
Phaeoacremonium	http://www.cbs.knaw.nl/phaeoacremonium	DMS	See CBS collections	28	Yes	No	Yes	Yes	No
Phoma	http://www.q-bank.eu/fungi	DEMS	149	95	Yes	No	Yes	Yes	No
Phytophthora	http://www.q-bank.eu/fungi	DEMS	2183	751	Yes	No	Yes	Yes	No
Qiime	http://qiime.org	D	?	?	Yes	Yes	Yes	Yes	Yes
Resupinate russulales	http://www.cbs.knaw.nl/russulales	MS	161	404	Yes	No	Yes	Yes	No
Russula	http://www.cbs.knaw.nl/russula	DMS	0	490	Yes	No	Yes	Yes	No
SILVA	http://www.arb-silva.de/	D	3194778 (SSU), 288717 (LSU), 0 (ITS) LSU	?	Yes	No	Yes	Yes	Yes
Stenocarpella	http://www.q-bank.eu/fungi	DEMS	32	6	Yes	No	Yes	Yes	No
Unite	http://unite.ut.ee	DSE	6816	1,977	Yes	No	Yes	Yes	Not yet working
Yeasts	http://www.cbs.knaw.nl/Collections/BiolomicsID.aspx?IdentScenario=Yeast2011ID	DEMS	2082	1,769	Yes	No	Yes	Yes	No

Type of data: D, DNA/molecular; E, ecological and source data; M, morphology and biological characteristics; S, species or specimen metadata.

sequence alignments from the central MycoBank hub. Results obtained from several websites are combined and ordered according to a decreasing rating index, summarizing the similarity between sequences, their overlap, the number of fragments, and the probabilistic score. At this stage, only single-sequence submissions are possible but this is likely to evolve and batch alignments should be available soon.

Another major, curated, and generalist source of ITS and other sequences is the CBS-KNAW culture collection (http://www.cbs.knaw.nl/collections), which is the largest and most diverse public fungal collection in the world. Almost all of the 75,000 living strains will be sequenced by the end of 2014 for the ITS and 26S (LSU) loci, making it one of most reliable and information-rich fungal reference databases. CBS-KNAW provides a pairwise sequence alignment hub facility similar to the one proposed by MycoBank.

As a result of a lack of reliable ITS sequence data for human and animal pathogenic fungi, the International Society of Human and Animal Mycology (ISHAM) working group for DNA barcoding has started an initiative to establish a well-curated quality-controlled ITS database for human and animal pathogenic fungi. This database is a result of a global collaboration of 11 leading medical mycology research institutes and currently contains 2195 ITS sequences corresponding to 378 species, 109 genera, and 54 families. The database can be searched at http://www.mycologylab.org/.

In addition to these central pairwise sequence alignments facilities, a number of specialized websites have also been developed by fungal taxonomists that provide group specific databases (i.e., *Ceratocystis*, *Colletotrichum*, Dermatophytes, *Fusarium* ID, indoor fungi, medical fungi, *Monilinia*, *Morchella*, *Mycosphaerella*, *Penicillium* subgenus *Penicillium*, *Phaeoacremonium*, *Phoma*, *Phytophthora*, *Resupinate russulales*, *Stenocarpella*, and yeasts). A selected list with their characteristics is provided in Table 4.3. These websites usually allow pairwise sequence alignments against curated lists of species, strains, and sequences but also allow for polyphasic identifications including a combination of morphological, physiological, and/or molecular criteria. As briefly discussed earlier, the ability to use more than one character (e.g., ITS, LSU, and some morphological data) at the same time for identification is a major advantage for the reliability of the identification.

Some very specialized databases and associated websites allow for the identification at the strain or sequence-type level. For example, the medical fungi MLST database of the University of Sydney (http://MLST.mycologylab.org/) or the MLST website of the Imperial College in London (http://mlst.net) allow the inclusion of several sequences from different very specific loci to perform strain sequence typing. Such tools are used frequently in epidemiological studies, where tracing the spread of diseases is important.

4.4 FUTURE

DNA barcoding and sequencing of specific loci were major steps in biology with a large number of applications in food quality control, biomonitoring, ecological studies, medical diagnostics, industrial microbiology, etc. Although very important, these are still small steps compared to metabolomics, functional genomics, and metagenomics [96,97]. Second- and third-generation sequencing methods have been developed recently and are likely to quickly replace Sanger or first-generation sequencing. This is exemplified by the rates at which these newly generated data are deposited at the NCBI sequence archive (http://www.ncbi.nlm.nih.gov/Traces/sra/). Recently, papers have been published on the use of second- and third-generation sequencing [98–101]. Such studies and many others demonstrate the huge potential of NGS methods in terms of new species discovery, environment monitoring, ecological studies, etc. They are evolving so fast that whatever is written today about them is almost obsolete tomorrow. Read lengths are still quite short for most existing systems ranging between 50 and 800 bp. There is no doubt that NGS will become one of the major tools for species identification, even for routine diagnostic when read lengths and accuracy improve, and the costs decrease significantly. With metagenomics studies using NGS, the following problems have to be carefully addressed before general adoption:

- Prices can remain high unless multiplexing.
- Short reads are problematic since they might not be sufficiently specific or informative.
- Accuracy of some NGS methods can be quite low—a serious problem for correct species or function identifications.
- Polyphasic or multiple locus sequencing on metagenomes samples is virtually impossible since DNA fragments belonging to different loci cannot be connected to each other and associated with a given individual strain or specimen. This is especially true when the different loci are located on different chromosomes. At this stage, it is unlikely that this will be solved by current and foreseeable technologies. This means that increases in read length and accuracy are much needed to obtain correct identifications.
- Short read lengths and low accuracy are creating huge bioinformatics challenges for the assembly in longer contigs that might carry more information and allow for better identifications. The risk of making chimeric sequences is very important in such cases.
- Multiple copy loci such as the ITS (official fungal barcode) will induce the amplification of all or most of the other nonidentical copies suggesting higher levels of diversity than really present. The choice of single-copy genes as DNA barcode would help avoid this problem.
- Although there are some bioinformatics pipelines capable of handling metagenomics datasets (GreenGenes, QIIME; [102–104]), however, none is capable of analyzing large amounts of samples in a way that would be fast enough for routine and high-throughput diagnostics. While it is straightforward to align single sequences against a relatively large reference database, it is quite another challenge to compare millions of sequences from a single sample against constantly updated reference databases.
- Database storage technologies have to be upgraded to allow gigantic datasets to be properly stored. Improved data compression technologies will have to be implemented to limit database growth.
- Much faster heuristic and efficient algorithms for identification will have to be developed in order to increase the speed of pairwise alignments.

Even with the aforementioned and current limitations, metagenomics using NGS methods remains an attractive technology and could induce significant cost reductions for diagnostics laboratories since a single method would potentially replace a large number of previously used technologies. Almost identical methods could be used to analyze samples containing different organisms (virus, bacteria, fungi, animals, plants, etc.).

ACKNOWLEDGMENT

The work of Felipe Borges dos Santos was supported by a CNPq scholarship, Brazil.

REFERENCES

1. Beijerinck, W.M. 1889. L'auxanographie, ou la méthode de l'hydrodiffusion dans la gelatine appliquée aux recherches biologiques. *Arch. Neerl. Sc. Ex. Nat.* 23: 367–372.
2. Hansen, E.C. 1888. Recherches sur la physiologie et la morphologie des ferments alcooliques. VII. Action des ferments alcooliques sur les diverses espèces de sucre. *Compt. Rend. Trav. Lab. Carlsberg* 2: 143–192.
3. Hansen, E.C. 1891. Recherches sur la physiologie et la morphologie des ferments alcooliques. *Compt. Rend. Trav. Lab. Carlsberg* 3: 44–66.

4. Hansen, E.C. 1898. Recherches sur la physiologie et la morphologie des ferments alcooliques. IX. Sur la vitalité des ferments alcooliques et leur variation dans les milieux nutritifs et a l'état sec. *Compt. Rend. Trav. Lab. Carlsberg* 4: 93–121.
5. Hansen, E.C. 1902. Recherches comparatives sur les conditions de la croissance végétative et le développement des organes de reproduction des levures et des moisissures de la fermentation alcoolique. *Compt. Rend. Trav. Lab. Carlsberg* 5: 68–107.
6. Lodder, J. 1934. Die anaskosporogenen Hefen, I. Halfte. *Verh. K. Ned. Akad. Wet., Afd. Natuurk.* Section II, 32: 1–256.
7. Lodder, J. and N.J.W. Kreger-van Rij. 1952. *The Yeasts: A Taxonomic Study*. North-Holland, Amsterdam, the Netherlands.
8. Lodder, J. 1970. *The Yeasts: A Taxonomic Study*, 2nd edn. North-Holland, Amsterdam, the Netherlands.
9. Kreger-Van Rij, N.J.W. 1984. *The Yeasts: A Taxonomic Study*, 3rd edn. Elsevier Scientific, Amsterdam, the Netherlands.
10. Wickerham, L.J. 1951. *Taxonomy of Yeasts*. United States Department of Agriculture, Technical Bulletin No. 1029, Washington, DC.
11. Wickerham, L.J. and K.A. Burton. 1948. Carbon assimilation tests for the classification of yeasts. *J. Bacteriol.* 56: 363–371.
12. Rohm, H., F. Lechner, and M. Lehner. 1990. Evaluation of the API ATB 32C system for the rapid identification of foodborne yeasts. *Int. J. Food Microbiol.* 11: 215–224.
13. Meyer, W. et al. 1997. Identification of pathogenic yeasts of the imperfect genus *Candida* by PCR-fingerprinting. *Electrophoresis* 18(9): 1548–1559.
14. Welsh, J. and M. McClelland. 1990. Fingerprinting genomes using PCR with arbitrary primers. *Nucl. Acids Res.* 18: 7213–7218.
15. Williams, J.G.K. et al. 1990. DNA polymorphisms amplified by arbitrary primers are useful as genetic markers. *Nucl. Acids Res.* 18: 6531–6535.
16. Williams, D.W. et al. 1995. Identification of Candida species by PCR and restriction fragment length polymorphism analysis of intergenic spacer regions of ribosomal DNA. *J. Clin. Microbiol.* 33(9): 2476–2479.
17. Burmeister, M. and L. Ulanovsky. 1992. Pulsed-field gel electrophoresis. *Methods in Molecular Biology* 12, Humana Press, Totowa, NJ, pp. 481.
18. Meyer, W. et al. 2003. Molecular typing of IberoAmerican *Cryptococcus* neoformans isolates. *J. Emerg. Infect. Dis.* 9(2): 189–195.
19. Meyer, W. et al. 1999. Molecular typing of global isolates of *Cryptococcus* neoformans var. neoformans by PCR-Fingerprinting and RAPD—A pilot study to standardize techniques on which to base a detailed epidemiological survey. *Electrophoresis* 20(8): 1790–1799.
20. Vos, P. et al. 1995. AFLP: A new technique for DNA fingerprinting. *Nucl. Acids Res.* 23: 4407–4414.
21. Boekhout, T. et al. 2001. Hybrid genotypes in the pathogenic yeast *Cryptococcus* neoformans. *Microbiology* 147: 891–907.
22. Price, C.W., G.B. Fuson, and H.J. Phaff. 1978. Genome comparison in yeast systematics: Delimitation of species within the genera *Schwanniomyces*, *Saccharomyces*, *Debaryomyces* and *Pichia*. *Microbiol. Rev.* 42(1): 161–193.
23. Kurtzman, C.P. 1984. Resolution of varietal relationships within the species *Hansenula anomala*, *Hansenula bimundalis* and *Pichia nakazawae* through comparisons of DNA relatedness. *Mycotaxon* 19: 271–279.
24. Tamaoka, M. and K. Komagata. 1984. Determination of DNA base composition by reversed phase high-performance liquid chromatography. *FEMS Microbiol. Lett.* 25: 125–128.
25. Nakase, T. and K. Komagata. 1968. Taxonomic significance of base composition of yeast DNA. *J. Gen. Appl. Microbiol.* 14: 345–357.
26. Seidler, R.J. and M. Mandel. 1971. Quantitative aspects of deoxyribonucleic acid renaturation: Base composition, state of chromosome replication, and polynucleotide homologies. *J. Bacteriol.* 106(2): 608–614.
27. Martini, A. and H.J. Phaff. 1973. The optical determination of DNA–DNA homologies in yeasts. *Ann. Microbiol.* 23: 59–68.
28. Meyer, S.A. et al. 1975. Physiological and DNA characterization of *Candida maltosa*, a hydrocarbon-utilizing yeast. *Arch. Microbiol.* 104(3): 225–231.
29. Kurtzman, C.P., M.J. Smiley, and C.J. Johnson. 1980. Emendation of the genus *Issatchenkia Kudriavzev* and comparison of species by deoxyribonucleic acid reassociation, mating reaction, and ascospore ultrastructure. *Int. J. Syst. Bacteriol.* 30(2): 503–513.

30. Kurtzman, C.P. 2011. Recognition of yeast species from gene sequence comparisons. *Open Appl. Inform. J.* 5(Suppl. 1-M4): 20–29.
31. White, T.J., T. Bruns, S. Lee, and J.W. Taylor. 1990. Amplification and direct sequencing of fungal ribosomal RNA genes for phylogenetics. In *PCR Protocols: A Guide to Methods and Applications*, Innis, M.A., D.H. Gelfand, J.J. Sninsky, and T.J. White (eds.). Academic Press, Inc., New York, pp. 315–322.
32. Kurtzman, C.P. 1993. DNA–DNA hybridization approaches to species identification in small genome organisms. In *Methods in Enzymology*, Vol. 224, Zimmer, E.A., T.J. White, R.L. Cann, and A.C. Wilson (eds.). Academic Press, Inc., New York, pp. 335–348.
33. Fell, J.W. 1995. rDNA targeted oligonucleotide primers for the identification of pathogenic yeasts in a polymerase chain reaction. *J. Ind. Microbiol.* 14(6): 475–477.
34. Fell, J.W. et al. 2000. Biodiversity and systematics of basidiomycetous yeasts as determined by large subunit rD1/D2 domain sequence analysis. *Int. J. Syst. Evol. Microbiol.* 50: 1351–1371.
35. Kretzer, A.M. and T.D. Bruns. 1999. Use of atp6 in fungal phylogenetics: An example from the Boletales. *Mol. Phylogen. Evol.* 3: 483–492.
36. Kurtzman, C.P. and C.J. Robnett. 1998. Identification and phylogeny of ascomycetous yeasts from analysis of nuclear large subunit (26S) ribosomal DNA partial sequences. *Antonie Van Leeuwenhoek* 73(4): 331–371.
37. Robert, V. 2000. BioloMICS web software for online publication and polyphasic identification of biological data. Version 1, BioAware, Hannut, Belgium, Germany.
38. Robert, V., P. Evrard, and G.L. Hennebert. 1997. BCCM/Allev 2.00 an automated system for the identification of yeasts. *Mycotaxon* 64: 455–463.
39. Robert, V., J.-E. de Bien, and G.L. Hennebert. 1994. ALLEV, a new program for computer-assisted identification of yeasts. *Taxon* 43: 433–439.
40. Abarenkov, K. et al. 2010. PlutoF—A web based workbench for ecological and taxonomic research, with an online implementation for fungal ITS sequences. *Evol. Bioinform.* 6: 189–196.
41. Kõljalg, U. et al. 2005. UNITE: A database providing web-based methods for the molecular identification of ectomycorrhizal fungi. *New Phytol.* 166: 1063–1068.
42. Nilsson, R.H. et al. 2011. Molecular identification of fungi: Rationale, philosophical concerns, and the UNITE database. *Open Appl. Inform. J.* 5(Suppl. 1-M9): 81–86.
43. Wang, Q., G.M. Garrity, J.M. Tiedje, and J.R. Cole. 2007. Naive Bayesian classifier for rapid assignment of rRNA sequences into the new bacterial taxonomy. *Appl. Environ. Microbiol.* 73: 5261–5267.
44. Liu, K.L., A. Porras-Alfaro, C.R. Kuske, S.A. Eichorst, and G. Xie. 2012. Accurate, rapid taxonomic classification of fungal large-subunit rRNA genes. *Appl. Environ. Microbiol.* 78: 1523–1533.
45. Schoch, C.L. et al. 2009. The Ascomycota tree of life: A phylum wide phylogeny clarifies the origin and evolution of fundamental reproductive and ecological traits. *Syst. Biol.* 58: 224–239.
46. James, T.Y. et al. 2006. Reconstructing the early evolution of the fungi using a six gene phylogeny. *Nature* 443: 818–822.
47. Lutzoni, F. et al. 2004. Assembling the fungal tree of life: Progress, classification, and evolution of subcellular traits. *Am. J. Bot.* 91: 1446–1480.
48. Kauff, F., C.J. Cox, and F. Lutzoni. 2007. WASABI: An automated sequence processing system for multigene phylogenies. *Syst. Biol.* 56(3): 523–531.
49. Robbertse, B. et al. 2011. Hal: An automated pipeline for phylogenetic analyses of genomic data. *PLoS Curr* 3: RRN1213.
50. Kurtzman, C.P. and C.J. Robnett. 2003. Phylogenetic relationships among yeasts of the *Saccharomyces* complex determined from multigene sequence analyses. *FEMS Yeast Res.* 3: 417–432.
51. Daniel, H.M. and W. Meyer. 2003. Evaluation of ribosomal RNA and actin gene sequences for the identification of ascomycetous yeasts. *J. Int. Food Microbiol.* 86: 61–78.
52. Tsui, C.K.M. et al. 2008. Reexamining the phylogeny of clinically relevant *Candida* species and allied genera based on multigene analyses. *FEMS Yeast Res.* 8: 651–659.
53. Dupuis, J.R., A.D. Roe, and F.A. Sperling. 2012. Multi-locus species delimitation in closely related animals and fungi: One marker is not enough. *Mol. Ecol.* 21(18): 4422–4436.
54. Jurjevic, Z. et al. 2012. Two novel species of *Aspergillus* section Nigri from indoor air. *IMA Fungus* 3(2): 159–173.
55. Martin, F. et al. 2011. Sequencing the fungal tree of life. *New Phytol.* 190: 818–821.
56. Floudas, D. et al. 2012. The Paleozoic origin of enzymatic lignin decomposition reconstructed from 31 fungal genomes. *Science* 336(6089): 1715–1719.
57. Zhiqiang, A. et al. 2010. China's fungal genomics initiative: A whitepaper. *Mycol. Int. J. Fungal Biol.* 1(1): 1–8.

58. Eberhardt, U. 2010. A constructive step towards selecting a DNA barcode for fungi. *New Phytol.* 187: 265–268.
59. Begerow, D. et al. 2010. Current state and perspectives of fungal DNA barcoding and rapid identification procedures. *Appl. Microbiol. Biotechnol.* 87: 99–108.
60. Seifert, K.A. 2009. Progress towards DNA barcoding of fungi. *Mol. Ecol. Res.* 9: 83–89.
61. Schoch, C.L. et al. 2012. Nuclear ribosomal internal transcribed spacer (ITS) region as a universal DNA barcode marker for Fungi. *Proc. Natl. Acad. Sci. U.S.A.* 109: 6241–6246.
62. Hebert, P.D.N., A. Cywinska, S.L. Ball, and J.R. deWaard. 2003. Biological identifications through DNA barcodes. *Proc. R. Soc. Lond. B* 270: 596–599.
63. Ratnasingham, S. and P.D.N. Hebert. 2007. BOLD: The barcode of life data system (www.barcodinglife.org). *Mol. Ecol. Notes* 7: 355–364.
64. McNeill, J. et al. 2006. International Code of Botanical Nomenclature (Vienna Code). Regnum Vegetabile 146. A.R.G. Gantner Verlag KG. http://www.iapt-taxon.org/icbn/main.htm.
65. Schindel, D.E. and S.E. Miller. 2005. DNA barcoding a useful tool for taxonomists. *Nature* 435: 17.
66. Hollingsworth, P.M. et al. 2009. CBOL Plant Working Group. A DNA barcode for land plants. *PNAS* 106(31): 12794–12797.
67. Hollingsworth, P.M. 2011. Refining the DNA barcode for land plants. *Proc. Natl. Acad. Sci. U.S.A.* 108: 19451–19452.
68. Li, D.Z. et al. 2011. Comparative analysis of a large dataset indicates that internal transcribed spacer (ITS) should be incorporated into the core barcode for seed plants. *Proc. Natl. Acad. Sci. U.S.A.* 108: 19641–19646.
69. Seifert, K.A. et al. 2007. Prospects for fungus identification using C01 DNA barcodes, with *Penicillium* as a test case. *Proc. Natl. Acad. Sci. U.S.A.* 104: 3901–3906.
70. Robideau, G.P. et al. 2011. DNA barcoding of oomycetes with cytochrome c oxidase subunit I and internal transcribed spacer. *Mol. Ecol. Res.* 11: 1002–1011.
71. Lachance, M.A. et al. 2003. The D1/D2 domain of the large subunit rDNA of the yeast species *Clavispora lusitaniae* is unusually polymorphic. *FEMS Yeast Res.* 4(3): 253–258.
72. Nilsson, R.H. et al. 2006. Taxonomic reliability of DNA sequences in public sequence databases: A fungal perspective. *PLoS One* 1(1): e59.
73. Kiss, L. 2012. Limits of nuclear ribosomal DNA internal transcribed spacer (ITS) sequences as species barcodes for Fungi. *Proc. Natl. Acad. Sci. U.S.A.* 109(27): E1811.
74. Simon, U.K. and M. Weiss. 2008. Intragenomic variation of fungal ribosomal genes is higher than previously thought. *Mol. Biol. Evol.* 25: 2251–2254.
75. James, S.A. et al. 2009. Repetitive sequence variation and dynamics in the ribosomal DNA array of Saccharomyces cerevisiae as revealed by whole–genome resequencing. *Gen. Res.* 19: 626–635.
76. Kovács, G.M., T. Jankovics, and L. Kiss. 2011. Variation in the nrDNA ITS sequences of some powdery mildew species: Do routine molecular identification procedures hide valuable information? *Eur. J. Plant Pathol.* 131: 135–141.
77. Alper, I., M. Frenette, and S. Labrie. 2011. Ribosomal DNA polymorphisms in the yeast *Geotrichum candidum*. *Fungal Biol.* 115: 1259–1269.
78. Lindner, D.L. and M.T. Banik. 2011. Intragenomic variation in the ITS rDNA region obscures phylogenetic relationships and inflates estimates of operational taxonomic units in genus *Laetiporus*. *Mycologia* 103: 731–740.
79. Harris, D.J. and K.A. Crandall. 2000. Intragenomic variation within ITS1 and ITS2 of freshwater crayfishes (Decapoda: Cambaridae): Implications for phylogenetic and microsatellite studies. *Mol. Biol. Evol.* 17: 284–291.
80. Leo, M.P. and S.C. Barker. 2002. Intragenomic variation in the ITS2 rDNA in the louse of humans, Pediculus humanus: ITS2 is not a suitable marker for population studies in this species. *Insect Mol. Biol.* 11: 651–657.
81. Elderkin, C.L. 2009. Intragenomic variation in the rDNA internal transcribed spacer (ITS1) in the freshwater mussel Cumberlandia monodonta (Say, 1828). *J. Molluscan. Stud.* 75: 419–421.
82. Thornhill, D.J., T.C. Lajeunesse, and S.R. Santos. 2007. Measuring rDNA diversity in eukaryotic microbial systems: How intragenomic variation, pseudogenes and PCR artifacts confound biodiversity estimates. *Mol. Ecol.* 16: 5326–5340.
83. Stewart, F.J. and C.M. Cavanaugh. 2007. Intragenomic variation and evolution of the internal transcribed spacer of the rRNA operon in bacteria. *J. Mol. Evol.* 65: 44–67.
84. Harpke, D. and A. Peterson. 2006. Non-concerted ITS evolution in Mammillaria (Cactaceae). *Mol. Phylogenet. Evol.* 41(2006): 579–593.

85. Bergeron, J. and G. Drouin. 2008. The evolution of 5S ribosomal RNA genes linked to the rDNA units of fungal species. *Curr. Genet.* 54: 123–131.
86. Mano, S. and H. Innan. 2008. The evolutionary rate of duplicated genes under concerted evolution. *Genetics* 180: 493–505.
87. Xiao, L.-Q., M. Möller, and H. Zhu H. 2010. High nrDNA ITS polymorphism in the ancient extant seed plant Cycas: Incomplete concerted evolution and the origin of pseudogenes. *Mol. Phyl. Evol.* 55: 168–177.
88. Daubin, V., M. Gouy, and G. Perriere. 2002. A phylogenomic approach to bacterial phylogeny: Evidence of a core of genes sharing a common history. *Genome. Res.* 12: 1080–1090.
89. Rokas, A. et al. 2003. Genome scale approaches to resolving incongruence in molecular phylogenies. *Nature* 425: 798–804.
90. Kurtzman, C.P. 2003. Phylogenetic circumscription of *Saccharomyces*, *Kluyveromyces* and other members of the Saccharomycetaceae, and the proposal of the new genera *Lachancea*, *Nakaseomyces*, *Naumovia*, *Vanderwaltozyma* and *Zygotorulaspora*. *FEMS Yeast Res.* 4: 233–245.
91. Taylor, J.W. and M.C. Fisher. 2003. Fungal multilocus sequence typing—It's not just for bacteria. *Curr. Opin. Microbiol.* 6(4): 351–356.
92. Kuramae, E. et al. 2006. Phylogenomics reveal a robust fungal tree of life. *FEMS Yeast Res.* 6(8): 1213–1220.
93. Robert, V. et al. 2011. The quest for a general and reliable fungal DNA barcode. *Open Appl. Inform. J.* 5: 45–61.
94. Lewis, C.T. et al. 2011. Identification of fungal DNA barcode targets and PCR primers based on Pfam protein families and taxonomic hierarchy. *Open Appl. Inform. J.* 5: 30–44.
95. Nilsson, R.H. et al. 2012. Five simple guidelines for establishing basic authenticity and reliability of newly generated fungal ITS sequences. *MycoKeys* 4: 37–63.
96. Rellini, P. et al. 2009. Direct spectroscopic (FTIR) detection of intraspecific binary contaminations in yeast cultures. *FEMS Yeast Res.* 9(3): 460–467.
97. Corte, L. et al. 2010. Development of a novel, FTIR (Fourier Transform InfraRed spectroscopy) based, yeast bioassay for toxicity testing and stress response study. *Anal. Chim. Acta.* 659(1–2): 258–265.
98. Buee, M. et al. 2009. 454 Pyrosequencing analyses of forest soils reveal an unexpectedly high fungal diversity. *New Phytol.* 184: 449–456.
99. Jumpponen, A. and K.L. Jones. 2009. Massively parallel 454 sequencing indicates hyperdiverse fungal communities in temperate Quercus macrocarpa phyllosphere. *New Phytol.* 184: 438–448.
100. Oepik, M. et al. 2009. Large-scale parallel 454 sequencing reveals host ecological group specificity of arbuscular mycorrhizal fungi in a boreonemoral forest. *New Phytol.* 184: 424–437.
101. Bellemain, E. et al. 2010. ITS as an environmental DNA barcode for fungi: An in silico approach reveals potential PCR biases. *BMC Microbiol.* 10: 189.
102. Caporaso, J.G. et al. 2010. QIIME allows analysis of high-throughput community sequencing data. *Nat. Methods.* 7(5): 335–336.
103. Caporaso, J.G. et al. 2011. Global patterns of 16S rRNA diversity at a depth of millions of sequences per sample. *Proc. Natl. Acad. Sci. U.S.A.* 15(108 Suppl. 1): 4516–4522.
104. Caporaso, J.G. et al. 2012. Ultra-high-throughput microbial community analysis on the Illumina HiSeq and MiSeq platforms. *ISME J.* 6(8): 1621–1624.

5 Metabolomics of Food- and Waterborne Fungal Pathogens

Danny Alexander, Adam D. Kennedy, Nalini Desai, Elizabeth Kensicki, and Kirk L. Pappan

CONTENTS

5.1	Introduction	57
5.2	Metabolomics and Metabolic Profiling Research Methods	57
5.3	High-Throughput Global Metabolic Profiling	58
5.4	Metabolomics in Fungal Biology	59
5.5	Metabolomics of Fungi	59
5.6	Metabolomics of Plant–Fungi Interactions	61
5.7	Metabolomics for Monitoring Fungal Contamination of Food	62
5.8	Metabolomics of Human Disease Caused by Mycotoxigenic Fungi and Mycotoxins	64
5.9	Conclusions	65
References		65

5.1 INTRODUCTION

Metabolomics, the nontargeted study of global changes in metabolic profiles, is a valuable technology for improving our understanding of physiology and biochemistry. It can be applied to many aspects of fungal pathogen research, including (1) the life cycle and pathogenicity of particular fungi, (2) plant–fungus interactions as related to crop yield and plant health, (3) transmission of pathogens and mycotoxins into the feed/food chain, and (4) the physiological interaction of fungi and their toxins with humans and other animals upon consuming tainted food, feed, and beverages. Biochemical reactions, pathways, and metabolic networks have been studied in detail for more than a century, but it has been only in the recent decade that technological systems with the capacity to simultaneously measure hundreds of biochemicals were developed—an innovation that has enabled the opportunity to view an organism's metabolic networks in a global, nontargeted manner and to relate that information to broader characteristics at the level of the genome, proteome, whole organism, and even populations. The successful implementation of metabolomics as a discovery platform entails the reproducible identification and measurement, in a medium- to high-throughput format, of a broad and useful range of metabolites in a given sample. The goal of metabolite screening is generation of experimentally testable biological hypotheses, which can provide the framework for directed and targeted validation studies.

5.2 METABOLOMICS AND METABOLIC PROFILING RESEARCH METHODS

Metabolomics, also known as global metabolic profiling, is the simultaneous measurement of all small metabolites or biochemicals in a biological sample. Its goal is to give a comprehensive static view of metabolites in a biological system in an unbiased manner. Such information, when placed in the context of biochemical pathways or reconstructed from an ontological vantage point, offers deep insight into the physiological and metabolic state of a system. The metabolome—all the small molecules in a sample—represents information created by the genome, transcriptome, and proteome of a biological system and serves as a key translator of that information into the observable phenotype of the system. Metabolomics is a systems biology technology that complements other *omics* technologies such as genomics, transcriptomics, and proteomics, which represent comprehensive approaches to measure changes in genes, transcripts, and proteins, respectively. Metabolomics provides insight into the metabolic state of a biological system across multiple areas of metabolism—such as energy production, amino and organic acids, lipids, carbohydrates, nucleotides, vitamins, cofactors, and xenobiotics—in response to genetic, physical, environmental, or biotic and abiotic perturbations.

Most metabolic profiling approaches rely on either nuclear magnetic resonance (NMR)- or mass spectrometry (MS)-based methods, which each has their distinct strengths and weaknesses. The tumult of data generated by both approaches necessitates the application of chemometric data handling techniques involving sophisticated software for complex data reduction, multivariate statistical analysis, and data visualization. NMR yields information on the chemical structure of biochemicals in solution by monitoring the chemical environment of every proton. Key advantages of NMR are that it is nondestructive, requires little or no special preparation for samples such as biological fluids, and has peak area directly proportional to compound abundance that allows direct quantification [1]. Limitations of NMR include (1) the complexity of its spectra, comprised of multiple overlapping peaks, (2) a limited capacity for automation, and (3) lower sensitivity and dynamic range that only permits detection of changes in the most abundant biochemicals in a sample.

MS with its better sensitivity compared to NMR, combined with the rich information, speed, and robustness it provides, has made it the preferred technology for metabolomic analysis. Biochemicals are ionized, sorted according to their mass to charge ratio (*m/z*) by a mass analyzer, and then measured by a mass detector [2]. The simultaneous measurement of all biochemicals requires MS to be coupled to an additional separation technique, such as liquid chromatography (LC) or gas chromatography (GC), to resolve the multitude of biochemicals present in most biological samples. However, biological samples can contain multiple biochemicals that share the same exact mass—restricting the ability to distinguish biochemicals on the basis of mass alone. Discrimination between biochemicals with the same mass, even if not resolved by chromatographic separation, can be enabled by performing multiple rounds of MS. The most basic form of this procedure, known as tandem MS (MS/MS), involves (1) sorting and isolating a subpopulation of molecules in one mass analyzer; (2) passing them to a second mass analyzer where they are subjected to fragmentation, sorted according to the *m/z* of the fragments; and (3) detected by a mass detector. The MS/MS ion fragment spectra often provide unique identifying chemical structural information that allows distinction between parent molecules of the same exact mass. Differences in mass, polarity, solubility, and ionization potential make it challenging to resolve all biochemicals with a single chromatographic method, but the use of combinations of a few robust and rapid analytical systems can form the basis of a metabolomic platform capable of achieving broad coverage ranging from small polar molecules, such as lactate, to large nonpolar ones such as cholesterol.

5.3 HIGH-THROUGHPUT GLOBAL METABOLIC PROFILING

Organic metabolites range from very simple one and two carbon molecules to (1) large molecules with complex stereochemistry, such as sterols, and (2) polymeric macromolecules such as the union of cellulose, hemicelluloses, and pectins to form a plant cell wall. Given the wide array of complexity and physical properties of metabolites, accurate identification of the metabolites and how they fit into metabolic pathways are essential for meaningful biological interpretation. The interpretation of data from complex biological interactions requires the unambiguous identification of metabolites. The MS-based platform developed at Metabolon uses a library-based method that matches the *m/z* of spectral *ion features* and other analytical characteristics, such as chromatographic retention time and fragmentation pattern, against a database of authentic standards to identify metabolites [2].

This platform and computer software also capture and track information for compounds of unknown structure that are not currently represented in the standard reference library [3]. Over time, as new molecules are added to the standard reference library, some of these *unnamed compounds* are identified through their match to new references. Unnamed compounds can also be pursued by traditional direct structural elucidation approaches—which can be important when an unnamed compound acts as a recurrent biomarker of a particular phenomenon.

Given the utility of MS platforms for producing high-quality biochemical data, the greatest challenges in practice involve dealing with large amounts of data in a timely and reproducible manner. Each sample yields thousands of ion features, the vast majority of which represent either redundant information or process artifacts [4]. Therefore, bioinformatic automation is essential to facilitate sample and data management, peak integration, chemical identification, data export, statistical analyses, and long-term archiving and data mining [3].

The metabolome is interrogated using analytical chemistry techniques. However, as an integrator of genetics, environment, nutrition, etc., it is fundamentally influenced by biology. The level of inherent biological and environmental variability fluctuates by the complexity of the system under study, so careful attention must be given to study design in terms of including sufficient controls and statistical power to ensure that legitimate differences between experimental groups are large enough to rise above the normal biological noise.

5.4 METABOLOMICS IN FUNGAL BIOLOGY

In nature, the metabolome of an organism is influenced by its physical environment, alterations in genetic composition and expression such as by mutations or epigenetics, and interaction with the biological environment. Thus, for a complex subject like the interaction of food- and waterborne fungal pathogens with their environment, there are multiple potential perspectives from which to view the application of global metabolic profiling as a tool to answer biological questions. In the case of a plant fungal pathogen, metabolomics could be used to ask questions such as *What differences in the metabolic profile of pathogenic strain X compared to nonpathogenic strain Y contribute to pathogenicity?* For the interaction of a fungal pathogen with its host plant, metabolomics could be used to ask questions such as *What changes occur in the metabolic phenotype of an infected plant relative to an uninfected plant, and can that information be used to identify a mechanism of pathology or identify traits that could be engineered into plants to confer resistance?* The metabolic profiles of food products made from adulterated materials could similarly be profiled, perhaps to look for biomarkers that serve as rapid, inexpensive means to detect fungal pathogens, mycotoxins, or nutrient degradation. Finally, metabolomics can be applied to understand the interaction between fungi or their toxins and humans and other animals that consume contaminated food or feed. For example, the procedure could be used to determine how consumption of contaminated material alters the animal metabolome, while looking for diagnostic biomarkers of foodborne illnesses or monitoring the impact of therapeutic treatments to better understand their efficacy or mode of action. The flexibility of metabolic profiling methods to extract and measure small biological chemicals from nearly any biological sample, and the close association between the metabolic phenotype of a biological system and important phenotypical qualities such as pathogenicity, disease susceptibility, and disease resistance, makes metabolomics a powerful discovery tool that complements other *omics* technologies. The next sections review the application of metabolomics to fungi, the plant–pathogen interaction, food–feed contamination, and the animal–pathogen interaction.

5.5 METABOLOMICS OF FUNGI

The focus of fungal metabolomics has largely been to address areas not related to food- and feed-borne pathogens. For example, metabolomics has been used extensively to examine industrially important fungi, such as *Saccharomyces cerevisiae*, to characterize different strains and optimize fermentation yields. However, the applicability of metabolomics to nearly all biological matrices and organisms, the close association between metabolomic profiles and phenotypes, and the growing awareness of metabolomics as a novel resource for discovery, hypothesis generation, and mechanistic insight into biological systems presage the wide adoption of this method for the study, monitoring, and control of food- and feed-borne pathogens. Metabolomic approaches have been used to (1) aid functional genomic analyses [5–7], (2) provide information for comparative genomic and taxonomic classification [8,9], (3) analyze secondary metabolism with its connections to primary metabolism and the growth environment [10], and (4) guide biotechnological engineering of fungi [11,12].

One early groundbreaking use of metabolomics in fungal biology research was to identify changes in NMR-deduced metabolic profiles that accompanied *silent* gene mutations that caused no overt phenotype in *S. cerevisiae* [5]. The concentrations of glycolytic intermediates were measured in midphase by growing wild-type and mutated strains bearing no observable phenotype differences. The authors demonstrated that although the wild-type and mutant strains did not differ in growth rate on minimal glucose media, the profiles of known glycolytic intermediates allowed for a clear discrimination between the strains at the metabolite level [5] and could serve as the basis for uncovering gene function. Since that report from the early part of the genomic era, metabolomic databases and mining tool techniques have improved greatly. For instance, the Yeast Metabolome Database (http://www.ymdb.ca) is a resource modeled after the human metabolomic database and serves as a highly annotated repository of metabolomic information for *S. cerevisiae*, which contains NMR- and MS-generated metabolite data [6]. The ability to interrogate the global metabolome of yeast has also advanced tremendously over the past decade. One recent evaluation of yeast grown in iron-replete and iron-deficient conditions measured 129 metabolites representing amino acids, peptides, carbohydrates, energy, lipid, nucleotide, and cofactor metabolism [7]. Yeast grown under iron-deficient conditions displayed altered glucose carbon partitioning that indicated a near-complete dependence on glycolysis for energy removed from iron-dependent enzymes of the mitochondria. The heavy dependence on glycolysis for energy came at the expense of lipid accumulation—which needs 3-phosphoglycerate to provide glycerol for triacylglycerol and glycerolipid synthesis. Although it was noted that yeast typically direct only 3% of glucose carbons to mitochondrial respiration, that small fraction supplies a large proportion of the energy to the cell.

Of particular relevance to fungal pathogen research, metabolomics has been used to understand secondary metabolism, including that leading to the synthesis of fungal toxins. Perhaps the earliest applications of metabolomic methods in fungal research were established with the recognition that secondary metabolites provide a unique chemical fingerprint that could form the basis of species identification and taxonomic classification [13–15]. These pioneering metabolic profiling approaches used thin-layer chromatography [16,17] and high-performance LC [18] for separation and UV/visible detection of compounds [19]. Today's MS- and NMR-based metabolomic profiling methods have continued the tradition of exploiting fungal chemodiversity for taxonomic classification. Thus, although metabolic profiling is often viewed as a useful route to identify new compounds that may serve as biotechnology products, it should be remembered that collecting chemical signatures in the absence of compound identification can provide information that can be used to efficiently identify and classify fungi [8]. Recently, cluster analysis of accurate direct infusion electrospray MS data was used to correctly classify 491 isolates from 57 species of *Penicillium* species from the *Penicillium* subgenus in accordance with the accepted classification. As an extension of this observation, Smith and Bluhm recently described a method for distinguishing two closely related fungal strains (i.e., wild type and mutant) by GC-MS metabolic fingerprinting [9]. They note that more than 200,000 unique fungal secondary metabolites—including toxins involved in host infection, antibiotics, UV-protective chemicals, and plant hormones—have been estimated to exist. Since most of these secondary metabolites are still structurally and functionally unknown, knowledge of a gene mutation may not provide information about the function of the gene or products associated with the change. Metabolic fingerprinting involves growing genetically related strains under identical conditions and then determining as wide of a range of metabolites as possible to pinpoint specific metabolic differences.

Metabolomics can also be used to assist functional genomics when coupled with routine techniques of molecular biology to alter the expression of specific genes or gene clusters. For example, differential analysis of the 2D NMR metabolic profiles of wild-type *Aspergillus fumigates* and the Δ*gliZ* knockout strain missing the gene encoding the transcriptional regulator for the *gli* nonribosomal peptide synthetase that produces gliotoxin was useful to identify heretofore unknown metabolic intermediates associated with the gene cluster [10]. The chemical structure information provided by 2D NMR also allowed for an understanding of the chemistry of gliotoxin synthesis and assigned functional roles to the identified pathway intermediates. Together, these examples demonstrate the utility of metabolomics for identifying and classifying new fungal isolates and for understanding the metabolic pathways leading to secondary metabolites such as mycotoxins.

Metabolomic approaches to studying metabolism and biochemical signatures of fungi could use a variety of study design strategies to:

- Identify and classify of fungal species by differential analysis of metabolic profiles (i.e., metabolites are identified) and fingerprints (i.e., differences are noted on the basis of different peaks in the spectra but metabolites are not necessarily identified)
- Elucidate of gene function and identification of metabolites and metabolic pathways—especially those of less well-characterized secondary metabolites
- Understand the effects of environmental factors such as temperature, light–dark cycles, and nutrient availability on metabolic profiles and production of mycotoxins
- Assist the rational metabolic engineering of fungal strains for biotechnological production of beneficial secondary metabolites such as antibiotics

5.6 METABOLOMICS OF PLANT–FUNGI INTERACTIONS

Fungi, through the production of mycotoxins, represent a major source of food and feed contamination worldwide [20,21]. Most of the approximately 100 fungal species that produce 300–400 known mycotoxin molecules emanate from the genera *Fusarium*, *Penicillium*, and *Aspergillus* [22,23]. Depending on the species and conditions, they can invade in the field or during storage (typically in grains), and the resulting contamination in feed and food can lead to a wide range of acute and chronic syndromes, including growth inhibition in farm animals, loss of fertility, immunosuppression, neurotoxicity, and cancers [24,25].

Managing the risks posed by mycotoxins involves various strategies to reduce fungal infection levels in the field and in postharvest storage locations. Proper crop rotation management and administration of fungicides presently play the most important roles in fighting infection in the field, and in many cases, the management of insects, through the use of insecticides or genetically modified crops (e.g., Bt toxin), provides significant reductions in mycotoxin levels because insect damage represents a primary route for fungal infection [24,26,27]. Modification (e.g., enzymatic detoxification) or absorption of mycotoxins in food/feed products has been used, but plays a relatively minor role. Beyond these approaches, plant geneticists have for many years studied the interactions of plants and their fungal pathogens, and genetic resistance to fungi has been a significant focus of research. Durable genetic resistance to some fungal pathogens has been widely incorporated into germplasm, for example, in wheat [28], but the range of these resistance genes often does not include the major mycotoxin-producing genera. While genetic resistance to *Aspergillus* and *Fusarium* species has been identified and is being incorporated into breeding programs, these often involve complex multigenic systems, and the levels of resistance are often suboptimal [29,30]. Also, various strategies have been explored to confer resistance through genetic engineering approaches [31–33], but it is unclear how effective these might eventually be, or how long it may take to commercialize.

Some form of genetic resistance to initial infections, whether acquired through conventional breeding or genetic engineering approaches, may represent the preferred route for the advancement of mycotoxin management in crops. In any case, a basic understanding of plant–pathogen interactions is essential for developing and validating novel approaches to resistance, and metabolomic analysis can play a key role in gaining a full picture of these complex biological systems [34–38]. Examples involving metabolic interactions between plants and fungi include rice blast disease [35]; the maize head smut pathogen *Ustilago* [39]; epiphytic fungi of rye grass [40,41]; *Magnaporthe* interactions with barley, rice, and *Brachypodium* [42]; and two studies involving mycotoxin-producing fungi—the interaction of *Aspergillus* and *Fusarium* in *Brassica* [43] and *Fusarium graminearum* of wheat [44]. Abdel-Farid et al. [43] used NMR metabolite profiling to show that a range of plant secondary metabolites, such as phenylpropanoids, flavonoids, and glucosinolates, were highly associated with mycotoxigenic fungal infection in *Brassica rapa* and that the identities and levels of induced metabolites followed pathogen-specific patterns. Chen et al. [44] also used NMR to study the wild-type wheat pathogen *F. graminearum* versus a mycotoxin-negative mutant (*Tri-5* gene deletion) and showed that deletion of the Tri-5 gene, which encodes the first committed enzyme for the synthesis of trichothecene mycotoxins, causes wide-scale metabolic alterations in the fungus, not just a simple ablation of the mycotoxin pathway. Such findings of unexpected metabolic complexity are not unusual in metabolomic studies and, when coupled with transcription data and other genomic information, can yield very powerful insights into biological regulatory mechanisms.

Metabolomic approaches to studying phytopathological systems could use a variety of study design strategies, some of which are listed in the following text:

- Direct studies of fungal pathogen metabolism in an in vitro culture system (if available) to understand metabolic profiles during growth and development. This would also be useful to compare genetic variants and mutants.
- Studies in vitro to test the effects of exogenous molecules (e.g., plant metabolites, antifungal compounds) on fungal metabolism.
- Effects of early fungal infection in host tissues, possibly through a time course postinfection, or at a fixed infection time, but initiated at varied stages of host plant development.
- Effects of different fungal genotypes on host plant metabolism during infection.
- Effects of a fungal pathogen on varied host plant genetic backgrounds during infection, for instance, in populations of near isogenic lines bearing resistance quantitative trait loci.
- Metabolic effects of exogenous treatments, such as fungicides, hormones, and environment, on the host plant in the presence or absence of pathogen.

5.7 METABOLOMICS FOR MONITORING FUNGAL CONTAMINATION OF FOOD

Throughout history, plant diseases and plant pathogens have had significant effects on human health and welfare (e.g., the Irish potato famine when the fungus-like oomycete *Phytophthora infestans* was instrumental in the deaths of a million residents of Ireland). Fungal contamination of the food supply is a threat that needs to be monitored at several points in the crop growing, harvesting, and food production processes. Mycotoxin-producing fungi can not only kill and contaminate crops but can also cause disease in humans that consume contaminated foods [45]. Identifying fungal contamination early in food processing is essential to control the spread of the contaminated crops. In addition to impacts on health, identifying infections and prescribing the most effective medication for those infections would ultimately have an economic impact [46]. Metabolomics can be leveraged to monitor crop health, examine nutritional impacts of contamination on crops during storage and food processing, identify contaminants, and characterize infection caused by fungal pathogens.

Biochemical profiling can be employed to monitor the growth of crops and to understand how crops respond to stress, including fungal infection. Molecular changes within a system occur prior to and drive the phenotypic changes of cells and tissues. For example, the ripening of strawberries was monitored by metabolomics, and several changes in amino acids, the shikimate pathway, and the tricarboxylic acid cycle occurred prior to the ripening and spoilage stages [47]. Metabolomic data can determine which factors drive variation between crop phenotypes by linking gene expression and allelic variation to changes in metabolite levels [48]. At the level of food consumption, metabolomic profiles can be correlated with sensory attributes to monitor how storage affects the stability of foods and which changes lead to less desirable tastes and smells [49].

Crops are processed for several reasons, including preservation during transportation, enrichment of nutritional value, and killing of contaminants that pose a danger to human consumption. Mycotic fungi infect a number of plants that are consumed by humans. Enzymatic, heat, and other sterilization procedures are necessary to kill the fungi and inactivate the toxins produced by these fungi. Processing of crops limits and prevents to some extent the transmission of mycotoxins and mycotoxigenic fungi from contaminated foods to humans and other animals. However, the processing procedures can also affect the levels of nutrients produced by grains, fruits, and vegetables. Biochemical profiling of different stages of food processing and comparison of the levels of the metabolites from each process stage have revealed that food processing enriches some metabolites, whereas it destroys others. Therefore, enrichment of essential nutrients is required after processing in order to preserve and deliver the most optimal nutritional value. Metabolomics can be employed as a tool for monitoring raw material quality and safety and can be further utilized for monitoring shelf life and postharvest processing.

Some biochemicals produced by plants serve important roles in host defense during plant growth. These include flavonoids, several amino acids, essential oils, and complex carbohydrates. Plant-produced essential oils such as terpinen-4-ol, eugenol, and carvone have antifungal activity against *Fusarium* and *Aspergillus* [50]. Many crops are treated during and after harvest in order to prevent the entry of fungal and bacterial pathogens into the food supply. For example, barley, wheat, and oats are prone to fungal contamination. Processing of plant materials through maceration, heat treatment, and solubilization can affect the stability of several different classes of biochemicals. Metabolomics can be utilized to detect these fungal pathogens and the changes in the mycotic fungi with treatment [51–53]. For example, conjugation of the major *Fusarium* mycotoxin deoxynivalenol to glucose to create deoxynivalenol-3-glucoside increased upon enzyme treatment and decreased upon baking [54]. The brewing process significantly decreased molecules from the mycotoxin class of enniatins [55]. Similarly, cleaning and sanitation of carrots affected the sugar content, carotene levels, and antioxidant capacity [56]. Genetic engineering has been employed in order to combat fungal infection in plants [57,58]. Thus, there are several ways to contest mycotic fungi at the level of agriculture and food processing, and a combination of these methods may be necessary in order to avoid ingestion and infection by these fungi.

Control of fungal growth is important in the management of raw feedstuffs, foodstuffs, condiments–spices, botanicals, and other consumable substances as they are grown, harvested, stored, and transported. Metabolomics has been applied to several areas within food and crop stability and in the field of mycotoxigenic fungi; it can be applied to the same extent as it is used to detect, control, and prevent bacterial contamination of meat and meat products [59–68]. These include the ability of metabolomics to monitor crops in the field and postharvest. Genetically modified plants have been utilized in the *commercial space* for several years. Genetic engineering has helped with such issues as production (e.g., increasing yield given limited areas of arable land), drought resistance, and resistance to pathogens (e.g., pests/insects and weeds). These genetic modifications are intended to bring beneficial traits and nutrients to consumers. Metabolomic analysis of genetically modified plants could support their commercialization by providing a comprehensive list of plant biochemical components and information necessary to show their benefit to the food supply. Metabolomics offers a unique method to identify these changes and discover biomarkers of mycotoxigenic-induced changes within these systems.

Metabolomic approaches to studying biochemical changes within foods and foodstuffs could use a variety of study design strategies, some of which are listed in the following text:

- Direct comparison of foods as they are processed
- Development of panels of biochemicals that can be utilized as biomarkers of fungal and mycotoxigenic contamination to identify products that need to be removed from market
- Correlation of genomic and metabolomic data from mycotoxigenic fungi cultures to identify driving factors for growth and identification of antifungal targets
- Treatment of plant and mammalian cells/tissues with mycotoxigenic fungi to catalog changes induced by toxin treatment and identify subsequent targets for therapeutic intervention

5.8 METABOLOMICS OF HUMAN DISEASE CAUSED BY MYCOTOXIGENIC FUNGI AND MYCOTOXINS

The use of metabolomics to examine the interaction between mycotoxic fungi/mycotoxins and human or animal health lags behinds its application to study fungal–plant interactions and fungal contamination of food. Nonetheless, there are examples of how metabolomic profiling can assist in the (1) identification of fungal pathogens, (2) development of antifungal medications, and (3) characterization of pathological effects and mechanisms of mycotoxins in animals—including humans. Metabolomics has been used to understand the relationship of opportunistic pathogens of *Saccharomyces* and strains commonly used in baking, wine making, and probiotics [69]. In this study, the relatively close clustering by principal component analysis of the metabolic profiles of two clinically isolated strains of *Saccharomyces* to these other *Saccharomyces* strains pointed to a common genetic background that was confirmed by phylogenetic analysis. These results led the authors to conclude that a food-related source was a high probability for the clinically isolated *Saccharomyces* strains. Metabolomics was among the techniques used to investigate the source of the toxin rhizoxin associated with *Rhizopus microsporus* and ultimately established the endosymbiotic bacteria *Burkholderia rhizoxinica* as the source of the toxin [70]. Rhizoxin poses a human health threat due to possible production during soy fermentation by *R. microsporus*. The authors also note that biotechnological production of this toxin could have beneficial applications as it is in the same family of macrolide antibiotics as erythromycin and is being clinically tested as a chemotherapy agent due to its ability to inhibit cytoskeleton formation and block mitosis [71].

The mechanisms by which aflatoxins contribute to liver fibrosis and hepatocellular carcinoma were investigated using ^{1}H-NMR to determine the metabolic changes that occurred in the liver, plasma, and urine of rats exposed to aflatoxin B1 (AFB1) in their chow [72]. AFB1 led to an elevation of plasma glucose, amino acids, and phospholipid metabolites such as choline, phosphocholine, and glycerophosphocholine. AFB1 exposure caused the accumulation of lipids in the liver (i.e., steatosis) and was accompanied by a reduction of lipids in plasma. Further evidence of altered liver function was seen by the reduction of liver glycogen and glucose. As expected, the most obvious signs of stress following AFB1 exposure were observed in the liver, but unexpectedly the elevation of gut microbiota cometabolites such as hippurate and phenylacetylglycine in urine at the later stages of recovery suggested that AFB1 exposure also has an effect on the digestive track and bacterial community during its absorption. A similar potential disruption of gut microbiota was identified by metabolomic profiling of urine from humans with known different levels of dietary exposure to the fungal toxin deoxynivalenol—where an accumulation of hippurate in urine from the high-exposure group was observed and proposed as a biomarker of deoxynivalenol exposure [73].

Metabolomic approaches to studying biochemical aspects of human and animal health in response to exposure to pathogenic fungi and mycotoxins could use a variety of study design strategies:

- Identification and classification of clinical isolates
- Understanding how secondary metabolism distinguishes between pathogenic and nonpathogenic strains
- Monitoring the biological effects of exposure in humans and animals to guide treatments to neutralize these effects

5.9 CONCLUSIONS

Current metabolomic approaches rely on MS- or NMR-based methods and offer biological insight into fungal pathogens distinct from other *omics* technologies. The metabolome—all the small molecules in a sample—represents information created by the genome, transcriptome, and proteome of a biological system and serves as a key translator of that information into the observable phenotype of the system. The flexibility of metabolic profiling methods to extract and measure small biological chemicals from nearly any biological sample type, and the close association between the metabolic phenotype of a biological system and important phenotypical qualities such as pathogenicity, disease susceptibility, and disease resistance, makes it a powerful discovery tool that complements other *omics* technologies. There are many potentially illuminating perspectives from which to study the interaction of food- and waterborne fungal pathogens with their environment using metabolomics, and we have reviewed examples of its application to fungi, the plant–pathogen interaction, food–feed contamination, and the animal–pathogen interaction. Growing awareness of this nascent technology and its innovative use in many areas of biological research ensures that it will find numerous applications in hypothesis-generating discovery research related to fundamental questions of fungal pathogen biology.

REFERENCES

1. Barding, G.A., Jr., R. Salditos, and C.K. Larive, Quantitative NMR for bioanalysis and metabolomics. *Anal Bioanal Chem*, 2012. **404**(4): 1165–1179.
2. Evans, A.M. et al., Integrated, nontargeted ultrahigh performance liquid chromatography/electrospray ionization tandem mass spectrometry platform for the identification and relative quantification of the small-molecule complement of biological systems. *Anal Chem*, 2009. **81**(16): 6656–6667.
3. Dehaven, C.D. et al., Organization of GC/MS and LC/MS metabolomics data into chemical libraries. *J Cheminform*, 2010. **2**(1): 9.
4. Evans, A.M., M.M., Dai, H., and DeHaven, C.D., Categorizing ion—Features in liquid chromatography/mass spectrometry metabolomics data. *Metabolomics*, 2012. **2**: 110.
5. Raamsdonk, L.M. et al., A functional genomics strategy that uses metabolome data to reveal the phenotype of silent mutations. *Nat Biotechnol*, 2001. **19**(1): 45–50.
6. Jewison, T. et al., YMDB: The yeast metabolome database. *Nucleic Acids Res*, 2012. **40**(Database issue): D815–D820.
7. Shakoury-Elizeh, M. et al., Metabolic response to iron deficiency in *Saccharomyces cerevisiae*. *J Biol Chem*, 2010. **285**(19): 14823–14833.
8. Smedsgaard, J. and J. Nielsen, Metabolite profiling of fungi and yeast: From phenotype to metabolome by MS and informatics. *J Exp Bot*, 2005. **56**(410): 273–286.
9. Smith, J.E. and B.H. Bluhm, Metabolic fingerprinting in *Fusarium verticillioides* to determine gene function. *Methods Mol Biol*, 2011. **722**: 237–247.
10. Forseth, R.R. et al., Identification of cryptic products of the gliotoxin gene cluster using NMR-based comparative metabolomics and a model for gliotoxin biosynthesis. *J Am Chem Soc*, 2011. **133**(25): 9678–9681.
11. Patnaik, R., Engineering complex phenotypes in industrial strains. *Biotechnol Prog*, 2008. **24**(1): 38–47.
12. Adrio, J.L. and A.L. Demain, Recombinant organisms for production of industrial products. *Bioeng Bugs*, 2010. **1**(2): 116–131.

13. Larsen, T.O. et al., Phenotypic taxonomy and metabolite profiling in microbial drug discovery. *Nat Prod Rep*, 2005. **22**(6): 672–695.
14. Frisvad, J.C., B. Andersen, and U. Thrane, The use of secondary metabolite profiling in chemotaxonomy of filamentous fungi. *Mycol Res*, 2008. **112**(Pt 2): 231–240.
15. Paterson, R.R. and C. Kemmelmeier, Neutral, alkaline and difference ultraviolet spectra of secondary metabolites from *Penicillium* and other fungi, and comparisons to published maxima from gradient high-performance liquid chromatography with diode-array detection. *J Chromatogr*, 1990. **511**: 195–221.
16. Frisvad, J.C., O. Filtenborg, and U. Thrane, Analysis and screening for mycotoxins and other secondary metabolites in fungal cultures by thin-layer chromatography and high-performance liquid chromatography. *Arch Environ Contam Toxicol*, 1989. **18**(3): 331–335.
17. Paterson, R.R., Standardized one- and two-dimensional thin-layer chromatographic methods for the identification of secondary metabolites in *Penicillium* and other fungi. *J Chromatogr*, 1986. **368**(2): 249–264.
18. Frisvad, J.C., High-performance liquid chromatographic determination of profiles of mycotoxins and other secondary metabolites. *J Chromatogr*, 1987. **392**: 333–347.
19. Frisvad, J.C. and U. Thrane, Standardized high-performance liquid chromatography of 182 mycotoxins and other fungal metabolites based on alkylphenone retention indices and UV-VIS spectra (diode array detection). *J Chromatogr*, 1987. **404**(1): 195–214.
20. Wu, F., Mycotoxin risk assessment for the purpose of setting international regulatory standards. *Environ Sci Technol*, 2004. **38**(15): 4049–4055.
21. Binder, E. et al., Worldwide occurrence of mycotoxins in commodities, feeds and feed ingredients. *Anim Feed Sci Technol*, 2007. **137**(3): 265–282.
22. Placinta, C., J. d'Mello, and A. Macdonald, A review of worldwide contamination of cereal grains and animal feed with *Fusarium mycotoxins*. *Anim Feed Sci Technol*, 1999. **78**(1): 21–37.
23. Wagacha, J. and J. Muthomi, Mycotoxin problem in Africa: Current status, implications to food safety and health and possible management strategies. *Int J Food Microbiol*, 2008. **124**(1): 1–12.
24. Binder, E.M., Managing the risk of mycotoxins in modern feed production. *Anim Feed Sci Technol*, 2007. **133**(1–2): 149–166.
25. Bryden, W.L., Mycotoxins in the food chain: Human health implications. *Asia Pac J Clin Nutr*, 2007. **16**(Suppl. 1): 95–101.
26. Bakan, B. et al., Fungal growth and *Fusarium* mycotoxin content in isogenic traditional maize and genetically modified maize grown in France and Spain. *J Agric Food Chem*, 2002. **50**(4): 728–731.
27. Wu, F., Mycotoxin reduction in Bt corn: Potential economic, health, and regulatory impacts. *Transgenic Res*, 2006. **15**(3): 277–289.
28. Krattinger, S.G. et al., A putative ABC transporter confers durable resistance to multiple fungal pathogens in wheat. *Science*, 2009. **323**(5919): 1360–1363.
29. Brown, R. et al., Advances in the development of host resistance in corn to aflatoxin contamination by *Aspergillus flavus*. *Phytopathology*, 1999. **89**(2): 113–117.
30. Munkvold, G.P., Cultural and genetic approaches to managing mycotoxins in maize. *Annu Rev Phytopathol*, 2003. **41**(1): 99–116.
31. Punja, Z.K., Genetic engineering of plants to enhance resistance to fungal pathogens—A review of progress and future prospects. *Can J Plant Pathol*, 2001. **23**(3): 216–235.
32. Gurr, S.J. and P.J. Rushton, Engineering plants with increased disease resistance: What are we going to express? *Trends Biotechnol*, 2005. **23**(6): 275–282.
33. Makandar, R. et al., Genetically engineered resistance to *Fusarium* head blight in wheat by expression of *Arabidopsis* NPR1. *Mol Plant Microbe Interact*, 2006. **19**(2): 123–129.
34. Jewett, M.C., G. Hofmann, and J. Nielsen, Fungal metabolite analysis in genomics and phenomics. *Curr Opin Biotechnol*, 2006. **17**(2): 191–197.
35. Allwood, J.W. et al., Metabolomic approaches reveal that phosphatidic and phosphatidyl glycerol phospholipids are major discriminatory non-polar metabolites in responses by *Brachypodium distachyon* to challenge by *Magnaporthe grisea*. *Plant J*, 2006. **46**(3): 351–368.
36. Berger, S., A.K. Sinha, and T. Roitsch, Plant physiology meets phytopathology: Plant primary metabolism and plant–pathogen interactions. *J Exp Bot*, 2007. **58**(15–16): 4019–4026.
37. Allwood, J.W., D.I. Ellis, and R. Goodacre, Metabolomic technologies and their application to the study of plants and plant–host interactions. *Physiol Plant*, 2008. **132**(2): 117–135.
38. Tan, K.A.R.C. et al., Assessing the impact of transcriptomics, proteomics and metabolomics on fungal phytopathology. *Mol Plant Pathol*, 2009. **10**(5): 703–715.
39. Doehlemann, G. et al., Reprogramming a maize plant: Transcriptional and metabolic changes induced by the fungal biotroph *Ustilago maydis*. *Plant J*, 2008. **56**(2): 181–195.

40. Rasmussen, S. et al., Metabolic profiles of *Lolium perenne* are differentially affected by nitrogen supply, carbohydrate content, and fungal endophyte infection. *Plant Physiol*, 2008. **146**(3): 1440–1453.
41. Cao, M. et al., Advanced data-mining strategies for the analysis of direct-infusion ion trap mass spectrometry data from the association of perennial ryegrass with its endophytic fungus, *Neotyphodium lolii*. *Plant Physiol*, 2008. **146**(4): 1501–1514.
42. Parker, D. et al., Metabolomic analysis reveals a common pattern of metabolic re-programming during invasion of three host plant species by *Magnaporthe grisea*. *Plant J*, 2009. **59**(5): 723–737.
43. Abdel-Farid, I. et al., Fungal infection-induced metabolites in *Brassica rapa*. *Plant Sci*, 2009. **176**(5): 608–615.
44. Chen, F. et al., Combined metabonomic and quantitative real-time PCR analyses reveal systems metabolic changes of *Fusarium graminearum* induced by Tri5 gene deletion. *J Proteome Res*, 2011. **10**(5): 2273–2285.
45. Madden, L.V. and M. Wheelis, The threat of plant pathogens as weapons against U.S. crops. *Annu Rev Phytopathol*, 2003. **41**: 155–176.
46. Cleveland, T.E. et al., United States Department of Agriculture-Agricultural Research Service research on pre-harvest prevention of mycotoxins and mycotoxigenic fungi in US crops. *Pest Manag Sci*, 2003. **59**(6–7): 629–642.
47. Zhang, J. et al., Metabolic profiling of strawberry (*Fragaria* × *ananassa* Duch.) during fruit development and maturation. *J Exp Bot*, 2011. **62**(3): 1103–1118.
48. Davies, H.V. et al., Metabolome variability in crop plant species—When, where, how much and so what? *Regul Toxicol Pharmacol*, 2010. **58**(3 Suppl.): S54–S61.
49. Sugimoto, M. et al., Metabolomic profiles and sensory attributes of edamame under various storage duration and temperature conditions. *J Agric Food Chem*, 2010. **58**(14): 8418–8425.
50. Morcia, C., M. Malnati, and V. Terzi, In vitro antifungal activity of terpinen-4-ol, eugenol, carvone, 1,8-cineole (eucalyptol) and thymol against mycotoxigenic plant pathogens. *Food Addit Contam Part A Chem Anal Control Expo Risk Assess*, 2012. **29**(3): 415–422.
51. Altomare, C. et al., Biological characterization of fusapyrone and deoxyfusapyrone, two bioactive secondary metabolites of *Fusarium semitectum*. *J Nat Prod*, 2000. **63**(8): 1131–1135.
52. Hughes, R.A., T. Cogan, and T. Humphrey, Exposure of *Campylobacter jejuni* to 6 degrees C: Effects on heat resistance and electron transport activity. *J Food Prot*, 2010. **73**(4): 729–733.
53. Sahgal, N. et al., Potential for detection and discrimination between mycotoxigenic and non-toxigenic spoilage moulds using volatile production patterns: a review. *Food Addit Contam*, 2007. **24**(10): 1161–1168.
54. Kostelanska, M. et al., Effects of milling and baking technologies on levels of deoxynivalenol and its masked form deoxynivalenol-3-glucoside. *J Agric Food Chem*, 2011. **59**(17): 9303–9312.
55. Vaclavikova, M. et al., "Emerging" mycotoxins in cereals processing chains: Changes of enniatins during beer and bread making. *Food Chem*, 2013. **136**(2): 750–757.
56. Ruiz-Cruz, S. et al., Sanitation procedure affects biochemical and nutritional changes of shredded carrots. *J Food Sci*, 2007. **72**(2): S146–S152.
57. Rommens, C.M. and G.M. Kishore, Exploiting the full potential of disease-resistance genes for agricultural use. *Curr Opin Biotechnol*, 2000. **11**(2): 120–125.
58. Wally, O. and Z.K. Punja, Genetic engineering for increasing fungal and bacterial disease resistance in crop plants. *GM Crops*, 2010. **1**(4): 199–206.
59. Brashears, M.M. et al., Microbial quality of condensation in fresh and ready-to-eat processing facilities. *Meat Sci*, 2012. **90**(3): 728–732.
60. Cevallos-Cevallos, J.M., M.D. Danyluk, and J.I. Reyes-De-Corcuera, GC-MS based metabolomics for rapid simultaneous detection of *Escherichia coli* O157:H7, Salmonella Typhimurium, Salmonella Muenchen, and Salmonella Hartford in ground beef and chicken. *J Food Sci*, 2011. **76**(4): M238–M246.
61. Liang, R. et al., Bacterial diversity and spoilage-related microbiota associated with freshly prepared chicken products under aerobic conditions at 4 degrees C. *J Food Prot*, 2012. **75**(6): 1057–1062.
62. Mace, S. et al., Characterisation of the spoilage microbiota in raw salmon (*Salmo salar*) steaks stored under vacuum or modified atmosphere packaging combining conventional methods and PCR-TTGE. *Food Microbiol*, 2012. **30**(1): 164–172.
63. Magwedere, K. et al., Brucellae through the food chain: The role of sheep, goats and springbok (*Antidorcus marsupialis*) as sources of human infections in Namibia. *J S Afr Vet Assoc*, 2011. **82**(4): 205–212.
64. Mellor, G.E., J.A. Bentley, and G.A. Dykes, Evidence for a role of biosurfactants produced by *Pseudomonas fluorescens* in the spoilage of fresh aerobically stored chicken meat. *Food Microbiol*, 2011. **28**(5): 1101–1104.

65. Oses, S.M. et al., Microbial performance of food safety management systems implemented in the lamb production chain. *J Food Prot*, 2012. **75**(1): 95–103.
66. Papadopoulou, O.S. et al., Transfer of foodborne pathogenic bacteria to non-inoculated beef fillets through meat mincing machine. *Meat Sci*, 2012. **90**(3): 865–869.
67. Podkowik, M., J. Bystron, and J. Bania, Prevalence of antibiotic resistance genes in staphylococci isolated from ready-to-eat meat products. *Pol J Vet Sci*, 2012. **15**(2): 233–237.
68. Sampers, I. et al., Performance of food safety management systems in poultry meat preparation processing plants in relation to *Campylobacter* spp. contamination. *J Food Prot*, 2010. **73**(8): 1447–1457.
69. MacKenzie, D.A. et al., Relatedness of medically important strains of *Saccharomyces cerevisiae* as revealed by phylogenetics and metabolomics. *Yeast*, 2008. **25**(7): 501–512.
70. Lackner, G. et al., Global distribution and evolution of a toxinogenic *Burkholderia-Rhizopus* symbiosis. *Appl Environ Microbiol*, 2009. **75**(9): 2982–2986.
71. Lackner, G. and C. Hertweck, Impact of endofungal bacteria on infection biology, food safety, and drug development. *PLoS Pathog*, 2011. **7**(6): e1002096.
72. Zhang, L. et al., Systems responses of rats to aflatoxin B1 exposure revealed with metabonomic changes in multiple biological matrices. *J Proteome Res*, 2011. **10**(2): 614–623.
73. Hopton, R.P. et al., Urine metabolite analysis as a function of deoxynivalenol exposure: An NMR-based metabolomics investigation. *Food Addit Contam Part A Chem Anal Control Expo Risk Assess*, 2010. **27**(2): 255–261.

6 Systems Biology in Fungi

Oscar Dias and Isabel Rocha

CONTENTS

- 6.1 Motivation 69
- 6.2 Metabolic Systems Biology 70
- 6.3 Functional Genomics 71
  - 6.3.1 Other Omics 72
- 6.4 Genome-Scale Metabolic Models 73
  - 6.4.1 Genome Annotation 76
  - 6.4.2 Assembling the Metabolic Network 77
    - 6.4.2.1 Genes, Proteins, and Reactions 78
    - 6.4.2.2 Spontaneous Reactions 79
    - 6.4.2.3 Stoichiometry 79
    - 6.4.2.4 Localization 79
    - 6.4.2.5 Manual Curation 79
  - 6.4.3 Converting the Metabolic Network to a Stoichiometric Model 80
  - 6.4.4 Validation of the Metabolic Model 83
- 6.5 Applications 84
- 6.6 Future Applications 86
  - 6.6.1 Health Applications 86
  - 6.6.2 Industrial Applications 87
- 6.7 Final Remarks 87
- References 88

6.1 MOTIVATION

Systems biology analyzes both the components and the interactions of organisms to understand their organization and to predict behavior.[1,2] Currently, systems biology has a variety of applications using industrial organisms and in medical problems. This holistic approach involves a combination of modeling and omics analyses and was naturally more rapidly and easily applied to prokaryotic organisms. Nevertheless, fungal systems biology started as a discipline quite soon, especially driven by the accumulated knowledge in the yeast *Saccharomyces cerevisiae* that is, simultaneously, a eukaryotic model organism and a widely used industrial organism.[3–5]

Within systems biology, integrated studies of metabolism (*metabolic systems biology*) emerge for two main reasons: availability of data and importance of applications. Indeed, accumulated knowledge on metabolism is vast and allows creating reliable models that allow simulation of microorganism behavior in a variety of conditions. Also, metabolism is directly related to valuable end products and a variety of diseases have a metabolic origin.[6,7] Knowledge on metabolism of a given organism is easily applied to other organisms, using simple bioinformatics tools such as the basic local alignment tool (BLAST),[8] while the same is not true for other functions such as transcription regulation and signaling. Therefore, it is easy to understand why a variety of metabolic models are available for organisms ranging from simple bacteria to filamentous fungi or even

Homo sapiens,[9] while only a few regulatory or signaling models have been constructed[10,11] and notably none for eukaryotes at a genome scale.

The general purpose of this chapter is detailing the state of the art of systems biology in fungi. Besides *S. cerevisiae*, several other fungi are being studied within metabolic systems biology, which are important for industrial or pharmaceutical purposes. Several applications of models are also described throughout this chapter besides detailing the methodologies to build a genome-scale metabolic model (GSMM) for fungal species.

6.2 METABOLIC SYSTEMS BIOLOGY

The first association between *systems* and *biology* was performed in 1915, by Walter Cannon, describing the human body as a control system.[12] In 1950, Ludwig von Bertalanffy introduced the general system theory,[13] and he was probably the first to declare that "… organismic conceptions have evolved in all branches of modern biology which assert the necessity of investigating not only parts but also relations of organization resulting from a dynamic interaction and manifesting themselves by the difference in behavior of parts in isolation and in the whole organism." Thus, the principle of analyzing the components, and the interactions between them, to understand and predict biological behavior is well established. The novelty arises from the availability of large-scale data sets,[5] which enables more complex analyses. The introduction of whole-genome high-throughput sequencing techniques, which promoted the completion of several whole-genome sequencing projects in the last decade, the advent of the Internet, and the development spree of various bioinformatics tools, motivated a paradigm shift in biology, bringing this field to the postgenomic era.[14]

Unlike traditional components biology, which sees cells as a set of individual components involved in biological processes, the study of biological systems, systems biology, similar to the study of any other type of system, involves quantifying the components in parallel and analyzing the interactions between them,[2,5,13,15,16] through the use of mathematical models. Thus, it is foreseeable that research will focus in the so-called emergent properties, which arise from the whole, instead of the individual parts and represent real biological properties.[16]

Leroy Hood provides a visual aid for this definition, declaring that cell systems can be seen as cars.[17] He proposes that understanding how a car works may be regarded as the formulation of a simple model. The determination of the car components would be performed by high-throughput technologies. Then removing a part of the car (i.e., perturbation) would allow comparing its behavior to normal cars. The integration of all this information would allow building a model of the car, and the model could be refined with integration of new data.

Although the major challenge of systems biology is to be able to represent the whole-cell behavior in a computer model, a good model is not just able to mimic cell comportment but rather predict such behavior.[18] Thus, these models should be able to foretell the phenotypical behavior of a cell, an organism, or an individual.

In the pregenomic era, strain development for industrial applications was carried out by random mutagenesis, followed by screening and selection of phenotypes of interest or by performing targeted modifications in genes known to be associated with the product of interest. This last reductionist approach, sustained by components biology, requires that biological events are related to only a few genes, or proteins,[5] and is often unsuccessful, because the identification of functional properties for specific cell components is insufficient to understand the biological systems as a whole.[19] Random mutagenesis, on the other hand, although successful, does not allow understanding of the cellular mechanisms leading to the end result.[20]

The models provided by systems biology approaches, together with an already high, yet growing, number of bioinformatics tools, allow identifying high-probability genetic targets for increasing yields, titers, productivities, and robustness in industrial biotechnology processes.[20–23] Thus, these systems-level models allow accelerating the development of biological processes, reducing the commitment of resources and commercialization time.[20]

6.3 FUNCTIONAL GENOMICS

The methodology for determining DNA sequences[24] was developed in 1975, and the *S. cerevisiae* budding yeast genome was completely sequenced,[25] through a worldwide collaboration in 1996. Yet only in the last dozen years has the process of sequencing genomes become a widespread and routine procedure.[5,20] From the full genome sequence of a single organism, it is possible, theoretically, to identify all gene products involved in complex biological processes.[26] Functional genomics is now a specific field focused on the determination of gene functions.[20] However, assigning functions to all genes in a sequenced genome is difficult, involving several steps (Figure 6.1).

The output of the high-throughput sequencing technologies is a set of short sequence reads that needs to be assembled. This process is called *de novo* genome assembly. The new sequencing technologies allow decoding microbial genomes at moderate costs employing only a small number of experiments.[27–30] However, the trade-off for these improvements is usually a much smaller read length,[31] which increases significantly the difficulty of the sequence reassembly process. The complexity of the sequence increases by a factor of 4 for each base added to the read, and the likelihood of detecting redundancy in a pool of sequences decreases drastically.[32] Thus, new sequencing approaches able to generate longer read lengths and improve data quality are welcome by those working on *de novo* genome assembly.[33] Even so, the cost for sequencing genomes has been decreasing, generating hundreds of gigabases of data per genome. Therefore, new algorithms, even more efficient, are needed to analyze and assemble a genome *de novo*.[34]

The next step in genome annotation is to find all genes in a given genomic sequence. This stage is called the genome structural annotation and consists of the identification of all protein-encoding genes, different types of RNA, and other DNA within the genome. This process is usually performed using bioinformatics resources, as experimental verification can be costly and time consuming.[14,35,36] In prokaryotes (and certain minor eukaryotes), it is fairly simple to pinpoint the frontiers of protein-encoding open reading frames (ORFs). Essentially, it is a question of identifying long ORFs within these genomes, by running a tool that identifies the ORFs longer than a given threshold within all six frames.[14] The accuracy of the tools, which predict ORFs in these organisms, is very high (over 90%).[35] In eukaryotes, the bigger genomes, a large number of introns and the alternative splicing pose a superior challenge for predicting the ORFs.[35,36] Moreover, gene finding is

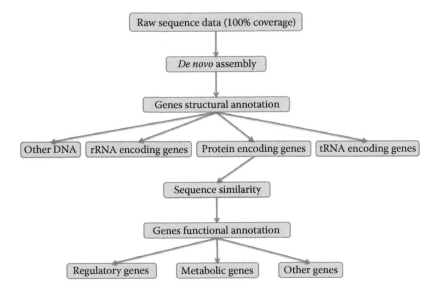

FIGURE 6.1 From genome to functional annotation. The raw genome contains virtually all information on the phenotypic potential of the organism. However, decoding such information is complex.

different in prokaryotic and eukaryotic genomes, since about 90% of the bacterial genome is coding sequences, whereas higher eukaryotes have less than 10% of coding sequences.[37] Several tools for gene prediction have been developed recently. Most of these use probabilistic methods, such as hidden Markov models (HMMs), to identify coding sequences within the ORFs (e.g., GLIMMER,[37] GenMark,[38] and EuGène[39]). Alternatively, some tools use methods other than HMM (e.g., Gismo[40]). Despite the availability of several software tools that perform ORF predictions, a clear winner has yet to emerge.[35] Thus, several programs should be used to check predictions and, when possible, the results confirmed by experimentation. Similarly, there are some software tools to predict genes that encode RNA instead of a functional protein product. More information in RNA gene prediction can be found in a review by Irmtraud Meyer.[41]

After knowing *Where* to find the protein-encoding genes, the question to ask is, *What* are the functions of those genes? This stage of the genome annotation is named genome functional annotation, and it consists of assigning putative functions to each protein-encoding gene. These functions are often identified by similarity to formerly characterized sequences kept in databases.[42] A classic genome functional annotation pipeline will start by performing similarity searches against databases of nucleotide or protein sequences, using an algorithm of the BLAST[8] family or HMMER.[43] The analysis of the homologous genes allows identifying the protein each ORF is most likely to encode, assigning each gene with specific functions. While performing the annotation, the product of a given gene may be unknown and dubbed as a hypothetical protein. Nevertheless, the assignment of a specific function to a given gene should be performed carefully because such function may not be the correct one, leading to a misclassification.[44] Finally, the genes can be grouped according to specific characteristics of interest. For example, if one is interested in building a hybrid metabolic-regulatory model, each gene within the genome will be classified either as a regulatory gene, a metabolic gene, or other type of gene.

6.3.1 Other Omics

The genome is a static entity that does not change significantly with time. In order to identify and characterize other cellular components, which provide context for utilization and regulate the expression levels of the genes, other *omics* technologies were developed. Unlike genomics, the other *omics* are susceptible to environmental and genetic perturbations.[5] Most of these technologies, namely, transcriptomics, proteomics, and metabolomics, provide *snapshots* of the physiological state of the system.

The transcriptome can be determined using next-generation sequencing RNA sequencing methods or microarray technology (DNA oligonucleotide and cDNA arrays). Both measure the mRNA expression levels for a given condition of virtually every ORF in the genome. These tools may also be useful for annotating genes.[5]

The proteome characterizes all proteins in the cell. It provides information on enzymes, transport proteins, regulatory proteins, signaling proteins, and others. Although proteins are encoded in the genome, no direct linear correlation has been found between the transcriptome and the proteome,[45,46] that is, between the quantity of mRNA molecules and the corresponding proteins. Traditionally, proteins are separated using 2D gel electrophoresis, followed by identification by mass spectrometry (MS). Recently, liquid chromatography (LC) or matrix-assisted laser ionization combined with MS proved to be more efficient methods.[47,48] Protein microarrays are also being increasingly used for protein identification and functional annotation.[5] Since protein amplification methods do not exist, as they do for DNA, the main problem in the development of this technology is the availability of sufficient amounts of proteins. Thus, for detailed functional and structural analysis, proteins have to be recombinantly produced and purified.

The metabolome represents the availability of metabolites in the system.[49] Metabolomics techniques were developed on the premise that cells control concentrations of intracellular metabolites very rigidly.[5] Metabolite profiling is very important in systems biology, as the connectivity of the networks is determined by the availability of metabolites. Although the number of metabolites is significantly lower than the number of proteins or genes, the full metabolic profile cannot be determined with the technology currently available. The analysis of the metabolome is typically determined by MS and nuclear magnetic resonance (NMR). Yet the combination of GC and LC with MS is probably one of the best techniques for metabolite profiling.

Though representing the physiological state of the cell, the fluxome cannot be quantified directly, in contrast to the *snapshot* of the system in a given moment provided by the concentrations measured with other omics.[50] Fluxes can be measured using stable isotope tracers, such as substrates labeled with C-13 markers[51,52] that can be analyzed by NMR or MS.[5] In either case, some or all metabolic fluxes are inferred from those isotopes, combined with measurements on the extracellular fluxes and biomass and with a metabolic model.[50]

Alternatively, metabolic fluxes can be estimated by performing flux balance analysis (FBA) on stoichiometric metabolic models. FBA is a mathematical approach, which applies linear programming, for analyzing the flow of metabolites through a metabolic network, maximizing or minimizing an objective function. Usually it is assumed that cells are under selective pressure and biomass precursor fluxes are favored.[53–55]

6.4 GENOME-SCALE METABOLIC MODELS

Metabolic reconstructions existed before genomic data were available where literature and biochemical characterization of enzymes were the main sources of information for these networks. Nowadays, the whole-genome sequences and the availability of well-studied biochemical reactions in several biological databases[26,56] allow generating metabolic networks at the genome scale, even for organisms less characterized in the literature.[57] Genome-scale metabolic networks (GSMNs) can be defined as the set of biological reactions retrieved from the enzymes encoded in the target organism's genome. A metabolic reconstruction process implies knowing, for each reaction in the network, which are the substrates and products, its stoichiometry, the reversibility of the reaction, and its location.[57] GSMNs characterize biochemical reactions, which produce compounds that are consumed by other reactions, and relations between the reactions. Although GSMNs allow determining some physiological and biochemical properties of the cells, only GSMMs can be used for predicting the capabilities of the metabolic system. These models may include reaction kinetics and regulatory information, although such information is currently only available for a few well-studied organisms.[56] Thus, the information contained in these models only include details on the biomass composition and energetic needs, apart from the network data. These models are currently used to predict, *in silico*, the response to perturbations of microorganisms and to identify candidate drug targets.[56]

The same process may, in theory, be applied for reconstructing eukaryotic and prokaryotic metabolic models.[58] Nevertheless, eukaryotic models are more demanding due to their larger knowledge base and genomes and the various compartments within the cells. The GSMM's reconstruction process currently involves widespread procedures, as several authors published guidelines and protocols for the reconstruction of these models.[56,58] Moreover, tools like *merlin 2.0*,[44,59] model SEED,[60] MicrobesFlux,[61] or Pathway tools,[62] developed specifically for model reconstruction, are becoming increasingly available. These tools are usually developed for assisting in the automation of some steps of the reconstruction of the model, although manual curation is always required.

Other tools, such as CellDesigner[63] or Cytoscape,[64] among others, allow visualizing networks within the models. Almost all tools developed for reconstructing, simulating, or visualizing metabolic models accept or export the models in the systems biology markup language (SBML)[65] format. This language was initially developed for representing dynamic models, yet it can also be used for stoichiometric models. The reconstruction of GSMMs is supported by information available in several online databases. These provide information about genome sequences and annotation and/or the functional capabilities of the proteins. Some of the most important data sources for developing GSMMs are listed in Table 6.1.

TABLE 6.1
Main Online Data Sources Used for the Reconstruction of Genome-Scale Metabolic Models

Database	Web Address	Description	Data Types	Curated	Reference
BioCyc	http://www.biocyc.org/	BioCyc is a collection of pathway/genome databases (PGDBs). Each PGDB in the BioCyc collection describes the genome and metabolic pathways of a single organism. These PGDBs contain additional features, including transport systems and gap fillers. Also, the BioCyc website contains tools for the visualization and analysis of the PGDBs.	Genomic, metabolic	•	Caspi et al.[66]
BKM	http://bkm-react.tu-bs.de/	BRENDA–KEGG–MetaCyc reactions (BKM-react) online. BKM-react is an integrated and nonredundant database containing known enzyme-catalyzed and spontaneous biological reactions collected from BRENDA, KEGG, and MetaCyc by aligning substrates and products.	Metabolic		Lang et al.[67]
BRENDA	http://www.brenda-enzymes.org/	BRaunschweig ENzyme DAtabase (BRENDA) is the main collection of enzyme functional data available to the scientific community. Contains functional and molecular information of enzymes, based on primary literature.	Metabolic	•	Schomburg et al.[68]
ExPASy	http://www.expasy.org/	Expert Protein Analysis System (ExPASy) is the Swiss Institute of Bioinformatics Resource Portal in different areas of life sciences including systems biology. Furthermore, ExPASy is one of the main bioinformatics resources for proteomics in the world.	Genomic, proteomic, metabolic		Artimo et al.[69]
GOLD	http://www.genomesonline.org	Genomes Online Database (GOLD) is a resource for comprehensive access to information regarding genome and metagenome sequencing projects.	Genomic		Pagani et al.[70]
KEGG	http://www.kegg.jp/	Kyoto Encyclopedia of Genes and Genomes (KEGG) is an online public repository that is, currently, the most extensive combined collection of information on genes, metabolites, reactions, and pathways.	Genomic, metabolic		Kanehisa et al.[71]
MetaCyc	http://www.metacyc.org/	MetaCyc is a database of nonredundant metabolic pathways. MetaCyc is curated from the scientific literature and contains pathways involved in primary and secondary metabolism and associated compounds, enzymes, and genes.	Metabolic	•	Caspi et al.[66]

(Continued)

Systems Biology in Fungi

TABLE 6.1 (*Continued*)
Main Online Data Sources Used for the Reconstruction of Genome-Scale Metabolic Models

Database	Web Address	Description	Data Types	Curated	Reference
NCBI	http://ncbi.nlm.nih.gov/	The National Center for Biotechnology Information (NCBI) is a repository of several databases that provides analysis, visualization, and retrieval resources for biomedical, genomic, and other biological data made available through the NCBI website.	Genomic		Sayers et al.[72]
SABIO-RK	http://sabio.villa-bosch.de/	SABIO-RK is a curated database that contains information about biochemical reactions, their kinetic rate equations with parameters, and experimental conditions.	Metabolic	•	Wittig et al.[73]
SGD	http://www.yeastgenome.org/	The Saccharomyces Genome Database (SGD) provides comprehensive integrated biological information for the budding yeast *Saccharomyces cerevisiae*.	Genomic	•	Cherry et al.[74]
TCDB	http://www.tcdb.org/	Transporter Classification Database (TCDB) comprehends a classification system for membrane transporter proteins known as the transporter classification system.	Genomic, metabolic	•	Saier et al.[75]
UniProt	http://www.uniprot.org/	Universal Protein Resource Knowledgebase (UniProtKB) is the central hub for the collection of accurate, rich, and consistent functional information on proteins. It consists of two sections: a section containing manually annotated records with information extracted from literature and computational analysis (referred to as *UniProtKB/Swiss-Prot*) and a section with computationally analyzed records waiting full manual annotation (*UniProtKB/TrEMBL*).	Genomic, metabolic	•	Apweiler et al.[76]

There are already several works that describe the reconstruction process.[56–58,77] In these, a concise description of the methodology for the reconstruction of GSMMs for unicellular eukaryotes is presented. The reconstruction process comprehends four fundamental stages, namely, genome annotation, assembling of a metabolic network from the genome, the conversion of the network to a stoichiometric model, and the validation of the metabolic model.

As depicted in Figure 6.2, the reconstruction of a GSMM is an iterative process in which the information retrieved from several data sources is compiled and used for assembling a draft of the GSMN. After obtaining the initial metabolic reconstruction from the genome annotation, the draft network is debugged and the network is converted to a GSMM by adding an equation that represents the biomass formation and other constraints. The biomass equation does not belong to the GSMN, since this reaction is not derived from the genome and is not a reaction that naturally occurs in the cell. It is a reaction that represents the cell, and inclusively, stoichiometry within this reaction is expressed in different units from the remaining reactions in the network (millimoles per gram of

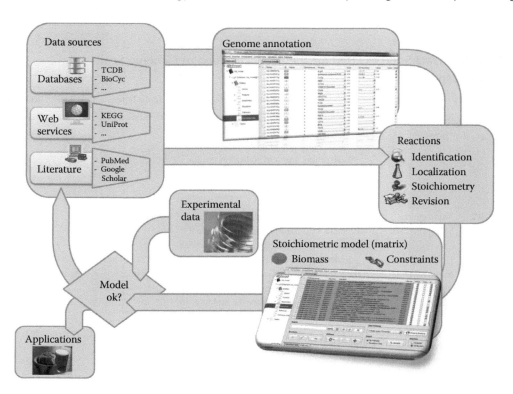

FIGURE 6.2 Description of the metabolic network reconstruction iterative process. This process starts with a thorough review of the current knowledge of the microorganism in multiple information sources. The construction and debugging of the reaction set is performed before building the steady-state metabolic model. Finally, the *in silico* simulation results are compared with experimental data. Once the *in silico* predictions comply with the experimental results, the model can be used for further applications. (Adapted from Rocha, I. et al., *Methods Mol. Biol.*, 416, 409, 2008.)

biomass). Thus, the flow through this reaction must be expressed in grams of biomass per time unit (i.e., the growth rate of a microorganism) or, in relative terms, in grams of biomass per grams of biomass per time unit (the specific growth rate), contrasting with fluxes through the other reactions (expressed in millimoles per time unit or millimoles per grams of biomass per time unit).[56]

The resulting model is validated by comparing experimental data to simulations performed with the GSMM. If the model does not comply, data sources can be used to revise the reactions set and improve the model. If it does comply, the model can be used in several applications such as gene deletion studies, or designing minimal media. A brief description of each of the stages, inferred from two reviews[56,58] of the reconstruction process, is provided next.

6.4.1 Genome Annotation

Every genome-scale reconstruction begins with the annotated genome of the target organism. The genome annotation stage is critical for developing high-quality GSMMs, because the annotation is assumed to be correct and it is performed only once throughout the whole reconstruction process. The genome annotation process assigns genes with functions, providing unique identifiers, such as the Enzyme Commission (EC)[78] and Transporter Classification (TC)[79] numbers, to the reconstruction. During the reconstruction process, subunits of protein complexes should also be identified, since more than one gene may be necessary to encode for an enzyme. Genes encoding enzymes or transport systems are labeled metabolic genes.

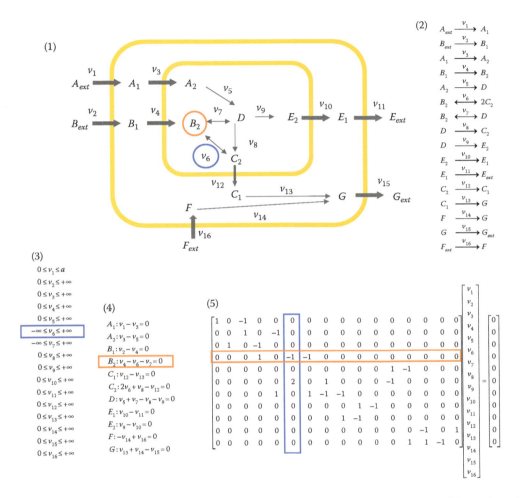

FIGURE 6.4 Example of a pseudo-metabolic network with seven metabolites (A–G) and 16 fluxes (v_1–v_{16}). The scheme of the reaction is described in (1), where the boundaries of the system are also outlined. Fluxes v_1–v_4, v_{10}–v_{12}, v_{15}, and v_{16} represent exchange fluxes of metabolic substrates (A, B, and F) and products (G and E). The reversibility of the reactions is indicated by the arrows, where double arrows represent reversible reactions and a forward arrow is used to characterize irreversible reactions. In Section (2), the stoichiometry of the network is represented. Section (3) shows the constraints around the flux values (where a represents the maximum uptake rate for the consumption of the limiting substrate A), and the steady-state mass balances are described in panel (4). A flux value may be negative for reversible reactions with unconstrained fluxes (e.g., v_6). Section (5) shows the stoichiometric matrix in which the mass balances are represented.

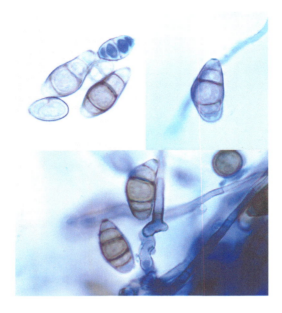

FIGURE 18.1 Lactophenol cotton blue preparation showing pigmented conidia of *Curvularia* spp. Notice that one of the central cells is larger than the others giving the conidia a curved appearance.

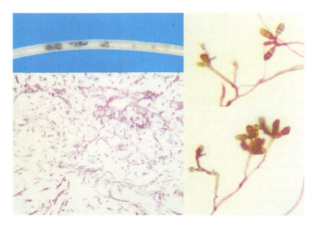

FIGURE 18.2 *C. lunata* in Tenckhoff catheter, histopathologic preparation of the pigmented material inside the catheter stained with hematoxylin and eosin and culture.

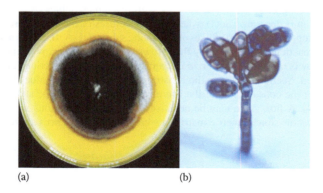

(a)   (b)

FIGURE 18.3 (a) *C. geniculata* culture and (b) lactophenol cotton blue preparation (Public Health Image Library, CDC).

Annotated genomes can be retrieved from several public repositories of genomic data, for example, National Center for Biotechnology Information (NCBI) or Kyoto Encyclopedia of Genes and Genomes (KEGG), in which a curation may have been performed, or more usually, curated organism-specific databases such as Saccharomyces Genome Database. However, if the genome annotation of the target organism is not available, the functional annotation of the genome can be performed, using specific tools.[80]

Since the quality of the curated genome annotation is critical for the reliability of the reconstructed model, the reannotation of previously annotated genomes may be required. Thus, some reannotations, with the purpose of developing GSMMs, have been performed.[80-82]

Performing the genome (re)annotation involves seeking specific data, namely, gene or ORF names, product names, and, if available, EC numbers. Only the so-called metabolic genes are mandatory for the development of GSMMs, which are the genes encoding enzymes and transport systems. Other genes involved in regulatory control or signaling are not included in GSMMs but may be useful for later integration of the model with these networks.

A Java application developed by the present authors for assisting in the reconstruction of GSMMs of organisms with sequenced genomes is *merlin 2.0*.[44,59] It performs several steps of the reconstruction process in a semiautomatic manner, including the functional genomic annotations of the whole genome. It utilizes two of the most used tools, BLAST and HMMER, for performing the (re) annotation of genomes. The similarity search results are then evaluated and an automatic annotation of the genome is presented. The tool assigns annotations to each gene of the target organism, using an internal scorer that weights both the frequency of functions and the taxonomy of every homologue of each gene of the target genome. Some parameters should be inputted for the computation of the annotation to be assigned to each gene. For fungi, the α (alpha) value should be set to 0.2. This parameter leverages the weight of each score (frequency and taxonomy), according to the following equation:

$$Score = \alpha \times Score_{frequency} + (1-\alpha) \times Score_{taxonomy} \qquad (6.1)$$

Usually, genes with annotation scores above 0.5 can be regarded as being always correct and a confidence level of 1 (highest) should be assigned to those annotations. On the other hand, gene annotation scores below 0.2 should be ignored. Gene assignments with intermediate scores should be manually curated, and this can be performed taking advantage of the graphical user interface (GUI). A concise pipeline for the manual curation of the metabolic annotations for fungal genomes is presented in Figure 6.3. It is based on *S. cerevisiae*, and if no information is available for the target fungus, curated information on the baker's yeast should be sought for inferring the gene's function.

BRaunschweig ENzyme DAtabase (BRENDA), as a primary source on enzymatic information, is used to confirm the annotations to be assigned to each gene, eliminating errors that originated by one of the following reasons: The EC number may have been transferred to another EC code, it may have been deleted, or there may be a mismatch between the EC number and the enzymatic function caused by errors in the annotation of the homologue gene. This procedure intends to limit the propagation of annotation errors. The present authors[80] used a similar schema to annotate the *Kluyveromyces lactis* genome. At the end of this stage, the metabolic annotation of the genome must be retrieved, so that the enzymatic reactions associated to the target organism's metabolism can be collected.

6.4.2 Assembling the Metabolic Network

The second stage of the reconstruction of metabolic models involves identifying and collecting biochemical reactions to form a network. This stage encompasses several steps. The first step is building the backbone of the network using reactions catalyzed by enzymes and transport systems encoded in the annotated genome.

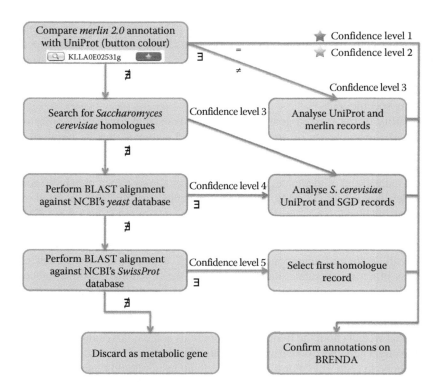

FIGURE 6.3 Annotation pipeline proposed for the assignment of enzymatic functions to fungal genes. Each gene annotation provided by *merlin 2.0* is assessed with UniProt. If the annotation cannot be inferred from this analysis, then a *S. cerevisiae* homologue is sought and its annotation is compared to merlin's annotation. If none of the previous strategies is satisfactory, then specific similarity searches should be performed against the UniProt's *Swiss-Prot* database. The information used to annotate the target genes should be always revised in BRENDA to verify the functions about to be assigned to such genes.

6.4.2.1 Genes, Proteins, and Reactions

The association between annotated genes, proteins, and reactions (the GPR associations) is usually performed by searching biological databases (Table 6.1) with the protein names, EC numbers, or other identifiers (e.g., KEGG reaction number) to which the reaction was associated.[77] TC numbers represent proteins that promote the relocation of metabolites. This system is analogous to the EC system, except that it incorporates functional and phylogenetic information. The TC code contains five elements, separated by four dots (#.*.#.#.#). The left most number represents one of the seven main divisions to which the transporters may belong to, namely, channels/pores, electrochemical potential-driven, primary active, group translocators, transmembrane electron carriers, accessory factors involved in transport, and incompletely characterized transport systems. The second element is a letter and the remainder elements are numbers. Each element to the right of the main class restricts the classification of the transporter. The Transporter Classification Database (TCDB) can be accessed for retrieving the metabolites and type of transport supported by a given carrier protein. These transport reactions should also be added to the draft network.

Therefore, the initial draft of the network can be reconstructed by associating enzymes and transporters, through the use of EC and TC numbers, with substrates and products using databases such as TCDB, BRENDA, and KEGG. These data should, preferably, be automatically retrieved using tools developed for that purpose, for example, *merlin 2.0*. Also, it should be noticed that proteins involved in DNA methylation and RNA modification, although commonly associated to EC numbers, are not usually incorporated into the model.

6.4.2.2 Spontaneous Reactions

The second step of this stage is the addition of nonenzymatic and spontaneous reactions to the network. These reactions can be found in published literature or in a few online data sources, such as KEGG. Information available in the latter can be retrieved automatically using tools that support such operation, for example, *merlin 2.0* or MicrobesFlux.[61]

6.4.2.3 Stoichiometry

After collecting the set of reactions, their stoichiometry should be revised. Information on this step may be found in databases such as BRENDA–KEGG–MetaCyc reactions (BKM-react), BRENDA, and MetaCyc.

6.4.2.4 Localization

The next step is the compartmentation of the reactions. The localization of enzymes inside compartments or outside the cell is important for the development of GSMMs, because it determines the organelles in which the enzymes operate. In prokaryotes, the compartments are typically limited to the cytosol, periplasmic space, and extracellular space. In fungi, and other eukaryotes, the reactions can occur in various compartments, including Golgi apparatus, lysosome, mitochondrion, endoplasmic reticulum, or glyoxysome. For higher eukaryotes, it may be further necessary to differentiate between tissues. The first GSMM reconstruction of *S. cerevisiae*[83] accounted for 3 compartments, the second[84] 8, and the consensus[85] 15. The existence of several *S. cerevisiae* GSMMs also demonstrates the dynamism of GSMMs, which are in continuous improvement.

It is important to distinguish similar reactions with the same metabolites and stoichiometry, but being held in different compartments, as distinct reactions. Likewise, it is critical to annotate one metabolite in different compartments as distinct metabolites. That is, the metabolite should be replicated in each compartment and its name and identifier should reflect each localization. For example, if glucose is present in the exterior of the cell and in cytosol, then two glucose species should be created and their names could be glucose$_{ext}$ and glucose$_{cyt}$.

Several bioinformatics tools were developed for predicting the localization of enzymes from the amino acid sequence of the proteins and physiological data of the organism. The most commonly used applications for this are tools from the PSort family[86,87] and TargetP.[88] Also, information about the localization of enzymes and reactions can be found in the literature and online databases such as UniProt. Nevertheless, when in doubt, reactions and metabolites are usually assigned to the cytosol. The compartmentalized draft GSMN serves as an input for the next step.

6.4.2.5 Manual Curation

Unfortunately, automatic methods, although very useful, are fallible.[77] An automatic draft reconstruction will be incomplete because it will have missing reactions and it may contain reactions irrelevant to the GSMM.[57] To obtain a metabolic network that reflects each organism's specificities, revision of all reactions added to the network is mandatory. Thus, the last step of the GSMN reconstruction is the thorough revision of the literature, including publications and textbooks, organism-specific databases, and consultation of expert researchers, for the validation of the reactions set—the so-called manual curation of the GSMN. This step may involve verifying the data sources used for building the network; hence, the decisions taken throughout the reconstruction process should be traceable. Unlike the automated generation of the draft GSMN, manual curation can be slow and tedious. Manual curation should deal with issues such as (1) the inconsistencies of the proteins and function identifiers, (2) the addition of new organism-specific reactions unavailable in the queried data sources, and (3) assessment of the assignment of reactions to ambiguous identifiers,[77] such as partial EC numbers (e.g., assignment of several reactions to EC number 3.6.1.).

The presence of each reaction of the model in the target organism metabolism should be confirmed in this step. Dubious reactions for which no evidence has been found in literature should be discarded.[56] For instance, various enzymes may potentially promote several reactions; however, only reactions specific to the target organism should be included in the GSMN. Accordingly, the charged formula of each metabolite should be determined, as the metabolites inside the cells may be protonated or deprotonated. For instance, the pH of organelles may alter the protonation state of the metabolites.

The reversibility of the reactions may be (1) assessed by biochemical studies of enzymes of the target or closely related organisms or (2) determined from the estimation of the standard Gibbs free energy of formation ($\Delta_f G'^0$) and of reaction ($\Delta_r G'^0$) as demonstrated in previous studies.[89,90]

Reconstructions can also be accelerated[91] and curated using comparative genomics, by paralleling the draft network with curated models from closely related organisms. A different approach, which can be combined with the previous, is comparing the draft network to known biological pathways and searching for gaps. A gap in a metabolic model refers to a reaction in a pathway uncoupled to a gene.[77] The lack of a reaction in a biological pathway would lead to accumulation of compounds produced by energetically favored reactions. Simultaneously, the downstream of the pathway would be halted because the substrate produced by the absent enzyme would be missing. Thus, gap-filling analysis should be performed so that missed reactions can be added. One of the major biological databases used for studying gaps in networks is KEGG pathways, which is a collection of manually drawn pathway maps representing molecular interaction and reaction networks.

Some enzymes may use several cofactors to convert substrates. However, substrate specificity and the directionality of the reaction may vary between organisms.[57] If it is known that the organism of interest only uses one of the cofactors in a reaction, such information should be taken into account when performing the manual curation.

Finally, a debugged GSMN is obtained, which is converted into a mathematical computational GSMM in the next stage.

6.4.3 Converting the Metabolic Network to a Stoichiometric Model

The third stage involves the conversion of the reactions set into a metabolic model encompassing the conversion of the network to a stoichiometric matrix and the addition of constraints to the model. Before converting the network to a GSMM, the biomass formation equation should be included in the reactions set. The biomass equation represents the macromolecular composition of the cell and the building blocks used to generate those molecules. To perform simulations it is necessary to include a reaction that denotes a drain of biomolecules (e.g., amino acids, nucleotides) into the biomass. The biomass formation reaction can be represented by the following equation:

$$\sum_{k=1}^{P} c_k X_k \rightarrow \text{biomass} \tag{6.2}$$

where c_k represents the coefficient of the metabolite X_k. The flux associated with this reaction represents the growth rate of an organism.[56] This equation should include growth-associated energy requirements in terms of ATP molecules per mass (grams) of biomass synthesized. Alternatively, if the biomass formation reaction cannot be determined, the biomass equation of a related organism is typically used. Previous studies[92] suggest that this alternative approach does not introduce significant errors in the model. Nevertheless, some studies to confirm this statement should be performed, as in some cases biomass composition can be significantly altered, such as in deletion mutants.[56]

When the metabolic network is complemented with the biomass equation and the nongrowth ATP requirements (represented simply by a drain of ATP into ADP and inorganic phosphate), the set of reactions can be represented in the form of a stoichiometric matrix. The classic principles of chemical engineering can be used to construct the matrix that represents the dynamic behavior of the metabolite concentration, by performing dynamic mass balances with ordinary differential equations, according to the following notation:

$$\frac{dX_i}{dt} = \sum_{j=0}^{N} S_{ij} \cdot v_j + \mu X, \quad i = 1, \ldots, M \tag{6.3}$$

Equation 6.3 represents the rate of change of the concentration of metabolite i with time t. X_i is then the concentrations of metabolite i, v_j is the rate of reaction j (i.e., its metabolic flux), and S_{ij} is the stoichiometric coefficient of metabolite i in reaction j. The growth rate of the system is represented by μX.

The development of models that predict all concentration profiles as functions of time would imply determining stoichiometry and kinetic rates of all biochemical reactions, at specific conditions in the model. However, at the present time it is virtually impossible to collect kinetic expressions and parameters at the genome scale, hindering the development of dynamic models. Thus, a steady-state approximation is applied where it is assumed that metabolite concentrations remain constant throughout time. Also, it is reasonable to assume that the values of the fluxes are several times greater than the specific growth rate. Thus, the rates of consumption become equal to the rates of production of the metabolites and Equation 6.3 is converted in the matrix format to

$$S \cdot v = 0 \tag{6.4}$$

where
   v is the flux vector
   S is the stoichiometric matrix, where columns represent reactions and rows the metabolites

Worthy of note is the fact that, in Equation 6.4, v also includes exchange fluxes.

Most metabolic networks are underdetermined systems, as the number of fluxes is much greater than the number of mass balance constraints. Therefore, an infinite number of solutions (flux distributions) may satisfy the mass balance constraints, the so-called null space of S.[2,56] It is therefore not possible to have detailed information on the cell behavior and compute a single solution.[16] Yet it is possible to establish constraints that limit such behavior. The imposition of these constraints (e.g., determining irreversibility of reactions, nongrowth ATP requirements, measuring of exchange flux values) can reduce the *null space of S* to a set of feasible solutions, the flux cone of solutions.

The main constraints that should be added to the mathematical model are related to the reversibility of the reactions. Usually, a reversible reaction is constrained between minus infinity and plus infinity; irreversible reactions should be constrained in the minimum (or the maximum depending on the directionality of the reaction) flux to zero.

Similarly, transport fluxes for most nutrients should be constrained between 0 and a maximum. It is mandatory that the limiting substrate maximum uptake rate is constrained to a specific uptake rate. The constraining of oxygen may be significant for simulating chemostat cultivations of Crabtree-positive yeasts,[93] as these yeasts exhibit fermentative metabolism in aerobic conditions at high-glucose concentrations and high growth rates. Metabolites unavailable in the medium should be constrained to zero, and metabolites that may be excreted should be left unconstrained in the outward direction.

As an example of the conversion of a GSMN to a GSMM, consider the pseudo metabolic system composed of metabolites A–G and compartments 1 and 2 (Figure 6.4).

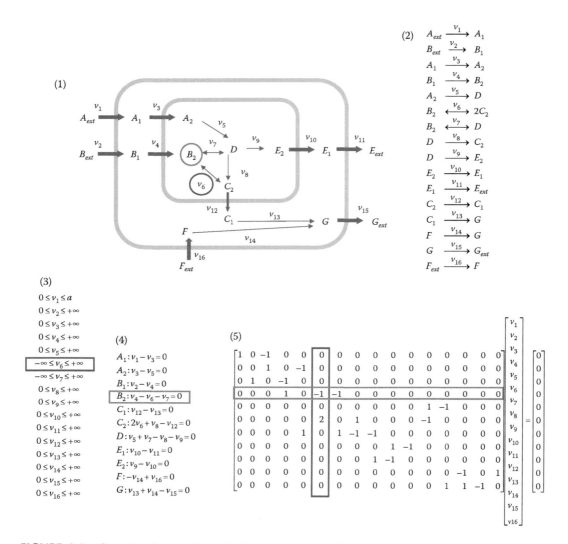

FIGURE 6.4 (See color insert) Example of a pseudo-metabolic network with seven metabolites (A–G) and 16 fluxes (v_1–v_{16}). The scheme of the reaction is described in (1), where the boundaries of the system are also outlined. Fluxes v_1–v_4, v_{10}–v_{12}, v_{15}, and v_{16} represent exchange fluxes of metabolic substrates (A, B, and F) and products (G and E). The reversibility of the reactions is indicated by the arrows, where double arrows represent reversible reactions and a forward arrow is used to characterize irreversible reactions. In Section (2), the stoichiometry of the network is represented. Section (3) shows the constraints around the flux values (where a represents the maximum uptake rate for the consumption of the limiting substrate A), and the steady-state mass balances are described in panel (4). A flux value may be negative for reversible reactions with unconstrained fluxes (e.g., v_6). Section (5) shows the stoichiometric matrix in which the mass balances are represented.

Several of the steps described thus far are illustrated in Figure 6.4. The network is described in Section (1), where metabolite A represents the limiting substrate available to the system. The flux for this metabolite is constrained to the maximum uptake rate, a, while metabolites B and F are freely uptaken into the system. Metabolites E and G exit the system as metabolic products through unconstrained fluxes. Section (2) provides the set of reactions, where the directionality and stoichiometry are shown. In Section (3), both the internal and exchange fluxes are bounded in a set of inequalities. Worthy of note is the fact that flux v_1 is restricted between 0 and a. Moreover, the fluxes for reversible reactions v_6 and v_7 are unconstrained. Section (4) shows the steady-state mass balances for each metabolite. The last section presents the stoichiometric matrix.

The mathematical representation of the model should then be saved in a computational-friendly format, for example, SBML,[65] so that simulations can be performed in specialized tools developed for such effect such as OptFlux[94] or COBRA.[95] Moreover, it is important to use standards, for example, MIRIAM,[96] when reconstructing GSMMs so that distinct reconstructions of the same organism can be compared and our understanding of such organism is further enhanced.[58]

6.4.4 Validation of the Metabolic Model

Once the mathematical representation of the model is created, it can be used to predict the behavior of the target organism and compare it to experimental data. Thus, increasing the knowledge on the physiology, biochemistry, and genetics of the target organism will improve the predictive capabilities of the model.[58] Nevertheless, if experimental data on the target organism are scarce or not available, data on phylogenetic neighbors can be of great help.[58]

FBA is currently the most used methodology to compute a solution from the flux cone of a GSMM. The linear programming of FBA, which can be resolved using one of several available solvers, corresponds to the maximization or minimization of a linear combination of metabolic fluxes, subject to the constraints imposed by the metabolic model and by upper and lower bounds on the fluxes. Usually the growth rate is the flux to be maximized, as various studies demonstrated that organisms tend to maximize the specific growth rate when exposed to limitations in the carbon source.[20,56] Therefore, FBA with maximization of biomass formation can be regarded as a simulation method. However, this approach can also be used in simulations with different objective functions such as maximization/minimization of ATP production, or to evaluate maximum production capabilities by maximizing a specific target compound.[97]

One of the first analyses that can be performed for model validation is using high-throughput growth phenotyping data, obtained, for instance, from Biolog, Inc.,[98] to assess the simulation results. These techniques allow testing several carbons, nitrogen, and other nutrient sources simultaneously. Thus, the model can be tested for growth in several limiting substrates and the results compared to the high-throughput data. If the GSMM predictions are not in accordance with the experimental results, the model should be examined and potentially missing reactions included and incorrect reactions removed.

Another assessment typically performed for the validation of GSMMs is the analysis of strategic fluxes, for example, specific growth rates and by-product formation, or the corresponding yields, for different growth conditions described in the literature. These fluxes or yields can be calculated by imposing the reported environmental conditions as model constraints and can be straightforwardly compared to the published data. Besides serving as model validation, this information can also be used to calibrate the model, for example, by fine-tuning the ATP (growth and nongrowth associated) parameters previously described. The study of active pathways under specific growth conditions (e.g., aerobic or anaerobic growth) can be performed for model validation. The nonzero fluxes should be analyzed, and if inconsistencies between the model and experimental data arise, the model should be reviewed and corrected.[56] Another approach for validating GSMMs is assessing simulation results to experiments performed with deletion mutants. This approach can provide valuable insights into the predictive capabilities of the model, and a good training set may be of great value for model debugging. The prediction of the phenotypical behavior of the microorganism when a deletion is simulated in the GSMM should replicate the experimental data. If it does not, the genome annotation should be revised and corrected.

Independent of the methodology used to validate the model, the fact is that if the model does not comply with the expectations, further debugging must be performed. The model must be thoroughly analyzed, so that the error(s) within it is (are) found. The data sources should be consulted subsequently and the reactions set and stoichiometric matrix corrected. When the model predictions are not in conformation with the experimental data, the process should be repeated from the second stage onward. In some specific infrequent cases, the genome annotation may also have to be reviewed, especially if the (re)annotation of the genome was not performed. Finally, the validation of the model may have to revisit some of the decisions taken in the manual curation step, where wrong conclusions may have been inferred.

6.5 APPLICATIONS

In this section, a brief overview of some of the most promising developments in this field will be performed. The rapid increase of fungal genome sequence availability affords the execution of comparative analysis within these organisms, and it provides insights into the genomic diversity that can unravel industrially relevant resources. Moreover, yeasts, particularly *S. cerevisiae*, have traditionally been regarded as model organisms for researching cellular physiology or biotechnological cell factories.

Table 6.2 shows the currently available fungi GSMM reconstructions. A total of eight fungal species have GSMMs available, and to our knowledge, two others are being finalized. Nevertheless, Table 6.2 exhibits 17 models, as some organisms, namely, *S. cerevisiae* and *Pichia pastoris*, have more than one model based on the same and different strains, respectively. Still, all models in the table are composed of a different number of genes, metabolites, and reactions. The first model was reconstructed in 2003, for *S. cerevisiae*, and a decade later, the reconstruction of the *Ashbya gossypii* and *K. lactis* models is being concluded within the research group of the present authors. After successfully validating the reconstruction of a GSMM, the model can be used to perform various tasks, including (1) the prediction of the phenotypical behavior of the target organism in different environmental and genetic perturbations, (2) the analysis of the robustness of the network when changing the flux levels of essential gene products,[99] and (3) performing *in silico* metabolic engineering.

The following examples demonstrate successful case studies where fungal metabolic models were used to accomplish several goals. In 2003, Famili et al.[112] computed the consequences of gene knockouts on growth phenotypes using the first GSMMs of *S. cerevisiae*. They observed that the results were consistent with experimental observations, proving that a constraint-based approach could be used to predict phenotypes. Meanwhile, several works have used GSMMs to predict such phenotypes. Bro et al.[113] in 2006 used the *iFF708 S. cerevisiae* GSMM to predict the best strategy for decreasing glycerol yield, while increasing ethanol yield, from glucose under anaerobic conditions. Four approaches were tested *in silico*, including the deletion and the insertion of several genes. The best strategy involved the heterologous expression of a D-glyceraldehyde-3-phosphate:NADP+ oxidoreductase (EC 1.2.1.9), which catalyzes the irreversible oxidation of glyceraldehyde-3-phosphate and NADP+ into 3-phosphoglycerate and NADPH during glycolysis. They were able to predict, *in silico*, the complete elimination of glycerol formation coupled with an increase of 10% in ethanol yield. The implementation of this strategy, *in vivo*, allowed engineering an *S. cerevisiae* strain that decreased the glycerol yield by about 40% on glucose, whereas the ethanol yield was increased by 3%, without any effect on the maximum specific growth rate. Asadollahi et al.[114] in 2009 used a GSMM to identify new target genes to enhance the biosynthesis of sesquiterpenes in *S. cerevisiae*. The effect of gene deletions on the flux distributions was assessed using the minimization of metabolic adjustments[115] (MOMA) as the objective function. The best target was the deletion of the NADPH-dependent glutamate dehydrogenase encoded by GDH1. *In vivo*, such deletion enhanced the availability of cytosolic NADPH, which could then be used by other enzymes such as HMG-CoA reductase. The cubebol yield was increased by approximately 85%, although a significant decrease in the maximum specific growth rate was also detected. Furthermore, in 2010, Brochado et al.[116] improved vanillin productivity in *S. cerevisiae* aiming at the development of an alternative to the chemical synthesis of this flavoring agent. Previous work[117] engineered the implementation

TABLE 6.2
Currently Available and Ongoing Fungal Reconstructions

Organism	Genes	Type	Model ID	Genes	Metabolites	Reactions	Compartments	Date	Reference
Ashbya gossipy ATCC 10895	4,726	f	iDG1137	1137	1285	1758	3 (c,e,m)	13	Gomes et al. unpublished data
Aspergillus nidulans	9,451	f	iHD666	666	732	794	4 (c,e,m,o)	08	David et al.[100]
Aspergillus niger CBS 513.88	14,165	f	iMA871	871	1045	1190	3 (c,e,m)	08	Andersen et al.[101]
Aspergillus oryzae RIB40	12,074	f	—	1314	1073	1053	3 (c,e,m,x)	08	Vongsangnak et al.[102]
Candida glabrata CCTCC M202019	6,885	b/f	iNX804	804	1025	1287	6 (c,m,p,v,g,e)	12	Xu et al.[103]
Kluyveromyces lactis CBS 2359	5,448	b	iOD907	907	1476	1867	4 (c,e,m,r)	13	Dias et al.[104]
Pichia pastoris DSMZ 70382	5,450	b	PpaMBEL1254	540	1147	1254	8 (c,e,g,m,n,r,v,x)	10	Sohn et al.[105]
Pichia pastoris GS115	5,313	b	iPP668	668	1177	1361	8 (c,e,g,m,n,r,v,x)	10	Chung et al.[106]
			iLC915	915	899	1423	7 (c,m,p,v,r,g,n)	12	Caspeta et al.[107]
Pichia stipitis CBS 6054	5,841	b	iSS884	884	992	1332	4 (c,m,p,v)	12	
Saccharomyces cerevisiae Sc288	6,183	b	iFF708	708	584	1175	3 (c,e,m)	03	Förster et al.[83]
			iND750	750	646	1149	8 (c,e,m,x,n,r,v,g)	04	Duarte et al.[84]
			iLL672	672	636	1038	3 (c,e,m)	05	Kuepfer et al.[108]
			iIN800	800	1013	1446	3 (c,e,m)	08	Nookaew et al.[109]
			iMH805/775	832	1168	1857	15 (c,e,g,m,n, r,v,x, gm,mm,nm,pm,rm,vm,xm,)	08	Herrgård et al.[85]
			iMM904	904	713	1412	8 (c,e,m,x,n,r,v,g)	09	Mo et al.[110]
Schizosaccharomyces pombe Sz-0205	4,940	f	SpoMBEL1693	605	1744	1693	8 (c,e,m,x,n,r,v,g)	12	Sohn et al.[111]

Compartment abbreviations: c, cytoplasm; e, extracellular; g, Golgi; m, mitochondrion; n, nucleus; o, glyoxysomes; r, ER; v, vacuole; x, peroxisome; gm, Golgi membrane; mm, mitochondrial membrane; nm, nuclear membrane; pm, plasma membrane; rm, ER membrane; vm, vacuolar membrane; xm, peroxisomal membrane.

Type of fungi: b, budding; f, filamentous.

of a *de novo* synthetic pathway for heterologous vanillin production from glucose in *S. cerevisiae*. The *S. cerevisiae* GSMM was revised to reflect the alterations performed in this study, which was enhanced with an *Arabidopsis thaliana* glycosyltransferase. The utilization of bioinformatics tools, that is, OptGene,[118] selected the deletion of two targets (PDC1 and GDH1) for experimental validation. The verification of the targets successfully led to overproducing strains, with vanillin yields increasing between 1.5- and 5-fold, when compared to the previous work on *de novo* vanillin biosynthesis in *S. cerevisiae*.

6.6 FUTURE APPLICATIONS

6.6.1 Health Applications

The potentialities of systems biology include the *in silico* design of novel drug targets, innovative therapies for the treatment of complex diseases, and developing cell factories for the production of drugs, biofuels, and chemicals of interest. Although, as seen in the previous segment, the latter two are already well implemented, the others are still in development.

In fact, it is expected that systems biology will facilitate the drug discovery process, using genome-scale models for simulations that may lead to the identification of optimal drug targets. The identification of these targets will allow developing more efficient drugs with fewer side effects. *Plasmodium falciparum* is the microorganism responsible for one of the world's most common and deadly diseases, specifically malaria. In 2004, Yeh et al.[119] initiated the *Plasmodium* genome project in which the PlasmoCyc network model was developed and used to identify 216 *chokepoint enzymes*, which are enzymes that catalyze a reaction that either uniquely consumes a specific substrate or uniquely produces a specific product. Interestingly, all three approved drugs for malaria and 87.5% of proposed drug targets with biological evidence in the literature are related to the so-called chokepoint enzymes. Hence, the identification of those chokepoint enzymes may represent one systematic way of identifying potential metabolic drug targets. Moreover, Fatumo et al.[120] in 2007 proceeded with this line of work and presented a refined list of 22 new potential candidate targets for *P. falciparum*, half of which had reasonable evidence to be valid targets against microorganisms and cancer. Since some yeasts such as *Candida albicans*[121] are major human pathogens, drug targets can also be predicted for pathogenic yeasts, and therefore for human diseases in the same way, they were predicted for *P. falciparum*.

Moreover, the genome sequences of fungi, especially yeasts, and humans revealed a high degree of conservation, which emphasizes the value of yeasts as a tool for drug discovery.[121] Thirty percent of the human genes involved in diseases had functional homologues in yeasts,[3,122] and yeasts are also currently seen as model systems for anticancer drug discovery.[123]

In 2010, Bordbar et al.[124] combined an alveolar macrophage submodel of the Recon 1[9] *H. sapiens* metabolic reconstruction, with the iNJ661 *Mycobacterium tuberculosis in silico* strain,[125] for constructing the iAB-AMØ-1410-Mt-661 host–pathogen model, with the premise that it is known that the pathogenesis of this bacterium is stimulated from its metabolic coupling to the host. The integrated host–pathogen network enabled the simulation of the metabolic changes during infection. This approach could be applied to better understand the human–pathogen interactions for several fungal infections, also helping to indicate potential drug targets. As an example, the fungus *Candida glabrata*, which is an emerging oral opportunistic pathogen, has a genome-scale reconstruction available, thus allowing the development and analysis of host–pathogen coupled models. The analysis of the metabolic interactions in these hybrid models should also provide valuable insights on the mechanisms that these type of pathogens use for infecting hosts.

These advances provide a view of the future in which systems biology will be used to analyze the interactions between pathogenic microorganisms and systems as complex as the human body for predicting diseases, from omics studies performed on each individual, starting a new scientific era.

The issue of mycotoxins produced by fungi may also be approached by using *in silico* models. For example, the *Aspergillus niger* model predicts the production of ochratoxin A, which is an abundant food-contaminating mycotoxin. The *Aspergillus nidulans* metabolic model includes the biosynthesis of sterigmatocystin, which is a carcinogenic mycotoxin precursor of aflatoxin B1. However, surprisingly, the *A. nidulans* model does not predict the excretion of this metabolite despite the fact that *A. nidulans* is a well-known sterigmatocystin producer.[126] Nevertheless, the study of this model should provide insights on the secondary metabolism for other fungi, which produce aflatoxins, namely, *Aspergillus flavus* and *Aspergillus parasiticus*,[127] facilitating the understanding and the eventual reconstruction of metabolic models, for the latter fungi. These models can be combined with data from other omics data, for instance, microarray analysis of *A. flavus* aflatoxin production,[128] for validation. Finally, the application of systems biology tools within these models may allow, in the future, to better understand the mycotoxin production pathways and their regulation and pinpoint ways to avoid food and feed contamination.

6.6.2 Industrial Applications

In the field of metabolic engineering, it is common to engineer fungi as cell factories. *Ashbya gossypii* is a filamentous fungus that had its genome sequenced in 2004 and presents one of the smallest genomes among the eukaryotes, with 4726 protein-coding genes. Despite presenting filamentous growth, a 95% genome similarity was reported to *S. cerevisiae*, even with the latter having a significantly higher number of genes. *A. gossypii* is an interesting target for metabolic engineering, as it benefits from the huge amount of information available for *S. cerevisiae*, which indirectly increases the knowledge regarding *A. gossypii* in addition to the low-complexity genome and the haploid nucleus, which are suitable for genetic manipulation. The ongoing reconstruction of a GSMM for *A. gossypii*, which has been intensively used for industrial riboflavin production, may be a critical step on the overall improvement of this fungus as an efficient cell factory system.

6.7 FINAL REMARKS

Systems biology may be regarded as an integrative discipline that aims at organizing the data provided from biology's reductionist approach. Large-scale data sets are converted into *in silico* models, allowing replication and prediction of the behavior of organisms. These data may come from several new *omics* technologies, namely, genomics, transcriptomics, fluxomics, or metabolomics.

A high-quality reconstruction is a combination of automatic retrieving of potentially relevant data and intensive manual curation.[57] A good metabolic reconstruction is the first step for understanding the genotype–phenotype association of a given organism. However, the metabolic reconstruction process is never concluded, since the knowledge on the metabolism of every organism is always growing.[77] The minimal requirement for considering that an operational model has been achieved is that the reconstructed metabolic network is consistent with the physiology of the organism it aims to model. For instance, it should include all pathways known to be present in the target organism, and reactions should be balanced and essential pathways complete if the purpose of the model is performing flux predictions. GSMMs represent a fraction (about 30%) of all genes, namely, genes encoding enzymes and transport systems. However, the absence of mechanistic details (kinetic rate constants) is a shortcoming of these models, since it restricts the model simulations to complete mutations (i.e., full gene additions or deletions). Metabolic models become truly useful when analyzing the solution space of a simulation where specific constraints, such as gene deletions and additions or other constraints, have been imposed.[5] Moreover, the integration of metabolic models with regulatory (gene regulation) and signaling (protein regulation) networks may improve the prediction capabilities of the models and the coverage of the genome.[77] Recent developments in systems biology demonstrate that this discipline will assume a crucial role in how biological and biomedical research is performed.

REFERENCES

1. Kitano, H. Systems biology: A brief overview. *Science (New York, N.Y.)* **295**, 1662–1664 (2002).
2. Palsson, B. Ø. *Systems Biology: Properties of Reconstructed Networks.* (Cambridge University Press: New York, 2006).
3. Mustacchi, R., Hohmann, S., and Nielsen, J. Yeast systems biology to unravel the network of life. *Yeast (Chichester, England)* **23**, 227–238 (2006).
4. Hohmann, S. The yeast systems biology network: Mating communities. *Current Opinion in Biotechnology* **16**, 356–360 (2005).
5. Petranovic, D. and Vemuri, G. N. Impact of yeast systems biology on industrial biotechnology. *Journal of Biotechnology* **144**, 204–211 (2009).
6. Zelezniak, A., Pers, T. H., Soares, S., Patti, M. E., and Patil, K. R. Metabolic network topology reveals transcriptional regulatory signatures of type 2 diabetes. *PLoS Computational Biology* **6**, e1000729 (2010).
7. Li, X. et al. RCM: A novel association approach to search for coronary artery disease genetic related metabolites based on SNPs and metabolic network. *Genomics* **100**, 282–288 (2012).
8. Altschul, S. F., Gish, W., Miller, W., Myers, E. W., and Lipman, D. J. Basic local alignment search tool. *Journal of Molecular Biology* **215**, 403–410 (1990).
9. Duarte, N. C. et al. Global reconstruction of the human metabolic network based on genomic and bibliomic data. *Proceedings of the National Academy of Sciences of the United States of America* **104**, 1777–1782 (2007).
10. Covert, M. W., Knight, E. M., Reed, J. L., Herrgard, M. J., and Palsson, B. O. Integrating high-throughput and computational data elucidates bacterial networks. *Nature* **429**, 92–96 (2004).
11. Faria, J. P. et al. Genome-scale bacterial transcriptional regulatory networks: Reconstruction and integrated analysis with metabolic models. *Briefings in Bioinformatics* **15**(4), 592–611 (2013). doi:10.1093/bib/bbs071.
12. Cannon, W. B. *Bodily Changes in Pain, Hunger, Fear, and Rage.* 311 (D. Appleton and Company: New York, 1915).
13. Von Bertalanffy, L. An outline of general system theory. *The British Journal for the Philosophy of Science* **1**, 134–165 (1950).
14. Stein, L. Genome annotation: From sequence to biology. *Nature Reviews. Genetics* **2**, 493–503 (2001).
15. Rizzetto, L. and Cavalieri, D. Friend or foe: Using systems biology to elucidate interactions between fungi and their hosts. *Trends in Microbiology* **19**, 509–515 (2011).
16. Palsson, B. The challenges of *in silico* biology. *Nature Biotechnology* **18**, 1147–1150 (2000).
17. Hood, L. Systems biology: Integrating technology, biology, and computation. *Mechanisms of Ageing and Development* **124**, 9–16 (2003).
18. Isalan, M. Systems biology: A cell in a computer. *Nature* **488**, 40–41 (2012).
19. Kitano, H. Computational systems biology. *Nature* **420**, 206–210 (2002).
20. Otero, J. M. and Nielsen, J. Industrial systems biology. *Biotechnology and Bioengineering* **105**, 439–460 (2010).
21. Patil, K. R., Akesson, M., and Nielsen, J. Use of genome-scale microbial models for metabolic engineering. *Current Opinion in Biotechnology* **15**, 64–69 (2004).
22. Vemuri, G. N. and Aristidou, A. A. Metabolic engineering in the -omics era: Elucidating and modulating regulatory networks. *Microbiology and Molecular Biology Reviews: MMBR* **69**, 197–216 (2005).
23. Stephanopoulos, G. Metabolic fluxes and metabolic engineering. *Metabolic Engineering* **1**, 1–11 (1999).
24. Sanger, F. and Coulson, A. R. A rapid method for determining sequences in DNA by primed synthesis with DNA polymerase. *Journal of Molecular Biology* **94**, 441–448 (1975).
25. Goffeau, A. et al. Life with 6000 genes. *Science* **274**, 546–567 (1996).
26. Palsson, B. Metabolic systems biology. *FEBS Letters* **583**, 3900–3904 (2009).
27. Mitra, R. D. and Church, G. M. In situ localized amplification and contact replication of many individual DNA molecules. *Nucleic Acids Research* **27**, e34 (1999).
28. Brenner, S. et al. Gene expression analysis by massively parallel signature sequencing (MPSS) on microbead arrays. *Nature Biotechnology* **18**, 630–634 (2000).
29. Margulies, M. et al. Genome sequencing in microfabricated high-density picolitre reactors. *Nature* **437**, 376–380 (2005).
30. Bentley, D. R. Whole-genome re-sequencing. *Current Opinion in Genetics & Development* **16**, 545–552 (2006).

31. Hernandez, D., François, P., Farinelli, L., Osterås, M., and Schrenzel, J. *De novo* bacterial genome sequencing: Millions of very short reads assembled on a desktop computer. *Genome Research* **18**, 802–809 (2008).
32. Warren, R. L., Sutton, G. G., Jones, S. J. M., and Holt, R. A assembling millions of short DNA sequences using SSAKE. *Bioinformatics (Oxford, England)* **23**, 500–501 (2007).
33. Chan, E. Y. Advances in sequencing technology. *Mutation Research* **573**, 13–40 (2005).
34. Simpson, J. T. and Durbin, R. Efficient *de novo* assembly of large genomes using compressed data structures. *Genome Research* **22**, 549–556 (2012).
35. Alioto, T. Gene prediction. *Methods in Molecular Biology (Clifton, N.J.)* **855**, 175–201 (2012).
36. Stanke, M. and Morgenstern, B. AUGUSTUS: A web server for gene prediction in eukaryotes that allows user-defined constraints. *Nucleic Acids Research* **33**, W465–W467 (2005).
37. Salzberg, S. L., Delcher, A. L., Kasif, S., and White, O. Microbial gene identification using interpolated Markov models. *Nucleic Acids Research* **26**, 544–548 (1998).
38. Borodovsky, M. and McIninch, J. GENMARK: Parallel gene recognition for both DNA strands. *Computers & Chemistry* **17**, 123–133 (1993).
39. Foissac, S. and Schiex, T. Integrating alternative splicing detection into gene prediction. *BMC Bioinformatics* **6**, 25 (2005).
40. Krause, L. et al. GISMO—Gene identification using a support vector machine for ORF classification. *Nucleic Acids Research* **35**, 540–549 (2007).
41. Meyer, I. M. A practical guide to the art of RNA gene prediction. *Briefings in Bioinformatics* **8**, 396–414 (2007).
42. Ouzounis, C. A. and Karp, P. D. The past, present and future of genome-wide re-annotation. *Genome Biology* **3**, 1–6 COMMENT2001 (2002).
43. Eddy, S. R. Profile hidden Markov models. *Bioinformatics (Oxford, England)* **14**, 755–763 (1998).
44. Dias, O., Rocha, M., Ferreira, E. C., and Rocha, I. Merlin: Metabolic models reconstruction using genome-scale information. *Proceedings of the 11th International Symposium on Computer Applications in Biotechnology (CAB 2010)*, Leuven, Belgium (J. R. Banga, P. Bogaerts, J. Van Impe, D. Dochain, I. Smets, eds.) pp. 120–125 (2010). doi:10.3182/20100707–3-BE-2012.0076.
45. Griffin, T. J. et al. Complementary profiling of gene expression at the transcriptome and proteome levels in *Saccharomyces cerevisiae*. *Molecular & Cellular Proteomics: MCP* **1**, 323–333 (2002).
46. Gygi, S. P., Rist, B., and Aebersold, R. Measuring gene expression by quantitative proteome analysis. *Current Opinion in Biotechnology* **11**, 396–401 (2000).
47. Shi, Y., Xiang, R., Horváth, C., and Wilkins, J. A. The role of liquid chromatography in proteomics. *Journal of Chromatography A* **1053**, 27–36 (2004).
48. Washburn, M. P., Wolters, D., and Yates, J. R. Large-scale analysis of the yeast proteome by multidimensional protein identification technology. *Nature Biotechnology* **19**, 242–247 (2001).
49. Nielsen, J. and Oliver, S. The next wave in metabolome analysis. *Trends in Biotechnology* **23**, 544–546 (2005).
50. Sauer, U. Metabolic networks in motion: 13C-based flux analysis. *Molecular Systems Biology* **2**, 62 (2006).
51. Christensen, B., Gombert, A. K., and Nielsen, J. Analysis of flux estimates based on 13C-labelling experiments. *European Journal of Biochemistry* **269**, 2795–2800 (2002).
52. Wiechert, W., Möllney, M., Petersen, S., and De Graaf, A. A. A universal framework for 13C metabolic flux analysis. *Metabolic Engineering* **3**, 265–283 (2001).
53. McCarthy, N. Systems biology: Lethal weaknesses. *Nature Reviews Cancer* **11**, 3109 (2011).
54. Papoutsakis, E. T. Equations and calculations for fermentations of butyric acid bacteria. *Biotechnology and Bioengineering* **26**, 174–187 (1984).
55. Savinell, J. M. and Palsson, B. O. Optimal selection of metabolic fluxes for in vivo measurement. I. Development of mathematical methods. *Journal of Theoretical Biology* **155**, 201–214 (1992).
56. Rocha, I., Förster, J., and Nielsen, J. Design and application of genome-scale reconstructed metabolic models. *Methods in Molecular Biology (Clifton, N.J.)* **416**, 409–431 (2008).
57. Feist, A. M., Herrgård, M. J., Thiele, I., Reed, J. L., and Palsson, B. Ø. Reconstruction of biochemical networks in microorganisms. *Nature Reviews Microbiology* **7**, 129–143 (2009).
58. Thiele, I. and Palsson, B. Ø. A protocol for generating a high-quality genome-scale metabolic reconstruction. *Nature Protocols* **5**, 93–121 (2010).
59. Dias, O., Rocha, M., Ferreira, E. C., and Rocha, I. Reconstructing genome-scale metabolic models with merlin 2.0. *Submitted* (2014).

60. Henry, C. S. et al. High-throughput generation, optimization and analysis of genome-scale metabolic models. *Nature Biotechnology* **28**, 977–982 (2010).
61. Feng, X., Xu, Y., Chen, Y., and Tang, Y. J. MicrobesFlux: A web platform for drafting metabolic models from the KEGG database. *BMC Systems Biology* **6**, 94 (2012).
62. Karp, P. D. et al. Pathway Tools version 13.0: Integrated software for pathway/genome informatics and systems biology. *Briefings in Bioinformatics* **11**, 40–79 (2010).
63. Funahashi, A., Morohashi, M., Kitano, H., and Tanimura, N. CellDesigner: A process diagram editor for gene-regulatory and biochemical networks. *BIOSILICO* **1**, 159–162 (2003).
64. Shannon, P. et al. Cytoscape: A software environment for integrated models of biomolecular interaction networks. *Genome Research* **13**, 2498–2504 (2003).
65. Hucka, M. et al. The systems biology markup language (SBML): A medium for representation and exchange of biochemical network models. *Bioinformatics (Oxford, England)* **19**, 524–531 (2003).
66. Caspi, R. et al. The MetaCyc database of metabolic pathways and enzymes and the BioCyc collection of pathway/genome databases. *Nucleic Acids Research* **40**, D742–D753 (2012).
67. Lang, M., Stelzer, M., and Schomburg, D. BKM-react, an integrated biochemical reaction database. *BMC Biochemistry* **12**, 42 (2011).
68. Schomburg, I., Chang, A., and Schomburg, D. B. Enzyme data and metabolic information. *Nucleic Acids Research* **30**, 47–49 (2002).
69. Artimo, P. et al. ExPASy: SIB bioinformatics resource portal. *Nucleic Acids Research* **40**, W597–W603 (2012).
70. Pagani, I. et al. The Genomes OnLine Database (GOLD) v.4: Status of genomic and metagenomic projects and their associated metadata. *Nucleic Acids Research* **40**, D571–D579 (2012).
71. Kanehisa, M. and Goto, S. KEGG: Kyoto encyclopedia of genes and genomes. *Nucleic Acids Research* **28**, 27–30 (2000).
72. Sayers, E. W. et al. Database resources of the National Center for Biotechnology Information. *Nucleic Acids Research* **37**, D5–D15 (2009).
73. Wittig, U. et al. SABIO-RK—Database for biochemical reaction kinetics. *Nucleic Acids Research* **40**, D790–D796 (2012).
74. Cherry, J. SGD: Saccharomyces genome database. *Nucleic Acids Research* **26**, 73–79 (1998).
75. Saier, M. H., Tran, C. V., and Barabote, R. D. TCDB: The Transporter classification database for membrane transport protein analyses and information. *Nucleic Acids Research* **34**, D181–D186 (2006).
76. Apweiler, R. et al. Ongoing and future developments at the universal protein resource. *Nucleic Acids Research* **39**, D214–D219 (2011).
77. Francke, C., Siezen, R. J., and Teusink, B. Reconstructing the metabolic network of a bacterium from its genome. *Trends in Microbiology* **13**, 550–558 (2005).
78. Barrett, A. J. et al. *Enzyme Nomenclature*. 862 (Academic Press: San Diego, CA, 1992).
79. Saier, M. H. A functional-phylogenetic classification system for transmembrane solute transporters. *Microbiology and Molecular Biology Reviews: MMBR* **64**, 354–411 (2000).
80. Dias, O., Gombert, A. K., Ferreira, E. C., and Rocha, I. Genome-wide metabolic (re-) annotation of *Kluyveromyces lactis*. *BMC Genomics* **13**, 517 (2012).
81. Gundogdu, O. et al. Re-annotation and re-analysis of the *Campylobacter jejuni* NCTC11168 genome sequence. *BMC Genomics* **8**, 162 (2007).
82. Haas, B. J. et al. Complete reannotation of the Arabidopsis genome: Methods, tools, protocols and the final release. *BMC Biology* **3**, 7 (2005).
83. Förster, J., Famili, I., Fu, P., Palsson, B. Ø., and Nielsen, J. Genome-scale reconstruction of the *Saccharomyces cerevisiae* metabolic network. *Genome Research* **13**, 244–253 (2003).
84. Duarte, N. C., Herrgård, M. J., and Palsson, B. Ø. Reconstruction and validation of *Saccharomyces cerevisiae* iND750, a fully compartmentalized genome-scale metabolic model. *Genome Research* **14**, 1298–1309 (2004).
85. Herrgård, M. J. et al. A consensus yeast metabolic network reconstruction obtained from a community approach to systems biology. *Nature Biotechnology* **26**, 1155–1160 (2008).
86. Yu, N. Y. et al. PSORTb 3.0: Improved protein subcellular localization prediction with refined localization subcategories and predictive capabilities for all prokaryotes. *Bioinformatics (Oxford, England)* **26**, 1608–1615 (2010).
87. Horton, P. et al. WoLF PSORT: Protein localization predictor. *Nucleic Acids Research* **35**, W585–W587 (2007).

88. Emanuelsson, O., Nielsen, H., Brunak, S., and Von Heijne, G. Predicting subcellular localization of proteins based on their N-terminal amino acid sequence. *Journal of Molecular Biology* **300**, 1005–1016 (2000).
89. Jankowski, M. D., Henry, C. S., Broadbelt, L. J., and Hatzimanikatis, V. Group contribution method for thermodynamic analysis of complex metabolic networks. *Biophysical Journal* **95**, 1487–1499 (2008).
90. Fleming, R. M. T., Thiele, I., and Nasheuer, H. P. Quantitative assignment of reaction directionality in constraint-based models of metabolism: Application to *Escherichia coli*. *Biophysical Chemistry* **145**, 47–56 (2009).
91. Notebaart, R. A., Van Enckevort, F. H. J., Francke, C., Siezen, R. J., and Teusink, B. Accelerating the reconstruction of genome-scale metabolic networks. *BMC Bioinformatics* **7**, 296 (2006).
92. Varma, A. and Palsson, B. O. Metabolic Capabilities of *Escherichia coli* II. Optimal Growth Patterns. *Journal of Theoretical Biology* **165**, 503–522 (1993).
93. Crabtree, H. G. The carbohydrate metabolism of certain pathological overgrowths. *The Biochemical Journal* **22**, 1289–1298 (1928).
94. Rocha, I. et al. OptFlux: An open-source software platform for *in silico* metabolic engineering. *BMC Systems Biology* **4**, 45 (2010).
95. Becker, S. A. et al. Quantitative prediction of cellular metabolism with constraint-based models: The COBRA toolbox. *Nature Protocols* **2**, 727–738 (2007).
96. Le Novère, N. et al. Minimum information requested in the annotation of biochemical models (MIRIAM). *Nature Biotechnology* **23**, 1509–1515 (2005).
97. Schuetz, R., Kuepfer, L., and Sauer, U. Systematic evaluation of objective functions for predicting intracellular fluxes in *Escherichia coli*. *Molecular Systems Biology* **3**, 119 (2007).
98. Bochner, B. R. Sleuthing out bacterial identities. *Nature* **339**, 157–158 (1989).
99. Edwards, J. S. and Palsson, B. O. Robustness analysis of the *Escherichia coli* metabolic network. *Biotechnology Progress* **16**, 927–939 (2000).
100. David, H., Ozçelik, I. S., Hofmann, G., and Nielsen, J. Analysis of *Aspergillus nidulans* metabolism at the genome-scale. *BMC Genomics* **9**, 163 (2008).
101. Andersen, M. R., Nielsen, M. L., and Nielsen, J. Metabolic model integration of the bibliome, genome, metabolome and reactome of *Aspergillus niger*. *Molecular Systems Biology* **4**, 178 (2008).
102. Vongsangnak, W., Olsen, P., Hansen, K., Krogsgaard, S., and Nielsen, J. Improved annotation through genome-scale metabolic modeling of *Aspergillus oryzae*. *BMC Genomics* **9**, 245 (2008).
103. Xu, N. et al. Reconstruction and analysis of the genome-scale metabolic network of *Candida glabrata*. *Molecular BioSystems* **9**, 205–216 (2013).
104. Dias, O. et al. iOD907, the first genome-scale metabolic model for the milk yeast *Kluyveromyces lactis*. *Biotechnology Journal* **9**, 776–790 (2014).
105. Sohn, S. B. et al. Genome-scale metabolic model of methylotrophic yeast *Pichia pastoris* and its use for *in silico* analysis of heterologous protein production. *Biotechnology Journal* **5**, 705–715 (2010).
106. Chung, B. K. et al. Genome-scale metabolic reconstruction and *in silico* analysis of methylotrophic yeast *Pichia pastoris* for strain improvement. *Microbial Cell Factories* **9**, 50 (2010).
107. Caspeta, L., Shoaie, S., Agren, R., Nookaew, I., and Nielsen, J. Genome-scale metabolic reconstructions of *Pichia stipitis* and *Pichia pastoris* and *in silico* evaluation of their potentials. *BMC Systems Biology* **6**, 24 (2012).
108. Kuepfer, L., Sauer, U., and Blank, L. M. Metabolic functions of duplicate genes in *Saccharomyces cerevisiae*. *Genome Research* **15**, 1421–1430 (2005).
109. Nookaew, I. et al. The genome-scale metabolic model iIN800 of *Saccharomyces cerevisiae* and its validation: A scaffold to query lipid metabolism. *BMC Systems Biology* **2**, 71 (2008).
110. Mo, M. L., Palsson, B. O., and Herrgård, M. J. Connecting extracellular metabolomic measurements to intracellular flux states in yeast. *BMC Systems Biology* **3**, 37 (2009).
111. Sohn, S. B., Kim, T. Y., Lee, J. H., and Lee, S. Y. Genome-scale metabolic model of the fission yeast *Schizosaccharomyces pombe* and the reconciliation of *in silico/in vivo* mutant growth. *BMC Systems Biology* **6**, 49 (2012).
112. Famili, I., Forster, J., Nielsen, J., and Palsson, B. O. *Saccharomyces cerevisiae* phenotypes can be predicted by using constraint-based analysis of a genome-scale reconstructed metabolic network. *Proceedings of the National Academy of Sciences of the United States of America* **100**, 13134–13139 (2003).
113. Bro, C., Regenberg, B., Förster, J., and Nielsen, J. *In silico* aided metabolic engineering of *Saccharomyces cerevisiae* for improved bioethanol production. *Metabolic Engineering* **8**, 102–111 (2006).

114. Asadollahi, M. A. et al. Enhancing sesquiterpene production in *Saccharomyces cerevisiae* through *in silico* driven metabolic engineering. *Metabolic Engineering* **11**, 328–334 (2009).
115. Segrè, D., Vitkup, D., and Church, G. M. Analysis of optimality in natural and perturbed metabolic networks. *Proceedings of the National Academy of Sciences of the United States of America* **99**, 15112–15117 (2002).
116. Brochado, A. R. et al. Improved vanillin production in baker's yeast through *in silico* design. *Microbial Cell Factories* **9**, 84 (2010).
117. Hansen, E. H. et al. *De novo* biosynthesis of vanillin in fission yeast (*Schizosaccharomyces pombe*) and baker's yeast (*Saccharomyces cerevisiae*). *Applied and Environmental Microbiology* **75**, 2765–2774 (2009).
118. Patil, K. R., Rocha, I., Förster, J., and Nielsen, J. Evolutionary programming as a platform for *in silico* metabolic engineering. *BMC Bioinformatics* **6**, 308 (2005).
119. Yeh, I., Hanekamp, T., Tsoka, S., Karp, P. D., and Altman, R. B. Computational analysis of *Plasmodium falciparum* metabolism: Organizing genomic information to facilitate drug discovery. *Genome Research* **14**, 917–924 (2004).
120. Fatumo, S. et al. Estimating novel potential drug targets of *Plasmodium falciparum* by analysing the metabolic network of knock-out strains *in silico*. *Infection, Genetics and Evolution: Journal of Molecular Epidemiology and Evolutionary Genetics in Infectious Diseases* **9**, 351–358 (2009).
121. Hughes, T. R. Yeast and drug discovery. *Functional and Integrative Genomics* **2**, 199–211 (2002).
122. Foury, F. Human genetic diseases: A cross-talk between man and yeast. *Gene* **195**, 1–10 (1997).
123. Simon, J. A. and Bedalov, A. Yeast as a model system for anticancer drug discovery. *Nature Reviews Cancer* **4**, 481–492 (2004).
124. Bordbar, A., Lewis, N. E., Schellenberger, J., Palsson, B. Ø., and Jamshidi, N. Insight into human alveolar macrophage and *M. tuberculosis* interactions via metabolic reconstructions. *Molecular Systems Biology* **6**, 422 (2010).
125. Jamshidi, N. and Palsson, B. Ø. Investigating the metabolic capabilities of *Mycobacterium tuberculosis* H37Rv using the *in silico* strain iNJ661 and proposing alternative drug targets. *BMC Systems Biology* **1**, 26 (2007).
126. Hajjar, J. D., Bennett, J. W., Bhatnagar, D., and Bahu, R. Sterigmatocystin production by laboratory strains of *Aspergillus nidulans*. *Mycological Research* **93**, 548–551 (1989).
127. Keller, N. P. and Adams, T. H. Analysis of a mycotoxin gene cluster in *Aspergillus Nidulans*. *SAAS Bulletin, Biochemistry and Biotechnology* **8**, 14–21 (1995).
128. Abdel-Hadi, A., Schmidt-Heydt, M., Parra, R., Geisen, R., and Magan, N. A systems approach to model the relationship between aflatoxin gene cluster expression, environmental factors, growth and toxin production by *Aspergillus flavus*. *Journal of the Royal Society, Interface/the Royal Society* **9**, 757–767 (2012).

7 Brief History of Fungal Genomics from Linkage Maps to Sequences

Kevin McCluskey and Scott E. Baker

CONTENTS

7.1 Linkage Mapping and the Origins of Fungal Genomics 93
    7.1.1 Fungal Chromosomes 93
    7.1.2 Linkage Maps 94
7.2 Genomic Libraries, Electrophoretic Karyotype Analysis, and Genome Sequencing 95
    7.2.1 Genomic Libraries 95
    7.2.2 Molecular Karyotyping and Reassociation Kinetics 96
    7.2.3 Sanger Sequencing and Fungal Genomics 96
    7.2.4 Next-Generation Fungal Genomics 97
        7.2.4.1 *De novo* Sequencing 97
        7.2.4.2 Resequencing Fungal Genomes 98
7.3 Preserving the Sequenced Fungi: A Crucial Role for the Biological Resource Centers 98
7.4 Conclusions 99
References 99

7.1 LINKAGE MAPPING AND THE ORIGINS OF FUNGAL GENOMICS

Genome analysis of fungi finds its foundational origins in the chromosome theory of biology. This theory was inherently observational and stated that inherited characteristics can be evaluated in terms of the presence or absence of chromosomes. It was based on the observation that traits on nonhomologous chromosomes assort independently [1]. At the time of these analyses, traits were being evaluated in terms of genetics for many organisms in addition to fungi. For some, such as bacteria and phages, the chromosome theory was not equivalent, but rather complementary. For most organisms, the traits that could be followed were physiological or morphological and included auxotrophies, temperature sensitivity, growth rate, and colony or cellular morphology [2]. The technology to study fungal genomes has changed over many years, and the information available from each technique is dependent on the resolving power (Table 7.1) and varies from a few data points per strain to millions.

7.1.1 FUNGAL CHROMOSOMES

Early analysis of the chromosome biology of fungi depended to a large extent on the ability to resolve fungal chromosomes using established cytological stains and was only practical for fungi such as *Neurospora* that have large chromosomes and exist in accessible tissue. McClintock first described the *Neurospora* karyotype as having seven chromosomes [3]. Similar descriptions were

TABLE 7.1
Approaches to Studying Fungal Genomes

Modality	Number of Data Points	Technical Difficulty	Limitations
Cytology	1–10	High	Species with large chromosomes
Genetics	1–1000	High	Genes with phenotypes
PFGE	2–30	Moderate	Variability among isolates
RFLP/AFLP	20–200	High	Markers are anonymous
RAPD	1–100	Moderate	Poorly reproducible
Whole-genome sequencing	$>10^6$	High	Computationally demanding

unavailable for other fungi that have many small chromosomes (such as *Ustilago*), and elucidation of chromosome numbers in many fungi occurred many years later.

Simultaneously, the study of fungal genetics relied on the ability to generate mutations and map them using the principles of Mendelian genetics [4]. These allowed the development of genetic linkage maps for some fungi including *Neurospora*. At these early stages, fungi had a significant advantage over other eukaryotic model systems, such as *Drosophila*, because fungi have haploid vegetative tissue, and even recessive mutations were immediately evident. Other fungi that were in use for the elucidation of the gene theory of inheritance included *Sordaria* [5], *Allomyces* [6], *Schizophyllum* [7], and *Aspergillus* [8]. In *Aspergillus*, sexual recombination was not discovered for many years, but genetic analysis was still possible by virtue of the ability to carry out mitotic or parasexual genetics [9].

For many fungi, the first character used in genetic mapping was the mating-type locus, and this was more readily apparent in heterothallic species. Originally designated as plus or minus (+ or −), the mating types of *Glomerella* were described as early as 1914 [10], and this was ultimately recapitulated in some fungi including *Neurospora* [11], *Sordaria* [5], *Schizophyllum* [12], and *Ustilago* [13].

7.1.2 LINKAGE MAPS

This emphasis on genetic analysis lasted for over 50 years, and while some systems such as *Allomyces* are no longer used for genetic analysis [14], the elucidation of genetic maps has been very important in genetic model systems and agriculturally and industrially important fungi. However, among the medically important fungi, sexual reproduction has not been widely used for analysis of genes and genomes. *Candida albicans* and *Aspergillus fumigatus* were considered to be asexual until mating-type genes could be characterized [15] or mobilized [16].

The genetic map of *Neurospora* was updated by Prof. D. Perkins of Stanford University [17] every other year and published by the Fungal Genetics Stock Center (FGSC) as part of their catalog [18]. The *Aspergillus nidulans* genetic map was maintained by Prof. J. Clutterbuck of Glasgow University and was originally published in the *Aspergillus Newsletter*. The *Aspergillus Newsletter* was merged with the *Neurospora Newsletter* in 1985 to form the *Fungal Genetics Newsletter* (FGN) and are available online in an historical archive at the FGSC (http://www.fgsc.net/ANL/ANLmain.html). The FGN was renamed Fungal Genetics Reports in 2008 to better reflect the fact that articles were peer-reviewed. The FGR is open access and adopted an online-only format after the 56th issue (2009). This remains an important source of genetic and technical information for working on filamentous fungi and provides an open archive for publishing the abstracts of meetings among other useful content.

A compendium of genetic loci was first published in 1982 because of the large number of markers on the *Neurospora* genetic map [19]. This included over 500 markers and for many years served as the main reference for *Neurospora* nomenclature and genetic information. Among these markers were nuclear and mitochondrial loci and other chromosomal landmarks including the nucleolus organizer, telomeres, and, building on the ability to isolate ordered tetrads for *Neurospora* [20], centromeres. The rapid expansion of the genetics of *Neurospora* meant that the compendium needed revision and in 2001 a second edition included over 1000 mapped loci [21].

Importantly, and in a manner that presaged the genome sequence, this version included genes identified by DNA sequences alone; the 1982 version required that a mutant allele was obtained prior to recognition of a gene location and assignment of a gene symbol. Modern technology, including restriction fragment length polymorphism (RFLP) mapping, allows genes to be identified without the identification of a loss or alteration of a function allele [22], although for functional analysis of these loci, reverse genetics is often employed to make targeted mutations [23]. The *Neurospora* compendium is continuously updated and presented online as *The Neurospora crassa e-Compendium* at the University of Leeds.

7.2 GENOMIC LIBRARIES, ELECTROPHORETIC KARYOTYPE ANALYSIS, AND GENOME SEQUENCING

Technological advances have had an impact on fungal genetics and genomics by increasing the amount of information available to researchers. Similarly, the sharing of molecular materials and information via open resources such as the FGSC and the FGN has been as an ocean tide lifting the entire field of fungal genetics. Some of these resources have enduring value, while others are only valuable for the information they provide and the physical resource loses its value after use. For example, arrayed cDNA clones have little value, while the cDNA sequence has been very valuable in annotation of whole-genome sequences.

7.2.1 GENOMIC LIBRARIES

For many years after the development of genome library technology and the identification of individual genes, the fungal genetics community shared libraries via the FGSC. The tremendous benefit that they realized came in terms of the ability to share gene locations in commonly used libraries, and these locations were published in the FGSC catalog [24] and online at the FGSC website (http://www.fgsc.net). Subsequently, it was possible to construct chromosome-specific sublibraries that greatly simplified screening for phenotypic complementation, or chromosome walking. Moreover, these libraries were used in the genome programs for *Neurospora* and for *Aspergillus*, as were cDNA libraries from the same fungi. The clones used in the construction of RFLP maps have also been useful in assembling genomes for *Fusarium graminearum* [25,26], *F. moniliforme* [25], *F. verticillioides* [27], and, to a lesser extent, *Magnaporthe grisea* [28] while not comprising full genome libraries. RFLP maps served an important role as they represented a significant advantage in terms of economics of scale compared to traditional forward genetic analysis. A lab (or even an ambitious student!) could generate a genetic map with multiple markers per linkage group in just a few years using RFLP mapping [25].

The FGSC received and still holds fosmid, bacterial artificial chromosome (BAC), and cosmid libraries associated with genome programs for core species in the FGSC collection including *N. crassa* [29], *A. nidulans* [30], and *F. graminearum* [31] as the genomes were completed. During the later years of the Sanger genome sequencing era, the FGSC also received genome-linked fosmid libraries for several fungi that (1) are impossible to culture away from the host (e.g., *Puccinia graminis*), (2) require USDA permits (e.g., *M. grisea* [32]), or (3) have significant community support (e.g., *Schizophyllum commune* [33], *Coprinopsis commune* [34], and *Batrachochytrium dendrobatidis* [35]).

The availability of genome-linked libraries has also enabled numerous experiments that otherwise would have been impossible. For example, using linked markers allows identification of *Neurospora* cosmids containing wild-type alleles of temperature-sensitive lethal alleles by complementation [36–38]. Similarly, cosmids have been used in the characterization of large gene clusters and may offer advantages over other approaches [39]. While not in high demand, the FGSC continues to provide small numbers of genome-linked cosmid clones to academia [40].

7.2.2 Molecular Karyotyping and Reassociation Kinetics

Contemporaneously with the development of genome libraries, a molecular approach that separated intact chromosomes in agarose gels [41] was being used to elucidate the number of chromosomes that comprise a haploid complement for many fungi. This technique verified the number of genetically determined chromosomes for *Aspergillus* [42] and the genetically and cytologically determined karyotype for *Neurospora* [3], although various translocation strains had to be used to allow resolution of each chromosome [43]. Electrophoretic karyotypes challenged cytologically determined karyotypes for many other fungi [44]. While earlier cytologically determined karyotypes were accurate for some fungi, for others, the large number of very small chromosomes, varying from 70 kb up to 1 Mb, made it impossible to resolve karyotypes accurately [45]. On the other hand, for some fungi, such as the rusts, the chromosomes are too large to resolve using electrophoretic technology, and suitable tissues for light microscopic determination of a karyotype are inaccessible. Electron microscopy was used to assemble a pachytene karyotype for *P. graminis* from serial sections of purified nuclei from teliospores [46]. The large size of this genome, carried on 18 chromosomes, has been confirmed by reassociation kinetics [47] and whole-genome analysis.

DNA reassociation kinetics provided another approach to characterize fungal genomes and for *Neurospora* predicted that the genome was comprised of 90% single-copy sequences [49]. Similar analyses for other model filamentous fungi including *Fusarium* [50], *Penicillium* [51], and *Schizophyllum* [52] were carried out and in *Schizophyllum*, for example, and suggested unique genome structure with the ribosomal DNA contributing less than 3% of the total nuclear DNA. Similarly, mitochondrial genomes of filamentous fungi were characterized by various techniques including physical analysis [53], RFLP analysis [54], and genome analysis [55], although the small size of mitochondrial genomes means that there are few genes to study.

7.2.3 Sanger Sequencing and Fungal Genomics

It became imperative to analyze the whole genome of important model fungi such as *Neurospora* and *Aspergillus*, and there was significant competition to be the first researchers to sequence these genomes subsequent to the publication of the *Saccharomyces cerevisiae* genome [56]. Each scientific community organized a significant effort to have a publicly available genome program, and a series of symposia were held to coordinate community efforts. The last of these was held in 2000 at the University of Georgia and included representatives of major groups including those involved in model, clinical, and agricultural fungi. This group effort culminated in the organization of a white paper called the Fungal Genome Initiative [57]. Fungal genetics researchers built upon similar structures put into place to sequence the human genome, guaranteeing that the fungal genome research was collaborative [58].

Early activity in parallel with the first years of the Fungal Genome Initiative included systematic cDNA sequencing for various fungi including *Neurospora* [59], *Aspergillus* [60], and *Fusarium* [61], among others. It became evident that as the sequences were publicly available, the physical clones were not valuable, and so they were de-accessed when the FGSC relocated in 2004 [40]. During this time, some genome sequences were generated in companies, and these included the first-pass (3X) coverage of *A. nidulans* and *F. verticillioides* that were sequenced at Monsanto [30] and Syngenta Co. [26], respectively.

As sequence data became available, it was soon recognized that filamentous fungi contain genes not found in the yeast, *Saccharomyces*, and so unique projects to analyze these genes were undertaken [62]. These included some of the first attempts at systematic classification of fungal genes by ontologies [62]. The *N. crassa* genome was the first to be completed by the Broad Institute [29], and this success was built upon by various historical efforts. *Neurospora* had myriad tools for anchoring the genome and the large number of characterized genes made assembly straight forward. Genome libraries had already been sorted into chromosome-specific sets, and the chromosomes were identified using genetic and molecular biological methods. Perhaps most significantly, *Neurospora* contains less repeated DNA than most fungi simplifying the informatic analysis [63]. Subsequent fungal genomes have proven more complicated and may never be fully assembled with contigs matching linkage groups [47].

However, the complete assembly into contigs that match the linkage groups is not always possible, and even for *Neurospora* with its robust tools, cosmid, and BAC libraries, optical mapping of DNA was necessary to generate the final robust assembly [29] despite the wealth of tools available for analysis of fungal genomes. It was not for many years that the ends of the *Neurospora* chromosomes were completely characterized [64] because of their underrepresentation in genome libraries.

Ultimately, there were several fungal genomes sequenced during the Sanger sequencing era, although publication of many of these genomes was not complete until multiple strains of related species were sequenced for comparative purposes [30].

7.2.4 NEXT-GENERATION FUNGAL GENOMICS

As novel DNA sequencing technologies have appeared, they have been applied to fungal genome analysis and each has different characteristics. Most fungal species do not have the preexisting tools used for genome assembly and annotation in the more widely studied model organisms, although their small genome sizes make them amenable to analysis by high-throughput technologies. These fungi are studied by comparative phylogenomics, and until more information on multiple isolates from each branch of the tree of life is available, these data will continue to be largely descriptive. Next-generation sequencing, whether short read, paired end, or long read with high error, is an area of rapid technological change. While the earliest techniques required infrastructure-level investment, new technologies promise that whole-genome sequence production can be as widely distributed as thermocyclers and the cost can be as low as the cost of sequencing a gene.

7.2.4.1 *De novo* Sequencing

High-depth/short-read next-generation sequencing technology has made low-cost *de novo* sequencing possible. The pace of fungal genome sequencing continues to increase rapidly. Perhaps the most impressive initiative associated with next-generation sequencing is the 1000 Fungal Genomes Initiative led by the DOE Joint Genome Institute. The rational for the 1000 fungal genomes has its roots in the exploration of fungal metabolic and enzymatic diversity and fungal systematics [65–67].

The enzymatic diversity encoded in the genomes of fungi has been exploited for centuries for production of food and drink. Enzymes to aid in digestion (e.g., Takamine) were developed at the close of the nineteenth century (reviewed in [68]). Over the last few decades, fungal enzymes have been produced for industrial uses such as the saccharification of plant biomass for biofuel and renewable chemical production and for food and fiber processing [65–67,69]. Fungal genomes also encode a diverse array of secondary metabolite biosynthetic pathways for the production of fatty acids, terpenes, and polyketides. These hydrocarbon backbone molecules are potential precursors to biofuels and renewable chemicals and as such may provide the processing potential to displace ethanol as the biofuel of choice [70]. For these reasons and for aiding taxonomic/systematic studies, sequencing a broad spectrum of fungi has been proposed as a way to increase the catalog of enzymes and biosynthetic pathways for use in metabolic engineering and synthetic biology approaches to production of biofuels and renewable chemicals.

7.2.4.2 Resequencing Fungal Genomes

Resequencing is defined as whole-genome analysis of genomes for which a high-quality reference genome has already been produced. Targets for resequencing of fungal genomes are numerous [71]. Next-generation sequencing has also allowed the rapid mapping of insertions, deletions, single-nucleotide variants, and other chromosomal aberrations within the genomes of mutated strains. For example, mutations from two strains of a lineage of *Trichoderma reesei* strains with increased cellulase secretion and decreased carbon catabolite repression were mapped using next-generation sequencing [72]. The findings indicated a number of mutations in genes encoding proteins involved in a wide variety of cellular processes. However, the complexities of the mutant phenotypes combined with the large number of mutations made it impossible to associate specific mutations with the enhanced enzyme secretion and carbon catabolite phenotypes.

Next-generation sequencing is a powerful tool for associating phenotypes with genetic mutations when used in systems with well-characterized genetic maps. *A. nidulans* and *N. crassa* are excellent targets for resequencing because of strong genetic foundations [71]. Recent studies in *N. crassa* have highlighted the power of next-generation sequencing, as physical loci (annotated genes in the DNA sequence) were rapidly associated with phenotypes in several strains [66,73]. Similarly, resequencing is being used to characterize and identify genes responsible for light perception (*Phycomyces*), meiosis (*Coprinopsis*), myriad virulence-related traits in *Fusarium*, and the characterization of agricultural and model plant species [74] and for human genome analysis [75].

7.3 PRESERVING THE SEQUENCED FUNGI: A CRUCIAL ROLE FOR THE BIOLOGICAL RESOURCE CENTERS

The publication of a complete fungal genome is more valuable when the actual biological material used is made available. The *Neurospora* community has benefitted from the widespread use of the reference genome strain. For example, identification of classical mutations in each of 18 strains subject to whole-genome sequence analysis was facilitated by the shared genetic lineage used by the community for many years. This was despite the fact that the strain originally used for DNA analysis was maintained in the lab that provided the DNA without documentation of the number of passages relative to the reference strain deposited at the FGSC (FGSC 2489, also known as 74-OR23–1VA). Because the spontaneous mutation rate for *N. crassa* has been measured as being on the order of one mutation per 10^5 or 10^6 bases per generation, it is impossible to ascertain whether this independent lineage is the source of the many polymorphisms detected in all of the resequenced strains relative to the so-called reference genome strain without further analysis [66]. Similarly, the *A. nidulans* scientific community has all used the same reference strain, FGSC A4, which was deposited in the FGSC collection in 1960 and which was used to generate the cDNA-, genome-, and chromosome-specific libraries [76].

The FGSC has endeavored to collect as many reference genome strains as is practical and holds 47 reference genome strains from 32 different species. These include ascomycete, basidiomycete, heterobasidiomycete, and zygomycete fungi.

While the FGSC is the acknowledged reference collection for whole-genome sequenced strains, some strains are not freely available because of material transfer agreement limitations placed upon them by depositors or other collections. However, this situation is made contradictory by the public availability of the genome sequence data. In the absence of the strain, access to the genome sequence means that through synthetic genetics, it should be possible to recapitulate any characteristic of interest. Similarly, the commercial use of genetic information may require compensation to the country of origin according to the terms of the Convention on Biological Diversity. The free availability of genome information does not imply that the information can be used without consideration of the source of the information. In an era when traits can be recapitulated using molecular technology, the ability to identify strains based on physiological traits becomes unreliable. The whole-genome sequence of a strain is the only reliable way to distinguish among related strains.

7.4 CONCLUSIONS

Fungi impact nearly every aspect of human life, often beneficially, but also negatively. The detrimental impacts include crop loss [77], both directly and through the production of mycotoxins on food crops, the degradation of fibers used for fabric and construction [78], and the production of copious allergenic spores [79] and through direct infection of susceptible humans and animals raised for food production [80]. The beneficial impacts fungi provide to human life include the direct use as food [81] in food processing [82], the production of pharmaceutical compounds such as immunomodulatory [83] or antibiotic molecules [84], and their use in myriad industrial processes [85] and as promising sources of biomass-degrading enzymes [86].

In addition to the impact it has had on the search for novel antifungal agents [87], fungal genome analysis has led to meaningful improvements in the use of fungi to produce enzymes for food and fiber processing [88]. Genome analysis, by virtue of the tremendous wealth of information it provides (Table 7.1), is finding application to every branch of scientific endeavor [71]. The generation of fungal genome sequences has enabled a breadth of global-level organismal analyses, and as a consequence, fungi are better understood, and their potential contributions to the future bioeconomy are well anticipated [89]. Similarly, the impact of culture collections is seen as a twofold increase in citations to materials that are shared openly [90]. As a foundational biological resource, culture collections provide the opportunity to build upon work by previous generations of scientists [40] and guarantee that important materials are available as reference for diagnostic purposes and also as the source of untapped diversity to provide novel solutions to industrial, food, health, and agricultural challenges as we move into the postgenomics era.

Genome analysis of fungi, from its beginnings in the chromosome theory of biology, and through the genetic, and molecular eras, has provided deep insight into the biology of this important group of organisms. It has provided the perspective that some fungi (such as *Neurospora*) are like crocodile with genomes that are stable and resist change, while others (such as *Fusarium oxysporum*) are like Darwin's finches [91] that change rapidly to exploit new environmental resources.

REFERENCES

1. Fincham, J.R. and P.R. Day, *Fungal Genetics*. 1963, Philadelphia, PA: F.A. Davis, Co.
2. Raper, K.B., Variation in molds; natural and induced. *J Bacteriol*, 1947. **53**(3): 379.
3. McClintock, B., Neurospora. I. Preliminary observations of the chromosomes of *Neurospora crassa*. *Am J Botany*, 1945. **32**(10): 671–678.
4. Beadle, G.W. and E.L. Tatum, Genetic control of biochemical reactions in Neurospora. *Proc Natl Acad Sci USA*, 1941. **27**(11): 499–506.
5. Carr, A.J. and L.S. Olive, Genetics of *Sordaria fimicola* III. Cross-compatibility among self fertile and self sterile cultures. *Am J Botany*, 1959. **46**: 81–91.
6. Emerson, R. and C.M. Wilson, The significance of meiosis in allomyces. *Science*, 1949. **110**(2847): 86–88.
7. Raper, J.R., Tetrapolar sexuality. *Q Rev Biol*, 1953. **28**(3): 233–259.
8. Pontecorvo, G. and J.A. Roper, Resolving power of genetic analysis. *Nature*, 1956. **178**(4524): 83–84.
9. Pontecorvo, G., The parasexual cycle in fungi. *Annu Rev Microbiol*, 1956. **10**: 393–400.
10. Edgerton, C.W., Plus and minus strains in the genus *Glomerella*. *Am J Botany*, 1914. **1**: 244.
11. Shear, C.L. and B.O. Dodge, Life histories and heterothallism of the red bread-mold fungi of the *Monilia sitophila* group. *J Agric Res*, 1927. **34**(11): 1019–1042.
12. Raper, J.R., M.G. Baxter, and R.B. Middleton, The genetic structure of the incompatibility factors in schizophyllum commune. *Proc Natl Acad Sci USA*, 1958. **44**(9): 889–900.

13. Perkins, D.D., Biochemical mutants in the smut fungus ustilago maydis. *Genetics*, 1949. **34**(5): 607–626.
14. Olson, L.W., *Allomyces*—A different fungus. *Opera Botanica*, 1984. **73**: 1–96.
15. Paoletti, M. et al., Evidence for sexuality in the opportunistic fungal pathogen *Aspergillus fumigatus*. *Curr Biol*, 2005. **15**(13): 1242–1248.
16. Berman, J. and L. Hadany, Does stress induce (para)sex? Implications for *Candida albicans* evolution. *Trends Genet*, 2012. **28**(5): 197–203.
17. Raju, N.B., David D. Perkins (1919–2007): A lifetime of *Neurospora* genetics. *J Genet*, 2007. **86**(2): 177–186.
18. McCluskey, K., From genetics to genomics: Fungal collections at the Fungal Genetics Stock Center. *Mycology*, 2011. **2**(3): 161–168.
19. Perkins, D.D. et al., Chromosomal loci of *Neurospora crassa*. *Microbiol Rev*, 1982. **46**(4): 426–570.
20. Bole-Gowda, B.N., D.D. Perkins, and W.N. Strickland, Crossing-over and interference in the centromere region of linkage group I of *Neurospora*. *Genetics*, 1962. **47**: 1243–1252.
21. Perkins, D.D., A. Radford, and M.S. Sachs, *The Neurospora Compendium: Chromosomal Loci*. 2001, San Diego, CA: Academic Press. ix, 325pp.
22. Nelson, M.A. and D.D. Perkins, Restriction polymorphism maps of *Neurospora crassa*: 2000 update. *Fungal Genet Newslett*, 2000. **47**: 23–39.
23. Selker, E.U. and P.W. Garrett, DNA sequence duplications trigger gene inactivation in *Neurospora crassa*. *Proc Natl Acad Sci USA*, 1988. **85**(18): 6870–6874.
24. McCluskey, K. and M. Plamann, Fungal Genetics Stock Center catalogue of strains, 11th edn, *Fungal Genet. Newslett*, 2006. **53S**: 1–221.
25. Xu, J. and J.F. Leslie, A genetic map of *Gibberella fujikuroi* mating population A (*Fusarium moniliforme*). *Genetics*, 1996. **143**(1): 175–189.
26. Ma, L.J. et al., Comparative genomics reveals mobile pathogenicity chromosomes in Fusarium. *Nature*, 2010. **464**(7287): 367–373.
27. Jurgenson, J.E., K.A. Zeller, and J.F. Leslie, Expanded genetic map of *Gibberella moniliformis* (*Fusarium verticillioides*). *Appl Environ Microbiol*, 2002. **68**(4): 1972–1979.
28. Dioh, W. et al., Mapping of avirulence genes in the rice blast fungus, *Magnaporthe grisea*, with RFLP and RAPD markers. *Mol Plant Microbe Interact*, 2000. **13**(2): 217–227.
29. Galagan, J.E. et al., The genome sequence of the filamentous fungus *Neurospora crassa*. *Nature*, 2003. **422**(6934): 859–868.
30. Galagan, J.E. et al., Sequencing of *Aspergillus nidulans* and comparative analysis with *A. fumigatus* and *A. oryzae*. *Nature*, 2005. **438**(7071): 1105–1115.
31. Cuomo, C.A. et al., The *Fusarium graminearum* genome reveals a link between localized polymorphism and pathogen specialization. *Science*, 2007. **317**(5843): 1400–1402.
32. Dean, R.A. et al., The genome sequence of the rice blast fungus *Magnaporthe grisea*. *Nature*, 2005. **434**(7036): 980–986.
33. Ohm, R.A. et al., Genome sequence of the model mushroom *Schizophyllum commune*. *Nat Biotechnol*, 2010. **28**(9): 957–963.
34. Stajich, J.E. et al., Insights into evolution of multicellular fungi from the assembled chromosomes of the mushroom *Coprinopsis cinerea* (*Coprinus cinereus*). *Proc Natl Acad Sci USA*, 2010. **107**(26): 11889–11894.
35. James, T.Y. et al., A molecular phylogeny of the flagellated fungi (Chytridiomycota) and description of a new phylum (Blastocladiomycota). *Mycologia*, 2006. **98**(6): 860–871.
36. Dieterle, M.G. et al., Characterization of the temperature-sensitive mutations un-7 and png-1 in *Neurospora crassa*. *PLoS One*, 2010. **5**(5): e10703.
37. Wiest, A., M. Plamann, and K. McCluskey, Identification of the *Neurospora crassa* mutation un-4 as the mitochondrial inner membrane translocase subunit tim16. *Fungal Genet Rep*, 2008. **56**: 37–39.
38. McCluskey, K. et al., Complementation of un-16 and the development of a selectable marker for transformation of *Neurospora crassa*. *Fungal Genet Newslett*, 2007. **54**: 9–11.
39. Nowrousian, M., A novel polyketide biosynthesis gene cluster is involved in fruiting body morphogenesis in the filamentous fungi *Sordaria macrospora* and *Neurospora crassa*. *Curr Genet*, 2009. **55**(2): 185–198.
40. McCluskey, K., A. Wiest, and M. Plamann, The Fungal Genetics Stock Center: A repository for 50 years of fungal genetics research. *J Biosci*, 2010. **35**(1): 119–126.
41. Schwartz, D.C. and C.R. Cantor, Separation of yeast chromosome-sized DNAs by pulsed field gradient gel electrophoresis. *Cell*, 1984. **37**(1): 67–75.

42. Brody, H. and J. Carbon, Electrophoretic karyotype of *Aspergillus nidulans*. *Proc Natl Acad Sci USA*, 1989. **86**(16): 6260–6263.
43. Orbach, M.J. et al., An electrophoretic karyotype of *Neurospora crassa*. *Mol Cell Biol*, 1988. **8**(4): 1469–1473.
44. Mills, D.I. and K. McCluskey, Electrophoretic karyotypes of fungi: The new cytology. *Mol Plant-Microb Interact*, 1990. **3**(6): 351–357.
45. Zolan, M.E., Chromosome-length polymorphism in fungi. *Microbiol Rev*, 1995. **59**(4): 686–698.
46. Boehm, E. et al., An ultrastructural pachytene karyotype for *Puccinia graminis* f. sp. tritici. *Can J Botany*, 1992. **70**(2): 401–413.
47. Backlund, J.E. and L.J. Szabo, Physical characteristics of the genome of the phytopathogenic fungus *Puccinia graminis*. *Curr Genet*, 1993. **24**(1): 89–93.
48. Duplessis, S. et al., Obligate biotrophy features unraveled by the genomic analysis of rust fungi. *Proceedings of the National Academy of Sciences*, 2011. **108**(22): 9166–9171.
49. Krumlauf, R. and G. Marzluf, Genome organization and characterization of the repetitive and inverted repeat DNA sequences in *Neurospora crassa*. *J Biol Chem*, 1980. **255**(3): 1138–1145.
50. Szecsi, A. and A. Dobrovolszky, Phylogenetic relationships among *Fusarium* species measured by DNA reassociation. *Mycopathologia*, 1985. **89**(2): 89–94.
51. Paterson, R., G. King, and P. Bridge, High resolution thermal denaturation studies on DNA from 14 *Penicillium strains*. *Mycol Res*, 1990. **94**(2): 152–156.
52. Dons, J. and J. Wessels, Sequence organization of the nuclear DNA of *Schizophyllum commune*. *Biochim Biophys Acta Nucleic Acids Protein Syn*, 1980. **607**(3): 385–396.
53. Taylor, J.W. and B.D. Smolich, Molecular cloning and physical mapping of the *Neurospora crassa* 74-OR23–1A mitochondrial genome. *Curr Genet*, 1985. **9**(7): 597–603.
54. Moody, S.F. and B.M. Tyler, Restriction enzyme analysis of mitochondrial DNA of the *Aspergillus flavus* group: *A. flavus, A. parasiticus*, and *A. nomius*. *Appl Environ Microbiol*, 1990. **56**(8): 2441–2452.
55. Bertrand, H. et al., Deletion mutants of Neurospora crassa mitochondrial DNA and their relationship to the "stop-start" growth phenotype. *Proc Natl Acad Sci USA*, 1980. **77**(10): 6032–6036.
56. Goffeau, A. et al., Life with 6000 Genes. *Science*, 1996. **274**(5287): 546–567.
57. Christina C.A. and B.W. Birren, The fungal genome initiative and lessons learned from genome sequencing. *Methods in e Enzymology*, 2010. **470**: 833–855.
58. Hall, D. et al., Using workflow to build an information management system for a geographically distributed genome sequencing initiative. *Mycol Ser*, 2003. **18**: 359–372.
59. Nelson, M.A. et al., Expressed sequences from *Conidial, Mycelial*, and sexual stages of *Neurospora crassa*. *Fungal Genet Biol*, 1997. **21**(3): 348–363.
60. Kupfer, D.M. et al., Multicellular ascomycetous fungal genomes contain more than 8000 genes. *Fungal Genet Biol*, 1997. **21**(3): 364–372.
61. Roe, B. et al., *Fusarium sporotrichioides* cDNA Sequencing Project. University of Oklahoma, Norman, OK and Texas A&M University, College Station, TX, 2000.
62. Schulte, U. et al., Large scale analysis of sequences from *Neurospora crassa*. *J Biotechnol*, 2002. **94**(1): 3–13.
63. Wöstemeyer, J. and A. Kreibich, Repetitive DNA elements in fungi (Mycota): Impact on genomic architecture and evolution. *Curr Genet*, 2002. **41**(4): 189–198.
64. Wu, C. et al., Characterization of chromosome ends in the filamentous fungus *Neurospora crassa*. *Genetics*, 2009. **181**(3): 1129–1145.
65. Baker, S.E. et al., Fungal genome sequencing and bioenergy. *Fungal Biol Rev*, 2008. **22**: 1–5.
66. McCluskey, K. et al., Rediscovery by whole genome sequencing: Classical mutations and genome polymorphisms in *Neurospora crassa*. *G3 Genes, Genomes Genet*, 2011. **1**(4): 303–316.
67. Martin, F. et al., Sequencing the fungal tree of life. *New Phytol*, 2011. **190**(4): 818–821.
68. Bennett, J.W. and S.E. Baker, An overview of the genus *Aspergillus*. In *The Aspergilli: Genomics, Medical Aspects, Biotechnology, and Research Methods*, G.H. Goldman and S.A. Osmani, eds. CRC Press, Taylor & Francis Group: Boca Raton, FL, 2008, pp. 3–13.
69. Olempska-Beer, Z.S. et al., Food-processing enzymes from recombinant microorganisms—A review. *Regul Toxicol Pharmacol*, 2006. **45**(2): 144–158.
70. Fortman, J. et al., Biofuel alternatives to ethanol: Pumping the microbial well. *Trends Biotechnol*, 2008. **26**(7): 375–381.
71. Baker, S.E., Selection to sequence: Opportunities in fungal genomics. *Environ Microbiol*, 2009. **11**(12): 2955–2958.

72. Le Crom, S. et al., Tracking the roots of cellulase hyperproduction by the fungus *Trichoderma reesei* using massively parallel DNA sequencing. *Proc Natl Acad Sci USA*, 2009. **106**(38): 16151–16156.
73. Pomraning, K.R., K.M. Smith, and M. Freitag, Bulk segregant analysis followed by high-throughput sequencing reveals the Neurospora cell cycle gene, ndc-1, to be allelic with the gene for ornithine decarboxylase, spe-1. *Eukaryot Cell*, 2011. **10**(6): 724–733.
74. Henry, R.J., Next-generation sequencing for understanding and accelerating crop domestication. *Briefings Funct Genomics*, 2012. **11**(1): 51–56.
75. Gonzaga-Jauregui, C., J.R. Lupski, and R.A. Gibbs, Human genome sequencing in health and disease. *Annu Rev Med*, 2012. **63**: 35–61.
76. Brody, H. et al., Chromosome-specific recombinant DNA libraries from the fungus *Aspergillus nidulans*. *Nucleic Acids Res*, 1991. **19**(11): 3105–3109.
77. Strange, R.N. and P.R. Scott, Plant disease: A threat to global food security. *Annu Rev Phytopathol*, 2005. **43**: 83–116.
78. Kubicek, C.P., The actors: Plant biomass degradation by fungi. In *Fungi and Lignocellulosic Biomass*, 2012, Oxford, UK: Wiley-Blackwell, pp. 29–44.
79. Madani, Y., A. Barlow, and F. Taher, Severe asthma with fungal sensitization: A case report and review of literature. *J Asthma*, 2010. **47**(1): 2–6.
80. Fisher, M.C. et al., Emerging fungal threats to animal, plant and ecosystem health. *Nature*, 2012. **484**(7393): 186–194.
81. Roman, M.D., The contribution of wild fungi to diet, income and health: A world review. *Prog Mycol*, 2010. 327–348.
82. van den Berg, M. et al., Fungal genes and their respective enzymes in industrial food, bio-based and pharma applications. *Biotechnology of Fungal Genes*, 2012, New Hampshire: Edenbridge Ltd, pp. 189–221.
83. Mishra, B.B. and V.K. Tiwari, Natural products: An evolving role in future drug discovery. *Eur J Med Chem*, 2011. **46**(10): 4769–4807.
84. Kawaguchi, M. et al., New method for isolating antibiotic-producing fungi. *J Antib*, 2012. **66**(1): 17–21.
85. Demain, A.L. and J.L. Adrio, Contributions of microorganisms to industrial biology. *Mol Biotechnol*, 2008. **38**(1): 41–55.
86. Demain, A.L., Biosolutions to the energy problem. *J Ind Microbiol Biotechnol*, 2009. **36**(3): 319–332.
87. Markovich, S. et al., Genomic approach to identification of mutations affecting caspofungin susceptibility in Saccharomyces cerevisiae. *Antimicrob Agents Chemother*, 2004. **48**(10): 3871–3876.
88. Pel, H.J. et al., Genome sequencing and analysis of the versatile cell factory *Aspergillus niger* CBS 513.88. *Nat Biotechnol*, 2007. **25**(2): 221–231.
89. Safferman, S., W. Liao, and C. Saffron, Engineering the Bioeconomy. *J Environ Eng*, 2009. **135**(11): 1085.
90. Furman, J.L. and S. Stern, Climbing atop the shoulders of giants: The impact of institutions on cumulative research. *Am Econ Rev*, 2011. **101**(5): 1933–1963.
91. Grant, P.R. and B.R. Grant, Unpredictable evolution in a 30-year study of Darwin's finches. *Science*, 2002. **296**(5568): 707–711.

8 Recommendations for Quantitative PCR *Aspergillus* Assays

Stéphane Bretagne, Odile Cabaret, and Jean-Marc Costa

CONTENTS

8.1 Introduction 103
8.2 Origin of *Aspergillus* DNA 104
8.3 PCR Format 105
    8.3.1 Prevention of False Positives 105
    8.3.2 Prevention of False Negatives 107
        8.3.2.1 Internal Control 107
        8.3.2.2 PCR Yield Optimization 108
8.4 Quality Control Design 109
8.5 Clinical Performance of a qPCR test 110
8.6 Perspectives 110
8.7 Conclusions 111
References 111

8.1 INTRODUCTION

Numerous polymerase chain reaction (PCR)-based assays have become routine in microbiology and are particularly effective for detecting and identifying infectious agents for which culture and microscopy methods are inadequate. In view of the success in viral diseases where PCR has surpassed culture and other diagnostic means, it is surprising to discover that this is not the case for several invasive fungal infections, specifically invasive aspergillosis (IA), one of the leading causes of mortality in patients with hematological malignancies receiving intensive chemotherapy and hematopoietic stem cell transplantation.[1] The diagnosis of invasive fungal diseases remains a multidisciplinary analysis of clinical, CT scan, and microbiological findings, including classical mycology and antigen detection.[2] However, the microbiological tools are (1) insensitive[3] and (2) hampered by false-positive results.[4,5] Hence, much hope has been placed on PCR. However, PCR data are not always included in the consensual criteria for definitions of invasive fungal diseases[2] although many PCR assays have been developed over the previous 20 years.[6,7] The problems are the large number of different PCR protocols published and the absence of consensus on the optimal PCR technique. For example, the choices include (1) the different DNA targets, although this is predominantly the rDNA gene,[8,9] (2) the chemistry of the probes (hydrolysis probes,[10–14] hybridization probes,[15–18] or molecular beacons[19]), and (3) the PCR apparatus used. These are the main reasons, but not the only ones, as different protocols are also published for viral diagnoses, but this does not hamper the acceptance of diagnostic PCR in this field.

    The main difference between viruses and fungi for PCR is the number of DNA targets in the clinical specimens when dealing with blood or serum. For viruses it is common to work with

>10^3 copies. This is not the case with IA where the number of copies is at the limit of detection of the PCR that decreases the limits and exacerbates the pitfalls of PCR. The problem is to detect minute concentrations of fungal DNA among huge amounts of undesirable human DNA for the detection of invasive fungal diseases.

The present chapter considers the diagnosis of IA, mainly caused by the species *Aspergillus fumigatus*, because it is by far the focus of most of the publications on the diagnosis of invasive fungal diseases. PubMed interrogation with *aspergillus*, *PCR*, and *diagnosis* over the last 5 years gives 228 publications and 21 reviews in English. IA is taken as an example to underline the difficulties when developing a PCR assay although some points are not necessarily relevant to the diagnosis of other invasive fungal infections.

The first focus for our attention is the lack of knowledge concerning the nature of the target to be amplified. The fungal element circulating in patients with different forms of IA is unknown: are the conidia and hyphae engulfed in macrophages or circulating as cell-free fungal DNA? These considerations affect protocols for sample selection, collection, and DNA extraction, which explain the difficulties in comparing results of studies using different methods. Some authors favored serum,[10,14,15,20] clot,[21] or a combination of plasma and whole blood,[22] although most used whole blood.[13,16,18,23]

The second factor is technical: PCR assay designs have evolved during these past 20 years. Internal controls[24] and enzymatic prevention of contamination,[24,25] although advocated, were absent for many years.[24,26,27] The technical issues are exacerbated by the presence of environmental fungi, which can contaminate every step of the PCR process. In view of insufficiencies in using experimental controls and optimization experimental details, some recommendations for valid PCR can be proposed.[28] If using the same quantitative PCR (qPCR) assay is unrealistic, one can nevertheless reach a consensual procedure based on the *Minimum Information for the Publication of Real-Time Quantitative PCR Experiments* (MIQE) guidelines.[28]

The third issue is the demonstration of the clinical relevance of such a test for patients at risk of IA. Beyond qPCR assay design, the challenge is to include the assay in a diagnostic strategy that is framed by clinical findings.

8.2 ORIGIN OF *ASPERGILLUS* DNA

Mammals are very resistant to invasive fungal diseases.[29] Consequently, they develop only in immunocompromised patients and are called opportunistic, with few exceptions. Therefore, the isolation of a fungus from clinical specimens that are open to the environment (the most frequent situation) does not necessarily reflect infection. A positive culture can be interpreted as true infection, colonization, or recovery from bystanders. To change from cultural techniques to molecular tools does not modify this seminal issue, and the final interpretation is often based on a somewhat subjective overall clinical assessment, hence the difficulties in establishing criteria for the definitions of invasive fungal diseases.[2] This is a major point to consider as it is not only the technical issues that are required to be improved for accurate diagnoses of invasive fungal diseases.

The second major issue is the origin of the DNA detected in blood samples. There is no definitive demonstration that whole blood or serum is more suitable for the diagnosis of IA,[30] which strongly affects the DNA extraction protocol. The mouse model with intravenous inoculation of conidia can provide clues, but it is too far removed from human IA.[31] Indeed, blood cultures are seldom positive for IA,[3] suggesting that living fungal particles are not the source of the positive results. Fungal DNA can come from (1) conidia or hyphae virtually engulfed in macrophages and/or (2) circulating cell-free fungal DNA. A stringent DNA extraction protocol should be followed for fungi in blood as suggested by the European *Aspergillus* PCR Initiative (EAPCRI).[9] In contrast, if the hypothesis of cell-free fungal DNA in serum is accepted, a softer DNA extraction procedure can be followed,[15,20,30,32,33] less damaging to DNA integrity and more amenable to automatization. On the other hand, cell-free DNA circulates as short DNA fragments, which require optimized amplification and specific DNA extraction procedures for efficient analysis of these from biologic fluids.[34] The absence of a pathogen-specific DNA enrichment procedure indirectly favors the hypothesis of circulating cell-free fungal DNA, over the cells of the pathogen being the source of the amplified DNA.[35] To add complexity, the rationale is to use stringent DNA extraction protocols to free the DNA from hyphae and conidia if respiratory specimens are used for diagnosis.[36–38] Therefore, the optimal DNA extraction for respiratory specimens has probably little in common with the equivalent extraction from serum. Each DNA extraction must be adapted to the clinical specimens and the initial hypotheses on the origin of the DNA.

To improve the interpretation of positive results for opportunistic diseases, defining thresholds is possible. Indeed, the real-time qPCR methods (see later) allow defining thresholds of positivity. Unfortunately, blood or serum *Aspergillus* DNA is always in low amounts in IA as already mentioned, and no definition of threshold values has been reported. This issue might be even more relevant in respiratory specimens, which is the central debate for the diagnosis of *Pneumocystis* pneumonia where the amount of DNA detected in BAL fluids serve to define the terms *infected* or *colonized* patients (i.e., with no clinical signs to confirm *Pneumocystis* pneumonia).[39–41]

8.3 PCR FORMAT

qPCR has become the most widely used PCR format to detect and quantify pathogens. Compared to the end point format of the 1990s, qPCR is a major breakthrough because of several features well suited to the routine laboratory such as speed, simplicity, robustness, and high throughput.[42]

Two major advantages must be underlined for fungal diagnoses.[6,7] qPCR dramatically reduces the risk of carryover contamination and the potential for false-positive results by avoiding overmanipulation of amplified PCR products. That is why the nested PCR method should be avoided even when including a qPCR step to enhance sensitivity.[43] Indeed, opening the tubes between the two runs of amplification of the nested PCR obviously ruins the main advantage of the close-tube format. Also, quantification addresses the issue of amplification yield in each clinical sample to avoid false-negative results (see later).

8.3.1 Prevention of False Positives

Every PCR assay for diagnosing invasive fungal infections is designed for detecting a low amount of fungus among a huge amount of human cells. Therefore, all the PCR parameters must be optimized. As a consequence, the risk of false-positive results is maximal from contamination with previously amplified products and ubiquitous environmental fungi.[6–8]

The carryover risk via aerosolization obliges workers to maintain physical prevention measures. The standard precautions include (1) working in laminar flow hoods; (2) wearing gloves and gown; (3) using a unidirectional workflow environment with physically separated laboratories for pre-, per-, and post-PCR analysis; and (4) using aerosol-resistant tips and specific pipettes. It should be noted that maintaining these precautions in a routine laboratory can be difficult. Physical measures should be completed by enzymatic prevention of contamination using uracil-*N*-glycosylase.[24,25,44] This enzyme is designed for cleaving previously amplified products, which then cannot serve as a molecular template for further amplifications.

However, the main breakthrough that improved the reliability of PCR is qPCR. This dramatically decreases the risk of amplified DNA aerosols contaminating the environment since amplification and detection take place in a close tube. However, qPCR format does not decrease, to the same extent, the risk of false positives related to microorganisms and/or fungal DNA in the environment.

Sample processing in laminar airflow cabinet can control contamination with airborne fungal spores but not with fungal DNA. Commercial enzymes used in PCR can be a source of residual fungal DNA since many are produced by fungi.[45,46] Thus, the simplest DNA extraction method is preferred to avoid unnecessary enzymes.[9] In-house and commercial reagents should be tested for contamination on a regular basis, and every qPCR run should include as many negative controls as possible, which should be as similar to the test samples as possible. However, these measures are inadequate when the contamination is in the sampling tubes[47] and are difficult to demonstrate since they are often at the limit of detection of the assay. As a consequence, results of the negative controls can be alternatively positive or negative according to the Poisson's law. Below 10 copies of target DNA, the alternatively positive and negative results can lead by chance to falsely validate a result. Therefore, regular tests of commercial and *in-house* reagents using negative extraction controls should be as numerous as possible although this is not a complete guarantee of detecting very-low-level contamination of initial tubes. Another possibility is the presence of fungal DNA in

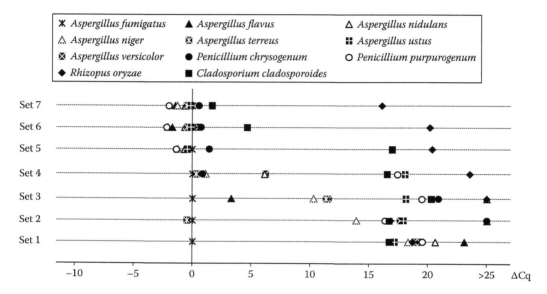

FIGURE 8.1 Specificity of seven primer sets to detect environmental molds. The results are expressed as ΔCq by subtracting the Cq value obtained for the *Aspergillus fumigatus* IP 2279.94 strain from each fungal species. The higher the difference, the more specific is the primer set for *A. fumigatus*. In contrast, a negative value means a higher specificity for the non-*fumigatus* species. Set 1 has the best specificity for *A. fumigatus*, that is, the highest ΔCq value between *A. fumigatus* and the first non-*fumigatus* species. The ΔCq value was >15, meaning that Set 1 amplified *A. fumigatus* 32,000 times more efficiently than the first non-*fumigatus* species, given a yield of the amplification close to 2. PCR assays using Set 2 and Set 3 include one additional species (*A. terreus* and *A. flavus* for Set 2 and Set 3, respectively) other than *A. fumigatus* within a ΔCq <5. Sets 4–7 amplify non-*fumigatus* species at least as efficiently as *A. fumigatus*. Reference strains list: *Aspergillus fumigatus* (IP 2279.94), *Aspergillus flavus* (ATCC 204304), *Aspergillus brasiliensis* (CBS 733.88), *Emericella nidulans* var. *nidulans* (CBS 121.35), *Aspergillus insuetus* (CBS 107.25), *Aspergillus versicolor* (CNRMA/F5-26), *Aspergillus terreus* (CNRMA/F07-91), *Penicillium chrysogenum* (CNRMA/F02-26), *Penicillium purpurogenum* (CNRMA/F08-52), *Rhizopus oryzae* (CBS 112.07), and *Cladosporium cladosporioides* (IP 1232). Fungi were grown on Sabouraud agar medium, and DNA was extracted from 7-day-old culture conidia using the MagNa Pure Compact Nucleic Acid Isolation Kit in a MagNa Pure Compact apparatus (Roche Diagnostics, Meylan, France) according to the manufacturer's instructions. DNA concentration was determined using a spectrophotometer (NanoDrop ND-1000, Thermo Scientific, Wilmington, USA). PCRs were carried out in duplicates in a 20 μL final volume containing 1X LightCycler® 480 SYBR Green I Master (Roche Diagnostics, Meylan, France), 500 nM of each primer (Set1[10], Set2[17], Set3[18], Set4[15], Set5[14], Set 6[16], and Set7[13]) and 5 ng of each fungal DNA species on a LightCycler® 480 instrument (Roche Diagnostics, Meylan, France) with an initial denaturation step at 95°C for 8 min followed by 50 cycles at 95°C for 10 s, 60°C for 10 s, and 72°C for 15 s. The melting curves of each PCR were analyzed for detecting primer dimer formation. The final results were expressed by subtracting to the Cq of a given species the Cq value obtained for the *A. fumigatus* IP 2279.94 strain. Each experiment was repeated twice.

products perfused to the patients that contain fungal DNA[48] that cannot be eliminated by any technical improvement. Detailed analysis of the clinical file is required containing all the medications taken by the patient, as are tests of any suspected products with the diagnostic PCR assay. The risk of sample cross-contamination between negative and positive samples handled and processed in parallel should be mentioned. Automated DNA extraction is a useful means to decrease this risk.[15,49]

To limit the risk of detecting environmental fungal DNA, one possibility is to design primers very specific for the main target, for example, *A. fumigatus* for IA. However, specificity is relative, and some primers designed for amplifying *A. fumigatus* DNA can also amplify other environmental mold DNAs (Figure 8.1). This is simply due to the possibility of mismatches between primers and DNA targets. Therefore, the use of *panfungal* primers is particularly prone to false positives due to environmental DNA in any step of the tube processing. On the other hand, when using very specific primers, one must accept not being able to detect other fungi that can be clinically relevant, such as *Fusarium* spp. or zygomycetes.[50]

For the clinical diagnostic assays, it is also mandatory to check the specificity of the amplified products by using specific probes.[6,7] The use of SYBR Green[51] lacks the required specificity for use in clinical diagnostic assays.[42] With the SYBR Green dye, the melting curve analysis is only indicative of the nature of the amplified fragment. Moreover, comparison from run to run is always very difficult and cannot be routinely used to assess the nature of the amplified fragment.

8.3.2 Prevention of False Negatives

Given the high sensitivity of PCR, less attention has been paid to the prevention of false negatives than to the prevention of false positives. Two main sources of false negatives can be prevented. First, PCR inhibitors can be detected using internal controls for the reaction[24] and by the quantification provided by qPCR.[6,7] Second, every step of the PCR, especially the first cycles of amplification, must be optimized to improve the limit of detection.

8.3.2.1 Internal Control

The DNA amplification yield is very sensitive to the presence of inhibitors that can be concentrated during sample processing and DNA extraction. This is even more relevant in fungal PCR where large volumes of blood are often tested. For example, the EAPCRI recommendations are to extract from at least 3 mL of whole blood.[9] However, the current authors demonstrated that to increase the volume of blood tested can be counterproductive in some diagnosis such as *Plasmodium* spp. detection,[52] and there is no reason to assume that it is different for fungal diagnoses. The optimal volume should be experimentally determined.

It is mandatory to have PCR inhibitor assessments at the DNA amplification stage when low copy number detection is required. The compulsory use of internal controls for the routine diagnoses of IFI has received acceptance only recently[9] despite publications supporting the use of an internal control in 1995 for *Aspergillus*[24] following other routine diagnoses for other pathogens.[44] A 2009 review reported that only half of the studies dealing with *Aspergillus* DNA detection use some control for PCR inhibitors[8] that consisted mainly of human DNA amplification. However, these controls were of no value to assess the presence of low amount of PCR inhibitors since human DNA is present in huge quantity in human specimens.[6,7]

An internal control is designed for controlling that the yield of the PCR in every clinical sample is the same as that determined during the different steps of optimization of the PCR assay. The internal control can be specific for each PCR assay,[24,44] but this is time-consuming when several routine diagnoses are run simultaneously. Therefore, generic internal controls are usually preferred such as mouse DNA when dealing with human samples,[53] plasmid,[43] seal virus,[54] or plant DNA.[55] Each has its own advantage, to mimic the eukaryotic target[11,55] or the ease of obtention.[54] With any qPCR assay, it is straightforward to obtain the quantification cycle (Cq) of the internal control and to check for every clinical sample that the Cq obtained has the expected value, that is, the value obtained when there is no PCR inhibitor. Any significant increase of this value means that a certain degree of inhibition occurred during amplification, alerting to the possibility of a false-negative result or, if considered, to the quantification of the fungal load.

The internal controls described earlier are not designed for evaluating the DNA extraction step. For fungal diagnoses, since the origin of the DNA is unknown, it is difficult to design a unique DNA extraction control. If the DNA is thought to come from conidia or hyphae, conidia can be a pertinent control if the DNA extraction is stringent enough to free DNA from the conidia. Of course, the spores must come from a fungal species not targeted by the diagnostic PCR assay. Some authors used a calibrated spore suspension of *Geotrichum candidum* for environmental studies, for instance, assuming that this species is not routinely collected from environmental sampling.[56] However, if the hypothesis is the presence of circulating cell-free fungal DNA in the samples, then fungal spores are not a valid control, and any of the generic internal controls listed earlier and added before the DNA extraction step can be relevant.

8.3.2.2 PCR Yield Optimization

It is difficult to summarize all the parameters to be controlled for optimization of a PCR assay, and a list of desirable and essential items to fulfil these has been proposed.[28] A hot start can be used to avoid unspecific amplification during the first cycle and then to improve the sensitivity of the whole amplification reaction. Primer sequences are now provided by software that considers the (1) length of the PCR product (the shorter the better to increase the PCR yield, usually <100 bp), (2) thermal properties of the primers, and (3) dimer formation among other parameters.[57] A positive point observed in the quality control performed using serum spiked with *A. fumigatus* DNA was that most PCR protocols generate satisfactory analytical performance.[33] This suggests that primer design and PCR platforms, which were different between the participants, were not critical in explaining differences between the laboratories and that preanalytical steps are probably more important for standardization.[9]

However, primer choice is a compromise between the constraints of the targeted sequence and the diagnostic strategy, and obviously the latter is not included in primer design software. To target a panfungal sequence for diagnosis allows the possibility of false-positive results (see earlier) and false-negative results. In fact, it is impossible to amplify each fungal species with a single set of primers.[58] Worse still, amplification can occur but without the same efficiency for each species[59] (Figure 8.1). Therefore, if contaminating environmental DNA is present in a clinical sample, the PCR can be deviated toward this irrelevant species, masking the presence of *A. fumigatus* DNA, for example. Indeed, with two fungi in the sample, only the more prevalent will be amplified, and not necessarily the most relevant one in terms of tissue damage or treatment (Figure 8.1).

The most specific primers possible should be used to overcome these pitfalls that improve assay sensitivity and allow direct species-level identification without additional steps, hence speeding diagnosis and therapeutic decision making when the etiological agents are in limited quantities. This is the case for endemic mycoses where the suspected species is well known but limited in quantity.[60,61] Since *A. fumigatus* is by far the predominant species and priority, accounting for more than 80% of the IAs.[1] However, extending primer specificity to other *Aspergillus* species such as *A. flavus* with similar antifungal susceptibility to *A. fumigatus* can be justified.[15] To extend further within the *Aspergillus* genus[62] harbors the risk of nondetection of *A. fumigatus* in mixed infection (see earlier) or, for example, to detect sample contamination with *Penicillium* DNA. Indeed, it is perhaps unsurprising that qPCR designed to amplify DNA of several *Aspergillus* spp. also amplifies *Penicillium* sp. DNA, given that *Aspergillus* and *Penicillium* have some similarity.[12] An additional step is then mandatory to identify at the species level to adapt the antifungal therapy, since all the *Aspergillus* spp. do not have the same susceptibility to antifungal drugs. For example, *A. terreus* is resistant to amphotericin B in contrast to *A. fumigatus*. Obviously, a qPCR assay to detect *A. fumigatus* will not necessarily recognize other clinically significant species such as the mucormycetes,[50] and qPCR assays will be required specifically for the other taxa.

Another limit of using panfungal primers is that fungi are eukaryotic cells with large convergences with mammal DNA in contrast to bacteria and viruses. For instance, the ITS1 primer[63] perfectly matches with *Homo sapiens* DNA. Therefore, although this primer can be used for amplifying fungal DNA from pure fungal culture, it will be inoperative when searching for fungal DNA in human samples.

8.4 QUALITY CONTROL DESIGN

In view of the difficulties in comparing the different qPCR assays, there is a need to develop quality controls.[64] This was the main objective of the EAPCRI working group, launched in 2006 on behalf of the International Society for Human and Animal Mycology (ISHAM). Several recommendations have been published since.[9,33,65] The most frequent variability depends on the preanalytical steps, rather than on the PCR itself, if the qPCR format is used. Some relevant points are as follows if comparisons between the different qPCR assays are performed.

The DNA target is usually a multicopy gene in order to improve sensitivity and is most frequently the multicopy rRNA genes, based on the assumption that all gene copies are identical between different isolates. However, this is not the case for *A. fumigatus* with variation in the copy number of amplifiable rDNA.[66] Some authors have targeted other multicopy genes such as the mitochondrial genome sequences.[15,18,67] Because 28S qPCR[10] and mitochondrial qPCR[11] target different loci, the current authors studied the mitochondrial/ribosomal DNA ratio variability according to the time of culture. The maximum ratio value (3.7 ± 0.6) was observed after 8 h of culture reflecting the expected production of mitochondria during fungal growth. This ratio then slowly decreased to achieve a value of 1.5 (± 0.3) at 72 h (Figure 8.2). Consequently, quality control must be designed with a well-characterized strain and the DNA used for control obtained under the same cultural conditions.

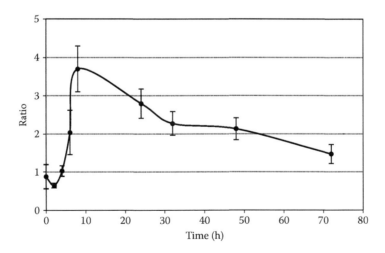

FIGURE 8.2 Mitochondrial DNA/ribosomal DNA ratio during growth of *Aspergillus fumigatus*. Quantification of mitochondrial and ribosomal DNA was determined from Cq using a standard curve derived from a dilution series of *A. fumigatus* DNA (IP 2279.94) using a mitochondrial[15] and a nuclear[10] target. Results were finally expressed as a ratio of the mitochondrial DNA to the ribosomal DNA. 10 mL of Dulbecco's Modified Eagle's Medium (Invitrogen Life Technologies, Cergy-Pontoise, France) with 10% heat-inactivated fetal bovine serum were seeded with 10^6 conidia/mL, cultured for 72 h at 37°C and 160 rpm. In these conditions, germination began at 4 h and filamentation at 6 h.[80] At time 2, 4, 6, 8, 24, 48, and 72 h, the mycelium was harvested by centrifugation at 3500 rpm for 7 min and DNA extracted as in Figure 8.1. Each growth experiment was repeated twice.

To simplify the qPCR issue, some advocate the use of commercially available kits.[68] However, commercial availability does not mean that every compulsory validation has been performed and the difficulty in obtaining details hampers checking crucial points (such as primer sequences). Hence, strategies developed for these kits can fail in fulfilling clinical requirements such as appropriate conditions for a screening test. Additionally, they can be technically very demanding and/or expensive.[68]

8.5 CLINICAL PERFORMANCE OF A qPCR TEST

Several elements complicate the clinical evaluation of surrogate markers for the diagnosis of IA.

First, there is no gold standard for the diagnosis of IA except the evidence of the fungus in biopsies, which is rarely obtained from at-risk patients.

Second, the probability of IA is a function of risk factors. The risk of IA evolves over time as a factor of neutropenia duration[69] where the risk decreases with the early administration of antifungal therapy before diagnosis.[70] Indeed, neutropenia is one of the main risk factors and the probability of IA increases with the neutropenia duration.

Third, the prognosis of IA is so poor[1] that empiric fungal treatment is the rule. This treatment is expected to impact on the growth of the fungus and, therefore, also on the positivity rate of the test. Consequently, the diagnostic value of any test strongly depends on the time when it is performed and whether the patient is given antifungal drugs or not. Indeed, when all serum samples collected during a study comparing two therapeutic strategies[71] were considered, the sensitivity of an evaluated qPCR test was 73%, in the range calculated from a meta-analysis (88% [95% CI, 75–94]).[23] However, when calculation was restricted to serum samples collected during the risk window of IA, that is, after the occurrence of neutropenia and before the administration of antifungal drugs, a sharp loss in sensitivity was observed. Therefore, it appears mandatory to report the timing of test results in diagnostic studies and the complete reporting of all possible confounders due to the management of IA.[23]

Finally, a qPCR assay must be validated in prospective clinical studies to be accepted in routine analysis. No study has demonstrated that a PCR-driven strategy improved the prognosis of the patients until now.[72–74]

8.6 PERSPECTIVES

If a consensus were to emerge on the procedure to be followed for using qPCR in a routine laboratory for microorganism detection,[26,28] the current technological developments are so large and rapid that other possibilities to qPCR emerge.

Digital polymerase chain reaction (dPCR) or droplet digital PCR (ddPCR) can overcome some difficulties encountered with qPCR. With ddPCR, the sample is partitioned into droplets so that a unique DNA molecule can be amplified. When all the droplets are PCR negative, the initial copy number in the sample can be calculated without reliance on the quantification based on the amplification yield. Indeed, with qPCR the initial copy number is proportional to the number of qPCR amplification cycles. However, with ddPCR the results are expressed as either positive or negative, and the number of copies is calculated according to Poisson's law. An absolute quantification can be provided without the need of exponential data analysis with an external standard curve. ddPCR could give reliable quantification since the number of *A. fumigatus* DNA molecules is always very low in IA.[75] This would avoid the potentially inaccurate positive or negative result when facing the limit of detection of qPCR. In addition, dPCR can detect copy number variations and relative gene expression in single cells.[76] Multiple commercial dPCR platforms are available and differ mainly in their method of individual reaction partitioning (i.e., chips designed with microfluidics channels or systems that divide diluted samples among many water-in-oil droplets).

dPCR is also the basis of clonal amplification for next-generation sequencing (NGS). This has the potential to explore (1) new variants of microorganisms (including fungi) for more effective disease management[77] and (2) commensal flora (the microbiome) and therefore mixed infections, among other applications.[78] However, to state that NGS improves molecular diagnoses in a routine lab is premature.

8.7 CONCLUSIONS

The large number of parameters to be controlled, from the design of the primers to the qPCR results,[6,28,79] should not restrain the use of qPCR in routine diagnostic laboratories. Some recommendations can be followed by any laboratory and there is no need to run exactly the same qPCR assay. When developing their own qPCR test, authors should compare the analytical performance observed with well-established qPCR.[12] This should save time for evaluating only qPCR assays with better analytical performances. In parallel, more information is required of DNA kinetics in humans to improve the preanalytical step of DNA extraction. A solution might be found with the emergence of dPCR[7] that allows an absolute quantification of *A. fumigatus* molecules, for the current limitations of the qPCR assays due to the low number of DNA targets in the clinical specimens that hampered reliable comparisons. Eventually, a qPCR assay in the diagnostic strategy for the specific issue of IA needs validation through well-designed prospective studies.

REFERENCES

1. Lortholary, O. et al. Epidemiological trends in invasive aspergillosis in France: The SAIF network (2005–2007). *Clin Microbiol Infect* **17**, 1882–1889 (2011).
2. De Pauw, B. et al. Revised definitions of invasive fungal disease from the European Organization for Research and Treatment of Cancer/Invasive Fungal Infections Cooperative Group and the National Institute of Allergy and Infectious Diseases Mycoses Study Group (EORTC/MSG) Consensus Group. *Clin Infect Dis* **46**, 1813–1821 (2008).
3. Hope, W.W., Walsh, T.J., and Denning, D.W. Laboratory diagnosis of invasive aspergillosis. *Lancet Infect Dis* **5**, 609–622 (2005).
4. Leeflang, M.M. et al. Galactomannan detection for invasive aspergillosis in immunocompromised patients. *Cochrane Database Syst Rev* **2008**, CD007394 (2008).
5. Pfeiffer, C.D., Fine, J.P., and Safdar, N. Diagnosis of invasive aspergillosis using a galactomannan assay: A meta-analysis. *Clin Infect Dis* **42**, 1417–1427 (2006).
6. Bretagne, S. Advances and prospects for molecular diagnostics of fungal infections. *Curr Infect Dis Rep* **12**, 430–436 (2011).
7. Bretagne, S. Molecular detection and characterization of fungal pathogens. In *Molecular Microbiology: Diagnostic Principles and Practice*, Vol. 42 (eds. D.H. Persing, F.C. Tenover, Y.-W. Tang, F.S. Nolle, R.T. Hayden, and A.V. Belkum), pp. 655–668 (ASM, Washington, DC, 2011).
8. Khot, P.D. and Fredricks, D.N. PCR-based diagnosis of human fungal infections. *Expert Rev Anti Infect Ther* **7**, 1201–1221 (2009).
9. White, P.L. et al. *Aspergillus* PCR: One step closer towards standardisation. *J Clin Microbiol* **48**, 1231–1240 (2010).
10. Challier, S., Boyer, S., Abachin, E., and Berche, P. Development of a serum-based Taqman real-time PCR assay for diagnosis of invasive aspergillosis. *J Clin Microbiol* **42**, 844–846 (2004).
11. Costa, C. et al. Development of two real-time quantitative TaqMan PCR assays to detect circulating *Aspergillus fumigatus* DNA in serum. *J Microbiol Methods* **44**, 263–269 (2001).
12. Johnson, G.L., Bibby, D.F., Wong, S., Agrawal, S.G., and Bustin, S.A. A MIQE-compliant real-time PCR assay for *Aspergillus* detection. *PLoS One* **7**, e40022.
13. Kami, M. et al. Use of real-time PCR on blood samples for diagnosis of invasive aspergillosis. *Clin Infect Dis* **33**, 1504–1512 (2001).
14. Pham, A.S. et al. Diagnosis of invasive mold infection by real-time quantitative PCR. *Am J Clin Pathol* **119**, 38–44 (2003).
15. Costa, C. et al. Real-time PCR coupled with automated DNA extraction and detection of galactomannan antigen in serum by enzyme-linked immunosorbent assay for diagnosis of invasive aspergillosis. *J Clin Microbiol* **40**, 2224–2227 (2002).
16. Loeffler, J. et al. Quantification of fungal DNA by using fluorescence resonance energy transfer and the light cycler system. *J Clin Microbiol* **38**, 586–590 (2000).

17. O'Sullivan, C.E. et al. Development and validation of a quantitative real-time PCR assay using fluorescence resonance energy transfer technology for detection of *Aspergillus fumigatus* in experimental invasive pulmonary aspergillosis. *J Clin Microbiol* **41**, 5676–5682 (2003).
18. Spiess, B. et al. Development of a LightCycler PCR assay for detection and quantification of *Aspergillus fumigatus* DNA in clinical samples from neutropenic patients. *J Clin Microbiol* **41**, 1811–1818 (2003).
19. Zhao, Y. et al. Rapid real-time nucleic acid sequence-based amplification-molecular beacon platform to detect fungal and bacterial bloodstream infections. *J Clin Microbiol* **47**, 2067–2078 (2009).
20. Suarez, F. et al. Detection of circulating *Aspergillus fumigatus* DNA by real-time PCR assay of large serum volumes improves early diagnosis of invasive aspergillosis in high-risk adult patients under hematologic surveillance. *J Clin Microbiol* **46**, 3772–3777 (2008).
21. McCulloch, E. et al. Don't throw your blood clots away: Use of blood clot may improve sensitivity of PCR diagnosis in invasive aspergillosis. *J Clin Pathol* **62**, 539–541 (2009).
22. Springer, J. et al. A novel extraction method combining plasma with a whole-blood fraction shows excellent sensitivity and reproducibility for patients at high risk for invasive aspergillosis. *J Clin Microbiol* **50**, 2585–2591.
23. Mengoli, C., Cruciani, M., Barnes, R.A., Loeffler, J., and Donnelly, J.P. Use of PCR for diagnosis of invasive aspergillosis: Systematic review and meta-analysis. *Lancet Infect Dis* **9**, 89–96 (2009).
24. Bretagne, S. et al. Detection of *Aspergillus* species DNA in bronchoalveolar lavage samples by competitive PCR. *J Clin Microbiol* **33**, 1164–1168 (1995).
25. Burkardt, H.J. Standardization and quality control of PCR analyses. *Clin Chem Lab Med* **38**, 87–91 (2000).
26. Hoorfar, J. et al. Making internal amplification control mandatory for diagnostic PCR. *J Clin Microbiol* **41**, 5835 (2003).
27. Paterson, R.R. Internal amplification controls have not been employed in fungal PCR hence potential false negative results. *J Appl Microbiol* **102**, 1–10 (2007).
28. Bustin, S.A. et al. The MIQE guidelines: Minimum information for publication of quantitative real-time PCR experiments. *Clin Chem* **55**, 611–622 (2009).
29. Casadevall, A. Fungi and the rise of mammals. *PLoS Pathog* **8**, e1002808.
30. Bernal-Martinez, L. et al. Analysis of performance of a PCR-based assay to detect DNA of *Aspergillus fumigatus* in whole blood and serum: A comparative study with clinical samples. *J Clin Microbiol* **49**, 3596–3599 (2011).
31. Morton, C.O. et al. Dynamics of extracellular release of *Aspergillus fumigatus* DNA and galactomannan during growth in blood and serum. *J Med Microbiol* **59**, 408–413.
32. Millon, L. et al. Use of real-time PCR to process the first galactomannan-positive serum sample in diagnosing invasive aspergillosis. *J Clin Microbiol* **43**, 5097–5101 (2005).
33. White, P.L. et al. Evaluation of *Aspergillus* PCR protocols for testing serum specimens. *J Clin Microbiol* **49**, 3842–3848 (2011).
34. Horlitz, M., Lucas, A., and Sprenger-Haussels, M. Optimized quantification of fragmented, free circulating DNA in human blood plasma using a calibrated duplex real-time PCR. *PLoS One* **4**, e7207 (2009).
35. Springer, J. et al. Pathogen-specific DNA enrichment does not increase sensitivity of PCR for diagnosis of invasive aspergillosis in neutropenic patients. *J Clin Microbiol* **49**, 1267–1273 (2011).
36. Avni, T. et al. Diagnostic accuracy of PCR alone compared to galactomannan in bronchoalveolar lavage fluid for diagnosis of invasive pulmonary aspergillosis: A systematic review. *J Clin Microbiol* **50**, 3652–3658 (2012).
37. Sun, W. et al. Evaluation of PCR on bronchoalveolar lavage fluid for diagnosis of invasive aspergillosis: A bivariate metaanalysis and systematic review. *PLoS One* **6**, e28467 (2011).
38. Tuon, F.F. A systematic literature review on the diagnosis of invasive aspergillosis using polymerase chain reaction (PCR) from bronchoalveolar lavage clinical samples. *Rev Iberoam Micol* **24**, 89–94 (2007).
39. Alanio, A. et al. Real-time PCR assay-based strategy for differentiation between active *Pneumocystis jirovecii* pneumonia and colonization in immunocompromised patients. *Clin Microbiol Infect* **17**(10), 1531–1537. doi:10.1111/j.1469-0691.2010.03400.x (2010).
40. Botterel, F. et al. Clinical significance of quantifying *Pneumocystis jirovecii* DNA by using real-time PCR in bronchoalveolar lavage fluid from immunocompromised patients. *J Clin Microbiol* **50**, 227–231 (2012).
41. Morris, A., Wei, K., Afshar, K., and Huang, L. Epidemiology and clinical significance of pneumocystis colonization. *J Infect Dis* **197**, 10–17 (2008).
42. Espy, M.J. et al. Real-time PCR in clinical microbiology: Applications for routine laboratory testing. *Clin Microbiol Rev* **19**, 165–256 (2006).

43. White, P.L., Linton, C.J., Perry, M.D., Johnson, E.M., and Barnes, R.A. The evolution and evaluation of a whole blood polymerase chain reaction assay for the detection of invasive aspergillosis in hematology patients in a routine clinical setting. *Clin Infect Dis* **42**, 479–486 (2006).
44. Bretagne, S. et al. Detection of *Toxoplasma gondii* by competitive DNA amplification of bronchoalveolar lavage samples. *J Infect Dis* **168**, 1585–1588 (1993).
45. Loeffler, J. et al. Contaminations occurring in fungal PCR assays. *J Clin Microbiol* **37**, 1200–1202 (1999).
46. Rimek, D., Garg, A.P., Haas, W.H., and Kappe, R. Identification of contaminating fungal DNA sequences in Zymolyase. *J Clin Microbiol* **37**, 830–831 (1999).
47. Harrison, E. et al. Fungal DNA contamination of blood collection tubes. In *48th Interscience Conference on Antimicrobial Agents and Chemotherapy* (American Society for Microbiology, Washington, DC, 2008).
48. Millon, L. et al. False-positive *Aspergillus* real-time PCR assay due to a nutritional supplement in a bone marrow transplant recipient with GVH disease. *Med Mycol* **48**, 661–4 (2010).
49. Costa, J.M. and Ernault, P. Automated assay for fetal DNA analysis in maternal serum. *Clin Chem* **48**, 679–680 (2002).
50. Bitar, D. et al. Increasing incidence of zygomycosis (mucormycosis), France, 1997–2006. *Emerg Infect Dis* **15**, 1395–1401 (2009).
51. Bu, R. et al. Monochrome LightCycler PCR assay for detection and quantification of five common species of *Candida* and *Aspergillus*. *J Med Microbiol* **54**, 243–248 (2005).
52. Farrugia, C. et al. Cytochrome b gene quantitative PCR for diagnosing *Plasmodium falciparum* infection in travelers. *J Clin Microbiol* **49**, 2191–2195 (2011).
53. Costa, J.M., Ernault, P., Gautier, E., and Bretagne, S. Prenatal diagnosis of congenital toxoplasmosis by duplex real-time PCR using fluorescence resonance energy transfer hybridization probes. *Prenat Diagn* **21**, 85–88 (2001).
54. van Doornum, G.J., Guldemeester, J., Osterhaus, A.D., and Niesters, H.G. Diagnosing herpes virus infections by real-time amplification and rapid culture. *J Clin Microbiol* **41**, 576–580 (2003).
55. Nolan, T., Hands, R.E., Ogunkolade, W., and Bustin, S.A. SPUD: A quantitative PCR assay for the detection of inhibitors in nucleic acid preparations. *Anal Biochem* **351**, 308–310 (2006).
56. Bellanger, A.P. et al. Indoor fungal contamination of moisture-damaged and allergic patient housing analysed using real-time PCR. *Lett Appl Microbiol* **49**, 260–266 (2009).
57. Zuker, M. Mfold web server for nucleic acid folding and hybridization prediction. *Nucleic Acids Res* **31**, 3406–3415 (2003).
58. Schoch, C.L. et al. Nuclear ribosomal internal transcribed spacer (ITS) region as a universal DNA barcode marker for Fungi. *Proc Natl Acad Sci USA* **109**, 6241–6246 (2012).
59. Bellemain, E. et al. ITS as an environmental DNA barcode for fungi: An in silico approach reveals potential PCR biases. *BMC Microbiol* **10**, 189 (2010).
60. Bialek, R., Gonzalez, G.M., Begerow, D., and Zelck, U.E. Coccidioidomycosis and blastomycosis: Advances in molecular diagnosis. *FEMS Immunol Med Microbiol* **45**, 355–360 (2005).
61. Maubon, D., Simon, S., and Aznar, C. Histoplasmosis diagnosis using a polymerase chain reaction method. Application on human samples in French Guiana, South America. *Diagn Microbiol Infect Dis* **58**, 441–444 (2007).
62. Torres, H.A. et al. *Aspergillosis* caused by non-fumigatus *Aspergillus* species: Risk factors and in vitro susceptibility compared with *Aspergillus fumigatus*. *Diagn Microbiol Infect Dis* **46**, 25–28 (2003).
63. White, T.J., Bruns, T.D., Lee, S.B., and Taylor, J.W. Amplification and sequencing of fungal ribosomal RNA genes for phylogenetics. In *PCR Protocols and Applications: A Laboratory Manual* (eds. M.A. Innis, D.H. Gelfand, J.J. Sninsky, and T.J. White), pp. 315–322 (Academic Press, New York, 1990).
64. White, P.L. et al. A consensus on fungal polymerase chain reaction diagnosis? A United Kingdom-Ireland evaluation of polymerase chain reaction methods for detection of systemic fungal infections. *J Mol Diagn* **8**, 376–384 (2006).
65. White, P.L. et al. Critical stages of extracting DNA from *Aspergillus fumigatus* in whole-blood specimens. *J Clin Microbiol* **48**, 3753–3755 (2010).
66. Herrera, M.L., Vallor, A.C., Gelfond, J.A., Patterson, T.F., and Wickes, B.L. Strain-dependent variation in 18S ribosomal DNA Copy numbers in *Aspergillus fumigatus*. *J Clin Microbiol* **47**, 1325–1332 (2009).
67. Florent, M. et al. Prospective evaluation of a polymerase chain reaction-ELISA targeted to *Aspergillus fumigatus* and *Aspergillus flavus* for the early diagnosis of invasive aspergillosis in patients with hematological malignancies. *J Infect Dis* **193**, 741–747 (2006).

68. Lehmann, L.E. et al. A multiplex real-time PCR assay for rapid detection and differentiation of 25 bacterial and fungal pathogens from whole blood samples. *Med Microbiol Immunol* **197**, 313–324 (2008).
69. Muhlemann, K., Wenger, C., Zenhausern, R., and Tauber, M.G. Risk factors for invasive aspergillosis in neutropenic patients with hematologic malignancies. *Leukemia* **19**, 545–550 (2005).
70. Gotzsche, P.C. and Johansen, H.K. Meta-analysis of prophylactic or empirical antifungal treatment versus placebo or no treatment in patients with cancer complicated by neutropenia. *BMJ* **314**, 1238–1244 (1997).
71. Cordonnier, C. et al. Empirical versus preemptive antifungal therapy for high-risk, febrile, neutropenic patients: A randomized, controlled trial. *Clin Infect Dis* **48**, 1042–1051 (2009).
72. Barnes, R.A. et al. Clinical impact of enhanced diagnosis of invasive fungal disease in high-risk haematology and stem cell transplant patients. *J Clin Pathol* **62**, 64–69 (2009).
73. Blennow, O. et al. Randomized PCR-based therapy and risk factors for invasive fungal infection following reduced-intensity conditioning and hematopoietic SCT. *Bone Marrow Transplant* **45**(12), 1710–1718 (2010).
74. Hebart, H. et al. A prospective randomized controlled trial comparing PCR-based and empirical treatment with liposomal amphotericin B in patients after allo-SCT. *Bone Marrow Transplant* **43**, 553–561 (2009).
75. Pinheiro, L.B. et al. Evaluation of a droplet digital polymerase chain reaction format for DNA copy number quantification. *Anal Chem* **84**, 1003–1011 (2011).
76. Zeng, Y., Novak, R., Shuga, J., Smith, M.T., and Mathies, R.A. High-performance single cell genetic analysis using microfluidic emulsion generator arrays. *Anal Chem* **82**, 3183–3190.
77. Le, T. et al. Low-abundance HIV drug-resistant viral variants in treatment-experienced persons correlate with historical antiretroviral use. *PLoS One* **4**, e6079 (2009).
78. Mardis, E.R. New strategies and emerging technologies for massively parallel sequencing: Applications in medical research. *Genome Med* **1**, 40 (2009).
79. Apfalter, P., Reischl, U., and Hammerschlag, M.R. In-house nucleic acid amplification assays in research: How much quality control is needed before one can rely upon the results? *J Clin Microbiol* **43**, 5835–5841 (2005).
80. Bellanger, A.P. et al. *Aspergillus fumigatus* germ tube growth and not conidia ingestion induces expression of inflammatory mediator genes in the human lung epithelial cell line A549. *J Med Microbiol* **58**, 174–179 (2009).

9 Acremonium

Richard C. Summerbell and James A. Scott

CONTENTS

9.1 Brief History of *Acremonium* 115
9.2 Scope of This Chapter 117
9.3 Molecular Analysis of *Acremonium* 118
9.4 *Acremonium* in Food 118
9.5 *Acremonium* in the Medical Context 120
9.6 Conclusions 125
References 126

All knowledge begins with epistemology, but in the twenty-first century, we are not accustomed to dealing with biological entities at this level. Apart from borderline situations like interspecies hybridization, we do not need to belabor *what is a horse* and *what is Pseudotsuga menziesii*. On the other hand, recent scientific advances have obliged us to go right back to square one and ask, *what is an Acremonium*? An upset in our biosystematic understanding has affected not only the genus but also nearly all of the individual species that have been placed in *Acremonium*. An applied question like *what is the significance of Acremonium in food?* cannot be framed until we know, in some intelligible form, what an *Acremonium* is.

9.1 BRIEF HISTORY OF *ACREMONIUM*

The genus *Acremonium* was described early in mycological history. Link[1] in 1809 coined the new genus name for a species he believed produced single spores at the ends of its fertile cells (hence the *mono*-root, for single, attached to the *acro*-root, situated at the top, as in acropolis). Gams[2,3] in 1968 examined Link's herbarium material and found that the type species, *Acremonium alternatum*, produced not single conidia, but rather conidia in chains, from thin, tapering phialides.

Many species closely resembling *A. alternatum* produce conidia in mucoid masses rather than in chains. These species were for many years placed scientifically in the genus *Cephalosporium* Corda 1839. Gams,[2,3] however, showed that the original *Cephalosporium* species appeared to be zygomycetous fungi, with aseptate mycelium, that would today be classified in the genera *Umbelopsis* and *Mortierella*. Therefore, he merged the septate, phylogenetically ascomycetous *Cephalosporium* species under the reinterpreted name *Acremonium* in his monumental monograph on that genus.[4] Also taken into *Acremonium* were morphologically similar species that had been disposed under other genus names, including *Gliomastix*, *Hyalopus*, *Mastigocladium*, *Paecilomyces pro parte* (selected monophialidic species only), *Oospora*, and *Monosporium*. The genus lost a number of species when the eurotialean genus *Sagenomella* was distinguished in 1978.[5] It then gained numerous, mostly unnamed, species when the *Acremonium*-like asexual states of *Chaetomium* and *Thielavia* species were added in 1982 as a new taxonomic section and asexual states related to grass endophyte fungi in the genus *Epichloe* were added as an additional new section.[6] This maximized circumscription of *Acremonium* prevailed for much of the rest of the twentieth century and remains highly influential today.

The first major molecular phylogenetic study of *Acremonium*, carried out by Glenn et al.,[7] in 1996, clearly showed that endophytes related to *Epichloe* formed a distinct clade. The genus *Neotyphodium* was described to remove these species from *Acremonium*. Literature searches on *Acremonium* to this day elicit large numbers of references on this economically important group of plant symbionts. The same phylogenetic study also showed that other *Acremonium* species were heterogeneous, but the very small numbers of species included prevented further revision of the genus. The *Acremonium* species related to *Chaetomium* were segregated into a new genus, *Taifanglania*, by Liang et al.[8] in 2009.

In 2011, Summerbell et al.[9] published a molecular phylogenetic overview in which more than 100 species in *Acremonium* and its related and similar taxa were sequenced for the first time. An epitype

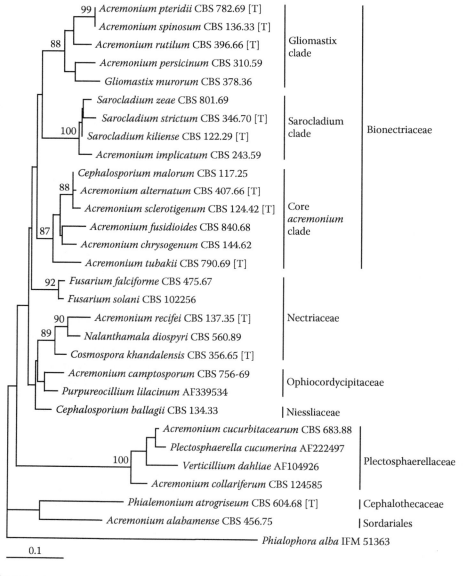

FIGURE 9.1 *Acremonium* and similar species important in food and medical mycology. Maximum likelihood tree based on 28S ribosomal DNA sequences.

was designated to root the type species, *A. alternatum*, and provide modern biosystematic mooring for the genus name *Acremonium*. Although this left most economically important *Acremonium* species associated with the genus, it excluded one well-defined group containing the economically and medically important names *Acremonium strictum* and *A. kiliense*. These fungi were transferred to the genus *Sarocladium*. In addition, the genus *Gliomastix* was reinstated for a phylogenetically distinct group of species. A general overview of the *Acremonium* species, based on 28S ribosomal sequences, is given in Figure 9.1.

Sarocladium strictum proved, in phylogenetic study, to be one of many species grouped under Gams'[4] concept of *Acremonium* that was highly heterogeneous, containing isolates from a wide diversity of haplotypes, often remotely related to one another. Of the 16 "*A. strictum*" isolates sequenced by Summerbell et al.[9] from the CBS Fungal Biodiversity Centre, Utrecht, the Netherlands collection (CBS), 8 were *S. strictum*; 3 were *Acremonium sclerotigenum*; 2 were *Acremonium zeae*; 1 was *Plectosphaerella melonis*, formerly *Acremonium cucurbitacearum*[10]; and 2 represented undescribed species. (Re "*A. strictum*" in the preceding sentence: in this chapter, as in Summerbell et al.,[9] historical morphological identifications that must be considered dubious or unsupported are cited in inverted commas, except in cases where their provisional status has already been made clear in the surrounding text. In addition, identifications that are taxonomically out of date but cannot yet be given a modern disposition are also cited in inverted commas, e.g., "*Cephalosporium malorum*".)

Phylogenetic analysis showed that the isolates identified as *A. strictum* in CBS spanned five phylogenetic genera from three families of fungi (Bionectriaceae, Nectriaceae, Plectosphaerellaceae) belonging to two distinct orders.[9]

Sequencing of a large number of *Acremonium* isolates, including all known ex-type isolates of species, revealed that most morphologically defined species concepts had accumulated heterogeneous collections of phylogenetic species.[9] For example, the three isolates in the CBS culture collection identified as *Acremonium blochii* belonged to three different species. Most species contained isolates belonging to more than one phylogenetic genus. The identifications, however, were by no means random, but were based on acutely precise morphological study. It appears that convergent evolution among these simply structured organisms was sufficient to defeat such morphological scrutiny. Just a few species, most notably *Sarocladium kiliense* and *Acremonium camptosporum*, were sufficiently distinct and unitary to be reliably morphologically identified, at least most of the time.

9.2 SCOPE OF THIS CHAPTER

The consequence of the biosystematic upheaval described earlier is that most existing literature on *Acremonium* is minimally scientifically reproducible, unless it was based on isolates that remain extant today and can be identified by sequencing. Therefore, a conventional review of *Acremonium* ss. W. Gams species associated with food and with biomedical risk would need to be based mainly on unreliable data and only minimally on reliable data. (The phrase *Acremonium* ss. W. Gams, where *ss.* abbreviates the Latin *sensu*, meaning *in the sense of*, is a taxonomic convention indicating the genus *Acremonium* as defined in the influential monograph of W. Gams.[4]) It is true that most isolates identified at the genus level as *Acremonium* in the food, industrial, and biomedical literatures were probably members of taxa that we can still legitimately assume would belong to the Gams[4] circumscription of the genus or its emendations up to 1978.[5,11] Since, however, that circumscription comprehends so many heterogeneous orders and families of fungi, the generic level identification is of little ecological significance. In essence, it is a vague hint at a certain range of morphological forms.

In this chapter, rather than laboriously summarizing the irreproducible material, we begin to build a reliable scientific literature on *Acremonium* and its segregate genera. In this task, we rely as much as possible on data attributable to genuinely identified—that is to say, in most cases, sequence identified—*Acremonium* species. Acremonioid species belonging to recently revised genera such as *Sarocladium* is included.

9.3 MOLECULAR ANALYSIS OF *ACREMONIUM*

As a preamble to the survey of *Acremonium* species in food and in medical mycology, the PCR methods that could be employed in producing well-confirmed *Acremonium* species identifications are surveyed briefly. Additional methods used by researchers producing multigene phylogenies for taxonomic purposes are not included in this survey.

In GenBank, the majority of published *Acremonium* sequences derived from authoritatively identified, internationally available culture collection isolates are currently 28S ribosomal DNA sequences. These sequences have the advantage of being alignable among a broad range of *Acremonium* groups, whereas the ribosomal internal transcribed spacer (ITS) sequences that are frequently used in mycology are very divergent and can only be validly aligned within restricted *Acremonium* subgroups.[9] Various authors[9,12,13] examining 28S sequences have used the NL1 and NL4 primers constructed by O'Donnell[14] or the V9G[15] and LR5 primers.[16] Both primer pairs amplify the D1 and D2 variable domains of the 28S rDNA.

Authors examining closely interrelated subgroups of *Acremonium* species[12,13,17] have used whole ITS region sequences, inclusive of the contiguous ITS1, 5.8S, and ITS2 subregions. These sequences have been obtained using primer combinations ITS5[18] and NL1R,[13] ITS5 and ITS4,[18] and ITS1-f and ITS4.[18,19] Much less commonly, especially to allow placement of very divergent organisms, whole 18S sequences have been amplified with primers NS1 and NS24 and sequenced using primers NS1–NS4, NS6, and NS24,[18,20] while partial sequences of the 5′ end of the 18S region have been performed using primers NS1 and NS2.[18]

9.4 *ACREMONIUM* IN FOOD

In the authoritative overviews of fungi in food mycology by Pitt and Hocking[21] and Samson et al.,[22] *Acremonium* species are mentioned as being among the food-spoiling microorganisms. Pitt and Hocking[21] cite "*A. strictum*" as the main organism of interest but acknowledge that the name is so frequently misapplied that the data are highly suspect. Moreover, their "*A. strictum*" records include those compiled for the even more indiscriminately applied, older name *Cephalosporium acremonium*. Nominally, then, this potentially wide range of fungi has been isolated from cereal grains, bananas, fresh vegetables, nuts, and preserved meats (salami, biltong). These records probably mostly apply at least to the genus *Acremonium* as defined by Gams.[4,5]

Samson et al.[22] mention *Acremonium butyri*, *Acremonium charticola*, and, once again, "*A. strictum*". *A. butyri* has recently been recombined as *Cosmospora butyri* by Gräfenhan et al.[23] It is only quasi-reliably known from its ex-type isolate, CBS 301.38. The isolation of CBS 301.38 was from butter, but the growth of *A. butyri* on this substrate has not been confirmed. It is a member of the *Cosmospora berkeleyana* (*Acremonium berkeleyanum*) species complex, and members of this group are often found growing, perhaps as mycoparasites, on other fungal species.[23] Upon our own request to obtain CBS 301.38 for sequencing, we were delivered an isolate that yielded the sequence of *Cadophora malorum*; possibly CBS 301.38 is a mixed culture containing a butter-colonizing *Cadophora* and its *Cosmospora* mycoparasite.

A. charticola is a name of uncertain application in modern systematics, since it is not rooted in living ex-type material. Our sequences of three isolates, so far, from CBS have yielded two widely divergent species. Samson et al.[22] say that *A. charticola* "probably causes apple- and pear-rot." The isolates that have been classified under this name are by no means commonly isolated from fruit, but one isolate from apple listed in CBS and Gams[4] as *A. charticola*, CBS 117.25, is the ex-type isolate of the synonymous name *C. malorum*. We have sequenced this isolate[9] and found that it is a close relative of *A. sclerotigenum*. *C. malorum* was described by Kidd and Beaumont[24] as having been isolated more than once from apples, but those authors noted that "*C. malorum* up to the present has always been isolated associated with other fungi." Hence, the species may be a mycoparasite or a decay organism growing on effete fungal material rather than a direct agent of apple spoilage.

To find some basis for determining which *Acremonium* ss. W. Gams species are important in foods, we checked the isolates in two major culture collections, CBS and the American Type Culture Collection (ATCC), Manassas, VA, USA. We looked for isolates of (1) sequence-confirmed identity that had obtained from food-related sources and (2) representatives of the few species that are reliably identified microscopically. CBS isolates identified by published and unpublished sequences produced by the current authors were included. The literature was also checked to the beginning of the 1990s in an attempt to find *Acremonium*-like fungi from food-related sources that had been identified by sequencing studies or deposited in a collection where such studies were done subsequently. No such studies were found, but in a few cases, identifications could be inferred with high probability based on the substrates and morphological identifications that were given.

Two members of *Acremonium* ss. W. Gams that are well documented in sequence-based studies are regularly associated with food materials. The former *A. zeae*, now *Sarocladium zeae*, is an endophyte of maize corn and is commonly isolated from healthy and diseased corn ears.[25] It produces pyrrocidine antifungal metabolites that deter the growth of the maize seed–rotting pathogens *Fusarium verticillioides* and *Aspergillus flavus*.[26] Numerous isolates were sequenced for studies by Wicklow et al.[26] and Summerbell et al.[9] Also, the former *Acremonium diospyri*, now *Nalanthamala diospyri*, is a wilt pathogen of the American persimmon, *Diospyros virginiana*. Isolates are typically from lesions formed beneath the tree bark, but isolation from inoculum deposited on surface-contaminated fruit is possible. Representative isolates, including the probable ex-type culture CBS 560.89, were sequenced by Schroers et al.[27] A related species, *Nalanthamala psidii*, is a similar wilt pathogen forming bark blisters on guava, but it was never classified in *Acremonium*.[27]

The other sequence-confirmed isolates of *Acremonium* ss. W. Gams that are accessed in CBS or ATCC are listed in Table 9.1. In addition, Table 9.1 includes isolates that have not yet been identified by sequencing studies, but belong to species that molecular biosystematic studies show are likely to be reliably identifiable by morphological examination alone. In these cases, morphological identification relies on truly distinctive characters; moreover, the biological entity referred to by the species name can be precisely specified. In particular, *A. sclerotigenum* isolates identified by confirming the formation of distinctive multicelled sclerotia are considered to be definitively identified. Apart from the isolates listed in Table 9.1, there is a small number of culture collection isolates of uncertain identity that derive from food-related sources. There are four such acremonioid

TABLE 9.1
Reliably Identified *Acremonium* Isolates Obtained from Food-Related Sources

Species	Current Identity in Collection	Accession Number	Substrate	Comment
Phialemonium atrogriseum	*Acremonium atrogriseum*	CBS 604.67	Noodles	Ex-type isolate
Cephalosporium malorum	*Acremonium charticola*	CBS 117.25	Rotten apple	Ex-type isolate, species in need of revision
Acremonium sp.	*Acremonium egyptiacum*	CBS 303.64	Stored wheat	Undescribed species
Acremonium recifei	*Acremonium recifei*	CBS 442.66	Brazil nut	
Acremonium sclerotigenum	*Acremonium sclerotigenum*	CBS 740.69	Fish meal	Morphological identification
Acremonium sclerotigenum	*Acremonium sclerotigenum*	CBS 343.64	Lenticel in apple peel	Morphological identification
Sarocladium strictum	*Acremonium strictum*	CBS 550.73	Stored apples	
Sarocladium strictum	*Acremonium strictum*	ATCC 10141	Pea	JX094783 (GenBank record)

isolates in CBS (365.64 "*A. egyptiacum*" from wheat, 109069 "*A. fusidioides*" from barley, 226.70 "*A. rutilum*" from apple scab, and 588.73A "*A. sordidulum*" from the *Annona reticulata* fruit) and five in ATCC (62726 "*A. implicatum*" from sunflower seed, 62827 "*A. strictum*" from maize, MYA-3363 and MYA-3384 "*A. strictum*" from sorghum, and MYA-3976 *Acremonium* sp. from barley). These isolates could in principle be definitively identified, adding to our small corpus of reliable knowledge of *Acremonium* in foods.

Any other information on *Acremonium* ss. W. Gams species associated with foods is, in terms of biosystematics, at the level of speculation. There are some tantalizing examples that require further investigation. Included are reports of "*A. strictum*" as a contributor to 2,4,6-trichloroanisole formation in corked wine,[28] "*A. strictum*" and "*A. implicatum*" as common isolations from rice,[29] *Acremonium* sp. as a common, visible contaminant in spoiled bottles of mineral water,[30] and "*A. recifei*" as a tomato pathogen.[31] The last study cited lists a culture collection number as a voucher, IMI 333806, but the isolate is not in the online catalogue of the IMI collection (CABI Biosciences, Egham, UK). Morphologically identified "*Acremonium tubakii*" and "*A. rutilum*" were demonstrated to be postharvest pathogens of peaches.[32] *A. tubakii* is a marine fungus,[33] so at least this peach pathogen is highly likely to have been misidentified.

In general, it appears that fungi with the appearance of the unrelated but morphologically similar *S. strictum* and *A. sclerotigenum* are among the most common acremonioid fungi isolated from many types of contaminated foods. Both species may also be plant endophytes coming into contact with foods as components of plant stems. For example, an *A. strictum*–like fungus is listed as an endophyte of oak[34] and may be the same species from wine corks.[28] The current authors possess *A. sclerotigenum* isolates as endophytes of various woody plants including grapevines (unpublished data). Future identifications made by sequence analysis should rapidly clarify the nature and origins of the common acremonioid fungi associated with food spoilage and related plant pathogenesis.

9.5   *ACREMONIUM* IN THE MEDICAL CONTEXT

The problem of *Acremonium* in medical mycology became firmly linked to food mycology when reports emerged of species infecting immunocompromised human patients via the gut. Schell and Perfect[35] reported "*A. strictum*" from a disseminated mycosis that began as a gastrointestinal colonization; however, the identification of the isolate was not confirmed molecularly at that time or subsequently. Other evidence suggests that *S. strictum per se* is unlikely to have been the etiologic agent. Perdomo et al.[17] undertook sequence identification of 14 isolates from U.S. medical sources that had been morphologically identified as *A. strictum* by a reputable reference laboratory, but none of them was a member of this species. Five were *A. sclerotigenum*, eight were unidentified species but not *S. strictum*, and one, from a contact lens, was *S. zeae*. Although Schell and Perfect's[35] identification of their etiologic agent as *A. strictum* is doubtful, illustrations confirm that the fungus was a member of *Acremonium* ss. W. Gams. From the epidemiological standpoint, the isolate's invasion of the patient via the gut strongly implies ingestion in food or water of the problematical inoculum, although swallowing of air spora deposited on the oropharyngeal mucous membranes cannot be ruled out.

A case report from Hong Kong documented "*Acremonium falciforme*" as an agent of upper gastrointestinal tract lesions in a bone marrow transplant patient with congenital severe combined immunodeficiency.[36] The isolate was not retained for later verification of its identity. The species *A. falciforme* itself was shown to be a member of the *Fusarium solani* complex and was transferred as *Fusarium falciforme* by Summerbell and Schroers.[37] Indeed, there are many reports of human opportunistic infections caused by members of the *F. solani* complex.[38]

S. kiliense, which is straightforward to identify correctly using morphological techniques,[17] has been recorded from esophagitis in an otherwise healthy 11-year-old boy.[39] Such an esophageal infection, like an infection starting in the lower gut, is suggestive of a food or water inoculum and remotely compatible with an airborne source. However, *S. kiliense* and *S. strictum* bear conidia in mucoid heads, a structure adapted to aqueous and arthropod propagule dispersal, and are uncommonly isolated from air. These records suggest that severely immunodeficient patients and, very rarely, immunocompetent patients can become infected by acremonioid fungi in ingested material.

In addition, *A. recifei* was isolated from the infected cornea of a man whose eye had been impaled by a flying shard of coconut shell.[40] Although deposition of the isolate in a culture collection was not mentioned in the publication, a correlation of provenance identified CBS 188.82 as the isolate, and the current authors have identified the isolate as *A. recifei* using molecular methods. This species, which can be confused with a small number of described and undescribed acremonioid species with curved conidia, has been regularly reported from coconut and Brazil nutshells.[4,38] A direct isolation from Brazil nutshell has been confirmed using molecular methods (Table 9.1).

Acremonium ss. W. Gams is a regularly occurring opportunistic pathogen of humans and animals.[38,41] The ecological evaluation of the roles played by acremonioid fungi in human and animal infections is complicated, however, by an epistemological problem that is just as problematical as the taxonomic identification problem mentioned earlier. That is, acremonioid isolates that are obtained from many types of medical specimens can be surface contaminants or the etiologic agents of opportunistic infections. In cases where these fungi are isolated from normally sterile sites and materials such as deep tissue biopsies and blood cultures, infectious status is all but guaranteed, but in isolations from samples like lung biopsy, skin scrapings, sputum, sinus biopsies, sinus swabs, and urine, harmless body surface contamination from normal air, dust, cosmetics, etc., may be the inoculum source.

Experienced clinical laboratorians tend to ensure etiologic agents are well defined rather than combining these with general contaminants. Nevertheless, physicians may assume incorrectly that laboratory-reported isolation of acremonioid species from nonsterile bodily materials straightforwardly signifies infectious status. The medical literature is replete with case reports where etiologic status for a fungus is incompletely confirmed or demonstrably incorrectly asserted. Species with maximum growth temperatures well below human body temperature are sometimes naively reported as pathogens. Many such dubious case reports are reported for acremonioids and other Eurotiomycetes fungi by Summerbell.[38]

The landmark molecular identification study of Perdomo et al.[17] revised existing concepts of *Acremonium* ss. W. Gams species in the clinical laboratory by identifying 75 clinical isolates using sequencing. The introduction to the study (in which the present senior author was a coauthor) made a statement that could be open to misinterpretation: "To assess the incidence of different species of *Acremonium* in human infections, we studied a large set of clinical isolates." To be stringent, one cannot study the incidence of any fungi in infections without rigorously determining which isolates are actually causing infections and which are merely contaminants isolated from diseases caused by something else. Later in the text, it was duly admitted that "we lack sufficient clinical information to ascertain whether any of these isolates was confirmed as a causal agent of infection." A reasonable rationalization was then given: "It is, however, likely that a significant proportion of them were causal agents, and the remainder may represent clinical contaminants, the identification of which will be a regular feature of clinical practice in the future."

Problematically, however, following a typical pattern, the first major citation of the paper, by Khan et al.,[42] stated that Perdomo et al.'s aims were "to clarify taxonomic uncertainties about the identification of *Acremonium* spp. and to assess their relative etiologic significance in human infections." In fact, the study included no methodology addressing the second aim attributed to it. The citing authors also said "this comprehensive study provided new insight into the etiologic spectrum of *Acremonium* spp. associated with human infections" and "another interesting finding of the study was the identification of several species that were not previously incriminated in human infections." In fact, the study did not implicate any previously undocumented species in the causation of infection. Some of the isolates identified may have been from as yet uncharacterized types of infections, but they were not distinguishable from species documented for the first time from biomedical contamination.

Although Perdomo et al.[17] have suggested the true spectrum of acremonioid species likely to be implicated in human infections, the actual number of definitively identified isolates confirmed as causing human infections is low. *S. kiliense* is an exception and can be identified readily by morphology. It is well documented and authenticated (e.g., via culture collection deposition of isolates) as an agent of eumycetoma.[38] In addition, it has been validly reported from a moderate number of posttraumatic or postsurgical ocular infections, occasional non-eumycetoma subcutaneous infections, and rare systemic or internal infections.[38,42] However, the review by Khan et al.[42] included some cases that Summerbell[38] found to be inadequately attested.

Another well-established eumycetoma agent is the tropical species *A. recifei* including the ex-type isolate CBS 137.35 that has been sequenced at several loci in order to anchor the phylogenetic identification of this species.[9] Apart from this isolate, only three of the *A. recifei* isolates in CBS are from well-demonstrated cases of human infection with two being published.[40,43] CBS 485.77 and CBS 110348 have been shown by molecular studies to come from a complex of undescribed species that are well separated from the true *A. recifei*.[9] The only sequence-identified (Summerbell, unpublished data) *real A. recifei* isolate from confirmed human disease, besides the ex-type, is CBS 188.82, the aforementioned isolate derived from keratomycosis caused by impalement of the eye with a shard of coconut shell.[40] CBS 485.77, representing an undescribed *A. recifei*–like species, is from confirmed eumycetoma of the hand.[43]

S. strictum, as *A. strictum*, was the subject of a substantial record of human infections until sequence identification threw all such cases into question. Unfortunately, few isolates from previous cases remained available to reinvestigate with modern techniques. Novicki et al.[44] studied a newly arising case of fatal disseminated infection apparently caused by *A. strictum* in a severely immunocompromised patient. Sequencing of the ribosomal ITS region revealed that the isolate was incompatible with the ex-type isolate anchoring the true *S. strictum*. The closest BLAST hit was to an isolate, CBS 223.70, identified as *A. alternatum*. Our studies[9] have shown this isolate, despite its formation of conidia in chains, to be conspecific with *A. sclerotigenum*, which normally forms conidia in mucoid heads. In fact, sequencing of multiple acremonioid isolates has disclosed that *A. sclerotigenum* includes a diversity of isolates with mucoid and catenulate conidia and that many of the latter were identified as *A. egyptiacum*.[9,17] The ex-type isolate of *A. egyptiacum* belongs to the same genetic group.[17] The name *A. egyptiacum* has nomenclatural priority over the later *A. sclerotigenum*, but we refrain from formally reducing *A. sclerotigenum* to synonymy because there are still earlier names in contention for the valid name of the species. Perdomo et al.[17] referred to the group as *A. sclerotigenum–A. egyptiacum*, a workable stopgap solution.

Novicki et al.,[44] having access only to a small number of comparison sequences in GenBank, elected to refer to the *A. sclerotigenum–A. egyptiacum* complex as *A. strictum* genogroup *II*. Regrettably, they also included a misidentified isolate of *S. kiliense* in their study, labeling it and depositing its sequence as *A. strictum* genogroup *III*. The isolate in the Novicki et al.[44] case appears to have been a typical *A. sclerotigenum* isolate based on its morphological description, but the special media needed to elicit the characteristic sclerotia were not used. One of the current authors (RCS) is acknowledged by Novicki et al. for assistance in the identification of the isolate, but was never sent the isolate itself and only commented on the possible significance of photos and other data.

Furthermore, an *A. strictum* genogroup *IV* sequence, not mentioned in the paper, was deposited by Novicki et al.[44] as GenBank ID AY138486. This sequence is of *Phialemoniopsis curvata*,[45] a fungus that was never classified in *Acremonium*. The erroneously labeled sequence has generated a biosystematically careless case report of vertebral osteomyelitis caused by *A. strictum*.[46] This otherwise medically excellent case report is illustrative of a general problem of unquestioningly performed GenBank searches.

Guarro et al.[47] identified an isolate from a heart valve vegetation as *A. strictum* based on Novicki et al.'s genogroup II. The sequence data indicate that the isolate is indeed *A. sclerotigenum*. A case of catheter-associated *A. strictum* peritonitis can also be reattributed to *A. sclerotigenum* as

Bibashi et al.[48] sent the isolate to CBS, where it has been ITS sequenced and is now indexed in the database as *A. sclerotigenum* (CBS 112783). An *S. kiliense* was incorrectly ascribed to *A. strictum* on the basis of Novicki et al.'s *genogroup III* by Foell et al.[49] Disseminated infection by *S. kiliense* co-occurred with pulmonary aspergillosis in a boy undergoing chemotherapy for acute lymphoblastic leukemia.

A. strictum had a modest track record as a causal agent of onychomycosis prior to the sequencing era.[38] In more recent times, where the available isolates from confirmed cases (as opposed to mere nail isolations and those from suggestive but unconfirmed cases) have been sequenced, they have proven to be *A. sclerotigenum*. A large number of well-confirmed onychomycosis isolates identified as *Acremonium* sp. in the studies of Gupta et al.[50] and Summerbell et al.[51] proved to be *A. sclerotigenum* upon sequencing; six of these isolates are now accessed in CBS (CBS 114224–114229). Zaias[52] confirmed numerous cases of onychomycosis that he ascribed to *Cephalosporium roseo-griseum*—a mysterious misidentification, since the dark-colored species indicated, now called *Gliomastix roseogriseum*, is completely unlike the isolates illustrated in the publication and equally unlike the two representatives later sent to CBS. Summerbell[38] pointed out that based on morphology, these isolates had been reidentified as *A. strictum* and *A. potronii* by Gams.[4] The "*A. strictum*" isolate CBS 395.70B has now been redetermined as *A. sclerotigenum*, while the "*A. potronii*" isolate CBS 781.69 belongs to one of several undescribed species still grouped under that name.

The consistent trend to reidentification of medically implicated "*A. strictum*" isolates as *A. sclerotigenum*[17] suggests that credence given to the name *A. strictum* in human disease contexts must be reevaluated. The report of Hitoto et al.[53] and those reviewed by Summerbell[38] should be considered taxonomically dubious unless authentic case isolates can be evaluated by sequencing.

It is possible that *A. sclerotigenum* may be responsible for the whole spectrum of infections heretofore attributed to *A. strictum*: these include keratomycosis, peritonitis, and onychomycosis in immunocompetent patients and (1) subcutaneous and other soft tissue mycoses and (2) pulmonary and disseminated infections in immunocompromised patients.[38,41,54]

It should be noted that there is one published case where a sequence genuinely pertaining to *S. strictum* (originally *A. strictum*) was attributed to the etiologic agent of pulmonary infection in a horse.[55] Bronchoalveolar lavage (BAL) fluid from the horse yielded four fungi in culture: *Aspergillus sp., Penicillium sp., Chaetomium sp., and Acremonium sp.* In that list, one already sees evidence of medical mycological unfamiliarity, in that, since some common *Aspergillus* species are potentially opportunistic and others are not, a report of *Aspergillus* sp. from an invasive sample such as BAL is inadequate, at least in human disease.[56] A Giemsa stain was made of centrifuged cells from the same BAL specimen; it showed subglobose "spores" 3–5 μm in diameter in monocytes. These spores were, without explanation, interpreted as potentially consistent with *Acremonium*. It should be noted that a conventional demonstration of positive histopathology in a systemic case caused by an *Acremonium*-like fungus consists of showing characteristically thin, septate filaments in a biopsy specimen. In the case at hand, studied by Pusterla et al.,[55] the veterinary surgery that would have been needed to obtain such a specimen was, understandably, not performed. The histopathology in this case, however, cannot be reinterpreted as excluding acremonioids, since hypocrealean fungi such as *Fusarium* and *Acremonium* are known to produce conidia in infected human tissue.[35] *S. strictum* conidia, even when produced submerged in medium, are generally cylindrical, not subglobose.[4] Schell and Perfect,[35] however, depict their acremonioid etiologic agent (called "*A. strictum*"; actual identity uncertain) producing yeastlike secondary budding in tissue, giving rise to some rounded cells. No actively budding cells were seen in Pusterla et al.'s histopathology, but the rounded spores they saw could conceivably have been products of a budding acremonioid fungus in tissue.

Pusterla et al.[55] attempted to clarify the etiology of their case[55] using a company providing real-time PCR–based identification based on a proprietary method and confirmed *S. strictum* in the isolated acremonioid culture and in the BAL fluid. Notwithstanding these findings, the causal agent in this case may not have been *S. strictum*.

First, BAL specimens may carry material of contaminant fungi from the air, aspirated saliva, or food. Noninfecting, airborne propagules are often inactivated on respiratory surfaces by pulmonary phagocytes, so the presence of spores in phagocytes in BAL does not directly signal a disease process. Pusterla et al.[55] noted that such spores occasionally are seen in BAL specimens from horses. Horses eat plant material, a common source of *S. strictum*, and may inhale or aspirate the fungus. The turnover time of such materials in mammalian respiratory tracts is seldom studied. In principle, any fungus that can be isolated in culture from BAL should be detectable by PCR, which does not distinguish surface contamination from etiology in medical mycological specimens.[57] Therefore, the PCR confirmation of *S. strictum* in the BAL specimen did not establish etiology.

Second, acremonioid fungi are opportunistic pathogens, usually requiring severe preexisting immunodeficiencies in their hosts before they can cause infection. This is especially predictable in organs such as the lungs that are well defended by the cellular immune system. Infections of otherwise healthy humans or animals are usually limited to cases involving traumatic or surgical penetration of especially vulnerable body sites such as corneas and peritoneal cavities. Third, the putatively *Acremonium*-infected horse[55] was apparently successfully treated with fluconazole, an antifungal drug now well known not to have any inhibitory effects on any acremonioid species tested.[17,58]

The pulmonary disease in an otherwise healthy horse from California, the small, round spores in sputum, and the fluconazole response all suggest acute coccidioidomycosis. The spores seen may be consistent with newly released *Coccidioides* endospores, which may be as small as 2 μm in diameter,[59] or, in the differential diagnostic, with phagocytosed airborne contaminants or aspirated yeasts. Even a nonfungal infection responding to concomitant therapy with antibacterials cannot be ruled out. Infection by *S. strictum* seems highly unlikely, especially now that most previous cases of disease ascribed to it have been reassigned to the *A. sclerotigenum* complex. The Pusterla et al.[55] study may provide an illustration of PCR results exacerbating medical confusion about fungal contaminants on environmentally exposed body surfaces.

There are a few other definitively identified acremonioid isolates convincingly linked to human disease although some are from unpublished cases and will not be discussed. The ex-type isolate of *Acremonium spinosum*, CBS 136.33, derives from a well-confirmed case of onychomycosis studied in Argentina and published in 1933.[60,61] Sequencing studies reveal that CBS 136.33 has ribosomal 28S[9] and ITS sequences (unpublished) identical to those of CBS 782.69, the ex-type isolate of *Acremonium pteridii*. The latter species is mainly encountered in temperate forests and soils and on decaying vegetation and fungal fruiting bodies. Typical *A. pteridii* isolates differ from the ex-type of *A. spinosum* by having very pale colonies and only "very finely warty" conidia,[4] in contrast to the potentially deeper yellowish, reddish, and purple colony reverse colors and "clearly finely warty"[4] conidia seen in *A. spinosum*. However, the differences do not appear to be at the species level, but rather at the level of the individual strain. Indeed, there are morphologically typical *A. pteridii* isolates with ITS sequences that diverge more from that of the ex-type than does the ITS sequence of *A. spinosum* (Summerbell, unpublished). Therefore, *A. pteridii* is here reduced to synonymy with *A. spinosum*.

> *Acremonium spinosum* (Negroni) W. Gams, *Cephalosporium*-artige Schimmelpilze (Stuttgart): 78. 1971[4]
> Basionym: *Cephalosporium spinosum* Negroni., *Rev. Soc. Argent. Biol.* 9(1):16 (1933) and C. R. Séanc. *Soc. Biol.* Paris, 113: 478, 1933[60,61]
> ≡ *Acremonium pteridii* W. Gams & Frankland, *Cephalosporium*-artige Schimmelpilze (Stuttgart): 81. 1971

Reports of *Acremonium atrogriseum* (now *Phialemonium atrogriseum*[45]), *A. blochii*, *A. curvulum*, *A. hyalinulum*, and *A. implicatum* (including the synonym, fide Gams,[4] *Fusidium terricola*) in human disease are refuted as incorrect or unsubstantiated by Summerbell.[38] Early reports of "*A. potronii*" in disease are also discussed by Summerbell[38]; some valid cases were described early in the twentieth century, but the identity of the long-lost case isolates is uncertain. The case attributed by Lehner,[62] Ballagi,[63] and Oomen[64] to the ex-type isolate of *Cephalosporium ballagii*, placed into synonymy by Gams[4] with *A. charticola*, was not confirmed as a genuine mycosis showing invasion of tissue and appears to have been an allergic, contact dermatitis response to the fungus and/or to geraniums and other flowers handled by the patient. An undiagnosed underlying condition cannot be excluded. No isolate currently placed in the heterogeneous species mixture called *A. charticola* is a confirmed etiologic agent of humans or animals. A case attributed to *A. alabamense* by Wetli et al.[65] cannot be confirmed, in that the case isolate was not preserved in a public collection and *A. alabamense* is now known to be part of a highly diverse complex of probably mostly undescribed species in the genus *Taifanglania*.[8] Even if correctly identified according to morphological criteria, Wetli et al.'s case isolate may be any one of an as yet undetermined number of morphologically similar species.

It should be noted that *Acremonium collariferum*, a species recently described from Panama after isolation from human skin and nails, was not confirmed as a human pathogen and is likely to represent surface contamination.[12] This species belongs to a loosely structured group of plesiomorphic, simply structured Plectosphaerellaceae that are phylogenetically distant from true *Acremonium* species. No confirmed human pathogens are known in this family. Diligent postmolecular morphotaxonomy in this group has given distinctive names to all the more elaborately structured apomorphs—*Verticillium* ss. str., *Acrostalagmus*, *Musicillium*, *Plectosporium*, and *Gibellulopsis*—but left the intervening simple plesiomorphs in the ancient Saccardoan form taxonomy as *Acremonium*. There is no satisfying conventional postmolecular solution for this problem.

We do not know of any case of mycotoxicosis that has been plausibly ascribed to acremonioid fungi. The best known secondary metabolites from acremonioids are the medically useful cephalosporins from *A. chrysogenum* and various species in the *Emericellopsis* clade.[33] Clarification of the biosystematics of the acremonioids should greatly aid in the study of potentially interesting and useful metabolites, such as the antifungal pyrrocidines found in *S. zeae*.[26]

9.6 CONCLUSIONS

Biosystematic confusion and meager use of culture collections have kept the study of acremonioid fungi in food and human pathology in a state of infancy. Thanks to the increasing accessibility of gene sequencing to stabilize identification, the roles played by various acremonioid species may now begin to emerge. Morphological identification of acremonioid species is not recommended. Only a small number of species with highly distinctive characters can be identified in this manner, for example, (1) *S. kiliense* isolates forming chlamydospores and, on Sabouraud glucose agar, visible diffusing melanin and (2) *A. sclerotigenum* isolates forming multicellular sclerotia. No isolate resembling *S. strictum* should be reported as such in a publication without sequence confirmation. Persons using sequences for identification must be aware of the many erroneously labeled sequences in public databases. They should be familiar with the concept of nomenclatural typification so that they can distinguish which isolates can truly be said to belong to a given species.

Published medical case reports should ensure that standard practices for demonstrating mycotic disease involvement are stringently followed. This entails culture isolation and visual direct microscopic demonstration of compatible elements in tissue in a normally sterile body site or in partly environmentally exposed deep tissue (e.g., lung tissue). In fungemia, it entails more than one consistently positive blood culture. In skin or nail infections, it entails the demonstration of positive culture and/or PCR, plus the demonstration of visible, morphologically compatible tissue elements too restricted in diameter to be dermatophyte filaments, at more than one independent sampling time from the same lesion.

Finally, isolates that are reported as (1) novel or interesting etiologic agents, (2) significant novel agents of food spoilage, or (3) postharvest plant pathogenesis should be deposited in an international public culture collection. A public collection is defined as a collection with full-time staff paid, as their primary responsibility, to receive, preserve, catalogue online, and, in expert compliance with international regulations, ship out fungal cultures.

REFERENCES

1. Link HF. Observationes in ordines plantarum naturales. Dissertatio I. *Mag Ges Naturf Fr Berlin* 1809;3:3–42.
2. Gams W. Was ist *Cephalosporium acremonium*? In: Lange- de la Camp M. Das Art- und Rassenproblem bei Pilzen. *Symposium Werningerode 1967*. Biologische Gesellschaft der DDR, Berlin, Germany, 1968, pp. 27–41.
3. Gams W. Typisierung der Gattung *Acremonium*. *Nova Hedwigia* 1968;16:141–145.
4. Gams W. *Cephalosporium-artige Schimmelpilze (Hyphomycetes)*. G. Fischer, Stuttgart, Germany, 1971.
5. Gams W. Connected and disconnected chains of phialoconidia and *Sagenomella* gen. nov. segregated from *Acremonium*. *Persoonia* 1978;10:97–112.
6. Morgan-Jones G, Gams W. Notes on hyphomycetes. XLI. An endophyte of *Festuca arundinacea* and the anamorph of *Epichloe typhina*, new taxa in one of two new sections of *Acremonium*. *Mycotaxon* 1982;15:311–318.
7. Glenn AE, Bacon CW, Price R, Hanlin RT. Molecular phylogeny of *Acremonium* and its taxonomic implications. *Mycologia* 1996;88:369–383.
8. Liang ZQ, Han YF, Chu HL, Fox RTV. Studies on the genus *Paecilomyces* in China V. *Taifanglania* gen. nov. for some monophialidic species. *Fungal Divers* 2009;34:69–77.
9. Summerbell RC, Gueidan C, Schroers HJ, de Hoog GS, Starink M, Arocha Rosete Y, Guarro J, Scott JA. *Acremonium* phylogenetic overview and revision of *Gliomastix*, *Sarocladium*, and *Trichothecium*. *Stud Mycol* 2011;68:139–162.
10. Carlucci A, Raimondo ML, Santos J, Phillips AJL. *Plectosphaerella* species associated with root and collar rots of horticultural crops in southern Italy. *Persoonia* 2012;28:34–48.
11. Gams W. *Cephalosporium*-like hyphomycetes: Some tropical species. *Trans Br Mycol Soc* 1975;64:389–404.
12. Weisenborn JLF, Kirschner R, Piepenbring M. A new darkly pigmented and keratinolytic species of *Acremonium* (Hyphomycetes) with relationship to the Plectosphaerellaceae from human skin and nail lesions in Panama. *Nova Hedwigia* 2010;90:457–468.
13. Bills GF, Platas G, Gams W. Conspecificity of the cerulenin and helvolic acid producing '*Cephalosporium caerulens*', and the hypocrealean fungus *Sarocladium oryzae*. *Mycol Res* 2004;108:1291–1300.
14. O'Donnell K. *Fusarium* and its near relatives. In: Reynolds DR, Taylor JW, eds., *The Fungal Holomorph: Mitotic, Meiotic and Pleomorphic Speciation in Fungal Systematics*. CAB International, Wallingford, U.K., 1993, pp. 225–238.
15. de Hoog GS, Gerrits van den Ende AHG. Molecular diagnostics of clinical strains of filamentous basidiomycetes. *Mycoses* 1998;41:183–189.
16. Vilgalys R, Hester M. Rapid genetic identification and mapping of enzymatically amplified ribosomal DNA from several *Cryptococcus* species. *J Bacteriol* 1990;172:4238–4246.
17. Perdomo H, Sutton DA, García D, Fothergill AW, Cano J, Gené J, Summerbell RC, Rinaldi MG, Guarro J. Spectrum of clinically relevant *Acremonium* species in the United States. *J Clin Microbiol* 2011;49:243–256.
18. White TJ, Bruns T, Lee S, Taylor J. Amplification and direct sequencing of fungal ribosomal RNA genes for phylogenetics. In: Innis MA, Gelfand DH, Sninsky JJ, White TJ, eds., *PCR-Protocols: A Guide to Methods and Applications*. Academic Press, Inc., London, U.K., 1990, pp. 315–322.
19. Gardes M, Bruns TD. ITS primers with enhanced specificity of Basidiomycetes: Application to the identification of mycorrhizae and rusts. *Mol Ecol* 1993;2:113–118.
20. Gargas A, Taylor JW. Polymerase chain reaction (PCR) primers for amplifying and sequencing 18S rDNA from lichenized fungi. *Mycologia* 1992;84:589–592.
21. Pitt JI, Hocking AD. *Fungi and Food Spoilage*, 2nd edn. Blackie Academic & Professional, London, U.K., 1997.
22. Samson RA, Hoekstra ES, Frisvad JC. *Introduction to Food-Borne Fungi*, 7th edn. Centraalbureau voor Schimmelcultures, Utrecht, the Netherlands, 2004.

23. Gräfenhan T, Schroers H-J, Nirenberg HI, Seifert KA. An overview of the taxonomy, phylogeny, and typification of nectriaceous fungi in *Cosmospora, Acremonium, Fusarium, Stilbella,* and *Volutella. Stud Mycol* 2011;68:79–113.
24. Kidd MN, Beaumont A. Apple rot fungi in storage. *Trans Br Mycol Soc* 1924;10:98–118.
25. Wicklow DT, Roth S, Deyrup SM, Gloer JB. A protective endophyte of maize: *Acremonium zeae* antibiotics inhibitory to *Aspergillus flavus* and *Fusarium verticillioides. Mycol Res* 2005;109:610–618.
26. Wicklow DT, Poling SM, Summerbell RC. Occurrence of pyrrocidine and dihydroresorcylide production among *Acremonium zeae* populations from maize grown in different regions. *Canad J Plant Pathol* 2008;30:425–433.
27. Schroers HJ, Geldenhuis MM, Wingfield MJ, Schoeman MH, Yen YF, Shen WC, Wingfield BD. Classification of the guava wilt fungus *Myxosporium psidii,* the palm pathogen *Gliocladium vermoesenii* and the persimmon wilt fungus *Acremonium diospyri* in Nalanthamala. *Mycologia* 2005;97:375–395.
28. Alvarez-Rodríguez ML, López-Ocaña L, López-Coronado JM, Rodríguez E, Martínez MJ, Larriba G, Coque JJ. Cork taint of wines: Role of the filamentous fungi isolated from cork in the formation of 2,4,6-trichloroanisole by o-methylation of 2,4,6-trichlorophenol. *Appl Environ Microbiol* 2002;68:5860–5869.
29. Abdel-Hafez SI, el-Kady IA, Mazen MB, el-Maghraby OM. Mycoflora and trichothecene toxins of paddy grains from Egypt. *Mycopathologia* 1987;100:103–112.
30. Fujikawa H, Wauke T, Kusunoki J, Noguchi Y, Takahashi Y, Ohta K, Itoh T. Contamination of microbial foreign bodies in bottled mineral water in Tokyo, Japan. *J Appl Microbiol* 1997;82:287–291.
31. Oladiran AO, Iwu LN. Studies on the fungi associated with tomato fruit rots and effects of environment on storage. *Mycopathologia* 1993;121:157–161.
32. Fernández-Trujillo JP, Martínez JA, Salmerón MC, Artés F. Isolation of *Acremonium* species causing postharvest decay of *Peaches* in Spain. *Plant Dis* 1997;81:958.
33. Zuccaro A, Summerbell RC, Gams W, Schroers H-J, Mitchell JI. A new *Acremonium* species associated with *Fucus* spp., and its affinity with a phylogenetically distinct marine *Emericellopsis* clade. *Stud Mycol* 2004;50:283–297.
34. Gonthier P, Gennaro M, Nicolotti G. Effects of water stress on the endophytic mycota of *Quercus robur. Fungal Divers* 2006;21:69–80.
35. Schell WA, Perfect JR. Fatal, disseminated *Acremonium strictum* infection in a neutropenic host. *J Clin Microbiol* 1996;34:1333–1336.
36. Lau YL, Yuen KY, Lee CW, Chan CF. Invasive *Acremonium falciforme* infection in a patient with severe combined immunodeficiency. *Clin Infect Dis* 1995;20:197–198.
37. Summerbell RC, Schroers H-J. Analysis of phylogenetic relationship of *Cylindrocarpon lichenicola* and *Acremonium falciforme* to the *Fusarium solani* species complex and a review of similarities in the spectrum of opportunistic infections caused by these fungi. *J Clin Microbiol* 2002;40:2866–2875.
38. Summerbell RC. *Aspergillus, Fusarium, Sporothrix, Piedraia* and their relatives. In: Howard DH, ed., *Pathogenic Fungi in Humans and Animals.* Marcel Dekker Press, New York, 2003, pp. 237–498.
39. Simon G, Rákóczy G, Galgóczy J, Verebély T, Bókay J. *Acremonium kiliense* in oesophagus stenosis. *Mycoses* 1991;34:257–260.
40. Simonsz HJ. Keratomycosis caused by *Acremonium recifei,* treated with keratoplasty, miconazole and ketoconazole. *Doc Ophthalmol* 1983;56:131–135.
41. de Hoog GS, Guarro J, Gené J, Figueras MJ. *Atlas of Clinical Fungi,* 2nd edn. Centraalbureau voor Schimmelcultures, Utrecht, the Netherlands, 2000.
42. Khan Z, Al-Obaid K, Ahmad S, Ghani AA, Joseph L, Chandy R. *Acremonium kiliense*: Reappraisal of its clinical significance. *J Clin Microbiol* 2011;49:2342–2347.
43. Koshi G, Padhye AA, Ajello L, Chandler FW. *Acremonium recifei* as an agent of mycetoma in India. *Am J Trop Med Hyg* 1979;28:692–696.
44. Novicki TJ, LaFe K, Bui L, Bui U, Geise R, Marr K, Cookson BT. Genetic diversity among clinical isolates of *Acremonium strictum* determined during an investigation of a fatal mycosis. *J Clin Microbiol* 2003;41:2623–2628.
45. Perdomo H, García D, Gené J, Cano J, Sutton DA, Summerbell RC, Guarro J. *Phialemoniopsis,* a new genus of Sordariomycetes, and new species of *Phialemonium* and *Lecythophora. Mycologia* 2013;105(2):398–421. doi: 10.3852/12-137.
46. Keynan Y, Sprecher H, Weber G. *Acremonium* vertebral osteomyelitis: Molecular diagnosis and response to voriconazole. *Clin Infect Dis* 2007;45:e5–e6.
47. Guarro J, Del Palacio A, Gené J, Cano J, González CG. A case of colonization of a prosthetic mitral valve by *Acremonium strictum. Rev Iberoam Micol* 2009;26:146–148.

48. Bibashi E, Kokolina E, Sigler L, Sofianou D, Tsakiris D, Visvardis G, Papadimitriou M, Memmos D. Three cases of uncommon fungal peritonitis in patients undergoing peritoneal dialysis. *Perit Dial Int* 2002;22:523–525.
49. Foell JL, Fischer M, Seibold M, Borneff-Lipp M, Wawer A, Horneff G, Burdach S. Lethal double infection with *Acremonium strictum* and *Aspergillus fumigatus* during induction chemotherapy in a child with ALL. *Pediatr Blood Cancer* 2007;49:858–861.
50. Gupta AK, Cooper EA, McDonald P, Summerbell RC. Inoculum counting (Walshe/English criteria) in the clinical diagnosis of onychomycosis caused by non-dermatophytic filamentous fungi. *J Clin Microbiol* 2000;39:2115–2121.
51. Summerbell RC, Cooper E, Bunn U, Jamieson F, Gupta AK. Onychomycosis: A critical study of techniques and criteria for confirming the etiologic significance of nondermatophytes. *Med Mycol* 2005;43:39–59.
52. Zaias N. Superficial white onychomycosis. *Sabouraudia* 1966;5:99–103.
53. Hitoto H, Pihet M, Weil B, Chabasse D, Bouchara JP, Rachieru-Sourisseau P. *Acremonium strictum* fungaemia in a paediatric immunocompromised patient: Diagnosis and treatment difficulties. *Mycopathologia* 2010;170:161–164.
54. Guarro J, Gams W, Pujol I, Gené J. *Acremonium* species: New emerging fungal opportunists–in vitro antifungal susceptibilities and review. *Clin Infect Dis* 1997;25:1222–1229.
55. Pusterla N, Holmberg TA, Lorenzo-Figueras M, Wong A, Wilson WD. *Acremonium strictum* pulmonary infection in a horse. *Vet Clin Pathol* 2005;34:413–416.
56. Summerbell RC. Mould identification. In: Isenberg H, ed., *Clinical Microbiology Procedures Handbook*, 2nd edn. ASM Press, Washington, DC, 2004, Vol. 2, pp. 8.9.1–8.9.59.
57. Kardjeva V, Kantardjiev T, Summerbell RC, Devliotou-Panagiotidou D, Sotiriou E, Gräser Y. 48-hour diagnosis of onychomycosis including subtyping of *Trichophyton rubrum* strains. *J Clin Microbiol* 2006;44:1419–1427.
58. Saldarreaga A, Garcia Martos P, Ruiz Aragón J, García Agudo L, Montes de Oca M, Puerto JL, Marín P. Antifungal susceptibility of *Acremonium* species using E-test and Sensititre. *Rev Esp Quimioter.* 2004;17:44–47.
59. Kwon-Chung KJ, Bennett JE. *Medical Mycology*. Lea & Febiger, Philadelphia, PA, 1992.
60. Negroni P. Onicomicosis por *Cephalosporium spinosus* n. sp. Negroni, 1933. *Rev Soc Argent Biol* 1933;9:16–22.
61. Negroni P. Onycomycose par *Cephalosporium spinosus* n. sp. Negroni 1933. *CR Séanc Soc Biol Buenos Aires* 1933;113:478–480.
62. Lehner E. Über einen Fall van Acremoniosis. *Arch Dermatol Syphilol* 1932;166:399–404.
63. Ballagi S. Mykologische Beschreibung der Acremoniosis. *Arch Dermatol Syphilol* 1932;166:405–407.
64. Oomen HAPC. Über *Cephalosporium ballagii* nov. spec. *Zbl Bakt, Parasitenkde, Infektionskr Hyg, I Abt, Orig, Reihe A* 1935;134:475–477.
65. Wetli CV, Weiss SD, Cleary TJ, Gyori E. Fungal cerebritis from intravenous drug use. *J Forensic Sci* 1984;29:260–268.

10 Alternaria Mycoses

Giuliana Lo Cascio and Marco Ligozzi

CONTENTS

10.1 Introduction 129
    10.1.1 Classification, Morphology, and Biology 129
    10.1.2 Clinical Presentation, Pathogenesis, and Epidemiology 130
        10.1.2.1 Clinical Presentation... 130
        10.1.2.2 Pathogenesis.. 131
        10.1.2.3 Epidemiology.. 132
    10.1.3 Laboratory Diagnosis 132
        10.1.3.1 Conventional Diagnosis 132
        10.1.3.2 Molecular Techniques... 133
10.2 Methods 135
10.3 Conclusion 135
References 135

10.1 INTRODUCTION

10.1.1 CLASSIFICATION, MORPHOLOGY, AND BIOLOGY

Fungi are ubiquitous microorganisms that are often associated with the spoilage and biodeterioration of a large variety of foods and feedstuffs [1]. Some molds can also adversely affect human and animal health, and they may produce mycotoxins that have been related to a range of pathologies, from gastroenteritis to cancer [2]. The genus *Alternaria* was established on the basis of the conidial morphology in 1933 to cover a group of filamentous fungi in the phylum Ascomycota, the family Dematiaceae, that had previously been known as *Alternaria tenuis* and *Torula alternata* [3,4]. *Alternaria* encompasses a complex group of saprophytic and pathogenic fungal species [5]. *Alternaria* spp. are frequently reported as allergenic, food spoilers, mycotoxicogenic, opportunistic fungi associated with mycosis in animals and humans and destructive plant pathogens [6].

The genus is assigned to the class Hyphomycetes. Many *Alternaria* spp. lack a sexual stage [7]. Fungi in this genus are anamorphs of Ascomycetes, including members of the genus *Pleospora*. *Alternaria* is a dematiaceous mold that includes, at present, about 80 species and varieties. However, only few species have been implicated as food or water contaminants: *Alternaria alternata*, *Alternaria brassicicola* [8], *Alternaria chartarum*, *Alternaria stemphylioides*, *Alternaria dianthicola*, *Alternaria infectoria*, *Alternaria pluriseptata*, *Alternaria malorum* [9], and *Alternaria tenuissima*. *Alternaria iridis* is mentioned only as causing allergic reactions [10,11].

In 1817, Nees described this telluric fungus characterized by chains of spores with apical beak under the name of *A. tenuis*. Subsequently, Fries (1832) described that *Alternaria* species are characterized by very distinctive large multicellular dictyospores that have a beak and are produced in chains.

Alternaria spp. grow rapidly and the colony size reaches a diameter of 3–9 cm following incubation at 25°C for 7 days on potato glucose agar. The optimal growth temperature for *Alternaria* is 25°C–28°C, and the maximal growth temperature is 31°C–32°C. The typical colony is flat and downy to woolly and is covered by grayish, short, aerial hyphae with the passage of time.

The surface is grayish white at the beginning, which later darkens and becomes greenish black or olive brown with a light border. The reverse side is typically brown to black due to pigment production [12,13].

On microscopic examination, *Alternaria* spp. have septate, brown hyphae. The conidiophores arise singly or in small groups, simple or branched, straight or flexible, sometimes geniculate, pale to mid-olivaceous or golden brown, smooth, up to 50 µm long, and 3–6 µm thick with one or several conidial scars. They bear simple or branched large conidia (7-10–23-34 µm) that have transverse and longitudinal septations. These conidia may be observed singly or in acropetal chains and may produce germ tubes. They are ovoid to obclavate, darkly pigmented, muriform, and smooth or roughened. The end of the conidium nearest the conidiophore is round while it tapers toward the apex. This gives the typical beak or club-like appearance of the conidia [12,13]. In addition, some species, such as *A. malorum*, show sterile hyphae, branched, sometimes forming strands, occasionally anastomosing, smooth to faintly rough walled, septate, occasionally constricted at the septa, subhyaline to pale olivaceous, slender, and usually 1–4 µm wide [9].

Species of *Alternaria* occur as parasites on a number of crop plants, causing early blight or leaf spot diseases, or as saprobes on a wide variety of organic substrates: they are also isolated from food such as honey [14]. This genus is prominent in the aerobiological literature because it is recognized as an important aeroallergen and plant pathogen [15,16].

10.1.2 Clinical Presentation, Pathogenesis, and Epidemiology

10.1.2.1 Clinical Presentation

The genus *Alternaria* comprises a large number of predominantly saprobic or plant-pathogenic species that infects mainly immunocompromised hosts. Infections of immunocompetent hosts rarely involve invasive diseases [17]. *Alternaria* can be detected on normal human and animal skin and conjunctiva. This fungus has been associated frequently with hypersensitivity pneumonitis, bronchial asthma, and allergic sinusitis and rhinitis. It can also cause several different types of human infections, for example, paranasal sinusitis, ocular infections, onychomycosis, cutaneous and subcutaneous infections, and in some cases, soft palate perforation, and disseminated disease [17,18]. The portal of entry of the infection is usually through corneal trauma or penetration of the skin barrier.

10.1.2.1.1 Cutaneous and Subcutaneous Infections

As mentioned, most clinical presentations are from direct trauma inoculations, and infections are predominantly in patients with immunosuppression and in transplant patients. Phaeohyphomycosis may be difficult to recognize because lesions are variable in size and aspect. They are located on exposed areas, mainly on upper or lower limbs. *Alternaria* infection is sometimes misdiagnosed as a yeast or blastomycosis, and in vitro distinction may be problematic. Cutaneous infection can also be from colonization of pathologically altered skin or from an exogenous inhalation of conidia and systematic spread resulting in secondary cutaneous infection. Colonization has been recorded from steroid-treated eczema of the face [18]. A very rare case occurred in an immunocompetent 18-year-old man who developed severe meningoencephalitis and arachnoiditis caused by *A. alternata*, which was diagnosed in the context of difficult-to-treat hydrocephalus. Etiological diagnosis was made based on fungal culture and histopathologic examination [19]. Another case was a 74-year-old man with well demarcated and elevated erythematous plaques with irregular scattered pustules on his right forearm. Furthermore, a 77-year-old woman had well demarcated and elevated erythematous nodules within an erythematous patch on her right forearm. In both cases, the lesions started at the senile purpura site [20]. Finally, a 53-year-old patient who had received a kidney transplant presented with multiple verrucous lesions on the distal extremities. Positive histopathology and cultures, in addition to ribosomal DNA (rDNA) internal transcribed spacer (ITS) region sequencing, identified the fungal isolate as *A. infectoria* [21].

10.1.2.1.2 Rhinosinusitis

Alternaria DNA has been found in surgical specimens from the mucosa of the middle meatus and the paranasal sinuses, or in ethmoid sinus mucosal specimens in almost all patients with chronic rhinosinusitis (CRS), with none being detected in the controls. Patients with CRS have greater cytokine and humoral immune responses to fungi and particularly to *Alternaria* [18].

10.1.2.1.3 Ocular Infections

The risk of oculomycosis from this fungus is probably related to the risk of trauma by organic matter and varies according to geographical location. Surgical trauma is also associated. The incidence ranges from 2.3% to 10.4% [18] and most cases are keratitis with endophthalmitis associated frequently. Causative agents were often *A. alternata* or *A. infectoria* although identifications to species were frequently not performed [18].

10.1.2.1.4 Onychomycosis

Alternaria spp. were the causative agent of onychomycosis only in the 0.08%–2.50% of culture-proven clinical cases, and a history of contact with soil or trauma in the nail was indicated [18]. Fingernails and toenails can be infected. Identification to species was often not reported although *A. alternata* is the most prevalently identified with some reports of *Alternaria humicola, A. pluriseptata,* and *Alternaria chlamydospora*.

10.1.2.2 Pathogenesis

Alternaria produce a dark melanin pigment that plays an important role in the evasion of host defense by fungal pathogens, including dematiaceous molds. Melanins are composed of polymerized phenolic and/or indolic compounds and are insoluble in aqueous or organic solvents, resistant to concentrated acid, and susceptible to bleaching by oxidizing agents [22]. Melanin in dematiaceous molds such as *A. alternata* is synthesized through the dihydroxynaphthalene polyketide pathway [23,24] and production may protect the fungi from radiation and adverse environmental conditions. Melanins contribute to virulence by various mechanisms: they provide protection against oxidants because they are highly effective scavengers of free radicals and have electron transfer properties that can facilitate redox cycling.

Species of the genus are ubiquitous and are reported to occur in different ecosystems and geographic regions, such as Antarctic soils, deserts, and the tropics [25,26]. Fungi are abundant in the troposphere and can be transported vast distances on prevailing winds. Even after traveling for 10 days across the Pacific Ocean in the free troposphere, diverse and viable microbial populations, including *A. infectoria*, were detected in Asian air samples [27]. Many pathogenic species of *Alternaria* have a worldwide distribution and are infective to a variety of plants including potato, tomato, onion, and members of the Brassicaceae [6].

In general, *Alternaria* attack plants under stress, especially those affected by drought, insect infestation, or senescence. *Alternaria* spores are passively dispersed from infected leaves by moderate to strong gusty wind, with velocities of 2–3 m/s required for spore release. As a component of the dry-air spora, dispersal typically occurs during dry weather that immediately follows periods of rain or heavy dew. *Alternaria* spores colonize plants via wounds to the pericarp layer. About 7–10 days after host colonization, *Alternaria* sporulates and conidia germinate for several weeks. The result of this colonization is the spot and cankers in stem, twigs, fruit, tuber, seeds, and root rots [28]. *Alternaria* are among the principal contaminating fungi in wheat, sorghum, and barley and have also been reported to occur in oilseeds such as sunflower and rapeseed, tomatoes, apples, citrus fruits, olives, and several other fruits and vegetables. *A. alternata* is the most common species isolated.

Alternaria species produce more than 70 phytotoxins, but only a small proportion of them have been chemically characterized and reported to act as mycotoxins to humans and animals.

Some toxins such as alternariol (AOH), alternariol monomethyl ether (AME), tenuazonic acid (TeA), and altertoxins are described to induce harmful effects in animals, including fetotoxic and teratogenic effects. Culture extracts of *A. alternata* and individual mycotoxins such as AOH and AME are mutagenic and clastogenic in various in vitro systems. In addition, it has been suggested that in certain areas in China, *Alternaria* toxins in grains might be responsible for esophageal cancer [29]. In samples containing *Alternaria*, toxins such as AOH, AME, TeA, and tentoxin were generally found in certain grains and grain-based products, tomato and tomato products, sunflower seeds and sunflower oil, fruits and fruit products including fruit juices, and beer and wine. Hence, due to their possible harmful effects, *Alternaria* toxins are of concern for public health, although they have not been reported to cause animal toxicosis as a result of exposure from feed. The presence of the toxins in foods indicates that the producing fungi were also present at some stage and so could be the source of infections of humans.

10.1.2.3 Epidemiology

Alternaria has a worldwide distribution as a saprophyte in soil, air, and a variety of environmental habitats; *Alternaria* species are present in soil, wood, and decomposing plants debris, especially in humid and warm climates [11]. The fungus can also be found on normal human and animal skin and conjunctiva [30]. During spring, summer, and autumn, *Alternaria* species produce abundant airborne spores and mycelium debris due to degradation of vegetation. Exposure to airborne spores has frequently been associated with hypersensitivity pneumonitis, bronchial asthma, and allergic sinusitis and rhinitis. Furthermore, some CRS patients have elevated populations of fungi in their sinuses [31]. Farmers cultivating peppermint, chamomile, and other herbs are exposed during processing of this plants to large concentrations of airborne microorganisms that create a respiratory risk for exposed workers [32] that may, in part, consist of *Alternaria*. The fungus can also cause several different types of human infections such as paranasal sinusitis, ocular infections, onychomycosis, and cutaneous and subcutaneous infections, the latter more frequently in immunosuppressed patients [33,34] and in Cushing's syndrome. Rarely, *Alternaria* species are responsible for disseminated diseases [35]. Most clinical isolates have been shown to be either *A. alternata* or *A. infectoria* [36]. However, ITS region sequences have demonstrated that *Alternaria longipes* and *A. tenuissima* cannot be distinguished from *A. alternata* [37].

10.1.3 Laboratory Diagnosis

10.1.3.1 Conventional Diagnosis

Because of the importance of fungi in mycosis, quick and accurate procedures to detect and enumerate these contaminants in, for example, food commodities, are essential. The same techniques can be used to detect the fungi from other sources that may be involved in pathogenesis. Traditional culture techniques for detection of foodborne fungi involve the use of selective microbiological media, followed by the isolation of pure cultures, and finally the application of confirmatory tests. Although effective, these procedures are extremely labor intensive and require several days or weeks. The competing presence of other fungi could be problematic for the isolation of *Alternaria* from environmental samples, such as food and water. To overcome this problem, it is recommended to use selective media in the detection procedures. Dichloran rose Bengal yeast extract sucrose agar (DRYES) is especially useful for analyzing samples containing *spreader molds* (e.g., *Mucor*, *Rhizopus*) since the added dichloran and rose Bengal effectively slow down the growth of fast-growing fungi, thus readily allowing detection of other yeasts and molds that have lower growth rates. The media dichloran rose Bengal chloramphenicol optimized fungi detection in terms of abundance and variety in water sources using culture-based methods [38]. Several authors recommend in addition potato carrot agar (PCA) and V-8 juice agar that are selective for *Alternaria* species and allow abundant sporulation [39]. DRYES agar plates are incubated at 25°C for 7 days in darkness, while the PCA plates are incubated at 25°C under an alternate light/dark cycle consisting of 8 h of cool-white daylight followed by 16 h darkness [40].

Colonies of *Alternaria* grow rapidly and appear flat and downy to woolly and are covered by grayish, short, aerial hyphae in time. The surface is grayish white at the beginning, which later darkens and becomes greenish black or olive brown with a light border. The reverse side is typically brown to black due to pigment production.

Until recently, the identification of *Alternaria* rests upon microscopic morphology, with the most significant characteristics being the morphology of the conidia and the formation of conidial chains. Morphology evaluation normally are based upon cultures that have been grown on a medium such as potato dextrose agar or cornmeal dextrose agar at 25°C–30°C for approximately 2 weeks. Slide culture preparation using cornmeal dextrose agar is ideal for conidiogenesis, because these nutritionally minimal media usually stimulate the formation of spores [41].

The morphological characteristics useful for distinction among the three species more frequently isolated are the following: *A. alternata* shows medium-brown conidia with a short, cylindrical beak, forming long and profusely branched chains, usually ten or more conidia; in *A. tenuissima*, the conidia are golden brown, frequently tapering gradually into a beak that is up to half the length of the conidium, and usually occur in unbranched chains of three to five conidia; and *A. infectoria* species-group comprises more than 30 named anamorph taxa. Morphologically, the *A. infectoria* species-group differs from other *Alternaria* species in the 3D sporulation pattern and has more scarce conidia, as this species usually sporulates poorly in common media, and its small conidia (up to 70 µm in length) occur in branched chains with long, geniculate multiseptate secondary conidiophores (up to 120 µm) between conidia [9].

10.1.3.2 Molecular Techniques

Identification of pathogenic dematiaceous fungi such as *Alternaria* spp. is typically undertaken by morphological and physiological procedures [42,43]. However, these procedures are time consuming, require technical expertise, and are ineffective for identification of species with poor conidia production and a wide diversity in anamorphic life cycles [44]. PCR-based methods have become very important and are widely used to detect *Alternaria* spp. in food. The diagnostic PCR approach for the detection of mycotoxinogenic fungi such as *Alternaria* is an indirect method. A positive PCR can, therefore, be taken as an indication that the sample potentially contains mycotoxins and should be analyzed further. The methods can also be used in cases of mycosis.

Some molecular methods based on immunological and genotypic techniques have been developed for revealing the presence of undesirable microorganisms, including fungi, in different food matrices [45–47]. Among these, PCR is one of the most promising analytical tools in food microbiology and food control because of its specificity and sensitivity [48,49]. However, conventional PCR methods do not distinguish between viable, viable but nonculturable (VBNC), and dead cells. VBNC bacterial cells are defined as those cells that have lost the ability to express genes but may return to a culturable state [50]. The presence of these cells limits the use of PCR for microbiological monitoring of food samples, where metabolically injured or nonviable cells are generally present after the stresses imposed during food processing, although assessments of fungi as VBNC are required.

In contrast to DNA, mRNA is turned over rapidly in viable cells; most mRNA species have half-lives measured in minutes [51]. Therefore, detection of mRNA by reverse transcriptase PCR (RT-PCR), as opposed to DNA-based methods, is considered a better indicator of cell viability [52]. Recently, some authors described a method for the detection of viable bacteria, molds, and yeasts by RT-PCR, with primers specific for an elongation factor gene in heat-treated milk samples, yogurts, and pasteurized food products [53]. Real-time RT-PCR allows the determination of the initial template concentration and, therefore, an accurate estimation of cell number offering an enormous potential for the quantification of a range of microorganisms of medical, alimentary, and environmental importance [54].

Since mycotoxin biosynthetic genes sequences are not available for *Alternaria* spp., two different approaches are applied for species identification: the use of species-specific primers [55] or the use of primers binding to conserved regions of a certain gene present in all species to be identified and subsequent identification of the species by sequencing [56]. To achieve a high sensitivity in the PCR assays, genes present in a high copy number within the genome are used as targets. These genes include rDNA. Similar to other eukaryotes, *Alternaria* ribosomal rDNA genes consist of three subunits (18S, 5.8S, and 28S) that are tandemly repeated. In one unit, the genes are separated by two ITSs (ITS1 and ITS2), and two rDNA units are separated by the intergenic sequences (IGS).

The 18S rDNA evolves relatively slowly and is useful for comparing distantly related organisms, whereas the noncoding regions (ITS and IGS) evolve faster and are useful for comparing fungal species within a genus or strains within a species. Some regions of the 28S rDNA are also variable between species. The development of PCR and the design of primers for the amplification of the various rDNA regions have considerably facilitated taxonomic studies of fungi [56]. These primers were designed from conserved regions, allowing the amplification of the fragment they flank in most fungi.

Since *Alternaria* species are capable of producing a large array of mycotoxin, some of which are powerful mycotoxins that have been implicated in the development of cancer in mammals [5], it is useful for the food industry to know if toxigenic fungi are present in raw vegetables, water, and animal products.

Two avenues are available for the selection of target DNA: the first is the selection of the target DNA using specific sequence information from databases, allowing primers to be designed across conserved and variable regions, and the second is cloning and sequencing of arbitrary parts of the fungal genome.

Ribosomal RNA (rRNA) is an essential component of protein synthesis, thus ubiquitous. The sequence of rRNA has highly conserved and variable regions. Examination of this sequence reveals the relatedness between the species or the genetic distance and the organisms in question. An additional advantage of this is that these genes are not horizontally transferred, like other prokaryotic genes, for example, drug resistance genes.

Nuclear rRNA is coded by rDNA, which is organized in ribosomal operons (usually 100–200 identical copies in fungi) located in chromosomes. The nuclear-encoded rRNA genes of fungi exist as a multiple-copy gene family comprised of highly similar DNA sequences (typically from 8–12 kb each) arranged in a head-to-toe manner. Each operon codes for the large subunit (LSU-rRNA; 28S rRNA), small subunit (SSU-rRNA; 18S rRNA), 5.8S rRNA, and 5S rRNA. The position of 5S rRNA varies, but the organization of the rest of the genes is the same in all fungi.

The fungal identification is based on detection of conserved sequences in 5.8S and 28S rDNA that enable the amplification of the ITS region between these two regions and detection of D1/D2 domain contained in 28S rDNA. Then, the tools for molecular investigation are represented by (1) DNA sequencing of the full length ITS region ITS1-2 and (2) DNA sequencing of the D1/D2 region of LSU gene in 28S rDNA.

The main area for the development of fungal diagnostics is ribosomal genes [56] present in all organisms and at high copy numbers aiding detection and the sensitivity of PCR. The fungal nuclear rDNA consists of three genes, the LSU gene (28S), the SSU gene (18S), and the 5.8S gene, separated by ITS regions, in a unit repeated many times. The ITS region is an area of particular importance to fungal diagnostics. It has areas of high conservation and areas of high variability and is an ideal starter for the development of specific PCR primers for identification of fungal species. Universal primers [56] are available for fungi that isolate the regions of the ITS; once cloned, these sequences can be compared to the wealth of other sequences in the sequence database and diagnostic primers developed for a particular fungus. The MicroSeq D2 LSU rDNA sequencing kit appears to be accurate and useful for the identification of filamentous fungi, even those that are relatively uncommon, that are seen in the clinical laboratory. However, the library includes more of the common *Alternaria* spp. and other environmental flora that cause disease in immunocompromised patients.

10.2 METHODS

The *Alternaria* molecular detection tests can be used indirectly on molds isolated from an environmental specimen or directly on biological samples. Several commercial kits are available for mold DNA extraction in a microbiology laboratory [57].

Alternaria DNA can be extracted from food or water samples using commercial kits such as PowerFood Ultraclean Microbial DNA Isolation kit (MO BIO Laboratories, Solana Beach, CA, USA) [58].

Pavón et al. [59] employed a pair of primers from the ITS1 and ITS2 regions of fungal DNA Dir1ITSAlt (5′-TGT CTT TTG CGT ACT TCT TGT TTC CT-3′) and Inv1ITSAlt (5′-CGA CTT GTG CTG CGC TC-3′) to achieve PCR amplification of a 370 bp fragment. After amplification, the PCR fragment needs to be analyzed by comparison with sequence in the National Center for Biotechnology Information GenBank or the European Molecular Biology Laboratory Nucleotide Sequence database using the Basic Local Alignment Search Tool [60] alignment program to find the most identical sequence.

10.3 CONCLUSION

Alternaria species are ubiquitous plant pathogens and saprobes and are often found on different food and water products. They can also cause disease in humans. Methods are available for the detection of these fungi using molecular biology techniques. The application of nucleic acid–based techniques such as PCR procedures makes rapid, sensitive, and specific identification and detection of *Alternaria* species. Most molecular assays have targeted *Alternaria* rRNA and ITS1 and ITS2. Because of the relative genetic conservation in these regions, the resulting PCR products are often of similar size, which requires extra steps such as DNA sequencing and restriction enzyme digest to distinguish among related species. Whatever PCR methods are used, it does not distinguish between all human pathogenic *Alternaria* species. Specific primers were developed for detection of *A. alternata*, *Alternaria radicina*, and *Alternaria dauci* plant pathogens [61], but does not cover all the known species. Moreover, molecular methods could help in the near future to understand more on the pathogenic role of environmental fungi, even in the era of a large-scale use of immunosuppressant, the main important risk factor predisposing to mycotic disease.

REFERENCES

1. Filtenborg, O., Frisvad, J. C., and Thrane, U. 1996. Moulds in food spoilage. *Int. J. Food Microbiol.* 33:85–102.
2. Hussein, H. S. and Brasel, J. M. 2001. Toxicity, metabolism and impact of mycotoxins on human and animals. *Toxicology* 167:101–134.
3. Simmons, E. G. 1992. *Alternaria* taxonomy: Current status view-point, challenge. In Chelkowski, J. and Visconti, A, (Eds.), *Alternaria: Biology, Plant Disease and Metabolites*. Elsevier Science Publishers BV, Amsterdam, the Netherlands, pp. 1–36.
4. Chelkowski, J. and Viscont, A. 1992. *Alternaria: Biology, Plant Diseases and Metabolites*. Elsevier, Amsterdam, the Netherlands.
5. Thomma, B. P. H. J. 2003. *Alternaria* spp.: From general saprophyte to specific parasite. *Mol. Plant Pathol.* 4:225–236.
6. Rotem, J. 1994. The genus *Alternaria*. In *Biology, Epidemiology, and Pathogenicity*. APS Press, American Phytopathological Society Press, St. Paul, MN.
7. Berbee, M. L., Payne, B. P., Zhang, G., Roberts, R. G., and Turgeon, B. G. 2003. Shared ITS DNA substitutions in isolates of opposite mating type reveal a recombining history for three presumed asexual species in the filamentous ascomycete genus *Alternaria*. *Mycol. Res.* 107:169–182.

8. Köhl, J., van Tongeren, C. A. M., Groenenboom-de Haas, B. H., van Hoof, R. A., Driessen, R., and van der Heijden, L. 2010. Epidemiology of dark leaf spot caused by *Alternaria brassicicola* and *A. brassicae* in organic seed production of cauliflower. *Plant Pathol.* 59(2):358–367.
9. Braun, U., Crous, P. W., Dugan, F., Groenewald, J. Z., and de Hoog, G. S. Phylogeny and taxonomy of *Cladosporium*-like hyphomycetes, including *Davidiella* gen. nov., the teleomorph of *Cladosporium s. str.* 2003. *Mycol. Prog.* 2(1):3–18.
10. De Bièvre, C. 1991. Les *Alternaria* pathogens pour l'homme: Mycologie èpidemiologique. *J. Mycol. Mèd.* 118:50.
11. de Hoog, G. S., Guarro, J., Genè, J., and Figueras, M. J. 2000. *Atlas of Clinical Fungi*, 2nd edn. Centraalbureau voor Schimmelcultures/Universitat Rovira I Virgili, Tarragona, Spain.
12. Collier, L., Balows, A., and Sussman, M. 1998. *Topley & Wilson's Microbiology and Microbial Infections*, 9th edn., Vol. 4. Arnold, London, U.K.
13. St-Germain, G. and Summerbell, R. 1996. *Identifying Filamentous Fungi—A Clinical Laboratory Handbook*, 1st edn. Star Publishing Company, Belmont, CA.
14. Kačániová, M., Kňazovická, V., Felšöciová, S., and Rovná, K. 2012. Microscopic fungi recovered from honey and their toxinogenity. *J. Environ. Sci. Health A Tox. Hazard. Subst. Environ. Eng.* 47(11):1659–1664.
15. Levetin, E. 2007. Aerobiology of agricultural pathogens. In Hurst, C.J. (Ed.), *Manual of Environmental Microbiology*, 3rd edn, ASM Press, Washington DC.
16. American Academy of Allergy, Asthma, and Immunology. 1999. 1998 Pollen and spore report. American Academy of Allergy, Asthma, and Immunology, Milwaukee, WI.
17. Pastor, F. J. and Guarro, J. 2008. *Alternaria* infections: Laboratory diagnosis and relevant clinical features. *Clin. Microbiol. Infect.* 14:734.
18. Lo Cascio, G. and Ligozzi, M. 2011. *Alternaria*. In *Molecular Detection of Human Fungal Pathogens*. Taylor & Francis CRC Press, Boca Raton, FL, pp. 27–36.
19. Silveira, C. J. C., Amaral, J., Gorayeb, R. P., Cabral, J., and Pacheco, T. 2013. Fungal meningoencephalitis caused by *Alternaria*: A clinical case. *Clin. Drug Invest.* 33(Suppl 1):S27–S31.
20. Chae, I. S., Kim, I. Y., Ko, D. K., Chung, K. H., Jun, J. B., Bang, Y. J., and Park, J. S. 2012. Two cases of cutaneous infection by *Alternaria alternata* on senile purpura site. *Korean J. Med. Mycol.* 17(3):183–188.
21. Cunha, D. et al. 2012. Phaeohyphomycosis caused by Alternaria infectoria presenting as multiple vegetating lesions in a renal transplant patient. *Rev. Iberoam. Micol.* 29(1):44–46.
22. Nosanchuk, J. D. and Casadevall, A. 2003. The contribution of melanin to microbial pathogenesis. *Cell Microbiol.* 5:203.
23. Walsh, T. J., Groll, A., Hiemenz, J., Fleming, R., Roilides, E., and Anaissie, E. 2004. Infections due to emerging and uncommon medically important fungal pathogens. *Clin. Microbiol. Infect.* 10:48.
24. Taylor, B., Wheeler, M. H., and Szaniszlo, P. J. 1987. Evidence for pentaketide melanin biosynthesis in dematiaceous human pathogenic fungi. *Mycologia* 79:320–322.
25. Malosso, E., Waite, I. S., English, L., Hopkins, D. W., and O'Donnell, A. G. 2006. Fungal diversity in maritime Antarctic soils determined using a combination of culture isolation, molecular fingerprinting and cloning techniques. *Polar Biol.* 29:552–561.
26. Grishkan, I., Beharav, A., Kirzhener, V., and Nevo, E. 2007. Adaptive spatiotemporal distribution of soil microfungi in "Evolution Canyon" III, Nahal Shaharut, extreme southern Negev desert, Israel. *Biol. J. Linn. Soc. Lond.* 90:263–277.
27. Smith, D. J., Jaffe, D. A., Birmele, M. N., Griffin, D. W., Schuerger, A. C., Hee, J., and Roberts, M. S. 2012. Free tropospheric transport of microorganisms from Asia to North America. *Microb. Ecol.* 64(4):973–985.
28. Logrieco, A., Bottalico, A., Mule, G., Moretti, A., and Perrone, G. 2003. Epidemiology of toxinogenic fungi and their associate mycotoxin for some Mediterranean crops. *Eur. J. Plant Pathol.* 109:645–667.
29. Liu, G., Quian, Y., Zhang, P., Dong, W., Qi, Y., and Guo, H. 1992. Etiological role of *Alternaria alternata* in human esophageal cancer. *Chinese Med. J.* 105:394–400.
30. Midgley, G., Moore, M. K., Cook, J. C., and Phan, Q. G. September 1994. Mycology of nail disorders. *J. Am. Acad. Dermatol.* 31:S68–S74.
31. Murr, A. H., Goldberg, A. N., Pletcher, S. D., Dillehay, K., Wymer, L. J., and Vesper, S. J. July 2012. Some chronic rhinosinusitis patients have elevated populations of fungi in their sinuses. *Laryngoscope* 122(7):1438–1445.

32. Skórska, C., Sitkowska, J., Krysińska-Traczyk, E., Cholewa, G., and Dutkiewicz, J. 2005. Exposure to airborne microorganisms, dust and endotoxin during processing of peppermint and chamomile herbs on farms. *Ann. Agric. Environ. Med.* 12(2):281–288.
33. Lo Cascio, G., Ligozzi, M., Maccacaro, L., and Fontana, R. November 2004. Utility of molecular identification in opportunistic mycotic infections: A case of cutaneous *Alternaria infectoria* infections in a cardiac transplant recipient. *J. Clin. Microbiol.* 42(11):5334–5336.
34. Lavergne, R. A. et al. December 2012. Simultaneous cutaneous infection due to *Paecilomyces lilacinus* and Alternaria in a heart transplant patient. *Transpl. Infect. Dis.* 14(6):E156–E160.
35. Brandt, M. E. and Warnock, D. W. November 2003. Epidemiology, clinical manifestations, and therapy of infections caused by dematiaceous fungi. *J. Chemother.* 15(Suppl 2):36–47.
36. Revankar, S. G. and Sutton, D. A. 2010. Melanized fungi in human disease. *Clin. Microbiol. Rev.* 23(4):884–928.
37. de Hoog, G. S. and Horre, R. 2002. Molecular taxonomy of the *Alternaria* and *Ulocladium* species from humans and their identification in the routine laboratory. *Mycoses* 45:259–276.
38. Pereira, V. J., Fernandes, D., Carvalho, G., Benoliel, M. J., San Romão, M. V., and Barreto Crespo, M. T. 2010. Assessment of the presence and dynamics of fungi in drinking water sources using cultural and molecular methods. *Water Res.* 44(17):4850–4859.
39. Kosiak, B., Torp, M., Skjerve, E., and Andersen, B. 2004. *Alternaria* and *Fusarium* in Norwegian grains of reduced quality—A matched pair sample study. *Int. J. Food Microbiol.* 93:51–62.
40. Simmons, E. G. 2007. *Alternaria: An Identification Manual*. CBS Biodiversity Series, Utrecht, the Netherlands, pp. 1–775.
41. Schell, W. A., Salkin, I. F., and McGinnis M. R. 2003. *Bipolaris, Exophiala, Scedosporium, Sporothrix,* and other dematiaceous fungi. In Murray, P. R. et al. (eds.) *Manual of Clinical Microbiology*, 8th edn. ASM Press, Washington, DC, pp. 1820–1847.
42. Steadham, J. E., Geis, P. A., and Simmank, J. L. 1986. Use of carbohydrate and nitrate assimilations in the identification of dematiaceous fungi. *Diagn. Microbiol. Infect. Dis.* 5:71.
43. Fothergill, A. W. 1996. Identification of dematiaceous fungi and their role in human disease. *Clin. Infect. Dis.* 22:179.
44. de Hoog, G. S. et al. 1994. Phaeoanamorphic life cycle of Exophiala (Wangiella) dermatitidis. *Antonie van Leeuwenhoek* 65:143.
45. Kappe, R., Okeke, N., Fauser, C., Mainwald, M., and Sonntag, H. G. 1998. Molecular probes for the detection of pathogenic fungi in the presence of human tissue. *J. Med. Microbiol.* 47:811–820.
46. Li, S., Marquardt, R. R., and Abramanson, D. 2000. Immunochemical detection of molds: A review. *J. Food Prot.* 63:281–291.
47. van der Vossen, J. M. and Hofstra, H. 1996. DNA based typing, identification and detection systems for food spoilage microorganisms: Development and implementation. *Int. J. Food Microbiol.* 33:35–49.
48. Swaminathan, B. and Feng, P. 1994. Rapid detection of food-borne pathogenic bacteria. *Annu. Rev. Microbiol.* 48:401–426.
49. Zur, G., Hallerman, E. M., Sharf, R., and Kashi, Y. 1999. Development of a polymerase chain reaction-based assay for the detection of *Alternaria* fungal contamination in food products. *J. Food Prot.* 62:1191–1197.
50. Del Mar Lleó, M., Pierobon, S., Tafi, M. C., Signoretto, C., and Canepari, P. 2000. mRNA detection by reverse transcription-PCR for monitoring viability over time in an *Enterococcus faecalis* viable but nonculturable population maintained in a laboratory microcosm. *Appl. Environ. Microbiol.* 66:4564–4567.
51. Ingle, C. A. and Kushner, S. R. 1996. Development of an in vitro mRNA decay system for *Escherichia coli*: Poly(A) polymerase I is necessary to trigger degradation. *Proc. Natl. Acad. Sci. USA* 93:12926–12931.
52. Sheridan, G. E. C., Masters, C. I., Shallcross, J. A., and Mackey, B. M. 1998. Detection of mRNA by reverse transcription-PCR as an indicator of viability in *Escherichia coli* cells. *Appl. Environ. Microbiol.* 64:1313–1318.
53. Vaitilingom, M., Gendre, F., and Brignon, P. 1998. Direct detection of viable bacteria, molds and yeasts by reverse transcriptase PCR in contaminated milk samples after heat treatment. *Appl. Environ. Microbiol.* 64:1157–1160.
54. Bleve, G., Rizzotti L., Dellaglio F., and Torriani S. 2003. Development of reverse transcription (RT)-PCR and real-time RT-PCR assays for rapid detection and quantification of viable yeasts and molds contaminating yogurts and pasteurized food products. *Appl. Environ. Microbiol.* 69:4116–4122.

55. Andersen, B., Smedsgaard, J., Jorring, I., Skouboe, P., and Pedersen, L. H. 2006. Real-time PCR quantification of the AM-toxin gene and HPLC qualification of toxigenic metabolites from Alternaria species from apples. *Int. J. Food Microbiol.* 111(2):105–111.
56. White, T., Burns, T., Lee, S., and Taylor, J. 1990. Amplification and direct sequencing of fungal ribosomal RNA genes for phylogenetics. In Innis, M. A., Gelfand, D. H., Sninsky, J. J., and White, T. J. (eds.) *PCR Protocols. A Guide to Methods and Applications*. Academic Press, Inc., San Diego, CA, pp. 315–322.
57. Fredricks, D. N., Smith, C., and Meier, A. 2005. Comparison of six DNA extraction methods for recovery of fungal DNA as assessed by quantitative PCR. *J. Clin. Microbiol.* 43:5122.
58. Tournas, V., Stack, M. E., Mislivec, P. B., Koch, H. A., and Bandler, R. 2001. Chapter 18. Yeasts, molds and mycotoxins. *Bacteriological Analytical Manual.* pp. 259–268.
59. Pavón, M. Á., González, I., Rojas, M., Pegels, N., Martín, R., and García, T. February 2011. PCR detection of *Alternaria* spp. in processed foods, based on the internal transcribed spacer genetic marker. *J. Food Prot.* 74(2):240–247.
60. Altschul, S. F., Gish, W., Miller, W., Myers, E. W., and Lipman, D. J. 1990. Basic local alignment search tool. *J. Mol. Biol.* 215:403–410.
61. Konstantinova, P., Bonants, P. J. M., van Gent-Pelzer, M. P. E., van der Zouwen, P., and van den Bulk, R. 2002. Development of specific primers for detection and identification of *Alternaria* spp in carrot material by PCR and comparison with blotter and plating assay. *Mycol. Res.* 106(1):23–33.

11 *Alternaria* spp. and Mycotoxins

Miguel Ángel Pavón, Isabel González,
Rosario Martín, and Teresa García

CONTENTS

11.1 Introduction 139
11.2 *Alternaria* sp. Mycotoxins 139
    11.2.1 Mycotoxin Synthesis.. 140
    11.2.2 Physiology of Mycotoxin Production 141
    11.2.3 Chemistry and Physical Properties.. 141
11.3 Toxicity of *Alternaria* sp. Mycotoxins.. 141
    11.3.1 Dibenzo-α-Pyrones 142
    11.3.2 Tenuazonic Acid 142
    11.3.3 Altertoxins 144
    11.3.4 AAL Toxins 144
11.4 Occurrence of *Alternaria* Toxins in Food and Feed.. 144
11.5 Human Exposure to *Alternaria* sp. Mycotoxins 146
11.6 Control and Prevention of Mycotoxins in Foods 146
References.. 147

11.1 INTRODUCTION

Alternaria is a cosmopolitan fungal genus that includes saprophytic, endophytic, and pathogenic species, widely distributed in soil and organic matter; the fungus is involved in decomposition in general. Plant pathogenic species affect cereals, vegetables, and fruit crops in the field and during storage. *Alternaria* sp. contamination is responsible for some of the world's most devastating plant diseases, causing serious reduction of crop yields and considerable economic losses. Certain species of the fungus are responsible for mycoses [1], which will be dealt with elsewhere in this book. The production of mycotoxins is the focus of the current chapter.

11.2 *ALTERNARIA* SP. MYCOTOXINS

Alternaria species produce more than 70 secondary metabolites toxic to plants (phytotoxins), but only a small fraction of them have been chemically characterized and reported to act as mycotoxins to humans and animals (Table 11.1) [2–4]. Based on their effect on plant crops, *Alternaria* sp. toxins are grouped in non-host-specific toxins and host-specific toxins. Non-host-specific toxins affect a broad range of food crops, including cereals, oilseeds, Apiaceae, Solanaceae, Cucurbitaceae, Brassicaceae, citrus, Rosaceae, olives, nuts, and leguminous crops, although the mode of action of most of these toxins has not been identified [5,6]. However, the phytotoxic activities of toxins such as zinniol, tenuazonic acid (TeA), and tentoxin (TEN) have been studied in detail. TeA inhibits protein synthesis and zinniol affects membrane permeabilization. TEN acts as a photophosphorylation inhibitor through specific binding to chloroplast ATP synthase, causing the inhibition of ATP hydrolysis and ATP synthesis [7]. Host-specific toxins have a limited host range causing severe

TABLE 11.1
Toxins Produced by the Major *Alternaria* Species

Species	Toxins
Alternaria alternata	AOH, AME, ALT, altenusin, ATX-I, ATX-II, ATX-III, TEN, AAL toxins, AS toxin, AF toxin, ACR toxin, ACT toxin
A. arborescens	AOH, AME, ALT, ATX-I, TeA
A. brassicae	AOH, AME, TeA
A. brassicicola	AOH, AME, TeA, ATX-I
A. citri	AOH, AME, ALT, TeA
A. cucumerina	AOH, AME, macrosporin
A. dauci	AOH, AME, macrosporin, zinniol
A. gaisen	AOH, AME, ALT, altenusin, TeA, ATX-I
A. infectoria	Infectopyrones, novae-zelandins
A. longipes	AME, TeA, ATX-I
A. mali	TeA, ATX-I, ATX-II, ATX-III, TEN, AM toxin
A. porri	AOH, AME, ALT, TeA, altersolanol, TEN, macrosporina, zinniol
A. radicina	ALT, TeA, ATX-I, radicinin, radicinol
A. solani	AOH, AME, altersolanol, macrosporin, zinniol
A. tenuissima	AOH, AME, ALT, TeA, ATX-I, TEN

Abbreviations: AOH, alternariol; AME, alternariol monomethyl ether; ALT, altenuene; TeA, tenuazonic acid; ATX, altertoxin; TEN, tentoxin.

alterations in crops infected with *Alternaria* species. The mode of action of some host-specific toxins has been determined and they play a critical role in plant pathogenicity. *Alternaria alternata* f. sp. *lycopersici* toxins (AAL toxins) inhibit the enzyme ceramide synthase that is involved in the sphingolipid biosynthetic pathway [8]. Toxins from *A. alternata* f. sp. *fragariae* (ACT toxin), *A. alternata* f. sp. *fragariae* (AF toxin), and *A. gaisen* pear pathotype (AK toxin) have in common an epoxyeicosatrienoic acid structure, and they exert their primary effect on the plasma membrane, causing permeabilization. *A. mali* toxin (AM toxin) not only affects the plasma membrane but also acts on chloroplasts, while the ACT toxin was found to affect mitochondria [7,9]. Toxin from *A. alternata* rough lemon pathotype (ACR toxin) induces changes in the mitochondria including swelling, decrease in the electron density of the matrix, increase in the rate of NADH oxidation, and inhibition of malate oxidation [9].

11.2.1 Mycotoxin Synthesis

Mycotoxin synthesis has been studied by culturing toxin-producing species under optimal growth conditions. However, mycotoxin production depends on the (1) fungal strain, (2) substrate on which it grows, and (3) environmental growth conditions. Toxins such as alternariol (AOH), alternariol monomethyl ether (AME), altenuene (ALT), TeA, altertoxins (ATXs), TEN, or altenusin can be synthesized by several species of the genus [10,11]. In contrast, altersolanol, macrosporin, or zinniol is only produced by the *A. porri* species-group, which includes the species *A. solani*, *A. porri*, *A. dauci*, and *A. cucumerina* [12,13], while the *A. infectoria* species-group produces the metabolites infectopyrones and novae-zelandins that are not found in other *Alternaria* species-groups [14,15]. There are toxins that can only be synthesized by certain pathogenic strains including AAL toxins (from *A. alternata* f. sp. *lycopersici*), AS toxin (from *A. alternata* pathogenic to sunflower), AF toxin (from *A. alternata* f. sp. *fragariae*), ACR toxin

Alternaria spp. and Mycotoxins 141

(from the *A. alternata* rough lemon pathotype), and ACT toxin (from the *A. alternata* tangerine pathotype). Finally, AK toxin and AM toxin have been detected in pears and apples contaminated with pathogenic strains of *A. gaisen* and *A. mali*, respectively [7,16,17]. However, strains belonging to these groups do not always produce those toxins, because their synthesis is influenced by environmental conditions.

11.2.2 Physiology of Mycotoxin Production

Alternaria species can produce mycotoxins within a very broad range of temperatures, but always at water activities (a_w) above 0.90. Young et al. [18] inoculated *A. tenuissima* into cottonseed samples and found that maximal TeA production occurred at 20°C and 37.5% grain moisture. Magan and Lacey [19] showed the effect of temperature (5°C, 15°C, 25°C, and 30°C) and water relations (0.98, 0.95, and 0.90 a_w) on AOH, AME, and ALT production by *A. alternata* on autoclaved wheat grains. All three toxins were produced optimally at 25°C and 0.98a_w. In another study, Pose et al. [20] reported that the optimum conditions for AOH, AME, and TeA in tomato pulp inoculated with five *A. alternata* strains were 21°C/0.954a_w, 35°C/0.954a_w, and 21°C/0.982a_w, respectively. Finally, Oviedo et al. [21] showed maximal AOH and AME production at 0.98a_w but at different temperatures, depending on the *A. alternata* strain inoculated into irradiated soybeans.

11.2.3 Chemistry and Physical Properties

Alternaria sp. toxins are divided into five different groups based on their chemical structures:

1. Dibenzo-α-pyrones: AOH, AME and ALT
2. Perylene quinones: ATX-I, ATX-II, ATX-III
3. Tetramic acids: TeA
4. AAL toxins: AAL-TA1, AAL-TA2, AAL-TB1, AAL-TB2
5. Miscellaneous structures: TEN

Table 11.2 and Figures 11.1 and 11.2 show the chemical properties and structures of the main *Alternaria* sp. toxins [3]. The study of the biosynthetic pathways and metabolism of AOH, AME, and ALT showed that they can be conjugated to other polar substances such as glucose, amino acids, and sulfates present in the vegetable substrates where they occur. These modified mycotoxins may hinder the development of analytical methods for their detection, due to differences in polarity between the native mycotoxins and their metabolites [22–25].

The information on metabolism of *Alternaria* sp. toxins in the mammalian organism is limited to AOH, AME, and ALT. In vitro studies indicate that they are hydroxylated mostly to catechol metabolites and conjugated with glucuronic acid and sulfate. Hence, whether they are true mycotoxins remains to be determined.

11.3 TOXICITY OF *ALTERNARIA* SP. MYCOTOXINS

Exposure to *Alternaria* sp. mycotoxins is associated with a great variety of adverse effects on the health of humans and animals. Cultures of *Alternaria* spp. were toxic in rats, chick embryos, and human cell cultures and also were teratogenic and fetotoxic in mice [26–30]. Moreover, culture extracts of *A. alternata* and individual mycotoxins such as AOH and AME were mutagenic and clastogenic in in vitro systems and carcinogenic in rats fed contaminated feed [31–34]. In addition, *Alternaria* sp. toxins in grains might be responsible for esophageal cancer in China [35]. The toxic effects caused by *Alternaria* sp. cultures in different in vitro and in vivo models are highly variable, as they depend on the mycotoxin(s) synthesized (Table 11.3).

TABLE 11.2
Chemical Names, Molecular Weights, and Formulas of the Main *Alternaria* sp. Toxins

Toxin	Chemical Name	Mw (Da)	Formula
Alternariol (AOH)	3,7,9-Trihydroxy-1-methyl-6H-dibenzo[b,d]pyran-6-one	258	$C_{14}H_{10}O_5$
Alternariol monomethyl ether (AME)	3,7-Dihydroxy-9-methoxy-1-methyl-6H-dibenzo[b,d]pyran-6-one	272	$C_{15}H_{12}O_5$
Altenuene (ALT)	(2R,3R,4aR)-rel-2,3,4,4a-Tetrahydro-2,3,7-trihydroxy-9-methoxy-4a-methyl-6H-dibenzo[b,d]pyran-6-one	292	$C_{15}H_{16}O_6$
Altertoxin I (ATX-I)	(1S,12aR,12bS)-1,2,11,12,12a,12b-Hexahydro-1,4,9,12a-tetrahydroxy-3,10-perylenedione	352	$C_{20}H_{16}O_6$
Altertoxin II (ATX-II)	(7aR,8aS,8bS,8cR)-7a,8a,8b,8c,9,10-Hexahydro-1,6,8c-trihydroxyperylo[1,2-b]oxirene-7,11-dione	350	$C_{20}H_{14}O_6$
Altertoxin III (ATX-III)	(1aR,1bS,5aR,6aR,6bS,10aR)-1a,1b,5a,6a,6b,10a-Hexahydro-4,9-dihydroxy-perylo[1,2-b:7,8-b']bisoxirene-5,10-dione	348	$C_{20}H_{12}O_6$
Tenuazonic acid (TeA)	(5S)-3-Acetyl-1,5-dihydro-4-hydroxy-5-[(1S)-1-methylpropyl]-2H-pyrrol-2-one	197	$C_{10}H_{15}O_3N$
Tentoxin (TEN)	Cyclo[N-methyl-L-alanyl-L-leucyl-(αZ)-α,β-didehydro-Nmethylphenylalanylglycyl]	414	$C_{22}H_{30}O_4N_4$
AAL TA1 toxin	(2R)-1,2,3-Propanetricarboxylic acid, 1-[(1S,3S,9R,10S,12S)-13-amino-9,10,12-trihydroxy-1-[(1R,2R)-1-hydroxy-2-methylbutyl]-3-methyltridecyl] ester	521	$C_{25}H_{47}O_{10}N$
AAL TA2 toxin	(2R)-1,2,3-Propanetricarboxylic acid, 1-[(1R,2S,4S,10R,11S,13S)-14-amino-2,10,11,13-tetrahydroxy-4-methyl-1-[(1R)-1-methylpropyl]tetradecyl] ester	521	$C_{25}H_{47}O_{10}N$
AAL TB1 toxin	(2R)-1,2,3-Propanetricarboxylic acid, 1-[(1S,3S,10R,12S)-13-amino-10,12-dihydroxy-1-[(1R,2R)-1-hydroxy-2-methylbutyl]-3-methyltridecyl] ester	505	$C_{25}H_{47}O_9N$
AAL TB2 toxin	(2R)-1,2,3-Propanetricarboxylic acid, 1-[(1R,2S,4S,11R,13S)-14-amino-2,11,13-trihydroxy-4-methyl-1-[(1R)-1-methylpropyl]tetradecyl] ester	505	$C_{25}H_{47}O_9N$

Mw, molecular weight.

11.3.1 DIBENZO-α-PYRONES

AOH and AME were mutagenic, genotoxic, carcinogenic, and cytotoxic against bacterial and animal cell cultures. AOH and AME were strongly mutagenic toward *Bacillus subtilis* [36] and *Escherichia coli* [37,38], and they were reported as not mutagenic or weakly mutagenic against *Salmonella* serovar Typhimurium [31,39]. Genotoxicity studies show that AOH had greater activity than AME [32,40–44]. However, there is a lack of comprehensive carcinogenic studies of these mycotoxins in animals. Precancerous changes were reported in esophageal mucosa of mice (groups of 10 animals) fed in drinking water with 50–100 mg/kg body weight (b.w.) per day of AME for 10 months [45], suggesting that progression to esophageal cancer might occur after prolonged exposure. The 50% lethal concentration dose of ALT was 375 g/mL in brine shrimp larvae (*Artemia salina*), while the doses for TeA and AOH were 75 and 200 mg/mL, respectively [46].

11.3.2 TENUAZONIC ACID

The toxicity of TeA has been reported in plants, chick embryos, and several animal species [47]. Two beagle dogs (male and female) received TeA orally at 10 mg/kg b.w. per day (gelatin capsules, four separate doses of 2.5 mg/kg b.w.) for 8–9 days. The animals became moribund by days 8 and 9.

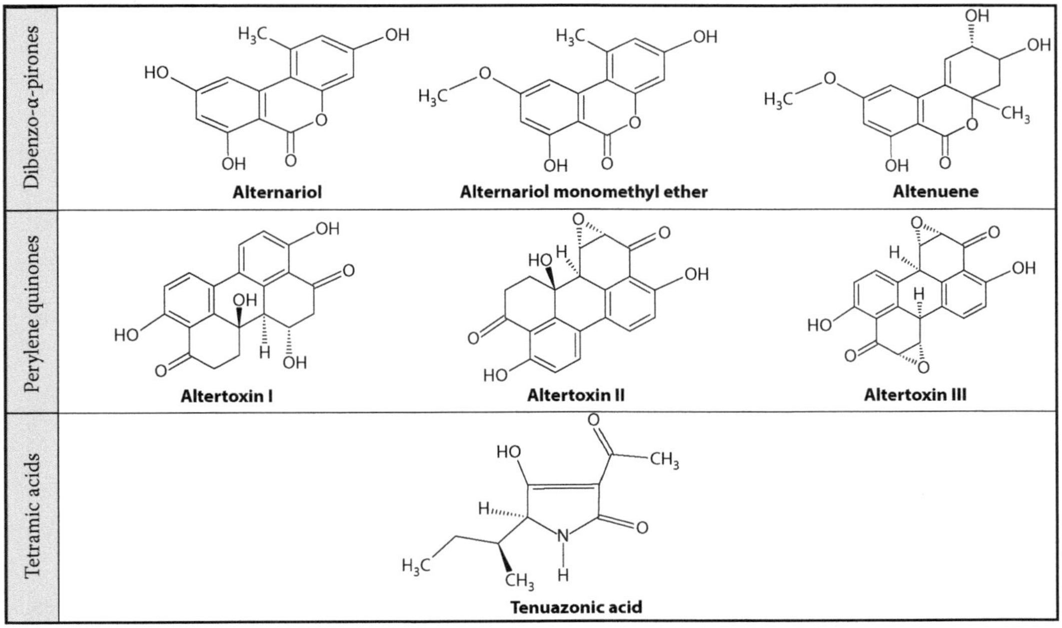

FIGURE 11.1 Chemical structures of AOH; AME; ALT; ATX-I, ATX-II, and ATX-III; and TeA.

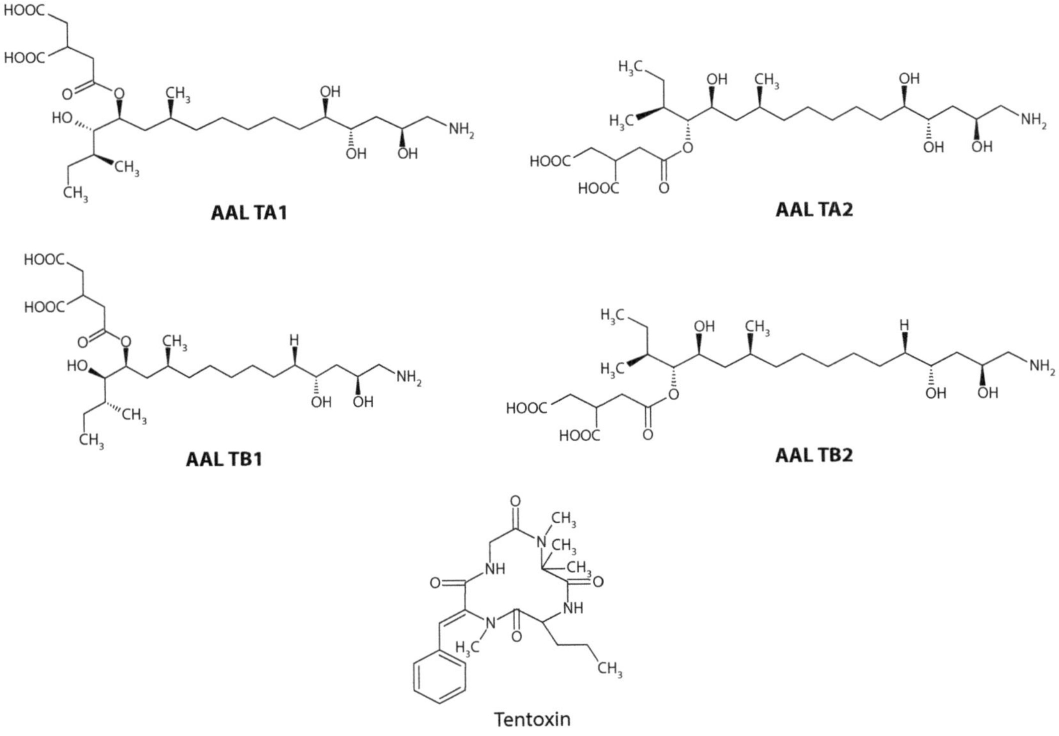

FIGURE 11.2 Chemical structures of AAL toxins and TEN.

TABLE 11.3
Main Toxic Effects Caused by *Alternaria* sp. Mycotoxins

Mycotoxin	Toxic Effect	References
AOH	Mutagenic activity in bacterial cultures	[31,36,38,39]
	Genotoxic activity in animal cultures	[32,40–43]
	Cytotoxic activity in brine shrimp larvae (*Artemia salina*)	[46]
AME	Mutagenic activity in bacterial cultures	[31,36,38,39]
	Genotoxic activity in animal cultures	[32,42,43]
	Precancerous changes in esophageal mucosa of mice	[45]
	Cytotoxic activity in brine shrimp larvae (*Artemia salina*)	[46]
ALT	Cytotoxic activity in brine shrimp larvae (*Artemia salina*)	[46]
TeA	Internal hemorrhages in dogs and chicks	[47]
	Precancerous changes in esophageal mucosa of mice	[45]
	Human hematological disorder (Onyalai)	[3,5]
ATX	Mutagenic activity in the Ames test	[31,34,48–51]
	Genotoxic activity in mouse cell cultures (ATX-I and ATX-III)	[53]

Clinical signs of toxicity were diarrhea, vomiting, and hemorrhages in the lungs and gastrointestinal tract. Microscopic analyses gave evidence of hemorrhage in many organs, particularly the zona fasciculata of the adrenal glands and degenerative changes in the liver [4]. In day-old chicks fed with TeA at levels of up to 10 mg/kg feed, decreased weight gain and lowered feed efficiency were observed during the second and third week of toxin administration. However, TeA tested negative in bacterial mutagenicity assays [31,34]. As with AME, precancerous changes were observed in esophageal mucosa of mice fed in the drinking water with TeA for 10 months, but in this case, the daily dose was 25 mg/kg b.w. per day [45]. Moreover, sorghum grain contaminated with TeA was associated with the human hematological disorder known as onyalai [3,5].

11.3.3 ALTERTOXINS

ATXs were mutagenic in the Ames test [31,48–50]. ATX-I, ATX-II, and ATX-III caused gene mutations in TA98, TA100, and TA1537 strains with or without metabolic activation in the following order: ATX-III > ATX-II > ATX-I. The potency of ATX-III was 10-fold lower than that of the aflatoxin B_1, which is highly mutagenic and a well-established human hepatocarcinogen [48,49]. Moreover, the mutagenic activity of ATX (ATX-I, II, III) in mice was higher than that of AOH and AME, with ATX-III being the most active [34,51,52]. It has also been shown that ATX-I and ATX-III are genotoxic in mouse cell cultures [53].

11.3.4 AAL TOXINS

AAL toxins are structurally related to sphinganine and closely resemble the mycotoxin fumonisin B_1 that was identified as a *Fusarium moniliforme* toxin. Like fumonisin B_1, AAL toxins inhibited the sphingolipid (ceramide) biosynthesis. These mycotoxins induced cell death and neoplastic events in mammalian cell cultures [7,54].

11.4 OCCURRENCE OF *ALTERNARIA* TOXINS IN FOOD AND FEED

Alternaria mycotoxins have been frequently isolated from raw and processed fruits, vegetables, and oilseeds infected with *Alternaria* (Table 11.4). The presence of *Alternaria* spp. in cereals is very common, because growth is favored in grains harvested during wet weather and stored in moist

TABLE 11.4
Concentrations of *Alternaria* sp. Mycotoxins Reported for Several Types of Food

Type of Food	Mycotoxins Detected[a]	Maximum Concentration (µg/kg or µg/L)	References
Grains and grain-based products	AOH	2320	[55]
	AME	7451	[57]
	ALT	1480	[55]
	TeA	8814	[57]
	TEN	38	[4]
Oilseeds	AOH	1800	[65]
	AME	440	[4]
	TeA	5400	[4]
	TEN	880	[61]
	ALT	9	[4]
	ATX-I	80	[4]
Seed oil	AOH	18	[4]
	AME	85	[4]
	TeA	390	[4]
	TEN	67	[4]
Legumes	AOH	290	[62]
	AME	1153	[63]
Vegetables and vegetable products	AOH	8756	[69]
	AME	1734	[69]
	ALT	22820	[68]
	TeA	4021	[69]
	TEN	9.2	[4]
Fruits and fruit products	AOH	151	[4]
	AME	42	[4]
	TeA	8700	[4]
Alcoholic beverages	AOH	7.59	[74]
	AME	1.5	[4]
	TeA	175	[59]
Infant food	AME	3.3	[4]
	TeA	1	[4]

[a] Minimum concentrations detected in all cases were 0.01 µg/kg for AOH and AME, 1 µg/kg for ALT, 2 µg/kg for TeA and TEN, and 10 µg/kg for ATX-I [4], with the exception of AME in legumes that was 16 µg/kg [63].

conditions. Hence, AOH, AME, ALT, TeA, and TEN have been detected in several cereal grains (e.g., wheat, barley, oats, rye, rice, and maize). Processing of cereal grains does not avoid the presence of mycotoxins, as AOH, AME, and TeA were found in breakfast cereals, flour, pasta, or beer [4,55–59]. Also, AME has been detected in baby food and TeA in infant food containing cereals [4], which are causes for particular concern.

AOH and AME were found in rapeseed, sunflower, sesame, and linseed although ALT was detected only in rapeseed and linseed and TeA in sunflower and sesame. TEN was detected only in sunflower and ATX-I in sesame. Furthermore, mycotoxins isolated from sunflower and sesame were also detected in oils obtained from these sources [4,60,61]. Finally, AOH and AME were also found in legumes such as lentils and soybeans [62,63].

AOH, AME, ALT, TeA, and TEN have been isolated from fruits (apple, pear, melon, apricot, grapes, raisins, strawberry, olive, citrus fruits, and dried figs) [6,64,65], vegetables (tomato, pepper, and carrot) [6,66–68], and tubers (potato) [6]. Direct consumption of moldy fruits and vegetables by the consumer is improbable, but the presence of moldy raw materials being included in processed food products is quite likely. Thus, *Alternaria* mycotoxins have been frequently detected in processed tomato products (tomato sauce, ketchup, and dried tomato) [6,66,68–70] and in wine or fruit juices (apple, tomato, grape, and orange) [6,58,71–75]. The presence of mycotoxins in processed foods could be due to the use of raw materials where symptoms affect only the internal part, as in the case of *Alternaria* rot of apples and citrus. Furthermore, *Alternaria* mycotoxins can be transferred into the surrounding tissues although visibly infected areas have been removed [76]. Food processing industries do not always have efficient methods to detect and completely remove raw materials affected by fungal spoilage.

The presence of *Alternaria* mycotoxins has also been studied in feedstuffs. In fact, high concentrations of AOH, AME, ALT, ATX-I, TeA, TEN, and AAL toxins were found in cereal grains and oilseeds intended for animal feed. Usually, samples containing higher amounts of mycotoxins showed some evidence of spoilage, associated with mold. The high incidence of *Alternaria* sp. mycotoxins in these samples could be due to the storage process, particularly during the early stages, when the temperature and high moisture are appropriate for growth and mycotoxin production [58,63,77–79].

11.5 HUMAN EXPOSURE TO *ALTERNARIA* SP. MYCOTOXINS

The European Food Safety Authority has recently published the results of a nutritional study carried out to evaluate daily exposure of AOH, AME, TeA, and TEN in population groups between 18 and 65 years of age [4]. In the adult population, the mean chronic dietary exposure to AOH across dietary surveys ranged from 1.9 to 39 ng/kg b.w. per day. The exposure estimates for AME were 0.8–4.7 ng/kg b.w. per day. The mean chronic dietary exposure for TeA ranged from 36 to 141 ng/kg b.w. per day, and the lowest estimates were for TEN (0.01–7 ng/kg b.w. per day).

The dietary exposure in children might be higher compared to adults by a factor of 2–3 due to the higher food consumption per kilogram body weight. The estimates in this survey took into account only vegetable foods, since there is no evidence for the presence of *Alternaria* mycotoxins in animal foods.

11.6 CONTROL AND PREVENTION OF MYCOTOXINS IN FOODS

Contamination of food with mycotoxins occurs naturally, and their concentration can increase as a result of environmental conditions or improper operations during collection, storage, and processing of foodstuffs. It is therefore necessary to establish control programs to prevent toxigenic mold contamination. To achieve these goals, the official authorities of food control are establishing preventive programs and procedures to minimize the risks associated with the presence of mycotoxins in foodstuffs that include good agricultural practices, good manufacturing practices, and the application of the hazard analysis and critical control point (HACCP) system. The HACCP system identifies and evaluates, through a structured and systematic approach, the hazards that must be controlled to ensure food safety from farm to table. To carry out the effective implementation of HACCP in relation to hazards arising from the presence of toxigenic molds in food, there is need for rapid techniques to detect toxigenic molds in raw and processed foods that could help industry to decide whether to use or refuse selected lots, thereby preventing the occurrence of mycotoxins in processed foods.

There are no specific regulations for any of the *Alternaria* toxins in foods despite *Alternaria* being one of the most commonly isolated fungal genera of vegetables [6]. However, considering their potential effects on human health and the frequency of their presence in foodstuffs, systematic testing for *Alternaria* mycotoxins in these commodities is desirable to evaluate consumer health risk [20,69].

REFERENCES

1. Lo Cascio, G. and Ligozzi, M. 2011. *Alternaria*. In *Molecular Detection of Human Fungal Pathogens*, ed. D. Liu, pp. 27–36. Boca Raton, FL: CRC Press.
2. Barkai-Golan, R. 2008. *Alternaria* mycotoxins. In *Mycotoxins in Fruits and Vegetables*, eds. R. Barkai-Golan and P. Nachman, pp. 185–203. San Diego, CA: Academic Press.
3. Ostry, V. 2008. *Alternaria* mycotoxins: An overview of chemical characterization, producers, toxicity, analysis and occurrence in foodstuffs. *World Mycotoxin J* 1: 175–188.
4. European Food Safety Authority (EFSA). 2011. Scientific opinion on the risks for animal and public health related to the presence of *Alternaria* toxins in feed and food. *EFSA J* 9: 2407.
5. Logrieco, A., Bottalico, A., Mulé, M., Moretti, A., and Perrone, G. 2003. Epidemiology of toxigenic fungi and their associated mycotoxins for some Mediterranean crops. *Eur J Plant Pathol* 109: 645–667.
6. Logrieco, A., Moretti, A., and Solfrizzo, M. 2009. *Alternaria* toxins and plant diseases: An overview of origin, occurrence and risks. *World Mycotoxin J* 2: 129–140.
7. Thomma, B. 2003. *Alternaria* spp.: From general saprophyte to specific parasite. *Mol Plant Pathol* 4: 225–236.
8. Abbas, H. K., Tanaka, T., Duke, S. O., Porter, J. K., Wray, E. M., Hodges, L., Sessions, A. E., Wang, E., Merrill, A. H., and Riley, R. T. 1994. Fumonisin- and AAL-toxin-induced disruption of sphingolipid metabolism with accumulation of free sphingoid bases. *Plant Physiol* 106: 1085–1093.
9. Otani, H., Kohmoto, K., and Kodama, M. 1995. *Alternaria* toxins and their effects on host plants. *Can J Bot* 73: S453–S458.
10. Andersen, B., Kroger, E., and Roberts, R. G. 2001. Chemical and morphological segregation of *Alternaria alternata*, *A. gaisen* and *A. longipes*. *Mycol Res* 105: 291–299.
11. Andersen, B., Kroger, E., and Roberts, R. G. 2002. Chemical and morphological segregation of *Alternaria arborescens*, *A. infectoria* and *A. tenuissima* species-groups. *Mycol Res* 106: 170–182.
12. Andersen, B., Dongo, A., and Pryor, B. M. 2008. Secondary metabolite profiling of *Alternaria dauci*, *A. porri*, *A. solani* and *A. tomatophila*. *Mycol Res* 112: 241–250.
13. Phuwapraisirisan, P., Rangsan, J., Siripong, P., and Tip-pyang, S. 2009. New antitumour fungal metabolites from *Alternaria porri*. *Nat Prod Res* 23: 1063–1071.
14. Christensen, K. B., van Klink, J. W., Weavers, R. T., Larsen, T. O., Andersen, B., and Phipps, R. K. 2005. Novel chemotaxonomic markers for the *Alternaria infectoria* species–group. *J Agric Food Chem* 53: 9431–9435.
15. Andersen, B., Sorensen, J. L., Nielsen, K. F., van den Ende, B. G., and de Hoog, S. 2009. A polyphasic approach to the taxonomy of the *Alternaria infectoria* species–group. *Fungal Genet Biol* 46: 642–656.
16. Montemurro, N. and Visconti, A. 1992. *Alternaria* metabolites—Chemical and biological data. In *Alternaria. Biology, Plant Diseases and Metabolites*, eds. J. Chełkowski, and A. Visconti, pp. 449–557. Amsterdam, the Netherlands: Elsevier.
17. Xu, L. and Du, L. 2006. Direct detection and quantification of *Alternaria alternata lycopersici* toxins using high-performance liquid chromatography-evaporative light-scattering detection. *J Microbiol Methods* 64: 398–405.
18. Young, A. B., Davis, N. D., and Diener, U. L. 1980. The effect of temperature and moisture on tenuazonic acid production by *Alternaria tenuissima*. *Phytopathology* 70: 607–609.
19. Magan, N. and Lacey, J. 1984. Effect of water activity and temperature on mycotoxin production by *Alternaria alternata* in culture and wheat grain. *Appl Environ Microbiol* 47: 1113–1117.
20. Pose, G., Patriarca, A., Kyanko, V., Pardo, A., and Fernández, V. 2010. Water activity and temperature effects on mycotoxin production by *Alternaria alternata* on a synthetic tomato medium. *Int J Food Microbiol* 142: 348–353.
21. Oviedo, M. S., Ramirez, M. L., Barros, G. G., and Chulze, S. N. 2011. Influence of water activity and temperature on growth and mycotoxin production by *Alternaria alternata* on irradiated soya beans. *Int J Food Microbiol* 149: 127–132.
22. Berthiller, F., Dall'Asta, C., Schuhmacher, R., Lemmens, M., Adam, G., and Krska, R. 2005. Masked mycotoxins: Determination of a deoxynivalenol glucoside in artificially and naturally contaminated wheat by liquid chromatography-tandem mass spectrography. *J Agric Food Chem* 53: 3421–3425.
23. Koch, K., Podlech, J., Pfeiffer, E., and Metzler, M. 2005. Total synthesis of alternariol. *J Org Chem* 70: 3275–3276.
24. Altemöller, M., Podlech, J., and Fenske, D. 2006. Total synthesis of altenuene and isoaltenuene. *Eur J Org Chem* 7: 1678–1684.

25. Saha, D., Fetzner, R., Burkhardt, B., Podlech, J., Metzler, M., Dang, H., Lawrence, D., and Fischer, R. 2012. Identification of a polyketide synthase required for alternariol (AOH) and alternariol-9-methyl ether (AME) formation in *Alternaria alternata*. *PLoS One* 7: e40564.
26. Pero, R. W., Posner, H., Blois, M., Harvan, D., and Spalding, J. W. 1973. Toxicity of metabolites produced by the '*Alternaria*'. *Environ Health Perspect* 4: 87–94.
27. Harvan, D. J. and Pero, R. W. 1976. The structure and toxicity of the *Alternaria* metabolites. In *Mycotoxins and Other Fungal Related Food Problems*, Advances in Chemistry Series 149, ed. J. V. Rodricks, pp. 344–355. Washington, DC: American Chemical Society.
28. Sauer, D. B., Seitz, L. M., Burroughs, R., Mohr, H. E., West, J. L., Milleret, R. J., and Anthony, H. D. 1978. Toxicity of *Alternaria* metabolites found in weathered sorghum grain at harvest. *J Agric Food Chem* 26: 1380–1383.
29. Griffin, G. F. and Chu, F. S. 1983. Toxicity of the *Alternaria* metabolites alternariol, alternariol monomethyl ether, altenuene and tenuazonic acid in the chicken embryo assay. *Appl Environ Microbiol* 46: 1420–1422.
30. Schwarz, C., Kreutzer, M., and Marko, D. 2012. Minor contribution of alternariol, alternariol monomethyl ether and tenuazonic acid to the genotoxic properties of extracts from *Alternaria alternata* infested rice. *Toxicol Lett* 214: 46–52.
31. Scott, P. M. and Stoltz, D. R. 1980. Mutagens produced by *Alternaria alternata*. *Mutat Res* 78: 33–40.
32. Liu, G., Qian, Y., Zhang, P. et al. 1991. Relationships between *Alternaria alternata* and oesophageal cancer. In *Relevance to Human Cancer of N-Nitroso Compounds, Tobacco Smoke and Mycotoxins*, eds. I. K. O'Neill, J. Chen, and H. Bartsch, pp. 258–262. Lyon, France: International Agency for Research on Cancer.
33. Combina, M., Dalcero, A., Varsavsky, E., Torres, A., Etcheverry, M., Rodríguez, M., and González, H. 1999. Effect of heat treatments on stability of alternariol, alternariol monomethyl ether and tenuazonic acid in sunflower flour. *Mycotoxin Res* 15: 33–38.
34. Schrader, T. J., Cherry, W., Soper, K., Langlois, I., and Vijay, H. M. 2001. Examination of *Alternaria alternata* mutagenicity and effects of nitrosylation using the Ames *Salmonella* test. *Teratog Carcinog Mutagen* 21: 261–274.
35. Liu, G., Qian, Y., Zhang, P., Dong, W., Qi, Y., and Guo, H. 1992. Etiological role of *Alternaria alternata* in human oesophageal cancer. *Chin Med J* 105: 394–400.
36. Kada, T., Sadaie, Y., and Sakamoto, Y. 1984. *Bacillus subtilis* repair test. In *Handbook of Mutagenicity Test Procedures*, 2nd edn., eds. B. J. Kilbey, M. Legator, W. Nichols, and C. Ramel, pp. 13–31. Amsterdam, the Netherlands: Elsevier.
37. An, Y., Zhao, T., Miao, J., Liu, G., Zheng, Y., Xu, Y., and Van Etten, R. L. 1989. Isolation, identification and mutagenicity of alternariol monomethyl ether. *J Agric Food Chem* 37: 1341–1343.
38. Zhen, Y. Z., Xu, Y. M., Liu, G. et al. 1991. Mutagenicity of *Alternaria alternata* and *Penicillium cyclopium* isolated from grains in an area of high incidence of oesophageal cancer—Linxian China. In *Relevance to Human Cancer of N-Nitroso Compounds, Tobacco Smoke and Mycotoxins*, eds. I. K. O'Neill, J. Chen, and H. Bartsch, pp. 253–257. Lyon, France: International Agency for Research on Cancer.
39. Davis, V. M. and Stack, M. E. 1994. Evaluation of alternariol and alternariol methyl ether for mutagenic activity in *Salmonella typhimurium*. *Appl Environ Microbiol* 60: 3901–3902.
40. Xu, D. S., King, T. Q. and Ma, J. Q. 1996. The inhibitory effect of extracts from *Fructus lycii* and *Rhizoma polygonati* on in vitro DNA breakage by alternariol. *Biomed Environ Sci* 9: 67–70.
41. Lehmann, L., Wagner, J., and Metzler, M. 2006. Estrogenic and clastogenic potential of the mycotoxin alternariol in cultured mammalian cells. *Food Chem Toxicol* 44: 398–408.
42. Marko, D. 2007. Mechanisms of the genotoxic effect of *Alternaria* toxins. In *Gesellschaft fur Mycotoxin Forschung. Proceedings of the 29th Mycotoxin Workshop*, p. 48, Stuttgart-Fellbach, Germany.
43. Pfeiffer, E., Eschbach, S., and Metzler, M. 2007. *Alternaria* Toxins: DNA strand-breaking activity in mammalian cells *in vitro*. *Mycotoxin Res* 23: 152–157.
44. Solhaug, A., Vines, L. L., Ivanova, L., Spilsberg, B., Holme, J. A., Pestka, J., Collins, A., and Eriksen, G. S. 2012. Mechanisms involved in alternariol-induced cell cycle arrest. *Mutat Res, Fundam Mol Mech Mutagen.* 738–739: 1–11.
45. Yekeler, H., Bitmis, K., Ozcelik, N., Doymaz, M. Z., and Calta, M. 2001. Analysis of toxic effects of *Alternaria* toxins on oesophagus of mice by light and electron microscopy. *Toxicol Pathol* 29: 492–497.
46. Panigrahi, S. and Dallin, S. 1994. Toxicity of the *Alternaria* spp metabolites, tenuazonic acid, alternariol, altertoxin-I and alternariol monomethyl ether to brine shrimp (*Artemia salina* L) larvae. *J Sci Food Agric* 66: 493–496.

47. Scott, P. M. 2004. Other mycotoxins. In *Mycotoxins in Food: Detection and Control*, eds. N. Magan and M. Olsen, pp. 406–440. Cambridge, England: Woodhead Publishing Ltd.
48. Stack, M. E., Mazzola, E. P., Page, S. W., Pohland, A. E., Highet, R. J., Tempesta, M. S., and Chorley, D. G. 1986. Mutagenic perylenequinones metabolites of *Alternaria alternata*: Altertoxins I, II, and III. *J Nat Prod* 49: 866–871.
49. Stack, M. E. and Prival, M. J. 1986. Mutagenicity of the *Alternaria* metabolites altertoxins I, II, and III. *Appl Environ Microbiol* 4: 718–722.
50. Schrader, T. J., Cherry, W., Soper, K., and Langlois, I. 2006. Further examination of the effects of nitrosylation on *Alternaria alternata* mycotoxin mutagenicity *in vitro*. *Mutat Res Genet Toxicol Environ Mutagen* 606: 61–71.
51. Chelkowski, J. and Visconti, A. 1992. *Alternaria. Biology, Plant Diseases and Metabolites*. Amsterdam, the Netherlands: Elsevier.
52. Fleck, S. C., Burkhardt, B., Pfeiffer, E., and Metzler, M. 2012. *Alternaria* toxins: Altertoxin II is a much stronger mutagen and DNA strand breaking mycotoxin than alternariol and its methyl ether in cultured mammalian cells. *Toxicol Lett* 214: 27–32.
53. Osborne, L. C., Jones, V. I., Peeler, J. T., and Larkin, E. P. 1988. Transformation of C3H/10T½ cells and induction of EBV-early antigen in Raji cells by altertoxins I and III. *Toxicol In Vitro* 2: 97–102.
54. Wang, W., Jones, C., Ciacci-Zanella, J., Holt, T., Gilchrist, D. G., and Dickman, M. B. 1996. Fumonisins and *Alternaria alternata lycopersici* toxins: Sphinganine analog mycotoxins induce apoptosis in monkey kidney cells. *Proc Natl Acad Sci USA* 93: 3461–3465.
55. Abd El-Aal, S. S. 1997. Effects of gamma radiation, temperature and water activity on the production of *Alternaria* mycotoxins. *Egypt J Microbiol* 32: 379–396.
56. Aresta, A., Cioffi, N., Palmisano, F., and Zambonin, C. G. 2003. Simultaneous determination of ochratoxin A and cyclopiazonic, mycophenolic and tenuazonic acids in cornflakes by solid-phase microextraction coupled to high-performance liquid chromatography. *J Agric Food Chem* 51: 5232–5237.
57. Azcarate, M. P., Patriarca, A., Terminiello, L., and Fernández, V. 2008. *Alternaria* toxins in wheat during the 2004 to 2005 Argentinean harvest. *J Food Prot* 71: 1262–1265.
58. Asam, S., Konitzer, K., and Rychlik, M. 2010. Precise determination of the *Alternaria* mycotoxins alternariol and alternariol monomethyl ether in cereal, fruit and vegetable products using stable isotope dilution assays. *Mycotoxin Res* 27: 23–28.
59. Siegel, D., Merkel, S., Koch, M., and Nehls, I. 2010. Quantification of the *Alternaria* mycotoxin tenuazonic acid in beer. *Food Chem* 120: 902–906.
60. Skarkova, J., Ostry, V., and Prochazkova, I. 2005. Planar chromatographic determination of *Alternaria* toxins in selected foodstuffs. *Proceedings of the International Symposium on Planar Separations,* pp. 29–31, *Planar Chromatography*, Milestones in Instrumental TLC, Siofok, Hungary.
61. Kralova, J., Hajšlová, J., Poustka, J., Hochman, M., Bjelková, M., and Odstrčilová, L. 2006. Occurrence of *Alternaria* toxins in fibre flax, linseed and peas grown in organic and conventional farms: Monitoring pilot study. *Czech J Food Sci* 24: 288–296.
62. Ostry, V., Skarkova, J., and Ruprich, J. 2004. Occurrence of *Alternaria* mycotoxins and *Alternaria* spp. In *Lentils and Human Health, 26*. Abstract of the Mycotoxin-Workshop, p. 87, Herrsching, Germany.
63. Barros, G. G., Oviedo, M. S., Ramirez, M. L. et al. 2011. Safety aspects in soybean food and feed chains: Fungal and mycotoxins contamination. In *Soybean: Biochemistry, Chemistry and Physiology*, ed. N. Tzi-Bun, pp. 7–20. Rijeka, Croatia: InTech.
64. Logrieco, A., Bottalico, A., Solfrizzo, M., and Mulé, G. 1990. Incidence of *Alternaria* species in grains from Mediterranean countries and their ability to produce mycotoxins. *Mycologia* 82: 501–505.
65. Bottalico, A. and Logrieco, A. 1998. Toxigenic *Alternaria* species of economic importance. In *Mycotoxins in Agriculture and Food Safety*, eds. K. K. Sinha, and D. Bhatnagar, pp. 65–108. New York: Marcel Dekker Inc.
66. da Motta, S. and Soares, L. M. V. 2001. Survey of Brazilian tomato products for alternariol, alternariol monomethyl ether, tenuazonic acid and cyclopiazonic acid. *Food Addit Contam* 18: 630–634.
67. Solfrizzo, M., de Girolamo, A., Vitti, C., Visconti, A., and van den Bulk, R. 2004. Liquid chromatographic determination of *Alternaria* toxins in carrot. *J AOAC Int* 87: 101–106.
68. Pavón, M. A., González, I., Luna, A., Martín, R., and García, T. 2012. PCR-based assay for the detection of *Alternaria* species and correlation with HPLC determination of altenuene, alternariol and alternariol monomethyl ether production in tomato products. *Food Control* 25: 45–52.
69. Terminiello, L., Patriarca, A., Pose, G., and Fernández, V. 2006. Occurrence of alternariol, alternariol monomethyl ether and tenuazonic acid in Argentinean tomato puree. *Mycotoxin Res* 22: 236–240.

70. Noser, J., Schneider, P., Rother, M., and Schmutz, H. 2011. Determination of six *Alternaria* toxins with UPLC-MS/MS and their occurrence in tomatoes and tomato products from the Swiss market. *Mycotoxin Res* 27: 265–271.
71. Delgado, T. and Gómez-Cordovés, C. 1998. Natural occurrence of alternariol and alternariol methyl ether in Spanish apple juice concentrates. *J Chromatogr A* 815: 93–97.
72. Scott, P. M., Lawrence, G. A., and Lau, B. P. Y. 2006. Analysis of wines, grape juices and cranberry juices for *Alternaria* toxins. *Mycotoxin Res* 22: 142–147.
73. Magnani, R. F., De Souza, G. D., and Rodrigues-Filho, E. 2007. Analysis of alternariol and alternariol monomethyl ether on flavedo and albedo tissues of tangerines (*Citrus reticulata*) with symptoms of *Alternaria* brown spot. *J Agric Food Chem* 55: 4980–4986.
74. Asam, S., Konitzer, K., Schieberle, P., and Rychlik, M. 2009. Stable isotope dilution assays of alternariol and alternariol monomethyl ether in beverages. *J Agric Food Chem* 57: 5152–5160.
75. Ackermann, Y., Curtui, V., Dietrich, R., Gross, M., Latif, H., Märtlbauer, E., and Usleber, E. 2011. Widespread occurrence of low levels of alternariol in apple and tomato products, as determined by comparative immunochemical assessment using monoclonal and polyclonal antibodies. *J Agric Food Chem* 59: 6360–6368.
76. Robiglio, A. L. and López, S. E. 1995. Mycotoxin production by *Alternaria alternata* strains isolated from Red Delicious apples in Argentina. *Int J Food Microbiol* 24: 413–417.
77. Nawaz, S., Scudamore, K. A., and Rainbird, S. C. 1997. Mycotoxins in ingredients of animal feeding stuffs: I. Determination of *Alternaria* mycotoxins in oilseed rape meal and sunflower seed meal. *Food Addit Contam* 14: 249–262.
78. Häggblom, P., Stepinska, A., and Solyakov, A. 2007. *Alternaria* mycotoxins in Swedish feed grain. In *Gesellschaft fur Mycotoxin Forschung. Proceedings of the 29th Mycotoxin Workshop*, p. 35, Stuttgart-Fellbach, Germany.
79. Mansfield, M. A., Archibald, D. D., Jones, A. D., and Kuldau, G. A. 2007. Relationship of sphinganine analog mycotoxin contamination in maize silage of seasonal weather conditions and to agronomic and ensiling practices. *Phytopathology* 97: 504–511.

12 Aspergillus and Aspergillosis

Malcolm D. Richardson and Riina Richardson

CONTENTS

12.1 Introduction 151
12.2 *Aspergillus* Species.. 151
12.3 The Spectrum of Aspergillosis... 152
12.4 Detection.. 154
    12.4.1 In Food... 154
    12.4.2 In Water 156
        12.4.2.1 Domestic Water Supplies 156
        12.4.2.2 Hospital Water Supplies.. 157
    12.4.3 In Biofilms 159
    12.4.4 In Humans 160
        12.4.4.1 ELISA and Lateral Flow Devices 160
        12.4.4.2 Molecular Platforms 160
        12.4.4.3 Diagnosis of Invasive Aspergillosis in Practice...................................... 161
12.5 Conclusions.. 161
References.. 162

12.1 INTRODUCTION

Aspergillosis is a term used to refer to infections caused by molds belonging to the genus *Aspergillus*.[1] These conditions are seen worldwide. In immunocompromised individuals, inhalation of conidia or hyphal fragments can give rise to life-threatening invasive infection of the lungs or sinuses: dissemination to other organs often follows. This condition is termed invasive aspergillosis (IA). In nonimmunocompromised persons, these molds can cause localized infection of the lungs, sinuses, and other sites. Human disease can also result from noninfectious mechanisms: inhalation of conidia of these ubiquitous organisms can exacerbate allergic symptoms in both atopic and non-atopic individuals. Several *Aspergillus* species produce mycotoxins that are harmful to humans and animals when ingested.

Molds of the anamorphic genus *Aspergillus* are widespread in the environment, growing in the soil, on plants, and on decomposing organic matter. These molds are often found in the outdoor and indoor air, in water, on food items, and in dust. Aspergilli are saprophytes that commonly grow on decaying plant material. They are able to utilize a wide range of organic substrates and adapt well to a broad range of environmental conditions. In contact with air, the mycelium forms specialized structures, the so-called conidiophores. These produce large numbers of conidia (asexual spores) that are efficiently dispersed through the air and inhaled by humans.

12.2 *ASPERGILLUS* SPECIES

Aspergillus includes over 185 species. Around 20 species have so far been reported as causative agents of opportunistic infections in man.[1] Among these, *Aspergillus fumigatus* is the most commonly isolated species, followed by *A. flavus* and *A. niger*, *A. clavatus*, *A. glaucus* group, *A. nidulans*, *A. oryzae*, *A. terreus*, *A. ustus*, and *A. versicolor* are other species less commonly isolated as opportunistic pathogens. Fewer than 40 species cause diseases in humans or animals, and some of these have been reported only once.[1] The fungus grows quite readily on ordinary laboratory culture media at room temperature, at 37°C, and higher. Czapek's solution agar or Sabouraud agar may be used. The colonies are green to bluish green at first and darken with age so as to appear almost black. The colonies vary from velvety to floccose. *Aspergillus* spores and hyphal fragments are commonly found in air, water, soil, plant debris, rotten vegetation, manure, sawdust litter, bagasse litter, animal feed, indoor air, and on animals.

12.3　THE SPECTRUM OF ASPERGILLOSIS

A. fumigatus still accounts for most human cases of aspergillosis, with *A. flavus* the second most frequent pathogen. *A. terreus* is resistant to amphotericin B and is increasingly been reported as a cause of invasive disease in immunocompromised patients in some hospitals. *A. fumigatus* is also important as a human allergen, while *A. clavatus* and *A. fumigatus* are among the occupational causes of extrinsic allergic bronchoalveolitis.[1]

Inhalation of *Aspergillus* conidia is the usual mode of infection in humans. The incubation period is unknown. Less frequently, infection follows the traumatic implantation of spores as in corneal infection or direct inoculation from contaminated dressings. Inhalation of contaminated water aerosols during patient showering has been suggested as an additional potential source of infection.[1] Inhalation can give rise to a number of different clinical forms of aspergillosis, depending on the immunological status of the host. In immunocompromised patients, there is widespread growth of the fungus in the lungs or sinuses, and dissemination to other organs often follows. It must, however, be emphasized that with early diagnosis and treatment, a significant number of patients can now be cured. In nonimmunocompromised individuals, *Aspergillus* can cause localized infection of the lungs, sinuses, and other sites, or act as a potent allergen. A wide range of subacute or chronic forms of aspergillosis are now recognized, which are distinguished from each other and from acute IA, by their radiological appearance, rate of clinical progression, and histopathological findings. However, there is considerable overlap between these forms of disease.[1]

The etiological agents of aspergillosis are ubiquitous in the environment, and the likelihood that infection will occur following inhalation or implantation of spores largely depends on host factors. IA has emerged as a major problem in several groups of immunocompromised patients. Those at greatest risk include persons with hematological malignancies, particularly acute leukemia, hematopoietic stem cell transplant (HSCT) recipients, solid organ transplant (SOT) recipients, individuals receiving high-dose corticosteroid treatment, and persons with neutrophil deficiencies or dysfunction, such as children with chronic granulomatous disease. Emerging at-risk populations include persons with the acquired immunodeficiency syndrome (AIDS), patients receiving intensive care, patients with chronic obstructive pulmonary disease, and patients receiving tumor necrosis factor blockers. The likelihood of aspergillosis developing in these individuals depends on a number of host factors, the most important of which is the level of immunosuppression, whether this manifests as profound or prolonged neutropenia, as graft-versus-host disease (GVHD) in HSCT recipients, or as rejection in SOT recipients. With new developments in the management of the underlying disorders, the patient groups who are at highest risk of developing IA are changing, and there is an ongoing need to reassess the factors that can affect different groups of susceptible hosts.

The occurrence rate of IA in neutropenic patients with hematological malignancies is variable and is dependent on the underlying disorder, the treatment regimen, and other supportive measures used. Most at risk are those patients who remain neutropenic for 2 weeks or longer. The use of corticosteroids in an antineoplastic regimen increases the risk for aspergillosis. Patients who have gone through several periods of profound neutropenia (during remission induction, consolidation, or relapse treatment) are at increasing risk of developing aspergillosis, and a history of infection is often predictive of relapse during a subsequent period of immunosuppression.

However, some high-risk individuals do not develop the disease, suggesting an unidentified protection mechanism.

Allogeneic HSCT recipients, particularly HLA-mismatched and unrelated donor recipients, are at high risk for IA for a prolonged period of time because of disruption of mucosal barriers, delayed engraftment, GVHD, and the use of corticosteroids. Active acute GVHD has been identified as a major risk factor for the early development of *Aspergillus* infection, while extensive chronic grade III–IV GVHD is a major risk factor for late infection. The risk of aspergillosis is further increased by concomitant cytomegalovirus (CMV) infection.

Among SOT recipients, those with renal allografts are least at risk. In contrast, lung transplantation carries a high risk of IA. Among the risk factors that have been identified are graft rejection, obliterative bronchiolitis, CMV infection, and increased immunosuppression. Liver transplant recipients are another high-risk group. Many of those who develop aspergillosis have signs of poor allograft function and/or renal dysfunction that required dialysis.

It has long been recognized that few patients with hematological malignancies and IA will survive unless their neutrophil count recovers. Earlier diagnosis and optimal treatment have also been associated with a better prognosis. Risk factors associated with mortality in IA include progression of an underlying malignancy, receipt of allogeneic HSCT, preexisting respiratory function impairment, renal impairment, diffuse lung involvement or disseminated infection, and prior corticosteroid treatment. Among the factors that have been associated with a poor outcome in allogeneic HSCT recipients with aspergillosis are active acute GVHD of grade II or more, or extensive chronic GVHD at the time of diagnosis, and a high cumulative dose of prednisolone.

The incidence of IA in high-risk patient groups is difficult to determine for a number of reasons. In the past, these included the lack of a consistent case definition, which made it difficult to compare the incidence rates reported in different studies. This situation has improved with the adoption of an agreed set of consensus definitions for proven and probable invasive fungal infection. However, because these definitions require the use of aggressive diagnostic procedures, which may not be performed in all cases, it is likely that incidence figures will underestimate the true burden of disease.

There is a widespread perception that there has been a marked increase in the prevalence of IA over the past several decades. Although this belief may be correct, the evidence most frequently cited in its support has been derived from longitudinal studies conducted among HSCT recipients in individual hospitals and may not be representative of the situation in other patient groups, institutions, or countries. However, there are other studies that support the perception that aspergillosis is increasing in incidence. Most of these have relied on retrospective analysis of large databases, such as hospital discharge and death records. Although helpful for investigating trends, these databases tend to underestimate the incidence of fungal infections, such as aspergillosis. Nonetheless, these data indicate that the incidence of aspergillosis as a hospital discharge diagnosis in the Unites States increased about eightfold between 1976 and 1996, from 4.8 to 38 cases per million population.

The incidence of IA differs from one high-risk patient group to another.[1] Among 8000 HSCT and SOT recipients at 19 hospitals in the Unites States during 2001 and 2002, the cumulative 12-month incidence of aspergillosis was 0.5% after autologous HSCT, and ranged from 2.3% after allogeneic HSCT from an HLA-matched-related donor to 3.9% after transplantation from an unrelated donor. Among SOT recipients, the incidence was 2.4% after lung transplantation, 0.8% after heart transplantation, 0.3% after liver transplantation, and 0.1% after kidney transplantation. These figures are lower than those published in earlier reports. Many factors could account for this, including changes in transplantation practices, such as the use of nonmyeloablative conditioning regimens and receipt of peripheral blood stem cells; improved diagnostic procedures, principally galactomannan antigen testing and PCR (see in the following text); and improvements in supportive care.

The mortality rate from IA remains high, ranging from 50% to 100% in almost all groups of immunocompromised patients. In a large review of cases reported up to 1995, case–fatality ratios for cerebral, pulmonary, and sinus infections were 99%, 86%, and 66%, respectively.

Some improvement has been achieved since 1995. Among 642 patients who developed IA at 23 hospitals in the Unites States between 2001 and 2005, 58% of HSCT recipients died within 12 weeks of transplantation, compared with 34% of SOT recipients. Other reports have also described an improvement in outcome over time: in a recent report from a group of Italian hospitals, the attributable mortality rate among patients with acute leukemia fell from 60% in 1987–1988 to 32% in 2002–2003.

The most important extrinsic risk factor for IA is the presence of *Aspergillus* conidia and viable hyphal fragments in the hospital environment. Health care–associated outbreaks of the disease have become a well-recognized complication of construction or renovation work in or near units in which high-risk patients are housed. These reports have served to highlight the fact that the hospital environment is often contaminated with *Aspergillus* spores and have contributed to the widespread perception that most cases of aspergillosis in immunocompromised persons are hospital acquired. However, despite the fact that most hospital outbreaks of aspergillosis go unpublished, it is clear that these events in general are uncommon.

Most cases of IA are sporadic in nature, and it is much more difficult to determine whether these infections are acquired inside or outside the hospital setting. It seems probable that some individuals are colonized before their admission to hospital and develop invasive disease when rendered neutropenic. Indeed, it has been estimated that up to 70% of cases of IA diagnosed over a 2-year period of surveillance during construction in one North American hospital were acquired outside the hospital setting.

There are other evidences that a significant number of sporadic cases of IA are now being acquired outside the hospital environment. In the 1970s and 1980s, most cases of *Aspergillus* infection among leukemic patients occurred during the first few weeks after transplantation, before engraftment had occurred, or while individuals were undergoing intensive remission induction treatment. However, recent reports have indicated that many HSCT recipients now develop aspergillosis some months after transplantation, usually in association with GVHD and its treatment.

12.4 DETECTION

12.4.1 IN FOOD

The gold standard *Aspergillus* diagnostic test used in hospital laboratories for circulating biomarkers in blood is based on the monoclonal antibody EB-A2 that binds to an epitope located on the galactofuranose-containing side chains of the galactomannans of *Aspergillus* species and immunologically similar molds.[1] Therefore, a number of commentators have suggested that the test should be named more exactly the mold reveal kit. The specificity of the Platelia *Aspergillus* ELISA for the detection of galactomannan is hampered by the occurrence of false-positive results. In order to prove whether or not the false-positive reactions may be caused by the uptake of the soluble galactomannan antigen from the environment, the presence of the antigen has been tested in foods, air samples, antibiotics for therapeutic use, and feces.[2] *Aspergillus* antigen was detected in 15 out of 19 (79%) samples of meals prepared in a hospital kitchen, in 5 out of 6 canned vegetables from a supermarket, and in all of 6 samples of pasta and rice bought in health shops. The presence of galactofuranose in food products is reflected by the detection of the antigen in the feces of bone marrow transplant (BMT) recipients.[2] The concentration of the antigen in fecal material was calculated to be in the range of 1.2–38.4 µg/g. The authors suggest that the fecal galactomannan antigen may reach the circulation in patients with dysfunction of the intestinal mucosal barrier, for example, BMT recipients, thus leading to diagnostically false-positive antigenemia.

A number of studies have shown that vegetable and cereal foods contain components that react in the Platelia *Aspergillus* ELISA.[3,4] Because of the specificity of the monoclonal antibody, these components are most probably galactomannans of molds. During growth and harvesting, molds that originate from soil, water, and air most likely contaminate the surface of fruit, vegetables and cereals. The galactomannans of the contaminants are heat stable and are not destroyed or eliminated by the production processes and by the preparation of meals. Consequently, galactomannan has been found in the feces of healthy subjects and hospitalized patients, demonstrating that galactomannan is not destroyed by digestion. The amount can be calculated on the basis of the sensitivity threshold of the Platelia *Aspergillus* kit, that is, 15 ng/mL, according to the manufacturer's specifications. Cell wall and exocellular galactomannans of *A. fumigatus* have molecular weights in the range of 20,000–75,000 Da. In subjects with an intact intestinal mucosa, molecules of such size are not absorbed in appreciable amounts. However, in BMT recipients, for example, the conditioning chemotherapy can damage the intestinal mucosa for up to 4 weeks after transplantation, with acute GVHD also compromising barrier function. Foodborne mold galactomannan may also pass through the damaged mucosa into the circulation. The amounts of antigen found in the feces appear to be sufficient to cause an antigenemia detectable with the Platelia *Aspergillus* ELISA. The occurrence of galactomannan in some batches of the antibiotics piperacillin and co-amoxiclav suggests a further possibility of false-positive galactomannan antigenemia. The penams piperacillin and amoxicillin, and the oxypenam clavulanic acid, combined with amoxicillin in co-amoxiclav, are semisynthetic drugs that are based on natural compounds derived from molds of the genus *Penicillium*. Therefore, positive test results in patients receiving piperacillin/tazobactam should be interpreted cautiously and confirmed by other diagnostic methods.

Different foods will have different associated mycobiotas, qualitatively and quantitatively. For *Aspergillus* species there are qualitative differences (Table 12.1). For example, *A. niger* and *A. carbonarius* are more common than *A. ochraceus* and *A. westerdijkiae* on grapes, while the reverse is the case for green coffee beans. Knowledge on the associated mold mycobiota on foods will add to our understanding of false-positives in the Platelia *Aspergillus* ELISA. Several foods can also be colonized by *Aspergillus* and other molds and can lead to primary gastrointestinal

TABLE 12.1
Aspergillus Species Associated with Crops and Their Products

Crop	Product	Species
Beans and peas	Black beans, cowpeas	*A. flavus, A. ochraceus, A. parasiticus*
	Rice	*A. flavus, A. niger*
	Sorghum	*A. flavus*
	Wheat bread	*A. flavus*
	Wheat, rye, barley, oat	*A. flavus, A. parasiticus*
Cheese	Hard cheese	*A. versicolor*
Coffee	Coffee: monsoon	*A. candidus, A. niger, A. tamari*
	Coffee: traditional	*A. carbonarious, A. steynii, A. westerdijkiae*
Fruit	Dried fruits	*A. carbonarius, A. flavus, A. niger, A. ochraceus*
	Grapes	*A. carbonarius, A. niger, A. tubingensis*
Nuts	Almonds, hazelnuts, pistachio, walnuts	*A. flavus, A. niger, A. tamari*
Oil crop	Olives	*A. versicolor*
	Peanuts	*A. flavus*
		A. niger
	Sunflower	*A. flavus, A. niger, A. Parasiticus*
Vegetables	Pepper: black	*A. flavus, A. parasiticus, A. tamari*
	Yam chips	*A. flavus, A. niger*

colonization and subsequent systemic infection. Massive *Aspergillus* contamination of foods was first reported in pepper, and subsequently in regular and herbal tea, corn, coconut, cashew nuts, coffee, beans, soy, cheese, smoked meat, downy-skinned fruits (apricots, kiwis, and peaches), smooth-skinned fruits (apples, bananas, lemons, and oranges), freeze-dried soups, and individual food wrappings. It has been previously recommended that immunocompromised patients should avoid such contaminated foods, and granulocytopenic patients should receive sterile diets or diets low in microbial content. However, such restrictions in diet may affect the well-being of these patients.

Several studies have examined the efficiency of disinfection procedures applicable to foods for eradication of *A. fumigatus*.[5] Boiling and microwave treatment appears to fully decontaminate contaminated liquid foods (e.g., reconstituted dried food, herbal tea). Pepper can be decontaminated when it is heated for 15 min at 220°C but not by microwaving. Fruit skin can be partially decontaminated by 70% ethanol. These studies show that *A. fumigatus* spores can be eradicated from food by heating to a temperature of at least 100°C. When foods cannot be exposed to high temperature or microwaving, ethanol only partially reduces the level of surface contamination.

12.4.2 IN WATER

Fungi can enter drinking water distribution systems through several contamination pathways, including treatment breakthrough, deficiencies in stored water facilities cross-connections, mains breaks and intrusions, and during mains installation and maintenance. Once introduced, fungal species can become established on the inner surfaces of pipes, including interaction and reaction with sealings and coatings, and biofilms within distribution systems, or can be suspended in the water.

Fungi have been reported from all types of water, from raw water to treated water, and from heavily polluted water to distilled or ultrapure water: they have also been reported from bottled drinking water.[6] Statistically, it has been established that the odds for fungal recovery are three times higher in surfaces-sourced water compared with ground-sourced water, and fungi are more commonly recovered from cold water and shower water than from hot tap water.[7]

Few studies have investigated the fungi found in treated drinking water.[6] The numbers of fungi found in the existing studies range from 1 to 5000 colony-forming unit (cfu)/L. Of the 65 genera that have been isolated in the studies analyzed during this review, the majority were filamentous fungi. The most commonly isolated genera were *Penicillium*, *Cladosporium*, *Aspergillus*, *Phialophora*, and *Acremonium*.

Due to the low level of literature on the topic of *Aspergillus* in drinking water, there are a number of aspects that remain poorly understood. Research needs to include the determination of the importance of drinking water as the environmental source of fungal infection in vulnerable or at-risk population groups. Greater knowledge on the importance of ingestion, as opposed to inhalation or skin contact, as exposure pathways for fungi in drinking water will ensure that mitigation measures for at-risk patients are appropriate. Finally, greater understanding of the effect of the analytical method on the results obtained and development of a standard method would facilitate further research into fungi in drinking water.

A number of different methods of analyzing drinking water samples for the presence of *Aspergillus* have been used, including culture, antigen detection, measurement of ergosterol, quantitative PCR (QPCR), gene markers and probes, protein probes, direct observation, and mass spectrometry. There is currently no international standard specifically for the measurement of fungi in drinking water, and there is no widespread adoption of other relevant standards. Therefore, differences in analysis methods limit the extent to which results can be compared between studies.

12.4.2.1 Domestic Water Supplies

The results of sample analysis from customer taps and other points within distribution systems often reveal higher numbers of fungi than the analysis of samples following treatment, prior to entry into the distribution system. Such increases through the distribution system could be due to

TABLE 12.2
Global Distribution of *Aspergillus* Species in Water Supplies

Country, Place	Period of Time	Type of Water	Main Isolation Method
United Kingdom	Autumn and spring	Surface water and network	Membrane filtration, direct plating, and baiting
Greece	One collection (126 samples)	Tap water (hospital and community)	Membrane filtration
Greece	One collection (255 samples)	Municipal water supplies of hemodialysis centers	Membrane filtration
Norway	December, June, and September	14 networks drinking water (surface and groundwater)	Membrane filtration
Pakistan	One collection (30 samples)	Water (and fruit juice)	Direct plating
Australia	18 months	Municipal water	Membrane filtration
Brazil	5 months	Water treatment plant: tap water	Membrane filtration
Portugal	4 months	Surface water; spring water; groundwater	Membrane filtration

Source: Adapted from Siqueira, V.M. et al., *Int. J. Environ. Res. Public Health*, 8, 456, 2011.

three reasons: fungi (1) that remain present after treatment and multiply within the system, (2) that were only partially inactivated later recover, and/or (3) enter the system via pathways of secondary contamination. Accumulation of fungi in stored water at the consumer end, such as in water tanks, has also been observed. For example, higher numbers of cfu of *Aspergillus* have been found in hospital water storage tanks than in the municipal water supply.[8] A high occurrence of *A. fumigatus* has been found in private domestic wells in Brazil.[9] A summary of studies from many areas of the globe show the wide distribution in *Aspergillus* species in water supplies (Table 12.2)

12.4.2.2 Hospital Water Supplies

Nosocomial aspergillosis can be waterborne, as shown in reports of the disease in otherwise healthy people who have near-drowning accidents. *Aspergillus* species share several characteristics with *Legionella* bacteria, which are known waterborne pathogens; these include amplification in water reservoirs, presence in the water system biofilm, certain growth requirements, and an association with construction activity.

There is little evidence that the pathogenicity of *Aspergillus* arises from its presence in drinking water. More severe invasive infections are limited to those with immune deficiency, due to, for example, HIV/AIDS, chemotherapy, immunosuppressive therapy following transplants, or other underlying health conditions, such as cystic fibrosis or diabetes mellitus. Such invasive infections carry a high mortality rate, estimated at between 50% and 100%. The extent to which infections arise from at-risk individuals is not well known. The continuing rise of *Aspergillus* infections in at-risk individuals, despite hospital-based measures to control airborne fungal spores, suggests that another environmental source exists. A small number of studies have linked the genotypes of *Aspergillus* species recovered from patients to those of fungi from hospital water supplies. An illustrative example is an outbreak of nosocomial *A. niger* infection in which an ice-making machine was implicated as a possible reservoir.[10] A 3-year prospective study of the air, environmental surfaces, and water distribution system of a hospital in which there were known cases of aspergillosis was conducted to determine other possible sources of infection.[11]

Aspergillus species were found in the hospital water system. Significantly higher concentrations of airborne *Aspergillus* propagules were found in bathrooms, where water use was highest (2.95 cfu/m^3) than in patient rooms (0.78 cfu/m^3) and in corridors (0.61 cfu/m^3). A correlation was found between the rank orders of *Aspergillus* species recovered from hospital water and air. Water from tanks yielded higher counts of cfu than did municipal water. An isolate of *A. fumigatus* recovered from a patient with aspergillosis was genotypically identical to an isolate recovered from the shower wall in the patient's room.

The significance of exposure via drinking the water, as opposed to washing with it, has not been specifically studied. Aerosolization of *Aspergillus* during showering or from running taps has received more attention; numbers of airborne fungi have been found to increase after running taps or showers. This notion has several arguments in its favor. First, the incidence of aspergillosis continues to increase nationwide, despite the widespread installation of expensive air filtration systems to prevent the entry of contaminated outside air into hospitals. Second, recent data suggest no correlation between air spore counts of aspergilli and the rate of nosocomial aspergillosis or colonization with *Aspergillus* species. Third, there are data that suggest other portals of entry, such as the gut or skin, for *Aspergillus* species and other molds. Fourth, counts of cfu of *Aspergillus* species were seen to increase in a cancer unit, despite the presence of a properly functioning HEPA filtration system; the source of the molds was determined to be a rotten sink cabinet with leaking plumbing lines. Finally, the reports that support the notion that nosocomial aspergillosis is primarily airborne are retrospective in nature and include limited or no information on such key factors as the number of patient-days at risk and the degree of risk. In addition, these studies did not sample water as a possible source of infection.

Collectively, the findings from a number of studies should not be generalized to all institutions for various reasons, including differences in the ages and types of water distribution systems at different institutions. Most cases of IA worldwide are not caused by *A. niger*, the principal species found in water samples, although this species is known to cause otomycosis, pneumonia, endocarditis, pericarditis, peritonitis, and disseminated disease.[1] However, there are well-known geographical variations in the *Aspergillus* species that cause disease. These differences may be the consequence of the frequency and the virulence of each of the *Aspergillus* species present in the environment. For example, *A. fumigatus* is the species most frequently recovered from a hospital and community water system in Norway,[12] whereas the predominant *Aspergillus* species recovered from hospital water in Little Rock, Arkansas, is *A. niger* and in Houston, Texas, it is *A. terreus*.[13] However, temperature may have an effect as *A. fumigatus* is thermotolerant whereas the others are less so, and may relate to air-conditioning in the United States or artificial heating in Norway, for example.

A study was carried out over a 4-month winter period in order to assess the presence of filamentous fungi in the water distribution system of the University Hospital of Liege.[14] A total of 197 hot and cold water samples were collected from the main water supply lines and from the taps at three different hospital sites. Overall, filamentous fungi were recovered from 55% to 50% of the main water distribution system and tap water samples, respectively, with a mean of 3.5 ± 1.5 cfu/500 mL water. Nine different genera were identified, all belonging to the Hyphomycetes class. *Aspergillus* spp. were recovered from 6% of the samples of the water distribution system, and *A. fumigatus* was the most frequently recovered species (66.6%). However, this species was not isolated from water taps. *Fusarium* spp. were predominant at one site, where it was found in 28% of tap water samples. No *Aspergillus*, but some *Fusarium* spp. were identified in samples collected from high-risk units. Filters were introduced at the point of use in the hematology unit after completion of the study. The findings of this study confirm the need for further documented studies to evaluate the safety of the hospital water system and to define new preventive measures.

Rapid QPCR has been applied to the measurement of *A. fumigatus*, *A. flavus*, *A. terreus*, and *A. niger* in home tap water and hospital water supplies.[17] Water samples were taken from the kitchen tap in the homes of 60 patients who were diagnosed with legionellosis. Water samples were also taken from three locations in a hospital that generated all of its hot water by flash heating. *A. terreus*

DNA was found in 16.7% and *A. fumigatus* DNA in 1.7% of the samples taken from the kitchen tap. None of the *Aspergillus* species was found in any of the hospital water samples. This study typifies the development of a simple DNA extraction method along with QPCR analysis that is suitable for rapid screening of tap water for opportunistic fungal pathogens.[15] Results are available in about 3 h, instead of waiting from days to weeks for cultural data.

Other precautionary measures in the case of high-risk patients have been suggested with respect to water quality.[13] It was recommended that hospital water should be tested for the presence of disease-related fungi. High-risk patients should avoid exposure to hospital water and use sterile water for drinking, not only because of the danger of fungal infections but also due to the presence of other serious waterborne pathogens. High-risk patients should avoid showers, and these patients should use sterile sponges for bed baths instead. It was also suggested that shower facilities should be thoroughly cleaned to prevent fungal aerosols, and that education of patients and health-care workers to the fungal risk should be emphasized.[13]

More recent studies have investigated the risk of fungal infections related to the water supply in a cross section of hospitals in an attempt to standardize the conditions and develop guidelines. A survey in France was conducted in 10 university hospital centers.[16] The study was conducted under the same conditions as for bacteria, that is, water filtration through a cellulose acetate membrane cultured on agar. Departments with the highest patient risk were selected, including hematology, organ transplantation, and burn units. The investigators selected 98 sites and sampled water and water-related surfaces at each, namely, three one-liter water samples (the first flow, cold, and hot water) and two or three surface samples (inside the tap, pommel of the shower, and siphon). Water from taps equipped with sterilized filtration was sampled without further filtration. There was a significant difference ($p=0.039$) in the number of positive cultures between the three types of water sampled: (unsurprisingly) hot water (>50°C) was colonized less often than first flow or cold water. Only 4% of the hot-water samples had positive cultures, compared to the 52% of the cold-water samples. Colonization was slight except in two hospitals with generalized contamination of the water pipes (one with *Exophiala* spp. and the other with *Fusarium* spp.). Cold water was more colonized than hot water, but 79% of the samples yielded fewer than 5 cfu/L. Dematiaceous hyphomycetes were isolated; *Aspergillus* spp. were rare. The number of cfu in surface samples (i.e., biofilms) was higher (mean = 15 cfu per sample) but surfaces were positive less often than water (13% compared with 43% of all water samples). Sampling from siphons was productive more often than from taps (23%), but the molds isolated differed from those in the related water. Relations to bacterial flora and *Pseudomonas aeruginosa* were also studied, together with the effects of chemical treatment. The authors concluded that the absence of knowledge about the threshold of contamination at which there is a risk of nosocomial invasive fungal infections makes it difficult to formulate and impose routine-monitoring guidelines. Mycological surveys of water are required during hospital renovation, plumbing work, pipe maintenance, and when air samples are negative during nosocomial infection investigations. Finally, the planning of new hospitals should include designs to take into account recent information on how to combat fungal contamination.

12.4.3 In Biofilms

Biofilms are important habitats for fungi in drinking water. Their development is influenced by many factors including temperature, nutrient concentration, pipe material, and water flow rate. However, exactly how such factors affect biofilm development and specifically the role of fungi in biofilms is poorly known.

Fungi have the ability to grow attached to a substrate, forming part of microbial biofilms on pipe surfaces, debris, or sediments. They are likely to become established where there are cracks, pitting, or dead ends.[17]. Several studies have detected fungi in biofilms on water and wastewater pipe surfaces.[18–20] Fungi established in biofilms in water systems may be protected and are more resistant to water treatment, and fungi may also colonize filters in treatment plants and consequently affect the water treatment. The occurrence and distribution of filamentous fungi was investigated in a Norwegian study, and the results indicated that several species of filamentous fungi have the ability to establish in biofilms or sediments in water distribution systems.[7,21,22] The ecology of fungi in biofilms has been studied only to a small degree, and further research should aim to investigate the character of fungi in biofilms.

The presence of filamentous fungi in drinking water has become an area worthy of investigation with various studies now being published and reviewed.[6,8] These studies highlight the problems associated with fungi include blockage of water pipes, organoleptic deterioration, pathogenic fungi and mycotoxins. Fungal biofilm formation is a less developed field of study. The review by Siqueira and colleagues update the topic and introduce novel methods on fungal biofilm analysis, particularly from work based in Brazil.[8] In the United Kingdom, the recent review of fungi in drinking water published by the Department for Environment, Food and Rural Affairs acknowledges the importance of fungal biofilms in water distribution systems.[6]

12.4.4 IN HUMANS

12.4.4.1 ELISA and Lateral Flow Devices

Establishing the diagnosis of IA in patients with hematological malignancies, and transplant recipients is often difficult to accomplish. In most cases, the diagnosis is based on a combination of clinical, radiological, microbiological, and histopathological findings.[1] These might include (1) a positive culture result for a specimen obtained from a normally sterile and clinically or radiologically abnormal site or (2) histopathological or cytopathological examination showing hyphae consistent with *Aspergillus* in a biopsy specimen or aspirate. Unfortunately, those patients at highest risk of invasive infection are also at highest risk for complications from the invasive procedures needed to make a definitive diagnosis. Therefore, in many clinical settings the diagnosis is presumptive, and based upon chest CT findings and clinical signs in a patient with recognized risk factors. Several non-culture-based methods of diagnosis have been the subject of extensive evaluation in recent years. These include (1) tests for detection of *Aspergillus* antigens in blood and other body fluids by ELISA and lateral flow devices and (2) molecular methods to detect circulating *Aspergillus* DNA.

Early reports on non-culture-based diagnosis using human immune serum to detect circulating *Aspergillus* antigens showed that immunoelectrophoresis and ELISA formats were suitable platforms, but the nature of the antigen(s) was unknown.[23–26] Over the past 25 years, the focus has turned to hybridoma technology and the use of monoclonal antibodies (MAbs) to detect signature molecules of infection.[27] The detection of one such signature molecule, galactomannan (and associated galactomannoprotein molecules), forms the basis of the commercial *Aspergillus* Platelia enzyme immunoassay, an assay that has found widespread use in IA diagnosis (see previous text). Nevertheless, concerns surrounding its accuracy and sensitivity mean that alternative strategies to diagnosis have been sought including detection of the fungal cell wall component β-1-3-D-glucan and the use of PCR. The poor specificity of *panfungal* β-1-3-D-glucan tests and current lack of standardization of PCR assays have led to the recent development of next-generation MAb-based assays that detect surrogate markers of infection and that have been incorporated into *point-of-care* diagnostic devices, for example, the *Aspergillus* lateral flow device (LFD).[27] In a number of validation studies, if used as a screening test (one positive serum required for test positivity), or to rule out invasive fungal disease, the *Aspergillus* LFD has shown a comparable diagnostic performance to the *Aspergillus* galactomannan ELISA.[28,29] In terms of practicability, the *Aspergillus* LFD has demonstrated to be a quick (15 min) and easy-to-use test for single-patient detection of *Aspergillus* antigens.

12.4.4.2 Molecular Platforms

Real-time (RT) PCR clearly represents an advance compared to the detection of *Aspergillus* by culture, which showed great lack of sensitivity and specificity and needs more time to achieve the diagnosis.[30] A study performed by Badiee and colleagues in 194 hematological patients with proven and probable IA verified values for the molecular methodology–specific *Aspergillus* spp. of 66%, 96%, 63%, and 97% for sensitivity, specificity, positive predictive value (PPV), and negative predictive value (NPV), respectively.[31] The *Aspergillus* galactomannan ELISA showed higher PPV for bronchoalveolar lavage samples obtained from lung transplant recipients, while pan-*Aspergillus* PCR showed higher NPV.[32] Sensitivity and specificity of *A. fumigatus*-specific PCR were 85% and 96%, respectively, and were comparable to diagnosis based on galactomannan detection. When comparing galactomannan and RT-PCR, PCR positivity preceded galactomannan by 2.8 ± 4.1 days.[33] Several studies suggest that the combination of RT-PCR and galactomannan may provide improved diagnosis of IA. In other studies, successful antifungal treatment was obtained following PCR negative results for 14 days; in patients with unsuccessful treatment, the PCR diagnostic results were positive until death. It is difficult to distinguish the asymptomatic carrier from true IA cases in most clinical samples, and this fact represents a major limitation for molecular approaches.[34] *Aspergillus* is ubiquitous in nature, and it is noteworthy that 25% of the healthy population has a positive PCR for *Aspergillus* in respiratory samples.[35] RT-PCR may decrease the number of false positive results compared to conventional PCR by allowing the definition of cutoff values.

A study performed by Schabereiter-Gurtner and colleagues compared the results of culture and histology with molecular diagnosis, which revealed that 20% of the samples showed no visible growth when RT-PCR was positive.[36] Moreover, RT-PCR has recently shown 100% sensitive and specific with a PPV of 100% and a NPV of 97% for the molecular diagnosis of IA.[37] The lower detection limit was 10 conidia per mL of blood. Molecular assays are useful to confirm/exclude mold disease in high-risk patients.[38] Numerous molecular in-house approaches have been described employing distinct primers located at genomic and/or mitochondrial sites and being genus and species specific. The results can be different depending on the tested primers, DNA extraction methods, and PCR conditions.[39]

Commercially available *Aspergillus* PCR assays available provide standardized methodology with quality controlled reagents. Myconostica MycAssay PCR (Lab21, Cambridge, U.K.) had a sensitivity of 65% and a specificity of 95% when testing serum, and the performance has been shown to be similar to the galactomannan ELISA.[40] The MycAssay *Aspergillus* PCR appears to be a sensitive and specific molecular test for the diagnosis of IA but further validation work is required.

12.4.4.3 Diagnosis of Invasive Aspergillosis in Practice

The diagnosis of aspergillosis is difficult but several improvements have been observed during the last decade. By exploring the full potential of current diagnostic strategies, early diagnosis of IA is achievable. Possible IA is usually diagnosed based on the presence of distinct host and clinical factors allowing empirical antifungal therapy to be initiated.[1] Nevertheless, confirmation of IA should always be performed and there is no doubt that a HR-CT scan can suggest early signs of IA. It is important that clinicians understand the potential, and limitations of each method employed for diagnosis. Molecular diagnosis, particularly strategies employing RT-PCR targeting a broad range of fungi, and the detection of biomarkers such as β-1-3-D-glucan are methods associated with a high NPV and may exclude patients with IA.

12.5 CONCLUSIONS

Inhalation of *Aspergillus* spores is the usual mode of infection in humans. The etiological agents of aspergillosis are ubiquitous in the environment, and the likelihood that infection will occur following inhalation or implantation of spores largely depends on host factors. The incubation period is unknown. Less frequently, infection follows the traumatic implantation of spores as in corneal infection, or direct inoculation from contaminated dressings. Inhalation of contaminated water aerosols during patient showering has been suggested as an additional potential source of infection. In addition, contaminated food items have been suggested as a source of *Aspergillus*. IA has emerged as a major problem in several groups of immunocompromised patients.

It has long been recognized that few patients with hematological malignancies and IA will survive unless their neutrophil count recovers. Earlier diagnosis and optimal treatment have also been associated with a better prognosis. Risk factors associated with mortality in IA include progression of an underlying malignancy, receipt of allogeneic HSCT, preexisting respiratory function impairment, renal impairment, diffuse lung involvement or disseminated infection, and prior corticosteroid treatment.

The most important extrinsic risk factor for IA is the presence of *Aspergillus* spores in the hospital environment. Health care–associated outbreaks of the disease have become a well-recognized complication of construction or renovation work in or near units in which high-risk patients are housed. These reports have served to highlight the fact that the hospital environment is often contaminated with *Aspergillus* spores and have contributed to the widespread perception that most cases of aspergillosis in immunocompromised persons are hospital acquired.

However, despite the fact that most hospital outbreaks of aspergillosis go unpublished, it is clear that these events in general are uncommon. In contrast, numerous sporadic cases of IA are acquired outside the hospital setting.[41,42] It is noteworthy that the influence of the outdoor filamentous fungal flora on the occurrence and impact of community-acquired IA (or non-hospital-acquired IA) has recently attracted interest.[43,44]

Establishing the diagnosis of IA in patients with hematological malignancies and transplant recipients is often difficult to accomplish. In most cases, the diagnosis is based on a combination of clinical, radiological, microbiological, and histopathological findings. These might include a positive culture result for a specimen obtained from a normally sterile and clinically or radiologically abnormal site or histopathological or cytopathological examination showing hyphae consistent with *Aspergillus* in a biopsy specimen or aspirate. Unfortunately, those patients at highest risk of invasive infection are also at highest risk for complications from the invasive procedures needed to make a definitive diagnosis. Therefore, in many clinical settings, the diagnosis is presumptive, and based upon chest CT findings and clinical signs in a patient with recognized risk factors. Several non-culture-based methods of diagnosis have been the subject of extensive evaluation in recent years. These include tests such as ELISA and lateral flow devices for detection of *Aspergillus* antigens in blood and respiratory samples, as well as molecular methods to detect circulating *Aspergillus* DNA.

REFERENCES

1. Richardson, M.D., Warnock, D.W. 2012. *Fungal Infection: Diagnosis and Management*, 4th edn. Wiley-Blackwell, Chichester, U.K.
2. Ansorg, R., van den Boom, R., Rath, P.-M. 1997. Detection of *Aspergillus*-galactomannan-antigen in foods and antibiotics. *Mycoses* 40: 353–357.
3. Gangneux, J.-P. et al. 2002. Transient *Aspergillus* antigenaemia: Think of milk. *Lancet* 359: 1251.
4. Murashige, N. et al. 2004. False-positive results of *Aspergillus* enzyme-linked immunosorbent assays for a patient with gastrointestinal graft-versus-host disease taking a nutrient containing soybean protein. *Clin Infect Dis* 40: 333–334.
5. Gangneux, J.P. et al. 2004. Experimental assessment of disinfection procedures for eradication of *Aspergillus fumigatus* in food. *Blood* 104: 2000–2002.
6. DEFRA. 2011. A review of fungi in drinking water and the implications for human health. Reference WD0906. Department for Environment, Food & Rural Affairs, Nobel House, U.K.
7. Hageskal, G., Lima, N., Skaar, I. 2009. The study of fungi in drinking water. *Mycol Res* 113: 165–172.
8. Siqueira, V.M., Oliveira, H.M.B., Santos, C., Paterson, R.R.M., Gusmão, N.B., Lima, N. 2011. Filamentous fungi in drinking water, particularly in relation to biofilm formation. *Int J Environ Res Public Health* 8: 456–469.
9. Arroyo, M.G. et al. 2012. Water from private domestic wells: High occurrence of *Aspergillus fumigatus*. Abstracts of the *18th International Society for Human and Animal Mycology Congress*, Berlin, Germany, 744p.
10. Loudon, K.W. et al. 1996. Kitchens as a source of *Aspergillus niger* infection. *J Hosp Infect* 32: 191–198.

11. Anaissie, E.J. et al. 2002. Pathogenic *Aspergillus* species recovered from a hospital water system: A 3-year prospective study. *Clin Infect Dis* 34: 780–790.
12. Warris, A. et al. 2001. Recovery of filamentous fungi from water in a paediatric bone marrow transplantation unit. *J Hosp Infect* 47: 143–148.
13. Anaissie, E.J., Penzak, S.R., Dignani, M.C. 2002. The hospital water supply as a source of nosocomial infections. A plea for action. *Arch Intern Med* 162: 1483–1492.
14. Hayette, M.P. 2010. Filamentous fungi recovered from the water distribution system of a Belgian university hospital. *Med Mycol* 48: 969–974.
15. Vesper, S.J. 2007. Opportunistic Aspergillus pathogens measured in home and hospital tapwater by quantitative PCR(QPCR). *J Water Health* 5: 427–431.
16. Kauffman-Lacroix, C. et al. 2008. Prevention of fungal infections related to the water supply in French hospitals: Proposal for standardization of methods. *Presse Med* 37: 751–759.
17. Paterson, R.R.M., Lima, N. 2005. Fungal contamination of drinking water. In *Water Encyclopedia*; Lehr, J. et al. eds. John Wiley & Sons, New York, pp. 1–7.
18. Doggett, M.S. 2000. Characterization of fungal biofilms within a municipal water distribution system. *Appl Environ Microbiol* 66: 1249–1251.
19. Hendrickx, T.L. 2002. Influence of the nutrient balance on biofilm composition in a fixed film process. *Water Sci Technol* 46: 7–12.
20. Gonçalves, A.B. 2006. Survey and significance of filamentous fungi from tapwater. *Int J Hyg Environ Health* 209: 257–264.
21. Hageskal, G. et al. 2007. Occurrence of moulds in drinking water. *J Appl Microbiol* 102: 774–780.
22. Hageskal, G. 2011. Emerging pathogen *Aspergillus calidoustus* colonizes water distribution systems. *Med Mycol* 49: 588–593.
23. White, L.O., Richardson, M.D., Gibb, E., Newham, H.C., Warren, R.C. 1977. Circulating antigen of *Aspergillus fumigatus* in cortisone-treated mice challenged with conidia: Detection by counter immuno electrophoresis. *FEMS Lett* 2: 153–156.
24. Warren, R., White, L., Mohan, S. Richardson, M. 1979. The occurrence and treatment of false positive reactions in enzyme-linked immunosorbent assays (ELISA) for the presence of fungal antigens in clinical samples. *J Immunol Methods* 28(1–2): 177–186.
25. Richardson, M., White, L., Warren, R. 1979. Detection of circulating antigen of Aspergillus fumigatus in sera of mice and rabbits by enzyme-linked immunosorbent assay. *Mycopathologia* 67(2): 83–88.
26. Mohan, S., Warren, R., Richardson, M. 1980. Sero-diagnosis of invasive aspergillosis: Attempts to determine antigen and antibody relevance to infection. *Mycopathologia* 70(1): 37–41.
27. Thornton, C.R. 2010. Detection of invasive aspergillosis. *Adv Appl Microbiol* 70: 187–216.
28. White, P.L., Parr, C., Thornton, C., Barnes, R.A. 2013. Evaluation of real-time PCR, galactomannan enzyme-linked immunosorbent assay (ELISA), and a novel lateral-flow device for diagnosis of invasive aspergillosis. *J Clin Microbiol* 51: 1510–1516.
29. Held, J., Schmidt, T., Thornton, C.R., Kotter, E., Bertz, H. 2013. Comparison of a novel *Aspergillus* lateral-flow device and the Platelia galactomannan assay for the diagnosis of invasive aspergillosis following haematopoietic stem cell transplantation. *Infection* 41: 1163–1169.
30. Beirão, F., Araujo, R. 2013. State of the art diagnostic of mold diseases: A practical guide for clinicians. *Eur J Clin Microbiol Infect Dis* 32: 3–9.
31. Badiee, S., Franc, B.L., Webb, E.M., Chu, B., Hawkins, R.A., Shakiba, E. 2008. Molecular detection of invasive aspergillosis in hematologic malignancies. *Infection* 36: 580–584.
32. Luong, M.L. et al. 2010. Clinical utility and prognostic value of bronchoalveolar lavage galactomannan in patients with hematologic malignancies. *Diagn Microbiol Infect Dis* 68: 132–139.
33. Kami, M. et al. 2001. Use of real-time PCR on blood samples for diagnosis of invasive aspergillosis. *Clin Infect Dis* 33: 1504–1512.
34. Kawazu, M. et al. 2003. Rapid diagnosis of invasive pulmonary aspergillosis by quantitative polymerase chain reaction using bronchial lavage fluid. *Am J Hematol* 72: 27–30.
35. Del Palacio, A., Cuétara, M.S., Pontón, J. 2003. Invasive aspergillosis. *Rev Iberoam Micol* 20: 77–78.
36. Schabereiter-Gurtner, C. et al. 2007. Development of novel real-time PCR assays for detection and differentiation of eleven medically important *Aspergillus* and *Candida* species in clinical specimens. *J Clin Microbiol* 45: 906–914.
37. Deshpande, P. et al. 2011. Standardisation of fungal polymerase chain reaction for the early diagnosis of invasive fungal infection. *Indian J Med Microbiol* 29: 406–410.
38. Chen, S.C., Kontoyiannis, D.P. 2010. New molecular and surrogate biomarker-based tests in the diagnosis of bacterial and fungal infection in febrile neutropenic patients. *Curr Opin Infect Dis* 23: 567–577.

39. Boudewijns, M., Verweij, P.E., Melchers, W.J. 2006. Molecular diagnosis of invasive aspergillosis: The long and winding road. *Future Microbiol* 1: 283–293.
40. White, P.L. et al. 2011. Evaluation of analytical and preliminary clinical performance of Myconostica MycAssay *Aspergillus* when testing serum specimens for diagnosis of invasive aspergillosis. *J Clin Microbiol* 49: 2169–2174.
41. Hajjeh, R.A., Warnock, D.W. 2001. Counterpoint: Invasive aspergillosis and the environment—Rethinking our approach to prevention. *Clin Infect Dis* 33: 1549–1552.
42. Nicolle, M.C., Benet, T., Vanhems, P. 2011. Aspergillosis: Nosocomial or community-acquired? *Med Mycol* 49: 524–529.
43. Panackal, A.A. et al. 2010. Geoclimatic influences on invasive aspergillosis after hematopoietic stem cell transplantation. *Clin Infect Dis* 50: 1588–1597.
44. Brenier-Pinchart, M.-P. et al. 2011. Community-acquired aspergillosis and outdoor filamentous fungal spore load: A relationship? *Clin Microbiol Infect* 17: 1387–1390.

13 *Aspergillus* Mycotoxins

*János Varga, Sándor Kocsubé, Gyöngyi Szigeti,
Nikolett Baranyi, and Beáta Tóth*

CONTENTS

13.1 Introduction 165
13.2 Aflatoxins 167
    13.2.1 Producers 167
    13.2.2 Occurrence 167
    13.2.3 Biological Effects 168
    13.2.4 Biosynthesis 169
    13.2.5 Molecular Detection 170
13.3 Ochratoxins 170
    13.3.1 Producing Organisms 171
    13.3.2 Occurrence 171
    13.3.3 Biological Effects 171
    13.3.4 Biosynthesis 172
    13.3.5 Molecular Detection 172
13.4 Fumonisins 173
    13.4.1 Producing Organisms 173
    13.4.2 Occurrence 174
    13.4.3 Biological Effects 174
    13.4.4 Biosynthesis 174
    13.4.5 Molecular Detection 174
13.5 Patulin 175
    13.5.1 Producers 175
    13.5.2 Occurrence 175
    13.5.3 Biological Effects 175
    13.5.4 Biosynthesis 176
    13.5.5 Molecular Detection 176
13.6 Other Mycotoxins Produced by Aspergilli 176
13.7 Potential Effects of Climate Change on the Occurrence of *Aspergillus*
    Mycotoxins in Temperate/Continental Climates 177
13.8 Conclusions 178
Acknowledgments 179
References 179

13.1 INTRODUCTION

The name *Aspergillus* was first used for filamentous fungi bearing conidial heads and stalks by the Italian priest–mycologist Pietro Antonio Micheli (1679–1737) in *Nova Plantarum Genera* of 1729. He called these fungi *Aspergillus* because the spore-bearing structures characteristic of the genus reminded him a device called an aspergillum or holy water sprinkler used in Catholic liturgy. The fungi began to be recognized as active agents of decomposition processes and causes of animal and human diseases only in the middle of the ninetieth century. Today, *Aspergillus* is one of the most widely distributed and economically most important fungal genera on Earth [1]. Taxonomically, the *Aspergillus* genus is divided into 8 subgenera and >20 sections consisting of about 300–350 species [2,3].

Aspergillus species can be harmful and beneficial for mankind. The greatest economic benefits of aspergilli have been the production of industrial enzymes including amylases, glycosidases, pectinases, proteases, and organic acids such as citric acid, gluconic acid, or itaconic acid [1]. *Aspergillus oryzae*, *Aspergillus sojae*, and *Aspergillus tamarii* are called *koji molds* and have been used for centuries to produce sake and soy sauce in oriental food fermentation processes [1]. Aspergilli are also able to produce various secondary metabolites that have useful biotechnological and pharmacological properties for mankind, including lovastatin, a cholesterol-lowering drug produced by *Aspergillus terreus*; penicillin, an antibiotic produced by *Aspergillus nidulans*; and echinocandins, antifungal compounds produced by *Aspergillus rugulosus* [4]. Aspergilli are also used as hosts for heterologous expression of various proteins [5].

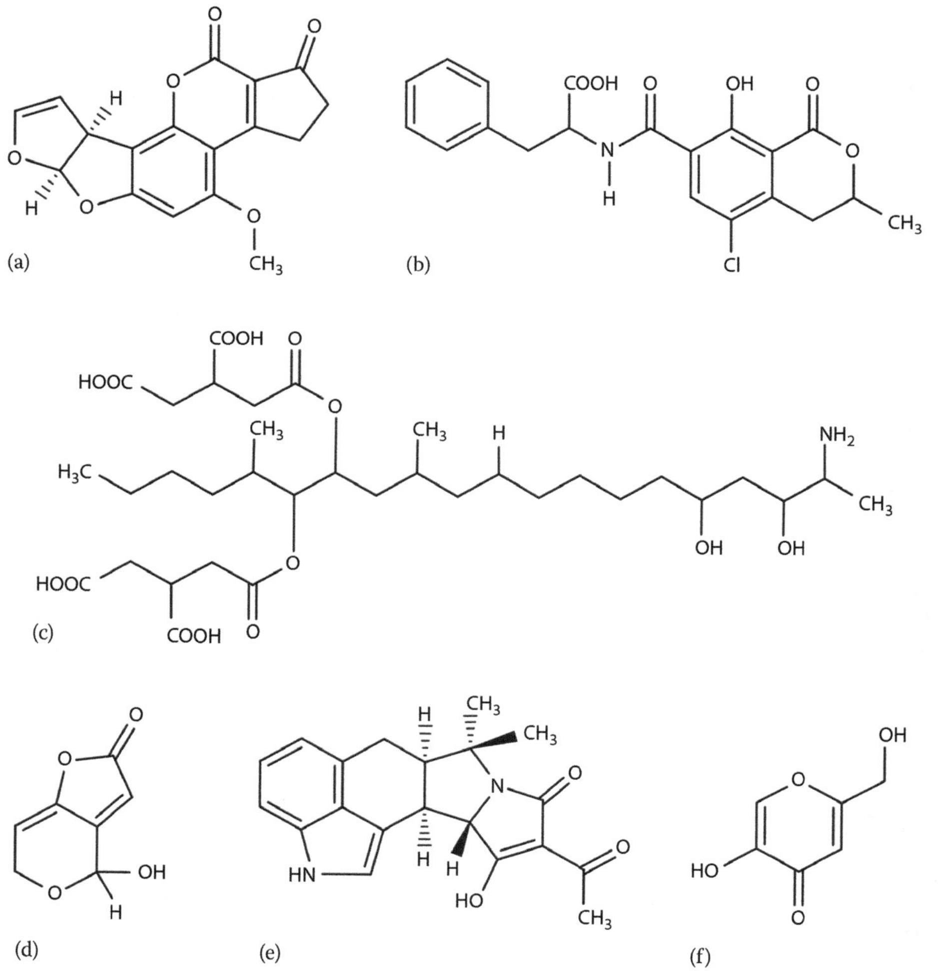

FIGURE 13.1 Chemical structures of some of the mycotoxins produced by Aspergilli: (a) aflatoxin B_1, (b) ochratoxin A, (c) fumonisin B_2, (d) patulin, (e) cyclopiazonic acid, and (f) kojic acid.

However, taxa of this genus can also be seriously deleterious, negatively impacting human and animal health causing diseases called aspergilloses. Although *Aspergillus fumigatus* is the most common human pathogen accounting for over 90% of invasive and noninvasive human aspergillosis, other species have also been identified to be able to cause infections [6].

Aspergillus species can also contaminate foods and feeds at different stages including pre- and postharvest, processing, and handling. The most important aspect of food and feed spoilage is the formation of mycotoxins, which may have harmful effects on human and animal health. Several *Aspergillus* mycotoxins have been identified as contaminants in foods and feeds, the economically most important of which are the aflatoxins, ochratoxin A (OTA), patulin, and fumonisins (Figure 13.1). Aflatoxins have also been identified in water [7]. Due to their importance and the extensive genome sequencing efforts carried out on aspergilli, the biosynthetic routes and the molecular genetic background of the biosynthesis of several *Aspergillus* mycotoxins have been studied in detail. In this review, we wish to give a general overview of aspergilli producing these mycotoxins, occurrence in foods and feeds, biosynthesis and molecular detection.

13.2 AFLATOXINS

Aflatoxins are the most thoroughly studied mycotoxins. They were discovered when the toxicity of animal feeds containing contaminated peanut meal led to the death of more than 100,000 turkeys from acute liver necrosis in the early 1960s [8,9]. *Aspergillus flavus* was identified as the aflatoxin-producing fungus, and the toxic agent was named after this fungus. Aflatoxins are a group of structurally related difuranocoumarins that were named as aflatoxins B_1, B_2, G_1, and G_2 based on their fluorescence under UV light (blue or green) and relative chromatographic mobility during thin-layer chromatography. Aflatoxin B_1 (Figure 13.1a) is the most potent natural carcinogen known [10] and is usually the major aflatoxin produced by toxigenic strains. Apart from the aforementioned, over a dozen of other structural analogues including aflatoxins P_1, Q_1, B_{2a}, and G_{2a} have been described, especially as mammalian biotransformation products of the major metabolites, while aflatoxin D_1 was detected in ammoniated corn and aflatoxin B_3 as a metabolite of *A. flavus* [11]. Aflatoxin M_1, a hydroxylated metabolite, is found primarily in animal tissues and fluids (milk and urine) as a metabolic product of aflatoxin B_1 [11].

13.2.1 PRODUCERS

Recent data indicate that aflatoxins are produced by 17 species assigned to sections *Flavi*, *Ochraceorosei*, and *Nidulantes* of the genus *Aspergillus* (Table 13.1) [11–14]. Several species claimed to produce aflatoxins have been synonymized with other aflatoxin producers, including *Aspergillus toxicarius* (= *A. parasiticus*), *A. flavus* var. *columnaris* (= *A. flavus*), or *Aspergillus zhaoqingensis* (= *Aspergillus nomius*). Although aflatoxin production was claimed for several other species and genera (and for bacteria), none of these observations were confirmed [11].

Sterigmatocystin (ST) is a penultimate precursor of aflatoxins and also a toxic and carcinogenic substance produced by many species belonging mainly to section *Versicolores* and *Nidulantes* of *Aspergillus*. ST production also occurs in the phylogenetically unrelated genera *Monocillium*, *Chaetomium*, *Humicola*, and *Bipolaris* [14].

13.2.2 OCCURRENCE

Aflatoxins are primarily produced by *A. flavus* and *A. parasiticus* on agricultural commodities including cereals (wheat, corn, rice), cotton, peanut, tree nuts, pepper, and spices [11]. Aflatoxins were detected and *A. flavus* was identified in water from a cold water storage tank for the first time by Paterson et al. [15]. *A. flavus* was also identified in the fungal flora of tap water from an Iranian university hospital [16] and Polish water distribution systems [17]. *A. nomius* is an important

TABLE 13.1
Aspergillus Species Able to Produce Aflatoxins and Other Mycotoxins

Species	Occurrence	Type of Aflatoxin Produced	Other Mycotoxins
Aspergillus* section *Flavi			
A. arachidicola	Argentina, Africa	Aflatoxins B_1, B_2, G_1, and G_2	Kojic acid, aspergillic acid
A. bombycis	Japan, Indonesia	Aflatoxins B_1, B_2, G_1, and G_2	Kojic acid, aspergillic acid
A. flavus	Worldwide	Aflatoxins B_1 and B_2	Cyclopiazonic acid, kojic acid, aspergillic acid
A. minisclerotigenes	Argentina, United States, Australia, Nigeria	Aflatoxins B_1, B_2, G_1, and G_2	Cyclopiazonic acid, kojic acid, aspergillic acid
A. nomius	United States, Japan, Thailand, India, Brazil	Aflatoxins B_1, B_2, G_1, and G_2	Kojic acid, aspergillic acid, tenuazonic acid
A. novoparasiticus	Colombia, Brazil	Aflatoxins B_1, B_2 and G_1, G_2	Kojic acid, aspergillic acid
A. parasiticus	United States, Japan, Australia, India, South America, Uganda	Aflatoxins B_1, B_2 G_1, and G_2	Kojic acid, aspergillic acid
A. parvisclerotigenus	Africa	Aflatoxins B_1, B_2, G_1, and G_2	Cyclopiazonic acid, kojic acid
A. pseudocaelatus	Argentina	Aflatoxins B_1, B_2, G_1, and G_2	Cyclopiazonic acid, kojic acid
A. pseudonomius	United States	Aflatoxin B_1	Kojic acid
A. pseudotamarii	Japan, Argentina, India	Aflatoxins B_1 and B_2	Cyclopiazonic acid, kojic acid
A. togoensis	Africa	Aflatoxin B_1	ST
Aspergillus* section *Ochraceorosei			
A. ochraceoroseus	Ivory Coast	Aflatoxins B_1 and B_2	ST
A. rambellii	Ivory Coast	Aflatoxins B_1 and B_2	ST
Aspergillus* section *Nidulantes			
Aspergillus astellatus (= Emericella astellata)	Ecuador	Aflatoxin B_1	ST, terrein
Aspergillus olivicola (= Emericella olivicola)	Italy	Aflatoxin B_1	ST, terrein
Aspergillus venezuelensis (= Emericella venezuelensis)	Venezuela	Aflatoxin B_1	ST, terrein

Sources: Varga, J. et al., *World Mycotoxin J.*, 2, 263, 2009; Varga, J. et al., *Stud. Mycol.*, 69, 57, 2011; Goncalves, S.S. et al., *Med. Mycol.*, 50, 152, 2012; Rank, C. et al., *Fungal Biol.*, 115, 406, 2011.

producer of aflatoxins in Brazil nuts [18]. The other aflatoxin-producing species cannot be regarded as potential health hazards because they produce only small amounts of aflatoxins or are encountered rarely in food products (e.g., aflatoxin-producing *Aspergillus ochraceoroseus*, *Aspergillus venezuelensis*, and *A. astellatus* isolates). ST producers are also frequently identified in various substrates including food, feed, and indoor air [19,20].

13.2.3 Biological Effects

Aflatoxins exhibit hepatocarcinogenic and hepatotoxic properties and are usually referred to as one of the most potent naturally occurring carcinogens [21,22]. The diseases caused by aflatoxin consumption are called aflatoxicoses. Acute aflatoxicosis is caused when moderate to high levels of aflatoxins are consumed. These episodes of disease symptoms may include hemorrhage, acute liver damage, edema, alteration in digestion, absorption, and/or metabolism of nutrients and may result in death. While death is uncommon in humans, several deaths were attributed to aflatoxicosis in tropical and subtropical regions of the world, mainly due to the consumption of contaminated corn [23]. Chronic aflatoxicosis results from prolonged ingestion of low to moderate levels of aflatoxins, although the effects are usually subclinical and difficult to recognize. Some of the common symptoms are impaired food conversion and slower rates of growth, with or without the production of an overt aflatoxin syndrome. Comprehensive studies have shown that aflatoxin is a risk factor for human hepatocellular carcinoma, especially in Asia and sub-Saharan Africa [24,25]. The liver is the primary target organ, with liver damage occurring when poultry, fish, rodents, and primates are fed with aflatoxin B_1 [26]. The International Agency for Research on Cancer (IARC) has classified aflatoxins as group 1 carcinogens [27]. Aflatoxin M_1, the metabolite of aflatoxin B_1 found in the milk of lactating mammals, was classified in group 1 as possibly carcinogenic to humans [27]. Over 100 countries restrict the content of aflatoxin in the food and feed supplies [28].

13.2.4 BIOSYNTHESIS

Molecular analysis of aflatoxin production in *A. flavus* and *A. parasiticus* led to the identification of an about 70 kb DNA cluster consisting two specific transcriptional regulators (*aflR* and *aflS*) and 28 coregulated downstream metabolic genes in the aflatoxin biosynthetic pathway [29–31]. The regulatory genes *aflR* and *aflS* are located adjacent to each other within the aflatoxin cluster and are involved in the regulation of aflatoxin or ST gene expression. These genes are divergently transcribed, but they have independent promoters. The precise role of AflS in aflatoxin biosynthesis remains unclear [30].

The biosynthesis of aflatoxin occurs through a series of highly organized oxidation–reduction reactions. Aflatoxin biosynthesis starts with conversion of hexanoyl-CoA and 7 to a condensed polyketide noranthrone by the products of two fatty acid synthase genes (*fas-1* = *aflB* and *fas-2* = *aflA*) and a polyketide synthase gene (*pksA* = *aflC*) [32]. Another gene *hypC* [33] encodes for an anthrone oxidase that is involved in the catalytic conversion of noranthrone to norsolorinic acid (NOR). NOR is the first stable metabolite that can be isolated. *Nor-1* (= *aflD*, a dehydrogenase), *norA* (= *aflE*, a reductase), and *norB* (= *aflF*, a dehydrogenase) are involved in the reduction from NOR to averantin (AVN) [31].

The next catalytic step is the conversation of AVN to hydroxyaverantin (HAVN) by a cytochrome P450 monooxygenase enzyme encoded by the gene *avnA* (= *aflG*). The alcohol dehydrogenase encoded by *adhA* (= *aflH*) [34] can catalyze the conversion from HAVN to averufin (AVR). AVR is converted to HVN by averufin monooxygenase encoded by *avfA* (= *aflI*) and to versiconal hemiacetal acetate (VHA) by a VHA synthase (*estA* = *aflJ*) [35]. The conversion of VHA to versicolorin B (VERB) is the key step in aflatoxin formation since it closes the bisfuran ring of aflatoxin. Several studies demonstrated the function of the VERB synthase gene (*vbs* = *aflK*) for the conversion of VHA to VERB in *A. parasiticus* [31]. The formation of versicolorin A (VERA) from VERB is a branch point in aflatoxin biosynthetic pathway where the biosynthesis of AFB_1 and AFG_1 is separated from that of AFB_2 and AFG_2. Similar to AFB_2 and AFG_2, VERB contains a tetrahydrobisfuran ring in its structure; and, like AFB_1 and AFG_1, VERA contains a dihydrobisfuran ring. The branching step between AFB_1 or AFG_1 and AFB_2 or AFG_2 is the desaturation reaction from VERB to VERA [31]. The *verB* (= *aflL*) gene encodes a cytochrome P450 monooxygenase/desaturase that is presumed to be involved in the conversion of VERB to VERA. The gene responsible for the conversion directly from VERB to demethyldihydrosterigmatocystin (DMDHST) and then to AFB_2 and AFG_2 has not been identified yet. It is possible that *verB* is involved in the conversion of both VERB to VERA and VERB to DMDHST [31]. The dihydrobisfuran ring in VERA and the tetrahydrobisfuran ring in VERB are maintained through the next steps. The intermediates after these versicolorins are demethylsterigmatocystin (DMST) for VERA and dihydrodemethylsterigmatocystin

(DHDMST) for VERB. *Ver-1* (= *aflM*, ketoreductase) and *verA* (= *aflN*, a cytochrome P-450 monooxygenase) are required for the conversion of VERA to DMST [31,36]. An O-methyltransferase I encoded by *omtB* (= *aflO*) catalyzes the transfer of the methyl groups from S-adenosylmethionine to the hydroxyl groups of DMST and DHDMST in order to produce ST and dihydrosterigmatocystine [36]. Another O-methyltransferase enzyme is also involved in aflatoxin biosynthesis [37]. The rule of O-methyltransferase II is the conversion of ST to O-methylsterigmatocystin (OMST) and DMST to dihydro-O-methylsterigmatocystin [36]. It is encoded by the *omtA* (= *aflP*) gene [31]. The absence of an omtA homolog in *A. nidulans* is responsible for ST as the final product of the biosynthetic pathway in this fungus.

The final step in the formation of aflatoxins is the conversion of OMST or DHOMST to aflatoxins B_1, B_2, G_1, and G_2, requiring the presence of a NADPH-dependent monooxygenase, *ordA* (= *aflQ*) [31]. The formation of the G-type aflatoxins involves an additional step, possibly involving the enzyme encoded by *ordB* (= *aflX*) and *cypA* (= *aflU*) encoding a cytochrome P450 monooxygenase [38]. More recently, the *nadA* gene [39,40] has also been found to play a role in AFG_1/AFG_2 formation [41]. This gene was previously suggested to be part of the *sugar cluster* adjacent to the aflatoxin biosynthesis gene cluster [31]. Only the G-type aflatoxin producer *A. parasiticus* has intact *nadA* and *norB* genes [41]. Another gene with nuclear function, *aflT*, encodes an ABC transporter protein that may be necessary for aflatoxin efflux from the cells [31].

13.2.5 MOLECULAR DETECTION

Several regions of the genome have been used for the identification of aflatoxin-producing species in various products. The most widely used DNA target regions for discriminating *Aspergillus* species are those of the rDNA gene cluster (mainly the internal transcribed spacer [ITS] regions 1 and 2 [ITS1 and ITS2]) [42,43] or protein-coding genes such as calmodulin or β-tubulin [12]. However, these regions are unrelated to the structural genes involved in aflatoxin biosynthesis; consequently, they can be used for species identification but do not confirm aflatoxin production. Several PCR-based approaches have been used for detection of aflatoxin-producing isolates targeting three to four genes involved in the biosynthesis of aflatoxins in multiplex PCR assays [44]. The targets used include various combinations of structural and regulatory genes [45–47]. Real-time PCR approaches have also been developed, which provided a tool for accurate and sensitive quantification of target DNA of aflatoxin producers using either SYBR Green or TaqMan probes [48,49]. However, in all these studies, the presence of the genes involved in aflatoxin biosynthesis was not in close correlation with aflatoxin-producing abilities of the isolates. Recently, multiplex reverse transcription polymerase chain reaction protocol has also been developed to discriminate aflatoxin-producing from aflatoxin-nonproducing strains of *A. flavus* [50]. Five genes of the aflatoxin gene cluster including two regulatory (*aflR* and *aflS*) and three structural genes (*aflD*, *aflO*, and *aflQ*, the synonyms of which are *nor-1*, *omtB*, and *ordA*) were targeted with specific primers to highlight their expression in mycelia cultivated under inducing conditions for aflatoxin production. However, most of these genes are also involved in the biosynthesis of ST, so false positives could be detected [44,51]. It was suggested that the application of primers targeting genes not involved in ST biosynthesis including *aflP* and *aflQ* would be a good choice to detect aflatoxin producers in various products [51]. A microarray has also been developed to detect the expression of genes taking part in the biosynthesis of various mycotoxins including aflatoxins [52] and used successfully to examine the effect of various environmental conditions on aflatoxin formation [53,54].

13.3 OCHRATOXINS

OTA (Figure 13.1b) was discovered in 1965 in the ferment broth of an *Aspergillus ochraceus* isolate (later identified as *A. westerdijkiae*) during the systematic examination of the metabolites of molds [55]. Ochratoxins are cyclic pentaketide dihydroisocoumarin derivatives linked to an

L-β-phenylalanine by an amide bond. The most toxic member of the group is OTA, which is a chlorinated pentaketide. According to the IUPAC nomenclature, the chemical name of this economically important mycotoxin is L-phenylalanine-N-[(5-chloro-3,4-dihydro-8-hydroxy-3-methyl-1-oxo-1H-2-benzopyrane-7-yl)carbonyl]-(R)-isocoumarin. Several OTA-derived metabolites have also been identified. Ochratoxin B is a dechlorinated analogue of OTA, while ochratoxin C is a chlorinated ethyl ester derivative. Ochratoxin α is the isocoumarin core of OTA, while ochratoxin β is a dechlorinated analogue of ochratoxin α. Other derivatives also exist such as 4-hydroxyochratoxin A, 10-hydroxyochratoxin A, OTA methyl ester, OTB ethyl and methyl esters, and several amino acid analogues [56,57].

13.3.1 Producing Organisms

Ochratoxins are produced mostly by *Penicillium* species in colder temperate climates and by a number of *Aspergillus* species in warmer and tropical regions. *Aspergillus* isolates usually produce OTA and ochratoxin B, while *Penicillium* species produce only OTA. In *Aspergillus* section *Circumdati*, *A. cretensis, A. flocculosus, A. pseudoelegans, A. roseoglobulosus, A. westerdijkiae, A. sulphureus*, and *Neopetromyces muricatus* produce consistently high amounts of OTA [58]. Isolates belonging to the *A. ochraceus* and *A. sclerotiorum* species produce either large or small amounts of OTA, while three further species produce OTA in trace amounts: *A. melleus, A. ostianus,* and *A. persii* [58]. The most important species for potential OTA production in coffee, rice, beverages, and other foodstuffs are *A. ochraceus, A. westerdijkiae,* and *A. steynii*. Recently, a new ochratoxin-producing species, *A. affinis*, has been described in section *Circumdati* [59]. In *Aspergillus* section *Flavi*, *A. albertensis* and *A. alliaceus* produce OTA in synthetic media [60], and in *Aspergillus* section *Nigri*, *A. carbonarius, A. niger, A. welwitschiae,* and *A. sclerotioniger* are proven to be potential OTA-producing organisms [61,62]. While most *A. carbonarius* isolates are able to produce large amounts of OTA, different surveys found that only 5%–15% of *A. niger* and *A. welwitschiae* isolates produce OTA in smaller quantities [62]. Among penicillia, *Penicillium verrucosum* and *Penicillium nordicum* are able to produce this mycotoxin [63,64].

13.3.2 Occurrence

Ochratoxins have been detected in several agricultural products including cereals, coffee beans, cocoa, spices, soybeans, peanut, rice, corn, figs, and grapes [65]. Recent studies clarified that *A. westerdijkiae* is important in OTA contamination of coffee beans in Thailand [66], while *A. alliaceus* is mostly responsible for OTA contamination of figs in California [67]. Black aspergilli, mainly *A. carbonarius* and *A. niger*, are responsible for OTA contamination of grapes and grape-derived products [68]. Regarding penicillia, in stored cereals, the main OTA producer is *P. verrucosum*, while in the case of meat products, *P. nordicum* is the causative agent of OTA contamination [64]. Ochratoxin-producing organisms are usually considered as causing postharvest spoilage. However, *P. verrucosum* is often isolated from surface-sterilized cereals examined at harvest indicating that preharvest infestation by OTA-producing fungi is at least partly responsible for OTA contamination [69].

13.3.3 Biological Effects

Several nephropathies affecting animals and humans have been attributed to OTA. It is the etiologic agent of Danish porcine nephropathy and renal disorders observed in other animals [70]. In humans, OTA is often cited as the possible causative agent of Balkan endemic nephropathy [71], although recently, aristolochic acid has been suggested to play a major role in the etiology of this disease [72]. OTA also proved to exhibit immunosuppressive, teratogenic, hepatotoxic, and carcinogenic properties [73]. Several studies suggested that OTA may cause genotoxic effects, but the role of this toxin in genotoxicity is still unclear due to the inconsistent results obtained in different mammalian tests [74]. However, evidence of OTA-mediated DNA-adduct formation has been provided by numerous authors [74]. Recently, it has been suggested that OTA has a role in chronic karyomegalic interstitial nephropathy and chronic interstitial nephropathy in Tunisia [75], urothelial tumors (end-stage renal disease) in Egypt [76], and testicular cancer [77]. The IARC classified OTA as a possible human carcinogen (group 2B) [27].

13.3.4 Biosynthesis

Unlike other economically important mycotoxins such as fumonisins and aflatoxins, the complete OTA biosynthetic pathway has not been elucidated in *Aspergillus* or *Penicillium*. Early studies on OTA biosynthesis showed that phenylalanine was incorporated into OTA, whereas ochratoxin α was constructed from five acetate units with one carbon at C-8 obtained from methionine [78]. Wei et al. [79] demonstrated incorporation of ^{36}Cl into OTA, possibly due to the action of a chloroperoxidase enzyme [80]. The isocoumarin group is a pentaketide skeleton formed from acetate and malonate via a polyketide pathway. Phenylalanine derived from the shikimic acid pathway is linked through the additional carboxyl group. Because of the structural similarity of mellein and the heterocyclic core of OTA, Huff and Hamilton [81] suggested that mellein is a precursor of OTA. However, precursor feeding experiments did not support an intermediary role for mellein in OTA biosynthesis [82].

Regarding the genes taking part on OTA biosynthesis, O'Callaghan et al. [83] successfully cloned and characterized a putative polyketide synthase gene (*pks*) from *A. ochraceus*. The involvement of this PKS in the biosynthesis of ochratoxin was demonstrated by the disruption of the corresponding gene. Later, two putative p450-type monooxygenase genes (p450-H11 and p450-B03) putatively taking part in OTA biosynthesis were also identified by this group [84]. A *pks aolc35-12* has been identified in *A. westerdijkiae* [85], and the authors proved its involvement in the biosynthesis of OTA by the disruption of the gene. In a recent study, a nonribosomal peptide synthetase (AcOTAnrps) taking part in the biosynthesis of OTA in *A. carbonarius* was identified [86]. The disruption of the gene stopped the production of OTA, ochratoxin B, and ochratoxin α, but the amount of ochratoxin β increased, indicating that the formation of ochratoxin B from phenylalanine and ochratoxin β occurs earlier than the chlorination step.

Regarding penicillia, several genes responsible for the biosynthesis of OTA in *P. nordicum* have been characterized. These genes include a nonribosomal peptide synthase (*otanps*), a pks gene (*otapks*), a putative transport protein-coding gene (*otatra*), and the gene of a putative chlorinating enzyme (*otachl*) [86]. Interestingly, the pks gene found in *P. nordicum* is different from that present in *P. verrucosum* [87].

13.3.5 Molecular Detection

Several genomic regions have been targeted to detect potential OTA producers in various products. The variable ITS region is applicable for developing species-specific primers for potentially OTA-producing species such as *A. niger* [88,89], *A. ochraceus* [89,90], *A. steynii*, or *A. westerdijkiae* [90]. Protein-coding genes, including β-tubulin and calmodulin, have also been used to design species-specific primer pairs for the detection of OTA producers [91–93]. AFLP-based sequence-characterized amplified region primers were applied successfully to detect the species *A. ochraceus* and *A. carbonarius*, but the discrimination between the producers and the atoxigenic isolates was not possible [94,95]. Application of RAPD-based PCR primer pairs were also tested; primer pairs based on a characteristic fragment obtained by the use of the OPX-07 decamer were designed and used successfully for detecting *A. niger* or *A. carbonarius* isolates [96,97]. However, none of these approaches allow the distinction between the OTA-producing strains and the nonproducers.

Species-specific PCR methods were also developed based on mycotoxin biosynthetic genes. Dao et al. [98] developed two primer sets based on the conserved ketosynthase (KS) region of a *pks* identified in *A. westerdijkiae*. One of the primer sets could be used to detect both OTA- (*A. carbonarius*, *A. melleus*, *A. westerdijkiae*, and *A. sulphureus*) and citrinin-producing fungi (*Monascus ruber* and *Penicillium citrinum*), while the other one could be used to detect only *A. ochraceus* [98]. Since several other fungi are able to produce citrinin (see Section 13.6), which has a similar polyketide backbone to OTA, the appearance of false positives during the application of PKS-based primers for the detection of OTA producers should be considered [51].

A real-time PCR method was also developed based on the sequence of a *pks* gene (*An15g07920*) identified in the total genomic sequence of *A. niger* [99] that showed strong similarity to the PKS involved in OTA biosynthesis in *A. ochraceus*. PCR products were only obtained with the potential ochratoxigenic *A. niger* and *A. welwitschiae* species. A specific primer pair has also been designed for an acyltransferase domain of a *pks* gene (*AC12RL3*) for the detection of *A. carbonarius* [100]. Using a quantitative real-time PCR method on grape samples, the authors found good correlation between the OTA content and the *A. carbonarius* DNA content of berries. Gallo et al. [101] designed a primer pair based on KS domain in *A. carbonarius* (*ACpks*) that expression correlated with the OTA content of the cultures by using RT-PCR approach. With these primers, the authors were able to detect not only *A. carbonarius* but two closely related species *Aspergillus ibericus* and *A. sclerotioniger*. By using the same KS domain, a specific PCR-based detection was designed, which yielded products only in the case of *A. carbonarius* [102]. In conclusion, none of the methods applied so far were able to distinguish OTA-producing isolates from nonproducing ones.

13.4 FUMONISINS

Several studies have indicated that some of these compounds are produced by *Aspergillus* species (see Section 13.4.1). Fumonisins were purified and chemically characterized in 1988 in South Africa from cultures of *Fusarium moniliforme* strain MRC 826 (= *F. verticillioides*, teleomorph *Gibberella moniliformis*) on corn [103,104]. These nonaketide-derived mycotoxins can be classified into four groups based on their chemical structure: series A, B, C, and P. Recent surveys identified about 100 isomers of fumonisins [105,106]. The most abundant and toxicologically most important fumonisins belong to the B-series (FB). In fusaria, usually FB_1 is predominant accounting for 70%–80% of the total fumonisin content [107]. The B-series fumonisins consist of a 19–20 carbon backbone with an amine group at C-2, two methyl groups at C-12 and C-16, two tricarballylic ester groups attached to C-14 and C-15, and one to four hydroxyl residues. The A-series fumonisins (FA) are *N*-acetyl analogues of the B-series and have less biological activity. The P-series fumonisins (FP) have a 3-hydroxypyridinium group instead of the C-2 amine group, and in fumonisin C analogues (FC), the terminal methyl group is missing [108].

13.4.1 Producing Organisms

Besides *F. verticillioides*, several other species of the genus *Fusarium* were found to be able to produce fumonisins. *Fusarium oxysporum* produces only C-series fumonisins [109]. Interestingly, during the sequencing of the genome of *Aspergillus niger*, a region homologous to the *Fusarium* fumonisin gene cluster was discovered [110]. Later, production of fumonisins in *A. niger* was confirmed by Frisvad et al. [111]. According to recent findings, fumonisins are also produced by another black *Aspergillus* species, *A. welwitschiae* [112]. Black aspergilli produce mainly fumonisins B_2 and B_4, although other analogs have also been identified in smaller quantities [113]. According to Mogensen et al. [114], several *Tolypocladium* species are also able to produce fumonisins B_2 (Figure 13.1c) and B_4.

13.4.2 Occurrence

Both *Fusarium* and black *Aspergillus* species are widely distributed saprotrophic fungi. Regarding fumonisin-producing aspergilli, *A. niger* is an important opportunistic pathogen of grapes, causing bunch rot or berry rot and raisin mold [115]. It also infects corn, peanuts, onions, mangoes, apples, and dried meat products [116]. In contrast with fusaria, which produce large amounts of fumonisins in media containing plant extracts (barley, malt, oat, rice, potatoes, and carrots), aspergilli produce fumonisins in high quantities on substrates with low water activities [111]. Fumonisins produced by black aspergilli have been detected in Thai coffee beans [117], wine [118], raisins [119], onions [120], dates, and figs (Varga et al., unpublished data). Potential fumonisin-producing black aspergilli have also been detected in indoor air and water [17,121–123].

13.4.3 Biological Effects

Fumonisins are structurally similar to sphinganine and disrupt the biosynthesis of sphingolipids by inhibition of ceramide synthase enzyme [124]. Ingestion is associated with several fatal diseases in domestic animals, for example, equine leukoencephalomalacia and porcine pulmonary edema. Fumonisins can cause nephrotoxicity, hepatotoxicity, and hepatocarcinogenicity in laboratory animals [125]. They also inhibit folic acid transport and cause neural tube defect in mouse embryos [126]. Fumonisins are the possible etiologic agents of esophageal cancer in several countries such as China and South Africa [127,128]. Fumonisin B_1 is treated as possibly carcinogenic to humans (group 2B) by the IARC [27].

13.4.4 Biosynthesis

The 42 kb fumonisin gene cluster includes 17 coregulated genes located on chromosome 1 in *F. verticillioides* [129]. In the genome of *A. niger*, the genes involved in fumonisin production are also clustered together. However, the fumonisin cluster in *A. niger* contains fewer genes: *fum11*, *fum12* (= *fum2*), *fum16*, *fum17*, and *fum18* are missing in this cluster, and some of the genes are relocated in different regions of the genome [110]. The *fum12* gene is essential for FB_1/FB_3 synthesis in *F. verticillioides* [130]. Its absence from the *A. niger* fumonisin gene cluster is in agreement with the observation that this species produces predominantly fumonisins B_2 and B_4 [111]. Another missing gene is *fum11* that is similar to the mitochondrial membrane-bound tricarboxylate transporters and presumably transports tricarboxylic acids from the mitochondria to make them available for fumonisin biosynthesis [131]. The *fum16–18* genes were found to be nonessential for fumonisin production in *F. verticillioides* [130].

13.4.5 Molecular Detection

Accensi et al. [132] distinguished two types (N and T) of the black aspergilli using a PCR-RFLP method. The substance of the method is the PCR amplification of the ITS region followed by the digestion of the amplicon with the restriction endonuclease *Rsa*I. The fumonisin-producing species (*A. welwitschiae* and *A. niger*) belong to the N type. Another possibility to detect the fumonisin producers is the amplification of the genes involved in fumonisin biosynthesis. In our previous study, we analyzed the sequence data of the *fum1* and *fum8* genes. Although the phylogenetic analysis indicated that there is no strict correlation between the phylogenetic trees based on these sequences and those of the partial calmodulin gene, this method could be useful to detect potential fumonisin producers [133]. Species-specific primers have also been developed for *A. welwitschiae* and *A. niger*, which could be used for fast detection of these potential fumonisin-producing species (Kocsubé et al., unpublished data).

13.5 PATULIN

Patulin (4-hydroxy-4*H*-furo[3,2-*c*]pyran-2(6*H*)-one, Figure 13.1d) is a water-soluble lactone that is produced through the polyketide pathway by several fungi. The compound was first isolated by Birkinshaw et al. [134] in 1943 from *Penicillium* species as part of a screening effort to find new antibiotics of fungal origin. It was discovered independently in several laboratories, and it was named variously (e.g., clavacin, expansine, claviformin, clavatin, gigantic acid, or myosin C) [135]. Patulin was the first compound tested in a controlled clinical trial under the brand name tercinin for the treatment of the common cold [136]. However, the trials were unsuccessful, besides patulin proved toxic to humans and animals [137].

13.5.1 PRODUCERS

Patulin was identified from up to 30 fungal genera [135,138]. However, recent studies clarified that patulin is produced only by a limited number of species belonging to the genera *Penicillium*, *Aspergillus*, *Paecilomyces*, and *Byssochlamys*. Among aspergilli, only three members of section *Clavati* are able to produce patulin (*Aspergillus clavatus*, *A. giganteus*, and *A. longivesica*) [139]. Although *A. terreus* and *A. candidus* are frequently cited in the literature as patulin producers, none of the isolates of these species have been found to be able to produce this mycotoxin [140,141]. Among penicillia, a recent survey clarified that 13 species are able to produce patulin, including *Penicillium carneum*, *P. clavigerum*, *P. concentricum*, *P. coprobium*, *P. dipodomyicola*, *P. expansum*, *P. glandicola*, *P. gladioli*, *P. griseofulvum*, *P. marinum*, *P. paneum*, *P. sclerotigenum*, and *P. vulpinum* [142]. Regarding *Paecilomyces* and *Byssochlamys*, a recent study showed that only *Byssochlamys nivea* and some strains of *Paecilomyces saturatus* produce patulin [143].

13.5.2 OCCURRENCE

Patulin-producing penicillia have been isolated from a variety of fruits and vegetables [135,144]. Apple soft rot and blue mold rot is most commonly caused by *P. expansum*, which also causes blue rot in other fruits including grapes, cherries, plums, blueberries, oranges, strawberries, and melons [135,144]. Patulin was first detected in rotted apples infected by *P. expansum* in 1956 [145]. Besides fruits and vegetables, patulin has also been isolated from other sources including cheese and grain products such as barley and wheat malts, feed silages, cereal stubbles, bread, and related flour/dough products [135]. The economically most important patulin-producing *Aspergillus* species *A. clavatus* is frequently isolated from various cereals. The other patulin-producing aspergilli are rare species and cannot be considered as health hazards in foods and feeds. Patulin has not yet been detected in water.

13.5.3 BIOLOGICAL EFFECTS

Patulin provokes congestion and edema of pulmonary, hepatic, and gastrointestinal blood vessels and tissues. Regarding its acute toxicity, patulin mainly induces gastrointestinal disorders with ulceration, distension, and bleeding [146]. It was suggested to be partly responsible for several animal intoxications in Germany, France, Hungary, and Japan [147]. Patulin has a strong affinity for sulfhydryl groups, which explains its inhibition of many enzymes [148]. Regarding the carcinogenicity of patulin, subcutaneous injection of patulin produced local sarcomas in rats [146]. However, the studies on long-term toxicity of patulin showed an absence of tumors in rats [149]. According to the IARC, patulin is classified in group 3 as *not classifiable as to its carcinogenicity to humans* [27].

13.5.4 BIOSYNTHESIS

Studies based on isotope labeling clarified most of the patulin metabolic pathway. The biosynthesis of patulin is initiated by the condensation of an acetyl CoA and 3 units of malonyl CoA to form 6-methylsalicylic acid (6-MSA) by 6-MSA synthetase, making it a tetraketide [150]. 6-MSA is converted to *m*-cresol and to *m*-hydroxybenzyl alcohol by the activities of 6-MSA decarboxylase and *m*-cresol-2 hydroxylase, respectively. This step is followed by a hydroxylation reaction that leads to gentisaldehyde formation. Once gentisaldehyde has been formed, it is converted to isoepoxydon, phyllostine, neopatulin, E-ascladiol, and finally patulin. The conversion of isoepoxydon to phyllostine is accomplished by a NADP-dependent isoepoxydon dehydrogenase [151].

Several genes taking part in the biosynthesis of patulin have been identified. The first identified gene was the 6-methylsalicylic acid synthase (6MSAS) gene from *P. griseofulvum* [152]. Recently, a cluster of 15 genes involved in patulin biosynthesis has been identified in the *A. clavatus* genome [153]. All the genes were found to be located in a 40 kb genomic region. The genes encode the enzymes necessary for the biosynthesis of the toxin, a transcription factor and three genes coding for transporters [148]. Related clusters have also been identified in other species not producing patulin including *Penicillium chrysogenum*, *Talaromyces stipitatus*, and *A. terreus* [148]. Regarding *A. terreus*, it was hypothesized that the cluster possibly encodes for enzymes taking part in terreic acid biosynthesis [148].

13.5.5 MOLECULAR DETECTION

For the molecular detection of patulin-producing organisms, a PCR-based approach was developed using sequences of the isoepoxydon dehydrogenase (*idh*) gene as target [154]. The primers designed were used successfully to identify potential patulin-producing *Penicillium* and *Aspergillus* species [155–157]. However, unpredictable results occurred on some occasions, that is, the *idh* gene was also detected in patulin nonproducers including *Penicillium brevicompactum*, and the gene was not observed in some *P. expansum* isolates [154]. It was suggested that the false-positive results might have been caused by the mutagenic effect of metabolites present in the growth medium [158], while the negative results could have been caused by the presence of PCR inhibitors [159].

13.6 OTHER MYCOTOXINS PRODUCED BY ASPERGILLI

Aspergillus species produce a wide range of other mycotoxins that may contaminate our foods and can be harmful to humans. Among these, cyclopiazonic acid (Figure 13.1e) is an indole-tetramic acid mycotoxin, which was first discovered in 1968 as a metabolite of *Penicillium cyclopium* in groundnuts [160]. Several other *Penicillium* and *Aspergillus* species are able to produce this mycotoxin, including *P. camemberti*, *P. commune*, *P. dipodomyicola*, *P. griseofulvum*, *P. palitans* [142], *A. flavus*, *A. oryzae*, *A. minisclerotigenes*, *A. pseudotamarii* [161], and *A. lentulus* [162]. These fungi can contaminate various agricultural products including cheese, meat products, and various grains and seeds. Cyclopiazonic acid is a specific inhibitor of calcium-dependent ATPase in the sarcoplasmic reticulum, resulting in increased muscle contraction. It also causes degenerative changes and necrosis in the liver, spleen, pancreas, kidney, salivary glands, myocardium, and skeletal muscles [163] and was probably also involved together with aflatoxins in the turkey X disease in the United Kingdom [164]. It has been detected in Kodo millet seed in India, where an association with poisoning of humans was suggested [165]. Four of the genes involved in cyclopiazonic acid biosynthesis form a mini gene cluster closely linked to the aflatoxin biosynthesis gene cluster in this fungus [166].

Citrinin is a pentaketide-derived mycotoxin structurally related to ochratoxins. The genus *Aspergillus* includes several citrinin-producing species mainly belonging to section *Terrei* such as *A. niveus*, *A. carneus*, *A. alabamensis*, *A. allahabadii*, *A. floccosus*, *A. terreus*, and *A. flavipes* [140].

Some penicillia including *P. citrinum* and other species in section *Citrina* [167], *P. expansum*, *P. radicicola*, and *P. verrucosum*, and *Monascus* species also produce this nephrotoxic compound [142]. Citrinin commonly contaminates grains, food, and feedstuffs, such as wheat, corn, oats, wheat bran, and fruit juices [168]. Citrinin can be ingested by animals and humans and causes chronic diseases. It is known primarily as a nephrotoxin and several studies have addressed its potential for immunotoxicity [169]. Although the whole genome sequence of a potentially citrinin-producing *A. terreus* isolate (NIH 2624) is known, the genes taking part in citrinin biosynthesis have not yet been identified in aspergilli. However, the observation that primer pairs developed based on KS domain sequences of a *pks* taking part in the biosynthesis of OTA also detected citrinin producers indicates that the polyketide backbone of citrinin is synthesized by a similar enzyme to that of OTA [98].

Numerous other mycotoxins produced by aspergilli can be detected in foods and feeds [22,170]. Due to the results of extensive genome sequencing efforts in aspergilli, the molecular background of the production of several *Aspergillus* mycotoxins has been clarified recently, including gliotoxin [171], kojic acid (Figure 13.1f) [172], naphthopyrones, asperfuranone, and orsellinic acid [173]. The clarification of the possible health effects of these mycotoxins on humans and animals needs further studies.

13.7 POTENTIAL EFFECTS OF CLIMATE CHANGE ON THE OCCURRENCE OF *ASPERGILLUS* MYCOTOXINS IN TEMPERATE/CONTINENTAL CLIMATES

The production of several *Aspergillus* mycotoxins including aflatoxins, OTA, and fumonisins is favored by moisture and high temperature. Several papers have dealt with the effects of global warming caused by climate change on the appearance of mycotoxin-producing fungi and mycotoxins in agricultural products [174–179]. All these studies emphasize that aflatoxin-producing fungi and consequently aflatoxins are specifically expected to become more prevalent with climate change. Indeed, in 2003, prolonged hot and dry weather caused an outbreak of *A. flavus* contamination of maize in northern Italy, with consequent problems of aflatoxin contamination that had previously been uncommon in Europe [174–176]. Several other recent reports have indicated the occurrence of aflatoxin-producing fungi and consequently aflatoxin contamination in agricultural commodities in several European countries that did not face with this problem before. In western Romania, Curtui et al. [180] found that all the examined maize samples were free from aflatoxins in 1997. However, more recently, Tabuc et al. [181] have found that about 30% of maize samples collected between 2002 and 2004 in southeastern Romania were contaminated with AFB_1, and in 20% of these samples, the level of toxin exceeded the European Union (EU) limit of 5 µg/kg. In Serbia, Jakic-Dimic et al. [182] isolated *A. flavus* from 18.7% of the maize samples analyzed, and aflatoxins were also detected in 18.3% of the samples, while Jaksic et al. [183] detected aflatoxins in 41.2% of the analyzed maize samples in the range of 2–7 µg/kg. Polovinski-Horvatovic et al. [184] detected aflatoxin M_1 in 30.4% of milk samples collected from small farms in Serbia in amounts exceeding the allowable legislation of the EU. Similarly, Torkar and Vengust [185] detected aflatoxin M_1 above the EU limit in 10% of the examined milk samples in Slovenia. Halt [186] detected aflatoxins in 9.4% of Croatian flour samples and could isolate *A. flavus* from 38% of the flour samples in 2004 [187]. Although Haberle [188] could not detect aflatoxins in Croatian milk samples, Bilandzic et al. [189] and Markov et al. [190] could detect aflatoxin M_1 above the EU limit in some milk samples collected in Croatia. Regarding Hungary, Richard et al. [191] examined the mycotoxin-producing abilities of 22 isolates collected from various sources in Hungary, and none of the isolates were found to produce aflatoxins. However, more recently, Borbély et al. [192] have examined mycotoxin levels in cereal samples and mixed feed samples collected in eastern Hungary and detected AFB_1 levels above the EU limit in 4.8% of the samples. Dobolyi et al. [193] identified aflatoxin-producing *A. flavus* isolates in several maize fields in Hungary. Aflatoxin contamination of maize (2003) and milk (2007, 2011, 2012) originating from Hungary, Serbia, Romania, and Slovenia has also been detected recently in the framework of the Rapid Alert System for Food and Feed of the EU (https://webgate.ec.europa.eu/rasff-window/portal/).

Recently, a microarray-based method has been developed to examine the effect of combinations of water activity and temperature on the activation of 30 aflatoxin biosynthetic genes in *A. flavus* [194]. This model could be used to predict the impact of climate change on toxin production in vitro.

The occurrence of other mycotoxins is also influenced by climatic conditions. For example, climate is considered as an important risk factor for the increase of the contamination of wines with OTA. In fact, the occurrence of OTA and black aspergilli is higher in grapes grown in southern European regions compared to those of the north. According to a geostatistical analysis performed by Battilani et al. [195], an increase of the incidence of black aspergilli on grapes is seen in the warmer Mediterranean regions of the south compared to the cooler regions of the north. Moreover, the main OTA producer *A. carbonarius* was shown to be isolated more frequently from grapes grown in the Mediterranean regions [196]. The higher recovery rates of black aspergilli in grapes grown in warmer weather conditions can be explained by their ability to grow at higher temperature than other filamentous fungi potentially present on grapes. Since the presence of black aspergilli on grapes and the subsequent contamination of grapes with OTA are favored by high temperatures, climate change is likely to cause the migration of OTA-producing black aspergilli to regions where these species had not been widely found previously [197]. The incidence of black aspergilli was found to be higher in years with very hot summers in Spain, while humidity was less relevant [198], while in recent surveys, ochratoxin-producing black aspergilli have been identified in Hungary, and low levels of OTA contamination was observed in some of the Hungarian wines tested recently [199,200]. OTA-producing black aspergilli have also been identified recently in onions and red pepper harvested in Hungary [120,201].

13.8 CONCLUSIONS

The mycotoxins produced by *Aspergillus* species are extremely important for humans. The pathways and genes taking part in the biosynthesis of the most important, including aflatoxins, ochratoxins, fumonisins, and patulin, have mostly been elucidated, partly due to the ongoing genome sequencing efforts. Based on these sequence data, several molecular detection methods have been developed. However, some of these approaches should be used with care, since there are potential pitfalls. Several genes involved in the biosynthesis of aflatoxins and ochratoxins also take part in the synthesis of ST and citrinin, respectively, potentially leading to false-positive results in PCR tests. The potential effects of mycotoxins as PCR inhibitors and as mutagens should also be considered during molecular detection approaches. However, such molecular approaches could be very useful for the establishment of critical control points in food production.

Most aspergilli prefer warmer climatic conditions, and with global warming, the mycotoxins produced by these species could occur in agricultural products in countries that did not face with this problem before. The most important case is the recent appearance of aflatoxins and producers in agricultural products in Central Europe. In addition, the occurrence of other *Aspergillus* mycotoxins (ochratoxins, fumonisins) has also been detected recently in countries with temperate climate. These observations strongly indicate that climate change seriously affects the distribution of thermotolerant aspergilli and their mycotoxins, leading to the migration of aflatoxin-, fumonisin-, and ochratoxin-producing aspergilli to areas with temperate climatic conditions. However, only limited data are available on the potential effect of climate change on the occurrence of other mycotoxins including patulin, cyclopiazonic acid, and citrinin. Further studies are needed to clarify the possible consequences of further global warming on these fungi.

ACKNOWLEDGMENTS

Parts of the studies cited were supported by OTKA grant Nos. K84077 and K84122 and by the János Bolyai Research Scholarship (B. Tóth) of the Hungarian Academy of Sciences. The project is cofinanced by the EU through the Hungary-Serbia IPA Cross-border Co-operation Programme (ToxFreeFeed, HU-SRB/1002/122/062) and by the European Social Fund (*Broadening the knowledge base and supporting the long term professional sustainability of the Research University Centre of Excellence at the University of Szeged by ensuring the rising generation of excellent scientists*; TÁMOP-4.2.2/B-10/1-2010-0012).

REFERENCES

1. Raper, K. B. and Fennell, D. I. 1965. *The Genus Aspergillus*. Baltimore, MD: Williams & Wilkins.
2. Peterson, S. W. 2008. Phylogenetic analysis of *Aspergillus* species using DNA sequences from four loci. *Mycologia* 100: 205–226.
3. Samson, R. A. and Varga, J. 2009. What is a species in *Aspergillus*? *Med. Mycol.* 47 (Suppl. 1): S13–S20.
4. Lubertozzi, D. and Keasling, J. D. 2009. Developing *Aspergillus* as a host for heterologous expression. *Biotechnol. Adv.* 23: 53–75.
5. Demain, A. L. 2007. The business of biotechnology. *Ind. Biotechnol.* 3: 269–283.
6. Hope, W. W., Walsh, T. J., and Denning, D. W. 2005. The invasive and saprophytic syndromes due to *Aspergillus* spp. *Med. Mycol.* 43 (Suppl. 1): S207–S238.
7. Gonçalves, A. B., Paterson, R. R. M., and Lima, N. 2006. Survey and significance of filamentous fungi from tap water. *Int. J. Hyg. Environ. Health* 209: 257–264.
8. Blout, W. P. 1961. Turkey "X" disease. *Turkeys* 9: 52–77.
9. Van der Zijden, A. S. M., Blanche Koelensmid, W. A. A., Boldingh, J. et al. 1962. *Aspergillus flavus* and Turkey X disease: Isolation in crystalline form of a toxin responsible for Turkey X disease. *Nature* 195: 1060–1062.
10. Squire, R. A. 1981. Ranking animal carcinogens: A proposed regulatory approach. *Science* 214: 877–880.
11. Varga, J., Frisvad, J. C., and Samson, R. A. 2009. A reappraisal of fungi producing aflatoxins. *World Mycotoxin J.* 2: 263–277.
12. Varga, J., Frisvad, J. C., and Samson, R. A. 2011. Two new aflatoxin producing species, and an overview of *Aspergillus* section *Flavi*. *Stud. Mycol.* 69: 57–80.
13. Goncalves, S. S., Stchigel, A. M., Cano, J. F. et al. 2012. *Aspergillus novoparasiticus*: A new clinical species of the section *Flavi*. *Med. Mycol.* 50: 152–160.
14. Rank, C., Nielsen, K. F., Larsen, T. O. et al. 2011. Distribution of sterigmatocystin in filamentous fungi. *Fungal Biol.* 115: 406–420.
15. Paterson, R. R. M., Kelley, J., and Gallagher, M. 1997. Natural occurrence of aflatoxins and *Aspergillus flavus* (Link) in water. *Lett. Appl. Microbiol.* 25: 435–436.
16. Hedayati, M. T., Mayahi, S., Movahedi, M. et al. 2011. Study on fungal flora of tap water as a potential reservoir of fungi in hospitals in Sari city, Iran. *J. Mycol. Med.* 21: 10–14.
17. Grabinska-Loniewska, A., Konillowicz-Kowalska, T., Wardzynska, G. et al. 2007. Occurrence of fungi in water distribution system. *Polish J. Environ. Stud.* 16: 539–547.
18. Olsen, M., Johnsson, P., Möller, T. et al. 2008. *Aspergillus nomius*, an important aflatoxin producer in Brazil nuts? *World Mycotoxin J.* 1: 123–126.
19. Versilovskis, A. and De Saeger, S. 2010. Sterigmatocystin: Occurrence in foodstuffs and analytical methods—An overview. *Mol. Nutr. Food Res.* 54: 136–147.
20. Nielsen, K. F. 2003. Mycotoxin production by indoor molds. *Fungal Genet. Biol.* 39: 103–117.
21. Chu, F. S. 1991. Mycotoxins: Food contamination, mechanisms, carcinogenic potential and preventive measures. *Mutation Res.* 259: 291–306.
22. Bennett, J. W. and Klich, M. 2003. Mycotoxins. *Clin. Microbiol. Rev.* 16: 497–516.
23. Nyikal, J., Misore, A., Nzioka, C. et al. 2004. Outbreak of aflatoxin poisoning—Eastern and Central Provinces, Kenya, January–July 2004. *Morbid. Mortal. Week. Rep.* 53: 790–793.
24. Groopman, J. D., Johnson, D., and Kensler, T. W. 2005. Aflatoxin and hepatitis B virus biomarkers: A paradigm for complex environmental exposures and cancer risk. *Cancer Biomark.* 1: 5–14.
25. Eaton, D. L. and Groopman, J. D. 1994. *The Toxicology of Aflatoxins: Human Health, Veterinary, and Agricultural Significance*. Academic Press, San Diego, CA.

26. Newman, S. J., Smith, J. R., Stenske, K. A. et al. 2007. Aflatoxicosis in nine dogs after exposure to contaminated commercial dog food. *J. Vet. Diagn. Invest.* 19: 168–175.
27. International Agency for Research on Cancer (IARC). 2012. *IARC Monographs on the Evaluation of Carcinogenic Risks to Humans.* Vol. 100F. Chemical agents and related occupations. A review of human carcinogens. IARC Press, Lyon, France.
28. van Egmond, H. P., Schothorst, R. C., and Jonker, M. A. 2007. Regulations relating to mycotoxins in food: Perspectives in a global and European context. *Anal. Bioanal. Chem.* 389: 147–157.
29. Bhatnagar, D., Ehrlich, K., and Cleveland, T. 2003. Molecular genetic analysis and regulation of aflatoxin biosynthesis. *Appl. Microbiol. Biotechnol.* 61: 83–93.
30. Georgianna, D. R. and Payne, G. A. 2009. Genetic regulation of aflatoxin biosynthesis: From gene to genome. *Fungal Genet. Biol.* 46: 113–125.
31. Yu, J., Chang, P. K., Ehrlich, K. C. et al. 2004. Clustered pathway genes in aflatoxin biosynthesis. *Appl. Environ. Microbiol.* 70: 1253–1262.
32. Cary, J. W., Linz, J. E., and Bhatnagar, D. 2000. Aflatoxins: Biological significance and regulation of biosynthesis. In *Microbial Foodborne Diseases: Mechanisms of Pathogenesis and Toxin Synthesi*, eds. J. W. Cary, J. E. Linz, and D. Bhatnagar, pp. 317–361. Lancaster, U.K.: Technomic.
33. Ehrlich, K. C. 2009. Predicted roles of the uncharacterized clustered genes in aflatoxin biosynthesis. *Toxins* 1: 37–58.
34. Chang, P. K., Yu, J., Ehrlich, K. C. et al. 2000. The aflatoxin biosynthesis gene adhA in *Aspergillus parasiticus* is involved in the conversion of 5′-hydroxyaverantin to averufin. *Appl. Environ. Microbiol.* 66: 4715–4719.
35. Yabe, K., Chihaya, N., Hamamatsu, S. et al. 2003. Enzymatic conversion of averufin to hydroxyversicolorone and elucidation of a novel metabolic grid involved in aflatoxin biosynthesis. *Appl. Environ. Microbiol.* 69: 66–73.
36. Yabe, K. and Nakajima, H. 2004. Enzyme reactions and genes in aflatoxin biosynthesis. *Appl. Microbiol. Biotechnol.* 64: 745–755.
37. Yabe, K., Ando, Y., Hashimoto, J. et al. 1989. Two distinct O-methyltransferases in aflatoxin biosynthesis. *Appl. Environ. Microbiol.* 55:2172–2177.
38. Ehrlich, K. C., Chang, P. K., Yu, J. et al. 2004. Aflatoxin biosynthesis cluster gene *cypA* is required for G aflatoxin formation. *Appl. Environ. Microbiol.* 70: 6518–6524.
39. Cai, J., Zeng, H., Shima, Y. et al. 2008. Involvement of the *nadA* gene in formation of G-group aflatoxins in *Aspergillus parasiticus*. *Fungal Genet. Biol.* 45: 1081–1093.
40. Ehrlich, K. C., Chang, P. K., Yu, J. et al. 2011. Control of aflatoxin biosynthesis in Aspergilli. In *Aflatoxins—Biochemistry and Molecular Biology*, ed. R. G. Guevara-Gonzalez, pp. 21–40. Rijeka, Croatia: Intech Open.
41. Yu, J. 2012. Current understanding on aflatoxin biosynthesis and future perspective in reducing aflatoxin contamination. *Toxins* 4: 1024–1057.
42. Sardiñas, N., Vázquez, C., Gil-Serna, J. et al. 2011. Specific detection and quantification of *Aspergillus flavus* and *Aspergillus parasiticus* in wheat flour by SYBR® Green quantitative PCR. *Int. J. Food Microbiol.* 145: 121–125.
43. Cruz, P. and Buttner, M. P. 2008. Development and evaluation of a real-time quantitative PCR assay for *Aspergillus flavus*. *Mycologia* 100: 683–690.
44. Levin, R. E. 2012. PCR detection of aflatoxin producing fungi and its limitations. *Int. J. Food Microbiol.* 156: 1–6.
45. Shapira, R., Paster, N., Eyal, O. et al. 1996. Detection of aflatoxigenic molds in grains by PCR. *Appl. Environ. Microbiol.* 62: 3270–3273.
46. Färber, P., Geisen, R., and Holzapfel, W. 1997. Detection of aflatoxigenic fungi in figs by a PCR reaction. *Int. J. Food Microbiol.* 36: 215–220.
47. Rahimi, K. P., Sharifnabi, B., and Bahar, M., 2008. Detection of aflatoxin in *Aspergillus* species isolated from pistachio in Iran. *J. Pathol.* 156: 15–20.
48. Passone, M. A., Rosso, L. C., Ciancio, A. et al. 2010. Detection and quantification of *Aspergillus* section *Flavi* spp. in stored peanuts by real-time PCR of nor-1 gene, and effects of storage conditions on aflatoxin production. *Int. J. Food Microbiol.* 138: 276–281.
49. Rodrigues, P., Soares, C., Kozakiewicz, Z. et al. 2007. Identification and characterization of *Aspergillus flavus* and aflatoxins. In *Communicating Current Research and Educational Topics and Trends in Applied Microbiology*, Vol. 2, ed. A. Méndez-Vilas, pp. 527–534. Badajoz, Spain: Formatex Publications.
50. Degola, F., Berni, E., Dall'Asta, C. et al. 2007. A multiplex RT-PCR approach to detect aflatoxigenic strains of *Aspergillus flavus*. *J. Appl. Microbiol.* 103: 409–417.

51. Paterson, R. R. M. 2006. Identification and quantification of mycotoxigenic fungi by PCR. *Proc. Biochem.* 41: 1467–1471.
52. Schmidt-Heydt, M. and Geisen, R. 2007. A microarray for monitoring the production of mycotoxins in food. *Int. J. Food Microbiol.* 117: 131–140.
53. Schmidt-Heydt, M., Abdel-Hadi, A., Magan, N. et al. 2009. Complex regulation of the aflatoxin biosynthesis gene cluster of *Aspergillus flavus* in relation to various combinations of water activity and temperature. *Int. J. Food Microbiol.* 135: 231–237.
54. Schmidt-Heydt, M., Magan, N., and Geisen, R. 2008. Stress induction of mycotoxin biosynthesis genes by abiotic factors. *FEMS Microbiol. Lett.* 284: 142–149.
55. van der Merwe, K. J., Steyn, P. S., Fourie, L. et al. 1965. Ochratoxin A, a toxic metabolite produced by *Aspergillus ochraceus* Wilh. *Nature* 205: 1112–1113.
56. Moss, M. O. 1996. Mode of formation of ochratoxin A. *Food Addit. Contam.* 13 (Suppl.): 5–9.
57. Xiao, H., Marquardt, R. R., Frohlich, A. A. et al. 1995. Synthesis and structural elucidation of analogs of ochratoxin A. *J. Agric. Food Chem.* 43: 524–530.
58. Frisvad, J. C., Frank, J. M., Houbraken, J. A. M. P. et al. 2004. New ochratoxin A producing species of *Aspergillus* section *Circumdati*. *Stud. Mycol.* 50: 23–43.
59. Davolos, D., Persiani, A. M., Pietrangeli, B. et al. 2012. *Aspergillus affinis* sp. nov., a novel ochratoxin A-producing *Aspergillus* species (section *Circumdati*) isolated from decomposing leaves. *Int. J. Syst. Evol. Microbiol.* 62: 1007–1015.
60. Varga, J., Kevei, É., Rinyu, E. et al. 1996. Ochratoxin production by *Aspergillus* species. *Appl. Environ. Microbiol.* 62: 4461–4464.
61. Samson, R. A., Houbraken, J. A. M. P., Kuijpers, A. F. A. et al. 2004. New ochratoxin or sclerotium producing species in *Aspergillus* section *Nigri*. *Stud. Mycol.* 50: 45–61.
62. Perrone, G., Stea, G., Epifani, F. et al. 2011. *Aspergillus niger* contains the cryptic phylogenetic species *A. awamori*. *Fungal Biol.* 115: 1138–1150.
63. Pitt, J. I. 1987. *Penicillium viridicatum*, *Penicillium verrucosum*, and production of ochratoxin A. *Appl. Environ. Microbiol.* 53: 266–269.
64. Cabañes, F. J., Bragulat, M. R., and Castellá, G. 2010. Ochratoxin A producing species in the genus *Penicillium*. *Toxins* 2: 1111–1120.
65. Varga, J., Rigó, K., Téren, J. et al. 2001. Recent advances in ochratoxin research I. Production, detection and occurrence of ochratoxins. *Cereal Res. Commun.* 29: 85–92.
66. Noonim, P., Mahakarnchanakul, W., Nielsen, K. F. et al. 2009. Isolation, identification and toxigenic potential of ochratoxin A-producing *Aspergillus* species from coffee beans grown in two regions of Thailand. *Int. J. Food Microbiol.* 128: 197–202.
67. Bayman, P., Baker, J. L., Doster, M. A. et al. 2002. Ochratoxin production by the *Aspergillus ochraceus* group and *Aspergillus alliaceus*. *Appl. Environ. Microbiol.* 68: 2326–2329.
68. Varga, J. and Kozakiewicz, Z. 2006. Ochratoxin A in grapes and grape-derived products. *Trends Food Sci. Technol.* 17: 72–81.
69. Miller, J. D. 1995. Fungi and mycotoxins in grain: Implications for stored product research. *J. Stored Prod. Res.* 31: 1–16.
70. Smith, J. E. and Moss, M. O. 1985. *Mycotoxins: Formation, Analysis and Significance*. John Wiley & Sons, Chichester, U.K.
71. Krogh, P., Hald, B., Plestina, R. et al. 1977. Balkan (endemic) nephropathy and food-borne ochratoxin A: Preliminary results of a survey of foodstuff. *Acta Pathol. Microbiol. Scand. Sect. B* 85: 238–240.
72. Grollman, A. P., Shibutani, S., Moriya, M. et al. 2007. Aristolochic acid and the etiology of endemic (Balkan) nephropathy. *Proc. Natl. Acad. Sci. USA* 104: 12129–12134.
73. O'Brien, E. and Dietrich, D. R. 2005. Ochratoxin A: The continuing enigma. *Crit. Rev. Toxicol.* 35: 33–60.
74. Pfohl-Leszkowicz, A. and Manderville, R. A. 2007. Ochratoxin A: An overview on toxicity and carcinogenicity in animals and humans. *Mol. Nutr. Food Res.* 51: 61–91.
75. Hassen, W., Abid-Essafi, S., Achour, A. et al. 2004. Karyomegaly of tubular kidney cells in human chronic interstitial nephropathy in Tunisia: Respective role of ochratoxin A and possible genetic predisposition. *Hum. Exp. Toxicol.* 23: 339–346.
76. Wafa, E. W., Yahya, R. S., Sobh, M. A. et al. 1998. Human ochratoxicosis and nephropathy in Egypt: A preliminary study. *Hum. Exp. Toxicol.* 17: 124–129.
77. Schwartz, G. G. 2002. Hypothesis: Does ochratoxin A cause testicular cancer? *Cancer Causes Control.* 13: 91–100.
78. Ferreira, N. P. and Pitout, M. J. 1969. The biosynthesis of ochratoxin. *J. South Afr. Chem. Inst.* 22: S1.

79. Wei, R. D., Strong, F. M., and Smalley, R. D. 1971. Incorporation of chlorine-36 into ochratoxin A. *Appl. Microbiol.* 22: 276–277.
80. Steyn, P. S. and Holzapfel, C. W. 1967. The isolation of the methyl and ethyl esters of ochratoxins A and B, metabolites of *Aspergillus ochraceus* Wilh. *J. South Afr. Chem. Inst.* 20: 186–189.
81. Huff, W. E. and Hamilton, P. B. 1979. Mycotoxins—Their biosynthesis in fungi: Ochratoxins—Metabolites of combined pathways. *J. Food Protect.* 42: 815–820.
82. Harris, J. P. and Mantle, P. G. 2001. Biosynthesis of ochratoxins by *Aspergillus ochraceus*. *Phytochemistry* 58: 709–716.
83. O'Callaghan, J., Caddick, M. X., and Dobson, A. D. 2003. A polyketide synthase gene required for ochratoxin A biosynthesis in *Aspergillus ochraceus*. *Microbiology* 149: 3485–3491.
84. O'Callaghan, J., Stapleton, P. C., and Dobson, A. D. 2006. Ochratoxin A biosynthetic genes in *Aspergillus ochraceus* are differentially regulated by pH and nutritional stimuli. *Fungal Genet. Biol.* 43: 213–221.
85. Bacha, N., Atoui, A., Mathieu, F. et al. 2009. *Aspergillus westerdijkiae* polyketide synthase gene "aoks1" is involved in the biosynthesis of ochratoxin A. *Fungal Genet. Biol.* 46: 77–84.
86. Gallo, A., Bruno, K. S., Solfrizzo, M. et al. 2012. New insight into the ochratoxin A biosynthetic pathway through deletion of a nonribosomal peptide synthetase gene in *Aspergillus carbonarius*. *Appl. Environ. Microbiol.* 78: 8208–8218.
87. Geisen, R., Schmidt-Heydt, M., and Karolewiez, A. 2006. A gene cluster of ochratoxin A biosynthetic genes in *Penicillium*. *Mycotoxin Res.* 22: 134–141.
88. Gonzalez-Salgado, A., Patiño, B., Vazquez, C. et al. 2005 Discrimination of *Aspergillus niger* and other *Aspergillus* species belonging to section *Nigri* by PCR assays. *FEMS Microbiol. Lett.* 245: 353–361.
89. Patiño, B., Gonzalez-Salgado, A., Gonzalez-Jaen, M. T. et al. 2005. PCR detection assays for the ochratoxin-producing *Aspergillus carbonarius* and *Aspergillus ochraceus* species. *Int. J. Food Microbiol.* 104: 207–214.
90. Gil-Serna, J., Vázquez, C., Sardiñas, N. et al. 2009. Discrimination of the main ochratoxin A-producing species in *Aspergillus* section *Circumdati* by specific PCR assays. *Int. J. Food Microbiol.* 136: 83–87.
91. Morello, L. G., Sartori, D., Martinez, A. L. O. et al. 2007. Detection and quantification of *Aspergillus westerdijkiae* in coffee beans based on selective amplification of β-tubulin gene by using real-time PCR. *Int. J. Food Microbiol.* 119: 270–276.
92. Perrone, G., Susca, A., Stea, G. et al. 2004. PCR assay for identification of *Aspergillus carbonarius* and *Aspergillus japonicus*. *Eur. J. Plant Pathol.* 110: 641–649.
93. Mulè, G., Susca, A., Logrieco, A. et al. 2006. Development of a quantitative real-time PCR assay for the detection of *Aspergillus carbonarius* in grapes. *Int. J. Food Microbiol.* 111: 28–34.
94. Schmidt, H., Ehrmann, M., Vogel, R. F. et al. 2003. Molecular typing of *Aspergillus ochraceus* and construction of species specific SCAR-primers based on AFLP. *Syst. Appl. Microbiol.* 26: 434–438.
95. Schmidt, H., Taniwaki, M. H., Vogel, R. F. et al. 2004. Utilization of AFLP markers for PCR-based identification of *Aspergillus carbonarius* and indication of its presence in green coffee samples. *J. Appl. Microbiol.* 97: 899–909.
96. Pelegrinelli-Fungaro, M. H., Vissoto, P. C., Sartori, D. et al. 2004. A molecular method for detection of *Aspergillus carbonarius* in coffee beans. *Curr. Microbiol.* 49: 123–127.
97. Sartori, D., Furlaneto, M. C., Martins, M. K. et al. 2006. PCR method for the detection of potential ochratoxin-producing *Aspergillus* species in coffee beans. *Res. Microbiol.* 157: 350–354.
98. Dao, H. P., Mathieu, F., and Lebrihi, A. 2005. Two primer pairs to detect OTA producers by PCR method. *Int. J. Food Microbiol.* 104: 61–67.
99. Castellá, G. and Cabañes, F. J. 2011. Development of a real time PCR system for detection of ochratoxin A-producing strains of the *Aspergillus niger* aggregate. *Food Control* 22: 1367–1372.
100. Atoui, A., Mathieu, F., and Lebrihi, A. 2007. Targeting a polyketide synthase gene for *Aspergillus carbonarius* quantification and ochratoxin A assessment in grapes using real-time PCR. *Int. J. Food Microbiol.* 115: 313–318.
101. Gallo, A., Perrone, G., Solfrizzo, M. et al. 2009. Characterisation of a pks gene which is expressed during ochratoxin A production by *Aspergillus carbonarius*. *Int. J. Food Microbiol.* 129: 8–15.
102. Spadaro, D., Patharajan, S., Kartikeyan, M. et al. 2010. Specific PCR primers for the detection of isolates of *Aspergillus carbonarius* producing ochratoxin A on grapevine. *Ann. Microbiol.* 61: 267–272.
103. Bezuidenhout, S. C., Gelderblom, W. C., Gorst-Allman, C. P. et al. 1988. Structure elucidation of the fumonisins, mycotoxins from *Fusarium moniliforme*. *J. Chem. Soc. Chem. Commun.* 11: 743–745.
104. Gelderblom, W. C., Jaskiewicz, K., Marasas, W. F. et al. 1988. Fumonisins—Novel mycotoxins with cancer-promoting activity produced by *Fusarium moniliforme*. *Appl. Environ. Microbiol.* 54: 1806–1811.

105. Bartók, T., Szécsi, Á., Szekeres, A. et al. 2006. Detection of new fumonisin mycotoxins and fumonisin-like compounds by reversed-phase high-performance liquid chromatography/electrospray ionization ion trap mass spectrometry. *Rapid Commun. Mass Spectrom.* 20: 2447–2462.
106. Bartók, T., Tölgyesi, L., Szekeres, A. et al. 2010. Detection and characterization of twenty-eight isomers of fumonisin B_1 (FB1) mycotoxin in a solid rice culture infected with *Fusarium verticillioides* by reversed-phase high-performance liquid chromatography/electrospray ionization time-of-flight and ion trap mass spectrometry. *Rapid Commun. Mass Spectrom.* 24: 35–42.
107. Marasas, W. F. (1996) Fumonisins: History, world-wide occurrence and impact. *Adv. Exp. Med. Biol.* 392: 1–17.
108. Branham, B. E. and Plattner, R. D. 1993. Isolation and characterization of a new fumonisin from liquid cultures of *Fusarium moniliforme*. *J. Nat. Prod.* 56: 1630–1633.
109. Seo, J. A., Kim, J. C., and Lee, Y. W. 1996. Isolation and characterization of two new type C fumonisins produced by *Fusarium oxysporum*. *J. Nat. Prod.* 59: 1003–1005.
110. Pel, H. J., de Winde, J. H., Archer, D. B. et al. 2007. Genome sequencing and analysis of the versatile cell factory *Aspergillus niger* CBS 513.88. *Nat. Biotechnol.* 25: 221–231.
111. Frisvad, J. C., Smedsgaard, J., Samson, R. A. et al. 2007. Fumonisin B_2 production by *Aspergillus niger*. *J. Agric. Food Chem.* 55: 9727–9732.
112. Hong, S. B., Lee, M., Kim, D. H. et al. 2013. *Aspergillus luchuensis*, an industrially important black *Aspergillus* in East Asia. *PLoS One* 8: e63769.
113. Varga, J., Kocsubé, S., Suri, K. et al. 2010. Fumonisin contamination and fumonisin producing black Aspergilli in dried vine fruits of different origin. *Int. J. Food Microbiol.* 143: 143–149.
114. Mogensen, J. M., Moller, K. A., von Freiesleben, P. et al. 2011. Production of fumonisins B_2 and B_4 in *Tolypocladium* species. *J. Ind. Microbiol. Biotechnol.* 38: 1329–1335.
115. Varga, J., Juhász, Á., Kevei, F. et al. 2004. Molecular diversity of agriculturally important *Aspergillus* species. *Eur. J. Plant Pathol.* 110: 627–640.
116. Pitt, J. I. and Hocking, A. D. 1997. *Fungi and Food Spoilage*, 2nd edn. London, U.K.: Blackie Academic & Professional Publisher.
117. Noonim, P., Mahakarnchanakul, W., Nielsen, K. F. et al. 2009. Fumonisin B_2 production by *Aspergillus niger* from Thai coffee beans. *Food Addit. Contam.* 26: 94–100.
118. Logrieco, A., Ferracane, R, Haidukowsky, M. et al. 2009. Fumonisin B_2 production by *Aspergillus niger* from grapes and natural occurrence in must. *Food Addit. Contam.* 26: 1495–1500.
119. Varga, J., Kocsubé, S., Koncz, Z. et al. 2006. Mycobiota and ochratoxin A in raisins purchased in Hungary. *Acta Aliment.* 35: 289–294.
120. Varga, J., Kocsubé, S., Szigeti, G. et al. 2012. Black Aspergilli and fumonisin contamination in onions purchased in Hungary. *Acta Aliment.* 41: 414–423.
121. Panagopoulou, P., Filioti, J., Farmaki, E. et al. 2007. Filamentous fungi in a tertiary care hospital: Environmental surveillance and susceptibility to antifungal drugs. *Infect. Control Hosp. Epidemiol.* 28: 60–67.
122. Kelley, J., Kinsey, G., Paterson, R. et al. 2003. *Identification and Control of Fungi in Distribution Systems*. Denver, CO: AWWA Research Foundation and American Water Works Association.
123. Hageskal, G., Knutsen, A. K., Gaustad, P. et al. 2006. Diversity and significance of mold species in Norwegian drinking water. *Appl. Environ. Microbiol.* 72: 7586–7593.
124. Wang, E., Norred, W. P., Bacon, C. W. et al. 1991. Inhibition of sphingolipid biosynthesis by fumonisins. Implications for diseases associated with *Fusarium moniliforme*. *J. Biol. Chem.* 266: 14486–14490.
125. Stockmann-Juvala, H. and Savolainen, K. 2008. A review of the toxic effects and mechanisms of action of fumonisin B_1. *Hum. Exp. Toxicol.* 27: 799–809.
126. Marasas, W. F., Riley, R. T., Hendricks, K. A. et al. 2004. Fumonisins disrupt sphingolipid metabolism, folate transport, and neural tube development in embryo culture and *in vivo*: A potential risk factor for human neural tube defects among populations consuming fumonisin-contaminated maize. *J. Nutr.* 134: 711–716.
127. Chu, F. S. and Li, G. Y. 1994. Simultaneous occurrence of fumonisin B_1 and other mycotoxins in moldy corn collected from the People's Republic of China in regions with high incidences of esophageal cancer. *Appl. Environ. Microbiol.* 60: 847–852.
128. Rheeder, J. P., Marasas, W. F., Thiel P. G. et al. 1992. *Fusarium moniliforme* and fumonisins in corn in relation to human esophageal cancer in Transkei. *Phytopathology* 82: 353–357.
129. Brown, D. W., Butchko, R. A., Busman, M. et al. 2007. The *Fusarium verticillioides* FUM gene cluster encodes a Zn (II) 2Cys6 protein that affects FUM gene expression and fumonisin production. *Eukaryot. Cell* 6: 1210–1218.

130. Butchko, R. A., Plattner, R. D., and Proctor, R. H. 2006. Deletion analysis of FUM genes involved in tricarballylic ester formation during fumonisin biosynthesis. *J. Agric. Food Chem.* 54: 9398–9404.
131. Proctor, R. H., Brown, D. W., Plattner, R. D. et al. 2003. Co-expression of 15 contiguous genes delineates a fumonisin biosynthetic gene cluster in *Gibberella moniliformis*. *Fungal Genet. Biol.* 38: 237–249.
132. Accensi, F., Cano, J., Figuera, L. et al. 1999. New PCR methods to differentiate species in the *Aspergillus niger* aggregate. *FEMS Microbiol. Lett.* 180: 191–196.
133. Varga, J., Frisvad, J. C., Kocsubé, S. et al. 2011. New and revisited species in *Aspergillus* section *Nigri*. *Stud. Mycol.* 69: 1–17.
134. Birkinshaw, J. H., Michael, S. E., Bracken, A. et al. 1943. Patulin in the common cold. Collaborative research on a derivative of *Penicillium patulum* Bainier. II. Biochemistry and chemistry. *Lancet* 245:5625–5630.
135. Moake, M. M., Padilla-Zakour, O. I., and Worobo, R. W. 2005. Comprehensive review of patulin control methods in foods. *Compr. Rev. Food Sci. Food Saf.* 1: 8–21.
136. Chalmers, I. and Clarke, M. 2004. The 1944 patulin trial: The first properly controlled multicentre trial conducted under the aegis of the British Medical Research Council. *Int. J. Epidemiol.* 33: 253–260.
137. Walker, K. and Wiesner, B. P. 1944. Patulin and clavicin. *Lancet* 246: 294.
138. Steinman, R., Seigle-Murandi, F., Sage, L., and Krivobok, S. 1989. Production of patulin by micromycetes. *Mycopathologia* 105: 129–133.
139. Varga, J., Due, M., Frisvad, J. C. et al. 2007. Taxonomic revision of *Aspergillus* section *Clavati* based on molecular, morphological and physiological data. *Stud. Mycol.* 59: 89–106.
140. Samson, R. A., Peterson, S. W., Frisvad, J. C. et al. 2011. New species in *Aspergillus* section *Terrei*. *Stud. Mycol.* 69: 39–55.
141. Varga, J., Frisvad, J. C., and Samson, R. A. 2007. Polyphasic taxonomy of *Aspergillus* section *Candidi* based on molecular, morphological and physiological data. *Stud. Mycol.* 59: 75–88.
142. Frisvad, J. C., Smedsgaard, J., Larsen, T. O. et al. 2004. Mycotoxins, drugs and other extrolites produced by species in *Penicillium* subgenus *Penicillium*. *Stud. Mycol.* 49: 201–242.
143. Samson, R. A., Houbraken, J., Varga, J. et al. 2009. Polyphasic taxonomy of the heat resistant ascomycete genus *Byssochlamys* and its *Paecilomyces* anamorphs. *Persoonia* 22: 14–27.
144. Frank, H. K. 1977. Occurrence of patulin in fruit and vegetables. *Ann. Nutr. Aliment.* 31: 459–465.
145. Brian, P. W., Elson, G. W., and Lowe, D. 1956. Production of patulin in apple fruits by *Penicillium expansum*. *Nature* 178: 263–264.
146. Joint FAO/WHO Expert Committee on Food Additives (JECFA). March 9–13, 1998. *Position Paper on Patulin*, pp. 1–8, The Hague, the Netherlands.
147. Scott, P. M. 1974. Patulin. In *Mycotoxins*, ed. I. F. H. Purchase, pp. 383–403. Amsterdam, the Netherlands: Elsevier Science Publishing.
148. Puel, O., Galtier, P., and Oswald, I. P. 2010. Biosynthesis and toxicological effects of patulin. *Toxins* 2: 613–631.
149. Becci, P. J., Hess, F. G., Johnson, W. D. et al. 1981. Long-term carcinogenicity and toxicity studies of patulin in the rat. *J. Appl. Toxicol.* 1: 256–261.
150. Steyn, P. 1992. The biosynthesis of polyketide-derived mycotoxins. *J. Environ. Pathol. Toxicol. Oncol.* 11: 47–59.
151. Sekiguchi, J. and Gaucher, G. M. 1979. Isoepoxydon, a new metabolite of the patulin pathway in *Penicillium urticae*. *Biochem. J.* 182: 445–453.
152. Beck, J., Ripka, S., Siegner, A., Schiltz, E. et al. 1990. The multifunctional 6-methylsalicylic acid synthase gene of *Penicillium patulum*. Its gene structure relative to that of other polyketide synthases. *Eur. J. Biochem.* 192: 487–498.
153. Artigot, M. P., Loiseau, N., Laffitte, J. et al. 2009. Molecular cloning and functional characterization of two CYP619 cytochrome P450s involved in biosynthesis of patulin in *Aspergillus clavatus*. *Microbiology* 155: 1738–1747.
154. Paterson, R. R. M., Archer, S., Kozakiewicz, Z. et al. 2000. A gene probe for the patulin metabolic pathway with potential for use in patulin and novel disease control. *Biocontr. Sci. Technol.* 10: 509–512.
155. Paterson, R. R. M. 2007. The isoepoxydon dehydrogenase gene PCR profile is useful in fungal taxonomy. *Rev. Iberoam. Micol.* 24: 289–293.
156. Paterson, R. R. M. 2004. The isoepoxydon dehydrogenase gene of patulin biosynthesis in cultures and secondary metabolites as candidate PCR inhibitors. *Mycol. Res.* 108: 1431–1437.
157. Varga, J., Rigó, K., Molnár, J. et al. 2003. Mycotoxin production and evolutionary relationships among species of *Aspergillus* section *Clavati*. *Antonie van Leeuwenhoek* 83: 191–200.

158. Paterson, R. R. M. and Lima, N. 2013. Biochemical mutagens affect the preservation of fungi and biodiversity estimations. *Appl. Microbiol. Biotechnol.* 97: 77–85.
159. Paterson, R. R. M. 2007. Internal amplification controls have not been employed in diagnostic fungal PCR hence potential false negative results. *J. Appl. Microbiol.* 102: 1–10.
160. Holzapfel, C. W. 1968. The isolation and structure of cyclopiazonic acid, a toxic metabolite of *Penicillium cyclopium* Westling. *Tetrahedron* 24: 2101–2119.
161. Varga, J., Frisvad, J. C., and Samson, R. A. 2011. Two new aflatoxin producing species, and an overview of *Aspergillus* section *Flavi*. *Stud. Mycol.* 69: 57–80.
162. Larsen, T. O., Smedsgaard, J., Nielsen, K. F. et al. 2007. Production of mycotoxins by *Aspergillus lentulus* and other medically important and closely related species in section *Fumigati*. *Med. Mycol.* 45: 225–232.
163. Riley, R. T., Goeger, D. E., Yoo, H. et al. 1992. Comparison of three tetramic acids and their ability to alter membrane function in cultured skeletal muscle cells and sarcoplasmic reticulum vesicles. *Toxicol. Appl. Pharmacol.* 114: 261–267.
164. Cole, R. J. 1986. Etiology of turkey "X" disease in retrospect: A case for the involvement of cyclopiazonic acid. *Mycotoxin Res.* 2: 3–7.
165. Rao, B. L. and Husain, A. 1985. Presence of cyclopiazonic acid in kodo millet (*Paspalum scrobiculatum*) causing "kodua poisoning" in man and its production by associated fungi. *Mycopathologia* 89: 177–180.
166. Chang, P. K., Horn, B. W., and Dorner, J. W. 2009. Clustered genes involved in cyclopiazonic acid production are next to the aflatoxin biosynthesis gene cluster in *Aspergillus flavus*. *Fungal Genet. Biol.* 46: 176–182.
167. Houbraken, J., Frisvad, J. C., and Samson, R. A. 2011. Taxonomy of *Penicillium* section *Citrina*. *Stud. Mycol.* 70: 53–138.
168. Li, Y., Zhou, Y. C., Yang, M. H. et al. 2012. Natural occurrence of citrinin in widely consumed traditional Chinese food red yeast rice, medicinal plants and their related products. *Food Chem.* 132: 1040–1045.
169. EFSA Panel on Contaminants in the Food Chain. 2012. Scientific opinion on the risks for public and animal health related to the presence of citrinin in food and feed. *EFSA J.* 10: 2605.
170. Klich, M. A. 2009. Health effects of *Aspergillus* in food and air. *Toxicol. Industr. Health* 25: 657–667.
171. Gardiner, D. M. and Howlett, B. J. 2005. Bioinformatic and expression analysis of the putative gliotoxin biosynthetic gene cluster of *Aspergillus fumigatus*. *FEMS Microbiol. Lett.* 248: 241–248.
172. Terabayashi, Y., Sano, M., Yamane, N. et al. 2010. Identification and characterization of genes responsible for biosynthesis of kojic acid, an industrially important compound from *Aspergillus oryzae*. *Fungal Genet. Biol.* 47: 953–961.
173. Klejnstrup M. L., Frandsen, R. J. N., Holm, D. K. et al. 2012. Genetics of polyketide metabolism in *Aspergillus nidulans*. *Metabolites* 2: 100–133.
174. Paterson, R. R. M. and Lima, N. 2010. How will climate change affect mycotoxins in food? *Food Res. Int.* 43: 1902–1914.
175. Paterson, R. R. M. and Lima, N. 2011. Further mycotoxin effects from climate change. *Food Res. Int.* 44: 2555–2566.
176. Tirado, M. C., Clarke, R., Jaykus, L. A. et al. 2010. Climate change and food safety: A review. *Food Res. Int.* 43: 1745–1765.
177. Miraglia, M., Marvin, H. J., Kleter, G. A. et al. 2009. Climate change and food safety: An emerging issue with special focus on Europe. *Food Chem. Toxicol.* 47: 1009–1021.
178. Cotty, P. J. and Jaime-Garcia, R. 2007. Influences of climate on aflatoxin producing fungi and aflatoxin contamination. *Int. J. Food Microbiol.* 119: 109–115.
179. Giorni, P., Magan, N., Pietri, A. et al. 2007. Studies on *Aspergillus* section *Flavi* isolated from maize in northern Italy. *Int. J. Food Microbiol.* 113: 330–338.
180. Curtui, V., Usleber, E., Dietrich, R. et al. 1998. A survey on the occurrence of mycotoxins in wheat and maize from western Romania. *Mycopathologia* 143: 97–103.
181. Tabuc, C., Marin, D., Guerre, P. et al. 2009. Molds and mycotoxin content of cereals in southeastern Romania. *J. Food Protect.* 72: 662–665.
182. Jakic-Dimic, D., Nesic, K., and Petrovic, M. 2009. Contamination of cereals with aflatoxins, metabolites of fungi *Aspergillus flavus*. *Biotechnol. Anim. Husbandry* 25: 1203–1208.
183. Jaksic, S. M., Prunic, B. Z., Milanov, D. S. et al. 2011. Fumonisins and co-occurring mycotoxins in North Serbian corn. *Proc. Natl. Sci. Matica Srpska Novi. Sad.* 120: 49–59.
184. Polovinski-Horvatovic, M., Juric, V., and Glamocic, D. 2009. Two year study of incidence of aflatoxin M_1 in milk in the region of Serbia. *Biotechnol. Anim. Husbandry* 25: 713–718.

185. Torkar, K. G. and Vengust, A. 2007. The presence of yeasts, moulds and aflatoxin M_1 in raw milk and cheese in Slovenia. *Food Control* 19: 570–577.
186. Halt, M. 1994. *Aspergillus flavus* and aflatoxin B_1 in flour production. *Eur. J. Epidemiol.* 10: 555–558.
187. Halt, M., Klapec, T., Subaric, D. et al. 2004. Fungal contamination of cookies and the raw materials for their production in Croatia. *Czech J. Food Sci.* 22: 95–98.
188. Haberle, V. 1988. Aflatoxin M_1 determination in samples of market milk produced in Croatia. *Hrana I Ishrana* 29: 195–196 (in Croatian).
189. Bilandzic, N., Varenina, I., and Solomun, B. 2010. Aflatoxin M_1 in raw milk in Croatia. *Food Control* 21: 1279–1281.
190. Markov, K., Frece, J., Cvek, D. et al. 2010. Aflatoxin M_1 in raw milk and binding of aflatoxin by lactic acid bacteria. *Mljekarstvo* 60: 244–251 (in Croatian).
191. Richard, J. L., Bhatnagar, D., Peterson, S. et al. 1992. Assessment of aflatoxin and cyclopiazonic acid production by *Aspergillus flavus* isolates from Hungary. *Mycopathologia* 120: 183–188.
192. Borbély, M., Sipos, P., Pelles, F. et al. 2010. Mycotoxin contamination in cereals. *J. Agroal. Proc. Technol.* 16: 96–98.
193. Dobolyi, C., Sebők, F., Varga, J. et al. 2011. Identification of aflatoxin-producing *Aspergillus flavus* strains originated from maize kernels. *Növényvédelem* 47: 125–133 (in Hungarian).
194. Abdel-Hadi, A., Schmidt-Heydt, M., Parra, R. et al. 2012. A systems approach to model the relationship between aflatoxin gene cluster expression, environmental factors, growth and toxin production by *Aspergillus flavus*. *J. R. Soc. Interf.* 9: 757–767.
195. Battilani, P., Barbano, C., Marin, S. et al. 2006. Mapping of *Aspergillus* section *Nigri* in Southern Europe and Israel based on geostatistical analysis. *Int. J. Food Microbiol.* 111: S72–S82.
196. Perrone, G., Susca, A., Cozzi, G. et al. 2007. Biodiversity of *Aspergillus* species in some important agricultural products. *Stud. Mycol.* 59: 53–66.
197. Magan, N., Medina, A., and Aldred, D. 2011. Possible climate-change effects on mycotoxin contamination of food crops pre- and postharvest. *Plant Pathol.* 60: 150–163.
198. Belli, N., Mitchell, D., Marin, S. et al. 2005. Ochratoxin A-producing fungi in Spanish wine grapes and their relationship with meteorological conditions. *Eur. J. Plant Pathol.* 113: 233–239.
199. Varga, J., Kiss, R., Mátrai, T. et al. 2005. Detection of ochratoxin A in Hungarian wines and beers. *Acta Aliment.* 34: 381–392.
200. Varga, J., Koncz, Z., Kocsubé, S. et al. 2007. Mycobiota of grapes collected in Hungarian and Czech vineyards in 2004. *Acta Aliment.* 36: 329–341.
201. Fazekas, B., Tar, A., and Kovács, M. 2005. Aflatoxin and ochratoxin A content of spices in Hungary. *Food Addit. Contam.* 22: 856–863.

14 Aureobasidium

Hasima Mustafa Bamadhaj, Giek Far Chan, and Noor Aini Abdul Rashid

CONTENTS

14.1 Introduction 187
14.2 Morphology and Identification 187
14.3 Taxonomy 189
14.4 Chemotaxonomy 189
14.5 Molecular Biology 189
14.6 *Aureobasidium* in Food 191
14.7 Epidemiology 192
14.8 Clinical Features 192
14.9 Pathogenesis 193
14.10 Treatment and Prevention 193
14.11 Conclusions 193
References 193

14.1 INTRODUCTION

Aureobasidium species are ubiquitous. They can be found almost everywhere, for example, soil, plants, young leaves, flowers, wood, water, textiles, wet and damp bathrooms, air-conditioning systems, hypersaline habitats, coastal waters, deep sea, and even in the Arctic [1–3]. Pathogenic strains of *Aureobasidium* species were previously isolated from human body parts, such as skin and nails, and are found commonly as contaminants in clinical laboratories [4]. It is common to isolate *Aureobasidium* species from food and water. However, the pathogenicity of *Aureobasidium* was considered uncommon and the species were recognized as neither primary human pathogens nor significant producers of mycotoxins [5]. Risks of fungal infections among immunocompromised patients and infection due to exposure of high concentrations of airborne *Aureobasidium* conidia were reported [1,6]. *Aureobasidium pullulans*, *A. proteae*, and *A. mansoni* are emerging as rare pathogens causing mycoses among immunocompromised patients [6–9].

14.2 MORPHOLOGY AND IDENTIFICATION

Aureobasidium is usually yeast-like initially. Colonies of *Aureobasidium* are mostly spreading, pinkish when young, becoming blackish at a later stage or upon exposure to light when cultured in media such as Sabouraud dextrose agar or potato dextrose agar [10]. The colony color of *A. pullulans* is due to melanin production that differs among strains where the expression is influenced by temperature and pH [11]. The colonies appear as low and mucoid and faintly pink, becoming gray to black in areas after 7–10 days with diameters of 25–35 mm on Czapek yeast extract agar and malt extract agar. Colonies can grow up to 10–12 mm in diameter on 25% glycerol nitrate agar [12]. A strain of *A. pullulans* isolated from a patient with persistent cutaneous infection showed morphological color changes depending on the age of the culture grown on Sabouraud dextrose agar, which was from pinkish (3 days) to brownish (1 week) and darkly pigmented (3–4 weeks) [13].

Zalar et al. reported that the main difference between isolates was due to pigmentation or darkening of the cultures among Arctic strains [3]. While some strains remained pinkish for at least 1 week, the majority became pigmented after 3 weeks. Additionally, the yeast-like morphology became entirely filamentous, with either marginal or central aerial mycelium [3].

Expanding hyphal ends of *Aureobasidium* showed irregular branching that bends in a distal direction with septa developing irregularly at later stages [10]. Small denticles (a projection between conidiogenous cells and conidium) can be formed from the hyphal walls or from short lateral protrusions on the hyphae. Approximately 2–4 denticles on one cell produce conidia synchronously [12]. *A. pullulans* is noted for a very complex polymorphic life cycle, consisting of various unicellular forms (Figure 14.1a and b) and multicellular filamentous mycelium (Figure 14.1c). The unicellular forms include budding blastospores and swollen cells (Figure 14.1a). Some of the swollen cells could convert to resting-stage chlamydospores or revert to blastospores [14,15]. Chlamydospores are usually spherical in shape with a homogeneous cytoplasm [14], which contrast to the younger cells that contain cytoplasmic granules of a birefringent character (Figure 14.1b). Additionally, hyphal development with lateral outgrowth of blastoconidia has been known to form chlamydospores with melanin in the cell walls. Chlamydospores are freed from the hyphae and bud

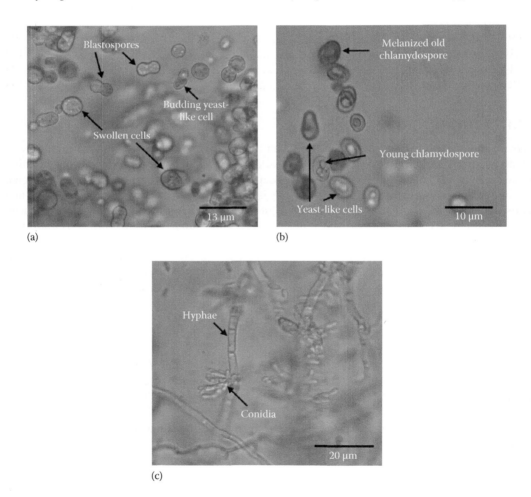

FIGURE 14.1 Morphology of *A. pullulans* grown in basal mineral medium: (a) budding yeast-like cell, blastospore, and swollen cell; (b) yeast-like cells, young chlamydospore, and melanized old chlamydospore; and (c) conidia and hyphae.

to form hyaline blastoconidia [16]. Two types of hyphae were observed from this fast-growing fungus (Figure 14.1c): (1) large thick-walled black cells that have small single-celled budding conidia and (2) thin-walled delicate hyphae [17].

Takeo and de Hoog proposed the identification of *Aureobasidium* based on karyology and hyphae maturation to differentiate the genus from other ascomycetous black yeasts [10]. Nutritional physiology was insufficient for unambiguous recognition of *Aureobasidium* from related *Hormonema* Lagerb. & Melin and *Kabatiella* Bubak. In addition, infraspecific variability occurred among strains of *A. pullulans* in the production of exopolysaccharides that were considered as possible useful characters [18].

14.3 TAXONOMY

Aureobasidium is in the phylum Ascomycota, order Dothideales, and family Dothioraceae [19]. The genus contains 31 species as listed in MycoBank (www.mycobank.org) [20], namely, *A. aleuritis, A. apocryptum, A. australiense, A. bolleyi, A. caulivorum, A. dalgeri, A. foliicola, A. harposporum, A. indicum, A. leucospermi, A. lilii, A. lini, A. mansoni, A. microstictum, A. microstromoides, A. nigricans, A. nigrum, A. oleae, A. proteae, A. prunicola, A. prunorum, A. pullulans, A. ribis, A. salmonis, A. sanguinariae, A. slovacum, A. thujae-plicatae, A. umbellulariae, A. vaccinii, A. vitis,* and *A. zeae*. Recently, newly identified *A. iranianum* and *A. thailandense* were added [21,22].

The most well-known species is *A. pullulans* (de Bary) G. Arnaud, which was termed as the *black yeast* and formerly described as a new fungal species, *Dermatium pullulans*, by de Bary in 1866. However, in 1910, Arnaud proposed *A. pullulans* as a new taxonomic combination of *D. pullulans* and *A. vitis* [23]. Zalar et al. defined the morphology and phylogenetics of four varieties of *A. pullulans*, namely, *pullulans*, *melanogenum*, *subglaciale*, and *namibiae* [3].

14.4 CHEMOTAXONOMY

The biochemical properties of *A. pullulans* are well studied, although a pronounced random variability in biochemical activity was observed among taxa [24]. The 43 tested strains could be divided into three major phenoms based on 77 biochemical tests. The first phenom of *A. pullulans* strains was characterized by proteases, amylases, cellulases, endo-1,4-β-xylanase, and exo-1,4-β-xylosidases. The growth of strains of the first phenom was not stimulated by riboflavin, biotin, and calcium pantothenate, and the strains could not utilize lignosulfonate and lignins from the wood of deciduous trees [24]. The second phenom showed high activity in the (1) decomposition of cellobiose, raffinose, and starch; (2) production of pullulan; (3) ability to utilize D-xylose, D-galacturonic acid, and D-glucuronic acid; and (4) significant endo-1,4-α-xylanase and exo-1,4-α-xylosidase activity [23]. The third phenom exhibited (1) monophenol monooxygenase activity, (2) high melanin production, (3) growth highly stimulated by riboflavin, and (4) metabolism of lactose, catechol, vanillin, the D component of lignosulfonate, and lignin from beech and hornbeam wood [24].

14.5 MOLECULAR BIOLOGY

Aureobasidium species possess eight chromosomes, as determined by CHEF agarose gel electrophoresis [25], although the genome size remains unknown. Recently, whole-genome shotgun sequencing data of a pathogenic strain of *A. pullulans* were included on the NCBI database. The genome contains 515 contigs of approximately 26.72 Mbp in total size and 49.8% GC content [26]. Additionally, a total of 10,288 putative open reading frames (ORFs) were predicted and 8,101 ORFs were annotated. The genome also revealed the presence of 261 tRNAs, 19 copies of 8S ribosomal RNA (rRNA), and a copy (each) of 28S and 18S rRNA [26].

Until now, the taxonomy of the genus *Aureobasidium* remained uncertain and overlapped with the genus *Kabatiella* [19]. However, several molecular methods were applied for the taxonomic delimitation of *Aureobasidium*. Yurlova et al. developed molecular markers for the distinction of a large number of closely related taxa using random fragment length polymorphism of ITS2 and partial LSU rRNA amplicons [18,27]. *Msp*I restriction proved sufficient for the differentiation of *Aureobasidium* from *Hormonema* and *A. pullulans* differed from *H. dematioides* by lacking a *Rsa*I restriction site. Differences between *Aureobasidium*, *Kabatiella*, and *Selenophoma* were found with *Dde*I (*Aureobasidium* vs. *Kabatiella*) and *Hha*I (*Aureobasidium* vs. *Selenophoma*) [27]. Sequence comparisons of *Aureobasidium* and related genera using (1) ITS1–ITS2 and (2) 5.8S and partial ITS2 rDNA domains were performed by Yurlova et al. [28] and de Hoog et al. [29]. The main subdivisions from ITS1 to ITS2 comparisons demonstrated that there was a good correlation with conidiogenesis prevalent to the species studied [27]. On the other hand, *Aureobasidium* species and related genera were unable to be delimited on the basis of 5.8S and partial ITS2 rDNA data [29].

PCR ribotyping [18], arbitrary primed PCR (ap-PCR) fingerprinting [30], random amplified polymorphic DNA PCR fingerprinting [31], and fluorescent amplified fragment length polymorphism (fAFLP) [32] were applied to assess infraspecific variability and strain differentiation

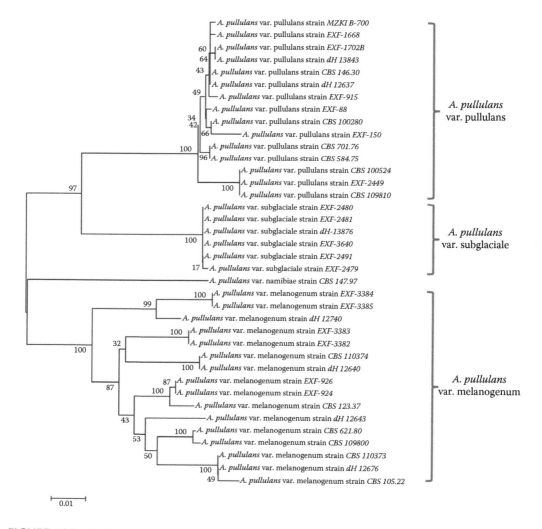

FIGURE 14.2 Phylogram of *ELO* genes from varieties of *A. pullulans* aligned using multiple sequence comparison by log-expectation (MUSCLE) and constructed with neighbor-joining analysis and bootstrapping test at 500 using molecular evolutionary genetic analysis (MEGA 5.1).

of *A. pullulans*. The molecular fingerprinting techniques demonstrated a wide diversity within *A. pullulans*. In addition, the genetic variability of Arctic *A. pullulans* strains and pan-global strains was characterized using internal transcribed spacer (ITS), partial 28S rRNA, partial introns and exons of β-tubulin (*TUB*), translation elongation factor (*EF1α*), and elongase (*ELO*) genes [3]. These multilocus analyses indicated that strains of *A. pullulans* from Arctic habitats could consistently be grouped into vars. *pullulans* and *melanogenum*. The partial *ELO* gene was discovered as a reliable phylogenetic marker at the infraspecific level [3], and Figure 14.2 shows the phylogram constructed based on the partial *ELO* genes of *A. pullulans* variants. This study suggests that identification of new *A. pullulans* strains at the infraspecific level could be attempted based on the partial sequence of *ELO* genes.

Molecular identification of new species of *Aureobasidium* has been frequently performed. For instance, *A. iranianum* from bamboo stem in Iran was distinguished not only by its abundant pigmented arthroconidia in culture but also by the ITS sequence that showed a low similarity of 97% with ITS sequences of *A. pullulans* from GenBank [21]. Another novel species, designated as *A. thailandense*, possessed a 500 bp type I intron in the 18S rRNA, which was lacking in *A. pullulans* [22]. Additionally, molecular detection of *A. pullulans* using specific primers and molecular beacon probes with varied quencher chemistry (BHQ-1 and Tamra) was attempted based on the fungal 18S rRNA gene [13]. The optimized quantitative real-time PCR assays could detect and quantify up to 1 pg concentration of *A. pullulans* genomic DNA [13].

14.6 *AUREOBASIDIUM* IN FOOD

Food contamination by *Aureobasidium* species is significant in general. Also, there is a risk of infection of immunocompromised patients by the fungus from foods [33]. Being ubiquitous in nature, *Aureobasidium* species are commonly isolated from various food sources even under extreme conditions. Kuehn and Gunderson revealed that *A. pullulans* was obtained from frozen pastries of blueberry, cherry, apple and raspberry, and chicken pies [34]. An *Aureobasidium* strain was isolated from blueberry and cherry pastries. The fungus was dominant in crust and was the most common psychrophile in the filling comprising 90% of the fungi [35]. The fungus could thrive under psychrophilic (0°C and 5°C) and mesophilic (10°C and 20°C) storage environment [34]. However, it rarely causes food spoilage [12].

A. pullulans has also been isolated from fresh fruits such as strawberries [36], sweet cherries [37], grapes [38], and apples [39]. Interestingly, *A. pullulans* is the most commonly found fungi on wine grapes and brewers' barley and was found to be almost 39% of the total microbial population of ice wine musts [40]. However, the fungus rarely presents a problem to the wine industry. *A. pullulans* is inactive during fermentation because it is an obligate aerobe and cannot grow at sugar concentration below 1% [40]. The fungus was also found in yoghurts, fresh vegetables, cabbage, citrus and pasteurized orange juice, shrimp, green olives, flour, oats, and nuts [12].

A. pullulans was also known to be part of the mycobiota colonizing cereal grain and grapes during ripening [41], wheat bran [42], and cheeses [43]. Since *A. pullulans* is halotolerant [44], the fungus has been isolated from the (1) brine solutions used in ripening of Taleggio PDO cheese [45] and (2) fermentation of black Conservolea olives [46]. Being moderate xerophiles [3], species of *Aureobasidium* were isolated from dried–salted fish [47].

Aureobasidium species are mostly contaminants derived from food processing environments. For example, *A. pullulans* was isolated from indoor environments, air control points, and other control points of equipment, plastic film, milk, and brine used in factories producing semihard cheeses [48]. Consequently, strict procedures are required to reduce mold contamination in production processes and food processing facilities [48]. *A. pullulans* strains were successful as antagonists against phytopathogens causing postharvest rots [30,36,49]. However, the safety of these should be thoroughly evaluated in view of the health risk associated with opportunistic *Aureobasidium* species.

14.7 EPIDEMIOLOGY

Aureobasidium species have contributed to the changing epidemiology of opportunistic cutaneous and invasive mycoses. The fungi can be present in the air in temperate zones throughout Europe [50] and the tropics [51]. Natural habitats for *A. pullulans* are in the Mediterranean ecosystems, where it was found, for example, to colonize popular bathing beaches in Greece [52]. The presence of *Aureobasidium* in the air samples from the homes of asthmatic children in seven cities in the United States was discovered [53]. An alarming emerging series of clinical case reports of *A. pullulans* has been published, especially in cases involving cutaneous and invasive infections of immunocompromised patients throughout Europe and North America, despite being considered an opportunistic microorganism [6,52].

14.8 CLINICAL FEATURES

A. pullulans remains the most reported species of *Aureobasidium* from the clinical viewpoint. *Aureobasidium* infections are associated with indwelling peritoneal dialysis catheter sites including the ones related to cerebrospinal fluid, splenic abscesses, bronchoalveolar lavage specimens, corneal ulcers, sclera, skin lesions (lower lip), mandibular abscesses, and systemic and invasive pulmonary infections [6]. *A. pullulans* has also been reported as a common allergen associated with upper and lower airway diseases [1]. Finally, *A. mansoni* and *A. proteae* have been reported to cause fungal meningitis [7–9].

The clinical manifestation of *Aureobasidium* infection remains varied. Chan et al. implicated persistent cutaneous infection to an immunocompromised patient suffering from primary aldosteronism with *A. pullulans*, which was coisolated with the yeast *Candida orthopsilosis* [54]. The infection resulted in the affected areas of the limbs to desiccate, harden, and crack (Figure 14.3). *A. pullulans* caused the lesions to start almost simultaneously at the distal edge of both thumbnails and gradually spread to the third and fourth fingers in a case involving a 25-year-old woman suffered from a 1-year history of painful discoloration on her hands. A red inflammatory linear area was observed to divide the onycholysis region from the healthy nail [4].

A case of lymphatic systemic infection on a 23-year-old patient caused by *A. pullulans* was reported by de Morais et al. [55]. The patient was being treated for erythema nodosum leprosum and presented a 60-day complaint of daily fever, hoarseness, odynophagia, and weight loss. Subsequent laboratory tests showed pancytopenia with severe neutropenia, cervical adenomegaly, and a solid contrast uptake lesion in the oropharyngeal region [55]. In another case, a 19-year-old man was

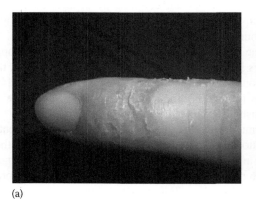
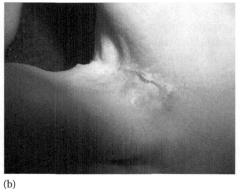

(a)   (b)

FIGURE 14.3 Skin lesion of fungal infection by *A. pullulans* and *C. orthopsilosis* on the (a) finger and (b) foot of an immunocompromised patient.

reported to have fever, dyspnea, and pleuritic chest pain that lasted for over 10 days. His past medical history was significant only for mild intermittent asthma, occasionally treated with salbutamol. The patient had severe eosinophilia, most probably caused by an allergic reaction to *A. pullulans*, which then caused a thrombotic diathesis and widespread thromboembolism [56].

14.9 PATHOGENESIS

There is no study on the pathogenesis of *A. pullulans* in human mycoses, and infection by this fungus was regarded as rare and positive identifications were suspected to be of laboratory contaminants. The prevalence of *A. pullulans* could be due to the prior use of antibiotics and corticosteroids among immunocompromised patients [17]. However, the importance of the fungus has increased as it contributes to the morbidity and mortality of patients [17]. Hence, the study on the pathogenesis of *A. pullulans* in mycoses could be a future direction for its treatment and prevention.

14.10 TREATMENT AND PREVENTION

A series of antifungal and antibiotic treatments has been reported for infection caused by *A. pullulans*. Oral itraconazole 200 mg once daily (which had shown MIC < 1 µg mL^{-1}) for 2 weeks and local bifonazole therapy 1% solution once daily empirically for 2 months have completely cured the symptoms of lesions in a case involving a patient with onychomycosis [4]. *A. pullulans* causing keratitis was found to be sensitive to fluconazole, itraconazole, and flucytosine [17]. Although in some cases, amphotericin B, either alone or in combination with other drugs, showed success [6,57–59]. However, there are very few clinical reports and there is no standard antifungal therapy against *A. pullulans* [17]. Furthermore, there is no preventive method or optimal standard procedure that exists to counteract *Aureobasidium*. Reuse of medical devices intended for single use is not advisable despite reprocessing and sterilization, as *Aureobasidium* species are lab contaminants and commonly found in medical devices. Indeed, after the pseudo-outbreak of *Aureobasidium* isolated from plastic bronchoscopy stopcocks, the reuse of these single-use stopcocks was halted and no further *Aureobasidium* was isolated during the 6-month follow-up tests [60].

Physicians are likely to encounter numerous opportunistic fungal infections in the future, of which *Aureobasidium* is one example, with increasing survival of chronically ill patients, and the common use of surgically implanted Silastic material [6]. *A. pullulans* was found to have high affinity for synthetic materials and surgically implanted Silastic devices; hence, it was frequently isolated from peritoneal dialysis catheters and central venous lines in hospitals [61,62]. It remains crucial to maintain good aseptic environments and avoid nosocomial infections, especially among immunocompromised patients whether in hospitals or homes.

14.11 CONCLUSIONS

As *Aureobasidium* species are frequently isolated from food and the environment, the ubiquitous polymorphic fungi are a potential source of mycoses, especially among immunocompromised patients. The clinical significance of the fungal species is emerging, although studies of fungal pathogenesis and treatment of infections remain lacking.

REFERENCES

1. Taylor, P. E., R. Esch, R. C. Flagan, J. House, L. Tran, and M. M. Glovsky. 2006. Identification and possible disease mechanisms of an under-recognized fungus, *Aureobasidium pullulans*. *Int. Arch. Allergy Immun.* 139:45–52.
2. Chi, Z. M., F. Wang, Z. Chi, L. Yue, G. Liu, and T. Zhang. 2009. Bioproducts from *Aureobasidium pullulans*, a biotechnologically important yeast. *Appl. Microbiol. Biotechnol.* 82:793–804.

3. Zalar, P., C. Gostincar, G. S. de Hoog, V. Ursic, M. Sudhadham, and N. Gunde-Cimerman. 2008. Redefinition of *Aureobasidium pullulans* and its varieties. *Stud. Mycol.* 61:21–38.
4. Arabatzis, M. 2012. Hypothyroidism-related onycholysis with *Aureobasidium pullulans* colonization successfully treated with antifungal therapy. *Clin. Exp. Dermatol.* 37:370–373.
5. Pritchard, R. C. and D. B. Muir. 1987. Black fungi: A survey of dematiaceous hyphomycetes from clinical specimens identified over a five-year period in a reference laboratory. *Pathology* 19:281–284.
6. Hawkes, M., R. Rennie, C. Sand, and W. Vaudry. 2005. *Aureobasidium pullulans* infection: Fungemia in an infant and a review of human cases. *Diagn. Microbiol. Infect. Dis.* 51:209–213.
7. Kutlesa, M., E. Mlinaric-Missoni, L. Hatvani, D. Voncina, S. Simon, D. Lepur, and B. Barsic. 2012. Chronic fungal meningitis caused by *Aureobasidium proteae*. *Diagn. Microbiol. Infect. Dis.* 73:271–272.
8. Krcmery, V. Jr., S. Spanik, A. Danisovicova, Z. Jesenska, and M. Blahova. 1994. *Aureobasidium mansoni* meningitis in a leukemia patient successfully treated with amphotericin B. *Chemotherapy* 40:70–71.
9. Huttova, M., K. Kralinsky, J. Horn, I. Marinova, K. Iligova, J. Fric, S. Spanik et al. 1998. Prospective study of nosocomial fungal meningitis in children—Report of 10 cases. *Scand. J. Infect. Dis.* 30:485–487.
10. Takeo, K. and G. S. de Hoog. 1991. Karyology and hyphal characters as taxonomic criteria in ascomycetous black yeasts and related fungi. *Antonie van Leeuwenhoek* 60:35–42.
11. Wickerham, L. J. and C. P. Kurtzman. 1975. Synergistic color variants of *Aureobasidium pullulans*. *Mycologia* 67:342–361.
12. Pitt, J. I. and A. D. Hocking. 2009. *Fungi and Food Spoilage*. New York: Springer.
13. Chan, G. F., M. S. A. Puad, C. F. Chin, and N. A. A. Rashid. 2011. Emergence of *Aureobasidium pullulans* as human fungal pathogen and molecular assay for future medical diagnosis. *Folia Microbiol.* 56:459–467.
14. Dominguez, J. B., F. M. Goni, and F. Uruburu. 1978. The transition from yeast-like to chlamydospore cells in *Pullularia pullulans*. *J. Gen. Microbiol.* 108:111–117.
15. Pechak, D. G. and R. E. Crang. 1977. An analysis of *Aureobasidium pullulans* developmental stages by means of scanning electron microscopy. *Mycologia* 69:783–792.
16. Slepecky, R. A. and W. T. Starmer. 2009. Phenotypic plasticity in fungi: A review with observations on *Aureobasidium pullulans*. *Mycologia* 101:823–832.
17. Panda, A., H. Das, M. Deb, B. Khanal, and S. Kumar. 2006. *Aureobasidium pullulans* keratitis. *Clin. Exp. Ophthalmol.* 34:260–264.
18. Yurlova, N. A., I. V. Mokrousov, and G. S. de Hoog. 1995. Intraspecific variability and exopolysaccharide production in *Aureobasidium pullulans*. *Antonie van Leeuwenhoek* 68:57–63.
19. Pitkaranta, M. and M. D. Richardson. 2011. *Aureobasidium*. In *Molecular Detection of Human Fungal Pathogens*, ed. D. Liu, pp. 37–47. Boca Raton, FL: CRC Press.
20. Robert, V., D. Vu, A. B. H. Amor et al. 2013. MycoBank gearing up for new horizons. *IMA Fungus* 4:371–379.
21. Arzanlou, M. and S. Khodaei. 2012. *Aureobasidium iranianum*, a new species on bamboo from Iran. *Mycosphere* 3:404–408.
22. Peterson, S. W., P. Manitchotpisit, and T. D. Leathers. 2012. *Aureobasidium thailandense*, a new species isolated from leaves and wooden surfaces. *Int. J. Syst. Evol. Microbiol.* doi: 10.1099/ijs.0.047613-0.
23. Matsumoto, T., A. A. Padhye, and L. Ajello. 1987. Medical significance of the so-called black yeasts. *Eur. J. Epidemiol.* 3:87–95.
24. Cernakova, M., A. Kockova-Kratochvilova, L. Suty, J. Zemek, and L. Kuniak. 1980. Biochemical similarities among strains of *Aureobasidium pullulans* (de Bary) Arnaud. *Folia Microbiol.* 25:68–73.
25. Thornewell, S. J., R. B. Peery, and P. L. Skatrud. 1995. Cloning and characterization of the gene encoding translation elongation factor 1α from *Aureobasidium pullulans*. *Gene* 162:105–110.
26. Chan, G. F., H. M. Bamadhaj, H. M. Gan, and N. A. A. Rashid. 2012. Genome sequence of *Aureobasidium pullulans* AY4, an emerging opportunistic fungal pathogen with diverse biotechnological potential. *Eukaryot. Cell* 11:1419–1420.
27. Yurlova, N. A., J. M. J. Uijthof, and G. S. de Hoog. 1996. Distinction of species in *Aureobasidium* and related genera by PCR-ribotyping. *Antonie van Leeuwenhoek* 69:323–329.
28. Yurlova, N.A., G. S. de Hoog, and A. H. G. Gerrits van den Ende. 1999. Taxonomy of *Aureobasidium* and allied genera. *Stud. Mycol.* 43:63–69.
29. de Hoog, G. S., P. Zalar, C. Urzi, F. de Leo, N. A. Yurlova, and K. Sterflinger. 1999. Relationships of dothideaceous black yeasts and meristematic fungi based on 5.8S and ITS2 rDNA sequence comparison. *Stud. Mycol.* 43:31–37.

30. Schena, L., A. Ippolito, T. Zahavi, L. Cohen, F. Nigro, and S. Droby. 1999. Genetic diversity and biocontrol activity of *Aureobasidium pullulans* isolates against postharvest rots. *Postharvest Biol. Technol.* 17:189–199.
31. Urzi, C., F. de Leo, C. Lo Passo, and G. Criseo. 1999. Intra-specific diversity of *Aureobasidium pullulans* strains isolated from rocks and other habitats assessed by physiological methods and by random amplified polymorphic DNA (RAPD). *J. Microbiol. Methods* 36:95–105.
32. de Curtis, F., L. Caputo, R. Castoria, G. Lima, G. Stea, and V. de Cicco. 2004. Use of fluorescent amplified fragment length polymorphism (fAFLP) to identify specific molecular markers for the biocontrol agent *Aureobasidium pullulans* strain LS30. *Postharvest Biol. Technol.* 34:179–186.
33. Tomsikova, A. 2002. Risk of fungal infection from foods, particularly in immunocompromised patients. *Epidemiol. Mikrobiol. Imunol.* 51:78–81.
34. Kuehn, H. H. and M. F. Gunderson. 1963. Psychrophilic and mesophilic fungi in frozen food products. *Appl. Microbiol.* 11:352–356.
35. Kuehn, H. H. and M. F. Gunderson. 1962. Psychrophilic and mesophilic fungi in fruit-filled pastries. *Appl. Microbiol.* 10:354–358.
36. Lima, G., A. Ippolito, F. Nigro, and M. Salerno. 1997. Effectiveness of *Aureobasidium pullulans* and *Candida oleophila* against postharvest strawberry rots. *Postharvest Biol. Technol.* 10:169–178.
37. Schena, L., F. Nigro, I. Pentimone, A. Ligorio, and A. Ippolito. 2003. Control of postharvest rots of sweet cherries and table grapes with endophytic isolates of *Aureobasidium pullulans*. *Postharvest Biol. Technol.* 30:209–220.
38. Prakitchaiwattana, C. J., G. H. Fleet, and G. M. Heard. 2004. Application and evaluation of denaturing gradient gel electrophoresis to analyse the yeast ecology of wine grapes. *FEMS Yeast Res.* 4:865–877.
39. Granado, J., B. Thurig, E. Kieffer, L. Petrini, A. Fließbach, L. Tamm, F. P. Weibel, and G. S. Wyss. 2008. Culturable fungi of stored "golden delicious" apple fruits: A one-season comparison study of organic and integrated production systems in Switzerland. *Microb. Ecol.* 56:720–732.
40. Subden, R. E., J. I. Husnik, R. van Twest, G. van der Merwe, and H. J. J. van Vuuren. 2003. Autochthonous microbial population in a Niagara Peninsula icewine must. *Food Res. Int.* 36:747–751.
41. Magan, N., D. Aldred, R. Hope, and D. Mitchell. 2010. Environmental factors and interactions with mycobiota of grain and grapes: Effects on growth, deoxynivalenol and ochratoxin production by *Fusarium culmorum* and *Aspergillus carbonarius*. *Toxins* 2:353–366.
42. Tancinova, D. and R. Labuda. 2009. Fungi on wheat bran and their toxinogenity. *Ann. Agric. Environ. Med.* 16:325–331.
43. Kure, C. F. and I. Skaar. 2000. Mould growth on the Norwegian semi-hard cheeses Norvegia and Jarlsberg. *Int. J. Food Microbiol.* 62:133–137.
44. Kogej, T., J. Ramos, A. Plemenitas, and N. Gunde-Cimerman. 2005. The halophilic fungus *Hortaea werneckii* and the halotolerant fungus *Aureobasidium pullulans* maintain low intracellular cation concentrations in hypersaline environments. *Appl. Environ. Microbiol.* 71:6600–6605.
45. Panelli, S., J. N. Buffoni, C. Bonacina, and M. Feligini. 2012. Identification of moulds from the Taleggio cheese environment by the use of DNA barcodes. *Food Control* 28:385–391.
46. Nisiotou, A. A., N. Chorianopoulos, G. J. Nychas, and E. Z. Panagou. 2010. Yeast heterogeneity during spontaneous fermentation of black Conservolea olives in different brine solutions. *J. Appl. Microbiol.* 108:396–405.
47. Atapattu, R. and U. Samarajeewa. 1990. Fungi associated with dried fish in Sri Lanka. *Mycopathologia* 111:55–59.
48. Kure, C. F., I. Skaar, and J. Brendehaug. 2004. Mould contamination in production of hard cheese. *Int. J. Food Microbiol.* 93:41–49.
49. Bencheqroun, S. K., M. Bajji, S. Massart, M. Labhilili, S. E. Jaafari, and M. H. Jijakli. 2007. In vitro and in situ study of postharvest apple blue mold biocontrol by *Aureobasidium pullulans*: Evidence for the involvement of competition for nutrients. *Postharvest Biol. Technol.* 46:128–135.
50. Mandelin, T. M. and M. F. Madelin. 1995. Biological analysis of fungi and associated molds. In *Bioaerosols Handbook*, eds. C. S. Cox and C. M. Wathes, pp. 361–386. Boca Raton, FL: Lewis Publishers.
51. Punnapayak, H., M. Sudhadham, S. Prasongsuk, and S. Pichayangkura. 2003. Characterization of *Aureobasidium pullulans* isolated from airborne spores in Thailand. *J. Ind. Microbiol. Biotechnol.* 30:89–94.
52. Efstratiou, M. A. and A. Velegraki. 2009. Recovery of melanized yeasts from Eastern Mediterranean beach sand associated with the prevailing geochemical and marine flora patterns. *Med. Mycol.* 9:1–3.

53. O'Connor, G. T., M. Walter, H. Mitchell, M. Kattan, W. J. Morgan, R. S. Gruchalla, J. A. Pongracic et al. 2004. Airborne fungi in the homes of children with asthma in low-income urban communities: The inner-city asthma study. *J. Allergy Clin. Immun.* 114:599–606.
54. Chan, G. F., M. S. A. Puad, and N. A. A. Rashid. 2011. *Candida orthopsilosis* and *Aureobasidium pullulans*—Rare fungal pathogens causing persistent skin infection. *Insight Infect. Dis.* 1:1–4.
55. de Morais, O. O., C. Porto, A. S. S. L. Coutinho, C. M. S. Reis, M. de Melo Teixeira, and C. M. Gomes. 2011. Infection of the lymphatic system by *Aureobasidium pullulans* in a patient with erythema nodosum leprosum. *Braz. J. Infect. Dis.* 15:288–292.
56. Sade, K., I. Schwartz, E. Lev, S. Kivity, and Y. Levo. 2001. Widespread thromboembolism in allergy. *Allergy* 56:253–254.
57. Huang, Y.-T., C.-H. Liao, S.-J. Liaw, J.-L. Yang, D.-M. Lai, Y.-C. Lee, and P.-R. Hsueh. 2008. Catheter-related septicemia due to *Aureobasidium pullulans*. *Int. J. Infect. Dis.* 12:e137–e139.
58. Pikazis, D., I. D. Xynos, V. Xila, A. Velegraki, and K. Aroni. 2009. Extended fungal skin infection due to *Aureobasidium pullulans*. *Clin. Exp. Dermatol.* 34:892–894.
59. Joshi, A., R. Singh, M. S. Shah, S. Umesh, and N. Khattry. 2010. Subcutaneous mycosis and fungemia by *Aureobasidium pullulans*: A rare pathogenic fungus in a post allogeneic BM transplant patient. *Bone Marrow Transplant.* 45:203–204.
60. Wilson, S. J., R. J. Everts, K. B. Kirkland, and D. J. Sexton. 2000. A pseudo-outbreak of *Aureobasidium* species lower respiratory tract infections caused by reuse of single-use stopcocks during bronchoscopy. *Infect. Control Hosp. Epidemiol.* 21:470–472.
61. Caporalel, N. E., L. Calegari, D. Perezl, and E. Gezuele. 1996. Peritoneal catheter colonization and peritonitis with *Aureobasidium pullulans*. *Periton. Dial. Int.* 16:97–98.
62. Ibanez, P., J. Chacon, A. Fidalgo, J. Martin, V. Paraiso, and J. L. Munoz-Bellido. 1997. Peritonitis by *Aureobasidium pullulans* in continuous ambulatory peritoneal dialysis. *Nephrol. Dial. Transpl.* 12:1544–1545.

15 Candida as Foodborne Pathogens

Sónia Silva, Cláudia Botelho, and Mariana Henriques

CONTENTS

15.1 *Candida* Virulence Factors 197
15.2 *Candida* Adhesion Ability 198
15.3 *Candida* Biofilm Formation 200
15.4 *Candida* Invasion and Damage of Host Tissues 202
      15.4.1 Filamentous Form Development 202
      15.4.2 Secreted Aspartyl Proteinases 203
      15.4.3 Phospholipases and Lipases 203
15.5 *Candida* Species Identification 204
      15.5.1 PCR-Based Molecular Methods 204
      15.5.2 Non-PCR-Based Molecular Methods 205
15.6 Concluding Remarks 206
References 206

Foodborne and waterborne pathogens can be divided into three large groups: molds, yeast, and bacteria [1] but also include protozoa and viruses. The yeast group has two different roles: they are (1) important in food and beverage production, such as bread and wine, and (2) extremely prejudicial to human health. *Saccharomyces cerevisiae* is the most important yeast species in the baking industry and in beer, wine, and ethanol production in general [2,3]. *Candida* can spoil wine, beer, and other fermented products and in some cases cause severe food poisoning [1]. In addition, significant contamination of mineral and tap water by *Candida parapsilosis*, *Candida glabrata*, and *Candida albicans* has been reported (Table 15.1) [4].

A study performed at different hematology units, in Paris, France, demonstrated that there was a high load of fungal contamination on the food served in these institutions, with *Candida norvegensis*, which causes dangerous invasive infections [5], being isolated from all the soft cheese samples, and the only species of *Candida* detected [6].

Candida species can be extremely prejudicial to human health, and one of the most common ways for *Candida* to enter the human body is by ingestion of food and/or water. Such infections may be (1) superficial, affecting the mucosal membranes, or (2) systemic, involving major body organs [7]. The virulence factors associated with pathogenic yeasts are described throughout this chapter, and laboratory techniques used in the identification are described.

15.1 *CANDIDA* VIRULENCE FACTORS

Candida is the most frequently recovered organism from human fungal infection. The genus contains over 150 species [8], but only minorities have been implicated in human candidoses. Additionally, approximately 65% of species are unable to grow at 37°C, which precludes them from being deep-seated pathogens of humans [8].

TABLE 15.1
Frequency of Yeast Detected on Bottled Mineral and Tap Water

Yeast	Bottled Mineral Water	Tap Water
Candida parapsilosis	10	4
Candida glabrata	6	1
Candida albicans	0	1
Candida spp.	3	1
Non-*Candida* spp.	1	0

Source: Adapted from Yamaguchi, M.U. et al., *Braz. Arch. Biol. Technol.*, 501, 2007.

TABLE 15.2
Most Relevant Virulence Factors Associated with *Candida*–Host Interactions

Virulence Factors	Host Interaction Mechanisms
Adhesion ability	*Promoting retention on biotic and abiotic surfaces*
Relative cell hydrophobicity	Nonspecific adherence process
Expression of cell surface proteins	Facilitates specific adherence mechanisms
Biofilm formation ability	*Promoting antifungal and host defense resistance*
Matrix production	Retention of antifungal molecules
	Protection from host responses
Expression of specific cell proteins	Facilitates specific adhesion and proliferation mechanisms
Invasion and damage of host tissues	
Hyphal development	Promotes invasion of host tissues
Protease production	Host cell and extracellular matrix damage
Phospholipase and lipase production	Damage of host cells' membrane

C. albicans is the most prevalent in healthy and diseased humans [8,9]. However, while *C. albicans* represents over 80% of isolates from all forms of human candidoses [8], the number of infections due to other *Candida* species has increased significantly in the previous two decades [10,11], which may partly be related to improvements in diagnostic methods, such as the use of agar media with the ability to differentiate species, and the introduction of molecular techniques in routine diagnoses [12]. Moreover, the pathogenesis of invasive candidoses is facilitated by virulence factors, including the ability to adhere to medical devices and/or host cells, biofilm formation, filamentous development (hyphae and/or pseudohyphae), and secretion of hydrolytic enzymes (proteases, phospholipases [PLs], and lipases) that allows the organisms to invade and destroy host tissues (Table 15.2).

Despite the intensive research to identify determinants of pathogenicity in fungi, and particularly in *Candida* species, little is known about the actual interactions with hosts.

15.2 *CANDIDA* ADHESION ABILITY

The primary event in *Candida* infection is its adherence to host epithelium promoting retention on biotic surfaces [13,14] (Figure 15.1). Thus, the adhesion phenomenon is considered extremely important as a virulence factor, which cannot be explained by a single specific event but by a combination of specific and nonspecific interactions. Adhesion extension has been related to the surface properties of *Candida*, such as hydrophobicity and electrostatic forces [15,16] (i.e., nonspecific factors). The colony morphology alterations and the expression of specific cell-wall proteins that in

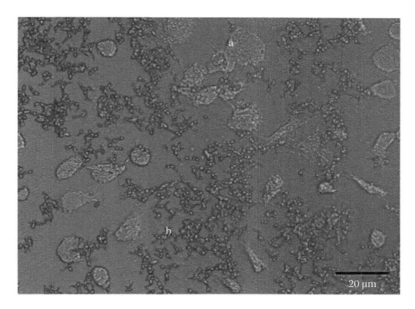

FIGURE 15.1 *C. albicans* invading urinary epithelial cells. ((a) human cell, (b) *Candida* cells; 40×, scale bar 100 μm).

turn can influence cell-to-cell, cell-to-surface, and cell-to-host tissue adherence are considering the most relevant specific factors [17].

The hydrophobicity of *C. albicans* is extremely sensitive to growth conditions although *C. glabrata* is insensitive to the same conditions [18]. In addition, Camacho et al. [19] found no correlation between the hydrophobicity and adherence of *Candida* on siliconized latex catheters, demonstrating that hydrophobicity was not a single predictor for adhesion. *C. glabrata* was described as having a twofold greater tendency to adhere to denture acrylic surfaces compared to *C. albicans* [20], and it was again shown that *C. glabrata* adhered more to urinary epithelial cells than other *Candida* [21]. Recently, Silva et al. [22] showed that oral isolates of *C. albicans* and *C. glabrata* were able to adhere to acrylic (Figure 15.2). *C. albicans* was the most adherent to vascular endothelium in vitro, while *C. glabrata* was the least, followed by *C. parapsilosis* and *C. tropicalis* [23]. Silva et al. [16,24,25]

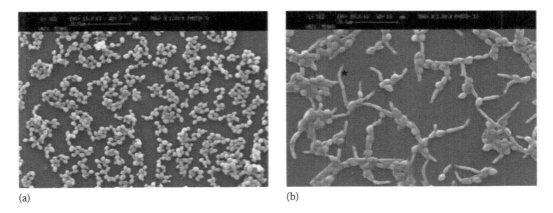

FIGURE 15.2 Scanning electron microscopy images of (a) *C. glabrata* and (b) *C. albicans* adhered to acrylic surfaces at 2 h. *Presence of filamentous forms. Magnification 1000×; bar 20 μm.

showed that *C. glabrata*, *C. parapsilosis*, and *C. tropicalis* presented different extents of colonization on oral epithelium, which is indicative of differences in cell-wall proteins among *Candida* species.

Specific proteins in the *Candida* cell wall, called adhesins, have been correlated with the ability to adhere. Contained within the *C. albicans* adhesins is the agglutinin-like sequence (Als) family, encoded by eight distinct genetic loci (*ALS1* to *ALS7, ALS9*) [26,27]. Als proteins have a similar structure, and mature Als molecules are large glycoproteins that are linked to β-1,6-glucan in the *C. albicans* cell wall [28]. Als3 makes an important contribution to *C. albicans* adhesion to TR146 oral epithelial cells, and subsequent epithelial damage: loss of Als3 results in the reduced capacity of *C. albicans* to induce epithelial cytokines [29].

In *C. glabrata*, a major group of adhesins is encoded by the epithelial adhesin (Epa) gene family [30]. The overall structure of Epa proteins is similar to that of the Als proteins of *C. albicans*. Furthermore, the deletion of merely *EPA1* reduces adherence in vitro despite the large number of *EPA* genes [30]. Although *EPA6* is not expressed in vitro, its expression increases during in vivo urinary infection, suggesting that *C. glabrata* is capable of adapting to different environmental conditions [31]. However, few studies of epithelial and endothelial adhesion have been undertaken with *C. parapsilosis*. Panagoda et al. [32] reported a greater ability (21%) of *C. parapsilosis* for buccal epithelial cell adherence compared to *C. albicans* with an increase of 14% in adhesion to acrylic. These studies suggest that the prevalence and severity of infection caused by *C. parapsilosis* associated with medical devices and/or host surfaces can be due to their high ability to colonize and survive on biomaterial surfaces, even in unfavorable conditions. Furthermore, a bioinformatics search of pathogen-specific gene families of *Candida* species revealed a number of genes for cell-wall proteins in *C. parapsilosis*. This study included genes for five Als proteins and six predicted GPI-anchored protein 30s (Pga 30) [33]. Unfortunately, there has been no further work on the roles these proteins play in *C. parapsilosis* adhesion.

C. tropicalis has the ability to colonize urinary epithelial cells, silicone, and latex catheters [34]; however, the extent of adhesion is strain dependent. At least 3 *ALS* genes were identified through southern and western blot analyses with anti-Als antibody concerning proteins from the *C. tropicalis* cell wall [27]. However, no further work has been undertaken in this area.

Data on adhesion mechanisms to medical devices and human cells of pathogenic fungi are still limited apart from those from *C. albicans*. However, the cell wall is thought to play a crucial role in colonization and infection. Elucidation of its structure and composition would lead to a better understanding of the pathogenesis of *Candida* infections and also to an improvement in treatment, since some cell-wall components constitute valuable targets for antifungal drugs.

15.3 *CANDIDA* BIOFILM FORMATION

The initial attachment of *Candida* to host and/or medical devices is followed by cell division, proliferation, and biofilm development [35]. Biofilms are described as surface-associated communities of microorganisms embedded within an extracellular matrix and represent the most prevalent growth forms of microorganisms [36]. Biofilm formation is a potent virulence factor for some *Candida* species, as it confers significant resistance to antifungal therapy by limiting the penetration of substances through the matrix and protecting cells from host immune responses [37]. Moreover, biofilms formed by *C. albicans*, *C. parapsilosis*, *C. tropicalis*, and *C. glabrata* isolates have been associated with higher mortality rates of patients compared to isolates incapable of forming biofilms [38]. *C. albicans* biofilm formation is associated with the dimorphic switch from yeast to hyphal growth, and the biofilm structure involves, generally, two distinct layers: a thin, basal yeast layer and a thicker, less compact hyphal layer [39]. It is assumed that the formation of mature biofilms and the subsequent production of extracellular matrix are strongly dependent upon species, strain, and environmental conditions (pH, medium composition, and oxygen concentrations) [35,40].

C. albicans biofilm matrix is mainly composed of carbohydrates, proteins, phosphorus, and hexosamines [41], and many studies have focused on *C. albicans* biofilms, due to the associated virulence, whereas only a few reports of other *Candida* species biofilms are available. For example, Shin et al. [42] showed that biofilm formation by *C. glabrata* was reduced compared to other *Candida* species when grown in rich culture media. Biofilms are readily formed by *C. parapsilosis* cells grown in media containing high glucose and lipid concentrations, which is associated with the increased prevalence of bloodstream infections in patients receiving parenteral nutrition. The preference of this species for plastic medical devices is of particular interest, as biofilm formation enhances the capacity of the organism to colonize catheters and intravascular cellular lines [34,43]. *C. parapsilosis* biofilms are thinner, less structured, and consist exclusively of aggregate blastospores in contrast to *C. albicans* [44]. It is known that biofilms are affected by farnesol, a quorum-sensing molecule produced by *C. albicans*. Martins et al. [45] reported that *C. parapsilosis* cells secrete farnesol, a finding that was in contrast to earlier studies by Trofa et al. [43]. However, according to Trofa et al. [43], treatment with extracellular farnesol did arrest growth without an apparent effect on *C. parapsilosis* morphology.

C. parapsilosis lipase knockout mutants have a decreased ability to form biofilms: *C. parapsilosis* mutants produced significantly less biofilm than the wild type [46], and the biofilm and cell-wall regulator (*BCR*) gene was necessary for complete biofilm formation [47]. Notably, the biofilm-deficient *C. parapsilosis* lipase mutants were less virulent in tissue culture infection models and in mice [46]. Lattif et al. [48] demonstrated that the two newly identified species, *C. orthopsilosis* and *C. metapsilosis*, were able to form biofilms similar to *C. parapsilosis*.

C. tropicalis clinical isolates are strong biofilm formers [21,49]. Al-Fattani et al. [36] showed that matrix material extracted from biofilms of *C. tropicalis* and *C. albicans* contained carbohydrates, proteins, hexosamine, phosphorus, and uronic acid. However, the major component in *C. tropicalis* biofilm matrices was hexosamine (27%). The same authors also reported that these biofilms were partially detached after treatment with lipase type VII and chitinase, which contrasted with biofilms of *C. albicans* that were detached after treatment with proteinase K, chitinase, DNase I, or β-*N*-aceytyglucosaminidase. While extensive work has been performed concerning the *C. albicans* genes involved on biofilm formation, little is known about equivalent controlling genes in other *Candida* species.

The studies described highlight the diversity found in terms of biofilm forming ability, structure, and matrix composition of *C. glabrata*, *C. parapsilosis*, and *C. tropicalis* (Figure 15.3). It is very important to continue these studies to elucidate such inherent differences and the possibility to identify and especially combat strains adapted to infection at particular body sites.

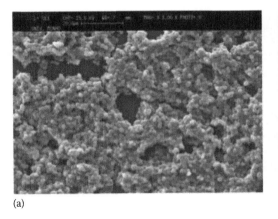
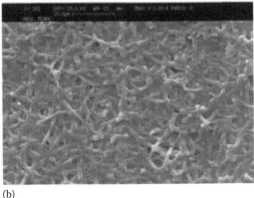

(a)   (b)

FIGURE 15.3 Scanning electron microscopy biofilm images of (a) *C. glabrata* and (b) *C. albicans* on acrylic surfaces formed on artificial saliva at 48 h. Magnification 1000×; bar 20 μm.

15.4 *CANDIDA* INVASION AND DAMAGE OF HOST TISSUES

Candida species can colonize host surfaces, invade deeper into host tissue, and evade host defenses. This section focuses on filamentous structure formation and enzyme secretion (i.e., proteases, PLs, and lipases), which are important prerequisites to pathogenicity.

15.4.1 Filamentous Form Development

Candida comprises an extremely heterogeneous group of fungi that have the ability to grow in several morphological states including yeast form (Figure 15.4 C2), pseudohyphae (Figure 15.4 C1), and/or true filamentous forms (Figure 15.4 C3). The distinction between hyphae (C3) and pseudohyphae (C1) is related to the way in which they are formed: pseudohyphae are formed from yeast cells or from hyphal budding, but the new growth remains attached to the parent cell and is elongate, resulting in filaments with constrictions at the cell–cell junctions. There are no internal cross walls (septa) associated with pseudohyphae. In comparison, true hyphae are formed from yeast cells or even as branches of existing hyphae. Generally, the development of true hyphae is initiated by a *germ tube*, which can elongate and then branch with defined septa that divide the hyphae into separate cells.

C. glabrata is not polymorphic, growing only as blastoconidia, while other *Candida* can develop pseudohyphae and/or hyphae. *C. parapsilosis* does not produce true hyphae but can generate pseudohyphae, which are characteristically large and curved, and often referred to as *giant cells*. True hyphal development is an attribute reserved for *C. albicans* and *C. dubliniensis*.

The ability of *C. albicans*, *C. tropicalis*, and *C. parapsilosis* to switch from yeast to a filamentous morphology promotes invasion of the epithelium and its damage once the yeasts are attached to the host surfaces. *C. glabrata* cells were able to colonize, but not to invade, reconstituted human oral epithelium, and mixed colonization with *C. albicans* enhanced the invasiveness [25]. Moreover, the hyphal morphology increases resistance to phagocytosis by host immune cells [50]. A number of genes correlate positively with hyphal production in *C. albicans* including

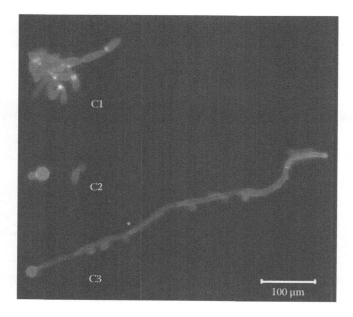

FIGURE 15.4 Photocomposition of the different morphological growth forms of *Candida species*: (C1) pseudohyphae formation, (C2) yeast form, (C3) hyphae formation. *Internal cross walls (septa).

hyphae-specific G1 cyclin (*HGC1*), which encodes for a cytoplasmatic protein without which the hyphal morphology does not occur. *C. albicans* mutants unable to express *HGC1* showed that nonfilamentous strains were less invasive and consequently less virulent [51]. However, *Candida* can respond rapidly to environmental changes, such as pH, glucose concentrations, and hormonal variations, and this could allow these microorganisms to take advantage of impaired immunity and facilitate disease. So one evolutionary advantage that *Candida* species has is the ability to rapidly expressed phenotype switching in response to specific environmental factors, helping them to invade and damage relevant host tissues.

15.4.2 SECRETED ASPARTYL PROTEINASES

The production of hydrolases facilitates invasion of pathogenic microorganisms. Hydrolytic enzymes fulfill a number of functions in addition to the simple role of digesting molecules for nutrient acquisition. Secreted aspartyl proteinases (Saps) facilitate colonization and invasion of host tissues by disruption of the host mucosal membranes [52] and by degrading important immunological and structural defense proteins [53].

The secretion of Sap1p to Sap10p is an important virulence determinant of *C. albicans* [54–56]. *C. glabrata* is capable of proteinase production, but the type has not been specified [57]. In addition, *C. parapsilosis* has a low Sap activity [58] compared to *C. albicans*. Only three *SAP* genes have been identified in *C. parapsilosis* (*SAPP1-3*), two of which remain largely uncharacterized [59]. However, vaginal and skin isolates of *C. parapsilosis* exhibit higher in vitro Sap activity than blood isolates [60,61].

As with *C. albicans*, in vitro studies revealed that *C. tropicalis* secreted higher levels of Saps in a medium containing bovine serum albumin (BSA) as the sole source of nitrogen. Furthermore, it is known that *C. tropicalis* possesses at least four genes encoding Saps (*SAPT1-4*) [62,63]. Sap1 is the only enzyme that has been purified, biochemically characterized, and crystallized [62,64]. The presence of aspartic proteinases secreted by *C. tropicalis* has also been reported on the surface of fungal elements penetrating tissues during disseminated infection and evading macrophages after phagocytosis of yeast cells [52,65]. Moreover, *C. tropicalis* is the most virulent of all *Candida* species apart from *C. albicans* and can colonize and penetrate the gastrointestinal mucosa of mice [66].

Naglik et al. [67] and Lerman et al. [68] have demonstrated that *C. albicans SAP* expression in an oral environment was not associated with invasion and tissue damage. Furthermore, proteinases are not involved in the invasion of reconstituted human oral epithelium by *C. tropicalis* [69] and *C. parapsilosis* [70], but a role for these enzymes in tissue damage caused by *C. parapsilosis* was indicated [70].

15.4.3 PHOSPHOLIPASES AND LIPASES

Pls are often considered as factors in *Candida* pathogenicity. Pls hydrolyze phospholipids into fatty acids, and according to the different and specific ester bonds cleaved, they have been classified into Pls A, B, C, and D [37]. The production of all classes of Pls has been described for *Candida* species, and their production could contribute to (1) host cell membrane damage promoting cell damage and/or (2) exposing receptors, hence facilitating adherence [71,72]. The most widely used diagnostic method for Pl determination is growth on egg yolk agar media. Several studies indicate that *Candida* species other than *C. albicans* are able to produce extracellular PLs [73–75] or significantly smaller amounts of Pls compared to *C. albicans* [71]. There have been contradictory findings, with some investigators reporting Pl activity in 51% of strains assayed [71] and others describing none [72]. While *C. tropicalis* appears to have a reduced ability to produce extracellular PLs, this production is strongly strain dependent [21,73–76]. Furthermore, no studies have been reported concerning *C. glabrata* Pl production and only a few on *C. tropicalis* and *C. parapsilosis* [77].

Lipases are involved in the hydrolysis and synthesis of triacylglycerols and are characterized by being stable at high temperatures and in organic solvents and being resistant to proteolysis [78]. Ten lipase genes have been identified in *C. albicans* [58]. Two lipase genes, *CpLIP1* and *CpLIP2*, were reported with CpLIP2 known to encode for an active protein in *C. parapsilosis* [78,79]. Recently, Gácser et al. [46] demonstrated that a lipase inhibitor significantly reduced tissue damage during *C. parapsilosis* infection of reconstituted human tissues and that *CpLIP1–CpLIP2* mutants formed thinner and less complex biofilms. They also (1) had reduced growth in lipid-rich media, (2) were more efficiently ingested and killed by macrophage-like cells, and (3) were less virulent in infections of reconstituted human oral epithelium. Recent genomic DNA sequencing projects suggest that two additional *CpLIP* genes may exist in *C. parapsilosis* [43]. Sequences similar to *C. albicans* (*LIP1-10*) were also detected in other pathogenic *Candida* species such as *C. tropicalis* but not *C. glabrata* [80]. However, no relevant studies have been performed concerning the role of these genes on virulence of this species.

15.5 *CANDIDA* SPECIES IDENTIFICATION

The laboratory identification and quantification of *Candida* species is essential for establishing a diagnosis of candidoses. Standard approaches to the laboratory identification of *Candida* depend on (1) recovery of samples in cultures of blood, body fluids, water, food, or specimens from other sites; (2) direct microscopic analyses; and (3) identification and quantification to the species level. Samples can be obtained by a variety of methodologies with most common approaches being swab, imprint culture from contaminants (e.g., tissues, food, water, and medical utensils). These methods have drawbacks, and the most appropriate method is largely governed by the *Candida* contamination and its source. Samples for the detection of *Candida* are generally cultured on Sabouraud dextrose agar (SDA). Moreover, in recent years, a differential medium, CHROMagar Candida, was developed that allows the identification of certain *Candida* species based on colony appearance and color following primary culture. Presumptive identification of yeasts based on primary culture media can be confirmed through a variety of supplemental tests traditionally based on morphological and physiological characteristics of the isolates. Moreover, molecular identification methods are becoming popular due to their accuracy, sensitivity, and specificity for the identification and differentiation of *C. albicans* from other *Candida* species. For molecular identification, several procedures have been proposed to detect and differentiate *Candida* species by polymerase chain reaction (PCR) or non-PCR molecular methods.

15.5.1 PCR-Based Molecular Methods

The invention of PCR was a landmark in the progress of molecular microbiology and had a substantial impact on the diagnosis of candidoses. The key strength of these techniques consists of the amplification and detection of microbial nucleic acid within the background of host DNA [81] or even directly from samples [82]. To ensure high sensitivity of PCR detection, primers should preferentially target multicopy genes. Also, specificity should be secured by targeting sequences found only in the pathogen of interest. The ribosomal RNA (rRNA) gene appears to meet the criteria. This consists of the small subunit rRNA gene (18S), the 5.8S gene, and the large subunit rRNA (25S) gene, separated by the internal transcribed spacer regions, ITS1 and ITS2. While rRNA genes are highly conserved in fungi, ITS regions involve highly variable and highly conserved areas, thus allowing the detection and identification of *Candida* species. The ITS regions have been used to detect *Candida* species from serum [83], whole blood [84,85], respiratory samples [86], and tissues [87]. Recent studies have suggested a multiplex PCR is more sensitive and specific to rapidly and simultaneously identify the most common pathogenic fungi using tandem multiplex PCR [88–90]. For example, Carvalho et al. [90] described a multiplex PCR strategy allowing the identification of eight clinically relevant *Candida* species, namely, *C. albicans*, *C. glabrata*, *C. parapsilosis*, *C. tropicalis*,

C. krusei, C. guilliermondii, C. lusitaniae, and *C. dubliniensis*. The multiplex PCR was based on the amplification of two fragments from the ITS1 and ITS2 regions by the combination of two yeast-specific and eight species-specific primers in a single PCR reaction. With this approach, it was possible to identify 231 clinical isolates with high specificity directly from clinical specimens, which attests to the laboratory applicability of the method. Importantly, this straightforward method presented a sensitivity of approximately 2 cells mL^{-1} and can be undertaken within 5 h.

In addition, nested PCR can be used to increase the sensitivity and specificity of PCR detection where two rounds of PCR are performed. In the first round, outer primers target a larger region for amplification. Amplicons from this round are then added as template into the second round reaction mixture, where inner primers target a fragment of the first round amplification. Specificity of the assay is increased, because four primers have to anneal in an arranged fashion instead of just two in a single PCR. It is important to emphasize that when the second round of primers are carefully designed to prevent interference, primer mixes can be used in a common reaction mixture reducing the costs of the multiplex PCR methodologies. However, a potential problem of this method is nontarget contamination from the greater degree of manipulation of samples required (see Bretagne et al. in this current book). Nested PCR approach was adapted for use with *Candida* species by Kanbe et al. [91]. Degenerate and specific primers based on the genomic sequences of DNA topoisomerase II of *C. albicans, C. dubliniensis, C. tropicalis* (genotypes I and II), *C. parapsilosis* (genotypes I and II), *C. krusei, C. kefyr, C. guilliermondii, C. glabrata,* and *C. lusitaniae* were designed, and their specificities tested in PCR-based identifications. Each of the specific primers selectively and exclusively amplified its own DNA fragment, not only from the corresponding genomic DNA of the *Candida* species but also from DNA mixtures containing other DNAs from several fungal species. Recently, Sugawara and coauthors [92] proposed a new method for the detection of fungal pathogens in hematological patients based also on PCR. The panfungal PCR assay system presented a broad range of PCR targets in the highly conserved sequence of the 18S rRNA gene in fungal DNA. According to the authors, the sensitivity and specificity of this assay was 100% and 92%, respectively.

The real-time PCR (RT-PCR) uses fluorescent reporter molecules to visualize the production of amplicons during each cycle of the PCR reaction. This is in contrast to endpoint detection in conventional PCR, where the amplicon is detected after completed amplification only. Moreover, the process of amplification can be monitored using labeled probes, which specifically hybridize to the newly formed amplicon molecules, or by staining newly formed double-stranded dyes (e.g., SYBR Green I, BEBO, or LC Green). The use of probes increases the specificity of PCR, because an additional sequence homology between the amplicon and the probe is necessary for successful amplification. Recent studies have used RT-PCR using different technologies to measure gene expression more rapidly and accurately [83,93–102]. This methodology provides a rapid automated combined PCR amplification and detection system with no postamplification manipulation of amplicons, thereby considerably reducing the risk of contamination. For example, TaqMan PCR was used for a rapid identification of clinically important *Candida* species (*C. albicans, C. glabrata, C. tropicalis, C. krusei, C. parapsilosis,* and *C. kefyr*) [103]. Primers and probe sets were shown to be 100% specific for their respective species. Similarly, Shin et al. [104] demonstrated that the fluorescent species-specific probes detected and correctly identified 95.1% of *C. albicans, C. glabrata, C. tropicalis, C. krusei,* and *C. parapsilosis* without false positives. RT PCR and consecutive high-resolution melting analysis was used for the detection and differentiation of various fungal pathogens [105] and discriminated between most relevant *Candida* species from *Aspergillus* and *Cryptococcus neoformans* with high sensitivity.

15.5.2 Non-PCR-Based Molecular Methods

Fluorescent in situ hybridization (FISH) with fluorescein-labeled oligonucleotide probes is a convenient way to detect yeasts without the need for pure culture. The employment of novel peptide nucleic acid (PNA) probes combines high affinity with the advantage of targeting highly structured rRNA regions, which has extended the utility of this method. The *Candida* PNA FISH assay has very high sensitivity and specificity [106,107]. The advantages of PNA FISH technology include the ability for it to be performed directly from positive blood culture bottles where the identifications are definitive due to the highly species-specific PNA probes. Shepard et al. [106] evaluated the use of PNA FISH for accurate real-time identification of *C. albicans* and *C. glabrata* from blood cultures newly assessed as positive for *Candida*. This study supported the use of the assay for rapid, simultaneous identifications of *C. albicans* and *C. glabrata* under these circumstances. Also, the probes can be added directly to smears made from the contents of the blood culture bottle. This method facilitates the accurate identification of clinical yeast isolates using two scoring techniques: flow cytometry and fluorescence microscopy. It is important to emphasize that the sensitivity of PNA FISH is similar to most results obtained by PCR-based assays. The entire PNA FISH takes only 2.5 h after a blood culture is designed positive for the presence of *Candida* due to the straightforward protocol that excludes DNA extraction.

15.6 CONCLUDING REMARKS

Candida species can be isolated from food and water, which increases their probability of causing infections. Moreover, the yeast presents a set of different virulence factors that make difficult treatment and control. The opportunistic pathogens utilize several genes and proteins that play important roles in adhesion, biofilm formation, filamentous development, and enzyme secretion. These determinants are involved in virulence. Given the increased incidence of candidoses and the unacceptable high morbidity and mortality levels, it is essential to increase our knowledge of the virulence determinants that will contribute toward the identification of new targets for future therapeutics against *Candida*. Furthermore, rapid and accurate identification of the disease-causing species of *Candida* is crucial for clinical treatment of localized and systemic candidiases. Molecular strategies, such as PCR or non-PCR molecular methods, have been used to complement conventional methods providing a more accurate and less time-consuming methodology. Given the high accuracy and speed with which molecular typing techniques can be carried out and rapid advances in technology, it is likely that most of these methods will improve routine clinical laboratory identification of *Candida* species.

REFERENCES

1. Lasztity, R. Food microbiology, p. 289. In *Food Quality and Standards*, Vol. III, Budapest, Hungary, 2004.
2. Bekatorou, A. et al. Production of food grade yeasts. *Food Technol. Biotechnol.* 44, 407, 2006.
3. Nevoigt, E. et al. Progress in metabolic engineering of *Saccharomyces cerevisiae*. *Microbiol. Mol. Biol. Rev.* 72, 379, 2008.
4. Yamaguchi, M.U. et al. Yeasts and filamentous fungi: In bottled mineral water and tap water from municipal supplies. *Braz. Arch. Biol. Technol.* 501, 2007.
5. Nielsen, H. et al. *Candida norvegensis* peritonitis and invasive disease in a patient on continuous ambulatory peritoneal dialysis. *J. Clin. Microbiol.* 28, 1664, 1990.
6. Bouakline, A. et al. Fungal contamination of food in hematology units. *J. Clin. Microbiol.* 11, 4272, 2000.
7. Rupping, M.J. et al. Patients at high risk of invasive fungal infections: When and how to treat. *Drugs* 68, 1941, 2008.
8. Calderone, R.A. Introduction and historical perspectives, pp. 15–25. In R. Calderone (ed.), *Candida and Candidiasis*. ASM Press, Washington, DC, 2002.
9. Samaranayake, L.P. et al. Fungal infections associated with HIV infection. *Oral Dis.* 8, 151, 2002.
10. Manzano-Gayosso, P. et al. Candiduria in type 2 diabetes mellitus patients and its clinical significance. *Candida* spp. antifungal susceptibility. *Rev. Med. Inst. Mex. Seguro. Soc.* 46, 603, 2008.
11. Ruan, S. et al. Invasive candidiasis: An overview from Taiwan. *J. Med. Assoc.* 108, 443, 2009.

12. Liguori, G. et al. Oral candidiasis: A comparison between conventional methods and multiplex polymerase chain reaction for species identification. *Oral Microbiol. Immunol.* 24, 76, 2009.
13. Crump, A. et al. Intravascular catheter-associated infections. *Eur. J. Clin. Microbiol. Infect. Dis.* 19, 1, 2000.
14. Chandra, J. et al. Biofilm formation by the fungal pathogen *Candida albicans*: Development, architecture, and drug resistance. *J. Bacteriol.* 183, 5385, 2001.
15. Henriques, M. et al. Experimental methodology to quantify *Candida albicans* cell surface hydrophobicity. *Biotechnol. Lett.* 24, 1111, 2002.
16. Silva, S. et al. Silicone colonisation by non-*Candida albicans Candida* species in the presence of urine. *J. Med. Microbiol.* 59, 747–754, 2010.
17. Chaffin, W.L. *Candida albicans* cell wall proteins. *Microbiol. Mol. Biol. Rev.* 72, 495, 2008.
18. Kikutani, H. et al. The murine autoimmune diabetes model: NOD and related strains. *Adv. Immunol.* 51, 285, 1992.
19. Camacho, D. et al. The effect of chlorhexidine and gentian violet on the adherence of *Candida* spp. to urinary catheters. *Mycopathalogia* 163, 261, 2007.
20. Luo, G. et al. Reverse transcriptase polymerase chain reaction (RT-PCR) detection of HLP gene expression in *Candida glabrata* and its possible role in vitro haemolysin production. *APMIS* 112, 283, 2004.
21. Negri, M. et al. Crystal violet staining to quantify *Candida* adhesion to epithelial cells. *Br. J. Biomed. Sci.* 67, 120, 2010.
22. Silva, S. et al. Co-colonization of *Candida glabrata* and *Candida albicans* on acrylic and the effect of silver nanoparticles and nystatin on pre-formed biofilms. *Med. Mycol.* DOI:10.3109/13693786.2012.700492.
23. Klotz, S.A. et al. Adherence and penetration of vascular endothelium by *Candida* yeast. *Infect. Immun.* 42, 955, 1985.
24. Silva, S. et al. Biofilms of non-*Candida albicans Candida* species: Quantification, structure and matrix composition. *Med. Mycol.* 47, 681–689, 2009.
25. Silva, S. et al. *Candida glabrata* and *Candida albicans* co-infection of an in vitro epithelium. *J. Oral Pathol. Med.* 40, 421, 2011.
26. Hoyer, L.L. et al. Discovering the secrets of the *Candida albicans* agglutinin-like sequence (*ALS*) gene family—A sticky pursuit. *Med. Mycol.* 46, 1, 2008.
27. Hoyer, L.L. et al. The *ALS* gene family of *Candida albicans*. *Trends Microbiol.* 9, 176, 2001.
28. Kapteyn, J.C. et al. The cell wall architecture of *Candida albicans* wild-type cells and cell wall-defective mutants. *Mol. Microbiol.* 35, 601, 2000.
29. Murciano, C. et al. Evaluation of the role of *Candida albicans* agglutinin-like sequence (*ALS*) proteins in human oral epithelial cell interactions. *PLoS One* 7(3), e33362, 2012.
30. De Las Penas, A. et al. Virulence-related surface glycoproteins in the yeast pathogen *Candida glabrata* are encoded in subtelomeric clusters and subject to RAP1- and SIR-dependent transcriptional silencing. *Genes Dev.* 17, 2245, 2003.
31. Domergue, R. et al. Nicotinic acid limitation regulates silencing of *Candida* adhesins during UTI. *Science* 308, 870, 2005.
32. Panagoda, G.J. et al. Adhesion of *Candida parapsilosis* to epithelial and acrylic surfaces correlates with cell surface hydrophobicity. *Mycoses* 44, 29, 2001.
33. Butler, G. et al. Evolution of pathogenicity and sexual reproduction in eight *Candida* genomes. *Nature* 459, 657, 2009.
34. Negri, M. et al. An in vitro evaluation of *Candida tropicalis* infectivity using human cell monolayers. *J. Med. Microbiol.* 60, 1275, 2011.
35. Ramage, G. et al. *Candida* biofilms on implanted biomaterials: A clinically significant problem. *FEMS Yeast Res.* 6, 979, 2006.
36. Al-Fattani, M.A. et al. Biofilm matrix of *Candida albicans* and *Candida tropicalis*: Chemical composition and role in drug resistance. *J. Med. Microbiol.* 55, 999, 2006.
37. Mukherjee, P.K. et al. *Candida* biofilm resistance. *Drug Resist. Update* 7, 301, 2004.
38. Kumamoto, C.A. *Candida* biofilms. *Curr. Opin. Microbiol.* 5, 608, 2002.
39. Donlan, R.M. et al. Biofilms: Survival mechanisms of clinically relevant microorganisms. *Clin. Microbiol. Rev.* 15, 167, 2002.
40. Jain, N. et al. Biofilm formation by and antifungal susceptibility of *Candida* isolates from urine. *Appl. Environ. Microbiol.* 73, 1697, 2007.
41. Baillie, G.S. et al. Matrix polymers of *Candida* biofilms and their possible role in biofilm resistance to antifungal agents. *J. Antimicrob. Chemother.* 46, 397, 2000.

42. Shin, J.H. et al. Biofilm production by isolates of *Candida* species recovered from nonneutropenic patients: Comparison of bloodstream isolates with isolates from other sources. *J. Clin. Microbiol.* 40, 1244, 2002.
43. Trofa, D. et al. *Candida parapsilosis*, an emerging fungal pathogen. *Clin. Microbiol. Rev.* 21, 606, 2008.
44. Kuhn, D.M. et al. Comparison of biofilms formed by *Candida albicans* and *Candida parapsilosis* on bioprosthetic surfaces. *Infect. Immun.* 70, 878, 2002.
45. Martins, M. et al. Presence of extracellular DNA in the *Candida albicans* matrix and its contribution to biofilms. *Mycopathologia* 5, 323, 2010.
46. Gácser, A. et al. Virulence of *Candida parapsilosis*, *Candida orthopsilosis*, and *Candida metapsilosis* in reconstituted human tissue models. *Fungal Genet. Biol.* 44, 1336, 2007.
47. Din, C. et al. Development of a gene knockout system in *Candida parapsilosis* reveals a conserved role of *BCR1* in biofilm formation. *Eukaryot Cell* 6, 1310, 2007.
48. Lattif, A.A. et al. Characterization of biofilms formed by *Candida parapsilosis*, *C. metapsilosis*, and *C. orthopsilosis*. *Int. J. Med. Microbiol.* 300, 265, 2009.
49. Negri, M. et al. Examination of potential virulence factors of *Candida tropicalis* clinical isolates from hospitalized patients. *Mycopathologia* 169, 175, 2010.
50. Svobodová, E. et al. Differential interaction of the two related fungal species *Candida albicans* and *Candida dubliniensis* with human neutrophils. *J. Immunol.* 1892, 502, 2012.
51. Zheng, X. et al. Hgc1, a novel hypha-specific G1 cyclin-related protein regulates *Candida albicans* hyphal morphogenesis. *EMBO J.* 23, 1845, 2004.
52. Rüchel, R. et al. *Candida* acid proteinases. *J. Med. Vet. Mycol.* 30, 123, 1992.
53. Pichová, I. et al. Secreted aspartic proteases of *Candida albicans*, *Candida tropicalis*, *Candida parapsilosis* and *Candida lusitaniae*: Inhibition with peptidomimetic inhibitors. *Eur. J. Biochem.* 268, 2669, 2001.
54. Naglik, J.R. et al. In vivo analysis of secreted aspartyl proteinases expression in human oral candidiases. *Infect. Immun.* 67, 2482, 1999.
55. Hube, B. et al. The role and relevance of phospholipase D1 during growth and dimorphism of *Candida albicans*. *Microbiology* 147, 879, 2001.
56. Monod, M. et al. Secreted aspartic proteases as virulence factors of *Candida* species. *Biol. Chem.* 383, 1087, 2002.
57. Chakrabarti, A. et al. in vitro proteinase production by *Candida* species. *Mycopathologia* 114, 163, 1991.
58. Kobayashi, C. et al. Candiduria in hospital patients: A study prospective. *Mycopathologia* 158, 49, 2004.
59. Merkerová, M. et al. Cloning and characterization of Sapp2p, the second aspartic proteinase isoenzyme from *Candida parapsilosis*. *FEMS Yeast Res.* 6, 1018, 2006.
60. Cassone, A. et al. Biotype diversity of *Candida parapsilosis* and its relationship to the clinical source and experimental pathogenicity. *J. Infect Dis.* 171, 967, 1995.
61. Dagdeviren, M. et al. Acid proteinase, phospholipase and adherence properties of *Candida parapsilosis* strains isolated from clinical specimens of hospitalized patients. *Mycoses* 48, 321, 2005.
62. Togni, G. et al. Isolation and nucleotide sequence of the extracellular acid protease gene (ACP) from the yeast *Candida* tropicalis. *FEBS Lett.* 286, 181, 1991.
63. Zaugg, C. et al. Secreted aspartic proteinase family of *Candida tropicalis*. *Infect. Immun.* 69, 405, 2001.
64. Symersky, J. et al. High-resolution structure of the extracellular aspartic proteinase from *Candida tropicalis* yeast. *Biochemistry* 36, 12700, 1997.
65. Borg, M. et al. Demonstration of fungal proteinase during phagocytosis of *Candida albicans* and *Candida tropicalis*. *J. Med. Vet. Mycol.* 28, 3, 1990.
66. Wingard, J.R. et al. Pathogenicity of *Candida tropicalis* and *Candida albicans* after gastrointestinal inoculation in mice. *Infect. Immun.* 29, 808, 1990.
67. Naglik, J.R. et al. Quantitative expression of the *Candida albicans* secreted aspartyl proteinase gene family in human oral and vaginal candidiasis. *Microbiology* 154, 3266, 2008.
68. Lermann, U. et al. Secreted aspartic proteases are not required for invasion of reconstituted human epithelia by *Candida albicans*. *Microbiology* 154, 3281, 2008.
69. Silva, S. et al. The role of secreted aspartyl proteinases in *Candida tropicalis* invasion and damage of oral mucosa. *Clin. Microbiol. Infect.* 17, 264, 2010.
70. Silva, S. et al. Characterization of *Candida parapsilosis* infection of an in vitro reconstituted human oral epithelium. *Eur. J. Oral Sci.* 117, 669, 2009.
71. Ghannoum, M.A. Potential role of phospholipases in virulence and fungal pathogenesis. *Clin. Microbiol. Rev.* 13, 122, 2000.

72. Kantarciŏlu, A.S. et al. Phospholipase and protease activities in clinical *Candida* isolates with reference to the sources of strains. *Mycoses* 45, 160, 2002.
73. Furlaneto-Maia, L. et al. in vitro evaluation of putative virulence attributes of oral isolates of *Candida* spp. obtained from elderly healthy individuals. *Mycopathologia* 166, 209, 2007.
74. Cafarchia, C. et al. Phospholipase activity of yeasts from wild birds and possible implications for human disease. *Med. Mycol.* 46, 429, 2008.
75. Galan-Ladero, M.A. et al. Enzymatic activities of *Candida tropicalis* isolated from hospitalized patients. *Med. Mycol.* 48, 207–210, 2010.
76. Kumar, V.G. et al. Phospholipase C, proteinase and hemolytic activities of *Candida* spp. isolated from pulmonary tuberculosis patients. *J. Mycol. Med.* 19, 3–10, 2009.
77. Brocherhoff, H. Model of interaction of polar lipids, cholesterol, and proteins in biological membranes. *Lipids* 9, 645–650, 1974.
78. Neugnot, V. et al. The lipase/acyltransferase from *Candida parapsilosis*: Molecular cloning and characterization of purified recombinant enzymes. *Eur. J. Biochem.* 269, 1734, 2002.
79. Brunel, L. et al. High-level expression of *Candida parapsilosis* lipase/acyltransferase in Pichia pastoris. *J. Biotechnol.* 111, 41, 2004.
80. Filler, S.G. et al. *Candida albicans* stimulates endothelial cell eicosanoid production. *J. Infect. Dis.* 164, 928–935, 1991.
81. Williams, D.W. et al. Identification of *Candida* species by PCR and restriction fragment length polymorphism analysis of intergenic spacer regions of ribosomal DNA. *J. Clin. Microbiol.* 33, 2476–2479, 1995.
82. Quiles-Melero, I. et al. Rapid identification of yeast from positive blood culture bottles by pyrosequencing. *Eur. J. Clin. Microbiol. Infect. Dis.* 30, 21–24, 2010.
83. McMullan, R. et al., A prospective clinical trial of a real-time polymerase chain reaction assay for diagnosis of candidemia in nonneutropenic, critically ill adults. *Clin. Infect. Dis.* 46, 890, 2008.
84. Skovberg, S. et al., Optimization of the detection of microbes in blood from immunocompromised patients with haematological malignancies. *Clin. Infect. Dis.* 15, 680, 2009.
85. Baddie, P. et al., Detection of systemic candidiasis in the whole blood of patients with hematologic malignancies. *Clin. Infect. Dis.* 62, 1, 2009.
86. Schabereiter-Gurtner, C. et al. Development of novel real time PCR assays for detection and differentiation of eleven medically important *Aspergillus* and *Candida* species in clinical specimens. *J. Clin. Microbiol.* 45, 906, 2007.
87. Hendolin, P.H. et al. Panfungal PCR and multiplex liquid hybridization for detection of fungi in tissue specimens. *J. Clin. Microbiol.* 38, 4186, 2000.
88. Lau, A. et al., Development and clinical application of a panfungal PCR assay to detect and identify fungal DNA in tissue specimens. *J. Clin. Microbiol.* 45, 380, 2007.
89. Lau, A. et al. Multiplex Tandem PCR: A Novel platform for rapid detection and identification of fungal pathogens from blood culture specimens. *J. Clin. Microbiol.* 46, 3021, 2008.
90. Carvalho, A. et al. Multiplex PCR identification of eight clinically relevant *Candida* species. *Med. Mycol.* 45, 619, 2007.
91. Kanbe, T. et al. PCR-based identification of pathogenic *Candida* species using primer mixes specific to Candida DNA topoisomerase II genes. *Yeast* 19, 973, 2002.
92. Yumiko, S. et al. Clinical utility of a Panfungal polymerase chain reaction assay for invasive fungal diseases in patients with haematologic disorders. *Eur. J. Haematol.* DOI: 10.1111/ejh.12078, 2013.
93. White, P.L. et al. Comparison of non-culture based methods for detection of systemic fungal infections with emphasis on invasive Candida infections. *J. Clin. Microbiol.* 43, 2181, 2005.
94. Klingspor, L. Molecular detection and identification of *Candida* and *Aspergillus* spp. from clinical samples using real-time PCR. *J. Clin. Microbiol.* 12, 745, 2006.
95. Wellinghausen, N. et al. Rapid diagnosis of candidemia by real-time PCR detection of Candida DNA blood samples. *J. Med. Microbiol.* 58, 1106, 2009.
96. Maaroufi, Y. et al. Rapid identification of *Candida albicans* in clinical blood samples by using a TaqMan-based PCR assay. *J. Clin. Microbiol.* 41, 3293, 2003.
97. Jordanides, N.E. et al., A prospective study of real time panfungal PCR test for the early diagnosis of invasive fungal infection in haemato-oncology patients. *Bone Marrow Transplant* 35, 389, 2005.
98. Basová, L. et al. The Pan-AC assay: A single-reaction real time PCR test for quantitative detection of a broad range of *Aspergillus* and *Candida* Species. *J. Med. Microbiol.* 56, 1167, 2007.
99. Innings, A. et al. Multiplex real-Time PCR targeting the RNase P RNA gene for detection and identification of *Candida* species in blood. *J. Clin. Microbiol.* 45, 1176, 2007.

100. Vollmer, T. et al. Evaluation of a novel broad-range real-time PCR assay for rapid detection of human pathogenic fungi in various clinical specimens. *J. Clin. Microbiol.* 46, 906, 2007.
101. Bizerra, F.C. et al. Characteristics of biofilm formation by *Candida tropicalis* and antifungal resistance. *FEMS Yeast Res.* 8, 442, 2008.
102. Nett, J.E. et al. Development and validation of an in vivo *Candida albicans* biofilm denture model. *Infect. Immun.* 78, 3650, 2010.
103. Xie, Z. et al. A quantitative real-time RT-PCR assay for mature *C. albicans* biofilms. *BMC Microbiol.* 11, 93, 2011.
104. Shin, J.H. et al. Rapid identification of up to three *Candida* species in a single reaction tube by a 5′ exonuclease assay using fluorescent DNA probes. *J. Clin. Microbiol.* 37, 165, 2010.
105. Somogyvari, F. et al. Detection of invasive fungal pathogens by real-time PCR and high-resolution melting analysis. *In Vivo* 26, 979.
106. Shepard, J.R. et al. Multicenter evaluation of the *C. albicans*/*C. glabrata* PNA FISH method for simultaneous dual color identification of *Candida albicans* and *Candida glabrata* directly from blood culture bottles. *J. Clin. Microbiol.* 46, 50, 2008.
107. Farina, C. et al. Evaluation of the peptide nucleic acid fluorescence in situ hybridisation technology for yeast identification directly from positive blood cultures: An Italian experience. *Mycoses* 55, 388, 2012.

16 Chaetomium*

Vit Hubka

CONTENTS

16.1 Introduction 211
16.2 Clinical Significance.. 212
      16.2.1 Key to Clinically Significant *Chaetomium/Achaetomium* Species....................... 217
16.3 Mycotoxins 217
16.4 Occurrence in Water.. 218
16.5 Occurrence in Foods.. 219
      16.5.1 Key to *Chaetomium* Species Relevant to Foods (at Least Twice
            Reported from Food Products) 220
16.6 Diagnosis 220
16.7 Conclusions 221
Acknowledgments... 221
References... 221

16.1 INTRODUCTION

Chaetomium is a member of the ascomycete family Chaetomiaceae (Sordariales, Sordariomycetes). This species-rich genus encompasses approximately 130 accepted species; most of them colonize cellulose-rich substrates, such as plant debris, herbivore dung, paper, seeds, soil, textiles, or timber.[1–3] *Chaetomium* spp. are generally referred to as soil saprophytes and are also commonly isolated from the soil by techniques for isolation of keratinophilic fungi.[4,5] The genus includes predominantly homothallic species producing ostiolate perithecia that are usually covered with hairs and attached to substrates by rhizoidal hyphae. The morphology of ascomata, hairs, and ascospores (brownish, grayish, olive brown; never opaque) and maximum growth temperature are the most important features in species differentiation.[2,3]

Some anamorphs are associated with *Chaetomium* species and are produced simultaneously with ascomata or are phylogenetically embedded within a robust Chaetomium clade, but teleomorphs have not been observed. *Botryotrichum*, *Humicola*, *Staphylotrichum*, *Trichocladium*, *Acremonium*-like, *Chrysosporium*-like, *Papulaspora*-like, *Paecilomyces*-like, *Scopulariopsis*-like, and *Histoplasma*-like anamorphs have been reported.[2,3,6–11] Many *Chaetomium* spp., including the majority of medically important species, lack anamorph and are known only by the sexual state. These species are exceptional in this aspect, because the infections due to fungi propagating strictly by ascospores are very rare in medical mycology.

Chaetomium spp. are uncommonly, however increasingly, reported as causal agents of various infections in humans. Many authors also include *Chaetomium* in the list of fungi responsible for allergic reactions and asthma.[12] *C. globosum* is distributed worldwide and frequently found in indoor environments particularly water-damaged buildings[13,14] and was also detected in the air of special-care units of a hospital.[15] *C. globosum* together with *Aspergillus fumigatus* and *Eurotium* spp. was found to be the major contributor to the higher Environmental Relative Moldiness Index[16] values in the homes of North Carolina asthmatic children compared to homes in the rest of the United States[17]

---

* This chapter covers literature sources published before February 2013 (accepted March 13, 2013).

and is also present on the list of species probably contributing to the development of childhood respiratory illness.[18,19] *Chaetomium* spp. are occasionally identified as contaminants of clinical material, in particular of specimens from skin and nails,[20] and their etiological significances have to be clarified. The majority of proven infections can be classified as phaeohyphomycosis based upon histopathological observation of dark-walled dematiaceous (syn. phaeoid) elements. The Masson-Fontana stain is the most appropriate technique for demonstrating dematiaceous fungal elements in tissues.[21] Only hyaline hyphae were observed in two described cases of onychomycosis due to *Chaetomium*,[22,23] and it is possible that *Chaetomium* spp. may, in some cases, manifest as hyaline or weakly pigmented hyphae in invaded tissue, although melanin in the walls can be accentuated using Masson-Fontana stain supporting the diagnosis of phaeohyphomycosis. A disadvantage of many other stains is that they make it impossible to determine whether the fungus is dematiaceous or hyaline—crucial information for the diagnosis of phaeohyphomycosis.

Controversial and not fully resolved is the position of the genus *Achaetomium* with respect to *Chaetomium*. *Achaetomium* spp. based on the Rodríguez et al.[24] concept have thicker ascomatal walls in comparison to *Chaetomium*, dark, opaque ascospores and ascomata covered with yellowish hyphae-like hairs (tomentose). *Achaetomium* spp. are thermotolerant in contrast to the majority of *Chaetomium* spp. that are mesophilic or psychrotolerant with only several thermophilic species.[3,25,26] However, the molecular support for distinctness of *Achaetomium* from *Chaetomium* is low,[24] and more detailed studies are needed. The medically important species *Achaetomium strumarium*, a causal agent of cerebral phaeohyphomycosis, can be designated as *Chaetomium* or *Achaetomium*, and due to the taxonomic ambiguity, this species is discussed in this chapter.

Chaetomium members produce a rich spectrum of secondary metabolites (including mycotoxins in a strict sense) with medical implications.[27] Some species are used as producers of thermostable cellulases[28] or as biological control agents of plant pathogens.[29] Species are commonly isolated as water and food contaminants or spoilage fungi of specific food products. However, their metabolites were detected mostly only in vitro or may be produced by more significant spoilage fungi (e.g., chaetoglobosins by penicillia or sterigmatocystin by aspergilli), and further investigations are needed to confirm health risks associated with *Chaetomium* members isolated from food and water.

16.2 CLINICAL SIGNIFICANCE

The infections due to *Chaetomium* manifest histopathologically as various forms of phaeohyphomycosis including superficial (cutaneous and ungual where only keratinized tissues are involved), subcutaneous, corneal, mucosal, systemic (including cerebral), and disseminated phaeohyphomycosis (multiple organs involved or the isolate is recovered from blood).[30–32] Cutaneous, ungual, subcutaneous, and corneal phaeohyphomycosis predominantly occurs in immunocompetent individuals. The predisposing factor is either unknown or typically the infection follows a trauma (Table 16.1). Cutaneous and subcutaneous phaeohyphomycosis occurs in farmers[33–35] or as a result of traumatic implantation of contaminated plant material or soil[36] into the skin. Subcutaneous phaeohyphomycosis manifests usually as a localized abscess that appears primarily as a nodule, and the infection may progress and be fatal for immunodeficient patients.[37] Granulomatous reaction may occur in subcutaneous infections, and the disease may progress to produce verrucose plaques reminiscent of chromoblastomycosis.[32] Consequently, a case of chronic skin infection due to *C. funicola* from Panama[34] showing tumorous masses on the hand dorsum was diagnosed incorrectly as a chromoblastomycosis. However, brown muriform bodies were lacking in histopathology, which are pathognomonic for chromoblastomycosis, and hence, the case should be reclassified as subcutaneous phaeohyphomycosis. The infections affecting the skin were predominantly caused by *C. globosum* and usually localized on extremities or face (Table 16.1); two cases were caused by *C. funicola*[34,38] and one by *C. murorum*.[33] The only case of *C. brasiliense* infection in human manifested as otitis externa.[20] The appropriate treatment for skin lesions is unknown: two patients

TABLE 16.1
Reported Cases of *Chaetomium* and *Achaetomium* Infections in Humans

Species	Patient Age/Sex of Patient	Site of Infection	Predisposing Factor	Treatment Outcome	Location	Year of Publication/ Reference
A. strumarium	25/M	Brain	Intravenous drug use	Death	Australia	1995[56]
	28/M	Brain	Intravenous drug use	Death	United States	1995[56,65]
	20/M	Brain	Intravenous drug use	Death	United States	1995[56]
C. atrobrunneum	32/M	Brain	Renal transplantation	Death	India	1989[66]
	15/F	Multiple organs including brain	ALL	Death	United States	1991[68]
	31/M	Lung, brain	Multiple myeloma, allogenic BMT, central venous catheter insertion	Death	United States	1998[63,64]
	12/M	CSF, brain	AML	Death	Saudi Arabia	2007[183]
	4 months/M	Lung	Hemophagocytic syndrome	Death	Saudi Arabia	2007[183]
	1/M	Cutaneous infection (eyelid)[a]	None	Cured	China	2010[184]
	11/M	Retina, brain	Hodgkin's lymphoma, allogenic BMT	Cured	Saudi Arabia	2010[67]
	44/M	Cornea	None	Cured	India	2012[40]
C. brasiliense	61/F	External auditory canal	Injury of head, surgeries, tumor	Death	Czech Republic	2011[20]
C. funicola	71/M	Cutaneous and subcutaneous lesion (buttocks, scrotum)	Age	NR	Germany	1965[38]
	83/M	Cutaneous and subcutaneous lesion (hand dorsum)	Age, farmer	Failed	Panama	2007[34]
C. globosum	57/M	Abdominal cavity (dialysis fluid)	Renal failure, peritoneal dialysis	Cured	France	1984[78]
	64/F	Cutaneous lesion (forearm)	Trauma	Cured	Brazil	1988[36]
	62/F	Fingernails and surrounding skin	Trauma	Failed	Brazil	1988[36]
	NR	Nails (three cases)	NR	NR	United States	1988[46]
	26/M	Fingernails	Trauma	NR	India	1991[42]
	83/F	Toenails	Age	NR	United States	1992[23]

(*Continued*)

TABLE 16.1 (*Continued*)
Reported Cases of *Chaetomium* and *Achaetomium* Infections in Humans

Species	Age/Sex of Patient	Site of Infection	Predisposing Factor	Treatment Outcome	Location	Year of Publication/Reference
	13/M	Cutaneous lesion (face, extremities, palmar and plantar surfaces, buttocks), fingernails, toenails	Farmer	Improvement	China	1998[35]
	19/F	Lung pleura[b]	Lymphoma, autologous BMT	Death	France	1999[185]
	57/M	Fingernails, toenail	Intensive traveling	Improvement	Japan	2000[43]
	34/M	Cervical and axillary lymph nodes	CML, allogenic BMT	Cured	Brazil	2003[76]
	14/M	Cutaneous and subcutaneous lesion (face, forehead, upper extremities)	Immunodeficiency, cardiomyopathy	Death	China	2006[37]
	23/M	Toenail	Football player	Cured	Spain	2007[22]
	25/M	Fingernails	Acid burn	NR	India	2010[45]
	46/M	Cutaneous lesion (interdigital region of the foot)	NR	Cured	Italy	2010[39]
	48/M	Toenails	Trauma, eczema	Cured	Czech Republic	2011[20]
	66/M	Toenail[c]	Often barefoot walking, poor hygiene, diabetes mellitus type 2	Cured	Canada	2012[47]
C. homopilatum[a,b]	31/M	Lung	AML	Death	Germany	1997[186]
C. murorum	25/F	Subcutaneous tissues (chest, abdominal region)	Farmer	Improvement	China	1995[33]
C. perlucidum	45/F	Lungs, brain, and myocardium	AML, umbilical cord blood transplant	Death	United States	2003[59]
	78/F	Lung	Asthma, chronic bronchiectasis	Cured	United States	2003[59]

(*Continued*)

TABLE 16.1 (Continued)
Reported Cases of *Chaetomium* and *Achaetomium* Infections in Humans

Species	Age/Sex of Patient	Site of Infection	Predisposing Factor	Treatment Outcome	Location	Year of Publication/Reference
Chaetomium sp.	8/F	Cutaneous lesion (inner thigh area)	NR	NR	Brazil	1944[187]
	24/M	Lung[d]	ALL, following *Pneumocystis* pneumonia	Death	United States	1983[188]
	15/M	Blood[e]	ALL, autologous BMT	Cured	France	1990[80]
	19/M	Lung	AML	Death	United Kingdom	1996[62]
	72/F	Paranasal sinuses[f]	Repeatedly punctured maxillary sinus	Cured	Denmark	1997[77]
	11/F	Toenails	Eczema	Cured	Spain	2009[44]
	44/M	Tongue (ulcer)	HIV infection	NR	India	2011[79]
	65/M	Cornea	Injury by vegetable matter	Cured	India	2011[41]
Papulaspora sp.[g]	39/F	Cornea	Injury by wire, contact lenses	Cured	Mexico	2012[7]

Abbreviations: NR, not reported; BMT, bone marrow transplant; AML, acute myelogenous leukemia; ALL, acute lymphoblastic leukemia; CML, chronic myelogenous leukemia.

a Mixed infection with *Clavispora lusitaniae*.
b The etiological significance of *Chaetomium* isolate was not clarified.
c Mixed infection with *Trichophyton mentagrophytes*.
d The isolated failed to produce fertile ascomata, and the identification as *C. globosum* is questionable due to well growth at 37°C.
e The identification as *C. globosum* was incorrect based on ascospore morphology.[189]
f Mixed infection with *A. fumigatus*.
g Anamorph phylogenetically included within *Chaetomium* spp. (the closest species based on ITS region was *C. cuniculorum*).

were successfully healed by topical oxyconazole[36] or terbinafine in combination with tioconazole.[39] However, in other patients, the therapy was unsuccessful or not started (Table 16.1). Three cases of keratitis were described recently[7,40,41]; two of them followed corneal injury, and patients were successfully treated with topical natamycine in monotherapy[41] or in combined therapy with oral fluconazole[7] or ketoconazole.[40]

Nail infections due to *Chaetomium* manifest as onychomycosis with usually strikingly brown to black[20,23,36,42] or yellow-brown discoloration.[22,43–45] These infections progress slowly, and patients often present several years after primary insult.[22,36,42] Toenails and fingernails may be affected (Table 16.1), and the lesion may spread to the surrounding skin.[36] The cases have been described worldwide including in tropical countries,[36,42,45] Japan,[43] United States,[23,46] Canada,[47] and Europe.[20,22] Despite the fact that a small number of cases were reported, the infection has been most effectively eradicated when treatment included terbinafine or itraconazole.[20] Similar intensive brown to brown-black nail discoloration has been reported in nail lesions caused by some other filamentous fungi such as *Trichophyton rubrum* var. *nigricans*, *Scytalidium dimidiatum*, *Alternaria alternata*, *Microascus desmosporus*, *Curvularia lunata*, *Exophiala dermatitidis*, *Fusarium oxysporum*, *Aspergillus* spp., *Onychocola canadensis*, and *Scopulariopsis brumptii*.[48–55] In differential diagnosis, it may be important to exclude melanonychia associated with subungual melanoma in some instances.[49]

C. globosum is clearly the predominant species in dermatomycotic infections and the only species from the genus *Chaetomium* that has thus far been isolated as an etiological agent of onychomycosis (Table 16.1). A previous report of *C. brasiliense* onychomycosis[36] (published under the synonymous name *C. perpulchrum*) was subsequently revised[56,57] and reidentified as *C. globosum*.

C. globosum together with two other species isolated as agents of subcutaneous phaeohyphomycosis, *C. funicola* and *C. murorum*, is mesophilic like the majority of *Chaetomium* spp. and has growth optima between 20°C and 30°C.[2,3] These species do not grow at 37°C or grow restrictedly in comparison with 25°C.[3,20,58,59] Even *C. murorum* was referred to as psychrotolerant with strongly reduced growth at 30°C.[3] This probably explains the predilection of these species for cooler body areas such as skin and nails. In contrast, the ability to cause systemic infections is typical for other *Chaetomium* spp. with higher growth optima.[59] *C. atrobrunneum*, *C. perlucidum*, and *A. strumarium* were reported as etiological agents of cerebral and disseminated phaeohyphomycosis in humans (Table 16.1). In contrast to *C. globosum*, these three species grow rapidly at 37°C and are able to grow at 42°C.[59] Also, *C. brasiliense* grows faster at 37°C than 25°C and in this aspect resembles the species of *Chaetomium* that cause systemic infections.[20] However, in contrast to these species, *C. brasiliense* is unable to grow at 42°C.

The invasive *Chaetomium* infections involved patients with hematological malignancies, immunosuppression after transplantation, or intravenous drug use (Table 16.1). Some of these infections are probably of iatrogenic origin (i.e., surgery, central venous catheter), from direct inoculation via drug injection and from spore inhalation. Airborne ascospores in a hospital environment,[15] for example, from a contaminated filter system,[60] or contaminated infusion fluid or fluid dialysis[61] bear risk for immunodeficient patients.

The pulmonary infections mostly resembled aspergillosis based on histopathological finding and imaging methods, whereas others manifested as pneumonia[62] or lung abscess.[63,64] Cerebral infections associated with *A. strumarium* were described only in patients with histories of intravenous drug use.[56,65] The neurotropism also was described from *C. atrobrunneum* and *C. perlucidum* (Table 16.1). The infections due to neurotropic *Chaetomium* species manifested either as solitary brain lesions[56,65–67] (single or multiple abscesses) or multiple organs were involved including the brain[59,63,64,68] (disseminated phaeohyphomycosis). The ability to cause phaeohyphomycosis in man and neurotropism was also reported in the related nonostiolate sordarialean genus *Thielavia*, namely, in *T. subthermophila*[69,70] and *Myceliophthora thermophila* (the anamorph of heterothallic *T. thermophila*).[71–75]

The reports of invasive infections due to *C. globosum* were often inadequately documented, and the etiological significance of the isolated fungus is questionable, or the etiological agent was reidentified[56,59] (Table 16.1). A case of systemic infection involving axillary and cervical lymph nodes was described in a chronic myeloid leukemia patient who underwent bone marrow transplantation.[76] The fungus resembled *C. globosum* based on micromorphology; however, it grew at 37°C. Additional physiological tests (e.g., colony diameters at 25°C and 37°C or ability to grow at 42°C), sequence data, and/or diagnostic PCR confirmation are necessary to verify the identity of the isolate. *Chaetomium* spp. were only once recorded as causative agents of sinusitis,[77] peritonitis,[78] and tongue ulcer.[79]

With several exceptions,[59,76,80] systemic and disseminated infections were fatal (Table 16.1), and the optimal management is unclear despite the use of several drugs. Susceptibility of clinically significant *Chaetomium* species to conventional antifungals was studied,[20,56,57,59,81] and all species showed resistance to 5-fluorocytosine and fluconazole. Some triazoles showed relatively high in vitro activity against *Chaetomium* spp., and interspecies differences were found in the activity of amphotericin B.[20,57,81] However, surgical therapy may be appropriate for localized infections.[59]

16.2.1 KEY TO CLINICALLY SIGNIFICANT *CHAETOMIUM*/*ACHAETOMIUM* SPECIES

(1a) Asci narrow, cylindrical; ascospores uniseriate	2
(1b) Asci clavate	3
(2a) Ascomatal hairs undulate or spirally coiled; peridium with *textura angularis*; ascospores 7–8.5 µm in long axis; no growth at 42°C	*C. brasiliense*
(2b) Ascomata covered with yellowish hyphae-like hairs; peridium with *textura intricata*; ascospores 10–13 µm in long axis; able to grow at 42°C	*A. strumarium*
(3a) Ascospores <9 µm in long axis	4
(3b) Ascospores >9 µm in long axis	5
(4a) Ascomatal hairs in part repeatedly dichotomously branched; ascospores ovoidal; anamorph absent	*C. funicola*
(4b) Ascomatal hairs predominantly straight and unbranched; ascospores lemon-shaped; chlamydospores (*Humicola* anamorph) usually present	*C. homopilatum*
(5a) Ascospores >12 µm in long axis	6
(5b) Ascospores <12 µm in long axis	7
(6a) Ascospores usually with three longitudinal bands and apical germ pore	*C. murorum*
(6b) Ascospores with subapical germ pore and without bands	*C. perlucidum*
(7a) No growth at 42°C; ascospores lemon-shaped	*C. globosum*
(7b) Able to grow at 42°C; ascospores fusiform	*C. atrobrunneum*

16.3 MYCOTOXINS

Chaetomium spp. are potent producers of secondary metabolites, many of which have considerable potential for medicine or plant protection as they exhibit various bioactivities including cytotoxic, immunomodulatory, enzyme inhibitory, and antimicrobial.[27] Chaetoglobosins, sterigmatocystin, chetomin, gliotoxin, and fumitremorgin are the most important mycotoxins in the strict sense.[82–87]

Chaetoglobosins belong to the general group of cytochalasins (macrocyclic polyketide alkaloids)[88] that are known to target cytoskeletal processes[89] and show various biological activities such as cytotoxic[90–92] (including antitumor[93,94]), antibacterial,[95] antifungal,[96] and phytotoxic.[97] Production of chaetoglobosins within *Chaetomium* is probably restricted to some species from *C. globosum* species group.[3] *C. globosum* is a remarkable producer of chaetoglobosins[92,98–100] and also the most frequently reported *Chaetomium* species from foods and water (see Sections 16.4 and 16.5). It is one of the most commonly isolated species in buildings[101] and associated with water-damaged building material.[102,103] The production of chaetoglobosin A and C was also demonstrated when cultured on building material.[104] Also, *C. subaffine*[98] and *C. elatum*[90] are able to produce chaetoglobosins. Chaetoglobosins are present in several nonrelated orders across fungi, and those detected in food are produced particularly by *Penicillium* species.[105,106]

The cancerogenic polyketide sterigmatocystin[107] is produced in significant amounts particularly in *Aspergillus* species from sections *Versicolores* and *Nidulantes*[108] and was detected under laboratory conditions in *C. brasiliense*,[109] *Humicola fuscoatra*,[110] *Botryotrichum piluliferum* (anamorph of heterothallic *C. piluliferum*),[111] and several rare *Chaetomium* species insignificant in terms of human health and occurrence in food and water.[84,108] Sterigmatocystin and O-methylsterigmatocystin were detected in *Chaetomium* spp. and are intermediate products of the sterigmatocystin–aflatoxin pathway in aflatoxigenic fungi.[108] However, aflatoxins have never been detected in *Chaetomium*.[108] Sterigmatocystin is produced by nonrelated fungal groups, and it was discovered that at least in some cases the sterigmatocystin gene cluster was horizontally transferred rather than independently developed.[112]

Chetomin, gliotoxin, and fumitremorgin C belong to the diketopiperazine class of compounds. Gliotoxin and fumitremorgin C were detected recently in *C. globosum*[87] and are produced by nonrelated fungal genera.[113,114] Gliotoxin shows strong antimicrobial[115–117] and immunosuppressive[118] activity; and fumitremorgin C is neurotoxic.[119] The possible impact of these mycotoxins on health was studied particularly in relation to *A. fumigatus* and related species.[113,119,120] Antimicrobial and cytotoxic chetomin bears a close similarity to the sporidesmin responsible for sheep facial eczema[100,121–123] and has been isolated from *C. globosum*, *C. funicola*,[121,124] and some other rare species.[100] The toxicity of moldy corn to animals was proved to be due to chetomin.[122,123]

The relevance of *Chaetomium* mycotoxins for water and food safety and human health in general is unknown; however, *Chaetomium* species should be included in the hazardous group of toxigenic fungi due to their ability to produce a number of mycotoxins and bioactive substances.

Other compounds detected in *Chaetomium* spp. are not considered as mycotoxins in the strict sense, but many of them show a wide range of biological activities. Chaetochromin, chaetocins, chaetocochins, chaetoconvosins, chaetocyclinones, chaetomanone, chaetominedione, chaetominine, chaetomugilins, chaetopyranin, chaetoquadrins, chaetoviridins, chaetoxanthones, chetoseminudins, chetracins, cochliodinols, echinulin, eugenitin, globosumones, heptelidic acid, mollicelins, oosporein, orsellides, radicicol, and tetramic acids are among the more significant compounds showing considerable biological activities (reviewed by Zhang et al.[27]).

16.4 OCCURRENCE IN WATER

Besides actinomycetes and cyanobacteria, *C. globosum* was identified as a producer of the volatile metabolites geosmin and 2-phenylethanol,[125] which contribute to earthy–musty odor and taste of public water supplies. It is probable that production of these odor metabolites by fungi takes place predominantly in soil (rather than in water) with respect to the ecology of potential producers[125,126] and is then washed into water sources.[127,128] However, *Chaetomium*[129–132] and anamorphic *Humicola*[130,133] and *Botryotrichum*[130] were repeatedly isolated from drinking water samples, and further investigations are needed to clarify the origin of odor metabolites and the significance of the fungal contribution to production in water systems. Similarly, the potential mycotoxin production by *Chaetomium* spp. in water and their concentration has not been evaluated. *C. globosum* was isolated in some studies from relatively high number of drinking water samples,[129,130] and such contamination may tend to increase in stored water. Some *Chaetomium* species have the potential to cause life-threatening infections in immunocompromised individuals, but these species have not been reported from drinking water, and other sources in the indoor environment are more likely the source of infection.

16.5 OCCURRENCE IN FOODS

Chaetomium spp. are frequently isolated from food products, but most frequently as simple contamination, whereas only in a limited number of cases do they act as spoilage fungi. The isolates are often not identified to species level, or the identification is questionable. However, this information is valuable with regard to species-specific secondary metabolite spectra. *Chaetomium* is the only ascomycetous genus producing brown to black ascomata, which is commonly encountered in food products[58] making them straightforward to identify.

Only a few *Chaetomium* species are commonly found in foods. *C. globosum* is the most frequently encountered species; additionally, *C. brasiliense* and *C. funicola* are common species in tropical commodities.[58,134] Relatively high infection levels by these species were encountered in soybeans, mung beans, black beans, rice, maize, barley, nuts overall, copra, and sorghum.[58,135–139] Significantly, these are also the most frequently implicated species in superficial human infections affecting the skin and nails (Table 16.1).

C. globosum has been isolated from a variety of commodities, particularly wheat,[137,140–143] barley,[135,137,141,144] maize (also as *C. cochliodes*),[135,136,138,143,145,146] oat grains (as *C. ochraceum*[143,146] and *C. cochliodes*[143]), rice (also as *C. olivaceum*),[134,142,146] sorghum,[137] millet,[143] beans (also as *C. cochliodes*),[134,137,138,143,147] soybeans (also as *C. olivaceum* and *C. cochliodes*),[134,137,138,142,143,148] mung beans,[137,138] copra,[136] peanuts,[136,138] kemiri nuts,[138] cashew nuts,[136,139] walnuts,[149] hazelnuts,[149] pea seeds (also as *C. cochliodes*),[143,146] tomato (as *C. cochliodes*),[146] margarine,[142] green tea,[150] black tea,[151] sugarcane (as *C. fibripilium*),[146] and spices[152–156] such as black pepper,[152,153,156] white pepper,[156] chili[154] and hot pepper seeds (as *C cochliodes*),[143] ginger,[152] cinnamon,[152] cumin,[152,153] fennel,[152] and Bishop's weed.[153] High level of infection by *C. globosum* was observed particularly in rice, copra, soybean, mung beans, maize, barley, cashew nuts, candle nuts, and nuts overall.[135,138,139] This species has also been implicated in causing disease in pears in Egypt.[157]

C. funicola has been recorded at relatively high frequencies from soybean samples, beans overall, and cashew nuts.[58,136,139] The species was further isolated from rice,[58,137,142] maize,[136,138,146,158] peanuts,[136] cashew nuts,[136,139] copra,[136,137] beans,[137] mung beans,[137,138] soybeans,[138,158] pea seeds (also as *C. dolichotrichum*),[142,146,158] sorghum,[137] radish,[142] eggplant,[142] and spices (as *C. dolichotrichum*).[155] *C. indicum* is phylogenetically close to *C. funicola*, but distinct.[3] This species also resembles *C. funicola* in morphology and is distinguished solely by the absence of stiff, unbranched ascomatal hairs.[2,142,158] *C. indicum* has been isolated from beans,[142,158] soybeans,[142] pea seeds,[158] rice,[142] and spices,[155] and it is possible that it is commonly misidentified with *C. funicola*. Another species clearly distinct from *C. funicola* based on ITS and LSU rDNA data, but morphologically similar, is *C. reflexum* that has been reported from pepper[158] and pea seeds.[158] *C. brasiliense* was isolated from a similar spectrum of tropical commodities as *C. funicola* including soybeans,[58,137,138] mung beans,[58,137,138] black beans,[137] rice,[58,137] maize,[58,137] cashew nuts,[136] peanuts,[58] sorghum,[137] and black pepper.[58]

Other species are isolated rarely from food products. *C. aureum*,[142] *C. atrobrunneum*,[142] and *C. murorum*[142,146] were reported from rice; *C. nigricolor*, *C. raii*, and *C. subaffine* from cereals[2]; *C. bostrychodes*,[143] *C. elatum*,[158] *C. murorum*,[143] and *C. succineum*[143] from pea seeds; *C. aureum* and *C. bostrychodes* from oat grains[143]; and *C. aureum* from snap beans[159] and butter.[143] *C. convolutum* (as *C. biapiculatum*)[155] was reported from spices; *C. carinthiacum* from thyme[154]; *Achaetomium globosum* from cumin[154]; *C. bostrychodes*,[143] *C. robustum*,[156] and *C. aureum*[143,146] from pepper; *Chaetomium* cf. *fusiforme*[151] and *C. aureum* from tea.[160] Webb and Mundt[161] isolated *C. fimeti* (usually classified in the genus *Chaetomidium*[2,162]) at relatively high frequencies from vegetables; *C. crispatum* was isolated from rotting potatoes,[163] *C. elatum* from rotting onion,[146] and *C. murorum* from Latundan banana.[2]

Chaetomium spp. were reported from various seeds, some of which may be used as food or for oil extraction. *C. funicola* (as *C. dolichotrichum*),[158] *C. globosum* (also as *C. cochliodes*),[143]

C. indicum,[158] *C. madrasense*,[2] and *C. murorum*[143] were isolated from linseed (*Linum usitatissimum*); *C. bostrychodes* and *C. globosum* from okra (*Hibiscus esculentus*)[143]; *C. aureum*,[143] *C. globosum*,[143] *C. funicola*,[146] and *C. murorum*[143] from cucumber seeds (*Cucumis sativus*); *C. elatum*,[158] *C. globosum*,[143] and *C. murorum*[143] from pumpkin seeds (*Cucurbita maxima*); *C. carinthiacum* from poppy seeds (*Papaver somniferum*)[2]; *C. funicola* from sesame (*Sesamum indicum*)[137]; and *C. globosum* from safflower (*Carthamus tinctorius*)[143] and vegetable marrow (*Cucurbita pepo*, also as *C. cochliodes*).[143] Unidentified *Chaetomium* spp. were found on meat products,[164] dried milk,[164] rice,[134,137,165–168] maize,[136,145,169] wheat,[169] sorghum,[137] soybeans,[137,170] beans,[137] mung beans,[137] copra,[136] tapioca,[137] black pepper,[171] peanuts,[136] cashew nuts,[136] sorghum,[172] sesame seeds,[137] Bishop's weed,[153] cumin,[153] lotus seeds,[137] pumpkin seeds,[173] and herbal drugs.[174]

16.5.1 Key to *Chaetomium* Species Relevant to Foods (at Least Twice Reported from Food Products)

(1a) Ascomatal hairs—dichotomously branched	2
(1b) Ascomatal hairs—unbranched (arcuate, coiled, straight, or flexed)	5
(2a) Ascospores >10 μm in long axis	*C. elatum*
(2b) Ascospores <10 μm in long axis	3
(3a) Terminal hairs—arcuate with reflexed branches	*C. reflexum*
(3b) Branches of terminal hairs—essentially straight	4
(4a) Ascomatal hairs—partly branched, partly straight, and unbranched	*C. funicola*
(4b) All or nearly all ascomatal hairs—dichotomously branched at maturity	*C. indicum*
(5a) Asci—cylindrical; ascospores uniseriate	*C. brasiliense*
(5b) Asci—clavate	6
(6a) Ascospores >13 μm in long axis	*C. murorum*
(6b) Ascospores <13 μm in long axis	7
(7a) Ascospores—unevenly fusiform to navicular	*C. aureum*
(7b) Ascospores—symmetrical	8
(8a) Ascospores—ellipsoidal, fusiform, width (shorter dimension) <6 μm	*C. carinthiacum*
(8b) Ascospores—lemon-shaped, width >6 μm	9
(9a) Terminal hairs—undulate	*C. globosum*
(9b) Terminal hairs—spirally coiled	*C. bostrychodes*

16.6 DIAGNOSIS

Molecular method for diagnosis of invasive infections caused by *Chaetomium* from paraffin wax embedded tissue using panfungal primers for ITS region was developed by Paterson et al.[175,176] The sequences of ITS rDNA are available for almost all species relevant to foods and medically important *Chaetomium* spp. They are sufficiently variable for species differentiation. Partial LSU rDNA and β-tubulin gene (*benA*) data are available for smaller number of species and are also informative. It is important to mention that sequences of reliably determined isolates or type strains should serve as standards for comparison[3,20,177] when BLAST similarity searches are undertaken.

Quantitative PCR assays targeted rDNA were developed for the detection of important indoor fungi including *C. globosum* from dust samples.[178] An oligonucleotide array targeted the ITS region, which was developed by Hung et al.[179] to identify common allergenic and pathogenic airborne fungi including *C. globosum* and *C. funicola*. The ITS region was amplified, and PCR products were hybridized with specific oligonucleotide probes of the array giving species-specific hybridization patterns.[179]

A PCR fingerprinting method using universal rice primers (URP, i.e., repeat sequences from the rice genome) was used for characterization of *Chaetomium* isolates.[180] The method was able to generate species-specific band patterns or was capable of being used for typification of isolates belonging to the same species. Random amplified polymorphic DNA (RAPD) is another method that was found suitable for distinguishing isolates within individual species and was tested in *C. globosum*.[181] Similarly, markers based on amplified fragment length polymorphisms (AFLP) have been demonstrated as a useful genetic approach to identify and quantitatively monitor fungal strains including *Chaetomium*.[182]

16.7 CONCLUSIONS

Chaetomium is a large genus distributed worldwide, and its members are commonly isolated from natural substrates, indoor environment, food, and water. Some species are increasingly reported as cause of phaeohyphomycosis in human with very variable clinical manifestations. Superficial infections occur predominantly in immunocompetent patients, typically presenting secondary to traumatic implantation. *C. globosum* is clearly the major cause of skin and nail infection. Systemic infections affect predominantly immunocompromised patients, or intravenous drug users, and are caused by species with higher growth maxima than *C. globosum*. Remarkable neurotropism was described for several species, and brain infections comprise important part of systemic infections that have been reported. Therapy strategy is not established due to the small number of cases with positive treatment outcome. However, superficial infections have been most effectively eradicated when treatment included terbinafine or itraconazole; corneal infections were successfully treated with topical natamycine in monotherapy or in combined therapy; and only a few number of systemic infections were cured after amphotericin B, voriconazole, or surgical therapy. *Chaetomium* is a rich source of structurally complex and bioactive compounds including mycotoxins. Chaetoglobosins, sterigmatocystin, chetomin, gliotoxin, and fumitremorgin C are the most hazardous mycotoxins detected in *Chaetomium*. Their relevance for human health with respect to pathogenicity of species and occurrence in food, water, and environment is, in general, unknown. The role of volatile metabolites detected in *Chaetomium* that contribute to earthy–musty odor of water is controversial as they are produced also by bacteria. Morphological, culture-dependent species identification may be time-consuming and requires expertise. Culture-dependent as well as culture-independent molecular methods have been developed and are increasingly applied for species identification and detection.

ACKNOWLEDGMENTS

I would like to thank Dr. Miroslav Kolařík and Dr. Ondřej Koukol for their valuable comments. This work was supported by the project "BIOCEV—Biotechnology and Biomedicine Centre of the Academy of Sciences and Charles University" (CZ.1.05/1.1.00/02.0109), from the European Regional Development Fund and by the Ministry of Education, Youth and Sports (SVV project).

REFERENCES

1. Morejón, K.C.R. Estudio taxonómico (morfológico y molecular) de especies del género Chaetomium y géneros afines, PhD thesis, Universitat Rovira i Virgili, Tarragona, Spain (2003).
2. Arx, J.A., Guarro, J., and Figueras, M.J. The ascomycete genus *Chaetomium*. *Beihefte zur Nova Hedwigia* **84**, 1–162 (1986).
3. Asgari, B. and Zare, R. The genus *Chaetomium* in Iran, a phylogenetic study including six new species. *Mycologia* **103**, 863–882 (2011).
4. Hubálek, Z. Keratinophilic fungi associated with free-living mammals and birds. In *Biology of Dermatophytes* (eds. Kushwaha, R.K.S. and Guarro, J.), pp. 93–103 (Revista Iberoamericana de Micología, Bilbao, Spain, 2000).
5. Marchisio, V.F. Keratinophilic fungi: Their role in nature and degradation of keratinic substrates. In *Biology of Dermatophytes and Other Keratinophilic Fungi* (eds. Kushwaha, R.K.S. and Guarro, J.), pp. 86–92 (Revista Iberoamericana de Micología, Bilbao, Spain, 2000).
6. Summerbell, R.C. et al. *Acremonium* phylogenetic overview and revision of *Gliomastix*, *Sarocladium*, and *Trichothecium*. *Studies in Mycology* **68**, 139–162 (2011).

7. Mootha, V.V. et al. Identification problems with sterile fungi, illustrated by a keratitis due to a non-sporulating *Chaetomium*-like species. *Medical Mycology* **50**, 361–367 (2012).
8. Morgan-Jones, G. and Gams, W. Notes on Hyphomycetes. XLI. An endophyte of *Festuca arundinacea* and the anamorph of *Epichloe typhina*, new taxa in one of two new sections of *Acremonium*. *Mycotaxon* **15**, 311–318 (1982).
9. Nonaka, K. et al. *Staphylotrichum boninense*, a new hyphomycete (Chaetomiaceae) from soils in the Bonin Islands, Japan. *Mycoscience* **53**, 312–318 (2012).
10. Hambleton, S., Nickerson, N.L., and Seifert, K.A. *Leohumicola*, a new genus of heat-resistant hyphomycetes. *Studies in Mycology* **53**, 29–52 (2005).
11. Guarro, J., Gene, J., Stchigel, A.M., and Figueras, M.J. *Atlas of Soil Ascomycetes*, p. 486 (CBS-KNAW Fungal Biodiversity Centre, Utrecht, the Netherlands, 2012).
12. Wilken-Jensen, K. and Gravesen, S. *Atlas of Moulds in Europe Causing Respiratory Allergy*, p.110 (Ask Publishing, Copenhagen, Denmark, 1984).
13. Samson, R.A., Houbraken, J., Thrane, U., Frisvad, J.C., and Andersen, B. *Food and Indoor Fungi*, p. 269 (CBS KNAW Biodiversity Center, Utrecht, the Netherlands, 2010).
14. Andersen, B. and Nissen, A.T. Evaluation of media for detection of *Stachybotrys* and *Chaetomium* species associated with water-damaged buildings. *International Biodeterioration & Biodegradation* **46**, 111–116 (2000).
15. Rainer, J., Peintner, U., and Pöder, R. Biodiversity and concentration of airborne fungi in a hospital environment. *Mycopathologia* **149**, 87–97 (2000).
16. Vesper, S. et al. Development of an environmental relative moldiness index for US homes. *Journal of Occupational and Environmental Medicine* **49**, 829–833 (2007).
17. Vesper, S. et al. Quantitative PCR analysis of molds in the dust from homes of asthmatic children in North Carolina. *Journal of Environmental Monitoring* **9**, 826–830 (2007).
18. Vesper, S.J. et al. Relative moldiness index as predictor of childhood respiratory illness. *Journal of Exposure Science and Environmental Epidemiology* **17**, 88–94 (2007).
19. Vesper, S.J. et al. Quantitative polymerase chain reaction analysis of fungi in dust from homes of infants who developed idiopathic pulmonary hemorrhaging. *Journal of Occupational and Environmental Medicine* **46**, 596–601 (2004).
20. Hubka, V., Mencl, K., Skorepova, M., Lyskova, P., and Zalabska, E. Phaeohyphomycosis and onychomycosis due to *Chaetomium* spp., including the first report of *Chaetomium brasiliense* infection. *Medical Mycology* **49**, 724–733 (2011).
21. Jensen, H.E. and Chandler, F.W. Histopathological diagnosis of mycotic disease. in *Topley and Wilson's Microbiology and Microbial Infections: Medical Mycology*, 10th edn. (eds. Merz, W.G. and Hay, R.J.) pp. 121–143 (Hodder Arnold, London, U.K., 2005).
22. Aspiroz, C., Gené, J., Rezusta, A., Charlez, L., and Summerbell, R.C. First Spanish case of onychomycosis caused by *Chaetomium globosum*. *Medical Mycology* **45**, 279–282 (2007).
23. Stiller, M.J., Rosenthal, S., Summerbell, R.C., Pollack, J., and Chan, A. Onychomycosis of the toenails caused by *Chaetomium globosum*. *Journal of the American Academy of Dermatology* **26**, 775–776 (1992).
24. Rodríguez, K., Stchigel, A.M., Cano, J.F., and Guarro, J. A new species of *Achaetomium* from Indian soil. *Studies in Mycology* **50**, 77–82 (2004).
25. Mouchacca, J. Thermotolerant fungi erroneously reported in applied research work as possessing thermophilic attributes. *World Journal of Microbiology and Biotechnology* **16**, 869–880 (2000).
26. Mouchacca, J. Heat tolerant fungi and applied research: Addition to the previously treated group of strictly thermotolerant species. *World Journal of Microbiology and Biotechnology* **23**, 1755–1770 (2007).
27. Zhang, Q., Li, H.Q., Zong, S.C., Gao, J.M., and Zhang, A.L. Chemical and bioactive diversities of the genus *Chaetomium* secondary metabolites. *Mini Reviews in Medicinal Chemistry* **12**, 127–148 (2012).
28. Umikalsom, M.S. et al. Production of cellulase by a wild strain of *Chaetomium globosum* using delignified oil palm empty-fruit-bunch fibre as substrate. *Applied Microbiology and Biotechnology* **47**, 590–595 (1997).
29. Reissinger, A., Winter, S., Steckelbroeck, S., Hartung, W., and Sikora, R.A. Infection of barley roots by *Chaetomium globosum*: Evidence for a protective role of the exodermis. *Mycological Research* **107**, 1094–1102 (2003).
30. Revankar, S.G., Patterson, J.E., Sutton, D.A., Pullen, R., and Rinaldi, M.G. Disseminated phaeohyphomycosis: Review of an emerging mycosis. *Clinical Infectious Diseases* **34**, 467–476 (2002).

31. Revankar, S.G. and Sutton, D.A. Melanized fungi in human disease. *Clinical Microbiology Reviews* **23**, 884–928 (2010).
32. Cooper, J.R. Deep phaeohyphomycosis. In *Topley and Wilson's Microbiology and Microbial Infections, Medical Mycology*, 10th edn. (eds. Merz, W.G. and Hay, R.J.) pp. 739–748 (Hodder Arnold, London, U.K., 2005).
33. Lin, Y. and Li, X. First case of phaeohyphomycosis caused by *Chaetomium murorum* in China. *Chinese Journal of Dermatology* **28**, 367–369 (1995).
34. Piepenbring, M., Mendez, O.A.C., Espinoza, A.A.E., Kirschner, R., and Schöfer, H. Chromoblastomycosis caused by *Chaetomium funicola*: A case report from Western Panama. *British Journal of Dermatology* **157**, 1025–1029 (2007).
35. Wang, J., Zhang, Q., Li, L., and Feng, W. Phaeohyphomycosis caused by *Chaetomium globosum*: First case report in China. *Chinese Journal of Dermatology* **31**, 273–275 (1998).
36. Costa, A.R. et al. Cutaneous and ungual phaeohyphomycosis caused by species of *Chaetomium* Kunze (1817) ex Fresenius, 1829. *Journal of Medical and Veterinary Mycology* **26**, 261–268 (1988).
37. Yu, J., Yang, S., Zhao, Y., and Li, R. A case of subcutaneous phaeohyphomycosis caused by *Chaetomium globosum* and the sequences analysis of *C. globosum*. *Medical Mycology* **44**, 541–545 (2006).
38. Koch, H.A. and Haneke, H. *Chaetomium funicolum* as possible aetiological agent of the deep mycosis. *Mykosen* **9**, 23–28 (1965).
39. Tullio, V. et al. Non-dermatophyte moulds as skin and nail foot mycosis agents: *Phoma herbarum*, *Chaetomium globosum* and *Microascus cinereus*. *Fungal Biology* **114**, 345–349 (2010).
40. Balne, P.K., Nalamada, S., Kodiganti, M., and Taneja, M. Fungal keratitis caused by *Chaetomium atrobrunneum*. *Cornea* **31**, 94–95 (2012).
41. Kaliamurthy, J., Kalavathy, C.M., Jesudasan, C.A.N., and Thomas, P.A. Keratitis due to *Chaetomium* sp. *Case Reports in Ophthalmological Medicine* **2011**, 696145 (2011) doi:10.1155/2011/740640.
42. Naidu, J., Singh, S.M., and Pouranik, M. Onychomycosis caused by *Chaetomium globosum* Kunze. *Mycopathologia* **113**, 31–34 (1991).
43. Hattori, N. et al. Case report. Onychomycosis due to *Chaetomium globosum* successfully treated with itraconazole. *Mycoses* **43**, 89–92 (2000).
44. Falcón, C.S. et al. Onychomycosis by *Chaetomium* spp. *Mycoses* **52**, 77–79 (2009).
45. Latha, R., Sasikala, R., Muruganandam, N., and Prakash, M.R.S. Onychomycosis due to ascomycete *Chaetomium globosum*: A case report. *Indian Journal of Pathology and Microbiology* **53**, 572–573 (2010).
46. Rippon, J.W. *Medical Mycology. The Pathogenic Fungi and the Pathogenic Actinomycetes*. p. 797 (Saunders, Philadelphia, PA, 1988).
47. Lagacé, J. and Cellier, E. A case report of a mixed *Chaetomium globosum/Trichophyton mentagrophytes* onychomycosis. *Medical Mycology Case Reports* **1**, 76–78 (2012).
48. Naidu, J., Singh, S.M., and Pouranik, M. Onychomycosis caused by *Scopulariopsis brumptii*. *Mycopathologia* **113**, 159–164 (1991).
49. Stuchlík, D., Mencl, K., Hubka, V., and Skořepová, M. Fungal melanonychia caused by *Onychocola canadensis*: First records of nail infections due to *Onychocola* in the Czech Republic. *Czech Mycology* **63**, 83–91 (2011).
50. Gupta, A.K., Horga-Bell, C.B., and Summerbell, R.C. Onychomycosis associated with *Onychocola canadensis*: Ten case reports and a review of the literature. *Journal of the American Academy of Dermatology* **39**, 410–417 (1998).
51. Lateur, N. and Andre, J. Melanonychia: Diagnosis and treatment. *Dermatologic Therapy* **15**, 131–141 (2002).
52. Gianni, C. and Romano, C. Clinical and histological aspects of toenail onychomycosis caused by *Aspergillus* spp.: 34 cases treated with weekly intermittent terbinafine. *Dermatology* **209**, 104–110 (2004).
53. Romano, C., Miracco, C., and Difonzo, E.M. Skin and nail infections due to *Fusarium oxysporum* in Tuscany, Italy. *Mycoses* **41**, 433–437 (1998).
54. Kamalam, A., Ajithadass, K., Sentamilselvi, G., and Thambiah, A.S. Paronychia and black discoloration of a thumb nail caused by *Curvularia lunata*. *Mycopathologia* **118**, 83–84 (1992).
55. Hata, Y., Naka, W., and Nishikawa, T. A case of melanonychia caused by *Exophiala dermatitidis*. *Japanese Journal of Medical Mycology* **40**, 231–234 (1999).
56. Abbott, S.P. et al. Fatal cerebral mycoses caused by the ascomycete *Chaetomium strumarium*. *Journal of Clinical Microbiology* **33**, 2692–2698 (1995).
57. Guarro, J., Soler, L., and Rinaldi, M.G. Pathogenicity and antifungal susceptibility of *Chaetomium* species. *European Journal of Clinical Microbiology and Infectious Diseases* **14**, 613–618 (1995).

58. Pitt, J.I. and Hocking, A.D. *Fungi and Food Spoilage*, 3rd edn. (Springer Verlag, Heidelberg, Germany, 2009).
59. Barron, M.A. et al. Invasive mycotic infections caused by *Chaetomium perlucidum*, a new agent of cerebral phaeohyphomycosis. *Journal of Clinical Microbiology* **41**, 5302–5307 (2003).
60. Woods, G.L., Davis, J.C., and Vaughan, W.P. Failure of the sterile air-flow component of a protected environment detected by demonstration of *Chaetomium* species colonization of four consecutive immunosuppressed occupants. *Infection Control and Hospital Epidemiology* **9**, 451–456 (1988).
61. Febré, N. et al. Contamination of peritoneal dialysis fluid by filamentous fungi. *Revista Iberoamericana de Micologia* **19**, 238–239 (1999).
62. Yeghen, T. et al. *Chaetomium* pneumonia in patient with acute myeloid leukaemia. *Journal of Clinical Pathology* **49**, 184–186 (1996).
63. Thomas, C., Mileusnic, D., Carey, R.B., Kampert, M., and Anderson, D. Fatal *Chaetomium* cerebritis in a bone marrow transplant patient. *Human Pathology* **30**, 874–879 (1999).
64. Guppy, K.H., Chinnamma, T., Kurian, T., and Douglas, A. Cerebral fungal infections in the immunocompromised host: A literature review and a new pathogen—*Chaetomium atrobrunneum*: Case report. *Neurosurgery* **43**, 1463–1468 (1998).
65. Aribandi, M., Bazan, C., and Rinaldi, M.G. Magnetic resonance imaging findings in fatal primary cerebral infection due to *Chaetomium strumarium*. *Australasian Radiology* **49**, 166–169 (2005).
66. Anandi, V. et al. Cerebral phaeohyphomycosis caused by *Chaetomium globosum* in a renal transplant recipient. *Journal of Clinical Microbiology* **27**, 2226–2229 (1989).
67. Tabbara, K.F., Wedin, K., and Al-Haddab, S. *Chaetomium* retinitis. *Retinal Cases & Brief Reports* **4**, 8–10 (2010).
68. Rinaldi, M.G. et al. Fatal *Chaetomium atrobrunneum* Ames, 1949, systemic mycosis in a patient with acute lymphoblastic leukemia. In *Abstracts of the XI Congress of the International Society for Human and Animal Mycology*, p. 107 (Montreal, Quebec, Canada, 1991).
69. Badali, H. et al. Fatal cerebral phaeohyphomycosis in an immunocompetent individual due to *Thielavia subthermophila*. *Journal of Clinical Microbiology* **49**, 2336–2341 (2011).
70. Theoulakis, P., Goldblum, D., Zimmerli, S., Muehlethaler, K., and Frueh, B.E. Keratitis resulting from *Thielavia subthermophila* Mouchacca. *Cornea* **28**, 1067–1069 (2009).
71. Destino, L. et al. Severe osteomyelitis caused by *Myceliophthora thermophila* after a pitchfork injury. *Annals of Clinical Microbiology and Antimicrobials* **5**, 21 (2006).
72. Bourbeau, P., McGough, D.A., Fraser, H., Shah, N., and Rinaldi, M.G. Fatal disseminated infection caused by *Myceliophthora thermophila*, a new agent of mycosis: Case history and laboratory characteristics. *Journal of Clinical Microbiology* **30**, 3019–3023 (1992).
73. Morio, F. et al. Invasive *Myceliophthora thermophila* infection mimicking invasive aspergillosis in a neutropenic patient: A new cause of cross-reactivity with the *Aspergillus* galactomannan serum antigen assay. *Medical Mycology* **49**, 883–886 (2011).
74. le Naourèsa, C., Bonhommeb, J., Terzic, N., Duhamelb, C., and Galateau-Salléa, F. A fatal case with disseminated *Myceliophthora thermophila* infection in a lymphoma patient *Diagnostic Microbiology and Infectious Disease* **70**, 267–269 (2011).
75. Farina, C., Gamba, A., Tambini, R., Beguin, H., and Trouillet, J.L. Fatal aortic *Myceliophthora thermophila* infection in a patient affected by cystic medial necrosis. *Medical Mycology* **36**, 113–118 (1998).
76. Teixeira, A.B.A. et al. Phaeohyphomycosis caused by *Chaetomium globosum* in an allogeneic bone marrow transplant recipient. *Mycopathologia* **156**, 309–312 (2003).
77. Aru, A., Munk-Nielsen, L., and Federspiel, B.H. The soil fungus *Chaetomium* in the human paranasal sinuses. *European Archives of Oto-Rhino-Laryngology* **254**, 350–352 (1997).
78. Barthez, J.P., Pierre, D., de Bievre, C., and Arbeille, M. Peritonitis due to *Chaetomium globosum* in a patient treated by C. A. P. D. *Bulletin de la Societe Francaise de Mycologie Medicale* **13**, 205–208 (1984).
79. Puri, A., Gupta, G., Kamal, R., and Dahiya, P. *Chaetomium* infection in a HIV positive patient—A case report. *Journal of the Indian Dental Association* **5**, 52–53 (2011).
80. Barale, T., Fumey, M.H., Reboux, G., and Mallea, M. Septicaemia due to *Chaetomium* sp. in a leukaemic child associated with a bone marrow transplant. *Bulletin de la Societe Francaise de Mycologie Medicale* **19**, 43–46 (1990).
81. Serena, C. et al. In vitro activities of new antifungal agents against *Chaetomium* spp. and inoculum standardization. *Antimicrobial Agents and Chemotherapy* **47**, 3161–3164 (2003).
82. Gravesen, S., Frisvad, J.C., and Samson, R.A. *Microfungi*. (Munksgaard International Publishers Ltd, Copenhagen, Denmark, 1994).

83. Frisvad, J.C. and Thrane, U. Mycotoxin production by common filamentous fungi. In *Introduction to Food- and Airborne Fungi*, 7th edn. (eds. Samson, R.A., Hoekstra, E.S., and Frisvad, J.C.) pp. 321–330 (Centraalbureau voor Schimmelcultures, Utrecht, the Netherlands, 2004).
84. Udagawa, S. Taxonomy of mycotoxin producing *Chaetomium*. In *Toxigenic Fungi: Their Toxins and Health Hazard*, pp. 139–147 (Elsevier, Amsterdam, the Netherlands, 1984).
85. Cole, R.J. and Cox, R.H. *Handbook of Toxic Fungal Metabolites* (Academic Press, New York, 1981).
86. Jen, W.C. and Jones, G.A. Effect of chaetomin on growth and acidic fermentation products of rumen bacteria. *Canadian Journal of Microbiology* **29**, 1399–1404 (1983).
87. Li, H.Q. et al. Antifungal metabolites from *Chaetomium globosum*, an endophytic fungus in *Ginkgo biloba*. *Biochemical Systematics and Ecology* **39**, 876–879 (2011).
88. Binder, M. and Tamm, C. The cytochalasans: A new class of biologically active microbial metabolites. *Angewandte Chemie International Edition in English* **12**, 370–380 (1973).
89. Scherlach, K., Boettger, D., Remme, N., and Hertweck, C. The chemistry and biology of cytochalasans. *Natural Product Reports* **27**, 869–886 (2010).
90. Thohinung, S. et al. Cytotoxic 10-(Indol-3-yl)-[13]cytochalasans from the fungus *Chaetomium elatum* ChE01. *Archives of Pharmacal Research* **33**, 1135–1141 (2010).
91. Sekita, S., Yoshihira, K., Natori, S., and Kuwano, H. Structures of chaetoglobosin A and B, cytotoxic metabolites of *Chaetomium globosum*. *Tetrahedron Letters* **14**, 2109–2112 (1973).
92. Jiao, W., Feng, Y., Blunt, J.W., Cole, A.L.J., and Munro, M.H.G. Chaetoglobosins Q, R, and T, three further new metabolites from *Chaetomium globosum*. *Journal of Natural Products* **67**, 1722–1725 (2004).
93. Tikoo, A., Cutler, H., Lo, S.H., Chen, L.B., and Maruta, H. Treatment of Ras-induced cancers by the F-actin cappers tensin and chaetoglobosin K, in combination with the caspase-1 inhibitor N1445. *The Cancer Journal from Scientific American* **5**, 293–300 (1999).
94. Ali, A., Sidorova, T.S., and Matesic, D.F. Dual modulation of JNK and Akt signaling pathways by chaetoglobosin K in human lung carcinoma and ras-transformed epithelial cells. *Investigational New Drugs*, doi:10.1007/s10637-012-9883-x (2012).
95. Namgoong, J. et al. Isolation and structural determination of anti-*Helicobacter pylori* compound from fungus 60686. *Sanop Misaengmul Hakhoechi* **26**, 137–142 (1998).
96. Kang, J.G., Kim, K.K., and Kang, K.Y. Antagonism and structural identification of antifungal compound from *Chaetomium cochliodes* against phytopathogenic fungi. *Agricultural Chemistry and Biotechnology* **42**, 146–150 (1999).
97. Lim, C.H., Kim, M.Y., Lee, J.W., Yun, B.S., and Baek, S.H. Phytotoxin isolated from the culture broth of *Chaetomium* sp. *Journal of the Korean Society for Applied Biological Chemistry* **50**, 316–320 (2007).
98. Udagawa, S. et al. The production of chaetoglobosins, sterigmatocystin, O-methylsterigmatocystin, and chaetocin by *Chaetomium* spp. and related fungi. *Canadian Journal of Microbiology* **25**, 170–177 (1979).
99. Ohtsubo, K., Saito, M., Sekita, S., Yoshihira, K., and Natori, S. Acute toxic effects of chaetoglobosin A, a new cytochalasan compound produced by *Chaetomium globosum*, on mice and rats. *Japanese Journal of Experimental Medicine* **48**, 105–110 (1978).
100. Sekita, S. et al. Mycotoxin production by *Chaetomium* spp. and related fungi. *Canadian Journal of Microbiology* **27**, 766–772 (1981).
101. Fogle, M.R., Douglas, D.R., Jumper, C.A., and Straus, D.C. Growth and mycotoxin production by *Chaetomium globosum*. *Mycopathologia* **164**, 49–56 (2007).
102. Jarvis, B.B. Chemistry and toxicology of molds isolated from water-damaged buildings. *Advances in Experimental Medicine and Biology* **504**, 43–52 (2002).
103. Nielsen, K.F. Mycotoxins from mould infested building materials. *Mycotoxin Research* **16**, 113–116 (2000).
104. Nielsen, K.F. et al. Production of mycotoxins on artificially and naturally infested building materials. *Mycopathologia* **145**, 43–56 (1999).
105. Larsen, T.O., Smedsgaard, J., Nielsen, K.F., Hansen, M.E., and Frisvad, J.C. Phenotypic taxonomy and metabolite profiling in microbial drug discovery. *Natural Product Reports* **22**, 672–695 (2005).
106. Barkai-Golan, R. *Penicillium* mycotoxins. In *Mycotoxins in Fruits and Vegetables*. (eds. Barkai-Golan, R. and Paster, N.) pp. 153–183 (Elsevier, San Diego, CA, 2008).
107. Fujii, K., Kurata, H., Odashima, S., and Hatsuda, Y. Tumor induction by a single subcutaneous injection of sterigmatocystin in newborn mice. *Cancer Research* **36**, 1615–1618 (1976).
108. Rank, C. et al. Distribution of sterigmatocystin in filamentous fungi. *Fungal Biology* **115**, 406–420 (2011).

109. Li, G.Y., Li, B.G., Yang, T., Liu, G.Y., and Zhang, G.L. Secondary metabolites from the fungus *Chaetomium brasiliense*. *Helvetica Chimica Acta* **91**, 124–129 (2008).
110. Joshi, B.K., Gloer, J.B., and Wicklow, D.T. Bioactive natural products from a sclerotium-colonizing isolate of *Humicola fuscoatra*. *Journal of Natural Products* **65**, 1734–1737 (2002).
111. Sy, A.A., Swenson, D.C., Gloer, J.B., and Wicklow, D.T. Botryolides A-E, decarestrictine analogues from a fungicolous *Botryotrichum* sp. (NRRL 38180). *Journal of Natural Products* **71**, 415–419 (2008).
112. Slot, J.C. and Rokas, A. Horizontal transfer of a large and highly toxic secondary metabolic gene cluster between fungi. *Current Biology* **21**, 134–139 (2011).
113. Scharf, D.H. et al. Biosynthesis and function of gliotoxin in *Aspergillus fumigatus*. *Applied Microbiology and Biotechnology* **93**, 467–472 (2012).
114. Samson, R.A., Hong, S., Peterson, S.W., Frisvad, J.C., and Varga, J. Polyphasic taxonomy of *Aspergillus* section *Fumigati* and its teleomorph *Neosartorya*. *Studies in Mycology* **59**, 147–203 (2007).
115. Waksman, S.A. and Geiger, W.B. The nature of the antibiotic substances produced by *Aspergillus fumigatus*. *Journal of Bacteriology* **47**, 391–397 (1944).
116. Rightsel, W.A. et al. Antiviral activity of gliotoxin and gliotoxin acetate. *Nature* **204**, 1333–1334 (1964).
117. Herrick, J.A. Effects of gliotoxin on *Trichophyton gypseum*. *The Ohio Journal of Science* **45**, 45–46 (1945).
118. Sutton, P., Newcombe, N.R., Waring, P., and Mullbacher, A. In vivo immunosuppressive activity of gliotoxin, a metabolite produced by human pathogenic fungi. *Infection and Immunity* **62**, 1192–1198 (1994).
119. Land, C.J., Hult, K., and Fuchs, R. Tremorgenic mycotoxins from *Aspergillus fumigatus* as a possible occupational health problem in sawmills. *Applied and Environmental Microbiology* **53**, 787–790 (1987).
120. Dorner, J.W., Cole, R.J., and Hill, R.A. Tremorgenic mycotoxins produced by *Aspergillus fumigatus* and *Penicillium crustosum* isolated from molded corn implicated in a natural intoxication of cattle. *Journal of Agricultural and Food Chemistry* **32**, 411–413 (1984).
121. Brewer, D. et al. The structure of chaetomin, a toxic metabolite of *Chaetomium cochliodes*, by nitrogen-15 and carbon-13 nuclear resonance spectroscopy. *Journal of the Chemical Society, Perkin Transactions 1*, **10**, 1248–1251 (1978).
122. Brewer, D. et al. Ovine ill-thrift in Nova Scotia. 5. The production and toxicology of chetomin, a metabolite of *Chaetomium* spp. *Canadian Journal of Microbiology* **18**, 1129–1137 (1972).
123. Christensen, C.M., Nelson, G.H., Mirocha, C.J., Bates, F., and Dorworth, C.E. Toxicity to rats of corn invaded by *Chaetomium globosum*. *Applied Microbiology* **14**, 774–777 (1966).
124. Brewer, D. and Taylor, A. The production of toxic metabolites by *Chaetomium* spp. isolated from soils of permanent pasture. *Canadian Journal of Microbiology* **24**, 1078–1081 (1978).
125. Kikuchi, T., Kadota, S., Suehara, H., Nishi, A., and Tsubaki, K. Odorous metabolites of a fungus, *Chaetomium globosum* Kunze ex Fr.: Identification of geosmin, a musty-smelling compound. *Chemical and Pharmaceutical Bulletin* **29**, 1782–1784 (1981).
126. Frisvad, J.C., Smedsgaard, J., Larsen, T.O., and Samson, R.A. Mycotoxins, drugs and other extrolites produced by species in *Penicillium* subgenus *Penicillium*. *Studies in Mycology* **2004**, 201–241 (2004).
127. Zaitlin, B. and Watson, S.B. Actinomycetes in relation to taste and odour in drinking water: myths, tenets and truths. *Water Research* **40**, 1741–1753 (2006).
128. Hageskal, G., Lima, N., and Skaar, I. The study of fungi in drinking water. *Mycological Research* **113**, 165–172 (2009).
129. Gashgari, R.M., Elhariry, H.M., and Gherbawy, Y.A. Molecular detection of mycobiota in drinking water at four different sampling points of water distribution system of Jeddah City (Saudi Arabia). *Geomicrobiology Journal* **30**, 29–35 (2013).
130. Nasser, L.A. Incidence of terrestrial fungi in drinking water collected from different schools in Riyadh region, Saudi Arabia. *Pakistan Journal of Biological Sciences* **7**, 1927–1932 (2004).
131. Gonçalves, A.B., Paterson, R.R.M., and Lima, N. Survey and significance of filamentous fungi from tap water. *International Journal of Hygiene and Environmental Health* **209**, 257–264 (2006).
132. Hageskal, G., Knutsen, A.K., Gaustad, P., de Hoog, G.S., and Skaar, I. Diversity and significance of mold species in Norwegian drinking water *Applied and Environmental Microbiology* **72**, 7586–7593 (2006).
133. Göttlich, E. et al. Fungal flora in groundwater-derived public drinking water. *International Journal of Hygiene and Environmental Health* **205**, 269–279 (2002).
134. Saito, M., Singh, R.B. and Saite, M. Reports on study of mycotoxins in foods in relation to liver diseases in Malaysia and Thailand, p. 85 (University of Tokyo, Tokyo, Japan, 1976).
135. Sumalan, R., Alexa, E., Pop, G., Dehelean, C., and Sumalan, R. The biodiversity and dissemination of mycotoxin-producing fungi in cereals and cereal products. In *Proceedings of the 46th Croatian and 6th International symposium on agriculture*, pp. 770–773 (Opatija, Croatia, 2011).

136. Pitt, J.I. et al. The normal mycoflora of commodities from Thailand. 1. Nuts and oilseeds. *International Journal of Food Microbiology* **20**, 211–226 (1993).
137. Pitt, J.I. et al. The normal mycoflora of commodities from Thailand. 2. Beans, rice, small grains and other commodities. *International Journal of Food Microbiology* **23**, 35–53 (1994).
138. Pitt, J.I. et al. The mycoflora of food commodities from Indonesia. *Journal of Food Mycology* **1**, 41–60 (1998).
139. Freire, F.C.O., Kozakiewicz, Z., and Paterson, R.R.M. Mycoflora and mycotoxins of Brazilian cashew kernels. *Mycopathologia* **145**, 95–103 (1999).
140. Pelhate, J. Inventaire de la mycoflore des blés de conservation. *Bulletin trimestriel de la Société mycologique* **84**, 127–143 (1968).
141. Flannigan, B. Comparison of seed-borne mycofloras of barley, oats and wheat. *Transactions of the British Mycological Society* **55**, 267–276 (1970).
142. Udagawa, S. A taxonomic study on the Japanese species of *Chaetomium*. *Journal of General and Applied Microbiology* **6**, 223–251 (1960).
143. Skolko, A.J. and Groves, J.W. Notes on seed-borne fungi VII. *Chaetomium*. *Canadian Journal of Botany* **31**, 779–809 (1953).
144. Abdel-Kader, M.I.A., Moubasher, A.H., and Abdel-Hafez, A.A.I. Survey of the mycoflora of barley grains in Egypt. *Mycopathologia* **69**, 143–147 (1979).
145. Farnworth, E.R. and Neish, G.A. Analysis of corn seeds for fungi and mycotoxins. *Canadian Journal of Plant Pathology* **60**, 727–731 (1980).
146. Seth, H.K. A monograph of the genus *Chaetomium*. *Beihefte zur Nova Hedwigia* **37**, 133 (1970).
147. Tseng, T.C., Tu, J.C., and Tzean, S.S. Mycoflora and mycotoxins in dry beans (*Phaseolus vulgaris*) produced in Taiwan and in Ontario, Canada. *Botanical Bulletin of Academia Sinica* **36**, 229–234 (1995).
148. El-Kady, I.A. and Youssef, M.S. Survey of mycoflora and mycotoxins in Egyptian soybean seeds. *Journal of Basic Microbiology* **33**, 371–378 (1993).
149. Abdel-Hafez, A.I.I. and Saber, S.M. Mycoflora and mycotoxin of hazelnut (*Corylus avellana* L.) and walnut (*Juglans regia* L.) seeds in Egypt. *Zentralblatt für Mikrobiologie* **148**, 137–147 (1993).
150. Kubátová, A., Váňová, M., Prášil, K., and Fassatiová, O. Microfungi contaminating foods with low water content. *Novitates Botanicae Universitatis Carolinae* **13**, 13–25 (2000).
151. Řezáčová, V. and Kubátová, A. Saprobic microfungi in tea based on *Camellia sinensis* and on other dried herbs. *Czech Mycology* **57**, 79–89 (2005).
152. Mandeel, Q.A. Fungal contamination of some imported spices. *Mycopathologia* **159**, 291–298 (2005).
153. Shrivastava, A. and Jain, P.C. Seed mycoflora of some spices. *Journal of Food Science and Technology* **29**, 228–230 (1992).
154. Udagawa, S. and Sugiyama, Y. Additions to the interesting species of ascomycetes from imported spices. *Transactions of the Mycological Society of Japan* **22**, 197–212 (1981).
155. Misra, N. Influence of temperature and relative humidity on fungal flora of some spices in storage. *Zeitschrift für Lebensmittel-Untersuchung und Forschung* **172**, 30–31 (1981).
156. Freire, F.C.O., Kozakiewicz, Z., and Paterson, R.R.M. Mycoflora and mycotoxins in Brazilian black pepper, white pepper and Brazil nuts. *Mycopathologia* **149**, 13–19 (2000).
157. Ismail, M.E. and Abdalla, H.M. The fungus *Chaetomium globosum* a new pathogen to pear fruits in Egypt. *Assiut Journal of Agricultural Science* **36**, 177–188 (2005).
158. Skolko, A.J. and Groves, J.W. Notes on seed-borne fungi V. *Chaetomium* species with dichotomously branched hairs. *Canadian Journal of Research* **26**, 269–280 (1948).
159. Watanabe, T. *Pictorial Atlas of Soil and Seed Fungi: Morphologies of Cultured Fungi and Key to Species*, 2nd edn., p. 504 (CRC Press, Boca Raton, FL, 2002).
160. Kubátová, A. *Chaetomium* in the Czech Republic and notes to three new records. *Czech Mycology* **58**, 155–171 (2006).
161. Webb, T.A. and Mundt, J.O. Molds on vegetables at the time of harvest. *Applied and Environmental Microbiology* **35**, 655–658 (1978).
162. Greif, M.D., Stchigel, A.M., Miller, A.N., and Huhndorf, S.M. A re-evaluation of genus *Chaetomidium* based on molecular and morphological characters. *Mycologia* **101**, 554–564 (2009).
163. Ames, L.M. *A Monograph of Chaetomiaceae*, Series 2. p. 125 (The United States Army Research and Development, Washington, DC, 1963).
164. Jesenská, Z. *Micromycetes in Foodstuffs and Feedstuffs* (Elsevier, Amsterdam, the Netherlands, 1993).
165. Park, J.W., Choi, S.Y., Hwang, H.J., and Kim, Y.B. Fungal mycoflora and mycotoxins in Korean polished rice destined for humans. *International Journal of Food Microbiology* **103**, 305–314 (2005).

166. Udagawa, S. Distribution of mycotoxin-producing fungi in foods and soil from New Guinea and Southeast Asia. *Proceedings of the Japanese Association of Mycotoxicology* **2**, 10–15 (1976).
167. Nusrath, M. and Ravi, V. Mycoflora, mycotoxins and toxigenic fungi from paddy. *Indian Journal of Botany* **7**, 94–97 (1984).
168. Tullis, E.C. Fungi isolated from discolored rice kernels. *US Department of Agriculture Technical Bulletin* **540**, 1–11 (1936).
169. Orton, C.R. Seed-borne parasites. A bibliography. *West Virginia Agricultural Experiment Station Bulletin* **245**, 1–47 (1931).
170. Tervet, I.W. The influence of fungi on storage, on seed viability, and seedling vigor of soybeans. *Phytopathology* **35**, 3–15 (1945).
171. Wada, Y. et al. Studies on the fungal contamination of spices in Japan. *Journal of the Food Hygienic Society of Japan* **19**, 128–132 (1978).
172. Hussaini, A.M., Timothy, A.G., Olufunmilayo, H.A., Ezekiel, A.S., and Godwin, H.O. Fungi and some mycotoxins found in mouldy *Sorghum* in Niger state, Nigeria. *World Journal of Agricultural Sciences* **5**, 5–17 (2009).
173. Weidenbörner, M. Pumpkin seeds—The mycobiota and potential mycotoxins. *European Food Research and Technology* **212**, 279–281 (2001).
174. Chourasia, H.K. Mycobiota and mycotoxins in herbal drugs of Indian pharmaceutical industries. *Mycological Research* **99**, 697–703 (1995).
175. Paterson, P.J. et al. Molecular confirmation of invasive infection caused by *Chaetomium globosum*. *Journal of Clinical Pathology* **58**, 334 (2005).
176. Paterson, P.J., Seaton, S., McLaughlin, J., and Kibbler, C.C. Development of molecular methods for the identification of *Aspergillus* and emerging moulds in paraffin wax embedded tissue sections. *Molecular Pathology* **56**, 368–370 (2003).
177. Untereiner, W.A., Debois, V., and Naveau, F.A. Molecular systematics of the ascomycete genus *Farrowia* (Chaetomiaceae). *Canadian Journal of Botany* **79**, 321–333 (2001).
178. Meklin, T. et al. Quantitative PCR analysis of house dust can reveal abnormal mold conditions. *Journal of Environmental Monitoring* **6**, 615–620 (2004).
179. Hung, W.T., Su, S.L., Shiu, L.Y., and Chang, T.C. Rapid identification of allergenic and pathogenic molds in environmental air by an oligonucleotide array. *BMC Infectious Diseases* **11**, 91 (2011).
180. Aggarwal, R., Sharma, V., Kharbikar, L.L., and Renu. Molecular characterization of *Chaetomium* species using URP-PCR. *Genetics and Molecular Biology* **31**, 943–946 (2008).
181. Ahammed, S.K., Aggarwal, R., and Renu. Use of PCR based RAPD technique for characterization of *Chaetomium globosum* isolates. *Acta Phytopathologica et Entomologica Hungarica* **40**, 303–314 (2005).
182. Hynes, S.S., Chaudhry, O., Providenti, M.A., and Smith, M.L. Development of AFLP-derived, functionally specific markers for environmental persistence studies of fungal strains. *Canadian Journal of Microbiology* **52**, 451–461 (2006).
183. Al-Aidaroos, A. et al. Invasive *Chaetomium* infection in two immunocompromised pediatric patients. *Pediatric Infectious Disease Journal* **26**, 456–458 (2007).
184. Zhang, H. et al. *Clavispora lusitaniae* and *Chaetomium atrobrunneum* as rare agents of cutaneous infection. *Mycopathologia* **169**, 373–380 (2010).
185. Lesire, V. et al. Possible role of *Chaetomium globosum* in infection after autologous bone marrow transplantation. *Intensive Care Medicine* **25**, 124–125 (1999).
186. Schulze, H., Aptroot, A., Grote-Metke, A., and Balleisen, L. *Aspergillus fumigatus* and *Chaetomium homopilatum* in a leukaemic patient. Pathogenic significance of *Chaetomium* species. *Mycoses* **40**(Suppl. 1), 104–109 (1997).
187. Almeida, F., Gomes, J.M., and Salles, C. A study of a strain of *Chaetomium* isolated from a cutaneous lesion. *Anais de Faculdade de Medicina da Universidade de São Paulo* **20**, 145–154 (1944).
188. Hoppin, E.C., McCoy, E.L., and Rinaldi, M.G. Opportunistic mycotic infection caused by *Chaetomium* in a patient with acute leukemia. *Cancer* **53**, 555–556 (1983).
189. Guarro, J. Comments on recent human infections caused by ascomycetes. *Medical Mycology* **30**, 349 (1998).

17 *Claviceps*
The Ergot Fungus

Janine Hinsch and Paul Tudzynski

CONTENTS

17.1 *Claviceps* spp.: The Ergot Fungi 229
17.2 Molecular Biology of *C. purpurea* 230
17.3 Molecular Diagnostics 235
17.4 Ergot Alkaloids 235
    17.4.1 Pharmacological Activities 235
    17.4.2 Biosynthesis: The Ergot Alkaloid Gene Cluster 236
17.5 Ergot Infection: A Sophisticated Strategy 239
    17.5.1 Life Cycle 239
    17.5.2 Pathogenic Lifestyle of *C. purpurea* 239
    17.5.3 Specializations to Flower Infection 240
        17.5.3.1 Pollen Mimicry 241
        17.5.3.2 Molecular Mechanisms of *C. purpurea* Infections 242
17.6 Perspectives 244
Acknowledgments 245
References 245

17.1 *CLAVICEPS* SPP.: THE ERGOT FUNGI

The ascomycete genus *Claviceps* comprises more than 30 species [1] causing disease in a broad range of grasses (e.g., Poaceae, Juncaceae, and Cyperaceae), including all economically important cereal crop plants [2]. While most of the species in this group have a defined, narrow host range, *Claviceps purpurea* (Fries ex Fries) Tulasne is a broad host range pathogen infecting more than 400 plant species. The common name *ergot fungus* (*argot*: French for spur) refers to the purple to dark sclerotia appearing on infected grass ears 2–3 weeks after infection (see Section 17.5).

These survival structures contain the ergot alkaloids, which, due to their structural similarity to neurotransmitters, can have significant effects on the central nervous system. The notorious medieval *St. Anthony's Fire* disease was caused by consumption of rye bread contaminated with alkaloid-containing sclerotia. The disease symptoms described in medieval texts vary, probably because, as we know today, there are different *chemical races* of *C. purpurea* containing different sets of alkaloids. Two major groups of symptoms occurred: (1) convulsive ergotism (*Ergotismus convulsivus*) causing spasms, paranoia, and hallucinations, and (2) gangrenous ergotism (*Ergotismus gangraenosus*) characterized by disturbed peripheral sensation, edema, and the loss of affected limbs [3]. The link between rye contaminated with ergot and disease was suspected in the seventeenth century (see [4]), but this was not confirmed before the nineteenth century [5]. This realization led to a significant reduction in ergotism epidemics as ergot contamination in rye became ever more strictly controlled. Still, there are reports of recent epidemics, for example, in Germany (1879–1881), Russia (1926–1927), and Ethiopia (1977–1978) [3,6,7].

The therapeutic value of these *Janus-faced* ergot alkaloids was already apparent in the middle ages: ergot sclerotia were used by midwives as an aid in childbirth or to induce abortion (see Adam Lonicer's *Kreuterbuch* from 1582 [8]). To the present day, ergot alkaloids have been widely used to prevent excessive bleeding during childbirth, one of the major reasons for maternal mortalities in developing countries [9]. In addition, ergot alkaloids were used for blood pressure control and treatment of migraine [3] and against degenerative diseases of the central nervous system, such as Parkinson's disease. An increased understanding of the chemistry and pharmacology of ergot alkaloids allowed the development of highly active derivatives of the natural products [10,11]. The ergot alkaloids produced by *C. purpurea* (mainly ergotamine, ergocryptine, and related ergopeptine alkaloids, plus simpler lysergyl amides; all derivatives of the basic ergoline ring structure; see Section 17.4) can be converted by alkaline or acid hydrolysis to D-lysergic acid (LA), as a starting material for pharmaceuticals. Many semisynthetic ergot alkaloids have been developed, one of them the most potent hallucinogen known, LA diethylamide [12].

The Janus-faced role of ergot remains valid. Ergot alkaloids belong to the most prominent mycotoxins in cereal crops, and the risk of intoxication remains high. Therefore, continuous control measures are necessary [13]. Hence, the ergot disease can cause severe economic losses as often the grain might be classified as too poisonous for use although the number of seeds is not severely reduced (5%–10%; [14]). Scientific interest in this fungus continues, because the development of new defense strategies against ergot disease and strain improvement programs for biotechnological purposes require a detailed understanding of the biology, physiology, and genetics of the fungus. Here, we will present the state of the art on the available molecular tools and the molecular background of alkaloid biosynthesis and pathogenicity.

17.2 MOLECULAR BIOLOGY OF *C. PURPUREA*

C. purpurea belongs to the few biotrophic filamentous fungi that can be cultivated saprophytically in axenic culture. Therefore, it is straightforward to establish analytical methods for the biology of this fungus at the molecular level. The most critical development was probably the establishment of a transformation technique in 1989 [15]; Van Engelenburg et al. used a bleomycin-resistance gene (bleo[R]) fused to the *Aspergillus nidulans* trpC promoter as a selectable marker because *Claviceps* is highly sensitive to the glycopeptide antibiotic phleomycin. The construct was transferred into the fungus via polyethylene glycol–mediated transformation of protoplasts. However, at first, only low transformation rates could be achieved (0.25/μg), so the construct was optimized gradually, for example, by utilization of different promoters and terminators. In addition, further selection markers were established (e.g., hygromycin) to expand the possible applications.

The ability to integrate exogenous DNA into *C. purpurea* strains allows functional analysis of genes by integrating constitutively active genes, overexpression constructs, reporter gene fusions, or targeted gene deletion. In general, the integration of exogenous DNA into the genome relies on the DNA repair and recombination mechanisms of organisms. Two main pathways have been described for fungi. The nonhomologous DNA end joining (NHEJ) repairs double-strand breaks (DSB) independent of sequence homology by simple ligation of two free ends, leading to ectopic integration of DNA constructs into the genome. This mechanism is preferentially used by higher eukaryotes [16]. In contrast, the homologous recombination (HR) pathway depends on homologous DNA sequences. This mechanism leads to the integration of exogenous DNA into homologous regions within the genome [17,18]. For targeted gene deletion in *C. purpurea*, one takes advantage of the homologous integration mechanism. Flanking sequences 5' and 3' of the targeted gene are fused to a resistance cassette and form the deletion fragment.

Recently, a codon-optimized phleomycin gene for *C. purpurea* was artificially created reducing the guanine–cytosine (GC)-content from 70% to 58% in the new *cpble* for improved amplification via PCR (B. Oeser, WWU Muenster, unpubl.). This allows the generation of deletion fragments via the rapid yeast recombinational cloning method. This high-throughput method relies on the recombination machinery of yeast and bypasses the classical restriction–ligation approach. All components are amplified individually via PCR, adding overlapping sequences to their ends and cotransformed into yeast with a linearized shuttle vector. The overlapping sequences enable the HR machinery to recircularize the vector by ligating all added PCR fragments in one step [19]. Finally, the vector containing the deletion fragment can be isolated, linearized, and used for *C. purpurea* transformation. After DNA uptake, a double crossover at the homologous 5′ and 3′ flanking sequences can lead to the replacement of the targeted gene by the resistance cassette within the *C. purpurea* genome (Figure 17.1). Unfortunately, *Claviceps*, like most higher eukaryotes, preferentially uses the NHEJ and therefore has a very low homologous integration rate of around 1%–2% (e.g., see [20]). This makes it excessively difficult to generate deletion mutants in *C. purpurea* and necessitates the screening of hundreds of transformants per knockout. Therefore, a Δku70 mutant of *C. purpurea* was generated to improve the gene-targeting efficiency [21]. Previously, it had been shown in other fungi that this deletion blocks the NHEJ, because Ku70 is part of the main complex responsible for this process [22–25]. The *C. purpurea* Δku70 showed an increased rate of HR of up to 50%–60% without negative side effects in early generations [21]. However, long-term analysis revealed a slow degeneration of the strain that finally leads to reduced growth rates and the complete loss of virulence on rye (J. Hinsch, P. Tudzynski, WWU Muenster, unpubl.). This observation corresponds to findings in other organisms where it could be shown

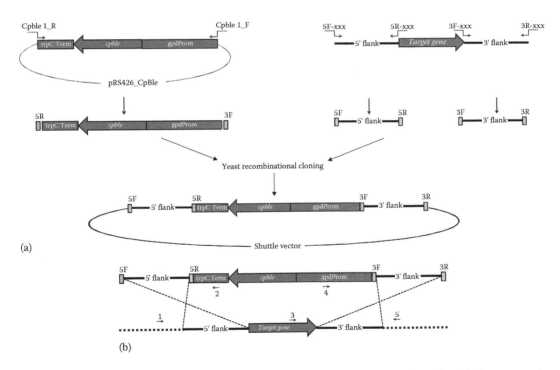

FIGURE 17.1 Scheme of a targeted gene deletion strategy. (a) For construction of a deletion vector via yeast recombinational cloning, the codon-optimized phleomycin resistance cassette (*cpble*) and the flanking sequences of the targeted gene (5′flank, 3′flank) are amplified via PCR adding overlapping sequences (5R/F, 3R/F). Together with a linearized shuttle vector, these fragments are transformed into competent yeast cells. The vector is assembled by homologous recombination at the overlapping sequences. (b) The linearized deletion fragment is transformed into *C. purpurea* protoplasts. Homologous recombination at the 5′ and 3′ flanking region of the targeted gene leads to the replacement of the gene by the resistance cassette. The correct integration can be verified via diagnostic PCR (using primers 1/2 for the 5′ flank and primers 4/5 for the 3′ flank).

that Ku70 is not only involved in DSB but also in telomere maintenance and thus in chromosome stability [26–28]. For this reason, *C. purpurea* Δku70 is no longer used for targeted gene deletions. Despite that, to date, around 30 knockout mutants of *C. purpurea* 20.1 from different kinds of gene families ranging from signaling components to degrading enzymes have been successfully generated and characterized (Table 17.1). In particular, the influence of targeted genes on virulence, polar growth, and alkaloid production has been the main focus and is discussed in detail in Sections 17.4 and 17.5.

The high ectopic integration rate (98%–99%) of *C. purpurea* can be turned to our advantage for the integration of nonhomologous DNA constructs such as reporter genes. One application is the labeling of strains with fluorescent proteins to visualize fungal development within plant tissue. For instance, a high and stable expression rate can be achieved with the vector pPgpd-DsRed_HygB containing a *dsRed* gene under the control of the *A. nidulans gpd* promoter and a hygromycin resistance cassette from Mikkelsen et al. [29], leading to strong and uniform fluorescence in the fungal cytoplasm (Figure 17.2). Moreover, the vector system created by Schumacher [30] for the related ascomycete *Botrytis cincerea* has successfully been used to generate fusion proteins in *C. purpurea*. With this system, a codon-optimized green fluorescent protein (oGFP) or mCherry protein, respectively, can be fused to proteins of interest using a constitutively active promoter or the native promoter of the gene. In this manner, the subcellular localization of *C. purpurea* proteins can be analyzed via live cell imaging giving an insight into their function. The fusion protein of oGFP with the small guanosine triphosphatase (GTPase), Cdc42, for example, leads to strong fluorescence at the hyphal tips (Figure 17.2) [31]. This localization pattern has been described for several fungal Cdc42 proteins [32–34] and strengthens the evidence suggesting that Cdc42 is involved in the maintenance of polar growth in *C. purpurea* [20]. For other proteins, for example, the transmembrane protein CpPls1, localization at the plasma membrane, and the endoplasmatic reticulum could be shown. Moreover, the GFP-CpPls1 fusion protein was the first fusion protein in *C. purpurea* proven to be functional as this construct complements a Δcppls1 deletion mutant [35].

The reporter gene fusion system will be especially useful for future research dealing with fungal effector proteins. These are small secreted proteins of fungal origin that are translocated into the host where they modulate the plant immune response [36]. It was already shown that *C. purpurea* is able to efficiently secrete mCherry fusion proteins containing an N-terminal secretion signal. This will allow the direct tracking of fungal secreted proteins within the plant tissue.

Further, potent applications of reporter genes in *C. purpurea* include bimolecular fluorescence complementation (BiFC) assays to study protein–protein interactions in situ. Split oGFP constructs were already used to prove the functionality of the BiFC system in *C. purpurea*; cotransformation of constructs containing putative proteins of the polarity complex led to fluorescence at the hyphal tips [31]. Due to the high autofluorescence of rye ovaries at these wavelengths, those strains cannot be analyzed in planta. Therefore, the generation of a split mCherry vector system is ongoing work.

Another major boost for the efficient molecular analysis of *C. purpurea* was the release of a genome assembly in 2013 as part of a comparative analysis of 15 genomes of different species belonging to the family of Clavicipitaceae [37]. All 15 species produce different kinds of psychoactive and bioprotective alkaloids during the interaction with their host species. Therefore, the comparison of the distribution and composition of ergot alkaloid biosynthesis clusters in the different strains was the major focus of this study. The genomic DNA of *C. purpurea* strain 20.1 was sequenced via 454 pyrosequencing using a mate-pair approach. The sequence is available online at http://www.ebi.ac.uk/ena/data/view/Project:76493 [37]. In total, the genome size is 32.1 Mb with 39-fold coverage of the sequence. The assembled genome consists of 1878 contigs that could finally be organized in 186 scaffolds. The predicted coding sequence length of 12.2 Mb is comparable to the other Clavicipitaceae and corresponds to 39.4% of the total genome. Notably, *C. purpurea* 20.1 has the lowest rate of repetitive DNA sequences of only 4.7% (compared to *Claviceps fusiformis* 45.7% and *Claviceps paspali* 17.5%) and the highest average GC content within the 15 analyzed

TABLE 17.1
Compilation of Knockout Mutants Available for *C. purpurea*

Gene Product	Gene	Impact on Pathogenicity	Other Defects	Reference
AP-transcription factor	*cpap1*	N/A	Increased sensitivity to menadione	D. Buttermann, P. Tudzynski, WWU Muenster, unpubl.
Cellobiohydrolase	*cpcel1*	None		[103]
Endo-1,4-β-xylanase	*cpxyl1*	None		[101]
Endo-1,4-β-xylanase	*cpxyl2*	Slight delay of colonization		J. Scheffer, S. Giesbert, P. Tudzynski, WWU Muenster, unpubl.
	cpxyl1/cpxyl2	Slight delay of colonization		J. Scheffer, S. Giesbert, P. Tudzynski, WWU Muenster, unpubl.
Polygalacturonase	*cppg1/cppg2*	Nonpathogenic		[104]
Cu/Zn superoxide dismutase	*cpsod1*	None	Reduced growth rate	[113]
Catalase	*cpcat1*	None		[114]
NADPH oxidase	*cpnox1*	Significant reduction of virulence	No ripe sclerotia	[121]
NADPH oxidase	*cpnox2*	Increase of honeydew production	No ripe sclerotia	[35]
Tetraspanin	*cppls1*	Significant reduction of virulence	No ripe sclerotia	[35]
Multicopper oxidase	*cpmco1*	None		S. Moore, P. Tudzynski, WWU Muenster, unpubl.
Transcription factor	*cptf1*	Significant reduction of virulence	Reduction of catalase activity	[115]
MAPK	*cpmk1*	Nonpathogenic		[108]
MAPK	*cpmk2*	Significant reduction of virulence	Defective in conidiation and cell wall structure	[109]
MAPK	*cpsak1*	Nonpathogenic	No sporulation, loss of septum formation	D. Buttermann, P. Tudzynski, WWU Muenster, unpubl.
Serine/threonine kinase	*cpcot1*	Nonpathogenic	Cell shape and branching strongly affected	[122]
Small GTPase	*cpcdc42*	Nonpathogenic (no colonization)	Hyperbranching, hypersporulation	[20]
Small GTPase	*cprac*	Nonpathogenic (no penetration)	Cell shape and branching strongly affected	[112]
PAK kinase	*cpcla4*	Nonpathogenic (no penetration)	Cell shape and branching strongly affected	[112]
Pentahydrophobin	*cpph1*	None		[123]
Histidine kinase	*cphk1*	N/A	Hypersporulation	D. Buttermann, P. Tudzynski, WWU Muenster, unpubl.
Histidine kinase	*cphk2*	Reduction of virulence (conidia only)	Significantly reduced conidial germination rate, increased resistance to menadione	[124]

(*Continued*)

TABLE 17.1 (Continued)
Compilation of Knockout Mutants Available for *C. purpurea*

Gene Product	Gene	Impact on Pathogenicity	Other Defects	Reference
Stretch-activated ion channel	*cpmid1*	Nonpathogenic	Increased sensitivity to cell wall stress	[125]
Subunit involved in nonhomologous end joining	*ku70*	None		[21]

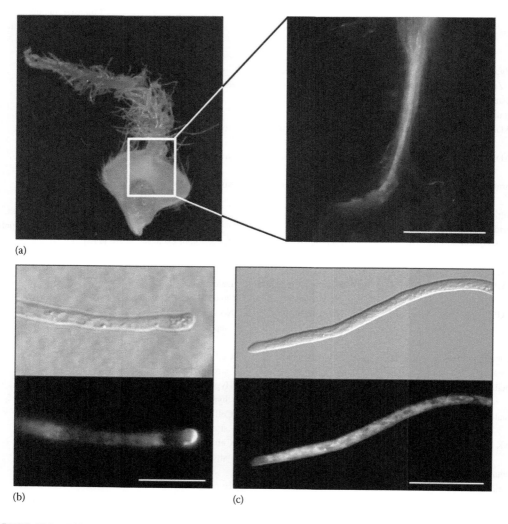

(a)

(b)   (c)

FIGURE 17.2 Life cell imaging of *C. purpurea*. (a) Epifluorescence image of hyphal growth of *C. purpurea* expressing DsRed within an ovary. The picture was taken 5 days after inoculation. Bar = 100 μm. (b) Epifluorescence image of *C. purpurea* hyphae expressing a GFP-Cdc42 fusion protein that is localized at the hyphal tip. Bar = 20 μm. (c) Epifluorescence image of *C. purpurea* hyphae expressing a GFP-Pls1 fusion protein that is most likely localized at the endoplasmatic reticulum and the plasma membrane. Bar = 20 μm.

Clavicipitaceae (52%; compared to *C. fusiformis* 37% and *C. paspali* 48%). By comparing the different chemical races of the analyzed fungi and their respective ergot alkaloid synthesis (EAS) gene clusters, a direct connection between cluster genes without experimentally determined roles in the pathway and the production of specific ergot alkaloid forms could be identified.

Apart from ergot alkaloids, a genome-wide bioinformatic analysis shows that *C. purpurea* 20.1 comprises several gene clusters potentially encoding for the production of additional secondary metabolites that might contaminate crops (B. Oeser, WWU Muenster, unpubl.). In general, the availability of a near-complete genome sequence facilitates practical work with *C. purpurea* as, for example, the time-consuming identification, amplification, and sequencing of genes is no longer necessary. Furthermore, it constitutes a suitable reference for RNA-sequencing (Seq) analyses of host and pathogen simultaneously.

A powerful set of tools for the molecular characterization of *C. purpurea* has been developed during the past years strengthening the use of this fungus as a model organism for the study of polar growth and the biosynthesis of pharmacologically active compounds. Finally, these methods will help better understand the close interaction between *C. purpurea* and host plants.

17.3 MOLECULAR DIAGNOSTICS

Claviceps purpurea—in contrast to most other phytopathogens—can be readily identified by its characteristic symptoms such as honeydew production and sclerotia formation. Since the main problem of *C. purpurea* infections is not the damage of host plants or crop loss but the contamination of seeds and flour with alkaloids, most diagnostic tools deal with the chemical identification of ergot alkaloids [38], or accompanying characteristic metabolites such as the fatty acid ricinoleate [39]. To distinguish different intraspecific groups (which might also differ in their alkaloid content), conidial morphology and ecological characteristics can be used [40]. However, molecular diagnostic tools are available; various PCR-based detection systems have been described, which allow the identification and (with semiquantitative or real-time PCR) quantification of fungal mycelium (for review, see [41]). So far, these techniques only find limited application for *C. purpurea*; however, they have considerable importance in other *Claviceps* species with high destructive potential, especially sorghum pathogens (*Claviceps africana, Claviceps sorghi, C. fusiformis*, and *C. paspali*). These species can be detected by direct PCR or quantified by real-time PCR using specific primers based, for example, on the beta tubulin gene [42–44]. The availability of genome data (see the previous text) and especially detailed sequence information on genes involved in EAS will help to develop new diagnostic tools to identify not only *C. purpurea* as such, but specific chemical races of the fungus, helping to improve the risk assessment.

17.4 ERGOT ALKALOIDS

17.4.1 PHARMACOLOGICAL ACTIVITIES

Ergot alkaloids are derivatives of prenylated tryptophan with the tetracyclic ergoline system as the basic structure. Their pharmacological activities are mostly due to their structural similarity to neurotransmitters: structures of noradrenaline, dopamine, and serotonin (5-hydroxytryptamine) fit well onto the d-LA ring structure. The strength and mode (agonistic or antagonistic) of interaction of ergot alkaloids with receptors for these neurotransmitters depend on the substituents attached to the carboxyl group of d-LA [45,46]. Natural isolates of *C. purpurea* can differ significantly in their alkaloid spectra (isolates with distinct spectra are sometimes referred to as *chemical races* or *chemotypes*); they can produce either the basic clavine alkaloids (intermediates/derivatives of the LA biosynthesis), simple LA derivatives such as ergometrine, or one or two of a large set of complex ergopeptines, which contain a tripeptide (with variable composition) attached to the LA carboxyl group (see Table 17.3). Many of these ergot alkaloids can probably interact with receptors for all three of these neurotransmitters as agonist, antagonist, or even in a dual role as partial agonist and antagonist [45]. This broad and variable specificity can cause highly complex (and unpredictable) intoxification symptoms when contaminated food is ingested. Even chemically pure natural alkaloids as applied in therapeutics can have unfavorable side effects; therefore, the improvement of natural ergot alkaloids by narrowing the specificity of the compounds by chemical modification represents a major challenge in pharmaceutical research [47]. However, a de novo chemical synthesis of ergot alkaloids is hampered by the stereospecificity of the pharmacological effects; derivatives of d-isolysergic acid, the stereoisomer of d-LA, show little or no pharmacological activity.

Natural ergopeptines mainly have vasoconstrictive and sympatholytic–adrenolytic effects because of their high affinity for adrenergic receptors [48]. However, modification of side chains can have drastic effects: the simple derivative dihydroergotamine, for example, has a much more dominant adrenolytic effect than ergotamine, with a parallel reduction of the vasoconstrictive effect; it is therefore preferentially used for the treatment of migraine [49–51]. Dihydroergotoxin, a mixture of several ergopeptines, is used for the treatment of high blood pressure and cerebral dysfunctions in the elderly [52,53]. Ergotoxines, including ergocryptine, act as inhibitors for the release of the peptide hormone prolactin. A semisynthetic derivative, 2-bromo-ergocryptine (bromocriptine), is used for the treatment of hyperprolactinemia, which can lead to reproductive disorders such as galactorrhea, prolactin-dependent mammary carcinoma, amenorrhea, acromegaly, or anovulation and long-term complications such as osteoporosis [54]. Bromocriptine is also used in the treatment of Parkinson's disease because of its high affinity to dopaminergic receptors [55]. Following a period of reduced interest in natural/semisynthetic ergot alkaloids due to their side effects, research has been intensified, and new and promising applications of ergot derivatives, especially in the field of degenerative diseases, have been described. Even clinical interest in the therapeutical potential of ergoline hallucinogens is growing, for example, for the treatment of autism [56,57].

17.4.2 Biosynthesis: The Ergot Alkaloid Gene Cluster

EAS [59–60] was originally studied by feeding experiments with radiolabeled putative precursors or intermediates added to fermentation cultures of *C. purpurea*, *C. fusiformis*, or *C. paspali* (the latter species produce only ergoclavine alkaloids or simple LA derivatives, respectively). The biosynthetic building blocks of the ergoline ring system were established as tryptophan and an isoprene unit derived from mevalonic acid. The pathway leading to LA (Figure 17.3b) starts with the formation of 4-dimethylallyltryptophan (DMAT) by a specific prenyltransferase, dimethylallyl tryptophan synthase (DMATS), the first enzyme of this pathway that was characterized in detail [61]. As typical for the first specific enzyme and hence the determinant step of the pathway, DMATS is strictly regulated; tryptophan serves as inducer, whereas later products of the pathway such as elymoclavine or agroclavine cause feedback regulation of the enzyme [62].

The *dmaW* gene encoding DMATS was the first gene from the EAS pathway to be cloned, originally from a *C. fusiformis* strain [63], and later from *C. purpurea* strain P1, a mutant of the ATCC strain 20102 producing alkaloids in axenic culture [64,65]. Using a chromosome-walking approach, the neighborhood of the *dmaW* gene was scanned for other potential EAS genes. This led to the detection of a cluster of 14 coregulated genes, predicted to encode ergot alkaloid biosynthetic enzymes (Figure 17.3a) [65–67]. Several of the genes have so far been analyzed functionally by gene disruption and the analysis of intermediates in *C. purpurea* and other fungi by different research groups (see the following text). To allow comparisons among ergot alkaloid–producing fungi, a systematic set of names for the genes of the ergot alkaloid pathway has been introduced [58]; EAS (*eas*) genes that have not yet been functionally characterized are designated *easA* through *easH*. Genes whose products have been biochemically characterized are named according to the enzyme activities of the encoded proteins. A list of the genes and the shown or predicted function is given in Table 17.2.

Besides *dmaW*, seven enzyme-encoding *eas* cluster genes have been functionally analyzed by gene disruption and analysis of intermediates in *C. purpurea* P1. These include four nonribosomal

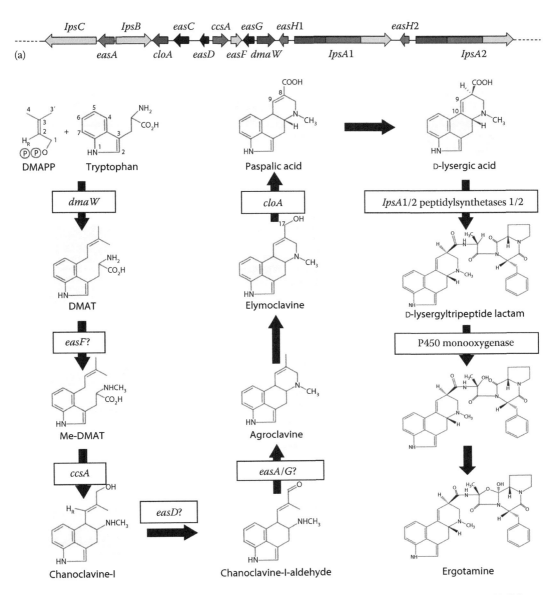

FIGURE 17.3 The ergot alkaloid cluster genes and their involvement in the biosynthetic pathway. (a) Scheme of the alkaloid biosynthesis cluster region in *C. purpurea* strain P1. For gene designations, see Table 17.2; direction of transcription is indicated by orientation of the arrows. (b) Biosynthetic pathway of the ergot alkaloids lysergic acid and ergotamine. DMAT, dimethylallyltryptophan; Me-DMAT, *N*-methyl-DMAT. (Modified after [66].)

peptide synthetase (NRPS) genes, *lpsA₁/A₂*, *lpsB*, and *lpsC*. Biochemical evidence had shown that the final steps of the ergopeptine synthesis in *C. purpurea* include a complex of two interacting NRPSs (a new finding in fungi): D-lysergyl peptide synthetase (LPS) 2, catalyzing the activation of LA, and LPS1, forming the tripeptide moiety [68]. Functional analysis showed that *lpsA₁* and *lpsA₂* both encode LPS1 enzymes, with LPS1-1 being necessary for synthesis of the major alkaloid of strain P1, ergotamine, and LPS1-2 for the synthesis of ergocryptine [21,65]. The gene *lpsC* encodes a monomodular NRPS enzyme that catalyzes the formation of ergonovine

TABLE 17.2
Genes of the Ergot Alkaloid Gene Cluster in *C. purpurea*

Gene	Enzyme	Function	Reference
dmaW	Prenyl transferase	DMATS	[63]
easF	Methyltransferase	Metylation of DMAT[a]	[74]
ccsA (*easE*)	FAD-oxidoreductase	Chanoclavine-I synthase	[71]
easD	Short-chain dehydrogenase	Chanoclavine-I aldehyde synthase[a]	[75]
easA	Oxidoreductase (old yellow enzyme)	Agroclavine synthase[b]	[73]
easG	Dehydrogenase	Agroclavine synthase	[72]
cloA	P450 oxidoreductase	Elymoclavine oxidase	[70]
easC	Catalase	Chanoclavine-I synthase[a]	[76]
lpsA1/2	NRPS	Assembly of tripeptide moiety of ergopeptines	[126]
lpsB	NRPS	Activation of lysergic acid	[126]
lpsC	NRPS	Ergonovine biosynthesis	[69]

[a] In *Aspergillus fumigatus*.
[b] In *Neotyphodium lolii*.

TABLE 17.3
Amino Acid Components of Ergopeptines in *Claviceps purpurea*

	Ergotamines	Ergoxines	Egotoxines
Position I → Position II ↓	Alanine	α-Amino Butyric Acid	Valine
Phenylalanine	Ergotamine	Ergostine	Ergocristine
Leucine	α-Ergosine	α-Ergoptine	α-Ergokyptone
Isoleucine	β-Ergosine	β-Ergoptine	β-Ergokryptine
Valine	Ergovaline	Ergonine	Ergocornine
α-Amino butyric acid	Ergobine	Ergobutine	Ergobutyrine

Note: The third position is always proline.

(ergometrine), an ergopeptine with a single amino acid side chain [69]. Thus, the NRPS enzymes encoded by the *C. purpurea* alkaloid gene cluster represent a highly flexible natural combinatorial system that is unique in eukaryotes; activated LA formed by LPS2 can serve as basis for the addition of several peptide moieties. This flexible biosynthesis scheme and the natural variability of the amino acid–binding domains of the LPS1 enzymes are the basis for the high variability of the ergot peptide alkaloid spectrum in the different natural chemical races of *C. purpurea* (see Table 17.3).

In recent years, the first part of the EAS pathway, that is, from DMAT to LA, was studied in detail. The function of two genes so far have been identified by a knockout approach in *C. purpurea*; the gene *cloA* encodes a cytochrome P450 oxidoreductase that catalyzes the conversion of elymoclavine to paspalic acid [70], and *ccsA* (*easE*) encodes a component of the chanoclavine-I synthase [71,72], which could show by heterologous expression of *easG* that its product (a dehydrogenase) is involved in biosynthesis of agroclavine. Additional information comes from studies in *Neotyphodium lolii*; in this endophytic fungus, the product of *easA* (an oxidoreductase of the *old yellow enzyme* type is necessary for agroclavine synthesis [73]. Since the first steps of the biosynthesis of the clavine alkaloid, fumigaclavine in the opportunistic human pathogen *Aspergillus fumigatus*, are identical to the EAS pathway in *C. purpurea*, the cluster contains several genes with high homology to the *C. purpurea* cluster. Functional analyses in this system showed that *easF* encodes a methyltransferase involved in biosynthesis of methylated DMAT [74]; *easD* codes for a chanoclavine-I aldehyde synthase [75]; and the product of *easC*, a catalase, is necessary for chanoclavine-I synthesis [76]. Thus, corresponding gene functions could be identified for most of the steps of the EAS pathway.

17.5 ERGOT INFECTION: A SOPHISTICATED STRATEGY

The ergot sclerotium, which is the only alkaloid-containing structure of *C. purpurea* field isolates, is exclusively formed during an infection of a host plant. It differentiates at the end of the infection process instead of a caryopsis, replacing the ovarian tissue. Therefore, *C. purpurea* is also defined as a tissue-replacement disease. The sclerotium resembles the shape of a grain but is normally bigger (2–25 mm; one to several times larger than the host seed), and its hard outer cortex is pigmented purple black, presumably to protect the inner plectenchymatous tissue from UV radiation. The main functions of the sclerotium for the fungus are overwintering and completion of the sexual cycle [77].

17.5.1 LIFE CYCLE

In spring, the sexual cycle of *C. purpurea* is induced in sclerotia; they germinate and differentiate stroma heads containing perithecia (Figure 17.4a). The needle-like ascospores are distributed via insects, wind, or rain. They land on the feathery stigmas of Poaceae, attach and germinate (Figure 17.4b). Like most Clavicipitaceae, *C. purpurea* enters the epidermal tissue directly without the formation of a specialized infection structure; however, the detailed mechanism remains unclear. After penetration of the cuticle, the colonization of the plant tissue proceeds mainly intercellularly. In rare cases, the hyphae even enter living host cells, presumably fulfilling haustoria-like functions [78]. During this early infection stage, the fungal hyphae grow in a strictly polar manner and are always directed to the base of the ovary—the rachilla—passing the interstylar region, the underlying transmitting tissue, and surrounding the ovule (Figure 17.4c). However, when the hyphae finally reach the rachilla and tap the vascular traces of the plant, frequently branched hyphae can be observed forming the so-called sphacelial tissue. Approximately 7 days postinfection, a persistent host–pathogen interface is established at this point as *C. purpurea* never enters the underlying tissue. The first typical macroscopic sign of an infection with *Claviceps*—the formation of honeydew—now becomes visible (Figure 17.4d). Honeydew is a sticky, sugar-rich fluid containing sphacelium-derived conidia exuded from infected grass spikelets (Figure 17.4e). This sweet exudate attracts insects, thus facilitating further spreading of the disease within a field. In addition, it serves as a preserving agent for the spores during dry and warm weather [79]. The production of honeydew ceases as the first sclerotial hyphae differentiate around 14 days postinfection. These are isodiametric storage cells that characteristically produce high levels of lipids (~30% (w/w)) and varying amounts of alkaloids (0.01%–0.5% (w/w)). The determination of the predominant fatty acid ricinoleate has even been used as a marker for ergot contamination in food [39]. The sclerotium ripens in ca. 4–5 weeks and falls to the ground for overwintering (Figure 17.4f and g).

17.5.2 PATHOGENIC LIFESTYLE OF *C. PURPUREA*

Claviceps is completely dependent on living host plants to fulfill its life cycle. Accordingly, the infection process needs to proceed without severe damage of plant material or the triggering of

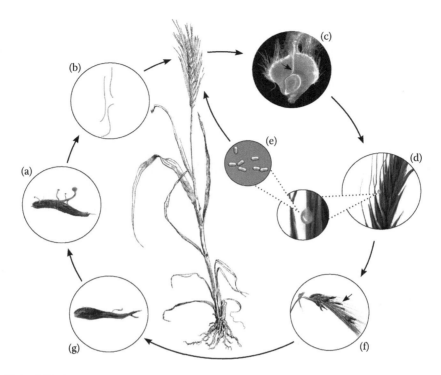

FIGURE 17.4 Life cycle of *C. purpurea*. (a) In spring, sclerotia germinate and differentiate stroma heads containing perithecia. (b) The needle-like ascospores land on the stigmatic hairs of flowering grass ears, attach, germinate, and penetrate the ovaries (c). In the early infection stages, the fungus is oriented toward the rachilla of the ovary and grows polar within interstylar region, the underlying transmitting tissue and surrounding the ovule (d). Around 7 days postinfection (dpi), it reaches the rachilla and a stable host–pathogen interface is established. At this time, the production of honeydew starts (e). It contains the conidia of the fungus and leads to secondary infections (f). Around 14 dpi, the formation of sclerotia is initiated that ripen in ca. 4–5 weeks (g) and finally fall to the ground for overwintering.

drastic host reactions like a hypersensitive response. Both might lead to host-cell and therefore also fungal death. Indeed, all microscopical analyses show that the cells surrounding fungal hyphae are intact during the colonization phase, proving that *C. purpurea* does not actively kill host cells to gain nutrients but somehow manages to keep them alive [80]. Only the separation of the ovarian cap from the nutrient supply during sclerotia formation results in an induction of plant cell death. However, this is regarded to be an inevitable process analogous to the induction of early host senescence by other biotrophic pathogens [81]. Finally, the replacement of the ovarian tissue leads to host sterilization. But as *C. purpurea* never infects all ovaries within one ear, it is only a partial sterilization that might even be beneficial for the host due to a protective effect against grazing animals [82]. Taken together, *C. purpurea* is an ecologically obligate parasite with an entirely biotrophic lifestyle.

17.5.3 Specializations to Flower Infection

Flowers represent an attractive infection target for phytopathogenic fungi: (1) they are naturally well supplied with nutrients as they serve to produce the plant embryo; (2) the stigmatic surface is relatively thin walled and therefore easily penetrable; and (3) they attract pollinators to distribute their male gametophytes, a feature that can be exploited by the pathogen to spread its spores. Indeed, fungi from different classes have evolved a wide variety of infection strategies to colonize all different flower parts [83]. The causal agent of one of the most important rice diseases in China, *Claviceps virens*, for example, is able to grow on the surface of stamens and pistils; thus, the primary infection sites are the stamen filaments [84,85]. In contrast, *C. purpurea* belongs to the small group of highly specialized gynoecial infecting fungi that solely infect the ovary of a plant via the stigma–style pathway. This pathway was considered as a natural opening through which pathogenic fungi can easily enter [86]. Nevertheless, it requires a precise adaption of the pathogen to the host. The formation of flowers, for example, is highly seasonal, and as *C. purpurea* can only infect ovaries of young unfertilized grasses, a synchronicity of ascospore formation and grass flowering is vital. Interestingly, sclerotia of *C. purpurea* and seeds of rye plants (the main cultivated host of *Claviceps*) require a vernalization period before germination can be induced [87,88]. Besides, the optimal germination temperature for ergot sclerotia varies for different isolates, leading to the suggestion that this results from adaption of ergots to the phonological development of their hosts in different habitats [89–91]. This phenomenon has not been analyzed in detail for *C. purpurea*; however, for other flower-infecting fungi such as *Monilinia vaccinii-corymbosi*, a synchronicity of spore formation and host-flower development could be proven [92].

17.5.3.1 Pollen Mimicry

The most outstanding feature of the gynoecial infection strategy is the polar and directed growth behavior of the pathogen. *C. purpurea* forms a hyphal network in axenic cultures, whereas during the colonization phase, mainly unbranched hyphae form a thick bundle of parallel hyphal strands within the style [93]. Fertilization experiments with *Secale cereale* reveal a high similarity between the fertilization and the infection process. Indeed, superficially, the attachment of pollen grains, penetration, and growth behavior within the ovarian tissue are similar to ergot infection that may be the result of a mimicry strategy whereby *C. purpurea* behaves exactly like a pollen tube in order to enter the pistil unrecognized. In fact, for *C. africana* and *C. sorghi*, field studies proved competition between fungal infection and fertilization [94,95]. It is also well established that male-sterile lines of *S. cereale* are more susceptible to infection with *C. purpurea* than fertile ones.

Despite the many similarities, some notable differences between the infection and the fertilization process have been discovered. First, pollen grains of *S. cereale* are around five times bigger than conidia of *C. purpurea*. Second, the fertilization process is significantly faster than an infection as it is completed approximately 75–90 min after pollen germination. The colonization of the same tissue by fungal hyphae takes 5–7 days, forming a denser structure within the female tissue compared to a single pollen tube. Most importantly, the infection route of *C. purpurea* is similar, but not identical, to the pollen tube pathway. After *C. purpurea* hyphae reach the transmitting tissue, they grow straight down until they come in physical contact with the ovule and completely surround it. In this infection phase, hyphae stay in close contact to the inner ovary wall until they reach the base of the ovary and grow toward the rachilla. In contrast, pollen tubes enter the ovary, grow toward the ovule but deviate before they touch its upper part, mainly growing in the middle of the ovarian mesophyll. Finally, they reach the ovule from the bottom at the micropyle to deliver the male gametes (Figure 17.5). These major differences suggest that the mere pathway mimicry is most likely not sufficient to avoid host defense reactions during the colonization process. Further aspects, for example, a molecular mimicry of specific pollen–stigma interaction signals by the fungus, might also play an important role. Likewise, evidence is growing that pathogenic fungi suppress host defense reactions by the secretion of small proteins into the fungus–host interface (reviewed in [36]). These effector proteins interact either directly with proteins of the defense machinery of the host (e.g., [96]) or indirectly via the interference with signaling pathways (e.g., [97]). Preliminary results of a transcriptome analysis of *C. purpurea* infecting *S. cereale* show that in this interaction a set of small secreted proteins is highly expressed in the fungus (B. Oeser, J. Schürmann, P. Tudzynski, WWU Muenster, unpubl.).

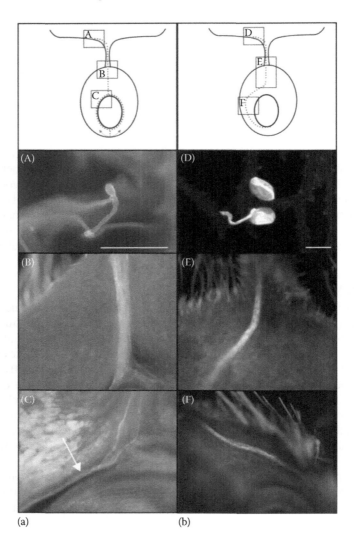

FIGURE 17.5 Comparison of the in planta growth of (a) *C. purpurea* hyphae and (b) pollen tubes. (a) Scheme of the infection route of *C. purpurea*. Boxes A–C show detailed microscopical images of indicated regions. Ovaries were inoculated with conidial suspensions, collected 7 dpi, and stained with anillin blue. (A) Germinated hypha on the stigmatic hairs. Bar = 50 μm, (B) hyphal growth within the transmitting tissue of the ovary, (C) hyphae growing in close contact to the inner ovary wall (indicated by arrow). (b) Scheme of the growth route of pollen tubes of *S. cereale* (cultivar *Conduct*) during fertilization. Boxes D–F show detailed microscopical images of indicated regions. The pollen grains were dryly applied to the stigmatic hairs, collected 3 h after inoculation and stained with anillin blue. (D) Germinated pollen grains on the stigmatic hairs. Bar = 50 μm, (E) pollen tube growth within the transmitting tissue of the ovary, (F) pollen tubes mainly growing within the ovarian mesophyll.

17.5.3.2 Molecular Mechanisms of *C. purpurea* Infections

To obtain a better understanding of the molecular mechanisms underlying the infection process of *C. purpurea*, knockout of putative virulence genes approaches have been performed. The analyzed genes can be categorized into three groups. They are either involved in polarity establishment, signaling cascades influencing fungal development, or the degradation of plant material.

The typical inter- and intracellular growth of *C. purpurea* within the plant tissue is accompanied by the degradation of cell wall material [98,99]. The fungus secretes limited amounts of degrading enzymes to facilitate the colonization of the tissue without destroying it completely. Monocotyledonous plants have evolved a special cell wall type mainly consisting of the polysaccharide cellulose, high amounts of glucuronoarabinoxylans, and only low amounts of pectins [100]. Accordingly, in axenic cultures of *C. purpurea* enzymatic activity of pectinases, xylanases, and cellulases (and a β-1,3-glucanase) have been detected [99–103]. Surprisingly, deletion of the genes for most of these cell wall–degrading enzymes does not have a severe effect on the virulence of the fungus. On the other hand, mutants lacking polygalacturonase genes (*cppg1*, *cppg2*) are drastically impaired in the colonization of plant tissue and only rarely produce honeydew and of limited amounts [104]. In lily, it has been shown that stylar pectin is necessary for pollen tube guidance [105]. Thus, it is possible that *C. purpurea* also follows a *pectin trail* within the ovary, again taking advantage of pollen tube mimicry. In the later infection stages, *cppg1* and *cppg2* might be necessary for the penetration of the sieve plates of the plant, hence facilitating honeydew formation. Further, carbohydrate-degrading enzymes are most likely involved in this process; a secreted β-1,3-glucanase is speculated to degrade phloem callose to keep the sieve plates unblocked for assimilate flow, whereas invertases that degrade sucrose, the most abundant transport sugars within plants, could keep the infected floret a sink tissue. Only recently, it was shown that *C. purpurea*, as the only analyzed ascomycete, contains seven homologous genes encoding for invertases. All of them were expressed in planta, strengthening the hypothesis that they are important for the pathogenesis of *C. purpurea* (S. Kind, J. Schürmann and P. Tudzynski, WWU Muenster, unpubl.).

The sensing of environmental stimuli is especially important for phytopathogenic fungi such as *C. purpurea*, as they need to adapt their development to their host plant (e.g., the formation of penetration structures, secretion of degrading enzymes). Members of the family of mitogen-activated protein kinases (MAPK) are part of signaling cascades in eukaryotes, transducing external signals to convert them into internal reactions. They are involved in various differentiation processes, such as sporulation and formation of fruiting bodies. In filamentous fungi, three MAPK cascades are conserved, corresponding to the yeast Fus3 (pheromone signaling), Slt2 (cell integrity pathway), and Hog1 (osmotic balance) cascades [106]. Deletion of the corresponding genes in various phytopathogenic fungi shows that MAPK signaling (especially the Fus3 cascades) has a negligible role in growth in axenic culture but controls key steps in pathogenesis and is therefore essential for full virulence of pathogens [107].

Accordingly, deletion mutants of *cpmk1* (encoding a Fus3 homologue) of *C. purpurea* are non-pathogenic on rye, and Δcpmk2 (slt2 pathway) mutants are affected negatively in virulence, as only 25% of inoculated ovaries are penetrated by the fungus [108,109]. Interestingly, the invading hyphae of Δcpmk2 are heavily branched and fail to colonize the whole ovary. This observation confirms the results of analyses of polarity complex components in *C. purpurea*, as all analyzed mutants with an altered polarity are apathogenic or at least drastically reduced in virulence. This proves that the typical polar growth of *C. purpurea* during the early infection stages is not only important but also mandatory for a successful infection. One example is the group of small GTPases. These highly conserved proteins are known to be connected to polarity establishment in various cell types ranging from plant root hairs to mammalian neurons [110,111]. Two classical members of the Rho subfamily, Cdc42 and Rac, have also been deleted in *C. purpurea*. Overall, the deletions led to different phenotypes but obviously both proteins are essential for normal polarity establishment. The phenotype of *C. purpurea* mutants lacking Cprac and its major downstream partner Cpcla4 involves a drastic change in colony morphology as the mutants completely lose polarity [112]. These strains are apathogenic because they are not able to penetrate the plant. Deletion of the second small GTPase Cdc42 induces hypersporulation and an increased branching frequency, but only in axenic culture. The morphology of hyphae growing within the stigmatic hairs is like the wild type, though with much lower density. Finally, the mutant stops prior to the ovarian cap, probably because plant defense reactions (including high H_2O_2 levels) hinder further progress of the fungus [20]. Usually, *C. purpurea* does not seem to face high oxidative stress in planta, as this kind of reaction has never been observed during wild-type infections. Moreover, the deletion of abundant reactive oxygen species (ROS)-scavenging enzymes (*cpsod, cpcat1*) has no significant influence on virulence of *C. purpurea* [113,114]. However, deletion of the basic region, leucine zipper (bZIP)-transcription factor *cptf1*, which regulates the expression of all catalase isoforms, results in a reduced virulence of the fungus and (comparable to Δcpcdc42) induces an oxidative burst of the plant [115]. Recent in planta transcriptome data from Δcptf1 show that this is most likely not only due to an influence of the transcription factor on the expression of the catalases but also on various not yet analyzed genes, including putative virulence factors that might be important to repress host defense responses (B. Oeser, J. Schürmann, P. Tudzynski, WWU Muenster, unpubl.).

The ROS that are produced by the plant during an oxidative burst are highly reactive molecules that damage macromolecules and therefore attack invading pathogens. However, despite these harmful effects, ROS can also act as signaling molecules between and within organisms, thus playing an ambivalent role during host–pathogen interactions. The best known ROS generating system is the NADPH-oxidase complex of mammalian phagocytes. Upon infection, the active complex, consisting of several cytosolic subunits and a membrane spanning NADPH-oxidase (gp91^{phox}), is formed catalyzing the reduction of oxygen to superoxide [116]. Plants and filamentous fungi contain similar proteins. In fungi, three subclasses of NADPH-oxidases have been described: the gp91^{phox} homologues NoxA and B are present in most filamentous fungi, whereas NoxC (characteristically containing an additional calcium-binding motif) is only rarely present in fungal genomes. Functional analyses show that a balanced ROS production is important to differentiate mainly fungal specific structures. For example, in *Botrytis cinerea*, NoxB is required to form functional appressoria, and in *Magnaporthe oryzae*, both Nox enzymes are indispensable for this differentiation process. Detailed analyses reveal an influence of Nox enzymes on actin and septin and an involvement of Cdc42, thus influencing cytoskeleton reorganization [117–119]. In *Epichloe festucae*, a close relative of *C. purpurea*, NoxA, regulates hyphal morphogenesis and the deletion of the gene renders the former endophytically growing fungus a pathogen [120]. Similarly, deletion mutants of *cpnox1* and *cpnox2* in *C. purpurea* are affected in the balanced infection of the host plant. Strains lacking *cpnox1* are reduced in virulence as they fail to fully colonize the ovarian tissue and only rarely produce immature sclerotia [121]. Contrarily, *C. purpurea* Δcpnox2 is rather hypervirulent, as this strain produces massive amounts of honeydew that could lead to a fungal outbreak in the field. This unique phenotype results from the fact that the morphological switch from spacelial (honeydew producing) to sclerotial growth is missing in the Δcpnox2 mutant [35]. Taken together, Nox-derived ROS are most likely not important for the host–pathogen signaling in *C. purpurea*, but rather are involved in the regulation of differentiation processes of the fungus *in planta*.

17.6 PERSPECTIVES

Claviceps sp., the ergot fungus, plays an important but ambivalent role in agriculture, pharmacology, and biotechnology. As a food contaminant, it represents a dangerous threat for consumers of cereal products, and not all toxic substances accompanying food contamination with *Claviceps* are known. On the other hand, it is the source of valuable pharmaceuticals. The fungus has caused recent dramatic and fatal epidemics, and yet it will help to establish new strategies for the treatment of important degenerative diseases that have enormous economic importance. Since no effective control strategy is available despite long-standing intensive research, *Claviceps* is a persistent threat to modern agriculture. The rapid developments in molecular technology open new perspectives for a better understanding of the infection strategies of this important pathogen. The fungus represents a fascinating model system to understand biotrophic host–pathogen interaction, with its special aspects of organ specificity, oriented in planta growth and suppression of host defense. Recent advances reported here give us good cause to hope that new control strategies can be achieved in the near future.

ACKNOWLEDGMENTS

We thank D. O'Sullivan for critical reading of the manuscript, B. Oeser and D. Buttermann for discussions and sharing of data prior to publication, and the Deutsche Forschungsgemeinschaft for financial support.

REFERENCES

1. Taber, W. A. 1985. Biology of *Claviceps*. In Biotechnology series, Vol. 6 *Biology of Industrial Microorganisms*, eds. A. L. Demain and A. S. Nadine, pp. 449–486. New York: The Benjamin Cummings Publishing Co. Inc.
2. Bové, F. J. 1970. *The Story of Ergot*. Basel, Switzerland: S. Karger.
3. Eadie, M. J. 2003. Convulsive ergotism: Epidemics of the serotonin syndrome? *Lancet Neurol.* 2:429–434.
4. Bauer, V. H. 1973. *Das Antoniusfeuer in Kunst und Medizin*. Berlin, Germany: Springer Verlag.
5. Tulasne, L.-R. 1853. Memoire sur l'ergot des glumacees. *Ann. Sci. Nat. (Parie Botanique)* 20:5–56.
6. Barger, G. 1931. *Ergot and Ergotism*. London, U.K.: Gurney & Jackson.
7. Urga, K., A. Debella, Y. N. A. W'Medihn, A. Bayu, and W. Zewdie. 2002. Laboratory studies on the outbreak of gangrenous ergotism associated with consumption of contaminated barley in Arsi, Ethiopian. *J. Health Dev.* 16:317–323.
8. van Dongen P. W. J. and A. N. J. A. de Groot. 1995. History of ergto alkaloids from ergotism to ergometrine. *Eur. J. Obstet. Gynecol. Reprod. Biol.* 60:109–116.
9. Li, X. F., J. A. Fortney, M. Kotelchuck, and L. H. Glover. 1996. The postpartum period: The key to maternal mortality. *Int. J. Gynaecol. Obstet.* 54:1–10.
10. Gröger, D. and H. G. Floss. 1998. Biochemistry of ergot alkaloids—Achievements and challenges. In *The Alkaloids: Chemistry and Biology*, ed. G. A. Cordell, Vol. 50, pp. 171–218. London, U.K.: Academic Press.
11. Sinz, A. 2008. Die Bedeutung der Mutterkorn-Alkaloide als Arzneistoffe. *Pharm. Unserer Zeit* 4:306–309.
12. Hofmann, A. 1978. Historical view on ergot alkaloids. *Pharmacology* 16:1–11.
13. Krska, R. and C. Crews. 2008. Significance, chemistry and determination of ergot alkaloids: A review. *Food Addit. Contam.* 25:722–731.
14. Alderman, S. C., D. D. Coats, F. J. Crowe, and M. D. Butler. 1998. Occurrence and distribution of ergot and estimates of seed loss in Kentuckey bluegrass grown for seed in central Oregon. *Plant Dis.* 82:89–93.
15. van Engelenburg, F., R. Smit, T. Goosen, H. van den Broek, and P. Tudzynski. 1989. Transformation of *Claviceps purpurea* using a bleomycin resistance gene. *Appl. Microbiol. Biotechnol.* 30:364–370.
16. Sonoda, E., H. Hochegger, A. Saberi, Y. Taniguchi, and S. Takeda. 2006. Differential usage of non-homologous end-joining and homologous recombination in double strand break repair. *DNA Repair* 5:1021–1029.
17. Mladenov, E. and G. Iliakis. 2011. Induction and repair of DNA double strand breaks: The increasing spectrum of non-homologous end joining pathways. *Mutat. Res./Fund. Mol. Mech. Mutagen.* 711:61–72. ISSN 0027-5107.
18. Neal, J. A. and K. Meek. 2011. Choosing the right path: Does DNA-PK help make the decision? *Mutat. Res.* 711:73–86. doi: 10.1016/j.mrfmmm.2011.02.010.
19. Oldenburg, K. R., K. T. Vo, S. Michaelis, and C. Paddon. 1997. Recombination-mediated PCR-directed plasmid construction in vivo in yeast. *Nucleic Acids Res.* 25:451–452.
20. Scheffer, J., C. Chen, P. Heidrich, M. B. Dickman, and P. Tudzynski. 2005. A CDC42 homologue in *C. purpurea* is involved in vegetative differentiation and is essential for pathogenicity. *Eukaryot. Cell* 4:1228–1238.
21. Haarmann, T., N. Lorenz, and P. Tudzynski. 2008. Use of a nonhomologous end joining deficient strain (Δku70) of the ergot fungus *Claviceps purpurea* for identification of the nonribosomal peptide synthetase gene involved in ergotamine biosynthesis. *Fungal Genet. Biol.* 45:35–44.
22. Kooistra, R. K., P. J. J. Hooykaas, and H. Y. Steensma. 2004. Efficient gene targeting in *Kluyveromyces lactis*. *Yeast* 21:781–792.

23. Ninomiya, Y., K. Suzuki, C. Ishii, and H. Inoue. 2004. Highly efficient gene replacements in *Neurospora* strains deficient for nonhomologous endjoining. *Proc. Natl. Acad. Sci. USA* 101:12248–12253.
24. Nayak, T., E. Szewczyk, C. E. Oakley et al. 2006. A versatile and efficient gene-targeting system for *Aspergillus nidulans*. *Genetics* 172:1557–1566.
25. Pöggeler, S. and U. Kück. 2006. Highly efficient generation of signal transduction knockout mutants using a fungal strain deficient in the mammalian ku70 ortholog. *Gene* 378:1–10.
26. Critchlow, S. E. and S. P. Jackson. 1998. DNA end-joining: From yeast to man. *Trends Biochem. Sci.* 23:394–398. doi: 10.1016/S0968-0004(98)01284-5.
27. Baumann, P. and T. R. Cech. 2000. Protection of telomeres by the Ku protein in fission yeast. *Mol. Biol. Cell* 11:3265–3275.
28. Bailey, S. M. and E. H. Goodwin. 2004. DNA and telomeres: Beginnings and endings. *Cytogenet. Genome Res.* 104:109–115. doi: 10.1159/000077474.
29. Mikkelsen, L., S. Sarrocco, M. Lübeck, and D. F. Jensen. 2003. Expression of the red fluorescent protein DsRed-Express in filamentous ascomycete fungi. *FEMS Microbiol. Lett.* 223:135–139.
30. Schumacher, J. 2012. Tools for *Botrytis cinerea*: New expression vectors make the gray mold fungus more accessible to cell biology approaches. *Fungal Gen. Biol.* 49:483–497.
31. Herrmann, A., B. A. Tillmann, J. Schürmann, M. Bölker, and P. Tudzynski, 2014. Small-GTPase-associated signaling by the guanine nucleotide exchange factors CpDock180 and CpCdc24, the GTPase effector CpSte20, and the scaffold protein CpBem1 in Claviceps purpurea. *Eukaryotic Cell* 13(4):470–482.
32. Hazan, I. and H. Liu. 2002. Hyphal tip-associated localization of Cdc42 is F-actin dependent in *Candida albicans*. *Eukaryot. Cell* 1:856–864. doi: 10.1128/EC.1.6.856-864.2002.
33. Virag, A., M. P. Lee, H. Si, and S. D. Harris. 2007. Regulation of hyphal morphogenesis by *cdc42* and *rac1* homologues in *Aspergillus nidulans*. *Mol. Microbiol.* 66:1579–1596.
34. Araujo-Palomares, C. L., C. Richthammer, S. Seiler, and E. Castro-Longoria. 2011. Functional characterization and cellular dynamics of the CDC-42 - RAC - CDC-24 module in *Neurospora crassa*. *PLoS One* 6:e27148. doi: 10.1371/journal.pone.0027148.
35. Schuermann, J., D. Buttermann, A. Herrmann, S. Giesbert, and P. Tudzynski. 2013. Molecular characterization of the NADPH oxidase complex in the Ergot fungus *Claviceps purpurea*: Cpnox2 and Cppls1 are important for a balanced host-pathogen interaction. *Mol. Plant Microbe Interact.* 26(10):1151–1164.
36. Dodds, P. N. and J. P. Rathjen. 2010. Plant immunity: Towards an integrated view of plant–pathogen interactions. *New Phytologist.* 183:993–1000. doi: 10.1111/j.1469-8137.2009.02922.x.
37. Schardl, C., C. Young, U. Hesse et al. 2013. Plant-symbiotic fungi as chemical engineers: Multi-genome analysis of the clavicipitaceae reveals dynamics of alkaloid loci. *PLoS Genetics* 9(2):e1003323.
38. Koleva, I. I., T. A. van Beek, A. E. M. F. Soffers, B. Dusemund, and I. M. C. M. Rietjens. 2012. Alkaloids in the human food chain—Natural occurrence and possible adverse effects. *Mol. Nutr. Food Res.* 56:30–52.
39. Franzmann, C., J. Wächter, N. Dittmer, and H.-U. Humpf. 2010. Ricinoleic acid as a marker for ergot impurities in rye and rye products. *J. Agric. Food Chem.* 58:4223–4229.
40. Fisher, A. J., J. M. DiTomaso, and T. R. Gordon. 2005. Conidial morphology and ecological characteristics as diagnostic tools for identifying *Claviceps purpurea* from salt-marsh habitats. *Can. J. Plant Pathol.* 27:389–395.
41. Munaut, F., F. van Hove, and A. Moretti. 2011. Molecular identification of mycotoxigenic fungi in food and feed. In *Woodhead Publishing in Food Science Technology and Nutrition. Determining Mycotoxins and Mycotoxigenic Fungi in Food and Feed*, ed. S. DeSaeger, pp. 298–331. Cambridge, U.K.: Woodhead Publ. LTD.
42. Tooley, P. W., E. D. Goley, M. M. Carras, R. D. Frederick, E. L. Weber, and G. A. Kuldau. 2001. Characterization of *Claviceps* species pathogenic on Sorghum by sequence analysis of the beta-tubulin gene intron 3 region and EF-1 alpha gene Intron 4. *Mycologia* 93:541–551.
43. Tooley, P. W., R. Bandyopadhyay, M. M. Carras, and S. Pažoutová. 2006. Analysis of *Claviceps africana* and *C. sorghi* from India using AFLPs, EF-1 alpha gene intron 4, and beta-tubulin gene intron 3. *Mycol. Res.* 110:441–451.
44. Tooley, P. W., M. M. Carras, A. Sechler, and A. H. Rajasab. 2010. Real-time PCR detection of Sorghum ergot pathogenesis *Claviceps africana*, *Claviceps sorghi* and *Claviceps sorghicola*. *J. Phytopathol.* 158:698–704.
45. Berde, B. and E. Stürmer. 1978. Introduction to the pharmacology of ergot alkaloids and related compounds. In *Ergot Alkaloids and Related Compounds*, eds. B. Berde and H. O. Schild, pp. 1–28. Berlin, Germany: Springer-Verlag.

46. Stadler, P. A. and R. Giger. 1984. Ergot alkaloids and their derivatives in medical chemistry and therapy. In *Natural Products and Drug Development*, eds. P. Krosgard-Larson, C. H. Christensen, and H. Kofod, pp. 463–485. Copenhagen, Denmark: Munksgaard.
47. Vendrell, M., E. Angulo, V. Casadó et al. 2007. Novel ergopeptides as dual ligands for adenosine and dopamine receptors. *J. Med. Chem.* 50:3062–3069.
48. Görnemann, T., S. Jähnichen, B. Schurad et al. 2008. Pharmacological properties of a wide array of ergolines at functional alpha$_1$-adrenoceptor subtypes. *Naunyn-Schmiedeberg's Arch. Pharmacol.* 376:321–330.
49. Villalon, C. M., P. de Vries, G. Rabelo, D. Centurion, A. Sanchez-Lopez, and P. R. Saxena. 1999. Canine external carotid vasoconstriction to methysergide, ergotamine and dihydroergotamine: Role of 5-HT$_{1B/1D}$ receptors and α_2-adrenoceptors. *Br. J. Pharmacol.* 126:585–594.
50. Willems, E. W., M. Trion, P. de Vries, J. O. C. Heiligers, C. M. Villalon, and P. R. Saxena. 1999. Pharmacological evidence that α_1- and α_2-adrenoceptors mediate vasoconstriction of carotid arteriovenous anastomoses in anaesthetized pigs. *Br. J. Pharmacol.* 127:1263–1271.
51. Tfelt-Hansen, P. C. and P. J. Koehler. 2008. History of the use of ergotamine and dihydroergotamine in migraine from 1906 and onward. *Cephalalgia* 28:877–886.
52. de Groot, A. N., P. W. van Dongen, T. B. Vree, Y. A. Hekster, and J. van Roosmalen, 1998. Ergot alkaloids. Current status and review of clinical pharmacology and therapeutic use compared with other oxytocics in obstetrics and gynaecology. *Drugs* 56:523–535.
53. Wadworth, A. N. and P. Crisp. 1992. Co-dergocrine mesylate. A review of its pharmacodynamic and pharmacokinetic properties and therapeutic use in age-related cognitive decline. *Drugs Aging* 2:153–173.
54. Crosignani, P. G. 2006. Current treatment issues in female hyperprolactinaemia. *Eur. J. Obstet. Gynaecol. Reprod. Biol.* 125:152–164.
55. Thobois, S. 2006. Proposed dose equivalence for rapid switch between dopamine receptor agonists in Parkinson's disease: A review of the literature. *Clin. Ther.* 28:1–12.
56. Fantegrossi, W. E., A. C. Murnane, and C. J. Reissig. 2008. The behavioral pharmacology of hallucinogens. *Biochem. Pharmacol.* 75:17–33.
57. Sigafoos, J., V. A. Green, C. Edrisinha, and G. E. Lancioni. 2007. Flashback to the 1960s: LSD in the treatment of autism. *Dev. Neurorehabil.* 10:75–81.
58. Schardl, C. L., D. G. Panaccione, and P. Tudzynski. 2006. Ergot alkaloids—Biology and molecular biology. *Alkaloids Chem. Biol.* 63:45–86.
59. Panaccione, D. G. 2010. Ergot alkaloids. In *The Mycota X*, ed. M. Hofrichter, pp. 195–214. Berlin, Germany: Springer-Verlag.
60. Wallwey, C. and S.-M. Li. 2011. Ergot alkaloids: Structure diversity, biosynthetic gene clusters and functional proof of biosynthetic genes. *Nat. Prod. Rep.* 28:496–510.
61. Gebler, J. C. and D. Poulter. 1992. Purification and characterization of dimethylallyl tryptophan synthase from *Claviceps purpurea*. *Arch. Biochem. Biophys.* 296:308–313.
62. Cheng, L. J., J. E. Robbers, and H. G. Floss. 1980. End-product regulation of ergot alkaloid formation in intact cells and protoplasts of *Claviceps* species, strain SD58. *J. Nat. Prod.* 43:329–339.
63. Tsai, H. F., H. Wang, J. C. Gebler, C. D. Poulter, and C. L. Schardl. 1995. The *Claviceps purpurea* gene encoding dimethylallyltryptophan synthase, the committed step for ergot alkaloid biosynthesis. *Biochem. Biophys. Res. Commun.* 216:119–125.
64. Keller, U. 1983. Highly efficient mutagenesis of *Claviceps purpurea* by using protoplasts. *Appl. Environ. Microbiol.* 46:580–584.
65. Tudzynski, P., K. Hölter, T. Correia, C. Arntz, N. Grammel, and U. Keller. 1999. Evidence for an ergot alkaloid gene cluster in *Claviceps purpurea*. *Mol. Gen. Genet.* 261:133–141.
66. Haarmann, T., C. Machado, Y. Lübbe et al. 2005. The ergot alkaloid gene cluster in *Claviceps purpurea*: Extension of the cluster sequence and intra species evolution. *Phytochemistry* 66:1312–1320.
67. Lorenz, N., E. V. Wilson, C. Machado, C. Schardl, and P. Tudzynski. 2007. Comparison of ergot alkaloid biosynthesis gene clusters in *Claviceps* species indicate loss of late pathway steps in evolution of *C. fusiformis*. *Appl. Env. Microbiol.* 73:7185–7191.
68. Riederer, B., M. Han, and U. Keller. 1996. D-Lysergyl peptide synthetase from the ergot fungus *Claviceps purpurea*. *J. Biol. Chem.* 271:27524–27530.
69. Ortel, I. and U. Keller. 2009. Combinatorial assembly of simple and complex D-lysergic acid alkaloid peptide classes in the ergot fungus *Claviceps purpurea*. *J. Biol. Chem.* 284:6650–6660.
70. Haarmann, T., I. Ortel, P. Tudzynski, and U. Keller. 2006. Identification of the cytochrome P450 monooxygenase that bridges the clavine and ergoline alkaloid pathways. *ChemBioChem.* 7:645–652.

71. Lorenz, N., J. Olsovska, M. Sulc, and P. Tudzynski. 2010. The alkaloid cluster gene ccsA of the ergot fungus *Claviceps purpurea* encodes the chanoclavine-I-synthase, an FAD-containing oxidoreductase mediating the transformation of N-methyl-dimethyltryptophan to chanoclavine I. *Appl. Environ. Microbiol.* 76:1822–1830.
72. Matuschek, M., C. Wallwey, X. Xie, and S.-M. Li. 2011. New insights into ergot alkaloid biosynthesis in *Claviceps purpurea*: An agroclavine synthase EasG catalyses, via a non-enzymatic adduct with reduced glutathione, the conversion of chanoclavine-I aldehyde to agroclavine. *Org. Biomol. Chem.* 9:4328–4335.
73. Cheng, J. Z., C. M. Coyle, D. G. Panaccione, and S. E. O'Connor. 2010. A role for old yellow enzyme in ergot alkaloid biosynthesis. *J. Am. Chem. Soc.* 132:1776–1777.
74. Rigbers, O. and S.-M. Li. 2008. Ergot alkaloid biosynthesis in *Aspergillus fumigatus*. Overproduction and biochemical characterization of A4-dimethylallyltryptophan N-methyltransferase. *J. Biol. Chem.* 283:26859–26868.
75. Wallwey, C., M. Matuschek, and S.-M. Li. 2010. Ergot alkaloid biosynthesis in *Aspergillus fumigatus*: Conversion of chanoclavine-I to chanoclavine-I aldehyde catalyzed by a short-chain alcohol dehydrogenase FgaDH. *Arch. Microbiol.* 192:127–134.
76. Goetz, K. E., C. M. Coyle, J. Z. Cheng, S. E. O'Connor, and D. G. Panaccione. 2011. Ergot cluster-encoded catalase is required for synthesis of chanoclavine-I in *Aspergillus fumigatus*. *Curr. Genet.* 57:201–211.
77. Alderman, S. C. 2003. Diversity and speciation in *Claviceps*. In *Clavicipitalean Fungi: Evolutionary Biology, Chemistry, Biocontrol and Cultural Impacts*, Mycology 19, eds. J. F. White, C. W. Bacon, N. L. Hywel-Jones, and J. W. Spatafora, pp. 195–246. New York: Marcel Dekker.
78. Tenberge, K. B. and P. Tudzynski. 1994. Early infection of rye ovaries by *Claviceps purpurea* is inter- and intracellular. *Bioeng. Sondernummer* 10:22.
79. Mower, R. L. and J. G. Hancock.1975. Sugar composition of ergot honeydews. *Can. J. Bot.* 53:2813–2825.
80. Shaw, B. I. and P. G. Mantle. 1980. Host infection by *Claviceps purpurea*. *Trans. Br. Mycol. Soc.* 75:77–90.
81. Parbery, D. G. 1996. Trophism and the ecology of fungi associated with plants. *Biol. Rev.* 71:473–527.
82. Haarmann, T., Y. Rolke, S. Giesbert, and P. Tudzynski 2009. Ergot: From witchcraft to biotechnology. *Mol. Plant Pathol.* 10:563–577.
83. Ngugi, H. K. and H. Scherm. 2006. Biology of flower-infecting fungi. *Annu. Rev. Phytopathol.* 44:261–282.
84. Ashizawa, T., M. Takahashi, M. Arai, and T. Arie. 2012. Rice false smut pathogen, *Ustilaginoidea virens*, invades through small gap at the apex of a rice spikelet before heading. *J. Gen. Plant Pathol.* 78:255.
85. Tang, Y. X., J. Jin, D.-W. Hu, M.-L. Yong, Y. Xu, and L.-P. He. 2012. Elucidation of the infection process of *Ustilaginoidea virens* (teleomorph: *Villosiclava virens*) in rice spikelets. *Plant Pathol.* doi: 10.1111/j.1365-3059.2012.02629.x.
86. Schönbeck, F. 1966. Untersuchungen über Blüteninfektionen. I. Allgemeine Untersuchungen zum Infektionsweg Narbe-Griffel. *Phytopathol. Z.* 59:157–182.
87. Behre, K.-E. 1992. The history of rye cultivation in Europe. *Veg. Hist. Archaeobot.* 1: 141–156.
88. Mitchell, D. T. and R. C. Cooke, 1968. Some effects of temperature on germination and longevity of sclerotia in *Claviceps purpurea*. *Trans. Br. Mycol. Soc.* 51:721–729.
89. Krebs, J. 1936. Untersuchungen über den Pilz des Mutterkorns *Claviceps purpurea*. *Tul. Ber, Schweiz. bot. Ges.* 65:71–165.
90. Kirchhoff, H. 1929. Beiträge zur Biologie und Physiologie des Mutterkornpilzes. *Zentbl. Bakt. Parasitkde.* 77:310–369.
91. Vladimirsky, S. V. 1939. Geographical distribution and zones of injurious influence of ergot on rye in the USSR. *Sovetskaya Botanika* 5:77–87. Translation by National Lending Library for Science and Technology, Boston Spa, Yorks.
92. Lehman, J. S. and P. V. Oudemans. 1997. Phenology of apothecium production in populations of *Monilinia vaccinii-corymbosi* from early- and late-maturing blueberry cultivars. *Phytopathology* 87:218–223.
93. Scheffer, J. and P. Tudzynski. 2006. In vitro pathogenicity assay for the ergot fungus *Claviceps purpurea*. *Mycol. Res.* 110:465–470.
94. Bandyopadhyay, R., D. E. Frederickson, N. W. McLaren, G. N. Odvody, and M. J. Ryley. 1998. Ergot: A new disease threat to sorghum, in the Americas and Australia. *Plant Dis.* 82:356–367.
95. Cisneros-López, E., L. E. Mendoza-Onofre, V. A. González-Hernández et al. 2010. Synchronicity of pollination and inoculation with *Claviceps africana* and its effects on pollen–pistil compatibility and seed production in sorghum. *Fungal Biol.* 114:285–292.
96. van Esse, H. P., J. W. van't Klooster, M. D. Bolton et al. 2008. The *Cladosporium fulvum* virulence protein Avr2 inhibits host proteases required for basal defense. *Plant Cell* 20:1948–1963.

97. Xiang, T., N. Zong Y. Zou et al. 2008. *Pseudomonas syringae* effector AvrPto blocks Innate immunity by targeting receptor kinases. *Curr. Biol.* 18:74–80.
98. Tudzynski, P., Tenberge, K. B., and Oeser, B. 1995. *Claviceps purpurea*. In *Pathogenesis and Host Specificity in Plant Diseases: Histopathological, Biochemical, Genetic and Molecular Bases*. Vol. II. *Eukaryotes*, eds. K. Kohmoto, U. S. Singh, and R. P. Singh, pp. 161–187. Oxford, U.K.: Pergamon Press, Elsevier Science.
99. Tenberge, K. B., V. Homann, B. Oeser, and P. Tudzynski. 1996. Structure and expression of two polygalacturonase genes of *Claviceps purpurea* oriented in tandem and cytological evidence for pectinolytic enzyme activity during infection of rye. *Phytopathology* 86:1084–1097.
100. Carpita, N. C. and D. M. Gibeaut. 1993. Structural models of primary cell walls in flowering plants: Consistency of molecular structure with the physical properties of the walls during growth. *Plant J.* 3:1–30.
101. Giesbert, S., H.-B. Lepping, K. B. Tenberge, and P. Tudzynski. 1998. The xylanolytic system of *Claviceps purpurea*: Cytological evidence for secretion of xylanases in infected rye tissue and molecular characterization of two xylanase genes. *Phytopathology*. 88:1020–1030.
102. Brockmann, B., R. Smit, and P. Tudzynski. 1992. Characterization of an extracellular β-1,3-glucanase of *Claviceps purpurea*. *Physiol. Mol. Plant Pathol.* 40:191–201.
103. Müller, U., K. B. Tenberge, B. Oeser, and P. Tudzynski. 1997. Ce11, probably encoding a cellobiohydrolase lacking the substrate binding domain, is expressed in the initial infection phase of *Claviceps purpurea* on *Secale cereale*. *Mol. Plant Microbe Interact.* 10:268–279.
104. Oeser, B., P. Heidrich, U. Müller, P. Tudzynski, and K. B.Tenberge. 2002. Polygalacturonase is a pathogenicity factor in the *Claviceps purpurea*/rye interaction. *Fungal Genet. Biol.* 36:176–186.
105. Mollet, J.-C., S.-Y. Park, E. A. Nothnagel, and E. M. Lord. 2000. A lily stylar pectin is necessary for pollen tube adhesion to an in vitro stylar matrix. *Plant Cell* 12:1737–1749.
106. Xu, J. R. 2000. Map kinases in fungal pathogens. *Fungal Gen. Biol.* 31:137–152.
107. Hamel, L.-P., M.-C. Nicole, S. Duplessis, and B. E. Ellisc. 2012. Mitogen-activated protein kinase signaling in plant-interacting fungi: Distinct messages from conserved messengers, *Plant Cell* 24:1327–1351.
108. Mey, G., B. Oeser, M. H. Lebrun, and P. Tudzynski. 2002. The biotrophic, non-appressoria forming grass pathogen *Claviceps purpurea* needs a *Fus3/Pmk1* homologous MAP kinase for colonization of rye ovarian tissue. *Mol. Plant Microbe Interact.* 15:303–312.
109. Mey, G., K. Held, J. Scheffer, K. B. Tenberge, and P. Tudzynski. 2002. CPMK2, a Slt2-homologous MAP-kinase, is essential for pathogenesis of *Claviceps purpurea* on rye: Evidence for a second conserved pathogenesis-related MAP-kinase cascade in phytopathogenic fungi. *Mol. Microbiol.* 46:305–318.
110. Cole, R. A. and J. E. Fowler. 2006. Polarized growth: Maintaining focus on the tip. *Curr. Opin. Plant Biol.* 9:579–588.
111. Jaffe, A. B. and A. Hall. 2005. Rho GTPases: Biochemistry and biology. *Annu. Rev. Cell Dev. Biol.* 21:247–269.
112. Rolke, Y. and P. Tudzynski. 2008. The small GTPase Rac and the PAK kinase Cla4 in *Claviceps purpurea*: Interaction and impact on polarity, development and pathogenicity. *Mol. Microbiol.* 68:405–423.
113. Moore, S., O. M. H. De Vries, and P. Tudzynski. 2002. The major Cu, Zn SOD of the phytopathogen *Claviceps purpurea* is not essential for pathogenicity. *Mol. Plant Pathol.* 3:9–22.
114. Garre, V., U. Müller, and P. Tudzynski. 1998. Cloning, characterization, and targeted disruption of cpcat1, coding for an *in planta* secreted catalase of *Claviceps purpurea*. *Mol. Plant Microbe Interact.* 11:772–783.
115. Nathues, E., S. Joshi, K. B. Tenberge et al. 2004. CPTF1 a CREB-like transcription factor, is involved in the oxidative stress response in the phytopathogen *Claviceps purpurea* and modulates ROS level in its host *Secale cereale*. *Mol. Plant Microbe Interact.* 17:383–393.
116. Lambeth, J. D. 2004. NOX enzymes and the biology of reactive oxygen. *Nat. Rev. Immunol.* 4:181–189.
117. Heller, J. and P. Tudzynski. 2011. Reactive oxygen species in phytopathogenic fungi: Signaling, development, and disease. *Annu. Rev. Phytopathol.* 49:369–390.
118. Tudzynski, P., J. Heller, and U. Siegmund. 2012. Reactive oxygen species generation in fungal development and pathogenesis. *Curr. Opin. Microbiol.* 15:653–659.
119. Ryder, L. S., Y. F. Dagdas, T. A. Mentlak et al. 2013. NADPH oxidases regulate septin-mediated cytoskeletal remodeling during plant infection by the rice blast fungus. PNAS, in press.
120. Tanaka, A., M. J. Christensen, M. Takemoto, P. Park, and B. Scott. 2006. Reactive oxygen species play a role in regulating a fungus–perennial ryegrass mutualistic interaction. *Plant Cell* 18:1052–1066.
121. Giesbert, S., T. Schürg, S. Scheele, and P. Tudzynski. 2008. The NADPH oxidase Cpnox1 is required for full pathogenicity of the ergot fungus *Claviceps purpurea*. *Mol. Plant Pathol.* 9:317–327.

122. Scheffer, J., C. Ziv, O. Yarden, and P. Tudzynski. 2005b. The COT1 homologue CPCOT1 regulates polar growth and branching and is essential for pathogenicity in *Claviceps purpurea*. *Fungal Genet. Biol.* 42:107–118.
123. Mey, G., T. Correia, and B. Oeser et al. 2003. Structural and functional analysis of an oligomeric hydrophobin gene from *Claviceps purpurea*. *Mol. Plant Pathol.* 4:31–41.
124. Nathues, E., C. Jörgens, N. Lorenz, and P. Tudzynski. 2007. The histidine kinase CpHK2 has impact on spore germination, oxidative stress and fungicide resistance, and virulence of the ergot fungus *Claviceps purpurea*. *Mol. Plant Pathol.* 8:653–665.
125. Bormann, J. and P. Tudzynski. 2009. Deletion of Mid1, a putative stretch-activated calcium channel in *Claviceps purpurea*, affects vegetative growth, cell wall synthesis and virulence. *Microbiology* 155:3922–3933.
126. Correia T., N. Grammel, I. Ortel, P. Tudzynski, and U. Keller. 2003. Molecular cloning and analysis of the ergopeptine assembly system in the ergot fungus *Claviceps purpurea*. *Chem. Biol.* 10:1281–1292.

18 Curvularia

Jeannette Guarner

CONTENTS

18.1 *Curvularia* in the Environment 251
18.2 Pathogenesis 253
    18.2.1 Melanin 253
    18.2.2 Allergens and Allergic Reactions 253
    18.2.3 Other Compounds 254
18.4 Clinical Presentations 255
    18.4.1 Eye Infections 255
    18.4.2 Skin Infections 256
    18.4.3 Allergic Fungal Rhinosinusitis 256
    18.4.4 Allergic Bronchopulmonary Disease 257
    18.4.5 *Curvularia* Infections Associated with Catheters 257
    18.4.6 Invasive Disease 258
18.5 Diagnostics 259
    18.5.1 Molecular Testing 261
18.6 Treatment 261
18.7 Conclusions 262
References 262

Curvularia is a dematiaceous or pigmented fungus found ubiquitously in plants and soil of tropical and subtropical areas.[1] It is a plant parasite that can cause blight in a variety of grasses. Human and animal infections with *Curvularia* have been described primarily as cutaneous or subcutaneous nodules, superficial eye infections, and respiratory allergic disease. Disseminated infections can also occur. The genus contains around 35 species, but most cases of human disease have been associated with only three species: *Curvularia lunata*, *Curvularia pallescens*, and *Curvularia geniculata*.[2] The characteristic pigmented and curved conidia of *Curvularia* species range in size from 8 to 14 μm by 21 to 35 μm (Figure 18.1). Features that help differentiate between species include the number of septa, shape and color of the conidia, existence of a dark median septum, and prominence of the geniculate growth pattern.[3]

This chapter describes the sites in the environment in which *Curvularia* spp. are found, pathogenesis, the clinical scenarios in which this mold occurs, how it is diagnosed, and the treatments that have been used in patients infected by this dematiaceous fungus.

18.1 *CURVULARIA* IN THE ENVIRONMENT

In general, it is accepted that humans and animals acquire the infection through exposure to environmental sources that contain the fungal spores such as air,[4,5] plants,[6,7] dirt,[1] metal,[8] fingernails,[9,10] and water.[11] For eye and skin infections, trauma with the spore-containing material is an important risk factor.[12–17] *Curvularia* has also been found in insects (ants)[18] and in horns and skin of ruminants (cattle, sheep, and others)[19] without causing disease. Only one human eye infection after surgery has been linked to animal exposure, a cat with skin *Curvularia* lesions.[20]

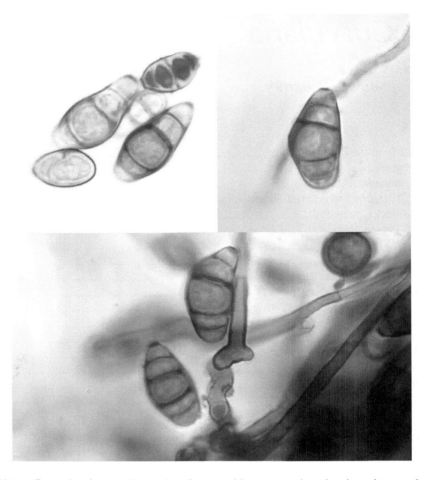

FIGURE 18.1 (See color insert.) Lactophenol cotton blue preparation showing pigmented conidia of *Curvularia* spp. Notice that one of the central cells is larger than the others giving the conidia a curved appearance.

Aspergillus, Fusarium, and *Curvularia* are the fungi that most frequently contaminate a variety of grains during the rainy season throughout the world.[21] *C. lunata* is particularly implicated in grain mold that is one of the major problems with sorghum productivity and profitability.[21] The amount of fungi and mycotoxins found in each crop will vary each year depending on the humidity and temperature.[7,22] Curvularin, a mycotoxin produced by *C. lunata*, has been found in a variety of grains and seeds including rice, maize, sorghum, millet, and sesame, as well as groundnuts,[23–25] although the presence of curvularin observed in these food items varies from country to country, for example, it was found in millet and feed in Mozambique but not in Burkina Faso.[24] Another product of *Curvularia* fermentation is curvarol, a protein synthesis and cell growth inhibitor.[26] Several mechanisms allow resistance of plants and grain to *C. lunata* infections including the sorghum proteins chitinase and sormatin.[21]

Some authors have suggested that *Curvularia* plays an important role in the life of plants. For example, the Indian medicinal plant *Tinospora cordifolia* is colonized by a variety of endophytic microbes. *Curvularia* was one of the endophytes present in the plant leaves particularly during the monsoon season and winter.[6] It was postulated that some of the medicinal substances present in *T. cordifolia* were produced by some of the endophytes.

Variations in humidity, temperature, and time of day are important to understand the fluctuations in the number of *Curvularia* spores in the environment. A study that correlated humidity and temperature in locations of the Gulf of Mexico with cases of *Curvularia* keratitis showed that infections were more common in the warm humid months as the molds dispersed during the harvest of sorghum and rice and proliferated on ensiled grains.[27] In Karachi, Pakistan, the concentration of *Curvularia* spores is highest during the evening hours and least at night,[5] while in Kolkata, India, the highest concentration of *Curvularia* spores is at noon.[4] The amount of *Curvularia* spores, inside and outside Florida homes without moisture problems during the rainy season, is very similar.[28] These environmental variations are important as they have an impact on the frequency of allergic reactions observed in humans and animals.

The presence of *Curvularia* in the environment correlates with the prevalence of the fungus from nasal cultures of asymptomatic individuals. A study performed after a woman was found to have *C. lunata* in a saline-filled silicone breast implant demonstrated that this fungus could be isolated from a sterile supply room that had a water-damaged ceiling and in the corridor to one of the operating rooms where the implants were stored.[11] The researchers investigating this outbreak studied nasal carriage in two facilities and found that the fungus was isolated in 35% of personnel working in the facility associated with the outbreak compared to 6.7% from another facility.[11] One of the workers had the fungus on the hands. Although *Curvularia* was isolated in one-third of the personnel from the facility where the initial case occurred, a recall of possible cases only identified three patients with implants who had symptoms. In addition, the fungus was also isolated from implants that had cloudy material in two asymptomatic persons. In this instance, the saline that was used to fill the silicon shells was probably contaminated while sitting in an open bowl before or during the surgery. A later study designed to identify fungi from electrodes and ultrasound transducers used in Brazilian clinics showed that *Curvularia* was found in two ultrasound transducers although infections associated with the devices were not documented.[8]

18.2 PATHOGENESIS

18.2.1 Melanin

As a dematiaceous fungus, *Curvularia* spp. produce melanin that is localized in the cell wall.[29] The melanin produced by dematiaceous fungi is derived from the dihydroxynaphthalene pathway, not the dihydroxyphenylalanine (L-dopa) pathway that is used by *Cryptococcus neoformans* to produce melanin. Melanin is considered an important virulence factor since it scavenges free radicals and hypochlorite produced by phagocytic cells thus protecting the mold from inflammatory cells.[29] Melanin also binds to hydrolytic enzymes, preventing their action on the fungal cell membrane.[29,30]

Since trihydroxynaphthalene reductase is an important step in producing melanin, some authors have studied the effect of inhibitors of this enzyme in the growth of *C. lunata*.[31] The inhibitors were searched from a National Cancer Institute bank of compounds using filtering procedures and a ligand-based 3D similarity search that focused in docking sites. The 28 selected compounds were tested to determine if they inhibited the enzyme and growth. Three of the compounds produced smaller hypopigmented colonies suggesting that these could be used for treatment of infections caused by dematiaceous molds. Other authors have studied plant flavonoids as inhibitors of *C. lunata* trihydroxynaphthalene reductase showing that apigenin and baicalein affect fungal pigmentation and growth.[32]

18.2.2 Allergens and Allergic Reactions

Several *Curvularia* allergens have been identified including alcohol dehydrogenase, serine protease, enolase, and cytochrome C.[33] The amino acids glycine, proline, and lysine of these molecules are frequently found in IgE B cell–binding sites thus suggesting these are major allergens. The three proteins that have been studied in more detail include Cur I 1 (31 kDa, serine protease), Cur I 2 (48 kDa, enolase), and Cur I 3 (12 kDa, cytochrome C). Cur I 1 showed IgE reactivity with sera of 80% of patients hypersensitive to *C. lunata* suggesting that this is a major allergen.[34] In addition, mice exposed to the serine protease Cur I 1 have demonstrated more inflammation with eosinophil infiltration, higher levels of IgE and IgG1, and higher production of IL-4 and IL-5 by splenocytes and in bronchoalveolar fluid.[35]

It is proposed that the protease cleavage of the IL-2 receptor (CD25) from T cells blocks Th1 cytokines and enhances Th2 activity. Elevated levels of Th2 cytokines have been observed when peripheral blood mononuclear cells are stimulated with Cur I 2 enolase.[36] A study of Cur l 2 found that residues important for activity and metal binding are similar to those found in other fungi including *Saccharomyces cerevisiae*.[36] The authors created a recombinant enolase and tested for IgE-binding activity using sera collected from patients with allergies against *Curvularia* in an ELISA and an immunoblot assay and found that all the selected patients had threefold or more binding against Cur I 2. The authors also demonstrated that patients allergic to *Curvularia* had increased lymphocyte proliferation capacity when stimulated by Cur I 2 and production of IL-4, IL-5, and IFNγ compared to controls. *Curvularia* enolase has been shown to be cross-reactive with that found in other fungi, bacteria, and latex enolase.[37]

The epitopes of Cur I 3 (cytochrome C) can be divided into those that have a higher binding capacity to IgE (B cell type) and those that elicit higher T cell lymphoproliferation with one of the latter having cryptic B cell function.[38] The Cur I 3 antigen and short peptides created from this antigen have been used to test immunotherapies in allergen-induced sensitization mouse models. The basic principle behind the immunotherapy includes the production of allergen-specific IgG antibodies that will reduce recruitment of eosinophils and induction of T cell anergy by shifting from a Th2 response to a Th1 response that is due to increased secretion of IL-10 and IFNγ and decreased secretion of IL-4.[39] Cur l 3 (cytochrome C) is cross-reactive with that found in other fungi and grasses.[40]

An animal study of the T cell response to different molds known to cause asthma showed that most molds cause a Th2-mediated allergic immune response that may switch to a Th1 response as the conidial burden increases.[41] Mice exposed to low doses of *C. lunata* had a combination of an eosinophilic response mixed with more neutrophils and macrophages secreting IFNγ compared to other molds that produced a primary eosinophilic response in bronchoalveolar lavage fluid. The switch to the Th1 response was seen in mice exposed to lower *Curvularia* conidial doses. Also, *Aspergillus fumigatus* and *C. lunata* elicited airway hyperactivity at lower conidia doses compared to other molds, and, interestingly, *C. lunata* responses tended to be unpredictable. In addition, *C. lunata*–exposed mice demonstrated significant weight loss and dishevelment, their lungs showed diffuse interstitial pneumonia with granuloma formation, and there was dissemination of the mold to the central nervous system.

In humans, allergic fungal rhinosinusitis is part of a spectrum of sinonasal disease derived from the presence of fungus and host factors. An association of allergic rhinosinusitis with class II human leukocyte antigen (HLA) DQB1*0301 and DQB1*0302 has been observed.[42] The spectrum of non-invasive fungal disease includes allergic rhinosinusitis and the presence of a fungus ball and minimally invasive diseases such as granulomatous indolent fungal sinusitis.[42] Through this spectrum, the affected nasal mucosa goes through cycles of chronic edema, mucous stasis, and bacterial superinfection, which results in improperly functioning sinuses.[43] The cycles may take years to manifest themselves; thus diagnostic criteria may not be all present initially.

18.2.3 Other Compounds

Several *Curvularia* species produce a variety of compounds that may be important for survival of the organism but that may also have other commercial applications. For example, *Curvularia verruculosa* produces a haloperoxidase that has antibacterial and antimycotic activity. Hansen et al. tested the haloperoxidase system against Gram-positive and Gram-negative bacteria, yeasts, and molds and demonstrated a significant reduction of microorganism counts and suggested that this could be an effective, less corrosive sanitizing system.[44] Several metabolites (modiolide and pyrone derivatives) of a marine-derived *Curvularia* species have antimicrobial activity against methicillin-resistant *Staphylococcus aureus*.[45] Five hybrid peptide–polyketides derived from *C. geniculata* have activity against *Candida albicans* and show synergism with the antifungal drug fluconazole.[46] The phytotoxins, phthalic acid, butyl isobutyl ester, and radicinin, recovered from *Curvularia* sp. FH01, which is found in the gut of the insect *Atractomorpha sinensis*, have been found to have herbicidal properties.[47]

Other products of importance include the production of hydrocortisone after 11β-hydrolyzation of steroids by *C. lunata*.[48] As the hydrolyzation of steroidal saponins occurs, sugar chains are produced. The sugar chains are then broken up by using glucoamylases that are similar to those produced by *Aspergillus niger*.[48] A study of dermatophytes and other fungi, including *Curvularia* isolated from the hair, wool, claws, horns, and skin scrapings of cattle, sheep, buffalo, and camel, demonstrated the ability of the fungi to produce proteinase, lipase, amylase, and keratinase.[19] In this study, *Curvularia* spp. produced the highest protease and amylase activity.

18.4 CLINICAL PRESENTATIONS

18.4.1 Eye Infections

Fungal corneal infections are an important cause of ocular morbidity that may lead to blindness if severe or ulcerated. Approximately 60% of fungal keratitis cases are caused by *Aspergillus* and *Fusarium* and 10% are caused by *Curvularia*[49]; however, this may vary from country to country. An Indian study of 1352 culture proven fungal keratitis cases demonstrated that *Curvularia* accounted for less than 3% of the isolates with *Fusarium* and *Aspergillus* accounting for more than 30% each. This same study showed that 54.4% of the fungal keratitis cases were preceded by trauma and that these infections were more frequent during the monsoon and winter season compared to the summer.[12] Most of the infections occurred in men who performed agricultural or outdoor work. A study from Nepal on corneal ulcerations isolated a fungus in 36% of patients with *Fusarium* and *Curvularia* being the most frequently identified fungi (32% and 18%, respectively) and found that in addition to trauma, the use of topical steroids was a risk factor.[13]

Wilhelmus and Jones reviewed all the cases that had *Curvularia* keratitis during a 30-year period in Houston, Texas.[14] Trauma with plants or dirt was a predisposing factor, and most of the infections occurred during the summer months. However, an allergic component to the inflammation has been implicated in the indolent behavior of these infections. The presentation of the 43 patients ranged from superficial, feathery infiltrates of the central cornea, to suppurative ulceration of the peripheral cornea. Although patients rarely developed hypopyon, its presence indicated increased risk of subsequent complications. The *Curvularia* species could not be identified in five cases. *Curvularia senegalensis*, *C. lunata*, *C. pallescens*, and *C. prasadii* were isolated from 11, 10, 4, and 2 cases, respectively. In addition, there is a report of keratitis coinfection with *Acanthamoeba* and *Curvularia*.[50]

Although keratitis is the most common clinical presentation seen in patients in whom *Curvularia* spp. has been isolated after penetrating injuries of the eye, fungal endophthalmitis can occur after trauma. A study from India demonstrated that if there was a positive fungal culture at the time of the trauma, there was a high possibility of having fungal endophthalmitis.[51] Three cases of *Curvularia* endophthalmitis have also been reported after surgical interventions to the eye[52–54] and one after a deep trauma to the eye.[55]

Eye infections due to *Curvularia* have been described in horses,[56,57] elephants,[58] dogs,[59] and tortoises.[60] The pathogenic mechanisms and symptoms are similar to those found in humans.

18.4.2 Skin Infections

The terms to describe cutaneous infections due to dematiaceous fungi have evolved through time. Phaeohyphomycosis describes all infections caused by dark-walled (dematiaceous) fungi, while chromoblastomycosis is reserved for distinctive skin and subcutaneous infections by pigmented fungi in which round, thick-walled, brown fungal cells called muriform or sclerotic bodies are found in microscopic examination of the lesion.[30,61,62] In addition, the term eumycotic mycetoma describes pigmented or nonpigmented fungal infections of the cutaneous and subcutaneous tissues that may involve adjacent bone and produce mycotic granules.

Independent of the challenging clinical nomenclature, defining the fungus present in the lesion is most important for treatment. A retrospective survey of 915 wounds demonstrated that 23% had a mold or yeast isolated from them.[63] The wound types included decubitus ulcers, diabetic foot ulcers, nonhealing surgical wounds, and venous leg ulcers. *Candida* species were the most frequently isolated fungi, followed by *Curvularia*, *Malassezia*, and others. Frequently, trauma or burns occurred before the *Curvularia* skin infections.[15–17] However, several cases have occurred without the histories of trauma, primarily in immunosuppressed patients.[64,65]

In immunocompetent individuals, the lesions in which *Curvularia* has been isolated have been described to have a variety of morphologies ranging from round to oval and brown to black, ulcerated, and having discharge.[15–17] In a patient that was pricked by a fallen tree, there was only a single lesion,[15] while in the patients with burns, several lesions were noted in the burned areas.[16,17] Descriptions of *Curvularia* skin lesions in immunosuppressed patients may show a different evolution: In a patient treated with prednisone and methotrexate for rheumatoid arthritis, the lesion started as a small papule in the leg that evolved in a 9-month period to an ulcerated eschar with an edematous undermined border.[65] In a heart transplant patient, a skin lesion started as a nodule in the upper right arm that disseminated to adjacent areas giving rise to nodules, vesicular lesions, or hyperchromic maculopapular skin lesions that extend to other areas of the body and mucosas including the mouth, esophagus, and lung.[66] *C. lunata* skin lesions have been identified from neonates: one with multiple cardiac malformations who was surgically treated and at postoperative day 13 was noted to have a black surface around the open sternotomy wound[67] and a set of twins who presented multiple hypochromic macules in the back, chest, and upper and lower extremities that rapidly progressed to centrally necrotic lesions.[68] None of these neonates survived the infections.

Onychomycosis caused by *Curvularia* are rare. Two studies from India of 218 patients with onychomycosis isolated *Curvularia* in only two patients.[9,10] In addition, *Curvularia* was isolated from a biopsy sample in a patient undergoing chemotherapy for a poorly differentiated lymphocytic lymphoma that presented invasive distal tinea unguium with a verrucous excrescence.[69]

Curvularia skin lesions in animals have been described as single lesions (mycetomas) in dogs[70] or disseminated skin infections in dogs[71,72] and cats.[20]

18.4.3 Allergic Fungal Rhinosinusitis

Allergic fungal rhinosinusitis is an immune-mediated clinical entity that is not completely understood. For the most part, the diagnostic criteria of Bent and Kuhn continue to be used.[73,74] Patients must meet the five major criteria (type 1 hypersensitivity, nasal polyposis, characteristic CT findings, eosinophilic mucin, and the presence of fungi in the mucous without invasion of tissues), while the minor criteria serve as support. The eosinophilic mucin and presence of fungi in mucous are determined by histopathologic examination of surgical specimens that are usually obtained as part of the treatment.[42,73] The surgical specimen should be divided with one portion sent for culture, while the other for histopathologic examination. Histopathologically, the mucous should show eosinophils and its breakdown products such as (1) Charcot–Leyden crystals, (2) lymphocytes and plasma cells, and (3) the mucosa that may be hypertrophic but should be intact and should not show necrosis or granulomas. Finding fugal elements can be difficult and special stains for fungi such as periodic acid–Schiff and silver stains should be used. Cultures help define the fungus present in the sinus so as to direct specific antifungal treatment if needed[42]; however, interpretation of fungal cultures can be problematic due to the (1) possibility of contamination since most of the fungi causing this allergic reaction are ubiquitous in the environment and (2) high frequency of negative cultures in patients with this clinical entity. Imaging studies demonstrate asymmetric involvement and erosion of bony structures that is due to the inflammation, rather than invasion, of the fungi.

Allergic fungal rhinosinusitis is a chronic condition that presents frequently in patients with atopia and asthma. It is estimated that mold allergies occur in 80% of patients with allergic asthma and 44% of those with atopy. This condition is analogous to allergic bronchopulmonary aspergillosis (ABPA) and as such can be elicited by a variety of molds. It appears that *C. lunata* is the sensitizing agent in 7%–16% of the population in India and 18%–28% in other countries.[33] As the patients become exposed to the fungus, they create an IgE-mediated allergic response with consequent edema of the mucosa that obstructs mucous flow and creates the characteristic allergic mucin. In these patients, it is common to find an IgE above 1000 U/mL.[73]

Curvularia has been identified as the most frequent culprit of allergic fungal rhinosinusitis in some studies,[75] while *Bipolaris* is believed to be the most common in others.[76] These differences may be due to the methods used for detection of the fungi that may under- or overestimate some agents. For example, a fungus could be found in patients in Egypt with allergic fungal rhinosinusitis in 62% of cases.[77] Dematiaceous molds were cultured or detected by using PCR in 71% of 42 cases. *Bipolaris* was the most frequent fungal genus, followed by *Curvularia* and *Aspergillus*. However, detection of *Aspergillus* and *Bipolaris* used PCR that is presumably more sensitive than detection of *Curvularia* that relied on cultures.[77]

18.4.4 Allergic Bronchopulmonary Disease

Allergic bronchopulmonary disease is most frequently due to *Aspergillus* and is commonly referred to as ABPA. However, in rare occasions, the syndrome is due to other fungi, and *Curvularia* has been reported in a few cases.[78–81] ABPA has a similar pathogenesis as allergic fungal rhinosinusitis with the production of allergic mucous in the lower respiratory tract and most of the patients having a previous history of allergic fungal rhinosinusitis. It is important to make the diagnosis of allergic bronchopulmonary fungal disease since treatment can prevent irreversible damage such as bronchiectasis. As would be expected, in the patients in whom the allergic reaction is not against *Aspergillus*, the usual tests (IgE against *Aspergillus* and prick skin tests) are negative; thus diagnosis can be delayed. Several methods have been used to help in diagnosing those cases not associated with *Aspergillus*, including culture of the allergic mucous obtained through endoscopy,[79,80] intradermal skin testing for *Curvularia*,[79] and detection of antibodies against *Curvularia* by immunodiffusion.[78,81] Lake et al.[78] identified 8 (2%) of 503 patients with allergic bronchopulmonary fungal disease due to *Bipolaris*/*Curvularia* in a 7-year period and found that most of these cases lived in remote, subtropical northern Australia. They used immunodiffusion and prick skin tests that could not differentiate between *Bipolaris* and *Curvularia*.

Similar to humans, horses can suffer from recurrent airway obstruction or *heaves* that correlate with increase in *Curvularia* spores during the hot humid months.[82,83]

18.4.5 Curvularia Infections Associated with Catheters

Several cases of *Curvularia* inside Tenckhoff catheters have been described. Darkly pigmented materials were seen floating inside the catheter or bags, attached to the catheter obstructing the flow (Figure 18.2) or, in one case, perforating the catheter.[67,84–87] However, not all the patients present signs and symptoms of peritonitis (abdominal pain accompanied by guarding, fever, and cloudy dialysate effluent that contains abundant leukocytes) suggesting that, in some cases, it is colonization of the catheter by the fungus rather than true peritonitis.[85] In one case, *C. lunata* was

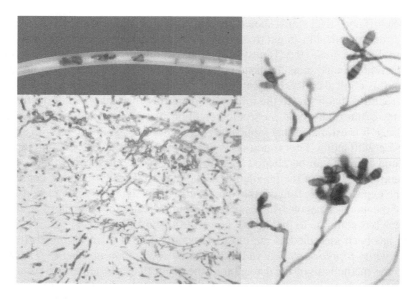

FIGURE 18.2 **(See color insert.)** *C. lunata* in Tenckhoff catheter, histopathologic preparation of the pigmented material inside the catheter stained with hematoxylin and eosin and culture.

accompanied by two Gram-negative bacteria in the peritoneal fluid; thus the peritoneal signs and symptoms could not be completely attributed to the fungus.[84] On the other hand, *C. geniculata* was cultured from the peritoneal fluid of a patient undergoing peritoneal dialysis with fever and diffusely tender abdomen, but in this case, the catheter did not show the dark material, and no other organisms were isolated from the peritoneal fluid.[88] There are two additional similar cases in which *Curvularia* sp. was present in the peritoneal dialysate. The patients had signs and symptoms of peritonitis, but no visible dark fungal material was found in the catheters, transfer sets, or bags.[89,90] Two risk factors have been identified for patients undergoing peritoneal dialysis and presenting with *Curvularia* in the peritoneal fluid, including having had previous episodes of either bacterial peritonitis[89,91] or bowel perforation.[86]

18.4.6 INVASIVE DISEASE

Eight cases of *Curvularia* in the central nervous system have been reported, and in all but one, the fungal infection led to death despite antifungal treatment. Five patients presented with chronic signs and symptoms of parenchymal masses in the brain; one was initially diagnosed as having a meningioma,[92] and the only patient that survived the infection had a pituitary mass that was called a mucocele.[93] Of those patients with parenchymal masses, three had lung[94–97] or skin[98] lesions that may have been the entry source, while one pathology was only found in the brain.[99] Except for the patient with meningeal involvement that had a plasma cell dyscrasia, the other patients were considered immunocompetent.

Curvularia endocarditis has been reported in two cases with one having the infection in a prosthetic valve[100] while the other was an aortic aneurysm.[101] The immune status of the patients with endocarditis was not specifically addressed in the publications. Finally, a child with acute monocytic leukemia was reported to have multiple liver abscesses from which *Curvularia* was isolated during one of the neutropenic episodes.[102] *Curvularia* has not been cultured from blood in any of the patients with disseminated disease.

Invasive *Curvularia* infection in animals is as rare as in humans and has only been described in a parrot with brain and lung disease.[103]

18.5 DIAGNOSTICS

As can be expected by the range of clinical presentations, the specimens sent for diagnosis of dematiaceous molds vary in type and quantity. Fluid specimens or eye scrapings can be placed on slides and stained with Gram, Giemsa, KOH, acridine orange, and calcofluor white. In patients with *Curvularia* keratitis, fungal elements were observed in 78% of corneal scrapings with similar detection sensitivities for Gram, Giemsa, acridine orange, and calcofluor white.[14] These stains are rapid and easily performed and give a presumptive diagnosis of the presence of hyphae.[29] With Giemsa and KOH stains, pigmentation of the cell wall may be evident, and a dematiaceous mold may be assumed. However, it is not advisable to homogenize tissue to make smears since any of the pauciseptated molds (e.g., *Mucor*) may be destroyed in the process that may not allow visualization of the hyphae and will prevent growth.[62] When plating tissue specimens, it is required to cut these into small pieces rather than homogenizing or pulverizing the specimen allowing growth of all fungi.

Curvularia grows rapidly producing blackish-brown, woolly colonies (Figure 18.3) that are also black on the back of Sabouraud, potato dextrose, cornmeal agars, and other media.[1] A study of *Curvularia* keratitis demonstrated growth of the fungi on blood or chocolate agar in approximately 2 days, while the growth on Sabouraud agar or in brain–heart infusion agar occurred in 4–5 days.[14] Visualization of the conidia once the fungus has grown is usually performed by fixing and staining the slide preparation with lactophenol cotton blue. Poroconidia develop from geniculate (zigzag shaped or bended) conidiophores (stalk). Conidia are pale brown with three or more transverse septa. As the fungi mature, one of the central cells in the conidia grows larger and darker than the lateral ones producing the characteristic bend or curve and gives the characteristic appearance of a *croissant* (Figure 18.1). Identification of species is based on morphological characteristics from cultured specimens and include the (1) size of the conidia, (2) number of cells and color of the conidia, and (3) degree of bending of the conidiophores.[89] For example, *C. lunata* usually has three septa and four cells, while *C. geniculata* usually has four septa and five cells.[3] The amount of sporulation varies between *Curvularia* species, for example, *C. lunata* sporulates easily in different media, while other species may require techniques such as cellulose-containing substrates (index card or filter paper) to enhance sporulation.[104] It should be noted that only 10% of the dematiaceous fungi grown in clinical laboratories are thought to be the cause of disease since these are ubiquitous fungi that are frequent contaminants.[29]

(a) (b)

FIGURE 18.3 (**See color insert.**) (a) *C. geniculata* culture and (b) lactophenol cotton blue preparation (Public Health Image Library, CDC).

Although *Curvularia* has not been isolated from the blood of patients with invasive disease, the use of an automated blood culture system, in which the peritoneal effluent was inoculated in blood culture bottles, showed growth of *C. inaequalis* in a patient with peritonitis associated with ambulatory peritoneal dialysis.[89] The growth of the peritoneal fluid in blood culture bottles took between 2 and 4 days, while the peritoneal fluid plated on blood agar and Sabouraud plates did not reveal fungal growth.

Histopathologic examination is extremely valuable since the pathologist may be able to determine the (1) reaction the host is mounting to the mold, (2) presence of invasion of tissues and vessels, and (3) presence of pigmented hyphae.[29] The host reaction in patients with invasive diseases is usually granulomatous inflammation, but occasionally there is an intense fibroblastic proliferation[92] with only few granulomas. Using hematoxylin and eosin stains in histopathologic preparations, *Curvularia* species show as variably pigmented and variably septated hyphae with dilated or bulbous and fragmented portions (Figure 18.4). In some instances, pigmentation may be difficult to identify,[93] and the septated hyphae can be confused with *Aspergillus*.[16,17] Special stains (e.g., Fontana Masson) to enhance the presence of melanin can be used, but nondematiaceous fungi may also stain positive.[105] Immunohistochemical assays using polyclonal and monoclonal anti-*Aspergillus* antibodies have shown cross-reactive staining of culture-confirmed *Curvularia* cases[106] suggesting that this technique is not useful for diagnosis. In summary, hyphae described as septate and pigmented in histopathologic preparations will indicate the patient is infected with a dematiaceous fungus; however, none of the aforementioned stains will specifically identify *Curvularia*.

In patients with allergic rhinitis, intradermal testing can be performed to define if the person has a specific allergy to *Curvularia*.[107] Allergen extracts are usually prepared from cultures and are heterogeneous in their potency and composition.[108] In addition, the intradermal antigens used in the tests appear to be cross-reactive with other fungi including *Alternaria alternata*, *Epicoccum nigrum*, *A. fumigatus*, and *Fusarium solani*.[37] The same allergens have been used for detection of IgE-binding activity[108] and serologic testing using some of the epitopes of Cur I 3 in an ELISA and have been able to demonstrate that 94% of the patients with positive intradermal reactions were positive with the ELISA.[38]

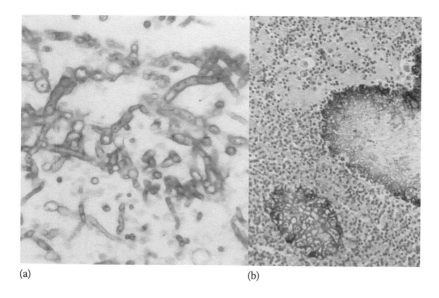

(a) (b)

FIGURE 18.4 *C. lunata* in Tenckhoff catheter (a) and *C. geniculata* from a mycetoma (b, image from the Public Health Image Library at CDC). Note the different pigment content in each.

18.5.1 Molecular Testing

Several molecular methods have been used to diagnose *Curvularia*. An in situ hybridization assay using tissue sections and a 24-base synthetic biotin-labeled oligonucleotide probe targeting abundant conserved fungal rRNA sequences of pigmented species has been used to define that the hyphae in the specimens tested corresponded to dematiaceous fungi.[109] However, this technique only defines that the hyphae present in the tissue are pigmented.

The use of conserved fungal primers followed by sequencing has proved useful for detecting different species of *Curvularia* after these have grown in cultures. Identification of *C. lunata* has been accomplished from isolates of cases having black grain mycetomas.[110] In these patients, universal fungal primers for the internal transcribed spacer 1 (ITS1)-5.8S-ITS2 DNA region were amplified and sequenced, and the sequences obtained were compared to published GenBank database where a homology greater than 90% was obtained. *C. lunata* and *C. senegalensis* were identified from four isolates of patients with keratitis by using panfungal highly conserved 5.8S ribosomal RNA, internal transcribed spacer 2 (ITS2) and 25S rRNA regions in a PCR.[14] Again, the amplified products were sequenced and compared to published databases. A similar methodology of sequencing the amplified conserved ITS region was used to identify *C. inaequalis* in an isolate from a patient in peritoneal dialysis with peritonitis.[89] In a patient with a pituitary mucocele, *C. geniculata* was detected by using a ribosomal deoxyribonucleic acid 28S D1/D2 region primer that generated a 600 base pair amplicon that was later sequenced and compared with Sequencher (Gene Codes Corp., Ann Arbor, MI) software and GenBank BLAST (National Center for Biotechnology Information, Bethesda, MD).[92] Theoretically, these methods may be able to be used in primary specimens or formalin-fixed, paraffin-embedded specimens[62]; however, publications describing this in such specimens from cases with *Curvularia* are not yet available. Whenever molecular techniques are used for identification of fungi, suitable controls are necessary, and establishing a causal relationship with the clinical symptoms is imperative to ensure the fungi identified are not contaminants.

18.6 TREATMENT

Keratitis should be treated with debridement of the infected tissue, discontinuation of topical corticosteroids, and a topical or oral antifungal agent.[61] A study comparing 5% topical natamycin to 1% topical itraconazole showed the same efficacy against *Curvularia* keratitis regardless of the severity of the corneal infection.[49] Topical suspension of 5% natamycin has been used for a median duration of 1 month, but a delay in diagnosis beyond 1 week can double the average length of topical antifungal treatment.[14] Topical clotrimazole and miconazole have also been used.[61] Oral treatment can be done with ketoconazole 200 mg twice daily.[61] In recalcitrant fungal keratitis cases that have not responded to topical and oral antifungals, injections of voriconazole around the abscess/ulcer (intrasomal) have been used successfully for *Curvularia* and other fungal keratitis.[111] However, fluconazole is ineffective against dematiaceous molds.[61]

Skin and subcutaneous infections should be treated with complete surgical excision,[61] and antifungal treatment is unnecessary if the lesion is excised completely. Oral treatment with amphotericin B, ketoconazole, or itraconazole should be used for those lesions incompletely resected; however, relapse of these incompletely removed lesions is common even after oral antifungal treatment.[61,69]

Treatment of allergic fungal sinusitis is primarily with surgical removal of the tenacious mucin that will promote drainage.[29,42] Other aspects of treatment include allergen avoidance and control of the allergic reaction with medications such as corticosteroids. Antifungal treatment may be used to decrease the amount of steroids used.

Treatment of *Curvularia* peritonitis consists of removal of the catheter and use of systemic amphotericin B and 5-flucytosine.[88,89,91] Nonetheless, resistance to amphotericin B was demonstrated in a 61-year-old man on peritoneal dialysis that presented with *C. lunata* peritonitis who had inadequate treatment response and a minimal inhibitory concentration (MIC) to amphotericin of 8 µg/mL (usually 0.125–1 µg/mL).[91] In this patient, the MIC of the *C. lunata* isolate for voriconazole was 1 µg/mL suggesting susceptibility to this agent; however, the patient developed septic shock and died soon after the susceptibility testing results were available.

Patients with invasive disease have a poor prognosis even with appropriate antifungal treatment and surgical removal of the mass when possible. Two of the patients with invasive disease described in the clinical section of this chapter have survived: the patient with a pituitary mucocele and one of the patients with endocarditis. The patient with a pituitary mucocele was treated with liposomal amphotericin B (4 mg/kg/day) for 1 month and then changed to oral voriconazole 200 mg BD for the second month and resulted in no residual disease by MRI and CT scan up to 8 months after initial presentation.[93] The patient with prosthetic valve endocarditis was initially treated with a combination of amphotericin B and ketoconazole while in the hospital; the patient was discharged on oral keotconazole.[100] After 2 months, the patient experienced gynecomastia and impotence; thus he was given terbinafine and continued with this treatment for 7 years. Revision of his surgery due to increase in symptoms revealed no fungi in the prosthesis by culture or histopathology.

MIC studies for *C. lunata* have been published, but the data supporting particular cutoff points are based on few and anecdotal experiences. In addition, susceptibility testing for molds is not frequently performed in clinical laboratories. The following MIC values have to be taken with the previous cautionary statements in mind. A study of 24 clinical isolates of *C. lunata* showed a range of MIC of (1) 0.5–2 µg/mL for amphotericin B, voriconazole, and posaconazole, (2) 0.5–4 µg/mL for itraconazole and ravuconazole, and (3) 1–4 µg/mL for isavuconazole suggesting good susceptibility to these agents. The MIC for fluconazole was 8–64 µg/mL suggesting resistance.[112] In addition, caspofungin was tested against four clinical isolates of *C. lunata* and showed a MIC range of 0.09–0.78 µg/mL indicating a strong in vitro activity of this echinocandin.[113] Finally, natamycin MIC was studied for some keratitis isolates and showed that these were inhibited by 4 µg/mL or less.[14]

18.7 CONCLUSIONS

Curvularia spp. are found ubiquitously in the environment. They can cause disease in humans and animals, with superficial infections (eye and skin) and respiratory allergic disease (allergic fungal rhinosinusitis and allergic bronchopulmonary disease) being more common than invasive disease. A variety of diagnostic methods have been used; some (culture and molecular sequencing) determine specifically that *Curvularia* is the causative agent, while others (cytology, histopathology) only define that a dematiaceous fungi is present. Tests that define allergy to *Curvularia* (prick skin tests, detection of antibodies) are not readily available or standardized. The rare cases of invasive disease have usually been fatal; however, treatment of superficial skin and eye infections has been successful using resection and/or antifungal agents. Treatment of respiratory allergic disease requires drainage of mucous (usually surgical removal) and immune modulation with steroids. Antifungals are rarely indicated in patients with respiratory allergic disease.

REFERENCES

1. Weber R. *Curvularia*. *Ann Allergy Asthma Immunol.* 2006;97:A4.
2. University-of-Adelaide. *Curvularia* sp. http://www.mycology.adelaide.edu.au/Fungal_Descriptions/Hyphomycetes_(dematiaceous)/Curvularia/. Accessed July 16, 2012.
3. Mycoses-Study-Group-DoctorFungus. *Curvularia* spp. http://www.doctorfungus.org/thefungi/curvularia.php. Accessed July 16, 2012.
4. Das S, Gupta-Bhattacharya S. Monitoring and assessment of airborne fungi in Kolkata, India, by viable and non-viable air sampling methods. *Environ Monit Assess.* 2012;184(8):4671–4684.
5. Hasnain S, Akhter T, Waqar M. Airborne and allergenic fungal spores of the Karachi environment and their correlation with meteorological factors. *J Environ Monit.* 2012;14:1006–1013.

6. Mishra A, Gond S, Kumar A et al. Season and tissue type affect fungal endophyte communities of the Indian medicinal plant *Tinospora cordifolia* more strongly than geographic location. *Microb Ecol.* 2012;64(2):388–398.
7. Jurjevic Z, Wilson J, Wilson D, Casper H. Changes in fungi and mycotoxins in pearl millet under controlled storage conditions. *Mycopathologia* 2007;164:229–239.
8. Mobin M, de-Moraes-Borba C, de-Moura-Filho O et al. The presence of fungi on contact electrical stimulation electrodes and ultrasound transducers in physiotherapy clinics. *Physiotherapy* 2011;97:273–277.
9. Veer P, Patwardhan N, Damle A. Study of onychomycosis: Prevailing fungi and pattern of infection. *Indian J Med Microbiol.* 2007;25:53–56.
10. Gupta M, Sharma N, Kanga A, Mahajan V, Tegta G. Onychomycosis: Clinico-mycologic study of 130 patients from Himachal Pradesh, India. *Indian J Dermatol Venereol Leprol.* 2007;73:389–392.
11. Kainer M, Keshavarz H, Jensen B et al. Saline-filled breast implant contamination with *Curvularia* species among women who underwent cosmetic breast augmentation. *J Infect Dis.* 2005;192:170–177.
12. Gopinathan U, Garg P, Fernandes M, Sharma S, Athmanathan S, Rao G. The epidemiological features and laboratory results of fungal keratitis: A 10-year review at a referral eye care center in South India. *Cornea* 2002;21:555–559.
13. Ganguly S, Salma K, Kansakar I, Sharma M, Bastola P, Pradhan R. Pattern of fungal isolates in cases of corneal ulcer in the western periphery of Nepal. *Nepal J Ophthalmol.* 2011;3:118–122.
14. Wilhelmus K, Jones D. *Curvularia* keratitis. *Trans Am Ophthalmol Soc.* 2001;99:111–132.
15. Hiromoto A, Nagano T, Nishigori C. Cutaneous infection caused by *Curvularia* species in an immunocompetent patient. *Br J Dermatol.* 2008;158:1371–1401.
16. Grieshop T, Yarbrough D-I, Farrar W. Phaeohyphomycosis due to *Curvularia lunata* involving skin and subcutaneous tissue after an explosion at a chemical plant. *Am J Med Sci* 1993;305:387–389.
17. Still-Jr J, Law E, Pereira G, Singletary E. Invasive burn wound infection due to *Curvularia* species. *Burns* 1993;19:77–79.
18. Guedes F, Attili-Angelis D, Pagnocca F. Selective isolation of dematiaceous fungi from the workers of *Atta laevigata* (Formicidae: Attini). *Folia Microbiol.* 2012;57:21–26.
19. Muhsin T, Salih T. Exocellular enzyme activity of dermatophytes and other fungi isolated from ruminants in Southern Iraq. *Mycopathologia* 2000;150:49–52.
20. Tuli S, Yoo S. *Curvularia* keratitis after laser in situ keratomileusis from a feline source. *J Cataract Refract Surg.* 2003;29:1019–1021.
21. Promo L, Waniska R, Kollo A, Rooney W, Bejosano F. Role of chitinase and sormatin accumulation in the resistance of sorghum cultivars to grain mold. *J Agric Food Chem.* 2005;53:5565–5570.
22. Ratnavathi C, Komala V, Kumar B, Das I, Patil J. Natural occurrence of aflatoxin B1 in sorghum grown in different geographical regions of India. *J Sci Food Agric.* 2012;92(12):2416–2420; doi: 10.1002/jsfa.5646.
23. Rout N, Nanda B, Gangopadhyaya S. Experimental pheohyphomycosis and mycotoxicosis by *Curvularia lunata* in albino rats. *Indian J Pathol Microbiol.* 1989;32:1–6.
24. Warth B, Parich A, Atehnkeng J et al. Quantitation of mycotoxins in food and feed from Burkina Faso and Mozambique using a modern LC-MS/MS multitoxin method. *J Agric Food Chem.* 2012;60:9352–9363.
25. Ezekiel C, Sulyok M, Warth B, Krska R. Multi-microbial metabolites in fonio millet (acha) and sesame seeds in Plateau State, Nigeria. *Eur Food Res Technol.* 2012;235:285–293.
26. Honda Y, Ueki M, Okada G et al. Isolation, and biological properties of a new cell cycle inhibitor, curvularol, isolated from *Curvularia* sp. RK97-F166. *J Antibiot (Tokyo).* 2001;54:10–16.
27. Wilhelmus K. Climatology of dematiaceous fungal keratitis. *Am J Ophthalmol.* 2005;140:1156–1157.
28. Codina R, Fox R, Lockey R, DeMarco P, Bagg A. Typical levels of airborne fungal spores in houses without obvious moisture problems during a rainy season in Florida, USA. *J Investig Allergol Clin Immunol.* 2008;18:156–162.
29. Revankar S, Sutton D. Melanized fungi in human disease. *Clin MIcrobiol Rev.* 2010;23:884–928.
30. Revankar S. Phaeohyphomycosis. *Infect Dis Clin N Am.* 2006;20:609–620.
31. Svegelj M, Turk S, Brus B, Rizner T, Stojan J, Gobec S. Novel inhibitors of trihydroxynaphthalene reductase with antifungal activity identified by ligand-based and structure-based virtual screening. *J Chem Inf Model.* 2011;51:1716–1724.
32. Brunskole M, Zorko K, Kerbler V et al. Trihydroxynaphthalene reductase of *Curvularia lunata*—A target for flavonoid action? *Chem Biol Interact.* 2009;178:259–267.
33. Nair S, Kukreja N, Singh B, Arora N. Identification of B cell epitopes of alcohol dehydrogenase allergen of *Curvularia lunata*. *PLoS One* 2011;6:1–9.
34. Gupta R, Sharma V, Sridhara S, Singh B, Arora N. Identification of serine protease as a major allergen of *Curvularia lunata*. *Allergy* 2004;59:421–427.

35. Tripathi P, Kukreja N, Singh B, Arora N. Serine protease activity of Cur l 1 from *Curvularia lunata* augments Th2 response in mice. *J Clin Immunol.* 2009;29:292–302.
36. Sharma V, Gupta R, Jhingran A et al. Cloning, recombinant expression and activity studies of a major allergen "enolase" from the fungus *Curvularia lunata. J Clin Immunol.* 2006;26:360–369.
37. Gupta R, Singh B, Sridhara S et al. Allergenic cross-reactivity of *Curvularia lunata* with other airborne fungal species. *Allergy* 2002;57:636–640.
38. Sharma V, Singh B, Gaur S, Pasha S, Arora N. Bioinformatics and immunologic investigation on B and T cell epitopes of Cur l 3, a major allergen of *Curvularia lunata. J Proteome Res.* 2009;8:2650–2655.
39. Sharma V, Singh B, Arora N. Cur l 3, a major allergen of *Curvularia lunata*–derived short synthetic peptides, shows promise for successful immunotherapy. *Am J Respir Cell Mol Biol.* 2011; 45:1178–1184.
40. Sharma V, Singh B, Gaur S, Arora N. Molecular and immunological characterization of cytochrome c: A potential cross-reactive allergen in fungi and grasses. *Allergy* 2008;63:189–197.
41. Porter P, Roberts L, Fields A et al. Necessary and sufficient role for T helper cells to prevent fungal dissemination in allergic lung disease. *Infect Immun.* 2011;79(11):4459–4471.
42. Schubert M. Allergic fungal sinusitis: Pathophysiology, diagnosis and management. *Med Mycol.* 2009;47:S324–S330.
43. Kuhn F, Swain R. Allergic fungal sinusitis: Diagnosis and treatment. *Curr Opin Otolaryngol Head Neck Surg.* 2003;11:1–5.
44. Hansen E, Albertsen L, Schafer T et al. *Curvularia* haloperoxidase: Antimicrobial activity and potential application as a surface disinfectant. *Appl Environ Microbiol.* 2003;69:4611–4617.
45. Trisuwan K, Rukachaisirikul V, Phongpaichit S, Preedanon S, Sakayaroj J. Modiolide and pyrone derivatives from the sea fan-derived fungus *Curvularia* sp. PSU-F22. *Arch Pharm Res.* 2011;34:709–714.
46. Chomcheon P, Wiyakrutta S, Aree T et al. Curvularides A–E: Antifungal hybrid peptide–polyketides from the endophytic fungus *Curvularia geniculata. Chem Eur J.* 2010;16:11178–11185.
47. Zhang Y, Kong L, Jiang D et al. Phytotoxic and antifungal metabolites from *Curvularia* sp. FH01 isolated from the gut of *Atractomorpha sinensis. Bioresour Technol.* 2011;102:3575–3577.
48. Feng B, Hu W, Ma B et al. Purification, characterization, and substrate specificity of a glucoamylase with steroidal saponin-rhamnosidase activity from *Curvularia lunata. Appl Microbiol Biotechnol.* 2007;76:1329–1338.
49. Kalavathy C, Parmar P, Kaliamurthy J et al. Comparison of topical itraconazole 1% with topical natamycin 5% for the treatment of filamentous fungal keratitis. *Cornea* 2005;24:449–452.
50. Gupta N, Samantaray J, Duggal S, Srivastava V, Dhull C, Chaudhary U. *Acanthamoeba* keratitis with *Curvularia* co-infection. *Indian J Med Microbiol.* 2012;28:67–71.
51. Bhala S, Narang S, Sood S, Mithal C, Arya S, Gupta V. Microbial contamination in open globe injury. *Nepal J Ophthalmol.* 2012;4:84–89.
52. Ehlers J, Chavala S, Woodward J, Postel E. Delayed recalcitrant fungal endophthalmitis secondary to *Curvularia. Can J Ophthalmol.* 2011;46:199–200.
53. Pathengay A, Shah G, Das T, Sharma S. *Curvularia lunata* presenting with a posterior capsular plaque. *Indian J Ophthalmol.* 2006;54:65–66.
54. Kaushik S, Ram J, Chakrabarty A, Dogra M, Brar G, Gupta A. *Curvularia lunata* endophthalmitis with secondary keratitis. *Am J Ophthalmol.* 2001;131:140–142.
55. Berbel R, Barbante-Casella A, de-Freitas D, Hofling-Lima A. *Curvularia lunata* endophthalmitis. *J Ocul Pharmacol Ther.* 2011;27:535–537.
56. Brooks D, Andrew S, Denis H et al. Rose bengal positive epithelial microerosions as a manifestation of equine keratomycosis. *Vet Ophthalmol.* 2000;3:83–86.
57. Weinstein W, Moore P, Sanchez S, Dietrich U, Wooley R, Ritchie B. In vitro efficacy of a buffered chelating solution as an antimicrobial potentiator for antifungal drugs against fungal pathogens obtained from horses with mycotic keratitis. *Am J Vet Res.* 2006;67:562–568.
58. Kodikara D, de-Silva N, Makuloluwa C, de-Silva N, Gunatilake M. Bacterial and fungal pathogens isolated from corneal ulcerations in domesticated elephants (*Elephas maximus maximus*) in Sri Lanka. *Vet Ophthalmol.* 1999;2:191–192.
59. Ben-Shlomo G, Plummer C, Barrie K, Brooks D. *Curvularia* keratomycosis in a dog. *Vet Ophthalmol.* 2010;13:126–130.
60. Myers D, Isaza R, Ben-Shlomo G, Abbott J, Plummer C. Fungal keratitis in a gopher tortoise (*Gopherus polyphemus*). *J Zoo Wildl Med.* 2009;40:579–582.
61. Brandt M, Warnock D. Epidemiologic, clinical manifestations, and therapy of infections caused by dematiaceous fungi. *J Chemother.* 2003;15:S36–S47.

62. Guarner J, Brandt M. Histopathologic diagnosis of fungal infections in the 21st century. *Clin Microbiol Rev.* 2011;24:247–280.
63. Dowd S, Delton-Hanson J, Rees E et al. Survey of fungi and yeast in polymicrobial infections in chronic wounds. *J Wound Care* 2011;20:40–47.
64. Vermeire S, de-Jonge H, Lagrou K, Kuypers D. Cutaneous phaeohyphomycosis in renal allograft recipients: Report of 2 cases and review of the literature. *Diag Microbiol Infect Dis.* 2010;68:177–180.
65. Berg D, Garcia J, Schell W, Perfect J, Murray J. Cutaneous infection caused by *Curvularia pallescens*: A case report and review of the spectrum of disease. *J Am Acad Dermatol.* 1995;32:375–378.
66. Tessari G, Forni A, Ferretto R et al. Lethal systemic dissemination from a cutaneous infection due to *Curvularia lunata* in a heart transplant recipient. *J Eur Acad Dermatol Venereol.* 2003;17:440–442.
67. Yau Y, de-Nanassy J, Summerbell R, Matlow A, Richardson S. Fungal sternal wound infection due to *Curvularia lunata* in a neonate with congenital heart disease: Case report and review. *Clin Infect Dis.* 1994;19:735–740.
68. Fernandez M, Noyola D, Rossmann S, Edwards M. Cutaneous phaeohyphomycosis caused by *Curvularia lunata* and a review of *Curvularia* infections in pediatrics. *Pediatr Infect Dis J.* 1999;18:727–731.
69. Safdar A. *Curvularia*-favorable response to oral itraconazole therapy in two patients with locally invasive phaeohyphomycosis. *Clin Microbiol Infect.* 2003;9:1219–1223.
70. Bridges C. Maduromycotic mycetomas in animals; *Curvularia geniculata* as an etiologic agent. *Am J Pathol.* 1957;33:411–427.
71. Herraez P, Rees C, Dunstan R. Invasive phaeohyphomycosis caused by *Curvularia* species in a dog. *Vet Pathol.* 2001;38:456–459.
72. Swift I, Griffin A, Shipstone M. Successful treatment of disseminated cutaneous phaeohyphomycosis in a dog. *Aust Vet J.* 2006;84:431–435.
73. Glass D, Amedee R. Allergic fungal rhinosinusitis: A review. *Ochsner J.* 2011;11:271–275.
74. Bent 3rd J, Kuhn F. Diagnosis of allergic fungal sinusitis. *Otolaryngol Head Neck Surg.* 1994;111:580–588.
75. deShazo R, Swain R. Diagnostic criteria for allergic fungal sinusitis. *J Allergy Clin Immunol.* 1995;96:24–35.
76. Manning S, Schaefer S, Close L, Vuitch F. Culture-positive allergic fungal sinusitis. *Arch Otolaryngol Head Neck Surg.* 1991;117:174–178.
77. El-Morsy S, Khafagy Y, El-Naggar M, Beih A. Allergic fungal rhinosinusitis: Detection of fungal DNA in sinus aspirate using polymerase chain reaction. *J Laryngol Otol.* 2012;124:152–160.
78. Lake F, Froudist J, McAleer R, Gillon R, Tribe A, Thompson P. Allergic bronchopulmonary fungal disease caused by *Bipolaris* and *Curvularia*. *Aust N Z J Med.* 1991;21:871–874.
79. Mroueh S, Spock A. Allergic bronchopulmonary disease caused by *Curvularia* in a child. *Pediatr Pulmonol.* 1992;12:123–126.
80. Travis W, Kwon-Chung K, Kleiner D et al. Unusual aspects of allergic bronchopulmonary fungal disease: Report of two cases due to *Curvularia* organisms associated with allergic fungal sinusitis. *Hum Pathol.* 1991;22:1240–1248.
81. Halwig J, Brueske D, Greenberger P, Dreisin R, Sommers H. Allergic bronchopulmonary curvulariosis. *Am Rev Respir Dis.* 1985;132:186–188.
82. Bowles K, Beadle R, Mouch S et al. A novel model for equine recurrent airway obstruction. *Vet Immunol Immunopathol.* 2002;87:385–389.
83. Costa L, Johnson J, Baur M, Beadle R. Temporal clinical exacerbation of summer pasture-associated recurrent airway obstruction and relationship with climate and aeroallergens in horses. *Am J Vet Res.* 2006;67:1635–1642.
84. Guarner J, del-Rio C, Williams P, McGowan J. Case report: Fungal peritonitis caused by *Curvularia lunata* in a patient undergoing peritoneal dialysis. *Am J Med Sci.* 1989;298:320–323.
85. Unal A, Sipahioglu M, Atalay M et al. Tenckhoff catheter obstruction without peritonitis caused by *Curvularia* species. *Mycoses* 2010;54:363–364.
86. Diskin C, Stokes T, Dansby L, Radcliff L, Carter T. Case report and review: Is the tendency for *Curvularia* tubular obstruction significant in pathogenesis? *Perit Dial Int.* 2008;28:678–679.
87. DeVault G-J, Brown S-I, King J, Fowler M, Oberle A. Tenckhoff catheter obstruction resulting from invasion by *Curvularia lunata* in absence of peritonitis. *Am J Kidney Dis.* 1985;6:124–127.
88. Vachharajani T, Zaman F, Latif S, Penn R, Abreo K. *Curvularia geniculata* fungal peritonitis: A case report with review of literature. *Int Urol Nephrol.* 2005;37:781–784.
89. Pimentel J, Mahadevan K, Woodgyer A et al. Peritonitis due to *Curvularia inaequalis* in an elderly patient undergoing peritoneal dialysis and a review of six cases of peritonitis associated with other *Curvularia* spp. *J Clin Microbiol.* 2005;43:4288–4292.

90. Lopes J, Alves S, Benevenga J, Brauner F, Castro M, Melchiors E. *Curvularia lunata* peritonitis complicating peritoneal dialysis. *Mycopathologia* 1994;127:65–67.
91. Varughese S, David V, Mathews M, Tamilarasi V. A patient with amphotericin-resistant *Curvularia lunata* peritonitis. *Perit Dial Int.* 2011;31:108–109.
92. Singh H, Irwin S, Falowski S et al. *Curvularia* fungi presenting as a large cranial base meningioma: Case report. *Neurosurgery* 2008;63:E177.
93. Smith T, Goldschlager T, Mott N, Robertson T, Campbell S. Optic atrophy due to *Curvularia lunata* mucocoele. *Pituitary* 2007;10:295–297.
94. Lampert H, Hutto J, Donnelly W, Shulman S. Pulmonary and cerebral mycetoma caused by *Curvularia pallescens.. J Pediatr.* 1977;91:603–605.
95. Friedman A, Campos J, Rorke L, Bruce D, Arbeter A. Fatal recurrent *Curvularia* brain abscess. *J Pediatr.* 1981;99:413–415.
96. de-la-Monte S, Hutchins G. Disseminated *Curvularia* infection. *Arch Pathol Lab Med.* 1985;109:872–874.
97. Pierce N, Millan J, Bender B, Curtis J. Disseminated *Curvularia* infection. Additional therapeutic and clinical considerations with evidence of medical cure. *Arch Pathol Lab Med.* 1986;110:959–961.
98. Rohwedder J, Simmons J, Colfer H, Gatmaitan B. Disseminated *Curvularia lunata* infection in a football player. *Arch Intern Med.* 1979;139:940–941.
99. Carter E, Boudreaux C. Fatal cerebral phaeohyphomycosis due to *Curvularia lunata* in an immunocompetent patient. *J Clin Microbiol.* 2004;42:5419–5423.
100. Bryan C, Smith C, Berg D, Karp R. *Curvularia lunata* endocarditis treated with terbinafine: Case report. *Clin Infect Dis.* 1993;16:30–32.
101. Kaufman S. *Curvularia* endocarditis following cardiac surgery. *Am J Clin Pathol.* 1971;56:466–470.
102. Shigemori M, Kawakami K, Kitahara T et al. Hepatosplenic abscess caused by *Curvularia boedijn* in a patient with acute monocytic leukemia. *Pediatr Infect Dis J.* 1996;15:1128–1129.
103. Clark F, Jones L, Panigrahy B. Mycetoma in a grand Eclectus (*Eclectus roratus roratus*) parrot. *Avian Dis.* 1986;30:441–443.
104. Pratt R. Enhancement of sporulation in species of *Bipolaris, Curvularia, Drechslera,* and *Exserohilum* by growth on cellulose-containing substrates. *Mycopathologia* 2006;162:133–140.
105. Kimura M, McGinnis M. Fontana-Masson—Stained tissue from culture-proven mycoses. *Arch Pathol Lab Med.* 1998;122:1107–1111.
106. Schuetz A, Cohen C. *Aspergillus* immunohistochemistry of culture-proven fungal tissue isolates shows high cross-reactivity. *Appl Immunohistochem Mol Morphol.* 2009;17:524–529.
107. Pumihirun P, Towiwat P, Mahakit P. Aeroallergen sensitivity of Thai patients with allergic rhinitis. *Asian Pac J Allergy Immunol.* 1997;15:183–185.
108. Gupta R, Singh B, Sridhara S, Gaur S, Chaudhary V, Arora N. Allergens of *Curvularia lunata* during cultivation in different media. *J Allergy Clin Immunol.* 1999;104:857–862.
109. Montone K, Livolsi V, Lanza D et al. Rapid in-situ hybridization for dematiaceous fungi using a broad-spectrum oligonucleotide DNA probe. *Diagn Mol Pathol.* 2011;20:180–183.
110. Desnos-Ollivier M, Bretagne S, Dromer F, Lortholary O, Dannaoui E. Molecular identification of black-grain mycetoma agents. *J Clin Microbiol.* 2006;44:3517–3523.
111. Sharma N, Agarwal P, Sinha R, Titiyal J, Velpandian T, Vajpayee R. Evaluation of intrastromal voriconazole injection in recalcitrant deep fungal keratitis: Case series. *Br J Ophthalmol.* 2011;95:1735–1737.
112. Gonzalez G. In vitro activities of isavuconazole against opportunistic filamentous and dimorphic fungi. *Med Mycol.* 2009;47(Special Issue):71–76.
113. del-Poeta M, Schell W, Perfect J. In vitro antifungal activity of pneumocandin L-743,872 against a variety of clinically important molds. *Antimicrob Agents Chemother.* 1997;41:1835–1836.

19 *Encephalitozoon*

*Carmen del Águila de la Puente,
Soledad Fenoy Rodríguez, and Nuno Henriques-Gil*

CONTENTS

19.1 Introduction	267
19.2 Systematics	268
19.3 Morphology and Biology	269
19.4 Genome of *Encephalitozoon*	272
19.5 Epidemiology	273
19.5.1 Human Infection	273
19.5.2 Source of Infection	275
19.5.2.1 Zoonotic Transmission	275
19.5.2.2 Waterborne Transmission	276
19.6 Intraspecific Variability	277
19.6.1 Intraspecific Variability in *E. cuniculi*	277
19.6.2 Intraspecific Variability in *E. hellem*	278
19.6.3 Intraspecific Variability in *E. intestinalis*	279
19.7 Clinical Features	279
19.7.1 Microsporidial Diarrhea and Biliary Pathology	280
19.7.2 Microsporidial Hepatitis and Peritonitis	280
19.7.3 Disseminated Infections	280
19.7.4 Ocular Microsporidiosis	281
19.8 Pathology and Immunology	281
19.9 Diagnosis and Detection	282
19.9.1 Electron Microscopy	282
19.9.2 Staining Methods	282
19.9.3 Fluorescent Staining Techniques	282
19.9.4 Immunofluorescence Test	282
19.9.5 PCR Methods	283
19.9.6 Cell Culture	283
19.9.7 Serology	283
19.9.8 Detection of Microsporidia in Water	283
19.10 Treatment	284
19.11 Prevention	285
19.12 Conclusions	285
Acknowledgments	286
References	286

19.1 INTRODUCTION

Microsporidia are obligate intracellular parasites considered as emerging pathogens. They are ubiquitous and capable of infecting all groups of animals. The organisms are characterized by the production of small and environmental resistant spores that have a unique mode of entering host cells via a polar tube. They are true eukaryotes with a nucleus, chromosome separation, and an intracytoplasmic membrane system. They lack a classical stacked Golgi apparatus, centrioles, peroxisomes, and mitochondria that have been substituted by a mitochondria-like organelle called a mitosome [1,2]. Three species belonging to the genus *Encephalitozoon* have been isolated from human infections: *Encephalitozoon cuniculi*, *Encephalitozoon hellem*, and *Encephalitozoon intestinalis* (synonym *Septata intestinalis*). Several reports in immunodeficient patients, especially AIDS patients, have been reported since the first case of human infection by *E. cuniculi*, described in 1959 in a 9-year-old Japanese boy with neurologic symptoms [1]. *E. cuniculi* infection often produces disseminated disease, although diarrhea or interstitial nephritis and cholecystitis produced by *E. intestinalis* have been described. However, keratoconjunctivitis, cystitis, ureteritis, and bronchiolitis have been associated with *E. hellem* infection. The epidemiology of this infection remains vague since the source of transmission of the parasites is obscure, and they have been identified in water, food, and animals.

19.2 SYSTEMATICS

Microsporidia were once considered primitive eukaryotes, living representatives of hypothetical nucleated cells, lacking mitochondria. However, their apparent simplicity is a secondary trait, derived from the adaptation to intracellular parasitism. It has now become clear that Microsporidia evolved from an ancestor that possessed mitochondria and are phylogenetically related to fungi [3–5]. Microsporidia are included in the first rank group Fungi, within the Eukaryota supergroup Opisthokonta (see the recent update of the classification of Eukaryotes: Adl et al. [6]). They have diverged widely, and the precise phylogenetic relationships among genera are under discussion. Ribosomal DNA-based phylogenies relate *Encephalitozoon* to the genera *Vairimorpha* and *Nosema* that infect insects such as silkworms and honeybees [7,8], although Keeling and Corradi [2] pointed out that *Encephalitozoon* and *Enterocytozoon* share a significant number of metabolic characteristics.

Five species have been referenced within the genus *Encephalitozoon*, differing in a number of morphological, immunological, and molecular features. Three of them—*E. cuniculi*, *E. intestinalis*, and *E. hellem*—are known to infect humans, while *E. romaleae* is a parasite of grasshoppers [9] and the less studied *E. lacertae* infects reptiles [10]. Genetic studies showed that they are clearly segregated entities and can be considered different species. However, the concept of species is necessarily arbitrary concerning clonal organisms. The similarity of a number of strains derives from selective constraints or recent divergence, rather than genetic flux due to sexual reproduction.

Indeed, a sexual cycle does not appear to exist in *Encephalitozoon* [11], despite the presence of genes in *E. cuniculi* that could be related to sexual transmission [12]. However, negative results do not mean that sexual reproduction is impossible, as a number of organisms, including many protozoa and fungi, also retain genes related to meiosis and sexual reproduction. For instance, in *Leishmania* species, the population distribution of genotypes indicates that sexual hybridization occurs occasionally, but this was only recently demonstrated in an intermediate mosquito host [13,14]. In Microsporidia, the detection of recombinants would be very difficult for technical reasons (there are no selectable markers such as different drug resistance, and Microsporidia cannot be transformed in vitro), and the intracellular environment in which two different strains must coincide and recombine may be very different from the cell cultures usually employed to multiply *Encephalitozoon* spp. Yet indirect data do not support sexual reproduction: when comparing strains that differ in a number of genetic markers (see later), the distribution of genotypes is compatible with strict clonal reproduction. A given mutation or gene reorganization seems to have occurred only in one lineage and is not found in other groups. There are no isolates that combine markers of two other strains [15].

19.3 MORPHOLOGY AND BIOLOGY

Microsporidia have a structure that is unique in nature, that is, the polar tube or polar filament involved in host cell invasion [16]. The infective stage is the spore (Figures 19.1 and 19.2), ranging from 1 to 20 µm long, with the species infecting mammals being small (1–3 µm). *Encephalitozoon* spp. spores measure 2.0–2.5 by 1.0–1.5 µm. They are Gram positive and environmentally resistant, with a thick wall composed of three layers: (1) an electron-dense proteinaceous outer layer, called the exospore, (2) an electron-lucent inner chitinous layer called the endospore, and (3) a plasma membrane enclosing an infective sporoplasm [1,17]. The content of the spores is composed of two functionally different parts: the sporoplasm and the extrusion apparatus. The sporoplasm is the infectious material of Microsporidia. It may contain a nucleus, as it is the case of *Encephalitozoon*, or be a diplokaryon as in *Nosema*, *Brachiola*, and *Vittaforma*, which consist of two closely apposed nuclei functioning as a single unit [16], with ribosomes and endoplasmic reticulum (ER) membranes [18]. The ribosomes (70S), ribosomal subunits (30S and 50S), and rRNAs (16S and 23S) are of prokaryotic size, and the rRNA has no separate 5.8S rRNA [1]. The extrusion apparatus is composed of (1) a polar tube, coiled around the posterior region of the spore; (2) an anchoring disk; (3) an anterior membrane-bounded organelle, termed a polaroplast; and (4) a vacuole at the posterior. The number and disposition of polar filament coils vary among Microsporidia, with five to seven coils in a single row in *Encephalitozoon* spp. Under certain conditions such as alkaline pH or increased concentrations of Na^+, K^+, Cl^-, and Ca^{2+} ions, the spore germinates, and an inflow of water dramatically increases the pressure inside the spore. This eventually ruptures the wall at the anterior end, where the exospore is less thick, forcing the polar filament to eject, turning inside out, to form a tube [18,19]. Germination is very quick: the polar tube acts as a projectile if it penetrates the host cell membrane, where it delivers the sporoplasm inside the host cell cytoplasm. Microsporidia can be taken up by phagocytosis, and they use the polar tube to escape from the vacuole [20].

Once inside the host cell cytoplasm, microsporidian proliferation (merogony) will occur within a parasitophorous vacuole, as in the case of *Encephalitozoon* spp. or direct contact with the host cell cytoplasm (e.g., *Enterocytozoon*). Inside the host cell, the sporoplasm undergoes extensive multiplication either by merogony (binary fission, e.g., *Encephalitozoon*) or schizogony (multiple fission).

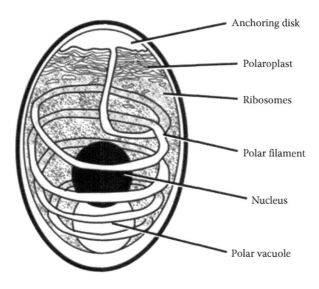

FIGURE 19.1 General representation of a Microsporidia spore.

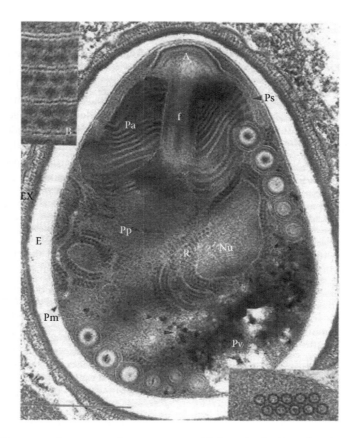

FIGURE 19.2 Electron microscopy of mature spore. (a) Longitudinal section (*Episeptum inversum*). (b) Polyribosomes attached to membranes (*Napamichum dispersus*). (c) Circularly arranged polyribosomes (*Flabelliforma magnivora*). *Abbreviations*: A, anchoring disk; E, endospore; EX, exospore; f, polar filament; Nu, nucleus; Pa, anterior polaroplast; Ps, polar sac; Pv, posterior vacuole; R, polyribosomes. (Reprinted from Vavra, J. and Larson, J.I.R., Structure of the microsporidia, in *The Microsporidia and Microsporidiosis*, eds. M. Witter and L.M. Weis, ASM Press, Washington, DC, 1999, pp. 7–84. With permission from ASM Press.)

The meronts are roundish cells encircled by a typical unit membrane. They have a large nuclear region. The cytoplasm contains poorly developed ER and many free ribosomes attached to vesicles of ER or the nuclear membrane [18]. The meronts undergo repeated divisions in the host cell, and nuclear division may occur without cell division, resulting in multinucleated plasmodial forms. Meronts develop into sporonts characterized by a dense surface coat and the increased presence of ER and ribosomes, which change organization. The ER becomes highly ordered, and the ribosomes increasingly form arrays attached to the ER, known as polyribosomes [21]. They will also multiply by binary or multiple fission and divide into sporoblasts that will finally develop into mature spores. At the end of the proliferative phase, the cytoplasm of the host cell is completely filled with spores, and the cell membrane will disrupt releasing the mature spores that can infect new cells to continue the life cycle [1,16,18] (Figure 19.3).

Simple life cycles, such as described here, are completed in a single host, as is the case of mammalian Microsporidia, such as *Encephalitozoon* in humans. Complex cycles requiring different hosts are also possible in nature, with the involvement of more than one generation of the parasite having different morphologies. These are mainly associated with invertebrate hosts [21].

Encephalitozoon

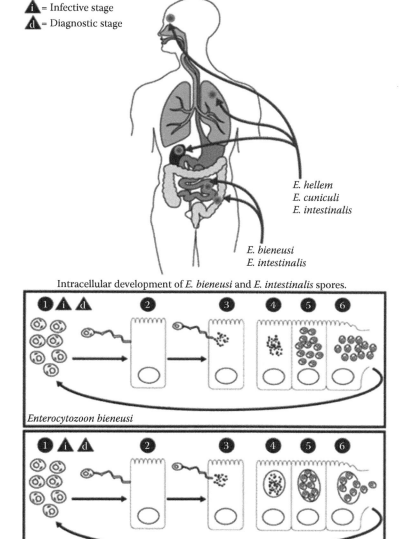

*Development inside parasitophorous vacuole also occurs in *E. hellem* and *E. cuniculi*.

FIGURE 19.3 Life cycle of microsporidia. (Reprinted from Microsporidiosis DPDx-CDC Parasitology Diagnostic Website, cited 2013, http://www.cdc.gov/dpdx/microsporidiosis/index.htlm. With permission.)

In human *Encephalitozoon* infections, the life cycle of the organism is mainly completed in the epithelial cells of the gastrointestinal tract, and the oral–fecal transmission of the spores is widely recognized, but the lung epithelium may also be colonized by inhaled Microsporidia spores [20,22]. Moreover, Microsporidia may infect and multiply in macrophages, contributing to the dissemination of the parasite in the organism. Sexual transmission of these organisms is based on the identification of Microsporidia in the urethra and prostate of several HIV+ patients. Infection through trauma or direct contact has been described for ocular colonization, and transplacental transmission of *Encephalitozoon* has been proved in rodents, carnivores, and nonhuman primates, but not yet in humans [22].

19.4 GENOME OF *ENCEPHALITOZOON*

Parallel to the reduction in size and intracellular organelles, the genomes of Microsporidia are also smaller. However, the amount of DNA in different microsporidian groups varies widely, from more than 25 Mb (which is equivalent to many free-living eukaryotic microbes) to 2.3 Mb [2]. The chromosomes of *Encephalitozoon* were first studied by pulsed-field gel electrophoresis (PFGE) by Sobottka et al. [23], and they are too small for observation under the microscope. The karyotype of *E. cuniculi* is composed by 11 chromosomes from approximately 200 kb to more than 300 kb. PFGE resolved only 10 DNA bands for *E. hellem* and *E. intestinalis*, although in reality they also have 11 chromosomes (two of them have similar size). These studies confirmed the very small genome of ca. three million base pairs (Mb), which is less than that of many prokaryotes (*Escherichia coli* has 4.6 Mb). The small size encouraged genome projects, and *E. cuniculi* was the first completely sequenced microsporidian [24].

The *E. cuniculi* genome project confirmed a total size of 2.9 Mb, and the 11 chromosomes (named I–XI) ranged from 217 to 317 Mb. The genome compaction derived from gene loss, together with gene and spacer shortening. *E. cuniculi* has approximately 2000 genes, and the entire metabolic functions have been lost, for which the microsporidian is entirely dependent from the host metabolism. In contrast, genes that codify transporters are overrepresented, which is in accordance with the intracellular parasitic way of life of these organisms [25]. The chromosomes of *E. cuniculi* have an additional number of peculiar features. Ribosomal genes 16S and 23S rRNA are located near the telomeres of all the chromosomes, and the distal segments are mainly composed by heterogeneous tandemly repeated sequences, including the telomeres [24,26]. Moreover, detailed analysis of chromosome I by Peyret et al. [27] showed that the two broad subtelomeric regions (about 45 kb) are symmetrically repeated and differ from the chromosome core in GC content, while the central 135 kb core of the chromosome contains mostly single-copy genes with a more or less typical organization as compared to the whole genome. Most of the genes located in these subtelomeric regions are of unknown functions, and their phylogenetic relationships to other genes have been difficult to establish.

After *E. cuniculi*, an important milestone in the genomics of Microsporidia was reached with the complete sequencing of three other genomes: *E. intestinalis*, *E. hellem*, and *E. romaleae* [25,28]. *E. intestinalis* has the smallest genome of 2.3 Mb that is the lowest amount of DNA ever found in a eukaryotic organism and seems to be near a limit of reduction [29]. *E. hellem* and *E. romaleae* have 2.5 Mb. The four genomes have common features—they all have 11 chromosomes, the genomes are similarly compacted, and genes are organized syntenically—demonstrating that they already existed in the common ancestor to the whole genus.

While this conservation of the genome structure is certainly remarkable, *Encephalitozoon* spp. also show relevant differences. Gene sequences have evolved rapidly, and any group of orthologue genes to be compared among the four species will show a wide variety of sequence differences. For instance, the histone H4 gene, a classic example of conservation among very different organisms, shows a mean number of almost 50 point substitutions between species, within a gene of 312 base pair (bp) in length. The rates of synonymous mutations are much higher than nonsynonymous mutations, which should be taken as evidence of purifying rather than directional selection. Keeling and Corradi [2] pointed out that spacers are often more conserved than coding sequences, suggesting that spacers have become as small as they can be, maintaining the ultimate elements necessary for the control of gene expression. Consequently, genome reorganization involving intergenic segments is very difficult without affecting gene expression and so would also explain why the general organization of the genome is constant [30].

E. hellem and *E. romaleae* are phylogenetically closer to each other, and they share unique acquisitions in the genome that do not exist in other Microsporidia. In contrast, the relationships

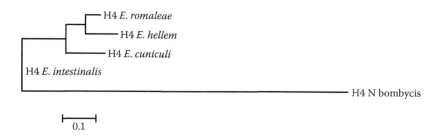

FIGURE 19.4 Maximum likelihood phylogeny of histone H4 gene of *Encephalitozoon* spp. *Nosema bombycis* was used as out-group. Note the basal position of *E. intestinalis*. Bar represents 0.1 substitutions per site.

with respect to *E. cuniculi* and *E. intestinalis* are less clear. They have all diverged strongly from a common ancestor, and the position of the root depends on the sequences employed in the analyses: *E. cuniculi* may cluster with *E. intestinalis* or not. The comparison of the histone H4 gene places *E. intestinalis* deeper in the tree (Figure 19.4).

Several authors have demonstrated recently that *E. hellem* and *E. romaleae* have a number of DNA sequences absent from all other Microsporidia, but show significant homology with phylogenetically distant prokaryotic and eukaryotic organisms [28,31]. Therefore, these Microsporidia must have acquired genetic material by horizontal gene transfer (HGT). Acquiring genes from phylogenetically distant organisms is known to occur occasionally in the evolution of eukaryotes, including human ancestors. But while those HGT are inferred by a more or less ancient branch in a phylogenetic tree, in *Encephalitozoon*, it is clearly recent and has stemmed from different origins on several occasions. This means that a microsporidium may occasionally coinfect a host cell together with different organisms, and free DNA (from the host or from the coinfecting parasite) may be incorporated into its genome.

Although the genomes are small, the relative differences are high: the 2.9 Mb of *E. cuniculi* represents approximately 25% more DNA than the 2.3 Mb of *E. intestinalis*, the function of which is unclear. Indeed, *E. intestinalis*, *E. cuniculi*, and *E. hellem* all infect vertebrate animals, including humans, with clinical symptoms not easily distinguishable: species identification requires molecular diagnoses. Moreover, those differences lie precisely in the extremes of the chromosomes mentioned earlier, that is, the subtelomeric regions not only differ in composition with respect to the central parts of the chromosomes but also differ from one *Encephalitozoon* sp. to another and may not even have equivalents.

In species that have reduced genomes to the absolute minimum, these heavy differences in DNA content are puzzling. Corradi and Slamovits [29] noticed that the gene loss in *E. intestinalis* compared to gene gain in *E. cuniculi* has no apparent impact on most metabolic functions, since it affects genes of hypothetical proteins with no known function [29]. Clearly, these genes deserve more thorough research, as they might play a relevant role in the biological properties that differentiate the species of *Encephalitozoon*.

19.5 EPIDEMIOLOGY

19.5.1 Human Infection

Information on the true level of human *Encephalitozoon* infections is difficult to discern, as the parasite is often misdiagnosed or overlooked since diagnoses are restricted to specialized laboratories with specific methods for detection and identification. The predominant information on encephalitozoonosis comes from AIDS patients and organ transplant recipients, although, more recently, the interest has also been focused in immunocompetent patients.

E. intestinalis has been described with variable presence in all continents depending on the country. In general, in HIV+ patients, it varies from 0.9% in Switzerland [32] to 12.8% in Russia, where the prevalence was unexpectedly higher than that observed for *Enterocytozoon bieneusi* [33] (Table 19.1). However, the data on immunocompetent individuals remain scarce. It has been associated mainly with the diarrhea of travelers, although serologic studies have

TABLE 19.1
Prevalence of *Encephalitozoon* sp. in HIV+ Patients

N° p/s	N°+	Species	Main Clinical Manifestation	Type of Sample	Country	References
240s	3	E. intestinalis	Diarrhea	Duodenal biopsies	Australia	[134]
68p	5	E. intestinalis	Diarrhea	Biopsy specimens	United States	[32]
48p	0	E. intestinalis	Nondiarrhea	—	—	—
50	2	E. intestinalis	Diarrhea	Feces	Germany	[23]
215p	49	E. intestinalis	—	Feces	Portugal	[135]
1p	1	E. cuniculi	Autopsy	—	Italia	[136]
2p	2	E. intestinalis	Intestinal infection, systemic infection	Sputum, urine, nasal lavage, and conjunctival cells	Sweden	[137]
1p	1	E. intestinalis	Diarrhea, abdominal pain, pansinusitis	—	Germany	[70]
120p	29	E. intestinalis	—	Sera	Slovakia	[138]
103p	4	E. bieneusi/E. intestinalis	Diarrhea	Feces	Colombia	[139]
80p	3	E. intestinalis	Diarrhea	Feces	Ethiopia	[140]
214p	6	E. intestinalis	Diarrhea	Feces	Ethiopia	[141]
110p	43	E. intestinalis	Diarrhea	Feces	EEUU	[74]
	2	E. hellem	—	Feces	—	—
42f	4	E. intestinalis	Diarrhea	—	Vietnam	[142]
51p	3	E. intestinalis	Diarrhea	Feces	Tunisia	[143]
—	1	E. hellem	—	—	—	—
46p	5	E. cuniculi	—	Sera	Russia	[144]
—	4	E. intestinalis	—	—	—	—
—	9	E. hellem	—	—	—	—
331	2	E. intestinalis	Diarrhea	Feces	India	[145]
247	3	E. intestinalis	—	Feces	Malaysia	[146]
119	9	E. intestinalis	—	Feces	Túnez	[143]
159	1	E. cuniculi I	—	Feces	Russia	[33]
	2	E. cuniculi II	—	—	—	—
	1	E. cuniculi	—	—	—	—
	1	E. hellem	—	—	—	—
193	2	E. intestinalis	Diarrhea	Feces	Nigeria	[38]
	1	E. cuniculi	—	—	—	—
50s	1	E. intestinalis	HIV+	Urine	Portugal	[147]
150s	2	E. cuniculi	—	Pulmonary specimens	—	—
363	20.8%	E. intestinalis	Diarrhea	Feces	India	[148]
—	3.8%	—	Nondiarrhea	—	—	—

p, patients; s, samples.

TABLE 19.2
Prevalence of *Encephalitozoon* sp. HIV-Negative Patients

N° p/s	N° +	Species	Characteristics of Patients	Type of Sample	Country	References
300	24	Encephalitozoon	Immunocompetent Dutch donors	Sera	Netherlands	[111]
—	13	—	Immunocompetents Pregnant French women	—	—	—
4	2	E. intestinalis	Travelers with diarrhea	Stool	France	[34]
406	11	Encephalitozoon	Immunocompetents Blood donors	Sera	Spain	[61]
21	21	E. intestinalis	Travelers with diarrhea	Stool	Austria	[35]
311	2	E. intestinalis	Cancer patients	Stool	Malaysia	[149]
—	1	E. hellem	—	—	—	—
360p	1	E. hellem	Risk with occupational exposure to animals	Urine, feces	Czech Republic	[39]
	10	E. hellem 1A	—	—	—	—
	1	E. cuniculi	—	—	—	—
	12	E. cuniculi II	—	—	—	—
265p	84	E. cuniculi II	Czechs, healthy	Stool	Czech Republic	[66]
	4	E. cuniculi I	—	—	—	—
	16	E. hellem 1A	—	—	—	—
117	2	E. intestinalis	Foreigners, healthy	—	—	—
	37	E. cuniculi II	—	—	—	—
	7	E. hellem 1A	—	—	—	—
55	20	E. intestinalis	CD4 < 200/µL	Stool	India	[148]
171	8	E. intestinalis	Chronic diarrhea	Stool	Pakistan	[150]
141	2	—	Healthy	—	—	—
18	—	—	Hepatocellular carcinoma	—	—	—

p, patients; s, samples.

shown high seroprevalence of up to 8% (Table 19.2) [34–36]. Human infections by *E. cuniculi* are traditionally described as less frequent than *E. intestinalis* infections in studies carried out with HIV+ and severe immunosuppression patients in Europe, the United States [37], and recently in Africa [38]. However, studies on immunocompetent individuals have shown a higher than expected prevalence, indicating that human exposure may be more common than previously suspected [39].

Reports of *E. hellem* infections in humans are rare. Approximately, 50 HIV+ patients infected with *E. hellem* have been described in Europe, the United States, and Africa [36]. However, recent studies of immunocompetent individuals have shown a higher prevalence [39].

19.5.2 Source of Infection

19.5.2.1 Zoonotic Transmission

The zoonotic potential of *Encephalitozoon* spp. has been recognized, since animals can be infected with human Microsporidia and animal contact has been described as a risk factor for microsporidiosis [36]. *E. intestinalis* has been described occasionally in a great variety of mammals, such as donkeys, dogs, pigs, cows, goats, cats, and avian hosts; the most likely source of this infection may be other infected humans and animals [22,40]. Molecular differences have not been detected among *E. intestinalis* from humans and animals, implying the absence of transmission barriers between host species [40,41]. On the other hand, molecular studies could not demonstrate intraspecies variability [36].

The zoonotic origin of *E. cuniculi* is evident. Molecular studies have shown that immunocompromised patients are mainly infected with genotypes I (rabbit strain) and III (dog strain) [36]. However, genotype II (mouse strain) was described recently in (1) HIV+ patients and (2) immunocompetent individuals [33].

E. hellem has been described mainly in avian hosts, such as parrots, ostriches, peach-faced lovebirds, and pigeons. The human infection is considered of zoonotic origin since some of these species are in close contact with humans [41–44]. Previous studies demonstrated the high presence of *E. hellem* 1A genotype in park pigeons, which is important since pigeons move together in groups and the flapping of their wings produces notable particle suspension in which earth mixes with debris and their feces [41]. Therefore, Microsporidia spores may penetrate through ocular mucosa, by inhalation, or by accidental ingestion after resting on the hands or toys [41]. This species produces ocular infections, so contaminated fingers can act as vehicles of transmission if they are used to touch the eyes [45].

19.5.2.2 Waterborne Transmission

Waterborne transmission must be considered, since the spores are environmentally resistant and survive in water [46–48]. In the case of *E. cuniculi*, under experimental conditions, the survival of spores has been demonstrated after exposure to different temperatures for months. Moreover, they can survive after freezing and release and acid and alkaline treatments [49]. Also, *E. hellem* and *E. intestinalis* can survive 10°C–30°C for months [50].

Different studies have demonstrated the relation between water and microsporidiosis infection especially from recreational water, drinking water, and wastewater [49] (Table 19.3) from which *E. intestinalis* and *E. hellem* are often found (Table 19.3). However, *E. cuniculi* (Genotypes I and III) has recently been detected in different types of water, in Spain for the first time [48]. Nowadays, the data about the extent of waterborne transmission is scarce and more studies must be done in order to standardize methods for detection and elimination of these pathogens.

TABLE 19.3

***Encephalitozoon* sp. Detection in Environmental Samples**

Type of Sample	Species Detected	Country	References
Tertiary sewage effluent	*E. intestinalis*	United States	[46]
Surface water	*E. intestinalis*		
Groundwater	*E. intestinalis*		
Irrigation water	*E. intestinalis*	United States	[117]
Water surface for public consumption	*E. intestinalis*	United States	[46]
Zebra mussels	*E. intestinalis*	Ireland	[124]
Feces of aquatic birds	*E. hellem*	Poland	[151]
	E. intestinalis		
Constructed subsurface flow and free-surface flow	*E. hellem*	Ireland	[152]
Wastewater	*E. intestinalis*	Ireland	[153]
	E. hellem		
Recreational rivers	*E. intestinalis*	Spain	[47]
Wastewater	*E. hellem*		
Wastewater treatment plants	*E. intestinalis*	Spain	[48]
Rivers	*E. cuniculi*		
	E. intestinalis		
	E. cuniculi		

19.6 INTRASPECIFIC VARIABILITY

It is fundamentally important to distinguish different strains of the same species. First, different strains may be related to different biological properties. Two genotypes may differ in host preferences, rate of dispersal, virulence, and/or drug resistance. Therefore, an accurate classification of strains is clinically relevant. Second, the distinction provides epidemiological information: two isolates that differ in several genetic markers clearly resulted from independent infections, while a perfect match would support a recent common origin. Third, the distribution of the variability geographically and among hosts may indicate reservoirs of the parasites.

PCR primers based on ribosomal sequences are currently used: it is possible to start with generic primers that amplify rDNA of all *Encephalitozoon* spp., followed by nested PCR using particular primers for each of the different species. A good example is the high (and apparently neutral) variability in the internal transcribed spacer (ITS) between the ribosomal genes in *E. bieneusi*: some genotypes circulate quite freely among different vertebrate hosts, while others are more restricted to a specific host and are only occasionally found in a different host species [51]. The ITS of the rDNA is commonly the focus to detect variability. The constant sequences of the ribosomal genes allow the design of efficient primers for PCR, and since the ITS is a noncoding sequence, the mutations can, in principle, be more easily maintained. However, the ITS has some limitations for the species of *Encephalitozoon* due to its short length that is approximately 28, 35, and 45 bp in *E. intestinalis*, *E. cuniculi*, and *E. hellem*, respectively. ITS has 243 bp in *E. bieneusi*.

19.6.1 Intraspecific Variability in *E. cuniculi*

The diversity in the ITS of *E. cuniculi* is due to the tetranucleotide GTTT, which may be repeated two, three, or four times [52–56]. Two repeats (type I) were found in strains isolated from mice and foxes, three repeats (type II) were detected in *E. cuniculi* from rabbits and mice, and four repeats (type III) were from dogs and humans. The ITS is used to classify strains of *E. cuniculi*, although the power of discrimination is limited.

Xiao et al. [15] described polymorphisms for the polar tube protein (PTP) and spore wall protein (SWP), which are useful for genotyping. *PTP1* and *SWP1* have intragenic repeated domains that include single-nucleotide polymorphisms (SNP), so different strains can be distinguished by the number of repeated domains and also in the order of such units. *SWP1* shows another source of variability, as some strains have two different copies of the gene instead of one. They both have the variable repeated domain but differently organized, revealing that such strains may express two SWPs [15]. These authors suggested that one of the two copies might be a pseudogene; however, both of them follow the same pattern, and the few point changes are synonymous, affecting the third base of the triplet. Therefore, either the duplication is very recent and the two copies have not diverged or they are both subject to the same purifying selection and, hence, are most probably expressed.

These markers have shown additional interest because the correspondent proteins are expressed in particular stages of the microsporidian life cycle and are located in the cell surface [18,57,58]. Although, data are still scarce, there is a strong possibility that these polymorphisms are directly involved in the biological properties of each strain. Genotyping with the ITS, *PTP1*, and *SWP1* is not routine for most isolates of *E. cuniculi*, apart from classification into the main types, such as the murine, rabbit, or human strains (the common names usually employed for the catalogued ATCC strains).

Dia et al. [59] analyzed the *InterB* multigenic family, a group of related genes existing in different species of Microsporidia including *E. cuniculi*. These genes, the function of which are unclear, are prone to reorganization and different copies, and variants are present in different chromosomes (unsurprisingly, they are located in the dynamic extremes of the chromosomes). *InterB* appears to be a good candidate for intraspecific characterization of different strains, but this has not been tested. The genome of *E. cuniculi* has a low number of short tandem repeats (STR) compared to other fungal genomes, and the authors in Refs. [11,60] reported variability in some unidentified loci that could be useful as genetic markers, but no further studies have been published concerning new genetic markers. In addition to being useful tools for genotyping, PTP1 and SWP1 (and PTP of *E. hellem*; see later) reveal that the regions with tandemly repeated domains are highly dynamic and prone to reorganizations, contrasting with the stability of the genomic structure in general.

19.6.2 INTRASPECIFIC VARIABILITY IN *E. HELLEM*

A parallel picture emerges in *E. hellem* with some relevant differences. First, genotype analyses have been on strains of human origin (particularly of immunosuppressed patients) and reveal considerable genetic variability, although *E. hellem* has been detected from nonhuman hosts.

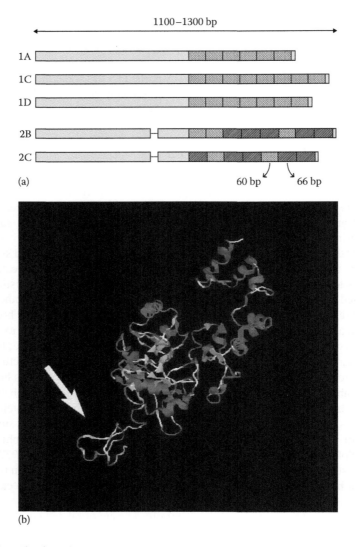

FIGURE 19.5 Organization of the polymorphic PTP gene of *E. hellem*. (a) The genotypes 1A, 1C, and 1D differ in the number of degenerated repeats of a 60 bp unit, while 2B and 2C differ in 60 and 66 bp units; these two also differ from the previous group in one 18 bp indel (dotted line). (b) Predicted model for the PTP of *E. hellem*; the polymorphic repeated domain occupies a peripheral location in the protein structure (arrowed).

Ribosomal DNA ITS is also variable due to insertion–deletions (indels) and SNP [54,61,62]. The PTPs appear to have identical roles in *E. cuniculi* and *E. hellem*. Additionally, they show a variable repeated domain, but, surprisingly, the repeated sequence is quite different in the two species: in *E. hellem* it is closer to the C-terminal, the repeats are of 60 or 66 bp in length, and the variability is much higher than in *E. cuniculi*. The prediction of structures suggests that the repeated domain of the PTP in *E. hellem* is located peripherally in an otherwise globular protein (Figure 19.5). In a comparative in vitro study, Haro et al. [63] observed that *E. hellem* strains differing in their *PTP* genotype showed differences in proliferative efficiency. It became clear that different strains do not behave alike, and it is possible that under in vivo conditions in different hosts, differences may also exist, although genetic differences other than the PTP could also be responsible for the different behavior.

Point substitutions in intergenic spacers (IGS) are also known to differ among strains of *E. hellem* [63]. In *E. hellem* and *E. cuniculi*, the classifications of strains are coherent although the panel of markers remains small. That is, if two *PTP* genotypes are found in different strains having the same ITS genotype, a different ITS type will appear associated to *PTP* types other than those two first mentioned. This hierarchical distribution of variants is expected after an exclusively clonal mode of reproduction and argues against a hypothetical sexual cycle in *Encephalitozoon*.

19.6.3 Intraspecific Variability in *E. intestinalis*

The case of *E. intestinalis* is peculiar even among the other *Encephalitozoon* Microsporidia. As mentioned earlier, it has the smallest genome ever found in a eukaryotic organism, and despite the fact that is has been detected worldwide from a number of hosts, its genetic variability is virtually nil [36,62]. A uniform ITS should not be surprising, since in this species it has only 28 bp, which provides little opportunity for variation. However, Sokolova et al. [33] reported two point differences in the ITS in isolates from Russia. Noncoding sequences cannot be considered irrelevant, as specific short sequences must be recognized particularly in the process of splicing, which separates the two ribosomal RNAs and eliminates the ITS. On the other hand, PTP and SWP that are polymorphic in the two other species are uniform in *E. intestinalis*. More strains of *E. intestinalis* require analysis as not many catalogued strains of *E. intestinalis* from different origins have been analyzed by molecular techniques.

Interestingly, one of the two *SWP* genes described by Hayman et al. [64] in *E. intestinalis* has 50 repeats of 12 or 15 amino acid units, and this is a good candidate for variability studies, but different strains have not been analyzed. Also, *PTP* has not revealed variability. Considering that the genomes of *E. cuniculi* and *E. intestinalis* tend to diverge even more in their coding sequences than in the correspondent IGS, the search for variability should be preferably focused on the genes. Following this notion, we have recently analyzed intragenic repeated sequences and found five variable loci (Galvan AL, Magnet A, Izquierdo F, Fenoy S, Henriques-Gil, N, del Aguila C; unpublished results). The use of this set of markers on isolates of *E. intestinalis* from as many origins as possible should provide a more accurate assessment of the genetic diversity of *E. intestinalis*.

19.7 CLINICAL FEATURES

Clinical features of microsporidiosis due to *Encephalitozoon* sp. are broad and include a variety of manifestations such as intestinal, ocular, renal, and pulmonary disorders, although the immune status of the host is important in the clinical course of the disease [22]. Clinically silent chronic infections usually developed in immunologically competent hosts, although clinical signs could develop after early infections [22,65,66]. In these cases, spore shedding has been described occasionally, but detectable amounts of microsporidial DNA may be excreted via urine that could indicate a capacity of Microsporidia to disseminate in this population [39,67]. This possibility requires further studies.

In immunocompromised hosts, microsporidial infections develop into serious diseases that may result in death. The most common clinical symptoms are chronic diarrhea and malabsorption although systemic diseases can develop. The first cases were recognized in AIDS patients with CD4+ T lymphocytes lower than 100 cells/mL. However, encephalitozoonosis has also been described in transplant recipients [37,48,67,68]. These parasites have been detected in numerous locations producing a great variety of manifestations due to their dissemination capacity. The more important are described as follows:

19.7.1 Microsporidial Diarrhea and Biliary Pathology

E. intestinalis is most implicated in intestinal and biliary pathology among *Encephalitozoon* spp. Chronic diarrhea and extreme weight loss are frequent in immunocompromised patients. This manifestation has been observed predominantly in HIV+ patients with CD4+ lymphocytes of less than 100 cells/mL [1]. Clinical manifestations of chronic diarrhea and weight loss are similar to *E. bieneusi* infection, the more common intestinal Microsporidia. The diarrhea is not bloody, there is no fever, and anorexia is associated with abdominal pain, sickness, and vomiting. Intestinal malabsorption is frequent but unspecific. Finally, a case with a small bowel perforation has been described [69].

Although rare, biliary microsporidiosis may be produced, causing cholangitis and cholecystitis in HIV+ patients. Systemic dissemination to the kidneys and other sites with no luminal connection to the intestine may occur, but intestinal symptoms appear to be predominant as described in HIV+ patients. The disease includes abdominal pain in the right hypochondrium, not always associated to jaundice [70,71].

E. cuniculi and *E. hellem* have only occasionally been found in the intestine of immunocompromised patients [67,72,73]. However, whether the presence of spores in fecal samples in these patients could be related to actual infection or be the result of inhaled or ingested spores passing through the intestine is under discussion [41,74]. It is important to note that spores are shed primarily in urine rather than feces [75].

19.7.2 Microsporidial Hepatitis and Peritonitis

The cases described for these are scarce, and isolates have been attributed to *E. intestinalis* and *E. hellem*. Only two cases have been reported with hepatic involvement: (1) the hepatitis was described at autopsy and produced by *E. cuniculi* [76] and (2) a fulminant hepatic failure was referenced in an HIV+ patient with previous diarrhea lasting 2 months [77]. Peritonitis is an unusual presentation of Microsporidia infections: only one case has been described and produced by *E. cuniculi* in an AIDS patient detected before autopsy [78].

19.7.3 Disseminated Infections

The three *Encephalitozoon* spp. have been described in immunocompromised patients, mainly in HIV+ patients with CD4+ below 50 cells/mL and in transplant recipients with multiple-organ failure, although they can also produce disseminated infections in the immunocompetent [79–82]. The clinical manifestations are variable. Beyond gastrointestinal manifestations, keratoconjunctivitis, bronchiolitis, sinusitis, nephritis, urethritis, cystitis, prostatitis, hepatitis, and peritonitis manifestations are frequent. Other organs can be involved such as the brain and genital tract [22,79,83,84]. *E. intestinalis* has been associated with enteric disease, although it can infect the kidneys, gallbladder, eyes, nasal mucosa, and skin. This microsporidium has been detected in saliva, urine, and bronchoalveolar lavage fluid. *E. hellem* has been described as the cause of pulmonary disease, keratitis/keratoconjunctivitis, kidney disease, and nasal polyps. Finally, *E. cuniculi* can infect the intestine, liver, peritoneum, kidneys, brain, and eyes [71,79,82,84,85].

19.7.4 Ocular Microsporidiosis

Ocular infections due to Microsporidia have been described in immunocompetent and immunocompromised patients and have recently received considerable attention [86]. This infection is mainly associated with *E. hellem* where a superficial punctate keratoconjunctivitis was described [87]. Although, these manifestations have been described in immunocompromised individuals or in contact lens wearers, recent studies indicated that this manifestation can occur in other immunocompetent individuals [88,89]. This infection is clinically characterized by multifocal, coarse, raised epithelial lesions with mild to moderate conjunctivitis, foreign-body sensation, blurred vision, and photophobia. The disease is most likely underdiagnosed or misdiagnosed as adenoviral keratitis [90].

19.8 PATHOLOGY AND IMMUNOLOGY

The pathology of encephalitozoonosis is not well known due to the severity of the disease and depends on the mode of infection, host factors (age and time of infection), the strain of the microsporidium, and, especially, the immune response of the host [79]. In immunocompetent and other healthy persons, the pathogenesis has not been evaluated, although observations in animal models showed the presence of clinical signs of acute stage of infection that disappeared despite the infection persisting [75]. There are no studies on how extensive this infection is. We cannot overlook the possible capacity of Microsporidia to disseminate in immunocompetent hosts since microsporidial DNA has been detected in urine sediments [66].

Studies on the pathology of *Encephalitozoon* spp. have been carried out in immunocompromised and especially AIDS patients. *E. intestinalis* may affect the perinuclear zone of intestinal lamina propria of macrophages, fibroblasts, and endothelial cells, although the parasite infects primarily small intestinal enterocytes [79,91]. It can disseminate to the liver, pancreas, colon, lungs, sinuses, kidneys, and conjunctiva via infected macrophages [79,92]. A severe ulceration of the small bowel with mucosal atrophy, along with acute and chronic inflammation, has been described in the intestine. Bile duct infections are related to papillary stenosis, bile duct dilatation, alithiasic cholecystitis, and sclerosing cholangitis [79].

Alterations produced by dissemination of *E. intestinalis*, *E. hellem*, and *E. cuniculi* external to the intestine are related to the tissue affected. An inflammatory reaction has been observed in the liver, pancreas, lung, kidneys, and eyes. In the case of respiratory tissues, infection has been detected in the lining epithelium of almost the entire length of the tracheobronchial tree [93].

Kidney lesions produce nephritis with an inflammatory infiltration of lymphocytes, plasma cells, macrophages, and neutrophils. Ureters may be affected, producing a granulomatous and lymphoplasmacytic inflammation and ulcerating cystitis, associated with a lymphohistiocytic inflammatory reaction [93]. The presence of parasites within vacuoles of the hepatic cells has been described. However, the pathologic features may vary from a granulomatous necrosis [76] to a nongranulomatous inflammatory reaction [78].

Parasites has been detected in the superficial epithelial of the conjunctiva or cornea in the case of ocular microsporidiosis. An inflammatory infiltration of neutrophils and mononuclear cells has been described [87,94,95].

The immunity against encephalitozoonosis depends mainly on cell-mediated immunity. This fact was suspected from the beginning, when clinicians observed that patients with severe AIDS and with low CD4+ T cells showed a more severe manifestation of infection. Recent studies in animal models have shown that protective immunity is mainly dependent on CD8+ T cells. Moreover, *E. cuniculi* infection induces a strong CD8+ cytotoxic T lymphocyte response, lysing the infected cells via a perforin mechanism and disabling the growth of the parasite. In contrast, the role of CD4+ cells does not appear necessary to control the infection, and the lack of CD4+ T cells does not affect induction of CD 8+ T cells [96,97]. Probably, other cells such as APC or NK and gamma delta provide a source of IFN-gamma that activates the cytotoxic mechanisms of CD 8+ T cells [97]. It is important to note that the presence of proinflammatory cytokines, such as IFN-gamma, tumor necrosis factor alpha, and interleukin-12, are determinants for resistance against infection [45].

In the case of humoral immunity, it is known that the infection activates antibody production and their presence is related to a latent infection. However, the presence alone does not appear to have a protective role, although a study carried out in mice experimentally infected with *E. cuniculi* has shown that antibodies can contribute to resistance [39,79].

19.9 DIAGNOSIS AND DETECTION

Methods developed for *Encephalitozoon* identification have been reviewed [71,79,98,99]. The diagnosis of *Encephalitozoon* is based on the observation of the characteristics spores in the cases of human infection and their detection in environmental samples. However, multiple and specific diagnostic methods are also required due to the small size of the spores. Microsporidia identifications were based on electron microscopic examination to determine the genus and species, although the species level was often not achieved. That method has been superseded by staining methods for light microscopy that is available more frequently in diagnostic laboratories, immunofluorescence procedures, and molecular tests.

The methods described for *Encephalitozoon* identifications require proper handling of the sample. Stool or duodenal aspirate can be analyzed in fresh samples or those conserved in 5%–10% formalin. In disseminated infections, fresh or preserved urine, along with other body fluids, is recommended for analyses. For cell culture and molecular studies, fresh samples should be used [99].

19.9.1 ELECTRON MICROSCOPY

As mentioned, electron microscopy was considered the gold standard to confirm the diagnosis of microsporidiosis and species identification, based in the characteristic of the polar tube. However, this method lacks sensitivity and is not available in most laboratories for clinical diagnosis. Nowadays, it is limited to taxonomic studies.

19.9.2 STAINING METHODS

Staining methods for *Encephalitozoon* detection include modified trichrome stains [100]. The spore wall stains pinkish to red, with the interior of the spore being clear or showing a horizontal or diagonal stripe, which represents the polar tube. This method can be used mainly for stools and body fluids. In the case of tissue sections, a Gram-chromotrope staining is recommended [101]. However, it does not allow species identification.

19.9.3 FLUORESCENT STAINING TECHNIQUES

Fluorescent staining techniques using optical fluorochrome is another method to confirm Microsporidia spores. It is based on the capacity of some chemofluorescent agents to stain chitin, the main component of the wall spores (calcofluor white, or Uvitex 2B) [99], and allows the detection of spores that could be missed by the trichrome stain.

19.9.4 IMMUNOFLUORESCENCE TEST

Species determination can be carried out by the indirect immunofluorescence test that has been improved since the development of monoclonal antibodies against the different species of *Encephalitozoon* spp. [97,102]. It combines high sensitivity and specificity and is comparable to PCR methods—the most used routinely despite high cost and complexity.

19.9.5 PCR Methods

Genetic diagnosis has been employed increasingly because it allows the detection of small numbers of spores or other stages of the parasite that may be undetected under the microscope. Molecular techniques provide species identification; although routinely used in clinical diagnosis, to date there are no commercial kits available, and their use is limited to specialized laboratories. PCR-based methods have been increasingly used in order to improve sensitivity and specificity since the first sequences of genes coding for rRNA of *Encephalitozoon* were described [103,104].

The application of PCR in clinical and environmental samples is subject to limitations such as the viable concentration of spores and the presence of inhibitors of the PCR. The latter is of special interest in the case of stool samples and water from wastewater sources. Furthermore, the sensitivity of this technique is much higher than optical microscopy [105]. Conventional PCR has been used with genes coding for the small subunit of ribosomal RNA, the polar tube gene, the ITS, and the IGS region [62,72,106]. However, conventional PCR has an important risk of contamination, and the parasitic burden cannot be estimated [105]. A review of primers to various rRNA genes used in clinical diagnosis was recently published [97].

In addition to PCR, an oligonucleotide microarray system has been developed for the simultaneous detection of the four human pathogenic microsporidian species from clinical samples [107]. However, this method, described as sensitive and specific, has not been used routinely in clinical diagnosis.

Recently, real-time PCR, a quantitative PCR assay that increases the efficiency and reduces the risk of contamination, is being used in the diagnosis of *Encephalitozoon* spp. [102,105,108]. This technique also allows the simultaneous detection of different species of human Microsporidia (multiplex real-time PCR) [109]. An SYBR Green RT-PCR has been developed for the detection and identification of the species mainly related to human infection, using a single primer set, and discriminates between microsporidian species by employing melting curves [110].

19.9.6 Cell Culture

All the species of *Encephalitozoon* can be cultured in vitro, but cell culture is not a routine method in diagnostic laboratories. Nevertheless, in vitro culture can provide useful information: the culture mass can be used in immunological, biochemical, physiological, and molecular characterization studies and for the development of immunological or molecular reagents. Furthermore, the efficacy of antimicrobial agents can be investigated, and a great variety of cell lines has been used from monkeys, rabbits, and humans [99].

19.9.7 Serology

Serology for the diagnosis of microsporidiosis due to *Encephalitozoon* spp. is not a useful tool. Studies on immunoglobulin G and M anti-*Encephalitozoon* have been carried out using a great variety of methods (such as ELISA and Western blot). However, a correlation between *Encephalitozoon* infection and symptomatic disease could not be established. It is important to note that these infections are mainly produced in immunodeficient patients with a low specific antibody response, and this methodology is limited to immunocompetent infections [61,111,112].

19.9.8 Detection of Microsporidia in Water

With regard to environmental samples, it is important to bear in mind that transmission of Microsporidia spores is a serious issue particularly to some American agencies concerned with the quality of drinking water. Microsporidia have been recognized as category B biodefense agents by the National Institutes of Health [113] and have been included in the Contaminant Candidate

List (CCL) of the U.S. Environmental Protection Agency (EPA) [114]. Spore identification, removal, and inactivation in drinking water are technologically challenging, and human infections are difficult to treat [74].

Different methods have been used to concentrate Microsporidia from water [50,102,115–117]. However, the use of IDEXX Filta-Max® system to concentrate *Cryptosporidium* and *Giardia*, published in method 1623 by the U.S. EPA [118], has been recommended recently as useful to detect Microsporidia in water [47]. Immunomagnetic separation (IMS), followed by real-time PCR, have shown a sensitivity of 78%–90% in ultrapure water [102]. Nevertheless, this method needs to be standardized for the analysis of water of high turbidity and especially regenerated wastewaters from urban, agricultural, industrial, recreative, and environmental use.

19.10 TREATMENT

The treatment of microsporidiosis is an unsolved problem, as it is an intracellular parasite and the spores are resistant. Various drugs have been used with variable effectiveness: albendazole, a benzimidazole that inhibits tubulin polymerization and has antihelminthic and antifungal activity, is the drug of choice in intestinal and disseminated (nonocular) *Encephalitozoon* infections, as demonstrated by the complete elimination of the parasite after 7 days of treatment [22,71,84]. In vitro studies have shown that parasite replication is inhibited and giant meronts, sporonts, and spores had malformed structures. Albendazole acts by preventing the assembly of microtubules, mainly by preventing the formation of the intranuclear spindle, the only known site of microtubule formation in Microsporidia [110]. The recommended dose is 400 mg, orally, twice daily [99]. It was found to be well tolerated, and absorption is improved when associated with fat-rich food [96].

Other benzimidazoles such as fenbendazole or oxfendazoles have shown activity against *E. intestinalis* and are nontoxic in vitro [119]. Although Mebendazole shows absorption problems, it has been used intravenously to access the brain [82]. Nitazoxanide, a broad-spectrum antiparasite, has proven its efficacy on cell cultures of *E. intestinalis*, but has not been used in human treatments as yet, although it is used in other human infections [42]. The drug inhibits pyruvate ferredoxin oxidoreductase of the electron transport system. Fumagillin has shown efficacy against *Encephalitozoon* sp.; however, its efficacy is counterbalanced by its adverse effects [71]. Fumagillin is an antibiotic from *Aspergillus fumigatus* and inhibits Microsporidia replication by blocking the site of action of a cellular metalloprotease, methionine aminopeptidase-2 (MetAP2), and by inhibition of RNA synthesis [119]. In the same way, TNP-470 (an analog of fumagillin) is on trial. The mechanism of action is through binding MetAP2, and it has shown activity in vitro and in vivo against *E. intestinalis* [91]. Fluoroquinolones, antimycotic compounds (pancrastistatin and 7-deoxynarciclasin), and polyamine analogs have also showed in vitro activity and appear to be promising, especially the latter. However, they have not been used in humans [22].

The use of cytokine therapy, especially IFN-gamma, has been evaluated in SCID mice infected with *E. cuniculi* and is suggesting to be potentially useful in CD4+-T-cell-deficient patients [120].

Since the use of highly active antiretroviral therapy (HAART), which is available in developed countries, a reduction in the prevalence of opportunistic infections, including microsporidiosis, has been observed without the use of a specific treatment, simply by induction of a progressive reconstitution of the immune system. However, the different situation in developing countries, where the rapid expansion of AIDS, together with limited access to HAART, has contributed to an increased incidence of this disease, which requires to be appreciated [38].

The treatment of eyes infected with *Encephalitozoon* spp. requires the use of various topical antimicrobial, lubrication, and anti-inflammatory agents. Itraconazole and propamidine isethionate have been used with or without albendazole in the case of *E. hellem* infection, with variable results [90]. Fumidil B, a soluble derivative of fumagillin applied topically, is the treatment of choice [121]. Other treatments have shown good efficacy in ocular microsporidiosis, such as the use of oral albendazole or topical voriconazole monotherapy, a topical antifungal agent [117,122].

19.11 PREVENTION

Since the sources of human infection are uncertain, specific measures of prevention are difficult to establish. Primary infection can occur by inhalation or ingestion of spores from environmental sources or by zoonotic transmission, suggesting the need of adequate hygiene rules, especially in the case of high-risk patients (HIV+ patients, transplant recipients, the elderly, and children) [123]. The measures included all those developed to prevent any infection of fecal–oral transmission through the hands and water and food ingestion [22,123]. Moreover, reconstitution of the immune system by the use of HAART is an important measure of prevention in the case of HIV+ patients since clinical encephalitozoonosis is related to the immune status of the host [71].

Measures related to animal contact should be developed, since one of the sources of human infection is animal to human [22]. Of interest is the description of the presence of *E. hellem* in park pigeons in their relation to humans—mainly children and the elderly—or to the immunosuppressed [41].

The presence of Microsporidia related to waterborne transmission has been documented. Studies in different types of water, including ditch and raw water, water treatment effluents, surface water and groundwater, irrigation and river water, recreational lakes, and wastewater have shown the presence of *Encephalitozoon* spp. [48,74], and their viability after treatment of drinking water and wastewater was demonstrated [124]. Wastewater requires special attention, since it can be discharged into a river or can be used for urban, agricultural, industrial, recreative, and environmental practices and may contribute to the contamination of the environment.

The reduction of the viability and potential infectivity of environmental Microsporidia have been achieved after boiling for at least 5 min or after the use of disinfectants: quaternary ammonium, peroxyacetic acid, 70% ethanol, formaldehyde, phenolic derivate, hydrogen peroxide, chloramine, sodium hydroxide, or amphoteric surfactants have been used to successfully kill *Encephalitozoon* spores [49,125–129]. Other treatments such as ozone, ultraviolet exposure, gamma irradiation, and chlorination are effective in reducing the viability and infectivity of *Encephalitozoon* spp. [130–133].

These observations suggest that the source of infection and modes of transmission require more definitive information in order to take measures to avoid health risks from Microsporidia.

19.12 CONCLUSIONS

The microsporidian genus *Encephalitozoon*, including *E. bieneusi*, has become notorious in the last decades due to the infection of immunocompromised patients (especially after the AIDS pandemia). This fact encouraged an increasing number of studies and the development of more efficient methods for diagnosis. It is now clear that these Microsporidia have a significant prevalence among immunocompetent individuals and also in many domestic and wild animals.

A few loci that vary among strains of a same species are available, but although some epidemiological relationships have been inferred, new markers and thorough studies, involving many different isolates, are still needed for a more accurate view of zoonotic, human-to-human, and environment-to-human pathways of infection, especially since Microsporidia have been included in the last two drinking water CCLs (CCL2 and CCL3) of the U.S. EPA. Finally, *Encephalitozoon* spp. have also challenged a number of concepts relevant not only from a clinical standpoint but also to basic biology: these Microsporidia represent the most extreme reduction among eukaryotes. Cellular organelles and structures seem to have reached the minimum for a peculiar and very efficient intracellular parasitic way of life, and the genome in extremely compacted. However, *Encephalitozoon* species are far from being simple organisms [154]. Moreover, a large part of the gene content are of unknown function, and no phylogenetic relationships could be established between such genes and any other sequence either from Microsporidia or any other organisms.

ACKNOWLEDGMENTS

We are grateful to L. Hamalainen for her help in the preparation of the manuscript. The authors' research on different aspects of *Encephalitozoon* infection included in this chapter was supported, partially by grants from the Fundación Universitaria San Pablo-CEU.

REFERENCES

1. Franzen, C. and A. Muller, Molecular techniques for detection, species differentiation, and phylogenetic analysis of microsporidia. *Clin Microbiol Rev*, 1999. **12**(2): 243–285.
2. Keeling, P.J. and N. Corradi, Shrink it or lose it. Balancing loss function with shrinking genomes in the microsporidia. *Virulence*, 2011. **2**: 67–70.
3. Franzen, C., Microsporidia: A review of 150 years research. *Open Parasitol J*, 2008. **2**(1): 34–35.
4. Lee, S.C. et al., Microsporidia evolved from ancestral sexual fungi. *Curr Biol*, 2008. **18**(21): 1675–1679.
5. Capella-Gutierrez, S., M. Marcet-Houben, and T. Gabaldon, Phylogenomics supports microsporidia as the earliest diverging clade of sequenced fungi. *BMC Biol*, 2012. **10**: 47.
6. Adl, S.M. et al., The revised classification of eukaryotes. *J Eukaryot Microbiol*, 2012. **59**(5): 429–493.
7. Nilsen, F., C. Endresen, and I. Hordvik, Molecular phylogeny of microsporidians with particular reference to species that infect the muscles of fish. *J Eukaryot Microbiol*, 1998. **45**(5): 535–543.
8. Dong, S. et al., Sequence and phylogenetic analysis of SSU rRNA gene of five microsporidia. *Curr Microbiol*, 2010. **60**(1): 30–37.
9. Johny, S. et al., Phylogenetic characterization of *Encephalitozoon romaleae* (Microsporidia) from a grasshopper host: Relationship to *Encephalitozoon* spp. infecting humans. *Infect Genet Evol*, 2009. **9**(2): 189–195.
10. Canning, E.U., *Encephalitozoon lacertae* n. sp., a microsporidian parasite of the lizard *Podarcis muralis*. In *Parasitological Topics*. Society of Protozoologists Special Publication, London, 1981, pp. 57–64.
11. Lee, S.C., L.M. Weiss, and J. Heitman, Generation of genetic diversity in microsporidia via sexual reproduction and horizontal gene transfer. *Commun Integr Biol*, 2009. **2**(5): 414–417.
12. Dyer, P.S., Evolutionary biology: Microsporidia sex—A missing link to fungi. *Curr Biol*, 2008. **18**(21): R1012–R1014.
13. Akopyants, N.S. et al., Demonstration of genetic exchange during cyclical development of *Leishmania* in the sand fly vector. *Science*, 2009. **324**(5924): 265–268.
14. Rougeron, V. et al., "Everything you always wanted to know about sex (but were afraid to ask)" in *Leishmania* after two decades of laboratory and field analyses. *PLoS Pathog*, 2010. **6**(8): e1001004.
15. Xiao, L. et al., Genotyping *Encephalitozoon cuniculi* by multilocus analyses of genes with repetitive sequences. *J Clin Microbiol*, 2001. **39**(6): 2248–2253.
16. Bigliardi, E. and L. Sacchi, Cell biology and invasion of the microsporidia. *Microbes Infect*, 2001. **3**(5): 373–379.
17. Vavra, J. and Larsson, J.I.R., Structure of the microsporidia. In *The Microsporidia and Microsporidiosis*, eds. M. Wittner and L.M. Weiss. Washington, DC: ASM Press, 1999, pp. 7–84.
18. Delbac, F. and V. Polonais, The microsporidian polar tube and its role in invasion. *Subcell Biochem*, 2008. **47**: 208–220.
19. Keeling, P., Five questions about microsporidia. *PLoS Pathog*, 2009. **5**(9): e1000489.
20. Franzen, C., How do microsporidia invade cells? *Folia Parasitol* (Praha), 2005. **52**(1–2): 36–40.
21. Keeling, P.J. and N.M. Fast, Microsporidia: Biology and evolution of highly reduced intracellular parasites. *Ann Rev Microbiol*, 2002. **56**: 93–116.
22. Didier, E.S., Microsporidiosis: An emerging and opportunistic infection in humans and animals. *Acta Trop*, 2005. **94**(1): 61–76.
23. Sobottka, I. et al., Inter- and intra-species karyotype variations among microsporidia of the genus *Encephalitozoon* as determined by pulsed-field gel electrophoresis. *Scand J Infect Dis*, 1999. **31**(6): 555–558.

24. Katinka, M.D. et al., Genome sequence and gene compaction of the eukaryote parasite *Encephalitozoon cuniculi*. *Nature*, 2001. **414**(6862): 450–453.
25. Corradi, N. et al., The complete sequence of the smallest known nuclear genome from the microsporidian *Encephalitozoon intestinalis*. *Nat Commun*, 2010. **1**: 77.
26. Brugere, J.F. et al., *Encephalitozoon cuniculi* (Microspora) genome: Physical map and evidence for telomere-associated rDNA units on all chromosomes. *Nucleic Acids Res*, 2000. **28**(10): 2026–2033.
27. Peyret, P. et al., Sequence and analysis of chromosome I of the amitochondriate intracellular parasite *Encephalitozoon cuniculi* (Microspora). *Genome Res*, 2001. **11**(2): 198–207.
28. Pombert, J.F. et al., Gain and loss of multiple functionally related, horizontally transferred genes in the reduced genomes of two microsporidian parasites. *Proc Natl Acad Sci USA*, 2012. **109**(31): 12638–12643.
29. Corradi, N. and C.H. Slamovits, The intriguing nature of microsporidian genomes. *Brief Funct Genomics*, 2011. **10**(3): 115–124.
30. Williams, B.A. et al., A high frequency of overlapping gene expression in compacted eukaryotic genomes. *Proc Natl Acad Sci USA*, 2005. **102**(31): 10936–10941.
31. Selman, M. and N. Corradi, Microsporidia: Horizontal gene transfers in vicious parasites. *Mob Genet Elements*, 2011. **1**(4): 251–255.
32. Coyle, C.M. et al., Prevalence of microsporidiosis due to *Enterocytozoon bieneusi* and *Encephalitozoon* (Septata) *intestinalis* among patients with AIDS-related diarrhea: Determination by polymerase chain reaction to the microsporidian small-subunit rRNA gene. *Clin Infect Dis*, 1996. **23**(5): 1002–1006.
33. Sokolova, O.I. et al., Emerging microsporidian infections in Russian HIV-infected patients. *J Clin Microbiol*, 2011. **49**(6): 2102–2108.
34. Raynaud, L. et al., Identification of *Encephalitozoon intestinalis* in travelers with chronic diarrhea by specific PCR amplification. *J Clin Microbiol*, 1998. **36**(1): 37–40.
35. Wichro, E. et al., Microsporidiosis in travel-associated chronic diarrhea in immune-competent patients. *Am J Trop Med Hyg*, 2005. **73**(2): 285–287.
36. Mathis, A., R. Weber, and P. Deplazes, Zoonotic potential of the microsporidia. *Clin Microbiol Rev*, 2005. **18**(3): 423–445.
37. Talabani, H. et al., Disseminated infection with a new genovar of *Encephalitozoon cuniculi* in a renal transplant recipient. *J Clin Microbiol*, 2010. **48**(7): 2651–2653.
38. Ojuromi, O.T. et al., Identification and characterization of microsporidia from fecal samples of HIV-positive patients from Lagos, Nigeria. *PLoS One*, 2012. **7**(4): e35239.
39. Sak, B., et al. Unapparent microsporidial infection among immunocompetent humans in the Czech Republic. *J Clin Microbiol*, 2011. **49**(3): 1064–1070.
40. Velasquez, J.N. et al., First case report of infection caused by *Encephalitozoon intestinalis* in a domestic cat and a patient with AIDS. *Vet Parasitol*, 2012. **190**(3–4): 583–586.
41. Haro, M. et al., First detection and genotyping of human-associated microsporidia in pigeons from urban parks. *Appl Environ Microbiol*, 2005. **71**(6): 3153–3157.
42. Black, S.S. et al., *Encephalitozoon hellem* in budgerigars (*Melopsittacus undulatus*). *Vet Pathol*, 1997. **34**(3): 189–198.
43. Gray, M.L., M. Puette, and K.S. Latimer, Microsporidiosis in a young ostrich (*Struthio camelus*). *Avian Dis*, 1998. **42**(4): 832–836.
44. Snowden, K.F., K. Logan, and D.N. Phalen, Isolation and characterization of an avian isolate of *Encephalitozoon hellem*. *Parasitology*, 2000. **121**(Pt 1): 9–14.
45. Didier, E.S. and L.M. Weiss, Microsporidiosis: Current status. *Curr Opin Infect Dis*, 2006. **19**(5): 485–492.
46. Dowd, S.E., C.P. Gerba, and I.L. Pepper, Confirmation of the human-pathogenic microsporidia *Enterocytozoon bieneusi*, *Encephalitozoon intestinalis*, and *Vittaforma corneae* in water. *Appl Environ Microbiol*, 1998. **64**(9): 3332–3335.
47. Izquierdo, F. et al., Detection of microsporidia in drinking water, wastewater and recreational rivers. *Water Res*, 2011. **45**(16): 4837–4843.
48. Galvan, A.L. et al., Molecular characterization of human-pathogenic microsporidia and *Cyclospora cayetanensis* isolated from various water sources in Spain: A year-long longitudinal study. *Appl Environ Microbiol*, 2013. **79**(2): 449–459.
49. Didier, E.S. et al., Epidemiology of microsporidiosis: Sources and modes of transmission. *Vet Parasitol*, 2004. **126**(1–2): 145–166.
50. Li, X. et al., Infectivity of microsporidia spores stored in water at environmental temperatures. *J Parasitol*, 2003. **89**(1): 185–188.

51. Henriques-Gil, N. et al., Phylogenetic approach to the variability of the microsporidian *Enterocytozoon bieneusi* and its implications for inter- and intrahost transmission. *Appl Environ Microbiol*, 2010. **76**(10): 3333–3342.
52. Deplazes, P. et al., Molecular epidemiology of *Encephalitozoon cuniculi* and first detection of *Enterocytozoon bieneusi* in faecal samples of pigs. *J Eukaryot Microbiol*, 1996. **43**(5): 93S.
53. Didier, E.S., Reactive nitrogen intermediates implicated in the inhibition of *Encephalitozoon cuniculi* (phylum microspora) replication in murine peritoneal macrophages. *Parasite Immunol*, 1995. **17**(8): 405–412.
54. Mathis, A. et al., Isolates of *Encephalitozoon cuniculi* from farmed blue foxes (*Alopex lagopus*) from Norway differ from isolates from Swiss domestic rabbits (*Oryctolagus cuniculus*). *Parasitol Res*, 1996. **82**(8): 727–730.
55. Vossbrinck, C.R., M.D. Baker, and E.S. Didier, Comparative rDNA analysis of Microsporidia including AIDS related species. *J Eukaryot Microbiol*, 1996. **43**(5): 110S.
56. Rinder, H., S. Katzwinkel-Wladarsch, A. Thomschke, and T. Löscher, Strain differentiation in microsporidia. *Tokai J Exp Clin Med*, 1999. **23**: 433–437.
57. Bohne, W. et al., Developmental expression of a tandemly repeated, glycine- and serine-rich spore wall protein in the microsporidian pathogen *Encephalitozoon cuniculi*. *Infect Immun*, 2000. **68**(4): 2268–2275.
58. Bouzahzah, B. et al., Interactions of *Encephalitozoon cuniculi* polar tube proteins. *Infect Immun*, 2010. **78**(6): 2745–2753.
59. Dia, N. et al., InterB multigenic family, a gene repertoire associated with subterminal chromosome regions of *Encephalitozoon cuniculi* and conserved in several human-infecting microsporidian species. *Curr Genet*, 2007. **51**(3): 171–186.
60. Karaoglu, H., C.M. Lee, and W. Meyer, Survey of simple sequence repeats in completed fungal genomes. *Mol Biol Evol*, 2005. **22**(3): 639–649.
61. del Aguila, C. et al., Seroprevalence of anti-Encephalitozoon antibodies in Spanish immunocompetent subjects. *J Eukaryot Microbiol*, 2001. Suppl.: 75S–78S.
62. Haro, M. et al., Intraspecies genotype variability of the microsporidian parasite *Encephalitozoon hellem*. *J Clin Microbiol*, 2003. **41**(9): 4166–4171.
63. Haro, M. et al., Variability in infection efficiency in vitro of different strains of the microsporidian *Encephalitozoon hellem*. *J Eukaryot Microbiol*, 2006. **53**(1): 46–48.
64. Hayman, J.R. et al., Developmental expression of two spore wall proteins during maturation of the microsporidian *Encephalitozoon intestinalis*. *Infect Immun*, 2001. **69**(11): 7057–7066.
65. Snowden, K.F. et al., Animal models of human microsporidial infections. *Lab Anim Sci*, 1998. **48**(6): 589–592.
66. Sak, B. et al., Latent microsporidial infection in immunocompetent individuals—A longitudinal study. *PLoS Negl Trop Dis*, 2011. **5**(5): e1162.
67. García, L.S., Intestinal protozoa (Coccidia and Microsporidia) and algae. In *Diagnostic Medical Parasitology*. L.S. García, Washington, DC: ASM Press, 2007, pp. 57–101.
68. George, B. et al., Disseminated microsporidiosis with *Encephalitozoon* species in a renal transplant recipient. *Nephrology* (Carlton), 2012. **17**(Suppl. 1): 5–8.
69. Soule, J.B. et al., A patient with acquired immunodeficiency syndrome and untreated *Encephalitozoon* (Septata) *intestinalis* microsporidiosis leading to small bowel perforation. Response to albendazole. *Arch Pathol Lab Med*, 1997. **121**(8): 880–887.
70. Lippert, U., J. Schottelius, and C. Menegold, Disseminated microsporidiosis (*Encephalitozoon intestinalis*) in a patient with HIV infection. *Dtsch Med Wochenschr*, 2003. **22**(128): 1769–1772.
71. Anane, S. and H. Attouchi, Microsporidiosis: Epidemiology, clinical data and therapy. *Gastroenterol Clin Biol*, 2010. **34**(8–9): 450–464.
72. del Aguila, C. et al., In vitro culture, ultrastructure, antigenic, and molecular characterization of *Encephalitozoon cuniculi* isolated from urine and sputum samples from a Spanish patient with AIDS. *J Clin Microbiol*, 2001. **39**(3): 1105–1108.
73. Chabchoub, N. et al., Genetic identification of intestinal microsporidia species in immunocompromised patients in Tunisia. *Am J Trop Med Hyg*, 2009. **80**(1): 24–27.
74. Graczyk, T.K. et al., Retrospective species identification of microsporidian spores in diarrheic fecal samples from human immunodeficiency virus/AIDS patients by multiplexed fluorescence in situ hybridization. *J Clin Microbiol*, 2007. **45**(4): 1255–1260.
75. Didier, E.S. and L.M. Weiss, Microsporidiosis: Not just in AIDS patients. *Curr Opin Infect Dis*, 2011. **24**(5): 490–495.

76. Terada, S. et al., Microsporidan hepatitis in the acquired immunodeficiency syndrome. *Ann Intern Med*, 1987. **107**(1): 61–62.
77. Sheth, S.G. et al., Fulminant hepatic failure caused by microsporidial infection in a patient with AIDS. *AIDS*, 1997. **11**(4): 553–554.
78. Zender, H.O. et al., A case of *Encephalitozoon cuniculi* peritonitis in a patient with AIDS. *Am J Clin Pathol*, 1989. **92**(3): 352–356.
79. Weber, R. et al., Human microsporidial infections. *Clin Microbiol Rev*, 1994. **7**(4): 426–461.
80. Cali, A., D.P. Kotler, and J.M. Orenstein, Septata intestinalis N. G., N. Sp., an intestinal microsporidian associated with chronic diarrhea and dissemination in AIDS patients. *J Eukaryot Microbiol*, 1993. **40**(1): 101–112.
81. Galvan, A.L. et al., First cases of microsporidiosis in transplant recipients in Spain and review of the literature. *J Clin Microbiol*, 2011. **49**(4): 1301–1306.
82. Ditrich, O. et al., *Encephalitozoon cuniculi* genotype I as a causative agent of brain abscess in an immunocompetent patient. *J Clin Microbiol*, 2011. **49**(7): 2769–2771.
83. Mohindra, A.R. et al., Disseminated microsporidiosis in a renal transplant recipient. *Transpl Infect Dis*, 2002. **4**(2): 102–107.
84. Robinson, T. et al., Drug-resistant genotypes and multi-clonality in *Plasmodium falciparum* analysed by direct genome sequencing from peripheral blood of malaria patients. *PLoS One*, 2011. **6**(8): e23204.
85. Kotler, D.P. and J.M. Orenstein, Clinical syndromes associated with microsporidiosis. *Adv Parasitol*, 1998. **40**: 321–349.
86. Loh, R.S. et al., Emerging prevalence of microsporidial keratitis in Singapore: Epidemiology, clinical features, and management. *Ophthalmology*, 2009. **116**(12): 2348–2353.
87. Didier, E.S. et al., Isolation and characterization of a new human microsporidian, *Encephalitozoon hellem* (n. sp.), from three AIDS patients with keratoconjunctivitis. *J Infect Dis*, 1991. **163**(3): 617–621.
88. Chan, C.M. et al., Microsporidial keratoconjunctivitis in healthy individuals: A case series. *Ophthalmology*, 2003. **110**(7): 1420–1425.
89. Sridhar, M.S. and S. Sharma, Microsporidial keratoconjunctivitis in a HIV-seronegative patient treated with debridement and oral itraconazole. *Am J Ophthalmol*, 2003. **136**(4): 745–746.
90. Das, S. et al., Diagnosis, clinical features and treatment outcome of microsporidial keratoconjunctivitis. *Br J Ophthalmol*, 2012. **96**(6): 793–795.
91. Conteas, C.N. et al., Therapy for human gastrointestinal microsporidiosis. *Am J Trop Med Hyg*, 2000. **63**(3–4): 121–127.
92. Orenstein, J.M., D.T. Dieterich, and D.P. Kotler, Systemic dissemination by a newly recognized intestinal microsporidia species in AIDS. *AIDS*, 1992. **6**(10): 1143–1150.
93. Schwartz, D.A. et al., Disseminated microsporidiosis (*Encephalitozoon hellem*) and acquired immunodeficiency syndrome. Autopsy evidence for respiratory acquisition. *Arch Pathol Lab Med*, 1992. **116**(6): 660–668.
94. Cali, A. et al., Corneal microsporidioses: Characterization and identification. *J Protozool*, 1991. **38**(6): 215S–217S.
95. Joseph, J. et al., Histopathological evaluation of ocular microsporidiosis by different stains. *BMC Clin Pathol*, 2006. **6**: 6.
96. Moretto, M. et al., Lack of CD4(+) T cells does not affect induction of CD8(+) T-cell immunity against *Encephalitozoon cuniculi* infection. *Infect Immun*, 2000. **68**(11): 6223–6232.
97. Ghosh, K. and L.M. Weiss, Molecular diagnostic tests for microsporidia. *Interdiscip Perspect Infect Dis*, 2009. **2009**: 926521.
98. Franzen, C. and A. Muller, Microsporidiosis: Human diseases and diagnosis. *Microbes Infect*, 2001. **3**(5): 389–400.
99. Garcia, L.S., Laboratory identification of the microsporidia. *J Clin Microbiol*, 2002. **40**(6): 1892–1901.
100. Weber, R. et al., Improved light-microscopical detection of microsporidia spores in stool and duodenal aspirates. The Enteric Opportunistic Infections Working Group. *N Engl J Med*, 1992. **326**(3): 161–166.
101. Visvesvara, G.S. and L.S. Garcia, Culture of protozoan parasites. *Clin Microbiol Rev*, 2002. **15**(3): 327–328.
102. Hoffman, R.M. et al., Development of a method for the detection of waterborne microsporidia. *J Microbiol Methods*, 2007. **70**(2): 312–318.
103. Zhu, X. et al., Nucleotide sequence of the small subunit rRNA of *Septata intestinalis*. *Nucleic Acids Res*, 1993. **21**(20): 4846.

104. Visvesvara, G.S. et al., Polyclonal and monoclonal antibody and PCR-amplified small-subunit rRNA identification of a microsporidian, *Encephalitozoon hellem*, isolated from an AIDS patient with disseminated infection. *J Clin Microbiol*, 1994. **32**(11): 2760–2768.
105. Menotti, J. et al., Development of a real-time PCR assay for quantitative detection of *Encephalitozoon intestinalis* DNA. *J Clin Microbiol*, 2003. **41**(4): 1410–1413.
106. Rossi, P. et al., Identification of a human isolate of *Encephalitozoon cuniculi* type I from Italy. *Int J Parasitol*, 1998. **28**(9): 1361–1366.
107. Wang, Z., P.A. Orlandi, and D.A. Stenger, Simultaneous detection of four human pathogenic microsporidian species from clinical samples by oligonucleotide microarray. *J Clin Microbiol*, 2005. **43**(8): 4121–4128.
108. Wolk, D.M. et al., Real-time PCR method for detection of *Encephalitozoon intestinalis* from stool specimens. *J Clin Microbiol*, 2002. **40**(11): 3922–3928.
109. Verweij, J.J. et al., Multiplex detection of *Enterocytozoon bieneusi* and *Encephalitozoon* spp. in fecal samples using real-time PCR. *Diagn Microbiol Infect Dis*, 2007. **57**(2): 163–167.
110. Polley, S.D. et al., Detection and species identification of microsporidial infections using SYBR Green real-time PCR. *J Med Microbiol*, 2011. **60**(Pt 4): 459–466.
111. van Gool, T. et al., High seroprevalence of *Encephalitozoon* species in immunocompetent subjects. *J Infect Dis*, 1997. **175**(4): 1020–1024.
112. Bouladoux, N., S. Biligui, and I. Desportes-Livage, A new monoclonal antibody enzyme-linked immunosorbent assay to measure in vitro multiplication of the microsporidium *Encephalitozoon intestinalis*. *J Microbiol Methods*, 2003. **53**(3): 377–385.
113. Nwachcuku, N. and C.P. Gerba, Emerging waterborne pathogens: Can we kill them all? *Curr Opin Biotechnol*, 2004. **15**(3): 175–180.
114. US Environmental Protection Agency, Announcement of the drinking water contaminant candidate list: Notice. *Fed Regist*, 1998. **63**, 10272–10287.
115. Sparfel, J.M. et al., Detection of microsporidia and identification of Enterocytozoon bieneusi in surface water by filtration followed by specific PCR. *J Eukaryot Microbiol*, 1997. **44**(6): 78S.
116. Fournier, S. et al., Detection of microsporidia in surface water: A one-year follow-up study. *FEMS Immunol Med Microbiol*, 2000. **29**(2): 95–100.
117. Thurston-Enriquez, J.A. et al., Detection of protozoan parasites and microsporidia in irrigation waters used for crop production. *J Food Prot*, 2002. **65**(2): 378–382.
118. US Environmental Protection Agency, *Method 1623: Cryptosporidium and Giardia in Water by Filtration/IMA/Fa*. 2005. (http://www.epa.gov/microbes/).
119. Schetinger, M.R. et al., New benzodiazepines alter acetylcholinesterase and ATPDase activities. *Neurochem Res*, 2000. **25**(7): 949–955.
120. Salat, J. et al., Efficacy of gamma interferon and specific antibody for treatment of microsporidiosis caused by *Encephalitozoon cuniculi* in SCID mice. *Antimicrob Agents Chemother*, 2008. **52**(6): 2169–2174.
121. Microsporidiosis DPDx-CDC Parasitology Diagnostic Website (http: //www.cdc.gov/dpdx/microsporidiosis/index.html). Cited 2013. dpd.cdc.gov/dpx.
122. Khandelwal, S.S. et al., Treatment of microsporidia keratitis with topical voriconazole monotherapy. *Arch Ophthalmol*, 2011. **129**(4): 509–510.
123. Garcia, L.S., Intestinal protozoa (Coccidia and Microsporidia) and algae. In *Diagnostic Medical Parasitology*, 5th edn., ed. L.S. García, Vols. 355–403. Washington, DC: ASM Press, 2007, pp. 57–101.
124. Graczyk, T.K. et al., Human waterborne parasites in zebra mussels (*Dreissena polymorpha*) from the Shannon River drainage area, Ireland. *Parasitol Res*, 2004. **93**(5): 385–391.
125. Shadduck, J.A. and M.B. Polley, Some factors influencing the in vitro infectivity and replication of *Encephalitozoon cuniculi*. *J Protozool*, 1978. **25**(4): 491–496.
126. Waller, T., Sensitivity of *Encephalitozoon cuniculi* to various temperatures, disinfectants and drugs. *Lab Anim*, 1979. **13**(3): 227–230.
127. Santillana-Hayat, M. et al., Effects of chemical and physical agents on viability and infectivity of *Encephalitozoon intestinalis* determined by cell culture and flow cytometry. *Antimicrob Agents Chemother*, 2002. **46**(6): 2049–2051.
128. Jordan, C.N., J.A. Dicristina, and D.S. Lindsay, Activity of bleach, ethanol and two commercial disinfectants against spores of *Encephalitozoon cuniculi*. *Vet Parasitol*, 2006. **136**(3–4): 343–346.
129. Ortega, Y.R. et al., Efficacy of a sanitizer and disinfectants to inactivate *Encephalitozoon intestinalis* spores. *J Food Prot*, 2007. **70**(3): 681–684.
130. Khalifa, A.M., M.M. El Temsahy, and I.F. Abou El Naga, Effect of ozone on the viability of some protozoa in drinking water. *J Egypt Soc Parasitol*, 2001. **31**(2): 603–616.

131. Li, X. et al., Effects of gamma radiation on viability of *Encephalitozoon* spores. *J Parasitol*, 2002. **88**(4): 812–813.
132. Johnson, C.H. et al., Chlorine inactivation of spores of *Encephalitozoon* spp. *Appl Environ Microbiol*, 2003. **69**(2): 1325–1326.
133. John, D.E. et al., Chlorine and ozone disinfection of *Encephalitozoon intestinalis* spores. *Water Res*, 2005. **39**(11): 2369–2375.
134. Field, A.S. et al., Microsporidia in the small intestine of HIV-infected patients. A new diagnostic technique and a new species. *Med J Aust*, 1993. **158**(6): 390–394.
135. Ferreira, F.M. et al., Intestinal microsporidiosis: A current infection in HIV-seropositive patients in Portugal. *Microbes Infect*, 2001. **3**(12): 1015–1019.
136. Tosoni, A. et al., Disseminated microsporidiosis caused by *Encephalitozoon cuniculi* III (dog type) in an Italian AIDS patient: A retrospective study. *Mod Pathol*, 2002. **15**(5): 577–583.
137. Svedhem, V. et al., Disseminated infection with *Encephalitozoon intestinalis* in AIDS patients: Report of 2 cases. *Scand J Infect Dis*, 2002. **34**(9): 703–705.
138. Halanova, M. et al., Detection of anti-microsporidial antibodies in patients with secondary immunodeficiency. *Epidemiol Microbiol Imunol*, 2004. **53**(2): 78–80.
139. Botero, J.H. et al., Frequency of intestinal microsporidian infections in HIV-positive patients, as diagnosis by quick hot Gram chromotrope staining and PCR. *Biomedica*, 2004. **24**(4): 375–384.
140. Endeshaw, T.K.A., J.J. Verweii, D. Wolday, A. Zewide, K. Tsige, Y. Abraham, T. Messele, A.M. Polderman, and B. Petros, Detection of intestinal microsporidiosis in diarrhoeal patients infected with the human immunodeficiency virus (HIV-1) using PCR and Uvitex-2B stain. *Ethiop Med J*, 2005. **43**(2): 97–101.
141. Endeshaw, T.K.A., J.J. Verweii, A. Zewide, K. Tsige, Y. Abraham, D. Wolday, T. Woldemichael, T. Messele, A.M. Polderman, and B. Petros, Intestinal microsporidiosis in diarrheal patients infected with human immunodeficiency virus-1 in Addis Ababa, Ethiopia. *Jpn J Infect Dis*, 2006. **59**(5): 306–310.
142. Espern, A. et al., Molecular study of microsporidiosis due to *Enterocytozoon bieneusi* and *Encephalitozoon intestinalis* among human immunodeficiency virus-infected patients from two geographical areas: Niamey, Niger, and Hanoi, Vietnam. *J Clin Microbiol*, 2007. **45**(9): 2999–3002.
143. Chabchoub, N. et al., Contribution of PCR for detection and identification of intestinal microsporidia in HIV-infected patients. *Pathol Biol* (Paris), 2012. **60**(2): 91–94.
144. Kucerova, Z. et al., Microsporidiosis and cryptosporidiosis in HIV/AIDS patients in St. Petersburg, Russia: Serological identification of microsporidia and *Cryptosporidium parvum* in sera samples from HIV/AIDS patients. *AIDS Res Hum Retroviruses*, 2011. **27**(1): 13–15.
145. Kulkarni, S.V.K.R., S.S. Sane, P.S. Padmawar, V.A. Kale, M.R. Thakar, S.M. Mehendale, and A.R. Risbud, Opportunistic parasitic infections in HIV/AIDS patients with diarrhoea by the level of immunosuppression. *Indian J Med Res*, 2011. **130**(1): 63–66.
146. Lono, A., S. Kumar, and T.T. Chye, Detection of microsporidia in local HIV-positive population in Malaysia. *Trans R Soc Trop Med Hyg*, 2011. **105**(7): 409–413.
147. Lobo, M.L. et al., Microsporidia as emerging pathogens and the implication for public health: A 10-year study on HIV-positive and -negative patients. *Int J Parasitol*, 2012. **42**(2): 197–205.
148. Saigal, K. et al., Intestinal microsporidiosis in India: A two year study. *Parasitol Int*, 2013. **62**(1): 53–56.
149. Lono, A.R., S. Kumar, and T.T. Chye, Incidence of microsporidia in cancer patients. *J Gastrointest Cancer*, 2008. **39**(1–4): 124–129.
150. Yakoob, J. et al., Microsporidial infections due to *Encephalitozoon intestinalis* in non-HIV-infected patients with chronic diarrhoea. *Epidemiol Infect*, 2012. **140**(10): 1773–1779.
151. Slodkowicz-Kowalska, A. et al., Microsporidian species known to infect humans are present in aquatic birds: Implications for transmission via water? *Appl Environ Microbiol*, 2006. **72**(7): 4540–4544.
152. Graczyk, T.K. et al., Human zoonotic enteropathogens in a constructed free-surface flow wetland. *Parasitol Res*, 2009. **105**(2): 423–428.
153. Cheng, H.W. et al., Municipal wastewater treatment plants as removal systems and environmental sources of human-virulent microsporidian spores. *Parasitol Res*, 2011. **109**(3): 595–603.
154. Keeling, P.J. and C.H. Slamovits, Simplicity and complexity of microsporidian genomes. *Eukaryotic Cell*, 2004. **3**(6): 1363–1369.

20 Enterocytozoon

Olga Matos and Maria Luisa Lobo

CONTENTS

- 20.1 Introduction 293
- 20.2 Taxonomy, Morphology, and Life Cycle of *Enterocytozoon bieneusi* 294
  - 20.2.1 Taxonomy 294
    - 20.2.1.1 Taxonomic Position of Microsporidia from Discovery to Present 295
    - 20.2.1.2 Biology and Life Cycle 296
- 20.3 Epidemiology of Human Microsporidiosis due to *Enterocytozoon bieneusi* 298
  - 20.3.1 Distribution of *Enterocytozoon bieneusi* in Humans 298
  - 20.3.2 Prevalence of *Enterocytozoon bieneusi* in Humans 299
  - 20.3.3 Sources of Infection and Modes of Transmission 299
    - 20.3.3.1 Direct Person-to-Person Transmission 299
    - 20.3.3.2 Zoonotic Transmission 300
    - 20.3.3.3 Waterborne Transmission 300
    - 20.3.3.4 Foodborne Transmission 303
    - 20.3.3.5 Airborne Transmission 303
  - 20.3.4 Molecular Epidemiology of *Enterocytozoon bieneusi* in Humans 303
  - 20.3.5 Prevention 306
- 20.4 Pathogenesis and Clinical Spectra of *Enterocytozoon bieneusi* in Humans 307
- 20.5 Diagnosis of Microsporidiosis by *Enterocytozoon bieneusi* 308
  - 20.5.1 Specimen Collection 308
  - 20.5.2 Transmission Electron Microscopy 308
  - 20.5.3 Histological Diagnosis 308
  - 20.5.4 Cytological Diagnosis 309
  - 20.5.5 Serologic Tests 310
  - 20.5.6 Cell Culture 310
  - 20.5.7 Molecular Methods 310
- 20.6 Therapy of *Enterocytozoon bieneusi* Disease in Humans 311
- 20.7 Conclusion 312
- References 312

20.1 INTRODUCTION

Microsporidia are a group of obligate intracellular eukaryotic parasites, ubiquitous in nature, and considered as emerging pathogens in humans and animals. Over the past three decades, microsporidia have risen from obscure organisms to well-recognized human pathogens. Several genera and species of microsporidia were found in humans, and diagnosis and clinical management of microsporidiosis cases have improved significantly. Despite progress, the epidemiology of diseases caused by microsporidia is still poorly known.

The first record of microsporidia comes from the first half of the nineteenth century, when a severe epidemic, known as pébrine (which in southern France means disease of pepper), affected the sericulture industry in Europe, particularly in France and Italy.[1,2] The agent of this epidemic was designated *Nosema bombycis* by microbiologist Karl Wilhelm von Nägeli (1857) of Switzerland.[3] In mammals, microsporidia were registered for the first time in 1922, when Wright and Craighead described the infection in laboratory rabbits with nervous signs,[4] and the first case of the microsporidial infection in humans was only registered in 1959. A case of microsporidiosis was described in a boy of 9 with neurological manifestations.[5] The agent involved was initially considered as belonging to the genus *Encephalitozoon*. Later, this child's infection was attributed to *Encephalitozoon cuniculi*.[1] Now more than 1200 species of microsporidia are described in the literature, belonging to 160 genera.[1,6] Most of these species infect arthropods and fish, and only 14 species have been identified as pathogenic for humans.[1,2] The pathogenic species for humans are *Enterocytozoon bieneusi*, the most prevalent, *Encephalitozoon intestinalis*, the second most prevalent, *Encephalitozoon hellem*, *E. cuniculi*, *Anncaliia algerae*, *Anncaliia connori*, *Anncaliia vesicularum*, *Microsporidium africanum*, *Microsporidium ceylonensis*, *Nosema ocularum*, *Pleistophora ronneafiel*, *Trachipleistophora antropophtera*, *Trachipleistophora hominis*, and *Vittaforma corneae*.[1,7–10]

Until the beginning of the 1980s, the records of microsporidiosis in humans were occasional and little valued by the scientific community. A new vision of these organisms, as emerging agents of opportunistic infections in humans, began to gain ground after the first records of the acquired immunodeficiency syndrome (AIDS) epidemic in 1981. In 1985 a new microsporidian, *E. bieneusi*, was described for the first time in AIDS patients, associated with persistent diarrhea and systemic infection.[11] Several hundred patients with chronic diarrhea attributed to this organism have been reported since. Cases of intestinal microsporidiosis due to *E. bieneusi* have also been increasingly reported in other immunocompromised individuals, such as organ transplant recipients, and in travelers, children, and the elderly.[12–16] Hepatobiliary and pulmonary involvement have also been observed in immunocompromised persons.[17,18] Some reports in humans suggested that this species can produce asymptomatic infections in immunocompromised and immunocompetent individuals.[19–23] In addition, *E. bieneusi* has been commonly identified in animals, especially mammals, and in water and food, raising public health concerns about its zoonotic, water, and foodborne transmission.[23–37]

20.2 TAXONOMY, MORPHOLOGY, AND LIFE CYCLE OF *ENTEROCYTOZOON BIENEUSI*

20.2.1 TAXONOMY

The phylum Microsporidia consists of a diverse group of obligate intracellular eukaryotic parasites that infects a broad range of invertebrates and humans. They were recognized as a causative agent for many hosts including insects, fish, rodents, and primates. The genus *Enterocytozoon* belongs to the family Enterocytozoonidae, suborder Apansporoblastina, phylum Microsporidia, and kingdom Fungi. The family Enterocytozoonidae includes five genera: *Desmozoon*, *Enterocytozoon*, *Microsporidium*, *Nucleospora*, and *Paranucleospora* and some unclassified Enterocytozoonidae (including Enterocytozoonidae gen. sp.). The genus *Enterocytozoon* consists of three recognized species: *Enterocytozoon bieneusi*, *Enterocytozoon hepatopenaei*, and *Enterocytozoon salmonis*, plus a few unassigned species: *Enterocytozoon* sp. IS2005R, *Enterocytozoon* sp. IS2005S, *Enterocytozoon* sp. IS2005T, *Enterocytozoon* sp. IS2005U, *Enterocytozoon* sp. IS2005V, *Enterocytozoon* sp. IS2005W, and *Enterocytozoon* sp. ST-2009a.[38]

Until recently, knowledge about these organisms was scarce and the microsporidian classification systems were accordingly simple and artificial. Traditionally, taxonomic studies and species classification of the microsporidia were mainly based on biological characters such as (1) size and morphology of spores, (2) number of coils of the polar filament, (3) development of spores, (4) microsporidium–host relationships, (5) life cycle, (6) parasitic site, and (7) method of transmission.[2]

The microsporidia are characterized at a structural level by a significant reduction in, or even absence of, certain constituents typical of eukaryotic organisms, such as mitochondria, peroxisomes, classically stacked Golgi apparatus, and 80S ribosomes. The microsporidia contain 70S ribosomes, ribosomal subunits (30S and 50S), and rRNAs (16S and 23S) of prokaryotic size, and they have no separate 5.8S rRNA. They also lack flagella, pili, and 9 + 2 microtubular structure.[39,40] The apparently simple cellular organization has impaired the establishment of the evolutionary relationships between the microsporidia and other eukaryotes, because these parasites do not have some of the characteristics usually used for such comparisons. Initially, these unusual features have led to the theory that microsporidia have diverged from a eukaryotic ancestor before eukaryotic cells have acquired mitochondria (endosymbiotic theory).[41] Early biochemical and molecular studies supported this ancient origin hypothesis. However, during the past decade, evidence has emerged, based on additional molecular phylogeny established from conserved protein genes, for the argument that microsporidia must be classified as fungi. It was suggested that features that were in the beginning interpreted as primitive are now considered as specialized adaptations inherent to their parasitic life cycle. Thus, in 1998 Cavalier-Smith classified the microsporidia into the kingdom Fungi. And in 2002, NCBI concluded a description of microsporidia as cellular organisms, eukaryote, fungi/metazoa group, fungi, and microsporidia.[42] There is no universal agreement on the evolution and classification of microsporidia because the Society of Classification has not approved the description.

20.2.1.1 Taxonomic Position of Microsporidia from Discovery to Present

When first described in 1857, the microsporidian species, *Nosema bombycis* identified by Nägeli, was depicted as being similar to yeasts and was included in the artificial group of schizomycete fungi.[2] Later, they were considered sporozoan protists and more specifically members of the subgroup Cnidosporidia, because of the introverted polar filament analogous to a cnidocyte, the stinging cell that fires a toxin projectile (the diagnostic feature of a cnidarian), a position favored for over 100 years.

In 1983, a new hypothesis radically departed from this idea, suggesting they were an ancient, primitive lineage that evolved before the origin of mitochondria. Microsporidia electron microscopy analysis not only allowed the discovery of the unique features of the spores but also revealed the absence of some typical characteristics of eukaryotes as mentioned.[43,44] Microsporidian genomes consist of multiple linear chromosomes with similarity to other eukaryotes. However, their genomes have reduced dimensions with an unusual disposition. *E. intestinalis*, with 2.3 Mpb, is one of the smallest eukaryotic genomes known to date. As *E. bieneusi* cannot be maintained indefinitely in vitro, sequencing its entire genome is a challenge, but it is thought to be about 6 Mbp.[45] Furthermore, the presence of ribosomal subunits similar to those of prokaryotes has been found in microsporidia.[46,47] Nevertheless and based on these arguments, the microsporidia were seen as a primitive group of eukaryotic organisms. This view was in agreement with inclusion of microsporidia in the Archezoa, proposed by Thomas Cavalier-Smith (1983), where it was postulated that the origin of the eukaryotic cell should have preceded the symbiotic origin of mitochondria.[39,44] It was assumed that microsporidia constitute a paraphyletic group, where these parasites were placed in the branch of the more ancient eukaryotes, because they are the only ones that had no 9 + 2 microtubule structure.[44,48]

Molecular data originally supported this hypothesis, but as data accumulated, it became clear that they were related to fungi, a conclusion broadly supported by the genomic data now available, although their exact relationship remains contentious. Initially, the microsporidia relationship to fungi was based on the occurrence of similarities in the life cycles, the presence of chitin in the wall of the spore, and the presence of the disaccharide trehalose in both groups of organisms. More recently, molecular evidence supported the proximity to kingdom Fungi, that is, phylogenetic analysis of α- and β-tubulin genes[49–51]; identification of chitinases similar to the fungal chitinases[52]; and phylogenetic analysis of genes encoding TATA-box-binding protein, large subunit (LSU) RNA polymerase II, heat shock protein of 70 kDa (HSP70), the glutamyl elongation synthetase, translocation factors EF-1 and EF-2, and the subunits α- and β-pyruvate dehydrogenase E1.[53–58] Gene sequences of *E. cuniculi* reinforce the link between the two groups.[47] Furthermore, the identification in microsporidia of more than a dozen genes encoding mitochondrial proteins, the location of HSP70 protein in the mitosome, and detection of Golgi-like membranes add support for this classification.[59–63]

At present, discussions are centered in deciding if these organisms share a common ancestor with fungi or if they derive from fungi. However, it is still not possible to reach a concrete answer to these questions, since the phylogenetic analyses are established from single genes and typically include only a small fraction of the diversity of microsporidia or fungi. The success of molecular methods in phylogenetic and systematic studies depends largely on the molecules chosen for analysis. Ideally, the use of a set of multigenic data with a broad taxonomic representation might overcome the problem. It has only been possible to study the genome of a small group of microsporidia, and characterization was obtained only for a limited number of genes (<2000). The high divergence of sequences determined not only makes it difficult to perform a molecular phylogenetic analysis but also makes this analysis extremely uncertain.[47,54,64,65] More recently, a different approach based on conservation of gene order was adopted. It was found that the gene order within the genome is highly conserved despite the gene sequences of microsporidia evolving very quickly.[64,66] These results suggest a link to zygomycetes, since both groups share a high degree of conservation in gene order. Other groups of fungi, such as basidiomycetes, showed a distinct rate of gene order.[63] On the other hand, an eight-locus genealogies study place the microsporidia as a sister group to a combined ascomycete and basidiomycete clade.[67] Finally, a genome-wide reanalysis of gene order conservation suggested that the phylogenetic placement of microsporidia as a sister to the zygomycetes needs to be reconsidered.[68]

The comparison of nucleic acid–based phylogeny analysis with the classic classification systems supported on phenotypic criteria alone allowed numerous contradictions in the classification of the phylum. The current trend of modern taxonomy is reassessing data on morphology, physiology, and ecology and integrating them with genomic information, in what is called by polyphasic taxonomy.[69] Molecular techniques definitely present an excellent means of identifying (1) species and (2) proposing evolutionary relatedness through phylogenetic analysis. Because sequence data are characters unique to an organism, species descriptions should whenever possible contain nucleic acid data (e.g., rRNA). On the other hand, morphological data (primarily obtained by electron microscopy for *Enterocytozoon*) also provide information that is often unique to a genus or even species, which should be expected since the morphology is determined by the genome. Nevertheless, all published classification systems reveal that morphology alone does not provide enough information for phylogenies. Several morphological characters of microsporidia, such as the number of nuclei, the number of spores or sporonts, the length, arrangement and structure of the polar tube, and other details of the life cycle, may change very rapidly during adaptation to different hosts or tissues. Indeed, some characters (e.g., being diplokaryotic and development in direct contact with the host cell cytoplasm) appear to have evolved several times simultaneously in different lineages of microsporidia. As morphology is the visual expression of the genome, careful evaluation of molecular and morphological data should lead to phylogenies that are based on two sets of characters and that correspond to, and are consistent with, each other.[2]

Clarification of microsporidia taxonomy is an essential step to understanding the epidemiology of these important parasites, allowing control of transmission. A large number of issues remain unresolved and continue to raise doubts among the scientific community, despite advances in recent years, to establish the evolutionary origin of microsporidia.

20.2.1.2 Biology and Life Cycle

Microsporidia are obligate intracellular, single-celled parasites with a unique ultrastructure and life cycle. The infectious stage of the microsporidia is the spore (Figure 20.1) and the parasites have no active stages outside their host cells. Despite microsporidia lacking some typical eukaryotic characteristics, they are true eukaryotes with a nucleus, an intracytoplasmatic membrane system,

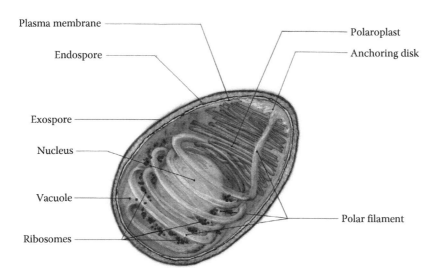

FIGURE 20.1 Diagram of the internal structure of an *Enterocytozoon bieneusi* spore. The spores have a thick wall, composed of an electron-dense outer region called the exospore, which is proteinaceous, and an electron-lucent inner region called the endospore, which is chitinous. A plasma membrane separates the spore coat from the spore contents and encloses the cytoplasm, the single nucleus, a posterior vacuole, ribosomes, the polaroplast membranes, and the unique extrusion apparatus. The extrusion apparatus consists of a coiled polar filament (five to six coils in a double row) and its anchoring disk. (Courtesy of Luís Filipe Marto, Museu - IHMT.)

and chromosome separation by mitotic spindles.[46,70] Polyadenylation occurs on mRNA (ribosomal ribonucleic acid) in microsporidia as in every other eukaryotic organism.[36]

Spores: Microsporidian spores are between 1 and 20 μm long and species that infect mammals are usually small, with diameters of 1–3 μm.[71] The spores have a thick wall, composed of three layers: (1) an electron-dense outer layer called the exospore, which is proteinaceous; (2) an electron-lucent inner layer called the endospore, which is chitinous; and (3) a plasma membrane enclosing the cytoplasm, the nucleus (sometimes two nuclei), a posterior vacuole, the polaroplast membranes, and the unique extrusion apparatus. The extrusion apparatus consists of a coiled polar filament and its anchoring disk, which is characteristic of all microsporidia (Figure 20.1). The number and arrangement of coils of the polar filament vary among genera and species. Upon ingestion or inhalation by a suitable host, the polar filament is discharged through the thin anterior end of the spore, thereby penetrating a new host cell and inoculating the infective sporoplasm into the host cell (Figures 20.1 and 20.2).[2]

The mature spores of *E. bieneusi* are the smallest of all spores of microsporidia and measure only 1.5 × 0.5 μm. They have a thick electron-lucent spore coat, a polar tube with five to six isofilar coils in a double row, a small electron-lucent inclusion in the sporoplasm, a single nucleus, and an anterior attachment complex extending down around the polaroplast.[72]

Life cycle: The general life cycle pattern of the microsporidia, while it may vary, can be divided into three phases (Figure 20.2): the infective phase, the proliferative phase (merogony), and the spore productive phase (sporogony). Following ingestion of the environmentally resilient spore in contaminated water or food, the coiled polar tube rapidly discharges, through the thin anterior end of the spore, thereby penetrating a new host cell and inoculating the infective contents (sporoplasm and nucleus) into the host cell (infective phase). Once inside a cell, the sporoplasm released from the polar tube initiates the proliferative phase of development called merogony. In suitable cells, the sporoplasms become meronts, which are rounded or irregular, sometimes elongated cells with little,

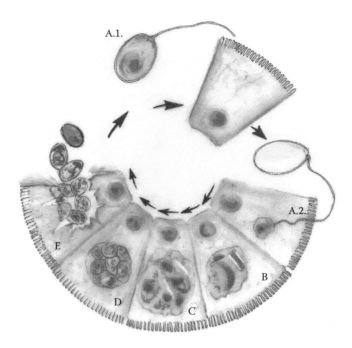

FIGURE 20.2 Diagram of *Enterocytozoon bieneusi* life cycle. (A.1.–A.2.) Following ingestion of *E. bieneusi* spore, the polar tube rapidly discharges, injecting the sporoplasm and nucleus into the cytoplasm of a host intestinal epithelial cell. (B to D) Intracellular developmental stages, in direct contact with the host cell cytoplasm. (B) Early proliferative plasmodial cell containing multiple elongated nuclei. (C) A late sporogonial plasmodium stage. (D) Several newly formed sporoblasts cells that will complete development into mature spores. (E) Mature spores are released from ruptured host cells. (Courtesy of Luís Filipe Marto, Museu - IHMT.)

or even without, endoplasmic reticulum. They are surrounded by a simple plasma membrane and proliferate by repeated binary or multiple fission or by plasmotomy. Early *E. bieneusi* proliferative plasmodial cells contain multiple elongated nuclei and electron dense inclusions. Meronts develop into sporonts, which is the stage that divides into sporoblasts. Later, sporogonial plasmodium has electron-dense disks, some in stacks or arcs in stages of polar tube formation. During sporogonial division individual nuclei with polar tube complexes segregate, maturing into separate sporoblast cells that complete development into mature spores.

During the intracellular part of the life cycle (merogony and sporogony), the *E. bieneusi* microsporidian lies in direct contact with the host cell cytoplasm, in the zone between the host cell nucleus, and the brush border. Spores are released from ruptured host cells and are excreted with feces into the environment from which they are ingested by the next host.[36,72]

20.3 EPIDEMIOLOGY OF HUMAN MICROSPORIDIOSIS DUE TO *ENTEROCYTOZOON BIENEUSI*

20.3.1 DISTRIBUTION OF *ENTEROCYTOZOON BIENEUSI* IN HUMANS

After the first reports of *E. bieneusi* infection in humans in the United States of America (USA) and some countries of Western Europe, an increasing interest in the diagnosis of these microorganisms was observed. Cases of *E. bieneusi* infection have been reported in the Americas (USA, Cuba, Colombia, Peru, Venezuela, Brazil, and Chile), the European continent (Portugal, Spain, France, Italy, Germany, Switzerland, United Kingdom, Holland, Czech Republic, and Russia), the

African continent (Tunisia, Mali, Niger, Nigeria, Cameroon, Gabon, Democratic Republic of the Congo, Ethiopia, Uganda, Tanzania, Zambia, Zimbabwe, Mozambique, and South Africa), the Asian continent (Iran, China, India, Thailand, and Vietnam), and the Australia/Oceania continent (Australia).[14–16,19–23,32,73–118]

20.3.2 Prevalence of *Enterocytozoon bieneusi* in Humans

Although microsporidial infection has already been reported on every continent, there are no reliable estimates of prevalence, in part by the scarcity of data based on accurate and precise sampling. Another factor affecting the determination of the prevalence is that routine diagnosis of microsporidiosis is not implemented in most countries. Also, *E. bieneusi* infections rates are difficult to compare because of the differences in the diagnostics methods employed, the specimens analyzed, the geographical locations, the patient groups, and the patient characteristics (sex, age, socioeconomic conditions, immune status, and clinical features).[119]

Despite the limitations, several studies show consistent results in some population groups. Human infection focuses on experience of human immunodeficiency virus (HIV)–infected patients. In developed countries in North America, in Europe, and in Australia, studies involving HIV-seropositive persons with diarrhea reported rates between 2% and 78%.[20,79,80,82,84,89,116] Lower infection rates in some cases (between 1.4% and 4.3%) were reported in HIV-seropositive persons without diarrhea.[20,80,116] *E. bieneusi* was the species detected most often in the majority of the studies performed in developed countries, followed by *E. intestinalis*.

In developing regions of the world where the implementation of highly active antiretroviral therapy (HAART) is still limited, microsporidiosis continues to be diagnosed in patients with AIDS, associated with risk factors such as low hygienic conditions and/or excessive exposure to animals. *E. bieneusi* prevalence rates were reported between 2.5% and 51% in HIV-seropositive adult patients with diarrhea and between 1.9% and 4.6% in patients without diarrhea.[73,74,94–96,100,101,103,108,110,111,113,114] In HIV-seronegative persons with and without diarrhea, *E. bieneusi* detection ranged from undetected to 58.1% of the fecal samples examined.[22,93,94,111,113,114] Studies conducted in children presented *E. bieneusi* prevalence rates that varied between 17.4% and 76.9% in HIV-seropositive children with diarrhea and 0.8% and 22.5% in immunocompetent or apparently immunocompetent children.[15,22,32,96,98–100] Some studies even indicate the existence of asymptomatic carriers of these microorganisms: for example, recovery of spores of *E. bieneusi* in (1) 0.8% of African children considered HIV seronegative, (2) 5.9% of healthy orphans, and (3) 1.9% of child-care workers in Asia.[96,120]

In industrialized countries, the records of cases of intestinal microsporidiosis due to *E. bieneusi* continue to increase in immunocompetent or apparently immunocompetent persons. Some studies reported *E. bieneusi* infection rates between 6.1% and 10% in travelers who returned from tropical destinations suffering from self-limiting diarrhea.[14] In addition, a prevalence of 17.0% of *E. bieneusi* infection in elderly persons was observed, leading to the speculation that age-related diminishment of immunity might predispose these persons to microsporidial infections.[15]

Prospective studies conducted in developed countries indicate that the prevalence of *E. bieneusi* in HIV-seropositive patients is progressively decreasing, probably due to the use of HAART, which restore the levels of patients' TCD4+ lymphocytes.[88] Similar observations are also made recently in some developing countries.[22,121]

20.3.3 Sources of Infection and Modes of Transmission

20.3.3.1 Direct Person-to-Person Transmission

Vertical transmission of microsporidiosis was reported in rodents, lagomorphs, carnivores, and nonhuman primates.[122] However, it has not been observed in humans.[123] The probability of horizontal transmission of microsporidia is high, given the well-documented presence of microsporidial spores in the gut, genitourinary, and respiratory tracts of the host.[124] Since the process of elimination of pathogens is usually closely associated with its mode of transmission, it is supposed that microsporidia infection is the result of fecal–oral, urinary–oral, oral–oral, sexual, and airborne transmission.[13] The hypothesis of direct transmission is reinforced by studies where homosexual relations and cohabitation between individuals with diarrhea are considered risk factors associated with intestinal microsporidiosis.[110,125] Also, person-to-person transmission has been suggested.[22,110,126]

It is possible to experimentally infect animals with *E. bieneusi*, *Encephalitozoon* spp., or *A. algerae* either inoculated orally, intrarectally, or through the ocular mucosa, respectively.[127–129] These data, once again, support the occurrence of horizontal transmission between animals and similarly, one can extrapolate to humans.

20.3.3.2 Zoonotic Transmission

In recent years, it was found that the majority of infectious diseases are caused by emerging or reemerging pathogens, 75% of which are considered zoonotic. Microsporidia infections have been identified in a wide range of animals where spores are discharged outdoors through urine, stools, or lung secretions or after the death of the host. Humans, pets, farm animals, and wild animals infected by microsporidia are potential sources of spread (Table 20.1).

It remains to be proven whether the microsporidia are transmitted from animals to humans, although several species are recognized as potential sources of infection for humans. Exposure to animals is reported in various studies on human microsporidiosis involving immunocompromised and immunocompetent persons. Studies showed an association of *E. bieneusi* infection in HIV-seropositive patients living in rural areas and from contact with cow dung or horses. Having been stung by a bee, hornet, or a wasp and contact with ducks or chicken droppings are other factors.[74,110,130]

The *E. bieneusi* genotypes that have been detected in humans have been identified in domestic, farm, and wild animals, strongly supporting the findings that the disease produced by this species can be of zoonotic origin. In Peru, an unusual *E. bieneusi* genotype (Peru 16), transmitted from guinea pigs, was found in a 2-year-old immunocompetent child complaining of diarrhea. The genetic uniqueness of the parasite, its wide occurrence in other guinea pigs in the community, and its absence in other children in the community suggest the possibility of zoonotic transmission of the infection to the child.[75] In a recent study conducted in China involving rural AIDS patients, the most prevalent *E. bieneusi* genotypes (EbpC, D and Type IV) detected were all potentially zoonotic. The common occurrence of EbpC was a feature of *E. bieneusi* transmission not seen in other areas. Despite the source of most *E. bieneusi* infections being still uncertain, contact with animals appears a risk factor for microsporidiosis.[113]

20.3.3.3 Waterborne Transmission

Water plays an essential role in the transmission of a significant number of infectious diseases.[131] Outbreaks due to contamination of public water systems occur worldwide and have the potential of causing disease in a large number of consumers. These outbreaks have economic consequences that go beyond the medical care costs of people affected, by causing loss of confidence in the quality of drinking water and in the water industry in general.[132] Hundreds of outbreaks of parasitic diseases have been associated with contaminated water supplies.[131,133]

The ability of transmitting infectious agents to humans from the environment is influenced by multiple factors.[134] The load of microorganisms that passes to the environment depends on the (1) prevalence of infection in humans or animals, (2) concentration of the agent in the stools or urine, and (3) duration of the elimination of the infectious agents by the host. Once in the environment, several critical points can affect the transmission capacity of the agent to humans. The survival of the organism in the environment is crucial for its transmission: the longer the survival time, the

TABLE 20.1
Enterocytozoon bieneusi Zoonotic Genotypes

Genotype Designation	Hosts	References
A	Humans, baboons, and birds	[22,34,73–75,86,97,102,158,165,166,168]
C (Type II)	Humans and mice	[23,87]
D (PigITS9, WL8, Peru9, PtEbVI, CEbC)	Humans, macaque, baboons, pigs, cattle, horse, dogs, fox, raccoon, beaver, muskrat, and falcons	[23,26,28,31,34,73,75,77,91,97,102,104,115, 148,158,159,206,210–212]
CAF1 (PebE)	Humans and pigs	[97,102,169]
EbpC (E, WL13, WL17, Peru4)	Humans, pigs, beaver, otter, muskrat, raccoon, and fox	[11,24,26,30,73–75,97,162,165,167,168,213]
Peru16	Humans and guinea pigs	[75]
Peru10	Humans and cats	[73–75,160]
Type IV (K, Peru2, PtEbIII, BEB5, BEB5-var)	Humans, cattle, cats, and dogs	[21,27–29,73–75,83,87,97,101,102,157–161,214]
WL11 (Peru5)	Humans, dogs, fox, and cats	[26,73–75,160,161]
O	Humans and pigs	[30,87,165,168]
PigEBITS7	Humans and pigs	[148,168]
Peru6 (PtEbI, PtEbVII)	Humans, cattle, dogs, birds	[28,73–75,80,162,164]
WL15 (WL16, Peru14)	Humans, beaver, fox, muskrat, and raccoon	[26,73,75,212]
EbpA (F)	Humans, cattle, pig, and birds	[23,30,87,149,163,166,167,170,212,213,215]
I (BEB2, CebE)	Humans and cattle	[27,32,87,164,211,213,214,216,217]
J (BEB1, CebB, PtEb X)	Humans, cattle, and birds	[27,28,32,87,164,213,214,218]
Peru7	Humans and baboons	[34,73–75,166]
Peru11 (Peru12)	Humans and baboons	[34,73–75,166,168]
BEB4	Humans and cattle	[23,27,31,164,211,216]
PigITS5 (PebA)	Humans and pigs	[23,148,167,169]
BFRmr2	Humans and pigs	[23,30]
CHN1	Humans, cattle, and pigs	[32]
CHN3	Humans and cattle	[32]
CHN4	Humans and cattle	[32]
PtEb II	Humans and birds	[80,163]
Peru 8	Humans and mice	[23,73–75]
CZ3	Humans and mice	[32,33]

Note: The primary genotype name according to *E. bieneusi* genotype terminology, based on that of Santin and Fayer, is shown in bold.[219]

greater the chance of the infectious agents to contact susceptible hosts. The time of survival in water depends on several physical factors, for example, pH, temperature, sun exposure, and the intrinsic characteristics of the microorganism, and may vary from hours to years. The characteristics of microsporidia make them extremely likely to survive in the environment and be spread by water. Their environmental developmental phase, the spore, is infectious immediately after being excreted to the exterior. Also, spores are resistant to adverse environmental conditions and processes of water treatment. Their small size and low density facilitate spread through water and increase their potential to be subsurface and groundwater contaminants.[135,136]

The main element in the transmission process is the infective dose, for example, the degree of exposure required to transmit the infection. In the case of microsporidia, the dose appears to be low, and inoculation with only 100 spores may cause lethal disease in athymic mice.[36,137]

In industrialized countries, protozoa belonging to the genera *Cryptosporidium* and *Giardia* are described as the most common waterborne parasites. But in recent years, several epidemiological studies describe contact with water also as being a risk factor associated with human intestinal microsporidiosis.[35,123,125,134,138,139] These observations led the U.S. Environmental Protection Agency (EPA) to consider them to be in the category B biodefense agents on the National Institutes of Health list and to place microsporidia spores on the list justifying the current concern of the occurrence of waterborne microsporidiosis.[140]

E. bieneusi spores have been detected in ground, surface, ditch, and crop irrigation water sources.[35,139,141–146] However, only one outbreak of microsporidiosis, associated with waterborne transmission, has been reported in the literature. This outbreak involved 200 patients, with and without HIV infection, and was recorded in mid-1995 in Lyon, France.[35] After this report, *E. bieneusi* spores were found in water samples from (1) the River Seine, France, from two rivers located near Paris where bathing and boating were frequent and (2) treated and untreated water samples from the water supply of Lisbon, Portugal.[138,142,144] Also, spores of *E. bieneusi* were reported in wastewater and sludge collected from 18 treatment plants located throughout Tunisia.[145] *E. bieneusi* genotype IV was identified in water samples from Portugal and Tunisia, and genotype D was also one of the most prevalent in Tunisia, demonstrating that the quality of public water was affected by humans, livestock, and rodents.[144,145] Genotype D was also the most prevalent in a study conducted in the sewer systems of four urban communities in China where anthroponotic transmission appeared to be important in the epidemiology of microsporidiosis.[146] The detection of microsporidia spores in tertiary sewage effluents indicates that they can survive wastewater treatment processes including mixed medium filtration and chlorination. *E. bieneusi* spores in surface water suggested the environmental reservoirs include domestic and wild animals.[139] All the *E. bieneusi* genotypes reported in water are summarized in Table 20.2.

Some studies have shown an association of *E. bieneusi* infection in HIV-seropositive patients with (1) consumption of nonpiped water; (2) contact with recreational water, hot tub, or spa; and

TABLE 20.2
Detection of *Enterocytozoon bieneusi* Genotypes in Water

Continent	Country	Water Source	Species/Genotype (Synonyms)	References
America	EUA	SW, MBW	*E. bieneusi*	[139,220]
Europe	Portugal	GW, WTP, FW	*E. bieneusi*/**Type IV** (K, Peru2, BEB5, BEB5-var, CMITS1, PtEb III)	[144]
	Spain	WTP, RW, FW	*E. bieneusi*/**C** (Type II), **D** (WL8, Peru9, PigEBITS9, PtEb VI, CEbc), **D-like genotypes**	[221]
	France	SW	*E. bieneusi*	[138,142]
	Ireland	WTP	*E. bieneusi*	[222]
Africa	Tunisia	WTP, DS, DHS	*E. bieneusi*/**D** (WL8, Peru9, PigEBITS9, PtEb VI, CEbc), **Type IV** (K, Peru2, BEB5, BEB5-var, CMITS1, PtEb III), **Peru8, WL1, WL2, WL4** (WL5), **WL12, BEB3, BEB6**	[145]
	China	WTP	*E. bieneusi*/**D** (WL8, Peru9, PigEBITS9, PtEb VI, CEbc), **EbpC** (E, WL13, Peru4, WL17), **EbpD, Peru6** (PtEb I, PtEb VII), **Peru8, Peru11** (Peru12), **C** (Type II), **EbpA** (F), **PtEbIV, BEB6, PigEBITS7, PigEBITS8, WL2, WL14, WL4, PtEbIX, WW1, WW2, WW3, WW4, WW5, WW6, WW7, WW8, WW9**	[146]

Notes: The primary genotype name according to *E. bieneusi* genotype terminology, based on that of Santin and Fayer, is shown in bold.[219] SW, surface water; GW, groundwater; WTP, waste treatment plants; MBW, marine beach water; FW, finished water; RW, raw water; DS, dry sludge; DHS, dehydrated sludge.

(3) occupational contact with water.[110,130] Another study reported an association between microsporidia transmission and consumption of foods irrigated with contaminated water.[143] Data supporting waterborne transmission of microsporidia, and especially *E. bieneusi*, are accumulating, but further studies are needed before appropriate public health initiatives can be considered in different regions by public health authorities.

20.3.3.4 Foodborne Transmission

Microsporidia transmission through contaminated food raises concerns due to several factors, including globalization (resulting in a high international traffic of cargo and people) and changes in food consumption.[134,147] Hutin et al. found an association between ingestion of undercooked meat at least once a month with microsporidiosis in HIV-seropositive patients.[125] *E. bieneusi* was identified in 32% of the samples analyzed from 176 slaughtered pigs where the genotypes were zoonotic D and F, suggesting the transmission of this organism through the ingestion of raw or undercooked meat.[148]

Vegetables and fruits were contaminated with microsporidia. In a study in Peru of HIV/AIDS patients, infection with *E. bieneusi* genotypes Peru-2 to Peru-11 were associated with watermelon consumption.[74] Another study demonstrated that fresh food produce, such as berries, sprouts, and green-leafed vegetables, sold at the retail level, can contain potentially viable microsporidian spores of human-virulent species, such as *E. bieneusi*, *E. intestinalis*, and *E. cuniculi*, at quantities representing a threat of foodborne infection.[149] Microsporidian spores were identified in lettuce, parsley, cilantro, and strawberries.[143] All products evaluated presented, at least once, *Cryptosporidium* sp., *Cyclospora* sp., and microsporidia, demonstrating the risk they represent to public health.

Recently, the first identified foodborne outbreak associated with *E. bieneusi* was reported in Sweden in 2009, involving more than 100 cases of gastrointestinal illness.[150] The investigations suggested that cucumber slices in cheese sandwiches and a salad contaminated with *E. bieneusi* were the most probable vehicle of transmission. All samples available for typing were infected with *E. bieneusi* genotype C. Although this genotype is potentially zoonotic, the fact that it has been rarely reported in animals made the authors suggest that the source of contamination in this outbreak was of human (fecal) origin.[150] All these findings indicate that *E. bieneusi* is capable of causing foodborne outbreaks of gastrointestinal illness and highlight the importance of introducing good agricultural practices in the countries/areas where there is a possibility of contamination of vegetables and/or fruits with pathogen-contaminated irrigation water.

20.3.3.5 Airborne Transmission

The occurrence of upper and lower airway infections by microsporidia species suggests that transmission can happen by this route through dust contaminated by microsporidia. This type of transmission has been indicated as the most likely vehicle in some cases of infection with *E. hellem* and *E. cuniculi*.[13] Also, respiratory tract involvement has been described, although rarely, in *E. bieneusi* infections suggesting that different modes of transmission are possible.[81,151] Besides the common fecal–oral route, the oral–oral route, and inhalation of aerosols are hypothetically possible.[152]

20.3.4 MOLECULAR EPIDEMIOLOGY OF *ENTEROCYTOZOON BIENEUSI* IN HUMANS

Enterocytozoon bieneusi is a complex species with multiple genotypes and diverse host range and pathogenicity. Since different strains cannot be discriminated morphologically, typing of this species relies on molecular methods.[25] PCR analysis of the internal transcribed spacer (ITS) of the rRNA gene, a hypervariable sequence of approximately 243 bp, followed by sequencing of the PCR products and comparison of obtained ITS sequences with those in a databases, is the standard method for genotyping *E. bieneusi* isolates from humans and animals. For several years these were the only genetic markers available.[25,73,100,103,113,128,153–156] A few researchers genotyped *E. bieneusi* based on the analysis of the ITS by PCR restriction fragment length polymorphism (RFLP).[83,103,157] Recently, two groups of researchers reported the analysis of the GenBank

E. bieneusi ITS sequence collection, using statistical methods, with the aim to obtain information about diversity, transmission, and evolution of this species.[155,156] These studies have identified the presence of host-adapted *E. bieneusi* genotypes in a variety of domestic animals and wild mammals and a large group of *E. bieneusi* genotypes that do not appear to have host specificity.[26,155,156] These latter genotypes, considered zoonotic, are responsible for most human infections. *E. bieneusi* genotypes in humans have been shown to differ from each other in virulence and geographic distribution.[74,156]

More than 90 different genotypes of *E. bieneusi*, identified by ITS sequences, have been described in humans and animals, and some genotypes have multiple names.[72] Approximately 64 of these genotypes have been found in humans (Table 20.3), and 27 genotypes have been reported in humans and animals (Table 20.1). All these observations were made in populations from different geographic regions, involving all continents.

In studies conducted in France involving HIV-seropositive and seronegative patients infected with *E. bieneusi*, Liguory et al. found five genotypes (Types I to V):[83,157]

Type I, synonym of genotype B, was the most prevalent genotype (75%).

Type II was designated as genotype C (10%). Genotypes B (Type I) and C (Type II) were also found in other populations.[86,87,102,118,158,159] Genotype B (Type I) even showed predisposition toward HIV/AIDS patients.[103,118,158]

Type III was the least prevalent genotype (4%), found only in HIV-seropositive patients.

Type IV, also named genotype K/Peru2/PtEbIII/BEB5/BEB5-var/CMITS1, was rare in France.[83] In contrast, this genotype was the only one in a population in Cameroon and one of the most frequent in AIDS patients in the Democratic Republic of Congo, China, Malawi, and the Netherlands.[101,103,113,158] Moreover, it was found in humans in other regions and in various animals.[21,27–29,63,73–75,87,97,102,158–161]

Type V was found only in one HIV-seropositive patient.

In Lima, Peru, 11 genotypes were identified (Peru-1 to Peru-11) in 2672 HIV/AIDS patients. The most prevalent genotypes were Peru-1 (39%), synonym of genotype A; Peru-2 (19%), synonym of Type IV; and Peru-9 (10%), also named genotype D. The remaining eight genotypes were found in the population at much lower frequencies (1%–9%).[73,74] Some of these genotypes were also found in animals, including genotype A (Peru-1), Type IV (Peru-2), Peru-4 (genotype EbpC/E/WL13/WL17), Peru-6 (PtEb1/PtEbVII, Peru-7), D (Peru-9), Peru-10, and Peru-11 (Peru-12).[18,24,26–30,34,63,75,87,101,160–167] Four genotypes (A/Peru-1, Peru-7, Peru 8, and Peru-11) were also identified in animals and found previously only in humans.[33,34,166] Genotype D was one of the most prevalent among patient groups from Malawi and the Netherlands, and it was found in one HIV-seropositive patient in Gabon.[102,158] This genotype, which was commonly detected in HIV-seropositive patients, was also found in HIV-seronegative individuals and in renal transplant recipients, confirming its widespread nature.[73,102,109,115,158,159,168] Genotype EbpC (synonym of Peru 4) that was found in low frequencies in Peru, Thailand, and Vietnam was the most frequent genotype in HIV-seropositive patients in Henan, China.[73,75,97,113,168]

Also in Peru, in a prospective pediatric-cohort study of enteric parasites, *E. bieneusi* infection was identified in 8% (31/388) of the children studied.[75] Thirty of these children had infections with genotypes Peru-1 (synonym of A) through to Peru-15 and Peru-17, and one child was infected with genotype Peru-16. Another genotype, Peru-14, also named genotype WL15/WL16, was also found in animals.[26] Genotype UG2145, identified for the first time in Uganda, was the only *E. bieneusi* detected in a population of 1779 children with diarrhea.[21]

Breton et al. revealed for the first time the presence of genotypes A and B in Africa.[102] These genotypes were also reported from HIV-seropositive and HIV-seronegative populations in Europe, Peru, Thailand, and Niger.[73,75,83,86,87,126,157–159,162,165] Three new genotypes CAF1, CAF2, and CAF3 that are close relatives of genotypes E, Type IV, and D, respectively, were found for

TABLE 20.3
Distribution of *Enterocytozoon bieneusi* Genotypes in Humans

Continent	Country	Genotype (Synonyms)	References
America	Peru	**A** (Peru1), **Type IV** (K, Peru2, BEB5, BEB5-var, CMITS1, PtEb III), **Peru3**, **EbpC** (E, WL13, Peru4, WL17), **WL11** (Peru5), **Peru6** (PtEb I, PtEb VII), **Peru7**, **Peru8**, **D** (WL8, Peru9, PigEBITS9, PtEb VI, CEbc), **Peru10**, **Peru11** (Peru12), **Peru13**, **WL15** (Peru14), **Peru15**, **Peru16**, **Peru17**	[73–75]
Europe	Portugal	**Type IV** (K, Peru2, BEB5, BEB5-var, CMITS1, PtEb III), **Peru6** (PtEb I, PtEb VII), **A** (Peru1), **C** (Type II), **D** (WL8, Peru9, PigEBITS9, PtEb VI, CEbc), **PtEb II**	[80]
	Spain	**D** (WL8, Peru9, PigEBITS9, PtEb VI, CEbc)	[206]
	France	**B** (Type I), **C** (Type II), **Type III**, **Type IV** (K, Peru2, BEB5, BEB5-var, CMITS1, PtEb III), **Type V**	[83,157]
	Germany	**A** (Peru1), **B** (Type I), **C** (Type II), **Q**	[86,87]
	Switzerland	**Q**	[223]
	United Kingdom	**B** (Type I), **D** (WL8, Peru9, PigEBITS9, PtEb VI, CEbc), **Type IV** (K, Peru2, BEB5, BEB5-var, CMITS1, PtEb III)	[159]
	The Netherlands	**A** (Peru1), **B** (Type I), **C** (Type II), **S7**, **S8**, **S9**, **D** (WL8, Peru9, PigEBITS9, PtEb VI, CEbc), **Type IV** (K, Peru2, BEB5, BEB5-var, CMITS1, PtEb III)	[158]
	Czech Republic	**EbpA**, **CZ3**, **PigEBITS5**, **BFRmr2**, **BEB4**, **CZ1**, **CZ2**	[23]
	Russia	**D** (WL8, Peru9, PigEBITS9, PtEb VI, CEbc)	[91]
Africa	Niger	**A** (Peru1), **D** (WL8, Peru9, PigEBITS9, PtEb VI, CEbc), **Type IV** (K, Peru2, BEB5, BEB5-var, CMITS1, PtEb III), **CAF1** (PEbE), **NIA1**	[97]
	Nigeria	**D** (WL8, Peru9, PigEBITS9, PtEb VI, CEbc), **Type IV** (K, Peru2, BEB5, BEB5-var, CMITS1, PtEb III), new genotype (similar to genotype K, with two nucleotide substitutions), **B**, new genotype (similar to P, Type IV, UG2145, Peru3, PtEb IV, PtEb V)	[98,99]
	Cameroon	**Type IV** (K, Peru2, BEB5, BEB5-var, CMITS1, PtEb III), **A** (Peru1), **B** (Type I), **D** (WL8, Peru9, PigEBITS9, PtEb VI, CEbc), **CAF4**	[101,102]
	Gabon	**A** (Peru1), **D** (WL8, Peru9, PigEBITS9, PtEb VI, CEbc), **Type IV** (K, Peru2, BEB5, BEB5-var, CMITS1, PtEb III), **CAF1** (PEbE), **CAF2**, **CAF3**, **CAF4**	[102]
	Uganda	**Type IV** (K, Peru2, BEB5, BEB5-var, CMITS1, PtEb III), **UG2145**	[21]
	Democratic Republic of Congo	**NIA1**, **D** (WL8, Peru9, PigEBITS9, PtEb VI, CEbc), **KIN1**, **KIN2**, **KIN3**	[103]
	Malawi	**Type IV** (K, Peru2, BEB5, BEB5-var, CMITS1, PtEb III), **D** (WL8, Peru9, PigEBITS9, PtEb VI, CEbc), **Peru8**, **UG2145**, **S1**, **S2**, **S3**, **S4**, **S5**, **S6**	[158]
Asia	Iran	**D** (WL8, Peru9, PigEBITS9, PtEb VI, CEbc)	[112]
	China	**I** (BEB2, CEbE), **J** (BEB1, CEbB, PtEb X), **CHN1**, **CHN2**, **CHN3**, **CHN4**	[32]
	Thailand	**D** (WL8, Peru9, PigEBITS9, PtEb VI, CEbc), **Peru 11** (Peru12), **EbpC** (E, WL13, Peru4, WL17), **PigEBITS7**, **O**, **R**, **S**, **T**, **U**, **V**, **W**	[16,22,115, 165,168]
	Vietnam	**D** (WL8, Peru9, PigEBITS9, PtEb VI, CEbc), **EbpC** (E, WL13, Peru4, WL17), **HAN1**	[97]
Oceania	Australia	**B** (Type I)	[118]

Note: The primary genotype name according to *E. bieneusi* genotype terminology, based on that of Santin and Fayer, is shown in bold.[219]

the first time in the HIV-seropositive patients from Gabon.[102] CAF1 was also detected in pigs in Korea.[169] Another new genotype, CAF4, a highly divergent genotype reported from humans, was equally present in HIV-seropositive patients in Gabon and HIV-seronegative individuals in Cameroon. Its high frequency (25%) in both countries may indicate that this genotype is common in Central Africa.[102]

Ten Hove et al. found nine new genotypes: six (S1–S6) were from Malawi and three (S7–S9) were from the Netherlands.[158] Genotype S7 showed no linkage to any of the described genotypes and seems to be more related to the genotypes isolated from cattle. Two other new unnamed genotypes were detected in this study, AF502396 and AY371283, and named as UE2145 and Peru-8, respectively.[158]

In a study conducted in Changchun City, China, involving stool samples from children with diarrhea and from pigs, dogs, and cows, and all collected in the same area around the city, Zhang et al. found for the first time in humans two genotypes previously reported in animals (genotypes I, also named BEB2/CEbE, and J, also named BEB1/CEbB, PtEb X) and 10 new genotypes (CHN1-10).[27,32] The novel genotype, CHN1, which was detected in children, cows, and pigs, was the most common genotype found in this study. Genotype CHN2 was found only in human samples. Genotypes CHN3 and CHN4 were also found in human and cow samples, and genotypes CHN5-10 were only detected in animals.[32]

The sequence analyses of *E. bieneusi* positive samples revealed 7 different genotypes in a study that investigated the occurrence and prevalences of *E. bieneusi* and *Encephalitozoon* spp. in 382 immunocompetent humans in the Czech Republic.[23] Genotypes that were previously reported only from cattle (BEB4 and EbpA) and pigs (PigEBITS5, EbpA, and BFRmr2) were identified in humans, and three new genotypes were detected (*E. bieneusi* CZ1 to CZ3). Genotype CZ3 was one of the most, and genotypes CZ1 and CZ2 were the least prevalent.[23]

Several studies indicate that the distribution of genotypes of *E. bieneusi* can vary by geographical locations, and it has been recently proposed that predominant genotypes in different geographical sites could be related to distinct sources of transmission.[87,126] Genotype D is considered the broadest host and geographic range, having been found in all continents and in several countries in Europe, United States, Peru, Malawi, Thailand, Vietnam, and Korea.[26,28,74,75,97,115,126,148,158,159,170]

All these findings suggest different transmission modes of *E. bieneusi* in diverse geographic regions and in special groups in the same region.

20.3.5 Prevention

Measures to prevent human microsporidiosis are difficult to establish since the potential sources of infection are unclear. Also, there are neither vaccines nor clinical trials being conducted in this context. Given these difficulties, the adoption of general preventive measures is the first step to avoid human contact with the parasites.[13] Some of the recommended actions to avoid microsporidiosis in humans are (1) adopt good hygiene practices—wash hands after going to the bathroom, after changing diapers for children, before preparing food, and before eating. People with diarrhea should also protect others from possible infection by avoiding the use of public water pools and water recreation; (2) avoid contact with water that may be contaminated—do not drink (a) recreational water; (b) water from wells, rivers, and lakes; and (c) untreated water and ice, especially in countries where the quality of the water supply is poor. Water should be boiled before consumption in cases of doubt; (3) avoid eating contaminated foods (wash or remove the rind of fruits and vegetables and avoid eating raw foods in countries where water treatment is poor), (4) avoid contact with potentially infected animals, and (5) avoid sexual practices that involve fecal–oral contact. The decrease in environmental contamination is also an important measure of prevention that may be achieved by reducing the transfer of infection between animals.[13] Several studies have been conducted to find strategies for reducing the viability and infectivity of microsporidia in the environment and to avoid the potential transmission through water and food. Successful disinfection of *E. intestinalis* in (1) water can be achieved by using chlorine and ozone disinfection and (2) in food by using high-pressure processing. Also, exposures of *E. cuniculi* to bleach, ethanol, HiTor, or Roccal were effective in reducing infectivity in a tissue culture model system.[6,171–173]

20.4 PATHOGENESIS AND CLINICAL SPECTRA OF *ENTEROCYTOZOON BIENEUSI* IN HUMANS

E. bieneusi is an obligate intracellular organism that invades, replicates, and matures within the intestinal epithelial cells and monocytes. The pathogenesis of intestinal disease is related to excess death of enterocytes from cellular infection. The invasion of enterocytes stimulates production of cytokines (e.g., interleukin-8), which in turn activates resident phagocytes and attracts new phagocytes into lamina propria from the blood.[174] The activated leukocytes release prostaglandins, leukotrienes, and platelet-activating factor to induce neurotransmitter-mediated intestinal secretion of chloride and water. They also generate histamine, serotonin, and adenosine to inhibit the absorption of epithelial cells. As a consequence of the inflammatory response, proteases and oxidants secreted by mast cells can negatively affect the growth of epithelial cells and produce villus atrophy and crypt hyperplasia.[174] Replication of microsporidia in epithelial villi of the small intestine, associated with a reduction in villus height and its surface area, appears to contribute to poor absorption of vitamin B12, D-xylose, and fat, leading to diarrhea.[17,124,174–177]

E. bieneusi infection is expressed clinically as a syndrome of malabsorption, leading to slow and progressive weight loss. The symptoms consist of chronic intermittent watery diarrhea, usually without blood or mucus.[17] Lesions are observed especially at the duodenum and jejunum, which are characterized microscopically by atrophy of the intestinal villi, with hyperplasia of the crypts of Lieberkühn and lymphocyte infiltration.[178] As the infection progresses, the epithelial cells undergo several changes, becoming progressively cuboidal, and at a later stage pleomorphic. The nuclei of the villus enterocytes lose basal orientation and turn hyperchromatic, and the cytoplasmic changes include vesiculation, vacuolization, and occasional lipid accumulation.[175,179] Although *E. bieneusi* propagates preferentially in the small intestine, on rare occasions extraintestinal infections have been described at the biliary tree, gall bladder, biliary tract and nonparenchymal liver cells, and pancreatic duct and in the lungs, which may provoke symptoms related to their specific localization.[18,81,124]

The pathogenicity of microsporidia is still not clearly defined, and the mechanism by which they induce diarrhea has not been determined. Some authors found that although HIV-seropositive patients infected by *E. bieneusi* had more diarrhea than those uninfected, they in fact had less inflammation than the uninfected HIV-seropositive persons, even though some genotypes of *E. bieneusi* can cause chronic diarrhea.[74,111]

E. bieneusi can be an important cause of persistent diarrhea, intestinal malabsorption, and wasting in HIV-seropositive adults. Although microsporidiosis is common in children <5 years of age, particularly those who live in developing countries or are HIV seropositive, the effects of infection on the nutritional health of these vulnerable populations were not well documented until recently.[16,21] Mor et al. found an association of *E. bieneusi* infection with lower rates of weight gain in children ≤60 months of age with persistent diarrhea. This relationship remained even after controlling for HIV and concurrent opportunistic diseases.[180]

Diarrhea and malabsorption seem to be the most common clinical problems associated with *E. bieneusi* infections reported in the literature.[17] However, in some studies there were no statistically significant associations between the presence of *E. bieneusi* in fecal specimens and patients with diarrhea.[80,101,105] The cases of *E. bieneusi* subclinical infection reported suggest that some persons may be asymptomatic carriers of these microorganisms.[22,96,106,120]

20.5 DIAGNOSIS OF MICROSPORIDIOSIS BY *ENTEROCYTOZOON BIENEUSI*

The signs and symptoms that accompany microsporidia infections are insufficiently clear to enable diagnosis based on clinical presentation. However, the presumptive diagnosis of human microsporidiosis is justified in HIV-seropositive patients and in those with other causes of immunosuppression, who present diarrhea, malabsorption, emaciation, and TCD4+ lymphocytes' count <100 cells/mm^3.[88]

The definitive diagnosis of enteric microsporidiosis in humans depends on identification of spores in stool or duodenal drainage specimens and/or intestinal biopsy specimens.[181] However, the identification of spores in these specimens is difficult because these organisms are small, stain penetration of the spore wall is low, and they cause weak inflammatory response in the infected tissues facilitating infection concealment. As such, the development of sensitive, specific, and rapid laboratory diagnostic methods has been the subject of recent studies.

20.5.1 SPECIMEN COLLECTION

Microsporidial spores are very resistant to environmental conditions and may remain infective for several years, especially if they are protected from dehydration. They can be stored at −20°C for years.[88,135,136] Because of the endurance of the spores, infected biological samples (unfixed) can be transported/mailed at room temperature.[88]

Spores of *E. bieneusi* can be identified after staining in (1) stools or (2) duodenal drainage specimens used fresh, preserved in 5%–10% formalin, or sodium acetate acetic acid formalin, and in (3) biopsy specimens. In cases with pulmonary involvement, *E. bieneusi* spores can be searched for in bronchoalveolar lavage and sputum. For (1) histologic examination, biopsy specimens should be fixed in formalin, and (2) for electron microscopy, it is preferable to fix the tissue biopsies in glutaraldehyde. Fresh specimens stored at 4°C are preferable for molecular analyses.[88,181]

20.5.2 TRANSMISSION ELECTRON MICROSCOPY

Definitive diagnosis of intestinal microsporidiosis is based on transmission electron microscopy (TEM) observation of microsporidial spores in stool specimens, tissue biopsies, duodenal aspirate, bile, or biliary aspirates. Visualization of the ultrastructure of the spore, with the characteristic polar filament, is diagnostic.[70,88,182] This technique allows the identification of microsporidia to the species level based on the morphological characteristics of the spores or their proliferative forms, the method of division, and the nature of the interface between parasite and host cell.[70,182] TEM remains the standard technique for confirming the diagnosis and identification of microsporidia.[70,88,182] All stages in the life cycle of the parasite can be found in infected tissues, whereas in body fluids and stools, in general, only the spores can be observed.[183]

Detection of microsporidia by TEM is a very specific, but low-sensitivity, technique, due to the small amount of sample that can be examined. Other disadvantages include the need for invasive procedures (biopsy) to observe the intracellular stages, and it is very laborious and costly. Furthermore, electron microscopes in routine laboratories are rare.[182,183]

20.5.3 HISTOLOGICAL DIAGNOSIS

Histological examination of biopsy specimens embedded in paraffin or fixed in formalin can be used for diagnosis of enteric microsporidiosis and to study physiopathological aspects of microsporidia infections inside infected cells. The small size of the spores, and the lack of an inflammatory response of infected tissues, makes it difficult to detect microsporidia by light microscopy. Their visualization depends on the contrast between the spores and other constituents of the host cells. Several staining methods are used in routine histological preparations of tissues. The most satisfactory ones are (1) Gram-derived and Warthin–Starry (silver staining) stains, (2) Giemsa and chromotrope 2R stains, and (3) the fluorescent staining technique, Uvitex 2B.[36,183–185]

20.5.4 CYTOLOGICAL DIAGNOSIS

The identification of microsporidia in samples by light microscopy obtained by noninvasive procedures posed a difficult challenge to researchers. Spores of microsporidia can be detected in several body fluids and stool samples as mentioned.[88] The spore wall consists of an outer proteinaceous layer (exospore) and an inner chitin layer (endospore) (Figure 20.1). Various stains, such as Gram, Giemsa, trichrome, or fluorescent, have been tested in microsporidia. In cytological diagnosis, Giemsa or Gram stains are not suitable since they do not allow the distinction between the microsporidia and other microorganisms present in biological specimens (e.g., enterobacteria with size, shape, and staining characteristics similar to microsporidia).[88,183] Only after the development of special fluorescent stains or stains based on chromotrope 2R can the microscopic examination of body fluids for diagnosis of microsporidia become routine.

Modified trichrome stains: The modified trichrome technique, developed by Weber et al., is the most common method used in diagnostic laboratories for the detection of microsporidia in stools, urine, and sputum.[70,88,151] The spores of microsporidia, at ×1000, show their cell wall colored pink, and in some cases, a pinkish band dividing spores diagonally or equatorially may be observed.[151] Small yeasts can stain in a similar manner but are distinguishable from microsporidia, as they are larger and more intensely blue. Bacteria stained with the same color can be distinguished from microsporidia by their nonoval shape and by their more uniform staining.[151] This technique is slow, and it is difficult to detect spores in specimens with low parasite burdens. However, some modifications accelerate the stain penetration and provide better contrast between the background and spores such as changes in temperature and time of staining, decreasing the level of phosphotungstic acid, and replacing the counterstain.[117,186] A positive control specimen should always be used. Of course, spore identification requires adequate illumination and magnification (oil immersion; total magnification ×1000).[181]

Gram-chromotrope stain: Moura et al. developed a *quick-hot Gram-chromotrope technique*.[187] This new and improved staining technique detects microsporidial spores in clinical specimens, such as stools, urine, saliva, nasopharyngeal fluid, and bronchoalveolar lavage fluid, and in formalin-fixed and paraffin-embedded tissue sections with the additional advantages of light microscopy. In this method, samples are stained with crystal violet solution and iodine, used in Gram staining, followed by a modified chromotrope solution The microsporidial spores are then stained dark violet on a pale-green background. This is a fast, reliable, and simple staining technique, easily adapted for use in routine laboratories.[187]

Fluorochrome stains: Chemofluorescent (optical brightening) agent stains may be used for detection of microsporidia in stools, urine, or sputum specimens using Uvitex 2B, Calcofluor, and Fungifluor.[88,90,188] The fluorochromes have high affinity for the chitin layer in spores (endospores), which are identified by their size, shape, and fluorescence characteristics (bluish-white or turquoise oval halos) under UV light.[88] The technique is straightforward and fast but requires a fluorescence microscope with excitation filter wavelength between 350 and 380 nm. The methods using fluorochromes are described in some studies as being more sensitive than those using trichrome; however, fluorochromes can stain yeasts, giving rise to false positives.[88,189,190] However, there are other studies that do not support these findings.[186] In one aspect, several authors are in agreement that the two techniques must be used simultaneously in order to maximize the correct diagnosis, especially in biological specimens with low parasite burden.[186,189,190]

Immunofluorescent reagents: The indirect immunofluorescence (IIF) technique associated with specific antimicrosporidia antibodies, labeled with fluorescein, has been used in the diagnosis and differentiation of microsporidia species. Poly- and monoclonal antibodies (MAbs) were also used for the analysis of various species of microsporidia by Western blot.[88] To date, the production of antibodies that recognize *E. hellem*, *E. intestinalis*, *E. cuniculi*, and *E. bieneusi* has been achieved.[191–194] Most of these MAbs are specific for antigens present in spores.[191] Other authors use MAbs that react with proteins of polar filament and MAbs that recognize surface antigens of spores.[192] Some of these MAbs are species specific, while others react with the spore wall or with the polar filament of several species of microsporidia. Antibody tests using monoclonal anti-*E. bieneusi* and anti-*E. intestinalis* antibodies were compared with species-specific PCR, demonstrating similar levels of performances, and the suitability of IIF-MAbs for diagnosis of enteric microsporidiosis.[95] Since *E. bieneusi* is clinically the most significant among the microsporidia causing diarrheic syndrome in HIV/AIDS patients, children, and elderly, the availability of commercial kits with MAbs for diagnostic purpose improves the ability to detect these cases.

20.5.5 Serologic Tests

Data on human microsporidiosis based on serological tests have focused almost exclusively on exposure to *Encephalitozoon* species. This is largely due to difficulty in obtaining pure *E. bieneusi* organisms.

Several serologic tests, such as indirect immunofluorescent antibody testing, enzyme-linked immunosorbent assay, carbon immunoassay, counterimmunoelectrophoresis, and Western blotting, have been applied in detecting IgG and IgM antibodies against microsporidial infections in humans.[195,196] However, the sensitivity and specificity of these methods are unknown since no comparative studies are available.[6,183] In HIV-seropositive and seronegative cases, antibodies against *E. cuniculi*, *E. intestinalis*, *E. hellem*, and *E. bieneusi* have been reported.[92,196] There is no certainty whether the antibodies are the result of a true infection, exposure to the parasite without establishment of infection, cross-reactions with other species of microsporidia, or even a nonspecific reaction. Serologic tests for serodiagnosis of *E. bieneusi* are not available because this parasite has not been propagated continuously in culture as a prerequisite for antigen production.

Given the increasing number of cases of microsporidial infection in immunocompetent persons, some serological methods have been developed, which utilize the whole parasite proteins or polar tube wall spore as antigens, in particular for microsporidia that do not grow in culture.[197,198] Having access to serologic tests for routine diagnosis would be very helpful but unfortunately they are not readily available.

20.5.6 Cell Culture

The isolation of microsporidia in vitro has no relevance as a routine diagnostic method, although it is an important tool for the development of immunologic reagents for diagnosis and determination of the efficacy of antimicrobial agents on several species, including *Encephalitozoon* spp.[88,182] Molecular, ultrastructural, biochemical, and antigenic analyses, combined with in vitro cultures, have served to confirm certain infections caused by species of microsporidia already identified and to define new species.[36,182] Regrettably, *E. bieneusi*, the most frequently identified microsporidia species in humans, has been propagated only in short-term cultures.[88,181] All efforts to establish a continuous culture of these organisms have failed.

20.5.7 Molecular Methods

Molecular methods have contributed to a deeper knowledge in many areas of biology. These methods are fundamental for identifications, taxonomic and phylogenetic analyses, and characterization of genotypes of *E. bieneusi* in immunocompromised and immunocompetent patient groups. There has been a gradual implementation of molecular methods to diagnose and characterize microsporidia. In routine laboratories, the first-line diagnosis is microscopy. However, in the last 15 years molecular methods have been mainly developed and applied in research centers. Several PCR methods based on the amplification of rRNA gene fragments have been described for use in detection and characterization of microsporidian species pathogenic to humans, which are considered the most sensitive and specific tools available for diagnosis of microsporidiosis.[109] The detection threshold of microsporidian spores in a stool sample by light microscopy ranges from 10^4 to 10^6 spores/g of stools, whereas the PCR detects concentrations of 10^2 spores/g of stools.[199] The first PCR assay developed for diagnosis of *E. bieneusi* was based on the small subunit rRNA (SSU-rRNA).[200] Primers used for diagnosis of *E. bieneusi* usually target the SSU and the LSU-rRNA.[182]

Several authors have described PCR-based techniques in order to detect the presence of microsporidia in biological and environmental specimens, overcoming the limitations of sensitivity and specificity presented by the microscopy techniques.[88,100,113,114,145,201–203] Although conventional PCR has been successfully applied to the detection of enteric microsporidia, their application in routine diagnosis has limitations due to difficulties in the DNA extraction from stool specimens and because it is prone to contamination, time consuming, and expensive. To overcome these difficulties, a real-time PCR was developed for routine diagnosis of enteric infection.[103,202] This technique allows the detection of amplicons in *real time* using fluorescent detection specific probes and permits quantification of parasites in stool specimens. Furthermore, it is standardized, can test a large number of samples simultaneously, and avoids post-PCR handling. Also, the use of a specific *Taq*Man probe guarantees specificity.[202] The detection limit for this technique is between 10^2 and 10^3 spores/mL of stools.[204] A multiplex real-time PCR was developed for the simultaneous detection of *E. bieneusi* and *Encephalitozoon* spp. in stool specimens.[202] This technique includes an internal control to detect inhibition of the amplification by components of the stool matrix. Also reported is a fluorescent in situ hybridization (FISH) technique developed to detect spores of *E. bieneusi*[203] that uses a fluorescently labeled probe that binds to the nucleic acid of the parasite.[205] Unlike PCR, it provides morphological and spatial information. Although the FISH assay appears an attractive tool as it provides extensive background information, two limitations may restrict its applicability to clinical diagnosis: it is (1) extremely laborious and (2) less sensitive than conventional PCR. An oligonucleotide microarray assay was also described for the simultaneous detection and discrimination of the four major species of microsporidia: *E. bieneusi*, *E. cuniculi*, *E. hellem*, and *E. intestinalis*.[201] The 18S SSU-rRNA gene was the amplification target, which was labeled with fluorescent dye, and hybridized to a series of species-specific oligonucleotide probes immobilized on a microchip. This microarray was successfully applied to investigate clinical specimens and was an attractive diagnostic tool for high-throughput detection and identification of microsporidian species in clinical and epidemiological investigations.[201]

Several authors argue that molecular methods should be used simultaneously with the colorimetric methods, to confirm the results obtained by the latter methods and to identify the species of microsporidia causing the infection. Until now, the enhanced diagnostic capacity obtained with the molecular methods has improved the identification of human cases of microsporidiosis due to *E. bieneusi*, not only in developed countries but especially in developing countries where *E. bieneusi* may be a particular public health problem due to the magnitude of the HIV epidemic and poor sanitary conditions.[14,15,20–23,32,73,74,77,79–81,83,91,93–95,101,102,105,110,111,115,118,,154,158,180,206] Nevertheless, despite the improvement described, the PCR-based methods (1) remain time consuming, (2) need considerable technician expertise, and (3) require a high level of equipment that translates into high cost and consequently are more difficult to support by laboratories, especially in developing countries.

20.6 THERAPY OF *ENTEROCYTOZOON BIENEUSI* DISEASE IN HUMANS

Currently, there is no effective treatment against infections caused by microsporidia. The difficulty of drug penetration into the thick wall of the parasite spore is a major obstacle, because even if other vegetative stages are destroyed and spore morphogenesis is disturbed, there is always the possibility of infection through the restoration of mature spores when treatment is interrupted. The few published studies on the treatment of microsporidiosis always fall into the context of therapeutic approaches to opportunistic infections in HIV-seropositive patients. The administration of HAART to these patients results in increased levels of TCD4+ lymphocytes that, with the concomitant reduction in HIV viral load, lead to a decrease in most opportunistic infections, including microsporidiosis.[207] Infections due to *Enterocytozoon* are very difficult to treat and there is currently no consensual treatment. Furthermore, the lack of tissue culture for *E. bieneusi* has limited in vitro pharmacological studies.[208] At present, albendazole, a benzimidazole that inhibits microtubule assembly, is the most effective drug in treating infections caused by most species of microsporidia in humans, although its effectiveness is reduced or even null against *E. bieneusi*.[208] Other drugs have been used in the treatment of microsporidiosis with varying results, among which are TNP-470, metronidazole, furazolidone, sinefungin, atovaquone, azithromycin, and sulfa drugs.[9,209] However, effectiveness needs to be confirmed by controlled trials. These studies utilized *Encephalitozoon* spp. because of the difficulty in working with *E. bieneusi*.

Since an effective treatment is not yet available for infections by *E. bieneusi*, while those induced by *Encephalitozoon* spp. in most cases respond to albendazole, it becomes necessary increasingly to differentiate the microsporidia taxa involved in the infection of patients with microsporidiosis.

20.7 CONCLUSION

Over the past 150 years, the classification of microsporidia in distinct taxonomic systems is unstable. Early studies indicated microsporidia as a primitive group of eukaryotic organisms, and currently, these parasites are classified as fungi. The recent application of PCR-based molecular methods to *E. bieneusi* identification and characterization has led to more reliable results compared to light microscopy of stained biological smears. PCR has (1) improved knowledge about animal reservoirs, modes of transmission, and risk factors, (2) allowed the study of intraspecific variability in *E. bieneusi*, and (3) increased knowledge on the distribution of genotypes by geographical location to enhance source tracking and calculate the pathogenic potential of an isolate. *E. bieneusi* is considered an opportunistic microorganism, being associated with different clinical spectra (symptomatic and asymptomatic infections in HIV-seropositive and seronegative persons), in different settings. Diarrhea and malabsorption are the most common clinical problems associated with this species. However, the implementation of microsporidia diagnosis in routine laboratories is still dependent on the development of more sensitive and specific techniques, requiring less processing, reading time, and expertise. These issues assume extreme importance in developing countries with a high rate of HIV-infected people who are not under HAART and consequently at risk of developing microsporidiosis. Few drug therapy studies on *E. bieneusi* have been published and, except in the case of infections caused by *Encephalitozoon* spp., treatment efficacy is poor. The best approach remains the improvement in immune function of persons infected by *E. bieneusi* as a result of effective antiretroviral therapy.

Despite the difficulties, our knowledge concerning microsporidia in general and *E. bieneusi* in particular has expanded remarkably. However, more research is needed to (1) elucidate their complex taxonomy and epidemiology, (2) develop and commercialize new and less expensive diagnostic techniques, and (3) find effective antimicrosporidial agents able to control *E. bieneusi* infections in humans.

REFERENCES

1. Wittner, M. Historic perspective on the microsporidia: Expanding horizons. In Wittner, M. and Weiss, L.M. (Eds.), *The Microsporidia and Microsporidiosis* (pp. 1–6). Washington, DC: American Society for Microbiology (1999).
2. Franzen, C. Microsporidia: A review of 150 years of research. *Open Parasitol J* 2, 1–34 (2008).
3. Nageli, K.W. Über die neue Krankheit der Seidenraupe und verwandte Organismen. *Bot Zeitung* 15, 760–761 (1857).

4. Wright, J.H. and Craighead, E.M. Infectious motor paralysis in young rabbits. *J Exp Med* 36, 135–140 (1922).
5. Matsubayashi, H. et al. A case of *Encephalitozoon*-like body infection in man. *A M A Arch Pathol* 67, 181–187 (1959).
6. Didier, E.S. and Weiss, L.M. Microsporidiosis: Current status. *Curr Opin Infect Dis* 19, 485–492 (2006).
7. Balbiani, G. Sur les microsporidies ou sporogspermies des articules. *C R Acad Sci* 95, 1168–1171 (1882).
8. Didier, E.S. Microsporidiosis: An emerging and opportunistic infection in humans and animals. *Acta Trop* 94, 61–76 (2005).
9. Didier, E.S. et al. Antimicrosporidial activities of fumagillin, TNP-470, ovalicin, and ovalicin derivatives in vitro and in vivo. *Antimicrob Agents Chemother* 50, 2146–2155 (2006).
10. Choudhary, M.M. et al. *Tubulinosema* sp. microsporidian myositis in immunosuppressed patient. *Emerg Infect Dis* 17(9), 1727–1730 (2011).
11. Desportes, I. et al. Occurrence of a new microsporidian: *Enterocytozoon bieneusi* n.g., n.sp., in the enterocytes of a human patient with AIDS. *J Protozool* 32, 250–254 (1985).
12. Sobottka, I. et al. Self-limited traveller's diarrhea due to a dual infection with *Enterocytozoon bieneusi* and *Cryptosporidium parvum* in an immunocompetent HIV-negative child. *Eur J Clin Microbiol Infect Dis* 14, 919–920 (1995).
13. Bryan, R.T. and Schwartz, D.A. Epidemiology of microsporidiosis. In Wittner, M. and Weiss L.M. (Eds.), *The Microsporidia and Microsporidiosis.* (pp. 502–516) Washington, DC: American Society of Microbiology (1999).
14. López-Vélez, R. et al. Microsporidiosis in travelers with diarrhea from the tropics. *J Travel Med* 6, 223–227 (1999).
15. Lores, B. et al. Intestinal microsporidiosis due to *Enterocytozoon bieneusi* in elderly human immunodeficiency virus-negative patients from Vigo, Spain. *Clin Infect Dis* 34, 918–921 (2002).
16. Leelayoova, S. et al. Intestinal microsporidiosis in HIV-infected children with acute and chronic diarrhea. *Southeast Asian J Trop Med Public Health* 32, 33–37 (2001).
17. Kotler, D.P. and Orenstein, J.M. Clinical syndromes associated with microsporidiosis. *Adv Parasitol* 40, 321–349 (1998).
18. Sodqi, M. et al. Unusual pulmonary *Enterocytozoon bieneusi* microsporidiosis in an AIDS patient: Case report and review. *Scand J Infect Dis* 36, 230–231 (2004).
19. Orenstein, J.M. et al. Intestinal microsporidiosis as a cause of diarrhea in human immunodeficiency virus-infected patients: A report of 20 cases. *Hum Pathol* 21, 475–481 (1990).
20. Coyle, C.M. et al. Prevalence of microsporidiosis due to *Enterocytozoon bieneusi* and *Encephalitozoon* (*Septata*) *intestinalis* among patients with AIDS-related diarrhea: Determination by polymerase chain reaction to the microsporidian small-subunit rRNA gene. *Clin Infect Dis* 23, 1002–1006 (1996).
21. Tumwine, J.K. et al. *Enterocytozoon bieneusi* among children with diarrhea attending Mulago Hospital in Uganda. *Am J Trop Med Hyg* 67, 299–303 (2002).
22. Pagornrat, W. et al. Carriage rate of *Enterocytozoon bieneusi* in an orphanage in Bangkok, Thailand. *J Clin Microbiol* 47, 3739–3741 (2009).
23. Sak, B. et al. Unapparent microsporidial infection among immunocompetent humans in the Czech Republic. *J Clin Microbiol* 49, 1064–1070 (2011).
24. Deplazes, P. et al. Molecular epidemiology of *Encephalitozoon cuniculi* and first detection of *Enterocytozoon bieneusi* in faecal samples of pigs. *J Eukaryot Microbiol* 43, 93S (1996).
25. Mansfield, K.G. et al. Identification of an *Enterocytozoon bieneusi*-like microsporidian parasite in simian-immunodeficiency virus-inoculated macaques with hepatobiliary disease. *Am J Pathol* 150, 1395–1405 (1997).
26. Sulaiman, I.M. et al. Molecular characterization of microsporidia indicates that wild mammals harbor host-adapted *Enterocytozoon* spp. as well as human-pathogenic *Enterocytozoon bieneusi*. *Appl Environ Microbiol* 69, 4495–4501 (2003).
27. Sulaiman, I.M. et al. Molecular characterization of *Enterocytozoon bieneusi* in cattle indicates that only some isolates have zoonotic potential. *Parasitol Res* 92, 328–334 (2004).
28. Lobo, M.L. et al. Genotypes of *Enterocytozoon bieneusi* in mammals in Portugal. *J Eukaryot Microbiol* 53, S61–S64 (2006).
29. Abe, N. et al. Molecular evidence of *Enterocytozoon bieneusi* in Japan. *J Vet Med Sci* 71, 217–219 (2009).
30. Reetz, J. et al. Identification of *Encephalitozoon cuniculi* genotype III and two novel genotypes of *Enterocytozoon bieneusi* in swine. *Parasitol Int* 58, 285–292 (2009).

31. Santín, M., Vecino, J.A., and Fayer R. A zoonotic genotype of *Enterocytozoon bieneusi* in horses. *J Parasitol* 96, 157–161 (2010).
32. Zhang, X. et al. Identification and genotyping of *Enterocytozoon bieneusi* in China. *J Clin Microbiol* 49, 2006–2008 (2011).
33. Sak, B. et al. The first report on natural *Enterocytozoon bieneusi* and *Encephalitozoon* spp. infections in wild East-European house mice (*Mus musculus musculus*) and West-European house mice (*M. m. domesticus*) in a hybrid zone across the Czech Republic-Germany border. *Vet Parasitol* 178, 246–250 (2011).
34. Li, W. et al. *Cyclospora papionis, Cryptosporidium hominis*, and human-pathogenic *Enterocytozoon bieneusi* in captive baboons in Kenya. *J Clin Microbiol* 49, 4326–4329 (2011).
35. Cotte, L. et al. Waterborne outbreak of intestinal microsporidiosis in persons with and without human immunodeficiency virus infection. *J Infect Dis* 180, 2003–2008 (1999).
36. Franzen, C. and Müller, A. Cryptosporidia and microsporidia–waterborne diseases in the immunocompromised host. *Diagn Microbiol Infect Dis* 34, 245–262 (1999).
37. Cheng, H.W. et al. Municipal wastewater treatment plants as removal systems and environmental sources of human-virulent microsporidian spores. *Parasitol Res* 109, 595–603 (2011).
38. Verweij, J.J. Enterocytozoon. In Liu, D. (Ed.), *Molecular Detection of Human Fungal Pathogens* (pp. 8131–8121). Boca Raton, FL: CRC Press (2011).
39. Cavalier-Smith, T. A 6-kingdom classification and a unified phylogeny. In Shenck, H.E.A. and Schwemmler, W.S. (Eds.), *Endocytobiology II. Intracellular Space as Oligogenetic* (pp. 1027–1034). Berlin, Germany: Walter de Gruyter & Co. (1983).
40. Dong, S. et al. Sequence and phylogenetic analysis of SSU rRNA gene of five microsporidia. *Curr Microbiol* 60, 30–37 (2010).
41. Li, J. et al. Tubulin genes from AIDS-associated microsporidia and implications for phylogeny and benzimidazole sensitivity. *Mol Biochem Parasitol* 78, 289–295 (1996).
42. Cavalier-Smith, T.A revised six-kingdom system of life. *Biol Rev Camb Philos Soc* 73, 203–266 (1998).
43. Vavra, J. and Larsson, J.I. Structure of the Microsporidia. In Wittner, M. and Weiss, L.M. (Eds.), *The Microsporidia and Microsporidiosis* (pp. 7–84). Washington, DC: American Society for Microbiology (1999).
44. Corradi, N. and Keeling, P.J. Microsporidia. A journey through radical taxonomical revisions. *Fungal Biol Rev* 23, 1–8 (2009).
45. Akiyoshi, D.E. et al. Genomic survey of the non-cultivatable opportunistic human pathogen, *Enterocytozoon bieneusi*. *PLoS Pathog* 5(1), 1–10 (2009).
46. Vossbrinck, C.R. et al. Ribosomal RNA sequence suggests microsporidia are extremely ancient eukaryotes. *Nature* 326, 411–414 (1987).
47. Katinka, M.D. et al. Genome sequence and gene compaction of the eukaryote parasite *Encephalitozoon cuniculi*. *Nature* 414, 450–453 (2001).
48. Patterson, D.J. Protozoa: Evolution and systematics. In Hausmann, K. and Hülsmann, N. (Eds.), *Proceeding of the Conference IX International Congress of Protozoology*, pp. 1–13. Stuttgart, Germany: Gustav Fischer-Verlag (1994).
49. Keeling, P.J. and Doolittle, W.F. Alpha-tubulin from early-diverging eukaryotic lineages and the evolution of the tubulin family. *Mol Biol Evol* 13, 1297–1305 (1996).
50. Edlind, T.D. et al. Phylogenetic analysis of beta-tubulin sequences from amitochondrial protozoa. *Mol Phylogenet Evol* 5, 359–367 (1996).
51. Müller, M. What are the microsporidia? *Parasitol Today* 13, 455–456 (1997).
52. Hinkle, G. et al. Genes coding for reverse transcriptase, DNA-directed RNA polymerase, and chitin synthase from the microsporidian *Spraguea lophii*. *Biol Bull* 193, 250–251 (1997).
53. Fast, N.M. Phylogenetic analysis of the TATA box binding protein (TBP) gene from *Nosema locustae*: Evidence for a microsporidia-fungi relationship and spliceosomal intron loss. *Mol Biol Evol* 16, 1415–1419 (1999).
54. Hirt, R.P. Microsporidia are related to Fungi: Evidence from the largest subunit of RNA polymerase II and other proteins. *Proc Natl Acad Sci USA* 96, 580–585 (1999).
55. Hirt, R.P. A mitochondrial Hsp70 orthologue in *Vairimorpha necatrix*: Molecular evidence that microsporidia once contained mitochondria. *Curr Biol* 7, 995–998 (1997).
56. Peyretaillade, E. et al. Microsporidia, amitochondrial protists, possess a 70-kDa heat shock protein gene of mitochondrial evolutionary origin. *Mol Biol Evol* 15, 683–689 (1998).
57. Brown, J.R. and Doolittle, W.F. Gene descent, duplication, and horizontal transfer in the evolution of glutamyl- and glutaminyl-tRNA synthetases. *J Mol Evol* 49, 485–495 (1999).

58. Fast, N.M. and Keeling, P.J. Alpha and beta subunits of pyruvate dehydrogenase E1 from the microsporidian *Nosema locustae*: Mitochondrion-derived carbon metabolism in microsporidia. *Mol Biochem Parasitol* 117, 201–209 (2001).
59. Williams, B.A. et al. Distinct localization patterns of two putative mitochondrial proteins in the microsporidian *Encephalitozoon cuniculi*. *J Eukaryot Microbiol* 55, 131–133 (2008).
60. Williams, B.A. et al. An ADP/ATP-specific mitochondrial carrier protein in the microsporidian *Antonospora locustae*. *J Mol Biol* 375, 1249–1257 (2008).
61. Tsaousis, A.D. et al. A novel route for ATP acquisition by the remnant mitochondria of *Encephalitozoon cuniculi*. *Nature* 453, 553–556 (2008).
62. Goldberg, A.V. et al. Localization and functionality of microsporidian iron-sulphur cluster assembly proteins. *Nature* 452, 624–628 (2008).
63. Lee, S.C. et al. Microsporidia evolved from ancestral sexual fungi. *Curr Biol* 18, 1675–1679 (2008).
64. Slamovits, C.H. et al. Genome compaction and stability in microsporidian intracellular parasites. *Curr Biol* 14, 891–896 (2004).
65. Thomarat, F. et al. Phylogenetic analysis of the complete genome sequence of *Encephalitozoon cuniculi* supports the fungal origin of microsporidia and reveals a high frequency of fast-evolving genes. *J Mol Evol* 59, 780–791 (2004).
66. Corradi, N. et al. Patterns of genome evolution among the microsporidian parasites *Encephalitozoon cuniculi*, *Antonospora locustae* and *Enterocytozoon bieneusi*. *PLoS One* 2(12), 1–8 (2007).
67. Gill, E.E. and Fast, N.M. Assessing the microsporidia-fungi relationship: Combined phylogenetic analysis of eight genes. *Gene* 375, 103–109 (2006).
68. Koestler, T. and Ebersberger, I. Zygomycetes, microsporidia, and the evolutionary ancestry of sex determination. *Genome Biol Evol* 3,186–94 (2011).
69. Rosselló-Mora, R. and Amann, R. The species concept for prokaryotes. *FEMS Microbiol Rev* 25, 39–67 (2001).
70. Weiss, L.M. and Vossbrinck, C.R. Molecular biology, molecular phylogeny, and molecular diagnostic approaches to the microsporidia. In Wittner, M. and Weiss, L.M. (Eds.), *The Microsporidia and Microsporidiosis* (pp. 129–171). Washington, DC: American Society for Microbiology (1999).
71. Cali, A. and Takvorian, P.M. Developmental morphology and life cycles of the microsporidia. In Wittner, M. and Weiss, L.M. (Eds.), *The Microsporidia and Microsporidiosis* (pp. 85–128). Washington, DC: American Society for Microbiology (1999).
72. Santín, M. and Fayer, R. Microsporidiosis: *Enterocytozoon bieneusi* in domesticated and wild animals. *Res Vet Sci* 90(3), 363–371 (2011).
73. Sulaiman, I.M. et al. A molecular biologic study of *Enterocytozoon bieneusi* in HIV-infected patients in Lima, Peru. *J Eukaryot Microbiol* 50, 591–596 (2003).
74. Bern, C. et al. The epidemiology of intestinal microsporidiosis in patients with HIV/AIDS in Lima, Peru. *J Infect Dis* 191, 1658–1664 (2005).
75. Cama, V.A. et al. Transmission of *Enterocytozoon bieneusi* between a child and Guinea pigs. *J Clin Microbiol* 45, 2708–2710 (2007).
76. Chacin-Bonilla, L. et al. Microsporidiosis in Venezuela: Prevalence of intestinal microsporidiosis and its contribution to diarrhea in a group of human immunodeficiency virus-infected patients from Zulia State. *Am J Trop Med Hyg* 74, 482–486 (2006).
77. Brasil, P. et al. Clinical and diagnostic aspects of intestinal microsporidiosis in HIV-infected patients with chronic diarrhea in Rio de Janeiro, Brazil. *Rev Inst Med Trop São Paulo* 42, 299–304 (2000).
78. Weitz, J.C., Botehlo, R., and Bryan, R. Microsporidiosis in patients with chronic diarrhea and AIDS, in HIV asymptomatic patients and in patients with acute diarrhea. *Rev Med Chil* 123, 849–856 (1995).
79. Ferreira, F.M. et al. Intestinal microsporidiosis: A current infection in HIV-seropositive patients in Portugal. *Microbes Infect* 3, 1015–1019 (2001).
80. Lobo, M.L. et al. Microsporidia as emerging pathogens and the implication for public health: A 10-year study on HIV-positive and -negative patients. *Int J Parasitol* 42, 197–205 (2012).
81. del Aguila, C. et al. Microsporidiosis in HIV-positive children in Madrid (Spain). *J Eukaryot Microbiol* 44, 84S–85S (1997).
82. Molina, J.M. et al. Intestinal microsporidiosis in human immunodeficiency virus-infected patients with chronic unexplained diarrhea: Prevalence and clinical and biologic features. *J Infect Dis* 167, 217–221 (1993).
83. Liguory, O. et al. Evidence of different *Enterocytozoon bieneusi* genotypes in patients with and without human immunodeficiency virus infection. *J Clin Microbiol* 39, 2672–2674 (2001).

84. Voglino, M.C. et al. Intestinal microsporidiosis in Italian individuals with AIDS. *Italian J Gastroenterol* 28, 381–386 (1996).
85. Franzen, C. et al. Polymerase chain reaction for microsporidian DNA in gastrointestinal biopsy specimens of HIV-infected patients. *AIDS* 10, F23–F27 (1996).
86. Rinder, H. et al. Evidence for the existence of genetically distinct strains of *Enterocytozoon bieneusi*. *Parasitol Res* 83, 670–672 (1997).
87. Dengjel, B. et al. Zoonotic potential of *Enterocytozoon bieneusi*. *J Clin Microbiol* 39, 4495–4499 (2001).
88. Weber, R. et al. Enteric infections and diarrhea in human immunodeficiency virus-infected persons: prospective community-based cohort study. Swiss HIV Cohort Study. *Arch Intern Med* 159, 1473–1480 (1999).
89. Kyaw, T. et al. The prevalence of *Enterocytozoon bieneusi* in acquired immunodeficiency syndrome (AIDS) patients from the north west of England: 1992–1995. *Br J Biomed Sci* 54, 186–191 (1997).
90. van Gool, T. et al. Diagnosis of intestinal and disseminated microsporidial infections in patients with HIV by a new rapid fluorescence technique. *J Clin Pathol* 46, 694–699 (1993).
91. Sokolova, O.I. et al. Emerging microsporidian infections in Russian HIV-infected patients. *J Clin Microbiol* 49, 2102–2108 (2011).
92. Kucerova, Z. et al. Microsporidiosis and cryptosporidiosis in HIV/AIDS patients in St. Petersburg, Russia: Serological identification of microsporidia and *Cryptosporidium parvum* in sera samples from HIV/AIDS patients. *AIDS Res Hum Retroviruses* 27, 13–15 (2011).
93. Anane, S. et al. Identification of *Enterocytozoon bieneusi* by PCR in stools of Tunisian immunocompromised patients. *Pathol Biol* 59, 234–239 (2011).
94. Maiga, I. et al. Human intestinal microsporidiosis in Bamako (Mali): The presence of *Enterocytozoon bieneusi* in HIV-seropositive patients. *Sante* 7, 257–262 (1997).
95. Alfa Cisse, O. et al. Evaluation of an immunofluorescent-antibody test using monoclonal antibodies directed against *Enterocytozoon bieneusi* and *Encephalitozoon intestinalis* for diagnosis of intestinal microsporidiosis in Bamako (Mali). *J Clin Microbiol* 40, 1715–1718 (2002).
96. Bretagne, S. et al. Prevalence of *Enterocytozoon bieneusi* spores in the stool of AIDS patients and African children not infected by HIV. *Bull Soc Pathol Exot* 86, 351–357 (1993).
97. Espern, A. et al. Molecular study of microsporidiosis due to *Enterocytozoon bieneusi* and *Encephalitozoon intestinalis* among human immunodeficiency virus-infected patients from two geographical areas: Niamey, Niger, and Hanoi, Vietnam. *J Clin Microbiol* 45, 2999–3002 (2007).
98. Ayinmode, A.B., Ojuromi, O.T., and Xiao L. Molecular identification of *Enterocytozoon bieneusi* isolates from Nigerian children. *J Parasitol Res* 2011, 129542 (2011).
99. Ojuromi, O.T. et al. Identification and characterization of microsporidia from fecal samples of HIV-positive patients from Lagos, Nigeria. *PLoS One* 7, 4 (2012).
100. Maikai, B.V. Molecular characterizations of *Cryptosporidium*, *Giardia*, and *Enterocytozoon* in humans in Kaduna State, Nigeria. *Exp Parasitol* 131, 452–456 (2012).
101. Sarfati, C. et al. Prevalence of intestinal parasites including microsporidia in human immunodeficiency virus-infected adults in Cameroon: A cross-sectional study. *Am J Trop Med Hyg* 74, 162–164 (2006).
102. Breton, J. et al. New highly divergent rRNA sequence among biodiverse genotypes of *Enterocytozoon bieneusi* strains isolated from humans in Gabon and Cameroon. *J Clin Microbiol* 45, 2580–2589 (2007).
103. Wumba, R. et al. Intestinal parasites infections in hospitalized AIDS patients in Kinshasa, Democratic Republic of Congo. *Parasite* 17, 321–328 (2010).
104. Wumba, R. et al. *Enterocytozoon bieneusi* identification using real-time polymerase chain reaction and restriction fragment length polymorphism in HIV-infected humans from kinshasa province of the democratic republic of Congo. *J Parasitol Res* 2012, 278028 (2012).
105. Endeshaw, T. et al. Intestinal microsporidiosis in diarrheal patients infected with human immunodeficiency virus-1 in Addis Ababa, Ethiopia. *Jpn J Infect Dis* 59, 306–310 (2006).
106. Cegielski, J.P. et al. *Cryptosporidium*, *Enterocytozoon*, and *Cyclospora* infections in pediatric and adult patients with diarrhea in Tanzania. *Clin Infect Dis* 28, 314–321 (1999).
107. Drobniewski, F.P. et al. Human microsporidiosis in African AIDS patients with chronic diarrhea. *J Infect Dis* 171, 515–516 (1995).
108. van Gool, T. et al. High prevalence of *Enterocytozoon bieneusi* infections among HIV-positive individuals with persistent diarrhoea in Harare, Zimbabwe. *Trans R Soc Trop Med Hyg* 89, 478–480 (1995).
109. Rinder, H. et al. Strain differentiation in microsporidia. *Tokai J Exp Clin Med* 23, 433–437 (1998).
110. Gumbo, T. et al. Intestinal parasites in patients with diarrhea and human immunodeficiency virus infection in Zimbabwe. *AIDS* 13, 819–821 (1999).

111. Samie, A. et al. Microsporidiosis in South Africa: PCR detection in stool samples of HIV-positive and HIV-negative individuals and school children in Vhembe district, Limpopo Province. *Trans R Soc Trop Med Hyg* 101, 547–554 (2007).
112. Agholi, M. et al. Microsporidia and coccidia as causes of persistence diarrhea among liver transplant children: Incidence rate and species/genotypes. *Pediatr Infect Dis J* 32(2), 185–187 (2013).
113. Wang, L. et al. Zoonotic *Cryptosporidium* species and *Enterocytozoon bieneusi* genotypes in HIV-positive patients on antiretroviral therapy. *J Clin Microbiol* 51, 557–563 (2013).
114. Saigal, K. et al. Intestinal microsporidiosis in India: A two year study. *Parasitol Int* 62, 53–56 (2013).
115. Saksirisampant, W. et al. Intestinal parasitic infections: Prevalences in HIV/AIDS patients in a Thai AIDS-care centre. *Ann Trop Med Parasitol* 103, 573–581 (2009).
116. Field, A.S. et al. Microsporidia in the small intestine of HIV-infected patients. A new diagnostic technique and a new species. *Med J* 158, 390–394 (1993).
117. Ryan, N.J. et al. A new trichrome-blue stain for detection of microsporidial species in urine, stool, and nasopharyngeal specimens. *J Clin Microbiol* 31, 3264–3269 (1993).
118. Stark, D. et al. Limited genetic diversity among genotypes of *Enterocytozoon bieneusi* strains isolated from HIV-infected patients from Sydney, Australia. *J Med Microbiol* 58, 355–357 (2009).
119. Matos, O., Lobo, M.L., and Xiao, L. Epidemiology of *Enterocytozoon bieneusi* infection in humans. *J Parasitol Res* 2012, 981424 (2012).
120. Mungthin, M. et al. Asymptomatic intestinal microsporidiosis in Thai orphans and child-care workers. *Trans R Soc Trop Med Hyg* 95, 304–306 (2001).
121. Bachur, T.P. et al. Enteric parasitic infections in HIV/AIDS patients before and after the highly active antiretroviral therapy. *Braz J Infect Dis* 12, 115–122 (2008).
122. Snowden, K.F. and Shadduck, J.A. Microsporidia in higher vertebrates. In Wittner, M. and Weiss, L.M. (Eds.), *The Microsporidia and Microsporidiosis*. (pp. 393–417) Washington, DC: American Society of Microbiology (1999).
123. Didier, E.S. et al. Epidemiology of microsporidiosis: Sources and modes of transmission. *Vet Parasitol* 126, 145–166 (2004).
124. Weber, R., Deplazes, P., and Schwartz, D. Diagnosis and clinical aspects of human microsporidiosis. *Contrib Microbiol* 6, 166–192 (2000).
125. Hutin, Y.J. et al. Risk factors for intestinal microsporidiosis in patients with human immunodeficiency virus infection: A case-control study. *J Infect Dis* 178, 904–907 (1998).
126. Leelayoova, S. et al. Transmission of *Enterocytozoon bieneusi* genotype a in a Thai orphanage. *Am J Trop Med Hyg* 73, 104–107 (2005).
127. Fuentealba, I.C. et al. Hepatic lesions in rabbits infected with *Encephalitozoon cuniculi* administered per rectum. *Vet Pathol* 29, 536–540 (1992).
128. Tzipori, S. et al. Transmission and establishment of a persistent infection of *Enterocytozoon bieneusi*, derived from a human with AIDS, in simian immunodeficiency virus-infected rhesus monkeys. *J Infect Dis* 175, 1016–1020. (1997).
129. Koudela, B. et al. The human isolate of *Brachiola algerae* (Phylum Microspora): Development in SCID mice and description of its fine structure features. *Parasitology* 123, 153–162 (2001).
130. Dascomb, K. et al. Microsporidiosis and HIV. *J Acquir Immune Defn Syndr* 24, 290–292 (2000).
131. Craun, G.F. et al. Causes of outbreaks associated with drinking water in the United States from 1971 to 2006. *Clin Microbiol Rev* 23, 507–528 (2010).
132. Corso, P.S. et al. Cost of illness in the 1993 waterborne *Cryptosporidium* outbreak, Milwaukee, Wisconsin. *Emerg Infect Dis* 9, 426–431 (2003).
133. Karanis, P., Kourenti, C., and Smith, H. Waterborne transmission of protozoan parasites: A worldwide review of outbreaks and lessons learnt. *J Water Health Rev* 5, 1–38 (2007).
134. Slifko, T.R., Smith, H.V., and Rose, J.B. Emerging parasite zoonoses associated with water and food. *Int J Parasitol* 30, 1379–1393 (2000).
135. Shadduck, J.A. and Polley, M.B. Some factors influencing the in vitro infectivity and replication of *Encephalitozoon cuniculi*. *J Protozool* 25, 491–496 (1978).
136. Waller, T. Sensitivity of *Encephalitozoon cuniculi* to various temperatures, disinfectants and drugs. *Lab Anim* 13, 227–230 (1979).
137. Schmidt, E.C. and Shadduck, J.A. Murine encephalitozoonosis model for studying the host-parasite relationship of a chronic infection. *Infect Immun* 40, 936–942 (1983).
138. Sparfel, J.M. et al. Detection of microsporidia and identification of *Enterocytozoon bieneusi* in surface water by filtration followed by specific PCR. *J Eukaryot Microbiol* 44, 78S (1997).

139. Dowd, S.E., Gerba, C.P., and Pepper, I.L. Confirmation of the human-pathogenic microsporidia *Enterocytozoon bieneusi*, *Encephalitozoon intestinalis*, and *Vittaforma corneae* in water. *Appl Environ Microbiol* 64, 3332–3335 (1998).
140. U.S. Environmental Protection Agency (EPA). Announcement of the drinking water contaminant candidate list: Notice. *Fed Regist* 63, 10272–10287 (1998).
141. Thurston-Enriquez, J.A. et al. Detection of protozoan parasites and microsporidia in irrigation waters used for crop production. *J Food Prot* 65, 378–382 (2002).
142. Coupe, S. et al. Detection of *Cryptosporidium*, *Giardia* and *Enterocytozoon bieneusi* in surface water, including recreational areas: A one-year prospective study. *FEMS Immunol Med Microbiol* 47, 351–359 (2006).
143. Calvo, M. et al. Prevalence of *Cyclospora* sp., *Cryptosporidium* sp, microsporidia and fecal coliform determination in fresh fruit and vegetables consumed in Costa Rica. *Arch Latinoam Nutr* 54, 428–432 (2004).
144. da Costa, M.L.. Epidemiologia e caracterização de espécies implicadas na microsporidiose humana em Portugal por análise parasitológica e molecular. PhD thesis, Instituto de Higiene e Medicina Tropical, Universidade Nova de Lisboa, Portugal, p. 280 (2010).
145. Ben Ayed, L. et al. Survey and genetic characterization of waste water in Tunisia for *Cryptosporidium* spp., *Giardia duodenalis*, *Enterocytozoon bieneusi*, *Cyclospora cayetanensis* and *Eimeria* spp. *J Water Health* 10, 431–444 (2012).
146. Li, N. et al. Molecular surveillance of *Cryptosporidium* spp., *Giardia duodenalis*, and *Enterocytozoon bieneusi* by genotyping and subtyping parasites in wastewater. *PLoS Negl Trop Dis* 6, e1809 (2012).
147. Orlandi, P.A. et al. Parasites and the food supply. *Food Technol* 56, 72–81 (2002).
148. Buckholt, M.A., Lee, J.H., and Tzipori, S. Prevalence of *Enterocytozoon bieneusi* in swine: An 18-month survey at a slaughterhouse in Massachusetts. *Appl Environ Microbiol* 68, 2595–2599 (2002).
149. Jedrzejewski, S. et al. Quantitative assessment of contamination of fresh food produce of various retail types by human-virulent microsporidian spores. *Appl Environ Microbiol* 73, 4071–4073 (2007).
150. Decraene, V. et al. First reported foodborne outbreak associated with microsporidia, Sweden. *Epidemiol Infect* 140, 519–527 (2012).
151. Weber, R. et al. Pulmonary and intestinal microsporidiosis in a patient with the acquired immunodeficiency syndrome. *Am Rev Respir Dis* 146, 1603–1605 (1992).
152. Rinder, H. Transmission of microsporidia to humans: Water-borne, food-borne, air-borne, zoonotic, or anthroponotic? *Southeast Asian J Trop Med Public Health* 35, 54–57 (2004).
153. Katzwinkel-Wladarsch, S. et al. Direct amplification and species determination of microsporidian DNA from stool specimens. *Trop Med Int Health* 1, 373–378 (1996).
154. Xiao, L. et al. Molecular epidemiology of human microsporidiosis caused by *Enterocytozoon bieneusi*. *Southeast Asian J Trop Med Public Health* 35(Suppl. 1), 40–47 (2004).
155. Widmer, G. and Akiyoshy, D.E. Host-specific segregation of ribosomal nucleotide sequence diversity in the microsporidian *Enterocytozoon bieneusi*. *Infect Genet Evol* 10, 122–128 (2010).
156. Henriques-Gil, N. et al. Phylogenetic approach to the variability of the microsporidian *Enterocytozoon bieneusi* and its implications for inter- and intrahost transmission. *Appl Environ Microbiol* 76, 3333–3342 (2010).
157. Liguory, O. et al. Determination of types of *Enterocytozoon bieneusi* strains isolated from patients with intestinal microsporidiosis. *J Clin Microbiol* 36, 1882–1885 (1998).
158. ten Hove, R.J. et al. Characterization of genotypes of *Enterocytozoon bieneusi* in immunosuppressed and immunocompetent patient groups. *J Eukaryot Microbiol* 56, 388–393 (2009).
159. Sadler, F. et al. Genotyping of *Enterocytozoon bieneusi* in AIDS patients from the north west of England. *J Infect* 44, 39–42 (2002).
160. Santín, M. et al. *Cryptosporidium*, *Giardia* and *Enterocytozoon bieneusi* in cats from Bogota (Colombia) and genotyping of isolates. *Vet Parasitol* 141, 334–339 (2006).
161. Santín, M. et al. *Enterocytozoon bieneusi* in dogs in Bogota, Colombia. *Am J Trop Med Hyg* 79, 215–217 (2008).
162. Breitenmoser, A.C. et al. High prevalence of *Enterocytozoon bieneusi* in swine with four genotypes that differ from those identified in humans. *Parasitology* 118, 447–453 (1999).
163. Lobo, M.L. et al. Identification of potentially human-pathogenic *Enterocytozoon bieneusi* genotypes in various birds. *Appl Environ Microbiol* 72, 7380–7382 (2006).
164. Santín, M., Trout, J.M., and Fayer, R. *Enterocytozoon bieneusi* genotypes in dairy cattle in the eastern United States. *Parasitol Res* 97, 535–538 (2005).

165. Leelayoova, S. et al. Genotypic characterization of *Enterocytozoon bieneusi* in specimens from pigs and humans in a pig farm community in Central Thailand. *J Clin Microbiol* 47, 1572–1574 (2009).
166. Kasicková, D. et al. Sources of potentially infectious human microsporidia: molecular characterisation of microsporidia isolates from exotic birds in the Czech Republic, prevalence study and importance of birds in epidemiology of the human microsporidial infections. *Vet Parasitol* 165, 125–130 (2009).
167. Abe, N. and Kimata, I. Molecular survey of *Enterocytozoon bieneusi* in a Japanese porcine population. *Vector Borne Zoonotic Dis* 10, 425–427 (2010).
168. Leelayoova, S. et al. Identification of genotypes of *Enterocytozoon bieneusi* from stool samples from human immunodeficiency virus-infected patients in Thailand. *J Clin Microbiol* 44, 3001–3004 (2006).
169. Jeong, D.K. et al. Occurrence and genotypic characteristics of *Enterocytozoon bieneusi* in pigs with diarrhea. *Parasitol Res* 102, 123–128 (2007).
170. Sak, B. et al. First report of *Enterocytozoon bieneusi* infection on a pig farm in the Czech Republic. *Vet Parasitol* 153, 220–224 (2008).
171. John, D.E. et al. Chlorine and ozone disinfection of *Encephalitozoon intestinalis* spores. *Water Res* 39, 2369–2375 (2005).
172. Jordan, C.N. et al. Effects of high-pressure processing on in vitro infectivity of *Encephalitozoon cuniculi*. *J Parasitol* 91, 1487–1488 (2005).
173. Jordan, C.N., Dicristina, J.A., and Lindsay, D.S. Activity of bleach, ethanol and two commercial disinfectants against spores of *Encephalitozoon cuniculi*. *Vet Parasitol* 136, 343–346 (2006).
174. Liu, D. and Didier, E.S. Encephalitozoon. In Liu, D. (Ed.), *Molecular Detection of Human Fungal Pathogens*. Chapter 92. (pp. 803–811) Boca Raton, FL: CRC Press (2011).
175. Kotler, D.P. and Orenstein, J.M. Clinical syndromes associated with microsporidiosis. In Wittner, M. and Weiss, L.M. (Eds.), *The Microsporidia and Microsporidiosis* (pp. 258–292). Washington, DC: American Society of Microbiology (1999).
176. Morpeth, S.C. and Thielman, N.M. Diarrhea in patients with AIDS. *Curr Treat Opt Gastroenterol* 9, 23–37 (2006).
177. Johnson, L.R. Regulation of gastrointestinal mucosal growth. *Physiol Rev* 68, 456–469 (1988).
178. Shadduck, J.A. and Orenstein, J.M. Comparative pathology of microsporidiosis. *Arch Pathol Lab Med* 117, 1215–1219 (1993).
179. Greenson, J.K. et al. AIDS enteropathy: Occult enteric infections and duodenal mucosal alterations in chronic diarrhea. *Ann Intern Med* 114, 366–372 (1991).
180. Mor, S.M. et al. Microsporidiosis and malnutrition in children with persistent diarrhea, Uganda. *Emerg Infect Dis* 15, 49–52 (2009).
181. Garcia, L.S. Laboratory identification of the microsporidia. *J Clin Microbiol* 40, 1892–1901 (2002).
182. Fedorko, D.P. and Hijazi, Y.M. Application of molecular techniques to the diagnosis of microsporidial infection. *Emerg Infect Dis* 2, 183–191 (1996).
183. Franzen, C. and Muller, A. Molecular techniques for detection, species differentiation, and phylogenetic analysis of microsporidia. *Clin Microbiol Rev* 12, 243–285 (1999).
184. Weber, R. et al. Intestinal *Enterocytozoon bieneusi* microsporidiosis in an HIV-infected patient: Diagnosis by ileo-colonoscopic biopsies and long-term follow up. *Clin Investig* 70, 1019–1023 (1992).
185. Franzen, C. et al. Tissue diagnosis of intestinal microsporidiosis using a fluorescent stain with Uvitex 2B. *J Clin Pathol* 48, 1009–1010 (1995).
186. Didier, E.S. et al. Comparison of three staining methods for detecting microsporidia in fluids. *J Clin Microbiol* 33, 3138–3145 (1995).
187. Moura, H. et al. A new and improved "quick-hot Gram-chromotrope" technique that differentially stains microsporidian spores in clinical samples, including paraffin-embedded tissue sections. *Arch Pathol Lab Med* 121, 888–893 (1997).
188. Conteas, C.N. et al. Fluorescence techniques for diagnosing intestinal microsporidiosis in stool, enteric fluid, and biopsy specimens from acquired immunodeficiency syndrome patients with chronic diarrhea. *Arch Pathol Lab Med* 120, 847–853 (1996).
189. DeGirolami, P.C. et al. Diagnosis of intestinal microsporidiosis by examination of stool and duodenal aspirate with Weber's modified trichrome and Uvitex 2B strains. *J Clin Microbiol* 33, 805–810 (1995).
190. Matos, O. et al. Diagnostic use of 3 techniques for identification of microsporidian spores among AIDS patients in Portugal. *Scand J Infect Dis* 34, 591–593 (2002).
191. Aldras, A.M. et al. Detection of microsporidia by indirect immunofluorescence antibody test using polyclonal and monoclonal antibodies. *J Clin Microbiol* 32, 608–612 (1994).

192. Beckers, P.J. et al. *Encephalocytozoon intestinalis*-specific monoclonal antibodies for laboratory diagnosis of microsporidiosis. *J Clin Microbiol* 34, 282–285 (1996).
193. Sak, B., Sakova, K., and Ditrich, O. Effects of a novel anti-exospore monoclonal antibody on microsporidial development in vitro. *Parasitol Res* 92, 74–80 (2004).
194. Accoceberry, I. et al. Production of monoclonal antibodies directed against the microsporidium *Enterocytozoon bieneusi*. *J Clin Microbiol* 37, 4107–4112 (1999).
195. Kucerova-Pospisilova, Z. and Ditrich, O. The serological surveillance of several groups of patients using antigens of *Encephalitozoon hellem* and *E. cuniculi* antibodies to microsporidia in patients. *Folia Parasitol (Praha)* 45, 108–112 (1998).
196. Sak, B. et al. Seropositivity for *Enterocytozoon bieneusi*, Czech Republic. *Emerg Infect Dis* 16, 335–337 (2010).
197. Peek, R. et al. Carbohydrate moieties of microsporidian polar tube proteins are targeted by immunoglobulin G in immunocompetent individuals. *Infect Immun* 73, 7906–7913 (2005).
198. Xu, Y. and Weiss, L.M. The microsporidian polar tube: A highly specialised invasion organelle. *Int J Parasitol* 35, 941–953 (2005).
199. Rinder, H. et al. Blinded, externally controlled multicenter evaluation of light microscopy and PCR for detection of microsporidia in stool specimens. The Diagnostic Multicenter Study Group on Microsporidia. *J Clin Microbiol* 36, 1814–1818 (1998).
200. Zhu, X. et al. Small subunit rRNA sequence of *Enterocytozoon bieneusi* and its potential diagnostic role with use of the polymerase chain reaction. *J Infect Dis* 168, 1570–1575 (1993).
201. Wang, Z., Orlandi, P.A., and Stenger, D.A. Simultaneous detection of four human pathogenic microsporidian species from clinical samples by oligonucleotide microarray. *J Clin Microbiol* 43, 4121–4128 (2005).
202. Verweij, J.J. et al. Multiplex detection of *Enterocytozoon bieneusi* and *Encephalitozoon* spp. in fecal samples using real-time PCR. *Diagn Microbiol Infect Dis* 57, 163–167 (2007).
203. Graczyk, T.K. et al. Retrospective species identification of microsporidian spores in diarrheic fecal samples from human immunodeficiency virus/AIDS patients by multiplexed fluorescence in situ hybridization. *J Clin Microbiol* 45, 1255–1260 (2007).
204. Wolk, D.M. et al. Real-time PCR method for detection of *Encephalitozoon intestinalis* from stool specimens. *Clin Microbiol* 40, 3922–3928 (2002).
205. Procop, G.W. Molecular diagnostics for the detection and characterization of microbial pathogens. *Clin Infect Dis* 45, S99–S111 (2007).
206. Galván, A.L. et al. First cases of microsporidiosis in transplant recipients in Spain and review of the literature. *J Clin Microbiol Rev* 49, 1301–1306 (2011).
207. Carr, A. et al. Treatment of HIV-1 associated microsporidiosis and cryptosporidiosis with combination antiretroviral therapy. *Lancet* 351, 256–261 (1998).
208. Molina, J.M. et al. Potential efficacy of fumagillin in intestinal microsporidiosis due to *Enterocytozoon bieneusi* in patients with HIV infection: results of a drug screening study. The French Microsporidiosis Study Group. *AIDS* 11, 1603–1610 (1997).
209. Didier, E.S. et al. Therapeutic strategies for human microsporidia infections. *Expert Rev Anti Infect Ther* 3, 419–434 (2005).
210. Chalifoux, L.V. et al. *Enterocytozoon bieneusi* as a cause of proliferative serositis in simian immunodeficiency virus-infected immunodeficient macaques (*Macaca mulatta*). *Arch Pathol Lab Med* 124, 1480–1484 (2000).
211. Samra, N.A. et al. *Enterocytozoon bieneusi* at the wildlife/livestock interface of the Kruger National Park, South Africa. *Vet Parasitol* 190(3–4), 587–590 (2012).
212. Wagnerová, P. et al. *Enterocytozoon bieneusi* and *Encephalitozoon cuniculi* in horses kept under different management systems in the Czech Republic. *Vet Parasitol* 190, 573–577 (2012).
213. Rinder, H. et al. Close genotypic relationship between *Enterocytozoon bieneusi* from humans and pigs and first detection in cattle. *J Parasitol* 86, 185–188 (2000).
214. Lee, J.H. Prevalence and molecular characteristics of *Enterocytozoon bieneusi* in cattle in Korea. *Parasitol Res* 101, 391–396 (2007).
215. Lallo M.A. et al. *Encephalitozoon* and *Enterocytozoon* (microsporidia) spores in stool from pigeons and exotic birds: Microsporidia spores in birds. *Vet Parasitol* 190, 418–22 (2012).
216. Fayer, R., Santín, M., and Trout, J.M. *Enterocytozoon bieneusi* in mature dairy cattle on farms in the eastern United States. *Parasitol Res Rev* 102, 15–20 (2007).
217. Fayer, R. et al. Detection of concurrent infection of dairy cattle with *Blastocystis, Cryptosporidium, Giardia*, and *Enterocytozoon* by molecular and microscopic methods. *Parasitol Res* 111, 1349–1355 (2012).

218. Reetz, J. et al. First detection of the microsporidium *Enterocytozoon bieneusi* in non-mammalian hosts (chickens). *Int J Parasitol* 32, 785–787 (2002).
219. Santin, M. and Fayer, R. *Enterocytozoon bieneusi* genotype nomenclature based on the internal transcribed spacer sequence: A consensus. *J Eukaryot Microbiol* 56, 34–38 (2009).
220. Graczyk, T.K. et al. Relationships among bather density, levels of human waterborne pathogens, and fecal coliform counts in marine recreational beach water. *Parasitol Res* 106(5), 1103–1108 (2010).
221. Galván, A.L. et al. Molecular characterization of human-pathogenic microsporidia and *Cyclospora cayetanensis* isolated from various water sources in Spain: A year-long longitudinal study. *Appl Environ Microbiol* 79(2), 449–459 (2013).
222. Graczyk, T.K. et al. Human zoonotic enteropathogens in a constructed free-surface flow wetland. *Parasitol Res* 105(2), 423–428 (2009).
223. Rinder, H. et al. Microsporidiosis of man: Where is the reservoir? *Mitt Osterr Ges Trop Med Parasitol* 22, 1–6 (2000).

21 Mycotoxins of *Fusarium* spp. Biochemistry and Toxicology

Anthony De Lucca and Thomas J. Walsh

CONTENTS

21.1 Introduction 323
21.2 Nonestrogenic Toxins 324
    21.2.1 Trichothecenes 324
        21.2.1.1 Trichothecenes: Type A 324
        21.2.1.2 Trichothecenes: Type B 326
    21.2.2 Fumonisins 329
21.3 Estrogenic Toxins: Zearalenone and Its Metabolites 331
21.4 Emerging Toxins 334
    21.4.1 Enniatins 334
    21.4.2 Beauvericin 336
    21.4.3 Moniliformin 336
    21.4.4 Fusaproliferin 337
21.5 *Masked Mycotoxins* Produced by Plants 338
21.6 Conclusion 339
References 339

21.1 INTRODUCTION

Fungi are ubiquitous in nature and are found on cereal grains before and after harvest. Among such fungi are members of the *Fusarium* species, which are septate, filamentous, and saprophytic though they can become pathogens of plants (especially cereal grains) or immunocompromised humans.* Climatic conditions in temperate regions, especially the north (e.g., Europe, America, Asia), favor the growth of *Fusarium* species and production of mycotoxins on contaminated grains [1]. Globally, these toxins contaminate an estimated 25% of cereal crops resulting in frequent exposure via foods in many populations and are considered among the most serious mycotoxins affecting humans and animals [2,3]. The trichothecenes (T-2, HT-2, deoxynivalenol [DON]), fumonisin B_1, and zearalenone (ZON) constitute the most relevant *Fusarium* toxins for human and animal health [4–7]. The *emerging toxin* group, comprised the enniatins, beauvericin, moniliformin, and fusaproliferin, are not directly linked to known human diseases. These toxins vary in structure, target organs, and/or their toxic dose level, as well as their mode of action. Finally, trichothecenes, ZON, and their metabolites are sometimes bound to plant constituents, thus forming less toxic *masked mycotoxins*, not detected by standard protocols, which can return to their toxic form after ingestion.

---

* For a review of *Fusarium* mycoses, see Manikandan et al. *Fusarium*. In Liu, D. (Ed.), *Molecular Detection of Human Fungal Pathogens*, CRC Press, Boca Raton, FL, pp. 417–433, 2011.

21.2 NONESTROGENIC TOXINS

21.2.1 TRICHOTHECENES

The trichothecenes, produced by *Fusarium* and several unrelated fungal genera, are a large family of low-molecular-weight compounds that share a tetracyclic sesquiterpenoid 12,13,-epoxytrichothec-9-ene ring [8,9] (Figure 21.1). They are toxic to animals and humans and are very stable molecules that withstand the conditions of cooking, baking, and irradiation [10–13]. In general, monogastric animals are more susceptible to trichothecenes than ruminants [14]. Toxicity is due to their binding to ribosomes leading to inhibition of protein synthesis [15,16]. Trichothecenes also have immunomodulatory properties on cellular immunity based upon dose, exposure frequency, and timing of functional immune assay [17]. Approximately 190 members of the trichothecene family have been identified with the most important being T-2, HT-2, DON, and nivalenol (NIV) [18].

21.2.1.1 Trichothecenes: Type A

21.2.1.1.1 T-2

Type A toxins, the most prominent of which are T-2 and HT-2, have cytotoxic, immunotoxic, and neurotoxic properties [19] (Figure 21.1). *Fusarium solani*, *Fusarium langsethiae*, and *Fusarium sporotrichioides* produce the water-soluble T-2 (also known as fusariotoxin, insariotoxin, NSC 138780) that is found in corn, barley, oats, wheat, sorghum, and peanuts [14,20]. It is considered the most toxic trichothecene and is about 10-fold more toxic than DON, a Type B trichothecene that is more prevalent in cereal grains than T-2 [15,19,21]. T-2 causes feed refusal in swine and is implicated as the causative agent of this disease in sheep [22,23]. No regulations (as of 2010) concerning only T-2's concentration in grains have been established [24]. However, in 2011, the European Food Safety Authority (EFSA) defined the combined allowable maximum daily intake of T-2 and HT-2 at 100 ng/kg body weight [25].

Ingestion of these toxins in food or feed is the most common exposure route. After ingestion by swine, T-2 is absorbed orally and then passes through the bloodstream to the liver where it inhibits P4503A and P450 activity [26,27]. Toxicokinetic studies suggest ingested T-2 does not accumulate in any particular organ but may attack all tissues [28]. T-2 causes oxidative damage resulting in depletion of the mitochondrial membrane potential reflecting mitochondrial dysfunction and plays a critical role in T-2-induced apoptosis [29,30]. Studies in pigs show that it inhibits protein synthesis by binding to cell ribosomes and prevents hepatic cytochrome P4503A activity [26,31–33]. In vitro, T-2 has the potential to act as an endocrine disruptor as observed by their effect on H295R (human adrenocortical carcinoma cell) viability, steroidogenesis, and alteration of gene expression [34].

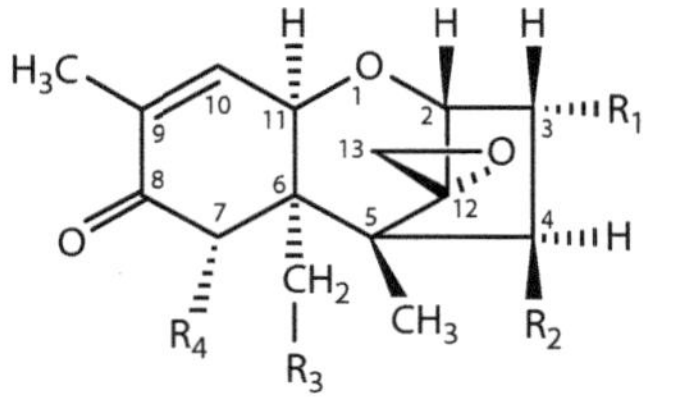

Type A trichothecenes
T-2 toxin; R = acetyl
H-2 toxin: R = H

Type B trichothecenes
Nonketone side chain or no side chain at C-8

FIGURE 21.1 Trichothecenes: general structure.

A massive European study showed that wheat and wheat-based products represent the major source of T-2 (and HT-2) in the diet of poultry and humans [35,36]. The ingestion of T-2-contaminated grains can cause lethal and sublethal toxicoses in farm animals and humans [37]. Pigs ingested with T-2 develop severe cutaneous ulceration and necrosis of the skin around the mouth and prepuce [38]. In chicken broilers and pigs, ingested T-2 induces reduced food intake, while in the former, it causes severe oral lesions and immunological dysfunction [39]. This feed refusal may be due to an increase in brain indoleamines induced by orally administered T-2 [40]. Ingestion of T-2-contaminated grain can be lethal to sheep due to rumenitis and ulcerative abomasitis, depletion of lymphocytes in lymphoid organs, necrosis of the exocrine pancreas, myocarditis, and intense edema of the skin and brain [41]. Lesions were observed in the sheep hearts, which suggest an additional cardiotoxic effect [41].

Lethal alimentary toxic aleukia (ATA) was reported in humans during World War II in Russia and was attributed to T-2 produced by *Fusarium poae* and *F. sporotrichioides* growing on overwintered wheat used for food [37,42]. In ATA, T-2 causes oral and alimentary tract necrotic lesions and reduction of the immune capacity due to leukopenia. Kashin–Beck disease is a chronic osteoarthritic disease that is endemic in parts of China. Reports suggest that this disease is caused by a dietary selenium deficiency combined with a high concentration of fulvic acid and ingestion of T-2 in contaminated grain, which together causes a reduction of the mRNA expression of antioxidants in cartilage tissue [43–45].

T-2 has immunotoxic properties, resulting in an increased threat of infections, as shown in vivo and in vitro studies. Subclinical concentrations (1324 or 2102 μg T-2/kg body weight) of T-2 fed to pigs reduce the humoral immune response [27]. In pigs, a nebulized dose of T-2 at 9 mg/kg (1.8–2.7 mg/kg retained by the animals) can be lethal to pigs that developed severe pneumonia as early as 8 h after exposure, while surviving pigs displayed mild pulmonary injury as well as transient impairment of pulmonary immunity [46]. T-2 ingested at low concentrations affects toll-like receptor activation by decreasing pattern recognition of pathogens and thus interferes with the initiation of the inflammatory immune response against bacteria and viruses [47]. In mice pretreated (intraperitoneally [IP]) with T-2 and then challenged with reovirus, an increase in lung viral burden, bronchopneumonia, and pulmonary cellular infiltration occur over the control group [48]. T-2 increases both the extent of gastrointestinal tract reovirus infection and fecal shedding with a dose as low as 50 μg/kg body weight, which also suppresses immunoglobulin and IFNγ responses [49]. An oral administration of T-2 in mice followed by a challenge with *Salmonella typhimurium* results in markedly larger and more bacterial-related lesions in the spleens, kidneys, and livers, leading to greater mortality when compared to the control group [50]. Histopathology studies show that chickens fed with T-2-contaminated feed developed moderate to severe necrosis to lymphoid organs and extensive lymphocyte depletion, resulting in a suppression of the immune system and aggravated pathology and pathogenesis with infectious bronchitis virus [51]. In vitro, T-2 exerts a high immunosuppressing effect in human peripheral blood mononuclear cells and causes apoptosis in human B and T lymphoid cell lines [52,53].

T-2 also has neurotoxic properties and can disrupt the blood–brain barrier [54,55]. In vitro tests with normal astrocytes in primary culture show that these cells were highly sensitive to the cytotoxic properties of T-2 [56]. T-2 also displays oxidative damage to the brain in a murine model [57].

Though ingestion is considered the main route of exposure, T-2 can also enter the system via the skin [57,58]. Dermal exposure to T-2 in mice induces alteration of the blood–brain barrier permeability that is mediated through oxidative stress and activation of matrix metalloproteinase-9 and proinflammatory cytokines, which indicates an earlier effect on the spleen than in the brain [54,57]. Topical exposure to T-2 results in skin inflammation and injury. These effects are mediated by a number of factors including oxidative stress, an increase in inflammatory cytokines, matrix metalloproteinase activity, myeloperoxidase and p38 MAPK activation, and apoptosis of epidermal cells [59].

Trichothecenes, including T-2, are present in grain dust and are a potential health risk through inhalation [60,61]. Besides its aforementioned immunotoxicity properties, an inhaled dose of T-2 in rats and guinea pigs was, respectively, approximately 20- and 2-fold as toxic as was the same dose given IP. Lesions in the major organs after inhalation were similar in both species to those described after systemic administration. However, when T-2 was inhaled, alterations of respiratory tract tissue were minimal and could not account for the increase in toxicity [62].

21.2.1.1.2 HT-2

Both T-2 and HT-2 are consistently produced by *F. langsethiae* and *F. sporotrichioides* isolates, while only a small percentage of *F. poae* isolates produce both [63,64]. In Europe as with T-2, wheat-containing products represent the major source of HT-2 in the human diet with the mean intake of T-2 plus HT-2 being 0.06 µg/kg body weight [35], which is lower than the EFSA combined maximum allowable intake (100 ng/kg body weight) for these two toxins in combination [25].

HT-2 is also produced as a degradation product of T-2 by the liver and gastric bacteria by the hydrolysis of the acetyl group at the C-4 position of T-2 [14]. In bovine and human liver homogenates (the latter being more rapid), the acetyl group at the C-4 position of T-2 is cleaved via hydrolysis resulting in the production of HT-2 [65]. Bacteria present in mammalian intestinal tracts and in the rumen, as well as in the environment, can also degrade T-2 into HT-2 [14,66–69].

In vitro, HT-2 is less cytotoxic than the parent compound, T-2, to human renal proximal tubule epithelial cells and normal lung fibroblasts [70]. When tested against human liver, epithelial, macrophage, and lymphocyte cell lines, the IC_{50} for HT-2 and T-2 values are 7.5–55.8 and 4.4–10.8 nmol/L, respectively, showing that the loss of toxicity exists in many different cell types [71]. Both toxins are also lethal for nonhuman cells with, for example, the IC_{50} of HT-2 and T-2 being 68 and 12 ng/mL, respectively, for mouse fibroblasts [72].

Neurotoxicity of HT-2 is also lower than that of T-2 as observed with porcine brain capillary cells (used to mimic the blood–brain barrier) where separate treatments with 10 nM HT-2 and T-2 are lethal to 4% and 35%, respectively, of these cells. Further, blood–brain barrier function impairment occurs with 200 nM of HT-2 but at the much lower level of 75 nM of T-2 [56]. Though both toxins inhibit V79 lung fibroblast cell growth in vitro, the IC_{50} value for HT-2 is greater than that for T-2, 0.0142 and 0.003 µM, respectively [73].

21.2.1.2 Trichothecenes: Type B

Type B trichothecenes differ structurally from the Type A group in that the latter have a nonketone side chain at the C-8 position, or no side chain at all, while, with the former, this position is occupied by a ketone functional group [74]. The major Type B trichothecenes are DON and NIV, which share an epoxy-sesquiterpenoid structure [75]. In wheat, these fungi can infect the developing kernels, causing *Fusarium* head blight.

21.2.1.2.1 DON

DON is produced by *Fusarium graminearum* and *Fusarium culmorum* (in some geographical locations). Of the mycotoxins, DON (also called vomitoxin) has the greatest impact on the safety of wheat and its derived products. It is responsible for feed refusal in swine, which are the most susceptible animal to this toxin, as well as toxicosis in other animals and humans [76]. In Europe, 57% of over 11,000 cereal grain samples contained DON, indicating that wheat and wheat-containing products are the major source of dietary intake of DON by humans. Similar results are reported in a smaller study of poultry feed [35,36].

The European Union governing body has set DON limits of 1250, 750, and 500 µg/kg in cereals, flour, and bread, respectively, with a temporary tolerable daily intake for DON of 1.0 µg/kg body weight per day [77,78]. Though DON is generally considered a problem in northern temperate climes [79], it is also found in maize growing in Nigeria and colonized by *Fusarium* species [80].

In the United States, DON is also a problem in barley, a major constituent in the making of beer, for which brewers set a DON limit of 0.0–0.5 ppm. The U.S. Food and Drug Administration advises a DON maximum concentration of 1 ppm in finished wheat products (e.g., flour, bran, and germ) intended for human consumption [81].

Animals are susceptible to the toxic effects of DON via the intestines. An in vitro intestinal absorption study based on healthy swine indicates that DON is absorbed at a 51% rate [82]. DON is known as *vomitoxin* due to its emetic properties, but it can also cause diarrhea, muscle weakness, tremors, and twilight coma in swine and is associated with human gastroenteritis [83,84]. Swine are the most susceptible with concentrations of 12 and 20 ppm in corn causing feed refusal and vomiting, respectively [85]. The plasma levels of two gut peptides, YY3-36 and 5-hydroxytryptamine, are significantly elevated within 30–60 min of IP exposure to DON (0.1 and 0.25 mg/kg body weight) and are mediators of DON-induced emesis [86].

As with the Type A trichothecenes, DON is cytotoxic, genotoxic, immunotoxic, and neurotoxic in both human and animal cell lines. Cytotoxicity occurs via apoptosis-associated rRNA cleavage through PKR/Hck-mediated p38 activation and subsequent p53-dependent induction of the extrinsic and intrinsic apoptosis pathways [87]. Using human erythroleukemia cells, cell cycle analysis indicates DON also causes apoptosis in human blood cells [88]. In vitro results show DON is cytotoxic to the primary cultures of human renal proximal tubule epithelial cells and lung fibroblast cells, with IC_{50} values of 2.4 and 0.7 µM, respectively [89]. Similarly, this toxin has an IC_{50} value of 6.0 µM in primary cultures of human hepatocytes but has a larger IC_{50} value of 41.4 µM for (nonprimary) HepG2 human hepatocellular liver carcinoma cells [90].

Grain dust, contaminated with DON, offers a possible route to initiate lung disease due to lung cell susceptibility. As with T-2 and HT-2 toxins, DON is cytotoxic (IC_{50} of 0.7 µM) to V79 lung fibroblast cells [73]. In a murine model, nasal exposure to DON (5 mg/kg body weight) induces a greater proinflammatory cytokine (e.g., IL-1β, IL-6, and TNFα) gene expression than the same amount in an oral dose [91].

DON displays genotoxicity toward DNA synthesis in Caco-2 cells (human intestinal cells) at concentrations (0.01–0.5 µM) likely encountered in the gastrointestinal tract [92]. It has a genotoxic modulatory effect on primary cultures of hepatocytes. DON, at 1 µg/mL, causes chromosomal aberrations in primary cultures of rat hepatocytes, which declined at a concentration of 10 µg/mL [93]. DON also causes oxidation damage to cells. For example, DON-caused oxidation increases the levels of superoxide anion in human embryonic kidney cells (Hek-293), which the cellular antioxidant system cannot counteract [94]. In Caco-2 cells, DON induces lipid peroxidation and at 5 µM inhibits DNA synthesis by 85%–90%, but inhibits protein synthesis by only 40%–45%, though at higher DON concentrations DNA synthesis seems to be restored suggesting promoter activity [95].

DON is cytotoxic to immune system cells. In vitro studies show DON causes apoptosis of human T lymphocytes, but is not as cytotoxic as T-2 and HT-2, which reduce mitochondrial activity in these cells at concentrations approximately 1000-fold lower than DON [96]. Similar effects are observed in vitro with human T and B lymphocyte activities, which are suppressed by DON [52]. However, as with human T cells, T-2 had a greater suppressive effect than DON on these cells. DON also inhibits human lymphocyte proliferation with an IC_{50} concentration of 216 ng/mL. At 400 ng/mL, DON causes a significant increase in IL-2 levels with a less pronounced increase of IL-4 and IL-6 [97]. The role of immature dendritic cells is to take up, process, and present protein antigens to T cells [98]. Though not cytotoxic as is T-2, DON (2 µM) reduces the in vitro viability of dendritic cells derived from human placenta by 40% after 24 h of incubation [99]. This activity of DON is related to its ability to interfere with phenotypic and functional characteristics of maturing dendritic cells [100,101].

Cytotoxicity associated with metabolic stress caused by DON in porcine IPEC-J2 intestinal cells is noted by a significant decrease in ATP levels after 48 h of incubation in a dose-dependent manner beginning at 2.5 µM [102]. Another study using IPEC-1 and IPEC-J2 cells shows DON has a dose- and time-dependent modulatory effect with the high dose (2000 ng/mL) being toxic while the low dose (200 ng/mL) producing cell proliferation after a 72 h incubation [103]. However, even a low in vitro dose (25.0 ng/mL) of DON renders these porcine epithelium cells more susceptible to *S. typhimurium* with a subsequent potentiation of the inflammatory response [104].

Porcine reproduction may also be adversely affected by DON in animal feed. Porcine oocytes are also affected by DON with a concentration of 2.0 µM or lower affecting developmental competence by interfering directly with microtubule dynamics during meiosis and by disturbing oocyte cytoplasmic maturation [105]. DON, at 0.337 µM, also inhibits progesterone production by porcine granulosa cells and may play a negative role in the pregnancy success in swine [106].

Since DON causes feed refusal and weight loss in animals, especially pigs, studies have centered on the neurotoxicity effects of DON on the central nervous system. In mice, DON reduces feeding and affects satiation and satiety by interfering with central neuronal systems that regulate food intake [107]. Also in mice, this toxin reduces leptin and AgRP levels leading to decreases in food intake and body fat as well as impaired lean weight increases [108]. In pigs, oral administration of *F. graminearum* extracts containing DON (1 mg/kg of body weight) shows that this toxin stimulates the main brain structures involved in food intake and suggests that catecholaminergic and NUCB2/nesfatin-1 neurons could contribute to the feed refusal effects of DON [109].

DON may also be a factor in other neurological diseases. Glial cells provide nutrients and oxygen, physical support, and insulation to brain neurons as well as remove dead neurons. Using animal and human glial cells in vitro, studies show that low doses of DON decrease glial cell viability and may cause modification of brain homeostasis and could play a role in neurological diseases in which alterations of glial cells are involved [110].

21.2.1.2.2 Nivalenol

NIV is produced by *Fusarium cerealis*, *F. poae*, *F. culmorum*, *Fusarium crookwellense*, and *F. graminearum* on wheat. Structurally, it is the same as DON except that NIV has a keto substitution in the C-8 position. NIV has a temporary total daily intake limit of 0.7 µg/kg body weight [78,111,112]. It is present in 16% of grain samples tested in Europe and has been implicated in esophageal cancer and gastric cardia [35,113]. As with T-2 and DON, NIV displays strong inhibition on protein synthesis by binding to ribosomes, inhibition of RNA and DNA synthesis, and toxic effects on cell membranes [78].

NIV is cytotoxic to several cell lines. When compared with DON, fumonisin B_1, and α-zearalenol (α-ZOL), NIV at a concentration as small as 0.0625 µM is the most potent inhibitor of porcine blood cell proliferation [114]. In a dose-respective manner, NIV shows cytotoxic effects against IPEC-J2 beginning at 1.0 µM and was more cytotoxic than fumonisin B_1, DON, and ZON [115].

In vitro, cells of the immune system are sensitive to NIV. Human T lymphocytes exhibit similar apoptosis activity with NIV as with DON but at a much lower levels than the Type A trichothecenes [96]. NIV has a modulatory effect on human and porcine lymphocytes, where lower concentrations (0.1–1.0 µg/mL) enhanced cellular proliferation, while higher doses (1.0–10.0 µg/mL) inhibit these cells [116]. In another study, the immunomodulatory effects of NIV and DON were determined for Jurkat cells (human lymphoblastoid T lymphocytes) and swine lymphocytes isolated from whole blood. NIV significantly inhibits Jurkat cell and porcine lymphocyte produced by *F. graminearum* and *F. culmorum* (in some geographical locations), respectively [117]. These results are similar to those for DON, which inhibits Jurkat and the porcine lymphocytes at 1.0 and 0.25 µM, respectively. Using an in vitro model simulating the gastrointestinal tract of healthy pigs, the intestinal absorption of NIV is estimated as 21% [82]. NIV is detected in blood samples taken 20 min after the start of feeding with the systemic peak concentration occurring 2.5–4.5 h after feeding [118]. NIV delivered per os to mice is transmitted in unchanged form to fetuses via the placenta or to suckling mice via milk [119].

NIV at high concentrations can affect the ability of the intestine and immune system to function properly. In vitro, high doses (5–80 µM) can significantly affect rat intestinal epithelial cells (IEC-6), while low doses (0.1–2.5 μM) affected epithelial cell migration that is needed for restitution and gastrointestinal wound healing [120]. In swine, exposure to NIV produces a reduced enzymatic capacity to utilize alpha-ketoglutarate in the tricarboxylic acid cycle suggesting an impaired energy supply in the epithelial tissue of the small intestine and colon [121]. In vivo studies show that rats fed NIV over a 90-day period display decreased body weight, affecting the immune function as observed by an increase of natural killer cell activity in a dose-dependent manner beginning at 0.4 mg/kg body weight [122]. At a dose of 100 mg/kg body weight, histopathology shows that NIV-related changes are seen in the hematopoietic and immune organs and the anterior pituitary in both sexes [123,124]. Studies in mice indicate that oral intake of NIV (30 ppm) results in significant anemia and slight leukopenia, while ultrastructural studies show polyribosomal breakdown of the bone marrow [125].

NIV, as does DON, affects reproduction in animals. Beginning at a concentration of 0.4 mg/kg body weight, NIV targets female reproductive organs as observed in an increase in atretic ovarian follicles and interstitial glands as well as a lack of corpora lutea development in severe cases [124]. Uterine atrophy with diestrus endometrial mucosal change and an increase in castration cells (an indicator of reduced sex hormone production in both sexes) occur at the much higher dose of 100 mg/kg body weight [124].

21.2.2 Fumonisins

Of the over 30 *Fusarium* species that produce fumonisins, only *Fusarium verticillioides* (formerly *F. moniliforme*) and *Fusarium proliferatum* produce significant amounts of these toxins [1,126]. Of this toxin family, only fumonisins B_1 (FB_1), B2 (FB_2), and B_3 (FB_3) are produced in substantial amounts. The Joint FAO/WHO Expert Committee on Food Additives rates the toxicological profiles of FB_2 and FB_3 as very similar to that of FB_1, which is the most common fumonisin found on grains [127]. Together, FB_1 and DON constitute the major *Fusarium* toxins occurring worldwide in corn, wheat, and other cereal grain intended for human and animal consumption [88,128]. Besides being isolated from plants, the aforementioned *Fusarium* species as well as *Fusarium subglutinans* can be present in clinical samples and produce FB_1, FB_2, and FB_3 when grown at 28°C in nonclinical media [129].

FB_1 is the diester of propane-1,2,3-tricarboxylic acid and 2S-amino-12S,16R-dimethyl-3S,5R,10R,14S,15R-pentahydroxyeicosane in which the C-14 and C-15 hydroxy groups are esterified with the terminal carboxyl group of propane-1,2,3-tricarboxylic acid [127]. The cancer-initiating ability and toxicity of the fumonisins are based on the amino group present and the intact molecule itself [130]. Due to its implications in human cancer, FB1 is designated a possible human carcinogen (group B carcinogen) by the International Agency for Research on Cancer [131].

FB_1 is the most toxic of the fumonisins and is known to cause, or is implicated in, disease in animals and humans. It causes equine leukoencephalomalacia (ELEM), pulmonary edema, and hydrothorax in swine and renal tubule tumors in rats [7,132–135]. FB_1 is implicated in humans as one of the factors involved in neural tube defects and esophageal cancer [135–138]. In sub-Saharan Africa, fumonisins, rather than aflatoxin, in maize are implicated in predisposing people to HIV [139].

In vitro studies show that FB_1 prevents sphingolipid biosynthesis in pig epithelial cells, rat hepatocytes, and neuronal cells [140–142]. FB_1 does so by blocking ceramide synthase conversion (IC_{50} dose, 0.1 μM FB_1) of sphingosine to ceramide with the tricarboxylic acid moiety of FB_1 required for maximum enzyme activity inhibition [140,142,143]. This inhibition leads to the accumulation of sphingolipid bases that affect cell growth, differentiation, and apoptosis [144]. FB_1 may affect renal tumor growth characteristics by inhibiting sphingolipid synthesis, which aids in the regulation of cell contact, growth, and differentiation, or possibly by interacting with the extracellular matrix [145]. In turn, the increase in sphingolipid bases stimulates DNA synthesis and mitogenesis and may explain FB_1 carcinogenicity [146]. Sphingolipid metabolism is inhibited by FB_1 leading to calmodulin expression, which in turn induces apoptosis [147].

FB_1 is hepatotoxic and nephrotoxic in horses, swine, chickens, turkeys, rats, and ducks and causes liver cancer in rats with this organ being the primary target [127]. In rats that die after 3 days of high dosing per os with FB_1, the liver has hydropic degeneration, scattered single-cell necrosis, an increase and enlargement of Kupffer cells, and hyaline droplet degeneration [148]. Dosing of FB_1 in rat diets over 6 months results in persistent oval cell proliferation and generation of hepatic adenomas and cholangiofibromas along with the development of chronic toxic hepatitis, liver fibrosis, and cirrhosis [149]. FB_1-induced apoptosis leads to constant regeneration of renal tube epithelial cells that may be partially responsible for initiation of renal tube carcinoma [150].

FB_1 affects the production of cytokines differently based on gender. Although female mice have higher basal expression of inflammatory cytokines (IL-Iβ, TNFα, IL-12, IFNγ, p40) than males, the induction of IL-6 by FB_1 is higher in females resulting in greater hepatotoxicity observed in females [151]. Gender sensitivity to FB_1 is observed in renal tubule tumors in males, while females developed hepatocellular tumors [150]. Besides inducing these cytokines in mice, FB_1 also induces TNFα, TNF receptor 55, receptor-interacting protein, and caspase 8 [152,153]. The increase in caspase 8, which is involved in the TNFα signaling pathway, only occurs in the liver and not in the kidney and may contribute to FB_1-induced apoptosis [153]. Overexpression of TNFα competitively prevents FB_1 binding to TNFα receptors in the liver, while the lack of TNFα receptors also protects against FB_1-induced hepatotoxicity [154–156]. Together, these studies indicate that FB_1-induced hepatotoxicity involves TNFα receptors in the liver.

Studies suggest FB_1 increases cell sensitivity to apoptosis caused by TNFα. As FB_1 increases the level of free sphingoid base, the populations of renal epithelial cells become arrested in the G_2/M phase making them susceptible to TNFα apoptosis. Additionally, phosphorylation of pro-apoptotic JNK, a mitogen-activated protein kinase that is activated by FB_1, possibly plays an important role [157].

Kupffer cells are macrophages present in the liver and lungs that are phagocytic and immunomodulatory. They also synthesize and secrete biologically active mediators, including cytokines such as TNFα [158]. In a murine model depleted of Kupffer cells by gadolinium chloride, animals challenged with FB_1 experienced less hepatotoxicity and reduced free sphinganine accumulation, while TNFα production increased [159]. Depletion of T cells in mice also has been shown to reduce FB_1-induced hepatotoxicity, possibly by eliminating proinflammatory cytokine production by T cells that would stimulate Kupffer cells [160]. The interaction of immune cells and their production of cytokines modulate FB_1-induced hepatotoxicity.

FB_1 can alter sphingolipid metabolism and TNFα expression in the heart and lungs of mice with females being more sensitive to sphingolipid metabolism disruption by this toxin than males [161]. However, no damage was evident in these murine tissues despite disruption to sphingolipid metabolism. The toxins FB_2 and FB_3 also disrupt sphingolipid metabolism, with the former being the more effective disruptor: both were implicated in liver injury development and ELEM in an equine study [162].

A dose of 20 mg of fumonisins (FB_1 + FB_2) per kg of feed is recommended by the European Commission as safe for turkeys, which are more resistant to this toxin than ducks and chickens [163]. However, though this dosage does not affect feed consumption, it does disrupt sphingolipid metabolism by altering the ratio of sphinganine to sphingosine (Sa/So) in serum and liver. This ratio serves as an early and sensitive biomarker for fumonisin exposure in turkeys [164]. In contrast to the acceptable dose in turkeys, broiler chickens fed a prolonged diet of 10 mg of FB_1/kg feed (half of the acceptable dose in turkeys) can develop liver lesions [165].

Primates also have a reduced tolerance to the fumonisins. Velvet monkeys fed a daily total fumonisin (FB_1 + FB_2) intake of about 0.3 or 0.8 mg/kg body weight over 60 weeks display liver damage as indicated by an increase in five liver enzymes and show disruption of sphingolipid biosynthesis as measured by the elevation of the Sa/So ratio in blood serum [166]. Similar buildup of sphinganine occurs in other cell types. In vitro studies with human epidermal keratinocytes suggest that FB_1 reduces ceramide levels with concomitant sphinganine accumulation that leads to cell apoptosis [167]. In swine fed low to moderate levels (0.1–10.0 mg FB_1/kg diet) of FB_1 over 8 weeks, the Sa/So ratio was increased in the liver, lungs, and kidney, and the animals experienced reduced growth with aberrations in blood biochemistry and in the weights of the pancreas and adrenals [168].

The intestinal epithelium is the initial barrier to ingested toxins and microorganisms and is susceptible to FB_1 toxicity. Cytotoxicity studies show FB_1 at 50 µM is cytotoxic to porcine proliferating IEC. The barrier function of intestinal cells also is compromised by chronic consumption of contaminated food or feed and induces intestinal damage in animals and possibly humans [169]. Ingestion of this toxin (0.5 mg FB_1/kg body weight/day for 7 days) by swine causes predisposition to pulmonary inflammation with an increase in expression of IL-8, IL-18, and IFNγ mRNA [170]. When exposed to an intratracheal challenge of the pathogen *Pasteurella multocida* after ingestion of FB_1, swine experience a delay in growth, an initiation of cough, and an increase in BALF cells, macrophages, and lymphocytes. Development of subacute interstitial pneumonia with macroscopic lung lesions also occurs, suggesting FB_1 ingestion may predispose animals to pulmonary disease [170].

Oral administration of FB_1 in piglets (dose, 0.5 mg/kg body weight/day for 7 days) decreases IL-8 mRNA expression in the porcine ileum due, at least in part, to the effect of FB_1 on the IEC. This weakening of the intestinal immune system may render it susceptible to colonization by pathogenic microorganisms [171]. An in vitro model using mucosa from the pig proximal jejunum in Ussing chambers was treated with a moderate dose (8–10 ppm) of FB_1 over 2 h [172]. After 60–90 min, an increase in transepithelial resistance and macromolecular passage occurs without a change in absorptive or secretory physiology, suggesting an increase in the small intestine of paracellular and transcellular permeability [172]. After the swine consumed FB_1 for 9 days, an increase in the colonic levels of αB crystalline and COX-1 occurs with smaller increases in various stress proteins along the gastrointestinal tract [173]. In a much longer (5 weeks) study, piglets ingested with FB_1 (260 µg/kg body weight per day) develop increased intestinal cytokine expression and adherent and tight junction proteins as well as morphological and histological changes as shown by atrophy and fusion of villi, decreased villi height, and cell proliferation in the jejunum [174]. These studies indicate that long-term ingestion of FB_1 alters the intestine and makes the animal more susceptible to intestinal pathogens.

Nixtamalization is an alkaline cooking process used for making masa and tortillas from corn. During this process, the levels of FB_1 are decreased with FB_1 hydrolyzed to HFB_1, effectively reducing the level of total fumonisin whether it is on the commercial or individual cooking level [175]. HFB_1, in contrast to FB_1, does not alter sphingolipid metabolism in a murine model indicating that HFB_1 does not pose a significant risk for neural tube defects [176,177]. The results indicate that nixtamalization could be employed to reduce exposure to FB_1 [177].

21.3 ESTROGENIC TOXINS: ZEARALENONE AND ITS METABOLITES

ZON is found in cereal crops worldwide and is produced mainly by *F. graminearum* (*Fusarium roseum graminearum*), though some isolates of *Fusarium tricinctum*, *F. culmorum*, and *Fusarium equiseti* also produce this compound [37,178,179]. ZON has a chemical structure of 6-(10-hydroxy-6-oxo-trans-1-undecenyl)-β-resorcyclic acid lactone and is noted for its estrogenic properties that affect reproduction, especially in pigs, which are highly susceptible to ZON and its metabolites [180,181]. Initial studies show *F. graminearum* produces a phase I metabolite of ZON called α-ZOL, but later both diastereomeric forms, α-ZOL and β-ZOL, were found to be produced by *F. culmorum* and *F. equiseti* in corn stalks as well as in commercially available corn [179,182–184].

ZON, a nonsteroidal lactone, is slightly soluble in water and is noted for its estrogenic properties and elicits production of the same 52 kDa uterine protein in vitro and in vivo as does 17β-estradiol, the main female sex hormone [185]. ZON and its metabolites bind to mammalian estrogen receptors (ERα or ERβ), which have affinity for 17β-estradiol, estriol, and estrone [186]. In animals (e.g., pig, rat, and chicken), differences exist in their sensitivity to ZON and its metabolites, which may be due to variations in their binding affinity for these estrogen receptors [187]. The European Commission Scientific Committee on Food has established a temporary daily intake maximum for ZON of 0.2 µg/kg body weight per day [188].

ZON is rapidly metabolized after ingestion through a two-phase pathway. In phase 1, a series of reduction metabolic steps occur, while the second phase concerns ZON and its reduced metabolites being conjugated into glucuronides or sulfates. ZON reduction occurs mainly in the liver where ZON is metabolized into α-ZOL and β-ZOL [181,189–191]. α-ZOL can be converted into 7-α-zearalanol (α-ZAL or zeranol), an anabolic growth promoter given to farm animals, which in turn can be converted in the liver to zearalanone (ZAN) with small amounts of 7-β-zearalanol (β-ZAL) also produced [192–194].

The conjugation (phase II metabolism) of ZON and its reduced metabolites occurs in the intestines, liver, and other organs and plays a major role in the excretion of these compounds by increasing their water solubility [195,196]. Caco-2 cells are epithelial colorectal adenocarcinoma cells, which upon culturing have the characteristics of mature small intestine enterocytes and are used as a model of the intestinal barrier [197]. Caco-2 cells can metabolically convert ZON into its phase I reduced metabolites as well as converting ZON and these metabolites into phase II glucuronide conjugates [198]. In another study, addition of ZON and α-ZAL to Caco-2 cells resulted in the formation of ZON and α-ZAL phase I metabolites as well as conversion of the parent compounds and their metabolites into phase II glucuronide and sulfate conjugates [199]. In vitro studies show that the microsomal fraction of sow intestinal mucosa, in the presence of NADPH, preferentially converts ZON to the phase I β-ZOL over the α-epimer, while the phase II conjugation of ZON occurs preferentially over its phase I reduction [200].

ZON and its metabolites are noted for their varying degrees of estrogenic properties, which can be ranked as follows: α-ZOL > α-ZAL > β-ZAL > ZAN > ZON > β-ZOL with their relative estrogenicity rankings calculated as 92, 18, 3.5, 2.5, 1.0, and 0.44, respectively [186,201]. Because they can adopt a structural conformation similar to 17β-estradiol, ZON and its metabolites also bind to estrogen receptors (ERα and ERβ) found in the various organs, which, depending on tissue type, can have both ERα and ERβ or either ERα or ERβ [202]. Estrogens with an affinity for ERα sites trigger proliferation, while those that bind to ERβ sites initiate signaling that inhibits proliferation but promotes apoptosis [203].

Of clinical importance is that the two major metabolites of ZON, α-ZOL and β-ZOL, have different estrogenic properties, with α-ZOL having significantly more estrogenic potency than β-ZOL or even the parent toxin, ZON [204,205]. Using in vitro assays with H295R human adrenocortical cells, α-ZOL exhibits the strongest estrogenic activity and is slightly less potent than 17β-estradiol [204]. In comparison, ZON has about 70-fold less estrogenic potency than α-ZOL but has twice that of β-ZOL [204]. One factor in the various susceptibilities of animals to ZON may be the differences in their conversion of ZON into these two metabolites. For example, in swine and sheep, conversion of ZON favors α-ZOL over β-ZOL, while in heifers, rats, horses, and chickens, this conversion greatly favors β-ZOL over α-ZOL [181,206–208]. ZON, α-ZOL, and β-ZOL have genotoxic properties. In mouse bone marrow and HeLa cells, they increase the percentage of chromosome aberrations with ZON and α-ZOL more genotoxic than β-ZOL in a dose-dependent manner [209].

Studies with rat liver microsomes indicate that ZON is converted into 2 reductive (α-ZOL and β-ZOL) and 14 oxidative (catechol) metabolites [210]. ZON and α-ZOL and the endogenous steroidal estrogens, 17β-estradiol (E2) and estrone (E1), can undergo aromatic hydroxylation via cytochrome P450 to catechol metabolites (15-hydroxy-ZON/α-ZOL, 13-hydroxy-ZON/α-ZOL, 2/4-hydroxy-E1/E2), which can induce oxidative damage to DNA [211]. Similarly, α-ZAL and ZAN can be hydroxylated by cytochrome P450 [212]. These catechol metabolites can undergo inactivation by catechol-O-methyltransferase also present in hepatic microsomes. However, since 15-hydroxy-ZEN/α-ZEL and 13-hydroxy-ZEN/α-ZEL are poorer substrates for methylation than 2/4-hydroxy-E1/E2, they are less susceptible to inactivation by this enzyme [213].

Hemopoietic and immune cells, such as T and B cells, dendritic cells, microglia, monocytes, and macrophages, found in different areas of the body, also express the estrogen receptors ERα and ERβ and therefore are targets of estrogenic endocrine disruption [214–218]. ZON causes a fivefold increase in the production of cytokines (IL-2, IL-6, and IFNγ) in splenic lymphocytes in chickens by a yet unknown mechanism of action [219]. This differed from the effect of ZON, α-ZOL, β-ZOL, and ZAN (10 μM) in a porcine progressive motor neuronopathy (PMN) model, whereby they significantly decreased IL-8 synthesis and PMN cell viability at concentrations between 53.1 and 73.4 μM, with ZON being less toxic than its metabolites [220]. However, these toxins at a much lower concentration (1 μM) significantly increased oxidative stress. In weaning piglets, the effect of ZON on lymphocytes differs from PMNs. ZON (316 ppb) increases the synthesis of inflammatory cytokines (TNFα, IL-1β, and IFNγ) and the respiratory burst of monocytes [221]. In contrast, in the liver, ZON decreases the synthesis of all inflammatory cytokines [221]. ZON also reduces, in vitro, the production of the antimicrobial peptides, β-defensin 1 and 2, by swine jejunal epithelial cells, which, in turn, could alter intestinal microbial populations and possibly impact the health of animals and humans [222].

ZON affects the reproductive process in animals, and it decreases litter size in small female ruminants. ZON, α-ZOL, and β-ZOL can act as endocrine disruptors of nuclear receptor signaling, altering hormone production [204]. For example, these three toxins at a concentration of 10 μM increase the production of the hormones progesterone, estradiol, testosterone, and cortisol. Evidence indicates they could affect normal placental activity resulting in pregnancy disorders or changes in fetal development by binding to estrogen receptors and increasing progesterone synthesis and progesterone receptors [204]. While ZEN promotes in vitro differentiation by HCG stimulation, α-ZOL and β-ZOL do not induce syncytialization [223]. In addition, exposure to these ZON metabolites causes variability in ATP-binding cassette (ABC) transporter expression by acting on the nuclear receptors [223].

ZON, at a high concentration of 2 mg/kg in the diet of prepubertal gilts, displays a greater level of ERα/ERβ mRNA expression than those animals fed diets of ZON with soybean isoflavone, which is believed to counteract the effects of ZON [224]. In pregnant rats exposed daily (during gestation days 7–20), ZON (1 mg/kg) modulates mRNA levels of ABC transporters in fetal liver and maternal tissues and modulates protein levels in the maternal uterus and fetal liver [225]. α-ZOL, possibly through pathways other than estrogen receptor binding, impairs the quality of preimplantation of swine embryos in a dose-dependent manner [226]. Ovarian cells may also be a target of these toxins. In vitro, ZON, α-ZOL, and β-ZOL are cytotoxic to ovarian Chinese hamster ovary (CHO)-K1 cells with IC_{50} values of 60.3, 30.0, and 55.0 μM, respectively, after 24 h of incubation indicating that α-ZOL is more cytotoxic to these cells than ZON or β-ZOL [227]. In CHO-K1 cells, evidence shows ZON causes reactive oxygen species (ROS) production in a dose- and time-dependent manner and that the presence of ROS, as well as its cytotoxicity properties, plays roles in the overall toxicity observed [228].

ZON and α-ZOL can affect male reproduction in animals. In vitro, incubation of diluted boar sperm with ZON or α-ZOL (10–30 μg/mL, 4 h incubation time) does not affect motility, while susceptibility to nuclear chromatin damage is dependent on the individual animal [229]. However, both mycotoxins at concentrations of 125, 187.5, and 250 μg/mL of diluted boar sperm significantly affected motility, viability, and spontaneous acrosome reaction in a time- and dose-dependent manner [230]. A longer (24 h) incubation period results in a greater sensitivity of boar sperm to these mycotoxins. At picomolar levels, ZON and α-ZOL affects chromatin structure stability and viability, while β-ZOL, at micromolar concentrations, adversely affects sperm motility [231].

A single dose of ZON (5 mg/kg, IP) causes germ cell degeneration about 12 h after dosing followed by apoptosis, especially of the spermatogonia and spermatocytes, which occurs in a time-dependent and stage-specific pattern [232]. Though it can bind to estrogen receptors, ZON causes testicular germ cell apoptosis in rats by involvement in the Fas signaling system with estrogen receptors not playing a major role [233]. In another study, male rat pups were injected with ZON (1, 50, or 100 μg/day) on postnatal days (PNDs) 1–5. Testes samples taken on PND90 show ZON induces both mRNA and protein expressions and significantly induced the expression of *Abcc5* mRNA at the injected doses. These results indicate that neonatal exposure to ZON could reduce spermatogenesis and male fertility [234].

Zeranol, the generic name for α-ZAL, is the active ingredient of anabolic growth promoters used in several countries, including the United States and Canada, to enhance fattening rates in cattle, but is banned in the European Union since 1985 [235,236]. Both ZON and α-ZAL are implicated as causative agents of idiopathic or central precocious puberty (CPP) in girls. In a study of 78 idiopathic precocious puberty (IPP) patients and 100 control subjects in China, the IPP patients have higher serum ZON concentrations and a higher positive rate of ZON than in control patients [237]. Girls in China affected with CPP had greater weight, height, and height velocity than in control subjects, indicating ZON could play a role in the development of CPP [238].

ZON or its metabolites may play roles in the development of hepatocellular and endometrial adenocarcinomas as well as hyperplasia [239]. In mice, a single dose of ZON (2 mg/kg either IP or orally) causes DNA adduct formation in the liver and kidneys in female mice and rats but not in the ovaries, though repeated doses of 1 mg/kg on days 1, 5, 7, 9, and 10 did induce ovarian DNA adducts [240]. The presence of α-ZOL (range, 0.1–1.7 μg/g) in all neoplastic endometrial tissues has been confirmed by liquid chromatography and its quantity correlated with tumor consistency [241]. Only trace quantities of ZON were found in just a few of these samples, suggesting α-ZOL, not ZON, may play a role in the development of this disease.

ZON, DON, and NIV are often found together in contaminated cereals. In assays (in vitro) using IPEC-J2, ZON at 40 μM is more cytotoxic than FB_1 at this same concentration, but not as cytotoxic as NIV or DON at 2 μM [115]. However, different combinations of ZON with the other *Fusarium* toxins at cytotoxic and noncytotoxic levels cause a decrease in cell viability, indicating the potential of interactive cytotoxic effects to these intestinal cells when ingested. Similarly, ZON, DON, NIV, and FB_1, individually at their cytotoxic concentrations, cause upregulation of proinflammatory cytokines (IL-1α, IL-1β, IL-6, IL-8, TNFα, and MCP-1). However, different combinations of two, three, or four of these toxins at noncytotoxic concentrations produce significant upregulation of these cytokines, suggesting that such combinations could be involved with intestinal inflammation [242].

21.4 EMERGING TOXINS

Though the major *Fusarium* produced toxins such as the trichothecenes, fumonisins, ZON, and their metabolites have been studied in great depth, there are several *emerging* or minor toxins that have not been thoroughly studied. Such toxins include the enniatins, beauvericin, fusaproliferin, and moniliformin, which are produced by many *Fusarium* species, found in contaminated grains (e.g., wheat, barley, oats, maize) and grain products [243–246].

21.4.1 ENNIATINS

Enniatins, produced by several fungal genera (e.g., *Halosarpheia* and *Verticillium*) besides *Fusarium* species, are six-membered cyclic hexadepsipeptides and have designations from *A* through *P*, with the major variants (enniatins A, A1, B, B1, B2, B3) produced by *Fusarium* species [247–252]. They are ionophores that form cation-selective pores in cell membranes because of their high affinity for Na^+, K^+, Ca^{2+}, and Mg^{2+} and interaction with the lecithin moieties in the cell phospholipid membrane [253,254] (Figure 21.2). In studies with human neural cells (Paju), mouse pancreatic beta-cell line (MIN 62), feline fetus lung cells, and boar spermatozoa, the ionophoric property of enniatins results in K^+ uptake from the cytoplasm. This results in positively charged enniatins binding with the

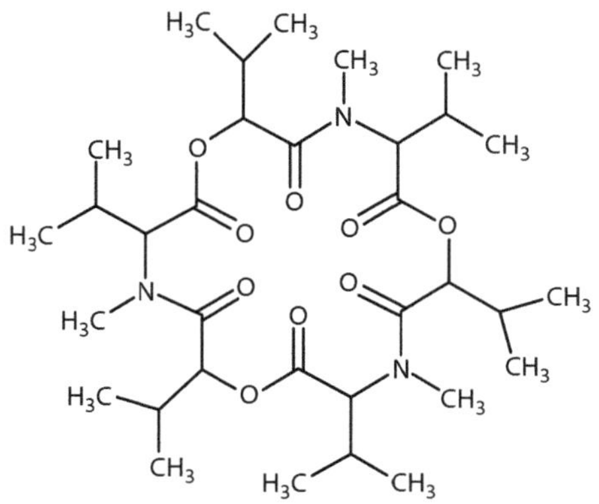

FIGURE 21.2 Enniatin general structure.

negatively charged mitochondrial matrix causing damage to mitochondrial ion homeostasis, volume regulation, and oxidative phosphorylation [255].

Enniatins A, A1, B, and B1 inhibit the growth of bacteria and fungi [256–258]. In vitro studies show they also have mammalian cytotoxic properties that could reduce the immune response during infections. For example, enniatin B has IC_{50} values for immature and mature dendritic cells of 1.6 and 2.6 µM, respectively, and for macrophages of 2.5 µM [259]. In RAW 267.4 murine macrophages, enniatin B arrests the normal cell developmental cycle in the G0/G1 stage and causes apoptosis along with inflammation and release of IL-1β [260]. Using the Alamar Blue assay, the enniatins A, A1, B, B1, B2, and B3 are cytotoxic for hepatocellular carcinoma cells, Hep G2, and fibroblast-like fetal lung cells (MRC-5) at concentrations of 7.3–206.7 and 1.9–5.9 µM, respectively [261]. Enniatin causes apoptosis by the mitochondrial pathway after only a short exposure time (≤8 h) at low levels (0.1–1.0 µM) to glioblastoma, melanoma, osteosarcoma, and human lung cancer cells [262].

Flow cytometry and cell viability shows that, at 1.5 and 3.0 µM, enniatins A and B induce apoptosis and necrosis in Hep G2 cells and that enniatin B is more toxic than enniatin A [263]. Besides kidney cells, enniatins are cytotoxic in vitro to other human cell types. Enniatin B has an IC_{50} value of 4.0 ± 1.5 µM for V79 lung fibroblast cells, which is much higher than that for the trichothecene, T-2, whose IC_{50} value for these cells is 0.003 ± 0.0016 µM [73]. Enniatin B probably causes apoptosis of V79 cells via nuclear fragmentation but does not induce mutagenicity or genotoxicity [264,265]. Human hematopoietic progenitors are also affected by the enniatins. Human granulocytes and monocytes are susceptible to enniatin B toxicity alone (IC_{50}, 4.4 µM), and, when combined with beauvericin, additive reduction occurs in viable cell numbers [266]. Enniatin B is cytotoxic for white blood cell, platelet, and red blood cell progenitors with IC_{50} values of 4.4, 1.3, and 3.3 µM, respectively [267]. In vitro studies show enniatins are toxic to intestinal cells. The enniatins A, A1, B, and B1, individually and in various concentrations of two, three, or four, are toxic to Caco-2 (colon carcinoma) cells in a dose-dependent manner [268]. Individual mycotoxin IC_{50} values range between 1.3 and greater than 15 µM, while the various combinations produce an additive effect on cell viability [268]. Damage to these cells begins after 3 h of exposure with cell death due to lysosomal destabilization, ROS production, and mitochondrial membrane permeabilization, which differs from the previous, described as pore-forming cellular death mechanism [269]. Hamster ovary cells (CHO-K1) are also targets of the enniatins A, A1, B, and B1 with IC_{50} values for the individual mycotoxins ranging from 1.65 to 11.0 µM and various combinations of having additive or synergistic effects [270]. Interestingly, antioxidant polyphenols (quercetin, rutin, myricetin, and t-pterostilbene) found in wine prevent enniatin cytotoxic effects on CHO-K1 cells [271].

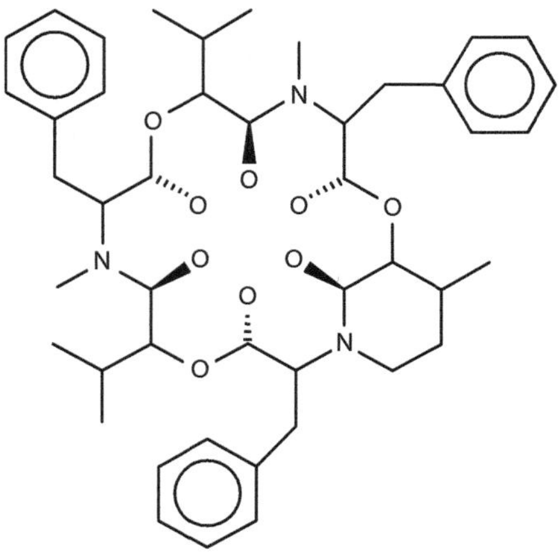

FIGURE 21.3 Beauvericin.

21.4.2 BEAUVERICIN

Beauvericin is a cyclic hexadepsipeptide produced by a number of *Fusarium* species and consists of alternating D-α-hydroxy-isovaleryl and aromatic-*N*-methyl-phenylalanine moieties [259,272] (Figure 21.3). This differs from those of the enniatins, which are also cyclic hexadepsipeptides, but which have three D-α-hydroxyvaleric acid moieties alternately attached to three L-*N* methyl amino acid molecules [259]. It is ionophoric, which affects ionic homeostasis and initiates pore formation in cell membranes and, once inside the cell, damages mitochondrial functions [255,273]. In Caco-2 and CHO-K1 cells, beauvericin causes ROS production and damage to mitochondria leading to apoptosis, suggesting that oxidative stress is another pathway for beauvericin-induced toxicity [228,268]. In contrast to the enniatins, beauvericin-related apoptosis is not induced by ROS production or DNA damage in epidermal carcinoma-derived cell line KB-3–1 and promyelocytic leukemia cell line HL-60 [274].

As with the enniatins, beauvericin inhibits the growth of bacterial pathogens such as *Salmonella enterica*, *Listeria monocytogenes*, *Clostridium perfringens*, and *Escherichia coli* at concentrations of 10, 100, 1, and 1000 ng, respectively, in zones of inhibition assays [275]. Beauvericin also shares in vitro cytotoxicity properties with the enniatins for a number of mammalian cell lines. For example, this toxin at 30 μM damages intracellular ionic homeostasis and mitochondrial activity in guinea pig ventricular myocytes [276]. It displays IC_{50} values of 1.0, 2.9, and 2.5 μM for immature dendritic cells, mature dendritic cells, and macrophages, respectively [259]. This is also true for human progenitors of white blood cells, platelets, and red blood cells for which this mycotoxin was cytotoxic at 32, 3.2, and 6.4 μM, respectively [266]. Beauvericin is cytotoxic for human granulocytes and macrophages with IC_{50} levels of 3.4 μM as well as an IC_{50} value of 1.6 μM for V79 lung fibroblast cells [73,267].

21.4.3 MONILIFORMIN

Moniliformin is produced by various *Fusarium* species, such as *F. proliferatum* and *Fusarium fujikuroi*, and is found worldwide in grains [277–280]. Its chemical structure is 3-hydroxycyclobut-3-ene-1,2-dione and occurs in nature as a potassium or sodium salt [259] (Figure 21.4). It is cytotoxic, but differs from the other emerging *Fusarium* toxins in its weak activity against various cell lines. In contrast to enniatin B and beauvericin, moniliformin, even at the highest concentration (80 μM) tested, is not lethal to macrophages and only weakly lethal to immature and mature dendritic cells [259]. At 100 μM, moniliformin is lethal to 5% and 31% for human white blood cell and platelet

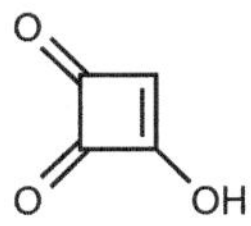

Mycotoxins of *Fusarium* spp.

FIGURE 21.4 Moniliformin.

progenitors, respectively, while enniatin B is significantly less toxic to these cells than DON and FB$_1$ [267]. It is essentially nontoxic (IC$_{50}$ estimated as 10,000 μM) to V79 lung fibroblast cells, in comparison to T-2, beauvericin, and enniatin B that have IC$_{50}$ values of 0.003, 0.7, and 1.6 μM, respectively, for this cell line [73]. Another report shows moniliformin to be weakly cytotoxic after 72 h to CHO-K1 (IC$_{50}$, 1020 μM), Caco-2 (IC$_{50}$, 315 μM), Balb/c mice keratinocyte (C5-O; IC$_{50}$, 349 μM), V79 (IC$_{50}$, >1020 μM), and Hep G2 (IC$_{50}$, 273 μM) cells [281].

Though moniliformin is weakly toxic to the aforementioned cell lines, it does affect cardiac cells. In Japanese quail, this toxin causes cardiomyocyte hypertrophy by day 7 after challenge as well as cardiomegaly, myocardial karyomegaly, myofibril disarray, and nuclear hyperchromasia by 28 days after dosing [282]. In mice orally challenged with moniliformin (29.46 mg/kg), the mitochondria of myocardial cells contain lesions with increased severity after 2–3 h as well as lesions appearing in the sarcolemma and myofibrils [283]. A daily dose of moniliformin (6 mg/kg) for nearly 2 months in rats produces mild lesions in myofibrils and myocardial mitochondria, while lesions in the sarcolemma are more pronounced after day 21 [283]. Young broilers and turkeys fed grain dosed with moniliformin (50 mg/kg) also develop cardiac lesions with losses in cardiomyocyte cross striations, increases in cardiomyocyte mitotic figures (in turkeys only), and greater cardiomyocyte nuclear size than in controls [284]. Besides cardiac tissue, moniliformin targets smooth muscle cells by impairing their contractile properties as observed in isolated guinea pig terminal ileum, aorta, pulmonary artery, and papillary muscle. However, the contractile effect was weaker in smooth muscle than cardiac tissue [285].

Experiments using the OECD Guideline 423 [286], whereby a single oral dose is given to rats, show that moniliformin is acutely toxic and causes weakness in the muscles, respiratory distress, and damage to the heart [287]. In this study, the LD$_{50}$ cutoff dose is 25 mg/kg body weight with pathology results showing the heart as the primary target with sudden arrhythmia as the cause of death [287]. In 6-week-old pigs fed moniliformin (50 mg, 100 mg, or 200 mg/kg), hematologic values are affected, while at the highest dose, death occurred in the test animals with heart weight gain and death attributed to cardiac failure [278].

Moniliformin is not directly linked to any human diseases. However, in vitro, it inhibits, in a dose- and time-dependent manner, chondrocyte viability and synthesis of aggrecan Type II collagen that suggests a possible role of this toxin in Kashin–Beck disease, a chronic and deformative osteoarthritis present in China [288]. However, experiments with combinations of DON, NIV, and T-2 toxins, in the presence of selenium deficiency, result in decreases of aggrecan and Type II collagen suggesting that selenium deficiency is needed for mycotoxins to play a role in this disease [289].

21.4.4 FUSAPROLIFERIN

This toxin is produced by a number of *Fusarium* species, with *F. proliferatum* and *F. subglutinans* being predominant [290,291]. First discovered as a product of *F. proliferatum* isolated from corn kernels, this toxin is a sesterterpene (C$_{27}$H$_{40}$O$_5$) described as 18-[1-(acetoxymethyl-2-methylethyl]-10-,17-dihydroxy-3,7,11,15-tetramethylbicyclo[13.3.0]octadeca-*trans,trans,trans,cis*-2,6,11,17-tetraen-16-one [292,293] (Figure 21.5). Though not as commonly found as the fumonisins, fusaproliferin is present in grapes, grain, and grain products from Europe, North America, and Africa [244,294–297]. Some information on its phytotoxicity, mammalian cytotoxicity, and lethality to brine shrimp is available. Fusaproliferin in combination with ZON, α-ZOL, FB$_1$, and DON,

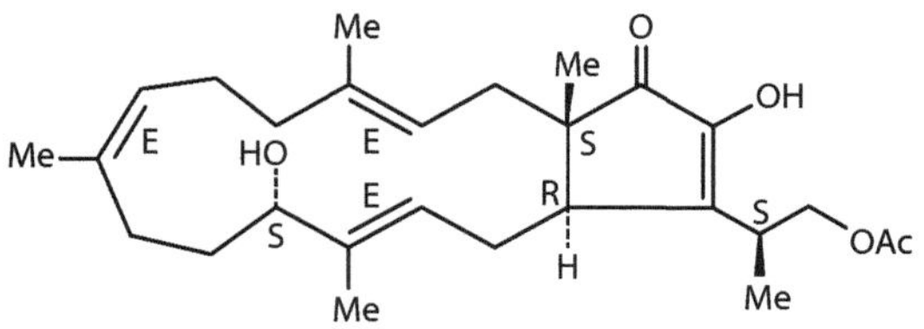

FIGURE 21.5 Fusaproliferin.

at a concentration of 3.5 μg/mL each, causes slight growth retardation (6%) and slightly lower chlorophyll concentrations with some disorganization of thylakoids in maize leaves by 3 days after inoculation with the toxin mixture [298].

Fusaproliferin is inactive in human colorectal adenocarcinoma (HT-29) and Caco-2 cells at up to 30 μM, while beauvericin has IC_{50} values of 15.0 and 24.6 μM, respectively [299]. In vitro, fusaproliferin is cytotoxic to human nonneoplastic B lymphocytes and lepidopteran cell line SF-9 with IC_{50} values of 55 and 70 μM, respectively, after a 48 h exposure [300]. Several studies indicate fusaproliferin toxicity to brine shrimp, *Artemia salina*. In one study, the LD_{50} of purified fusaproliferin is 53.4 μM after 24 h incubation, while toxicity of crude extracts of fusaproliferin depends on the species of *Fusarium* [291,300,301]. A much higher concentration (5 mM) added to the air sac of chicken eggs produces severe teratogenic effects in 20% of the embryos [308]. These malformations include the absence of the head, cephalic dichotomy, abnormal abdominal development, macrocephaly, and anomalous development of the legs and feet [301].

21.5 *MASKED MYCOTOXINS* PRODUCED BY PLANTS

The *Fusarium* toxins such as DON, ZON, T-2, and HT-2 are phytotoxic and aid in the virulence of the fungi that produce these mycotoxins [302–306]. Plants have developed the capacity to partly metabolize and detoxify these mycotoxins by transforming them into more polar compounds with the resulting conjugate, or *masked mycotoxin*, being undetectable by conventional analytical methods [307,308]. In plants, the transformation process is comprised three phases: (1) in phase 1, the mycotoxin is either reduced, acetylated, or oxidized, which adds reactive groups to the mycotoxins; (2) in phase II, the conjugation step, plant enzymes transform these mycotoxins by glycosylation or sulfation into more hydrophilic, less toxic conjugates that, (3) in phase III, are removed and bound to cell walls or placed in plant vacuoles or the apoplast [309–315].

Conjugated, *masked mycotoxins* are less toxic than the parent mycotoxins and are present in food and feed along with the parent toxin. For example, DON, ZON, T-2, HT-2, α-ZOL, β-ZOL, and their conjugated forms have been detected in oats, maize, wheat, bread and breakfast cereals, popcorn, and oatmeal [309,316]. These conjugates include 15-acetylDON, DON-3-glucoside, T-2-3-glucoside, HT-2-3-glucoside, HT-2-4-glucoside, ZON-4-glucoside, α-ZOL-4-glucoside, β-ZOL-4-glucoside, and β-ZON-4-sulfate [309,316,317]. Other conjugates found in cereal-based food include DON-3-glucopyranoside, 3-acetyl-DON, α-ZOL-4-glucopyranoside, and β-ZOL-4-glucopyranoside [318]. Of concern to the brewery industry is that DON and its metabolites 3-acetyl-DON, DON-3-glucoside, DON-3-diglucoside, DON-3-triglucoside, and DON-3-tetraglucoside are present in barley and both light and dark beers [316,319–322].

Concern exists about their effects on animal and human health after ingestion. Studies show that these conjugates can be cleaved thus releasing the toxic aglycone forms. For example, after ingestion of ZON-4-β-D-glucopyranoside for 2 weeks, the intestinal flora in swine can convert this *masked mycotoxin* into ZON and α-ZOL, which are found in the urine and feces [323]. Intestinal microflora in gavaged rats given DON-3-glucoside (3.1 mg/kg body weight) convert this DON conjugate into DON, some of which undergoes deep oxidation to the less toxic deepoxy-deoxynivalenol (DOM-1) and are excreted [324].

In vitro, human colonic microbiota deconjugate DON and ZON thereby releasing the toxic aglycone forms [325]. Results indicate that some humans have intestinal flora that can convert DON into a less toxic form. While microbes in human fecal matter from five volunteers metabolize DON-3-glucoside into DON, those from one volunteer cause deepoxidation of DON producing DOM-1 that is excreted in the urine [326].

21.6 CONCLUSION

Under the appropriate conditions, many *Fusarium* species produce one or more toxins on grains that enter into human or animal foods. Upon ingestion, these toxins vary in their target organ, toxic activity, and sensitivity of the host. Detection and elimination of these mycotoxins from the food chain are complicated by plants converting most of the *Fusarium* toxins into the *masked forms*, which can regain their toxigenic properties after ingestion following enzyme removal of the conjugated group. Continued research is needed to elucidate these multiple complex interactions among fungi, plants, grains, and grain-derived foods as well as to improve detection of both the natural and *masked* toxin forms.

REFERENCES

1. Creppy E. Update of survey, regulation and toxic effects of mycotoxins in Europe. *Toxicol Lett*. 2002; 127: 19–28.
2. Turner PC, Flannery B, Isitt C, Ali M, Peska J. The role of biomarkers in evaluating human health concerns from fungal contaminants in foods. *Nutr Res Rev*. 2012; 25: 162–179.
3. Pitt JI. Toxigenic fungi: Which are important? *Med Mycol*. 2000; 38(Suppl. 1): 17–22.
4. van Egmond HP, Schothorst RC, Jonker MA. Regulations relating to mycotoxins in food. Perspectives in a global and European context. *Anal Bioanal Chem*. 2007; 389: 147–157.
5. Bouaziz C, Martel C, Sharaf el Dein O, Abid-Essefi S, Brenner C, Lemaire C, Bacha H. Fusarial toxin-induced toxicity in cultured cells and in isolated mitochondrial involves PTPC-dependent activation of the mitochondrial pathway of apoptosis. *Toxicol Sci*. 2009; 110: 363–375.
6. Thiel PG, Marasas WFO, Syndenham EW, Shephard GS, Gelderblom WCA. The implications of naturally occurring levels of fumonisins in corn for human and animal health. *Mycopathology* 1992; 117: 3–9.
7. Müller S, Dekant W, Mally A. Fumonisin B1 and the kidney: Modes of action for renal tumor formation by fumonisin B1 in rodents. *Food Chem Toxicol*. 2012; 50: 3833–3846. Doi:10.1016/j.fct.2012.06.053.
8. Cole RA, Jarvis BB, Schweikert MA. *Handbook of Secondary Metabolites*. Academic Press, New York, 2003, pp. 199–560.
9. Grove JF. The trichothecenes and their biosynthesis. *Prog Chem Org Nat Prod*. 2007; 88: 63–130.
10. Lauren DR, Smith WA. Stability of the *Fusarium* mycotoxins nivalenol, deoxynivalenol and zearalenone in ground maize under typical cooking environments. *Food Addit Contam*. 2001; 18: 1011–1016.
11. Samar MM, Neira MS, Resnick SL, Pacin, A. Effect of fermentation on naturally occurring deoxynivalenol (DON) in Argentinean bread processing technology. *Food Addit Contam*. 2001; 18: 1004–1010.
12. O'Neill K, Damaglou AP, Patterson MF. The stability of deoxynivalenol and 3-acetyl deoxynivalenol to gamma irradiation. *Chem Res Toxicol*. 1993; 6: 524–529.
13. Kottapalli B, Wolf-Hall CE, Schwarz P. Effect of electron-beam irradiation on the safety and quality of *Fusarium*-infected malting barley. *Int J Food Microbiol*. 2006; 110: 224–231.
14. Dohnal V, Jezkova A, Jun D, Kuca K. Metabolic pathways of T-2 toxin. *Curr Drug Metab*. 2008; 9: 77–82.
15. Ueno Y. Toxicological features of T-2 toxin and related trichothecenes. *Fundam Appl Toxicol*. 1984; 4: S124–S132.
16. Pestka JJ, Smolinski AT. Deoxynivalenol: Toxicology and potential effects on humans. *J Toxicol Environ Health B Crit Rev*. 2005; 8: 39–69.
17. Peska JJ, Zhou H-R, Moon Y, Chung YJ. Cellular and molecular mechanisms for immune modulation by deoxynivalenol and other trichothecenes: Unraveling a paradox. *Toxicol Lett*. 2004; 153: 61–73.

18. Li Y, Wang Z, Beier RC, Shen J, De Smet D, De Saeger S, Zhang S. T-2 toxin, a trichothecene mycotoxin: Review of toxicity, metabolism, and analytical methods. *J Agric Food Chem*. 2011; 59: 3441–3453.
19. van der Fels-Klerx HJ, Stratakou I. T-2 toxin and HT-2 toxin in grain and grain-based commodities in Europe: Occurrence, factors affecting occurrence, co-occurrence and toxicological effects. *World Mycotoxin J*. 2010; 3: 349–367.
20. Ueno Y, Sawano M, Ishii K. Production of trichothecene mycotoxins by *Fusarium* species in shake culture. *Appl Microbiol*. 1975; 30: 4–9.
21. Canady R, Coker R, Egan S, Krska R, Kuiper-Goodman T, Olsen M, Pestka J, Resnik S, Schlatter J. Deoxynivalenol. In *Safety Evaluation of Certain Mycotoxins in Foods*, WHO Food Additives Series 47 and Joint FAO/WHO Expert Committee on Food Additives, International Programme on Chemical Safety, World Health Organization, Geneva, Switzerland, 2001, pp. 419–550.
22. Rafai P, Bata A, Ványi A, Papp Z, Brydl E, Jakab L, Tuboly S, Túry E. Effect of various levels of T-2 toxin on the clinical status, performance and metabolism of growing pigs. *Vet Rec*. 1995; 136: 485–489.
23. Tan DC, Flematti GR, Ghisalberti EL, Sivasithamparam K, Chakraborty S, Obanor F, Barbetti MJ. Mycotoxins produced by *Fusarium* species associated with annual legume pastures and 'sheep feed refusal disorders' in Western Australia. *Mycotoxin Res*. 2011; 27: 123–135.
24. Capriotti AL, Foglia P, Gubbiotti R, Roccia C, Samperi R, Laganà A. Development and validation of a liquid chromatography/atmospheric pressure photoionization-tandem mass spectrophotometric method for the analysis of mycotoxins subjected to commission regulation (EC) No. 1881/2006 in cereals. *J Chromatogr*. 2010; A 1217: 6044–6051.
25. EFSA Panel on Contaminants in the Food Chain (CONTAM). Scientific opinion on the risks for animal and public health related to the presence of T-2 and HT-2 toxin in food and feed. *EFSA J*. 2011; 12: 2481.
26. Goossens J, De Bock L, Osselaere A, Verbrugghe E, Devreese M, Boussery K, Van Bocxlaer J, De Backer P, Croubels S. The mycotoxin T-2 inhibits hepatic cytochrome P4503A activity in pigs. *Food Chem Toxicol*. 2013; 57C: 54–56.
27. Meissonnier GM, Laffitte J, Raymond I, Benoit E, Cossalter A-M, Pinton P, Bertin G, Oswald IP, Galtier P. Subclinical doses of T-2 toxin impair acquired immune response and liver P450 in pigs. *Toxicology* 2008; 247: 46–54.
28. Swanson SP, Corley RA. The distribution, metabolism and excretion of trichothecene mycotoxins. In *Trichothecene Mycotoxicosis Pathophysiologic Effects* (V.R. Beasley, Ed.) CRC Press, Boca Raton, FL, 1989, Vol. I, pp. 37–61.
29. Bouaziz C, Abid-Essenfi A, Bouslimi A, El Golli E, Bacha H. Cytotoxicity and related effects of T-2 toxin on cultured Vero cells. *Toxicon* 2006; 48: 343–352.
30. Bouaziz C, Martel C, el dein OS, Abid-Essefi S, Brenner C, Lemaire C, Bacha H. Fusarial toxin-induced toxicity in cultured cells and in isolated mitochondria involves PTPC-dependent activation of the mitochondrial pathway of apoptosis. *Toxicol Sci*. 2009; 110: 363–375.
31. Chaudhari M, Jayaraj R, Bhaskar ASB, Rao PVL. Oxidative stress induction by T-2 toxin causes DNA damage and triggers apoptosis via caspase pathway in human cervical cancer cells. *Toxicology* 2009; 262: 153–161.
32. Middlebrook JL, Leatherman DL. Binding of T-2 toxin to eukaryotic cell ribosomes. *Biochem Pharmacol*. 1989; 38: 3103–3110.
33. Middlebrook JL, Leatherman DL. Specific association of T-2 toxin with mammalian cells. *Biochem Pharmacol*. 1989; 38: 3093–3102.
34. Ndossi DG, Frizzell C, Tremoen NH, Fæste CK, Verhaegen S, Dahl E, Eriksen GS, Sørlie M, Connolly L, Ropstad E. An in vitro investigation of endocrine disrupting effect of trichothecenes deoxynivalenol (DON), T-2 and HT-2 toxins. *Toxicol Lett*. 2012; 214: 268–278.
35. Schothorst R, van Egmond HP. Report from SCOOP task 3.2.10 "Collection of occurrence data of *Fusarium* toxins in food and assessment of dietary intake by the population of EU member states" Subtask: Trichothecenes. *Toxicol Lett*. 2004; 153: 133–143.
36. Cegielska-Radziejewska R, Stuper K, Szablewski T. Microflora and mycotoxin contamination in poultry feed mixtures from western Poland. *Ann Agric Environ Med*. 2013; 20: 30–35.
37. Bennett JW, Klich M. Mycotoxins. *Clin Microbiol Rev*. 2003; 16: 497–516.
38. Harvey RB, Kubena LF, Corrier DE, Huff WE, Rottinghaus GE. Cutaneous ulceration and necrosis in pigs fed aflatoxin- and T-2 toxin-contaminated diets. *J Vet Diagn Invest*. 1990; 2: 227–229.
39. Devegowda G, Murthy TNK. Mycotoxins: Their effect in poultry and some practical solutions. In *The Mycotoxin Blue Book* (D.E. Diaz, Ed.) Nottingham, U.K., 2005, pp. 25–56.
40. MacDonald EJ, Cavan KR, Smith TK. Effect of acute oral doses of T-2 toxin on tissue concentrations of biogenic amines in the rat. *J Anim Sci*. 1988; 66: 434–441.

41. Ferreras MC, Benavides J, García-Pariente C, Delgado L, Fuertes M, Muñoz M, García-Marín F, Pérez V. Acute and chronic disease associated with naturally occurring T-2 mycotoxicosis in sheep. *J Comp Pathol.* 2013; 148: 236–242.
42. Jaffe AZ, Yagen B. Comparative study of the yield of T-2 toxin produced by *Fusarium poae*, *F. sporotrichioides* and *F. sporotrichioides* var. tricinctum strains from different sources. *Mycopathology* 1977; 60: 93–97.
43. Chen JH, Xue S, Li S, Qang ZL, Yang H, Wang W, Song D, Zhou X, Chen C. Oxidant damage in Kashin-Beck disease and a rat Kashin-Beck disease model by employing T-2 toxin treatment under selenium deficient conditions. *J Orthop Res.* 2012; 30: 1229–1237.
44. Peng A, Yang C, Rui H, Li H. Study on the pathogenic factors of Kashin-Beck disease. *J Toxicol Environ Health* 1992; 35: 79–90.
45. Sun LY, Li Q, Meng FG, Fu Y, Zhao ZJ, Wang LH. T-2 toxin contamination in grains and selenium concentration in drinking water and grains in Kashin-Beck disease endemic areas of Qinghai province. *Biol Trace Elem Res.* 2012; 150: 371–375.
46. Pang VF, Lambert RJ, Felsburg PJ, Beasley VR, Buck WB, Haschek WM. Experimental T-2 toxicosis in swine following exposure: Effects on pulmonary and systemic immunity, and morphologic changes. *Toxicol Pathol.* 1987; 15: 308–319.
47. Seeboth J, Solinhac R, Oswald IP, Guzylack-Pirlou L. The fungal T-2 toxin alters the activation of primary macrophages induced by TLR-agonists resulting in a decrease of the inflammatory response in the pig. *Vet Res.* 2008; 43: 35. Doi:10.1186/1297-9716-43-35.
48. Li M, Harkema JR, Islam Z, Cuff CF, Pestka JJ. T-2 toxin impairs murine response to respiratory reovirus and exacerbates viral bronchiolitis. *Toxicol Appl Pharmcol.* 2006; 217: 76–85.
49. Li M, Cuff CF, Pestka JJ. T-2 toxin impairment of enteric reovirus clearance in the mouse associated with suppressed immunoglobulin and IFN-γ responses. *Toxicol Appl Pharmacol.* 2006; 214: 318–325.
50. Tai JH, Pestka JJ. T-2 toxin impairment of murine response to *Salmonella typhimurium*: A histopathological assessment. *Mycopathology* 1990; 109: 149–155.
51. Yohannes T, Sharma AK, Singh SD, Goswami TK. Immunopathological effects of experimental T-2 mycotoxicosis in broiler chicken co-infected with infectious bronchitis virus (IBV). *Vet Immunol Immunopathol.* 2012; 146: 245–253.
52. Berek L, Petri IB, Mesterházy Á, Téren J, Molnár J. Effect of mycotoxins on human immune functions *in vitro*. *Toxicol In Vitro* 2001; 15: 25–30.
53. Minervini F, Fornelli F, Lucivero G, Romano C, Viscounti A. T-2 toxin immunotoxicity on human B and T lymphoid cell lines. *Toxicology* 2005; 210: 81–91.
54. Ravindran J, Agrawal M, Gupta N, Rao PVL. Alteration of blood brain barrier permeability by T-2 toxin: Role of MMP-9 and inflammatory cytokines. *Toxicology* 2011; 280: 44–52.
55. Weidner M, Hüwel S, Ebert F, Schwerdtle T, Galla H-J, Humpf HU. Influence of T-2 and HT-2 toxin on the blood-brain barrier *in vitro*: New experimental hints for neurotoxic effects. *PLoS One* 2013; 8(3): e60484. Doi:10.1371/journal.pone.0060484.
56. Weidner M, Lenczyk M, Schwerdt G, Gekle M, Humpf H-U. Neurotoxic potential and cellular uptake of T-2 toxin in human astrocytes in primary culture. *Chem Res Toxicol.* 2013; 26: 347–355.
57. Chaudhary M, Rao PVL. Brain oxidative stress after dermal and subcutaneous exposure of T-2 toxin in mice. *Food Chem Toxicol.* 2010; 48: 3436–3442.
58. Boonen J, Malysheva SV, Taevernier L, Diana Di Mavungu JD, De Saeger S, Spiegeleer, B. Human skin penetration of selected model mycotoxins. *Toxicology* 2012; 301: 21–32.
59. Agrawal M, Yadav P, Lomash V, Bhaskar ASB, Rao PVL. T-2 toxin induced skin inflammation and cutaneous injury in mice. *Toxicology* 2012; 302: 255–265.
60. Nordby K-C, Halstensen AS, Elen O, Clasen P-E, Langseth W, Kristensen P, Eduard W. Trichothecene mycotoxins and their determinants in settled dust related to grain production. *Ann Agric Environ Med.* 2004; 11: 75–83.
61. Halstensen AS, Nordby K-C, Elen O, Clasen PE, Eduard W. Toxigenic *Fusarium* spp. as determinants of trichothecene mycotoxins in settled grain dust. *J Occup Environ Hyg.* 2006; 3: 651–659.
62. Creasia DA, Thurman JD, Wannemacher RW Jr., Bunner DL. Acute inhalation toxicity of T-2 mycotoxin in the rat and guinea pig. *Fundam Appl Toxicol.* 1990; 14: 54–59.
63. Thrane U, Adler A, Clasen P-E, Galvano F, Langseth W, Lew H, Logrieco A, Nielsen KF, Ritieni A. Diversity in metabolite production by *Fusarium langsethiae*, *Fusarium poae*, and *Fusarium sporotrichioides*. *Int J Food Microbiol.* 2004; 95: 257–266.
64. Edwards SG, Imathiu SM, Ray RV, Back M, Hare MC. Molecular studies to identify the *Fusarium* species responsible for HT-2 and T-2 mycotoxins in UK oats. *Int J Food Microbiol.* 2012; 156: 168–175.

65. Ellison R, Kotsonis FN. In vitro metabolism of T-2 toxin. *Appl Microbiol.* 1974; 27: 423–424.
66. Swanson SP, Helaszek C, Buck WB, Rood HD Jr., Haschek WM. The role of intestinal microflora in the metabolism of trichothecene mycotoxins. *Food Chem Toxicol.* 1988; 26: 823–829.
67. Swanson SP, Nicoletti J, Rood HD, Jr., Buck WB, Cote LM, Yoshizawa T. Metabolism of three trichothecene mycotoxins, T-2 toxin, diacetoxyscirpenol and deoxynivalenol, by bovine rumen microorganisms. *J Chromatogr.* 1987; 414: 335–342.
68. Wu Q, Engemann A, Cramer B, Welsch T, Yuan Z, Humpf H-U. Intestinal metabolism of T-2 toxin in the pig cecum model. *Mycotox Res.* 2012; 28: 191–198.
69. Beeton S, Bull AT. Biotransformation and detoxification of T-2 toxin by soil and freshwater bacteria. *Appl Environ Microbiol.* 1989; 55: 190–197.
70. Königs M, Mulac D, Schwerdt G, Gekle M, Humpf H-U. Metabolism and cytotoxic effects of T-2 toxin and its metabolites on human cells in primary culture. *Toxicology* 2009; 258: 106–115.
71. Nielsen C, Casteel M, Didier A, Dietrich R, Märtlbauer E. Trichothecene-induced cytotoxicity on human cell lines. *Mycotox Res.* 2009; 25: 77–84.
72. Widestrand J, Lundh T, Pettersson H, Lindberg JE. Cytotoxicity of four trichothecenes evaluated by three colorimetric bioassays. *Mycopathology* 1999; 147: 149–155.
73. Behm C, Föllman W, Degen GH. Cytotoxic potency of mycotoxins in cultures of V79 lung fibroblast cells. *J Toxicol Environ Health, Part A* 2012; 75: 1226–1231.
74. Garvey GS, McCormick SP, Rayent I. Structural and functional characterization of the TRI101 trichothecene 3-*O*-acetyltransferase from *Fusarium sporotrichioides* and *Fusarium graminearum*. *J Biol Chem.* 2008; 283: 1660–1669.
75. Richard JL. Mycotoxins—an overview. In *Romer Labs' Guide to Mycotoxins* (J.L. Richard, Ed.) Vol. 1, Anytime Publishing Ltd, Leicestershire, England, 2000, pp.1–48.
76. Hussein HS, Brasel JM. Toxicity, metabolism and impact of mycotoxins on humans and animals. *Toxicology* 2001; 167: 101–134.
77. European Commission. Commission Regulation (EC) No. 1881/(2006) of 19 December (2006). Setting maximal levels for certain contaminants in foodstuffs. *Off J Eur Union.* 2006; 364: 7–8.
78. European Commission, Scientific Committee on Food. Opinion of the Scientific Committee on Food on Fusarium toxins, part 6: Group evaluation of T-2 toxin, HT-2 toxin, nivalenol, and deoxynivalenol (DON), 2002. http://ec.europa.eu/food/fs/sc/scf/out123_en.pdf, accessed April 8, 2015.
79. van der Fels-Klerx HJ, Goedhardt PW, Elen O, Börjesson T, Hietaniemi V, Booij CJH. Modeling deoxynivalenol contamination of wheat in northwestern Europe for climate change assessments. *J Food Protect.* 2012; 75: 1099–1106.
80. Adejumo TO, Hettwer U, Karlovsky P. Occurrence of *Fusarium* species and trichothecenes in Nigerian maize. *Int J Food Microbiol.* 2007; 116: 350–357.
81. Guidance for Industry and FDA. Advisory levels for Deoxynivalenol (DON) in finished wheat products for human consumption and grains and grain by-products used for animal feed, U.S. Food and Drug Administration, College Park, MD, June 2010. http://www.fda.gov/downloads/Food/GuidanceRegulation/UCM217558.pdf.
82. Avantaggiato G, Havenaar R, Visconoti A. Evaluation of the intestinal absorption of deoxynivalenol and nivalenol by an in vitro gastrointestinal model, and the binding efficiency of activated carbon and other adsorbent materials. *Food Chem Toxicol.* 2004; 42: 817–824.
83. Coppock RW, Swanson SP, Gelberg HB, Koritz GD, Hoffman WE, Buck WB, Vesonder RF. Preliminary study of the pharmacokinetics and toxicopathy of deoxynivalenol (vomitoxin) in swine. *Am J Vet Res.* 1985; 46: 169–174.
84. Bhat RV, Beedu SR, Ramakrishna Y, Munshi KL. Outbreak of trichothecene mycotoxicosis associated with consumption of mould-damaged wheat production in Kashmir Valley, India. *Lancet* 1989; 1(8628): 35–37.
85. Young LG, McGirr L, Valli VE, Lumsden JH, Lun A. Vomitoxin in corn fed to young pigs. *J Anim Sci.* 1983; 57: 655–664.
86. Wu W, Bates MA, Bursian SJ, Flannery B, Zhou H-R, Link JE, Zhang H, Pestka JJ. Peptide YY3–36 and 5-hydroxytryptamine mediate emesis induction by trichothecene deoxynivalenol (vomitoxin). *Toxicol Sci.* 2013; 133: 186–195.
87. He K, Zhou H-R, Peska JJ. Mechanisms for ribotoxin-induced ribosomal RNA cleavage. *Toxicol Appl Pharmacol.* 2012; 265: 10–18.
88. Minervini F, Fornelli F, Flynn KM. Toxicity and apoptosis induced by the mycotoxins nivalenol, deoxynivalenol and fumonisin B_1 in a human erythroleukemia cell line. *Toxicol In Vitro* 2004; 18: 21–28.

89. Königs M, Lenczyk M, Schwerdt G, Holzinger H, Gekle M, Humpf H-U. Cytotoxicity, metabolism and cellular uptake of the mycotoxin deoxynivalenol in human proximal tubule cells and lung fibroblasts in primary culture. *Toxicology* 2007; 240: 48–59.
90. Königs M, Scwerdt G, Gekle M, Humpf H-U. Effects of the mycotoxin deoxynivalenol on human hepatocytes. *Mol Nutr Food Res*. 2008; 52: 830–839.
91. Amuzie CJ, Harkema JR, Pestka JJ. Tissue distribution and proinflammatory cytokine induction by the trichothecene deoxynivalenol in the mouse: Comparison of nasal vs. oral exposure. *Toxicology* 2008; 248: 39–44.
92. Bony S, Carcelen M, Olivier L, Devaux A. Genotoxicity assessment of deoxynivalenol in the Caco-2 cell line model using the Comet assay. *Toxicol Lett*. 2006; 166: 67–76.
93. Knasmüller S, Bresgen N, Kassie F, Mersch-Sundermann V, Gelderblom W, Zöhrer E, Eckl PM. Genotoxic effects of three *Fusarium* mycotoxins, fumonisin B_1, moniliformin and vomitoxin in bacteria and in primary cultures of rat hepatocytes. *Mutat Res*. 1997; 391: 39–48.
94. Dinu D, Bodea GO, Ceapa CD, Munteanu MC, Roming FI, Serban AI, Hermenean A, Costache M, Zarnescu O, Dinischiotu A. Adapted response of the antioxidant defense system to oxidative stress induced by deoxynivalenol in Hek-293 cells. *Toxicon* 2011; 37: 1023–1032.
95. Kouadio JH, Mobio TA, Baudrimont I, Moukha S, Dano SD, Creppy EE. Comparative study of cytotoxicity and oxidative stress induced by deoxynivalenol, zearalenone or fumonisin B1 in human intestinal cell line Caco-2. *Toxicology* 2005; 213: 56–65.
96. Nasri T, Bosch RR, Voorde S, Fink-Gremmels J. Differential induction of apoptosis by Type A and B trichothecenes in Jurkat T-lymphocytes. *Toxicol In Vitro* 2006; 20: 832–840.
97. Meky FA, Hardie LJ, Evans SW, Wild CP. Deoxynivalenol-induced immunomodulation of human lymphocyte proliferation and cytokine production. *Food Chem Toxicol*. 2001; 39: 827–836.
98. Ni K, O'Neill HC. The role of dendritic cells in T cell activation. *Immunol Cell Biol*. 1997; 75: 223–230.
99. Hymery N, Sibiril Y, Parent-Massin D. In vitro effects of trichothecenes on human dendritic cells. *Toxicology* 2006; 20: 899–909.
100. Bimczok D, Döll S, Rau H, Govarts T, Wundrack N, Naumann M, Dänicke S, Rothkötter HJ. The *Fusarium* toxin deoxynivalenol disrupts phenotype and function of monocyte-derived dendritic cells in vivo and *in vitro*. *Immunobiology* 2007; 212: 655–666.
101. Luongo D, Severino L, Bergamo P, D'Arienzo R, Rossi M. Trichothecenes NIV and DON modulate the maturation of murine dendritic cells. *Toxicon* 2010; 55: 73–80.
102. Awad WA, Aschenbach JR, Zentek J. Cytotoxicity and metabolic stress induced by deoxynivalenol in the porcine intestinal IPEC-J2 cell line. *J Anim Physiol Anim Nutr*. 2011; 96: 709–716.
103. Diesing A-K, Nossol C, Panther P, Walk N, Post A, Kluess J, Kreutzmann P, Dänicke S, Rothkötter H-J, Kahlert S. Mycotoxin deoxynivalenol (DON) mediates biphasic cellular response in intestinal porcine epithelial cell lines IPC-1 and IPEC-J2. *Toxicol Lett*. 2011; 200: 8–18.
104. Vandenbroucke V, Croubles S, Martel A, Verbrugghe E, Goossens J, Van Deun K, Boyen F et al. The mycotoxin deoxynivalenol potentiates intestinal inflammation by *Salmonella typhimurium* in porcine ileal loops. *PLoS One* 2011; 6: e23871. Doi:10.1371/journal.pone.0023871.
105. Schoevers EJ, Fink-Gremmels J, Colenbrander B, Roelen BA. Porcine oocytes are most vulnerable to the mycotoxin deoxynivalenol during formation of the meiotic spindle. *Theriogenology* 2010; 74: 968–978.
106. Ranzenigo G, Caloni F, Cremonesi F, Aad PY, Spicer LJ. Effects of *Fusarium* mycotoxins on steroid production by porcine granulosa cells. *Anim Reprod Sci*. 2008; 107: 115–130.
107. Girardet C, Bonnet MS, Jdir R, Sadoud M, Thirion S, Tardivel C, Roux J et al. The food-contaminant deoxynivalenol modifies eating by targeting anorexigenic neurocircuitry. *PLoS One* 2011; 6: e26134. Doi10.1371/journal.pone.0026134.
108. Kobayashi-Hattori K, Amuzie CJ, Flannery BM, Pestka JJ. Body composition and hormonal effects following exposure to mycotoxin deoxynivalenol in the high-fat diet-induced obese model. *Mol Nutr Food Res*. 2011; 55: 1070–1078.
109. Gaigé S, Bonnet MS, Tardivel C, Pinton P, Trousard J, Jean A, Guzylack L, Troadec J-D, Dallaporta M. c-Fos immunoreactivity in the pig brain following deoxynivalenol intoxication: Focus on NUCB2/nesfatin-1 expressing neurons. *Neurotoxicology* 2013; 34: 135–149.
110. Razafimanjato H, Benzaria A, Taïeb N, Guo XJ, Vidal N, Di Scala C, Varini K, Maresca M. The ribotoxin deoxynivalenol affects the viability and functions of glial cells. *Glia* 2011; 59: 1672–1683.
111. European Commission, Scientific Committee on Food. Opinion of the Scientific Committee on Food on *Fusarium* Toxins, Part 4: Nivalenol, 2000. http://ec.europa.eu/food/fs/sc/scf/out74_en.pdf, accessed April 8, 2015.

112. Anon. Toxins derived from *Fusarium graminearum*, *F culmorum* and *F. crookwellense*: Zearalenone, deoxynivalenol, and fusarenone X. *IARC Monogr.* 1993; 56: 397–444.
113. Hsia CC, Wu ZY, Li YS, Zhang F, Sun ZT. Nivalenol, a main *Fusarium* toxin in dietary foods from high-risk areas of cancer of esophagus and gastric cardia in China, induced benign and malignant tumors in mice. *Oncol Rep.* 2004; 449–456.
114. Luongo D, De Luna R, Russo R, Severino L. Effects of four *Fusarium* toxins (fumonisins B1, α-zearalenol, nivalenol, and deoxynivalenol) on porcine whole-blood cellular proliferation. *Toxicon* 2008; 52: 156–162.
115. Wan LYM, Turner PC, El-Nezami H. Individual and combined effects of *Fusarium* toxins (deoxynivalenol, nivalenol, zearalenone and fumonisins B1 on swine jejunal epithelial cells. *Food Chem Toxicol.* 2013; 57: 276–283.
116. Taranu I, Marin DE, Burlacu R, Pinton P, Damian V, Oswald IP. Comparative aspects of in vitro proliferation of human and porcine lymphocytes exposed to mycotoxins. *Arch Anim Nutr.* 2010; 64: 383–393.
117. Servino L, Russo R, Luongo D, De Luna R, Ciarcia R, Rossi M. Immune effects of four *Fusarium*-toxins (FB_1, ZEA, NIV, DON) on the proliferation of Jurkat cells and porcine lymphocytes: In vitro study. *Vet Res Commun.* 2008; 32: S311–S313.
118. Hedman R, Pettersson H, Lindberg JE. Absorption and metabolism of nivalenol in pigs. *Arch Tierernahr.* 1987; 50: 13–24.
119. Poapolathep A, Sugita-Konishi Y, Phitsanu T, Doi K, Kumagai S. Placental and milk transmission of trichothecene mycotoxins, nivalenol and fusarenon-X, in mice. *Toxicon* 2004; 44: 111–113.
120. Bianco G, Fontanella B, Severino L, Quaroni A, Autore G, Marzocco S. Nivalenol and deoxynivalenol affect rat intestinal epithelial cells: A concentration related study. *PLoS One* 2012; 7: e52051. Doi:101371/journal.pone.0052051.
121. Madej M, Lundh T, Lindberg JE. Effect of exposure to dietary nivalenol on activity of enzymes involved in glutamine catabolism in the epithelium along the gastrointestinal tract of growing pigs. *Arch Tierernahr.* 1999. 52: 275–284.
122. Kubosaki A, Aihara M, Park BJ, Sugiura Y, Shibutani M, Kirose M, Suzuki Y, Takatori K, Sugita-Konishi Y. Immunotoxicity of nivalenol after subchronic dietary exposure to rats. *Food Chem Toxicol.* 2008; 46: 253–258.
123. Takahashi M, Shibutani M, Sugita-Konishi Y, Aihara M, Inoue K, Woo G-H, Fujimoto H, Hirose M. A 90-day subchronic toxicological study of nivalenol, a trichothecene mycotoxin, in F344 rats. *Food Chem Toxicol.* 2008; 46: 125–135.
124. Sugita-Konishi Y, Kubosaki A, Takahashi M, Park BJ, Tanaka T, Takatori K, Hirose M, Shibutani, M. Nivalenol and the targeting of the female reproductive system as well as haematopoietic and immune systems in rats after 90-day exposure through the diet. *Food Addit Contam Part A* 2008; 25: 1118–1127.
125. Ryu J-C, Ohtsubo K, Izumiyama N, Mori M, Tanaka T, Ueno Y. Effects of nivalenol on the bone marrow in mice. *J Toxicol Sci.* 1987; 12: 11–21.
126. Rheeder JP, Marasas WFO, Vismer HF. Production of fumonisins analogs by *Fusarium* species. *Appl Environ Microbiol.* 2002; 68: 2101–2105.
127. WHO Joint Technical Report Series. Evaluation of certain mycotoxins in foods, Fifty-sixth report of the Joint FAO/WHO Expert Committee on Food Additives, WHO Technical Report Series No. 906, 1-62, WHO, Geneva, Switzerland, 2002.
128. European Commission, Scientific Committee on Food. Updated opinion of the Scientific Committee on Food on Fumonisin B1, B2 and B3, 2003. http://ec.europa.eu/food/fs/sc/scf/out185_en.pdf, accessed April 8, 2015.
129. Sugiura Y, Barr JR, Barr DB, Brock JW, Elie CM, Ueno Y, Patterson G, Potter ME, Reiss E. Physiological characteristics and mycotoxins of human clinical isolates of *Fusarium* species. *Mycol Res.* 1999; 11: 1462–1468.
130. Gelderblom WCA, Cawood ME, Snyman SD, Vleggaar R, Marasas WFO. Structure–activity relationships of fumonisins in short-term carcinogenesis and cytotoxicity assays. *Food Chem. Toxicol.* 1993; 31: 407–414.
131. IARC, International Agency for Research on Cancer. Fumonisin B. In *IARC Monographs on the Evaluation of Carcinogenic Risks to Humans, Some Traditional Herbal Medicines, Some Mycotoxins, Naphthalene and Styrene*, IARC, Lyon, France, Vol. 82, 2002, pp. 301–366.
132. Marasas WF, Kellerman TS, Gelderblom WC, Coetzer JA, Thiel PG, van der Ludt JJ. Leukoencephalomalacia in a horse induced by fumonisins B1 isolated from *Fusarium moniliforme*. *Onderstepoort J Vet Res.* 1988; 55: 197–203.

133. Kellerman TS, Marasas WFO, Thiel PG, Gelderblom WC, Cawood M, Coetzer JA. Leukoencephalomalacia in two horses induced by oral dosing of fumonisins B1. *Onderstepoort J Vet Res*. 1990; 57: 269–275.
134. Harrison LR, Colvin BM, Greene JT, Newman LE, Cole JR Jr. Pulmonary edema and hydrothorax in swine produced by fumonisins B1, a toxic metabolite of *Fusarium moniliforme*. *J Vet Diagn Invest*. 1990; 2: 217–221.
135. Missmer SA, Suarez L, Felkner M, Wang E, Merrill AH, Jr., Rothman KJ, Hendricks KA. Exposure to fumonisins and the occurrence of neural tube defects along the Texas-Mexico border. *Environ Health Perspect*. 2006; 114: 237–241.
136. Suarez L, Felkner M, Brender JD, Canfield M, Zhu H, Hendricks KA. Neural tube defects on the Texas-Mexico border: What we've learned in the 20 years since the Brownsville cluster. *Birth Defects Res A Clin Mol Teratol*. 2012; 94: 882–892.
137. Chu FS, Li GY. Simultaneous occurrence of fumonisin B-1 and other mycotoxins in moldy corn collected from the People's Republic of China in regions with high incidences of esophageal cancer. *Appl Environ Microbiol*. 1994; 60: 847–852.
138. Sun G, Wang S, Hu X, Su J, Huang T, Yu J, Tan L, Gao W, Wang JS. Fumonisin B_1 contamination of home-grown corn in high-risk areas for esophageal and liver cancer in China. *Food Addit Contam*. 2007; 24: 181–185.
139. Williams JH, Grubb JA, Davis JW, Wang J-S, Jolly PE, Ankrah N-A, Elli WO, Afriyie-Gyawu E, Johnson NM, Robinson AG, Phillips TD. HIV and hepatocellular and esophageal carcinomas related to consumption of mycotoxin-prone foods in sub-Saharan Africa. *Am J Clin Nutr*. 2010; 92: 154–160.
140. Wang E, Norred WP, Bacon CW, Riley RT, Merrill AH, Jr. Inhibition of sphingolipid biosynthesis by fumonisins by fumonisins: Implications for diseases associated with *Fusarium moniliforme*. *J Biol Chem*. 1991; 266: 14486–14490.
141. Norred WP, Wang E, Yoo H, Riley RT, Merrill AH, Jr. In vitro toxicology of fumonisins and the mechanistic implications. *Mycopathology* 1992; 117: 73–78.
142. Merrill AH, Jr., van Echten G, Wang E, Sandhoff K. Fumonisin B_1 inhibits sphingosine (Sphinganine) *N*-acyltransferase and *de novo* sphingolipid biosynthesis in cultured neurons in situ. *J Biol Chem*. 1993; 268: 27299–27306.
143. van der Westhuizen L, Shephard GS, Snyman SD, Abel S, Swanevelder S, Gelderblom WCA. Inhibition of sphingolipid biosynthesis in rat primary hepatocyte cultures by fumonisin B1 and other structurally related compounds. *Food Chem Toxicol*. 1998; 36: 497–503.
144. Merrill AH, Jr., Schmeltz E-M, Dillehay DL, Spiegel S, Shayman JA, Schroeder JJ, Riley RT, Voss KA, Wang E. Sphingolipids—The enigmatic lipid class: Biochemistry, physiology, and pathology. *Toxicol Appl Pharmacol*. 1997; 142: 208–225.
145. Hard GC, Howard PC, Kovatch RM, Bucci TJ. Rat kidney pathology induced by chronic exposure to fumonisin B1 included rare variants of renal tubule tumor. *Environ Pathol*. 2001; 29: 379–386.
146. Schroeder JJ, Crane HM, Xia J, Liotta DC, Merrill AH, Jr. Disruption of sphingolipid metabolism and stimulation of DNA synthesis by fumonisin B1. *J Biol Chem*. 1994; 269: 3475–3481.
147. Kim MS, Lee D-Y, Wang T, Schroeder JJ. Fumonisin B_1 induced apoptosis in LLC-PK_1 renal epithelial cells via sphinganine- and calmodulin-dependent pathway. *Toxicol Appl Pharmacol*. 2001; 176: 118–126.
148. Gelderblom WCA, Jaskiewicz K, Marasas WFO, Thiel PG, Horak RM, Vleggar R, Kriek NPJ. Fumonisins-novel mycotoxins with cancer-promoting activity produced by *Fusarium moniliforme*. *Appl Environ Microbiol*. 1988; 54: 1806–1811.
149. Lemmer ER, Vessey CJ, Gelderblom WCA, Shephard EG, Van Schalkwyk DJ, Van Wijk RA, Marasas WFO, Kirsch RE, Hall PM. Fumonisin B_1-induced hepatocellular and cholangiocellular tumors in male Fischer 344 rats: Potentiating effects of 2-acetylaminofluorene on oval cell proliferation and neoplastic development in a discontinued feeding study. *Carcinogen* 2004; 25: 1257–1264.
150. Howard PC, Eppley RM, Stack ME, Warbrittion A, Voss KA, Lorentzen RJ, Kovatch RM, Bucci TJ. Fumonisin B1 carcinogenicity in a two-year feeding study using F344 rats and B6C3F1 mice. *Environ Health Perspect*. 2001; 109: 277–282.
151. Bhandari N, He Q, Sharma RP. Gender-related differences in subacute fumonisin B1 hepatotoxicity in BAL/c mice. *Toxicology* 2001; 165: 195–204.
152. Dugyala RR, Sharma RP, Tsunoda M, Riley RT. Tumor necrosis factor-α as a contributor in fumonisin B1 toxicity. *J Pharm Exp Ther*. 1997; 285: 317–323.
153. Bhandari N, Sharma RP. Fumonisin B_1-induced alterations in cytokine expression and apoptosis signaling genes in mouse liver and kidney after an acute exposure. *Toxicology* 2002; 172: 81–92.
154. Sharma RP, Bhandari N, Tsunoda M, Riley RT, Voss KA. Fumonisin hepatotoxicity is reduced in mice carrying the human tumor necrosis factor alpha transgene. *Life Sci*. 2000; 74: 238–248.

155. Sharma RP, Bhandari N, Riley RT, Voss KA, Meredith FI. Tolerance to fumonisin toxicity in a mouse strain lacking the P75 tumor necrosis factor receptor. *Toxicology* 2000; 143: 183–194.
156. Sharma RP, He Q, Meredith FI, Riley RT, Voss KA. Paradoxical role of tumor necrosis factor α in fumonisin-induced hepatotoxicity in mice. *Toxicology* 2002; 180: 221–232.
157. Johnson VJ, He Q, Kim SH, Kanti A, Sharma R. Increased susceptibility of renal epithelial cells to TNF-α-induced apoptosis following treatment with fumonisin B1. *Chem Biol Interact*. 2003; 145: 297–309.
158. Laskin DL, Weinburger B, Laskin JD. Functional heterogeneity in liver and lung macrophages. *J Leukoc Biol*. 2001; 70: 163–170.
159. He Q, Kim J, Sharma RP. Fumonisin B1 hepatotoxicity in mice is attenuated by depletion of Kupffer cells by gadolinium chloride. *Toxicology* 2005; 207: 137–147.
160. Sharma N, He Q, Sharma RP. Amelioration of fumonisin B1 hepatotoxicity in mice by depletion of T cells with anti-Thy-1.2. *Toxicology* 2006; 223: 191–201.
161. He Q, Bhandari N, Sharma RP. Fumonisin B (1) alters sphingolipid metabolism and tumor necrosis factor alpha expression in heart and lung of mice. *Life Sci*. 71: 2002; 2015–2023.
162. Riley RT, Showker Owens D.L., Ross PF. Disruption of sphingolipid metabolism and induction of equine leukoencephalomalacia by Fusarium proliferatum culture material containing fumonisin B2 and B3. *Environ Toxicol Pharmacol*. 1997; 3: 221–228.
163. European Union. 2006. Journal official UE, 23/8/2006. L229/7.
164. Tardieu D, Bailly JD, Skiba F, Métayer JP, Grosjean F, Guerre P. Chronic toxicity of fumonisins in turkeys. *Poult Sci*. 2007; 86: 1887–1893.
165. Del Bianchi M, Oliveira CAF, Albuquerque R, Guerra JL, Correa B. Effects of prolonger oral administration of aflatoxin B1 and fumonisin B1 in broiler chickens. *Poult Sci*. 2005; 84: 1835–1840.
166. Shepard GS, van der Westhuizen L, Thiel PG, Gelderblom WCA, Marasas WFO, Van Schalkwyk DJ. Disruption of sphingolipid metabolism in non-human primates consuming diets of fumonisin-containing *Fusarium moniliforme* culture material. *Toxicon* 1996; 34: 527–534.
167. Tolleson WH, Couch LH, Melchior WB, Jr, Jenkins GR, Muskhelishvili M, Muskhelishvili L, McGarrity LJ, Domon O, Morris SM, Howard PC. Fumonisin B1 induces apoptosis in cultured human keratinocytes through sphinganine accumulation and ceramide depletion. *Int J Oncol*. 1999; 14: 833–843.
168. Rotter BA, Thompson BK, Prelusky DB, Trenholm HL, Stewart B, Miller JD, Savard ME. Response of growing swine to dietary exposure to pure fumonisin B_1 during an eight-week period: Growth and clinical parameters. *Nat Toxins* 1996; 4: 42–50.
169. Bouhet S, Hourcade E, Loiseau N, Fikry A, Martinez S, Roselli M, Galtier P, Mengheri E, Oswald IP. The mycotoxin fumonisin B_1 alters the proliferation and barrier function of porcine intestinal cells. *Toxicol Sci*. 2004; 77: 165–171.
170. Halloy DJ, Gustin PG, Bouhet S, Oswald IP. Oral exposure to cultural material extract containing fumonisins predisposes swine to the development of pneumonitis. *Toxicology* 2005; 213: 34–44.
171. Bouhet S, Le Dorze E, Peres S, Fairbrother JM, Oswald IP. Mycotoxin fumonisin B_1 selectively down-regulates the basal IL-8 expression in pig intestine: In vivo and in vitro studies. *Food Chem Toxicol*. 2006; 44: 1768–1773.
172. Lallès JP, Lessard M, Boudry G. Intestinal barrier function is modulated by short-term exposure to fumonisin B_1 in Ussing chambers. *Vet Res Commun*. 2009; 33: 1039–1043.
173. Lallès J-P, Lessard M, Oswald IP, David J-C. Consumption of fumonisin B_1 for 9 days induces stress proteins along the gastrointestinal tract of pigs. *Toxicon* 2010; 55: 244–249.
174. Bracarense A-P FI, Lucioli J, Grenier B, Pacheco GD, Moll W-D, Schatzmayr G, Oswald IP. Chronic ingestion of deoxynivalenol and fumonisin, alone or in interaction, induces morphological and immunological changes in the intestine of piglets. *Br J Nutr*. 2012; 107: 1776–1786.
175. Palencia E, Torres O, Hagler W, Meredith FI, Williams L, Riley RT. Total fumonisins are reduced in tortillas using the traditional nixtamalization method of Mayan communities. *J Nutr*. 2003; 133: 3200–3203.
176. Voss KA, Riley RT, Snook M, Gelineau-van Waes J. Reproductive and sphingolipid metabolic effects of fumonisin B1 and its alkaline hydrolysis product in LM/Bc mice: Hydrolyzed fumonisin B1 did not cause neural tube defects. *Toxicol Sci*. 2009; 112: 459–467.
177. Grenier B, Bracarense A-PFL, Schwartz HE, Trumel C, Cossalter A-M, Schatzmayr G, Kolf-Clauw M, Moll W-D, Oswald IP. The low intestinal and hepatic toxicity of hydrolyzed fumonisin B1 correlates with its inability to alter the metabolism of sphingolipids. *Biochem Pharmacol*. 2012; 83: 1465–1473.
178. Caldwell RW, Tuite J, Stob M, Baldwin R. Zearalenone production by *Fusarium* species. *Appl Microbiol*. 1970; 20: 31–34.

179. Bottalico A, Visconti A, Logrieco A, Solfrizzo M, Mirocha CJ. Occurrence of Zearalenols (diastereomeric mixture) in corn stalk rot and their production by associated *Fusarium* species. *Appl Environ Microbiol*. 1985; 49: 547–551.
180. Urry WH, Wehrmeister HL, Hodge EB, Hidy PH. The structure of zearalenone. *Tetrahedron Lett*. 1966; 27: 3109–3114.
181. Malekinejad H, Maas-Bakker R, Fink-Gremmels J. Species differences in the hepatic biotransformation of zearalenone. *Vet J*. 2006; 172: 96–102.
182. Stipanovic RD, Schroeder HW. Zearalenol and 8'hydroxyzearalenone by *Fusarium roseum*. *Mycopathology* 1975; 57: 77–78.
183. Hagler WM, MIrocha CJ, Pathre SV, Behrens JC. Identification of the naturally occurring isomer of zearalenol produced by *Fusarium roseum* "Gibbosum" in rice culture. *Appl Environ Microbiol*. 1979; 37: 849–853.
184. Cerveró MC, Castillo MA, Montes R, Hernández E. Determination of trichothecenes, zearalenone and zearalenols in commercially available corn-based foods in Spain. *Rev Iberoam Micol*. 2007; 24: 52–55.
185. Kawabata Y, Tashiro F, Ueno Y. Synthesis of a specific protein induced by zearalenone and its derivatives in rat uterus. *J Biochem*. 1982; 91: 801–808.
186. Shier WT, Shier AC, Xie W, Mirocha CJ. Structure-activity relationships for human estrogenic activity in zearalenone mycotoxins. *Toxicon* 2001; 3: 1435–1438.
187. Fitzpatrick DW, Picken CA, Murphy LC, Buhr MM. Measurement of the relative binding affinity of zearalenone, alpha-zearalenol and beta-zearalenol for uterine and oviduct estrogen receptors in swine, rats and chickens: An indicator of estrogenic potencies. *Comp Biochem Physiol C* 1989; 94: 691–694.
188. SCF, Scientific Committee on Food. Opinion of the scientific committee of food on Fusarium toxins Part 2: Zearalenone (ZEA). SCF/CS/CNTM/MYC/22 Rev 3 Final, 2000.
189. Ueno Y, Tashiro F, Kobayashi T. Species differences in zearalenone-reductase activity. *Food Chem Toxicol*. 1983; 21: 167–173.
190. Olsen M, Kiessling KH. Species differences in zearalenone-reducing activity in subcellular fractions of liver from female domestic animals. *Acta Pharmacol Toxicol*. (*Copenh*) 1983; 52: 287–291.
191. Dong M, Tulayakul P, Li J-Y, Dong K-S, Manabe N, Kumagai S. Metabolic conversion of zearalenone to α-zearalenol by goat tissues. *J Vet Med Sci*. 2010; 72: 307–312.
192. Heitzman RJ. The absorption, distribution and excretion of anabolic agents. *J Anim Sci*. 1983; 57: 233–238.
193. Lindsay DG. Zeranol—A 'nature-identical' oestrogen? *Food Chem Toxicol*. 1985; 23: 767–774.
194. Pompa G, Montesiss, C, Di Lauro FM, Fadini L, Capua C. Zearanol metabolism by subcellular fractions from lamb liver. *J Vet Pharmacol Ther*. 1988; 11: 197–203.
195. Biehl ML, Prelusky DB, Koritz GD, Hartin KE, Buck WB, Trenholm HL. Biliary excretion and enterohepatic cycling of zearalenone in immature pigs. *Toxicol Appl Pharmacol*. 1993; 121: 152–159.
196. Pfeiffer E, Hildebrand A, Mikula H, Metzler M. Glucuronidation of zearalenone, zeranol and four metabolites *in vitro*: Formation of glucuronides by various microsomes and human UDP-glucuronosyltransferase isoforms. *Mol Nutr Food Res*. 2010; 54: 1468–1476.
197. Sambuy Y, De Angelis I, Ranaldi G, Scarino ML, Stammati A, Zucco F. The Caco-2 line as a model of the intestinal barrier: Influence of cell and culture-related factors on Caco-2 cell functional characteristics. *Cell Biol Toxicol*. 2005; 21: 1–26.
198. Schaut A, De Saeger S, Sergent T, Schneider Y-J, larondelle Y, Pussemier L, Van Peteghem C. Study of the gastrointestinal biotransformation of zearalenone in a Caco-2 cell culture system with liquid chromatographic methods. *J Appl Toxicol*. 2008; 28: 966–973.
199. Pfeiffer E, Kommer A, Dempe JS, Hildebrand AA, Metzler M. Absorption and metabolism of the mycotoxin zearalenone and the growth promoter zeranol in Caco-2 cells *in vitro*. *Mol Nutr Food Res*. 2011; 55: 560–567.
200. Olsen M, pettersson H, Sandholm K, Visconti A, Kiessling KH. Metabolism of zearalenone by sow intestinal mucosa *in vitro*. *Food Chem Toxicol*. 1987; 25: 681–683.
201. Malekinejad H, Maas-Bakker RF, Fink-Gremmels J. Bioactivation of zearalenone by porcine hepatic biotransformation. *Vet Res*. 2005; 36: 799–810.
202. Pearce ST, Jordan VC. The biological role of estrogen receptors α and β in cancer. *Crit Rev Oncol/Hematol*. 2004; 50: 3–22.
203. Yakimchuk K, Jondal M, Okret S. Estrogen receptor α and β in the normal immune system and in lymphoid malignancies. *Mol Cell Endocrinol*. 2013; 375: 121–129.

204. Frizzell C, Ndossi D, Verhaegen S, Dahl E, Erikson G, Sørlie M, Ropstad E, Muller M, Elliott C, Connolly L. Endocrine disrupting effects of zearalenone, alpha- and beta-zearalenol, at the level of nuclear receptor binding and steroidogenesis. *Toxicol Lett*. 2011; 206: 210–217.
205. Celius T, Haugeen TB, Grotmol T, Walther BT. A sensitive zonagenetic assay for rapid in vitro assessment of estrogenic potency of xenobiotics and mycotoxins. *Environ Health Perspect*. 1999; 107: 63–68.
206. Zöllner P, Jodlbauer J, Kleinova M, Kahlbacher H, Kuhn T, Hochsteiner W, Lindner W. Concentration levels of zearalenone and its metabolites on urine, muscle tissue, and live samples of pigs fed with mycotoxin-contaminated oats. *J Agric Food Chem*. 2002; 50: 2494–2501.
207. Kleinova M, Zöllner P, Kahlbacher H, Hochsteiner W, Linder W. Metabolite profiles of mycotoxin zearalenone and of the growth promoter zeranol in urine, liver and muscles of heifers. *J Agric Food Chem*. 2002; 371: 469–476.
208. Songsermsakul P, Böhm J, Aurich C, Zentek J, Razzazi-Fazel, E. The levels of zearalenone and its metabolites in plasma, urine, and faeces of horses fed with naturally *Fusarium* toxin-contaminated oats. *J Anim Physiol Anim Nutr*. 2013; 97: 155–161.
209. Ayed Y, Ayed-Boussema I, Ouanes Z, Bacha H. In vitro and in vivo induction of chromosome aberrations by alpha- and beta-zearalenols: Comparison to zearalenone. *Mut Res*. 2011; 726: 42–46.
210. Hildebrand AA, Pfieffer E, Rapp A, Metzler M. Hydroxylation of the mycotoxin zearalenone at aliphatic positions: Novel mammalian metabolites. *Mycotoxin Res*. 2012; 28: 1–8.
211. Fleck SC, Hildebrand AA, Müller E, Pfieffer E, Metzler M. Genotoxicity and inactivation of catechol metabolites of the mycotoxin zearalenone. *Mycotoxin Res*. 2012; 28: 267–273.
212. Hildebrand A, Pfeiffer E, Metzler M. Aromatic hydroxylation and catechol formation: A novel metabolic pathway of the growth promoter zeranol. *Toxicol Lett*. 2010; 12: 379–386.
213. Fleck SC, Hildebrand AA, Pfieffer E, Metzler M. Catechol metabolites of zeranol and 17 β-estradiol: A comparative in vitro study on the induction of oxidative DNA damage and methylation by catechol-O-methyltransferase. *Toxicol Lett*. 2012; 210: 9–14.
214. Cutolo M, Accardo S, Villaggio B, Clerico P, Bagnasco M, Coviello DA, Carruba G, Io Casto M, Castagnetta L. Presence of estrogen-binding sites on macrophage-like synoviocytes and CD8+, CD29+, CD45RO+ lymphocytes in normal and rheumatoid synovium. *Arthritis Rheum*. 1993; 36: 1087–1097.
215. Cutolo M, Sulli A, Seriolo B, Accard, S, Masi AT. Estrogens, the immune response and autoimmunity. *Clin Exp Rheumatol*. 1995; 13: 217–226.
216. Kovats, S. Estrogen receptors regulate an inflammatory pathway of dendritic cell differentiation: Mechanisms and implications for immunity. *Horm Behav*. 2012; 62: 254–262.
217. Wu WF, Tan XJ, Dai YB, Krishnan V, Warner M, Gustafsson JA. Targeting estrogen receptor β in microglia and T cells to treat experimental autoimmune encephalomyelitis. *Proc Natl Acad Sci USA* 2013; 110: 3543–3548.
218. Igarashi H, Kouro T, Yokota T, Comp PC, Kincade PW. Age and stage dependency of estrogen receptor expression by lymphocyte precursors. *Proc Natl Acad Sci USA* 2001; 98: 15131–15136.
219. Wang YC, Deng JL, Xu SW, Peng X, Zuo ZC, Cui HM, Wang Y, Ren ZH. Effects of zearalenone on IL-2, IL-6, and IFN-γ mRNA levels in the splenic lymphocytes of chickens. *Sci World J*. 2012. Article ID 567327. Doi:10.1100/2012/567327.
220. Marin DE, Taranu I, Burlac, R, Tudor DS. Effects of zearalenone and its derivatives on the innate immune response of swine. *Toxicon* 2010; 56: 956–963.
221. Marin DE, Pistol GC, Neagoe IV, Calin L, Taranu I. Effects of zearalenone on oxidative stress and inflammation in weanling piglets. *Food Chem Toxicol*. 2013; 58: 408–415.
222. Wan ML-Y, Woo C-JJ, Allen KJ, Turner PC, El-Nezami H. Modulation of porcine β-defensins 1 and 2 upon individual and combined *Fusarium* toxin exposure in a swine jejunal epithelial cell line. *Appl Environ Microbiol*. 2013; 79: 2225–2232.
223. Prouillac C, Koraichi F, Videmann B, Mazallon M, Rodriguez F, Baltas M, Lecoeur S. In vitro toxicological effects of estrogenic mycotoxins on human placental cells: Structure activity relationships. *Toxicol Appl Pharmacol*. 2012; 259: 366–375.
224. Wang DF, Zhang NY, Peng YZ, Qi DS. Interaction of zearalenone and soybean isoflavone on the development of reproductive organs, reproductive hormones and estrogen receptor expression in prepubertal gilts. *Anim Reprod Sci*. 2010; 122: 317–323.
225. Koraichi F, Videman B, Mazallon M, Benahmed M, Prouillac C, Lecoeur S. Zearalenone exposure modulates the expression of ABC transporters and nuclear receptors in pregnant rats and fetal liver. *Toxicol Lett*. 2012; 211: 246–256.
226. Wang H, Camargo Rodriguez O, Memili E. Mycotoxin alpha-zearalenol impairs the quality of preimplantation porcine embryos. *J Reprod Dev*. 2012; 58: 338–343.

227. Tatay E, Meca G, Font G, Ruiz M-J. Interactive effects of zearalenone and its metabolites on Cytotoxicity and metabolization in ovarian CHO-K1 cells. *Toxicol In Vitro* 2014; 28: 95–103.
228. Ferrer E, Juan-Garcia A, Font G, Ruiz MJ. Reactive oxygen species induced by Beauvericin, patulin and zearalenone in CHO-K1 cells. *Toxicol In Vitro* 2009; 23: 1504–1509.
229. Taskmakidis IA, Lymberopoulos AG, Khalifa TAA, Boscos CM, Saratsi A, Alexopoulos C. Evaluation of zearalenone and α-zearalenol toxicity on boar sperm DNA integrity. *J. Appl Toxicol.* 2008; 28: 681–688.
230. Tsakmakidis IA, Lymberopoulos AG, Alexopoulos C, Boscos CM, Kyriakis SC. In vitro effect of zearalenone and α-zearalenol on boar sperm characteristics and acrosome reaction. *Reprod Dom Anim.* 2006; 41: 394–401.
231. Benzoni E, Miniervini F, Giannoccaro A, Fornelli F, Vigo D, Visconti A. Influence of in vitro exposure to mycotoxin zearalenone and its derivatives on swine sperm quality. *Reprod Toxicol.* 2008; 25: 461–467.
232. Kim I-H, Son H-Y, Cho S-W, Ha C-S, Kang B-H. Zearalenone induces male germ cell apoptosis in rats. *Toxicol Lett.* 2003; 138: 185–192.
233. Jee Y, Eun-Mi N, Cho ES, Son H-Y. Involvement of the Fas and Fas ligand in testicular germ cell apoptosis by zearalenone in rat. *J Vet Sci.* 2010; 11: 115–119.
234. Koraïchi F, Inoubli L, Lakhdari N, Meunier L, Vega A, Mauduit C, Benahmed M, Prouillac C, Lecoeur S. Neonatal exposure to zearalenone induces long term modulation of ABC transporter expression in testis. *Toxicol.* 2013; 310: 29–38.
235. EU Council Directive 85/649/EEC, 1985. http://eur-lex.europa.eu/legal-content/EN/TXT/PDF/?uri=CELEX:31985L0649&from=en, accessed April 8, 2015.
236. Directive of the European Parliament and of the Council 96/22/EC, 1996. http://ec.europa.eu/food/food/chemicalsafety/residues/council_directive_96_22ec.pdf, accessed April 8, 2015.
237. Deng F, Tao F-B, Liu, D-Y, Xu Y-Y, Hao J-H, Sun Y, Su P-Y. Effects of growth environments and two environmental endocrine disruptors on children with idiopathic precocious puberty. *Eur J Endocrinol.* 2012; 166: 803–809.
238. Massart F, Meucci V, Saggese G, Soldani G. High growth rate of girls with precocious puberty exposed to estrogenic mycotoxins. *J Pediatr.* 2008; 152: 690–695.
239. Tomaszewski J, Miturski R, Semczuk A, Kotarski J, Jakowicki J. Tissue zearalenone concentration in normal, hyperplastic and neoplastic human endometrium. *Ginekol Pol.* 1998; 69: 363–366.
240. Phohl-Leszkowicz A, Chekir-Ghedira L, Bacha H. Genotoxicity of zearalenone, an estrogenic mycotoxin: DNA adduct formation in female mouse tissues. *Carcinogenesis.* 1995; 16: 2315–2320.
241. Gadzala-Kopciuch R, Cendrowski K, Cesarz A, Kielbasa P, Buszewski B. Determination of zearalenone and its metabolites in endometrial cancer by coupled separation techniques. *Anal Bioanal Chem.* 2011; 401: 2069–2078.
242. Wan L-Y, Woo C-SJ, Turner PC, Wan JM-F, El-Nezami H. Individual and combined effects of *Fusarium* toxins on the mRNA expression of pro-inflammatory cytokines in swine jejuna epithelial cells. *Toxicol Lett.* 2013; 220: 238–246.
243. Jestoi M, Rokka M, Yli-Mattila T, Parikka P, Rizzo A, Peltonen K. Presence and concentrations of the *Fusarium*-related mycotoxins beauvericin, enniatins and moniliformin in Finnish grain samples. *Food Addit Contam.* 2004; 21: 794–802.
244. Serrano AB, Font G, Mañes J, Ferrer E. Emerging *Fusarium* mycotoxins in organic and conventional pasta collected in Spain. *Food Chem Toxicol.* 2013; 51: 259–266.
245. Blesa J, Marin R, Lino CM, Mañes J. Evaluation of enniatins A, A1, B, B1 and beauvericin in Portuguese cereal-based foods. *Food Addit Contam Part A* 2012; 29: 1727–1735.
246. Vaclavikova M, Malachova A, Veprikova Z, Dzuman Z, Zachariasova M, Hajslova J. 'Emerging' mycotoxins in cereals processing chains: Changes of enniatins during beer and bread making. *Food Chem.* 2013; 15: 750–757.
247. Lin YC, Wang J, Wu XY, Zhou SN, Vrijmoed LLP, Jones EBG. A novel compound enniatin G from the mangrove fungus *Halosarpheia* sp. (strain 732) from the South China sea. *Aust J Chem.* 2002; 55: 225–227.
248. Nilanonta C, Isaka M, Chanphen D, Thong-Orn N, Tanticharoen M, Thebtaranonth Y. Unusual enniatins produced by the insect pathogenic fungus *Verticillium hemipterigenum*: Isolation and studies on precursor-directed biosynthesis. *Tetrahedron* 2003; 59: 1015–1020.
249. Tomoda H, Nishida H, Huang XH, Masuna R, Kim YK, Smura S. New cyclodepsipeptides, enniatin D, E, and F produced by *Fusarium* sp. FO-1305. *J Antibiot* (Tokyo). 1991; 45: 1207–1215.
250. Visconti A, Blais LA, Apsimon JW, Greenhalgh E, Miller JD. Production of Enniatins by *Fusarium acuminatum* and *Fusarium compactum* in liquid culture: Isolation and characterization of three new enniatins B2, B3, and B4. *J Agric Food Chem.* 1992; 40: 1076–1082.

251. Pohanka A, Capieau K, Broberg A, Stenlid J, Stenström E, Kenne L. Enniatins of *Fusarium* sp. strain F31 and their inhibition of *Botrytis cinerea* spore germination. *J Nat Prod*. 2004; 67: 851–857.
252. Song H-H, Lee H-S, Jeong J-H, Park HS, Lee C. Diversity in beauvericin and enniatins H,I, and MK1688 by *Fusarium oxysporum* isolated from potato. *Int J Food Microbiol*. 2008; 122: 296–301.
253. Levy D, Bluzat A, Seigneuret M, Rigaud J-L. Alkali cation transport through liposomes by the antimicrobial fusafungine and its constitutive enniatins. *Biochem Pharmacol*. 1995; 50: 2105–2107.
254. Kamyar M, Rawnduzi P, Studenik CR, Kouri K, Lemmens-Gruber R. Investigation of the electrophysiological properties of enniatins. *Arch Biochem Biophys*. 2004; 429: 215–223.
255. Tonshin AA, Teplova VV, Andersson MA, Salkinoja-Salonen MS. The *Fusarium* mycotoxins enniatins and beauvericin cause mitochondrial volume regulation, oxidative phosphorylation and ion homeostasis. *Toxicology* 2010; 276: 49–57.
256. Firáková S, Sturdiková M, Liptaj T, Prónayová N, Bezáková L, Proksa B. Enniatins produced by *Fusarium dimerum*, an endophytic fungal strain. *Pharmazie* 2008; 63: 539–541.
257. Meca G, Soriano JM, Gaspari A, Ritieni A, Moretti A Mañes J. Antifungal effects of the bioactive compounds enniatins A, A_1, B, B_1. *Toxicon* 2010; 56: 480–485.
258. Meca G, Sospedra I, Valero MA, Mañes J, Font G, Ruiz MJ. Antibacterial activity of the enniatin B, produced by *Fusarium tricinctum* in liquid culture, and cytotoxic effects on Caco-2 cells. *Toxicol Mech Methods*. 2011; 21: 503–512.
259. Ficheux AS, Sibiril Y, Parent-Massin D. Effects of beauvericin, enniatin b and moniliformin on human dendritic cells and macrophages: An in vitro study. *Toxicon* 2013; 71: 1–10.
260. Gammelsrud A, Solhuag A, Dendelé B, Sandberg WJ, Ivanova L, Kocbach Bølling A, Lagadic-Gosssmann D, Refsnes M, Becher R, Eriksen G, Holme JA. Enniatin B-induced cell death and inflammatory responses in RAW 267.8 murine macrophages. *Toxicol Appl Pharmacol*. 2012; 261: 74–87.
261. Ivanova L, Skjerve E, Eriksen GS, Uhli, S. Cytotoxicity of enniatins A, A1, B, B1, B2, and B3 from *Fusarium avenaceum*. *Toxicon* 2006; 47: 868–876.
262. Dornetshuber R, Heffete, P, Kamyar M-R, Peterbauer T, Berger W, Lemmens-Gruber R. Enniatin exerts p53-dependent cytostatic and p53-independent cytotoxic activities against human cancer cells. *Chem Res Toxicol*. 2007; 20: 465–473.
263. Juan-Garcia A, Manyes L, Ruiz M-J, Font G. Involvement of enniatins-induced cytotoxicity in human HepG2 cells. *Toxicol Lett*. 2013; 218: 166–173.
264. Behm C, Degen GH, Föllman W. The *Fusarium* toxin enniatin B exerts no genotoxic activity, but pronounced cytotoxicity in vitro. *Mol Nutr Food Res*. 2009; 53: 423–230.
265. Föllmann W, Behm C, Degen GH. The emerging *Fusarium* toxin enniatin B: In vitro studies on its genotoxic potential and cytotoxicity in V79 cells in relation to other mycotoxins. *Mycotoxin Res*. 2009; 25: 11–19.
266. Ficheux AS, Sibri, Y, Parent-Massin D. Co-exposure of *Fusarium* mycotoxins: In vitro myelotoxicity assessment on human hematopoietic progenitors. *Toxicon* 2012; 60: 1171–1179.
267. Ficheux AS, Sibiril Y, Le Garrec R, Parent-Massin D. In vitro myelotoxicity assessment of the emerging mycotoxins beauvericin, enniatin B and moniliformin on human hematopoietic progenitors. *Toxicon* 2012; 59: 182–191.
268. Prosperini A, Font G, Ruiz MJ. Interaction effects of *Fusarium* enniatins (A, A_1, B, and B_1) combinations on in vitro cytotoxicity of Caco-2 cells. *Toxicol In Vitro* 2014; 28: 88–94.
269. Ivanova L, Egge-Jacobson WM, Solhuag A, Thoen E, Fæste CK. Lysosomes as a possible target of enniatin B-induced toxicity in Caco-2 cells. *Chem Res Toxicol*. 2012; 25: 1662–1674.
270. Lu H, Fernández-Frazon M, Font G, Ruiz MJ. Toxicity evaluation of individual and mixed enniatins using and in vitro method with CHO-K1 cells. *Toxicol In Vitro* 2013; 27: 672–680.
271. Lombardi G, Prosperini A, Font G, Ruiz MJ. Effect of polyphenols on enniatin-induced cytotoxic effects in mammalian cells. *Toxicol Mech Methods* 2012; 22: 687–695.
272. Logrieco A, Moretti A, Castella G, Kostecki M, Golinski P, Ritieni A, Chelkowski J. Beauvericin production by *Fusarium* species. *Appl Environ Microbiol*. 1998; 64: 3084–3088.
273. Kouri K, Lemmens M, Lemmens-Gruber R. Beauvericin-induced channels in ventricular myocytes and liposomes. *Biochem Biophys Acta* 2003; 1609: 203–210.
274. Dornetshuber R, Heffeter P, Lemmens-Gruber R, Elbling L, Marko D, Micksche M, Berger W. Oxidative stress and DNA interactions are not involved in Enniatin- and Beauvericin-mediated apoptosis induction. *Mol Nutr Food Res*. 2009; 53: 1112–1122.
275. Meca G, Sospedra I, Soriano JM, Ritieni A, Moretti A, Mañes J. Antibacterial effect of the bioactive compound beauvericin produced by *Fusarium proliferatum* on solid medium of wheat. *Toxicon* 2010; 56: 349–354.

276. Kouri K, Duchen MR, Lemmens-Gruber R. Effects of beauvericin on the metabolic state and ionic homeostasis of ventricular myocytes of the guinea pig. *Chem Res Toxicol.* 2005; 18: 1661–1668.
277. Lambuda R, Tančinova D. Fungi recovered from Slovakian poultry feed mixtures and their toxigenicity. *Ann Agric Environ Med.* 2006; 13: 193–200.
278. Harvey RB, Edrington TS, Kubena LF, Rottingham GE, Turk JR, Genovese KJ, Nisbet DJ. Toxicity of moniliformin from *Fusarium fujikuroi* culture material to growing barrows. *J Food Protect.* 2001; 64: 1780–1784.
279. Morrison E, Kosiak B, Ritieni A, Aastveit AH, Uhlig S, Bernhoft A. Mycotoxin production by *Fusarium avenaceum* strains isolated from Norwegian grain and the cytotoxicity of rice culture extracts to porcine kidney epithelial cells. *J Agric Food Chem.* 2002; 50: 3070–3075.
280. Van der fels-Klerx HJ, de Rijk TC, Booij CJ, Goedhart PW, Boers EA, Zhao C, Waalwijk C, Mol HG, van der Lee TA. Occurrence of *Fusarium* Head Blight species and *Fusarium* mycotoxins in winter wheat in the Netherlands in 2009. *Food Addit Contam Part A* 2012; 29: 1716–1726.
281. Cetin Y, Bullerman LB. Cytotoxicity of *Fusarium* mycotoxins to mammalian cell cultures as determined by the MTT bioassay. *Food Chem Toxicol.* 2005; 43: 755–764.
282. Sharma D, Asrani RK, Ledoux DR, Rottinghaus GE, Gupta VK. Toxic interaction between fumonisin B1 and moniliformin for cardiac lesions in Japanese quail. *Avian Dis.* 2012; 56: 545–554.
283. Zhao D, Feng Q, Yan X, Li C, Pan Y, Cui Q. Ultrastructural study of moniliformin induced lesions of myocardium in rats and mice. *Biomed Environ Sci.* 1993; 6: 37–44.
284. Broomhead JN, Ledoux DR, Bermudez AJ, Rottinghaus GE. Chronic effects of moniliformin in broiler and turkeys fed dietary treatments to market age. *Avian Dis.* 2002; 46: 901–908.
285. Kamyar MR, Kouri K, Rawnduzi P, Studenik C, Lemmens-Gruber R. Effects of moniliformin in presence of cyclohexadepsipeptides on isolated mammalian tissue and cells. *Toxicol In Vitro* 2006; 20: 1284–1291.
286. OECD Guideline for Testing of Chemicals. Guideline 423: Acute Oral Toxicity—Acute Toxic Class Method, 2001. https://ntp.niehs.nih.gov/iccvam/suppdocs/feddocs/oecd/oecd_gl423.pdf, accessed April 8, 2015.
287. Jonsson M, Jestoi M, Nathanail AV, Kokkonen U-M, Anttila M, Koivsto P, Karhunun P, Peltonen K. Application of OECD Guideline 423 in assessing the acute oral toxicity of moniliformin. *Food Chem Toxicol.* 2013. 53: 27–32.
288. Zhang A, Cao J-L, Yang B, Zhang Z-T, Li S-Y, Fu Q, Hugnes CE, Caterson B. Effects of moniliformin and selenium on human articular cartilage metabolism and their potential relationships to the pathogenesis of Kashin-Beck disease. *J Zhejiang Univ Sci B (Biomed & Biotechnol)* 2010; 11: 200–208.
289. Lu M, Cao J, Liu F, Li S, Chen J, Fu Q, Zhang Z, Liu J, Luo M, Wang J, Li J, Caterson B. The effects of mycotoxins and selenium deficiency on tissue-engineered cartilage. *Cells Tissues Organs* 2012; 196: 241–250.
290. Munkvold G, Stahr HM, Logrieco A, Moretti A, Ritieni A. Occurrence of fusaproliferin and beauvericin in *Fusarium*-contaminated livestock feed in Iowa. *Appl Environ Microbiol.* 1998; 64: 3923–3926.
291. Moretti A, Mulé G, Ritieni A, Logrieco A. Further data on the production of Beauvericin, Enniatins and Fusaproliferin and toxicity to *Artemia salina* by *Fusarium* species of *Gibberella fujikuroi* species complex. *Int J Food Toxicol.* 2007; 118: 158–163.
292. Ritieni A, Fogliano V, Randazzo G, Scarallo A, Logrieco A, Moretti A, Mannina L, Bottalico A. Isolation and characterization of fusaproliferin, a new toxic metabolite from *Fusarium proliferatum*. *Nat Toxins* 1995; 17–20.
293. Santini A, Ritieni A, Fogliano V, Randazzo G, Mannina L, Logrieco A, Ettore Benedetti E. Structure and absolute stereochemistry of fusaproliferin, a toxic metabolite from *Fusarium proliferatum*. *J Nat Prod.* 1996; 59: 109–112.
294. Mikušova P, Šrobárová A, Sulyok M, Santini A. *Fusarium* fungi and associated metabolites presence on grapes from Slovakia. *Mycotoxin Res.* 2013; 29: 97–102.
295. Morgensen JM, Sørensen SM, Sulyok M, van der Westhuizen L, Shephard GS, Frisvad, JC, Thrane U, Krska R, Nielsen KF. Single-kernel analysis of fumonisins and other fungal metabolites in maize from South African subsistence farmers. *Food Addit Contam.* 2011; 28: 1724–1734.
296. Shephard GS, Sewram V, Nieuwoudt TW, Marasas WFO, Ritieni A. Production of the mycotoxins fusaproliferin and beauvericin by South African isolates in the *Fusarium* Section Liseola. *J Agric Food Chem.* 1999: 47: 5111–5115.
297. Reyes-Velázquez WP, Figueroa-Gómez RM, Barberis M, Reynoso MM, Rojo FG, Chulze SN, Torres AM. *Fusarium* species (section Liseola) occurrence and natural incidence of beauvericin, fusaproliferin and fumonisins in maize hybrids harvested in Mexico. *Mycotoxin Res.* 2011; 27: 187–194.

298. Nadubinska M, Ciamporova M. Toxicity of *Fusarium* mycotoxins in maize. *Mycotoxin Res*. 2001; 17 (Suppl. 1): 82–86.
299. Prosperini A, Meca G, Font G, Ruiz MJ. Study of the cytotoxic activity of Beauvericin and fusaproliferin and bioavailability in vitro on Caco-2 cells. *Food Chem Toxicol*. 2012; 50: 2356–2361.
300. Logrieco A, Moretti A, Fornelli F, Fogliano V, Ritieni A, Caiaffa MF, Randazzo G, Bottalico A, Macchia L. Fusaproliferin production by *Fusarium subglutinans* and its toxicity to *Artemia salina*, SF-9 insect cells, and IARC/LCL 171 human B lymphocytes. *Appl Environ Microbiol*. 1996; 62: 3378–3384.
301. Ritieni A, Monti SM, Randazzo G, Logrieco A, Moretti A, Peluso G, Ferracane R, Fogliano V. Teratogenic effects of fusaproliferin on chicken embryos. *J Agric Food Chem*. 1997; 45: 3039–3043.
302. Masuda D, Ishida M, Yamaguchi K, Yamaguchi I, Kimura M, Nishiuchi T. Phytotoxic effects of trichothecenes on the growth and morphology of *Arabidopsis thaliana*. *J Exp Bot*. 2007; 58: 1617–1626.
303. Snijders CHA. Resistance in wheat to *Fusarium* infection and trichothecene formation. *Toxicol Lett*. 2004; 153: 37–46.
304. Bukočáková M, Srobárová A, Vozár I. Zearalenone production in wheat cultivars infected with the fungus *Fusarium graminearum* Schwabe. *Mycotoxin Res*. 1991; 7 (Suppl. 1): 84–90.
305. McLean M. The phytotoxicity of selected mycotoxins on mature, germinating *Zea mays* embryos. *Mycopathology* 1995; 132: 173–183.
306. Nakagawa H, Sakamoto S, Sago Y, Nagashima H. Detection of type A trichothecene di-glucosides produced in corn by high-resolution liquid chromatography-orbitrap mass spectrometry. *Toxins* 2013; 5: 590–604.
307. Berthiller F, Sulyok M, Krska R, Schuhmacher R. Chromatographic methods for the simultaneous determination of mycotoxins and their conjugates in cereals. *Int J Food Microbiol*. 2007; 119: 33–37.
308. Berthiller F, Crews C, Dall'Asta C, De Saeger S, Haesaert G, Karlovsky P, Oswald IP, Seefelder W, Speijers G, Stroka J. Masked mycotoxins: A review. *Mol Nutr Food Res*. 2013; 57: 165–186.
309. De Boevre M, Jacxsens L, Lachat C, Eeckhout M, Diana Di Mavungu J, Audenaert K, Maene P, Haesaert G, Kolsteren P, De Meulenaer B, De Saeger S. Human exposure to mycotoxins and their masked forms through cereal-based foods in Belgium. *Toxicol Lett*. 2013; 218: 281–292.
310. Engelhardt G, Zill G, Wohner B, Wallnöfer PR. Transformation of the *Fusarium* mycotoxin zearalenone in maize suspension cultures. *Naturwissenschaften* 1988; 75: 309–310.
311. Sewald N, vn Gleissenthall JL, Schuster M, Müller G, Aplin RT. Structure elucidation of a plant metabolite of 4-desoxynivalenol. *Tetrahedron* 1992; 3: 953–960.
312. Poppenberger B, Berthiller F, Lucyshyn D, Sieberer T, Schuhmacher R, Krska R, Kuchler K, Glössl J, Luschnig C. Detoxification of the *Fusarium* mycotoxin deoxynivalenol by a UDP-glucosyltransferase from *Arabidopsis thaliana*. *J Biol Chem*. 2003; 278: 47905–47914.
313. Gardiner S, Boddu J, Berthiller F, Hametner C, Stuper RM, Adam G, Muehlbauer GJ. Transcriptome analysis of the barley-deoxynivalenol interaction: Evidence for a role of glutathione in deoxynivalenol detoxification. *Mol Plant Microbe Interact*. 2010; 23: 962–976.
314. Schweiger W, Boddu J, Shin S, Poppenberger B, Berthiller F, Lemmens M, Muehlbauer GJ, Adam G. Validation of a candidate deoxynivalenol-inactivating UDP-glucosyltransferase from barley by heterologous expression in yeast. *Mol Plant Microbe Interact*. 2010; 23: 977–986.
315. Coleman JOD, Blake-Kalff MMA, Davies TGE. Detoxification of xenobiotics by plants: Chemical modification and vacuolar compartmentation. *Trends Plant Sci*. 1997; 2: 144–151.
316. Lattanzio VMT, Visconti A, Haidukowski M, Pascale M. Identification and characterization of new *Fusarium* masked mycotoxins, T2 and HT2 glucosyl derivatives, in naturally contaminated wheat and oats by liquid chromatography-high-resolution mass spectrometry. *J Mass Spectrom*. 2012; 47: 466–475.
317. Zachariasova M, Vaclavikova M, Lacina O, Vaclavik L, Hajslova J. Deoxynivalenol oligosaccharides: New "masked" *Fusarium* toxins occurring in malt, beer and breadstuff. *J Agric Food Chem*. 2012; 60: 9280–9291.
318. Vendl, O, Crews C, MacDonald S, Krska R, Berthiller F. Occurrence of free and conjugated *Fusarium* mycotoxins in cereal-based food. *Food Addit Contam*. 2010; 27: 1148–1152.
319. Lancova K, Hajslova J, Poustka J, Krplova A, Zachariasova M, Dostalek P, Sachambula L. Transfer of *Fusarium* mycotoxins and "masked" deoxynivalenol (deoxynivalenol -3-glucoside) from field barley through malt to beer. *Food Addit Contam*. 2008; 25: 732–744.
320. Rasmussen PH, Nielsen KF, Ghorbani F, Spliid NH, Nielsen GC, Jørgensen LN. Occurrence of different trichothecenes and deoxynivalenol-3-β-D-glucoside in naturally and artificially contaminated Danish cereal grains and whole maize plants. *Mycotoxin Res*. 2012; 28: 181–190.

321. Kostelanska M, Haajslova J, Zachariasova M, Malachova A, Kalachova K, Poustka J, Fiala J, Scott PM, Berthiller F, Krska R. Occurrence of deoxynivalenol-3-glucoside, in beer and some brewing intermediates. *J Agric Food Chem*. 2009; 57: 3187–3194.
322. Varga E, Malachova A, Schwartz H, Krska R, Berthiller F. Survey of deoxynivalenol and its conjugates deoxynivalenol-3-glucoside and 3-acetyl-deoxynivalenol in 374 beer samples. *Food Addit Contam*. 2013; 30: 137–146.
323. Gareis M, Bauer J, Thiem J, Plank G, Grabley S, Gedek B. Cleavage of zearalenone-glycoside, a "masked" mycotoxin, during digestion in swine. *Zentralbl Veterinarmed B* 1990; 37; 236–240.
324. Nagl V, Schwartz H, Krska R, Moll W-D, Knasmüller S, Ritzmann M, Adam G, Berthiller F. Metabolism of the masked mycotoxin deoxynivalenol-3-glucoside in rats. *Toxicol Lett*. 2012: 367–373.
325. Dell'Erta A, Cirline M, Dall'Asta M, Del Rio D, Galavema G, Dall'Asta C. Masked mycotoxins are efficiently hydrolyzed by human colonic microbiota releasing their aglycones. *Chem Res Toxicol*. 2013; 26: 305–312.
326. Gratz SW, Duncan G, Richardson AJ. The human fecal microbiota metabolizes deoxynivalenol and deoxynivalenol-3-glycoside and may be responsible for urinary deepoxy-deoxynivalenol. *Appl Environ Microbiol*. 2013; 79: 1821–1825.

22 Lichtheimia (ex Absidia)

Volker U. Schwartze and Kerstin Kaerger

CONTENTS

22.1 Systematics 355
22.2 Morphology 356
22.3 Biology..359
22.4 *Lichtheimia* Species in Food Production and Agriculture 360
    22.4.1 Agriculture..360
    22.4.2 Food Production 361
22.5 *Lichtheimia* in Human Infections and Related Diseases...361
    22.5.1 Epidemiology and Risk Factors...361
    22.5.2 Clinical Features and Pathogenesis 362
    22.5.3 Farmer's Lung Disease 363
22.6 Detection and Identification...364
    22.6.1 Morphology and Physiology-Based Identification...364
    22.6.2 PCR-Based Identification...364
    22.6.3 Hybridization-Based Identification 366
    22.6.4 Mass Spectrometry–Based Identification 366
22.7 Antifungal Susceptibility and Treatment...366
    22.7.1 Antifungal Susceptibility..366
    22.7.2 Clinical Therapy 367
22.8 Conclusions...367
Acknowledgments..368
References..368

22.1 SYSTEMATICS

The genus *Lichtheimia* belongs to the order Mucorales within the subphylum Mucoromycotina (former part of the phylum Zygomycota) and comprises five species, namely, *Lichtheimia corymbifera*, *Lichtheimia ramosa*, *Lichtheimia ornata*, *Lichtheimia hyalospora*, and *Lichtheimia sphaerocystis* [1]. Together with the genus *Dichotomocladium*, *Lichtheimia* forms the family Lichtheimiaceae, which represents one of the most basal mucoralean families. *L. corymbifera* was originally described by Cohn in 1884 as *Mucor corymbifer* and was the first agent found to be involved in mucormycosis [2,3]. For a long time *Lichtheimia* species were placed within the genus *Absidia* because of its morphological similarities (pyriform sporangia with a funnel-shaped apophysis, branched sporangiophores originating from stolons) [4]. Yet physiological, morphological, and phylogenetic data have led to a wide acceptance of the polyphyly of *Absidia* [5–7]. The currently accepted classification of *Absidia*-like taxa comprises three different genera: (1) *Absidia* s.str. (syn. *Tieghemella*, *Proabsidia*), (2) *Lichtheimia* (syn. *Mycocladus*, *Pseudoabsidia*, *Protoabsidia*), and (3) *Lentamyces*. For a detailed chronological change of names and discussion, see Hoffmann [8], or about *Lichtheimia*, see Table 22.1 for a brief summary. The main features to distinguish *Lichtheimia* species from other *Absidia*-like species are as follows: (1) *Lichtheimia* spp. possess a higher optimal growth temperature at or above 37°C, while *Absidia* species are mesophilic with

TABLE 22.1
Chronological Overview about the Description and Name Changes for Accepted Species in the Genus *Lichtheimia* Vuill[a]

Year	Original Description (Basionym)	Currently Accepted Name	Type Strain	MycoBank No.
1903	*Mucor corymbifer* Cohn	*L. corymbifera* (Cohn 1884) Vuill.	CBS429.75 (NT)	MB#416447
1903	*Rhizopus ramosus* Zopf	*L. ramosa* (Zopf 1890) Vuill.	CBS582.65 (NT)	MB#416448
1906	*Tieghemella hyalospora* Saito	*L. hyalospora* (SAITO 1906) K. Hoffm., G. Walther & K. Voigt	CBS173.67 (T)	MB#512830
1965	*Absidia ornata* A.K. Sarbhoy	*L. ornata* (A.K. Sarbhoy 1965) A. Alastruey-Izquierdo & G. Walther	CBS291.66 (T)	MB#516506
2010	*L. sphaerocystis* A. Alastruey-Izquierdo & G. Walther	*L. sphaerocystis* A. Alastruey-Izquierdo & G. Walther	CBS420.70 (T)	MB#516505

MB, MycoBank, http://www.mycobank.org; CBS, Centraalbureau voor Schimmelcultures, http://www.cbs.knaw.nl; T, type NT, neotype.

[a] 1903 [MB#20308] (type species of this genus is *L. corymbifera* (Cohn) Vuill.; *Lichtheimia* belongs to the mucoralean family Lichtheimiaceae [MB#508680]).

optimal growth temperatures below 37°C; (2) in contrast to *Absidia* species, *Lichtheimia* species form nonappendaged zygospores (those sexual structures where meiosis occurs are developed in mating between compatible mating types of a species; zygospores are dark brownish with a wall possessing equatorial rings; the suspensors are opposed and nonappendages); and (3) phylogenetic analysis revealed a monophyletic origin of the thermotolerant species of *Absidia* (later *Lichtheimia*) placed near the base of the Mucorales. Based on these data, thermophilic *Absidia* species were first reclassified into the genus *Mycocladus*, comprising *Mycocladus hyalosporus*, *Mycocladus blakesleeanus*, and *Mycocladus corymbifer* [9]. In 2009, the name *Mycocladus* was corrected to *Lichtheimia* in accordance with the International Code of Botanical Nomenclature [10]. Within this genus, a diverse number of species were described over the years, including several name changes (synonyms). Obligate and facultative synonyms are listed at MycoBank (http://www.mycobank.org) [11]. One of the most important changes affected *L. ramosa*. *L. ramosa* was originally described as *Rhizopus ramosus* by Zopf [12] but was taken as a synonym for *L. corymbifera* [13,14]. In 2009, *L. ramosa* was given back species rank based on phenotypic and phylogenetic data [15]. The current classification of *Lichtheimia* with its five species was proposed in 2010, based on combined morphological, physiological, and phylogenetic data [1] (Table 22.1). A key to all accepted species is published in Alastruey-Izquierdo et al. [1]. In addition, *Lichtheimia hongkongensis* was proposed as an additional species but is not accepted as valid [16]. To date, only *L. corymbifera*, *L. ramosa*, and *L. ornata*, which form a monophyletic group, were found to be clinically relevant [1].

22.2 MORPHOLOGY

All species of *Lichtheimia* grow well on artificial media, for example, 3% malt extract agar, supplemented minimal medium according to Wöstemeyer [17], or potato dextrose agar, producing high amounts of aerial hyphae and stolons on these media. Colonies of *Lichtheimia* appear white to gray, later becoming dark gray or brownish. Mucoralean fungi are distinguishable by the appearance/features of the asexual spores. All *Lichtheimia* species produce large amounts of multispored sporangia, which are hyaline to gray (Figures 22.1 and 22.2). The sporangia are spherical to pyriform and 100–150 μm in diameter. Sporangiophores are simple or branched and arise from the substrate mycelium or aerial hyphae. *Lichtheimia* species can be distinguished from other mucoralean fungi by the funnel-shaped apophysis (Figures 22.1 and 22.2). Only the genus *Absidia* shares this feature, which led to the

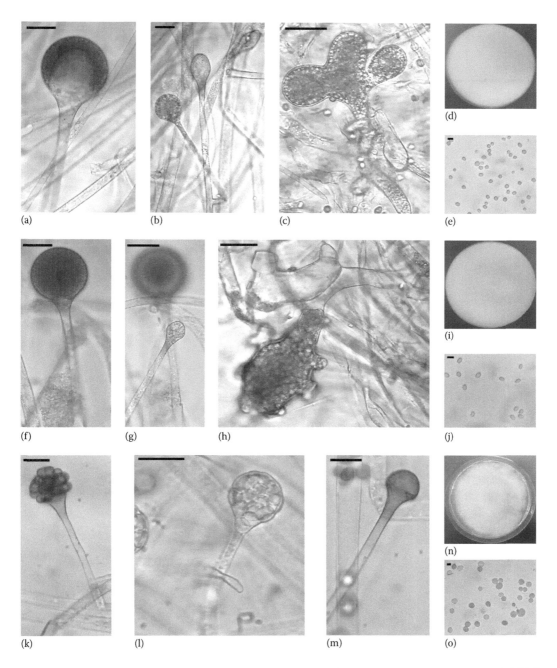

FIGURE 22.1 Main morphological features of *Lichtheimia* species. *L. corymbifera* FSU9682 (CBS429.75) (a–e), *L. ramosa* FSU10166 (CBS582.65) (f–j), and *L. hyalospora* FSU10163 (CBS173.63) (k–o). Sporangia are multispored and apophysate (a, f, k, l) and, in the case of *L. hyalospora*, spores (o) are larger and sporangia bear a smaller amount of asexual spores compared to that of other *Lichtheimia* species (e, j). Matured sporangium (k) and young sporangium (l). The columella (a bulbous vesicle at the sporangiophore apex) is protruding into the sporangium and remains after spore release (b, g, m). Young colonies of *Lichtheimia* (d, i, n) are in general white, turning to shades of gray and brown with age. In older cultures, variously shaped giant cells could be observed (c, h) (species specific, size, and complexity are dependent on culture conditions). This feature is not as reliable in axenic cultures as sporangia and depends on culture conditions and duration of culturing. Age of culture: 2 days (a, b, e–g, i–o), 5 days (d), 1 week (h), and 2 weeks (c). Scale bars, 20 μm, except spores, 5 μm.

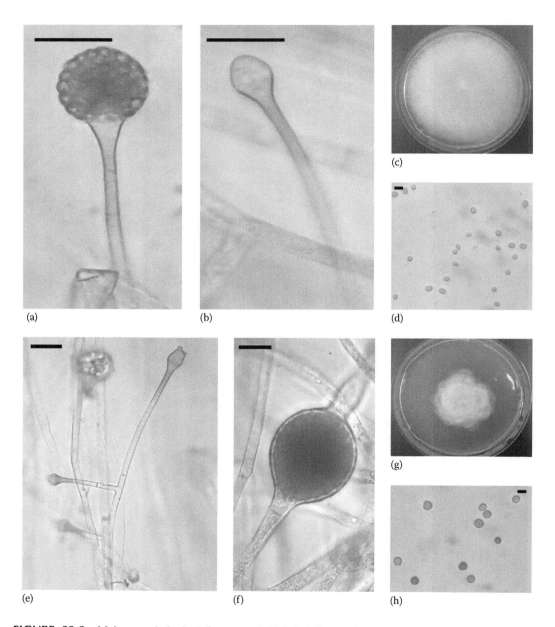

FIGURE 22.2 Main morphological features of *Lichtheimia* species, continued. *L. ornata* FSU10165 (CBS291.66) (a–d) and *L. sphaerocystis* FSU10079 (CBS420.70) (e–h). Sporangia are multispored and apophysate (a). The columella is protruding into the sporangium and remains after spore release (b, e). Young colonies of *Lichtheimia* (c, g) are in general white, turning to shades of gray with age. In older cultures, variously shaped giant cells could be observed (f) (species specific, size, and complexity are dependent on culture conditions). This feature is not as reliable in axenic cultures as sporangia and depends on culture conditions and duration of culturing. Age of culture: 2 days (a–e, g, h), 2 weeks (f). Scale bars, 20 μm, except spores, 5 μm.

previous assumption that *Lichtheimia* and *Absidia* belong to the same genus. Nevertheless, *Absidia* and *Lichtheimia* could be distinguished by their different (1) optimal and maximal growth temperatures, (2) phylogenetic relationships, and (3) morphological features (e.g., appearance of subsporangial septae, whorls of sporangia, and the zygospores; for a key to species, see Alastruey-Izquierdo et al. [1]). Moreover, most *Lichtheimia* species produce giant cells (Figures 22.1 and 22.2), a rare feature in *Absidia*. Sexually formed zygospores are brown with equatorial rings; the suspensors are free of appendages. The equatorial rings are not produced in *Absidia* but suspensors are appendaged.

While identification to the genus level is straightforward, species identification is more complicated since *Lichtheimia* species share most of the morphological markers. Pathogenic and nonpathogenic species *of Lichtheimia* are distinguishable mainly by the lower maximum growth temperature of the latter. While pathogenic *L. corymbifera*, *L. ramosa*, and *L. ornata* exhibit an optimal growth temperature at 37°C and are able to grow well at 43°C, nonpathogenic *L. hyalospora* and *L. sphaerocystis* show no or only slight growth at 43°C and have a growth optimum below 37°C. In addition, *L. hyalospora* and *L. sphaerocystis* produce dark gray to black or brown-colored sporangia containing sporangiospores with a size of greater than 6.5 μm. Both species are distinguishable by the differences in the form of the giant cells, which are globose in *L. sphaerocystis* and hyphal like in *L. hyalospora*. In contrast, the sporangia of the other three species are only light gray to brownish with sporangiospores less than 6.5 μm. However, no clear morphological features are available to distinguish the clinical species. Only the higher growth rate of *L. ramosa* at 43°C in comparison to *L. ornata* and *L. corymbifera* can be used for identification of this species. Despite Garcia-Hermoso et al. stating that *L. ramosa* and *L. corymbifera* can be distinguished by the form of the sporangiospores and the texture of the spores [15], these results could not be confirmed by other studies [1]. While *L. ramosa* produces more cylindrical spores and *L. corymbifera* produces more spherical ones, intermediate spore shapes are found in both species [1].

22.3 BIOLOGY

Most fungi of the order Mucorales, including the genus *Lichtheimia*, are saprophytic soil fungi with a worldwide distribution. Ecological data to determine the distribution of these fungi in the environment are missing although soil is believed to be the most important habitat. In addition, Mucorales can be found on a variety of organic substrates including fruits, vegetables, and dung, or anthropogenetic/man-made substrates such as processed food, or in farming environments on hay, straw, grain, and silage [18,19]. Some mucoralean species are also opportunistic pathogens of plants (e.g., *Choanephora cucurbitarum*, *Blakeslea trispora*, *Rhizopus stolonifer*, *Rhizopus oryzae*, *Mucor racemosus*), fungi (e.g., *Syzygites megalocarpus*, *Spinellus fusiger*, *Dicranophora fulva*, *Lentamyces parricida*, *Parasitella parasitica*, *Chaetocladium brefeldii*), and animals or humans (e.g., *Rhizopus microsporus*, *R. oryzae*, *L. corymbifera*, *L. ornata*, *L. ramosa*, *Mucor circinelloides*, *Apophysomyces elegans*). Thermotolerant or thermophilic species can also be found in environments with elevated temperature such as compost piles or silage silos. Knowledge on nutritional requirements including pH and humidity in Mucorales, especially in *Lichtheimia*, is sparse.

Despite their different lifestyles and different habitats, mucoralean fungi propagate by the same method (Figure 22.3). Mucorales are generally fast-growing and sporulate in large quantities. This formation of asexual, nonmotile sporangiospores released from variously shaped sporangia is the primary propagation form. Haploid spores can germinate and form haploid, vegetative mycelia, which produce sporangia with new sporangiospores. However, besides this asexual mechanism, mucoralean fungi also are able to reproduce via sexual zygospores. In heterothallic Mucorales such as *Lichtheimia* spp., two compatible mating partners (designated as + and −) are necessary to form the zygospore, where meiosis occurs. Mucorales display different sexual modalities of either homo- or heterothally, a feature discovered more than 100 years ago, with most species found to be heterothallic [20]. For successful zygospore formation, recognition of potential compatible mating types is necessary. The carotene derivate trisporic acid and its precursors are used as pheromones

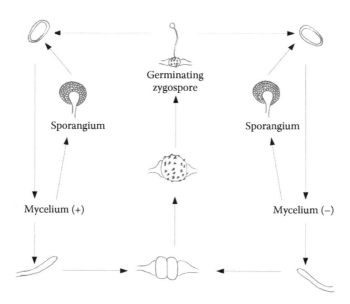

FIGURE 22.3 General life cycle of mucoralean fungi. Asexual sporangiospores are released from sporangia, forming unseptated or irregularly septated mycelia. If two compatible mating types (termed + and −) encounter each other, specialized hyphae (zygophores) are fused, producing a zygospore where meiosis occurs. In homothallic species, zygospores are produced within a single mycelium. Zygospores germinate into haploid mycelia, which produce asexual sporangia.

for recognition between the two mating partners [21,22]. Thereby both partners are necessary to process the precursors into trisporic acid [23]. Specialized hyphae (gametangia) are formed by both partners. The fusion of the gametangia and zygospore development is accompanied by the fusion of the cytoplasm and karyogamy. The matured zygospore eventually germinates with a sporangium, which releases spores and develop into a haploid mycelia. The importance of zygospores in natural environments remains unknown, whereas under laboratory conditions, germination of zygospores is a rare feature [24,25]. Depending on the species-specific compositions of the trisporoids, zygospores are generally observed only in intraspecific reactions [26]. Nevertheless, zygospores between species of different genera or species are possible [27,28] and also between different species of the genus *Lichtheimia* [1]. Yet interspecific zygospores display a (slightly) different morphology compared to intraspecific ones, which can be applied to biological species recognition [1].

22.4 *LICHTHEIMIA* SPECIES IN FOOD PRODUCTION AND AGRICULTURE

22.4.1 AGRICULTURE

Several mucoralean species are important postharvest pathogens of fruit, for example, bananas and sweet potatoes [29–33]. In addition, several species (e.g., *R. stolonifer*) are well-known plant pathogens and can lead to massive financial losses by causing diseases [34]. However, there are no reports of *Lichtheimia* species associated with fruit or plant rot.

Lichtheimia spp. cause severe infections in farm animals. *L. corymbifera* is a well-known cause of mastitis and abortions in cattle. Piancastelli et al. reported the isolation of *L. corymbifera* from lesions on an aborted bovine fetus in Italy [35], and additional isolates of *L. corymbifera* from bovine fetuses are available in public strain collections. In addition, *L. corymbifera* and *L. ramosa* were isolated from the rumen fluid of healthy cattle [36]. In a study regarding gastrointestinal mycoses in cattle, mucoralean fungi were found to be the causative agent in more than 60% of the cases [37]. *L. corymbifera* was isolated from several animals and was also found in coinfections with *Aspergillus fumigatus*. Since high amounts of molds, including *Lichtheimia* species, are present in cattle feed [38], this seems to be the most likely route of infection. Additional studies in slaughtered feedlot cattle showed that mucoralean fungi may also cause lymphadenitis [39]. Invasion of the lymph nodes by fungal hyphae resulted in necrosis with dystrophic calcification and the accumulation of immune cells including macrophages, lymphocytes, and plasma cells.

Lichtheimia infections also occur frequently in equine hosts. Guillot et al. documented two cases of mucormycoses in ponies from the same paddock [40]. One of the animals showed only a cutaneous infection with necrotic ulceration in the nostrils. The animal recovered after treatment with appropriate antifungals. In contrast, the second pony developed hyperthermia and neurological signs. After the first symptoms, the animal died a few days later due to a systemic mucormycosis. Lesions in the lungs, stomach, liver, and the digestive tract were found postmortem. In addition, a large infarct was present in the brain [40]. Coinfections with *A. fumigatus* have been reported in ponies with comparable clinical symptoms and pathological signs [41].

22.4.2 Food Production

Although species of *Lichtheimia* are well-known pathogens, they are widely used in food production. *L. corymbifera* is regularly isolated from soy fermentation products such as meju and nuruk in Asia [33,42,43] among other mucoralean fungi (e.g., species of *Rhizopus* and *Mucor*). Recently, Hong et al. surveyed the presence of zygomyceteous fungi in meju-related products and found *Lichtheimia* as the second most common genus in meju after *Mucor* species. Three hundred and sixty-four strains were isolated from more than 100 different spots of isolation including 87 strains of *Lichtheimia* [44]. The majority of *Lichtheimia* strains were *L. ramosa* and *L. corymbifera*, which, interestingly, are also the most important species in human infections regarding *Lichtheimia*. However, *L. ornata* and *L. hyalospora* were also isolated in this comprehensive study [44]. Other Asian fermentation products contain large amounts of mucoralean fungi besides meju and nuruk. Rice wine starter cultures contained members of the Mucorales as the most common fungal isolates. *L. corymbifera* and *L. ramosa* were found to be the most frequently isolated fungi, followed by *Aspergillus oryzae* and *Rhizomucor pusillus* [45]. *Lichtheimia* spp. can also be found on diverse fruits and food products. Interestingly, a recent study identified species of *Lichtheimia* as one of the most important contaminants of cocoa [46]. *Lichtheimia* was isolated from different processing states of cacao including sun drying, fermentation, storage, and in industrial cocoa products [46]. Fungi could be detected on 98% of the samples of peanuts with mucoralean fungi representing the majority of isolates (41%), especially *L. corymbifera* and *R. stolonifer* [47]. While most reports of *Lichtheimia* in food products are from Asian countries, studies from Europe show that mucoralean species are associated with a wide variety of products. For example, olive paste was found to be contaminated with different mucoralean fungi including *Rhizopus arrhizus*, *M. circinelloides*, and *L. corymbifera* [48]. However, while mucoralean fungi represented the majority of isolates in Asian food products, they were only rarely isolated in the European study (~5% of the samples). Although only *L. corymbifera* are identified in most studies, other species might have been present but were not designated correctly since several *Lichtheimia* species have been taken as synonyms of each other.

22.5 *LICHTHEIMIA* IN HUMAN INFECTIONS AND RELATED DISEASES

22.5.1 Epidemiology and Risk Factors

Infections with fungi formerly belonging to the *zygomycota* are called zygomycosis. However, a modern classification divides infections into entomophthoromycosis or mucormycosis depending on the causative agent of infection [49]. Entomophthoromycoses are cutaneous or subcutaneous infections with low mortality caused by members of the order Entomophthorales. In contrast, mucormycoses are life-threatening infections from mucoralean species and comprise the majority of zygomycosis cases [50]. These infections occur worldwide and have a wide range of clinical presentations depending on the causative agent. The first described case of mucormycosis was reported in 1885 with *L. corymbifera* as the causative agent [3]. Although mucormycoses are rare infections compared to aspergillosis and candidiasis, the number of clinical cases increased over recent decades [50], which may be due to the prolonged survival of transplantation patients or patients with malignancies. Additionally, better detection and increased awareness by the physicians might also play a role [51]. The main route of infections is inhaled sporangiospores, but fungal material from soil may also play a role in wound infections after accidents [52].

A variety of risk factors are associated with the development of mucormycoses in patients and one of the most important predispositions is diabetes mellitus [50,51,53]. Roden et al. found diabetes as a predisposition in 36% of the reviewed cases [50]. This was even more striking in a study from India where 70% of all reviewed cases involved patients with diabetes [54]. In contrast, diabetes was only the second most important predisposition in European studies at 17% [55] and 23% [56]. This may be due to differences in diagnosis and therapy of diabetes in different geographical regions and the resulting differences in the number of patients with uncontrolled diabetes. Animal infection experiments also demonstrate the high susceptibility of mice suffering from diabetic ketoacidosis [55–57]; it was shown that macrophages from diabetic mice were less able to clear spores of *R. oryzae* [58]. The endothelial cell receptor GRP78 is essential for the adhesion of *R. oryzae* spores, leading to phagocytosis of the spores and subsequent damaging of the cells [57,59]. This receptor is upregulated in patients with diabetes and may explain their high susceptibility for mucormycosis [57].

However, in European studies, immunosuppression such as occurs in patients with hematological malignancy was found to be the most important risk factor to develop mucormycosis in the case of almost 50% of the cases reviewed [60,61]. A study from a clinic in Belgium showed that there was a correlation between the incidence of mucormycoses and the number of patients with hematological malignancy [62]. Pulmonary and disseminated infections are the most common infections in patients with hematological malignancy [60]. However, several cases include patients without any classical predispositions.

Penetrating trauma has been found as the route of infection in healthy individuals in almost 20% of the reviewed cases [50,60,61], and infections after trauma are mainly cutaneous infections [60]. Additional risk factors are the application of voriconazole and the iron chelator deferoxamine as they increase mortality in murine infection models [63,64]. While it is established that deferoxamine can be used as a xenosiderophore by mucoralean fungi and therefore support growth, it is still unclear how voriconazole contributes to infections [63,64]. HIV patients, which represent an important risk group for other fungal pathogens, play a minor role in mucormycoses [50,60,61]. Mucormycoses are also found as nosocomial infections mainly causing superficial or gastrointestinal infections [65,66]. Surgical dressings, catheters, and bandages were found to be typical routes of infection for nosocomial mucormycoses [65–67].

To date more than 20 species of the Mucorales are known to cause infections in humans [68]. The majority of cases are caused by *Rhizopus* spp. with *R. oryzae* as the most important species representing 34% of the cases in Europe, 61% in India, and more than 47% worldwide [50,53,60,61,69]. In contrast, there are significant geographical differences in the abundance of infections with *Lichtheimia* species. Whereas *Lichtheimia* was identified in only a small number of mucoralean infections (~5% of all cases) in a worldwide and a U.S.-wide study [50,69], *Lichtheimia* appears as the second most common infectious agent in mucormycoses from Europe with 19%–29% of all cases [60,61].

22.5.2 Clinical Features and Pathogenesis

Mucoralean fungi can cause different infections in humans. The most common sites of infections in adult patients are pulmonary, rhinocerebral (sinuses, brain, nose passage, oral cavity), and cutaneous infections [50,61,70]. Pulmonary infections have been reported after organ transplantation and leukemia [70,71]. In general, pulmonary infections result in unspecific signs such as inflammation and worsening of breathing, but also in endobronchial bleeding [70,71]. Superficial infections include not only cutaneous but also subcutaneous infections and necrotizing fasciitis/cellulitis [72–75]. These infections occur not only in immunocompromised (e.g., patients suffering from leukosis or AIDS) but also in immunocompetent patients after injuries (e.g., penetrating traumata or burnings) and are associated with severe tissue necrosis [75–78]. However, cutaneous infections with mucoralean fungi tend to spread into deeper tissue and may disseminate [79,80]. Dissemination of *Lichtheimia* infections is reported frequently with involvement of the heart (endocarditis) and the central nervous system [81–83]. Gastrointestinal and cutaneous infections are the most common infections in neonates, representing 54% and 36% of the cases [84]. Also dissemination of *Lichtheimia* from cutaneous infections, with pulmonary, gastrointestinal, and cerebral involvement, has been reported from premature infants [80]. Infections with *Lichtheimia* could be associated with other fungal infections such as coinfections of *L. corymbifera* and *A. fumigatus* in pulmonary or disseminated infections [85,86] or cutaneous coinfection with *L. corymbifera* and *Candida albicans* in a case caused by traumatic wound after a car accident [87].

All types of mucormycoses are associated with rapid destruction of tissue and a high mortality [50,61]. Mortality rates of patients with superficial mucoralean infections are around 30%, while pulmonary and rhino-orbital infections show even higher rates (>60% mortality), increasing up to 100% in disseminated infections [50,88]. An important hallmark of mucormycosis is the rapid invasion of blood vessels, resulting in a combination of infarction, thrombosis, and tissue necrosis [89]. A secreted subtilisin protease of *R. microsporus* may be involved in the development of thrombosis [90,91]. This protease partially hydrolyses fibrinogen and an elastase activity of leukocytes activates the coagulation factor XIII, which finally results in blood clotting [90,91]. Moreover, the protease could be found in samples from *Rhizopus*-infected patients [90,91]. However, there are no comparable studies on *Lichtheimia* available.

Massive invasion of blood vessels, thrombosis, and necrosis is also observed in animal infections with *Lichtheimia* species [92,93]. Recent studies with *Lichtheimia* spp. revealed that the destruction of blood vessels was also present in infection experiments of the embryonated hen egg model, which is a suitable infection model for filamentous fungi [94–96].

Unsurprisingly, the virulence in the clinically relevant *Lichtheimia* species, namely, *L. corymbifera*, *L. ramosa*, and *L. ornata*, was higher compared to that of the nonclinically relevant species *L. hyalospora* and *L. sphaerocystis*. Moreover, destruction of blood vessels was more prone in the clinically relevant species of *Lichtheimia* [94]. However, in vitro interactions of *Lichtheimia* with endothelial cells did not result in damage to the cells, which is comparable to the situation in *Mucor* (Schwartze VU unpublished data and [61]). Thus, thrombosis seems a more likely explanation for the destruction of the blood vessels than direct damage to the cells.

Unfortunately, no data for virulence-associated genes and pathogenicity mechanisms of *Lichtheimia* are available. Nevertheless, all known virulence-associated genes of *R. oryzae* (e.g., proteases and iron uptake gene [51,85]) have been found in the genome of *L. corymbifera* [97]. Further research is necessary on the role of these genes in the infection process.

22.5.3 Farmer's Lung Disease

Mucoralean fungi do not only cause infections in humans but can be responsible for additional diseases. Mold-contaminated plant material such as hay especially can cause severe health problems. Farmer's lung disease (FLD) is one of the most important forms of occupational hypersensitivity pneumonitis, which is caused by recurrent exposure to several microorganisms. In the acute form such diseases can lead to influenza-like symptoms including nausea, chills, fever, sweating, and headache, while the (sub)chronic form is associated with coughing and dyspnea for several days or weeks [98]. A study from France found *L. corymbifera* as an important agent for FLD [99]. In addition, in vitro experiments on the cytokine production of lung epithelial cells after exposure to different extracts of microorganisms showed that *L. corymbifera* induced high expression of proinflammatory and allergic mediators (IL-8, IL-13), supporting the role of *Lichtheimia* [100]. Several studies investigated the impact of environmental factors and agricultural practice on the concentration of potential harmful organisms [18,38]. Moisture is a main factor supporting growth of microorganisms in hay. Since humidity is higher in mountain regions than in the plains, a gradient of *L. corymbifera* contamination was found. In addition, hay-packing strategies (including drying times) play an important role in the prevention of fungal contaminations [38]. Other factors such as the amount of rain during harvest influenced the remaining moisture in the hay and therefore microbial growth [18,38]. Thus, harvesting techniques, storage, and handling of agricultural products need to be optimized to prevent diseases in animals and farm workers.

22.6 DETECTION AND IDENTIFICATION

22.6.1 MORPHOLOGY AND PHYSIOLOGY-BASED IDENTIFICATION

Lichtheimia species are straightforwardly distinguished from other mucoralean fungi based on morphological examinations. Although morphological features (Figures 22.1 and 22.2) are accessible, morphological identification to species level requires considerable expertise since Mucorales display only a few differentiating morphological features. Morphological characteristics applied in species descriptions and identifications concerns appearance of the colony, mycelium, mode of septation, and morphology of sexual (zygospores) and asexual (sporangia and sporangiospores) reproductive structures [101,102]. Yet all morphological features are prone to environmental changes (e.g., temperature, nutrition, variations pH, salinity, and humidity). Therefore, all species descriptions refer to specific conditions where features can be observed (temperature, medium, etc.) [101,102]. Additional techniques have to be applied in order to perform reliable species identification. Although some morphological features can be readily applied to distinguish pathogenic from nonpathogenic species (e.g., temperature profiles), it is impossible to differentiate between the pathogenic species based solely on morphology [1]. Although the formation of sexual zygospores was proposed as a tool for proper species identification, based on the biological species concept [103], there are several (practical) problems that occur for *Lichtheimia*. In general, *Lichtheimia* species rarely produce zygospores. In addition, interspecies breeding is possible between several species of *Lichtheimia* and even between *Lichtheimia* and some *Mucor* species [1]. Although the interspecies zygospores are slightly different in morphology (i.e., smaller and lighter in color) [1], this method requires experience, is time consuming, and is not a routine method.

In addition to morphology, physiological and biochemical data can be applied for the identification of fungi. However, there are only few data on the use of such data for the identification of mucoralean fungi. One possibility is different carbon source assimilation profiles. Commercial available test kits (e.g., Api50 and ID32C strips, bioMerieux diagnostics) can be applied to differentiate clinically relevant zygomycetes including *L. corymbifera* [104]. However, it is not known if the carbon source assimilation pattern overlaps with different phyla or even genera. Thus, previous knowledge about the designation of the strains to a higher taxonomic level is still important and must be acquired by morphological examination.

22.6.2 PCR-BASED IDENTIFICATION

Molecular-based techniques allow fast and accurate species identification for almost all mucoralean fungi. If axenic cultures of the fungi are available, amplification and sequencing of molecular markers using universal primers is a possibility. DNA can be prepared after short incubation times since most mucoralean fungi show fast growth on artificial media. Usually loci of the ribosomal

DNA clusters are used for the identification depending on the desired taxonomic level of identification [105]. While the conserved small subunit of ribosomal DNA (18S ribosomal subunit) and the large subunit of ribosomal DNA (28S ribosomal subunit) allows reliable identification down to the genus level, the variable internal transcribed spacer (ITS) region can be used for species identification. The ITS combines high variability that is necessary to distinguish species and comes with the possibility to design universal primers since it is flanked by the highly conserved 18S and 28S rDNA. Also a recent publication on the identification of fungi based on DNA bar codes evaluated six different regions and proposed the ITS region as the most reliable bar code marker for fungi [105]. Table 22.2 includes available primers and their sequences for the amplification and sequencing of these regions. Reference sequences for most mucoralean species, for at least one of these regions, are available in public databases (e.g., GenBank, http://www.ncbi.nlm.nih.gov/). One major disadvantage in exclusively sequence-based identification is the inability to identify a fungal specimen unambiguously. Nevertheless, with a deliberate choice of the molecular marker in combination with a comprehensive and reliable reference database, even a single marker could result in a high approximation to the identity of the fungus. However, the discovery of new, undescribed or unsequenced specimens is impossible without the combination of additional data (e.g., morphology).

Furthermore, solely sequence-based methods require axenic cultures to prevent, for example, the amplification of host DNA or of contaminants with the given primers. However, a recent publication evaluated the use of universal ITS primers (ITS1/ITS2) and sequencing on the basis of paraffin-embedded tissue sections from infected mice. The authors were able to identify several mucoralean pathogens based on their ITS1 sequences [106], and since mucoralean fungi sometimes cannot be isolated from the patient material, direct amplification from specimens is valuable for diagnosis.

To avoid sequencing and thereby reduce the time necessary for identification, taxon-specific primers can be applied. Specific primers can be developed for different taxonomic levels ranging from phylum to species specificity. Positive results can be detected by the presence of a DNA band in an agarose gel electrophoresis since the primers only amplify the corresponding fragments of the target species. For specificity on a species level in mucoralean fungi, the highly variable ITS region is used in most cases to design the primers. Species-specific primers for pathogenic zygomycetes including many mucoralean fungi were developed and tested for specificity in a study by Voigt et al. [5]. Although such specific primers allow a rapid and easy identification of the organisms, they are restricted to the known fungi and separate primer pairs are necessary for each species. To date only one primer pair for *Lichtheimia* species is available with specificity for *L. corymbifera* (see Table 22.2).

TABLE 22.2
Oligonucleotide Primers for the Detection of *Lichtheimia*

Primer	Specificity	Target	Primer Sequence (5'-3')	Amplicon (bp)	References
Acy1	*L. corymbifera*	28S rDNA	CGGATTGTAAACTAAAGAGCG	577	Voigt et al. [5]
Acy2	*L. corymbifera*	28S rDNA	CCAAAGTAGATTACAGTTCTAG		Voigt et al. [5]
ITS1	Pan-eukaryotic	ITS	TCCGTAGGTGAACCTGCGG	~700–800	White et al. [137]
ITS4	Pan-eukaryotic	ITS	TCCTCCGCTTATTGATATGC		White et al. [137]
NL1	Pan-eukaryotic	28S rDNA	GCATATCAATAAGCGGAGGAAAAG	~640	O'Donnell [138]
NL4	Pan-eukaryotic	28S rDNA	GGTCCGTGTTTCAAGACGG		O'Donnell [138]
NS1	Pan-eukaryotic	18S rDNA	GTAGTCATATGCTTGTCTC	~1000–1200	White et al. [137]
NS41	Pan-eukaryotic	18S rDNA	CCCGTGTTGAGTCAAATTA		O'Donnell et al. [139]

22.6.3 Hybridization-Based Identification

To overcome time-consuming and cost-intensive sequencing steps and to avoid multiple PCR steps using specific primers, several hybridization-based methods for different fungal pathogens have been developed. All hybridization probes described so far are based on the ITS region of the rDNA cluster since it is highly variable and thus useful for generation of species-specific probes. The protocols rely on the generation of PCR primers labeled with biotin and subsequent detection with streptavidin coupled to fluorescent markers or peroxidase for chemiluminescence detection [107–109]. Zhao et al. described a method combining ITS PCR with a broad-range primer combined with a subsequent reverse line blot hybridization (PCR/RLB) [107]. Species-specific probes are blotted to a membrane and afterward incubated with the biotin-labeled PCR fragments. After washing steps, probe-bound PCR fragments are detectable using streptavidin–peroxidase conjugate and commercially available chemiluminescence detection solutions. Alternatively, species-specific probes can be bound to carboxylated beads and incubated with biotin-labeled PCR products [109,110]. After washing, probe-bound PCR fragments are detectable using fluorescence-labeled streptavidin. The described assays were also evaluated for mixed samples of different fungi (e.g., *Aspergillus*/Mucorales species) with high accuracy [107–109].

22.6.4 Mass Spectrometry–Based Identification

Matrix-assisted laser desorption/ionization (MALDI) time-of-flight (TOF) mass spectrometry (MS)–based identification have gained an importance recently. PCR-based methods can be time and labor consuming due to the DNA extraction, PCR reaction, purification, and sequencing. In contrast, MALDI-TOF-MS-based methods require only minutes of on-hand time (around 5 min) starting with an axenic culture if appropriate reference spectra are available. Also the costs for the MALDI-TOF identification are very low (ca. 50¢ U.S. per sample) compared to DNA sequencing (ca. $5 U.S.) [111]. Fast and reliable identification protocols have been described for several bacteria and yeasts [112–114]. In addition, MALDI-TOF MS was used accurately for the identification of filamentous fungi, for example, dermatophytes, *Aspergillus*, and *Penicillium* [115,116]. Recently, MS-based identification was successfully applied to *Lichtheimia*. *Lichtheimia* species could be unambiguously discriminated from other mucoralean pathogens and more than 90% of the strains were correctly identified to species [117]. However, all protocols described for mucoralean species still rely on the availability of axenic cultures. It was shown that MALDI-TOF MS can be used for the identification of bacteria directly from patient material (e.g., blood samples) [118]. Thus, the method might be capable of culture-independent identification of *Lichtheimia* with further development, which would be a great advantage.

22.7 ANTIFUNGAL SUSCEPTIBILITY AND TREATMENT

22.7.1 Antifungal Susceptibility

Mucoralean fungi are resistant to various antifungal agents including voriconazole, which is used in transplantation patients for prophylaxis [119]. The in vitro activity of several antifungal has been tested against members of the Mucorales revealing high differences in the susceptibility of different species. *Lichtheimia* species were found to be resistant to 5-fluorocytosine, voriconazole, and the echinocandins micafungin and caspofungin. In contrast, they were susceptible to posaconazole and terbinafine [120–122]. While several mucoralean strains such as *Rhizopus* spp. and *Mucor* spp. showed a high tolerance for itraconazole and even amphotericin B, all strains of *Lichtheimia* were highly susceptible to these drugs [120]. There are no significant differences in the susceptibility patterns between the three clinically relevant species *L. corymbifera*, *L. ramosa*, and *L. ornata* [122]. In vivo studies on the activity of antifungals against mucoralean fungi showed that posaconazole (10 mg/kg/day) and amphotericin B (1 mg/kg/day) significantly prolonged survival of *Lichtheimia*-infected mice [123]. Studies on the sequence of CYP51 protein (the main target of azole antifungal) in *L. corymbifera* and *R. oryzae* revealed several mutations, which result in resistance to voriconazole while the affinity for the binding of posaconazole was high. Finally, cellular influx and/or reduced efflux of posaconazole may explain its higher activity against mucoralean fungi [124].

Combination therapy with antifungal substances from different classes can increase the efficiency of the treatment due to synergistic effects. While echinocandins (e.g., micafungin and caspofungin) alone are not effective in therapy of mucormycoses, combination therapy with amphotericin B and caspofungin prolonged survival of *R. oryzae*–infected mice compared to mice treated with amphotericin B alone [125]. Comparable in vivo data are missing for *Lichtheimia*. However, in vitro data show that combination of caspofungin with immunosuppressive substances (calcineurin inhibitors) displayed synergistic effects in *L. ramosa* [126]. In addition, posaconazole was shown to enhance the activity of amphotericin B against hyphae of *L. corymbifera* but not against sporangiospores [127].

Iron overload is an important risk factor in the development of mucormycosis [128]. Several treatments include the application of iron chelators to reduce iron overload in addition to antifungals [129]. Synergistic effects of amphotericin B and the iron chelator deferasirox were shown in vitro, and combination therapy prolonged survival of *R. oryzae*–infected mice in vivo due to iron starvation of the fungus [130,131]. In contrast, the widely used iron chelator deferoxamine is associated with the development of mucormycosis since it can be used as xenosiderophore by mucoralean species [63]. In addition to the previously mentioned synergistic effects of antifungals, combinations of amphotericin B, micafungin, and deferasirox improved survival and reduced fungal burden in *R. oryzae*–infected mice [132]. However, to date no data about the application of iron chelators in the treatment of *Lichtheimia* infections exist.

22.7.2 Clinical Therapy

Treatment of mucormycoses is complicated and often involves an additional surgical removal of infected tissue [50,76,83]. Several successful treatments of *Lichtheimia* infections have been described in the literature in different types of infections. Subcutaneous infections were successfully treated with amphotericin B in combination with early removal of debris [76]. Patients with sinus infections of *Lichtheimia* responded also to debridement in combination with amphotericin B and itraconazole [54]. Successful combination therapies with amphotericin B and caspofungin or micafungin (echinocandines) have been reported for patients suffering from rhino-orbital *Lichtheimia* infections [133,134]. Although mortality in disseminated mucormycoses is very high, successful treatments are also reported for such severe cases. A triple therapy with high doses of liposomal amphotericin B, posaconazole, and caspofungin together with surgical treatment resulted in the survival of a patient who suffered from disseminating *Lichtheimia* infection after chemotherapy [135]. However, all infections needed therapy for several months. Despite the good results with iron chelators in laboratory studies, results from clinical trials could not support the positive effect of iron chelators on amphotericin B treatment [136].

22.8 CONCLUSIONS

Mucoralean fungi, including the genus *Lichtheimia*, represent a variety of ubiquitous, mostly saprophytic microorganisms. A variety of mucoralean species have been used for food production for a long time, especially in Asian countries. Yet several species are also known as plant pathogens and agents causing postharvest losses. Some of these species have pathogenic potential against animals, and under appropriate conditions, infections in humans are also possible. Although mucormycoses are rare compared to infections with *Aspergillus* or *Candida*, the number of clinical cases is increasing. This might be due to the prolonged survival of transplantation patients or patients with malignancies. Additionally, better detection and increased awareness by the physicians might also play a role. An important hallmark of mucormycoses is the rapid invasion of blood vessels, resulting in a combination of infarction, thrombosis, and tissue necrosis. All types of mucormycoses are associated with rapid destruction of tissue and a high mortality. Fast identification and proper treatment is crucial for a good outcome of such infections. However, the rapid progress of mucormycoses combined with its various manifestations means a sufficiently rapid identification of the causing organism is difficult. Nevertheless, secure protocols and methods for identification are constantly developed making survival chances of patients more likely, along with susceptibility studies for effective treatments.

ACKNOWLEDGMENTS

The authors are thankful to Stefanie Ponert for the figure of the general life cycle. After finishing the manuscript, the genus Lichtheimia was extended by one additional species isolated from Brazilian soil—Lichtheimia brasiliensis. So far, no impact on food production and agriculture is known for this species. Furthermore, L. brasiliensis is without any pathogenic potential [140,141].

REFERENCES

1. Alastruey-Izquierdo A, Hoffmann K, De Hoog GS, Rodriguez-Tudela JL, Voigt K et al. (2010) Species recognition and clinical relevance of the zygomycetous genus *Lichtheimia* (syn. *Absidia* pro parte, *Mycocladus*). *Journal of Clinical Microbiology* 48: 2154–2170.
2. Cohn FJ. (1884) *Mucor corymbifer*. In: Lichtheim L, ed. *Über pathogene Mucoineen und die durch sie erzeugten Mykosen des Kaninchens*. Zeitschrift für Klinische Medicin. Verlag von August Hirschwald, Berlin, p. 149.
3. Platauf AP. (1885) Mycosis mucorina. *Vierchows Archive* 102: 543.
4. van Tieghem, P. (1876) Troisiéme mémoire sur les Mucorinées. *Annales des Sciences Naturelles Botanique* 4: 312–399.
5. Voigt K, Cigelnik E, O`Donnell K. (1999) Phylogeny and PCR identification of clinically important zygomycetes based on nuclear ribosomal-DNA sequence data. *Journal of Clinical Microbiology* 37: 3957–3964.
6. Voigt K, Wöstemeyer J. (2001) Phylogeny and origin of 82 zygomycetes from all 54 genera of the Mucorales and Mortierellales based on combined analysis of actin and translation elongation factor EF-1alpha genes. *Gene* 270: 113–120.
7. O`Donnell K, Lutzoni FM, Ward TJ, Benny GL. (2001) Evolutionary relationships among mucoralean fungi (Zygomycota): Evidence for family polyphyly on a large scale. *Mycologia* 93: 286–296.
8. Hoffmann K. (2010) Identification of the genus *Absidia* (Mucorales, Zygomycetes): A comprehensive taxonomic revision. In: Gherbawy Y, Voigt K, eds. *Molecular Identification of Fungi*. Springer Verlag Berlin Heidelberg, pp. 439–460.
9. Hoffmann K, Discher S, Voigt K. (2007) Revision of the genus *Absidia* (Mucorales, Zygomycetes) based on physiological, phylogenetic, and morphological characters: Thermotolerant *Absidia* spp. form a coherent group, Mycocladiaceae fam. nov. *Mycological Research* 111: 1169–1183.
10. Hoffmann K, Walther G, Voigt K. (2009) *Mycocladus* vs. *Lichtheimia*: A correction (Lichtheimiaceae fam. nov., Mucorales, Mucoromycotina). *Mycological Research* 113: 277–278.
11. Crous PW, Gams W, Stalpers JA, Robert V, Stegehuis G. (2004) MycoBank: An online initiative to launch mycology into the 21st century. *Studies in Mycology* 50: 19–22. http://www.mycobank.org/, accessed December, 2012.
12. Vuillemin P. (1903) Le genre Tieghemella et la série de Absidées. *Bulletin de la Société Mycologique de France* 19: 119–127.
13. Nottebrock H, Scholer HJ, Wall M. (1974) Taxonomy and identification of mucormycosis-causing fungi. I. Synonymity of *Absidia ramosa* with *A. corymbifera*. *Sabouraudia* 12: 64–74.
14. Schipper MAA. (1990) Notes on Mucorales-I. Observations on *Absidia*. *Persoonia* 14: 133–149.

15. Garcia-Hermoso D, Hoinard D, Gantier J-C, Grenouillet F, Dromer F et al. (2009) Molecular and phenotypic evaluation of *Lichtheimia corymbifera* (formerly *Absidia corymbifera*) complex isolates associated with human mucormycosis: Rehabilitation of *L. ramosa*. *Journal of Clinical Microbiology* 47: 3862–3870.
16. Woo PCY, Lau SKP, Ngan AHY, Tung ETK, Leung S-Y et al. (2010) *Lichtheimia hongkongensis* sp. nov., a novel *Lichtheimia* spp. associated with rhinocerebral, gastrointestinal, and cutaneous mucormycosis. *Diagnostic Microbiology and Infectious Disease* 66: 274–284.
17. Wöstemeyer J. (1985) Strain-dependent variation in ribosomal DNA arrangement in *Absidia glauca*. *European Journal of Biochemistry* 146: 443–448.
18. Reboux G, Reiman M, Roussel S, Taattola K, Millon L et al. (2006) Impact of agricultural practices on microbiology of hay, silage and flour on finnish and french farms. *Annals of Agricultural and Environmental Medicine* 13: 267–273.
19. Kotimaa MH, Oksanen L, Koskela P. (1991) Feeding and bedding materials as sources of microbial exposure on dairy farms. *Scandinavian Journal of Work, Environment & Health* 17: 117–122.
20. Blakeslee A. (1904) Sexual reproduction in the Mucorineae. *Proceedings of the National Academy of Arts and Sciences* 40: 205–319.
21. Van den Ende H. (1967) Sexual factor of the Mucorales. *Nature* 215: 211–212.
22. Gooday G. (1968) Hormonal control of sexual reproduction in *Mucor mucedo*. *New Phytologist* 67: 815–821.
23. Werkman B. (1976) Localization and partial characterization of a sex-specific enzyme in homothallic and heterothallic mucorales. *Archives of Microbiology* 109: 209–213.
24. Michailides TJ, Spotts R. (1988) Germination of zygospores of *Mucor piriformis* on the life history of *Mucor piriformis*. *Mycologia* 80: 837–844.
25. Yi MQ, Ko W. (1997) Factors affecting germination and mode of germination of zygospores of *Choanephora cucurbitarum*. *Journal of Phytopathology* 145: 357–361.
26. Sutter RP, Dadok J, Bothner-By AA, Smith RR, Misha P. (1989) Cultures of separated mating types of *Blakeslea trispora* make D and E forms of trisporic acids. *Biochemistry* 28: 4060–4066.
27. Blakeslee AF, Cartledge J. (1927) Sexual dimorphism in Mucorales II. Interspecific reactions. *Botanical Gazette* 84: 51–57.
28. Stalpers JA, Schipper M. (1980) Comparison of zygospore ornamentation in intra- and interspecific matings in some related species of *Mucor* and *Backusella*. *Persoonia* 11: 39–52.
29. Webb TA, Mundt JO. (1978) Molds on vegetables at the time of harvest. *Applied and Environmental Microbiology* 35: 655–658.
30. Eseigbe DA, Bankole SA. (1997) Fungi associated with post-harvest rot of black plum (*Vitex doniana*) in Nigeria. *Mycopathologia* 136: 109–114.
31. Ray RC, Ravi V. (2005) Post harvest spoilage of sweetpotato in tropics and control measures. *Critical Reviews in Food Science and Nutrition* 45: 623–644.
32. Kwon JH, Ryu JS, Chi TTP, Shen SS, Choi O. (2012) Soft Rot of *Rhizopus oryzae* as a postharvest pathogen of banana fruit in Korea. *Mycobiology* 40: 214–216.
33. Kim H, Kim J, Bai D, Ahn B. (2011) Identification and characterization of useful fungi with α-amylase activity from the korean traditional nuruk. *Microbiology* 39: 278–282.
34. Shtienberg D. (1997) *Rhizopus* head rot of confectionery sunflower: Effects on yield quantity and quality and implications for disease management. *Phytopathology* 87: 1226–1232.
35. Piancastelli C, Ghidini F, Donofrio G, Jottini S, Taddei S et al. (2009) Isolation and characterization of a strain of *Lichtheimia corymbifera* (ex *Absidia corymbifera*) from a case of bovine abortion. *Reproductive Biology and Endocrinology* 7: 138.
36. Lund A. (1973) Yeasts and moulds in the bovine rumen. *Journal of General Microbiology* 81: 453–462.
37. Jensen HE, Olsen SN, Aalbaek B. (1994) Gastrointestinal aspergillosis and zygomycosis of cattle. *Veterinary Pathology* 31: 28–36.
38. Gbaguidi-Haore H, Roussel S, Reboux G, Dalphin J, Piarroux R. (2009) Multilevel analyis of the impact of environmental factors. *Annals of Agricultural and Environmental Medicine* 16: 219–225.
39. Ortega J, Uzal FA, Walker R, Kinde H, Diab SS et al. (2010) Zygomycotic lymphadenitis in slaughtered feedlot cattle. *Veterinary Pathology* 47: 108–115.
40. Guillot J, Collobert C, Jensen HE, Huerre M, Chermette R. (2000) Two cases of equine mucormycosis caused by *Absidia corymbifera*. *Equine Veterinary Journal* 32: 453–456.
41. Thirion-Delalande C, Guillot J, Jensen HE, Crespeau FL, Bernex F. (2005) Disseminated acute concomitant aspergillosis and mucormycosis in a pony. *Journal of Veterinary Medicine A* 52: 121–124.

42. Lee J-H, Kim T-W, Lee H, Chang HC, Kim H-Y. (2010) Determination of microbial diversity in meju, fermented cooked soya beans, using nested PCR-denaturing gradient gel electrophoresis. *Letters in Applied Microbiology* 51: 388–394.
43. Yeun J, Yeo S, Baek SY, Choi HS. (2011) Molecular and morphological identification of fungal species isolated from bealmijang meju. *Journal of Microbiology and Biotechnology* 21: 1270–1279.
44. Hong S, Kim D, Lee M, Baek S, Kwon S et al. (2012) Zygomycota associated with traditional meju, a fermented soybean starting material for soy sauce and soybean paste. *The Journal of Microbiology* 50: 386–393.
45. Yang S, Lee J, Kwak J, Kim K, Seo M et al. (2011) Fungi associated with the traditional starter cultures used for rice wine in Korea. *Journal of the Korean Society of Applied Biological Chemistry* 54: 933–943.
46. Copetti MV, Iamanaka BT, Frisvad JC, Pereira JL, Taniwaki MH. (2011) Mycobiota of cocoa: From farm to chocolate. *Food Microbiology* 28: 1499–1504.
47. Mphande FA, Siame BA, Taylor JE. (2004) Fungi, aflatoxins, and cyclopiazonic acid associated with peanut retailing in Botswana. *Journal of Food Protection* 67: 96–102.
48. Baffi MA, Romo-Sánchez S, Ubeda-Iranzo J, Briones-Pérez AI. (2012) Fungi isolated from olive ecosystems and screening of their potential biotechnological use. *New Biotechnology* 29: 451–456.
49. Kwon-Chung KJ. (2012) Taxonomy of fungi causing mucormycosis and entomophthoramycosis (zygomycosis) and nomenclature of the disease: Molecular mycologic perspectives. *Clinical Infectious Diseases* 54(S1): S8–S15.
50. Roden MM, Zaoutis TE, Buchanan WL, Knudsen TA, Sarkisova TA et al. (2005) Epidemiology and outcome of zygomycosis: A review of 929 reported cases. *Clinical Infectious Diseases* 41: 634–653.
51. Lanternier F, Lortholary O. (2009) Zygomycosis and diabetes mellitus. *Clinical Microbiology and Infection* 15(S5): 21–25.
52. Richardson M. (2009) The ecology of the Zygomycetes and its impact on environmental exposure. *Clinical Microbiology and Infection* 15(S5): 2–9.
53. Chakrabarti A, Das A, Mandal J, Shivaprakash MR, George VK et al. (2006) The rising trend of invasive zygomycosis in patients with uncontrolled diabetes mellitus. *Medical Mycology* 44: 335–342.
54. Kindo AJ, Shams NR, Srinivasan V, Kalyani J, Mallika M. (2007) Multiple discharging sinuses: An unusual presentation caused by *Absidia corymbifera*. *Indian Journal of Medical Microbiology* 25: 9–12.
55. Ibrahim AS, Gebremariam T, Lin L, Luo G, Husseiny MI et al. (2010) The high affinity iron permease is a key virulence factor required for *Rhizopus oryzae* pathogenesis. *Molecular Microbiology* 77: 587–604.
56. Waldorf AR, Ruderman N, Diamond RD. (1984) Specific susceptibility to mucormycosis in murine diabetes and bronchoalveolar macrophage defense against *Rhizopus*. *The Journal of Clinical Investigation* 74: 150–160.
57. Liu M, Spellberg B, Phan QT, Fu Y, Fu Y et al. (2010) The endothelial cell receptor GRP78 is required for mucormycosis pathogenesis in diabetic mice. *The Journal of Clinical Investigation* 120: 1914–1924.
58. Waldorf AR, Levitz SM, Diamond RD. (1984) In vivo bronchoalveolar macrophage defense against *Rhizopus oryzae* and *Aspergillus fumigatus*. *The Journal of Infectious Diseases* 150: 752–760.
59. Ibrahim AS, Spellberg B, Avanessian V, Fu Y, Edwards JE Jr. (2005) *Rhizopus oryzae* adheres to, is phagocytosed by, and damages endothelial cells in vitro. *Infection and Immunity* 73: 778–783.
60. Lanternier F, Dannaoui E, Morizot G, Elie C, Garcia-Hermoso D et al. (2012) A global analysis of mucormycosis in France: The RetroZygo study (2005–2007). *Clinical Infectious Diseases* 54(S1): S35–S43.
61. Skiada A, Pagano L, Groll A, Zimmerli S, Dupont B et al. (2011) Zygomycosis in Europe: Analysis of 230 cases accrued by the registry of the European Confederation of Medical Mycology (ECMM) Working Group on Zygomycosis between 2005 and 2007. *Clinical Microbiology and Infection* 17: 1859–1867.
62. Saegeman V, Maertens J, Meersseman W, Spriet I, Verbeken E et al. (2010) Increasing incidence of mucormycosis in University Hospital, Belgium. *Emerging Infectious Diseases* 16: 1456–1458.
63. Boelaert JR, De Locht M, Van Cutsem J, Kerrels V, Cantineaux B et al. (1993) Mucormycosis during deferoxamine therapy is a siderophore-mediated infection. In vitro and in vivo animal studies. *The Journal of Clinical Investigation* 91: 1979–1986.
64. Lamaris GA, Ben-Ami R, Lewis RE, Chamilos G, Samonis G et al. (2009) Increased virulence of Zygomycetes organisms following exposure to voriconazole: A study involving fly and murine models of zygomycosis. *The Journal of Infectious Diseases* 199: 1399–1406.
65. Perlroth J, Choi B, Spellberg B. (2007) Nosocomial fungal infections: Epidemiology, diagnosis, and treatment. *Medical Mycology* 45: 321–346.

66. Rammaert B, Lanternier F, Zahar J, Dannaoui E, Bougnoux M-E et al. (2012) Healthcare-associated mucormycosis. *Clinical Infectious Diseases* 54(S1): S44–S54.
67. Christiaens G, Hayette, MP, Jacquemin D, Melin P, Mutsers J, De Mol P. (2005) An outbreak of *Absidia corymbifera* infection associated with bandage contamination in a burns unit. *The Journal of Hospital Infections* 61: 88.
68. De Hoog GS, Guarro J, Gene J, Figueras M. (2000) *Atlas of Clinical Fungi*, 2nd edn. American Society of Microbiology, Washington DC, pp. 60–114.
69. Alvarez E, Sutton DA, Cano J, Fothergill AW, Stchigel A et al. (2009) Spectrum of zygomycete species identified in clinically significant specimens in the United States. *Journal of Clinical Microbiology* 47: 1650–1656.
70. Barr A, Nolan M, Grant W, Costello C, Petrou MA. (2006) Rhinoorbital and pulmonary zygomycosis post pulmonary aspergilloma in a patient with chronic lymphocytic leukaemia. *Acta Bio-Medica* 77: 13–18.
71. Mattner F, Weissbrodt H, Strueber M. (2004) Two case reports: Fatal *Absidia corymbifera* pulmonary tract infection in the first postoperative phase of a lung transplant patient receiving voriconazole prophylaxis, and transient bronchial *Absidia corymbifera* colonization in a lung transplant patient. *Scandinavian Journal of Infectious Diseases* 36: 312–314.
72. Thami GP, Kaur S, Bawa AS, Chander J, Mohan H et al. (2003) Post-surgical zygomycotic necrotizing subcutaneous infection caused by *Absidia corymbifera*. *Clinical Dermatology* 28: 251–253.
73. Almaslamani M, Taj-Aldeen SJ, Garcia-Hermoso D, Dannaoui E, Alsoub H et al. (2009) An increasing trend of cutaneous zygomycosis caused by *Mycocladus corymbifer* (formerly *Absidia corymbifera*): Report of two cases and review of primary cutaneous *Mycocladus* infections. *Medical Mycology* 47: 532–538.
74. Pasticci MB, Terenzi A, Lapalorcia LM, Giovenale P, Pitzurra L et al. (2008) *Absidia corymbifera* necrotizing cellulitis in an immunocompromised patient while on voriconazole treatment. *Annals of Hematology* 87: 687–689.
75. Shakoor S, Jabeen K, Idrees R, Jamil B, Irfan S. (2011) Necrotising fasciitis due to *Absidia corymbifera* in wounds dressed with non sterile bandages. *International Wound Journal* 8: 651–655.
76. Blazquez D, Ruiz-Contreras J, Fernández-Cooke E, González-Granado I, Delgado MD et al. (2010) *Lichtheimia corymbifera* subcutaneous infection successfully treated with amphotericin B, early debridement, and vacuum-assisted closure. *Journal of Pediatric Surgery* 45: e13–e15.
77. Liu ZH, Lv GX, Chen J, Sang H, She XD et al. (2009) Primary cutaneous zygomycosis due to *Absidia corymbifera* in a patient with cutaneous T cell lymphoma. *Medical Mycology* 47: 663–668.
78. Leong KW, Crowley B, White B, Crotty GM, Briain DSO et al. (1997) Cutaneous mucormycosis due to *Absidia corymbifera* occurring after bone marrow transplantation. *Bone Marrow Transplantation* 19: 513–515.
79. Skiada A, Petrikkos G. (2009) Cutaneous zygomycosis. *Clinical Microbiology and Infection* 15(S5): 41–45.
80. Amin SB, Ryan RM, Metlay LA, Watson WJ. (1998) *Absidia corymbifera* infections in neonates. *Clinical Infectious Diseases* 26: 990–992.
81. Ritz N, Ammann RA, Aebischer CC, Gugger M, Jaton K et al. (2005) Failure of voriconazole to cure disseminated zygomycosis in an immunocompromised child. *European Journal of Pediatrics* 164: 231–235.
82. Skiada A, Vrana L, Polychronopoulou H, Prodromou P, Chantzis A et al. (2009) Disseminated zygomycosis with involvement of the central nervous system. *Clinical Microbiology and Infection* 15(S5): 46–49.
83. Mitchell ME, McManus M, Dietz J, Camitta BM, Szabo S et al. (2010) *Absidia corymbifera* endocarditis: Survival after treatment of disseminated mucormycosis with radical resection of tricuspid valve and right ventricular free wall. *The Journal of Thoracic and Cardiovascular Surgery* 139: e71–e72.
84. Roilides E, Zaoutis TE, Walsh TJ. (2009) Invasive zygomycosis in neonates and children. *Clinical Microbiology and Infection* 15: 50–54.
85. McLintock LA, Gibson BES, Jones BL. (2005) Mixed pulmonary fungal infection with *Aspergillus fumigatus* and *Absidia corymbifera* in a patient with relapsed acute myeloid leukaemia. *British Journal of Haematology* 128: 737.
86. Hasan D, Med C, Fleischhack G, Gillen J, Bialek R et al. (2010) Successful management of a simultaneous *Aspergillus fumigatus* and *Absidia corymbifera* invasive fungal infection. *Journal of Pediatric Hematology/Oncology* 32: 22–24.
87. Horré R, Jovanić B, Herff S, Marklein G, Zhou H et al. (2004) Wound infection due to *Absidia corymbifera* and *Candida albicans* with fatal outcome. *Medical Mycology* 42: 373–378.

88. Mantadakis E, Samonis G. (2009) Clinical presentation of zygomycosis. *Clinical Microbiology and Infection* 15(S5): 15–20.
89. Sugar AM. (1992) Mucormycosis. *Clinical Infectious Diseases* 14(S1): S126–S129.
90. Rüchel R, Elsner C, Spreer A. (2004) A probable cause of paradoxical thrombosis in zygomycosis. *Mycoses* 47: 203–207.
91. Spreer A, Rüchel R, Reichard U. (2006) Characterization of an extracellular subtilisin protease of *Rhizopus microsporus* and evidence for its expression during invasive rhinoorbital mycosis. *Medical Mycology* 44: 723–731.
92. Sodhi MP, Khanna RN, Sadana JR, Chand P. (1998) Experimental *Absidia corymbifera* infection in rabbits: Sequential pathological studies. *Mycopathologia* 143: 25–31.
93. Corbel, MJ, Eades S. (1975) Factors determining the susceptibility of mice to experimental phycomycosis. *Journal of Medical Microbiology* 8: 551–564.
94. Schwartze VU, Hoffmann K, Nyilasi I, Papp T, Vágvölgyi C et al. (2012) *Lichtheimia* species exhibit differences in virulence potential. *PloS one* 7: e40908.
95. Härtl A, Hillesheim HG, Künkel W, Schrinner EJ. (1995) The Candida infected hen's egg. An alternative test system for systemic anticandida activity. *Arzneimittelforschung* 45: 926–928.
96. Jacobsen ID, Große K, Slesiona S, Hube B, Berndt A et al. (2010) Embryonated eggs as an alternative infection model to investigate *Aspergillus fumigatus* virulence. *Infection and Immunity* 78: 2995–3006.
97. Schwartze VU, Winter S, Shelest E, Marcet-Houben M, Horn F et al. (2014) Gene expansion shapes genome architecture in the human pathogen Lichtheimia corymbifera: An evolutionary genomics analysis in the ancient terrestrial mucorales (Mucoromycotina). *PLoS Genetics* 10(8): e1004496.
98. Lacasse Y, Cormier Y. (2006) Hypersensitivity pneumonitis. *Orphanet Journal of Rare Diseases* 1: 25.
99. Reboux G, Piarroux R, Mauny F, Madroszyk A, Millon L et al. (2001) Role of molds in farmer's lung disease in eastern France. *American Journal of Critical Care Medicine* 163: 1534–1539.
100. Bellanger AP, Reboux G, Botterel F, Candido C, Roussel S et al. (2010) New evidence of the involvement of *Lichtheimia corymbifera* in farmer's lung disease. *Medical Mycology* 48: 981–987.
101. Zycha H, Siepmann R, Linnemann G. (1969) Mucorales. Eine Beschreibung aller Gattungen und Arten dieser Pilzgruppe. Verlag von J. Cramer, Lehre, Germany, 355pp.
102. O'Donnell K. (1979) Zygomycetes in culture. Palfrey Contributions in Botany no. 2, Department of Botany, University of Georgia, Athens, Georgia, p. 257.
103. Weitzman I, Whittier S, Mckitrick JC, Della-Latta P. (1995) Zygospores: The last word in identification of rare or atypical zygomycetes isolated from clinical specimens. *Journal of Clinical Microbiology* 33: 781–783.
104. Schwarz P, Lortholary O, Dromer F, Dannaoui E. (2007) Carbon assimilation profiles as a tool for identification of zygomycetes. *Journal of Clinical Microbiology* 45: 1433–1439.
105. Schoch CL, Seifert KA, Huhndorf S, Robert V, Spouge JL et al. (2012) Nuclear ribosomal internal transcribed spacer (ITS) region as a universal DNA barcode marker for Fungi. *Proceedings of the National Academy of Sciences of the United States of America* 109: 6241–6246.
106. Dannaoui E, Schwarz P, Slany M, Loeffler J, Jorde AT et al. (2010) Molecular detection and identification of zygomycetes species from paraffin-embedded tissues in a murine model of disseminated zygomycosis: A collaborative European Society of Clinical Microbiology and Infectious Diseases (ESCMID) Fungal Infection Study Group (EFISG) evaluation. *Journal of Clinical Microbiology* 48: 2043–2046.
107. Zhao Z, Li L, Wan Z, Chen W, Liu H et al. (2011) Simultaneous detection and identification of *Aspergillus* and mucorales species in tissues collected from patients with fungal rhinosinusitis. *Journal of Clinical Microbiology* 49: 1501–1507.
108. Imhof A, Schaer C, Schoedon G, Schaer DJ, Walter RB et al. (2003) Rapid detection of pathogenic fungi from clinical specimens using LightCycler real-time fluorescence PCR. *European Journal of Clinical Microbiology & Infectious Diseases* 22: 558–560.
109. Landlinger C, Preuner S, Willinger B, Haberpursch B, Racil Z et al. (2009) Species-specific identification of a wide range of clinically relevant fungal pathogens by use of Luminex xMAP technology. *Journal of Clinical Microbiology* 47: 1063–1073.
110. Liao M-H, Lin J-F, Li S-Y. (2012) Application of a multiplex suspension array for rapid and simultaneous identification of clinically important mold pathogens. *Molecular and Cellular Probes* 26: 188–193.
111. Dhiman N, Hall L, Wohlfiel SL, Buckwalter SP, Wengenack NL. (2011) Performance and cost analysis of matrix-assisted laser desorption ionization-time of flight mass spectrometry for routine identification of yeast. *Journal of Clinical Microbiology* 49: 1614–1616.

112. Bader O, Weig M, Taverne-Ghadwal L, Lugert R, Groß U et al. (2010) Improved clinical laboratory identification of human pathogenic yeasts by MALDI-TOF MS. *Clinical Microbiology and Infection* 17: 1359–1365.
113. Shah HN, Keys CJ, Schmid O, Gharbia SE. (2002) Matrix-assisted laser desorption/ionization time-of-flight mass spectrometry and proteomics: A new era in anaerobic microbiology. *Clinical Infectious Diseases* 35: S58–S64.
114. Seng P, Drancourt M, Gouriet F, La Scola B, Fournier PE et al. (2009) Ongoing revolution in bacteriology: Routine identification of bacteria by matrix-assisted laser desorption ionization time-of-flight mass spectrometry. *Clinical Infectious Diseases* 49: 543–551.
115. Hettick JM, Green BJ, Buskirk AD, Kashon ML, Slaven JE et al. (2008) Discrimination of *Penicillium* isolates by matrix-assisted laser desorption/ionization time-of-flight mass spectrometry fingerprinting. *Rapid Communications in Mass Spectrometry* 22: 2555–2560.
116. Santos C, Paterson RRM, Venâncio A, Lima N. (2010) Filamentous fungal characterizations by matrix-assisted laser desorption/ionization time-of-flight mass spectrometry. *Journal of Applied Microbiology* 108: 375–385.
117. Schrödl W, Heydel T, Schwartze VU, Hoffmann K, Grosse-Herrenthey A et al. (2012) Direct analysis and identification of pathogenic *Lichtheimia* species by matrix-assisted laser desorption ionization-time of flight analyzer-mediated mass spectrometry. *Journal of Clinical Microbiology* 50: 419–427.
118. Prod'hom G, Bizzini A, Durussel C, Bille J, Greub G. (2010) Matrix-assisted laser desorption ionization-time of flight mass spectrometry for direct bacterial identification from positive blood culture pellets. *Journal of Clinical Microbiology* 48: 1481–1483.
119. Pongas GN, Lewis RE, Samonis G, Kontoyiannis DP. (2009) Voriconazole-associated zygomycosis: A significant consequence of evolving antifungal prophylaxis and immunosuppression practices? *Clinical Microbiology and Infection* 15(S5): 93–97.
120. Vitale RG, De Hoog GS, Schwarz P, Dannaoui E, Deng S et al. (2012) Antifungal susceptibility and phylogeny of opportunistic members of the order mucorales. *Journal of Clinical Microbiology* 50: 66–75.
121. Dannaoui E. (2002) In vitro susceptibilities of zygomycetes to conventional and new antifungals. *Journal of Antimicrobial Chemotherapy* 51: 45–52.
122. Alastruey-Izquierdo A, Cuesta I, Walther G, Cuenca-Estrella M, Rodriguez-Tudela JL. (2010) Antifungal susceptibility profile of human-pathogenic species of *Lichtheimia*. *Antimicrobial Agents and Chemotherapy* 54: 3058–3060.
123. Spreghini E, Orlando F, Giannini D, Barchiesi F. (2010) In vitro and in vivo activities of posaconazole against zygomycetes with various degrees of susceptibility. *Journal of Antimicrobial Chemotherapy* 65: 2158–2163.
124. Chau AS, Chen G, McNicholas PM, Mann PA. (2006) Molecular basis for enhanced activity of posaconazole against *Absidia corymbifera* and *Rhizopus oryzae*. *Antimicrobial Agents and Chemotherapy* 50: 3917–3919.
125. Spellberg B, Fu Y, Edwards JE Jr, Ibrahim AS. (2005) Combination therapy with amphotericin B lipid complex and caspofungin acetate of disseminated zygomycosis in diabetic ketoacidotic mice. *Antimicrobial Agents and Chemotherapy* 49: 830–832.
126. Thakur M, Revankar SG. (2011) In vitro interaction of caspofungin and immunosuppressives against agents of mucormycosis. *Journal of Antimicrobial Chemotherapy* 66: 2312–2314.
127. Perkhofer S, Locher M, Cuenca-Estrella M, Ru R, Wu R et al. (2008) Posaconazole enhances the activity of amphotericin B against hyphae of zygomycetes in vitro. *Antimicrobial Agents and Chemotherapy* 52: 2636–2638.
128. Symeonidis AS. (2009) The role of iron and iron chelators in zygomycosis. *Clinical Microbiology and Infection* 15(S5): 26–32.
129. Spellberg B, Ibrahim A, Roilides E, Lewis RE, Lortholary O et al. (2012) Combination therapy for mucormycosis: Why, what, and how? *Clinical Infectious Diseases* 54: 73–78.
130. Ibrahim AS, Gebermariam T, Fu Y, Lin L, Husseiny MI et al. (2007) The iron chelator deferasirox protects mice from mucormycosis through iron starvation. *The Journal of Clinical Investigation* 117: 2649–2657.
131. Ibrahim AS, Gebremariam T, French SW, Edwards JE Jr, Spellberg B. (2010) The iron chelator deferasirox enhances liposomal amphotericin B efficacy in treating murine invasive pulmonary aspergillosis. *Journal of Antimicrobial Chemotherapy* 65: 289–292.
132. Ibrahim AS, Gebremariam T, Luo G, Fu Y, French SW et al. (2011) Combination Therapy of murine mucormycosis or aspergillosis with iron chelation, polyenes, and echinocandins. *Antimicrobial Agents and Chemotherapy* 55: 1768–1770.

133. Ogawa T, Takezawa K, Tojima I, Shibayama M, Kouzaki H et al. (2012) Successful treatment of rhino-orbital mucormycosis by a new combination therapy with liposomal amphotericin B and micafungin. *Auris, Nasus, Larynx* 39: 224–228.
134. Reed C, Bryant R, Ibrahim AS, Edwards J Jr, Filler SG et al. (2009) Combination polyene-caspofungin treatment of rhino-orbital-cerebral mucormycosis. *Clinical Infectious Diseases* 47: 364–371.
135. Osteosarcoma W, Roux G. (2010) Successful triple combination therapy of disseminated *Absidia corymbifera* infection in an adolescent. *Journal of Pediatric Hematology/Oncology* 32: 131–133.
136. Spellberg B, Ibrahim AS, Chin-Hong PV, Kontoyiannis DP, Morris MI et al. (2012) The Deferasirox-AmBisome Therapy for Mucormycosis (DEFEAT Mucor) study: A randomized, double-blinded, placebo-controlled trial. *The Journal of Antimicrobial Chemotherapy* 67: 715–722.
137. White TJ, Bruns T, Lee S, Taylor J. (1990) Amplification and direct sequencing of fungal ribosomal RNA genes for phylogenetics. In: Innis MA, Gelfand DH, Sninsky JJ, White T, eds. *PCR Protocols: A Guide to Methods and Applications*. Academic Press, Inc., New York, pp. 315–322.
138. O'Donnell K. (1993) Fusarium and its near relatives. In: Reynolds DR, Taylor J, eds. *The Fungal Holomorph: Mitotic, Meiotic and Pleomorphic Speciation in Fungal Systematics*. CAB International, Wallingford, U.K., p. 225.
139. O'Donnell K, Cigelnik E, Benny G. (1998) Phylogenetic relationships among Harpellales and Kickxellales. *Mycologia* 90: 624.
140. de A. Santiago ALCM, Hoffmann K, Lima DX, de Oliveira RJV, Vieira HEE, Malosso E, Maia LC, da Silva GA. (2014) A new species of Lichtheimia (Mucoromycotina, Mucorales) isolated from Brazilian soil. *Mycological Progress* 13: 343–352.
141. Schwartze VU, de A. Santiago ALCM, Jacobsen ID, Voigt K. (2014) The pathogenic potential of the Lichtheimia genus revisited: Lichtheimia brasiliensis is a novel, non-pathogenic species. *Mycoses* 57(S3): 128–131.

23 Microascus/Scopulariopsis

Sean P. Abbott

CONTENTS

23.1 Introduction 375
    23.1.1 Taxonomy and Biology 375
23.2 Incidence in Water 376
23.3 Incidence in Food 377
    23.3.1 *Microascus* and Cheese 378
23.4 Detection and Identification 379
23.5 Mycotoxins 380
23.6 Mycoses 380
    23.6.1 Onychomycosis and Infection of Keratinous Tissue 380
    23.6.2 Deep-Tissue Infections and Systemic Disease 381
    23.6.3 Diagnosis and Treatment 381
23.7 Conclusions 382
References 382

23.1 INTRODUCTION

23.1.1 TAXONOMY AND BIOLOGY

The genus *Microascus* comprises a group of saprobic ascomycetes in the family Microascaceae and includes the anamorphic (asexual) stage *Scopulariopsis*. *Microascus* and *Scopulariopsis* are found primarily on cellulosic and protein-rich substrata (soil, plant litter, wood, dung, and animal remains) and have a global distribution in tropical, temperate, and polar regions. They are prevalent in human environments and can be important as agents of biodeterioration, indoor contamination, and opportunistic infection.

Microascus was established by Zukal in 1885 and is circumscribed today to comprise about 50 species, including related anamorph taxa [1–10]. These perithecial ascomycetes produce black, smooth (rarely sparsely setose) ascomata with an indistinct to prominent neck (Figure 23.1). The single-celled, light reddish orange, asymmetrical (reniform, lunate, lenticular, or triangular) ascospores (Figure 23.2), released from deliquescent asci, are typically extruded from the perithecial ostiole as a cirrus at maturity (rarely remaining enclosed resembling cleistothecia). The anamorphic stage, *Scopulariopsis*, is a filamentous mold producing dry chains of spores from annellidic conidigenous cells (annellides) on branched (simple to penicillate) conidiophores (Figure 23.3). Conidia can be smooth to verrucose, subglobose to ellipsoidal or apically pointed, and frequently exhibit a broad flattened base where they are produced from the annellide. Most species of *Microascus* are known to produce a *Scopulariopsis* anamorph, but some species described in *Scopulariopsis* are still known only from the asexual state. One *Microascus* species is known to produce a *Wardomyces* anamorph [10]. The type species of anamorph genera *Wardomyces* (darkly pigmented condia with a germ slit produced solitary or in stellate clusters) and *Cephalotrichum* (= *Doratomyces*) (dry chains of annelloconidia produced on synnematous conidiophores) also belong in the Microascaceae and are closely related to *Microascus*, as confirmed by molecular data [1,7]. It is this broad concept of *Microascus* that is adopted here. The complete life cycle, including the teleomorph (sexual)

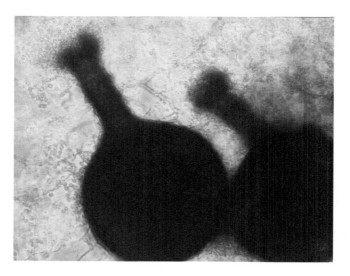

FIGURE 23.1 *Microascus cirrosus*, perithecium.

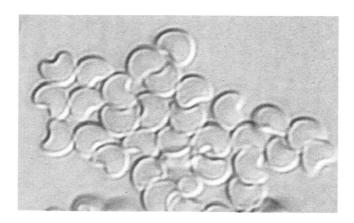

FIGURE 23.2 *Microascus longirostris*, ascospores.

and anamorph (asexual), is referred to as the holomorph [4–6], and each species is referred to by its holomorph name *Microascus* where known or retained in *Scopulariopsis*, *Cephalotrichum*, or *Wardomyces* if the sexual stage is unknown or uncertain. Several species with asexual states referable to as *Scopulariopsis* and *Cephalotrichum* are linked to *Kernia*, another genus of Microascaceae [1,7].

The most common and best-known species is *Scopulariopsis brevicaulis*, described by Saccardo in 1881 as *Penicillium brevicaule*. Despite being a ubiquitous mold found worldwide in soil, plant and animal matter, and air, it was not until over a century later that the *Microascus* sexual stage was discovered and described as *Microascus brevicaulis* [4]. Proteolytic, cellulolytic, and keratinolytic abilities allow *M. brevicaulis* to be an agent of biodeterioration of food and other organic products and as an occasional agent of opportunistic human infection.

23.2 INCIDENCE IN WATER

Microascus and *Scopulariopsis* species are unknown as contaminants of drinking water, and proliferation in environmental water sources has not been reported. However, given the

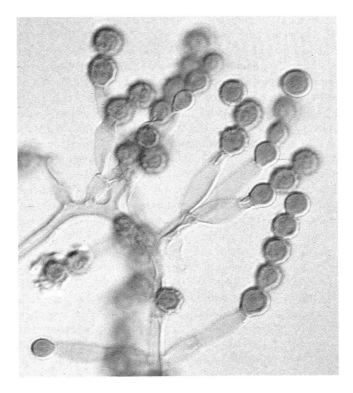

FIGURE 23.3 *Microascus brevicaulis*, conidiophore and conidia.

widespread occurrence of *M. brevicaulis* in air and soil samples, occasional isolation of this or other species from water might be expected at a very low frequency. A potential source of contamination by *Microascus* species is associated with sewage treatment processes and materials. *M. brevicaulis* and *Microascus manginii* have been isolated from sewage sludge, and *M. brevicaulis* has been recorded from polluted water [11]. Another member of the Microascaceae known to cause opportunistic disease, *Pseudallescheria boydii*, has also been repeatedly isolated from sewage and polluted water [11–17]. Ulfig and associates demonstrated that the incidence of these fungi colonizing sewage sludge increased as sewage dried [12,14]. Given the mesophilic nature of *Microascus/Scopulariopsis* species (*M. brevicaulis* minimum water activity for growth is $0.90a_w$) [18], it is unlikely that species contribute to significant biodeterioration in saturated environments.

23.3 INCIDENCE IN FOOD

Microascus and *Scopulariopsis* species are generally of low incidence in food. One common species, *M. brevicaulis*, is isolated as a contaminant on a variety of food sources, and a number of other species have been recorded. Compared with the ubiquity of these species in the environment, they are considered as uncommon in foods [18]. Pitt and Hocking [18] also stated that *Microascus*, the teleomorph of *Scopulariopsis*, does not occur in foods, in reference to the fact that isolates recovered from food are identified by the rapidly growing anamorphic *Scopulariopsis* state and frequently do not produce ascomata under typical isolation conditions using routine isolation media and a short incubation period.

Nevertheless, the proteolytic and cellulolytic abilities of *Microascus* species allow for growth on plant- and animal-based foodstuffs, and their presence in food can be significant. In particular,

Microascus species have an affinity for high protein seeds (e.g., rice, corn, and cereal grains) and animal products (e.g., dairy products, eggs, and meat). Twenty species of *Microascus,* including *Scopulariopsis, Wardomyces,* and *Cephalotrichum,* have been isolated from foodstuffs.

M. brevicaulis (including the anamorph *S. brevicaulis* or its synonym *Scopulariopsis koningii*) is by far the most frequently reported species and has been recorded from a wide variety of foodstuffs, including plant matter: rice and rice flour, barley, wheat, oats, corn, soy beans, meju (Korean fermented soy beans), mung beans, peanuts and groundnuts, pine seeds, sunflower seeds, black pepper, red pepper, caraway, hops, apples and fresh apple juice [4,8,11,18–23]; and from animal matter: nonfat dried milk, butter, cheese, salami, bacon, ham, biltong, eggs, and fishmeal used as poultry feed [8,18,20,21,24]. *M. manginii* (including anamorph *Scopulariopsis candida*) has been reported from milled rice, cereal grains, corn, buckwheat, and cheese [5,7,20,25–27]. *Microascus niger* (including anamorph *Scopulariopsis asperula* or its synonyms *S. fusca* and *Stemonitis roseola*) has been isolated from wheat flour and fermented wheat, cacao beans and leaves, cheese, and chicken eggs [5,8,19,20,26]. *Microascus soppii* (including anamorph synonym *Scopulariopsis flava*) has been reported from cheese, salami, and coffee [6,26,28,29]. *Microascus trigonosporus* has been isolated from milled rice, wheat and cereal grains, sorghum, legume seeds, pecans, and onions [8,9,11,30,31]. *Microascus cinereus* has been reported from wheat flour, oats, corn, peanuts, pepper, and coffee [7,9,22,30,32]. *Microascus cirrosus* has been isolated from milled rice, cereal grains, corn, flax, and cumin [9,30,33]. *Microascus longirostris* has been isolated from corn and pea seeds [9,27]. *Microascus intermedius* (including synonym *Pithoascus intermedius*) is reported from stored corn and blueberry seeds [9,34]. *Microascus schumacheri* has also been isolated from stored corn [9]. *Microascus senegalensis* has been isolated from stored grains in Venezuela [7]. *Microascus caviariformis* was originally described from rotting meat in a cave [35]. The environmentally common species *Scopulariopsis brumptii* has been reported infrequently from food and feed but has been isolated from milled rice in the United States and from hay in Australia [7,8,11]. *Scopulariopsis coprophila,* long known as *Scopulariopsis fimicola* or *white plaster mold* is associated with substrata colonization and poor crop development during commercial mushroom-growing processes [8,36]. *Scopulariopsis gracilis* (as synonym *Paecilomyces fuscatus*) is known from the original description based on a culture isolated from wheat flour in Japan [19,37]. Three species of *Wardomyces* were originally isolated and described from food. *Wardomyces anomalus* was described from rabbit meat in cold storage and is also known from cold-stored oiled eggs [38,39]. *Wardomyces columbinus* was isolated from plum jelly and *Wardomyces simplex* from milled rice [11,40]. *Cephalotrichum* species have been reported only occasionally from food. *Cephalotrichum stemonitis* is recorded from oats, soybeans, broad beans, sugar beet, potatoes, apples, cheese, and frozen meat [8,11].

23.3.1 MICROASCUS AND CHEESE

The occurrence of *Microascus/Scopulariopsis* species in cheese deserves special attention as at least five species have been reported [5,6,8,25,26,41,42]. *M. brevicaulis* has been isolated from moldy cheese in North America and Europe, and appears to be an occasional agent of biodeterioration. It is described as imparting an ammoniacal odor and taste [21,43]. Ropars et al. [26] obtained isolates from cheese production and starter cultures, and identified several species that occur as an integral component of the microflora used to culture these cheeses. However, they did not find any strains of *M. brevicaulis* associated with cheese production, and their findings support the contention that *M. brevicaulis* is present as a common contaminant. Strains used in the production of two French cheeses, Tomme des Pyrénées (cow's milk) and Ossau-Iraty (sheep's milk), were isolated and identified as three species of *Scopulariopsis,* all closely related to *M. brevicaulis.* The nomenclature and identification of this group is complex [4–6,8]. The cheese isolates were assigned to *M. manginii* (as *S. candida*), *M. niger* (as *S. fusca*), and *M. soppii* (as *S. flava*). Molecular data suggests that environmental and cheese isolates examined as *S. flava*, *S. candida*, and *S. fusca* may represent several species [26], and the nomenclature of cheese isolates attributable to *S. flava* warrants further investigation. Abbott et al. [6] recommended that the name *S. flava* be treated as a *nomen dubium* since there is no extant-type material and the name has been variously applied since. They described *M. soppii* after examination of isolates from environmental sources, and documented heterothallism wherein strictly anamorphic strains exhibiting only the *Scopulariopsis* state were mated to produce the teleomorphic stage. The identity of other strains described in the literature as *S. flava* was mixed. A number of isolates were determined to be atypical strains of *M. brevicaulis*, often with pale colony colors, including authentic material of *S. brevicaulis* var. *alba* [6,26,43]. One isolate from cheese was examined and compared to *M. soppii* and *M. brevicaulis*, although the strain was somewhat degenerate, strictly anamorphic and did not produce ascomata in mating trials [6]. Ropars et al. [26] provide molecular evidence that cheese strains identified by them as *S. flava* are clearly distinct from *M. brevicaulis*, *M. manginii*, and *M. niger* and suggest that this taxon may be restricted to cheese. Whether their *S. flava* corresponds to the anamorph of *M. soppii* or another unrecognized species of *Microascus* is uncertain. Another name in the literature, *Scopulariopsis casei*, was described from cheese. Its identity remains unclear, but it is likely based on the anamorph of *M. brevicaulis* or *M. soppii* [6,8]. Morton and Smith [8] additionally report *Scopulariopsis acremonium* from cheese based on older literature reports using other names of uncertain application, but the presence of this species in cheese has not been substantiated by recent isolation reports. *C. stemonitis* has also been reported as a contaminant of cheese [8]. Further molecular work including teleomorphic and strictly anamorphic isolates from cheese and environmental sources could provide a better understanding of the species involved with cheese production and cheese spoilage.

23.4 DETECTION AND IDENTIFICATION

Detection of species of *Microascus/Scopulariopsis* in food has relied on traditional isolation methods [e.g., 18,20], and in general, the direct or dilution-plating methods using general purpose media (e.g., malt extract agar, potato dextrose agar, and rose bengal dichloran agar) are effective for isolation of *M. brevicaulis* and other species. Specialized media may also be employed to enhance recovery or aid in identification of *Microascus* species. Some isolates produce ascocarps on oatmeal agar after extended incubation to aid in species identification [4]. A medium developed for selective isolation of the proteolytic fungi creatine sucrose dichloran agar [20] may be useful when used for primary isolation. Tolerance to benomyl and cycloheximide allows addition of these fungal inhibitors to other media for selective recovery of *Microascus* species [4,7].

Identification of *Microascus* species is often quickly and accurately accomplished using morphological features of colonies and microscopic morphological features of spores and sporulating structures. Typically, the fast-growing anamorph state (*Scopulariopsis*) produces abundant conidia within 1 week, and isolates are identified on this basis. The ubiquitous *M. brevicaulis* can be rapidly screened in food and identified by these traditional means. Some species may require extended incubation on specialized media to produce the sexual stage required for confirmation of species identity.

Use of molecular tools for detection of molds in food does not include specific tests developed to target *Microascus* species. Currently, molecular methods provide supplemental data for positive identification, and sequences obtained from cultures can be compared with other strain data in genetic databases (e.g., BLAST comparison in GenBank). Although further molecular work is required to delimit species and relationships within the Microascaceae, the species most frequently encountered in association with the human environment are readily identified. Culture collections provide an invaluable resource for identification, allowing comparisons with morphological and molecular characters of well-documented isolates. For example, the extensive collection of living strains of Microascaceae preserved at the University of Alberta Microfungus Collection and Herbarium (UAMH, Edmonton, Canada) represents many described species and includes most of the available type cultures. Despite recent advances in understanding the holomorph relationships between the various anamorphic taxa and their teleomorphic stages [1,2,4–7], literature reports continue to use anamorph names and synonyms [e.g., 26,44].

23.5 MYCOTOXINS

Microascus and *Scopulariopsis* species are not known to produce mycotoxins [45]. With the exception of *M. brevicaulis*, most species have likely not been extensively investigated for secondary metabolite production. Species of *Microascus*, especially *M. brevicaulis*, are encountered infrequently to fairly frequently in the human environment. They have been documented growing on water-damaged building materials and isolated from air and dust in mold contaminated environments, as well as in agricultural settings [4,5,7,11,20].

    M. brevicaulis has long been known for its ability to release arsenic compounds into the environment as trimethylarsine gas from methylation of arsenic-containing materials [46–48]. Another metalloid compound, antimony, can also be degraded by *M. brevicaulis* [47–49]. While these chemicals represent a toxicological potential, recent studies have demonstrated low toxicity for trimethylarsine, and the environmental health impact of arsenic degradation by *Microascus* species appears to be negligible [50].

23.6 MYCOSES

Incidence of human infectious disease by *Microascus* species is well documented [51–53], and detailed modern reviews are provided by Issakainen and Liu [2], Iwen et al. [3], and Sandoval-Denis et al. [54]. Ten species of *Microascus* have been implicated in human disease [3,51,52,54,55]. Infections caused by *Microascus* (often reported as species of *Scopulariopsis*) can be categorized as agents of noninvasive onychomycosis of nails and infection of other keratinous tissues, or invasive mycotic disease in deep tissues. While *Microascus* is considered an infrequent, but regular, nondermatophytic agent of onychomycosis, by slight contrast, it is only rarely involved as an agent of opportunistic systemic infection [2,3,53]. The incidence of infection reported and species involved have been increasing with the expansion of immunosuppressive therapies and immunocompromised conditions in the human population [3].

23.6.1 ONYCHOMYCOSIS AND INFECTION OF KERATINOUS TISSUE

Chronic nail infections are the most common clinical conditions associated with *Microascus*. Infections are primarily seen on the nail of the big toe (rarely other nails) as distal or proximal subungual onychomycosis in middle aged to older individuals. Various symptoms of tissue damage include fragility, thickening, discoloration, deformation, and loss of nails. Approximately 1%–6% of all nail infections are due to *Microascus* species [2,53,56]. *M. brevicaulis* (as *S. brevicaulis* and synonym *S. koningii*) is responsible for the majority of reported infections, and is one of the most prevalent nondermatophytic agents of onychomycosis. Much lower incidence is attributable to *M. manginii* (as *S. candida*), *M. niger* (as *S. fusca*), *M. cirrosus*, *M. cinereus*, and *Scopulariopsis brumptii* [2,44,52,57,58]. *Microascsus* infection may be secondary to dermatophytosis or the primary infectious agent. Keratinolytic abilities of *Microascus* species indicate a significant potential for superficial infection of nails and other keratinous materials. Infections of skin and ears by *Microascus* species are also reported and previous debilitating factors such as prolonged skin moisture, or immunosuppression are typically involved [2]. Veterinary cases of hair loss and hyperkeratosis due to *M. brevicaulis* are also known [24,59].

23.6.2 DEEP-TISSUE INFECTIONS AND SYSTEMIC DISEASE

Mycoses of deep tissues and disseminated disease are rare, but serious, opportunistic infections. Deep mycoses are typically associated with immunosuppression, other underlying disease, and/or physical trauma. These infections can be difficult to treat and several fatalities are known. Similar to the situation with onychomycosis, cases of deep mycoses report involvement of a wide variety of species, although there is somewhat less dominance by the ubiquitous *M. brevicaulis* [2,3]. In the respiratory tract, *Microascus* species are known from cases of invasive sinusitis and lung infections (fungus ball, pneumonia). A variety of species are reported as causative agents for respiratory infections, including *M. brevicaulis, M. cinereus, M. cirrosus, M. manginii, M. trigonosporus*, and *Scopulariopsis acremonium* [2,3,55,60–62]. Sandoval-Denis et al. [54] document a variety of *Microascus* and *Scopulariopsis* isolates recovered from respiratory sources such as bronchoalveolar lavage and sputum samples, additionally including *S. brumptii* and *S. gracilis* from respiratory sites. Infections of the central nervous system (CNS abscess, brain lesion, and meningitis) have involved *M. cinereus, S. brumptii* and *S. acremonium* [2,3,63]. Invasive disease in other deep sites includes involvement of internal organs (endocarditis, peritonitis, and kidney infection), eye infection and, most frequently, deep cutaneous infections of other soft tissues. Causal agents of these deep infections include *M. brevicaulis, M. cinereus, M. cirrosus, S. acremonium*, and *S. gracilis* [2,3,54,64–69]. Isolation of *Microascus* from blood is very rare [3,70,71], and the absence of positive culture from blood does not indicate absence of systemic infection. Iwen et al. [3] present a case of fatal disseminated *M. brevicaulis* infection and list 10 other cases of disseminated *Microascus* infections, additionally including *M. cirrosus, S. acremonium*, and *S. brumptii*. Fatal infections have been attributed to *M. brevicaulis, M. cirrosus, M. trigonosporus, S. acremonium*, and *S. gracilis* [3,54,55,62,71].

23.6.3 DIAGNOSIS AND TREATMENT

Diagnosis of *Microascus* infection relies on evidence from histopathology (inflammatory response and direct microscopic examination of fungal elements in tissue) and positive repeat culture for species identification by morphological and genetic techniques. Iwen et al. [3] emphasize the diagnostic methodology adopted by a recent consensus group [72] to meet their criteria of *proven invasive fungal disease* in their review of literature cases involving *Microascus* and *Scopulariopsis*. Traditional micromorphological and physiological characters used for culture identification can be supplemented by molecular techniques, including genomic sequencing of the nuclear ribosomal large subunit, internal transcribed spacer, β-tubulin, and translation elongation factor 1-alpha (tef1) [1,2,26,54]. Currently, there are no commonly used *Microascus*-specific molecular tests to detect infection in tissues. Issakainen and Liu [2] provide details of potential molecular methods that may be developed for purposes of specifically targeting detection of (1) species of Microascaceae at the family level, (2) *Microascus* species, including anamorph taxa *Scopulariopsis, Cephalotrichum*, and *Wardomyces* (genus level), and (3) *M. brevicaulis* (species level).

Rapid detection and identification may be particularly important for this group of fungi because they demonstrate significant resistance to many antifungal drugs and are known for their resilience of infection, with a potentially fatal outcome. Antifungal resistance has been well documented for other species in the Microascaceae, including well-known opportunistic human pathogens in *Pseudallescheria* (including *P. boydii* and anamorph *Scedosporium apiospermum*) and *Petriella* (including anamorph *Scedosporium prolificans*) [51,73–76], and it appears that *M. brevicaulis* and other species of *Microascus* exhibit similar intrinsic resistance [54,58,64,77,78]. Recent studies on *M. brevicaulis* have shown it to be multiresistant to currently available broad-spectrum antifungal compounds and documented high in vitro or in vivo resistance to amphotericin B, flucytosine, the azole drugs (fluconazole, itraconazole, ketoconazole, miconazole, posaconazole, and voriconazole), the echinocandins (anidulafungin, caspofungin, micafungin), terbinafine, mycostatin, and griseofulvin [54,64,77–79]. Cases reporting in vivo administration of antifungals (especially amphotericin B) for *Microascus* infections demonstrate the potential difficulties encountered in providing effective treatment.

23.7 CONCLUSIONS

The saprobic ascomycete genus *Microascus* includes anamorphic taxa in *Scopulariopsis*, *Wardomyces*, and *Cephalotrichum*, as confirmed by cultural and molecular studies. Species demonstrate proteolytic, cellulolytic, and keratinolytic abilities and are known as agents of biodeterioration and human infection. The most commonly encountered species, *M. brevicaulis*, is a ubiquitous mold with a global distribution. Twenty species of *Microascus* have been isolated from a variety of foodstuffs and have a particular affinity for high-protein seeds and animal products. *Microascus* species are not known to produce mycotoxins. A variety of species are known as agents of human disease, including noninvasive infection of keratinous tissue and invasive mycotic disease in deep tissues. *Microascus* species exhibit significant resistance to many antifungal drugs, and fatal infections have been reported. Molecular tools, including genomic sequencing, can aid in rapid detection and identification of *Microascus* species and will help elucidate relationships among species in the family Microascaceae.

REFERENCES

1. Issakainen, J., J. Jalva, J. Hyvönen, N. Sahlberg, T. Pirnes, and C.K. Campbell. 2003. Relationships of *Scopulariopsis* based on LSU rDNA sequences. *Med. Mycol.* 41:31–42.
2. Issakainen, J. and D. Liu. 2011. *Microascus*, including *Scopulariopsis*. In: *Molecular Detection of Human Fungal Pathogens* (Liu, D. ed.). Boca Raton, FL: CRC Press.
3. Iwen, P.C., S.D. Schutte, D.F. Floresco, R.K. Noel-Hurst, and L. Sigler. 2012. Invasive *Scopulariopsis brevicaulis* infection in an immunocompromised patient and review of prior cases caused by *Scopulariopsis* and *Microascus* species. *Med. Mycol.* 50:561–569.
4. Abbott, S.P., L. Sigler, and R.S. Currah. 1998. *Microascus brevicaulis* sp. nov., the teleomorph of *Scopulariopsis brevicaulis*, supports placement of *Scopulariopsis* with the Microascaceae. *Mycologia* 90:297–302.
5. Abbott, S.P. and L. Sigler. 2001. Heterothallism in the Microascaceae demonstrated by three species in the *Scopulariopsis brevicaulis* series. *Mycologia* 93:1211–1220.
6. Abbott, S.P., T.C. Lumley, and L. Sigler. 2002. Use of holomorph characters to delimit *Microascus nidicola* and *M. soppii* sp. nov., with notes on the genus *Pithoascus*. *Mycologia* 94:362–369.
7. Abbott, S.P. 2000. Holomorph studies of the Microascaceae. PhD dissertation, University of Alberta, Edmonton, Alberta, Canada.
8. Morton, F.J. and G. Smith. 1963. The genera *Scopulariopsis* Bainer, *Microascus* Zukal, and *Doratomyces* Corda. *Mycol. Pap.* 86:1–96.
9. Barron, G.L., R.F. Cain, and J.C. Gillman. 1961. The genus *Microascus*. *Can. J. Bot.* 39:1609–1631.
10. Malloch, D. 1970. New concepts in the Microascaceae illustrated by two new species. *Mycologia* 62:727–740.
11. Domsch, K.H., W. Gams, and T.-H. Anderson. 1993. *Compendium of Soil Fungi*, 2nd edn. London, U.K.: Academic Press.
12. Ulfig, K., G. Plaza, M. Terakowskip, and K. Janda-Ulfig. 2006. Sewage sludge open air drying affects on keratinolytic, keratinophilic and actidione-resistant fungi. *Rocz. Panstw. Zakl. Hig.* 57:371–379.
13. Ulfig, K., G. Plaza, and K. Janda-Ulfig. 2007. The growth of proteolytic microorganisms affects ketatinophylic fungi in sewage sludge. *Rocz. Panstw. Zakl. Hig.* 58:471–479.
14. Ulfig, K., G. Plaza, A. Markowska-Szczupak, K. Janda, and S. Kirkowska. 2010. Keratinolytic and nonkeratinolytic fungi in sewage sludge. *Polish J. Environ. Stud.* 19:635–642.
15. Muhsin, T.M. and R.B. Hadi. June 2011. Incidence of keratinlytic and dermatophytic fungi in sewage sludge in Basrah, Iraq. *J. Basrah Res. (Sci.)* 37(3A):1–9.
16. Musial, C.E., F.R. Cockerill, and G.D. Roberts. 1988. Fungal infections of the immunocompromised host: Clinical and laboratory aspects. *Clin. Microbiol. Rev.* 1:349–364.

17. Cooke, W.B. and P. Kabler. 1955. Isolation of potentially pathogenic fungi from polluted water and sewage. *Public Health Rep.* 70:689–694.
18. Pitt, J.I. and A.D. Hocking. 1999. *Fungi and Food Spoilage*, 2nd edn. Gaithersburg, MD: Aspen Publishers.
19. Inagaki, N. 1962. On some fungi isolated from food. *Trans. Mycol. Soc. Jpn.* 4:1–5.
20. Samson, R.A., E.S. Hoekstra, J.C. Frisvad, and O. Filtenborg (eds.). 2000. *Introduction to Food- and Airborne Fungi*, 6th edn. Utrecht, the Netherlands: Centraalbureau voor Schimmelcultures.
21. Bothast, R.J., E.B. Lancaster, and C.W. Hesseltine. 1975. *Scopulariopsis brevicaulis*: Effect of pH and substrate on growth. *Eur. J. Appl. Microbiol.* 1:55–66.
22. Alwakeel, S.S. and L.A. Nasser. 2011. Microbial contamination and mycotoxins from nuts in Riyadh, Saudi Arabia. *Am. J. Food Technol.* 6:613–630.
23. Peter, K.V. (ed.). 2006. *Handbook of Herbs and Spices*, Volume 3. Cambridge, U.K.: Woodhead Publishing Limited.
24. Miljkovic, B., Z. Pavlovski, D. Jovicic, O. Radanovic, and B. Kureljusic. 2011. Fungi on feathers of common clinically healthy birds in Belgrade. *Biotechnol. Animal Husbandry* 27:45–54.
25. Filtenborg, O., J.C. Frisvad, and R.A. Samson. 2000. Specific association of fungi to foods and influence of physical environmental factors. In: *Introduction to Food- and Airborne Fungi*, 6th edn. (Samson et al. eds.). Utrecht, the Netherlands: Centraalbureau voor Schimmelcultures.
26. Ropars, J., C. Cruaud, S. Lacoste, and J. Dupont. 2012. A taxonomic and ecological overview of cheese fungi. *Int. J. Food Microbiol.* 155:199–210.
27. Udagawa, S. 1963. Microascaceae in Japan. *J. Gen. Appl. Microbiol.* 9:137–148.
28. Food and Agriculture Organization of the United Nations (FAO). 2005. Geographical distribution of fungi in coffee. http://www.fao.org/fileadmin/user_upload/agns/pdf/coffee/Annex-C.2.xls (accessed January 1, 2013).
29. Iacumin, L., L. Chiesa, D. Boscolo et al. 2009. Moulds and ochratoxin A of surfaces of artisanal and industrial dry sausages. *Food Microbiol.* 26:65–70.
30. Udagawa, S. 1962. *Microascus* species new to the mycoflora of Japan. *J. Gen. Appl. Microbiol.* 8:39–51.
31. Thaung, M.M. 2009. Additions (and annotations) to fungi of Burma. *Aust. Mycol.* 28:68–69.
32. Freire, F.C., Z. Kozakiewicz, and R.R. Paterson. 2000. Mycoflora and mycotoxins in Brazilian black pepper, white pepper and Brazil nuts. *Mycopathologia* 149:13–19.
33. Bhatnagar, K., K. Mather, and B.S. Sharma. 1994. *Microascus cirrosus* on cumin—New host record. *Indian Phytopathol.* 47:214.
34. Gourley, C.O. and L. Nickerson. 1979. *Pithoascus intermedius* from seeds of *Vaccinium angustifolium*. *Can. J. Bot.* 57:1218–1219.
35. Malloch, D. and J.-M. Hubart. 1987. An undescribed species of *Microascus* from the cave of Ramioul. *Can. J. Bot.* 65:2384–2388.
36. Charles, V.K. and E.B. Lambert. 1933. Plaster molds occurring in beds of the cultivated mushroom. *J. Ag. Res.* 46:1089–1098.
37. Samson, R.A. and A. von Klopotek. 1972. *Scopulariopsis murina*, a new fungus from self-heated compost. *Arch. Mikrobiol.* 85:175–180.
38. Brooks, F.T. and C.G. Hansford. 1923. Mould growths upon cold-store meat. *Trans. Brit. Mycol. Soc.* 8:113–142.
39. Hennebert, G.L. 1962. *Wardomyces* and *Asteromyces*. *Can. J. Bot.* 40:1203–1216.
40. Hennebert, G.L. 1968. Echinobotryum, Wardomyces and Mammaria. *Trans. Brit. Mycol. Soc.* 51:749–762.
41. Bourdichon, F., S. Casaregola, C. Farrokh et al. 2012. Food fermentations: Microorganisms with technological beneficial use. *Int. J. Food Microbiol.* 154:87–97.
42. Moreau, C. 1979. Nomenclature des *Penicillium* utiles à la preparation du Camembert. *Le Lait* 59:219–233.
43. Thom, C. 1930. *The Penicillia*. Baltimore, MD: The Williams & Wilkins Company.
44. Lee, M.H., S.M. Hwang, M.K. Suh, G.Y. Ha, H. Kim, and J.Y. Park. 2012. Onychomycosis caused by *Scopulariopsis brevicaulis*: Report of two cases. *Ann. Dermatol.* 24:209–213.
45. Frisvad, J. and U. Thrane. 2000. Mycotoxin production by common filamentous fungi. In: *Introduction to Food- and Airborne Fungi*, 6th edn. (Samson et al. eds.). Utrecht, the Netherlands: Centraalbureau voor Schimmelcultures.
46. Challenger, F. and C. Higginbottom. 1935. The production of trimethylarsine by *Penicillium brevicaule* (*Scopulariopsis brevicaulis*). *Biochem. J.* 29:1757–1778.

47. Andrewes, P., W.R. Cullen, and E. Polishchuk. 2000. Arsenic and antimony biomethylation by *Scopulariopsis brevicaulis*: Interaction of arsenic and antimony compounds. *Environ. Sci. Technol.* 34:2249–2253.
48. Bentley, R. and T.G. Chasteen. 2002. Microbial methylation of metalloids: Arsenic, antimony, and bismuth. *Microbiol. Mol. Biol. Rev.* 66:250–271.
49. Jenkins, R.O., P.J. Craig, W. Goessler, D. Miller, N. Ostah, and K.J. Irgolic. 1998. Biomyethylation of antimony compounds by an aerobic fungus: *Scopulariopsis brevicaulis*. *Environ. Sci. Technol.* 32:882–885.
50. Cullen, W.R. and R. Bentley. 2005. The toxicity of trimethylarsine: An urban myth. *J. Environ. Monit.* 7:11–15.
51. de Hoog, G., J.G. Guarro, and M.J. Figueras (eds.). 2000. *Atlas of Clinical Fungi*, 2nd edn. Utrecht, the Netherlands: Centraalbureau voor Schimmelcultures.
52. Kane J., R.C. Summerbell, L. Sigler, S. Krajden, and G. Land (eds.). 1997. *Laboratory Handbook of Dermatophytes. A Clinical Guide and Laboratory Manual of Dermatophytes and Other Filamentous Fungi from Skin, Hair, and Nails*. Belmont, CA: Star Publishing Co.
53. Summerbell, R.C., J. Kane, and S. Krajden. 1989. Onychohmycosis, tinea pedis and tinea manuum caused by non-dermatophytic filamentous fungi. *Mycoses* 32:609–619.
54. Sandoval-Denis, M., D.A. Sutton, A.W. Fothergill et al. 2013. *Scopulariopsis*, a poorly known opportunistic fungus: Spectrum of species in clinical samples and *in vitro* responses to antifungal drugs. *J. Clin. Microbiol.* 51:3937–3943.
55. Mohammedi, I., M.A. Piens. C. Audigier-Valette et al. 2004. Fatal *Microascus trigonosporus* (anamorph *Scopulariopsis*) pneumonia in a bone marrow transplant recipient. *Eur. J. Clin. Microbiol. Infect. Dis.* 23:215–217.
56. Issakainen, J., H. Heikkila, E. Vainio et al. 2007. Occurrence of *Scopulariopsis* and *Scedosporium* in nails and keratinous skin. Finland. A 5-year retrospective multi-center study. *Med. Mycol.* 45:201–209.
57. De Vroey, C., A. Lasagni, E. Tosi, F. Schroeder, and M. Song. 1992. Onychomycoses due to *Microascus cirrosus* (syn. *M. desmosporus*). *Mycoses* 35:193–196.
58. Naidu, J., S.M. Singh, and M. Pouranik. 1991. Onychomycosis caused by *Scopulariopsis brumptii*: A case report and sensitivity studies. *Mycopathologia* 113:159–164.
59. Ogawa, S.S., T. Shibahara, A. Sano, K. Kadota, and M. Kubo. 2008. Generalized hyperkeratosis caused by *Scopulariopsis brevicaulis* in a Japanese Black calf. *J. Comp. Pathol.* 138:145–150.
60. Wheat, L.J., M. Bartlett, M. Ciccarelli, and J.W. Smith. 1979. Opportunistic *Scopulariopsis* pneumonia in an immunocompromised host. *South. Med. J.* 77:1608–1609.
61. Aznar, G., C. de Bievre, and C. Guiguen. 1989. Maxillary sinusitis from *Microascus cinereus* and *Aspergillus repens*. *Mycopathologia* 105:93–97.
62. Beltrame, A., L. Sarmati, L. Cudillo et al. 2009. A fatal case of invasive sinusitis by *Scopulariopsis acremonium* in a bone marrow transplant recipient. *Int. J. Infect. Dis.* 13:488–492.
63. Baddley, J.W., S.A. Moser, D.A. Sutton, and P.G. Pappas. 2000. *Microascus cinereus* (anamorph *Scopulariopsis*) brain abscess in a bone marrow transplant recipient. *J. Clin. Microbiol.* 38:395–397.
64. Sekhon, A.S., D.J. Willans, and J.H. Harvey. 1974. Deep scopulariopsosis: A case report and sensitivity studies. *J. Clin. Pathol.* 27:837–843.
65. Phillips, P., W.S. Wood, G. Phillips, and M.G. Rinaldi. 1989. Invasive hyalohyphomycosis caused by *Scopulariopsis brevicaulis* in a patient undergoing allogeneic bone marrow transplant. *Diagn. Microbiol. Infect. Dis.* 12:429–432.
66. Krisher, K.A., N.B. Holdridge, M.M. Mustafa, M.G. Rinaldi, and D.A. McGough. 1995. Disseminated *Microascus cirrosus* infection in pediatric bone marrow transplant patient. *J. Clin. Microbiol.* 33:735–737.
67. Marques, A.R., K.J. Kwon-Chung, S.M. Holland, M.L. Turner, and J.I. Gallin. 1995. Suppurative cutaneous granulomata caused by *Microascus cinereus* in a patient with chronic granulomatous disease. *Clin. Infect. Dis.* 20:110–114.
68. Anandan, V., V. Nayak, S. Sundaram, and P. Srikanth. 2008. An association of *Alternaria alternata* and *Scopulariopsis brevicaulis* in cutaneous phaeohyphomycosis. *Indian J. Dermatol. Venereol. Leprol.* 74:244–247.
69. Issakainen, J., J.H. Salonen, V.J. Anttila et al. 2010. Deep, respiratory tract and ear infections caused by *Pseudallescheria* (*Scedosporium*) and *Microascus* (*Scopulariopsis*) in Finland. A 10-year multi-center study. *Med. Mycol.* 48:458–465.
70. Neglia, J.P., D.D. Hurd, P. Ferrieri, and D.C. Snover. 1987. Invasive *Scopulariopsis* in the immunocompromised host. *Am. J. Med.* 83:1163–1166.

71. Miossec, C., F. Morio, T. Lepoivre et al. 2011. Fatal invasive infection with fungemia due to *Microascus cirrosus* after heart and lung transplantation in a patient with cystic fibrosis. *J. Clin. Microbiol.* 49:2743–2747.
72. De Pauw, B., T.J. Walsh, J.P. Donnelly et al. 2008. Revised definitions of invasive fungal disease from the European Organization for Research and Treatment of Cancer/Invasive Fungal Infections Cooperative Group and the National Institute of Allergy and Infectious Diseases Mycoses Study Group (EORTC/MSG) Consensus Group. *Clin. Infect. Dis.* 46:1813–1821.
73. Salkin, I.F., M.R. McGinnis, M.J. Dykstra, and M.G. Rinaldi. 1988. *Scedosporium inflatum*, an emerging pathogen. *J. Clin. Microbiol.* 26:498–503.
74. Issakainen, J., J. Jalava, E. Eerola, and C.K. Campbell. 1997. Relatedness of *Pseudallescheria*, *Scedosporium* and *Graphium pro parte* based on SSU rDNA sequences. *J. Med. Vet. Mycol.* 35:389–398.
75. Guarro, J., A.S. Kantarcioglu, R. Horré et al. 2006. *Scedosporium apiospermum*: Changing clinical spectrum of a therapy-refractory opportunist. *Med. Mycol.* 44:295–327.
76. Cortez, K.J., E. Roilides, F. Quiroz-Telles et al. 2008. Infections caused by *Scedosporium* species. *Clin. Microbiol. Rev.* 21:157–197.
77. Cuenca-Estrella, M., A. Gomez-Lopez, E. Mellado, M.J. Buitrago, A. Monzon, and J.L. Rodriguez-Tudela. 2003. *Scopulariopsis brevicaulis*, a fungal pathogen resistant to broad-spectrum antifungal agents. *Antimicrob. Agents Chemother.* 47:2339–2347.
78. Aguilar, C., I. Pujol, and J. Guarro. 1999. in vitro antifungal susceptibilities of *Scopulariopsis* isolates. *Antimicrob. Agents Chemother.* 43:1520–1522.
79. Gupta, A.K. and T. Gregurek-Novak. 2001. Efficacy of itraconazole, terbinafine, fluconazole, griseofulvin and ketoconazole in the treatment of *Scopulariopsis brevicaulis* causing onychomycosis of the toes. *Dermatology* 202:235–238.

24 Mucormycosis

Luis Zaror, Patricio Godoy-Martínez, and Eduardo Álvarez

CONTENTS

24.1 Introduction..387
24.2 Taxonomic Position and Characteristics of the Genus 388
24.3 Most Common Clinical *Mucor* Species..389
24.4 Mechanisms of Pathogenicity and Risk Factors 391
24.5 Epidemiology...392
24.6 Sites and Patterns of Infection 393
24.7 Mycological Diagnosis...393
24.8 Molecular Identification...394
24.9 Therapy and Antifungals 394
24.10 Combined Therapy 395
24.11 *Mucor* in Foods and Industry 395
24.12 Conclusions 395
References..396

24.1 INTRODUCTION

Fungi pertaining to *Mucor* are classified into one of the most variable groups, which is the subdivision Mucoromycotina. This replaced the widely known division Zygomycota, which is now considered obsolete for not constituting a monophyletic group and was recently substituted in accordance with Hibbett et al.[1] The taxa that were located before in Zygomycota are included in the new divisions Glomeromycota (without clinical interest) and Entomophthoromycota and various other subdivisions of uncertain position such as Mucoromycotina, Kickxellomycotina, and Zoopagomycotina[1] (Figure 24.1). In this sense, the subdivision Mucoromycotina includes the orders Mucorales, Endogonales, and Mortierellales. In Entomophthoromycota,[2] several genera are found of clinical interest such as *Basidiobolus* and *Conidiobolus*, which are extremely deformative, limited to the tropics, and associated with chronic cutaneous or subcutaneous mycoses. Other genera located in Entomophthoromycota are, for example, *Batkoa*, *Entomophthora*, and *Zoophthora*, which have been isolated from insects and other soil invertebrates.

Mucoromycotina, also called Mucorales or mucormycetes, have common coenocytic, nontabicated hyphae (diameter 2–20 μm) that produce long sporangiophores, at the ends of which are columellae that are totally or partially covered by sporangia or similar (sporangioles or merosporangia). In these, a mitosporic process occurs that produces asexually numerous sporangiospores, whereas zygospores are produced sexually.

These fungi contain ubiquitous, saprotrophic, and cosmopolitan species found on the ground, in water, and on vegetation.

The following genera are found in the Mucorales, *Rhizopus*, *Rhizomucor*, *Mucor*, *Lichtheimia* (formerly *Absidia*), *Apophysomyces*, *Cunninghamella*, and *Saksenaea*, that produce the following types of mycosis: rhinocerebral, pulmonary, cutaneous, gastrointestinal, angioinvasive mycosis, and other less frequent types, or they infect immunocompromised or predisposed patients.

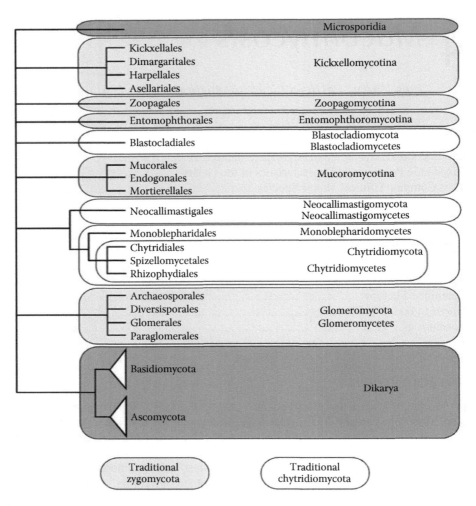

FIGURE 24.1 Taxonomic proposal for kingdom fungi.[1] Zygomycota disappears in order to accept new divisions such as Glomeromycota and the subphyla with uncertain position such as Mucoromycotina, Entomophthoromycotina (recently described as phylum Entomophthoromycota), Kickxellomycotina, and Zoopagomycotina.

The following key permits the identification of Mucoromycotina with clinical interest to the genus level (Table 24.1)

24.2 TAXONOMIC POSITION AND CHARACTERISTICS OF THE GENUS

The genus *Mucor* constitutes one of the largest groups in the subdivision Mucoromycotina, encompassing approximately 52 taxonomically valid species. These fungi are of rapid growth, saprotrophic, and ubiquitous. Some species are known to cause infections in humans and animals.[3–5] *Mucor* species have been isolated from food raw materials and the final processed foods. There are no mycotoxin-producing species. However, they are responsible for the deterioration of foods, fruits, and vegetables.[6,7] Additionally, these microorganisms are the cause of a series of alterations of the organoleptic properties in foods, given that they are capable of producing different enzymes such as lipase or protease.[6,8,9] Also, many *Mucor* species are used in the fermentation process of international culinary products, especially those of Oriental origin such as sufu, ragi, tempeh, and mureha.[10,11]

TABLE 24.1
Synoptic Identification Key at the Genus Level of Mucoromycotina

1. Abundant mycelium, large quantity of asexual spores that form sporangia and are released passively (Mucorales)	2
1'. Dense, waxy, and low-growing mycelium, single one-celled asexual spores formed in sporangiophores within mycelium, actively released (Entomophthorales)	9
2. Sporangia with only one or few spores (sporangiola), forming in swollen areas of the apex of the sporangiophore or the branches	3
2'. Sporangium with a large number of spores that form on the apex of the sporangiophore or on one of the branches	4
3. Spores in a single-spored sporangium	*Cunninghamella*
3'. Spores in rows in cylindrical sporangia (merosporangia)	*Syncephalastrum*
4. Sporangia with apophysis (swelling at the base of the sporangium)	5
4'. Sporangia without apophysis	7
5. Sporangia with an elongated upper part, sporangiophores generally simple	*Saksenaea*
5'. Spherical, subspherical, or pyriform sporangia, sporangiophores in whorls	6
6. Spherical sporangia, spores with surface striation usually very dark	*Rhizopus*
6'. Pyriform sporangia, smooth spores, pale	*Absidia/Lichtheimia/Lentamyces*
7. Presence of columella	8
7'. Absence of columella	*Mortierella*
8. Rhizoids present, thermophilic	*Rhizomucor*
8'. Rhizoids absent	*Mucor*
9. Mycelium colonies, with hyphae that can be septate, elongated sporangiophores, rounded spores on the apex	*Conidiobolus*
9'. Yeast colonies, short sporangiophores, conical spores	*Basidiobolus*

Only a small number of species grow at 37°C and, for that reason, are capable of producing deep infections in humans. In this sense, five species are historically reported from clinical samples: *Mucor circinelloides*, *Mucor indicus*, *Mucor ramosissimus*, and less frequently, *Mucor hiemalis* and *Mucor racemosus*.[5] It is important to add the recently described *Mucor ellipsoideus* and *Mucor velutinosus*[12] and the relocated *Mucor irregularis* (*Rhizomucor variabilis* var. *variabilis*)[12] that produce a series of infections.[13–16]

Mucor grow well on Sabouraud agar, malt extract agar (MEA), lactrimel agar, potato agar, and others. The colony in 2–5 days can occupy all of the Petri dish. Hyaline or colored nonseptate (coenocytic) hyphae are observed, but septa are observed in older cultures. The species are characterized by sporangiophores, which can be branched or unbranched, and nonapophysated. They have spherical, subspherical, or pyriform sporangia, with smooth or lightly ornamented sporangiospores. They do not produce rhizoids and in general reproduce asexually by the production of internal asexual spores contained in the sporangia. Rarely chlamydospores are produced, especially in aged cultures; however, their production is characteristic of some species.

Zygospore formation occurs when hyphae of different mating sexuality are attracted to one another forming the progametangia. These fuse to produce a gametangium, which undergoes plasmogamy and karyogamy. The wall becomes thickened and forms the zygosporangium. The zygospores can be heterothallic or homothallic.

24.3 MOST COMMON CLINICAL *MUCOR* SPECIES

M. circinelloides: This is the most frequently isolated species from clinical samples.[12] Also, it is isolated from bovine, pork, turkey, and platypus mycoses. Occasionally, it is found in diabetic humans with ketoacidosis and is described as producing subcutaneous mycosis.[17,18] Colonies of

M. circinelloides are usually grey and approximately 6 mm high. Long, branched, and sometimes circinate sporangiophores are observed. Sporangia can reach a length of 80 µm, with the columella being spherical or subspherical, of about 55 × 50 µm; ellipsoidal sporangiospores are 4.4–6.8 × 3.7–4.7 µm. At 37°C, growth is slow and the optimum temperature is between 30°C and 37°C. Xiaoli et al.[19] confirmed the first case of mucormycosis by *M. circinelloides* in catfish in China by molecular and morphological identification. This infectivity study showed 100% mortality.

M. indicus: *M. indicus* has been described in a case of human gastritis and in one of pulmonary mycosis.[20] The colonies are clear yellow, 10 mm high, and aromatic. The sporangiophores show sympodial branching, with long branches, and sporangia up to 75 µm diameter. Columellae are subglobose and 56 × 53 µm: sporangiospores are subglobose to cylindrical–ellipsoidal and 5.4–5.7 × 4.4–4.8 µm. *M. indicus* (1) grows and sporulates between 25°C and 37°C, (2) grows but does not sporulate at 40°C, and (3) does not grow at 45°C.

M. hiemalis: *M. hiemalis* has been described in two cases of cutaneous mycosis.[21] It presents as light yellow or grayish, with a height of 15 mm. The sporangiophores are branching and sometimes a yellow pigmentation on the whole sporangiophore can be observed. The sporangium measures 70 µm in diameter, with an ellipsoidal columella of 38 × 30 µm; the sporangiospores are usually ellipsoidal and sometimes have a flat side, measuring 5.7–8.7 × 2.7–5.4 µm. Good growth at 25°C and no growth at 30°C are shown. Given the growth temperatures, its virulence appears to be limited to cutaneous mycoses and not systemic.

M. racemosus: The pathogenicity of this species is dubious. Colonies are grey, reaching 20 mm in height. Sporangiophores are simple or branched. The sporangia are up to 90 µm in diameter. Columellae are subspheric, ellipsoidal, or cylindrical–ellipsoidal, slightly pyriform, from 55 × 377 µm; the sporangiospores are ellipsoidal or subglobose at 5.5–8.5 × 4–7 µm: chlamydospores are frequently observed. The growth temperature is up to 35°C and there is poor growth at 37°C.

M. ramosissimus: *M. ramosissimus* has been described as producing chronic cutaneous mycosis, with tissue destruction to the face and cutaneous lesions. Rhinocerebral mycosis and septic arthritis in a newborn child are recorded.[12,22] The colonies are yellow grey and lightly aromatic, with a height of 2 mm. The sporangiophores are branched, and the sporangia are globose with diameters of 75 µm. Columellae reach 40 × 50 µm, and sporangiospores are subglobose to ellipsoidal at 5–8 × 4–7 µm. Growth is very slow at 36°C and is quicker between 30°C and 35°C.

M. ellipsoideus: This species was isolated from a patient with peritoneal dialysis. The colonies were white, 2–5 mm in height, and occupy the whole Petri dish after 9 days of incubation. The sporangiophores are simple or branched. The sporangia are subglobose with a diameter of 16–48 × 18–50 µm. The columellae are globose to conical, with a diameter of 10–45 µm. Sporangiospores largely ellipsoidal, sometimes flat on one side, measuring 4–8 × 2–3 µm, hyaline, and smooth walled. The chlamydospores are abundant, oidia-like, terminal or intercalar, and solitary or in large chains. The fungus presents an optimum growth temperature of 25°C and grows at 37°C. At 42°C, no growth is observed.

M. irregularis: *M. irregularis* was described for the first time from a chronic cutaneous lesion from a patient in China.[15] Recently, cases were again reported from China and the United States.[16] It was first described as a species of *Rhizomucor*; however, due to the genotype and phenotype characters, it was relocated in *Mucor*.[12] The colonies are grayish-yellow on MEA and 6–10 mm in height. The sporangiophores can be simple or branched. The sporangium is subspherical, approximately 100 µm in diameter; columellae are spherical to subspherical and 40 µm in diameter. Hyaline sporangiospores are smooth and variable in shape and 2–7 × 3–11 µm. Chlamydospores are produced. The formation of zygospores has not been described and it grows until 30°C, with an optimum of 25°C.

M. velutinosus: *M. velutinosus* was isolated from hemocultures and the skin biopsy from a patient with acute myelogenous leukemia.[13] It presents velvet-like grey colonies, approximately 0.5–2 mm in height in MEA. The sporangiophores show sympodial branching, and sporangia are globose or

TABLE 24.2
Key to Distinguish the Most Relevant Clinical Species of *Mucor*

1. Sporangiophores unbranched or weakly branched	2
1′. Sporangiophores repeatedly branched	3
2. Sporangiospores spherical, 3.4–5.4 μm in diameter	*Mucor amphibiorum*
2′. Sporangiospores ellipsoidal, 5.7–8.7 × 2.7–5.4 μm	*Mucor hiemalis*
3. Growth at 40°C	*M. indicus*
3′. No growth at 40°C	4
4. Sporangiospores globose or nearly so	5
4′. Sporangiospores otherwise	8
5. Sporangiospores verrucose	6
5′. Sporangiospores smooth walled	7
6. Grows at 37°C: sporangiospores coarsely ornamented	*M. velutinosus*
6′. No growth at 37°C: sporangiospores finely ornamented	*Mucor plumbeus*
7. Grows at 35°C: sporangiophores with several swellings; sporangia on short lateral branches; sporangiospores 5.0–8.0 × 4.5–6.0 μm	*M. ramosissimus*
7′. No growth at 35°C: sporangiophores lack swellings; sporangia on long lateral branches; sporangiospores 5.5–10.0 × 4.0–7.0 μm	*M. racemosus*
8. Sporangiospores ellipsoidal, 4.4–6.8 × 3.7–4.7 μm	*M. circinelloides*
8′. Sporangiospores very variable in shape	9
9. Columellae irregular in shape; sporangiospores 2.5–16.5 × 2.0–7.0 μm	*M. irregularis*
9′. Columellae globose; sporangiospores 5.5–17.6 × 3.4–12.8 μm	*Mucor lusitanicus*

Source: Alvarez, E. et al., *Med. Mycol.*, 49(1), 62, 2011.

subglobose, approximately 15–60 μm in diameter. Columellae are globose to conical, 10–50 μm in diameter. The sporangiospores are globose, subglobose, ovoid, or irregular with thick walls and lightly verruculose, measuring 4–15 μm in diameter. The fungus presents terminal or intercalary chlamydospores. *M. velutinosus* grows up to 37°C (Table 24.2).

24.4 MECHANISMS OF PATHOGENICITY AND RISK FACTORS

Mucormycosis is the third leading cause of invasive mycosis after aspergillosis and candidiasis. This represents between 8.3% and 13% of all the cases of mycosis found in autopsies of patients with hematological diseases.[23–25] Mycosis is acquired by the inhalation of spores, by ingestion of contaminated food, or by exposed cutaneous lesion expressed in immunocompetent individuals.

Why do fungi such as *Mucor* express pathogenicity in immunocompromised individuals and not in those who are immunocompetent? Mucorales produce various proteins and metabolites that are toxic to humans and animals. The capacity to capture iron in the plasma and tissues is considered an important factor in pathogenesis. Boelaert et al.[26,27] demonstrated that Mucorales use desferrioxamine (DFO) as a xenosiderophore by which it captures iron. This was confirmed by Ibrahim et al.[28] in a study carried out on the FTR1 gene (high-affinity iron permease) in *Rhizopus oryzae* and its products. When these investigators were able to silence the gene, *R. oryzae* was unable to capture iron in vitro and showed a low virulence when inoculated in rats. On the other hand, anti-FTR1 antibodies protected the rats from infection. For this reason, an important protective factor against mucormycosis is a low concentration of iron in the plasma. Iron is essential for the Mucorales in order to boost development and hyphae growth in vitro or increasing its pathogenicity.[26] It is relevant that patients in hemodialysis under treatment with DFO present an elevated risk for the development of mucormycosis. Boelaert et al.[29] reported an increased mortality (89%) in 46 patients that developed mucormycosis during therapy with DFO.

Angioinvasion and neurotropism are essential factors in the dissemination of mucormycosis, with the subsequent infarction of annexed tissues. Another important strategy of the Mucorales is adhesion to endothelial cells as they grow in cells.[30] Liu et al.[31] identified a glucose-regulated protein 78 (GRP78) as a novel host receptor that mediates invasion and damage of human endothelial cells by *R. oryzae*. When an elevated concentration of glucose and iron is observed, as seen in diabetics, the expression of GRP78 is incremented with an increase in the endothelial damage in diabetic rats.

Frater et al.[32] in their evaluation of histopathological cuts from 20 patients with invasive mycosis found a high percentage of perineural invasion. Chamilos et al.[33] showed a possible endosymbiotic relationship with a toxin-producing bacteria. However, these researchers showed afterward that this relationship was unrelated to the pathogenesis of Mucorales. Other potential virulence factors could be proteolytic, lipolytic, and glycosidic enzymes and secondary metabolites. However, their direct involvement in human cases of mucormycosis is unrecorded.[34]

Transplant patients with immunosuppression, antirejection therapy, and graft dysfunction allow the growth of Mucorales spores or propagules for the development of clinical manifestation. The use of some antimycotics, especially azoles, may contribute to the appearance of mucormycosis in patients that receive hematopoietic transplant.[35]

Risk factors for invasive diseases include patients with unbalanced diabetes mellitus, oncohematological patients, organ and bone marrow transplant patients, patients receiving DFO therapy or therapy with corticoids, or patients with other conditions that induce immunosuppression.[30] The limited activity of some of the principal antifungal drugs (echinocandins and azole derivatives), vascular invasion and neurotrophic activity, could explain the elevated mortality observed in mucormycosis.[36,37]

Prolonged neutropenia is the principal risk factor for the development of this disease. Therapeutic interventions (i.e., corticosteroid therapy) cause functional defects in macrophages and neutrophils and represent an additional factor for mucormycosis. Diabetes, itself, causes damage to the neutrophil function, contributing to the seriousness of mucormycosis in patients with ketoacidosis.[38,39]

Recently, the use of voriconazole was related to confirmed cases of mucormycosis in different institutions around the world. This is primarily from its use in aspergillosis in high-risk patients with hematological diseases or bone marrow transplants. The use of voriconazole as prophylaxis, or for the treatment of aspergillosis in immunocompromised patients, could be considered a risk for the acquisition of emerging fungal pathogens that are intrinsically resistant to azoles, as is the case for mucormycosis.[35,40–47]

24.5 EPIDEMIOLOGY

Mucormycosis have different clinical presentations related to the predisposing factors. The rhino-orbital-cerebral form is the most common, representing between 44% and 49% of the cases. Cutaneous presentation has 10%–16%, pulmonary 10%–11%, disseminated 6%–11%, and gastrointestinal 2%–11% of the cases. In oncohematological patients, the relationship changes; the most common manifestation is pulmonary, followed by the disseminated form and the rhino-orbital-cerebral. In diabetics, the principal presentation is the rhino-orbital-cerebral and pulmonary. Finally, mucormycosis effects 0.9%–1.9% of patients with allogeneic bone marrow transplants.[48]

Skiada and Petrikkos,[21] in a revision of 78 cases of cutaneous mucormycosis between 2004 and 2008, found that the principal etiological agents identified were *Rhizopus* spp. 43%, *Lichtheimia (Absidia) corymbifera* 21%, *Apophysomyces elegans* 11%, *Saksenaea vasiformis* 7%, *Mucor* spp. 5%, *Rhizomucor* spp. 5%, and *Cunninghamella bertholletiae* 5%. On the other hand, of the 78 cases of cutaneous mucormycosis, 36% was acquired on hospital grounds. The authors show that penetrating trauma is the principal cause of cutaneous presentations, which was true in 61% of the cases.

The revision of Roden et al.[48] showed that the cutaneous presentation was the third leading cause of mucormycosis, after the rhinocerebral and pulmonary forms. However, in children, the cutaneous form was the most common. Kubak et al.,[49] in a revision of 2410 kidney transplants (1988–2008), observed that the mucormycosis was the principal mycosis in 11 of the 21 patients with invasive mycosis. The point of entry included the sinuses, lungs, gastrointestinal tract, and skin, with the presentations of rhino-sino-orbital and pulmonary as the principal clinical manifestations after solid organ transplant.

Almyroudis et al.[50] examined 116 cases of mucormycosis in transplant recipients of solid organs: 73 kidneys, 16 heart, 4 pulmonary, 2 coronary–pulmonary, 19 liver, and 2 kidney–pancreas. The use of corticoids was the most relevant risk factor and other predisposing factors being diabetes 31.8%, hyperglycemia 21.5%, renal failure 13.7%, acidosis 5.1%, hyperglycemia and acidosis 3.4%, neutropenia 0.8%, and the use of DFO 0.8%.

Among 116 patients, localized zygomycosis was present in 101 (87.1%) and disseminated in 15 (12.9%). The most frequent presentation of zygomycosis was rhinosinusitis with/without local extension to brain, followed by pulmonary zygomycosis. Finally, the largest number of clinical samples was investigated by Alvarez et al.,[12] who demonstrated that the agents involved in mucormycosis were *R. oryzae* (44.7%), *Rhizopus microsporus* (22.1%), *M. circinelloides* (9.5%), *L. corymbifera* (5.3%), *Rhizomucor pusillus* (3.7%), *C. bertholletiae* (3.2%), *M. indicus* (2.6%), *Cunninghamella echinulata* (1%), and *A. elegans* (0.5%). At the same time, the most frequent clinical presentations were sinus/parasinus (25.8%), pulmonary (26.8%), and cutaneous (28%).

24.6 SITES AND PATTERNS OF INFECTION

The agents of mucormycosis are acquired via the respiratory tract, skin from trauma, and less frequently, the intestine. These fungi have a tendency to invade the blood vessels and cause thrombosis and necrosis in infected tissues in patients in critical conditions.[51,52] The infection acquired by trauma is confined to the dermis and epidermis. Fascia, muscles, and bones in immunocompetent individuals are rarely affected, but if the individual is immunocompromised, the fungus can rapidly disseminate (see Table 24.3).

24.7 MYCOLOGICAL DIAGNOSIS

Histopathology shows an inflammatory reaction with the presence of neutrophils, angioinvasion, with posterior infarct, or perineural invasion. Histology or direct microscopic examinations are fundamental to document a case of mucormycosis. The analysis of β-1–3 glucan is not important in these fungi, due to the low concentration in Mucorales. Molecular tools are being developed as diagnostic tests but are not commercially available.[53,54] An early and rapid diagnostic is urgent,

TABLE 24.3
Mode of Transmission of 179 Cutaneous Mucormycosis Cases

Mode of Transmission	Cases N (%)
Penetrating traumas	60 (34%)
Dressings	26 (15%)
Surgery	26 (15%)
Burns	11 (6%)
Motor vehicle accidents	5 (3%)
Falls	5 (3%)
Not well described	42 (24%)

Source: Roden, M. et al., *Clin. Infect. Dis.*, 41(5), 634, 2005.

due to the rapid extension and necrosis produced by the fungus. In culture, precaution should be taken not to excessively macerate the tissue sample because it can affect the viability of the fungus due to its coenocytic character.

The presence of wide, nonseptate hyphae with right-angled branches are diagnostic or suggestive of a mucormycosis, and culture will permit the identification of the agent. *Mucor* develop well in Sabouraud agar, potato agar, and lactrimel agar, among others, at 3–5 days of incubation on average. Once the structures are obtained, they can be identified following the key in Table 24.2.[12] In the laboratory, the diagnosis should be based principally on the evaluation of the structures and the physiological characters. Alvarez et al.[12] demonstrated that the correct use of these tools allowed for the identification at the species level in the majority of the species studied. A correlation of 92% was obtained between morphological identification and molecular identification.[12]

24.8 MOLECULAR IDENTIFICATION

In recent decades, molecular methods have contributed to a better understanding of the taxonomy and phylogeny of mucormycetes. In the laboratory routine, the identification of *Mucor* species traditionally has been carried out by the use of morphological features. However, several authors have reported that the use of morphology is a difficult and time-consuming task. Hence, the use of molecular techniques, especially the sequence-based methods, appears as a reliable tool in specific identification. For a specific identification, both an informative marker and an accurate sequence database should be available. Recently, Alvarez and colleagues[55] have showed that the correct use of morphological features leads to a high positive correlation with molecular identification. These authors obtained a high percentage (92%) of correlation between two identification methods. Ideally, the marker or DNA target should be sufficiently informative in order to be applicable for all the species of the study. Several DNA targets, such as ribosomal DNA cluster,[12,56–59] actin,[57,60] elongation factor 1 – α,[57,58,59,61,62] and lactate dehydrogenase[63] genes, have been employed for phylogenetic species recognition and for taxonomical reorganization. The most used marker in the identification of mucormycetes, and consequently in Mucor, is the ribosomal DNA cluster, the internal transcribed spacer (ITS) region being the most informative target. Using this marker, it was possible to study the phylogenetic relationships within the *Mucor* genus. Additionally, it was possible to elucidate the species' boundaries, and describing and relocating several species was possible. Several new species have been proposed, such as *Mucor renisporus*, *M. ellipsoideus*, and *M. velutinosus*.[12,64] In addition, based on these studies, it has been possible to elucidate and relocate, as *M. irregularis*, strains previously identified as *Rhizomucor variabilis*.[12]

The emergence of mucormycetes as etiological agents enhances the need for better and reliable identification methods. Recent studies have shown promising results in the reliable identification of mucormycetes by the use of sequence-based tools as a complement of classical taxonomical identification.

24.9 THERAPY AND ANTIFUNGALS

Currently, mucormycosis presents an elevated rate of mortality with values approximating 50%.[48,65] This could be explained in part by the large variability in the sensitivity of Mucorales to antifungals.[66–68] Diverse studies show that amphotericin B is the antifungal of choice, being the most active in vitro and with an acceptable in vivo activity.[69,70] A recent study showed good activity observed against 17 strains of *M. circinelloides*.[70] However, species of *Saksenaea* and of *Apophysomyces* show a high variability in their sensitivity.[66,67,71,72]

On the other hand, posaconazole can be considered as satisfactory.[67,68,73] In vitro, studies suggest the use of posaconazole as the second option for mucormycosis.[12,67,68,73] Notwithstanding, a recent study demonstrated the lack of activity of posaconazole in vivo.[70] The results of Salas et al. are interesting, in that they demonstrate the effectiveness of posaconazole in *Apophysomyces variabilis* and *S. vasiformis* in a mouse model,[71,72] demonstrating the great variability in different species and genera of mucormycetes.

Itraconazole is the only marketed azole drug that has in vitro activity against Mucorales. There are case reports of successful therapy with itraconazole alone. Furthermore, animal studies revealed that itraconazole was completely ineffective against *Rhizopus* and *Mucor* spp. even though the isolates were susceptible in vitro.[37,73] Furthermore, in vivo studies showed the good activity of itraconazole against *Lichtheimia* spp.,[74] but it was not effective against *Mucor* spp.,[67] *R. microsporus*, and *R. oryzae*.[68,75] Therefore, given the limited activity of the azoles against mucormycetes, the combination of 2 or 3 antifungal agents is a possibility.[76–78]

24.10 COMBINED THERAPY

The high mortality rate of mucormycosis with currently available monotherapy, particularly in hematology patients, has stimulated interest in studying novel combinations of antifungal agents to determine whether superior outcomes might be achieved.[79] Gomez-Lopez et al.[80] evaluated the combination in vitro of terbinafine with itraconazole or AmB against 17 clinical strains of mucormycetes and found that terbinafine + itraconazole exhibited a synergistic effect in 82% of the strains, especially for *R. microsporus*, *L. corymbifera*, and *C. bertholletiae*. The echinocandins did not present activity in vitro but could be an option when utilized in combination with other drugs in the treatment of mucormycosis. Animal models show that the interaction between amphotericin B and caspofungin or posaconazole improved the survival in rats with disseminated mucormycosis, indicating a synergistic effect between these drugs.[81–83] Sugar and Liu[84] reported the synergistic effect of the combination of azolic and quinolone drugs in rats with pulmonary mucormycosis. On the other hand, caspofungin combined with AmB presented a better response than AmB alone in patients in treatment with rhino-orbital-cerebral mucormycosis,[78] and this combination was also satisfactory for the treatment of a case of rhinocerebral mucormycosis in a hematological cancer patient.[85]

Some promising therapies consist of the use of iron chelators as mentioned previously, despite the treatment of DFO being associated with a high risk of acquiring mucormycosis. New iron chelators (deferiprone and deferasirox) are not associated with the increment of risk in acquiring mucormycosis and are used as therapeutic agents in cases of experimental mucormycosis.[39,86,87]

Case reports have suggested that hyperbaric oxygen may be a beneficial adjunct to the standard surgical and medical antifungal therapy of mucormycosis, particularly for patients with rhinocerebral disease. It is hypothesized that hyperbaric oxygen might be useful for treating mucormycosis in conjunction with standard therapy because higher oxygen pressure improves the ability of neutrophils to kill the organism.[88,89]

24.11 *MUCOR* IN FOODS AND INDUSTRY

Mucor and *Rhizopus* are important in the production of fermented foods such as tempeh, sufu, and lao-chao. They are used in the production of organic acids and they are a cause of the postharvest deterioration and rotting of fruits and vegetables.[90–92]

The genera of Mucoromycotina of interest in industry are *Mucor*, *Absidia*, *Rhizopus*, *Zygorhynchus*, *Phycomyces*, *Mortierella*, *Syncephalastrum*, *Thamnidium*, and *Helicostylum*. *M. racemosus* and other Mucorales are responsible for the softness or hardness of cheeses. Soft cheese included camembert, brie, and ricotta.

24.12 CONCLUSIONS

The *Mucor* genus, belonging to subphylum Mucoromycotina, is a widespread cosmopolitan fungus, being reported as the etiological agent of mucormycosis in immunocompromised people with high mortality rates (ca. 50%). Several publications have shown the increasing incidence of mucormycosis around the world as the cause of invasive diseases from AIDS, hematological malignancies, and immunosuppressive therapies.

The diagnosis has been performed mainly on the basis of the morphological features, histological and radiological findings, and the symptoms of the infection. However, incorrect or incomplete identification is observed commonly. In order to resolve these, molecular tools have been proposed for identifications. Several markers have been proposed, such as those of the ribosomal cluster. In addition, new species of *Mucor* have been proposed, for example, from mucormycosis by *M. velutinosus*.

Historically, amphotericin B has been the first-line antifungal drug in the treatment of mucormycosis. However, recent studies have shown the high intra-/interspecific variability to these. In addition, several toxicity effects have been associated with the use of amphotericin B. Some investigations have indicated posaconazole as a second-line antifungal drug. The use of voriconazole has been associated as a risk factor for mucormycosis. Also, several investigations demonstrate the promise of combined treatments of two or more antifungal drugs, representing a real alternative to decrease the mortality rates associated with mucormycosis. Finally, the importance of mucormycosis as an emerging infectious disease in immunocompromised people demands the standardization and development of better methods to diagnose and treat these infections. However, more clinical studies are needed to elucidate the species boundaries and their susceptibility patterns.

REFERENCES

1. Hibbett DS, Binder M, Bischoff JF et al. A higher-level phylogenetic classification of the fungi. *Mycol Res.* 2007;111:509–547.
2. Gryganskyi AP, Humber RA, Smith ME, Miadlikovska J, Wu S, Voigt K, Walther G, Anishchenko IM, and Vilgalys R. Molecular phylogeny of the Entomophthoromycota. *Mol Phylogenet Evol.* 2012;65(2):682–694.
3. Medoff G and Kobayashi GS. Pulmonary mucormycosis. *N Engl J Med.* 1972;286(2):86–87.
4. Oliver MR, Van Voorhis WC, Boeckh M, Mattson D, and Bowden RA. Hepatic mucormycosis in a bone marrow transplant recipient who ingested naturopathic medicine. *Clin Infect Dis.* 1996;22(3):521–524.
5. De Hoog GS, Guarro J, Figueras MJ, and Gené J. *Atlas of Clinical Fungi*, 2nd edn. Centraalbureau voor Schimmelcultures, Baarn, the Netherlands, and Universitat Rovira i Virgili, Reus, Spain, 2000.
6. Pitt J and Hocking A. *Fungi and Food Spoilage*, 3rd edn. Springer, New York, 2009.
7. Raudoniene V and Lugauskas A. *Micromycetes* on imported fruit and vegetables. *Botanica Lithuanica.* 2005;7:55–64.
8. Foschino A, Garzaroli C, and Ottogali G. Microbial contaminants cause swelling and inward collapse of yoghurt packs. *Le Lait* 1993;73(4):395–400.
9. Hayaloglu AA, Kirbag S. Microbial quality and presence of moulds in Kuflu cheese. *Int J Food Microbiol.* 2007;115(3):376–380.
10. Golbitz P. Traditional soyfoods: Processing and products. *J Nutr.* 1995;125(3):570s–572s.
11. Liu K. Expanding soybean food utilization. *Food Technol.* 2000;54:46–58.
12. Alvarez E, Cano J, Stchigel A, Sutton D, Fothergill A, Salas V, Rinaldi M, and Guarro J. Two new species of *Mucor* from clinical samples. *Med Mycol.* 2011;49(1):62–72.
13. Sugui JA, Christensen JA, Bennett JE, Zelazny AM, and Kwon-Chung KJ. Hematogenously disseminated skin disease caused by *Mucor velutinosus* in a patient with acute myeloid leukemia. *J Clin Microbiol.* 2011;49(7):2728–2732.
14. Hemashettar BM, Patil RN, O'Donnell K, Chaturvedi V, Ren P, and Padhye AA. Chronic rhinofacial mucormycosis caused by *Mucor irregularis* (*Rhizomucor variabilis*) in India. *J Clin Microbiol.* 2011;49(6):2372–2375.
15. Li DM and Lun LD. *Mucor irregularis* infection and lethal midline granuloma: A case report and review of published literature. *Mycopathologia.* 2012;174(5–6):429–439.
16. Schell WA, O'Donnell K, and Alspaugh JA. Heterothallic mating in *Mucor irregularis* and first isolate of the species outside of Asia. *Med Mycol.* 2011;49(7):714–723.

17. Khan ZU, Ahmad S, Brazda A, and Chandy R. *Mucor circinelloides* as a cause of invasive maxillofacial zygomycosis: An emerging dimorphic pathogen with reduced susceptibility to posaconazole. *J Clin Microbiol*. 2009;47:1244–1248.
18. Palacio AD, Ramos MJ, Perez A, Arribi A, Amondarian I, Alonso S, and Cruz Ortiz YM. Zigomicosis. A propósito de cinco casos. *Rev Iberoam Micol*. 1999;16:50–56.
19. Xiaoli K, Jianguo W, Ming Li, Zemao G, and Xiaoning G. First report of *Mucor circinelloides* occurring on yellow catfish (*Pelteobagrus fulvidraco*) from China. *FEMS Microbiol Lett*. 2010;302(2): 144–150.
20. Deja M, Wolf S, Weber-Carstens S, Lehmann TN, Adler A, Ruhnke M, and Tintelnot K. Gastrointestinal zygomycosis caused by *Mucor indicus* in a patient with acute traumatic brain injury. *Med Mycol*. 2006;44(7):683–687.
21. Skiada A and Petrikkos G. Cutaneous zygomycosis. *Clin Microbiol Infect*. 2009;15(5):41–45.
22. Weitzman I, Della-Latta P, Housey G, and Rebatta G. *Mucor ramosissimus* Samutsevitsch isolated from a thigh lesion. *J Clin Microbiol*. 1993;31(9):2523–2525.
23. Eucker J, Sezer O, Graf B, and Possinger K. Mucormycoses. *Mycoses*. 2001;44(7–8):253–260.
24. Nosari A, Oreste P, Montillo M et al. Mucormycosis in hematologic malignancies: An emerging fungal infection. *Haematologica*. 2000;85(10):1068–1071.
25. Funada H and Matsuda T. Pulmonary mucormycosis in a hematology ward. *Intern Med*. 1996; 35(7):540–544.
26. Boelaert JR, De Locht M, Van Cutsem J et al. Mucormycosis during deferoxamine therapy is a siderophore-mediated infection: In vitro and in vivo animal studies. *J Clin Invest*. 1993;91(5): 1979–1986.
27. Boelaert JR, Van Cutsem J, De Locht M, Schneider YJ, and Crichton RR. Deferoxamine augments growth and pathogenicity of *Rhizopus*, while hydroxypyridinone chelators have no effect. *Kidney Int*. 1994;45(3):667–671.
28. Ibrahim AS, Gebremariam T, Lin L et al. The high affinity iron permease is a key virulence factor required for *Rhizopus oryzae* pathogenesis. *Mol Microbiol*. 2010;77(3):587–604.
29. Boelaert JR, Fenves AZ, and Coburn JW. Deferoxamine therapy and mucormycosis in dialysis patients: Report of an international registry. *Am J Kidney Dis*. 1991;18(6):660–667.
30. Morace G and Borghi E. Invasive mold infections: Virulence and pathogenesis of mucorales. *Int J Microbiol*. 2012;2012:1–5.
31. Liu M, Spellberg B, Phan QT et al. The endothelial cell receptor GRP78 is required for mucormycosis pathogenesis in diabetic mice. *J Clin Invest*. 2010;120(6):1914–1924.
32. Frater JL, Hall GS, and Procop GW. Histologic features of zygomycosis. *Arch Pathol Lab Med*. 2001;125(3):375–378.
33. Chamilos G, Lewis RE, and Kontoyiannis DP. Multidrug resistant endosymbiotic bacteria account for the emergence of zygomycosis: A hypothesis. *Fungal Genet Biol*. 2007;44(2):88–92.
34. Ribes JA, Vanover-Sams CL, and Baker JK. Zygomycetes in human disease. *Clin Microbiol Rev*. 2000;13(2):236–301.
35. Trifilio SM, Bennett CL, Yarnold PR et al. Breakthrough zygomycosis after voriconazole administration among patients with hematologic malignancies who receive hematopoietic stem-cell transplants or intensive chemotherapy. *Bone Marrow Transplant*. 2007;39(7):425–429.
36. Roden MM, Zaoutis TE, Buchanan WL et al. Epidemiology and outcome of zygomycosis: A review of 929 reported cases. *Clin Infect Dis*. 2005;41(5):634–653.
37. Spellberg B, Edwards Jr J, and Ibrahim A. Novel perspectives on mucormycosis: Pathophysiology, presentation, and management. *Clin Microbiol Rev*. 2005;18(3):556–569.
38. Chinn RY and Diamond RD. Generation of chemotactic factors by *Rhizopus oryzae* in the presence and absence of serum: Relationship to hyphal damage mediated by human neutrophils and effects of hyperglycemia and ketoacidosis. *Infect Inmun*. 1982;38(3):1123–1129.
39. Kontoyiannis DP and Lewis RE. How I treat mucormycosis. *Blood*. 2011;118(5):1216–1224.
40. Imhof A, Balajee SA, Fredricks DN, Englund JA, and Marr KA. Breakthrough fungal infections in stem cell transplant recipients receiving voriconazole. *Clin Infect Dis*. 2004;39(5):743–746.
41. Marty FM, Cosimi LA, and Baden LR. Breakthrough zygomycosis after voriconazole treatment in recipients of hematopoietic stem-cell transplants. *N Engl J Med*. 2004;350(9):950–952.
42. Siwek GT, Dodgson KJ, De Magalhaes-Silverman M et al. Invasive zygomycosis in hematopoietic stem cell transplant recipients receiving voriconazole prophylaxis. *Clin Infect Dis*. 2004;39(4):584–587.
43. Kobayashi K, Kami M, Murashige N, Kishi Y, Fujisaki G, and Mitamura T. Breakthrough zygomycosis during voriconazole treatment for invasive aspergillosis. *Haematologica*. 2004;89(11):ECR42.

44. Kontoyiannis DP, Lionakis MS, Lewis RE et al. Zygomycosis in a tertiary-care cancer center in the era of *Aspergillus*-active antifungal therapy: A case-control observational study of 27 recent cases. *J Infect Dis*. 2005;191(8):1350–1360.
45. Lionakis MS and Kontoyiannis DP. Sinus zygomycosis in a patient receiving voriconazole prophylaxis. *Br J Haematol*. 2005;129(1):2.
46. Oren I. Breakthrough zygomycosis during empirical voriconazole therapy in febrile patients with neutropenia. *Clin Infect Dis*. 2005;40(5):770–771.
47. Vigouroux S, Morin O, Moreau P et al. Zygomycosis after prolonged use of voriconazole in immunocompromised patients with hematologic disease: Attention required. *Clin Infect Dis*. 2005;40(4):e35–e37.
48. Roden M, Zaoutis T, Buchanan W et al. Epidemiology and outcome of zygomycosis: A report of 929 reported cases. *Clin Infect Dis*. 2005;41(5):634–653.
49. Kubak BM and Huprikar SS. AST infectious diseases community of practice. Emerging and rare fungal infections in solid organ transplant recipients. *Am J Transplant*. 2009;9(4):S208–S226.
50. Almyroudis NG, Sutton DA, Linden P, Rinaldi MG, Fung J, and Kusne S. Zygomycosis in solid organ transplant recipients in a tertiary transplant center and review of the literature. *Am J Transplant*. 2006;6(10):2365–2374.
51. Hajdu S, Obradovic A, Presterl E, and Vécsei V. Invasive mycoses following trauma. *Injury*. 2009;40(5):548–554.
52. Mantadakis E and Samonis G. Clinical presentation of zygomycosis. *Clin Microbiol Infect*. 2009;15(5):15–20.
53. Jimenez C, Lumbreras C, Paseiro G et al. Treatment of mucor infection after liver or pancreas-kidney transplantation. *Transplant Proc*. 2002;34(1):82–83.
54. Dannaoui E. Molecular tools for identification of Zygomycetes and the diagnosis of Zygomycosis. *Clin Microbiol Infect*. 2009;15(5):66–70.
55. Alvarez E, Sutton DA, Cano J et al. Spectrum of zygomycete species identified in clinically significant specimens in the United States. *J Clin Microbiol*. 2009;47(6):1650–1656.
56. Abe A, Oda Y, Asano K et al. The molecular phylogeny of the genus Rhizopus based on rDNA sequences. *Biosci Biotechnol Biochem*. 2006;70(10):2387–2393.
57. Abe A, Oda Y, Asano K et al. Rhizopus delemar is the proper name for *Rhizopus* oryzae fumaric-malic acid producers. *Mycologia*. 2007;99(5):714–722.
58. Abe A, Asano K, Sone T et al. A molecular phylogeny-based taxonomy of the genus *Rhizopus*. *Biosci Biotechnol Biochem*. 2010;74(7):1325–1331.
59. Alastruey-Izquierdo A, Hoffmann K et al. Species recognition and clinical relevance of the zygomycetous genus *Lichtheimia* (syn. Absidia pro parte, Mycocladus). *J Clin Microbiol*. 2010;48(6):2154–2170.
60. Hoffmann K, Discher S et al. Revision of the genus *Absidia* (Mucorales, Zygomycetes) based on physiological, phylogenetic, and morphological characters; thermotolerant *Absidia* spp. form a coherent group, Mycocladiaceae fam. nov. *Mycol Res*. 2007;111(10):1169–1183.
61. Garcia-Hermoso D, Hoinard D and Voigt K. Molecular and phenotypic evaluation of *Lichtheimia corymbifera* (formerly *Absidia corymbifera*) complex isolates associated with human mucormycosis: Rehabilitation of L. ramosa. *J Clin Microbiol*. 2009;47(12):3862–3870.
62. O'Donnell KL, Lutzoni F, Ward TJ et al. Evolutionary relationships among mucoralean fungi (Zygomycota): Evidence for family polyphyly on a large scale. *Mycologia*. 2001;93:286–296.
63. Saito K, Saito A, Ohnishi M et al. Genetic diversity in *Rhizopus oryzae* strains as revealed by the sequence of lactate dehydrogenase genes. *Arch Microbiol*. 2004;182(1):3036.
64. Jacobs K and Botha A. *Mucor renisporus* sp. nov., a new coprophilous species from Southern Africa. *Fungal Div*. 2008;29:27–35.
65. Skiada A, Pagano L, Groll A, Zimmerli S, Dupont B, Lagrou K, Lass-Florl C et al. European Confederation of Medical Mycology Working Group on Zygomycosis. Zygomycosis in Europe: Analysis of 230 cases accrued by the registry of the European Confederation of Medical Mycology (ECMM) Working Group on Zygomycosis between 2005 and 2007. *Clin Microbiol Infect*. 2011;17(12):1859–1867.
66. Almyroudis NG, Sutton DA, Fothergill AW, Rinaldi MG, and Kusne S. In vitro susceptibilities of 217 clinical isolates of zygomycetes to conventional and new antifungal agents. *Antimicrob Agents Chemother*. 2007;51(7):2587–2590.
67. Sun QN, Najvar LK, Bocanegra R, Loebenberg D, and Graybill JR. In vivo activity of posaconazole against *Mucor* spp. in an immunosuppressed-mouse model. *Antimicrob Agents Chemother*. 2002;46(7):2310–2312.

68. Dannaoui E, Meis JF, Loebenberg D, and Verweij PE. Activity of posaconazole in treatment of experimental disseminated zygomycosis. *Antimicrob Agents Chemother*. 2003;47(11):3647–3650.
69. Takemoto K, Yamamoto Y, and Kanazawa K. Comparative study of the efficacy of liposomal amphotericin B and amphotericin B deoxycholate against six species of Zygomycetes in a murine lethal infection model. *J Infect Chemother*. 2010;16(6):388–395.
70. Salas V, Pastor FJ, Calvo E, Alvarez E, Sutton DA, Mayayo E, Fothergill AW, Rinaldi MG, and Guarro J. In vitro and in vivo activities of posaconazole and amphotericin B in a murine invasive infection by *Mucor circinelloides*: Poor efficacy of posaconazole. *Antimicrob Agents Chemother*. 2012;56(5):2246–2250.
71. Salas V, Pastor FJ, Calvo E, Sutton D, García-Hermoso D, Mayayo E, Dromer F, Fothergill A, Alvarez E, and Guarro J. Experimental murine model of disseminated infection by *Saksenaea vasiformis*: Successful treatment with posaconazole. *Med Mycol*. 2012;50(7):710–715.
72. Salas V, Pastor FJ, Calvo E, Sutton DA, Chander J, Mayayo E, Alvarez E, and Guarro J. Efficacy of posaconazole in a murine model of disseminated infection caused by *Apophysomyces variabilis*. *J Antimicrob Chemother*. 2012;67(7):1712–1715.
73. Sabatelli F, Patel R, Mann PA, Mendrick CA, Norris CC, Hare R, Loebenberg D, Black TA, and McNicholas PM. In vitro activities of posaconazole, fluconazole, itraconazole, voriconazole, and amphotericin B against a large collection of clinically important molds and yeasts. *Antimicrob Agents Chemother*. 2006;50(6):2009–2015.
74. Mosquera J, Warn PA, Rodriguez-Tudela JL, and Denning DW. Treatment of *Absidia corymbifera* infection in mice with amphotericin B and itraconazole. *J Antimicrob Chemother*. 2001;48(4):583–586.
75. Odds FC, Van Gerven F, Espinel-Ingroff A, Bartlett MS, Ghannoum MA, Lancaster MV, Pfaller MA, Rex JH, Rinaldi MG, and Walsh TJ. Evaluation of possible correlations between antifungal susceptibilities of filamentous fungi in vitro and antifungal treatment outcomes in animal infection models. *Antimicrob Agents Chemother*. 1998;42(2):282–288.
76. Dannaoui E, Afeltra J, Meis J, Verweij P, and The Eurofung Network. In vitro susceptibilities of zygomycetes to combinations of antimicrobial agents. *Antimicrob Agents Chemother*. 2002;46(8):2708–2711.
77. Ibrahim AS, Gebremariam T, Fu Y, Edwards Jr JE, and Spellberg B. Combination echinocandin-polyene treatment of murine mucormycosis. *Antimicrob Agents Chemother*. 2008;52(4):1556–1558.
78. Reed C, Bryant R, Ibrahim AS et al. Combination polyene–caspofungin treatment of rhino-orbital-cerebral mucormycosis. *Clin Infect Dis*. 2008;47(3):364–371.
79. Spellberg B, Ibrahim A, Roilides E, Lewis RE, Lortholary O, Petrikkos G, Kontoyiannis DP, and Walsh TJ. Combination therapy for mucormycosis: Why, what, and how? *Clin Infect Dis*. 2012;54(S1):73–78.
80. Gómez-López A, Cuenca-Estrella M, Mellado E, and Rodríguez-Tudela JL. In vitro evaluation of combination of terbinafine with itraconazole or amphotericin B against Zygomycota. *Diagn Microbiol Infect Dis*. 2003;45(3):199–202.
81. Ibrahim AS, Gebremariam T, Luo G, Fu Y, French SW, Edwards Jr JE, and Spellberg B. Combination therapy of murine mucormycosis or aspergillosis with iron chelation, polyenes, and echinocandins. *Antimicrob Agents Chemothe* 2011;55(4):1768–1770.
82. Spellberg B, Fu Y, Edwards Jr JE, and Ibrahim AS. Combination therapy with amphotericin B lipid complex and caspofungin acetate of disseminated zygomycosis in diabetic ketoacidotic mice. *Antimicrob Agents Chemother*. 2005;49(2):830–832.
83. Rodriguez MM, Serena C, Marine M, Pastor FJ, and Guarro J. Posaconazole combined with amphotericin B, an effective therapy for a murine disseminated infection caused by *Rhizopus oryzae*. *Antimicrob Agents Chemother*. 2008;52(10):3786–3788.
84. Sugar AM and Liu XP. Combination antifungal therapy in treatment of murine pulmonary mucormycosis: Roles of quinolones and azoles. *Antimicrob Agents Chemother*. 2000;44(7):2004–2006.
85. Vazquez L, Mateos JJ, Sanz-Rodriguez C, Perez E, Caballero D, and San Miguel JF. Successful treatment of rhinocerebral zygomycosis with a combination of caspofungin and liposomal amphotericin B. *Haematologica*. 2005;90(12):ECR39.
86. Ibrahim AS, Spellberg B, and Edwards Jr J. Iron acquisition: A novel perspective on mucormycosis pathogenesis and treatment. *Curr Opin Infect Dis*. 2008;21(6):620–625.
87. Spellberg B, Walsh TJ, Kontoyiannis DP, Edwards Jr J, and Ibrahim AS. Recent advances in the management of mucormycosis: From bench to bedside. *Clin Infect Dis*. 2009;48(12):1743–1751.
88. Couch L, Theilen F, and Mader JT. Rhinocerebral mucormycosis with cerebral extension successfully treated with adjunctive hyperbaric oxygen therapy. *Arch Otolaryngol Head Neck Surg*. 1988;114:791–794.
89. Garcia-Covarrubias L, Barratt DM, Bartlett R, and Van Meter K. Treatment of mucormycosis with adjunctive hyperbaric oxygen: Five cases treated at the same institution and review of the literature. *Rev Investig Clin*. 2004;56:51–55.

90. Guermani L, Villaume C, Bau HM, Nicolas JP, and Mejean L. Modification of soyprotein hypocholesterolemic effect after fermentation by *Rhizopus oligosporus* sp. T3. *Sci Aliments*. 1993;13:317–324.
91. Han BZ and Nout MJR. Effects of temperature, water activity and gas atmosphere on mycelia growth of tempe fungi *Rhizopus microsporus* var. *microsporus* and *R. microsporus* var. *oligosporus*. *World J Microbiol Biotechnol*. 2000;16:853–858.
92. Han BZ, Rombouts FM, and Nout MJR. A Chinese fermented soybean food. *Int J Food Microbiol*. 2001;65:1–10.

25 *Paecilomyces*: Mycotoxin Production and Human Infection

Cintia de Moraes Borba and Marcelly Maria dos Santos Brito

CONTENTS

25.1 The Genus *Paecilomyces* 401
25.2 Taxonomy and Morphological Characteristics of *Paecilomyces* 401
25.3 *Paecilomyces* in Food, Feed, and Drinking Water 402
25.4 *Paecilomyces* spp.–Producing Mycotoxins and Other Biological Compounds 403
25.5 Hyalohyphomycosis Caused by Paecilomyces spp. 404
       25.5.1 Human Model 405
       25.5.2 Animal Model 408
25.6 Laboratory Diagnosis for *Paecilomyces* spp. Infections 408
25.7 Treatment for Hyalohyphomycosis Caused by *Paecilomyces* spp. 411
25.8 In Vitro Antifungal Susceptibility 411
25.9 Experimental Models 412
25.10 Concluding Remarks 413
References 413

25.1 THE GENUS *PAECILOMYCES*

The genus *Paecilomyces* was firstly introduced by Bainier [1] as being closely related to *Penicillium* Link ex Fr. with differences in the color of colonies and shape of phialides. Subsequently, the genus was studied by Hughes [2], Brown and Smith [3], Morris [4], Onions and Barron [5], and de Hoog [6]. In 1974, Samson [7] monographed and redefined this genus. Based on morphological characters in pure culture, he discussed the delimitation of the genus and the relationships to other genera and proposed a subdivision into two sections: Section *Paecilomyces* (mesophilic to thermophilic species with yellow-brown to brown colonies) and Section *Isarioidea* (mesophilic species with white or other *bright* colonies that included several entomopathogenic or nematophagous species). Additionally, it was affirmed that the known perfect states belong to the Ascomycetes, *Byssochlamys* Westl., *Talaromyces* C.R. Benjamin, and *Thermoascus* Miehe.

25.2 TAXONOMY AND MORPHOLOGICAL CHARACTERISTICS OF *PAECILOMYCES*

Samson [7] described and illustrated 31 species that has been increased to 40 [8]; the classification of which remains based on morphological characteristics, which can be imprecise [9]. Fortunately, molecular tools have been used for identifications and phylogenetic studies. Analysis of the 18S ribosomal DNA (rDNA) demonstrated that *Paecilomyces* can be considered polyphyletic across two Ascomycetes orders, the Eurotiales and the Hypocreales and into two subclasses,

Sordariomycetidae and Eurotiomycetidae [8]. These authors indicated that the type of species, *Paecilomyces variotii*, and thermophilic relatives belong to the order Eurotiales (Trichocomaceae), and the mesophilic species related to *Paecilomyces farinosus* are in the order Hypocreales (Clavicipitaceae and Hypocreaceae). In addition, *Paecilomyces inflatus* had affinities to the order Sordariales. The genus is monophyletic and polyphyletic in orders Eurotiales and Hypocreales, respectively.

Luangsa-ard et al. [10] analyzed partial β-tubulin and ITS1-5.8S-ITS2 rDNA sequences of 34 strains of *Paecilomyces* and their associated hypocrealean teleomorphs *Torrubiella* and *Cordyceps* to better understand the relationship of *Paecilomyces* Section *Isarioidea*. They concluded that Section *Isarioidea* is not monophyletic, and there is not a monophyletic grouping for the invertebrate-associated species.

Samson et al. [11] studied the genus *Byssochlamys* and its *Paecilomyces* anamorphs using a polyphasic approach to find characters that differentiate species. They analyzed the internal transcribed spacer (ITS) region, parts of the β-tubulin and calmodulin genes, macro- and micromorphology, and extrolite profiles, and revealed that *Byssochlamys* includes nine species, five of which form a teleomorph (*Byssochlamys fulva*, *Byssochlamys lagunculariae*, *Byssochlamys nivea*, *Byssochlamys spectabilis*, *Byssochlamys zollerniae*), while four are strictly anamorphic (*Paecilomyces brunneolus*, *Paecilomyces divaricatus*, *Paecilomyces formosus*, *Paecilomyces saturatus*). The *P. variotii* complex was divided into four species, *P. divaricatus*, *P. formosus*, *P. saturatus*, and *P. variotii*.

Paecilomyces produce colonies that grow moderately well on culture media. Microscopic examination shows hyphae being hyaline to yellowish, septate, mostly smooth walled, with conidiogenous structures, synnematous or mononematous. These consist of verticillate or irregularly branched conidiophores, bearing terminally on each branch whorl of conidiogenous cells, which also may be solitary on the fertile hyphae. Phialides comprise a cylindrical or swollen basal portion, tapering often abruptly into a long distinct neck. One-celled conidia are in divergent or tangled basipetal chains, hyaline or slightly pigmented, smooth-walled or echinulate, and of various shapes. Chlamydospores are usually thick walled, borne singly or in short chains, smooth walled or ornamented, or absent [7,11].

Paecilomyces species are often recovered from soil and air, and can cause deterioration of grain, food, and paper [12]. Furthermore, they can be found in acidic habitats and tolerate microaerophilic conditions [11]. Certain species have a clinical importance because they produce infections in immunocompromised and immunocompetent patients. *P. variotii* and *Paecilomyces lilacinus* are the two species associated most frequently with human and animal diseases. *Paecilomyces marquandii* and *Paecilomyces javanicus* are also reported to infect humans [13].

A new genus name, *Purpureocillium*, was proposed for *P. lilacinus*. Molecular results showed that *P. lilacinus* is not related to *Paecilomyces*, which is represented by the thermophilic and often pathogenic, *P. variotii*. Thus, a new combination *Purpureocillium lilacinum* was made [14]. In addition, *P. javanicus* has been returned to the genus *Isaria* in the Hypocreales [8], and *P. marquandii* is considered not to be related to *P. variotii* and should be transferred to a new genus [14]. However, we will include these species in this chapter because the new nomenclature is not well known.

25.3 *PAECILOMYCES* IN FOOD, FEED, AND DRINKING WATER

Fungi of importance to the food and beverage industry include heat-resistant fungi [15], and *Byssochlamys* and *Neosartorya* are the heat-resistant fungal genera most implicated in the spoilage of fruit juices and foods. However, *Eupenicillium*, *Talaromyces*, and *Paecilomyces* have also been isolated [16]. The heat resistance is probably due to the presence of structures such as asci and ascospores, sclerotia, chlamydospores, and aleurospores [17]. *P. variotii* is the most frequent species encountered in relation to spoiled food- and feedstuffs, and its presence in pasteurized beverages causes great economic losses [18]. It has been reported as contaminant of foods, fruits, raw materials, especially those containing oils (cereals, nuts, meat products, margarine, edible oils, cheese, dried fruits, and seeds), silages, rye, and poultry feed among others [17,19–23].

Species have been reported in drinking water, although the frequencies are variable [24]. For example, *P. variotii, Paecilomyces carneus, P. farinosus*, and *P. lilacinus* (=*Purpureocillium lilacinum*) have been reported from different sources of water, such as drinking water, bottled water, and hospital water distribution system [25–33]. The problems associated with fungi in water include blockage of water pipes, organoleptic deterioration, pathogenic fungi, and mycotoxins [34]. However, the significance of drinking water as an exposure pathway to pathogenic, allergenic, or toxic fungal species or their metabolites remains obscure [32].

The method of detection and quantification of fungi in juice, on fruits, and other natural materials is often the dilution plate technique that consists in mixing part of sample with saline solution containing Tween 80 and shake for 30 min [22]. Siqueira et al. [34] state that one of the most serious problems for studying fungi is the use of isolation media that tends to vary among researchers and recommended Dichloran Rose Bengal chloramphenicol agar and Dichloran 18% glycerol agar for the isolation and quantification of fungi in foods of high water activity and for water analyses.

The identification of *Paecilomyces* spp. from natural products has been based on morphological characteristic and tends to be subjective. Hageskal et al. [35] emphasize that molecular techniques such as DNA sequencing are not dependent on sporulating or even viable fungi and is a good supplement to morphology. Kanzler et al. [30] recommend polymerase chain reaction (PCR) of the gene coding for the ribosomal ITS with the enclosed 5.8S rDNA, and subsequent sequencing can be employed with good results. They isolated DNA from *Paecilomyces* species using a purification kit, and after, the ITS region was amplified by PCR using the primer set: ITS 4 (TCCTCCGCTT ATTGA TATCG) and ITS5 (GGAAG TAAAAGTCGT AACAA GG). The ITS sequences were compared with entries in genomic data banks to identify the specific fungi.

Besides the undesirable deterioration of food-, feedstuffs, and water by *Paecilomyces*, several species can produce mycotoxins in these products. *P. variotii* produce patulin and viriditoxin, and *Byssochlamys*, the teleomorph of *Paecilomyces*, also produces patulin, byssotoxin A, asymmetrin, variotin, and byssochlamic acid, all of them toxic to many biological systems [15,36]. Therefore, these fungi can be classified not only as spoilage microorganisms but also as potential source of public health problems [16,37].

There are various types of mycotoxins detection methods [36,37]. However, molecular-based methods have been applied as companion assays for microbiological and chemical methods. Thus, PCR-based diagnosis has been employed for detection of producers of aflatoxins, trichothecenes, fumonisins, and patulin (*Aspergillus, Fusarium*, and *Penicillium*) [38]. In the case of *Paecilomyces* and patulin producers, Paterson [39] designed a primer pair specific to amplify 600 base pairs of the *idh* gene (IDH1 sequence: 5′-CAATGTGTCGTACTGTGCCC-3′ and IDH2 sequence: 5′-ACCTTCAGTCGCTGTTCCTC-3′), a gene coding for isoepoxydon dehydrogenase that participates in patulin biosynthesis, and showed the presence of this gene in, inter alia, *P. variotii, B. nivea*, and *B. fulva*. In addition, the isoepoxydon dehydrogenase gene PCR profile of *Byssochlamys* and *Paecilomyces* may be useful in classification and identification [40].

25.4 *PAECILOMYCES* SPP.–PRODUCING MYCOTOXINS AND OTHER BIOLOGICAL COMPOUNDS

Many studies have been demonstrating the ability of different species of this genus to produce mycotoxins and other biological compounds.

According to Samson et al. [11], *P. variotii* sensu lato produces the genotoxic mycotoxin patulin [41] and sphingofungin E and F that are sphingosine-like compounds that inhibit serinepalmitoyl transferase, an enzyme essential in the biosynthesis of sphingolipids [42]. In addition, the fungus produces viriditoxin [43] that has broad-spectrum antibacterial activity against clinically relevant Gram-positive pathogens [44]. *P. variotii* also produces variotin (mycotoxin), cornexistins, SCH 643432, and penicillin-like compound among others [45–57].

P. lilacinus (=*Purpureocillium lilacinum*) produces an antibiotic called leucinostatin active against some Gram-positive bacteria and a wide range of pathogenic and nonpathogenic fungi; it was cytotoxic to HeLa cell culture and showed some inhibitory effect on Ehrlich subcutaneous solid tumor in mice [58]. Later, the same research group described that leucinostatin was found to be a mixture of several components that were separated in leucinostatins A and B [59]. Mikami et al. [60] obtained leucinostatin from two *P. lilacinus* strains (from soil and a human oculomycosis) and affirmed that it had high toxicity to experimental animals. Furthermore, data have demonstrated that leucinostatins are indicators of nematicidal activity [61]. Screening of strains collected from various sources including clinical isolates showed that among 20 strains tested, 19 were found to produce the leucinostatins A and B [62]. On the other hand, there was a strain of *P. lilacinus 251* used in biological control, which did not produce leucinostatin [63].

P. lilacinus also produces enzymes such as serine proteases and chitinases that may be relevant to pathogenicity in a general sense. Bonants et al. [64] described that *P. lilacinus* CBS 143.75 produced an extracellular serine protease possibly involved in penetration through the egg shell of nematodes. *P. lilacinus 251* produced proteases and chitinases in the presence of egg yolk and chitin, respectively [65]. An extracellular thiol-dependent serine protease was isolated from the culture medium of *P. lilacinus* [66], and this group of enzymes can digest the host cuticle during invasion of insects or nematodes serving as an important virulence factor [67].

P. marquandii is also a leucinostatins producer. Casinovi et al. [68] and Rossi et al. [69] reported the isolation and structure elucidation of two peptide antibiotics produced by submerged cultures of *P. marquandii*. Later, they described a new peptide metabolite called leucinostatin D with biological activity against Gram-positive bacteria and several fungi, including *Cryptococcus neoformans* and *Candida albicans* [70]. Additionally, two novel peptides were described called leucinostatin H and K showing biological activity against Gram-positive bacteria and fungi, as do leucinostatins D and leucinostatin A from *P. lilacinus* [71].

P. carneus, strain P-177 was identified to produce secondary metabolites such as paeciloquinones A, B, C, D, and F, as inhibitors of protein tyrosine kinase. These quinones inhibited epidermal growth factor receptor protein tyrosine kinase in the micromolar range, and A and C were the most potent [72].

Paecilomyces tenuipes has been used in the Orient as a herbal medicine to treat allergic diseases, asthma, cancer, and tuberculosis [73]. It is known to produce cytotoxic components called ergosterol peroxide and acetoxyscirpenediol active against several tumor cells lines [74]. Nilanonta et al. [75] isolated two antimycobacterial and antiplasmodial cyclodepsipeptides from *P. tenuipes* BCC 1314 called beauvericin and beauvericin A. Later, trichotecane derivatives were isolated and identified as spirotenuipesine A and B, with potent activity in neurotrophic factor biosynthesis in glial cells, besides trichotecane mycotoxins [76,77]. In addition, alkaloids-denominated isariotins A, B, C, and D from *Isaria tenuipes* BCC 7831 (*P. tenuipes* was designated by Luangsa-ard et al. [10] one of the species belonging the *Isaria* clade) was described, but they did not show activities in the assays against malarial parasite and fungi nor cytotoxicity to cancer cell lines and Vero cells [78]. On the other hand, a bioactive compound from this fungus was isolated and has an inhibitory effect on monoamine oxidase A and B in vitro and in vivo, which indicates it may be a potential antidepressant [79].

25.5 HYALOHYPHOMYCOSIS CAUSED BY PAECILOMYCES SPP.

Hyalohyphomycosis is a term proposed by Ajello and McGinnis [80] to describe mycotic infections in which the tissue forms of the etiological agents are hyaline septate hyphae. The genera that cause this mycosis are *Acremonium* sp., *Beauveria* sp., *Fusarium* sp., *Paecilomyces* sp., *Pseudallescheria* sp., and *Scopulariopsis* sp. [81,82]. The most important clinical species of *Paecilomyces* are *P. variotii* and *P. lilacinus* (=*Purpureocillium lilacinum*) followed by *P. marquandii* and *P. javanicus*. Because of that, the clinical manifestations in human and animal models will be described only for these species.

25.5.1 HUMAN MODEL

The genus *Paecilomyces* has been recognized as etiological agent of hyalohyphomycosis, mainly in immunocompromised patients, although many cases were reported without species identifications [83–95]. Furthermore, most clinical manifestations were associated with peritoneal, sinus, cutaneous, and ocular infections. *P. variotii* has been reported to infect humans, especially the immunocompromised. The first case of *P. variotii* infection was an endocarditis following cardiac surgery with isolation of the fungus from blood cultures [96]. Novel cases were reported since 1971 including those related to endocarditis, pyelonephritis, sinusitis, pneumonia, endophthalmitis, osteomyelitis, peritonitis, cutaneous and disseminated infection, and fungemia among others (Table 25.1) where peritonitis was the most common clinical manifestation. This species is a saprobe in soil, peat, silage, and water. Until 1990, there had been no association with continuous ambulatory peritoneal dialysis peritonitis [97]. The most recent case reported in the literature was an endophthalmitis following an intraocular lens implantation in an immunocompetent man [98]. *P. variotii* has rarely been implicated in ocular infections [99], although it is known to be resistant to most sterilizing techniques including formaldehyde [100,101].

P. marquandii also has medical relevance as, although it is a saprophytic fungus, it can infect immunosuppressed patients. Harris et al. [102] reported the first case of cellulitis in a kidney transplant patient caused by *P. marquandii*, and it was cured with miconazole intravenous therapy. Another case of cellulitis by this fungus was described in a man who had undergone renal transplant 15 years before the cutaneous symptoms [103]. However, it is important to emphasize that the last case found in the literature was of an immunocompetent female with symptoms of irritation, photophobia, and reduced vision but without a history of known trauma, contact-lens wearing, or herpes simplex infection. This patient was diagnosed with keratitis due to *P. marquandii* and susceptibility tests demonstrated that this fungus was resistant to all antifungals, except voriconazole [104].

P. javanicus has been described as etiological agent of endocarditis. Allevato et al. [105] reported the first case of *P. javanicus* endocarditis of a native and prosthetic aortic valve with consequent cerebral fungal embolism and leptomeningeal vasculitis [106,107].

P. lilacinus (=*Purpureocillium lilacinum*) is the species with the highest number of reported cases (Table 25.2). It is widely considered cosmopolitan, saprophytic, and frequently detected in environmental soil samples and can cause deterioration of grains, food, and paper. The fungus can be recovered from contaminated skin creams and lotions used clinically. Catheters and plastic implants can be sources [108]. It is an important opportunistic pathogen in immunocompromised and immunocompetent hosts. The portal of entry of this fungus is attributed to breakdown of the skin barrier, indwelling catheters, or inhalation [13]. Most clinical manifestations are associated with ocular, cutaneous, and subcutaneous infections, but other associations have also been reported. The fungus is considered resistant to common sterilizing methods. Volná and Maderová [109] isolated one *P. lilacinus* strain from a case of ophthalmological infection. During sensitivity tests to disinfectants and chemical sterilizing agents, it was shown that the strain was not killed even by the action of higher concentrations of disinfectants solutions. In addition, the authors demonstrated that the strain resisted 10 min treatment with 0.5% peroxyacetic acid (Persteril®) and was killed only after 30 min treatment.

Pastor and Guarro [13] reviewed human cases of *P. lilacinus* (=*Purpureocillium lilacinum*) infection since 1964 to 2004 and affirmed this fungus has a special tropism to ocular structures and showed the highest incidence of ocular infections. They pointed to the risk factors of this infection: intra-ocular lens implantation, wearing of contact lenses, nonsurgical trauma with or without a foreign body, and ophthalmic surgery. Furthermore, they connected the cutaneous and subcutaneous infections to organ transplantation, corticosteroid therapy, primary immunodeficiency, diabetes mellitus, AIDS, and malignancies. Antas et al. [108] reviewed the cases from 2006 to 2011 including cases of fungemia, keratitis, cutaneous infection, bursitis, endocarditis, rhinitis, cellulitis, onychomycosis, eumycetoma, and one case associated with dog bites. Most recently, (1) maxillary

TABLE 25.1
Reported Human Cases of Hyalohyphomycosis Caused by *Paecilomyces variotii*

Infection	Number of Cases	Reference
Cutaneous infection	7	Naidu and Singh [162]
		Williamson et al. [181]
		Athar et al. [182]
		Abbas et al. [183]
Disseminated infection	1	Chamilos and Kontoyiannis [164]
		Vasudevan et al. [217]
Endocarditis	5	Uys et al. [96]
		Senior and Saldarriaga [155]
		Silver et al. [184]
		McClellan et al. [185]
		Kalish et al. [186]
Endophthalmitis	3	Anita et al. [98]
		Lam et al. [99]
		Tarkkanen et al. [158]
Fungemia	2	Shing et al. [187]
		Salle et al. [188]
Osteomyelitis	1	Cohen-Abbo and Edwards [189]
Peritonitis	15	Bibashi et al. [97]
		Marzec et al. [151]
		Kovac et al. [152]
		Rinaldi et al. [157]
		Crompton et al. [190]
		Eisinger and Weinstein [191]
		Nankivell et al. [192]
		Chan et al. [193]
		Wright et al. [194]
Pneumonia	6	Dharmasena et al. [150]
		Polat et al. [218]
		Byrd et al. [156]
		Akhunova and Shustnova [195]
		Das et al. [196]
		Yepes et al. [197]
Pyelonephritis	2	Sriram et al. [154]
		Steiner et al. [219]
Sinusitis	3	Sherwood and Dansky [198]
		Otcenasek et al. [199]
		Thompson et al. [200]
Other infection sites	8	Eloy et al. [201]
		Lee et al. [153]
		Wang et al. [159]
		Fagerburg et al. [202]
		Dhindsa et al. [203]
		Young et al. [204]
		Niazi et al. [205]
		Arenas et al. [206]
		Kantarcioglu et al. [207]

TABLE 25.2
Reported Human Cases of Hyalohyphomycosis Caused by *Paecilomyces lilacinus* (=*Purpureocillium lilacinum*)

Infection	Number of Cases	Reference
Cutaneous and subcutaneous	64	Reviewed by Pastor and Guarro [13]
		Reviewed by Antas et al. [108]
		Keshtkar-Jahromi et al. [112]
		Ezzedine et al. [114]
		Rosmaninho et al. [170]
		Shin et al. [208]
		Kitami et al. [209]
		Bassiri-Jahromi [220]
		Nagamoto et al. [225]
		Rimawi et al. [226]
		Lavergne et al. [228]
		Saegeman et al. [229]
Disseminated	4	Reviewed by Pastor and Guarro [13]
		Reviewed by Antas et al. [108]
Endocarditis	1	Reviewed by Antas et al. [108]
Endophthalmitis	38	Reviewed by Pastor and Guarro [13]
		Trachsler et al. [113]
		Shukla et al. [210]
		Pflugfelder et al. [211]
		Watanabe et al. [212]
		Gupta et al. [213]
		Chakrabarti et al. [214]
Fungemia	7	Reviewed by Pastor and Guarro [13]
		Reviewed by Antas et al. [108]
		Ding et al. [221]
Keratitis	55	Pastor and Guarro [13]
		Reviewed by Pastor and Guarro [13]
		Reviewed by Antas et al. [108]
		Maier et al. [111]
		Minogue et al. [215]
		Deng et al. [216]
		Todokoro et al. [222]
		Arnoldner et al. [223]
		López-Medrano et al. [224]
		McLintock et al. [227]
		Monden et al. [230]
Onychomycosis	2	Reviewed by Pastor and Guarro [13]
		Reviewed by Antas et al. [108]
Pneumonia	3	Reviewed by Pastor and Guarro [13]
		Khan et al. [115]
Sinusitis	7	Reviewed by Pastor and Guarro [13]
		Permi et al. [110][a]
		Wong et al. [116]
Other infection sites	9	Reviewed by Pastor and Guarro [13]
		Reviewed by Antas et al. [108]
		Schweitzer et al. [231]
		Wolley et al. [232]

[a] We suggest rechecking the classification of the species.

sinusitis in an immunocompetent patient [110],* (2) keratitis associated with contact lens in a female patient [111], (3) hand cutaneous involvement with synovitis in an immunocompetent patient [112], (4) endophthalmitis following cataract surgery [113], (5) cutaneous infection in a patient who had undergone uncontrolled courses of corticosteroid therapy [114], and (6) cavitary pulmonary disease [115] were reported. Also, the first report was published of *P. lilacinus* (=*Purpureocillium lilacinum*) causing sinusitis in a fit young woman with no identified predisposing factors in the United Kingdom [116].

25.5.2 ANIMAL MODEL

Some reports on animal infections are provided as they can be models for the human diseases. Hyalohyphomycosis by *Paecilomyces* spp. has been described in several animals. *Paecilomyces* belongs to mycobiota on the fur of sheep, hamsters, and dogs [117–119], and they have been found in organs of apparently healthy free-living rodents, the nails of ducks, and the guts of mullet [120–122]. It is difficult to identify a predisposing disease or immune-system deficiency in the affected animals. Captivity, high-population density, and temperature fluctuations are among the factors that may compromise the immune functions of the animals, predisposing them to the opportunistic pathogens [108].

The most common port of entry for *Paecilomyces* spp. is primary skin inoculation. However, disseminated disease may develop via spread from the primary site to various tissues [123]. The common clinical manifestations are skin lesions, white spots or soft tissue swelling on animal bodies, depression, anorexia, and eventually death. Beyond the skin in a disseminated infection, many organs can be largely affected, such as liver, spleen, lymph nodes, kidneys, and lungs [108].

Several animal cases were reported without identification of species. Fleischman and McCracken [124] described focal granulomas caused by *Paecilomyces* sp. at subcutaneous and laryngeal sites in a rhesus monkey (*Macaca mulatta*). Novel cases have been described in green turtles with pneumonia, a dog with granulomas in different organs, and disseminated infection in dogs and horses [125–130].

As is the case for humans, *P. variotii* and *P. lilacinus* (=*Purpureocillium lilacinum*) are of higher veterinarian interest [108]. *P. variotii* was first described causing infection in pigeons [131], and more cases have been documented as mastitis in a goat [132], systemic infection in a dog [133], pulmonary and cutaneous infections in laboratory rats [134], pneumonia in laboratory mice [135], disseminated infection in a German shepherd dog [136], calcinosis cutis associated with disseminated infection [123], and cutaneous infection in an African pygmy hedgehog [137]. *P. lilacinus* has been associated with systemic infection in (1) armadillos [138], (2) a tortoise [139], (3) a crocodile [140], (4) cutaneous and pulmonary infections in cats [141,142], (5) cutaneous infection in turtles [143,144], and (6) disseminated infection in captive sharks [145].

25.6 LABORATORY DIAGNOSIS FOR *PAECILOMYCES* SPP. INFECTIONS

Paecilomyces species are differentiated on the basis of conidial color and growth rates [7]. However, the diagnosis of *Paecilomyces* infections is based on morphological characteristics and histology of the lesions [108]. Identifications on the basis of fungal specific structures and histopathology may be difficult, and requires specialists because *Paecilomyces* may be confused with *Penicillium* and *Paecilomyces* species in tissue is commonly confused with that of other fungi [146].

Hence, molecular-based methods such as sequencing of amplified fragments containing internal spacer regions are a good alternative. Castelli et al. [147] amplified and analyzed the DNA

---

* The authors identified the etiological agent as *P. lilacinus* based on morphological characteristics. However, we suggest rechecking the classification because this species, according to Samson [7] and Luangsa-ard et al. [14], does not produce yellow green or yellow-brown colonies and chlamydospores as described in the case report.

sequences comprising the ITS1 and ITS2 regions from 58 clinical isolates of *Paecilomyces* using the primers ITS1 (5′-TCAGTGAACCTGCG-3′) and ITS4 (5′-TCCTCCGCTTATTGATATGC-3′). The method was useful for distinguishing species. Recently, Houbraken et al. [148] analyzed a set of 34 clinical isolates identified morphologically as of *P. variotii* and *P. lilacinus* (=*Purpureocillium lilacinum*) by sequencing ITS regions (ITS1 and ITS2) including 5.8S rDNA and a part of β-tubulin gene. They showed discordant results between the two identification methods. They concluded that both loci of the genes used exhibited sufficient interspecific variation for identification of these species.

Molecular identification is useful when atypical isolates are found on culture media: A *Paecilomyces* strain was isolated from a case of pulmonary fungus ball, and the diagnosis was established by isolation of the fungus in culture, although it failed to produce fruiting structures. Identification was achieved by sequencing of the ITS regions of the rDNA using primers ITS1 (5′TCCGTAGGTGAACCTGCGG3′) and ITS4 (5′TCCTCCGCTTATTGATATGC3′). However, after comparing sequences with reference strains, the clinical isolate was a *Paecilomyces* but a species identification was impossible [91]. Another case of pulmonary disease caused by an atypical isolate was related by Khan et al. [115]. *P. lilacinus* (=*Purpureocillium lilacinum*) was identified by sequencing of multiple conserved loci using primers to amplify the (1) divergent domains (D1/D2) of the 28S rRNA gene, (2) ITS region (ITS1, 5.8S rRNA, and ITS2) of rDNA, (3) variable region of the β-tubulin gene, and (4) variable region of the calmodulin gene. They showed that the atypical isolate exhibited 100% identity with the type strain of *P. lilacinus* (=*Purpureocillium lilacinum*).

Nevertheless, culturing remains the most common method for identification of *Paecilomyces* species, and a summary of morphological characteristics are described in the following paragraphs [7]:

P. variotii on malt extract agar at 25°C grows rapidly, developing floccose or funiculose, or tufted colonies. The colony color depend on the strain, ranging from deep olive buff to dark olive buff or light yellowish olive, yellowish olive to ecru olive. Reverse can be yellow to yellow brown, and in some strains, dark brown to almost black. Microscopic examination shows vegetative hyphae, hyaline, smooth walled or sometimes encrusted with yellow granules and conidiophores consisting of dense whorls of verticillately or irregularly arranged branches with two to seven phialides. Phialides in whorls or solitary consisting of a cylindrical or ellipsoidal basal portion, tapering abruptly into a long cylindrical neck. Conidia smooth walled, subglobose to ellipsoidal, and in some strains, ellipsoidal to cylindrical or clavate, hyaline to yellow, yellow brown in mass (Figure 25.1). Chlamydospores are present normally.

P. marquandii on malt extract agar at 25°C grows moderately fast, developing floccose, and pale vinaceous to violet colonies. Reverse is bright yellow to orange yellow. Microscopic structures are vegetative hyphae, hyaline, smooth walled, and conidiophores consisting of verticillate branches with whorls of two to four phialides. Phialides show a short cylindrical to ellipsoidal basal portion, tapering into a distinct neck. Conidia hyaline or pale vinaceous in mass, smooth walled to roughened, ellipsoidal to fusiform in dry divergent chains. Chlamydospore-like structures usually present.

P. javanicus on malt extract agar at 25°C grows slowly, developing floccose to cottony, at first white and after cream colonies. Reverse is uncolored to yellow. Microscopic examination shows vegetative hyphae smooth, hyaline, and erect conidiophores consisting of phialides in whorls of two to three. Phialides present a cylindrical basal portion, tapering into a thin neck. Conidia are hyaline, smooth walled, and cylindrical to fusiform. Chlamydospores are absent.

P. lilacinus (=*Purpureocillium lilacinum*) on malt extract agar at 25°C grows fast, developing floccose colonies, firstly white and after sporulation vinaceous. Microscopic examination shows vegetative hyphae, hyaline, smooth walled, and conidiophores with verticillate branches with whorls of two or four phialides. Phialides have a swollen basal portion tapering into a short distinct neck. Conidia arise from phialides in divergent chains, ellipsoidal to fusiform, smooth walled to slightly roughened, hyaline, or purple in mass (Figure 25.2). Chlamydospores are absent.

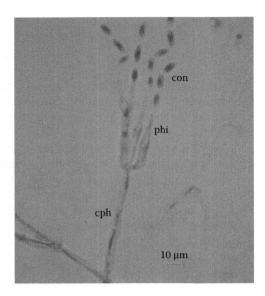

FIGURE 25.1 Micromorphology of *Paecilomyces variotii* showing the typical reproductive structures. Conidiophores (cph) with phialides (phi) consisting of a cylindrical basal portion, tapering abruptly into a long cylindrical neck and conidia (con) subglobose to ellipsoidal. Magnification ×1000.

FIGURE 25.2 Micromorphology of *Purpureocillium lilacinum* showing the reproductive typical structures. Conidiophores (cph) with whorls of phialides (phi) with a swollen basal part, tapering into a thin distinct neck and conidia (con) ellipsoidal to fusiform. Magnification ×1000.

Although definitive identification of the etiological agents of hyalohyphomycosis has been done with fungal culture, they can be identified provisionally, especially *Fusarium*, *Acremonium*, and *Paecilomyces* species, in tissue sections by a combination of histologic characteristics in a period of 24 h, whereas culture identification may require several days [149]. These authors emphasize the presence of adventitious sporulation, consisting of phialides and phialoconidia, and the presence of irregular hyphae with both 45° and 90° branching as diagnostic clues for hyalohyphomycosis-etiological agents.

25.7 TREATMENT FOR HYALOHYPHOMYCOSIS CAUSED BY *PAECILOMYCES* SPP.

Treatment of *Paecilomyces* infections has been disappointing in several cases. The management of infection is quite difficult because of the (1) immunological condition of patients, (2) presence of biological devices, (3) issues of infection sites, and (4) different susceptibility to antifungal agents that may determine the outcome of the disease.

Various *Paecilomyces* species have entirely different susceptibility to antifungal agents. *P. variotii* is considered susceptible to amphotericin B and is the agent of choice [99,150–155]. In addition, ketoconazole and fluconazole have been demonstrated to be effective alone or associated with amphotericin B against *P. variotii* [97,151,156–159].

Two cases of cellulitis caused by *P. marquandii* demonstrated in vivo susceptibility to miconazole and itraconazole with clearance of infection after therapy [102,103]. Lee et al. [104] showed that therapy with itraconazole in an immunocompetent patient with keratitis caused by this fungus did not solve the problem. It was necessary to change the antifungal therapy to topical and oral voriconazole and surgery to obtain fungal clearance.

Unfortunately, data concerning treatment of *P. javanicus* infection are unavailable as the patient from the case reported died before the fungi were identified.

P. lilacinus (=*Purpureocillium lilacinum*) is resistant to many antifungal agents, and radical treatments are necessary in some cases [108]. Pastor and Guarro [13] described that older antifungal agents, such as amphotericin B, flucytosine, fluconazole, miconazole, and itraconazole, have been used in many cases of cutaneous, subcutaneous, and ocular infections but generally with poor outcomes. However, there were patients with hyalohyphomycosis by this fungus who were successfully treated with itraconazole [160,161]. However, the novel azoles, voriconazole, and posaconazole have been demonstrated to be efficient against this fungus [108,112,114,115]. In many cases, the combination of antifungal agents and surgery was required to elicit remission [108].

25.8 IN VITRO ANTIFUNGAL SUSCEPTIBILITY

Information about in vitro susceptibility test for *Paecilomyces* spp. is limited and some reports show results of susceptibility for the genus but not species, and the use of different methods that make results difficult to compare [147]. Moreover, susceptibility profiles among *Paecilomyces* spp. are different and only on rare occasions are there in vitro and in vivo correlation of results [162]. Despite these disadvantages, in vitro susceptibility test of the isolate is important because it points to susceptibility of the fungus and helps to define the line of therapy. Therefore, it is extremely important to have the correct identification of *Paecilomyces* species.

P. variotii is susceptible in vitro and in vivo to amphotericin B but not to fluconazole, voriconazole, and ravuconazole [163–165]. In addition, it is susceptible in vitro to most of the azoles, for example, minimum inhibitory concentrations (MICs) of <2 µg/mL were found to amphotericin B, itraconazole, posaconazole, terbinafine, and echinocandins [147].

In vitro resistance to amphotericin B and flucytosine with MICs ranging from 25 to 50 µg/mL, respectively, and susceptibility to miconazole were found for the only strain of *P. marquandii* isolated from the first-reported case of cellulitis [102]. Moreover, Aguilar et al. [163] tested nine *P. marquandii* isolates, and amphotericin B, miconazole, itraconazole, ketoconazole, fluconazole, and flucytosine had poor activity against this species. According to Lee et al. [104], the treatment of *P. marquandii* infection is a challenge as the species is resistant as mentioned. These authors described the isolation of resistant strains to all antifungals, except voriconazole, from a keratitis case.

Unfortunately, there is very little information about the in vitro antifungal activities against *P. javanicus*. Aguilar et al. [163] is the most detailed regarding antifungal susceptibility and that includes *P. javanicus*. They tested four *P. javanicus* isolates and demonstrated high resistance to fluconazole and flucytosine, and MICs of amphotericin B, miconazole, and ketoconazole were moderately low. Finally, itraconazole was effective against this species.

P. lilacinus (=*Purpureocillium lilacinum*) is much more resistant to amphotericin B than species described earlier. Pastor and Guarro [13] summarized the results of in vitro activity of antifungal agents against this species from 1978 to 2004. They showed that (1) amphotericin B, flucytosine, and fluconazole had poor in vitro activity; (2) ketoconazole, miconazole, clotrimazole, and itraconazole demonstrated contrasting data; (3) posaconazole, voriconazole, and ravuconazole showed variable activities; and (4) echinocandins, caspofungin, and micafungin had different MICs. Previous findings have shown good activities of voriconazole, posaconazole, and ravuconazole against *P. lilacinus* (=*Purpureocillium lilacinum*) and different activities of amphotericin B, terbinafine, and older and new azoles [114,147,161,166–169]. Recently, Wong et al. [116] isolated a multidrug-resistant strain from an immunocompetent patient with sinusitis, which was susceptible only to caspofungin and voriconazole. However, several reports have demonstrated that voriconazole alone or combined with other antifungal agent presents an in vitro activity against this fungus and has clinical efficacy [108,170].

25.9 EXPERIMENTAL MODELS

Despite several limitations, in vivo and in vitro experimental models remain essential to knowledge of host–fungal interactions, virulence of fungal strains, mechanisms of antifungal agent action, and its toxicity. However, few studies have been done with *Paecilomyces* species. Experimental pathogenicity of *P. variotii*, *P. lilacinus* (=*Purpureocillium lilacinum*), and *P. javanicus* showed that the former two species are more virulent than *P. javanicus* [171]. In addition, liposomal amphotericin B was more efficient to treat *P. variotii* experimental murine infection than amphotericin B deoxycholate and itraconazole [172,173]. *P. marquandii* caused splenomegaly and several nodules on liver with abundant spores, short hyphae, and yeast-like cells in immunocompetent mice [174]. *P. lilacinus* (=*Purpureocillium lilacinum*) is considered more virulent in experimental models than the other species [13,138,172,174–178]. Some authors state that the virulence of this species is generally low, and it is necessary to use a large inoculum and cause immunosuppression of the host for infection to take place [108]. On the other hand, Brito et al. [179] using immunocompetent and immunosuppressed BALB/c mice inoculated, intravenously, with a low inoculum (4.4×10^4 conidia) that showed an establishment of infection with observation of fungal structures in different organs and the recovery of fungal cells. Furthermore, immunocompetent and immunosuppressed C57BL/6 mice model were also infected with this species where a small inoculum was employed (D. Sequeira and C. Borba, personal communication) and IgG antibodies were readily detected by conventional indirect immunofluorescence and non-conventional flow cytometry approaches [233]. Study of antifungal efficiency using immunosuppressed murine model demonstrated that voriconazole reduced the fungal load in the spleen, kidneys, and liver of infected mice, otherwise amphotericin B was not able to reduce the tissue burden [178].

In vitro experimental models have been performing to show the pathogenicity of different *P. lilacinus* (=*Purpureocillium lilacinum*) isolates. Peixoto et al. [180] studied by optical microscopy the interaction of isolates from clinical forms of hyalohyphomycosis with C57BL/6 peritoneal and J774 lineage-mice macrophages and verified that, after 30 min of interaction, conidia was adhered to the host cells. The number of conidia inside the macrophages gradually increased after 1-day incubation. Germ tubes were produced and septate-branched hyphae were seen inside the macrophages that destroyed the host cells within 48 h. The three isolates studied by these authors were equally capable of surviving inside the phagocytes. Similarly, in vitro interactions of *P. lilacinus* (=*Purpureocillium lilacinum*) with peritoneal macrophages from inducible nitric oxide synthase knock-out mice (C. Ponte and C. Borba, personal communication) and human macrophages and dendritic cells have shown conidia and hyphae capable of surviving inside the phagocytes [234].

25.10 CONCLUDING REMARKS

Currently, some species of the genus *Paecilomyces* are included as less common fungal pathogen. Their identification is based mainly on morphological characteristics that is difficult and requires experienced professionals. This genus is similar to *Penicillium* but differs in the absence of greenish-colored colonies and by the short cylindrical phialides that taper into long necks. Molecular-based methods are gaining ground. Some *Paecilomyces* species are responsible for food- and feedstuffs spoilage and water deterioration. Some species produce mycotoxins in these products, and they can be considered as a potential source of public health problems.

P. variotii, *P. lilacinus* (=*Purpureocillium lilacinum*), *P. marquandii*, and *P. javanicus* are able to infect immunocompromised hosts with negative outcome in most cases including death, eye enucleation, and functional impairment. These species have been described to cause infection in immunocompetent patients except for *P. javanicus*.

Considering that human infections caused by *Paecilomyces* spp. constitute a devastating event, a better comprehension of the infection mechanisms may provide essential insights for potential novel treatments.

REFERENCES

1. Bainier, G. Mycothèque de l'école de Pharmacie. XI. *Paecilomyces*, genre nouveau de Mucédinées. *Bull. Soc. Mycol. Fr.* **23**, 26–27 (1907).
2. Hughes, S.J. Studies on micro-fungi. XI. Some Hyphomycetes, which produce phialides. *Mycol. Pap.* **45**, 1–36 (1951).
3. Brown, A.H.S. and Smith, G. The genus *Paecilomyces* Bainier and its perfect stage *Byssochlamys* Westling. *Trans. Br. Mycol. Soc.* **40**, 17–89 (1957).
4. Morris, E.F. *The Synnematous Genera of the Fungi Imperfecti: Western Illinois University Series in the Biological Sciences No. 3*, Western Illinois University, Macomb, IL (1963).
5. Onions, A.H.S. and Barron, G.L. Monophialidic species of *Paecilomyces*. *Mycol. Pap.* **107**, 1–25 (1967).
6. De Hoog, G.S. The genera *Beauveria*, *Isaria*, *Tritirachium* and *Acrodontium* gen. nov. *Stud. Mycol.* **1**, 1–41 (1972).
7. Samson, R.A. *Paecilomyces* and some allied Hyphomycetes. *Stud. Mycol.* **6**, 1–119 (1974).
8. Luangsa-ard, J.J., Hywel-Jones, N.L., and Samson, R.A. The polyphyletic nature of *Paecilomyces sensu lato* based on 18S-generated rDNA phylogeny. *Mycologia* **96**, 773–780 (2004).
9. Inglis, P.W. and Tigano, M.S. Identification and taxonomy of some entomopathogenic *Paecilomyces* spp. (Ascomycota) isolates using rDNA-ITS sequences. *Gen. Mol. Biol.* **29**, 132–136 (2006).
10. Luangsa-ard, J.J., Hywel-Jones, N.L., Manoch, L., and Samson, R.A. On the relationships of *Paecilomyces* sect. *Isarioidea* species. *Mycol. Res.* **109**, 581–589 (2005).
11. Samson, R.A., Houbraken, J., Varga, J., and Frisvad, J.C. Polyphasic taxonomy of the heat resistant ascomycete genus *Byssochlamys* and its *Paecilomyces* anamorphs. *Persoonia* **22**, 14–27 (2009).
12. De Hoog, G.S., Guarro, J., Gené, J., and Figueras, M.J. *Atlas of Clinical Fungi*, 2nd edn. Centraalbureau voor Schimmelcultures, Baarn, the Netherlands (2000).
13. Pastor, F.J. and Guarro, J. Clinical manifestations, treatment and outcome of *Paecilomyces lilacinus* infections. *Clin. Microbiol. Infect.* **12**, 948–960 (2006).
14. Luangsa-ard, J.J. et al. *Purpureocillium*, a new genus for the medically important *Paecilomyces lilacinus*. *FEMS Microbiol. Lett.* **321**, 141–149 (2011).
15. Tournas, V. Heat-resistant fungi of importance to the food and beverage industry. *Crit. Rev. Microbiol.* **20**, 243–263 (1994).
16. Tribst, A.A.L., Sant'Ana, A.S., and Massaguer, P.R. Review: Microbiological quality and safety of fruit juices—Past, present and future perspectives. *Crit. Rev. Microbiol.* **35**, 310–339 (2009).

17. Piecková, E. and Samson, R.A. Heat resistance of *Paecilomyces variotii* in sauce and juice. *J. Ind. Microbiol. Biotechnol.* **24**, 227–230 (2000).
18. Houbraken, J., Vargas, J., Rico-Munoz, E., Johnson, S., and Samson, R.A. Sexual reproduction as the cause of heat resistance in the food spoilage fungus *Byssochlamys spectabilis* (anamorph *Paecilomyces variotii*). *Appl. Environ. Microbiol.* **74**, 1613–1619 (2008).
19. Splittstoesser, D.F., Kuss, F.R., Harrison, W., and Prest, D.B. Incidence of heat-resistant molds in eastern orchards and vineyards. *Appl. Microbiol.* **21**, 335–337 (1971).
20. Ismail, M.A. Deterioration and spoilage of peanuts and desiccated coconuts from two sub-Saharan tropical East African countries due to the associated mycobiota and their degradative enzymes. *Mycopathologia* **150**, 67–84 (2000).
21. Lugauskas, A. Potential toxin producing micromycetes on food raw material and products of plant origin. *Bot. Lithuanica* **7**, 3–16 (2005).
22. Labuda, R. and Tancinová, D. Fungi recovered from Slovakian poultry feed mixtures and their toxinogenity. *Ann. Agric. Environ. Med.* **13**, 193–200 (2006).
23. Biro, D. et al. Occurrence of microscopic fungi and mycotoxins in conserved high moisture corn from Slovakia. *Ann. Agric. Environ. Med.* **16**, 227–232 (2009).
24. Hageskal, G., Gaustad, P., Heier, B.T., and Skaar, I. Ocurrence of moulds in drinking water. *J. Appl. Microbiol.* **102**, 774–780 (2007).
25. Nagy, L.A. and Olson, B.H. The occurrence of filamentous fungi in drinking water distribution systems. *Can. J. Microbiol.* **28**, 667–671 (1982).
26. Doggett, M.S. Characterization of fungal biofilms within a municipal water distribution system. *Appl. Environ. Microbiol.* **66**, 1249–1251(2000).
27. Anaissie, E.J. et al. Pathogenic molds (including *Aspergillus* species) in hospital water distribution systems: A 3-year prospective study and clinical implications for patients with hematologic malignancies. *Blood* **101**, 2542–2546 (2003).
28. Hageskal, G., Knutsen, A.K., Gaustad, P., Sybren de Hoog, G., and Skaar, I. Diversity and significance of mold species in Norwegian drinking water. *Appl. Environ. Microbiol.* **72**, 7586–7593 (2006).
29. Ribeiro, A. et al. Fungi in bottled water: A case study of a production plant. *Rev. Iberoam. Micol.* **23**, 139–144 (2006).
30. Kanzler, D. et al. Occurrence and hygienic relevance of fungi in drinking water. *Mycoses* **51**, 165–169 (2007).
31. Hayette, M. et al. Filamentous fungi recovered from the water distribution system of a Belgium university hospital. *Med. Mycol.* **48**, 969–974 (2010).
32. Defra (Department for Environment, Food & Rural Affairs). *A Review of Fungi in Drinking Water and the Implications for Human Health*. Reference: WD 0906. Final Report. Bio Intelligence Service, Paris, France, pp. 1–107 (2011).
33. Hedayati, M.T., Mayahi, S., Movahedi, M., and Shokohi, T. Study on fungal flora of tap water as a potential reservoir of fungi in hospitals in Sari city, Iran. *J. Mycol. Med.* **21**, 10–14 (2011).
34. Siqueira, V.M. et al. Filamentous fungi in drinking water, particularly in relation to biofilm formation. *Int. J. Environ. Res. Public Health* **8**, 456–469 (2011).
35. Hageskal, G., Lima, N., and Skaar, I. The study of fungi in drinking water. *Mycol. Res.* **113**, 165–172 (2009).
36. Bhat, R., Rair, R.V., and Karim, A.A. Mycotoxins in food and feed: Present status and future concerns. *Compr. Rev. Food. Sci. Food Saf.* **9**, 57–81 (2010).
37. Cast (Council for Agricultural Science and Technology). *Mycotoxins: Risks in Plant, Animal, and Human Systems*. Task Force Report, Ames, IA, Vol. 139, pp. 1–196 (2003).
38. Niessen, L. PCR-based diagnosis and quantification of mycotoxin producing fungi. *Int. J. Food Microbiol.* **119**, 38–46 (2007).
39. Paterson, R.R.M. The isoepoxydon dehydrogenase gene of patulin biosynthesis in cultures and secondary metabolites as candidate PCR inhibitors. *Mycol. Res.* **108**, 1431–1437 (2004).
40. Paterson, R.R.M. The isoepoxydon dehydrogenase gene PCR profile is useful in fungal taxonomy. *Rev. Iberoam. Micol.* **24**, 289–93 (2007).
41. Hopmans, E.C. Patulin: A mycotoxin in apples. *Perish. Hand Q.* **91**, 5–6 (1997).
42. Horn, W.S. et al. Sphingofungins E and F: Novel serinepalmitoyl transferase inhibitors from *Paecilomyces variotii*. *J. Antibiot. (Tokyo)* **45**, 1692–1696 (1992).
43. Jiu, J. and Mizuba, S. Metabolic products from *Spicaria divaricata* NRRL 5771. *J. Antibiot. (Tokyo)* **27**, 760–765 (1974).

44. Wang, J. et al. Discovery of a small molecule that inhibits cell division by blocking FtsZ, a novel therapeutic target of antibiotics. *J. Biol. Chem.* **278**, 4424–4428 (2003).
45. Sakaguchi, K., Inoue, T., and Tada, S. On the production of ethyleneoxide-α, β-dicarboxylic acid by moulds. *Zbl. Bakt. Paras-Kunde (Abt. II)* **100**, 302–307 (1939).
46. von Loesecke, H.W. A review of information on mycological citric acid production. *Chem. Eng. News* **23**, 1952–1959 (1945).
47. Burton, H.S. Antibiotics from Penicillia. *Br. J. Exp. Pathol.* **30**, 151–158 (1949).
48. Takeuchi, S., Yonehara, H., and Umezawa, H. Studies on variotin, a new antifungal antibiotic 1. Preparations and properties of variotin. *J. Antibiot. (Tokyo)* **12**, 195–200 (1959).
49. Takeuchi, S., Yonehara, H., and Shoji, H. Crystallized variotin and its revised chemical structure. *J. Antibiot. (Tokyo)* **17**, 267–268 (1964).
50. Bakalinerov, D. Indole derivatives in soil microscopic fungi. *Pochv. Agrokhim.* **3**, 81–86 (1968).
51. Aldridge, D.C., Carman, R.M., and Moore, R.B. A new tricarboxylic acid anhydride from *Paecilomyces variotii*. *J. Chem. Soc. Perkin. Trans.* **1**, 2134–2135 (1980).
52. Domsch, K.H., Gams, W., and Anderson, T.H. *Compendium of Soil Fungi*. Academic Press, London, U.K. (1980).
53. Suzuki, K., Nozawa, K., Nakajima, S., and Kawai, K. Structure revision of mycotoxin, viriditoxin, and its derivatives. *Chem. Pharm. Bull.* **38**, 3180–3181 (1990).
54. Nakajima, M. et al. Cornexitin: A new fungal metabolite with herbicidal activity. *J. Antibiot. (Tokyo)* **44**, 1065–1072 (1991).
55. Fields, S.C., Mireles-Lo, L., and Gerwick, B.C. Hydroxycornexitin: A new phytotoxin from *Paecilomyces variotii*. *J. Nat. Prod.* **59**, 698–700 (1996).
56. Omolo, J.O., Anke, H., Chhabra, S., and Sterner, O. New variotin analogues from *Aspergillus viridinutans*. *J. Nat. Prod.* **63**, 975–977 (2000).
57. Hedge, V.R. et al. Novel fungal metabolites as cell wall active antifungals: Fermentation, isolation, physico-chemical properties, structure and biological activity. *J. Antibiot. (Tokyo)* **56**, 437–447 (2003).
58. Arai, T., Mikami, Y., Fukushima, K., Utusumi, T., and Yazawa, K. A new antibiotic, leucinostatin, derived from *Penicillium lilacinum*. *J. Antibiot. (Tokyo)* **26**, 157–161 (1973).
59. Mori, Y., Tsuboi, M., Makoto, S., Fukushima, K., and Arai, T. Isolation of leucinostatin A and one of its constituents, the new amino acid, 4-metthyl-6-(2-oxobutyl)-2 piperidinecarboxylic acid, from *Paecilomyces lilacinus* A-267. *J. Antibiot. (Tokyo)* **35**, 543–544 (1982).
60. Mikami, Y. et al. Leucinostatins, peptide mycotoxins produced by *Paecilomyces lilacinus* and their possible roles in fungal infection. *Zentralbl. Bakteriol. Mikrobiol. Hyg. A* **257**, 275–283 (1984).
61. Park, J.O. et al. Production of leucinostatins and nematicidal activity of Australian isolates of *Paecilomyces lilacinus* (Thom) Samson. *Lett. Appl. Microbiol.* **38**, 271–276 (2004).
62. Mikami, Y. et al. Paecilotoxin production in clinical or terrestrial isolates of *Paecilomyces lilacinus*. *Mycopathologia* **108**, 195–199 (1989).
63. Khan, A., Williams, K., and Nevalainen, H. Testing the nematophagous biological control strain *Paecilomyces lilacinus* 251 for paecilotoxin production. *FEMS Microbiol. Lett.* **227**, 107–111 (2003).
64. Bonants, P.J.M. et al. A basic serine protease from *Paecilomyces lilacinus* with biological activity against *Meloidogyne hapla* eggs. *Microbiology* **141**, 775–784 (1995).
65. Khan, A., Williams, K., Molloy, M.P., and Nevalainen, H. Purification and characterization of a serine protease and chitinases from *Paecilomyces lilacinus* and detection of chitinase activity on 2D gels. *Protein Expr. Purif.* **32**, 210–220 (2003).
66. Kotlova, E.K. et al. Thiol-dependent serine proteinase from *Paecilomyces lilacinus*: Purification and catalytic properties. *Biochemistry (Mosc)* **72**, 117–123 (2007).
67. Ye, F. et al. Preliminary crystallographic study of two cuticle-degrading proteases from the nematophagous fungi *Lecanicillium psalliotae* and *Paecilomyces lilacinus*. *Acta Crystallogr. Sec. F: Struct. Biol. Cryst Commun.* **65**, 271–274 (2009).
68. Casinovi, C.G., Tuttobello, L., Rossi, C., and Benciari, Z. Structural elucidation of the phytotoxic antibiotic peptides produced by submerged cultures of *Paecilomyces marquandii* (Massee) Hughes. *Phytopathol. Mediterr.* **22**, 103–106 (1983).
69. Rossi, C., Benciari, Z., Casinovi, C.G., and Tuttobello, L. Two phytotoxic antibiotic peptides produced by submerged cultures of *Paecilomyces marquandii* (Massee) Hughes. *Phytopathol. Mediterr.* **22**, 209–211 (1983).
70. Rossi, C., Casinovi, C.G., and Radics, L. Leucinostatin D, a novel peptide antibiotic from *Paecilomyces marquandii*. *J. Antibiot. (Tokyo)* **40**, 130–33 (1987).

71. Radics, L. et al. Leucinostatins H and K, two novel peptide antibiotics with tertiary amine-oxide terminal group from *Paecilomyces marquandii* isolation, structure and biological activity. *J. Antibiot. (Tokyo)* **40**, 714–716 (1987).
72. Petersen, F. et al. Paeciloquinones A, B, C, D, E and F: New potent inhibitors of protein tyrosine kinases produced by *Paecilomyces carneus* I. Taxonomy, fermentation, isolation and biological activity. *J. Antibiot. (Tokyo)* **48**, 191–198 (1995).
73. Chung, E.J., Choi, K., Kim, H.W., and Lee, D.H. Analysis of cell cycle gene expression responding to acetoxyscirpendiol isolated from *Paecilomyces tenuipes*. *Biol. Pharm. Bull.* **26**, 32–36 (2003).
74. Nam, K.S., Jo, Y.S., Kim, Y.H., Hyun, J.W., and Kim, H.W. Cytotoxic activities of acetoxyscirpenediol and ergosterol peroxide from *Paecilomyces tenuipes*. *Life Sci.* **69**, 229–237 (2001).
75. Nilanonta, C. et al. Antimycobacterial and antiplasmodial cyclodepsipeptides from the insect pathogenic fungus *Paecilomyces tenuipes* BCC 1614. *Planta Med.* **66**, 756–758 (2000).
76. Kikuchi, H. et al. Novel spirocyclic trichothecanes, spirotenuipesine A and B, isolated from entamopathogenic fungus, *Paecilomyces tenuipes*. *J. Org. Chem.* **69**, 352–356 (2004).
77. Kikuchi, H. et al. A novel carbon skeletal trichothecane, tenuipesine A, isolated from an entomopathogenic fungus, *Paecilomyces tenuipes*. *Org. Lett.* **6**, 4531–4533 (2004).
78. Haritakun, R., Srikitikulchai, P., Khoyaiklang, P., and Isaka, M. Isariotins A–D, alkaloids from the insect pathogenic fungus *Isaria tenuipes* BCC 7831. *J. Nat. Prod.* **70**, 1478–1480 (2007).
79. Yin, Y.Y. et al. Bioactive compounds from *Paecilomyces tenuipes* regulating the function of the hypothalamo-hypophyseal system axis in chronic unpredictable stress rats. *Chin. Med. J. (Engl)* **120**, 1088–1092 (2007).
80. Ajello, L. and McGinnis, M.R. Factors of the microbe colonization. In *Nomenclature of Human Pathogenic Fungi, Basis of the Antiseptic*. Part 4, ed. Krasilnikow, A.P. Peoples Publishing House and Health, Berlin, Germany, pp. 363–377 (1984).
81. Matsumoto, T., Ajello, L., Matsuda, T., Szaniszlo, P.J., and Walsh, T.J. Developments in hyalohyphomycosis and phaeohyphomycosis. *J. Med. Vet. Mycol.* **32**, 329–349 (1994).
82. Das, S., Saha, R., Dar, S.S., and Ramachandran, V.G. *Acremonium* species: A review of the etiological agents of emerging hyalohyphomycosis. *Mycopathologia* **170**, 361–375 (2010).
83. Haldane, E.V., McDonald, J.L., Gittens, W.O., Yuce, K., and van Rooyen, C.E. Prosthetic valvular endocarditis due to the fungus *Paecilomyces*. *Can. Med. Assoc. J.* **111**, 963–968 (1974).
84. Rowley, S.D. and Strom, C.G. *Paecilomyces* fungus infection of the maxillary sinus. *Laryngoscope* **92**, 332–334 (1982).
85. Wilhelmus, K.R., Robinson, N.M., Font, R.A., Hamill, M.B., and Jones, D.B. Fungal keratitis in contact lens wearers. *Am. J. Ophthalmol.* **106**, 708–714 (1988).
86. Lye, W.C. *Paecilomyces* peritonitis in a patient on continuous ambulatory peritoneal dialysis. *Nephrol. Dial. Transplant.* **5**, 1053–1054 (1990).
87. Leigheb, G. et al. Sporotrichosis-like lesions caused by a *Paecilomyces* genus fungus. *Int. J. Dermatol.* **33**, 275–276 (1994).
88. Mizunoya, S. and Watanabe, Y. *Paecilomyces* keratitis with corneal perforation salvaged by a conjunctival flap and delayed keratoplasty. *Br. J. Ophthalmol.* **78**, 157–158 (1994).
89. Castro, L.G. Sporotrichosis-like lesions caused by a *Paecilomyces* genus fungus. *Int. J. Dermatol.* **34**, 364 (1995).
90. Alscher, D.M. et al. Moulds in containers with biological wastes as a possible source of peritonitis in two patients on peritoneal dialysis. *Perit. Dial. Int.* **18**, 643–646 (1998).
91. Gutiérrez, F. et al. Pulmonary mycetoma caused by an atypical isolate of *Paecilomyces* species in an immunocompetent individual: Case report and literature review of *Paecilomyces* lung infections. *Eur. J. Clin. Microbiol. Infect. Dis.* **24**, 607–611 (2005).
92. Veillette, M., Cormier, Y., Israël-Assayaq, E., Meriaux, A., and Duchaine, C. Hypersensitivity pneumonitis in a hardwood processing plant related to heavy mold exposure. *J. Occup. Environ. Hyg.* **3**, 301–307 (2006).
93. Yildiz, E.H. et al. *Alternaria* and *Paecilomyces* keratitis associated with soft contact lens wear. *Cornea* **29**, 564–568 (2010).
94. Kim, J.E. et al. Synchronous infection with *Mycobacterium chelonae* and *Paecilomyces* in a heart transplant patient. *Transpl. Infect. Dis.* **13**, 80–83 (2011).
95. Rai, S., Tiwari, R., Sandhu, S.V., and Rajkumar, Y. Hyalohyphomycosis of maxillary antrum. *J. Oral. Maxillofac. Pathol.* **16**, 149–152 (2012).
96. Uys, C.J., Don, P.A., Schrire, V., and Barnard, C.N. Endocarditis following cardiac surgery due to the fungus *Paecilomyces*. *S. Afr. Med. J.* **37**, 1276–1280 (1963).

97. Bibashi, E. et al. Three cases of uncommon fungal peritonitis in patients undergoing peritoneal dialysis. *Perit. Dial. Int.* **22**, 523–525 (2002).
98. Anita, K.B., Fernandez, N., and Rao, R. Fungal endophthalmitis caused by *Paecilomyces variotii*, in an immunocompetent patient, following intraocular lens implantation. *Indian J. Med. Microbiol.* **28**, 253–254 (2010).
99. Lam, D.S.C. et al. Endogenous fungal endophthalmitis caused by *Paecilomyces variotii*. *Eye (Lond)* **13**, 113–116 (1999).
100. Kondo, T., Morikawa, Y., and Hayashi, N. Purification and characterization of alcohol oxidase from *Paecilomyces variotii* isolated as a formaldehyde-resistant fungus. *Appl. Microbiol. Biotechnol.* **77**, 995–1002 (2008).
101. Fukuda, R. et al. Purification and properties of S-hydroxymethylglutathione dehydrogenase of *Paecilomyces variotii* a formaldehyde-degrading fungus. *AMB Exp.* **2**, 32 (2012).
102. Harris, L.F., Dan, B.M., Lefkowitz, L.B. Jr., and Alford, R.H. *Paecilomyces* cellulitis in a renal transplant patient: Successful treatment with intravenous miconazole. *South Med. J.* **72**, 897–898 (1979).
103. Naldi, L., Lovati, S., Farina, C., Gotti, E., and Cainelli, T. *Paecilomyces marquandii* cellulitis in a kidney transplant patient. *Br. J. Dermatol.* **143**, 647–649 (2000).
104. Lee, G.A., Whitehead, K., and McDougall, R. Management of *Paecilomyces* keratitis. *Eye (Lond)* **21**, 262–264 (2007).
105. Allevato, P.A., Ohorodnik, J.M., Mezger, E., and Eisses, J.F. *Paecilomyces javanicus* endocarditis of native and prosthetic aortic valve. *Am. J. Clin. Pathol.* **82**, 247–252 (1984).
106. Ho, K.L. and Allevato, P.A. Hirano body in an inflammatory cell of leptomeningeal vessel infected by fungus *Paecilomyces*. *Acta Neuropathol.* **71**, 159–162 (1986).
107. Ho, K.L., Allevato, P.A., King, P., and Chason, J.L. Cerebral *Paecilomyces javanicus* infection. An ultrastructural study. *Acta Neuropathol.* **72**, 134–141 (1986).
108. Antas, P.R.Z., Brito, M.M.S., Peixoto, E., Ponte, C.G.C., and Borba, C.M. Neglected and emerging fungal infections: Review of hyalohyphomycosis by *Paecilomyces lilacinus* focusing in disease burden, in vitro antifungal susceptibility and management. *Microbes Infect.* **14**, 1–8 (2011).
109. Volná, F. and Máderová, E. *Paecilomyces lilacinus*—Sensitivity to disinfectants. *Cesk. Epidemiol. Mikrobiol. Imunol.* **39**, 315–317 (1990).
110. Permi, H.S. et al. A rare case of fungal maxillary sinusitis due to *Paecilomyces lilacinus* in an immunocompetent host, presenting as a subcutaneous swelling. *J. Lab. Phys.* **3**, 46–48 (2011).
111. Maier, A.K., Reichenbach, A., and Rieck, P. *Paecilomyces lilacinus* keratitis. *Ophthalmologe* **108**, 966–968 (2011).
112. Keshtkar-Jahromi, M., McTighe, A.H., Segalman, K.A., Fothergill, A.W., and Campell, W.N. Unusual case of cutaneous and synovial *Paecilomyces lilacinus* infection of hand successfully treated with voriconazole and review of published literature. *Mycopathologia* **174**, 255–258 (2012).
113. Trachsler, S., Eberhard, R., Kocher, C., and Fleischhauer, J. *Paecilomyces lilacinus* endophthalmitis following cataract surgery: A therapeutic challenge. *Klin. Monbl. Augenheilkd.* **229**, 441–442 (2012).
114. Ezzedine, K. et al. Cutaneous hyphomycosis due to *Paecilomyces lilacinus*. *Acta Derm. Venereol.* **92**, 156–157 (2012).
115. Khan, Z. et al. *Purpureocillium lilacinum* as a cause of cavitary pulmonary disease: A new clinical presentation and observations on atypical morphologic characteristics of the isolate. *J. Clin. Microbiol.* **50**, 1800–1804 (2012).
116. Wong, G., Nash, R., Barai, K., Rathod, R., and Singh, A. *Paecilomyces lilacinus* causing debilitating sinusitis in an immunocompetent patient: A case report. *J. Med. Case Rep.* **6**, 86 (2012).
117. Ali-Shtayeh, M.S., Arda, H.M., Hassouna, M., and Shaheen, S.F. Keratinophilic fungi on sheep hairs from the West Bank of Jordan. *Mycopathologia* **106**, 95–101 (1989).
118. Cabanes, F.J., Abarca, M.L., Bragulat, M.R., and Castella, G. Seasonal study of the fungal biota of the fur of dogs. *Mycopathologia* **133**, 1–7 (1996).
119. Bagy, M.M., el-Shanawany, A.A., and Abdel-Mallek, A.Y. Saprophytic and cycloheximide resistant fungi isolated from golden hamster. *Acta Microbiol. Immunol. Hung.* **45**, 195–207 (1998).
120. Hubálek, Z., Rosicky, B., and Otcenasek, M. Fungi from interior organs of free-living small mammals in Czechoslovakia and Yugoslavia. *Folia Parasitol. (Praha)* **27**, 269–279 (1980).
121. Abdel-Gawad, K.M. and Moharram, A.M. Keratinophilic fungi from the duck nails in Egypt. *J. Basic Microbiol.* **29**, 259–263 (1989).
122. Mountfort, D.O. and Rhodes, L.L. Anaerobic growth and fermentation characteristics of *Paecilomyces lilacinus* isolated from mullet gut. *Appl. Environ. Microbiol.* **57**, 1963–1968 (1991).

123. Holahan, M.L., Loft, K.E., Swenson, C.L., and Martinez-Ruzafa, I. Generalized calcinosis cutis associated with disseminated paecilomycosis in a dog. *Vet. Dermatol.* **19**, 368–372 (2008).
124. Fleischman, R.W. and McCracken, D. Paecilomycosis in a nonhuman primate (*Macaca mulatta*). *Vet. Pathol.* **14**, 387–391 (1977).
125. Jacobson, E.R., Gaskin, J.M., Shield, R.P., and White, F.H. Mycotic pneumonia in mariculture-reared green sea turtles. *J. Am. Vet. Med. Assoc.* **175**, 929–933 (1979).
126. Nakagawa, Y. et al. A canine case of profound granulomatosis due to *Paecillomyces* fungus. *J. Vet. Med. Sci.* **58**, 157–159 (1996).
127. March, P.A., Knowles, K., Dillavou, C.L., Lakowski, R., and Freden, G. Diagnosis, treatment, and temporary remission of disseminated paecilomycosis in a vizsla. *Am. Anim. Hosp. Assoc.* **32**, 509–514 (1996).
128. García, M.E. et al. Disseminated mycoses in a dog by *Paecilomyces* sp. *J. Vet. Med. A: Physiol. Pathol. Clin. Med.* **47**, 243–249 (2000).
129. Foley, J.E., Norris, C.R., and Jang, S.S. Paecilomycosis in dogs and horses and a review of the literature. *J. Vet. Intern. Med.* **16**, 238–243 (2002).
130. Cheatwood, J.L. et al. An outbreak of fungal dermatitis and stomatitis in a free-ranging population of pigmy rattlesnakes (*Sistrurus miliarius barbouri*) in Florida. *J. Wildl. Dis.* **39**, 329–337 (2003).
131. Laclaire, M.C., Morand, M., and Euzeby, J. Investigation mycologique. Note II. Recherche sur quelques syndromes pseudo-aspergillaires. Paecilomycose des sacs aériens des poumons. *Bull. Soc. Sci. Vét. Méd. Comp. Lyon* **76**, 317–319 (1974).
132. Pepin, G.A. and Pritchard, G.C. Fungal mastitis in a goat due to infection with *Paecilomyces variotii*. *Vet. Med. J.* **5**, 12 (1984).
133. Littman, M.P. and Goldschmidt, M.H. Systemic paecilomycosis in a dog. *J. Am. Vet. Med. Assoc.* **191**, 445–447 (1987).
134. Kunstyr, I., Jelinek, F., Bitzenhofer, U., and Pittermann, W. Fungus *Paecilomyces*: A new agent in laboratory animals. *Lab. Anim.* **31**, 45–51 (1997).
135. France, M.P. and Muir, D. An outbreak of pulmonary mycosis in respiratory burst-deficient (gp91(phox-/-) mice with concurrent acidophilic macrophage pneumonia. *J. Comp. Pathol.* **123**, 190–194 (2000).
136. Booth, M.J., Van der Lugt, J.J., Van Heerden, A., and Picard, J.A. Temporary remission of disseminated paecilomycosis in a German shepherd dog treated with ketoconazole. *J. S. Afr. Vet. Assoc.* **72**, 99–104 (2001).
137. Han, J.I. and Na, K.J. Cutaneous paecilomycosis caused by *Paecilomyces variotii* in an african pygmy hedgehog (*Atelerix albiventris*). *J. Exot. Pet. Med.* **19**, 309–312 (2010).
138. Gordon, M.A. *Paecilomyces lilacinus* (Thom) Samson, from systemic infection in an armadillo (*Dasypus novemcinctus*). *Sabouraudia* **22**, 109–116 (1984).
139. Heard, D.J. et al. Hyalohyphomycosis caused by *Paecilomyces lilacinus* in an *Aldabra* tortoise. *J. Am. Vet. Med. Assoc.* **189**, 1143–1145 (1986).
140. Maslen, M., Whitehead, J., Forsyth, W.M., McCracken, H., and Hocking, A.D. Systemic mycotic disease of captive crocodile hatchling (*Crocodylus porosus*) caused by *Paecilomyces lilacinus*. *J. Med. Vet. Mycol.* **26**, 219–225 (1988).
141. Rosser, E.J. Jr. Cutaneous paecilomycosis in a cat. *J. Am. Anim. Hosp. Assoc.* **39**, 543–546 (2003).
142. Pawloski, D.R., Brunker, J.D., Singh, K., and Sutton, D.A. Pulmonary *Paecilomyces lilacinus* infection in a cat. *J. Am. Anim. Hosp. Assoc.* **46**, 197–202 (2010).
143. Posthaus, K. et al. Systemic paecilomycosis in a hawksbill turtle (*Eretmochelys imbricate*). *J. Mycol. Méd.* **7**, 223–226 (1997).
144. Li, X.L., Zhang, C.L., Fang, W.H., and Lin, F.C. White-spot disease of Chinese soft-shelled turtles (*Trionyx sinens*) caused by *Paecilomyces lilacinus*. *J. Zhejiang. Univ. Sci. B.* **9**, 578–581 (2008).
145. Marancik, D.P. et al. Disseminated fungal infection in two species of captive sharks. *J. Zoo Wildl. Med.* **42**, 686–693 (2011).
146. Saberhagen, C., Klotz, S.A., Bartholomew, W., Drews, D., and Dixon, A. Infection due to *Paecilomyces lilacinus*: A challenging clinical identification. *Clin. Infect. Dis.* **25**, 1411–1413 (1997).
147. Castelli, M.V. et al. Susceptibility testing and molecular classification of *Paecilomyces* spp. *Antimicrob. Agents Chemother.* **52**, 2926–2928 (2008).
148. Houbraken, J., Verweij, P.E., Rijs, A.J.M.M., Borman, A.M., and Samson, R.A. Identification of *Paecilomyces variotii* in clinical samples and settings. *J. Clin. Microbiol.* **48**, 2754–2761 (2010).
149. Liu, K., Howell, D.N., Perfect, J.R., and Schell, W.A. Morphologic criteria for the preliminary identification of *Fusarium*, *Paecilomyces*, and *Acremonium* species by histopathology. *Am. J. Clin. Pathol.* **109**, 45–54 (1998).

150. Dharmasema, F.M.C., Davies, G.S.R., and Catovsky, D. *Paecilomyces varioti* pneumonia complicating hairy cell leukaemia. *Br. Med. J. (Clin. Res. Ed.)* **290**, 967–968 (1985).
151. Marzec, A. et al. *Paecilomyces variotii* in peritoneal dialysate. *J. Clin. Microbiol.* **31**, 2392–2395 (1993).
152. Kovac, D. et al. Treatment of severe *Paecilomyces variotii* peritonitis in a patient on continuous ambulatory peritoneal dialysis. *Nephrol. Dial. Transplant.* **13**, 2943–2946 (1998).
153. Lee, J. et al. Delayed sternotomy wound infection due to *Paecilomyces variotii* in a lung transplant recipient. *J. Heart Lung Transplant.* **21**, 1131–1134 (2002).
154. Sriram, K., Mathews, M.S., and Goplakrishnan, G. *Paecilomyces* pyelonephritis in a patient with urolithiasis. *J. Urol.* **23**, 195–197 (2007).
155. Senior, J.M. and Saldarriaga, C. Endocarditis infecciosa por *Paecilomyces variotii*. *Biomedica* **29**, 177–180 (2009).
156. Byrd, R.P. Jr., Roy, T.M., Fields, C.L., and Lynch, J.A. *Paecilomyces varioti* pneumonia in a patient with diabetes mellitus. *J. Diabetes Complications* **6**, 150–153 (1992).
157. Rinaldi, S., Fiscarelli, E., and Rizzoni, G. *Paecilomyces variotii* peritonitis in an infant on automated peritoneal dialysis. *Pediatr. Nephrol.* **14**, 365–366 (2000).
158. Tarkkanen, A. et al. Fungal endophthalmitis caused by *Paecilomyces variotii* following cataract surgery: A presumed operating room air-conditioning system contamination. *Acta Ophthalmol. Scand.* **82**, 232–235 (2004).
159. Wang, S.M., Shieh, C.C., and Liu, C.C. Successful treatment of *Paecilomyces variotii* splenic abscesses: A rare complication in a previously unrecognized chronic granulomatous disease child. *Diagn. Microbiol. Infect. Dis.* **53**, 149–152 (2005).
160. Lin, W.L, Lin, W.C., and Chiu, C.S. *Paecilomyces lilacinus* cutaneous infection associated peripherally inserted central catheter insertion. *J. Eur. Acad. Dermatol. Venerol.* **22**, 1236–1278 (2008).
161. Motswaledi, H.M., Mathekga, K., Sein, P.P., and Nemutayhanani, D.L. *Paecilomyces lilacinus* eumycetoma. *Int. J. Dermatol.* **48**, 858–861 (2009).
162. Naidu, J. and Singh, S.M. Hyalohyphomycosis caused by *Paecilomyces variotii*: A case report, animal pathogenicity and in vitro sensitivity. *Antonie Van Leeuwenhoek* **62**, 225–230 (1992).
163. Aguilar, C., Pujol, I., Sala, J., and Guarro, J. Antifungal susceptibilities of *Paecilomyces* species. *Antimicrob. Agents Chemother.* **42**, 1601–1604 (1998).
164. Chamilos, G. and Kontoviannis, D.P. Voriconazole-resistant disseminated *Paecilomyces variotii* infection in a neutropenic patient with leukaemia on voriconazole prophylaxis. *J. Infect.* **51**, e225–e228 (2005).
165. Lamagni, T.L., Campell, C., Pezzoli, L., and Johnson, E. Unexplained increase in *Paecilomyces variotii* blood culture isolates in the UK. *Euro Surveill.* **11**, E061116.2 (2006).
166. Gonzalez, G.M., Fothergill, A.W., Sutton, D.A., Rinaldi, M.G., and Loebenberg, D. In vitro activities of new and established triazoles against opportunistic filamentous and dimorphic fungi. *Med. Mycol.* **43**, 281–284 (2005).
167. Sponsel, W. et al. Topical voriconazole as a novel treatment for fungal keratitis. *Antimicrob. Agents Chemother.* **50**, 262–268 (2006).
168. Wessolossky, M., John, P.H., and Bagchi, K. *Paecilomyces lilacinus* olecranon bursitis in an immunocompromised host: Case report and review. *Diagn. Microbiol. Infect. Dis.* **61**, 354–357 (2008).
169. Rosmaninho, A. et al. *Paecilomyces lilacinus* in transplant patients: An emerging infection. *Eur. J. Dermatol.* **20**, 643–644 (2010).
170. Chang, B.P. et al. *Paecilomyces lilacinus* peritonitis complicating peritoneal dialysis cured by oral voriconazole and terbinafine combination therapy. *J. Med. Microbiol.* **57**, 1581–1584 (2008).
171. Pujol, I. et al. Experimental pathogenicity of three opportunistic *Paecilomyces* species in murino model. *J. Micol. Méd.* **12**, 86–89 (2002).
172. Pujol, I., Ortoneda, M., Aguilar, C., Pastor, F.J., and Guarro, J. Experimental treatment of murine infection by *Paecilomyces variotii*. *Mycoses* **45**(Suppl. 2), 48 (2002).
173. Pujol, I., Aguilar, C., Ortoneda, M., Pastor, J., and Guarro, J. Experimental treatment of murine disseminated infection by *Paecilomyces variotii*. *J. Mycol. Méd.* **13**, 141–143 (2003).
174. Hubálek, Z. and Hornich, M. Experimental infection of white mouse with *Chrysosporium* and *Paecilomyces*. *Mycopathologia* **62**, 173–78 (1977).
175. Agrawal, P.K., Lal, B., Wahab, S., Srivastrava, O.P., and Misra, S.C. Orbital paecilomycosis due to *Paecilomyces lilacinus* (Thom) Samson. *Sabouraudia* **17**, 363–370 (1979).
176. Yuan, X. et al. Pathogenesis and outcome of *Paecilomyces* keratitis. *Am. J. Ophthalmol.* **147**, 691–696 (2009).
177. Rodríguez, M.M., Pastor, F.J., Serena, C., and Guarro, J. Efficacy of voriconazole in a murine model of invasive paecilomycosis. *Int. J. Antimicrob. Agents* **35**, 362–365 (2010).

178. Rodríguez, M.M., Pastor, F.J., Serena, C., and Guarro, J. Posaconazole efficacy in a murine disseminated infection caused by *Paecilomyces lilacinus*. *J. Antimicrob. Chemother.* **63**, 361–364 (2009).
179. Brito, M.M.S. et al. Characteristics of *Paecilomyces lilacinus* infection comparing immuncompetent with immunosuppressed murine model. *Mycoses* **54**, e513–e521 (2011).
180. Peixoto, E., Oliveira, J.C., Antas, P.R.Z., and Borba, C.M. In vitro study of the host-parasite interactions between mouse macrophages and the opportunistic fungus *Paecilomyces lilacinus*. *Ann. Trop. Med. Parasitol.* **104**, 529–534 (2010).
181. Williamson, P.E., Kwon-Chung, K.J., and Gallin, J.I. Successful treatment of *Paecilomyces variotii* infection in a patient with chronic granulomatous disease and a review of *Paecilomyces* species infections. *Clin. Infect. Dis.* **14**, 1023–1026 (1992).
182. Athar, M.A., Sekhon, A.S., Mcgrath, J.V., and Malone, R.M. Hyalohyphomycosis caused by *Paecilomyces variotii* in an obstetrical patient. *Eur. J. Epidemiol.* **12**, 33–35 (1996).
183. Abbas, S.Q. et al. A report of *Paecilomyces variotii* on human from Pakistan. *Pak. J. Bot.* **41**, 467–472 (2009).
184. Silver, M.D., Tuffnell, P.G., and Bigelow, W.G. Endocarditis caused by *Paecilomyces variotii* affecting an aortic valve allograft. *J. Thorac. Cardiovasc. Surg.* **61**, 278–281 (1971).
185. McClellan, J.R., Hamilton, J.D., Alexander, J.A., Wolfe, W.G., and Reed, J.B. *Paecilomyces variotii* endocarditis on a prosthetic aortic valve. *J. Thorac. Cardiovasc. Surg.* **71**, 472–475 (1976).
186. Kalish, S.B. et al. Infective endocarditis caused by *Paecilomyces variotii*. *Am. J. Clin. Pathol.* **78**, 249–252 (1982).
187. Shing, M.M., Ip, M., Li, C.K., Chik, K.W., and Yuen, P.M. *Paecilomyces varioti* fungemia in a bone marrow transplant patient. *Bone Marrow Transplant.* **17**, 281–283 (1996).
188. Salle, V. et al. *Paecilomyces variotii* fungemia in a patient with multiple myeloma: Case report and literature review. *J. Infect.* **51**, e93–e95 (2005).
189. Cohen-Abbo, A. and Edwards, K.M. Multifocal osteomyelitis caused by *Paecilomyces varioti* in a patient with chronic granulomatous disease. *Infection* **23**, 55–57 (1995).
190. Crompton, C.H., Balfe, J.W., Summerbell, R.C., and Silver, M.M. Peritonitis with *Paecilomyces* complicating peritoneal dialysis. *Pediatr. Infect. Dis. J.* **10**, 869–871 (1991).
191. Eisinger, R.P. and Weinstein, M.P. A bold mold? *Paecilomyces variotii* peritonitis during continuous ambulatory peritoneal dialysis. *Am. J. Kidney Dis.* **18**, 606–608 (1991).
192. Nankivell, B.J., Pacey, D., and Gordon, D.L. Peritoneal eosinophilia associated with *Paecilomyces variotii* infection in continuous ambulatory peritoneal dialysis. *Am. J. Kidney Dis.* **18**, 603–605 (1991).
193. Chan, T.H., Koehler, A., and Li, P.K. *Paecilomyces varioti* peritonitis in patients on continuous ambulatory peritoneal dialysis. *Am. J. Kidney Dis.* **27**, 138–142 (1996).
194. Wright, K. et al. *Paecilomyces* peritonitis: Case report and review of the literature. *Clin. Nephrol.* **59**, 305–310 (2003).
195. Akhunova, A.M. and Shutova, V.I. *Paecilomyces* infection. *Probl. Tuberk.* **8**, 38–42 (1989).
196. Das, A. et al. *Paecilomyces variotii* in a pediatric patient with lung transplantation. *Pediatr. Transplant.* **4**, 328–332 (2000).
197. Yepes, M.S., Richard, J.M., and Gadea, M.C.S.J. Infección pulmonary por *Paecilomyces variotii* en una paciente con neoplasia de mama. *Med. Clin. (Barc)* **129**, 438 (2007).
198. Sherwood, J.A. and Dansky, A.S. *Paecilomyces* pyelonephritis complicating nephrolithiasis and review of *Paecilomyces* infections. *J. Urol.* **130**, 526–528 (1983).
199. Otcenasek, M., Jirousek, Z., Nozicka, Z., and Mencl, K. Paecilomycosis of the maxillary sinus. *Mykosen* **27**, 242–251 (1984).
200. Thompson, R.F., Bode, R.B., Rhodes, J.C., and Gluckman, J.L. *Paecilomyces variotii*. An unusual cause of isolated sphenoid sinusitis. *Arch. Otolaryngol. Head Neck Surg.* **114**, 567–569 (1988).
201. Eloy, O. et al. Sinusite à *Paecilomyces variotii*. À propôs d'un cas. *J. Mycol. Méd.* **8**, 30–31 (1998).
202. Fagerburg, R., Suh, B., Buckley, H.R., Lorber, B., and Karian, J. Cerebrospinal fluid shunt colonization and obstruction by *Paecilomyces variotii*. Case report. *J. Neurosurg.* **54**, 257–260 (1981).
203. Dhindsa, M.K., Naidu, J., Singh, S.M., and Jain, S.K. Chronic suppurative otitis media caused by *Paecilomyces variotii*. *J. Med. Vet. Mycol.* **33**, 59–61 (1995).
204. Young, V.L., Hertl, M.C., Murray, P.R., and Lambros, V.S. *Paecilomyces variotii* contamination in the lumen of a saline-filled breast implant. *Plast. Reconstr. Surg.* **96**, 1430–1434 (1995).
205. Niazi, Z.B., Salzberg, C.A., and Petro, J.A. *Paecilomyces variotii* contamination in the lumen of a saline-filled breast implant. *Plast. Reconstr. Surg.* **98**, 1323 (1996).
206. Arenas, R., Arce, M., Munoz, H., and Ruiz-Esmenjaud, J. Onychomycosis due to *Paecilomyces variotii*. Case report and review. *J. Mycol. Méd.* **8**, 31–33 (1998).

207. Kantarcioglu, A.S., Hatemi, G., Yücel, A., de Hoog, G.S., and Mandel, N.M. *Paecilomyces variotii* central nervous system infection in a patient with cancer. *Mycoses* **46**, 45–50 (2003).
208. Shin, S.B. et al. Cutaneous abscess caused by *Pacilomyces lilacinus* in a renal transplant patient. *Korean J. Med. Mycol.* **3**, 185–189 (1998).
209. Kitami, Y., Kagawa, S., and Iijima, M. A case of cutaneous *Paecilomyces lilacinus* infection on the face. *Nihon Ishinkin Gakkai Zasshi* **46**, 267–272 (2005).
210. Shukla, P.K., Khan, Z.A, Bal, B., Agrawal, P.K., and Srisvastava, O.P. A study on the association of fungi in human corneal ulcers and their therapy. *Mykosen* **27**, 385–390 (1984).
211. Pflugfelder, S.C. et al. Exogenous fungal endophthalmitis. *Ophthalmology* **95**, 19–30 (1988).
212. Watanabe, K., Yamaha, T., and Inomata, H. Postoperative fungal endophthalmitis caused by *Paecilomyces lilacinus*. *Jpn. J. Clin. Ophthalmol.* **39**, 1141–1144 (1985).
213. Gupta, A., Srinivasan, R., Kaliaperumal, S., and Saha, I. Post-traumatic fungal endophthalmitis—A prospective study. *Eye* **22**, 13–17 (2008).
214. Chakrabarti, A. et al. Fungal endophthalmitis fourteen years' experience from a Center in India. *Retina* **28**, 1400–1407 (2008).
215. Minogue, M.J. et al. Successful treatment of fungal keratitis caused by *Paecilomyces lilacinus*. *Am. J. Ophthalmol.* **98**, 626–627 (1984).
216. Deng, S.X., Kamal, K.M., and Hollander, D.A. The use of voriconazole in the management of post-penetrating keratoplasty *Paecilomyces* keratitis. *J. Ocul. Pharmacol. Ther.* **25**, 175–177 (2009).
217. Vasudevan, B. et al. First reported case of subcutaneous hyalohyphomycosis caused by *Paecilomyces variotii*. *Int. J. Dermatol.* **52**, 711–713 (2013).
218. Polat, M. et al. Successful treatment of *Paecilomyces variotii* peritonitis in a liver transplant patient. *Mycopathologia* (published on line) (2014).
219. Steiner, B. et al. *Paecilomyces variotii* as an emergent pathogenic agent of pneumonia. *Case Rep. Infect. Dis.* (published on line) (2013).
220. Bassiri-Jahromi, S. Cutaneous *Paecilomyces lilacinus* infections in immunocompromised and immunocompetent patients. *Indian J. Dermatol. Venereol. Leprol.* **80**, 331–334 (2014).
221. Ding, C.H. et al. *Paecilomyces lilacinus* fungaemia in an AIDS patient: the importance of mycological diagnosis. *Pak J. Med. Sci.* **30**, 914–916 (2014).
222. Todokoro, D., Yamada, N., Fukuchi, M., and Kishi, S. Topical voriconazole therapy of *Purpureocillium lilacinum* keratitis that occurred in disposable soft contact lens wearers. *Int. Ophthalmol.* **34**, 1159–1163 (2014).
223. Alnoldner, M.A., Kheirkhah, A., Jakobiec, F.A., Durand, M.L., and Hamrah, P. Successful treatment of *Paecilomyces lilacinus* keratitis with oral posaconazole. *Cornea* **33**, 747–749 (2014).
224. López-Medrano, R., Pérez Madera, A., and Fuster, F.C. Eye infections caused by *Purpureocillium lilacinum*: a case report and literature review. *Rev. Iberoam. Micol.* **2**, S1130–S1406 (2014).
225. Nagamoto, E. et al. A case of *Paecilomyces lilacinus* infection occurring in necrotizing fasciitis-associated skin ulcers on the face and surrounding a tracheotomy stoma. *Med. Mycol. J.* **55**, E21–E27 (2014).
226. Rimawi, R.H., Carter, Y., Ware, T., Christie, J., and, Siraj, D. Use of voriconazole for the treatment of *Paecilomyces lilacinus* cutaneous infections: case presentation and review of published literature. Mycopathologia *175*, 345–349 (2013).
227. McLintock, C.A., Lee, G.A., and Atkinson, G. Management of recurrent *Paecilomyces lilacinus* keratitis. *Clin. Exp. Optom.* **96**, 343–345 (2013).
228. Lavergne, R.A. et al. Simultaneous cutaneous infection due to *Paecilomyces lilacinus* and *Alternaria* in a heart transplant patient. *Transpl. Infect. Dis.* **14**, E156–E160 (2012).
229. Saegeman, V.S., Dupont, L.J., Verleden, G.M., and Lagrou, K. *Paecilomyces lilacinus* and *Alternaria infectoria* cutaneous infections in a sarcoidosis patient after double-lung transplantation. *Acta Clin. Belg.* **67**, 219–221 (2012).
230. Monden, Y., Sugita, M., Yamakawa, R., and Nishimura, K. Clinical experience treating *Paecilomyces lilacinus* keratitis in four patients. *Clin. Ophthalmol.* **6**, 949–953 (2012).
231. Schweitzer Jr K.M., Richard, M.J., Leversedge, F.J., and Ruch, D.S. *Paecilomyces lilacinus* septic olecranon bursitis in an immunocompetent host. *Am. J. Orthop. (Bekke Nead NJ)* **41**, E74–E75 (2012).
232. Wolley, M., Collins, J., and Thomas, M. *Paecilomyces lilacinus* peritonitis in a peritoneal dialysis patient. *Perit. Dial. Int.* **32**, 364–365 (2012).
233. de Sequeira, D.C. et al. Detection of antibody to *Purpureocillium lilacinum* by immunofluorescent assay and flow cytometry in serum of infected C57BL/6 mice. *J. Immunol. Methods* **396**, 147–154 (2013).
234. Peixoto, M. L. et al. Interaction of an opportunistic fungus *Purpureocillium lilacinum* with human macrophages and dendritic cells. *Rev. Soc. Bras. Med. Trop.* **47**, 613–617 (2014).

26 Penicillium Mycotoxins Physiological and Molecular Aspects

Rolf Geisen

CONTENTS

26.1 Taxonomic Aspects 423
26.2 Biological Aspects 424
26.3 Aspects of Pathogenicity 425
26.4 Aspects of Toxin Biosynthesis 426
26.5 Toxicological Aspects 427
    26.5.1 Molecular Detection and Monitoring Methods of
           Mycotoxin-Producing Penicillia 427
    26.5.2 Regulation of Mycotoxin Biosynthesis by Food-Relevant External Factors 429
    26.5.3 Ochratoxin A and Citrinin Production by *Penicillium* Species 431
    26.5.4 Patulin Production by *Penicillium* Species 438
    26.5.5 Cyclopiazonic Acid–Producing *Penicillium* 440
    26.5.6 PR-Toxin Production by *Penicillium* 442
26.6 Conclusion 442
References 443

26.1 TAXONOMIC ASPECTS

The genus *Penicillium* consists of about 250 species. Most species are anamorphic because their sexual cycle is unknown. It can however be expected that many of these anamorphic species do carry mating type genes and are able to mate, as was recently shown for *Penicillium roqueforti* [1]. The sexual reproductive species corresponding to *Penicillium* are *Eupenicillium* and *Talaromyces*. Species of both genera are nutritionally undemanding and can be found especially in low-water activity foods. *Penicillium* belongs to the order *Eurotiales* and the family *Trichocomaceae* and to this family belong also *Aspergillus* and *Paecilomyces*. Whereas the morphology of the conidiophore of *Aspergillus* is quite different to *Penicillium*, *Paecilomyces* has a similar conidiophore structure; however, it is less densely organized and the conidia are more ellipsoidal than *Penicillium*. Because of these morphological similarities, *Paecilomyces* was formerly grouped with *Penicillium*.

The conidiophores of *Penicillium* form the typical brush-like structure, which gave this genus its name (penicillus = brush). Based on the micromorphology of the conidiophores, which can be monoverticillate, biverticillate, or terverticillate, depending on the number of branching levels, the species are grouped into the subgenera *Aspergilloides*, *Biverticillium*, *Furcatum* (also biverticillate), and *Penicillium* [2]. Subgenus *Penicillium* contains species with terverticillate conidiophores, and many food-related and important mycotoxin-producing species belong to this subgenus such as (1) *P. roqueforti*, *Penicillium carneum*, and *Penicillium paneum* (section *Roqueforti*); (2) *Penicillium chrysogenum* and *Penicillium nalgiovense* (section *Chrysogena*);

(3) *Penicillium expansum* and *Penicillium italicum* (section *Penicillium*); (4) *Penicillium digitatum* (section *Digitatum*); and (5) *Penicillium verrucosum*, *Penicillium nordicum*, *Penicillium allii*, *Penicillium solitum*, *Penicillium camemberti*, *Penicillium commune*, *Penicillium caseifulvum*, and *Penicillium crustosum* (section *Viridicata*) [3]. The subgenus *Biverticillium* contains no common food-related *Penicillium* species. However *Penicillium marneffei*, the only true human pathogenic *Penicillium* species, belongs to this subgenus [4]. *Penicillium citrinum*, which is an important toxin-producing and ubiquitous fungus, is the only important food-related species that belongs to subgenus *Furcatum*. In the subgenus *Aspergillioides*, *Penicillium citreonigrum* and *Penicillium glabrum* are grouped, which can occasionally be found in foods such as rice, peanuts, meats, and fresh cabbage and water [2,5]. As mentioned, most *Penicillium* species are nonpathogenic against plants and humans and are saprophytes occurring in soil or decaying plant materials. For this reason, they are important spoilage organisms of plant-derived foods and some are able to produce important mycotoxins. Most Penicillia grow slowly and produce compact, blue to green colonies with abundant conidiophores.

26.2 BIOLOGICAL ASPECTS

Most food-relevant Penicillia have growth optima between 20°C and 25°C. They are adapted to moderate climates and are a minor problem in tropical countries where *Aspergillus* species play a more important role as food spoilage organisms. Among the exceptions is *Penicillium oxalicum*, which has a growth optimum of 37°C and can be found on tropical food products [2]. A typical feature of many *Penicillium* species is their capacity to grow at low temperatures and many can even be regarded as psychrophilic. For example, *P. roqueforti* grows well at refrigeration temperatures (≤5°C) [2]. Some species are xero/halophilic, and furthermore, some species of *Penicillium* are adapted to extreme environments in relation to temperature and osmotic constraints. Gunde-Cimerman et al. [6] analyzed the relationship between adaptation to low temperature and the water activity of fungi occurring in arctic ice. They identified several *Penicillium* species, including food species such as *P. chrysogenum*, *P. nalgiovense*, *P. commune*, *Penicillium palitans*, *Penicillium brevicompactum*, *Penicillium olsonii*, and *P. nordicum*. Interestingly, this species composition is related to the population occurring in certain dry-cured foods. Similar species cope with salt stress (e.g., ionic stress) and can be found in NaCl-rich environments, such as salt lakes [7–9]. Some Penicillia (e.g., *P. nordicum*) even favor the addition of NaCl to growth media and have a growth optimum at between 30 and 40 g/L [10], and it has been suggested that NaCl used in the production of foods is a vector for the introduction of *P. nordicum* into dry-cured food products [11]. These specialized conditions, as with most environmental conditions, have a profound influence on the regulation of mycotoxins as will be discussed later. Penicillia are typical *storage fungi* that grow and multiply on plant food and feed material mainly after harvest and during storage. For example, *P. citrinum* grows at a water activity (a_w) of 0.775 in wheat [12]. Species such as *P. roqueforti* can grow at high CO_2 and low O_2 concentrations [13].

Penicillia may also be found in water. Nevarez et al. [5] characterized a *P. glabrum* strain, which was isolated from aromatized mineral water. Also, Gonçalves et al. [14] isolated a *P. glabrum* strain from tap water; however, many more *P. expansum* and *P. brevicompactum* strains were found during this analysis. Also, *Penicillium corylophilum*, *P. crustosum*, *Penicillium griseofulvum*, *Penicillium raistrikii*, and *Penicillium waksmanii* were isolated. Overall, some species have a very broad habitat range such as *P. citrinum*, *P. aurantiogriseum*, and *P. chrysogenum*, enabling them to occur as contaminants in many foods. Other species are well adapted to specific environments (Table 26.1).

P. digitatum and *P. italicum* are perfectly adapted to citrus fruits and are very important spoilage organisms in this habitat [15] and produce green (*P. digitatum*) or blue rot (*P. italicum*) on lemons or oranges, respectively. They are nonpathogenic organisms, because infection and spoilage mainly takes place after harvest and during storage and transport. However, due to heavy sporulation, they

TABLE 26.1
Important Spoilage and Mycotoxin-Producing Penicillia

Species	Occurrence/Food Relevance	Pathogenic/Spoilage Effect/Characteristics	Toxin
P. verrucosum	Cereals (wheat), olives, dry-cured meats	Negligible	Ochratoxin A, citrinin, depending on the external conditions.
P. citrinum	Ubiquitous in foods, soil, extreme environments like saline environments	Rapid distribution because of strong sporulation	Citrinin, production quite consistent under various environmental conditions.
P. expansum	Apples and other fruits like pears, peaches, or grapes	Green rot of apples	Patulin, citrinin, generally less citrinin compared to *P. citrinum*.
P. nordicum	Dry-cured NaCl-rich foods like fermented meats or cheeses	Slow growth but very consistent growth on substrates with high concentrations of sodium chloride, high competitiveness under these conditions	Ochratoxin A, production increases with increasing NaCl concentrations, at higher NaCl concentration, still high production of ochratoxin A, albeit after a prolonged lag phase.
P. commune, *P. camemberti*	*P. camemberti* domesticated form of *P. commune*, used as starter culture for cheeses, *P. commune* ubiquitous, but common spoilage organism of cheeses and dry-cured meats	*P. commune*, spoilage because of discoloration	Cyclopiazonic acid can also be found in cheeses, especially when the storage temperature was high, lower toxicity, so that the amounts produced in cheeses are regarded to have no measurable effect.
P. roqueforti	Starter culture for cheeses but also important spoilage organism for bread and for silages	Very competitive during growth in silages, because it can grow under very low oxygen concentrations, highly distributive because of its capacity to produce masses of spores	Produces various toxic secondary metabolites, among them, PR toxin that exhibits the highest toxic activity, it is produced in cheeses too, but it is unstable in this environment and it is converted to the much less toxic PR imine.

can rapidly colonize whole batches of fruits, once an initial colonization has taken place. *P. expansum* is the typical blue rot fungus of apples and *P. verrucosum* occurs on cereals. *P. commune* can be found as a spoilage fungus of cheeses.

26.3 ASPECTS OF PATHOGENICITY

There are only a few reports about human pathogenic activities beside *P. marneffei*, which is a true pathogenic *Penicillium* species with dimorphic growth, growing yeast-like and in filaments in its pathogenic and nonpathogenic states, respectively [16]. Allergic reactions of persons working in cheese factories (cheese workers' lung) have been described. These reactions are triggered by the inhalation of large number of conidia during handling of mold-ripened cheeses and can be attributed to *P. camemberti*, *P. roqueforti*, and *P. commune* [17]. Mainly immune-compromised persons were involved in cases when *Penicillium* species other than *P. marneffei* have been found in human infection. The *Penicillium* species were only identified as being present in the infectious environment in most cases, and a clear correlation to the infection could not be made. Surprisingly, some of the species involved are of food-borne origin [18]. These authors reported three rare cases of patients with different symptoms, where *Penicillium purpurogenum* (subgenus *Biverticillium*, non-food related), *P. chrysogenum* (subgenus *Penicillium*, food related), or *Penicillium decumbens* (subgenus *Aspergilloides*, non-food related) could be identified. Furthermore, *P. citrinum* was involved in urinary tract infections [19]. *P. chrysogenum* was a cause of endophthalmitis (Eschete et al. [20]) and unidentified *Penicillium* species invaded artificial heart valves [21]. Fiocchi et al. [21] described a case of anaphylaxis that was traced to *Penicillium lanosoceruleum*, a species that is seldom found on foods. As mentioned, *P. marneffei* is able to grow at 37°C and this feature maybe a prerequisite for human infection. Pitt [22] lists the *Penicillium* species that grow at this temperature: *Penicillium chermesinum*, *P. decumbens* (subgenus *Aspergilloides*), *Penicillium janthinellum*, *P. oxalicum*, *Penicillium simplicissimum* (subgenus *Furcatum*), *P. chrysogenum* (subgenus *Penicillium*), *P. marneffei*, *Penicillium minioluteum*, *Penicillium pinophilum*, *Penicillium funiculosum*, *P. purpurogenum*, *Penicillium islandicum*, and *Penicillium verruculosum* (subgenus *Biverticillium*). Most of these species occur in subgenus *Biverticillium* whose member species play only a minor role in the food environment. Several *Penicillium* species were isolated from human sources [22] and they also mainly belonged to subgenus *Biverticillium*, in particular *P. marneffei*, *Penicillium piceum*, *P. purpurogenum*, *P. minioluteum*, and *P. funiculosum*.

26.4 ASPECTS OF TOXIN BIOSYNTHESIS

Many of the food-borne *Penicillium* species are able to produce mycotoxins, toxic products synthesized during secondary metabolism. It is these compounds that can cause toxicoses in humans. Primary and secondary metabolisms are historical terms and are not clearly distinguishable. Secondary metabolite biosynthesis always starts with a primary metabolite and results in chemical substances that are not necessarily needed for growth of the fungus. Some hypotheses about the importance of these metabolites have been described, which mainly raise ecological aspects as reasons for their biosynthesis (see Section 26.5.2).

Important mycotoxins produced by members of the genus *Penicillium* are ochratoxin A, citrinin, patulin, α-cyclopiazonic acid (CPA), PR toxin, mycophenolic acid, and roquefortine C. They can occur in a variety of different foods due to the presence of the producing fungus. Some of them are not *Penicillium* specific such as ochratoxin A, patulin, CPA, and citrinin. Moreover, *Penicillium* is able to produce a set of less well-known mycotoxins, for example, chaetoglobosins, citreoviridin, penitrem A, rubratoxin, and verrucosidin [23]; however, comprehensive data about the toxicities of these are unavailable.

P. expansum is the most important spoilage organism in apples and is a prominent patulin-producing fungus, which is responsible for patulin in this commodity. *P. verrucosum* is another important fungus and produces ochratoxin A, mainly in cereal products. However, *P. verrucosum* is also able to produce citrinin, a related mycotoxin, and both toxins are mutually regulated in this organism. *P. camemberti* and *P. roqueforti* are used for food biotechnology purposes and are able to produce CPA, and *P. roqueforti* is further able to produce PR toxin. Molecular knowledge for these mycotoxins is available, which will be presented later.

Despite the tremendous progress made on the genomics of *Aspergillus*, and *Fusarium*, which are very important mycotoxigenic genera, much less information about *Penicillium* is known. Only the genome of *P. chrysogenum* has been published, at the time of writing, but only in a raw and unannotated form. *P. chrysogenum* produces penicillin as its most important secondary metabolite, and roquefortine C and CPA [24]. So the genome can perhaps be used as a blueprint to identify the biosynthesis genes of these two mycotoxins from Penicillia. It can be assumed that additional mycotoxin biosynthesis gene clusters are present, because usually, several are present in fungi [25]. The molecular bases of mycotoxin biosynthesis in *Penicillium* are not advanced; however, the background is partially available. Definitive molecular tools have been established to specifically identify and quantify mycotoxin-producing Penicillia in foods or to monitor the expression of mycotoxin biosynthesis directly in the food system. With these tools, clear statements about the occurrence of mycotoxin-producing *Penicillium* species and the influence of environmental parameters upon the regulation of mycotoxin biosynthesis in the food system can be given.

26.5 TOXICOLOGICAL ASPECTS

By definition, mycotoxins are secondary metabolites from food-related filamentous fungi and may be found as contaminants in foods, when the conditions in the food allow the biosynthesis of the toxins by the fungus. Intoxications by mycotoxins are called mycotoxicosis. Of course, the toxic activity of these secondary metabolites against humans and animals is only a side activity. The primary role for their biosynthesis by the fungus is of ecological nature and in most instances will increase the competiveness of the fungus in the environment. Mycotoxins are a diverse group of chemical compounds. The most important groups are polyketides (aflatoxins, citrinin, patulin), terpenes/isoprenoids (CPA), or composite compounds, for example, ochratoxin A, which consists of a polyketide and phenylalanine. Zearalenone has a lactone structure and nonribosomal peptides or depsipeptides, such as the beauvericins and enniatins, exist, although these are not produced by the Penicillia. The toxic activity can be differentiated into acute (mycotoxicosis) and chronic (often carcinogenesis). Higher amounts of mycotoxins must be ingested to observe acute effects, compared to chronic effects that may occur after prolonged continuous intake of minor concentrations with food. Acute toxicity due to mycotoxins is rarely observed; nevertheless, several cases concerning the fatal action of high amounts of aflatoxins in food have been described [26]. This information is provided to indicate that mycotoxin ingestion can be very serious although aflatoxins are not *Penicillium* mycotoxins of course. The last case was an intoxication of several hundred people in Kenya, who had consumed aflatoxin-contaminated corn where more than 100 people died.

The etiology of chronic diseases caused by mycotoxins is sometimes difficult to prove. However, there seems to be a clear correlation between the increased intake of aflatoxins and the increase in the rate of liver cancer in some Asian and African countries due to the consumption of foods of low quality [27]. Also, a correlation between the ingestion of ochratoxin A and the occurrence of endemic Balkan nephropathy (BEN) has been assumed [28]. Many of the important mycotoxins such as aflatoxins, ochratoxin A, and fumonisins are potentially carcinogenic. The European Union has set regulatory limits for many mycotoxins, for example, aflatoxins, ochratoxin A, fumonisins, trichothecenes, zearalenone, and patulin, where foods with concentrations above the limits are unacceptable for trade. Trichothecenes, zearalenone, patulin, and CPA are not carcinogenic, but show various other toxic activities.

26.5.1 MOLECULAR DETECTION AND MONITORING METHODS OF MYCOTOXIN-PRODUCING PENICILLIA

The identification of toxigenic fungi in foods and feeds is an important aspect in food safety as it is crucial to know which species may be present in a given food system. Also, knowledge about the change in species composition over time, for example, during storage or ripening, is necessary. The identification of toxigenic Penicillia is a time-consuming task and can only be fulfilled accurately by experts. There are numerous reports about misidentified *Penicillium* strains, which lead to the wrong assessments about the possible occurrence of mycotoxins [23,29], although further analyses are required, such as chromatographic, to determine if particular mycotoxins are present. In general, misidentifications are reduced when molecular detection methods are applied [30].

Molecular detection methods are mainly PCR or real-time PCR, which are specific for the mycotoxin-producing fungus or fungi. They are targeted against a specific sequence that might be identified in variable regions, for example, the ITS regions or any other specific arbitrary region or toward mycotoxin biosynthesis genes. Several specific PCR systems for mycotoxin-producing Penicillia have been described, for example, for patulin-producing *P. expansum* [31,32]. Whereas Marek et al. [31] used the general polygalacturonase gene, Paterson et al. [32] used the patulin specific *idh* gene as target. Pedersen et al. [33] described the development of two primer pairs for the characterization of species of the former *P. roqueforti* species complex. The first primer pair was specific for all tested *Penicillium* species from the subgenus *Penicillium*; the second primer set was specific for *P. roqueforti* and *P. carneum* of the same subgenus: both primer pairs were targeted against the ITS regions. Geisen et al. [34] described a real-time PCR system for detection and quantification of ochratoxin A–producing *P. nordicum* strains based on the polyketide synthase gene. Recently, a real-time PCR system for the detection of verrucosidin-producing Penicillia, such as *P. polonicum* and *P. aurantiogriseum*, in dry-cured foods has been described [35]. The same group also used formerly described primer pairs from the ochratoxin *nps*PN gene (nonribosomal peptide synthetase) to detect various ochratoxin A–producing species in a multiplex reaction together with patulin- and aflatoxin-producing fungi [36].

Conventional PCR methods are suitable to decide if a toxigenic species is present in a sample. To obtain information about the amount of fungal biomass in a sample, real-time PCR systems are necessary, which allow quantification of the DNA and thereby the quantification of the fungal genomes. DNA-based real-time PCR systems have been used to study the growth of ochratoxin A–producing *P. nordicum* in wheat [34]. The genome copy numbers of *P. nordicum* determined correlated with the colony forming units (cfus) counts determined in parallel, indicating that the quantification of the fungal biomass by real-time PCR is a useful approach. Rodriguez et al. [35] found a good correlation between real-time PCR *genome data* and the cfu data of verrucosidin-producing Penicillia. Haugland et al. [37] quantified various *Penicillium* species isolated from house dust by real-time PCR. The authors found a good agreement of the expected results of samples spiked with spores of *Penicillium* or species of the other two genera tested. The described methods give a positive or negative result about the presence of a mycotoxin-producing *Penicillium* species or provide information about the amount of fungal biomass in a sample. However, the biosynthesis of mycotoxins is a tightly regulated process (see Section 26.5.2), which means that mycotoxins are not produced under all conditions, which is especially the case in food systems. The growth conditions for fungi in foods are usually suboptimal compared to what can be arranged in vitro. That means that data obtained in vitro may not reflect the conditions in food systems. Nevertheless, the data obtained from laboratory media about the regulation of mycotoxin biosynthesis are highly important, because they demonstrate and model the influence of the respective analyzed parameter on the biosynthesis of the mycotoxin. By analyzing the expression of certain key genes directly in the food habitat, statements about the activation of mycotoxin biosynthesis genes and thereby the subsequent production of the respective mycotoxin can be made. For this approach, methods such as northern blotting, RT-PCR (reverse transcriptase PCR), RT real-time PCR, or microarray technology are highly suitable. Northern blots were used to follow the expression of aflatoxin biosynthesis genes in *Aspergillus parasiticus* [38]. Doohan et al. [39] used RT-PCR to study the expression of the *tri*5 gene of *Fusarium* species in planta and showed that suboptimal conditions of fungicides (such as prochloraz or tebuconazole) resulted in stress of the fungus and activation of mycotoxin biosynthesis genes. Also, suboptimal concentrations of preservatives [40] or fungicides [41] can activate ochratoxin biosynthesis in *P. verrucosum* and *P. nordicum*. This was also shown for patulin biosynthesis by *P. expansum* [42].

Ochiai et al. [43] developed a reporter gene strain of *Fusarium graminearum* to analyze the influence of environmental factors on the expression of trichothecene biosynthesis genes. They cloned the promotor of the *tri*5 gene in front of a reporter gene, which was a green fluorescent protein (GFP). It was demonstrated that certain fungicides such as ipconazole, metconazole, and tebuconazole lead to a high intensification of the green fluorescence of the reporter strains, indicating and confirming the fact that suboptimal concentrations of fungicides lead to an activation of the trichothecene biosynthesis genes. This was also shown with ochratoxin A biosynthesis by *Penicillium*. Suboptimal concentrations of preservatives led to an activation of the ochratoxin A biosynthesis genes and to a later increased biosynthesis of ochratoxin [40]. A GFP reporter strain has also been constructed for *P. nordicum* [44], a strong and consistent ochratoxin A–producing species. This strain was used to analyze the influence of substrate composition on the regulation of ochratoxin A biosynthesis genes. Minimal medium containing glucose as carbon source strongly inhibited the expression of the GFP protein, indicating that ochratoxin A biosynthesis in *Penicillium* is subjected

Penicillium Mycotoxins

to catabolite repression. In addition, microarray technology has been used to study the expression of ochratoxin A biosynthetic genes in *Penicillium*. A microarray with (1) putative ochratoxin A biosynthesis genes and (2) associated genes was activated during ochratoxin A biosynthesis. These were identified by differential display PCR, which was used for expression analysis of the ochratoxin A–related genes under various conditions. This microarray was called MycoChip and consists of different subarrays with the biosynthetic pathway genes of important food-relevant mycotoxins, for example, the aflatoxin pathway, the fumonisin pathway, the trichothecene type A and B pathway, the ochratoxin A–related genes, and some genes of the patulin biosynthetic pathway [45]. By using it to study the effect of systematically changed environmental parameters on ochratoxin A biosynthesis in *Penicillium*, a typical stress-related expression and ochratoxin A production profile could be elucidated (see later) [46]. Microarray technology has been used (1) to analyze the influence of temperature on the expression of aflatoxin biosynthetic genes [47], (2) to perform expression profile analysis of the fumonisin biosynthetic genes of *Fusarium verticillioides* [48], and (3) for genomic mining of novel secondary metabolites in *Aspergillus* and *Fusarium* [49,50]. Moreover, microarrays can be used for identification purposes [51] by spotting with oligonucleotides specific for a certain species, for example, those from ITS or other variable regions. The generation of a definitive spot hybridization pattern indicates a certain species or group of species, depending on the specificity. However, no microarray for the detection and identification of mycotoxin-producing *Penicillium* species has been described, although Harris et al. [52] used an Affymetrix whole-genome array of *P. chrysogenum* to study penicillin G biosynthesis at the transcriptional level. These examples show how versatile these approaches are and that much more information about the conditions that allow mycotoxins biosynthesis in foods can be gained. This more detailed information can be used to improve the control of mycotoxins biosynthesis and reduce the exposure of animals and humans to these toxic fungal metabolites.

26.5.2 Regulation of Mycotoxin Biosynthesis by Food-Relevant External Factors

Mycotoxin production is tightly regulated and they are produced only under specific conditions. Several environmental parameters especially play a role as regulatory factors: the nature of the substrate, temperature, water activity, and, perhaps less importantly, pH. Surprisingly, light is an important modulator of secondary metabolite biosynthesis [53–56]. It is considered that (especially) nitrogen limitation in the substrate leads to activation of mycotoxin biosynthesis [57–59]. The influence of external parameters on the regulation of mycotoxin biosynthesis genes is carried out at the transcriptional level [43,46,58,60]. There might be a direct coupling between expression of a certain gene and the phenotypic biosynthesis of the mycotoxin, but this is not a necessity. Scherm et al. [61], for example, showed that the expression of only some genes of the aflatoxin biosynthetic pathway is correlated with phenotypic aflatoxin biosynthesis, in particular *afl*D, *afl*O, and *afl*P, whereas others are not. Also Jurado et al. [60] demonstrated a direct correlation between the expression of the *fum*1 gene and fumonisin biosynthesis by two strains of *F. verticillioides*. For *Penicillium*, a direct correlation between expression of the ochratoxin A polyketide synthase gene (*otapks*) and the biosynthesis of ochratoxin A has been established [34]. These and other reports clearly show that the regulation of mycotoxin biosynthesis by external factors in the food is being exerted at the transcriptional level. In order for the fungus to react toward changes in environmental parameters, it has to be able to sense these changes. This is accomplished by the action of signal transduction pathways, which can sense changes of the environment and can transmit this signal to transcription. This leads to an activation or repression of mycotoxin biosynthesis genes, which, in a second step, results in activation or inhibition of toxin biosynthesis. One important signal transduction cascade, which is related to the biosynthesis of secondary metabolites, is the high-osmolarity glycerol (HOG) cascade. This is a MAP kinase cascade, which means that starting from a sensor protein in the cell membrane, a consecutive phosphorylation chain is initiated and ends at the level of transcription factors, either positive or negative, which in turn activates or represses mycotoxin biosynthesis.

The involvement of the HOG signal cascading pathway in the regulation of mycotoxin biosynthesis has been reported for *Fusarium proliferatum* [59] and for *Alternaria alternata* [62]. An inactivation of the HOG MAP kinase leads to reduced or abolished biosynthesis of the mycotoxins. Another signal transduction pathway involved in the regulation of mycotoxin biosynthesis is the G-protein/cAMP/PKA pathway, in which a heterotrimeric G-protein, c-AMP, and protein kinase A are active. This signaling pathway participates in the regulation of aflatoxin in *Aspergillus flavus* [63,64], citrinin in *Monascus ruber* [65], and roquefortine C in *P. chrysogenum* [66]. These two signal cascading pathways also seem to play a role in the regulation of ochratoxin and citrinin biosynthesis in *Penicillium* (Schmidt-Heydt et al., 2015). Li et al. [65] identified *mga*1, a gene for an α-subunit of a heterotrimeric G-protein involved in the regulation of citrinin and pigment production (monacolin) in *M. ruber*. The amino acid sequence of the gene product has 96% identity to FadA, a Gα-subunit involved in sterigmatocystin biosynthesis in *Aspergillus nidulans* and a gene with homology has recently been identified in *P. verrucosum* (unpublished, Markus Schmidt-Heydt, Dominik Stoll, Rolf Geisen). The concentration of c-AMP plays a role in signal transduction by the G-protein/c-AMP/PKA pathway. Furthermore, Miyake et al. [67] analyzed the influence of the c-AMP concentration on secondary metabolite production in *Monascus*. By adding exogenous c-AMP, they could repress the biosynthesis of certain secondary metabolites, in particular lovastatin, red pigments (monacolin), and citrinin. This indicates the importance of c-AMP for this type of regulation. According to new data, the interplay between the HOG- and the G-protein/c-AMP/PKA signal transduction pathways regulates the ratio of ochratoxin to citrinin produced by *P. verrucosum*, which seems to be important for the adaptation to certain environments. (Markus Schmidt-Heydt, Dominic Stoll, Rolf Geisen unpublished results) (Figure 26.1).

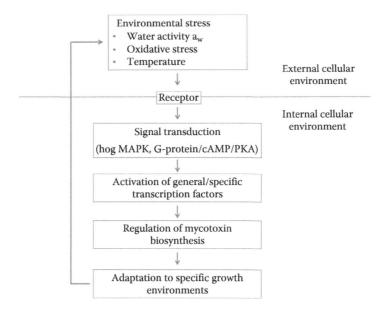

FIGURE 26.1 Scheme of the influence of external parameters/external stress on the transcription of mycotoxin biosynthesis gene via signal transduction pathways. Signal transduction pathways can activate or inhibit transcription factors according to the external signal. Two classes of transcription factors are involved in the regulation of secondary metabolite biosynthesis gene clusters, for example, global transcription factors (e.g., *lae*A [62] or *ve*A [61]), which have an influence on a set of genes/gene clusters and specific transcription factors, which are specific for the respective gene cluster (e.g., *afl*R/*afl*J [74]) in the case of the aflatoxin biosynthesis gene cluster. In certain cases the biosynthesis of mycotoxins can be regarded as an adaptive response to changes in the environment.

The organization of the mycotoxin biosynthesis genes usually follows the general rule that they are clustered. The clustering of the genes might facilitate transfer of the genes to other species [68,69] or the coordinated regulation of the whole cluster by the specific regulatory gene or by epigenetic mechanisms [70]. Usually, the cluster contains *typical* secondary metabolite-specific genes such as the key enzymes, polyketide synthases (pks), nonribosomal peptide synthases (nrps), or hybrid proteins of both (nrps-pks). These genes lead to the formation of polyketides. The other important key genes are the terpene cyclases, which in turn lead to terpenoids such as the sesquiterpenoids including the trichothecenes. Usually, the cluster contains further genes, for example, dehydrogenases, cytochromoxidases, MFS or ABC transporters, and finally genes for specific transcription factors. These specific regulatory factors are in turn regulated by higher ordered global regulatory factors, for example, the catabolite repressing protein, CREA [71], the light-regulated inductor of secondary metabolite expression, *ve*A [72], or the general inductor of secondary metabolism biosynthesis, *lae*A [73]. The clusters of aflatoxin [74], trichothecene [75], fumonisin [76], and CPA [77] biosynthesis are well described. Information about putative ochratoxin biosynthesis genes in *P. nordicum* [78] or *Aspergillus carbonarius* [79] are also available. Also, in these cases, genes for a putative polyketide synthase and a nonribosomal peptide synthase have been identified. With the advent of the genomic era, it became clear that many species contain high numbers of putative gene clusters. For example, Amaike and Keller [80] reported that *A. flavus* contains 55 putative secondary metabolite clusters; however, most of these metabolites are unknown. Similar numbers were also reported for other species for which the genome is known.

The issues described earlier show that environmental conditions have a profound influence of mycotoxin biosynthesis, which is very tightly regulated by signaling cascades and feedback regulatory mechanisms. By systematically analyzing the expression behavior of the ochratoxin A biosynthetic genes in *Penicillium* [46] under changing environmental conditions, a clear and representative expression profile became obvious. This expression profile was correlated to the biosynthesis of ochratoxin A. The general expression profile was independent of the parameter changes, for example, temperature, water activity, or pH [46]. Interestingly, the expression of the ochratoxin A biosynthetic genes react accordingly when one of the parameters was changed continuously. A change of the parameter from marginal to optimum growth lead to the formation of at least two expression peaks of the *otapks*PN gene. A major high peak was near the conditions where growth was optimal and a second minor, lower peak was at conditions that inhibits growth [46]. At both conditions, higher amounts of ochratoxin A were produced (Figure 26.2). This typical behavior, which could also be confirmed during an expression analysis of the *fum*1 gene, important for fumonisin biosynthesis in *F. verticillioides* [60], suggests a stress-induced activation of the mycotoxin biosynthesis genes. At the margins of growth, the stress (temperature, water activity, or pH) is obvious because of the strongly reduced growth rate. However, under optimal growth conditions, a high peak of biosynthesis also could be observed, which was interpreted as nutritional stress. Under these conditions, the growth rate is high and high amounts of nutrients are needed. In a solid medium, these nutrients have to diffuse toward the growing colony, which may lead to nutrient depletion. As mentioned earlier, limited nitrogen supply is a major signal for initiating mycotoxin biosynthesis and this might be the case under these circumstances. These expression profiles, which were obtained by RT real-time analysis of the *otapks*PN gene, could also be confirmed by microarray analysis.

26.5.3 Ochratoxin A and Citrinin Production by *Penicillium* Species

Ochratoxin A is an important mycotoxin produced by various fungal species first described by van der Merwe et al. [81], who isolated it from *Aspergillus ochraceus* (it is now considered that it was *Aspergillus westerdijkiae*). Ochratoxin A is a polyketide mycotoxin that is coupled to the amino acid phenylalanine and the polyketide part is a dihydroisocoumarin derivative. Ochratoxin A is the chlorinated form of the metabolite (Figure 26.3). The chlorine is coupled to C-atom 5 of

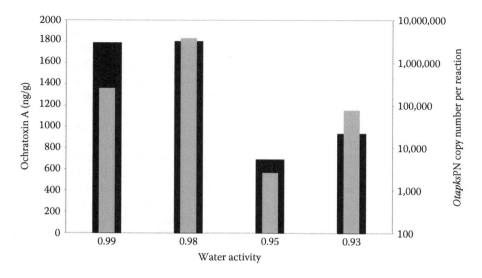

FIGURE 26.2 Correlation between expression of the *otapks*PN gene (narrow gray columns) and biosynthesis of ochratoxin A (wide black columns) under changing environmental conditions. A very similar behavior at all levels could be observed with various different parameters [31] but shown here with changing water activity as an example. Generally, the *otapks*PN gene is highly expressed at conditions near the growth optimum. Consequently, high amounts of ochratoxin A are produced under these conditions. However, a second minor peak of gene expression and of production of ochratoxin A is obvious at conditions where growth is at its margins, due to the extreme values of the external parameter (in this case, a water activity of 0.93).

the polyketide part of the molecule. Ochratoxin B has the same structure, except that the chlorine is exchanged by hydrogen (Figure 26.3). Ochratoxin B is less toxic than ochratoxin A by a factor of about 100 [82]. Depending on the species, ochratoxin B can be coproduced with ochratoxin A as an end product of the metabolism, and the ratio of ochratoxin B to ochratoxin A is also highly dependent on the environmental conditions [83].

Ochratoxin A is strongly nephrotoxic [84]. It is regarded by the WHO/FAO as a type 2 carcinogen for the kidney, which means that it is assumed that it is carcinogenic, but the epidemiological proof is not completely demonstrated. The mycotoxin is considered to be associated with BEN, the symptoms of which resemble those seen in animal experiments, where cancer of the kidneys is observed [85,86]. A clear correlation between the occurrence of ochratoxin A and nephrological disorders in swine has been described as Danish porcine nephropathy [87]. Also, citrinin is a polyketide mycotoxin whose target organ is the kidney. The structure of citrinin resembles the polyketide part of ochratoxin A, except that it contains no chlorine (Figure 26.3). The available toxicological data about citrinin are much sparser compared to ochratoxin A. However, it is assumed that the toxins might act synergistically [88]. Interestingly, Vrabcheva et al. [28] found 2–200 times more citrinin in Bulgarian cereal samples than ochratoxin A, indicating that under certain circumstances, the biosynthesis of citrinin is favored over the biosynthesis of ochratoxin A. Because of the well-proven toxicity of ochratoxin A, concentration limits have been set in various countries (Commission Regulation [EU] No. 1881/2006), although no regulations exist for citrinin.

At the molecular level, ochratoxin A is an inhibitor of protein biosynthesis. It is able to competitively inhibit the t-RNA synthetase for the amino acid phenylalanine [89], which stops protein biosynthesis. This toxic activity is toward higher eukaryotic cells and bacteria, for example, *Streptococcus faecalis* [90].

Some of the genes responsible for ochratoxin A biosynthesis in *Penicillium* have been characterized [78]: these include a polyketide synthase (*otapks*PN) and a nonribosomal peptide synthase, which is responsible for the formation of the peptide bond between the phenylalanine and the

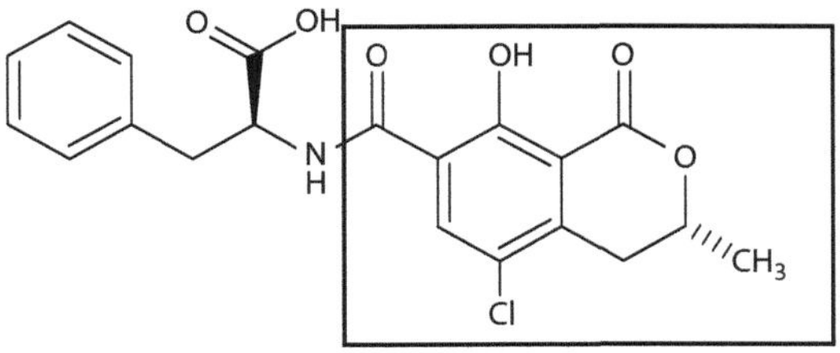

Penicillium Mycotoxins

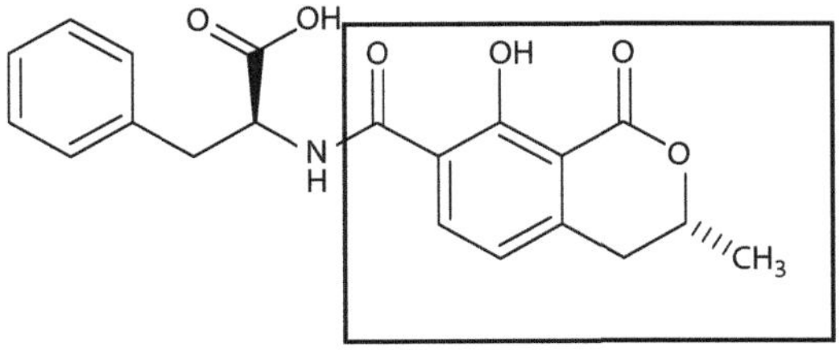

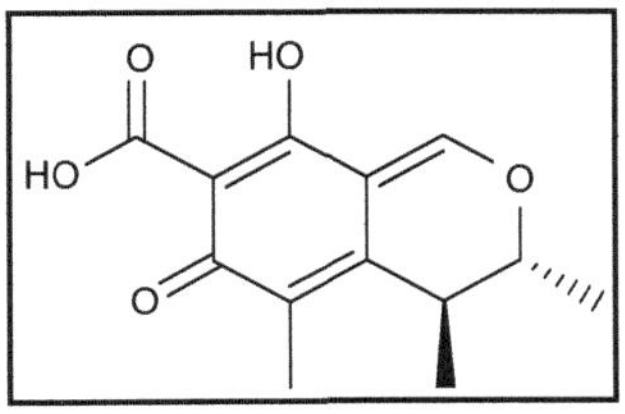

FIGURE 26.3 Structures of the three mycotoxins ochratoxin A, ochratoxin B, and citrinin produced by *P. verrucosum*. The similarity of the polyketide part, for example, the dihydroisocoumarin between ochratoxin and citrinin, is obvious. However, an important difference is the fact that in ochratoxin A chlorine is ligated to position 5.

dihydroisocoumarin parts. Interestingly, analysis of the expression profiles of these genes suggests that the biosynthesis of ochratoxin A seems to be organized in such a way that the much less toxic ochratoxin B is chlorinated only in the last step of the biosynthetic pathway, just before excretion from the cell [78,91]. This putative late chlorination of ochratoxin B to ochratoxin A in the biosynthesis pathway can be regarded as a self-protection mechanism. This view is supported by recent results of Gallo et al. [68] who analyzed the biosynthesis of ochratoxin A in *A. carbonarius* and came to the same conclusions. The gene cluster for citrinin has also been evaluated, although only in *Monascus*. Shimizu et al. [92] identified a polyketide synthase gene, which is responsible for the biosynthesis of citrinin in *Monascus purpureus*. The authors showed that the transcription of this gene is correlated with the biosynthesis of citrinin. Furthermore, disruption of this gene leads to a mutant strain, which was unable to produce citrinin indicating its involvement in citrinin biosynthesis. The same group identified an activator gene (*ctnA*) involved in citrinin biosynthesis. This gene is located adjacent to the polyketide synthase gene and after gene inactivation by disruption; the biosynthesis of citrinin was ceased. Other genes, namely, a dehydrogenase, an oxygenase, an oxidoreductase, and a transporter, were identified. However, nothing is known about the citrinin biosynthesis genes in *P. verrucosum*, which require future studies.

Only two species of the genus *Penicillium* are described as ochratoxin A producers, *P. verrucosum* and *P. nordicum*. They are morphologically similar species, which each have distinct chemotypes, including the physiology and regulation of the secondary metabolites produced under particular conditions. This has implications with respect to adaption to specific habitats (Figure 26.4).

P. verrucosum is a typical contaminant in stored wheat and is responsible for the occurrence of ochratoxin A and citrinin in that commodity [93]. This is especially true in northern European

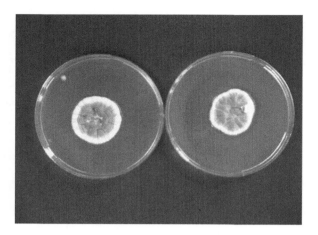

Penicillium nordicum	*Penicillium verrucosum*
• High amounts of ochratoxin A, no citrinin	• Moderate amounts of ochratoxin A, citrinin
• Chemotype I	• Chemotype II
• RAPD type I	• RAPD type II
• AFLP type I	• AFLP type II
• Habitat: NaCl rich foods	• Habitat: cereals, occasionally on NaCl rich foods

FIGURE 26.4 Phenotypical differences between the two closely related ochratoxin A–producing fungal species *P. verrucosum* and *P. nordicum*. Both species were grown on malt extract medium. (The data were collated according to Larsen, T.O. et al. [166] and Castella, G. et al. [100].)

countries where temperatures are moderate and generally humidity is high. Wheat appears to be an optimal substrate for *P. verrucosum* to biosynthesize ochratoxin A. Lund and Frisvad [94] showed that contamination of individual kernels of >7% indicates the presence of ochratoxin A if the relative humidity of the sample was above 14%–15% [95]. To be able to monitor the activation of the ochratoxin A biosynthesis genes, in situ real-time PCR and microarray technology were developed [96]. The real-time PCR system was directed against a polyketide synthase gene, which was coregulated with the production of ochratoxin A by *P. verrucosum*, indicating that this gene is involved in its biosynthesis. The DNA of *P. verrucosum* was targeted by using the real-time system to follow the growth of the fungus in wheat, where the cfu of *P. verrucosum* were also determined. A near-absolute congruence between the cfu and the real-time PCR data (given in copy numbers pks gene/g) was achieved. The wheat was artificially inoculated with 10^6 spores/g cfu indicated a reduction of the viable spores by 10^6–10^4 spores/g within the first 8 days. Then the spores replicated and reached 10^8–10^9 at approximately day 20. Interestingly, real-time PCR data were only produced after day 8, that is, after the spores had started to replicate. The unexpected reduction in the cfu value after inoculation on wheat indicates adaptation from the YES medium used to produce the spores to the wheat. Furthermore, these results demonstrate that the fungal spores and/or cells must be in an active growth phase; otherwise, they would not be detectable by real-time PCR. The real-time PCR was also used in a reverse transcriptase real-time PCR approach to study the activation of the ochratoxin A biosynthesis genes directly in wheat [96]. Wheat was inoculated with *P. verrucosum* and stored at ambient temperatures for up to 60 days. Samples were withdrawn and the ochratoxin A was analyzed by HPLC, and the expression of the polyketide synthase was determined by real-time PCR. This experiment revealed a good correlation between the expression of the *otapks*PN gene and the phenotypic biosynthesis. However, expression of the *otapks*PN gene could be measured before detectable ochratoxin A was produced. A clear induction of the *otapks*PN gene arose after incubation for 22 days and the optimum was reached after 35 days. In contrast to the early monitoring of *otapks*PN gene expression, phenotypically produced ochratoxin A began only to be detected at day 42 at elevated levels. Optimum ochratoxin A production was reached after 58 days (whereas the *otapks*PN expression optimum was on day 35), indicating that the induction of the *otapks*PN gene expression can be taken as an early alert for ochratoxin A biosynthesis and hence *otapks*PN expression can be used to predict ochratoxin A production, allowing time for preventative measures to be taken such as immediate drying to a water contend below 15%.

As discussed earlier, external stress conditions can activate ochratoxin biosynthesis and this could be demonstrated in the wheat system. Wheat was inoculated with *P. verrucosum* and stored in silos at the natural ambient temperature. Interestingly, when the ambient temperature dropped during autumn, an induction of ochratoxin A biosynthesis genes could be observed by real-time PCR, which was paralleled with an increase in ochratoxin A biosynthesis [97]. The wheat was stored in silos for several months when the average temperatures of the silos dropped from approximately 20°C in summer to approximately 6°C in winter. In month 3 of the experiment, the average temperature was 11°C, which is the temperature at which the growth of *P. verrucosum* drops but marginal growth is still possible, which in turn leads to an induction of ochratoxin A biosynthesis genes as demonstrated in previous challenge experiments [97]. As mentioned earlier, stress conditions induced by suboptimal concentrations of preservatives/fungicides can induce the biosynthesis of ochratoxin A [98]. In the analysis of Arroyo et al. [98], calcium propionate and potassium sorbate were used as preservatives and 3000 ppm led to complete inhibition of the growth of *P. verrucosum*. However, at suboptimal concentrations of 300 ppm, marginal growth was possible and an activation of ochratoxin A biosynthesis was observed [98]. This was confirmed in a subsequent analysis [40] where the activation of a polyketide synthase gene was also studied and the expression of the *otapks*PV gene was induced.

Despite the fact that *P. nordicum* and *P. verrucosum* are two related species, the regulation of ochratoxin A biosynthesis differs completely. Whereas *P. nordicum* is a very consistent at producing (detectable) ochratoxin A [99], *P. verrucosum* is more variable. Four groups of *P. verrucosum* strains have been reported that produce only (1) ochratoxin A, (2) citrinin, (3) both toxins, or (4) none that was confirmed at the molecular level by Frisvad et al. [93]. They analyzed 321 isolates of *P. verrucosum* by thin layer chromatography and 236 (74%) produced ochratoxin A under the conditions used. Of this group, 185 were analyzed by RFLP fingerprinting, which revealed that 138 strains showed unique AFLP patterns. This suggests a high genotypic variability of the single strains within the species *P. verrucosum*. In contrast, *P. nordicum* showed a very homogenous RAPD and AFLP type [100], indicating that the genetic variation is much more restricted.

This high and consistent biosynthesis of ochratoxin A by *P. nordicum*, in contrast to *P. verrucosum*, has an ecological basis. *P. nordicum* nearly exclusively can be isolated from protein and NaCl-rich foods such as dry-cured meat products and cheeses [99,101–103]. Interestingly, it can also be isolated from saline solutions or pure salt [11]. This indicates the adaptation of *P. nordicum* to high NaCl concentrations. High ionic concentrations impose stress on the fungus, and Gunde-Cimerman et al. [6] showed during the isolation of extremophilic fungi that increased concentrations of NaCl lead to a reduction in the number of species isolated. Highest numbers were found in media with 5% NaCl, which corresponds to a water activity value of 0.951. A concentration of 24% NaCl in the medium, corresponding to an a_w of 0.828 and was the upper limit for the isolation of adapted species. Interestingly, many food-related *Penicillium* species were found, for example, *P. chrysogenum*, *P. nalgiovense*, *P. commune*, *Penicillium lanosum*, *Penicillium echinulatum*, *P. palitans*, and, importantly for the current discussion, *P. nordicum*. A similar species population was found in NaCl-rich dry-cured foods [99–103].

The production of ochratoxin A and citrinin in *P. verrucosum* is regulated in a mutually exclusive manner: either high amounts of ochratoxin A and low amounts of citrinin are produced or vice versa [10] and this regulation is employed to adapt to different environments. *P. verrucosum* is adapted to cereals as the main habitat. It is regarded as a storage fungus, but spores are probably present on cereal plants during growth of the field. Under these conditions, they are subjected to bright sunlight that supports citrinin biosynthesis [55]. However, *P. verrucosum* can also be found in low frequencies in NaCl-rich products such as dry-cured ham or olives [104–107] and indicates that *P. verrucosum* can also adapt to these environments.

The influence of NaCl on the regulation of ochratoxin A biosynthesis in both species has been analyzed [10] by growth on a medium with increasing amounts of NaCl (0–100 g/L). Growth and secondary metabolites were determined and also the expression of the *otapks*PN gene in *P. nordicum*. Interestingly, there was a clear difference in the growth and regulation of the ochratoxin A biosynthesis between both species at increasing concentrations of NaCl. *P. nordicum* showed a steady high ochratoxin A biosynthesis over a broad NaCl concentration range (0–100 g/L) and biosynthesis in *P. verrucosum* is induced only at high NaCl concentrations. The secondary metabolite profile of *P. verrucosum* changed from citrinin at low NaCl to ochratoxin A at high NaCl concentrations indicating that NaCl induced ochratoxin A biosynthesis in *P. verrucosum*. This situation points to the intriguing hypothesis that ochratoxin A might be a shuttle for chloride transport out of the cell. Under high NaCl, the fungal cells are exposed to high Na^+ and Cl^- concentrations, which are partly taken up by the cell. High ionic concentrations in the cell however impose stress or are even toxic to the fungus and Cl^- ions appears to be especially disadvantageous to fungi. Samapundo [108] analyzed the influence of salts with different anion/cation combinations on the growth of various fungi. Chloride-containing salts had a much higher inhibiting activity than other salts. Also, Ayodele and Ojoghoro [109] showed that chloride salts were more toxic toward *Pleurotus tuberregium* compared to other salts, as the Na^+ can be effectively transported out of the cell. Kumar et al. [110] demonstrated that the halotolerant yeast *Debaryomyces nepalensis* accumulates K^+ but not Na^+ after growth at either high KCl or NaCl conditions and the Na^+ is generally more toxic than the K^+ [111]. A balanced influx and efflux of cations ensure cation homeostasis. Two effectively regulated Na^+ transporters are described for *Cryptococcus neoformans* [111]. These transporter systems are, for example, HOG signal cascade regulated, and ensure Na^+ homeostasis. On the other hand, chloride is effectively taken up by the fungus, but obvious excretion systems are less effective. Simkovich et al. [112] demonstrated that the anion transporter that imports anions (specifically Cl^-) into the cell is very effective, but an equally effective efflux of Cl^- could not be detected. This situation indicates that a potential imbalance between import and excretion of Cl^- may exist in fungi, which is especially problematic at high NaCl conditions. So in this respect, it is interesting to note that high NaCl conditions induce the biosynthesis of ochratoxin A at the expense of citrinin in *P. verrucosum*. Ochratoxin A carries a chloride and is excreted out of the cell leading to a permanent efflux of chloride. There are some reports that the amount of the produced and excreted ochratoxin A is not constant but oscillates around a constant upper level. This oscillation suggests that a part of the excreted ochratoxin A is degraded outside the cell and is subsequently resynthesized [83]. This situation leads to a permanent efflux of chloride out of the cell, which would ensure an intracellular chloride homeostasis, even under high external Cl^- concentrations. Degradation of ochratoxin A during prolonged incubation is a common phenomenon and has been described repeatedly [113].

This intriguing hypothesis could be confirmed in accumulation experiments [10]: *P. nordicum*, the NaCl-adapted species, was grown on YES medium with increasing amounts of NaCl (0, 20, and 80 g/L). For this experiment, two strains were used, one that strongly produces ochratoxin A and the other, a mutant strain, which was not able to produce ochratoxin A at high NaCl concentrations. After a growth of 7 days, the mycelium was harvested and the chloride content of the mycelium was determined. Interestingly, at 0 and 20 g/L NaCl, no difference in the cellular chloride

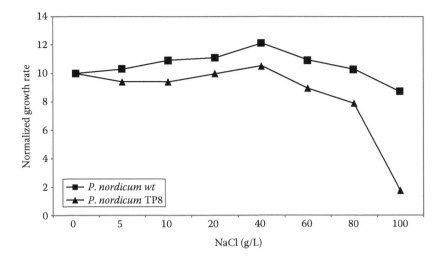

FIGURE 26.5 Comparison of the relative growth rates of the *P. nordicum* wild type (wt) with that of the *P. nordicum* mutant TP8. At low NaCl concentrations, when the mutant was still able to produce ochratoxin A, the growth rates between both strains were very similar. However at higher NaCl concentrations, when the mutants strain was no longer able to produce ochratoxin A, the growth rate of the mutant was drastically reduced compared to the wild type.

content could be observed in both strains; however, at 80 g/L, the nonproducing strain showed significant higher cellular chloride content compared to the producing strain. This clearly indicates that the biosynthesis and excretion of ochratoxin A contribute to the cellular chloride homeostasis under high NaCl conditions and apparently ensure competitiveness of the ochratoxin A–producing strains under these conditions. Interestingly, the relative growth rate at high NaCl concentrations was drastically different between the *P. nordicum* wild type wt and the *P. nordicum* mutant TP8 (Figure 26.5).

An increasing concentration of NaCl (or other chloride salts) in the medium shifts secondary metabolism in *P. verrucosum* from citrinin toward ochratoxin A, which is an adaptation of the fungus to this particular environment and increases the competitiveness of the fungus by ensuring partial chloride homeostasis. However, the secondary metabolite production of *P. verrucosum* can also be shifted from ochratoxin A to citrinin by incubating the fungus under differing light conditions [55]. Certain light conditions induce oxidative stress [114] and it is an indication that the biosynthesis of citrinin may be induced by oxidative stress. It was shown recently that citrinin represents an effective antioxidant [115]. Hence, the production of citrinin by *P. verrucosum* under conditions of oxidative stress at the expense of ochratoxin A is again an adaptation to a particular environment.

The production of ochratoxin A and citrinin are adaptive processes that increases the competitiveness of the fungi under stressful conditions. Under high NaCl concentrations, the biosynthesis of ochratoxin A is induced. In cases when the oxidative stress is increasing, for example, in *P. verrucosum* spores in the field from exposure to bright day light, citrinin is produced at the expense of ochratoxin A that again increases the competitiveness of *P. verrucosum* by reducing oxidative damages. That citrinin may function as a sun protectant was already suggested by Stormer et al. [116]. The light-induced activation of citrinin biosynthesis is further confirmed by Vrabcheva et al. [28] who found that Bulgarian wheat samples have up to 200 times more citrinin than ochratoxin A. Also, the citrinin concentration in cereal products may be higher compared to the ochratoxin concentration in cases where both toxins occur [117,118].

26.5.4 Patulin Production by *Penicillium* Species

As with ochratoxin A, patulin is a polyketide mycotoxin. In fact, the patulin polyketide synthase of *Penicillium patulum* was one of the first identified pks genes responsible for the biosynthesis of a mycotoxin [119]. Patulin is produced by a range of different food-relevant fungi, for example, *P. expansum*, *P. patulum*, *P. crustosum*, *P. griseofulvum* (synonym *P. patulum*, *Penicillium urticae*), *P. paneum*, *Paecilomyces variotii*, *Byssochlamys nivea*, *Aspergillus giganteus*, *Aspergillus terreus*, and *Aspergillus clavatus* [120,121]. Chemically, patulin is 4-hydroxy-4H-furo[3,2c]pyran-2[6H]-one. It was originally isolated as an antibacterial antibiotic. However, various toxic side effects against animals and humans were detected [121] and there are several described negative health effects of patulin. These are, among others symptoms, convulsions, GI tract distension, nausea, intestinal inflammation, genotoxicity, neurotoxicity, teratogenicity, plasma membrane disruption, inhibition of protein and DNA biosynthesis, and inhibition of RNA polymerase [121]. It also induces DNA double-strand breaks [122]. It is regulated in several countries and levels have been set because of these considerable toxicological activities. The upper limit of patulin is set at 50 mg/L for European countries. Patulin is especially problematic in pomaceous fruits, in particular in apples. *P. expansum* is the main apple-rotting fungus and leads to the typical blue green rot of this fruit. When *P. expansum* has colonized the fruit tissues, it can produce patulin and the toxin is able to diffuse at least 1 cm into the apple and pear tissue [123,124]. Patulin has been found on other fruits such as grapes [125,126], blueberries, raspberries, strawberries, cherries, peaches plums, and black currants [127]. Interestingly, like ochratoxin A, patulin can also be found coproduced with citrinin [126], and both have been identified in grapes or grape juices [128,129], which is unsurprising in that *P. expansum* produces both mycotoxins. *P. expansum* co occurs also very often with ochratoxin A–producing Penicillia [6] as does *P. citrinum*. Martins et al. [130] detected patulin and citrinin in 19.6% of 351 analyzed apple samples. They generally found much higher concentrations of patulin (up to 80.5 mg/kg) than citrinin (0.92 mg/kg). As mentioned, patulin is a polyketide mycotoxin, that is, it is biosynthesized from acetyl-CoA and malonyl-CoA precursors. The first intermediate produced by the polyketide synthase, the 6-MSA synthase, is 6-methylsalicylic acid, which is converted to *m*-cresol by 6 MSA decarboxylase and further to *m*-hydroxybenzyl alcohol by *m*-cresol 2-hydroxylase. One intermediate is isoepoxydon, which is converted to phyllostine by the isoepoxydon dehydrogenase [121].

Based on this information, molecular detection systems for *P. expansum* in fruits have been developed. Marek et al. [31] described a PCR system specific for *P. expansum*, which was based on the polygalacturonase gene. It gave a 404 bp fragment with all analyzed *P. expansum* strains, but it produced no amplicon with *P. roqueforti*, *P. solitum*, *P. echinulatum*, *P. crustosum*, *P. commune*, *Penicillium notatum*, or *P. chrysogenum*. With this system, the authors could detect down to 25 spores per reaction. Paterson et al. [131] described primer pairs that were directed against the *idh* gene (isoepoxy dehydrogenase gene). They only gave a fragment of 600 bp with *P. expansum* and *P. brevicompactum*. All of the analyzed *P. expansum* strains were able to produce patulin; however, from none of the three analyzed *P. brevicompactum* strains could patulin be detected. However, the results demonstrated that *P. brevicompactum* strains obviously carry the gene or even the whole cluster in a silent form, which may be activated under specific conditions. An alternative interpretation is that the strains were mutated as discussed in [132]. All other tested *Penicillium* species such as *P. citrinum*, *P. corylophilum*, *P. glabrum*, *P. simplicissimum*, *brevicompactum janczewskii*, and *Penicillium spinulosum* were negative, although an internal amplification control was not used and so the results could have been false. Hoorfar et al. [133] demanded a mandatory internal amplification control for food testing PCR reactions. Paterson [134] indicated for the first time how PCR inhibitors could be present in the culture medium of food-related fungi. According to the review by Paterson [135], internal amplification controls are not yet routinely used. In a further analysis, Paterson et al. [32] showed that the *idh* gene is obviously present in strains of various *Penicillium* species such as *P. brevicompactum*, *Penicillium paxilli*, and *P. roqueforti*. There was however a clear correlation between the presence of the *idh* gene and patulin biosynthesis. All 50 analyzed strains that were negative for the *idh* gene did not produce detectable patulin, whereas 66% of the positive strains where able to produce the toxin. This was also true for the positive strains of *P. roqueforti* and *P. paxilli*. In contrast, some of the *P. brevicompactum* strains of this analysis gave ambiguous results in that they partly showed a weak positive PCR reaction in correlation with very low but possible patulin biosynthesis under the conditions used. The work in penicillia was extended to include results indicating *idh* positive strains of *Aspergillus terreus*, *A. clavatus*, *A. giganteus*, *Byssochlamys fulva*, *B. nivea*, and *Paecilomyces varioti* [134]. In addition, a wider range of penicillia were also demonstrated to possess the gene within subgenera *Furcatum* and *Aspergilloides* apart from subgenus *Penicillium*. The same author [136] tested the possibility to use the *idh* primers for HACCP purposes within apple orchards. He found positive PCR reactions for the presence of the *idh* gene in soils around the apple trees and in various parts of the plant, such as bark or twigs. Varga et al. [113] detected the *idh* gene in various *Aspergillus* species, for example, *A. clavatus*, *Aspergillus pallidus*, *Aspergillus clavatonanica*, and *Aspergillus longivesica* by using the primers described by Paterson [32]. The presence of the gene correlated well with the ability of the strains to produce patulin. However, in some cases, a positive reaction was found but the respective strain did not produce detectable patulin. That was the case for the ex-type strain of *A. clavatus* (however, not for the other *A. clavatus* strains) and a strain of *Aspergillus carneus* and *A. terreus*. One of the primers used by Paterson [32] (idh1) was used by Luque et al. [137] to develop a real-time PCR system to detect and quantify patulin-producing fungi. With their system, they achieved a detection limit of 10 conidia per gram of inoculated food matrices, that is, peach or dry fermented sausage. In their list of strains used to demonstrate the specificity of the real-time PCR reaction, they found a low C_t value in strains or species that were shown to be able to produce patulin. However, some of the species, which were patulin positive, are not typical patulin-producing species such as *A. flavus*, *Aspergillus oryzae*, and *P. verrucosum* and some others [138]. Paterson [139] further used the *idh* primer pair for taxonomical purposes. By using these primers at a hybridization temperature of 52°C, the author could generate the expected 620 bp fragment with most of the *P. expansum* and *P. griseofulvum* strains analyzed; however, also additional bands occurred. This was true also for other species! Based on the bands produced, it was possible to group the species, which fits well with the morphological characterization. Hence, the author suggests that this approach may be used for taxonomical purposes. Dombrink-Kurtzman et al. [140] determined the sequence of the *idh* gene from various different *Penicillium* species able to produce patulin. These species include *P. carneum*, *Penicillium clavigerum*, *Penicillium concentricum*, *Penicillium coprobium*, *Penicillium dipodomyicola*, *P. expansum*, *Penicillium gladioli*, *Penicillium glandicola*, *P. griseofulvum*, *P. paneum*, *Penicillium sclerotigenum*, and *Penicillium vulpinum*. The authors compared the sequences and found eight of the nine conserved nucleotides of the *idh* sequence were also conserved in this study, whereas the ninth expected amino acid, lysine at position 46, was present in all *P. griseofulvum* (from which the sequence was originally generated) and in the *P. dipodomyicola* strains, but lysine was exchanged by threonine in all other strains. The strains having lysine at position 46 were the strongest producers of patulin. For this reason, it was suggested that the amino acid at this position is responsible for the optimal function of the enzyme and that it is responsible for the binding of the cofactor $NADP^+$. There were additional amino acid substitutions in *P. expansum*, which could not be identified in the other species. White et al. [127] identified further genes of the patulin biosynthetic gene cluster from *P. expansum*. They cloned a 470 bp fragment of the 6-methylsalicylic acid synthase (the patulin polyketide synthase) and showed its homology to the originally identified homologue from *P. urticae* [119]. At the nucleotide level, this gene showed high homology to the respective gene of *P. urticae* [119], whereas at the amino acid level, it resembles the proteins of *Byssochlamys nivea* and *A. terreus*. The *idh* gene of *P. expansum* was cloned by using primers deduced from the *idh* gene of *P. urticae* [141]. The gene of *P. expansum* showed 85% similarity to the gene of *P. urticae*. The authors identified a further gene upstream of the *idh* gene by the use of heterologous primers. After sequence comparison, it resembles an ABC transporter gene of *Botryotinia fuckeliana*, *C. albicans*, and *P. digitatum*. In a further suppression subtractive hybridization PCR under conditions permissive for patulin biosynthesis, they identified two additional genes with similarity to p-450 monooxygenase genes, in particular p-450-1 and p450-2. In an expression analysis, a high level of expression was found for all five genes under conditions, which supported patulin biosynthesis indicating the involvement of these genes in patulin biosynthesis and the possible regulation of patulin biosynthesis at the transcriptional level. The p450-2 transcript was upregulated 1127 fold and the p450-1 transcript 250 fold. Dombrink-Kurtzman [142] compared the sequence of a 600 bp fragment of the *idh* gene of *P. griseofulvum* with that of the respective fragment of *P. expansum*. The author found a difference of 12 amino acids of these two fragments. This difference corresponds with the ability of both species to produce patulin. Generally, *P. griseofulvum* produced greater amounts of patulin than *P. expansum*. An alcohol isoamyl oxidase gene from *P. griseofulvum* was described by the same author [143]. The gene was isolated by primer walking starting from the known *idh* gene and so it is in direct vicinity of this gene. Transcripts of this gene could be identified; hence, it was actively transcribed; however, a correlation to patulin biosynthesis was not shown. Puel et al. [144] analyzed the presence of the 6-methylsalicylic acid synthase gene and the *idh* gene in 8 *Byssochlamys nivea* and in 11 *B. fulva* strains. Interestingly, these authors found both genes in all 8 *B. nivea* strains, but not in any of the 11 *B. fulva* strains. This was correlated with the ability of all *B. nivea* strains to produce patulin, but none of the *B. fulva* strains did produce the toxin under the given conditions. The authors therefore suggest that *B. fulva* is not a source of patulin, because of the lack of the respective biosynthetic genes. However, in other studies, the *idh* gene was detected in *B. fulva* [134,139].

The growth of *P. expansum* as the most important apple spoilage organism is partly being controlled during storage by the use of fungicides [145]. Karaoglanidis et al. [145] analyzed 236 strains of *P. expansum* isolated from decayed apple fruits for their sensitivity against the fungicides tebuconazole, fludioxonil, iprodione, and cyprodinil. They identified a subpopulation of strains (7.5%), which were less sensitive toward these fungicides; some of them even showed a multiresistance that indicates the risk of increased patulin occurrence in apple tissue. Cabanas et al. [146] analyzed the β-tubulin gene of 71 *P. expansum* strains. Thirty-seven of these strains were sensitive toward the fungicide thiabendazole, whereas 34 were resistant. The β-tubulin gene of these strains was screened for mutations because this type of fungicide binds to β-tubulin and inhibits microtubule polymerization [147]. Strains resistant to this type of fungicides have mutations at certain positions in the β-tubulin gene. These mutations altered the amino acid sequence of the benzimidazole (a derivative of thiabendazole) binding site. In Cabanas et al. [146], all of the sensitive strains lacked mutations in this particular region, except for one, whereas all strains with a mutation at position 198 belonged to the resistant group. However, the occurrence of this mutation was not consistently present in the resistant group. Many of them had no mutation in the analyzed region indicating that other mechanisms are also involved in thiabendazole resistance.

Recently, Sanzani et al. [148] could show that the production of patulin is obviously important for the colonization of apples by the fungus and hence it constitutes a pathogenicity factor. The patulin pks gene was knocked out by integrative transformation, and the resulting strains that were no longer able to produce the toxin demonstrated a considerably reduced capacity to colonize apples.

26.5.5 Cyclopiazonic Acid–Producing *Penicillium*

CPA is produced by several *Aspergillus* and *Penicillium* species. It is named after *Penicillium cyclopium* from which the toxin was first isolated [149]. However, it was later stated that *P. cyclopium* is not able to synthesize CPA but the correct producer is *P. griseofulvum* [29]. Certain strains of *A. flavus* can produce CPA and aflatoxin. The genetic background of CPA biosynthesis in *Aspergillus* is well elucidated, but much less data about Penicillia are available. About 77% of the aflatoxin-producing *A. flavus* strains are able to synthesize CPA. Also *A. oryzae*, which is closely related to *A. flavus*, but which cannot produce aflatoxin, is able to synthesize CPA. However, only a subset of *A. oryzae* strains, as with the *A. flavus* strains, is able to produce this secondary metabolite. *Penicillium commune* and *P. camemberti* are known CPA-producing organisms and are closely related, where *P. camemberti* is regarded as a domesticated form of *P. commune*. Because of the production of CPA by *P. camemberti*, it can be found on mold-ripened cheeses such as camembert, especially in the outer layers at up to 4 µg/g, and it can be found on dry-cured meat products [150–152]. Bailly et al. [153] detected CPA at up to 50 mg/kg on dry-cured ham. CPA was found on peanuts at up to 6 µg/g [154], often in combination with aflatoxin, indicating that the occurrence of CPA in these commodities was due to the growth of *A. flavus*.

CPA is an indole tetramic acid that is synthesized via a polyketide pathway and it shows a moderate toxicity in various animal models. At the molecular level, it is a specific inhibitor of the Ca^{2+}-ATPase and thereby inhibiting Ca^{2+} flux through membranes. In rat experiments, various toxic effects could be observed after administration, and among these were neurological problems, decreased weight gain, diarrhea, dehydration, depression, hyperesthesia, hypokinesis, convulsion, and death, depending on the dose applied.

Le Bars [150] showed that all of 20 strains of *P. camemberti* were able to produce CPA, indicating that this is a common feature of this species. This is confirmed by the results of El Banna et al. [24] who showed that all 61 *P. camemberti* strains were able to produce this toxin. Because no known natural non-CPA-producing variants of *P. camemberti* exist, Geisen et al. [155] mutagenized *P. camemberti* conidia and performed a screening for non-CPA-producing strains. It was possible to isolate two strains with a mutant phenotype; one strain had completely lost the ability to produce CPA, whereas the other strain produced 50- to 100-fold less CPA than the wild type. The former strain however had a changed morphology compared to the wild type and had a tendency to form revertants toward the wild-type morphology. These revertants were able to produce CPA, which suggests that the biosynthesis of CPA is an important feature for *P. camemberti*. Interestingly, the latter mutant, which was more stable, produced a novel unknown compound. In a subsequent analysis, Geisen [156] located the mutation, which was obtained by mutagenizing *P. camemberti* with nitrous acid, in the CPA pathway. According to Holzapfel [149], the synthesis of CPA starts from the amino acid tryptophan and progresses with the inclusion of acetyl-CoA and dimethyl allyl pyrophosphate, which is also synthesized by acetyl-CoA units. By using either ^{3}H-labeled tryptophan or ^{3}H-labeled acetate, it could be demonstrated that the biosynthesis of acetate precursors was affected in this mutation rather than the biosynthesis of tryptophan [156]. Also, Benkhemmar et al. [157] mutated *A. oryzae* strains to obtain non-CPA-producing strains for possible application in the food industry. They used *N*-methyl-*N'*-nitro-*N*-nitrosoguanidine for mutation and were able to obtain the desired mutants. In a further experiment, the possibility of the transfer of the CPA production feature between strains by anastomosis and heterokaryosis was proven. The conclusion was that CPA production can be transferred rapidly after anastomosis between a producing and a nonproducing strain, but not if two nonproducing strains were crossed. This result again indicates that the production of CPA is advantageous to the fungus.

For *A. flavus*, the genome of which is sequenced, the CPA biosynthesis gene cluster is almost known [77]. In congruence with the results of Tokuoka et al. [158], the CPA gene cluster was located in a telomeric region near the aflatoxin biosynthesis gene cluster. Also, Chang et al. [77] identified pks-nrps, monoamine oxidase, dimethylallyl tryptophan synthase, and possible transcription factor genes. The first three genes were connected to the CPA biosynthesis pathway by a gene disruption approach and it was verified that they connect to the CPA gene cluster [159]. The genes could be located in an 87 kb subtelomeric region adjacent to the aflatoxin gene cluster on chromosome 3. The adjacent location of the aflatoxin and CPA gene cluster is demonstrated by the fact that many of the *A. flavus* strains, which do not produce aflatoxin, are also not able to produce CPA [148]. A current representation of the biosynthetic pathway of CPA starting from acetyl-CoA by the activity of the multicomponent PKS-NRPS enzyme is shown in Chang et al. [159].

Furthermore, Chang et al. [159] proposed an ecological reason for the biosynthesis of CPA related to the fact that CPA is a good chelator of iron. So it is possible that CPA-bound iron can serve as a stockpile for growth under iron-limiting conditions. This is, as in the case of the production of ochratoxin A, citrinin or patulin, an ecological explanation for the production of mycotoxins. However, it is not clear yet if the molecular background described earlier is the same in *Penicillium*.

26.5.6 PR-Toxin Production by *Penicillium*

Most strains of *P. roqueforti* are able to produce PR toxin, a secondary metabolite of quite high toxicity [160]. One of the precursors of PR toxin is the metabolite aristolochene. A gene (*ari*1) that codes for an aristolochene synthetase was cloned from *P. roqueforti*. It is assumed that aristolochene is a precursor of various sequiterpenoid toxins produced by the fungus, among them PR toxin [161]. The gene *ari*1 codes for a sesquiterpene cyclase, the aristolochene synthase from *P. roqueforti*. The gene was also heterologously expressed in *E. coli* and it could be demonstrated that it has sesquiterpene cyclase activity [162]. The same gene was isolated from *A. terreus*, which has 66% identity at the nucleotide level to that of *P. roqueforti* [163]. The gene product catalyzes the cyclization of farnesyl diphosphate to the sesquiterpene aristolochene. Jelen et al. [164] determined the influence of octanoic acid on the formation of PR toxin. *P. roqueforti* is used during ripening of Roquefort cheese, and during this process, octanoic acid may be present influencing the biosynthesis of PR toxin. These authors demonstrated that octanoic acid drastically reduced the biosynthesis of PR toxin, and they concluded that the free fatty acids in cheeses maybe a factor that inhibits the biosynthesis of PR toxin in this environment. Furthermore PR toxin is degraded by *P. roqueforti* itself to PR-acid, which is assumed to be further degraded to PR-amide [165]. Scott [160] indicated that the production of PR toxin in cheeses is unlikely, because the conditions such as aeration, low carbohydrate concentration, and the presence of NaCl are not supportive for its production. Furthermore, PR toxin is not stable in blue cheese, but is converted to the much less toxic PR imine. As mentioned, PR toxin is one of the most acutely toxic metabolites from *P. roqueforti* and it can induce degenerative changes in the liver of rats. On the molecular level, PR toxin is an inhibitor of nucleic acid and protein biosynthesis and showed mutagenic activity in *Saccharomyces cerevisiae* and in *Neurospora crassa* [161].

26.6 CONCLUSION

The *Penicillium* toxins for which the most molecular data are available are ochratoxin A, citrinin, patulin, CPA, and PR toxin. Ochratoxin A and citrinin are nephrotoxic compounds, which often occur simultaneously and synergistically. Ochratoxin A is considered as a human carcinogen and its maximum concentration in foods is regulated. Also, the maximum concentration of patulin is regulated, because of the potent toxicity of this secondary metabolite. PR toxin has a quite high toxicity, but its stability in cheese, which is the most important environment for its occurrence, is limited, so that it is unlikely that the consumer is exposed to toxic concentrations of this metabolite. CPA showed only moderate toxicity in various animal model experiments; nevertheless, this metabolite can be produced on cheeses under certain conditions, and novel *P. camemberti* starter cultures without CPA capacity should be investigated in selection programs.

Despite the reason for fungi producing secondary metabolites is not yet known, it is clear that they are not produced because of their toxic activity against humans and animals. It is important to know under which food-related conditions the mycotoxins biosynthesis genes are activated. The approaches described earlier point toward this understanding and may enable novel strategies to reduce the negative influence of the mycotoxins on human health. Because of the described molecular knowledge about mycotoxins regulation in food, these conditions during the storage of certain foods must be prevented. This may be reached by adjusting the parameters (e.g., temperature, a_w, pH) in a way that no growth, instead of low growth, is achieved.

New approaches to control mycotoxins biosynthesis also involves understanding the reasons why mycotoxins are being produced by the fungus. Possible ecological functions for the biosynthesis of many mycotoxins could be elucidated during molecular analysis. In each case, this putative ecological function increases the viability or competitiveness of the fungal strain under specific conditions. These conditions are mainly related to stress of the fungus. A stressful environment is recognized by the fungus by specialized signal transduction cascades, which in turn lead to fine-tuned regulation of the activity of the mycotoxin biosynthesis genes in relation to changing environmental parameters. The production of mycotoxins better adapts the fungus to a particular environment. It appears that the biosynthesis of mycotoxins has most importance under these specialized, mainly stressful, conditions and that may be a reason why nontoxin-producing mutants or producing strains show the same growth rate on complete laboratory media where they both cope equally well. Furthermore, the new information obtained by the transcriptomic approaches demonstrates that stressful conditions, which do not completely abolish the growth of the fungus and which often occur during the storage of certain foods, can lead to increased mycotoxin biosynthesis, which in turn is counter active with respect to food safety. The new results may help to overcome this problem by more carefully adjusting these conditions.

REFERENCES

1. Ropars, J., Dupont, J., Fontanillas, E., Rodríguez de la Vega, R. C., Malagnac, F., Coton, M., Giraud, T., and Lópet-Villavicencio, M. (2012). Sex in cheese: Evidence for sexuality in the fungus *Penicillium roqueforti*. *PLoS One* 7:1–9.
2. Pitt, J. I. and Hocking, A. D. (1999). *Fungi and Food Spoilage*. Aspen Publishers, Gaithersburg, MD.
3. Samson, R. A. and Frisvad, J. C. (2004). *Penicillium Subgenus Penicillium: New Taxonomic Schemes, Mycotoxins and Other Extrolites*. Studies in Mycology 49, Centraalbureau voor Schimmelcultures, Utrecht, the Netherlands.
4. Samson, R. A., Yilmaz, N., Spierenburg, H., Seifert, K. A., Peterson, S. W., Varga, J., and Frisvad J. C. (2011). Phylogeny and nomenclature of the genus Talaromyces and taxa accommodated in *Penicillium* subgenus Biverticillium. *Studies Mycol* 70:159–183.
5. Nevarez, L., Vasseur, V., Le Madec, A., Le Bras, M. A., Coroller, L., Leguérinel, I., and Barbier, G. (2009). Physiological traits of *Penicillium glabrum* strain LCP 08.5568, a filamentous fungus isolated from bottled aromatized mineral water. *Int J Food Microbiol* 130:166–171.
6. Gunde-Cimerman, N., Sonjak, S., Zalar, P., Frisvad, J. C., Diderichsen, B., and Plemenitas, A. (2003). Extremophilic fungi in arctic ice: A relationship between adaptation to low temperature and water activity. *Phys Chem Earth* 28:1273–1278.
7. Lu, Z. Y., Lin, Z. J., Wang, W. L., Du, L., Zhu, T. J., Fang, Y. C., Gu, Q. Q., and Zhu, W. M. (2008). Citrinin dimers from the halotolerant fungus *Penicillium citrinum* B-57. *J Nat Prod* 71(543):546.
8. Abbas, A. S. and Ali, N. (2011). Isolated of halotolerant *Penicillium* strains form the Howz Soltan Lake to produce a-amylase. *Middle East J Sci Res* 7:407–412.
9. Yadav, J., Verma, J. P., Yadav, S. K., and Tiwari, K. N. (2011). Effect of salt concentration and pH on soil inhabiting fungus *Penicillium citrinum* Thom. for solubilization of tricalcium phosphate. *Microbiol J* 1:25–32.
10. Schmidt-Heydt, M., Graf, E., Stoll, D., and Geisen, R. (2012). The biosynthesis of ochratoxin A by *Penicillium* as one mechanism for adaptation to NaCl rich foods. *Food Microbiol* 29:233–241.
11. Sonjak, S., Licen, M., Frisvad, J. C., and Gunde-Cimerman, N. (2011). Salting of dry-cured meat—A potential cause of contamination with the ochratoxin A-producing species *Penicillium nordicum*. *Food Microbiol* 28:1111–1116.
12. Comerio, R., Fernandez Pinto, V. E., and Vaamonde, G. (1998). Influence of water activity on *Penicillium citrinum* growth and kinetics of citrinin accumulation in wheat. *Int J Food Microbiol* 42:219–223.
13. Taniwaki, M. H., Hocking, A. D., Pitt, J. I., and Fleet, G. H. (2009). Growth and mycotoxin production by food spoilage fungi under high carbon dioxide and low oxygen atmospheres. *Int J Food Microbiol* 132:100–108.

14. Gonçalves, A. B., Paterson, R. R. M., and Lima, N. (2006). Survey and significance of filamentous fungi from tap water. *Int J Hyg Environ Health* 209:257–264.
15. Holmes, G. J. and Eckert, J. W. (1999). Sensitivity of *Penicillium digitatum* and *P. italicum* to postharvest citrus fungicides in California. *Phytopathology* 89:716–721.
16. Duong, T. A. (1996). Infection due to *Penicillium marneffei*, an emerging pathogen: Review of 155 reported cases. *Clin Infect Dis* 23:125–130.
17. Gravesen, S., Frisvad, J. C., and Samson, S. A. (1994). *Microfungi*. Munksgaard, Copenhagen, Demark.
18. Lyratzopoulos, G., Ellis,. M., Nerringer, R., and Denning, D. W. (2002). Invasive infection due to *Penicillium* species other than *P. marneffei*. *J Infect* 45:184–207.
19. Gilliam, J. S. and Vest, S. A. (1951). *Penicillium* infection of the urinary tract. *J Urol* 65:484–489.
20. Eschete, M. L., King, J. W., West, B. C., and Oberle, A. (1981). *Penicillium chrysogenum* endophthalmitis. First reported case. *Mycopathologia* 74:125–127.
21. Upshaw, C. B. 1974. *Penicillium* endocarditis of aortic valve prosthesis. *J Thoracic Cardiovasc Surg* 68:428–431.
21 Fiocchi, A., Mirri, G. P., Santini, I., Bernado, L., Ottoboni, F., and Riva, E. (1997). Exercise-induced anaphylaxis after food contaminant ingestion in double-blinded, placebo-controlled, food-exercise challenge. *J Allergy Clin Immunol* 100:425–427.
22. Pitt, J. I. (1994). The current role of *Aspergillus* and *Penicillium* in human and animal health. *J Med Veter Mycol* 32:17–32.
23. Frisvad, J. C., Thrane, U., Samson, R. A. and Pitt, J. I. (2006). Important mycotoxins and the fungi which produce them. In: *Advances in Food Mycology*, A. D. Hocking, J. I. Pitt, R. A. Samson, and U. Thrane, Eds. Springer, New York, pp. 3–31.
24. El Banna, A., Pitt, J. I., and Leistner, L. (1987). Production of mycotoxins by *Penicillium* species. *Syst Appl Microbiol* 10:42–46.
25. Bergmann, S., Schümann, J., Scherlach, K., Lange, C., Brakhage, A. A., and Hertweck, C. (2007). Genomics-driven discovery of PKS-NRPS hybrid metabolites from *Aspergillus nidulans*. *Nat Chem Biol Lett* 3:213–217.
26. Wagacha, J. M. and Muthomi, J. W. (2008). Mycotoxin problem in Africa: Current status, implications to food safety and health and possible management strategies. *Int J Food Microbiol* 124:1–12.
27. Liu, Y. and Wu, F. (2010). Global burden of aflatoxin-induced hepatocellular carcinoma: A risk assessment. *Environ Health Perspect* 118:818–824.
28. Vrabcheva, T., Usleber, E., Dietrich, R., and Märtlbauer, E. (2000). Co-occurrence of ochratoxin A and citrinin in cereals from Bulgarian villages with a history of balkan endemic nephropathy. *J Agric Food Chem* 48:2483–2488.
29. Frisvad, J. (1989). The connection between the Penicillia and Aspergilli and mycotoxins with special emphasis on misidentified isolates. *Arch Environ Contam Toxicol* 18:452–467.
30. González-Salgado, A., Patino, B., Vázquez, C., and González-Jaén, M. T. (2005). Discrimination of *Aspergillus niger* and other *Aspergillus species* belonging to section Nigri by PCR assays. *FEMS Microbiol Lett* 245:353–361.
31. Marek, P., Annamalai, T., and Venkitanarayanan, K. (2003). Detection of *Penicillium expansum* by polymerase chain reaction. *Int J Food Microbiol* 89:139–144.
32. Paterson, R. R. M., Kozakiewicz, Z., Locke, T., Brayford, D., and Jones, S. C. B. (2003). Novel use of the isoepoxydon dehydrogenase gene probe of the patulin metabolic pathway and chromatography to test penicillia isolated from apple production systems for the potential to contaminate apple juice with patulin. *Food Microbiol* 20:359–364.
33. Pedersen, H. L., Skouboe, P., Boysen, M., Soule, J., and Rossen, L. (1997). Detection of *Penicillium* species in complex food samples using the polymerase chain reaction. *Int J Food Microbiol* 35:169–177, 1997.
34. Geisen, R., Mayer, Z., Karolewiez, A., and Färber, P. (2004). Development of a Real Time PCR system for detection of *Penicillium nordicum* and for monitoring ochratoxin A production in foods by targeting the ochratoxin polyketide synthase gene. *Syst Appl Microbiol* 27:501–507.
35. Rodriguez, A., Cordoba, J. J., Werning, M. L., Andrade, M. J., and Rodriguez, M. (2012). Duplex-real-time PCR method with international amplification control for quantification of verrucosidin producing molds in dry-ripened foods. *Int J Food Microbiol* 153:85–91.
36. Rodriguez, A., Rodriguez, M., Andrade, M. J., and Cordoba, J. J. (2012). Development of a multiplex real-time PCR to quantify aflatoxin, ochratoxin A and patulin producing molds in foods. *Int J Food Microbiol* 155:10–18.

37. Haugland, R. A., Varma, M., Wymer, L. J., and Vesper, S. J. (2004). Quantitative PCR analysis of selected Aspergillus, *Penicillium* and *Paecilomyces* species. *Syst Appl Microbiol* 27:198–210.
38. Sweeney, M. J., Pamies, P., and Dobson, A. D. W. (2000). The use of reverse transcription-polymerase chain reaction (RT-PCR) for monitoring aflatoxin production in *Aspergillus parasiticus* 439. *Int J Food Microbiol* 56:97–103.
39. Doohan, F. M., Weston, G., Rezanoor, H. N., Parry, D. W., and Nicholson, P. (1999). Development and use of a reverse transcription-PCR assay to study expression of *Tri*5 by *Fusarium* species in vitro and *in planta*. *Appl Environ Microbiol* 65:3850–3854.
40. Schmidt-Heydt, M., Baxter, E., Geisen, R., and Magan, N. (2007). Physiological relationship between food preservatives, environmental factors, ochratoxin and *otapks*PV gene expression by *Penicillium verrucosum*. *Int J Food Microbiol* 119:277–283.
41. Sandra, L. (2012). Molekulare Grundlagen des Einflusses von Licht auf lebensmittelrelevante Schimmelpilze im natürlichen Habitat. Master thesis, KIT, Karlsruhe, Germany.
42. Paterson, R. R. M. (2007). Some fungicides and growth inhibitor/biocontrol-enhancer 2-deoxy-D-glucose increase patulin from *Penicillium expansum* strains in vitro. *Crop Protect* 26:543–548.
43. Ochiai, N., Tokai, T., Takahashi-Ando, N., Fujimura, M., and Kimura, M. (2007). Genetically engineered *Fusarium* as a tool to evaluate the effects of environmental factors on initiation of trichothecene biosynthesis. *FEMS Microbiol Lett* 275:53–61.
44. Schmidt-Heydt, M., Schunck, T., and Geisen, R. (2009). Expression of a *gfp* gene in *Penicillium nordicum* under control of the promoter of the ochratoxin A polyketide synthase gene. *Int J Food Microbiol* 133:161–166.
45. Schmidt-Heydt, M. and Geisen, R. (2007). A microarray for monitoring the production of mycotoxins in food. *Int J Food Microbiol* 117:131–140.
46. Schmidt-Heydt, M., Magan, N., and Geisen, R. (2008). Stress induction of mycotoxin biosynthesis genes by abiotic factors. *FEMS Microbial Lett* 284:142–149.
47. O'Brian, G. R., Georgianna, D. R., Wilkinson, J. R., Yu, J., Abbas, H. K., Bhatnagar, D., Cleveland, T. E., Nierman, W., and Payne, G. A. (2007). The effect of elevated temperature on gene transcription and aflatoxin biosynthesis. *Mycologia* 99:232–239.
48. Pirttilä, A. M., McIntyre, L. M., Payne, G. A., and Woloshuk, C. P. (2004). Expression profile analysis of wild-type and *fcc*1 mutant strains of *Fusarium verticillioides* during fumonisin biosynthesis. *Fungal Genet Biol* 41:647–656.
49. Bok, J. W., Maggio-Hall, L. A., Murillo, R., Glasner, J. D., and Keller, N. P. (2006). Genomic mining for *Aspergillus* natural products. *Chem Biol* 13:31–37.
50. Brown, D. W., Butchko, R. A. E., Busman, M., and Proctor, R. H. (2012). Identification of gene clusters associated with fusaric acid, fusarin and perithecial pigment production in *Fusarium verticillioides*. *Fungal Genet Biol* 49:521–532.
51. Nicolaisen, M., Justesen, A., Thrane, U., Skouboe, P., and Holmström, K. (2005). An oligonucleotide microarray for the identification and differentiation of trichothecene producing and non-producing *Fusarium* species occurring in cereal grain. *J Microbiol Method* 62:57–69.
52. Harris, D. M., van der Krogt, Z. A., Klaassen, P., Raamsdonk, L. M., Hage, S., van den Berg, M., Bovenberg, R. A. L., Pronk, J. T., and Daran, J. M. (2009). Exploring and dissecting genome-wide gene expression responses of *Penicillium chrysogenum* to phenylacetic acid consumption and penicillin G production. *BMC Genom* 10:75–95.
53. Bayram, Ö., Sari, F., Braus, G. H., and Irniger, S. (2009). The protein kinase ImeB is required for light-mediated inhibition of sexual development and for mycotoxin production in *Aspergillus nidulans*. *Mol Microbiol* 71:1278–1295.
54. Atoui, A., Kastner, C., Larey, C. M., Thokala, R., Etxebeste, O., Espeso, E. A., Fischer, R., and Calvo, A. M. (2010). Cross-talk between light and glucose regulation controls toxin production and morphogenesis in *Aspergillus nidulans*. *Fungal Genet Biol* 47:962–972.
55. Schmidt-Heydt, M., Rüfer, C. E., Raupp, F., Bruchmann, A., Perrone, G., and Geisen, R. (2011). Influence of light on food relevant fungi with emphasis on ochratoxin producing species. *Int J Food Microbiol* 145:229–237.
56. Fanelli, F., Schmidt-Heydt, M., Haidukowski, M., Geisen, R., Logrieco, A., and Mulè, G. (2012). Influence of light on growth, fumonisin biosynthesis and FUM1 gene expression by *Fusarium proliferatum*. *Int J Food Microbiol* 153:148–153.
57. Kim, H. and Woloshuk, C. P. (2008). Role of AREA, a regulator of nitrogen metabolism, during colonization of maize kernels and fumonisin biosynthesis in *Fusarium verticillioides*. *Fungal Genet Biol* 45:947–953.

58. Georgianna, D. R. and Payne, G. A. (2009). Genetic regulation of aflatoxin biosynthesis: From gene to genome. *Fungal Genet Biol* 46:125.
59. Kohut, G., Ádám, A. L., Fazekas, B., and Hornok, L. (2009). N-starvation stress induced FUM gene expression and fumonisin production is mediated via the HOG-type MAPK pathway in *Fusarium proliferatum. Int J Food Microbiol* 130:65–69.
60. Jurado, M., Marín, P., Magan, N., and González-Jaén, M. T. (2008). Relationship between solute and matric potential stress, temperature, growth, and FUM1 gene expression in two *Fusarium verticillioides* strains from Spain. Appl Environ Microbiol 74:2032–2036.
61. Scherm, B., Palomba, M., Serra, D., Marcello, A., and Migheli, Q. (2005). Detection of transcripts of the aflatoxin genes *aflD*, *aflO*, and *aflP* by reverse-transcription-polymerase chain reaction allows differentiation of aflatoxin-producing and non-producing isolates of *Aspergillus flavus* and *Aspergillus parasiticus. Int J Food Microbiol* 98:201–210.
62. Graf, E., Schmidt-Heydt, M., and Geisen, R. (2012). HOG MAP kinase regulation of alternariol biosynthesis in *Alternaria alternata* is important for substrate colonization. *Int J Food Microbiol* 157:353–359.
63. Shimizu, K., Hicks, J. K., Huang, T. P., and Keller, N. P. (2003). Pka, Ras and RGS protein interactions regulate activity of AflR, a Zn(II)2Cys6 transcription factor in *Aspergillus nidulans. Genetics* 165:1095–1104.
64. Brodhagen, M. and Keller, N. P. (2006). Signalling pathways connecting mycotoxin production and sporulation. *Mol Plant Pathol* 285–301.
65. Li, L., Shao, Y., Li, Q., Yang, S., and Chen, F. (2010). Identification of Mga1, a G-protein alpha-subunit gene involved in regulating citrinin and pigment production in *Monascus ruber* M7. *FEMS Microbiol Lett* 308:108–114.
66. Garcia-Rico, R. O., Fierro, F., Mauriz, E., Gómez, A., Fernández-Bodega, M. A., and Martín, J. F. (2008). The heterotrimeric alpha protein Pga1 regulates biosynthesis of penicillin, chrysogenin and roquefortine in *Penicillium chrysogenum. Microbiology* 154:3567–3578.
67. Miyake, T., Zhang, M. Y., Kono, I., Nozaki, N., and Sammoto, H. (2006). Repression of secondary metabolite production by exogenous cAMP in *Monascus. Biosci Biotechnol Biochem* 70:1521–1523.
68. Khaldi, N., Collemare, J., Lebrun, M. H., and Wolfe, K. H. (2008). Evidence for horizontal transfer of a secondary metabolite gene cluster between fungi. *Genom Biol* 9:R18.
69. Slot, J. C. and Rokas, A. (2011). Horizontal transfer of a large and highly toxic secondary metabolic gene cluster between fungi. *Curr Biol* 21:134–139.
70. Roze, L. V., Arthur, A. E., Hong, S. Y., Chanda, A., and Linz, J. E. (2007). The initiation and pattern of spread of histone H4 acetylation parallel the order of transcriptional activation of genes in the aflatoxin cluster. *Mol Microbiol* 66:713–726.
71. Bhatnagar, D., Carry, J. W., Ehrlich, K., Yu, J., and Cleveland, T. E. (2006). Understanding the genetics of regulation of aflatoxin production and *Aspergillus flavus* development. *Mycopathologia* 162:155–166.
72. Bayram, Ö., Krappmann, S., Ni, M., Bok, J. W., Helmstaedt, K., Valerius, O., Braus-Stromeyer, S. et al. (2008). VelB/VeA/LaeA complex coordinates light signal with fungal development and secondary metabolism. *Science* 320:1504–1506.
73. Keller, N. P., Turner, G., and Bennett, J. W. (2005). Fungal secondary metabolism—From biochemistry to genomics. *Nat Rev Microbiol* 3:937–947.
74. Yu, J., Bhatnagar, D., and Cleveland, T. E. (2004). Completed sequence of aflatoxin pathway gene cluster in *Aspergillus parasiticus. FEBS Lett* 564:126–130.
75. Brown, D. W., Proctor, R. H., Dyer, R. B., and Plattner, R. D. (2003). Characterization of a *Fusarium* 2-gene cluster involved in trichothecene C-8 modification. *J Agric Food Chem* 51:7936–7944.
76. Brown, D. W., Butchko, R. A. E., Busman, M., and Proctor, R. H. (2007). The *Fusarium verticillioides* FUM gene cluster encodes a Zn (II)2Cys6 protein that affects *fum* gene expression and fumonisin production. *Eukaryot Cell* 6:1210–1218.
77. Chang, P. K., Horn, B. W., and Dorner, J. W. (2009). Clustered genes involved in α-cyclopiazonic acid production are next to the aflatoxin biosynthesis gene cluster in *Aspergillus flavus. Fungal Genet Biol* 46:176–182.
78. Geisen, R., Schmidt-Heydt, M., and Karolewiez, A. (2006). A gene cluster of the ochratoxin A biosynthetic genes in *Penicillium. Mycot Res* 22:134–141.
79. Gallo, A., Bruno, K., Solfrizzo, M., Perrone, G., Mulè, G., Visconti, A., and Baker, S. E. (2012). New insight in the ochratoxin A biosynthetic pathway by deletion of an *nrps* gene in *Aspergillus carbonarius. Appl Environ Microbiol* doi: 10. 1128/AEM.02508-12.
80. Amaike, S. and Keller, N. P. (2011). *Aspergillus flavus. Annu Rev Phytopathol* 49:107–133.

81. van der Merwe, K. J., Steyn, P. S., and Fourie, L. (1965). Ochratoxin A, a toxic metabolite produced by *Aspergillus ochraceus* Wilh. *Nature* 205:1112–1113.
82. Stander, M. A., Steyn, P. S., Lübben, A., Miljkovic, A., Mantle, P. G., and Marais G. J. (2000). Influence of halogen salts on the production of the ochratoxins by *Aspergillus ochraceus* Wilh. *J Agric Food Chem* 48:1865–1871.
83. Schmidt-Heydt, M., Bode, H., Raupp, F., and Geisen, R. (2010). Influence of light on ochratoxin biosynthesis by *Penicillium*. *Mycot Res* 26:1–8.
84. Petzinger, E. and Ziegler, K. (2000). Ochratoxin A from a toxicological perspective. *J Vet Pharmacol Therap* 23:91–98.
85. Bozic, Z., Duancic, V., Belicza, M., Kraus, O., and Skljarov, I. (1995). Balkan endemic nephropathy: Still a mysterious disease. *Eur J Epidemiol* 11:235–238.
86. Pfohl-Leszkowicz, A., Petkova-Bocharova, T., Chernozemsky, I. N., and Castegnaro, M. (2002). Balkan endemic nephropathy and associated urinary tract tumors: A review on aetiological causes and the potential role of mycotoxins. *Food Addit Contam* 19(3):282–302.
87. Krogh, P. (1987). Ochratoxins in food. In: *Mycotoxins in Food*, P. Krogh, Ed. Academic Press, London, U.K., pp. 97–121.
88. Braunberg, R. C., Barton, C. N., Gantt, O. O., and Friedman, L. (1994). Interaction of citrinin and ochratoxin A. *Nat Toxins* 2:124–131.
89. Dirheimer, G. (1996). Mechanistic approaches to ochratoxin toxicity. *Food Addit Contam* 13:45–48.
90. Heller, K. and Röschenthaler, R. (1978). Inhibition of protein synthesis in *Streptococcus faecalis* by ochratoxin A. *Can J Microbiol* 24:466–472.
91. Geisen, R., and Schmidt-Heydt, M. (2009). Physiological and molecular aspects of ochratoxin A biosynthesis. In: *The Mycota XV*, Anke, T. and Weber, D. Eds. Springer, Berlin, Germany, pp. 353–376.
92. Shimizu, T., Kinoshita, H., Ishihara, S., Sakai, K., Nagai, S., and Nihira, T. (2005). Polyketide synthase gene responsible for citrinin biosynthesis in *Monascus purpureus*. *Appl Environ Microbiol* 71:3453–3457.
93. Frisvad, J. C., Lund, F., and Elmholt, S. (2005). Ochratoxin A producing *Penicillium verrucosum* isolates from cereals reveal large AFLP fingerprinting variability. *J Appl Microbiol* 98:684–692.
94. Lund, F. and Frisvad, J. C. (2003). *Penicillium verrucosum* in wheat and barley indicates presence of ochratoxin A. *J Appl Microbiol* 95:1117–1123.
95. Lindblad, M., Johnsson, P., Jonsson, N., Lindqvist, R., and Olsen, M. (2004). Predicting noncompliant levels of ochratoxin A in cereal grain from *Penicillium verrucosum* counts. *J Appl Microbiol* 97:609–616.
96. Schmidt-Heydt, M., Richter, W., Michulec, M., Buttinger, G., and Geisen, R. (2008). A comprehensive molecular system to study presence, growth and ochratoxin A biosynthesis of *Penicillium verrucosum* in wheat. *Food Addit Contam* 25:989–996.
97. Schmidt-Heydt, M., Richter, W., Michulec, M., Buttinger, G., and Geisen, R. (2007). Molecular and chemical monitoring of growth and ochratoxin A biosynthesis of *Penicillium verrucosum* in wheat stored at different moisture conditions. *Mycot Res* 23:138–146.
98. Arroyo, M., Aldred, D., and Magan, N. (2005). Environmental factors and weak organic acid interactions have differential effects on control of growth and ochratoxin A production by *Penicillium verrucosum*. *Int J Food Microbiol* 98:223–231.
99. Bogs, C., Battilani, P., and Geisen, R. (2006). Development of a molecular detection and differentiation system for ochratoxin A producing *Penicillium* species and its application to analyse the occurrence of *Penicillium nordicum* in cured meats. *Int J Food Microbiol* 107:39–47.
100. Castella, G., Larsen, T. O., Cabanes, J., Schmidt, H., Alboresi, A., Niessen, L., Färber, P., and Geisen, R. (2002). Molecular characterization of ochratoxin A producing strains of the genus *Penicillium*. *Syst Appl Microbiol* 25:74–83.
101. Comi, G., Orlic, S., Redzepovic, S., Urso, R., and Iacumin, L. (2004). Moulds isolated from Istrian dried ham at the pre-ripening and ripening level. *Int J Food Microbiol* 96:29–34.
102. Battilani, P., Pietri, A., Giorni, P., Formenti, S., Bertuzzi, T., Toscani, T., Virgili, R., and Kozakiewicz, Z. (2007). *Penicillium* populations in dry-cured ham manufacturing plants. *J Food Protect* 70:975–980.
103. Sonjak, S., Licen, M., Frisvad, J. C., and Gunde-Cimerman, N. (2011). The mycobiota of three dry-cured meat products from Slovenia. *Food Microbiol* 28:373–376.
104. Sorensen, L. M., Jacobsen, T., Nielsen, P. V., Frisvad, J. C., and Granly Koch, A. (2008). Mycobiota in the processing areas of two different meat products. *Int J Food Microbiol* 124:58–64.
105. Peintner, U., Geiger, J., and Pöder, R. (2000). The Mycobiota of speck, a traditional tyrolean smoked and cured ham. *J Food Protect* 63:1399–1403.

106. El Adlouni, C., Tozlovanu, M., Naman, F., Faid, M., and Pfohl-Leszkowicz, A. (2006). Preliminary data on the presence of mycotoxins (ochratoxin A. citrinin and aflatoxin B1) in black table olives "Greek style" of Moroccan origin. *Mol Nutr Food Res* 50:507–512.
107. Heperkan, D., Dazkir, G. S., Kansu, D. Z., and Güler, F. K. (2009). Influence of temperature on citrinin accumulation by *Penicillium citrinum* and *Penicillium verrucosum* in black table olives. *Toxin Rev* 28:180–186.
108. Samapundo, S., Deschuyffeleer, N., Van Laere, D., De Leyn, I., and Devlieghere, F. (2010). Effect of NaCl reduction and replacement on the growth of fungi important to the spoilage of bread. *Food Microbiol* 27:749–756.
109. Ayodele, S. M. and Ojoghoro, O. J. (2007). Salt tress effects on the vegetative growth of *Pleurotus tuberregium* (FR) sing. *J Biol Sci* 7:1278–1281.
110. Kumar, S. and Gummadi, S. N. (2009). Osmotic adaptation in halotolerant yeast, *Debaryomyces nepalensis* NCYC 3413: Role of osmolytes and cation transport. *Extremophiles* 13:793–805.
111. Jung, K. W., Strain, A., Nielsen, K., Jung, K. H., and Bahn, Y. S. (2012). Two cation transporters Ena1 and Nha1 cooperatively modulated ion homeostasis, antifungal drug resistance, and virulence of *Cryptococcus neoformans* via the HOG pathway. *Fungal Genet Biol* 49:332–345.
112. Simkovic, M., Pokorny, R., Hudecová, D. and Varecka, L. (2004). Chloride transport in the vegetative mycelia of filamentous fungus *Trichoderma viride*. *J Basic Microbiol* 44:122–128.
113. Varga, J., Rigó, K., and Téren, J. (2000). Degradation of ochratoxin A by *Aspergillus* species. *Int J Food Microbiol* 59:1–7.
114. Miyazaki, J., Yamasaki, M., Mishima, H., Mansho, K., Tachibana, H., and Yamada, K. (2001). Oxidative stress by visible light irradiation suppresses immunoglobulin production in mouse spleen lymphocytes. *Biosci Biotechnol Biochem* 65:593–598.
115. Heider, E. M., Harper, J. K., Grant, D. M., Hoffmann, A., Dugan, F., Tomer, D. P., and O'Neil, K. O. (2012). Exploring unusual antioxidant activity in a benzoic acid derivative: A proposed mechanism for citrinin. *Tetrahedron* 62:1199–1208.
116. Stormer, F. C., Sandven, P., Huitfeldt, S., Eduard, W., and Skogstad, A. (1998). Does the mycotoxin citrinin function as a sun protectant in conidia from *Penicillium verrucosum*? *Mycopathologia* 142:43–47.
117. Moulinié, A., Faucet, V., Castegnaro, M., and Pfohl-Lezkowicz, A. (2005). Analysis of some breakfast cereals on the French market for their contents of ochratoxin A, citrinin and fumonisin B_1: Development of a method for simultaneous extraction of ochratoxin A and citrinin. *Food Chem* 92:391–400.
118. Meister, U. (2004). New method of citrinin determination by HPLC after polyamide column clean-up. *Eur Food Res Technol* 218:394–399.
119. Beck, J., Ripka, S., Siegner, A., Schiltz, E., and Schweizer, E. (1990). The multifunctional 6-methylsalicylic acid synthase gene of *Penicillium patulum*. *Eur J Biochem* 192:487–498.
120. Frisvad, J. (1988). Fungal species and their specific production of mycotoxins. In: *Introduction to Food Borne Fungi*, Samson, R. A. and van Reenen-Hoekstra E. S. Eds. Centraalbureau voor Schimmelcultures, Institute of the Royal Netherlands Academy of Arts and Sciences, Utrecht, the Netherlands, pp. 239–249.
121. Moake, M. M., Padilla-Zakour, O. I., and Worobo, R. W. (2005). Comprehensive review of patulin control methods in foods. *Comp Rev Food Sci Food Safety* 1:8–21.
122. Lee, K. S. and Roschenthaler, R. J. (1986). DNA-damaging activity of patulin in *Escherichia coli*. *Appl Environ Microbiol* 52:1046–1054.
123. Taniwaki, M. H., Hoenderboom, C., Vitali, A., and Eiroa, M. (1992). Migration of patulin in apples. *J Food Protect* 55:902–904.
124. Laidou, I. A., Thanassoulopoulos, C. C., and Liakopoulou-Kyriakides, M. (2001). Diffusion of patulin in the flesh of pears inoculated with four post-harvest pathogens. *J Phytopathol* 149:457–461.
125. Harwig, J., Blanchfield, B. J., and Scott, P. M. (1978). Patulin production by *Penicillium roqueforti* Thom from grape. *Can Inst Food Sci Technol J* 11:149–151.
126. Bragulat, M. R., Abarca, M. L., and Cabanes, F. J. (2008). Low occurrence of patulin- and citrinin-producing species isolated from grapes. *Lett Appl Microbiol* 47:286–289.
127. Drusch, S. and Ragab, W. (2003). Mycotoxins in fruits, fruit juices and dried fruits. *J Food Protect* 66:1514–1527.
128. Rychlik, M. and Schieberle, P. (1999). Quantification of the mycotoxin patulin by a stable isotope dilution assay. *J Agric Food Chem* 47:3749–3755.
129. Aziz, N. H. and Moussa, L. A. A. (2002). Influence of gamma radiation on mycotoxin producing moulds and mycotoxins in fruits. *Food Control* 13:281–288.
130. Martins, M. L., Gimeno, A., Martins, H. M., and Bernardo, F. (2002). Co-occurrence of patulin and citrinin in Portuguese apples with rotten spots. *Food Addit Contam* 19:568–574.

131. Paterson, R. R. M., Archer, S., Kozakiewicz, Z., Lea, A., Locke, T., and O'Grady, E. (2000). A gene probe for the patulin metabolic pathway with potential for use in patulin and novel disease control. *Biocontrol Sci Technol* 10:509–512.
132. Paterson, R. R. M. and Lima, N. (2013). Biochemical mutagens affect the preservation of fungi and biodiversity estimations. *Appl Microbiol Biotechnol* 97:77–85.
133. Hoorfar, J., Malorny, N., Wagner, M., De Medici, D., Abdulmawjood, A., and Fach, P. (2004). Diagnostic PCR: Making internal amplification control mandatory. *J Appl Microbiol* 96:221–222.
134. Paterson, R. R. M. (2004). The isoepoxydon dehydrogenase gene of patulin biosynthesis in cultures and secondary metabolites as candidate PCR inhibitors. *Mycol Res* 108:1431–1437.
135. Paterson, R. R. M. (2007). Internal amplification controls have not been employed in fungal PCR hence potential false negative results. *J Appl Microbiol* 102:1–10.
136. Paterson, R. R. M. (2006). Primers from the isoepoxydon dehydrogenase gene of the patulin biosynthetic pathway to indicate critical control points for patulin contamination of apples. *Food Control* 17:741–744.
137. Luque, M. L., Rodriguez, A., Andrade, M. J. Gordillo, R., Rodriguez, M., and Cordoba, J. J. (2011). Development of real-time PCR methods to quantify patulin-producing molds in food products. *Food Microbiol* 28:1190–1199.
138. Paterson, R. M. M. (2012). Idh PCR not only for *Penicillium*. *Food Control* 25:421.
139. Paterson, R. R. M. (2007). The isoepoxydon dehydrogenase gene PCR profile is useful in fungal taxonomy. *Rev Iberoam Micol* 24(289):293.
140. Dombrink-Kurtzman, M. A. (2007). The sequence of the isoepoxydon dehydrogenase gene of the patulin biosynthetic pathway in *Penicillium* species. *Antonie van Leeuwenhoek* 91:179–189.
141. White, S., O'Callaghan, J., and Dobson A. D. W. (2006). Cloning and molecular characterization of *Penicillium expansum* genes upregulated under conditions permissive for patulin biosynthesis. *FEMS Microbiol Lett* 255:17–26.
142. Dombrink-Kurtzman, M. A. (2006). The isoepoxydon dehydrogenase gene of the patulin metabolic pathway differs for *Penicillium griseofulvum* and *Penicillium expansum*. *Antonie van Leeuwenhoek* 89:1–8.
143. Dombrink-Kurtzman, M. A. (2008). A gene having sequence homology to isoamyl alcohol oxidase is transcribed during patulin production in *Penicillium griseofulvum*. *Curr Microbiol* 56:224–228.
144. Puel, O., Tadrist, S., Delaforge, M., Oswald, I. P., and Lebrihi, A. (2007). The inability of *Byssochlamys fulva* to produce patulin is related to absence of 6-methylsalicylic acid synthase and isoepoxydon dehydrogenase genes. *Int J Food Microbiol* 115:131–139.
145. Karaoglanidis, G. S., Markoglou, A. N., Bardas, G. A., Doukas, E. G., Konstantinou, S., and Kalampokis, J. F. (2011). Sensitivity of *Penicillium expansum* field isolates to tebuconazole, iprodione, fludioxonil and cyprodinil and characterization of fitness parameters and patulin production. *Int J Food Microbiol* 145:195–204.
146. Cabanas, R., Castellá, G., Abarca, M. L., Bragulat, M. R., and Cabanes, J. F. (2009). Thiabendazole resistance and mutations in the β-tubulin gene of *Penicillium expansum* strains isolated from apples and pears with blue mold decay. *FEMS Microbiol Lett* 297:189–195.
147. Davidse, L. C. (1986). Benzimidazole fungicides: Mechanisms of action and biological impact. *Annu Rev Phytopathol* 24:43–65.
148. Sanzani, S. M., Reverberi, M., Punelli, M., Ippolito, A., and Fanelli, C. (2012). Study on the role of patulin on pathogenicity and virulence of *Penicillium expansum*. *Int J Food Microbiol* 153:323–331.
149. Holzapfel, C. W. (1968). The isolation and structure of cyclopiazonic acid, a toxic metabolite of *Penicillium cyclopium* Westling. *Tetrahedron* 24:2101–2119.
150. LeBars, J. (1979). Cyclopiazonic acid production by *Penicillium camemberti* Thom and natural occurrence of this mycotoxin in cheese. *Appl Environ Microbiol* 38:1052–1055.
151. Leistner, L. (1984). Toxigenic penicillia occurring in feeds and foods: A review. *Food Technol Aus* 36:404–406.
152. Finoli, C., Vecchio, A., Galli, A., and Franzetti, L. (1999). Production of cyclopiazonic acid by molds isolated from Taleggio cheese. *J Food Protect* 62:1198–1202.
153. Bailly, J. D., Tabuc, C., Querin, A., and Guerre, P. (2012). Production and stability of patulin, ochratoxin A, citrinin and cyclopiazonic acid on dry cured ham. *J Food Protect* 68:1516–1520.
154. Lansden, J. A. and Davidson, J. I. (1983). Occurrence of cyclopiazonic acid in peanuts. *Appl Environ Microbiol* 45:766–769.
155. Geisen, R., Glenn, E., and Leistner, L. (1990). Two *Penicillium camemberti* mutants affected in the production of cyclopiazonic acid. *Appl Environ Microbiol* 56:3587–3590.

156. Geisen, R. (1992). Characterization of a mutation in a strain of *P. camemberti* affecting the production of cyclopiazonic acid. *Fungal Genet Newslett* 39:20–22.
157. Benkhemmar, O., Gaudemer, F., and Bouvier-Fourcade, I. (1985). Heterokaryosis between Aspergillus oryzae cyclopiazonic acid-defective strains: Method for estimating the risk of inducing toxin production among cyclopiazonic acid-defective industrial strains. *Appl Environ Microbiol* 50:1087–1093.
158. Tokuoka, M., Seshime, Y., Fujii, I., Kitamoto, K., Takahashi, T., and Koyama, Y. (2008). Identification of a novel polyketide synthase-nonribosomal peptide synthetase (PKS-NRPS) gene required for the biosynthesis of cyclopiazonic acid in *Aspergillus oryzae*. *Fungal Genet Biol* 45:1608–1615.
159. Chang, P. K., Ehrlich, K. C., and Fuji, I. (2009). Cyclopiazonic acid biosynthesis of *Aspergillus flavus* and *Aspergillus oryzae*. *Toxins* 1:74–99.
160. Scott, P. M. (1981). Toxins of *Penicillium* species used in cheese manufacture. *J Food Protect* 44:702–710.
161. Proctor, R. H. and Hohn, T. M. (1993). Aristolochene synthase. Isolation, characterization, and bacterial expression of a sesquiterpenoid biosynthetic gene (ARI1) from *Penicillium roqueforti*. *J Biol Chem* 268:4543–4548.
162. Cane, D. E., Wu, Z., Proctor, R. H., and Hohn, T. M. (1993). Overexpression in *Escherichia coli* of soluble aristolochene synthase from *Penicillium roqueforti*. *Arch Biochem Biophys* 304:415–419.
163. Cane, D. E. and Kang, I. (2000). Aristolochene synthase: Purification, molecular cloning, high-level expression in *Escherichia coli*, and characterization of the *Aspergillus terreus* cyclase. *Arch Biochem Biophys* 376:354–364.
164. Jelen, H. H., Mildner, S., and Czaczyk, K. (2002). Influence of octanoic acid addition to medium on some volatile compounds and PR-toxin biosynthesis by *Penicillium roqueforti*. *Lett Appl Microbiol* 35:37–41.
165. Chang, S. C., Yeh, S. F., Li, S. Y., Lei, W. Y., and Chen, M. Y. (1996). A novel secondary metabolite relative to the degradation of PR toxin by *Penicillium roqueforti*. *Curr Microbiol* 32:141–146.
166. Larsen, T. O., Svendsen, A., and Smedsgaard, J. (2001). Biochemical characterization of ochratoxin A-producing strains of the genus *Penicillium*. *Appl Environ Microbiol* 67:3630–3635.

27 *Phoma* spp. as Opportunistic Fungal Pathogens in Humans

Mahendra Rai, Vaibhav V. Tiwari, and Evangelos Balis

CONTENTS

27.1 Introduction 451
27.2 *Phoma* Infection in Humans 452
    27.2.1 Phaeohyphomycosis 453
    27.2.2 Keratitis 454
    27.2.3 Cutaneous Infection 454
    27.2.4 Pulmonary Infection 455
27.3 Identifications 456
    27.3.1 Morphological Characters 456
    27.3.2 Molecular Markers 456
27.4 Conclusion 457
References 457

27.1 INTRODUCTION

There has been a consistent rise in opportunistic fungal infections (OFIs) over the previous two decades [1]. These are mainly due to impairments in host defense mechanisms as a consequence of viral infections (especially the human immunodeficiency virus), premature neonates, age, major surgery, hematological disorders such as different types of leukemia, organ transplants, and more intensive and aggressive medical practices. Many clinical procedures and treatments, such as surgery, use of catheters, injections, radiation, chemotherapy, antibiotics, and steroids, are risk factors for fungal infections [2–7]. These types of infections are commonly known as mycoses. Nevertheless, there have been incidences of unusual OFI from different fungi that were previously presumed to be plant pathogens [8–13]. Thus, indications suggest that the epidemiological features of fungal infections are changing and that this group is becoming serious.

The biodiversity of the fungi is immense. It was estimated that approximately 1.5 million species are present worldwide [14]. Fungi are heterotrophs and obtain their nutrients by absorption. Saprophytic fungi secrete enzymes and breakdown dead organic matter, while parasitic fungi obtain their nutrients from living hosts [15]. The cell wall of pathogenic fungi is a complex structure principally composed of polysaccharides with 1,3-β-glucan being the most abundant [16–18], which serves as a skeleton for other polysaccharides of the cell wall to become attached [19,20]. Biogenesis of fungal cell walls encompasses a highly compound set of procedures that require many cell functions. Numerous methods have been categorized at the molecular level to analyze the role of many genes and proteins that are identified in relation to cell wall functions [21]. The main application of fungal cell wall biogenesis research is the identification of the potential targets for antifungal agents. For example, caspofungin, a derivative of echinocandin, is a cell-wall-targeted antifungal agent [20]. Thus, continuing efforts to study the fungal cell wall and its biogenesis are important in the study of fungal pathogenicity and related aspects. Furthermore, these fungi produce spores that are the causal agents of infections. Primarily, the infection can be caused either by inhalation of the spores or through the contact with the body surface, that is, skin. However, in the case of opportunistic infections, where the immune system is weakened, they invade deeply causing serious disorders in the host [15,22–25].

Phoma consists of over 2000 known species, found all over the world. They are common inhabitants of soil [26–32] and many are phytopathogens, periodically infecting plants through the roots [33–41]. *Phoma* belongs to the order Pleosporales of the Ascomycota. These fungi can reproduce asexually and are designated as dematiaceous, which means they often possess dark pigments in the cell walls, one of which may be melanin. Typically, the colonies of the genus have a velvety texture that can be slightly powdery, depending on the species. The colony may be white to gray with pink, yellow, and reddish purple colorations. *Phoma* species produce distinctive flask-shaped fruiting cellular bodies known as pycnidia, which consist of single-celled masses of spores called pycnidiospores. With the effect of different factors such as pH, temperature, and light, the culture characteristics and morphologies differ [42–44]. They are also known to produce a variety of secondary metabolites in the form of dyes and antibiotics, etc. [45–47].

Phoma species are also reported as causative organisms of human infections [48–51] and fungal infections in general contribute significantly to patient morbidity and mortality. The incidence of mycotic infection, especially opportunistic infections, has been increasing alarmingly in the past decade. However, the mortality rate has decreased with the advance in clinical treatment and therapeutics but we are still exposed to serious underlying diseases. The main reason behind the increased disease is the use of cytotoxic and immunosuppressive drugs, corticosteroids and broad-spectrum antibiotics, which are preferentially used for neoplastic disorders, organ transplants, and patients with HIV infection, etc. [2,52]. Such mycotic infections caused by opportunistic pathogens are known as opportunistic invasive fungal infections (OIFIs).

The immunocompromised individuals are less capable of battling infections because of a poor immune response. Immunocompromised individuals are more prone to contract opportunistic infections than healthy individuals [53]. Overreaction to the agent of infection while overcoming immunosuppression can be problematic, which is known as immune reconstitution syndrome and concerns the restoration of host immunity. In a previously immunosuppressed patient, the infection becomes dysregulated and overly robust, resulting in host damage and sometimes death [54]. The causes for immunocompromisation are as follows:

- Genetic—inherited genetic defects
- Acquired infections, such as HIV, and cancers, including leukemia, lymphoma, or multiple myeloma
- Chronic diseases such as end stage renal disease and dialysis, diabetes, and cirrhosis
- Medications such as steroids, chemotherapy, radiation, and immunosuppressive posttransplant medications
- Physical state such as pregnancy

Infections caused by *Phoma* in human beings are very rare; hence, this chapter will focus on OIFI.

27.2 *PHOMA* INFECTION IN HUMANS

Incidences of *Phoma* infections have been increased in the last 5 years and it has emerged as a new human pathogen (Table 27.1). *Phoma* covers a broad range of infections including skin [54], facial tissue [48] and eyes [49,55], foot [56–58], and internal organs such as lungs [59,60]. The infections caused (Figure 27.1) can be categorized as

- Phaeohyphomycosis
- Keratitis
- Cutaneous infection
- Pulmonary infection

TABLE 27.1
List of the Rare Infections Caused by Different *Phoma* Species in Humans

Phoma Species	Infections Caused	References
Phoma hibernica	Lesion on a leg	[56]
Phoma species	Patient with a transplanted kidney	[69]
Phoma oculohominis	Infection from corneal ulcer	[66]
Phoma eupyrena	Lesion on skin	[54]
Phoma minutispora	Over the face and around neck region	[70]
Phoma minutella	Subcutaneous phaeohyphomycosis on foot	[57]
Phoma sorghina	Skin infection	[48]
Phoma species	Infection to lung mass	[59]
Pleurophoma (Phoma)	Cutaneous lesion	[49]
Phoma cava	Subcutaneous infection	[75]
Phoma species	Subcutaneous phaeohyphomycosis	[64]
Phoma species	Subcutaneous phaeohyphomycosis	[65]
Phoma species	Cutaneous infection in solid organ transplantation	[74]
Phoma species	Renal transplant recipient	[73]
Phoma species	Infection within ulcer	[55]
Phoma exigua	Infection to lung mass	[60]
Phoma herbarum	Infection to skin and nail of foot	[58]
Phoma species	Cutaneous, subcutaneous, and deep tissue infection	[50]

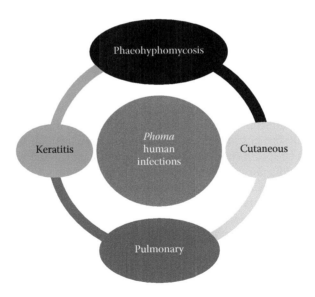

FIGURE 27.1 Types of infections caused by different *Phoma* species.

27.2.1 Phaeohyphomycosis

The term phaeohyphomycosis was coined by Ajello et al. [61] and is a group of mycotic infections characterized by the presence of dematiaceous (dark-walled) septate hyphae and/or yeasts. The infection often occurs on the skin via a cut where it forms nodules and cysts but can also invade deeper tissues including the brain. Phaeohyphomycosis tend to be opportunistic and immunocompromised people are particularly susceptible [62]. McGinnis [63] classified four forms of phaeohyphomycosis: (1) superficial (e.g., black piedra, tinea nigra), (2) cutaneous and corneal (e.g., dermatomycoses, mycotic keratitis, onychomycosis), (3) subcutaneous, and (4) systemic.

Phoma hibernica and *Phoma eypyrena* were reported from a human lesion [54,56] and are considered as being a phaeohyphomycosis. The former was isolated from a leg while the latter was isolated from the skin of an 18-month-old boy. He had a crusting, erythematous, perioral eruption for 1 month. Treatment with the antifungal drugs, clotrimazole with 15% zinc oxide paste and dimethicone, resulted in complete healing of the lesion [55,57]. Baker et al. reported a subcutaneous phaeohyphomycosis on the foot of a farmer caused by the dematiaceous hyphomycete, *Phoma minutella* [57]. The subject was undergoing corticosteroid therapy for myasthenia gravis and the microscopic study revealed the presence of fungal elements consistent with *P. minutella*. In addition, *Phoma sorghina* was reported for the first time as an opportunistic fungal pathogen [48] and Hirsh and Schiff [64] and Oh et al. [65] reported cases of subcutaneous phaeohyphomycoses caused by *Phoma* species.

27.2.2 Keratitis

Keratitis is a condition in which the cornea and the front part of the eye become inflamed. The condition is often marked by moderate to intense pain and usually involves impaired eyesight. It can be of the following types:

- Superficial punctate keratitis, in which the cells on the surface of the cornea die
- Interstitial keratitis, a condition present at birth
- Herpes simplex viral keratitis, caused by the sexually transmitted herpes virus
- Traumatic keratitis, which results when a corneal injury leaves scar tissue

Eyes become painful, watery, bloodshot, and sensitive to light. The condition is often accompanied by blurred or hazy vision.

Punithlingam reported an infection of *Phoma oculohominis* from the corneal ulcer [66]. However, a clinical history and histopathology was not given. A case of *Phoma* keratitis was reported by Rishi and Font [55] where the clinical study revealed brownish pigmentation within the ulcer, which failed to respond to medical treatment. The keratectomic study suggested large septate hyphae aligned along the plane of the Descemet's membrane, a basement membrane that lies between the corneal proper substance [67]. The specimen was cultured on Sabouraud agar and was identified as *Phoma* based on the typical pycnidia present in the microscopical examination. Furthermore, PCR analysis was done with panfungal primers [55], which amplifies a broad spectrum of different fungal DNA without cross amplification of other genomes [68]. This represents the first well-documented case of *Phoma* keratitis with histopathological and molecular diagnosis.

27.2.3 Cutaneous Infection

Cutaneous infections affect the skin and skin structures and may occur due to bacteria, filamentous fungi or yeast. However, invasive fungal infections are gaining much impetus and consist of the following forms:

Dermatophytosis is a classic skin and hair infection that characterizes the focus of much medical dermatology. Dermatomycosis is an infection of the skin and hair that in many ways parallel those caused by the dermatophytes. Onychomycosis is used to refer to nondermatophyte nail infections and to any fungal nail infection. Finally, there is a group of miscellaneous infections.

Generally, *Phoma* species are common plant pathogens and geophilic in nature. But there are reports of it being a potent causative invasive fungal infections in humans and animals as discussed earlier. Young et al. [69] reported subcutaneous infection of *Phoma* in a patient with a transplanted kidney. The patient was in an immunosuppressive state prior to the infection. The fungus closely resembles *Pyrenochaeta romeroi* but the brief microbiological study of the morphology of both organisms signified the presence of *Phoma* species [69]. Infection by *Phoma minutispora* was reported from an 18-year-old woman and a 20-year-old male farmer [70]. The fungus was isolated by Mathur from Indian soil and has not been reported from elsewhere [71]. The woman possessed furfuraceous spots covering the face, while the man had a rapid development of polycyclic and polymorphic patches around the neck. The cultural studies were performed on Sabouraud dextrose agar supplemented with chloramphenicol and cycloheximide. Pathogenicity test on rabbits was carried out, which revealed branched filaments and ostiolate pycnidia [70]. Furthermore, Gordon et al. [72] reported infection of *Phoma cava* from a tissue section of the ear fragments of another patient, which demonstrated apparent pycnidial walls and many septate hyphae. Another case study reports the infection by *P. sorghina* of a 24-year-old lecturer and a 19-year-old college student. The pathogen was identified by morphological characteristics. Moreover, pathogenicity test were carried on rabbits that exhibited symptoms of the disease readily. The patients were treated with miconazole nitrate (1%) twice a day and the infection was cured in 1 month [48].

A cutaneous lesion caused by *Pleurophoma* (*Phoma*) was reported by Rosen et al. [49] from facial eruptions. Physical examination revealed erythematous, tender facial plaques covering from the zygoma to the proximal jawline on the right cheek. A case of invasive *Phoma* species infection was reported from a 50-year-old female renal transplant recipient [73]. She experienced pain, warmth, and swelling in the dorsum of the left wrist. An open surgical synovial biopsy was performed, which encountered a cloudy fluid that demonstrated hyphal elements. The patient was subjected to amphotericin B antifungal therapy, and as the therapy was discontinued, the symptoms recurred. However, the organism was sensitive to amphotericin B and the symptoms resolved quickly with its therapy with no further recurrence.

Furthermore, similar kind of primary cutaneous fungal infections from solid organ transplant were reported by Miele et al. [74]. Transplantations comprised one cardiac, two renal, and one renal pancreatic transplant. Fungal infections were restricted to the skin and there was no indication of disseminated disease in any case. Individuals were efficaciously treated with surgical debridement, antifungal agents, and reduction of immunosuppressive therapy.

27.2.4 Pulmonary Infection

Pulmonary infection occurs when normal lung or systemic defense mechanisms are impaired. Infection begins with colonization of the upper respiratory tract by pathogens followed by aspiration into the lower respiratory tract. The pathogenic organisms gain access through (1) the airways, (2) blood stream, (3) traumatic implantation, and/or (4) direct spread across the diaphragm from a subphrenic source, probably through the lymphatics. The most likely route for fungi is through the airway where fungal spores reach lung tissue and produce disease. In the immunocompromised host, many fungi, including normally nonpathogenic ones, have the potential to cause serious morbidity and mortality.

Rare cases of invasive fungal infections to the lungs by *Phoma* species were reported [59,60,75]. The first report of invasive *Phoma* infection of lung mass was attributed to Morris et al. [60]. Zaitz et al. [75] reported subcutaneous infection in a 63-year-old male patient with pulmonary sarcoidosis. Histopathological examinations revealed the presence of a *Phoma* species as a causal agent, which was later confirmed as *P. cava* based on morphological studies. Amphotericin B was administered in a cumulative dose (610 mg) and itraconazole (400 mg/day) was maintained. Balis et al. reported lung infection by *Phoma exigua* where the patient was suffering from myeloid leukemia and diabetes. High-resolution computed tomography demonstrated a mass surrounded by an area of pneumonitis and the fungus was isolated from bronchoalveolar lavage specimens. PCR was performed to confirm the presence of *P. exigua* in a tissue biopsy material and pycnidia were revealed peculiar to *Phoma* species from morphological examination. Furthermore, the specific internal transcribed spacers (ITS) primers for *P. exigua* were designed from NCBI (National Center for Biotechnology Information), GenBank and the PCR product was sequenced, which showed 99% homology with the standard *P. exigua* sequences [60].

27.3 IDENTIFICATIONS

The major objective of any taxonomic study is systematic grouping of taxa of interest through generation of robust natural classifications based on constant characteristics, which reveal their factual evolutionary record, and development of trustworthy identification key(s) for uncomplicated taxon determination [76]. Morphological characteristics are often used by mycologists in traditional methods of identification. Generation of vast morphological and anatomical data over the years had built a strong base for taxonomic studies of *Phoma*. However, this often lacked satisfactory resolution and should be compared with molecular data sources. Many fungal taxonomic studies applied molecular tools for resolving relationships at the genus and species level [77–81]. Thus, with the use of morphological and molecular characters, an accurate identification can be obtained more readily.

27.3.1 MORPHOLOGICAL CHARACTERS

Identification by morphological characters forms the basis of classification of an organism, including fungi. However, nutritional and environmental conditions are known to affect some morphological characters. Morphological variability exists in *Phoma* and closely allied genera and among the different species [82]. The fungus can be stable at one condition but as the conditions change, the morphology can vary. This was supported by Boerema and Howeler who studied different varieties of *P. exigua* [83]. Also, White and Morgan-Jones studied variability in morphological characteristics in *P. sorghina* [84] and suggested that a high degree of genetic variability may exist within the species. Pazoutova assessed the morphological and genetic variability of *P. sorghina* isolates from Southern Africa and Texas [85] and showed that the haplotypes were separated on the basis of 53 markers from banding patterns obtained with rep-PCR (primers: M13 core, ERIC IR). The haplotype phylogram also partially reflected the shared geographic origin in the clades.

Boerema et al. [86] published a key to the identification of *Phoma* species. The method used was based on morphological characters that include in vitro culture on oatmeal agar (OA) and malt agar (MA). The following were employed: (1) colony characteristics, (2) pigment and crystal formation, (3) NaOH spot test, and (4) shape and size of pycnidia, conidia, and chlamydospores.

For identification of *Phoma*, comparative culture studies are carried out on OA and MA. The production of pycnidia is abundant on OA while MA stimulates pigment and crystal formation. After 7 days, the colony characters are studied that include diameter, outlines, and color of the colonies. The NaOH spot test and pigment and crystal formation play an important role in identifications.

However, the most important morphological characteristics are the formation of pycnidia, conidia, and chlamydospores. Species variation has been reported corresponding with the shape, size, and color of the pycnidia. Although, size and septation of the conidia are also important criteria, the conidia dimensions may differ between strains of the same species. This generates ambiguity in identifications and hence different molecular markers are exploited for more authentic identifications.

27.3.2 MOLECULAR MARKERS

Identification by morphological methods is time consuming and expensive and requires skilled taxonomists. In contrast, molecular methods are fast and can be reliable [87]: several molecular methods have been generated but the validities have been questioned as the species concepts are not yet fully understood [88–96].

When morphological and molecular markers are considered, the problems in the identification and differentiation of *Phoma* are often resolved. Unfortunately, many species have been misidentified or even placed in other genera. These ambiguities can be overcome with the judicious use of various molecular tools, and researchers have now generated methods that are very reliable and efficient in the identification of the *Phoma* species [26]. The introduction of ITS markers has begun a new era in the identification of species of *Phoma* [86,96–98]. In addition, two protein-encoding genes, namely, tef1 and β-tubulin, are used [96]. Davidson et al. [98] reported a new species of *Phoma*, isolated from *Ascochyta* blight lesion of field peas named *Phoma koolunga* sp. Nov using such molecular methods.

The genetic diversity within strains of the same species can also be studied with molecular markers. Sorensen et al. analyzed the sequences of *Phoma pomorum*, which were further studied for the production of the secondary metabolites, isocoumarins, which represented the first report of these compounds from *Phoma* [41]. Recently, phylogenetic studies carried out by Gruyter et al. led to the conclusion that phylogeny based on molecular markers such as sequence data of large subunit 28S nrDNA (LSU) and ITS regions 1 and 2 and 5.8S nrDNA, represents the importance of molecular systematics in differentiation and identification of the species. Thus, it can be demarcated that molecular taxonomic markers can be efficiently utilized for the correct identification and differentiation of this polyphyletic genus.

27.4 CONCLUSION

Phoma are normally plant pathogenic or soilborne. Nevertheless, some species are now emerging as opportunist pathogens of immunocompromised human beings. The frequency of OIFI has increased exponentially although our understanding of the epidemiological features is incomplete [99]. The incidence of infection is seen predominately in transplant patients and the patients undergoing immunosuppressive drug therapy are more prone to these infections. The identification of these fungi is of utmost importance in deciding the course of therapy.

In the past, the identification of these fungi was based on morphological and cultural characteristics, which tends to be time consuming and require high expertise. Moreover, identification based only on morphological characteristics is unreliable and there is a pressing need to develop accurate and rapid molecular tools for the identification of these pathogens.

REFERENCES

1. Nucci, M. et al., Epidemiology of opportunistic fungal infections in Latin America, *Clin. Infect. Dis.*, 51(5), 561, 2010.
2. Soll, D.R., The ins and outs of DNA fingerprinting of infectious fungi, *Clin. Microbiol. Rev.*, 13, 332, 2000.
3. Kojic, E.M. and Darouiche, R.O., *Candida* infections of medical devices, *Clin. Microbiol. Rev.*, 17(2), 255, 2004.
4. Warnock, D.W., Trends in the epidemiology of invasive fungal infections, *Jpn. J. Med. Mycol.*, 48, 1, 2007.
5. Walsh, T.J. et al., Treatment of aspergillosis: Clinical practice guidelines of the Infectious Diseases Society of America, *Clin. Infect. Dis.*, 46, 327, 2008.
6. Tiwari, V.V., Dudhane, M.N., and Rai, M.K., Molecular tools for identification and differentiation of different human pathogenic *Candida* species, in *Current Advances in Molecular Mycology*, Gherbawy, Y., Mach, R.L., and Rai, M.K., Eds., Nova Science Publishers, New York, 2009, p. 349.
7. Behiry, I.K., Hedeki, S.K.E., and Mahfouz, M., *Candida* infection associated with urinary catheter in critically ill patients. Identification, antifungal susceptibility and risk factors, *Res. J. Med. Med. Sci.*, 5(1), 79, 2010.
8. Abbott, S.P. et al., Fatal cerebral mycoses caused by the Ascomycete *Chaetomium strumarium*, *J. Clin. Microbiol.*, 33(10), 2692, 1995.

9. Hennequin, C. et al., Identification of *Fusarium* species involved in human infections by 28S rRNA gene sequencing, *J. Clin. Microbiol.*, 37(11), 3586, 1999.
10. Barron, M.A. et al., Invasive mycotic infections caused by *Chaetomium perlucidum*, a new agent of cerebral phaeohyphomycosis, *J. Clin. Microbiol.*, 41(11), 5302, 2003.
11. Mansoory, D. et al., Chronic *Fusarium* infection in an adult patient with undiagnosed chronic granulomatous disease, *Clin. Infect. Dis.*, 37, 107, 2003.
12. Thompson, G.R. and Patterson, T.F., Pulmonary aspergillosis, *Semin. Respir. Crit. Care. Med.*, 29, 103, 2008.
13. Yegneswaran, P.P. et al., *Colletotrichum graminicola* keratitis: First case report from India, *Ind. J. Ophthalmol.*, 58(5), 415, 2010.
14. Hawksworth, D.L., The fungal dimension of biodiversity: Magnitude, significance and conservation, *Mycol. Res.*, 95, 641, 1991.
15. Alexopolus, C.J., Mims, C.W., and Blackwell, M., *Introductory Mycology*, 4th edn. John Wiley & Sons, New York, 2002.
16. Latgé, J.P. et al., Galactomannan and the circulating antigens of *Aspergillus fumigatus*, in *Fungal Cell Wall and Immune Response*, Latgé, J.P. and Baocias, D., Eds., NATO ASI Series 53. Springer Verlag, Heidelberg, Germany, 1991, p. 143.
17. Hearn, V.M. and Sietsma, J.H., Chemical and immunological analysis of the *Aspergillus fumigatus* cell wall, *Microbiology*, 140, 789, 1994.
18. Mouyna, I. et al., Identification of the catalytic residues of the first family of β(1–3) glucanosyltransferases identified in fungi, *Biochem. J.*, 347, 741, 2000.
19. Fontaine, T. et al., From the surface to the inner layer of the fungal cell wall, *Biochem. Soc. Trans.*, 25, 194, 1997.
20. Duran, A. and Nombela, C., Fungal cell wall biogenesis: Building a dynamic interface with the environment, *Microbiology*, 150, 3099, 2004.
21. Schulze-Lefert, P., Knocking on the heaven's wall: Pathogenesis of and resistance to biotropic fungi at the cell wall, *Curr. Opin. Plant Biol.*, 7, 377, 2004.
22. Gutzmer, R. et al., Rapid identification and differentiation of fungal DNA in dermatological specimens by Light Cycler PCR, *J. Med. Microbiol.*, 53, 1207, 2004.
23. Alexander, B.D. and Pfaller, M.A., Contemporary tools for the diagnosis and management of invasive mycoses, *Clin. Infect. Dis.*, 43, S15, 2006.
24. Pauw, B.D. et al., Revised definitions of invasive fungal disease from the European Organization for Research and Treatment of Cancer/Invasive Fungal Infections Cooperative Group and the National Institute of Allergy and Infectious Diseases Mycoses Study Group (EORTC/MSG) Consensus Group, *Clin. Infect. Dis.*, 46, 1813, 2008.
25. Ostrosky-Zeichner, L., Invasive mycoses: Diagnostic challenges, *Am. J. Med.*, 125, S14, 2012.
26. Aveskamp, M.M., De Gruyter, J., and Crous, P.W., Biology and recent developments in the systematics of *Phoma*, a complex genus of major quarantine significance, *Fungal Divers.*, 31, 1, 2008.
27. Pellegrino, C. et al., Detection of *Phoma valerianellae* in lamb's lettuce seeds, *Phytoparasitica*, 38, 159, 2010.
28. Garibaldi, A., Gilardi, G., and Gullino, M.L., First report of leaf spot caused by *Phoma multirostrata* on *Fuchsia × hybrida* in Italy, *Plant Dis.*, 94(3), 382.1, 2010.
29. Strobel, G. et al., An endophytic/pathogenic *Phoma* sp. from creosote bush producing biologically active volatile compounds having fuel potential, *FEMS Microbiol. Lett.*, 320(2), 87, 2011.
30. Wang, X. et al., First report of leaf spot disease on *Schisandra chinensis* caused by *Phoma glomerata* in China, *Plant Dis.*, 96(2), 289, 2012.
31. Li, Y.P. et al., First report of *Phoma herbarum* on Tedera (*Bituminaria bituminosa* var. *albomarginata*) in Australia, *Plant Dis.*, 96(5), 769, 2012.
32. Patil, V.B. et al., First report of leaf spot caused by *Phoma costarricensis* on *Delphinium malabaricum* in Western Ghats of India, *Plant Dis.*, 96(7), 1074, 2012.
33. Rajak, R.C. and Rai, M.K., Species of *Phoma* from legumes, *Indian Phytopathol.*, 35(4), 609, 1982.
34. Rai, M.K., Identity and taxonomy of hitherto unreported pathogen causing leaf-spot disease of Ginger in India, *Mycotaxon*, XLVI, 329, 1993.
35. Rai, M.K. and Rajak, R.C., Distinguishing characteristics of some *Phoma* species, *Mycotaxon*, 48, 389, 1993.
36. Boerema, G.H. and Gruyter, J., Contributions towards a monograph of *Phoma* (Coelomycetes)—VII Section Sclerophomella: Taxa with thick-walled pseudoparenchymatous pycnidia, *Persoonia*, 17, 81, 1998.

37. Boerema, G.H. and Gruyter, J., Contributions towards a monograph of *Phoma* (Coelomycetes) III—Supplement Additional species of section Plenodomus, *Persoonia*, 17, 273, 1999.
38. Kövics, G.J., Gruyter, J., and van der Aa, H.A., *Phoma sojicola* comb nov, and other hyaline-spored coelomycetes pathogenic on soybean, *Mycol. Res.*, 103(8), 1065, 1999.
39. Kovics, G.J., Pandey, A.K., and Rai, M.K., *Phoma* Saccardo and related genera: Some new perspectives in taxonomy and biotechnology, in *Biodiversity of Fungi: Their Role in Human Life*, Deshmukh, S.K. and Rai, M. K., Eds., Science Publishers, Inc., Enfield, NH, 2005, p. 129.
40. Davari, M. and Hajieghrari, B., *Phoma negriana*, a new invasive pathogen for Moghan's vineyards in Iran, *Afr. J. Biotechnol.*, 7(6), 788, 2008.
41. Sorensen, J.L. et al., Chemical characterization of *Phoma pomorum* isolated from Danish maize, *Int. J. Food Microbiol.*, 136(3), 310, 2010.
42. Rajak, R.C. and Rai, M.K., Effect of different factors on the morphology and cultural characters of 18-species and 5-varieties of *Phoma*. I. Effect of different media, *Bibliogr. Mycol.*, 91, 301, 1983.
43. Rajak, R.C. and Rai, M.K., Effect of different factors on the morphology and cultural characters of 18-species and 5 varieties of *Phoma*. II. Effect of different pH, *Nova Hedwig.*, 40, 299, 1984.
44. Irinyi, L. et al., Studies of evolutionary relationships of *Phoma* species based on phylogenetic markers, in *Recent Developments of IPM*, Kovics, G.J. and David, I., Eds., *4th International Plant Protection Symposium*. Debrecen University, Debrecen University Centre for Agricultural Science, Debrecen, Hungary, October 18–19, 2006, p. 99.
45. Rai, M.K., Diversity and biotechnological applications of Indian species of *Phoma*, in *Frontiers of Fungal Diversity in India* (Prof. Kamal Festscrift), Rao, G.P., Manoharachari, C., Bhat, D.J., Rajak, R.C., and Lakhanpal, T.N., Eds., International Book Distributing Co., Lucknow, India, 2002, p. 179.
46. Rai, M.K. et al., *Phoma* Saccardo: Distribution, secondary metabolite production and biotechnological applications, *Crit. Rev. Microbiol.*, 35(3), 182, 2009.
47. Qin. S. et al., Two new metabolites, Epoxydine A and B, from *Phoma* sp., *Helv. Chim. Acta*, 93, 169, 2010.
48. Rai, M.K., *Phoma sorghina* infection in human being, *Mycopathologia*, 105, 167, 1989.
49. Rosen, T. et al., Cutaneous lesions due to *Pleurophoma (Phoma)* complex, *S. Med. J.*, 89(4), 431, 1996.
50. Roehm, C.E. et al., *Phoma* and *Acremonium* invasive fungal rhinosinusitis in congenital acute lymphocytic leukemia and literature review, *Int. J. Pediatr. Otorhinolaryngol.*, 76(10), 1387, 2012.
51. Rees, J.R. et al., The epidemiological features of invasive mycotic infections in the San Francisco bay area, 1992–1993: Results of population-based laboratory active surveillance, *Clin. Infect. Dis.*, 27, 1138, 1998.
52. Arnold, T.M., Sears, C.R., and Hage, C.A., Invasive fungal infections in the era of biologics, *Clin. Chest Med.*, 30, 279, 2009.
53. Elston, J.W. and Thaker, H., Immune reconstitution inflammatory syndrome, *Int. J. STD AIDS*, 20, 221, 2009.
54. Bakerspigel, A., Lowe, D., and Rostras, A., The isolation of *Phoma eupyrena* from a human lesion, *Arch. Dermatol.*, 117, 362, 1981.
55. Rishi, K. and Font, R.L., Keratitis caused by an unusual fungus, *Phoma* species, *Cornea*, 22(2), 166, 2003.
56. Bakerspigel, A., The isolation of *Phoma hibernica* from a lesion on a leg, *Sabouraudia*, 7, 261, 1970.
57. Baker, J.G. et al., First report of subcutaneous phaeohyphomycosis of the foot caused by *Phoma minutella*, *J. Clin. Microbiol.*, 25(12), 2395, 1987.
58. Tullio, V. et al., Non-dermatophyte moulds as skin and nail foot mycosis agents: *Phoma herbarum*, *Chaetomium globosum* and *Microascus cinereus*, *Fungal Biol.*, 114(4), 345, 2010.
59. Morris, J.T. et al., Lung mass caused by *Phoma* species, *Infect. Dis. Clin. Pract.*, 4, 58, 1995.
60. Balis, E. et al., Lung mass caused by *Phoma exigua*, *Scand. J. Infect. Dis.*, 38(6), 552, 2006.
61. Ajello, L. et al., A case of phaeohyphomycosis caused by a new species of *Phialophora*, *Mycologia*, 66, 490, 1974.
62. Rai, M.K. and Varma, A., New spectrum of fungal infections in immunocompromised hosts, in *Innovative Approaches in Microbiology*, Maheshwari, D.K. and Dubey, R.C., Eds., BSMPS Publishers, Dehradun, India, 2001, p. 299.
63. McGinnis, M.R., Chromoblastomycosis and phaeohyphomycosis: New concepts, diagnosis and mycology, *J. Am. Acad. Dermatol.*, 8, 1, 1983.
64. Hirsh, A.H. and Schiff, T.A., Subcutaneous phaeohyphomycosis caused by an unusual pathogen: *Phoma* species, *J. Am. Acad. Dermatol.*, 34(4), 679, 1999.

65. Oh, C.K. et al., Subcutaneous phaeohyphomycosis caused by *Phoma* species, *Int. J. Dermatol.*, 38(11), 874, 1999.
66. Punithlingam, E., *Phoma oculohominis* sp. nov. from corneal ulcer, *Trans. Br. Mycol. Soc.*, 67, 142, 1976.
67. Johnson, D.H., Bourne, W.M., and Campbell, R.J., The ultrastructure of Descemet's membrane I changes with age in normal cornea, *Arch. Ophthalmol.*, 100, 1942, 1982.
68. Kercher, L. et al., Molecular screening of donor corneas for fungi before excision, *Invest. Ophthalmol. Vis. Sci.*, 42, 2578, 2001.
69. Young, N.A., Kwon-Chung, K.J., and Freeman, J., Subcutaneous abscess caused by *Phoma* sp. resembling *Pyrenochaeta romeroi*: Unique fungal infection occurring in immunosuppressed recipient of renal allograft, *Am. J. Clin. Path.*, 59(6), 810, 1973.
70. Shukla, N.P. et al., *Phoma minutispora* as a human pathogen, *Mykosen*, 27(5), 255, 1983.
71. Mathur, P.N., Studies on members of Sphaeropsidales among Indian soil fungi and morphological and cultural studies on some members of plectascales from soils in India. PhD thesis, University of Agra, Agra, India, 1965.
72. Gordon, M.A., Salkia, I.K., and Stone, W.B., *Phoma (Peyronellaea)* as zoopathogen, *Sabouraudia*, 13, 329, 1975.
73. Everett, J.E. et al., A deeply invasive *Phoma* species infection in a renal transplant recipient, *Transplant. Proc.*, 35, 1387, 2003.
74. Miele, P.S. et al., Primary cutaneous fungal infections in solid organ transplantation: A case series, *Am. J. Transplant.*, 2, 678, 2002.
75. Zaitz, C. et al., Subcutaneous phaeohyphomycosis caused by *Phoma cava*: Report of a case and review of the literature, *Rev. Inst. Med. Trop. S. Paulo*, 39(1), 43–48, 1997.
76. Tiwari, V.V., Gade, A.K., and Rai, M.K., A study of phylogenetic variations among Indian *Phoma tropica* species by RAPD-PCR and ITS-rDNA sequencing, *Indian J. Biotechnol.*, 12(2), 187–194, 2013.
77. Gottlieb, A.M. and Lichtwardt, R.W., Molecular variation within and among species of *Harpellales*, *Mycologia*, 93, 66, 2001.
78. Nugent, K.G. and Saville, B.J., Forensic analysis of hallucinogenic fungi: A DNA based approach, *Forensic Sci. Int.*, 140, 147, 2004.
79. Morocko, I. and Fatehi, J., Molecular characterization of strawberry pathogen *Gnomonia fragariae* and its genetic relatedness to other *Gnomonia* species and members of Diaporthales, *Mycol. Res.*, 111, 603, 2007.
80. Padamsee, M. et al., The mushroom family Psathyrellaceae: Evidence for large-scale polyphyly of the genus *Psathyrella*, *Mol. Phylogenet. Evol.*, 46, 415, 2008.
81. Takamatsu, S. et al., Comprehensive molecular phylogenetic analysis and evolution of the genus *Phyllactinia* (Ascomycota: Erysiphales) and its allied genera, *Mycol. Res.*, 112, 299, 2008.
82. Rai, M.K., The genus *Phoma*: Identity and taxonomy, in *Advances in Plant Science Research*, Vols. VII and VIII, Rai, M. K., Ed., IBD Publisher & Distributors, Dehradun, India, 1998.
83. Boerema, G.H. and Höweler, L.H., *Phoma* exigua Desm, and its varieties, *Persoonia*, 5, 15, 1967.
84. White, J.F. and Morgan-Jones, G., Studies in genus *Phoma* II. Concerning *Phoma sorghina*, *Mycotaxon*, 18, 5, 1983.
85. Pazoutova, S., Genetic variation of *Phoma sorghina* isolates from Southern Africa and Texas, *Folia Microbiol.*, 54(3), 217, 2009.
86. Boerema, G.H. et al., *Phoma Identification Manual: Differentiation of Specific and Infra-Specific Taxa in Culture.* CABI, Noordwijkerhout, the Netherlands, 2004.
87. Miric, E., Aitken, E.A.B., and Goulter, K.C., Identification in Australia of the quarantine pathogen of sunflower *Phoma macdonaldii* (Teleomorph: *Leptosphaeria lindquistii*), *Aust. J. Agric. Res.*, 50, 325, 1999.
88. MacDonald, J.E., White, G.P., and Coté, M.J., Differentiation of *Phoma foveata* from *P. exigua* using a RAPD generated PCR-RFLP marker, *Eur. J. Plant Pathol.*, 106, 67, 2000.
89. Somai, B.M. et al., Internal transcribed spacer regions 1 and 2 and Random Amplified Polymorphic DNA analysis of *Didymella bryoniae* and related *Phoma* species isolated from cucurbits, *Phytopathology*, 92, 997, 2002.
90. Somai, B.M., Keinath, A.P., and Dean, R.A., Development of PCR-ELISA for detection and differentiation of *Didymella bryoniae* from related *Phoma* species, *Plant Dis.*, 86, 710, 2002.
91. Koch, C.A. and Utkhede, R.S., Development of a multiplex classical polymerase chain reaction technique for detection of *Didymella bryoniae* in infected cucumber tissues and greenhouse air samples, *Can. J. Plant Pathol.*, 26, 291, 2004.

92. Pethybridge, S.J., Scott, J.B., and Hay, F.S., Genetic relationships among isolates of *Phoma ligulicola* from pyrethrum and chrysanthemum based on ITS sequences and its detection by PCR, *Aust. Plant Pathol.*, 33, 173, 2004.
93. Balmas, V. et al., Characterization of *Phoma tracheiphila* by RAPD-PCR, microsatellite-primed PCR and ITS rDNA sequencing and development of specific primers for *in planta* PCR detection, *Eur. J. Plant Pathol.*, 111, 235, 2005.
94. Licciardella, G. et al., Identification and detection of *Phoma tracheiphila*, causal agent of Citrus Mal Secco disease, by real-time polymerase chain reaction, *Plant Dis.*, 90, 1523, 2006.
95. Cullen, D.W. et al., Development and validation of conventional and quantitative polymerase chain reaction assays for the detection of storage rot potato pathogens, *Phytophthora erythroseptica*, *Pythium ultimum*, *Phoma foveata*, *J. Phytopathol.*, 155, 309, 2007.
96. Irinyi, L., Kovics, G.J., and Sandor, E., Taxonomical re-evaluation of *Phoma*-like soybean pathogenic fungi, *Mycol. Res.*, 113, 249, 2009.
97. de Gruyter, J. et al., Redisposition of *Phoma*-like anamorphs in Pleosporales, *Stud. Mycol.*, 75, 1, 2012.
98. Davidson, J.A. et al., A new species of *Phoma* causes Ascochyta blight symptoms on field peas (*Pisum sativum*) in South Australia, *Mycologia*, 101(1), 120, 2009.
99. Ascioglu, S. et al., Defining opportunistic invasive fungal infections in immunocompromised patients with cancer and hematopoietic stem cell transplants: An international consensus, *Clin. Infect. Dis.*, 34, 7, 2002.

28 Yeasts Previously Included in the Genus *Pichia*

Volkmar Passoth

CONTENTS

28.1 Introduction 463
28.2 Molecular Methods for Species Identification and Determination of Phylogenetic
      Relationships 464
28.3 Current Taxonomy of the Former Genus *Pichia* 466
28.4 *Pichia, Kregervanrija, Saturnispora, Ambrosiozyma, Kuraishia, Ogataea,
      Peterozyma*, and *Nakazewaea* (Clade 5) 466
28.5 Clade 2: *Wickerhamomyces, Cyberlindnera, Barnettozyma, Starmera*,
      and *Pfaffomyces* 469
28.6 Clade 6: *Babjeviella, Yamadazyma, Hyphopichia, Pricomyces, Millerozyma,
      Meyerozyma, Scheffersomyces*, and *Kodamaea* 471
28.7 Clade 3: *Komagatella* and Clade 9: *Zygoascus* 472
28.8 Examples of Tools for Molecular Manipulation 473
28.9 Impact of Former and Recent *Pichia* Yeasts on Food Quality 474
    28.9.1 Coffee Fermentation 474
    28.9.2 Cocoa Fermentation 474
    28.9.3 Olive Fermentation 474
28.10 Conclusions 475
Acknowledgments 476
References 476

28.1 INTRODUCTION

Yeasts classified as *Pichia* have frequently been identified in food-related systems, where they can belong to the microflora necessary for the production of the food or act as spoilage organisms. Isolation from water samples has been reported less frequently.[1] In contrast to filamentous fungi, yeasts are not known to produce mycotoxins that can affect humans or vertebrates. Although they can impact the quality of food, health problems are rarely caused by these contaminations.[2] None of the isolates were explicitly pathogenic, with the possible exception of *Candida guilliermondii*, the anamorph of *Pichia guilliermondii* (now named *Meyerozyma guilliermondii*; see the following text). However, several were classified as opportunistic pathogens in immunocompromised patients; often, infections were connected to the use of devices such as catheters. Nevertheless, there is no evidence for outbreaks related to occurrence in food or water.[3–5]

Yeasts earlier or currently classified as *Pichia* spp. are designated as nonconventional yeasts. The term *nonconventional* refers to the amount of knowledge and methods for manipulation available for these yeasts. Nonconventional yeasts comprise, in principle, all the almost 2000 described yeast species apart from *Saccharomyces cerevisiae, Candida albicans*, and *Schizosaccharomyces pombe*. In most of these species, there are no, or only poorly developed, tools for molecular manipulation, and in spite of the increasing number of sequenced genomes, information about the genomes of nonconventional yeasts is often limited. However, molecular methods are widely used for species identification.

In the standard taxonomic yeast reference book from 1998, *Pichia* contained 91 species.[1] However, these species do not form a phylogenetic unit. The genus *Pichia*, as defined at that time, indicated that physiological characteristics were insufficient to build a taxonomic system reflecting phylogeny. This was already demonstrated 14 years earlier when Kurtzman showed that the only criterion, which differentiated *Hansenula* and *Pichia*, the ability of nitrate assimilation, was insufficient.[6] As a consequence, *Hansenula* was transferred to *Pichia*. Subsequently, sequencing the D1D2 region of the 26S rRNA genes proved that *Pichia* consisted of several, in some cases quite distantly related, groups,[7] and recently, efforts were intensified to establish a system of yeast taxonomy reflecting phylogeny.

Multigene sequencing resulted in splitting of *Pichia* into more than 20 genera.[4] However, the genus *Pichia* still exists, containing the type species *Pichia membranifaciens* (E.C. Hansen) and its relatives.[8] In addition, application of molecular tools will probably result in the identification of new genera. Moreover, following the recent decision to abolish the use of different names for teleomorphs and anamorphs,[9] additional species that were, up to now, named as *Candida* spp. will be included into *Pichia* and other genera (e.g., Urbina & Blackwell regrouped *Candida shehatae*, *Candida lignosa* and *Candida insectosa* to *Scheffersomyces*[10]).

At first sight, all these regroupings seem to be troublesome, and it may take some time until the new nomenclature is completely accepted by the scientific community. However, basing the nomenclature on evolution will finally provide a tool to better understand the role of the different species in their environments.

This chapter gives a survey about the actual genera that earlier were grouped into *Pichia*, with a specific focus on species that have been isolated from food or water and/or are of medical importance.

28.2 MOLECULAR METHODS FOR SPECIES IDENTIFICATION AND DETERMINATION OF PHYLOGENETIC RELATIONSHIPS

Sequencing the D1D2 domains of the large subunit rRNA gene (LSU-rRNA-gene, 26S rDNA) has become a standard for yeast identification, and databases exist for the D1D2 sequences of all known asco- and basidiomycetous yeast species.[7,11] The flanking sequences of the D1D2 region are highly conserved among yeasts, so that one set of primers can be used for all known yeasts. Using these primers, a PCR-product of about 600 bp is generated, which can be sequenced and compared with EMBL or Genbank databases by Blast search as a first step. A precise determination of phylogeny and relatedness requires more advanced alignment methods (e.g., Kurtzman and Robnett[12]). Usually, strains belonging to different species differ by more than 1% (6 bp) in their D1D2 sequences.[7] Using this technique, clinical isolates were identified that could not be identified by other (physiological) methods.[13] However, not all closely related species can be resolved by this technique, for instance *M. guilliermondii* differs from the closely related species *Meyerozyma caribbica* and *Candida carpophila* by only three and one base substitutions, respectively.[14] Multigene sequencing is therefore in some cases required for species identification. Other genes of the RNA cluster are frequently used in such analyses, including internally transcribed spacers (ITS) and small subunit rRNA genes (SSU, 18S rRNA genes). Besides these, the sequences of several non-RNA genes are involved, including actin, transcription elongation factor 1α (TEF-1α), *RPB1*, and *RPB2*. Sequences of ITS1 and ITS2, the 18S rRNA-gene, and TEF-1α were, for instance, sequenced to differentiate a clinical isolate of *Cyberlindnera fabianii* (syn. *Lindnera fabianii, Pichia fabianii, Hansenula fabianii*) from its close relative *Cyberlindnera mississippiensis*.[15] Currently, multilocus sequencing is the basis to establish a taxonomic system based on phylogeny (see the following text). Primer sequences are, for instance, listed by Kurtzman and Robnett.[12]

Other techniques include DNA–DNA reassociation, which was probably the first DNA technique that has been used to some extent for investigation of relatedness. DNA–DNA reassociation has been used for comparison of the genera *Pichia* and *Hansenula*, which finally resulted in the transfer of *Hansenula* to *Pichia*.[6] The reassociation technique has the disadvantage that only closely related species can be investigated. In contrast, gene sequence comparison can resolve both closely and distantly related species.[4]

Apart from sequencing and DNA–DNA reassociation, several other techniques have been applied for species/strain identification. Pulsed-field gel electrophoresis can provide a survey about the chromosome sizes of strains. However, in many species, a considerable variability in chromosome sizes has been observed among strains of the same species (e.g., for *Wickerhamomyces anomalus* (syn. *Hansenula anomala*, *Pichia anomala*, anamorph *Candida pelliculosa*; see the succeeding text).[16] Moreover, this method requires some optimization and is usually time consuming, so it is not the general method of choice for identification. In contrast, PCR-based methods provide rapid results. Species-specific primers can enable identification within hours, which is especially attractive in a clinical setting or in spoilage control. For example, clinical isolates of *M. guilliermondii* were difficult to differentiate from other isolates by standard physiological tests. Using specific primers or TaqMan® probes, respectively, they could be safely and more rapidly identified.[17,18] Specific primers for amplifying the rRNA intergenic spacer region of *W. anomalus* have been developed, and this method even allowed differentiating between clinical and nonclinical isolates.[19] In another approach, species-specific primers targeted to the D1D2 region have been developed for a group of yeasts, including *Pichia kudriavzevii* (synonym *Issatchenkia orientalis*, anamorph *Candida krusei*; see the following text) of clinical relevance isolated from dairy products.[20] Recently, a rapid identification method for clinical isolates based on real-time PCR and high-resolution melting analysis has been developed. Yeasts and filamentous fungi were differentiated using specific primers. Different yeasts, including *P. kudriavzevii* were identified by their HRM profile.[21] If equipment is available, PCR amplification can also be combined with pyrosequencing. By amplifying a region of the 18S rRNA gene and subsequent pyrosequencing, yeasts from blood cultures could be identified within a few hours.[22] Similarly, pyrosequencing of an ITS2 region in combination with a rapid Whatman FTA filter-based DNA isolation method enabled rapid identification of a number of clinical isolates including *W. anomalus* and *C. fabiani*.[23] Villa-Carvajal and colleagues[24] developed a method for differentiating species belonging to the polyphyletic genus *Pichia*, according to the former definition, based on restriction enzyme fragment length polymorphism of the ITS1-5.8S rDNA-ITS2 region. This technique has frequently been applied to analyzing yeast populations present during the fermentation of table olives,[25] and in fruits or juices.[26] For a first grouping of unknown isolates, fingerprinting methods can be applied, such as randomly amplified polymorphic DNA, arbitrary primed PCR, or DNA amplification based on microsatellite sequences.[27–30] For instance, microbial populations from feed fermentations and airtight storage systems of cereal grains were analyzed using microsatellite profiling using GTG_5 primers.[31,32]

The methods described earlier are usually based on template DNA isolated from cultivated yeasts. Culture-independent molecular methods have also been applied to identify yeasts, although the discrepancy between cultured and noncultured species does not seem to be as large as described for bacteria.[33] Denaturing gradient gel electrophoresis analysis of PCR products of the 26S rDNA region or analysis of length polymorphism of ITS1-5.8S rDNA-ITS2 provided results similar to culture-dependent investigations; dominating species were identified with culture-dependent and -independent methods. The culture-independent diversity was in some cases lower, which might be due to PCR inhibitors in food systems.[34–37] Brezná et al.[38] investigated the yeast diversity of Slovakian wine-related communities by f-ITS PCR, where one of the primers was FAM labeled. This was combined with capillary electrophoresis. Amplified ITS regions were cloned and sequenced. Using this method, several yeast species were identified besides the dominating *Hanseniaspora uvarum* and *S. cerevisiae*, including *P. membranifaciens*, *Pichia kluyveri*, *Pichia fermentans*, and *W. anomalus*.

28.3 CURRENT TAXONOMY OF THE FORMER GENUS *PICHIA*

The polyphyletic character of *Pichia* was recognized many years ago, and several attempts were made to divide it into monophyletic genera.[39–43] However, these efforts only recently gained wider acceptance after performing multilocus sequence analyses.[12,44] According to the current edition of the standard taxonomic book,[4] the revised genus (RG) *Pichia* contains 20 species, and some of them were previously included into the polyphyletic genus *Issatchenkia*. The other species previously assigned to *Pichia* are now associated with *Kregervanrija*, *Saturnispora*, *Ambrosiozyma*, *Kuraishia*, *Ogataea*, *Peterozyma*, *Nakazewaea*, *Wickerhamomyces*, *Cyberlindnera*, *Barnettozyma*, *Starmera*, *Pfaffomyces*, *Komagatella*, *Babjeviella*, *Yamadazyma*, *Hyphopichia*, *Pricomyces*, *Millerozyma*, *Meyerozyma*, *Scheffersomyces*, *Kodamaea*, and *Zygoascus* (Table 28.1).

According to recent investigations based on the sequences of large subunit rRNA (LSU rRNA), SSU rRNA, TEF-1α, RPB1, and RPB2, former *Pichia* species belong to five of the eleven identified clades within the hemiascomycetes,[12] illustrating the diversity of the former genus *Pichia*.

28.4 *PICHIA, KREGERVANRIJA, SATURNISPORA, AMBROSIOZYMA, KURAISHIA, OGATAEA, PETEROZYMA,* AND *NAKAZEWAEA* (CLADE 5)

Multilocus sequencing defined a large clade containing the RG *Pichia* and related genera. The family Pichiaceae as defined by Kurtzman et al. (2011),[4] contains *Pichia*, *Kregervanrija*, and *Saturnispora*. Clade 5 contains four more genera with species previously classified as *Pichia*. However, the clade also includes species never associated with *Pichia* such as *Dekkera (Brettanomyces) bruxellensis*.[12] Interestingly, the RG *Pichia* contains a variety of species associated with food. Strains of the type species, *P. membranifaciens* have been isolated during the fermentation of table olives,[45] wine and other alcoholic and nonalcoholic beverages.[46] It has also been found in fermented pig feed[31] and was used as a biocontrol agent.[47] *P. membranifaciens* strains do not grow at 37°C, therefore they have limited clinical importance.[8] *Pichia manshurika* is closely related to *P. fermentans* and was isolated from grape must and olives and probably plays an important role in food fermentations, but this role might have been underestimated because it is difficult to distinguish from *P. membranifaciens*. The yeast has also been isolated from human feces and grows at 37°C, thus it may have clinical importance.[8,46] *P. fermentans* was found in the spoilage of orange juice,[26] in dairy products[48] and brined green olives.[45] It can act as a biocontrol agent against *Monilinia fructicola* and *Botrytis cinerea* on apples, but it is a postharvest pathogen of peas.[49] Significantly, it has also been isolated from human and animal sputum and may thus have clinical importance.[8] *P. kluyveri* has been found in fermented food, such as olives[25] or wine[38] and in coffee fermentations, where it may play an important role in removing pectinaceous mucilage due to its pectinolytic activity.[50] *P. kudriavzevii* is frequently part of natural food fermentations, where it can improve the nutritional value because of its high-phytase activity.[51,52] It is not regarded as a common food spoilage agent.[8] The yeast is also an important opportunistic pathogen; it is among the six most frequently found clinical isolates (often mentioned under its anamorph name *C. krusei*). However, its frequency is rather low compared to the pathogens *C. albicans*, *Candida glabrata*, *Candida parapsilosis*, and *Candida tropicalis*; its proportion among clinical isolates is rarely higher than 2%–3%.[3,53]

Kregervanrija forms a distinct clade basal to *Pichia* and *Saturnispora*.[7,44,54] All three known species of the genus (i.e., *Kregervanrija fluxuum*, *Kregervanrija delftensis*, and *Kregervanrija pseudodelftensis*) are potential food spoilage yeasts. However, none are of clinical importance, and since they are not able to grow at 37°C, their ability to survive on the human body may be restricted. Strains of the type species *K. fluxuum* have been isolated from fluxes of wood sap of several trees. It can also form *films* on wine, beer, and vegetable brines exposed to air.[55] There was some discussion about the correct taxonomic designation of the species; D1D2 and ITS sequence analyses demonstrated that *Zygopichia chiantigiana* was conspecific with *K. fluxuum*, and since this name was established earlier, it should have priority.[46] However, the name *K. fluxuum* was retained since the

TABLE 28.1
Survey about Previous and Recent Taxonomy of the Genus *Pichia*

Genus	Important Species	Synonyms[a]	Anamorph Names[a]
Pichia	*P. membranifaciens*[T]	*P. membranaefaciens*	*Candida valida*
	P. fermentans	*Candida krusei* var. *transitoria*	*Candida lambica*
	P. kluyveri	*Hansenula kluyveri*	None
	P. kudriavzevii	*Issatchenkia orientalis*	*Candida krusei*
		Pichia orientalis	
Kregervanrija	*K. fluxuum*[T]	*Pichia fluxuum*	*Candida vini*
		Zygopichia chiantigiana	
Saturnispora	*S. dispora*[T]	*Saccharomyces disporus*	None
		Pichia dispora	
Ambrosiozyma	*A. monospora*[T]	*Endomycopsis fibuligera* var. *monospora*	None
		Pichia monospora	
	A. angophorae	*Pichia angophorae*	None
Kuraishia	*K. capsulata*[T]	*Hansenula capsulata*	None
		Pichia capsulata	
Ogataea	*O. minuta*[T]	*Hansenula minuta*	None
	O. polymorpha	*Hansenula polymorpha*	None
		Pichia angusta	
Peterozyma	*P. toletana*[T]	*Debaryomyces toletanus*	None
		Pichia toletana	
Nakazewaea	*N. holstii*[T]	*Hansenula holstii*	*Candida silvicola*
		Pichia holstii	
Wickerhamomyces	*W. canadensis*[T]	*Hansenula canadensis*	*Candida melinii*
		Pichia canadensis	
	W. anomalus	*Hansenula anomala*	*Candida pelliculosa*
		Pichia anomala	
Cyberlindnera[b]	*C. americana*[T]	*Hansenula americana*	
		Pichia americana	
		Lindnera americana	
	C. jadinii	*Hansenula jadinii*	*Candida utilis*
		Pichia jadinii	*Torulopsis utilis*
		Lindnera jadinii	
	C. fabiani	*Hansenula fabiani*	*Candida fabiani*
		Pichia fabiani	
		Lindnera fabiani	
Barnettozyma	*B. populi*[T]	*Hansenula populi*	None
		Pichia populi	
Starmera	*S. amethionina*[T]	*Pichia amethionina*	None
Pfaffomyces	*P. opuntiae*[T]	*Pichia opuntiae*	None
Komagataella	*K. pastoris*[T]	*Pichia pastoris*	
Babjeviella	*B. inositovora*[T]	*Pichia inositovora*	None
Yamadazyma	*Y. philogaea*[T]	*Pichia philogaea*	None
Hyphopichia	*H. burtonii*[T]	*Pichia burtonii*	*Candida variabilis*
Priceomyces	*P. haplophilus*[T]	*Pichia haplophila*	None
Millerozyma	*M. farinosa*[T]	*Pichia farinosa*	None
Meyerozyma	*M. guilliermondii*[T]	*Pichia guilliermondii*	*Candida guilliermondii*

(*Continued*)

TABLE 28.1 (*Continued*)
Survey about Previous and Recent Taxonomy of the Genus *Pichia*

Genus	Important Species	Synonyms[a]	Anamorph Names[a]
Scheffersomyces	*S. stipitis*[T]	*Pichia stipitis*	None
		Yamadazyma stipitis	
Kodamaea	*K. ohmeri*[T]	*Pichia ohmeri*	None
		Yamadazyma ohmeri	
Zygoascus	*Z. hellenicus*[T]	*Candida steatolytica*	*Candida steatolytica*

Source: Kurtzman, C.P. et al., *The Yeasts: A Taxonomic Study*, 5th edn., Elsevier, Amsterdam, the Netherlands, 2011.
T, type species.
[a] Only the most common synonyms and anamorph names are given. For a more comprehensive survey, see Kurtzman et al.[4]
[b] According to Minter.[84]

type strain of *Z. chiantigiana* is no longer available and the epithet *fluxuum* has been established for 50 years.[4] Only two strains of *K. delftensis* (syn. *Pichia delftensis*) have been isolated, one each from naturally fermented apple juice and Italian wine.[4,56] The one known strain of *K. pseudodelftensis* was isolated from an apple, which may indicate that this species has some potential to spoil fruits,[54] although more data are required.

Several *Saturnispora* are associated with trees, such as the type *Saturnispora dispora* or *Saturnispora saitoi* (syn. *Pichia saitoi*). The only known food-associated species is *Saturnispora mendoncae*, and all of its known strains were isolated from sauerkraut, indicating adaptation to highly acidic environments.[54] However, its role in sauerkraut fermentation has not yet been explored. There are no reports about the clinical importance of *Saturnispora*. Since no *Saturnispora* spp. is able to grow at 37°C its ability to infect humans or warm-blooded animals may be restricted.

Yeasts of *Ambrosiozyma* are frequently associated with bark beetles. The *Kuraishia* cluster contains three closely related species, *Kuraishia capsulata*, *Kuraishia molischiana* (syn. *Torulopsis molischiana*), and *Candida molischiana*. *C. molischiana* has earlier been regarded as the anamorph of *K. molischiana*, but this was disproven by biochemical and molecular tests.[7,57] The yeasts can grow on methanol,[57] and like other methylotrophic yeasts, its strains are frequently wood associated.[4] Extracellular phosphomannan of *K. capsulata* could reduce plaque formation in rats, when applied with the drinking water.[58] β-glucosidase of *K. molischiana* has been used to hydrolyze monoterpenol glycosides, improving the quality of Muscat wines.[59] The β-glucosidase and anthocyanase genes of the yeast have also been used to generate a recombinant wine strain.[60] Two strains of *K. molischiana* have been isolated from the mouth or skin of Cook Islanders. However, it has been concluded that these strains were rather randomly associated with human surfaces.[61]

Most of the more than 30 known species of *Ogataea* are able to grow on methanol as the sole carbon source. These species have a great potential for heterologous protein production due to the strong and inducible promoter of the alcohol oxidase gene (*AOX*). Most of the species have been isolated from wood or wood-associated insects, and *Ogataea polymorpha* is the sole species for which possible clinical importance has been reported. One strain, with the synonym *Candida thermophila*, was isolated from an infected human knee (i.e., catheter-related fungemia),[62] and another strain was from the intestine of swine,[63] which may indicate some potential as an emerging pathogen. The systematics of this yeast is complicated. When the genus *Hansenula* was transferred to *Pichia*, the name *P. polymorpha* could not be used, as it had been applied before for another species. Therefore, the name *Pichia angusta* was suggested, since *Hansenula angusta* was regarded as a synonym of *Hansenula polymorpha*. The type strain of *H. angusta* was defined as the type strain of *P. angusta*. However, multigene analysis revealed that the former type strain of *H. polymorpha* and the type strain of *P. angusta* are closely related, but not conspecific. The species *O. polymorpha* is now based on the type strain of the former *H. polymorpha*.[63] The name *P. angusta* was poorly accepted as a species name and *H. polymorpha* predominates in recent publications. A PubMed search on December 6, 2012, gave 636 hits for the combination *Hansenula polymorpha*, compared to 20 for *P. angusta* and only 2 for *O. polymorpha*. This resulted in the paradoxical situation that *O. polymorpha* is frequently seen as the *true Hansenula*, although it is only distantly related to the type species of the former genus *Hansenula*, *W. anomalus*.

Ogataea naganishii (syn. *Pichia naganishii*) has some importance in food production as it is the major component in *ersho*, a starter for *teff* (grain-based food in Ethiopia).[64]

The two currently recognized species of the genus *Peterozyma*, *P. toletana* and *Pichia xylosa* (syn. *P. xylosa*), are not known to have any clinical importance. The one currently known species of *Nakazawaea*, *Nakazawaea holstii*, has mainly been isolated from wood. Utilization for food production is not known and there is no evidence of clinical importance.

28.5 CLADE 2: *WICKERHAMOMYCES, CYBERLINDNERA, BARNETTOZYMA, STARMERA,* AND *PFAFFOMYCES*

The Wickerhamomycetaceae include species belonging to the earlier *Hansenula* clade, related to the former type species of *Hansenula*, *Hansenula anomala* (now *W. anomalus*).[65] As mentioned earlier, assimilation of nitrate as sole nitrogen source was not sufficient to distinguish the genera *Pichia* and *Hansenula*.[6] On the other hand, molecular methods demonstrated that most species of *Hansenula* formed a separate clade.[7,42] This was probably a reason that the transfer of *Hansenula* to *Pichia* was not completely accepted by the scientific community.[16,66] Also, the definition of the new genus *Wickerhamomyces*,[44] which includes the closest relatives of the former type species *H. anomala*, generated some debates.[66] Finally, an agreement was reached about using the genus name *Wickerhamomyces*, with the new type species *Wickerhamomyces canadensis*.[67]

Wickerhamomyces contains a variety of species connected with food and of clinical importance. *W. anomalus* belongs to the microflora of a variety of fermented foods, including wine, sourdough, or yogurt (in the latter it is mainly regarded as a spoilage organism). It is a very robust yeast, with resistance against extreme pH values, high osmotic pressure, and low-oxygen availability.[68–70] The yeast is a common spoilage yeast in dairy and baking products, wine, fruit and fruit juice concentrates, or high-salted foods.[68] It has also been identified as a potential beer spoilage organism that can contribute to biofilm formation in bottling plants.[71] On the other hand, it can also be a production organism contributing to the aroma and quality of wine and other fermented food by the production of volatile compounds and enzymes.[68] *W. anomalus* has an extensive antimicrobial potential in several environments, and its application as a biocontrol agent on cereal grain and apples has been explored.[32,72] One, but not the sole, component of this antimicrobial activity is the strong killer activity, which acts against a variety of other yeasts and even bacteria.[73] The yeast was also reported to act as opportunistic pathogen in heavily immunocompromised patients (frequently under its anamorph name *C. pellicolosa*) and is sometimes regarded as an emerging pathogen.[5,74] However, its frequency among clinical isolates is rather low; in a recent review of clinical *Candida* isolates, *C. pellicolosa* was even not specifically listed, and in a 2-year survey of British clinical isolates, the proportion of the total number of isolates was 0.3%.[13,53] The yeast may also have a medical potential by inhibiting Enterobacteriaceae.[75,76] Anti-idiotypic antibodies generated from killer proteins of *W. anomalus* have been shown to act against an amazingly broad range of pathogens, including yeasts and filamentous fungi, and even protists.[73] Recently the genome of this yeast has been sequenced,[77] enabling more systematic studies of its strong competitiveness in different habitats.

Wickerhamomyces subpelliculosus (syn. *Hansenula subpelliculosa*, *Pichia subpelliculosa*) is closely related to *W. anomalus*. The species shares some physiological characteristics with its relative, enabling it to grow on high-osmotic foods, reducing their quality. Spoilage of pickles and other brined vegetables has been reported,[78] and it has been found in fermentations of cachaca, a Brazilian alcoholic beverage.[79] There are no reports about clinical isolates of *W. subpelliculosa*. Only two strains are available for *Wickerhamomyces bisporus*: one has been isolated from insect tunnels in spruce; the other from an infected human scalp.[65] *W. bisporus* produces lipases active at pH 5.5 and 7.5.[80] However, little is known about the type species *Wickerhamomyces canadensis*. Isolates have mainly been associated with insect frass from different tree species and are reported to be found in fermented food and feed.[65] There are no reports about the clinical importance of the species, although it is listed among emerging pathogens.[81] *Wickerhamomyces lynferdii* (syn. *Hansenula lynferdii*, *P. lynferdii*) has been isolated from coffee fermentation,[82] and was identified as spoilage organism in butter.[83] *Wickerhamomyces onychis* (syn. *Pichia onychis*) has been isolated from a variety of habitats, including infected human nail and sputum, which may indicate a role as an emerging pathogen.[13]

Cyberlindnera was first described as *Lindnera* on the basis of multigene sequencing analysis of the polyphyletic *Pichia*.[44] However, since the name *Lindnera* was already in use to designate a plant genus, the name *Cyberlindnera* was proposed.[84] The genus consists of two subclades, one containing the type species *Cyberlindnera americana* and the other *Cyberlindnera jadinii* and the former *Williopsis* species *Cyberlindnera saturnus* and *Cyberlindnera mrakii*. Within the first subcluster, only *C. fabianii* is known to play roles in food, water, and as an opportunistic pathogen. It has initially been isolated as a contaminant from starch-based fermentations,[85] but it also has potential for waste water treatment of starch-based industrial processes.[86,87] In addition, *C. fabianii* has in a few cases been involved in yeast infections.[13,15] All strains of the type species *C. americana* have been isolated from frass in insect tunnels in pine trees in the United States. No association with food or clinical impact is known. Among the second cluster, *C. jadinii*, better known under its anamorph name *Candida utilis*, has a great biotechnological importance. This yeast has a long history of industrial application, mainly as fodder yeast,[88] but also as a food or food additive.[89] On the other hand, several strains of *C. jadinii* have been isolated from clinical samples or animals, for instance, the type strain has been isolated from a human abscess.[90] Recently, the genome and transcriptome in different growth phases have been determined for strain CBS 5609. The yeast showed several interesting features, including high-affinity glucose transporters and a specific regulation of alcohol dehydrogenase expression by antisense RNA.[91] Strains of *Cyberlindnera misumaiensis* have frequently been isolated from apples or apple products.[92] *C. mrakii* (syn. *Hansenula mrakii*, *Williopsis mrakii*, *Williopsis saturnus* var. *mrakii*) produces one of the most active killer proteins among yeasts. The gene encoding the killer protein has been expressed in *Aspergillus niger* and was active against spoilage yeasts in maize silage.[93] *C. saturnus* (syn. *Hansenula saturnus*, *Williopsis saturnus*) is very closely related to *C. mrakii*, and it also produces efficient killer proteins.[47] The yeast has been isolated from several habitats, including soil and freshwater: No clinical importance has been reported.

There is no clear evidence for the clinical importance of yeasts belonging to the genus *Barnettozyma*, however, most of the species do not grow at 37°C. The species are also very rarely associated with food production or storage. The type species *Barnettozyma populi* has been isolated from cottonwood trees. *Barnettozyma californica* (syn. *Hansenula californica*, *Williopsis californica*) produces lipases that can degrade the quality of olive oil.[94] *Barnettozyma salicaria* (syn. *Pichia salicaria*) is potentially the sole example of a clinical episode in this genus, since one strain (CBS 4208) was isolated from a human toe. However, identity of this strain was not confirmed by gene sequence analysis.[4]

Starmera are often associated with cacti or trees, however, clinical isolates and impact on food have not been reported. Similarly, the three currently described species of the closely related *Phaffomyces*, *Phaffomyces opuntiae*, *Phaffomyces antillensis*, and *Phaffomyces thermotolerans* are all cactus-associated yeasts. No utilization in food production or involvement in clinical processes has been reported.

28.6 CLADE 6: *BABJEVIELLA, YAMADAZYMA, HYPHOPICHIA, PRICOMYCES, MILLEROZYMA, MEYEROZYMA, SCHEFFERSOMYCES,* AND *KODAMAEA*

In several genera of this clade, there is an alternative translation of the CTG codon to serine instead of leucine, which is similar to the pathogenic *C. albicans*, also related to this clade.[95] However, it has not been completely investigated whether all yeasts of this clade alternatively translate the CTG codon. In *C. albicans*, about 97% of these codons are translated to Ser, but 3% to Leu.[96] However, for many species the mode of codon usage has not been determined. For example, the NCBI database states that *M. guilliermondii* uses the alternative translation, whereas the closely related *M. caribbica* is assumed to use the standard translation, but it is unclear if this has been confirmed by experimentation.

Babjeviella contains only *Babjeviella inositovora* and it is well separated from other genera within clade 6. Strains have been isolated from soil or arboreal habitats. No clinical or food-borne isolates are known; the species obviously has killer activity, which is connected with three linear DNA plasmids.[4,97]

The genus *Yamadazyma* was introduced in 1989 based on morphological and biochemical characteristics and contained 16 species formerly classified as *Pichia*.[98] Several of these species are now associated with *Scheffersomyces, Meyerozyma,* or *Millerozyma*. Sequence analysis showed that *Yamadazyma* was polyphyletic,[7] and thus the genus name was not widely accepted by the scientific community. The current genus, based on multigene sequencing, contains six species. Most of the species have been associated with trees or other plants, with a few exceptions. Strains of *Yamadazyma mexicana* have also been isolated from soil and grape must. The species is lipolytic; lipase production has been demonstrated,[80] and it is not inhibited by fatty acids such as hexanoic, octanoic, decanoic, or dodecanoic acid, which usually are inhibitory to yeasts.[99] *Yamadazyma triangularis* has been isolated from saline environments, including tamari soya, pickled Chinese mushrooms, and seawater. One isolate of this species represents the only potential clinical episode of the genus, as it was isolated from lung tissue. However, since the yeast is unable to grow at 37°C,[98] its clinical occurrence may be restricted to tissues with lower temperatures.

Hyphopichia contains two species. However, several imperfect species such as *Candida fennica* and *Candida ragii* will probably be included, as they are closely related to *Hyphopichia burtonii* or *Hyphopichia heimii*, respectively. *H. burtonii* is an amylase-producing yeast[100] and has been isolated from a variety of environments, including corn, wheat,[101] and fish ponds.[102] It is involved in starch degradation for murcha, a traditional starter for producing alcoholic beverages in the Himalaya region[100] and belongs to the yeast community, which develops during the curing of Italian ham.[103] One *H. burtonii* strain has been isolated from skin and another from sputum, but the clinical importance of this species seems to be restricted to tissues with lower temperature, because of its missing or weak ability to grow at 37°C. No correlation to clinical or food samples for *H. heimii* has been reported.[4]

Pricomyces have been previously classified as *Pichia, Debaryomyces, Torulaspora,* or *Yamadazyma*, mainly based on spore morphology. No case of pathogenicity is known. *Pichia carsonii* (syn. *Pichia carsonii, Debaryomyces carsonii, Torulaspora carsonii*) showed a strong killer activity against the human pathogens *Cryptococcus neoformans* and *C. parapsilosis*.[104] *P. carsonii* has been isolated from several sources, including spoiled wine or agricultural processing wastes. The yeast has been reported to produce styrene from the food preservative *trans*-cinnamic acid in chikuwa, a Japanese fish-based food, causing a petroleum-like odor.[4] *Pricomyces melissophilus* (syn. *Pichia melissophila, Torulaspora melissophila, Debaryomyces melissophilus*) belongs to the species commonly found on strawberry fruits, where it can produce volatile compounds that cause off-flavors of stored strawberries.[105]

Millerozyma is closely related to *Pricomyces* and contains *Millerozyma farinosa* and *Millerozyma acaciae* (syn. *Pichia acaciae*). *M. farinosa* is a rare opportunistic pathogen, isolated from cancer patients[106] and from a bloodstream infection.[107] The yeast belongs to the microbial consortium during cacao fermentation (under one of its synonyms, *Candida cacaoi*),[108] and it was among yeasts that inhibited the mold *Penicillium roquefortii* in airtight stored cereal grain.[109] Strain CBS 7064 has also been designated as *Pichia sorbitophila*, and there was a debate whether *P. sorbitophila* should be regarded as conspecific to *M. farinosa* as the strain shows substantial physiological and gene sequence differences. A recent investigation demonstrated that *M. farinosa* consists of at least four clades differing from each other in their genome by 8%–15%. However, hybridization between the clades does occur and strains of the species represent haploids or interclade hybrids. Diploids are able to sporulate forming a high proportion of viable spores. These are mainly diploid vegetative spores and the decrease of ploidy occurred by loss of heterozygosity. Strain 7064 was a hybrid of clade 2 and clade 4[110] and (under its synonym *P. sorbitophila*) has been partially sequenced during the Genolevure project.[111]

In the current yeast taxonomy book,[4] *Meyerozyma* is considered to contain two species. However, there are several nonsexual species, which will be incorporated into this genus, for example, *C. carpophila*, which is very closely related to the type species *M. guilliermondii*.[7] *Candida elateridarum* is also of note which forms true hyphae.[112] The type species *M. guilliermondii* is an opportunistic pathogen and it represents about 2% of clinical yeast isolates.[13,53,74] On the other hand, it has great biotechnological potential for the production of riboflavin[113] or xylitol[114] and was tested as a biocontrol yeast against several postharvest pathogens on a variety of fruits.[47] *M. caribbica* (syn. *Pichia caribbica*, anamorph *Candida fermentati*) performed well in cachaca fermentation.[115] One isolate has been associated with a human infection[14] and since the yeast is very closely related to *M. guilliermondii*, it is likely that this yeast can also act as an opportunistic pathogen.

Most of the species belonging to the *Scheffersomyces* clade are able to ferment xylose to ethanol; in fact, *Scheffersomyces stipitis* and the closely related *Scheffersomyces shehatae* (syn. *C. shehatae*)[10] are the most efficient xylose fermenters among all known organisms. Clinical importance associated with *Scheffersomyces* is unknown. *S. stipitis* has a high biotechnological potential to produce ethanol from lignocelluloses and the genome of this yeast has been determined.[116] However, it is not especially tolerant of ethanol and inhibitors are released during pretreatment of biomass. Therefore, transfer of the genes of its xylose metabolism to the established fermentation yeast *S. cerevisiae* may be a more promising strategy.[117] Recently, it has been shown that *S. stipitis* has a mold-inhibiting activity and can be used for a combined biopreservation and pretreatment of airtight stored wheat straw.[118] The yeast is also a model yeast for the response to low-oxygen concentration.[119,120] *Scheffersomyces spartinae* (syn. *Pichia spartinae*) has been isolated from water environments; it was isolated, for instance, from a biofilm clogging a water treatment plant in Australia.[121] The yeast has also a comparatively high-phytase activity with a very high optimal reaction temperature (75°C–80°C).[122] There are no reports about clinical isolates belonging to the genus.

Kodamaea contains five species, but only the type species, *Kodamaea ohmeri* was previously included in *Pichia*. This species is also the only one with clinical importance. Infections by *K. ohmeri* are rare and are associated with conditions, such as diabetes, hepatitis, cancer, or treatment with antibiotics. Identification was not always based on sequence analysis, so the true frequency of *K. ohmeri* infections is uncertain. Cases in premature newborns have also been reported, some of them very recently, and this yeast is regarded as an emerging pathogen.[123]

28.7 CLADE 3: *KOMAGATELLA* AND CLADE 9: *ZYGOASCUS*

These two former *Pichia* genera belong to a single clade and are thus more distantly related to all other species of the aforementioned clades. *Komagataella* is separated from all other known hemiascomycetous genera and forms its own clade and contains methylotrophic species. *Komagataella pastoris* is better known under its previous name *Pichia pastoris*, and *P. pastoris* is frequently believed to be the *true Pichia*. A PubMed search (7/12/2012) resulted in 3854 hits for

P. pastoris, and only 9 for *Komagatella pastoris*, although the yeast is more closely related to the *Wickerhamomyces* (*Hansenula*) clade.[12] Due to its extensive use as an expression host, there is a variety of genetic tools available for this yeast (see the succeeding text). The three species of the genus *K. pastoris*, *Komagatella phaffii*, and *Komagatella pseudopastoris* cannot be distinguished by physiological methods, which may explain some uncertainties about the identities of patented strains (see the succeeding text). All species have, like most methylotrophic species, been isolated from trees or rotten wood. Clinical importance has not been reported. Recent taxonomic investigations using multilocus sequencing revealed that the strains patented for single-cell protein production and of the expression system are not necessarily *K. pastoris*. One of the two strains mentioned in the patent was identified as *K. phaffii*, while the other belongs to *K. pastoris*. A strain isolated from the Invitrogen expression system (NRRL Y-11430) belongs most probably to *K. phaffii*. A genome sequence of *K. pastoris* is also available, but since no strain number is provided, it is uncertain which of the three species was sequenced.[4]

The genus *Zygoascus* belongs to clade 9 in the classification of Kurtzman and Robnett[12] and is more closely related to, for example, *Yarrowia lipolytica*, than to *P. membranifaciens*. Apart from the type species *Zygoascus hellenicus*, all described species of this genus were previously classified as *Pichia*.[4] *Z. hellenicus* has been mainly isolated from human samples or beverages,[4] for example, it was once isolated from human blood.[124] The closely related species *Zygoascus meyerae* (syn. *Pichia hangzhouana*)[125] has frequently been isolated from must or rotten fruits.[4] *Zygoascus ofunaensis* (syn. *Hansenula*, *Pichia ofunaensis*) was found to be one of the major yeasts in dry coffee fermentation.[82]

28.8 EXAMPLES OF TOOLS FOR MOLECULAR MANIPULATION

As stated earlier, former and current *Pichia* species are nonconventional yeasts, implying that tools for molecular manipulation are far less developed than those for *S. cerevisiae*. Nevertheless, especially for several methylotrophic species that are used as expression hosts and model organisms, a considerable diversity of manipulation tools has been developed. In *K. pastoris*/*K. phaffii*, mainly integrative plasmids are used, targeted to the *AOX1* or *HIS4*-genes, or to rDNA. There are also episomal plasmids available, but these are only stable under selective conditions. For protein production, promoters of the highly expressed and regulated genes involved in methanol metabolism or elements of these promoters are used. The genes include the *AOX1* and *AOX2* gene of *K. pastoris*, and the *AUG1* and *AUG2* (alcohol utilization gene) of *Ogataea methanolica*, which are repressed by glucose and induced by methanol. The methanol oxidase promoter of *O. polymorpha* is derepressed in the absence of glucose, which implies that it can be expressed even if methanol is absent in the medium. Some constitutive promoters, mainly from glycolytic genes, are also used in methylotrophic yeasts.[126] Selection markers in *K. pastoris* and *O. polymorpha* include markers complementing auxotrophies (for *K. pastoris HIS4*, *ARG4*, *ADE1*, and *URA3* from *K. pastoris* or *S. cerevisiae* and for *O. polymorpha LEU2*, *URA3*, *TRP3*, and *ADE11* either from *O. polymorpha*, *S. cerevisiae*, or *C. albicans*) or dominant resistance markers, for example, against zeocin, G418 or hygromycin B.[127,128] A broad host range vector system (CoMed™) has been constructed. This system can be used in *O. polymorpha* or *K. pastoris*, and in other yeasts, such as *S. cerevisiae* and *Blastobotrys* (*Arxula*) *adeninivorans*.[129]

For *C. jadinii* (*C. utilis*), a transformation system with a reusable marker has been established, based on hygromycin B resistance and the Cre-*loxP* system. This is especially valuable as strains of this species have a high ploidy.[13] For *C. fabiani*, a transformation system using a *URA3* selection marker has been established.[86] In *W. anomalus*, gene disruption using *URA3* as selection marker proved that β-1,3-glucanases are involved in biocontrol on apples.[72]

Some molecular tools have also been developed for the xylose-fermenting yeast *S. stipitis*. Most frequently, auxotrophic markers are used, complementing *ura3 leu2-* or *his3 trp1* mutants.[131,132] The latter has been used to establish a system for random and targeted integration mutagenesis, which may enable functional genomics in the future.[133] Establishing resistance markers has been a problem in this yeast because of its alternative CTG translation; however, a gene encoding for bleomycin resistance and a Cre-*loxP* site with exchanged CTG codons is now available for transformation.[134] Similarly, a modified bleomycin resistance gene was used in *M. farinosa* that also belongs to the CTG yeasts.[135]

28.9 IMPACT OF FORMER AND RECENT *PICHIA* YEASTS ON FOOD QUALITY

28.9.1 Coffee Fermentation

Yeasts in these processes are mainly of interest for their pectinolytic activity; however, they may act as spoilage agents in the later phases of fermentation.[136] Most investigations focused on wet fermentation processes. High-pectinolytic activity was observed in yeast species involved in *Coffea arabica* fermentation, especially with *W. anomalus* and *P. kluyveri*.[50] In dry fermentations, a broad spectrum of species were found, with *Blastobotrys adeninivorans* and *Z. (Pichia) ofunaensis* as the major isolates (about 15%). Other former *Pichia* species were also isolated, including *W. lynferdii*, *W. anomalus*, and *Wickerhamomyces cifferii*.[82] In a screening experiment, a variety of yeast strains were tested for pectinase activity and *M. (Pichia) guilliermondii* was one of the most promising.[137] As mentioned earlier, few strains of these yeast species have been involved in pathogenic processes, with *M. guilliermondii* and *W. anomalus* being most prominent. No correlation between coffee fermentation and yeast infections has been established.

28.9.2 Cocoa Fermentation

Similar to coffee, fruits of the cacao tree (*Theobroma cacao*) are fermented before processing, which removes the pectinaceous pulp surrounding the beans. Cocoa fermentation lasts 7 days, during which a microbial succession of yeasts, lactic acid bacteria (LAB), acetic acid bacteria, spore-forming bacteria, and molds occurs. During the first 4 days, when yeasts, LAB, and acetic acid bacteria are dominant, alcohols and acids are formed and the temperature increases to 50°C. As a result, the beans are killed and flavor precursors are formed. The fruit tissues degrade, which makes it easier to dry the bean. The yeast flora apparently varies considerably between different fermentations, and species dominating fermentations on one site were not found in another region. It is not clear whether these variations reflect differences due to geographic peculiarities or specific fermentation conditions. In most fermentations, *S. cerevisiae* was identified, which probably is the main producer of ethanol. During the high-temperature phase, thermotolerant yeasts such as *Kluyveromyces thermotolerans* have been observed. Several current and former *Pichia* species have been isolated from cocoa fermentations, including *W. anomalus* (*C. pelliculosa*), *P. fermentans*, *P. membranifaciens*, *P. kudriavzevii* (*C. krusei*), and *M. farinosa* (*C. cacoai*).[108] In fermentations of West African cocoa beans, *P. kudriavzevii* (*C. krusei*) was dominant in one type of fermentation (heap fermentation), and *P. membranifaciens* and *P. kluyveri* also played a role. In tray fermentations, *S. cerevisiae* and *P. membranifaciens* were dominant, and low cell numbers of *P. kudriavzevii* and *P. kluyveri* were also found.[138] Experiments with starter cultures have also been performed, confirming that pectinase activity is an important factor for cacao fermentation.[108] Most of these species may potentially act as pathogens (see previous text); however, an increase of yeast infections correlated to cocoa fermentation has not been reported.

28.9.3 Olive Fermentation

Yeasts can positively contribute to the taste of table olives or act as spoilage organisms. The fermentation is principally performed by LAB, and formation of lactic acid and decrease in pH is one major factor conserving the material. However, yeasts are always present, and in some cases, they dominate the process. The resulting product is milder in taste, but has a reduced shelf life. In some cases, yeast can even positively contribute to the growth of LAB by providing vitamins and other growth factors. Yeasts can also substantially contribute to the taste of the final product by generating ethanol, glycerol, and volatile compounds and by degrading polyphenolic compounds that are responsible for the bitter taste. They can even inhibit undesirable microbes and, by this, contribute to the stability of the product.[45,139] Negative effects of yeasts include increase in pH and thus susceptibility to contaminants, vigorous production of CO_2 and because of this, damage to the fruits, and production of enzymes such as proteases, xylanases, and pectinases occurs that soften the fruits. Yeast growth in olive packaging can result in swollen containers due to CO_2 production and off-flavors and odors, resulting in economic losses.[45,139] The yeast species composition in different fermentations varies, and the factors determining composition remain unknown. *S. cerevisiae*, *W. anomalus*, and *P. membranifaciens* are frequently (but not exclusively) present.[45,139] In fermentations of green table olives, differently treated against the olive fruit fly, *W. anomalus* was dominant in the beginning of the fermentation, while *S. cerevisiae* succeeded.[36] In contrast, *P. kluyveri* was the dominant species in a fermentation of green Silician table olives.[34] No correlation between the occurrence of potentially opportunistic strains and human yeast infection has been observed.

28.10 CONCLUSIONS

Yeasts are dominant organisms in many fermented foods. This includes the baking and brewing yeast *S. cerevisiae* and its relatives, but also (among others) the highly diverse group of yeasts previously associated with the genus *Pichia*. Molecular analyses demonstrated that these yeasts are distributed to almost all branches of the phylogenetic tree of the Hemiascomycetes. A wide variety of characteristics connect these yeasts with foods, from the production of volatiles, acids or alcohols, to the formation of enzymes such as pectinases, lipases, or proteases. Tolerance of extreme pH values and temperatures enable those yeasts to survive even in foods with added preservatives with antimicrobial activity. This makes them also potent spoilage organisms. However, consumption of food spoiled by yeasts usually does not have negative health impacts.[2] Many of the food-related yeasts, such as *P. kudriavzevii* or *W. anomalus*, also act as opportunistic pathogens or are at least occasionally involved in human yeast infections. However, there is no evidence for a correlation between occurrence in food and clinical outbreaks. *C. jadinii* (*C. utilis*) has been produced for almost 100 years for food and feed purposes, but there are no reports of enhanced yeast infections among process personnel. Other strains of the same species have been isolated from clinical samples. It is still largely unknown which factors make a strain pathogenic. The ability to act as opportunist does not seem to be restricted to any phylogenetic group. *M. guilliermondii*, one of the six most pathogenic yeast species, belongs to the same clade as *C. albicans*. On the other hand, the nonpathogenic *S. stipitis* also belongs to this clade. All three yeasts use the alternative translation of CTG, so this speciality does not seem to be involved in pathogenesis. *C. glabrata*, another pathogenic yeast, belongs to the *Saccharomyces* clade, whereas *P. kudriavzevii* (*C. krusei*) belongs to *Pichia sensu strictu*. Both of these latter yeasts do not use the alternative CTG translation. One could hypothesize that the major factor of becoming a pathogen is to survive in the host. This can happen when the immune system is suppressed or when a great number of living yeasts is ingested, such as the probiotic *S. cerevisiae* (syn. *Saccharomyces boulardii*).[74] From this background, the suggestion to use *P. kudriavzevii* as a probiotic yeast to decrease cholesterol concentration[48] should be viewed with some skepticism. Other characteristics like attachment to surfaces, stress tolerance, and production of lipases and/or proteases that are advantageous for food processing are also virulence factors in pathogenic yeasts.[140] Therefore, consumption of large numbers of living yeasts should probably be avoided. More research is required to understand under which circumstances harmless organisms turn into pathogens.

ACKNOWLEDGMENTS

I acknowledge the support for my previous and current research on *Pichia*, obtained from the European Commission (Marie-Curie Grant No. BIO4-CT98-5019), Deutsche Forschungsgemeinschaft, the Swedish Research Council for Environment, Agricultural Sciences and Spatial Planning (FORMAS), the Swedish Energy Authority (STEM), the Swedish Institute (Visby Programme), and the *MicroDrivE* program at the NL-faculty of the Swedish University of Agricultural Sciences. I also thank Dr. Su-Lin Leong for critically reading the manuscript and linguistic advice.

REFERENCES

1. Kurtzman, C.P. *Pichia* E.C. Hansen emend. Kurtzman. In *The Yeasts: A Taxonomic Study*, 4th edn., (eds. Kurtzman, C.P. and Fell, J.W.) pp. 273–352 (Elsevier, Amsterdam, the Netherlands, 1998).
2. Fleet, G.H. Spoilage yeasts. *Crit Rev Biotechnol* **12**, 1–44 (1992).
3. Kurtzman, C.P. and Fell, J.W. *The Yeasts, a Taxonomic Study*, 4th edn. (Elsevier, Amsterdam, the Netherlands, 1998).
4. Kurtzman, C.P., Fell, J.W., and Boekhout, T. *The Yeasts, a Taxonomic Study*, 5th edn. (Elsevier, Amsterdam, the Netherlands, 2011).
5. Hazen, K.C. New and emerging yeast pathogens. *Clin Microbiol Rev* **8**, 462–478 (1995).
6. Kurtzman, C.P. Synonymy of the yeast genera *Hansenula* and *Pichia* demonstrated through comparisons of deoxyribonucleic acid relatedness. *Antonie van Leeuwenhoek* **50**, 209–217 (1984).
7. Kurtzman, C.P. and Robnett, C.J. Identification and phylogeny of ascomycetous yeasts from analysis of nuclear large subunit (26S) ribosomal DNA partial sequences. *Antonie van Leeuwenhoek* **73**, 331–371 (1998).
8. Kurtzman, C.P. *Pichia* E.C. Hansen 1904. In *The Yeasts, a Taxonomic Study*, 5th edn., Vol. 2, (eds. Kurtzman, C.P., Fell, J.W., and Boekhout, T.) pp. 685–707 (Elsevier, Amsterdam, the Nethelands, 2011).
9. Hawksworth, D.L. et al. The Amsterdam declaration on fungal nomenclature. *IMA Fungus* **2**, 105–112 (2011).
10. Urbina, H. and Blackwell, M. Multilocus phylogenetic study of the *Scheffersomyces* yeast clade and characterization of the N-terminal region of xylose reductase gene. *PLoS One* **7**, e39128 (2012).
11. Fell, J.W., Boekhout, T., Fonseca, A., Scorzetti, G., and Statzell-Tallman, A. Biodiversity and systematics of basidiomycetous yeasts as determined by large-subunit rDNA D1/D2 domain sequence analysis. *Int J Syst Evol Microbiol* **50**(Pt 3), 1351–1371 (2000).
12. Kurtzman, C.P. and Robnett, C.J. Relationships among genera of the *Saccharomycotina* (Ascomycota) from multigene phylogenetic analysis of type species. *FEMS Yeast Res* **13**(1), 23–33 (2013).
13. Linton, C.J. et al. Molecular identification of unusual pathogenic yeast isolates by large ribosomal subunit gene sequencing: 2 years of experience at the United Kingdom mycology reference laboratory. *J Clin Microbiol* **45**, 1152–1158 (2007).
14. Vaughan-Martini, A., Kurtzman, C.P., Meyer, S.A., and O'Neill, E.B. Two new species in the *Pichia guilliermondii* clade: *Pichia caribbica* sp. nov., the ascosporic state of *Candida fermentati*, and *Candida carpophila* comb. nov. *FEMS Yeast Res* **5**, 463–469 (2005).
15. Gabriel, F., Noel, T., and Accoceberry, I. *Lindnera* (*Pichia*) *fabianii* blood infection after mesenteric ischemia. *Med Mycol* **50**, 310–314 (2012).
16. Naumov, G.I., Naumova, E.S., and Schnürer, J. Genetic characterization of the nonconventional yeast *Hansenula anomala*. *Microbiol Res* **152**, 551–562 (2001).
17. Mota, A.J., Back-Brito, G.N., and Nobrega, F.G. Molecular identification of *Pichia guilliermondii*, *Debaryomyces hansenii* and *Candida palmioleophila*. *Genet Mol Biol* **35**, 122–125 (2012).
18. Yamamura, M. et al. Polymerase chain reaction assay for specific identification of *Candida guilliermondii* (*Pichia guilliermondii*). *J Infect Chemother* **15**, 214–218 (2009).
19. Bhardwaj, S., Sutar, R., Bachhawat, A.K., Singhi, S., and Chakrabarti, A. PCR-based identification and strain typing of *Pichia anomala* using the ribosomal intergenic spacer region IGS1. *J Med Microbiol* **56**, 185–189 (2007).
20. Makino, H., Fujimoto, J., and Watanabe, K. Development and evaluation of a real-time quantitative PCR assay for detection and enumeration of yeasts of public health interest in dairy products. *Int J Food Microbiol* **140**, 76–83 (2010).

21. Goldschmidt, P. et al. New strategy for rapid diagnosis and characterization of fungal infections: The example of corneal scrapings. *PLoS One* **7**, e37660 (2012).
22. Quiles-Melero, I., Garcia-Rodriguez, J., Romero-Gomez, M.P., Gomez-Sanchez, P., and Mingorance, J. Rapid identification of yeasts from positive blood culture bottles by pyrosequencing. *Eur J Clin Microbiol Infect Dis* **30**, 21–24 (2011).
23. Borman, A.M., Linton, C.J., Miles, S.J., and Johnson, E.M. Molecular identification of pathogenic fungi. *J Antimicrob Chemother* **61**(Suppl. 1), i7–i12 (2008).
24. Villa-Carvajal, M., Querol, A., and Belloch, C. Identification of species in the genus *Pichia* by restriction of the internal transcribed spacers (ITS1 and ITS2) and the 5.8S ribosomal DNA gene. *Antonie van Leeuwenhoek* **90**, 171–181 (2006).
25. Botta, C. and Cocolin, L. Microbial dynamics and biodiversity in table olive fermentation: Culture-dependent and -independent approaches. *Front Microbiol* **3**, 245 (2012).
26. Las Heras-Vazquez, F.J., Mingorance-Cazorla, L., Clemente-Jimenez, J.M., and Rodriguez-Vico, F. Identification of yeast species from orange fruit and juice by RFLP and sequence analysis of the 5.8S rRNA gene and the two internal transcribed spacers. *FEMS Yeast Res* **3**, 3–9 (2003).
27. Lieckfeldt, E., Meyer, W., and Borner, T. Rapid identification and differentiation of yeasts by DNA and PCR fingerprinting. *J Basic Microbiol* **33**, 413–425 (1993).
28. Hadrys, H., Balick, M., and Schierwater, B. Application of random amplified polymorphic DNA (RAPD) in molecular ecology. *Mol Ecol* **1**, 55–63 (1992).
29. Welsh, J. and McClelland, M. Fingerprinting genomes using PCR with arbitrary primers. *Nucleic Acids Res* **18**, 7213–7218 (1990).
30. Williams, J.G.K., Kubelik, A.R., Livak, K.J., Rafalski, J.A., and Tingey, S.V. DNA polymorphism amplified by arbitrary primers are useful as genetic markers. *Nucleic Acids Res* **18**, 6531–6535 (1990).
31. Olstorpe, M., Lyberg, K., Lindberg, J.E., Schnürer, J., and Passoth, V. Population diversity of yeasts and lactic acid bacteria in pig feed fermented with whey, wet wheat distillers' grains, or water at different temperatures. *Appl Environ Microbiol* **74**, 1696–1703 (2008).
32. Olstorpe, M. and Passoth, V. *Pichia anomala* in grain biopreservation. *Antonie van Leeuwenhoek* **99**, 57–62 (2011).
33. Amann, R.I., Ludwig, W., and Schleifer, K.H. Phylogenetic identification and in situ detection of individual microbial cells without cultivation. *Microbiol Rev* **59**, 143–69 (1995).
34. Aponte, M. et al. Study of green Sicilian table olive fermentations through microbiological, chemical and sensory analyses. *Food Microbiol* **27**, 162–170 (2010).
35. Di Maro, E., Ercolini, D., and Coppola, S. Yeast dynamics during spontaneous wine fermentation of the Catalanesca grape. *Int J Food Microbiol* **117**, 201–210 (2007).
36. Muccilli, S., Caggia, C., Randazzo, C.L., and Restuccia, C. Yeast dynamics during the fermentation of brined green olives treated in the field with kaolin and Bordeaux mixture to control the olive fruit fly. *Int J Food Microbiol* **148**, 15–22 (2011).
37. Masoud, W., Cesar, L.B., Jespersen, L., and Jakobsen, M. Yeast involved in fermentation of *Coffea arabica* in East Africa determined by genotyping and by direct denaturating gradient gel electrophoresis. *Yeast* **21**, 549–556 (2004).
38. Brezná, B. et al. Evaluation of fungal and yeast diversity in Slovakian wine-related microbial communities. *Antonie van Leeuwenhoek* **98**, 519–529 (2010).
39. Yamada, Y., Kawasaki, H., Nagatsuka, Y., Mikata, K., and Seki, T. The phylogeny of the cactophilic yeasts based on the 18S ribosomal RNA gene sequences: The proposals of *Phaffomyces antillensis* and *Starmera caribaea*, new combinations. *Biosci Biotechnol Biochem* **63**, 827–832 (1999).
40. Yamada, Y., Suzuki, T., Matsuda, M., and Mikata, K. The phylogeny of *Yamadazyma ohmeri* (Etchells et Bell) billon-grand based on the partial sequences of 18S and 26S ribosomal RNAs: The proposal of *Kodamaea* gen. nov. (Saccharomycetaceae). *Biosci Biotechnol Biochem* **59**, 1172–1174 (1995).
41. Yamada, Y., Matsuda, M., Maeda, K., and Mikata, K. The phylogenetic relationships of methanol-assimilating yeasts based on the partial sequences of 18S and 26S ribosomal RNAs: The proposal of *Komagataella* gen. nov. (Saccharomycetaceae). *Biosci Biotechnol Biochem* **59**, 439–444 (1995).
42. Yamada, Y., Maeda, K., and Mikata, K. The phylogenetic relationships of the hat-shaped ascospore-forming, nitrate-assimilating *Pichia* species, formerly classified in the genus *Hansenula* Sydow et Sydow, based on the partial sequences of 18S and 26S ribosomal RNAs (Saccharomycetaceae): The proposals of three new genera, *Ogataea*, *Kuraishia*, and *Nakazawaea*. *Biosci Biotechnol Biochem* **58**, 1245–1257 (1994).

43. Yamada, Y., Matsuda, M., Maeda, K., Sakakibara, C., and Mikata, K. The phylogenetic relationships of the saturn-shaped ascospore-forming species of the genus *Williopsis* Zender and related genera based on the partial sequences of 18S and 26S ribosomal RNAs (Saccharomycetaceae): The proposal of *Komagataea* gen. nov. *Biosci Biotechnol Biochem* **58**, 1236–1244 (1994).
44. Kurtzman, C.P., Robnett, C.J., and Basehoar-Powers, E. Phylogenetic relationships among species of *Pichia*, *Issatchenkia* and *Williopsis* determined from multigene sequence analysis, and the proposal of *Barnettozyma* gen. nov., *Lindnera* gen. nov. and *Wickerhamomyces* gen. nov. *FEMS Yeast Res* **8**, 939–954 (2008).
45. Arroyo-López, F.N. et al. Yeasts in table olive processing: Desirable or spoilage microorganisms? *Int J Food Microbiol* **160**, 42–49 (2012).
46. Wu, Z.W., Robert, V., and Bai, F.Y. Genetic diversity of the *Pichia membranifaciens* strains revealed from rRNA gene sequencing and electrophoretic karyotyping, and the proposal of *Candida californica* comb. nov. *FEMS Yeast Res* **6**, 305–311 (2006).
47. Passoth, V. and Schnürer, J. Non-conventional yeasts in antifungal application. In *Functional Genetics of Industrial Yeasts*, Vol. 2, (ed. de Winde, J.H.) pp. 297–329 (Springer Verlag, Berlin, Germany, 2003).
48. Chen, L.S. et al. Screening for the potential probiotic yeast strains from raw milk to assimilate cholesterol. *Dairy Sci Technol* **90**, 537–548 (2010).
49. Fiori, S. et al. Identification of differentially expressed genes associated with changes in the morphology of *Pichia fermentans* on apple and peach fruit. *FEMS Yeast Res* **12**, 785–795 (2012).
50. Masoud, W. and Jespersen, L. Pectin degrading enzymes in yeasts involved in fermentation of *Coffea arabica* in East Africa. *Int J Food Microbiol* **110**, 291–296 (2006).
51. Daniel, H.M., Moons, M.C., Huret, S., Vrancken, G., and De Vuyst, L. *Wickerhamomyces anomalus* in the sourdough microbial ecosystem. *Antonie van Leeuwenhoek* **99**, 63–73 (2011).
52. Hellström, A.M., Almgren, A., Carlsson, N.G., Svanberg, U., and Andlid, T.A. Degradation of phytate by *Pichia kudriavzevii* TY13 and *Hanseniaspora guilliermondii* TY14 in Tanzanian togwa. *Int J Food Microbiol* **153**, 73–77 (2012).
53. Falagas, M.E., Roussos, N., and Vardakas, K.Z. Relative frequency of albicans and the various non-albicans *Candida* spp. among candidemia isolates from inpatients in various parts of the world: A systematic review. *Int J Infect Dis* **14**, e954–e966 (2010).
54. Kurtzman, C.P. New species and new combinations in the yeast genera *Kregervanrija* gen. nov., *Saturnispora* and *Candida*. *FEMS Yeast Res* **6**, 288–297 (2006).
55. Walker, H.W. and Ayres, J.C. Yeasts as spoilage organisms. In *The Yeasts*, Vol. 3, (eds. Rose, A.H. and Harrison, J.S.) pp. 463–527 (Academic Press, London, U.K., 1971).
56. Beech, F.W. *Pichia delftensis* sp. n. *Antonie van Leeuwenhoek* **31**, 81–83 (1965).
57. Lee, J.D. and Komagata, K. Further taxonomic study of methanol-assimilating yeasts with special references to electrophoretic comparison of enzymes. *J Gen Appl Microbiol* **29**, 395–416 (1983).
58. Shimotoyodome, A. et al. Reduction of saliva-promoted adhesion of *Streptococcus mutans* MT8148 and dental biofilm development by tragacanth gum and yeast-derived phosphomannan. *Biofouling* **22**, 261–268 (2006).
59. Gueguen, Y., Chemardin, P., Pien, S., Arnaud, A., and Galzy, P. Enhancement of aromatic quality of Muscat wine by the use of immobilized β-glucosidase. *J Biotechnol* **55**, 151–156 (1997).
60. Sanchez-Torres, P., Gonzalez-Candelas, L., and Ramon, D. Heterologous expression of a *Candida molischiana* anthocyanin-beta-glucosidase in a wine yeast strain. *J Agric Food Chem* **46**, 354–360 (1998).
61. Marples, M.J. and Somervil, D.A. Oral and cutaneous distribution of *Candida albicans* and other yeasts in Rarotonga, Cook Islands. *Trans R Soc Trop Med Hyg* **62**, 256–262 (1968).
62. Bar-Meir, M. et al. Catheter-related fungemia due to *Candida thermophila*. *J Clin Microbiol* **44**, 3035–3036 (2006).
63. Kurtzman, C.P. *Ogataea*. In *The Yeasts: A Taxonomic Study*, 5th edn., Vol. 2, (eds. Kurtzman, C.P., Fell, J.W., and Boekhout, T.) pp. 645–671 (Elsevier, Amsterdam, the Netherlands, 2011).
64. Ashenafi, M. Microbial flora and some chemical properties of Ersho, a starter culture for Teff (*Eragrostis tef*) fermentation. *World J Microbiol Biotechnol* **10**, 69–73 (1994).
65. Kurtzman, C.P. *Wickerhamomyces* Kurtzman, Robnett & Basehoar-Powers (2008). In *The Yeasts, a Taxonomic Study*, 5th edn., Vol. 2, (eds. Kurtzman, C.P., Fell, J.W., and Boekhout, T.) pp. 899–917 (Elsevier, Amsterdam, the Netherlands, 2011).
66. Passoth, V., Olstorpe, M., and Schnürer, J. Past, present and future research directions with *Pichia anomala*. *Antonie van Leeuwenhoek* **99**, 121–125 (2011).

67. Daniel, H.M., Redhead, S.A., Schnürer, J., Naumov, G.I., and Kurtzman, C.P. (2049–2050) Proposals to conserve the name *Wickerhamomyces* against *Hansenula* and to reject the name *Saccharomyces sphaericus* (Ascomycota: Saccharomycotina). *Taxon* **61**, 459–461 (2012).
68. Passoth, V., Fredlund, E., Druvefors, U.Ä., and Schnürer, J. Biotechnology, physiology and genetics of the yeast *Pichia anomala*. *FEMS Yeast Res* **6**, 3–13 (2006).
69. Walker, G.M. *Pichia anomala*: Cell physiology and biotechnology relative to other yeasts. *Antonie van Leeuwenhoek* **99**, 25–34 (2011).
70. Fredlund, E., Druvefors, U., Boysen, M.E., Lingsten, K.-J., and Schnürer, J. Physiological characteristics of the biocontrol yeast *Pichia anomala* J121. *FEMS Yeast Res* **2**, 395–402 (2002).
71. Timke, M., Wang-Lieu, N.Q., Altendorf, K., and Lipski, A. Identity, beer spoiling and biofilm forming potential of yeasts from beer bottling plant associated biofilms. *Antonie van Leeuwenhoek* **93**, 151–161 (2008).
72. Haïssam, J.M. *Pichia anomala* in biocontrol for apples: 20 years of fundamental research and practical applications. *Antonie van Leeuwenhoek* **99**, 93–105 (2011).
73. Polonelli, L., Magliani, W., Ciociola, T., Giovati, L., and Conti, S. From *Pichia anomala* killer toxin through killer antibodies to killer peptides for a comprehensive anti-infective strategy. *Antonie van Leeuwenhoek* **99**, 35–41 (2011).
74. Miceli, M.H., Diaz, J.A., and Lee, S.A. Emerging opportunistic yeast infections. *Lancet Infect Dis* **11**, 142–151 (2011).
75. Olstorpe, M., Schnürer, J., and Passoth, V. Growth inhibition of various Enterobacteriaceae species by the yeast *Hansenula anomala* during storage of moist cereal grain. *Appl Environ Microbiol* **78**, 292–294 (2012).
76. Olstorpe, M., Borling, J., Schnürer, J., and Passoth, V. *Pichia anomala* yeast improves feed hygiene during storage of moist crimped barley grain under Swedish farm conditions. *Anim Feed Sci Technol* **156**, 47–56 (2010).
77. Schneider, J. et al. Genome sequence of *Wickerhamomyces anomalus* DSM 6766 reveals genetic basis of biotechnologically important antimicrobial activities. *FEMS Yeast Res* **12**, 382–386 (2012).
78. Etchells, J.L. and Bell, T.A. Film yeasts on commercial cucumber brines. *Food Technol* **4**, 77–83 (1950).
79. Schwan, R.F., Mendonca, A.T., da Silva, J.J., Rodrigues, V., and Wheals, A.E. Microbiology and physiology of Cachaca (Aguardente) fermentations. *Antonie van Leeuwenhoek* **79**, 89–96 (2001).
80. Hou, C.T. pH-dependence and thermostability of lipases from cultures from the ARS culture collection. *J Ind Microbiol* **13**, 242–248 (1994).
81. Espinel-Ingroff, A. et al. Comparison of RapID yeast plus system with API 20C system for identification of common, new, and emerging yeast pathogens. *J Clin Microbiol* **36**, 883–886 (1998).
82. Silva, C.F., Schwan, R.F., Dias, E.S., and Wheals, A.E. Microbial diversity during maturation and natural processing of coffee cherries of *Coffea arabica* in Brazil. *Int J Food Microbiol* **60**, 251–260 (2000).
83. Mushtaq, M., Faiza, I., and Sharfun, N. Detection of yeast mycoflora from butter. *Pak. J Bot.* **39**, 887–896 (2007).
84. Minter, D.W. *Cyberlindnera*, a replacement name for *Lindnera* Kurtzman et al., nom. illegit. *Mycotaxon* **110**, 473–476 (2009).
85. Kurtzman, C.P. *Lindnera* Kurtzman, Robnett & Basehoar-Powers (2008). In *The Yeasts: A Taxonomic Study*, 5th edn., Vol. 2, (eds. Kurtzman, C.P., Fell, J.W., and Boekhout, T.) pp. 521–543 (Elsevier, Amsterdam, the Netherlands, 2011).
86. Kato, M., Iefuji, H., Miyake, K., and Iimura, Y. Transformation system for a wastewater treatment yeast, *Hansenula fabianii* J640: Isolation of the orotidine-5′-phosphate decarboxylase gene (*URA3*) and uracil auxotrophic mutants. *Appl Microbiol Biotechnol* **48**, 621–625 (1997).
87. Sato, S., Otani, M., Shimoi, H., Saito, K., and Tadenuma, M. Selection of flocculent mutants of yeasts for waste water treatment. Utilization of wild yeasts (part VIII). *J Brew Soc Jpn* **82**, 515–519 (1987).
88. Thaysen, A.C. Value of micro-organisms in nutrition (food yeast). *Nature* **151**, 406–408 (1943).
89. Bekatorou, A., Psarianos, C., and Koutinas, A.A. Production of food grade yeasts. *Food Technol Biotechnol* **44**, 407–415 (2006).
90. Sartory, A., Sartory, R., Weill, J., and Meyer, J. Un cas de blastomycose inveteree transmissible an cobaye, due a tin *Saccharomyces* pathogene (*Saccharomyces jadini* n. sp.). *Compte Rendu Hebdomadaire des Seances de l'Academie des Sciences* **194**, 1688–1700 (1932).
91. Tomita, Y., Ikeo, K., Tamakawa, H., Gojobori, T., and Ikushima, S. Genome and transcriptome analysis of the food-yeast *Candida utilis*. *PLoS One* **7**, e37226 (2012).
92. Coton, E., Coton, M., Levert, D., Casaregola, S., and Sohier, D. Yeast ecology in French cider and black olive natural fermentations. *Int J Food Microbiol* **108**, 130–135 (2006).

93. Lowes, K.F. et al. Prevention of yeast spoilage in feed and food by the yeast mycocin HMK. *Appl Environ Microbiol* **66**, 1066–1076 (2000).
94. Ciafardini, G., Zullo, B.A., Cloccia, G., and Iride, A. Lipolytic activity of *Williopsis californica* and *Saccharomyces cerevisiae* in extra virgin olive oil. *Int J Food Microbiol* **107**, 27–32 (2006).
95. Wang, H., Xu, Z., Gao, L., and Hao, B.L. A fungal phylogeny based on 82 complete genomes using the composition vector method. *BMC Evol Biol* **9**, 195 (2009).
96. Santos, M.A.S., Gomes, A.C., Santos, M.C., Carreto, L.C., and Moura, G.R. The genetic code of the fungal CTG clade. *Comptes Rendus Biol* **334**, 607–611 (2011).
97. Hayman, G.T. and Bolen, P.L. Linear DNA plasmids of *Pichia inositovora* are associated with a novel killer toxin activity. *Curr Genet* **19**, 389–393 (1991).
98. Kurtzman, C.P. *Yamadazyma* Billon-Grand (1989). In *The Yeasts: A Taxonomic Study*, 5th edn., Vol. 2, (eds. Kurtzman, C.P., Fell, J.W., and Boekhout, T.) pp. 919–925 (Elsevier, Amsterdam, the Netherlands, 2011).
99. Miranda, M., Holzschu, D.L., Phaff, H.J., and Starmer, W.T. *Pichia mexicana*, a new heterothallic yeast from cereoid cacti in the North-American Sonoran desert. *Int J Syst Bacteriol* **32**, 101–107 (1982).
100. Kato, S. et al. Molecular cloning and characterization of an alpha-amylase from *Pichia burtonii* 15-1. *Biosci Biotechnol Biochem* **71**, 3007–3013 (2007).
101. Kurtzman, C.P., Wickerham, L.J., and Hesseltine, C.W. Yeasts from wheat and flour. *Mycologia* **62**, 542–547 (1970).
102. Slavikova, E. and Vadkertiova, R. Yeasts and yeast-like organisms isolated from fish-pond waters. *Acta Microbiol Polonica* **44**, 181–189 (1995).
103. Simoncini, N., Rotelli, D., Virgili, R., and Quintavalla, S. Dynamics and characterization of yeasts during ripening of typical Italian dry-cured ham. *Food Microbiol* **24**, 577–584 (2007).
104. Criseo, G., Gallo, M., and Pernice, A. Killer activity at different pHs against *Cryptococcus neoformans* var. *neoformans* serotype A by environmental yeast isolates. *Mycoses* **42**, 601–608 (1999).
105. Ragaert, P. et al. Metabolite production of yeasts on a strawberry-agar during storage at 7 degrees C in air and low oxygen atmosphere. *Food Microbiol* **23**, 154–161 (2006).
106. Paula, C.R., Sampaio, M.C.C., Birman, E.G., and Siqueira, A.M. Oral yeasts in patients with cancer of the mouth, before and during radiotherapy. *Mycopathologia* **112**, 119–124 (1990).
107. Adler, A. et al. *Pichia farinosa* bloodstream infection in a lymphoma patient. *J Clin Microbiol* **45**, 3456–3458 (2007).
108. Schwan, R.F. and Wheals, A.E. The microbiology of cocoa fermentation and its role in chocolate quality. *Crit Rev Food Sci* **44**, 205–221 (2004).
109. Druvefors, U.Ä. and Schnürer, J. Mold-inhibitory activity of different yeast species during airtight storage of wheat grain. *FEMS Yeast Res* **5**, 373–378 (2005).
110. Mallet, S. et al. Insights into the life cycle of yeasts from the CTG clade revealed by the analysis of the *Millerozyma* (*Pichia*) *farinosa* species complex. *PLoS One* **7**, e35842 (2012).
111. de Montiguy, J. et al. Genomic exploration of the hemiascomycetous yeasts: 15. *Pichia sorbitophila*. *FEBS Lett* **487**, 87–90 (2000).
112. Suh, S.O. and Blackwell, M. Three new beetle-associated yeast species in the *Pichia guilliermondii* clade. *FEMS Yeast Res* **5**, 87–95 (2004).
113. Abbas, C.A. and Sibirny, A.A. Genetic control of biosynthesis and transport of riboflavin and flavin nucleotides and construction of robust biotechnological producers. *Microbiol Mol Biol Rev* **75**, 321–360 (2011).
114. Mussatto, S.I., Silva, C., and Roberto, I.C. Fermentation performance of *Candida guilliermondii* for xylitol production on single and mixed substrate media. *Appl Microbiol Biotechnol* **72**, 681–686 (2006).
115. Duarte, W.F., Amorim, J.C., and Schwan, R.F. The effects of co-culturing non-*Saccharomyces* yeasts with *S. cerevisiae* on the sugar cane spirit (cachaca) fermentation process. *Antonie van Leeuwenhoek* **103**, 175–194 (2013).
116. Jeffries, T.W. et al. Genome sequence of the lignocellulose-bioconverting and xylose-fermenting yeast *Pichia stipitis*. *Nat Biotechnol* **25**, 319–326 (2007).
117. Hahn-Hägerdal, B., Karhumaa, K., Fonseca, C., Spencer-Martins, I., and Gorwa-Grauslund, M.F. Towards industrial pentose-fermenting yeast strains. *Appl Microbiol Biotechnol* **74**, 937–953 (2007).
118. Passoth, V. et al. Enhanced ethanol production from wheat straw by integrated storage and pre-treatment (ISP). *Enzyme Microb Technol* **52**, 105–110 (2013).
119. Klinner, U., Fluthgraf, S., Freese, S., and Passoth, V. Aerobic induction of respiro-fermentative growth by decreasing oxygen tensions in the respiratory yeast *Pichia stipitis*. *Appl Microbiol Biotechnol* **67**, 247–253 (2005).

120. Passoth, V., Cohn, M., Schäfer, B., Hahn-Hägerdal, B., and Klinner, U. Analysis of the hypoxia-induced *ADH2* promoter of the respiratory yeast *Pichia stipitis* reveals a new mechanism for sensing of oxygen limitation in yeast. *Yeast* **20**, 39–51 (2003).
121. Gillings, M.R., Holley, M.P., and Selleck, M. Molecular identification of species comprising an unusual biofilm from a groundwater treatment plant. *Biofilms* **3**, 19–24 (2006).
122. Nakamura, Y., Fukuhara, H., and Sano, K. Secreted phytase activities of yeasts. *Biosci Biotechnol Biochem* **64**, 841–844 (2000).
123. Al-Sweih, N. et al. Kodamaea ohmeri as an emerging pathogen: A case report and review of the literature. *Med Mycol* **49**, 766–770 (2011).
124. Brandt, M.E. et al. Fungemia caused by *Zygoascus hellenicus* in an allogeneic stem cell transplant recipient. *J Clin Microbiol* **42**, 3363–3365 (2004).
125. Smith, M.T., Robert, V., Poot, G.A., Epping, W., and de Cock, A. Taxonomy and phylogeny of the ascomycetous yeast genus *Zygoascus*, with proposal of *Zygoascus meyerae* sp. nov. and related anamorphic varieties. *Int J Syst Evol Microbiol* **55**, 1353–1363 (2005).
126. Böer, E., Steinborn, G., Kunze, G., and Gellissen, G. Yeast expression platforms. *Appl Microbiol Biotechnol* **77**, 513–523 (2007).
127. Saraya, R. et al. Novel genetic tools for *Hansenula polymorpha*. *FEMS Yeast Res* **12**, 271–278 (2012).
128. van der Heide, M., Veenhuis, M., and van der Klei, I.J. The methylotrophic yeasts *Hansenula polymorpha* and *Pichia pastoris*: Favourable cell factories in various applications. In *Functional Genetics of Industrial Yeasts*, Vol. 2, (ed. de Winde, J.H.) pp. 207–225 (Springer, Berlin, Germany, 2003).
129. Steinborn, G. et al. Application of a wide-range yeast vector (CoMed) system to recombinant protein production in dimorphic *Arxula adeninivorans*, methylotrophic *Hansenula polymorpha* and other yeasts. *Microb Cell Fact* **5**, 33 (2006).
130. Ikushima, S., Fujii, T., and Kobayashi, O. Efficient gene disruption in the high-ploidy yeast *Candida utilis* using the Cre-*loxP* system. *Biosci Biotechnol Biochem* **73**, 879–884 (2009).
131. Lu, P., Davis, B.P., Hendrick, J., and Jeffries, T.W. Cloning and disruption of the beta-isopropylmalate dehydrogenase gene (*LEU2*) of *Pichia stipitis* with *URA3* and recovery of the double auxotroph. *Appl Microbiol Biotechnol* **49**, 141–146 (1998).
132. Piontek, M., Hagedorn, J., Hollenberg, C.P., Gellissen, G., and Strasser, A.W.M. Two novel gene expression systems based on the yeasts *Schwanniomyces occidentalis* and *Pichia stipitis*. *Appl Microbiol Biotechnol* **50**, 331–338 (1998).
133. Maassen, N., Freese, S., Schruff, B., Passoth, V., and Klinner, U. Nonhomologous end joining and homologous recombination DNA repair pathways in integration mutagenesis in the xylose-fermenting yeast *Pichia stipitis*. *FEMS Yeast Res* **8**, 735–743 (2008).
134. Laplaza, J.M., Torres, B.R., Jin, Y.S., and Jeffries, T.W. Sh ble and Cre adapted for functional genomics and metabolic engineering of *Pichia stipitis*. *Enzyme Microb Technol* **38**, 741–747 (2006).
135. Wang, X.X., Li, G., Deng, Y.T., Yu, X.W., and Chen, F. A site-directed integration system for the non-universal CUG^{Ser} codon usage species *Pichia farinosa* by electroporation. *Arch Microbiol* **184**, 419–424 (2006).
136. Jones, K.L. and Jones, S.E. Fermentations involved in the production of cocoa, coffee and tea. *Prog Ind Microbiol* **19**, 411–456 (1984).
137. Silva, C.F. et al. Evaluation of a potential starter culture for enhance quality of coffee fermentation. *World J Microbiol Biotechnol* (2012).
138. Jespersen, L., Nielsen, D.S., Honholt, S., and Jakobsen, M. Occurrence and diversity of yeasts involved in fermentation of West African cocoa beans. *FEMS Yeast Res* **5**, 441–453 (2005).
139. Arroyo-Lopez, F.N., Querol, A., Bautista-Gallego, J., and Garrido-Fernandez, A. Role of yeasts in table olive production. *Int J Food Microbiol* **128**, 189–196 (2008).
140. Calderone, R.A. and Fonzi, W.A. Virulence factors of *Candida albicans*. *Trends Microbiol* **9**, 327–335 (2001).

29 Medically Important *Rhodotorula* Species

Sahar Yazdani, Audrey N. Schuetz, Ruta Petraitiene, Malcolm D. Richardson, and Thomas J. Walsh

CONTENTS

29.1 Epidemiology 483
29.2 Clinical Manifestations 487
    29.2.1 Central Venous Catheter–Associated Fungemia and Sepsis 487
        29.2.1.1 Rhodotorula in Preterm Neonates 488
        29.2.1.2 Fungemia after Transplantation 488
        29.2.1.3 Rhodotorula Scleritis and Keratitis 488
    29.2.2 Lymphadenitis 488
    29.2.3 Meningitis 488
    29.2.4 Postoperative Infection 488
    29.2.5 CAPD-Related Peritonitis 488
    29.2.6 Peritonitis in a Liver Transplant Recipient 488
29.3 Laboratory Detection and Diagnosis 489
    29.3.1 Identification Methods 489
        29.3.1.1 Conventional Identification 489
    29.3.2 Molecular Identification 489
29.4 Treatment 490
References 491

29.1 EPIDEMIOLOGY

Rhodotorula is a member of basidiomycetous yeasts, which produces mucoid colonies with a characteristic carotenoid pigment ranging from yellow to red. *Rhodotorula* is a heterogeneous group of yeasts with currently more than 50 species described in the literature. The most frequent human pathogen is *Rhodotorula mucilaginosa* (formerly *Rhodotorula rubra*) followed by *Rhodotorula glutinis* and *Rhodotorula minuta*. *Rhodotorula* species have emerged as opportunistic pathogens that have the ability to colonize and infect susceptible patients. Studies have demonstrated that the incidence of fungemia caused by *Rhodotorula* spp. was between 0.5% and 2.3% in the United States and Europe [1].

    R. mucilaginosa is commonly isolated from foods and beverages. This ubiquitous yeast has a strong affinity for plastic, having been isolated from various kinds of medical equipment, such as dialysis equipment, fiberoptic bronchoscopes, and other environmental sources, including shower curtains, bathtubs, and toothbrushes [2,3]. In humans, this yeast has been isolated from the skin, nails, respiratory tract, gastrointestinal tract, and genital tract, usually representing harmless colonization.

    Isolation of this fungus from different ecosystems, including sites with unfavorable conditions, has been described, such as the depths of the Baltic Sea, the high-altitude Lake Patagonia, the soil and vegetation of Antarctica and aquatic, hypersaline and high-temperature environments such as the Dead Sea (Israel), Lake Enriquilli (Dominican Republic), the Great Salt Lake (United States), and beaches located in northern Brazil. Some studies have reported the occurrence of *Rhodotorula* spp. in marine waters polluted by household waste [4,5] (Table 29.1).

Compared to *R. mucilaginosa*, *R. glutinis* and *R. minuta* are less frequently isolated in natural environments. These species have been detected in air, seawater (including deep environments), freshwater, and goat's milk [6,7]. Environmental studies have documented the presence of *Rhodotorula* spp. in tropical fruits, sugarcane, and shrimp in the waters of Sepetiba Bay in Brazil [8,9]. Contamination of food by *Rhodotorula* spp. that is provided to immunocompromised patients in hospitals has been reported. As a direct consequence of the wide exposure to *Rhodotorula* in the hospital environment, patients who have a depressed immune system can develop *Rhodotorula* infection, causing a variety of systemic infections. Indeed, *Rhodotorula* spp. are the most common microorganism isolated from the hands of hospital employees and patients [10]. It has also been isolated from stool samples, indicating that these yeasts can survive in the extreme conditions of the gastrointestinal tract, but it is still unclear whether it is capable of translocating from the gastrointestinal tract into the bloodstream [11].

Regarding the pathogenicity of *Rhodotorula* spp. in animals, there are reports of an outbreak of skin infections in chickens and a report of a lung infection in sheep, both caused by *R. mucilaginosa* [12,13]. Also, it has been found to be the causative agent of epididymitis and cutaneous lesions in a sea lion and dermatitis in a cat that had crusted lesions and mastitis. Some authors have reported the *Rhodotorula* genus as a colonizing agent in the oropharynx and cloaca of ostriches, in fecal samples and the cloaca of wild birds and pigeons in urban and suburban areas, in the ear canals of adult cattle with parasitic otitis, in healthy rhesus monkeys, and the genital tract of healthy female camels [14,15]. Animal models have been used to study the mechanisms of pathogenesis of different human fungal diseases. The low pathogenicity of *Rhodotorula* spp. is probably related to its reduced ability to grow at 37°C, an attribute typically enhancing virulence of pathogenic strains. Animal experimentation is a valuable tool to study the pathogenesis of unusual human mycosis, such as *Rhodotorula* infection. The first experimental model of disseminated *Rhodotorula* infection was described in literature by some authors along with the histopathologic aspects of the infection. The results showed that the most affected organs by *R. mucilaginosa* were the lungs, spleen, and especially the liver, which was severely infected. Considering the animals were highly immunocompromised, histopathology of the involved affected organs revealed few epithelioid cells and multinuclear giant cells in association with abundant yeast forms with occasional granuloma formation [16].

Most cases of infection are associated with central vascular catheters in patients with hematological malignancies [17–19]. Localized infections without fungemia including endophthalmitis, onychomycosis, meningitis, prosthetic joint infections, and peritonitis (usually associated with continuous peritoneal dialysis) have been reported in immunocompromised and immunocompetent patients.

The first report of fungemia caused by *Rhodotorula* was made by Louria et al. in 1960 [17]. From 1970 until 1985, no cases of *Rhodotorula* infection were reported in hematological patients, but the number of cases of *Rhodotorula* infection in these patients increased after 1985. In all cases listed, the patients were using central venous catheters (CVCs), short- or long-term continuous ambulatory peritoneal dialysis (CAPD) catheters, and umbilical vein catheters. The most prevalent species was *R. mucilaginosa*, followed by *R. glutinis*. Most of the patients had an underlying disease, such as congenital heart disease, AIDS, cancer, or chronic intestinal disease, and two patients were transplant recipients (one lung and one hematopoietic stem cell). Two patients were neutropenic at the time of the development of fungemia, and five patients were receiving parenteral nutrition. All the patients received antifungal treatment. The increase of *Rhodotorula* fungemia related to catheters was associated with an increase of more aggressive treatment modalities, which included intensive care unit admissions, use of CVCs, short- and long-term parenteral nutrition, broad-spectrum antibiotics, organ transplants, and chemotherapy [20].

TABLE 29.1
Environmental Microbiology of *Rhodotorula* Species

Species	Environmental Location
R. acheniorum	Strawberries *Fragaria vesca* in the United Kingdom; leaf of bottlebrush plant *Callistemon viminalis* in Australia
R. acuta	Sulphited grape must and sponge cake in Japan
R. araucariae	Rotting monkey puzzle tree *Araucaria araucana* in Chile, wood in Portugal
R. armeniaca	Leaf of bottlebrush plant *Callistemon viminalis* in Australia
R. aurantiaca	Atmosphere in Japan; soil and Bantu beer in South Africa; bottlebrush plant *Callistemon viminalis* in Australia; bark beetle *Dendroctonus jeffreyi* in *Pinus jeffreyi* in the United States; brine bath in cheese factory in the Netherlands; sea water; colostrum from women
R. auricularia	Fruiting body of the Jew's ear fungus *Auricularia auricula* in Japan
R. bacarum	Skin of dolphin in the Netherlands, black currants *Ribes nigrum in* the United Kingdom; contaminant on cornmeal agar plate; leaf of *Platanus* spp. in Portugal
R. bogoriensis	Leaves of *Randia malleifera* in Indonesia; leaf litter in Portugal
R. buffonii	On the fungus *Boletus edulis*
R. creatinovora	Permafrost soil in Siberia
R. cresolica	Soil contaminated with ortho-cresol in the Netherlands
R. diffluens	Leaves of the *Tillandsia usneoides* in Paramaribo
R. ferulica	Polluted river and stagnant water in Portugal
R. foliorum	Leaf of *Drymoglossum pilloselloides* in Indonesia; leaf litter in Portugal
R. fragaria	Strains of this species grow on propan-1-ol or butan-1-ol as sole source of carbon
R. fujisanensis	Droppings of hare and wild grapes *Vitis coignetiae* in Japan; leaves in France; dry leaf, root of tree, and wood of *Quercus* sp. in Portugal; leaf of *Arcacia* sp. in New Zealand
R. futronensis	Rotting *Laurelia sempervirens* in Chile
R. glutinis (syn *R. rubra* and 27 others)	Atmosphere in Poland; tree in Thailand; leaf of peach tree *Prunus persica* in Italy; river water in Kazakhstan; atmosphere and soil in Japan; spoiled leather in France; leaf of *Desmodium repens* and garden soil in the Netherlands; sea water from Biscayne Bay and soil in the United States; fruit of *Phyllodendron* sp. in Italy; boracic lotion in Indonesia; water supply of brewery in Germany; sputum from case of bronchopneumonia
R. graminis	Leaf of *Citrus* sp. in Indonesia; grass in New Zealand; Atlantic ocean off the Bahamas; atmosphere in Japan; baneberry *Actaea spicata* in Canada; cacao
R. hinnulea	*Banksia collina* in Australia
R. hordea	Leaf of barley *Hordei hexastichi*
R. hylophila	Tunnels of pine-borer beetles *Xyleborus aemulus* in South Africa
R. ingeniosa	Grass in New Zealand; water of Pike Lake in Canada
R. javanica	Leaves of *Ixora coccinea* on Java, Indonesia
R. lactose	Atmosphere in Japan
R. lignophila	Rotten wood of *Drimys winteri* in Chile
R. marina	Shrimp from Gulf of Mexico off Texas
R. minuta	Atmosphere in Japan; sputum in the Netherlands; throat swabs in Norway; sputum in Germany; cucumbers pickled in 10% brine; sea water off Florida and shrimp from Gulf of Mexico off Texas in the United States; throat swab and feces of child in Hungary; gut of sheep in France; sea water off Sweden; decaying tooth in South Africa; blood from patient with enlarged lymph nodes; mycotic nodule in white rat; oat leaves; lorry
R. mucilaginosa	Trees in Thailand; exudate of aspen *Populus tremuloides* and water of Pike Lake in Canada; soil in Spain; rotten *Eucryphia cordifolia* in Chile; squash *Curcurbita melopepo* in the Philippines; case of emphysema; hair and lymph nodes in the Netherlands, red ibis *Guara rubra* in France; fermenting Kentucky tobacco and lung abscess in Italy; water of Rio Moca; larva of fruit fly *Drosophila pilimanae* on Hawaii; boracic lotion in Indonesia; sediment from Biscayne Bay in the United States; hair in France; atmosphere on Bonaire; culture contaminant and feces in Dominican Republic; atmosphere in Japan; pasteurized beer in Germany; water in Belgium; nail of a 12-year-old girl in Austria; ulcer in Madagascar; jam in France; malt syrup; powdered Spanish red pepper *Capsicum frutescens*; *Tremella mesenterica*

(Continued)

TABLE 29.1 (Continued)
Environmental Microbiology of *Rhodotorula* Species

Species	Environmental Location
R. muscorum	Decaying sphagnum moss in New Zealand
R. nothofagi	Rotten *Nothofagus oblique* in Chile
R. philyla	Wood in Portugal; unknown source from Thailand; tunnel of *Xyleborus ferrugeineum* in South Africa
R. phylloplana	Leaf of *Banksia collina* in Australia
R. pilatii	Litter of conifer *Abies alba* in France; rotten tree trunk in Portugal
R. pustule	Black currents *Ribes nigrum* in the United Kingdom
R. sonckii	Bathing pool and arm of patient in Finland
R. vanillica	Stagnant pond in Portugal
R. yakutica	Permafrost soil in Siberia

Source: Compiled from Barnett, J.A. et al., *Yeasts: Characteristics and Identification,* 3rd edn.

Four cases of CVC-related fungemia by *R. mucilaginosa* in a neonatal intensive care unit (NICU) in an Italian hospital were reported by Perniola et al., which showed that all newborns infected with fungemia were low birth weight. Three had bacteremia prior to fungemia. All four neonates had venous access since birth. Blood cultures performed at the end of antifungal therapy were negative [21].

Another retrospective study reviewed the demographics, risk factors, treatment, and outcome of seven patients with *Rhodotorula* fungemia over the years from 2002 to 2005 in a Brazilian hospital [1]. Risk factors included solid and hematologic malignancies in patients who were receiving corticosteroids and cytotoxic drugs, the presence of CVCs, and the use of broad-spectrum antibiotics. The result was favorable for patients who had just had the CVC removed. Duboc de Almeida et al. described 25 cases of fungemia by *R. mucilaginosa* [22]. The majority of the patients had a CVC, and 10% had undergone one hematopoietic stem cell transplantation (HSCT).

A recent literature review published in 2010 by Garcia-Suarez et al. analyzed 29 cases of *Rhodotorula* fungemia in patients with hematological disorders [20]. This study showed that 100% of patients who developed fungemia by *Rhodotorula* had some form of central venous access, such as a Hickman catheter.

The most common underlying condition associated with *Rhodotorula* infection was the use of CVCs. Unlike fungemia, some of the other infections caused by *Rhodotorula* were not necessarily linked to the use of CVCs or an underlying disease. Meningitis and endophthalmitis by *Rhodotorula* species have been reported as nosocomial infections especially in HIV-infected persons. Prosthetic joint infections caused by *Rhodotorula* also have been reported in an HIV-infected patient. Peritonitis caused by *Rhodotorula* species also has been reported in patients undergoing CAPD. Most of these patients were successfully treated with amphotericin B and fluconazole following peritoneal dialysis catheter removal.

A case report showed onychomycosis caused by *R. mucilaginosa* in an immunocompetent patient. In addition to aortic homograft endocarditis, dermatitis, oral ulcers, and lymphadenitis caused by *Rhodotorula* species have been reported in the literature. *R. mucilaginosa* rarely causes keratitis in immunocompromised individuals.

The literature of the epidemiology of infection caused by *Rhodotorula* spp. should be interpreted with caution as there are more than 50 species described in the literature with changes occurring frequently as the result of advances in molecular taxonomy [23–25]. For example, the most frequently reported human pathogen among the *Rhodotorula* spp. is *R. rubra*. However, due to improved understanding of the *Rhodotorula* taxon, *R. rubra* has been changed in nomenclature to *R. mucilaginosa*.

29.2 CLINICAL MANIFESTATIONS

Rhodotorula species must be considered as a potential pathogen in patients with immunosuppression and CVCs. The most frequent infections caused by these yeasts are bloodstream infections, which occur mainly in immunosuppressed patients and in the presence of CVCs [2,16,18,19,26]. To date, most infections caused by *Rhodotorula* species have been associated with intravenous catheters and with patients who have solid tumors, lymphoproliferative diseases, chronic renal failure, diabetes, endocarditis, pulmonary diseases, and AIDS [2,16,18,19,26].

R. mucilaginosa has been reported to cause fungemia, sepsis, endocarditis, meningitis, ventriculitis, endophthalmitis, keratitis, oral ulcers, chronic dacryocystitis, and peritonitis in immunocompromised hosts [18,27–35]. Associated with the widespread use of broad-spectrum antibiotics and corticosteroids in many chronic diseases and with prolonged survival in chronically debilitated patients, there has been a widening awareness of pathogenic fungi. *Rhodotorula* commonly cultured from skin and excreta of man and recovered from many sources in environment can manifest itself in various clinical forms.

29.2.1 CENTRAL VENOUS CATHETER–ASSOCIATED FUNGEMIA AND SEPSIS

Most of the cases of infection caused by *Rhodotorula* in humans are fungemia associated with CVC use. *R. mucilaginosa* was responsible for most cases followed by *R. glutinis*. The major putative risk factors were prolonged use of CVCs in patients with hematological and solid malignancies, who are taking corticosteroids or cytotoxic drugs [18]. Some studies have shown a correlation between *Rhodotorula* fungemia and the duration of catheter use [27,28]. Previous antibiotic use, parenteral nutrition, and abdominal surgery were also found to be risk factors of *Rhodotorula* fungemia. These factors might explain why most of the reported cases also entailed intensive care unit admission [2,19,29]. Even immunocompetent hosts are susceptible to these infections during long-term venous catheterization and broad-spectrum antibiotic therapy.

In 1960, a case was reported by Louria et al., which documented fungemia caused by *Rhodotorula* spp. manifested by fever and hypotension in a patient with subacute bacterial endocarditis being treated with penicillin and streptomycin [17]. At postmortem examination, *Rhodotorula* was readily cultured from a scarred aortic valve. Fifty-four years later, a case involving a patient who developed bioprosthetic aortic valve endocarditis due to *R. mucilaginosa* and *Staphylococcus epidermidis*, proven by culture and histopathology was reported from the same institution underscoring the continuing diagnostic and therapeutic challenges of this pathogen [30].

According to a retrospective study conducted at Hospital das Clinicas, a 2200-bed tertiary teaching complex, located in the city of Sao Paulo, Brazil [22], from January 1, 1996, to December 31, 2004, *Rhodotorula* spp. were isolated from blood cultures of 28 patients who were predominantly adults with a median age of 43 years (range, 22 days to 70 years). Seven of these (28%) patients were younger than 15 years of age including a newborn. The majority of patients had cancer as the underlying disease (72%). Of these, 94.4% had a diagnosis of hematological malignancy with non-Hodgkin's lymphoma being the most frequent type. CVC was present in 88% of the patients. The average time that the catheter was inserted prior to isolation of *Rhodotorula* in blood culture was 86.5 days (range, 4–261 days). Neutropenia was present in 3 patients and 11 patients were on previous antibiotic therapy: the third generation cephalosporin and glycopeptides were the most commonly used drugs. The mean duration of antibiotic therapy was 9.5 days. Notably, 40% of patients were HSCT recipients, which heretofore has been considered a rare complication in this population. Other relevant findings included the use of broad-spectrum antibiotics, exposure to steroids, other immunosuppressive agents, and chemotherapy in approximately half of the patients. The use of antibiotics and exposure to cytotoxic agents probably contributed to the increase in gastrointestinal colonization.

29.2.1.1 Rhodotorula in Preterm Neonates

Rhodotorula fungemia has been diagnosed in preterm neonates in the NICU who presented with respiratory failure and sepsis [21].

29.2.1.2 Fungemia after Transplantation

Rhodotorula species have been increasingly recognized as emerging pathogens particularly in immunocompromised patients after HSCT, as well as liver and kidney transplants.

29.2.1.3 Rhodotorula Scleritis and Keratitis

In several cases, exploration, pus aspiration, and culture from the scleral nodule showed the presence of yeast that was identified as *Rhodotorula* [31]. A 30-year-old male with history of minor trauma presented with cotton wool–like stromal infiltration and hypopyon in the left eye. Molecular identification of the organism was performed, which showed 100% homology with *R. mucilaginosa*.

29.2.2 Lymphadenitis

R. mucilaginosa was cultured from the lymph nodes of an HIV-infected patient with the diagnosis confirmed microbiologically and clinically by responses of lymphadenitis to antifungal therapy [32].

29.2.3 Meningitis

Rhodotorula has been reported as a causative agent for meningitis in an immunocompromised HIV-infected patient who presented with long-standing fever. Gram staining and culture of cerebrospinal fluid grew *R. glutinis* [27].

29.2.4 Postoperative Infection

R. mucilaginosa can cause postoperative infections especially in cases of nonunion fracture femur. The patient underwent repeated surgical debridement and received intensive antibiotic therapy before the diagnosis was made [36].

29.2.5 CAPD-Related Peritonitis

Cases have been reported showing *R. mucilaginosa* as a causative agent for CAPD-related peritonitis [37–41]. A 51-year-old female patient with a 2-year history of end-stage renal disease had been on CAPD treatment for 12 months when she was admitted to the hospital with chief complaints of abdominal pain, nausea, vomiting, and cloudy dialysate for 2 days. The exit site and tunnel of the CAPD catheter were found to be normal. Gram stain of the peritoneal fluid did not show any microorganisms. She was empirically started on an antibiotic regimen consisting of cefazolin and amikacin, but the patient's clinical status did not improve. On the eighth day after her admission to the hospital, yeast growing in the peritoneal fluid culture was identified as *R. mucilaginosa*. The catheter was removed immediately and intravenous antifungal therapy with amphotericin B was started. If CAPD-related peritonitis does not respond to standard antibiotic treatment, fungal peritonitis should be considered.

29.2.6 Peritonitis in a Liver Transplant Recipient

Rhodotorula has been cultured from ascitic fluid of patients posttransplantation after having fever for a prolonged period of time.

29.3 LABORATORY DETECTION AND DIAGNOSIS

29.3.1 Identification Methods

29.3.1.1 Conventional Identification

Colonies of *Rhodotorula* spp. are usually pink, reddish, or salmon colored due to production of carotenoid pigment and appear moist to slimy [42]. Variants of *Rhodotorula* can appear yellowish in color, and some strains may be dry. *Rhodotorula* spp. do not grow in the presence of cycloheximide. Microscopic morphology on cornmeal agar with Tween 80 demonstrates spherical to ellipsoidal cells with narrow budding, when present. Rudimentary pseudohypha may be present in *R. mucilaginosa* or *R. glutinis* but are often difficult to appreciate [43]. *Rhodotorula* spp. do not ferment sugars but are positive for urease by 4 days at 25°C incubation [44]. They assimilate glucose and sucrose. *R. mucilaginosa* will grow at 37°C, while *R. glutinis* and *R. minuta* will not. Other biochemical reactions that aid in speciation include assimilation of potassium nitrate (positive) for *R. glutinis* (negative for *R. mucilaginosa* and *R. minuta*) and assimilation of maltose (positive) for *R. glutinis* (variable for *R. mucilaginosa* and negative for *R. minuta*).

Ballistoconidia are not produced by *Rhodotorula* spp., as opposed to the other red-pigmented yeast, *Sporobolomyces*. In tissue sections or spinal fluid, *Rhodotorula* may not be readily differentiated from *Cryptococcus* spp., which also forms budding cells of similar size [45]. The presence of capsule of *Cryptococcus* spp. can be discerned by special histological stains, such as mucicarmine and Alcian blue.

Conventional identification techniques that rely upon a combination of morphologic and biochemical patterns can be labor intensive and time consuming. In addition, differentiation among the species via such techniques can be imperfect, particularly within the *R. glutinis* complex, which has been often misidentified in the older literature [46]. Difficulties in identification occur when using conventional biochemical testing systems or commercial kits, with occasional misidentifications among species within the *Rhodotorula* genus and even occasionally misidentification of *Rhodotorula* as *Candida* species [47]. Additionally, commercial databases often do not include all *Rhodotorula* spp. When correctly performed, molecular identification provides a more definitive and accurate characterization of *Rhodotorula* species.

29.3.2 Molecular Identification

A variety of molecular methods have been used to identify yeast isolates resembling *Rhodotorula* spp. [48]. Currently, gene sequencing is the most robust and most commonly used technique for accurate molecular identification of *Rhodotorula* spp. DNA extraction can be performed by a rapid, simple method described by Sampaio et al., which is suitable for many PCR-based methods [42].

Sequencing can be performed of the highly discriminative rRNA gene in noncoding internal transcribed sequence 1 (ITS) and/or ITS2 regions or in coding D1/D2 variable domains of the large subunit rDNA [49,50]. The sequences can then be compared to reference sequence data available at the GenBank database by using basic local alignment search tool. Sequencing of the ITS regions may allow for correct identification of *R. mucilaginosa* or *R. minuta* [51]. Nunes and colleagues successfully sequenced 51 clinical and 8 environmental *Rhodotorula* isolates using ITS sequencing [47]. However, discrimination between species of the *R. glutinis* complex may be difficult when solely using ITS sequencing [46,52]. Linton and colleagues demonstrated use of sequencing of the D1/D2 region of the large ribosomal subunit gene for identification of various yeasts, including *Rhodotorula* spp. [52]. However, D1/D2 sequence analysis also does not provide enough resolution to discriminate between various species in the *R. glutinis* complex. One problem with use of the GenBank database includes the incorrect sequence data submitted as inappropriate species designations. Thus, having well-known sequence submissions for comparison is important [53].

Universal primers useful for rDNA sequencing include the following:

ITS region: ITS1 (forward): (5′-TCC GTA GGT GAA CCT GCG G-3′) and ITS4 (reverse): (5′-TCC TCC GCT TAT TGA TAT GC-3′)
ITS1 region: ITS1 (forward): (5′-TCC GTA GGT GAA CCT GCG G-3′) and ITS2 (reverse): (5′-GCT GCG TTC TTC ATC GAT GC-3′)
ITS2 region: ITS3 (forward): (5′-GCA TCG ATG AAG AAC GCA GC-3′) and ITS4 (reverse): (5′-TCC TCC GCT TAT TGA TAT GC-3′)
D1/D2 region: NL1 (forward): (5′-GCA TAT CAA TAA GCG GAG GAA AAG-3′) and NL4 (reverse): (5′-GGT CCG TGT TTC AAG ACG G-3′)

Species-specific oligonucleotide primers can also be used to identify *Rhodotorula* species and includes two universal external limiting primers, either D1/D2 region or ITS, as well as a species-specific internal primer. Primers that can be used include [54] the following oligonucleotides:

External universal primers NL1 (forward): 5′GCA TAT CAA TAA GCG GAG GAA AAG-3′ and NL4 (reverse): 5′-GGT CCG TGT TTC AAG ACG G-3′. Internal specific primer (forward) of the D1/D2 region: 5′-TCA GAC TTG CTT GCC GAG CAA TCG-3′.

Other molecular approaches to identify *Rhodotorula* isolates include random amplification of polymorphic DNA and micro/minisatellite-primed PCR, and restriction fragment length polymorphism analysis of the ITS region, but such techniques are generally limited to research laboratories as opposed to most clinical laboratories. At this point in time, rDNA sequence analysis offers the most rapid and accurate diagnostic tool for definitive molecular identification of *Rhodotorula* spp. for clinical laboratories.

29.4 TREATMENT

Rhodotorula species are capable of causing invasive infections, especially in patients with impaired host defenses and/or in the presence of CVC. *Rhodotorula* infection remains as infrequent cause of fungemia. Correct identification is mandatory as *Rhodotorula* species are resistant to antifungal agents such as fluconazole and echinocandins.

The optimal therapy for these infections remains to be defined. Based on previous studies and reports, a definitive treatment recommendation cannot be made. Variable treatment regimens with different results have been described in literature. In some cases, most of the patients were subjected to a combined approach (CVC removal and systemic antifungal therapy) with large variations related to treatment initiation, dose, and duration of therapy. While there are no evidence-based guidelines for the management of this uncommon infection, previous cases have been managed successfully with the removal of indwelling vascular catheters and the initiation of appropriate antifungal agents.

Susceptibility studies of *Rhodotorula* species have shown that all tested isolates were susceptible to amphotericin B (minimum inhibitory concentration [MIC] range, 0.125–1.6 μg/mL), itraconazole (MIC range, 0.25–12.8 μg/mL) and resistant to fluconazole (MICs of ≥32 μg/mL) [55]. Furthermore, it has poor activity of caspofungin and micafungin [10]. Because of high-level fluconazole resistance and high-itraconazole MICs, these drugs are not recommended as treatment options for *Rhodotorula* infections. In some cases, administration of fluconazole may be discontinued and amphotericin B deoxycholate instituted after confirming the fugal pathogen as *R. mucilaginosa* [18].

Rhodotorula species infections are frequently associated with the presence of CVCs and other implantable medical devices. These devices provide the necessary surfaces for biofilm formation and are currently responsible for a significant percentage of human infections caused by *Rhodotorula* spp. In contrast to the extensive literature dealing with *Candida* species biofilms, little attention has been paid to biofilms caused by emerging fungal pathogens such as *Rhodotorula* species.

Rhodotorula species are able to form biofilms, which may play a role in the pathogenesis of infections by this species [47]. This study evaluated the ability of different *Rhodotorula* species to form biofilm and the putative differences between clinical and environmental isolates. Using crystal violet (CV) staining, it was demonstrated that clinical isolates of *R. mucilaginosa* were better at forming biofilms than the environmental isolates. Moreover, *R. mucilaginosa* and *R. minuta* were the best biofilm producers. Overall, the CV staining was a useful method to indirectly measure bulk biofilm, and results of biofilm production were confirmed by SEM.

Multiple antifungal susceptibility profiles consistently demonstrate that the genus *Rhodotorula* is not a target for fluconazole or caspofungin [18,19,22,55–58]. Other triazole agents also showed poor activity (≥2 μg/mL) against the majority of the isolates tested. In contrast, amphotericin B demonstrated the best activity in vitro [59]. Overall, the results obtained here are in agreement with previous studies and show that it is more appropriate to read the susceptibility test of *Rhodotorula* species after 72 h of incubation.

In conclusion, the importance of molecular methods to correctly identify *Rhodotorula* spp. isolates and non-*R. mucilaginosa* spp. in particular should be emphasized. *R. mucilaginosa* was the most prevalent species among the clinical and environmental samples. All *Rhodotorula* species isolates tested are resistant to fluconazole and caspofungin but susceptible to amphotericin B, suggesting that it should be considered the antifungal drug of choice for the treatment of *Rhodotorula* species invasive infections.

REFERENCES

1. Lunardi LW, Aquino VR, Zimerman RA, Goldani LZ. Epidemiology and outcome of *Rhodotorula* fungemia in a tertiary care hospital. *Clin Infect Dis*. 2006;43(6):e60–e63.
2. Kiehn TE, Gorey E, Brown AE, Edwards FF, Armstrong D. Sepsis due to *Rhodotorula* related to use of indwelling central venous catheters. *Clin Infect Dis*. 1992;14(4):841–846.
3. Pfaller MA, Diekema DJ. Rare and emerging opportunistic fungal pathogens: Concern for resistance beyond *Candida albicans* and *Aspergillus fumigatus*. *J Clin Microbiol*. 2004;42(10):4419–4431.
4. Ekendahl S, O'Neill AH, Thomsson E, Pedersen K. Characterisation of yeasts isolated from deep igneous rock aquifers of the Fennoscadian Shield. *Microb Ecol*. 2003;46(4):416–428.
5. Hagler AN, Mendonça-Hagler LC. Yeasts from marine and estuarine waters with different levels of pollution in the state of Rio de Janeiro, Brazil. *Appl Environ Microbiol*. 1981;41(1):173–178.
6. Nagahama T, Hamamoto M, Horikoshi K. *Rhodotorula pacifica* sp. nov., a novel yeast species from sediment collected on the deep-sea floor of the north-west Pacific Ocean. *Int J Syst Evol Microbiol*. 2006;56(1):295–299.
7. Callon C, Duthoit F, Delbès C, Delbès C, Ferrand M, Le Frileux Y, De Crémoux R, Montel MC. Stability of microbial communities in goat milk during a lactation year: Molecular approaches. *Syst Appl Microbiol*. 2007;30(7):547–560.
8. Trindade RC, Resende MA, Silva CM, Rosa CA. Yeasts associated with fresh and frozen pulps of Brazilian tropical fruits. *Syst Appl Microbiol*. 2002;25(2):294–300.
9. Pagnocca FG, Mendonça-Hagler LC, Hagler AN. Yeasts associated with the white shrimp *Penaeus schmitti*, sediment, and water of Sepetiba Bay, Rio de Janeiro, Brasil. *Yeast*. 1989;5:S479–S483.
10. Strausbaugh LJ, Sewell DL, Tjoelker RC, Heitzman T, Webster T, Ward TT, Pfaller MA. Comparison of three methods for recovery of yeasts from hands of health-care workers. *J Clin Microbiol*. 1996;34(2):471–473.
11. Silva JO, Franceschini SA, Lavrador MAS, Candido RC. Performance of selective and differential media in the primary isolation of yeasts from different biological samples. *Mycopathologia*. 2004;157(1):29–36.
12. Aruo SK. Necrotizing cutaneous rhodotorulosis in chickens in Uganda. *Avian Dis*. 1980;24(4):1038–1043.
13. Monga DP, Garg DN. Ovine pulmonary infection caused by *Rhodotorula rubra*. *Mykosen*. 1980;23(4):208–211.
14. Melville PA, Cogliati B, Mangiaterra MBBCD et al. Determinação da microbiota presenten a cloaca e orofaringe de avestruzes (*Struthiocamelus*) clinicamente sadios. *Ciência Rural*. 2004;34(6):1871–1876.
15. Amaral RC, Ibanez JF, Mamizuka EM, Gambale W, de Paula CR, Larsson CE. Microbiota indígena do meato acústico externo de gatos hígidos. *Ciência Rural, Santa Maria*. 1998;28(3):4441–4445.
16. Anatoliotaki M, Mandatadakis E, Galanakis E, Samonis G. *Rhodotorula* species fungemia: A threat to the immunocompromised host. *Clin Lab*. 2003;49:49–55.

17. Louria DB, Greenberg SM, Molander DW. Fungemia caused by certain nonpathogenic strains of the family *Cryptococcaceae*. *N Engl J Med*. 1960;263:1281–1284.
18. Tuon FF, Costa SF. *Rhodotorula* infection. A systematic review of 128 cases from literature. *Rev Iberoam Micol*. 2008;25(3):135–140.
19. Zaas AK, Boyce M, Schell W, Lodge BA, Miller JL, Perfect JR. Risk of fungemia due to *Rhodotorula* and antifungal susceptibility testing of *Rhodotorula* isolates. *J Clin Microbiol*. 2003;41(11):5233–5235.
20. García-Suárez J, Gómez-Herruz P, Cuadros JA, Burgaleta C. Epidemiology and outcome of *Rhodotorula* infection in haematological patients. *Mycoses*. 2011;54(4):318–324.
21. Perniola R, Faneschi ML, Manso E, Pizzolante M, Rizzo A, Sticchi Damiani A, Longo R. *Rhodotorula mucilaginosa* outbreak in neonatal intensive care unit: Microbiological features, clinical presentation, and analysis of related variables. *Eur J Clin Microbiol Infect Dis*. 2006;25(3):193–196.
22. Duboc De Almeida GM, Costa SF, Melhem M. et al. *Rhodotorula* spp. isolated from blood cultures: Clinical and microbiological aspects. *Med Mycol*. 2008;46(6):547–556.
23. Biswas SK, Yokoyama K, Nishimura K, Miyaji M. Molecular phylogenetics of the genus *Rhodotorula* and related basidiomycetous yeasts inferred from the mitochondrial cytochrome b gene. *Int J Syst Evol Microbiol*. 2001;51:1191–1199.
24. CBS yeast database [database on the internet]. The Centraalbureau voor Schimmelcultures, Utrecht, The Netherlands. http://www.cbs.knaw.nl/collections/Biolomics.aspx?Table=Yeasts%202011, accessed April 8, 2015.
25. Fell JW, Statzell-Tallman A, Harrisson FC. Rhodotorula. In: Kurtzman CP, Fell JW (eds.). *The Yeasts, a Taxonomic Study*, 4th edn. Elsevier: New York.
26. Braun DK, Kauffman CA. *Rhodotorula* fungaemia: A life-threatening complication of indwelling central venous catheters. *Mycoses* 1992;35:305–308.
27. Shinde RS, Mantur BG, Patil G, Parande MV, Parande AM. Meningitis due to *Rhodotorula glutinis* in an HIV infected patient. *Indian J Med Microbiol*. 2008;26:375–377.
28. Villar JM, Velasco CG, Delgado JD. Fungemia due to *Rhodotorula mucilaginosa* in an immunocompetent, critically ill patient. *J Infect Chemother*. 2012;18:581–583.
29. Spiliopoulou A, Anastassiou ED, Christofidou M. *Rhodotorula fungemia* of an intensive care unit patient and review of published cases. *Mycopathologia*. 2012;174:301–309.
30. Simon MS, Somersan S, Singh HK, Hartman B, Jenkins SG, Walsh TJ, Schuetz AN. *Rhodotorula* endocarditis: Case report and review of the literature. *J Clin Microbiol*. 2014;52:374–378.
31. Pradhan ZS, Jacob P. Management of *Rhodotorula* scleritis. *Eye (Lond)*. 2012;26(12):1587.
32. Fung HB, Martyn CA, Shahidi A, Brown ST. *Rhodotorula mucilaginosa* lymphadenitis in an HIV-infected patient. *Int J Infect Dis*. 2009;13(1):e27–e29.
33. Alothman A. *Rhodotorula* species peritonitis in a liver transplant recipient: A case report. *Saudi J Kidney Dis Transpl*. 2006;17(1):47–49.
34. Duggal S, Jain H, Tyagi A, Sharma A, Chugh TD. *Rhodotorula* fungemia: Two cases and a brief review. *Med Mycol*. 2011;49(8):879–882.
35. Mori T, Nakamura Y, Kato J, Sugita K, Murata M, Kamei K, Okamoto S. Fungemia due to *Rhodotorula mucilaginosa* after allogeneic hematopoietic stem cell transplantation. *Transpl Infect Dis*. 2012;14(1):91–94.
36. Goyal R, Das S, Arora A, Aggarwal A. *Rhodotorula mucilaginosa* as a cause of persistent femoral nonunion. *J Postgrad Med*. 2008;54(1):25–27.
37. Unal A, Koc AN, Sipahioglu MH, Kavuncuoglu F, Tokgoz B, Buldu HM, Oymak O, Utas C. CAPD-related peritonitis caused by *Rhodotorula mucilaginosa*. *Perit Dial Int*. 2009;29(5):581–582.
38. Piraino B, Bailie GR, Bernardini J et al. Peritoneal dialysis-related infections recommendations: 2005 update. *Perit Dial Int*. 2005;25:107–131.
39. Szeto CC, Chow KM, Wong TYH et al. Feasibility of resuming peritoneal dialysis after severe peritonitis and Tenckhoff catheter removal. *J Am Soc Nephrol*. 2002;13:1040–1045.
40. Cox SD, Walsh SB, Yaqoob MM, Fan SLS. Predictors of survival and technique success after reinsertion of peritoneal dialysis catheter following severe peritonitis. *Perit Dial Int*. 2007;27:67–73.
41. Perez Fontan M, Rodriguez-Carmona A. Peritoneal catheter removal for severe peritonitis: landscape after a lost battle. *Perit Dial Int*. 2007;27:155–158.
42. Sampaio JP, Gadanho M, Santos S, Duarte FL, Pais C, Fonseca A, Fell JW. Polyphasic taxonomy of the basidiomycetous yeast genus *Rhodosporidium*: *Rhodosporidium kratochvilovae* and related anamorphic species. *Int J Syst Evol Microbiol*. 2001;51(Part 2):687–697.
43. Kwon-Chung KJ, Bennett JE. *Medical Mycology*. Media, PA: Williams and Wilkins; 1992; pp. 770–772.

44. Larone DH. Medically important fungi: A guide to identification. 5th edn. Washington, DC: ASM Press; 2011.
45. Pore RS, Chen J. Meningitis caused by *Rhodotorula*. *Sabouraudia*. 1976;14(3):331–335.
46. Libkind D, Gadanho M, van Broock M, Sampaio JP. Studies on the heterogeneity of the carotenogenic yeast *Rhodotorula mucilaginosa* from Patagonia, Argentina. *J Basic Microbiol*. 2008;48(2):93–98.
47. Nunes JM, Bizerra FC, Ferreira RC, Colombo AL. Molecular identification, antifungal susceptibility profile, and biofilm formation of clinical and environmental *Rhodotorula* species isolates. *Antimicrob Agents Chemother*. 2013;57(1):382–389.
48. Libkind D. Rhodotorula. In: Liu, D (ed). *Molecular Detection of Human Fungal Pathogens*. Boca Raton, FL: Taylor & Francis, Inc.; 2011; pp. 653–667.
49. Chen YC, Eisner JD, Kattar MM, Rassoulian-Barrett SL, Lafe K, Bui U, Limaye AP, Cookson BT. Polymorphic internal transcribed spacer region 1 DNA sequences identify medically important yeasts. *J Clin Microbiol*. 2001;39(11):4042–4051.
50. Chen YC, Eisner JD, Kattar MM, Rassoulian-Barrett SL, LaFe K, Yarfitz SL, Limaye AP, Cookson BT. Identification of medically important yeasts using PCR-based detection of DNA sequence polymorphisms in the internal transcribed spacer 2 region of the rRNA genes. *J Clin Microbiol*. 2000;38(6):2302–2310.
51. Leaw SN, Chang HC, Sun HF, Barton R, Bouchara JP, Chang TC. Identification of medically important yeast species by sequence analysis of the internal transcribed spacer regions. *J Clin Microbiol*. 2006;44(3):693–699.
52. Linton CJ, Borman AM, Cheung G, Holmes AD, Szekely A, Palmer MD, Bridge PD, Campbell CK, Johnson EM. Molecular identification of unusual pathogenic yeast isolates by large ribosomal subunit gene sequencing: 2 years of experience at the United Kingdom mycology reference laboratory. *J Clin Microbiol*. 2007;45(4):1152–1158.
53. De Hoog GS, Guarro J, Figueras MJ. Atlas of clinical fungi, electronic version 3.1. Tarragona, Spain: Universitat Rovira i Virgili, Centraalbureau voor Schimmelcultures, 2011.
54. Fell JW. rDNA targeted oligonucleotide primers for the identification of pathogenic yeasts in a polymerase chain reaction. *J Ind Microbiol*. 1995;14(6):475–477.
55. Diekema DJ, Petroelje B, Messer SA, Hollis RJ, Pfaller MA. Activities of available and investigational antifungal agents against *Rhodotorula* species. *J Clin Microbiol*. 2005;43:476–478.
56. Pfaller MA, Diekema DJ, Gibbs DL et al. Results from the ARTEMIS DISK Global Antifungal Surveillance Study, 1997 to 2007: 10.5-year analysis of susceptibilities of noncandidal yeast species to fluconazole and voriconazole determined by CLSI standardized disk diffusion testing. *J Clin Microbiol*. 2009;47:117–123.
57. Gomez-Lopez A, Mellado E, Rodriguez-Tudela JL, Cuenca-Estrella M. Susceptibility profile of 29 clinical isolates of *Rhodotorula* spp. and literature review. *J Antimicrob Chemother*. 2005;55:312–316.
58. Guinea J, Recio S, Escribano P, Peláez T, Gama B, Bouza E. In vitro antifungal activities of isavuconazole and comparators against rare yeast pathogens. *Antimicrob Agents Chemother*. 2010;54:4012–4015.
59. Galán-Sánchez F, García-Martos P, Rodríguez-Ramos C, Marín-Casanova P, Mira-Gutiérrez J. Microbiological characteristics and susceptibility patterns of strains of *Rhodotorula* isolated from clinical samples. *Mycopathologia*. 1999;145:109–112.

30 Saccharomyces and Kluyveromyces Infections

Firas A. Aswad,* Vibhati V. Kulkarny,*
Kingsley Asare, and Samuel A. Lee

CONTENTS

30.1 Saccharomyces 495
    30.1.1 Introduction 495
    30.1.2 Microbiology 497
        30.1.2.1 Taxonomic Classification 497
        30.1.2.2 Genetics and Biochemistry 498
        30.1.2.3 Morphology and Features 498
        30.1.2.4 Uses: Fermentation, Bread, Nutritional Supplements, and Probiotics 499
    30.1.3 Epidemiology 500
        30.1.3.1 Ecology and Etiology 500
        30.1.3.2 Human Population 500
    30.1.4 Clinical Syndromes 501
        30.1.4.1 Fungemia 502
        30.1.4.2 Endocarditis 503
        30.1.4.3 Gastrointestinal Tract Infections 503
        30.1.4.4 Vaginitis 505
        30.1.4.5 Respiratory Tract Infections 505
        30.1.4.6 Genitourinary Tract Infections 507
        30.1.4.7 Keratitis and Endophthalmitis 507
    30.1.5 Diagnosis 507
    30.1.6 Treatment 508
    30.1.7 Conclusion 509
30.2 Kluyveromyces 509
    30.2.1 Introduction 509
    30.2.2 Classification and Microbiology 510
    30.2.3 Industrial Uses 510
    30.2.4 Opportunistic Infections 510
    30.2.5 Conclusion 511
References 511

30.1 SACCHAROMYCES

30.1.1 INTRODUCTION

The ascomycetous and the common plant saprophyte *Saccharomyces cerevisiae* is a yeast with a widespread distribution in nature and can be found on plants and fruits and in soil [1].

---

* These authors contributed equally to this work.

The industrial production and commercial use of yeasts started at the end of the nineteenth century after their identification and isolation by Pasteur as he showed that fermentation was a process caused by living yeast cells. Exposure to this organism is widespread and occurs mostly through foods, beverages, and nutritional supplements. Because of its extensive use in the brewing (alcoholic beverages) and baking (production of bread) industries, *S. cerevisiae* has been designated the brewer's and baker's yeast, respectively. It is employed to prevent food contamination by mycotoxins like aflatoxin B1 through inhibiting growth of *Aspergillus flavus* and binding aflatoxin B1 (a carcinogenic and toxigenic compound produced by *A. flavus* and *Aspergillus parasiticus*) [2].

Saccharomyces boulardii, a subtype of *S. cerevisiae*, is also used in probiotic preparation for the prevention and treatment of various diarrheal disorders including antibiotic-associated diarrhea (AAD) [3,4], recurrent *Clostridium difficile* disease [5–8], *C. difficile*–associated enterocolopathies in infants and children [9,10], acute diarrhea in children [11] and in adults [12], traveler's diarrhea [13,14], diarrhea in tube-fed patients [15,16], AIDS-related diarrhea [17], and relapses of Crohn's disease [18] or ulcerative colitis (UC) [19]. Its efficacy has been demonstrated in a number of double-blind, placebo-controlled studies [3,6,20]. For instance, Surawicz et al. [3] performed a prospective double-blind controlled study to evaluate the effect of *S. boulardii* to prevent AAD, and 180 patients completed the study; 22% of patients receiving placebo experienced diarrhea compared with 9.5% of the patients receiving *S. boulardii* ($p=0.038$). *S. boulardii* had been used as an antidiarrheal drug in France since the 1950s. It was discovered by the French microbiologist Henri Boulard while in Indochina in 1923 [1]. He visited the subcontinent during a cholera outbreak and observed that some people drinking a special tea made from cooking the outer skin of a tropical fruit did not develop cholera. He successfully isolated the responsible agent and named this strain of yeast *S. boulardii*.

S. cerevisiae has become increasingly important in biotechnology and is one of the most examined and well-characterized microorganism. It has been used as a model eukaryote for many years and is excellent for studies in eukaryotic cell biology. *S. cerevisiae* has been used extensively in the biotechnology sector as a cloning vector (single copy or multicopy) for protein production. For example, *S. cerevisiae* is used to produce small hepatitis B surface proteins for use in hepatitis B vaccines [21]. *S. cerevisiae* is also used to produce the interferon class of immune cytokines, which are important in the regulation of immune responses to certain diseases. There are many advantages for the utilization of *S. cerevisiae* in biotechnology. Additionally, culturing yeast is low cost and straightforward. *S. cerevisiae* is regarded as safe for human consumption and is commonly used in food and pharmaceutical production [22]. However, the clinical literature suggests that the incidence of serious *Saccharomyces* infections is increasing, likely due to greater utilization of immunosuppressive therapeutic regimens, long-term indwelling catheterization, broad-spectrum antibiotic use, and also the prolonged survival of highly immunocompromised patients [23]. There are some instances where infections due to *Saccharomyces* have occurred in patients showing no obvious predisposition or risk factors [24–26]. Among these infections, there are a growing number of reports regarding invasive infections with *S. boulardii*. This has led some authors in Europe to propose that *S. cerevisiae* is *losing its innocence* [27] and has been upgraded from *generally regarded as safe* to biosafety level 1 [27], indicating the ability to cause superficial or systemic infections under certain circumstances. Nonetheless, despite widespread exposure and occasional colonization, the incidence of infections caused by *S. cerevisiae* and other *Saccharomyces* strains are negligible compared to those caused by other yeast species, most notably *Candida* spp. *S. cerevisiae* accounts for 17% of clinical isolates of non-*Candida*, non-*Cryptococcus* yeasts in North America [28,29]. Therefore, *S. cerevisiae* should not be considered a strictly nonpathogenic yeast, as it has the ability to cause invasive disease, albeit infrequently.

30.1.2 Microbiology

30.1.2.1 Taxonomic Classification

The genus *Saccharomyces* is classified under the kingdom of Fungi, exhibiting chitin in its cell wall and the ability to form a community of organisms. In the phylum of Ascomycota, the defining feature of *Saccharomyces* is the formation of ascospores, or nonmotile spores. Due to the unorganized nature of these asci and the lack of a fruiting body, *Saccharomyces* is categorized within the class Hemiascomycetes in the order Saccharomycetales, according to its sexual reproduction via multilateral budding on a narrow base. The genus description of *Saccharomyces* is defined by vegetative reproduction (budding). In a multigene study, Kurtzman and Robnett [30] demonstrated that the genus *Saccharomyces* is restricted to members of the *S. cerevisiae* clade, including seven species, six heterothallic biological species, and the hybrid species *Saccharomyces pastorianus* [31]. Definition of the genus of this clade is based upon those species involved in alcoholic fermentation, subsequently termed a sensu stricto species [32,33]. The seven species (Table 30.1) are regarded as genetically isolated from one another on the basis of genetic crosses [31] and from biochemical and molecular comparisons [30,34]. *Saccharomyces cariocanus*, *S. cerevisiae*, and *S. paradoxus* appear to be separate biological species from genetic crosses [34], although they show relatively little gene sequence divergence [30].

The most well-known species from the *Saccharomyces* clade is *S. cerevisiae*, or *baker's or brewer's yeast*. It is widespread in the environment and found on fruit and vegetation and in the soil [1]. *S. boulardii* (nom. inval.) [35], used in the prevention and treatment of intestinal disorders such as various diarrheal disorders [36], is now considered to be an invalid taxon and has been reclassified molecularly as a subtype [37] of *S. cerevisiae*. Currently, it is only possible to discriminate *S. boulardii* from *S. cerevisiae* strains by determining microsatellite-containing locus length [38]. Microsatellites are short DNA (1–10 base pairs) sequence repeats, motifs, which exhibit a substantial level of polymorphism in numerous eukaryotic genomes [39]. Due to their abundance and high polymorphism levels, microsatellites are useful as markers in association studies, population genetics, and forensics. Identification of microsatellite DNA motifs in *S. cerevisiae* and *S. boulardii* therefore permits a better understanding of the evolution of mutations in clinical isolates and the epidemiology of *Saccharomyces* infections.

S. cerevisiae has been widely used in genetics and cell biology due to its simple eukaryotic structure, serving as a model for human cell biology including the study of cell cycle events, DNA

TABLE 30.1
Accepted Species of the *Saccharomyces* Clade

Accepted Species	Strain Discovery
Saccharomyces bayanus	Saccardo (1895)
Saccharomyces cariocanus	Naumov, James, Naumova, Louis, and Roberts (2000)
Saccharomyces cerevisiae[a]	E.C. Hansen (1883)
Saccharomyces kudriavzevii	Naumov, James, Naumova, Louis, and Roberts (2000)
Saccharomyces mikatae	Naumov, James, Naumova, Louis, and Roberts (2000)
Saccharomyces paradoxus	Bachinskaya (1914)
Saccharomyces pastorianus	E.C. Hansen (1904)

Notes: Each strain is regarded as genetically isolated from one another on the basis of genetic crosses [31] as well as from molecular comparisons [30,34].

[a] Type of species of *Saccharomyces*.

replication and recombination, cell division, and metabolism. The ease of yeast genetic manipulation (e.g., DNA transformation), rapid growth, replication, and mutant creation in the laboratory has enabled the development of powerful scientific tools and methods [40]. In 1996, *S. cerevisiae* was the first eukaryotic organism to have its genome of 12 million base pairs, completely sequenced as part of the genome project [41].

30.1.2.2 Genetics and Biochemistry

Yeasts are unicellular eukaryotes, existing both as diploids and haploids, which can reproduce by mitosis (sporulation or asymmetric budding). Diploid cells can undergo meiosis, thereby generating one to four ascospores encapsulated inside the mother cell or ascus, so as not to be released into the environment [42,43]. Haploid cells (4 µm spheroids) mate with other haploid cells to produce stable diploid cells (5–6 µm ellipsoids) or may elongate into pseudohyphae, but true hyphae are not produced. Under stressful conditions, such as nutrient depletion, diploid cells undergo meiosis to produce haploid spores [44,45]. Specific yeast strains are used in brewing and wine industries and other ingestible items such as bread, nutritional supplements, and probiotics.

In the normal diploid cell state under good nutritional conditions, recessive mutations can be isolated and grown in haploid strains. Genetic variation is high within the *Saccharomyces* species, both phenotypically and molecularly. The high variability is due to spontaneous mutations, mitotic crossing-over, and genome renewal [42,45]. This quick and efficient genomic renewal enhances the versatility of *S. cerevisiae* for use in gene cloning and genetic engineering. Genes corresponding to eukaryotic genetic traits (phenotypes) can be identified by complementation from plasmid or phage libraries; via homologous recombination, exogenous DNA can be directed to specific locations in the yeast genome. Thus, wild-type yeast DNA can be replaced with altered alleles. The resultant mutant phenotype contributes to defining gene function, structure and function of the proteins produced, and whole cell structure, *in vivo*. DNA manipulation in *S. cerevisiae* has also been used to improve the strain in the fermentation industry [46,47].

S. cerevisiae contains a haploid set of 16 well-characterized chromosomes, ranging in size from 200 to 2200 kb. The whole genome sequence of chromosomal DNA, consisting of 12,157,105 base pairs, was released in 1996 [48], including a total of 6,607 open reading frames (ORFs) [49]. Subsequently, comparisons to sequences of other *Saccharomyces* species in the clade indicated that *S. cerevisiae* contains 5,773 or 5,726 protein-coding genes [50,51]. Approximately 75.48% of these genes have been characterized experimentally, and, of the remaining 24.52% with uncharacterized or dubious function, approximately half either contain a motif of a characterized class of proteins or correspond to proteins that are structurally similar to already characterized gene products from yeast or from other organisms [49].

Yeasts are chemoorganotrophs, utilizing organic compounds as a source of energy in which carbon is obtained mostly from hexose sugars or disaccharides [52]. Yeast strains are obligate aerobes requiring oxygen for aerobic cellular respiration, or are facultative anaerobes and grow best in neutral or slightly acidic pH environments at varying temperatures. Yeast cells can survive freezing under certain conditions, with viability decreasing over time [53]. In laboratories, growth of yeast is readily maintained on solid media or in liquid broth. Common media used for the cultivation of yeasts include potato dextrose agar (PDA) or potato dextrose broth, agar, yeast peptone dextrose (YPD) agar, Sabouraud dextrose agar, and/or yeast mold (YM) agar or broth. Home brewers who cultivate yeast frequently use dried malt extract (DME) and agar as a solid growth medium [45,54].

30.1.2.3 Morphology and Features

S. cerevisiae are unicellular fungi that tend to organize themselves into multicellular aggregates or communities. Individual cells have a prominent central vacuole and a small nucleus surrounded by a cell wall containing glucan and mannoproteins. Ascospores, seen during budding, contain chitosan and dityrosine. Blastoconidia or cell buds are observed during growth and are unicellular, globular, and ellipsoid to elongated in shape. The species is known to have multipolar budding, and if pseudohyphae are present, they are not fully formed [55]. Also, some diploid strains of *S. cerevisiae* can form pseudohyphae when grown on agar. Pseudohyphal cells are significantly elongated and cause growth of branched chains outward from the center of the colony. The circular colonies of *Saccharomyces* grow rapidly, maturing in 3 days and form flat, smooth, moist, glistening, creamy mounds, usually 2–8 mm in diameter [55,56]. Community growth of yeast is diverse including ascospores, colonies, flocs, flors, mats, and biofilms [45].

30.1.2.4 Uses: Fermentation, Bread, Nutritional Supplements, and Probiotics

Brief history: Yeast, from Old English *gist, gyst*, meaning boil or foam [57], has been used for fermentation and baking throughout history. The Dutch naturalist Antonie van Leeuwenhoek microscopically observed yeast in 1680, and in 1857, French microbiologist Louis Pasteur in a paper, entitled *Mémoire sur la fermentation alcool*, demonstrated that alcoholic fermentation was conducted by living yeasts and not by chemical catalysis [58,59]. By the late eighteenth century, two yeast strains had been identified and were used in brewing: *S. cerevisiae* and *Saccharomyces carlsbergensis*. In 1883, Emil C. Hansen isolated brewing yeast and propagated the idea leading to the importance of yeast in brewing [60]. Charles L. Fleischmann who exhibited yeast blocks and usage in bread making marketed commercial yeast at the Centennial Exposition in 1876 in Philadelphia [61].

30.1.2.4.1 Fermentation

The first archeological evidence of production of fermented beverages dates back to 7000 BC in a Neolithic village in Jiahu, China [62], and the earliest evidence of winemaking is traced to Iran at the Hajji Firuz Tepe site (circa 5400–5000 BC) [63]. By 500 BC, wine was being produced in Italy, Sicily, Southern France, and the Iberian Peninsula. Further expansion of winemaking occurred during the European colonization of America (sixteenth century), South Africa (seventeenth century), and Australia and New Zealand (eighteenth to nineteenth centuries) [59,64]. Beer fermentation was first recorded in the Mesopotamian region and in Egypt. Brewing of beer diverged into two processes mainly differentiated by temperature: ale, acquired from the Middle East by Germanic and Celtic tribes around the first century AD, and lager, which appeared during the late middle ages in Europe [65,66]. There are hundreds of genetic strains of yeast and two types of beer yeast: ale yeast (the *top-fermenting* type, *S. cerevisiae*) and lager yeast (the *bottom-fermenting* type, *Saccharomyces uvarum*, formerly known as *S. carlsbergensis*). Yeast strains used in the beer and wine industry form flocs (bottom fermenting) and flors (surface fermenting), which separate themselves from the fermentation product [67]. *S. cerevisiae* converts fermentable sugars into alcohol and other by-products. As a result of recent reclassification of the *Saccharomyces* species, both ale and lager yeast strains are considered to be members of *S. cerevisiae* [66,68].

The life cycle of yeast consists of four phases—lag period, growth phase, fermentation, and sedimentation. During the lag phase, stored glycogen is broken down to glucose and used by the yeast cell for reproduction. Yeast cells in the growth or respiration phase use oxygen to oxidize acid compounds in their environment. Once available oxygen has been depleted, the anaerobic process of fermentation begins in which carbon dioxide and ethanol are produced. Flocculation and sedimentation begin 3–7 days later depending on the strain. During sedimentation, yeasts prepare themselves for dormancy by producing glycogen and collect at the bottom or float on top of the fermented product [60,68].

30.1.2.4.2 Baking

Description of bread making and analysis of bread remains from ancient Egypt suggest a simple composition made of coarsely milled emmer wheat (*Triticum dicoccum* Schubl.) [68]. Yeasts were observed in some breads of this time period, indicating that at least some leavened breads were being produced [66,69]. In modern times, *S. cerevisiae* is used extensively in baking as a leavening agent, converting the fermentable sugars present in dough into carbon dioxide. As the temperature increases, the yeast dies and leaves air pockets, giving the baked product a softer spongelike texture. In the process of bread making, the yeast initially respires aerobically, producing carbon dioxide and water. When the oxygen is depleted, fermentation begins, producing ethanol as a waste product that evaporates during baking [70].

30.1.2.4.3 Nutritional Supplements and Probiotics

Saccharomyces strains have been produced and marketed commercially as nutritional supplements. Also known as *nutritional yeast* when sold as a dietary supplement, it consists of deactivated *S. cerevisiae*. It is a rich source of protein and vitamins, such as B complex vitamins (B1 [thiamine], B2 [riboflavin], B3 [niacin], B5 [pantothenic acid], B6 [pyridoxine], B9 [folic acid], and H or B7 [biotin]), chromium, niacin, and selenium. Due to a cheese-like flavor, *Saccharomyces* has also been used as a cheese substitute and as a topping for snack foods. It is commonly sold in block form, flakes, or yellow powder. Probiotic supplement use of *S. boulardii*, a patented yeast preparation, to maintain and restore the natural flora in the gastrointestinal (GI) tract, has been used for the prevention of AAD in randomized, double-blind studies [71,72]. In addition, *S. boulardii* has been shown to reduce the symptoms of acute diarrhea in children [73]. *S. boulardii* possesses probiotic agent properties: it survives passage through the GI tract; it grows well at 37°C, both *in vitro* and *in vivo*; and it inhibits the growth of a number of microbial pathogens [74]. *S. boulardii* has been licensed for use in Europe as a probiotic agent for reequilibration of the intestinal tract. However, there are reports of patients given *S. boulardii* that have developed *Saccharomyces* fungemia, suggesting that *S. boulardii* can travel from the intestinal tract to the bloodstream [75].

30.1.3 Epidemiology

30.1.3.1 Ecology and Etiology

The ecology of *S. cerevisiae*, other than domesticated strains used for fermentation and food production, has a wide range in the environment. Wild strains have been isolated from mushroom fruiting bodies, oak tree–associated soils and fluxes [76–78], the skins of fruits and berries, and exudates from plants and in salt water [79–82]. Wild isolates of *S. cerevisiae* are also a major cause of spoilage of mango fruit [83] and peach puree [84], and it has recently been identified in surveys of the fungal diversity in beetle guts [85–87].

30.1.3.2 Human Population

S. cerevisiae is a commonly used industrial microorganism, is ubiquitous in the environment, and is widely used for human consumption. It can colonize the respiratory, genitourinary, and GI tracts of humans in an innocuous way and grow without appearing to cause any disease [1]. Industrial workers and the general public come into contact with *S. cerevisiae* via inhalation and ingestion. *S. cerevisiae* is an organism with an extensive history of safe use in food. Despite this and its use in research, there are increasing reports in the literature that the yeast can become pathogenic in debilitated and/or severely immunocompromised patients [88,89]. Very rarely, severe infections due to *S. cerevisiae* have been recorded in patients showing no obvious predisposing factors [90]. For instance, Smith et al. [91] report a case of *S. cerevisiae* fungemia as the etiology of an aortic graft infection in an immunocompetent adult. Individuals may ingest large quantities of *S. cerevisiae* on a daily basis, for example, ingestion of nutritional and/or probiotic supplements. Studies have been conducted to ascertain whether the ingestion of large quantities of yeast may result in either colonization or colonization and secondary spread to other organs of the body. Related to this, results indicate that *S. cerevisiae* can be found colonizing the colons of animals, with passage to draining lymph nodes [92].

Since reports of human infections due to *S. cerevisiae* have increased in the past several years [93], this yeast species is now considered in a group of emerging pathogens [88,89]. The epidemiologic characteristics of human infections from *S. cerevisiae* are not well defined, and it is not clearly established whether *S. cerevisiae* is a persistent commensal of the digestive tract or whether it is only transiently present after ingestion of contaminated food. Several reports have documented the occurrence of nosocomial acquisition, especially catheters with contamination through hand transmission [94,95]. The low concern for the pathogenicity of *S. cerevisiae* is illustrated in a series of surveys conducted at hospitals over two decades ago in which *S. cerevisiae* accounted for less than 1% of all invasive yeast infections isolated at a cancer hospital, and in most of the cases, the organism had been isolated from the respiratory system [96]. From 1991 to 1996, at Yale–New Haven Hospital (New Haven, CT, USA), 50 isolates of *S. cerevisiae* were recovered from patients, but all were considered nonpathogenic contaminants [97].

In the last decade, *S. cerevisiae* has been implicated in a number of human infections, including fungemia, pneumonia, endocarditis, peritonitis, and vaginitis [98]. It accounts for more than 5% of vaginal infections [99,100]. Studies involving *S. cerevisiae* and ingestion suggest that it is more of the exposure to high doses of yeast than the identity of the strain that is associated with infection [101]. *S. cerevisiae* does produce virulence factors with the ability to cause disease, including phospholipase A and lysophospholipase. These enzymes enhance the ability of the yeast to adhere to the mammalian cell wall surface and result in colonization of tissue as an initial step of infection. Nonpathogenic strains in the *Saccharomyces* clade had considerably lower phospholipase activities. Of a wide range of fungi assayed for phospholipase production, *S. cerevisiae* was found to have the lowest level of activity [102]. Based on these pathogenic attributes, in comparison to other fungi, *S. cerevisiae* has been considered to be of very low virulence. Nonetheless, *Saccharomyces* has been increasingly recognized as a cause of human infection.

30.1.4 CLINICAL SYNDROMES

S. cerevisiae has been isolated from persons with underlying comorbidities and/or immunosuppression and is recognized as an infrequent cause of invasive fungal infections (IFIs) [93,103–105]. More than 80% of reports of *S. cerevisiae* as the etiologic agents of serious IFIs have been published since 1990, and 40% of those infections have identified subtype *boulardii* as the causative agent [93,103]. The spectrum of clinical syndromes associated with *Saccharomyces* is diverse, ranging from allergic responses to the organism, asymptomatic colonization, superficial infection, and subacute or acute invasive disease. Invasive infection is often related to foreign bodies, prosthetic medical devices, and/or use of probiotic preparations intended for the prevention or treatment of various diarrheal disorders [106]. *S. cerevisiae* can also be a saprophytic colonizer of human mucosal surfaces and has been isolated from the digestive, respiratory, and genitourinary tracts and from the skin of healthy hosts where their presence is benign and asymptomatic [107]. In a study of patients with hematological disease, *S. cerevisiae* was isolated from throat, stool, urine, and perineal samples of patients at rates of 16%, 23%, 10%, and 20%, respectively [107]. In patients with human immunodeficiency virus (HIV) infection, *S. cerevisiae* has been isolated from periodontal lesions [108] and from plaques of oral leukoplakia [109]. Greer et al. [110] identified *Saccharomyces* as a colonizing isolate in 7% of expectorated sputum samples obtained from patients with pulmonary tuberculosis. Kiehn et al. [111] speciated 3340 yeast specimens isolated from 2208 cancer patients; *S. cerevisiae* was recovered from 24 patients, of which 19 were from sputum.

Patients with true IFI due to *Saccharomyces* may present clinically with nonspecific symptoms such as fever and systemic signs of illness, and the diagnosis of *Saccharomyces* is often unexpected. The presence of *Saccharomyces* in normally sterile sites has been attributed to rupture of local barriers; portals of entry include translocation of ingested organisms from the GI tract and oropharynx [112] and contamination of intravenous catheter sites [93,107,113]. Generally, the clinical presentation reflects the underlying immune defects and risk factors associated with each patient group, with greater immune suppression correlating with increased risk for invasive disease. *Saccharomyces* infections often clinically resemble invasive candidiasis, notably because chorioretinitis (fluffy yellow exudates) may be seen in both conditions in cases where fungemia is present.

Virulent isolates of *S. cerevisiae* are capable of growth at 42°C [114]; this is considered an important characteristic that would provide an inherent advantage for pathogenesis. Virulent isolates may also grow as pseudohyphae under certain conditions, and pseudohyphae have been seen to penetrate agar, which may give an indication of their role in vivo [115]. Pseudohyphal growth was not observed in nonclinical isolates of *S. cerevisiae*, and this ability may be important in the penetration of host tissue by virulent isolates and could play a role in the blocking of capillaries, which has been associated with animal mortality [116].

30.1.4.1 Fungemia

The most important clinical syndrome caused by *S. cerevisiae* is fungemia. The precise incidence of *S. cerevisiae* fungemia is unknown; however, population-based studies indicate that *S. cerevisiae* accounts for up to 0.1%–3.6% of all episodes of fungemia [94,117]. In one study, Munoz et al. [103] identified 60 cases of *S. cerevisiae* fungemia in the literature, of which 17 patients died, representing a mortality rate of 28%. Risk factors and underlying comorbidities are similar to those of candidemia including use of indwelling catheters, treatment with corticosteroids, parenteral nutrition, hemodialysis, broad-spectrum antibiotics, diabetes, neutropenia, immunosuppression due to HIV or neoplasm, and transplantation [94]. In addition, an isolated and exclusive risk factor is the ingestion of *S. boulardii* as a probiotic, which has been widely used for treatment of AAD [8,93,118]; therefore, a history of intake of health food supplements or probiotics should be ascertained. Lherm et al. [94] reported a series of 1395 ICU patients, who were treated prophylactically with *S. boulardii*. In this study, seven patients developed a systemic fungal infection with *S. boulardii*, and the authors concluded that the incidence of fungemia was approximately five per thousand patients who used *S. boulardii* in their unit.

Saccharomyces infection has been mostly found in immunosuppressed patients and critically ill patients; however, there have been a low number of relatively healthy persons infected as well [28]. *S. cerevisiae* should not be dismissed as a nonpathogenic microorganism when recovered from the blood or other normally sterile site. The clinical manifestations of *Saccharomyces* fungemia are similar to those of systemic candidiasis and candidemia. Fever of unknown etiology is the most frequent symptom associated with fungemia, and according to Enache-Angoulvant et al. [93], it is present in 75% of patients. Nosocomial transmission can occur through contamination from health-care workers to patients via indwelling central catheters and may be considered a health-care-related infection [93]. A cluster of isolates with identical genotypes among patients concurrently hospitalized in the same unit has been demonstrated [119]. In addition, fungemia by *S. cerevisiae* subtype *boulardii* has occurred in patients who have not been treated with a probiotic preparation of this organism when they share a room with treated patients [94,120]. Viable yeasts may be detected at a 1 m distance as a result of aerial transmission, and the yeast can survive for 2 h on room surfaces [113]. Hennequin et al. [113] demonstrated persistent hand contamination of the health-care workers that opened packets of *S. boulardii* despite careful hand washing. Contamination of a central venous catheter (CVC) insertion site is likely to be one of the main mechanisms of *S. cerevisiae* subtype *boulardii* fungemia [113]. Thus, it is recommended that health-care workers wear gloves when opening packages of *S. boulardii* and must do so outside the patient's room.

The potential therapeutic benefit of *S. boulardii* should be carefully evaluated in critically ill patients. It is important to note that probiotics are not completely safe and therefore must be used with caution in certain patient populations, particularly immunocompromised patients and those with indwelling CVCs. A specific protocol concerning the indications for the use of probiotics should be developed and implemented to minimize the risk to patient safety. If contamination of vascular catheters is suspected, removal of the central catheter should be considered. Smith et al. [91] reported a case of *S. cerevisiae* fungemia and aortic graft infection in an immunocompetent adult. The patient was found to have an aortic-enteric fistula between the proximal end of the aorta–bifemoral graft and proximal jejunum; it was suspected that *S. cerevisiae* infected the graft after the aorta-enteric fistula developed, as *S. cerevisiae* is a common colonizer of duodenal mucosa.

30.1.4.2 Endocarditis

Data from the International Collaboration on Endocarditis–Prospective Cohort Study (ICE-PCS) reported that less than 1% of more than 2700 patients with definite infective endocarditis (IE) of all microbiologic etiologies were infected with *Candida* species [121]. Thus, IE due to *Candida* species remains a rare, but potentially devastating infection. Contact with the health-care system has now emerged as the primary risk factor for most patients as 51% of cases of *Candida* IE have been attributed to health-care contact, whereas 12% had a history of injection drug use [121]. Fungal endocarditis is often a complication of medical and/or surgical interventions. The uses of cardiovascular devices, particularly prosthetic valves, and CVCs are predisposing factors. Of all causes of fungal IE, *Candida* species are by far the most common cause, with up to 41% of cases caused by non-*albicans* spp. Fungal prosthetic valve endocarditis (PVE) is an uncommon but extremely serious disease with historically high fatality rates [122]. Even with appropriate therapy, hospital survival rates remain lower than 40% [122], and late recurrences are frequent [123]. Para-annular abscesses are frequently associated with fungal PVE. Treatment results improve greatly with surgical replacement of the infected prosthetic valves in conjunction with prolonged antifungal therapy [122].

Patients diagnosed with fungal endocarditis require close follow-up over an extended period. In a detailed review of the literature, we found only three reported cases of IE due to *S. cerevisiae* (Table 30.2). Two cases involved PVE [124], and one case was native valve endocarditis in a preterm infant at 30 weeks of gestation, in the setting of broad-spectrum antimicrobial therapy and prolonged hospitalization [125].

30.1.4.3 Gastrointestinal Tract Infections

Saccharomyces species are common colonizers of human mucosal surfaces [107]. However, clinical experience suggests that *S. cerevisiae* may cause illness of the GI system in the immunocompromised patient. *S. cerevisiae* has been associated with esophageal malignancy [126], UC-related diarrhea, and HIV/AIDS. *Saccharomyces* esophagitis is a rare disease; Eng et al. [88] reported a patient with Waldenstrom's macroglobulinemia presenting with a mixed esophageal infection caused by *S. cerevisiae* and *Candida albicans*. Another report described a patient with HIV coinfection presenting with *Saccharomyces* esophagitis [127]. *Saccharomyces* esophagitis, duodenitis, jejunitis, and colitis are rare conditions that have been described in medical literature in a patient with HIV coinfection [128]. The clinical presentation may be indistinguishable from that of *Candida* spp. *S. cerevisiae* has been associated with chronic diarrhea in a German shepherd dog [129]. In several cases of UC unresponsive to therapy, diarrhea has been associated with opportunistic pathogens including cytomegalovirus and *C. albicans* during corticosteroid or immunosuppressant therapy [130]. Candelli et al. [131] have reported a case of *S. cerevisiae* intestinal infection presenting as acute diarrhea in a patient with a 10-year history of UC who was not previously treated with immunosuppressants, thus demonstrating the potential role for *S. cerevisiae* of causing infectious colitis and mimicking an exacerbation of UC.

Peritonitis is a leading complication of peritoneal dialysis (PD) and one of the most common factors contributing to technical failure and hospitalizations in patients on PD. Fungal peritonitis, although rare, is associated with substantial morbidity and mortality in patients on PD [132]. *Candida* species, especially *C. albicans*, are most often involved in fungal PD-related peritonitis [133]. The latest International Society of Peritoneal Dialysis (ISPD) guidelines recommend

TABLE 30.2
Clinical and Microbiologic Features of Three Patients with *Saccharomyces cerevisiae* IE

Age, Sex	Primary Isolate	Additional Isolates	Underlying Conditions	Central Intravenous Catheter	Previous Antibiotic Therapy	Treatment	Outcome	Reference
Not specified	PVE	Blood	Prosthetic valve	Yes	Yes	Amphotericin B and surgery	Cured	[124]
9 days, male	Tricuspid valve endocarditis	Blood	Premature birth, ductus arteriosus, necrotizing enterocolitis	Yes	Yes	Amphotericin B, fluconazole, and no surgery	Cured	[125]
57 years, female	Blood	Aortic PVE	Prosthetic valve	No	Yes	Ketoconazole and surgery	Death	[124]

removal of the PD catheter immediately after the diagnosis of fungal peritonitis [134]. There have been four published cases of peritonitis caused by *S. cerevisiae* in continuous ambulatory PD patients [135–138]. No reference is made regarding the source of the infection, except in one case where the patient's wife was in daily contact with *Saccharomyces* in her occupation as a baker [138].

Additionally, *S. cerevisiae* is used in food production and can rarely induce allergic symptoms such as mild to moderate systemic reactions (hives and/or cough) to more severe symptoms (e.g., bloody and purulent stools) and anaphylactic reactions to foods containing industrial *S. cerevisiae* extracts [139].

30.1.4.4 Vaginitis

Vaginal symptoms are extremely common, and vaginal discharge is among the 25 most common reasons for consulting physicians in private office practice in the United States [140]. Vaginitis is found in approximately 15% of women who visit sexually transmitted disease (STD) clinics [141]. Candidiasis is responsible for approximately 15%–30% cases of vaginitis, and 70%–75% of women are expected to experience at least one episode of the disease during their lifetime [142]. More than 90% of yeast species isolated from the vagina belong to *C. albicans*, followed by *Candida glabrata* [142–145]. The increasing number of patients at risk for debilitating and chronic diseases; the increasing need for and use of immunosuppressive treatments, broad-spectrum antibiotic therapies, and parenteral nutrition; the use of intravenous catheters; and the relative fungal resistance to azoles have resulted in an increased incidence of non-*albicans* and non-*Candida* yeast infections, including *S. cerevisiae* [119,146]. While vaginal colonization and symptomatic vaginitis by *S. cerevisiae* have been reported, fungal vaginitis other than candidiasis is very uncommon, representing less than 1% of the total episodes [133,143,147]. In contrast, Agatensi et al. [148] reported a higher incidence of 5.5% in Italian women. It has been reported that *S. cerevisiae* is more likely to be involved in chronic vaginal infections [25].

The first instance of transmission of *S. cerevisiae* between individuals was reported by Wilson et al. [149]. However, this finding was not supported by molecular epidemiological tools. The only other report of transmission of *S. cerevisiae* isolates between different individuals was demonstrated by Nyirjesy et al. [150]. Those investigators presented evidence that the same strain of *S. cerevisiae* was isolated from the finger of the husband of one of their four patients and from the dough used in his pizza shop. This study would appear to be the first documentation of an industrial isolate of *S. cerevisiae* being the etiological agent of a human disease.

The clinical presentation of vaginal infection due to *S. cerevisiae* is indistinguishable from *Candida* vaginitis, which complicates the etiological diagnosis [144]. Vaginal soreness, irritation, vulvar burning, dyspareunia, and external dysuria are commonly present. Odor, if present, is minimal. Examination frequently reveals erythema and swelling of the labia and vulva. Absence of hyphal elements on direct microscopy of vaginal specimens is often the first clue to the presence of *S. cerevisiae* [145].

30.1.4.5 Respiratory Tract Infections

Pulmonary infections due to *Saccharomyces* spp. are extremely rare (Table 30.3). *S. cerevisiae* has been implicated in pneumonia in an HIV/AIDS patient, and the presence of this organism in a number of organs at autopsy was recorded, suggesting that the yeast first colonized the oropharynx of the immunocompromised host and later was aspirated to the lungs, followed by hematogenous spread [151,152]. Sporadic cases of pulmonary infection have been attributed to occupational exposure; Ren et al. [153] reported a case of an otherwise healthy male employed in a bakery presenting with a single lung nodule, who underwent investigations to evaluate for pulmonary carcinoma. Lung biopsy was positive for yeast cells that were identified as *S. cerevisiae* by polymerase chain reaction (PCR) and nucleotide sequencing. The etiology of the lung nodule was

TABLE 30.3
Clinical and Microbiologic Features of Four Patients with *Saccharomyces cerevisiae* Respiratory Tract Infections

Age, Sex	Primary Isolate	Additional Isolates	Underlying Conditions	Central Intravenous Catheter	Previous Antibiotic Therapy	Treatment	Outcome	Reference
39 years, male	Lung, pneumonia	Spleen, digestive tract	HIV/AIDS	No	No	None	Death	[151]
40 years, male	Lung, nodule	None	None, occupational exposure as a baker	No	No	Surgery	Cured	[153]
17 years, male	Perihilar mass	Bronchoscopic specimens, esophageal lesions	HIV/AIDS	Not specified	Not specified	Amphotericin B	Death	[152]
60 years, female	Pleural fluid, empyema	Sputum	Cirrhosis, esophagopleural fistula	Not specified	No	Amphotericin B	Death	[154]

attributed to the occupational inhalation of dry baking yeast powder. Additional blood tests did not reveal any immune dysfunction. In another case, a patient developed *S. cerevisiae* empyema (pleural infection) that was caused by an esophagopleural fistula following variceal sclerotherapy. The patient had prepared a chicken potpie using brewer's yeast 2 weeks before presentation. *S. cerevisiae* was isolated from the lungs, and the ingestion of baker's yeast was identified as the source of infection [154].

30.1.4.6 Genitourinary Tract Infections

There have been four reported cases of urinary tract infections due to *S. cerevisiae*, two with renal abscesses associated with fungemia [88]. Fungus balls rarely result in acute renal failure, and the majority of cases have been caused by *C. albicans* [155]. Senneville et al. [156] reported one patient with bilateral ureteral obstruction due to *S. cerevisiae* fungus balls.

30.1.4.7 Keratitis and Endophthalmitis

Fungal keratitis is an uncommon syndrome. The main inciting cause of fungal keratitis worldwide is traumatic inoculation. However, in the developed world, risk factors include protracted corneal epithelial ulceration, use of soft contact lenses, corneal transplant, and topical ocular corticosteroid therapy. With increasing distance from the equator, the relative incidence of yeasts as causative organisms for the infection of the corneal ulcers increases [157]. Overall, fungi are present as part of the external ocular flora in 6%–8% of healthy patients [158]. A single case of *Saccharomyces* keratitis and endophthalmitis has been reported, which occurred following therapy with topical and systemic corticosteroid therapy for corneal graft rejection [159].

30.1.5 Diagnosis

The diagnosis of *Saccharomyces* infection is usually not suspected due to its low incidence and virulence. *Saccharomyces* infection clinically resembles candidiasis. When *Saccharomyces* is recovered from body sites that are normally nonsterile, especially in the absence of overt symptoms of infection, it may be difficult to clinically differentiate between colonization and infection. *Saccharomyces* grows in blood cultures and on Sabouraud dextrose media; detection generally relies on traditional clinical microbiologic cultures with biochemical identification and examination of histopathology. Standard commercially available automated systems (e.g., BD Phoenix™ or VITEK 2®) can identify *Saccharomyces* based on carbohydrate fermentation profiling [160]. CHROMagar™ Candida is a commercially available chromogenic agar that distinguishes between major *Candida* species and *Saccharomyces* based on differential coloration of growing colonies [161,162]. No specific serologic diagnostic tests are available to assist in the diagnosis.

PCR and sequencing may be helpful for definitive identification of *Saccharomyces* from clinical samples and for molecular typing for epidemiologic purposes, although these approaches are not routinely available from the standard clinical microbiology laboratory. It should be noted that molecular identification still primarily relies upon initial isolation via traditional subculturing from already-growing samples, and given that biochemical identification is generally straightforward, there is a limited need for molecular identification from a purely clinical standpoint unless identification can be achieved more rapidly. In contrast, nonculture-based molecular or other means of early diagnosis of IFIs (primarily *Candida* species, and molds such as *Aspergillus*, or rare fungi including of *Saccharomyces*) would be an important advance, but current approaches are limited in availability and/or utility that is restricted to specific clinical scenarios.

Nonculture-based molecular methods for early identification of IFIs remain in various stages of development and refinement. Detection of (1,3)-*beta*-D-glucan (BG) in the fungal cell wall is a bioassay based upon the horseshoe crab (*Tachypleus* or *Limulus*) clotting cascade [163]. The BG test can detect *Candida*, *Trichosporon*, *Saccharomyces*, *Aspergillus*, *Fusarium*, and other fungi, but not *Cryptococcus* or Zygomycetes [164–166]. Clinical utility of the BG test as an adjunctive means of diagnosis has been validated in patients with hematologic malignancies; however, its role in general intensive care patients and other settings remains unproven [167]. Major shortcomings of the BG test include its specificity, as false-positive reactions can occur in a variety of settings [168]. Overall, detection of fungal BG is potentially useful for early diagnosis of a variety of IFIs in specific patient populations, but it does not distinguish between fungal species.

A promising approach for nonculture-based detection of IFIs is identification of fungal signatures using matrix-assisted laser desorption ionization–time of flight mass spectrometry (MALDI-TOF MS) directly from blood or other fluids [169,170]. In principle, this approach can detect emerging yeasts, including *Saccharomyces*, although performance characteristics remain to be defined in the clinical setting [171].

The predominant approach for molecular diagnosis of IFIs cultured from clinical samples relies upon PCR amplification of fungal DNA, with a wide range of technical formats, including PCR–ELISA, nested PCR, real-time PCR approaches, multiplex PCR, pyrosequencing, microarrays, restriction fragment size analysis, and probe hybridization [172]. These approaches have yet to be standardized for use in the clinical setting. Several methods have been specifically devised for species-specific, nonculture detection of *Saccharomyces* in wine [173,174] and fecal material [175]. For instance, Martorell et al. [173] developed a random amplified polymorphic DNA (RAPD) approach using real-time PCR for the detection and quantification of *S. cerevisiae* in wine; this PCR product differentiated this species from other species associated with fermented foods. Limitations of nucleic amplification–based methods include issues with DNA extraction due to low circulating amounts of fungal DNA and a relatively tough cell wall, the presence of nonspecific inhibitors in fluid or tissue samples, false-positive reactions due to environmental contamination, and the inability to distinguish colonizing organisms from true pathogens. Because of the very low incidence of invasive *Saccharomyces* infections, any clinically useful nonculture-based fungal detection method would likely be panfungal in approach, and not limited to detection of only *Saccharomyces*.

The primary approach for specific molecular identification of cultured fungal species relies upon PCR amplification of conserved ribosomal DNA (rDNA) derived from samples growing in culture. Specific sequences of rDNA (18S, 5.8S, and 28S) from the internal transcribed spacer *ITS1* and *ITS2* variable regions are amplified, followed by sequencing and comparison using against fungal genomic databases or other method of downstream analysis [176,177]. This powerful molecular approach has been used to definitively identify common and unusual causative agents of IFIs [178,179], including *Saccharomyces* [176]. Ren et al. [153] used PCR assay of paraffin-embedded tissue and nucleotide sequencing with ribosomal *ITS1–ITS2* universal primers for identification of *S. cerevisiae* in a male patient employed in a bakery who presented with a single lung nodule.

Numerous genetic fingerprinting methods are readily available for determining the genotypes of *S. cerevisiae* in the research setting for molecular epidemiology purposes. These approaches include karyotyping using pulse field gel electrophoresis, restriction fragment polymorphism analysis, restriction enzyme analysis of mitochondrial DNA, and numerous and PCR-based methods including microsatellite amplification [160]. However, there is limited clinical data on the clinical molecular epidemiology of *S. cerevisiae* infections [25]. Using rDNA hybridization and Ty917 hybridization for Southern blotting, Posteraro et al. [180] demonstrated that most of 40 *S. cerevisiae* vaginal isolates from women of childbearing age were very heterogeneous, with the exception of those collected from individual patients with recurrent vaginitis. Although molecular approaches for the diagnosis and/or identification of *Saccharomyces* are not routinely available in the clinical setting, they may be available from specialized clinical mycology centers or research laboratories.

30.1.6 Treatment

A range of antifungal agents have been used to treat *S. cerevisiae* infections with mixed results. Underlying diseases and conditions such as severe immunosuppression are a major factor decreasing the likelihood of successful therapy [88]. Importantly, the *S. cerevisiae* genome contains multidrug resistance genes. Anderson et al. [181] reported that under conditions of low drug concentration, mutations in the genes *PDR1* and *PDR3*, which regulate ABC transporters associated with resistance to fluconazole, can develop. Moreover, under conditions of high drug concentration, recessive mutations in *ERG3* can occur, leading to resistance through altered sterol synthesis.

Although *Saccharomyces* appears to be susceptible to most antifungal drugs in vitro, including amphotericin B, 5-flucytosine, and the azoles, varying resistance of *S. cerevisiae* to antifungal agents has been reported [25,182,183]. In a survey of 32 *Saccharomyces* clinical isolates (mostly oropharyngeal, several urinary and abdominal, and one bloodstream source) obtained from 19 patients, there were increased minimal inhibitory concentrations (MICs) to fluconazole compared to amphotericin B and flucytosine [184]. Espinel-Ingroff et al. [185] noted an MIC_{50} value for fluconazole of 2 μg/mL for *S. cerevisiae* compared to an MIC_{50} of 0.5 μg/mL for *C. albicans*. Voriconazole has been tested in vitro against *S. cerevisiae* and had an MIC_{50} value of 0.12 μg/mL [185]. A recent publication has reported the successful use of voriconazole therapy for a patient with sepsis due to *S. boulardii* who did not respond to fluconazole [186].

Echeverría-Irigoyen et al. [147] tested the in vitro antifungal susceptibilities of eight *S. cerevisiae* isolates against multiple agents: for 5-flucytosine, (1) the MIC for susceptibility was ≤4 μg/mL, (2) intermediate susceptibility was 8–16 μg/mL, and (3) resistance was ≥32 μg/mL. For fluconazole, (1) the MIC for susceptibility was ≤8 μg/mL, (2) susceptible dose-dependent response was 16–32 μg/mL, and (3) resistance was ≥64 μg/mL. For amphotericin B, the MIC for susceptibility was ≤1 μg/mL. For posaconazole and voriconazole, (1) susceptibility was ≤1 μg/mL, (2) susceptible dose-dependent response was 2 μg/mL, and (3) resistance was ≥4 μg/mL. An MIC ≤2 μg/mL was considered susceptible to caspofungin.

Overall, most *S. cerevisiae* isolates appear to be moderately susceptible to fluconazole; however, inherent resistance to fluconazole has been described [107,186], which highlights the importance of the correct identification of the yeast species and antifungal susceptibility testing in clinical cases in order to provide appropriate therapy. Additional surveys of antifungal resistance in uncommon opportunistic yeast species including *Saccharomyces* are needed to better define antifungal resistance patterns. For these rare yeasts, determination of antifungal susceptibilities should be strongly considered. In patients with prosthetic valve infections and infected indwelling catheters, most experts advocate the removal of the indwelling catheter in addition to the use of antifungal agents [182].

30.1.7 CONCLUSION

Although widely used for the production of foods, beverages, and nutritional supplements for thousands of years, *S. cerevisiae* should now be regarded as a potential opportunistic pathogen of low virulence rather than as a completely nonpathogenic yeast. The use of *S. cerevisiae* subtype *boulardii* probiotics should be carefully reassessed for efficacy and safety, particularly in immunosuppressed or critically ill patients.

30.2 KLUYVEROMYCES

30.2.1 INTRODUCTION

Yeasts have a long tradition of being employed in biotechnology and a more recent history of use as research models for biochemistry, metabolism, genetics, and cell biology. *S. cerevisiae* has been the dominant representative in this field. There is tremendous diversity among yeasts, however, and the application of modern microbiological and molecular approaches has resulted in renewed focus on the biology and industrial potential of other yeasts. Previously, the yeast *Kluyveromyces lactis* was named *Saccharomyces lactis*, but since it appeared to be unable to undergo hybridization with the yeast *S. cerevisiae*, it was grouped under the genus *Kluyveromyces* and renamed *K. lactis* in 1965.

The dairy yeast *Kluyveromyces marxianus* is of particular interest in this regard because of traits that render it especially suitable for industrial application. These useful properties include the fastest growth rate of most eukaryotic microbes, thermotolerance, the capacity to assimilate a wide variety of sugars, secretion of lytic enzymes, and production of ethanol by fermentation. Despite the importance of these attributes and significant exploitation by the biotechnology sector, fundamental research with *K. marxianus* is just emerging from the shadow of its counterpart species, *Kluyveromyces lactis* [187,188].

30.2.2 Classification and Microbiology

The genus was composed originally on the basis of *Kluyveromyces polysporus* and *Kluyveromyces africanus*, which had been isolated from soil by van der Walt [189]. Van der Walt later amended the diagnosis of the genus to include organisms producing fewer spores, usually four per ascus. With this change, he transferred many former *Saccharomyces* species to the genus *Kluyveromyces*. An additional species, *K. africanus*, was also assigned to the genus. The ascospore shape was considered to be a generic characteristic, and thus the diagnosis of the genus *Kluyveromyces* was altered to include yeasts producing fewer reniform spores (generally four per ascus). As a result, many former *Saccharomyces* species and the genera *Fabospora*, *Zygofabospora*, *Dekkeromyces*, and *Guilliermondella* were transferred to the genus *Kluyveromyces*. However, the irregularity of ascosporic shape in *K. lactis* or *K. marxianus* did not support the view that ascospore shape underlies meaningful lines of development; later studies redefined the concept of the genus as more features were taken into account [190].

Reviews by Lachance [191,192] have established the current basis for the classification of the species of the genus *Kluyveromyces*. The genus currently consists of 18 species, including *Kluyveromyces sinensis* [193], *Kluyveromyces hubeiensis* [194], *Kluyveromyces piceae* [195], and *Kluyveromyces bacillisporus* [191], some of which have been the subject of extensive investigations concerning their genetics, ecology, and evolution.

Despite their close evolutionary relationship, *K. lactis* and *S. cerevisiae* have adapted their carbon utilization systems such that both yeasts differ in the GAL genetic switch according to their preference for the utilization of galactose as a carbon source [196–199]. Whereas glucose is the main carbon and energy source for *S. cerevisiae* within its ecological niche, *K. lactis* has instead adapted to the utilization of the main sugar present in the milk, the disaccharide lactose and galactose [196–199].

30.2.3 Industrial Uses

In the 1950s and 1960s, *K. lactis* biomass was manufactured on whey, where it utilizes lactose as a source of carbohydrate to produce alimentary yeast with high nutritional value in food and feed [200,201]. The ability of *K. lactis* to utilize lactose is also used for the production of lactase. This *K. lactis* enzyme, which hydrolyses milk lactose into its constituent monosaccharide glucose and galactose, is the most widely used commercial preparation for the production of low-lactose milk for intolerant populations [202–206]. The lactase preparation from *K. lactis* has been affirmed as generally recognized as safe (GRAS) by the U.S. Food and Drug Administration (FDA) as a direct ingredient of food [207,208]. It has also been found economically feasible to use *K. lactis* for microbiological recycling of whey in countries where millions of tons of this product are still regarded as industrial waste [209]. Finally, in the recent past, *K. lactis* has proven to be efficacious as a host for the production of heterologous proteins [206].

30.2.4 Opportunistic Infections

Fungemia, or infection of the bloodstream, is one of the main pathological effects caused by IFIs caused by yeasts. Because yeasts in general are disseminated via the bloodstream, infection of almost any organ system or anatomic site is possible, which can lead to substantial morbidity and mortality. Yeasts can colonize the skin and membranous areas, and infections of the oral cavity, respiratory system, vagina, and anal region may occur. *Kluyveromyces* species have been evaluated in a corticosteroid-treated mouse model and shown to have a very low potential for virulence [210]. Isolation of this fungus from human sources is extremely rare. *Kluyveromyces fragilis* has been isolated from clinical samples such as from tubercular lung, tonsillar lesion, feces, sputum, lung tissue, and thoracic cavity [211]. Diagnosis would typically rely on traditional microbiologic methods, although in principle, amplification and sequencing of small subunit18S rRNA, D1/D2 domains of large subunit 28S rRNA, and ITS regions should be able to precisely identify *Kluyveromyces* species [212].

The first documented case of IFI due to the yeast *K. fragilis* involved an immunosuppressed cardiac transplant patient with pulmonary infection [213]. The patient was treated with amphotericin B and recovered without sequel. After treatment, the patient had many other opportunistic infections over the next 5 years, but no recurrence of *K. fragilis* infection. In vitro susceptibility studies of several *Kluyveromyces* strains indicate inhibition by 5-fluorocytosine and miconazole and borderline susceptibility to amphotericin B [214].

30.2.5 Conclusion

The safety of microbial food enzymes can be established on the basis of a history of safe use of the production organism, supplemented with studies of the enzyme preparation [215,216]. The safety of the production organism *K. lactis* was reviewed extensively by Bonekamp and Oosterom in 1994 [217] and the European Food Safety Authority [218]. They concluded that *K. lactis* is generally recognized as a safe organism for use in lactase and chymosin production. As a lactose-fermenting yeast, *Kluyveromyces* is mainly associated with milk and dairy products. The ability of *Kluyveromyces* to survive in the environment is limited, and concern for infection consequently will be mainly associated with highly immunocompromised hosts. Consequently, IFI due to *Kluyveromyces* appears to be exceptionally rare.

REFERENCES

1. Kwon-Chung KJ, Bennett JE. (1992) *Medical Mycology*. Lea & Febiger, Philadelphia, PA.
2. Kusumaningtyas E, Widiastuti R, Maryam R. (2006) Reduction of aflatoxin B1 in chicken feed by using *Saccharomyces cerevisiae*, Rhizopus oligosporus and their combination. *Mycopathologia* 162(4):307–311.
3. Surawicz CM, Elmer GW, Speelman P, McFarland LV, Chinn J, van Belle G. (1989) Prevention of antibiotic-associated diarrhea by *Saccharomyces boulardii*: A prospective study. *Gastroenterology* 96:981–988.
4. McFarland LV, Surawicz CM, Greenberg RN. (1995) Prevention of β-lactam associated diarrhea by *Saccharomyces boulardii* compared with placebo. *Am J Gastroenterol* 90:439–448.
5. McFarland LV, Surawicz CM, Greenberg RN. (1994) A randomised placebo-controlled trial of *Saccharomyces boulardii* in combination with standard antibiotics for Clostridium difficile disease. *J Am Med Assoc* 271:1913–1918.
6. Surawicz CM et al. (2000) The search for a better treatment for recurrent *Clostridium difficile* disease: Use of high-dose vancomycin combined with *Saccharomyces boulardii*. *Clin Inf Dis* 31:1012–1017.
7. Tung JM, Dolovich LR, Lee CH. (2009) Prevention of *Clostridium difficile* infection with *Saccharomyces boulardii*: A systematic review. *Can J Gastroenterol* 23(12):817–821.
8. Hof H. (2010) Mycoses in the elderly. *Eur J Clin Microbiol Infect Dis* 29:5–13.
9. Ooi CY, Dilley AV, Day AS. (2009) *Saccharomyces boulardii* in a child with recurrent *Clostridium difficile*. *Pediatr Int* 51(1):156–158.
10. Buts JP, Corthier G, Delmée M. (1993) *Saccharomyces boulardii* for *Clostridium difficile*-associated enterocolopathies in infants. *J Pediatr Gastroenterol Nutr* 16:1497–1504.
11. Höchter W, Chase D, Hegenhoff G. (1990) *Saccharomyces boulardii* in treatment of acute adult diarrhea. Efficacy and tolerance of treatment. *Münch Med Wochenschr* 132:188–192.

12. Cetina-Sauri G, Sierra Basto G. (1994) Evaluation thérapeutique du *Saccharomyces boulardii* chez des enfants souffrant de diarrhée aiguë. *Ann Pediatr* 41:397–400.
13. Kollaritsch H, Holst H, Grobara P. (1993) Prophylaxe der Reisediarrhoëmith *Saccharomyces boulardii*. *Fortschr Med* 111:152–156.
14. Buts JP. (2002) Exemple d'un médicament probiotique: *Saccharomyces boulardii* lyophilisé. In: Rambaud JC et al. (eds.). *Flore Microbienne Intestinale: Physiologie et Pathologie Digestives*, pp. 221–244. John Libbey, Montrouge, France.
15. Bleichner G, Bléhaut H, Mentec H, Moyse D. (1997) *Saccharomyces boulardii* prevents diarrhea in critically ill tube-fed patients. A multi-center, randomised, double-blind, placebo-controlled trial. *Intensive Care Med* 23:517–523.
16. Lolis N et al. (2008) *Saccharomyces boulardii* fungaemia in an intensive care unit patient treated with caspofungin. *Crit Care* 12:414–415.
17. Saint-Marc T et al. (1991) Efficacité de *Saccharomyces boulardii* dans le traitement des diarrhées du SIDA. *Ann Med Intern* 142:64–65.
18. Guslandi M, Mezzi G, Sorghi M, Testoni PA. (2001) *Saccharomyces boulardii* in maintenance treatment of Crohn's disease. *Dig Dis Sci* 45:1462–1464.
19. Guslandi M, Giollo P, Testoni PA. (2003) A pilot trial of *Saccharomyces boulardii* in ulcerative colitis. *Eur J Gastroenterol Hepatol* 15:697–698.
20. Riaz M, Alam S, Malik A. (2012) Efficacy and safety of *Saccharomyces boulardii* in acute childhood diarrhea: A double blind randomised controlled trial. *Ind J Pediatr* 79(4):478–482.
21. Shouval D, Ilan Y, Adler R, Deepen R, Panet A, Even-Chen Z, Gorecki M, Gerlich WH. (1994) Improved immunogenicity in mice of a mammalian cell-derived recombinant hepatitis B vaccine containing pre-S1 and pre-S2 antigens as compared with conventional yeast derived vaccines. *Vaccine* 12:1453–1459.
22. Schreuder MP, Mooren TA, Toschka HY, Verrips CT, Klis FM. (1996) Immobilizing proteins on the surface of yeast cells. *Trends Biotechnol* 14:115–120.
23. Hanzen KC. (1995) New and emerging yeast pathogens. *Clin Microbiol Rev* 8:462–478.
24. Byron JK, Clemons KV, McCusker JH, Davis RW, Stevens DA. (1995) Pathogenicity of *Saccharomyces cerevisiae* in complement factor five-derived mice. *Infect Immun* 63:478–485.
25. Sobel JD, Vazquez J, Lynch M, Meriwether C, Zervos MJ. (1993) Vaginitis due to *Saccharomyces cerevisiae*: Epidemiology, clinical aspects and therapy. *Clin Infect Dis* 16:93–99.
26. Hamoud S, Keidar Z, Hayek T. (2011) Recurrent *Saccharomyces cerevisiae* fungemia in an otherwise healthy patient. *Isr Med Assoc J* 13:575–576.
27. De Hoog GS. (1996) Risk assessment of fungi reported from humans and animals. *Mycoses* 39:407–417.
28. Pfaller MA, Diekema DJ, Gibbs DL, Newell VA, Meis JF, Gould IM, Fu W, Colombo AL, Rodriguez-Noriega E, and the Global Antifungal Surveillance Study. (2007) Results from the ARTEMIS DISK Global Antifungal Surveillance Study, 1997 to 2005: An 8.5 year analysis of susceptibilities of *Candida* species and other yeast species to fluconazole and voriconazole determined by CLSI standardized disk diffusion testing. *J Clin Microbiol* 45:1735–1745.
29. Pfaller MA et al.; for the Global Antifungal Surveillance Group. (2009) Results from the ARTEMIS DISK Global Antifungal Surveillance Study, 1997 to 2007, 10.5 year analysis of susceptibilities of noncandidal yeast species to fluconazole and voriconazole determined by CLSI standardized disk diffusion testing. *J Clin Microbiol* 47:117–123.
30. Kurtzman CP, Robnett CJ. (2003) Phylogenetic relationships among yeasts of the '*Saccharomyces* complex' determined from multigene sequence analyses. *FEMS Yeast Res* 3:417–432.
31. Naumov GI, James SA, Naumova ES, Louis EJ, Roberts IN. (2000) Three new species in the *Saccharomyces sensu stricto* complex: *Saccharomyces cariocanus*, *Saccharomyces kudriavzevii* and *Saccharomyces mikatae*. *Int J Syst Evol Microbiol* 50:1931–1942.
32. Van der Walt JP. (1970) *Saccharomyces* Meyen emend. Reess. In: Lodder J (ed.). *The Yeasts, A Taxonomic Study*, 2nd edn., pp. 555–718. North-Holland, Amsterdam, the Netherlands.
33. Vaughan-Martini A, Martini A. (1998) *Saccharomyces* Meyenex Reess. In: Kurtzman CP, Fell JW (eds.). *The Yeasts, A Taxonomic Study*, 4th edn., pp. 358–371. Elsevier Science, Amsterdam, the Netherlands.
34. Vaughan-Martini A, Kurtzman CP. (1985) Deoxyribonucleic acid relatedness among species of the genus *Saccharomyces sensu stricto*. *Int J Syst Bacteriol* 35:508–511.
35. Hogenauer C, Hammer HF, Krejs GJ, Reisinger EC. (1998) Mechanisms and management of antibiotic-associated diarrhea. *Clin Infect Dis* 27:702–710.
36. Marteau PR, de Vrese M, Cellier CJ, Schrezenmeir J. (2001) Protection from gastrointestinal disease with the use of probiotics. *Am J Clin Nutr* 73(S2):430S–436S.

37. McCullough MJ, Clemons KV, McCusker JH, Stevens DA. (1998) Species identification and virulence attributes of *Saccharomyces boulardii* (nom. inval.). *J Clin Microbiol* 36:2613–2617.
38. Hennequin et al. (2001) Microsatellite typing as a new tool for identification of *S. cerevisiae* strains. *J Clin Microbiol* 39:551–559.
39. Richard G F, Hennequin C, Thierry A, Dujon B. (1999) Trinucleotide repeats and other microsatellites in yeasts. *Res Microbiol* 150:589–602.
40. Broach J R, Jones EW, Pringle JR. (eds.) (1991) The molecular and cellular biology of the yeast *Saccharomyces, Vol. 1. Genome Dynamics, Protein Synthesis, and Energetics.* Cold Spring Harbor Laboratory Press, Cold Spring Harbor, New York.
41. Williams N. (1996) Genome projects: Yeast genome sequence ferments new research. *Science* 272 (5261):481.
42. Rainieri S, Zambonelli C, Kaneko Y. (2003) *Saccharomyces sensu stricto*: Systematics, genetic diversity and evolution. *J Biosci Bioeng* 96:1–9.
43. Taxis C, Keller P, Kavagiou Z, Jensen LJ, Colombelli J, Bork P, Stelzer EHK, Knop M. (2005) Spore number control and breeding in *Saccharomyces cerevisiae. J Cell Biol* 171(4):627–640.
44. Scott MP, Matsudaira P, Lodish H, Darnell J, Zipursky L, Kaiser CA, Berk A, Krieger M. (2004) *Molecular Cell Biology*, 5th edn. WH Freeman and Co, New York.
45. Honigberg SM. (2011) Cell signals, cell contacts, and the organization of yeast communities. *Eukaryot Cell* 10(4):466–473.
46. Rainieri S, Pretorius IS. (2000) Selection and improvement of wine yeasts. *Ann Microbiol* 50:15–31.
47. Benitez T, Gasent-Ramirez JM, Castrejon F, Codon AC. (1996) Development of new strains for the food industry. *Biotechnol Prog* 12:149–163.
48. Goffeau A et al. (1996) Life with 6000 genes. *Sci Gen* 274(5287):563–567.
49. Cherry JM et al. (2012) *Saccharomyces* genome database: The genomics resource of budding yeast. *Nucleic Acids Res* 40(Database issue):D700–D705.
50. Cliften P, Sudarsanam P, Desikan A, Fulton L, Fulton B, Majors J, Waterston R, Cohen BA, Johnston M. (2003) Finding functional features in *Saccharomyces* genomes by phylogenetic footprinting *Science* 301:71.
51. Kellis M, Patterson N, Endrizzi M, Birren B, Lander ES. (2003) Sequencing and comparison of yeast species to identify genes and regulatory elements. *Nature* 423:241–254.
52. Barnett JA. (1975) The entry of D-ribose into some yeasts of the genus Pichia. *J Gen Microbiol* 90(1):1–12.
53. Arthur H, Watson K. (1976) Thermal adaptation in yeast: Growth temperatures, membrane lipid, and cytochrome composition of psychrophilic, mesophilic and thermophilic yeasts. *J Bacteriol* 128(1):56–68.
54. Ostergaard S, Olsson L, Nielsen J. (2000) Metabolic engineering of *Saccharomyces cerevisiae. Microbiol Mol Bio Rev* 64(1):34–50.
55. Sutton DA, Fothergill AW, Rinaldi MG (eds.). (1998) *Guide to Clinically Significant Fungi*, 1st edn. Williams & Wilkins, Baltimore, MD.
56. Larone DH. (1995) *Medically Important Fungi—A Guide to Identification*, 3rd edn. ASM Press, Washington, DC.
57. Appendix I: Indo-European Roots. (2000) *The American Heritage Dictionary of the English Language*, 4th edn. Houghton Mifflin Harcourt.
58. Barnett JA. (2003) Beginnings of microbiology and biochemistry: The contribution of yeast research. *Microbiology* 149(3):557–567.
59. McGovern PE. (2010) *Uncorking the Past: The Quest for Wine, Beer, and Other Alcoholic Beverages.* University of California Press, Berkeley, CA.
60. Boulton C, Quain D. (2001) *Brewing Yeast and Fermentation*, 1st edn. Blackwell Science, Oxford, U.K.
61. Snodgrass ME. (2004) *Encyclopedia of Kitchen History*. Fitzroy Dearborn, New York.
62. McGovern PE et al. (2004) Fermented beverages of pre- and proto-historic China. *Proc Natl Acad Sci USA* 101:17593–17598.
63. This P, Lacombe T, Thomas MR. (2006) Historical origins and genetic diversity of wine grapes. *Trends Genet* 22:511–519.
64. Pretorius IS. (2000) Tailoring wine yeast for the new millennium: Novel approaches to the ancient art of winemaking. *Yeast* 16:675–729.
65. Corran, HS. (1975) *A History of Brewing*. David & Charles, Newton Abbot, London, U.K.
66. Sicard D, Legras JL. (2011) Bread, beer and wine: Yeast domestication in the *Saccharomyces sensu stricto* complex. *CR Biol* 334:229–236.

67. Dengis PB, Rouxhet PG. (1997) Surface properties of top and bottom fermenting yeast. *Yeast* 13:931–943.
68. Pretorius IS, Curtain CD, and Chambers PJ. (2012) The winemaker's bug: From ancient wisdom to opening new vistas with frontier yeast science. *Bioeng Bugs* 3(3):147–156.
69. Delwen S. (1996) Investigation of ancient Egyptian baking and brewing methods by correlative microscopy. *Science* 273:488–490.
70. Legras JL, Merdinoglu D, Cornuet J-M, Karst F. (2007) Bread, beer and wine: *Saccharomyces cerevisiae* diversity reflects human history. *Mol Ecol* 16(10):2091–2102.
71. Sazawal S et al. (2006) Efficacy of probiotics in prevention of acute diarrhea: A meta-analysis of masked, randomized, placebo-controlled trials. *Lancet Inf Dis* 6:374–382.
72. McFarland LV. (2010) Systematic review and meta-analysis of *Saccharomyces boulardii* in adult patients. *World J Gastroenterol* 16(18):2202–2222.
73. Kurugol Z, Koturoglu G. (2005) Effects of *Saccharomyces boulardii* in children with acute diarrhea. *Acta Paediatrica* 94(1):44–47.
74. Czerucka D, Piche T, Rampal P. (2007) Review article—Yeast as Probiotics—*Saccharomyces boulardii*. *Alimentary Pharm Thera* 26:767–778.
75. Skovgaard N. (2007) New trends in emerging pathogens. *Int J Food Microbiol* 120:217–224.
76. Capriotti A. (1954) Yeasts in some Netherlands soils. *Antonie van Leeuwenhoek* 21:145–156.
77. Capriotti A. (1967) Yeasts from U.S.A. soils. *Archiv für Mikrobiologie* 57:406–413.
78. Naumov GI, Naumova ES, Korhola M. (1992) Genetic identification of natural *Saccharomyces sensu stricto* yeasts from Finland, Holland and Slovakia. *Antonie van Leeuwenhoek* 61:237–243.
79. Naumov GI, Naumova ES, Sniegowski PD. (1998) *Saccharomyces* paradoxus and *Saccharomyces cerevisiae* are associated with exudates of North American oaks. *Can J Microbiol* 44:1045–1050.
80. Sniegowski PD, Dombrowski PG, Fingerman E. (2002) *Saccharomyces cerevisiae* and *Saccharomyces paradoxus* coexist in a natural woodland site in North America and display different levels of reproductive isolation from European conspecifics. *FEMS Yeast Res* 1:299–306.
81. Boekhout T, Herman JP. (eds.). (2003) Yeast biodiversity. In: *Yeasts in Food, Beneficial and Detrimental Aspects*, pp. 1–38. Behr's Verlag, Hamburg, Germany.
82. Goddard MR, Burt A. (1999) Recurrent invasion and extinction of a selfish gene. *Proc Natl Acad Sci USA* 96:13880–13885.
83. Toraskar MV, Modi VV. (1989) Studies on spoilage of Alphonso Mangoes by *Saccharomyces cerevisiae*. *Acta Horticulturae* 231:697–708.
84. Garza S, Teixidó JA, Sanchis V, Viñas I, Condón S. (1994) Heat resistance of *Saccharomyces cerevisiae* strains isolated from spoiled peach puree. *Int J Food Microbiol* 23:209–213.
85. Sláviková E, Vadkertiová R. (2003) The diversity of yeasts in the agricultural soil. *J Basic Microbiol* 43(5):430–436.
86. Vega FE, Dowd PF. (2005) The role of yeasts as insect endosymbionts In: Vega FE, Blackwell M (eds.). *Insect-Fungal Associations: Ecology and Evolution*, pp. 211–243. Oxford University Press, New York.
87. Suh SO, McHugh JV, Pollock DD, Blackwell M. (2005) The beetle gut: A hyperdiverse source of novel yeasts. *Mycological Res* 109(3):261–265.
88. Eng RH., Drehmel R, Smith SM, Goldstein EJC. (1984) *Saccharomyces cerevisiae* infections in man. *J Med Vet Mycol* 22:403–407.
89. Murphy A, Kavanagh K. (1999) Emergence of *Saccharomyces cerevisiae* as a human pathogen. *Impl Biotech Enzyme Microbiol Tech (Rev)* 25:551–557.
90. Jensen DP, David MD, Smith MD. (1976) Fever of unknown origin secondary to brewer's yeast ingestion. *Arch Intern Med* 136:332–333.
91. Smith D, Metzgar D, Wills C, Fierer J. (2002) Fatal *Saccharomyces cerevisiae* Aortic Graft Infection. *J Clin Microbiol* 40(7):2691–2692.
92. Wolochow H, Hildegrand GJ, Lamanna C. (1961) Translocation of microorganisms across the intestinal wall of the rat: Effect of microbial size and concentration. *J Infect Dis* 116:523–528.
93. Enache-Angoulvant A, Hennequin C. (2005) Invasive *Saccharomyces* infection: A comprehensive review. *Clin Infect Dis* 41:1559–1568.
94. Lherm T et al. (2002) Seven cases of fungemia with *S. boulardii* in critically ill patients. *Inten Care Med* 28:797–801.
95. Cassone M et al. (2003) Outbreak of *S. cerevisiae* subtype *S. boulardii* fungemia in patients neighboring those treated with a probiotic preparation of the organism. *J Clin Microbiol* 41:5340–5343.
96. Kiehn TE, Edwards FF, Armstrong D. (1980) The prevalence of yeasts in gastrointestinal inoculation in antibiotic treated mice. *Sabouraudia* 21:27–33.

97. Dynamac. (1991) Human health assessment for *Saccharomyces cerevisiae*. Unpublished, U.S. Environmental Protection Agency, Washington, DC.
98. Miceli MH, Diaz JA, Lee SA. (2011) Emerging opportunistic yeast infections. *Lanet Infect Dis* 11:142–151.
99. McCuster J. (2006) *Saccharomyces cerevisiae*: An emerging and model pathogenic fungus. In: Heitman J, Filler SG, Edwards JE, Mitchell AP. (eds.). *Molecular Principles of Fungal Pathogenesis*, pp. 245–260. CRC Press, Boca Raton, FL.
100. Posteraro B, Sanguinetti M, Romano L, Torelli R, Novarese L, Fadda G. (2005) Molecular tools for differentiating probiotic and clinical strains of *Saccharomyces cerevisiae*. *Int J Food Microbiol* 103:295–304.
101. Hoheisel C, Jacq M, Johnston EJ, Louis HW, Mewes Y, Murakami P, Philippsen H, Jacques N, Casaregola S. (2008) Safety assessment of dairy microorganisms: The hemiascomycetous yeasts. *Int J Food Microbiol* 126:321–326.
102. Barrett Bree Y, Hayes Y, Wilson RG, and Ryley JF. (1985) A comparison of phospholipase activity, cellular adherence and pathogenicity of yeasts. *J Gen Microbiol* 131:1217–1221.
103. Munoz P, Bouza E, Cuenca-Estrella M, Eiros JM, Perez MJ, Sanchez-Somolinos M, Rincon C, Hortal J, Pelaez T. (2005) *Saccharomyces cerevisiae* fungemia: An emerging infectious disease. *Clin Infect Dis* 40:1625–1634.
104. Olver WJ, James SA, Lennard A, Galloway A, Roberts IN, Boswell TC, Russell NH. (2002) Nosocomial transmission of *Saccharomyces cerevisiae* in bone marrow transplant patients. *J Hosp Infect* 52:268–272.
105. Viggiano M, Badett C, Bernini V, Garabedian M, Manelli JC. (1995) *Saccharomyces boulardii* fungemia in a patient with severe burns. *Ann Fr Anesth Reanim* 14:356–358.
106. Herbrecht R, Niviox Y. (2005) *Saccharomyces cerevisiae* fungemia: An adverse effect of *Saccharomyces boulardii* probiotic administration. *Clin Infect Dis* 40:1635–1637.
107. Salonen HJ et al. (2000) Fungal colonization of haematological patients receiving cytotoxic chemotherapy: Emergence of azole-resistant *Saccharomyces cerevisiae*. *J Hosp Infect* 45:293–301.
108. Jabra-Rizk MA, Ferreira SM, Sabet M, Falkler WA, Merz WG, Meiller TF. (2001) Recovery of Candida dubliniensis and other yeasts from human immunodeficiency virus associated periodontal lesions. *J Clin Microbiol* 39:4520–4522.
109. Krogh P, Holmstrup P, Vedtofte P, Pindborg JJ. (1986) Yeast organisms associated with human oral leukoplakia. *Acta Derm Venereol* 121:51–55.
110. Greer AE, Gemoets HN. (1943) The coexistence of pathogenic fungi in certain chronic pulmonary diseases; with especial reference to pulmonary tuberculosis. *Dis Chest* 9:212–224.
111. Kiehn TE, Edwards FE, Armstrong D. (1980) The prevalence of yeasts in clinical specimens from cancer patients. *Am J Clin Pathol* 73:518–521.
112. Debelian GJ, Olsen I, Tronstad L. (1997) Observation of *Saccharomyces cerevisiae* in blood of patient undergoing root canal treatment. *Int Endo J* 30:313–317.
113. Hennequin C et al. (2000) Possible role of catheters in *Saccharomyces boulardii* fungemia. *Eur J Clin Microbiol Infect Dis* 19:16–20.
114. McCusker JH, Clemons KV, Stevens DA, Davis RW. (1994) Genetic characterization of pathogenic *Saccharomyces cerevisiae* isolates. *Genetics* 136:1261–1269.
115. McCusker JH, Clemons KV, Stevens DA, Davis RW. (1994) *Saccharomyces cerevisiae* virulence phenotype as determined with CD-1 mice is associated with the ability to grow at 42°C and form pseudohyphae. *Infect Immun* 62:5447–5455.
116. Clemons KV, McCusker JH, Davis RW, Stevens DA. (1994) Comparative pathogenesis of clinical and nonclinical isolates of *Saccharomyces cerevisiae*. *J Infect Dis* 169:859–867.
117. Rees JR, Pinner RW, Hajjeh RA, Brandt ME, Reingold AL. (1998) The epidemiological features of invasive mycotic infections in the San Francisco Bay Area, 1992–1993: Results of population-based laboratory active surveillance. *Clin Infect Dis* 27:1138–1147.
118. Stefanatou E et al. (2011) Probiotic sepsis due to *Saccharomyces* fungaemia in critically ill burn patient. *Mycoses* 54:e643–e646.
119. Zerva L, Hollis RJ, Pfaller MA. (1996) in vitro susceptibility testing and DNA typing of *Saccharomyces cerevisiae* clinical isolates. *J Clin Microbiol* 34:3031–3034.
120. Perapoch J, Planes AM, Querol A, Lopez V, Martinez-Bendayan I, Tormo R, Fernandez F, Peguero G, and Salcedo S. (2000) Fungemia with *Saccharomyces cerevisiae* in two newborns, only one of whom had been treated with Ultra-Levura. *Eur J. Clin Microbiol Infect Dis* 19:468–470.
121. Baddley JW et al. (2008) Candida infective endocarditis. *Eur J Clin Microbiol Infect Dis* 27:519.

122. Durack DT. (1986) Infective and non-infective endocarditis. In: Hurst JW (ed.). *The Heart: Arteries and Veins*, pp. 1130–601. McGraw-Hill, New York.
123. Johnston PG et al. (1991) Late recurrent *Candida* endocarditis. *Chest* 99:1531–1533.
124. Ubeda P, Viudes A, Pérez-Belles C, Marqués JL, Pemán J, Gobernado M. (2000) Endocarditis caused by *Saccharomyces cerevisiae* on the prosthetic valve. *Enferm Infect Microbiol Clin* 18:142.
125. Ruiz-Esquide F, Díaz MC, Wu E, Silva V. (2002) Verrucous endocarditis secondary to *Saccharomyces cerevisiae*. A case report. *Rev Med Chil* 130(10):1165–1169.
126. Kliemann DA, Antonello VS, Severo LC, Pasqualotto AC. (2011) *Saccharomyces cerevisiae* oesophagitis in a patient with oesophageal carcinoma. *J Infect Dev Ctries* 5(6):493–495.
127. Konecny P, Drummond FM, Tish KN, Tapsall JW. (1999) *Saccharomyces cerevisiae* oesophagitis in an HIV-infected patient. *Int J Std AIDS* 10:821–822.
128. Wadhwa A, Kaur R, Bhalla P. (2010) *Saccharomyces cerevisiae* as a cause of oral thrush and diarrhea in an HIV/AIDS patient. *Trop Gastroenterol* 3:227–229.
129. Milner RJ, Picard J, Tustin R. (1997) Chronic episodic diarrhea associated with apparent intestinal colonisation by the yeasts *Saccharomyces cerevisiae* and *Candida famata* in a German shepherd dog. *J S Afr Vet Assoc* 68:147–149.
130. Begos DG, Rappaport R, Jain D. (1996) Cytomegalovirus infection masquerading as an ulcerative colitis flare-up: Case report and review of the literature. *Yale J Biol Med* 69:323–328.
131. Candelli M et al. (2003) *Saccharomyces cerevisiae*—Associated diarrhea in an immunocompetent patient with ulcerative colitis. *J Clin Gastroenterol* 36(1):39–40.
132. Wang AY et al. (2000) Factors predicting outcome of fungal peritonitis in peritoneal dialysis: Analysis of a 9-year experience of fungal peritonitis in a single center. *Am J Kidney Dis* 36:1183–1192.
133. Garcia-Martos P et al. (2009) Fungal peritonitis in ambulatory continuous peritoneal dialysis: Description of 10 cases. *Nefrologia* 29:534–539.
134. Matuszkiewicz-Rowinska J. (2009) Update on fungal peritonitis and its treatment. *Perit Dial Int* 29(S2):S161–S165.
135. Snyder S. (1992) Peritonitis due to *Saccharomyces cerevisiae* in a patient on CAPD. *Perit Dial Int* 12(1):77–78.
136. Mydlik M, Tkacova E, Szovenyova K, Mizla P, Derzsiova K. (1996) *Saccharomyces cerevisiae* peritonitis complicating CAPD. *Perit Dial Int* 16(2):188.
137. Mocan H, Murphy AV, Beattie TJ, McAllister TA. (1989) Fungal peritonitis in children on continuous ambulatory peritoneal dialysis. *Scott Med J* 34(4):494–496.
138. Gomila Sard B, Téllez-Castillo C.J, García Pérez H, Moreno Muñoz R. (2009) Peritonitis caused by *Saccharomyces cerevisiae* in an ambulatory peritoneal dialysis patient. *Nefrología* 29(4):371–372.
139. Hwang J, Kang K, Kang Y, Kim A. (2009) Probiotic gastrointestinal allergic reaction caused by *Saccharomyces boulardii*. *Ann Allergy Asthma Immunol* 103(1):87–80.
140. Swedberg J et al. (1985) Comparison of single dose vs one week course of metronidazole for symptomatic bacterial vaginosis. *J Am Med Assoc* 254:1046.
141. Rivers CA, Adaramola OO, Schwebke JR. (2011) Prevalence of bacterial vaginosis and vulvovaginal candidiasis mixed infection in a southeastern American STD clinic. *Sex Transm Dis*. 38(7): 672–674.
142. Sobel JD. (2007) Vulvovaginal candidiasis. *Lancet* 369:1961–1971.
143. Richter SS, Galask RP, Messer SA, Hollis RJ, Diekema DJ, Pfaller MA. (2005) Antifungal susceptibilities of *Candida* species causing vulvovaginitis and epidemiology of recurrent cases. *J Clin Microbiol* 43:2155–2162.
144. Siccardi D, Rellini P, Corte L, Bistoni F, Fatichenti F, Cardinali G. (2006) General evidence supporting the hypothesis that *Saccharomyces cerevisiae* vaginal isolates originate from food industrial environments. *New Microbiol* 29:201–206.
145. Savini V et al. (2008) Two cases of vaginitis caused by itraconazole-resistant *Saccharomyces cerevisiae* and a review of recently published studies. *Mycopathologia* 166:47–50.
146. Passos XS, Costa CR, Araujo CR, Nascimento ES, e Souza LK, Fernandes Ode F, Sales WS, Silva Mdo R. (2007) Species distribution and antifungal susceptibility patterns of *Candida* spp. bloodstream isolates from a Brazilian tertiary care hospital. *Mycopathologia* 163(3):145–151.
147. Echeverrı́a-Irigoyen M et al. (2011) *Saccharomyces cerevisiae* vaginitis: Microbiology and in vitro antifungal susceptibility. *Mycopathologia* 172:201–205.
148. Agatensi L, Franchi F, Mondello F, Ceddia T, De Bernardis F, Cassone A. (1991) Vaginopathic and proteolytic Candida species in outpatient attending a gynecology clinic. *J Clin Pathol* 44:826e30.

149. Wilson JD, Jones BM, Kinghorn GR. (1988) Bread-making as a source of vaginal infection with *Saccharomyces cerevisiae*: Report of a case in a woman and apparent transmission to her partner. *Sex Transm Dis* 15:35–36.
150. Nyirjesy P, Vazquez JA, Ufberg DD, Sobel JD, Biokov DA, Buckley HR. (1995) *Saccharomyces cerevisiae* vaginitis: Transmission from yeast used in baking. *Obstet Gynecol* 86:326–329.
151. Tawfik OW, Papasian CJ, Dixon AY, Potter LM. (1989) *Saccharomyces cerevisiae* pneumonia in a patient with acquired immune deficiency syndrome. *J Clin Microbiol* 27:1689–1691.
152. Doyle MG, Pickering LK, O'Brien N, Hoots K, Benson JE. (1990) *Saccharomyces cerevisiae* infection in a patient with acquired immunodeficiency syndrome. *Pediatr Infect Dis J* 9:850–851.
153. Ren P, Sridhar S, Chaturvedi V. (2004) Use of paraffin-embedded tissue for identification of *Saccharomyces cerevisiae* in a Baker's lung nodule by fungal PCR and nucleotide sequencing. *J Clin Microbiol* 42(6):2840–2842.
154. Chertow GM, Marcantonio ER, Wells RG. (1991) *Saccharomyces cerevisiae* empyema in a patient with esophago-pleural fistula complicating variceal sclerotherapy. *Chest* 99:1518–1519.
155. Irby PH, Stoller ML, McAninich JW. (1990) Fungal bezoars of the upper urinary tract. *J Urol* 143:447–451.
156. Senneville E et al. (1996) Bilateral ureteral obstruction due to *Saccharomyces cerevisiae* fungus balls. *Clin Infect Dis* 23:636–637.
157. Koening SB. (1986) Fungal keratitis. In: Tabbara KF, Hyndiuk RA (eds.). *Infections of the Eye*, pp. 331–342. Little, Brown and Co, Boston, MA.
158. Ando N, Takatori K. (1982) Fungal flora of the conjunctival sac. *Am J Ophthalmol* 94:67–74.
159. Kirsch LS et al. (1999) *Saccharomyces* keratitis and endophthalmitis. *Can J Ophthalmol* 34:229–232.
160. Rossi F, Torriani S. (2011) *Saccharomyces cerevisiae*. In: Liu D (ed.). *Molecular Detection of Human Fungal Pathogens*, pp. 603–614. CRC Press, Boca Raton, FL.
161. Beighton D et al. (1995) Use of CHROMagar Candida medium for isolation of yeasts from dental samples. *J Clin Microbiol* 33:3025–3027.
162. Ghelardi E et al. (2008) Efficacy of Chromogenic Candida agar for isolation and presumptive identification of pathogenic yeast species. *Clin Microbiol Infect* 14(2):141–147.
163. Iwanaga S, Miyata T, Tokunaga F, Muta T. (1992) Molecular mechanism of hemolymph clotting system in Limulus. *Thromb Res* 68(1):1–32.
164. Miyazaki T et al. (1995) Plasma (1,3)-beta-D-glucan and fungal antigenemia in patients with candidemia, aspergillosis, and cryptococcosis. *J Clin Microbiol* 33(12):3115–3118.
165. Yoshida M et al. (1997) Detection of plasma (1,3)-beta-D-glucan in patients with Fusarium, Trichosporon, *Saccharomyces* and *Acremonium* fungaemias. *J Med Vet Mycol* 35(5):371–374.
166. Odabasi Z, Paetznick VL, Rodriguez JR, Chen E, McGinnis MR, Ostrosky-Zeichner L. (2006) Differences in beta-glucan levels in culture supernatants of a variety of fungi. *Med Mycol* 44(3):267–272.
167. Presterl E et al. (2009) Invasive fungal infections and (1,3)-beta-D-glucan serum concentrations in long-term intensive care patients. *Int J Infect Dis* 13(6):707–712.
168. Digby J, Kalbfleisch J, Glenn A, Larsen A, Browder W, Williams D. (2003) Serum glucan levels are not specific for presence of fungal infections in intensive care unit patients. *Clin Diagn Lab Immunol* 10(5):882–885.
169. Tan KE, Ellis BC, Lee R, Stamper PD, Zhang SX, Carroll KC. (2012) Prospective evaluation of a matrix-assisted laser desorption ionization-time off light mass spectrometry system in a hospital clinical microbiology laboratory for identification of bacteria and yeasts: A bench-by-bench study for assessing the impact on time to identification and cost-effectiveness. *J Clin Microbiol* 50(10):3301–3308.
170. Sendid B, Ducoroy P, François N, Lucchi G, Spinali S, Vagner O, Damiens S, Bonnin A, Poulain D, Dalle F. (2013) Evaluation of MALDI-TOF mass spectrometry for the identification of medically-important yeasts in the clinical laboratories of Dijon and Lille hospitals. *Med Mycol* 51(1):25–32.
171. Qian J, Cutler JE, Cole RB, Cai Y. (2008) MALDI-TOF mass signatures for differentiation of yeast species, strain grouping and monitoring of morphogenesis markers. *Anal Bioanal Chem* 392(3):439–449.
172. Bašková L, Buchta V. (2012) Laboratory diagnostics of invasive fungal infections: An overview with emphasis on molecular approach. *Folia Microbiol (Praha)* 57(5):421–430.
173. Martorell P et al. (2005) Rapid identification and enumeration of *Saccharomyces cerevisiae* cells in wine by real-time PCR. *Appl Environ Microbiol* 71(11):6823–6830.
174. Salinas F et al. (2009) Taqman real-time PCR for the detection and enumeration of *Saccharomyces cerevisiae* in wine. *Food Microbiol* 26(3):328–332.

175. Chang HW et al. (2007) Quantitative real time PCR assays for the enumeration of *Saccharomyces cerevisiae* and the *Saccharomyces* sensu stricto complex in human feces. *J Microbiol Methods* 71(3):191–201.
176. Linton CJ, Borman AM, Cheung G, Holmes AD, Szekely A, Palmer MD, Bridge PD, Campbell CK, Johnson EM. (2007) Molecular identification of unusual pathogenic yeast isolates by large ribosomal subunit gene sequencing: 2 years of experience at the United Kingdom mycology reference laboratory. *J Clin Microbiol* 45(4):1152–1158.
177. Landlinger C, Basková L, Preuner S, Willinger B, Buchta V, Lion T. (2009) Identification of fungal species by fragment length analysis of the internally transcribed spacer2 region. *Eur J Clin Microbiol Infect Dis* 28(6):613–622.
178. Vollmer T, Störmer M, Kleesiek K, Dreier J. (2008) Evaluation of novel broad-range real-time PCR assay for rapid detection of human pathogenic fungi in various clinical specimens. *J Clin Microbiol* 46:1919–1926.
179. Landlinger C et al. (2010) Diagnosis of invasive fungal infections by a real-time panfungal PCR assay in immunocompromised pediatric patients. *Leukemia* 24(12):2032–2038.
180. Posteraro B et al. (1999) Molecular and epidemiological characterization of vaginal *Saccharomyces cerevisiae* isolates. *J Clin Microbiol* 37(7):2230.
181. Anderson JB, Sirjusingh C, Ricker N. (2004) Haploidy, diploidy and evolution of antifungal drug resistance in *Saccharomyces cerevisiae*. *Genetics* 168:1915–1923.
182. Auscott JN, Fayen J, Grossnicklas H, Morrissey A, Ledermann MM, Salatta R. (1990) Invasive infection with *Saccharomyces cerevisiae*: Report of three cases and review. *Rev Infect Dis* 12:406–411.
183. Chitasombat MN et al. (2012) Rare opportunistic (non-*Candida*, non-*Cryptococcus*) yeast bloodstream infections in patients with cancer. *J Infect* 64:68e75.
184. Tiballi RN, Spiegel JE, Zarins LT, Kauffmann CA. (1995) *Saccharomyces cerevisiae* infections and antifungal susceptibility studies by colorimetric and broth macrodilution methods. *Diag Microbiol Infect Dis* 23:135–140.
185. Espinel-Ingroff A. (1998) in vitro activity of the new triazole Voriconazole (UK-109-496) against opportunistic filamentous and dimorphic fungi and common emerging yeast pathogens. *J Clin Microbiol* 36:198–202.
186. Burkhardt O, Kohnlein T, Pletz M, Welte T. (2005) *Saccharomyces boulardii* induced sepsis: Successful therapy with voriconazole after treatment failure with fluconazole. *Scand J Infect Dis* 37:69–72.
187. Fonseca GG. (2008) The yeast Kluyveromyces marxianus and its biotechnological potential. *Appl Microbiol Biotechnol* 79:339–354.
188. Lane MM et al. (November 2011) Physiological and metabolic diversity in the yeast *Kluyveromyces marxianus*. *Antonie van Leeuwenhoek* 100(4):507–519.
189. Van der Walt, JP. (1956) The yeast *Kluyveromyces africanus* nov. spec. and its phylogenetic significance. *Antonie van Leeuwenhoek* 22:321–326.
190. Lachance MA. (2007) Current status of *Kluyveromyces* systematics. *FEMS Yeast Res* 7:642–645.
191. Lachance MA. (1993) *Kluyveromyces* systematics since 1970. *Antonie Leeuwenhoek* 63:95–104.
192. Lachance MA. (1998) Kluyveromyces van der Walt emend. Van derWalt. In: Kurtzman CP, Fell JW (eds.). *The Yeasts, a Taxonomic Study*, 4th edn., pp. 227–247. Elsevier, Amsterdam, the Netherlands.
193. Li M, Fu X, Tang R. (1990) The yeasts in Shennongjia, China, and a new species of *Kluyveromyces*. *Acta Microbiol Sinica* 30:94–97.
194. Li M, Fu X, Tang R. (1992) *Kluyveromyces hubeiensis*, new species from Shennongjia. *Acta Microbiol Sinica* 30:94–97.
195. Weber G, Spaaij F, vander Walt JP. (1992) *Kluyveromyces piceae* sp. nov., a new yeast species isolated from the rhizosphere of *Picea abies* (L.) Karst. *Antonie van Leeuwenhoek* 62:239–244.
196. Johnston M. (1987) A model fungal gene regulatory mechanism: The GAL genes of *Saccharomyces cerevisiae*. *Microbiol Rev* 51:458–476.
197. Lohr D, Venkov P, Zlatanova J. (1995) Transcriptional regulation in the yeast GAL gene family: A complex genetic network. *Fed Am Soc Exp Biol* 9:777–787.
198. Schaffrath R, Breunig KD. (2000) Genetics and molecular physiology of the yeast *Kluyveromyces lactis*. *Fungal Genet Biol* 30:173–190.
199. Van Ooyen AJ, Dekker P, Huang M, Olsthoorn MM, Jacobs DI, Colussi PA, Taron CH. (2006) Heterologous protein production in the yeast Kluyveromyces lactis. *FEMS Yeast Res* 6:381–392.
200. Vrignaud Y. (1971) Levure lacliquo. *Rev Illst Pasteur Lyo* 4:147–165.
201. Valli S et al. (2012) A Study on the Lactase Potential of *Kluyveromyces lactis* Grown in Whey. *Int J Pharma Biol Arch* 3(4):877–883.

202. Woychik JH, Holsinger VI-J. (1977) Use of lactase in the manufacture of dairy products. In: Ory RL, Angelo AJ (eds.). *Enzymes in Foods and Beverage Processing*, pp. 30–31. San Francisco, CA, August 1976. [ACS (American Chemical, Society) series 47]. American Chemical Society, Washington, DC, pp. 67–79.
203. Nijpels HH. (1981) Lactases and their applications. In: Birch GG, Blakcbrough N, Parker KJ (eds.). Ind Univ Coop Symp (1980) (Publ 1981). *Enzymes and Food Processing*, pp. 89–103. Applied Science Publisher, London, U.K.
204. Bacqucs C, Simonpicri V. (1987) Lactose fermentation by *Kluyveromyces lactis* in half-skimmed milk. *Microbiol Aliments Nutr* 5:219–225.
205. Hussein L, Elasyed S, Foda S. (1989) Reduction of lactose in mil by purified lactase produced by *Kluyveromyces lactis. J Food Prot* 52:30–34.
206. Krijger et al. (2012) A novel, lactase-based selection and strain improvement strategy for recombinant protein expression in Kluyveromyces lactis. *Microbial Cell Factories* 11:112.
207. Randolph WF. (1984) Direct food substance affirmed as generally recognized as safe; lactase enzyme preparation from *Kluyveromyces lactis. Fed Reg* 49(234):47384–47387.
208. Fukuhara H. (2005) Kluyveromyces lactis—A retrospective. *FEMS Yeast Res* 6:323–324.
209. Foda MS, Mohammed S, Hussein L. (1988) Production of lactase from *Kluyveromyces lactis* propagated in media with different sodium chloride concentrations. *Zentralbl Mikrobil* 143:583–590.
210. Holzschu DL, Chandler FW, Ajello L, Ahearn DG. (1979) Evaluation of industrial yeasts for pathogenicity. *Sabouraudia* 17(1):71–78.
211. Van Der Walt JP. (1970) *Kluyveromyces*, Van der Walt emend, van der Walt, In: Lodder J (eds.). *The Yeasts—A Taxonomic Study*, pp. 34–113. North-Holland Publishing Company, Amsterdam, the Netherlands.
212. Liu D. (2011) "Kluyveromyces," In: Liu, D. (ed.). *Molecular Detection of Human Fungal Pathogens*, pp. 581–584. CRC Press, Boca Raton, FL.
213. Lutwick LI, Phaff HJ, Stevens DA. (1980) *Kluyveromyces fragilis* as an opportunistic fungal pathogen in man. *Sabouraudia* 18(1):69–73.
214. Stevens DA. (1977) Miconazole in the treatment of systemic fungi infections. *Am Rev Resp Dis* 116:801–806.
215. Battershill JM. (1993) Guidelines for the safety assessment of microbial enzymes in food. *Food Additives Contaminants* 10:479–488.
216. SCF. (1992) Guidelines for the presentation of data on food enzymes. Report of the Scientific Committee for Food 27th Report series (EUR14181); Office for Official Publications of the EEC, pp. 13–22.
217. Bonekamp FJ, Oosterom J. (1994) On the safety of *Kluyveromyces lactis*-a review. *App Microbiol Biotech* 41:1–3.
218. Introduction of a Qualified Presumption of Safety (QPS) approach for assessment of selected microorganisms referred to EFSA. *EFSA J* (2007) 587:1–16.

31 Trichoderma Mycoses and Mycotoxins

Christian P. Kubicek and Irina S. Druzhinina

CONTENTS

31.1 Introduction 521
31.2 Taxonomy and Identification of *Trichoderma* 522
31.3 Clinical Aspects of *Trichoderma* 523
31.4 *Trichoderma* in Food and Water... 524
31.5 Toxic Secondary Metabolites Produced by *Trichoderma* 524
    31.5.1 Peptaibols.. 524
    31.5.2 Epidithiodioxopiperazines 526
    31.5.3 Trichothecenes.. 528
31.6 Concluding Remarks 532
Acknowledgments.. 532
References.. 532

31.1 INTRODUCTION

Trichoderma (teleomorph *Hypocrea*, Hypocreales, Ascomycota, Dikarya) species are among the most common mitosporic fungi recovered in cultivation-based surveys of soil samples. They have also been isolated from an innumerable diversity of natural and artificial substrata, thus illustrating a high opportunistic potential and adaptability to various ecological conditions [1,2]. Some are applied in biotechnology due to their abilities to (1) produce enzymes such as cellulases and hemicellulases for conversion of plant biomass into soluble sugars used for biorefinery processes [3,4], (2) combat phytopathogenic fungi [5] and nematodes [6,7], and (3) stimulate plant defenses against plant pathogens [2,5]. Thus they are of considerable importance to mankind.

Most *Trichoderma* species are generally considered to be harmless fungi, because they are not known to produce toxins when growing on food or feedstocks. The industrially used cellulase producer *Trichoderma reesei* has even GRAS status (http://www.accessdata.fda.gov/scripts/fcn/gras_notices/grn000372.pdf). Yet a set of four closely related species (the *Trichoderma brevicompactum* clade) has been shown to produce trichothecene mycotoxins [8]. In addition, some *Trichoderma* spp. have become a member of opportunistic fungal pathogens against immunocompromised humans and mammals, although almost all reported cases are mainly from two genetically related species: the strictly anamorphic *Trichoderma longibrachiatum* and the sexually propagating *Hypocrea orientalis* [9,10]. In addition, some *Trichoderma* spp., including *T. longibrachiatum*, are frequently found as indoor contaminants of buildings [11,12]. Fungal indoor contaminants can lead to psychological responses in the occupants such as fatigue and nausea and lead to respiratory complaints and mucous membrane irritation [13]. While the mechanisms by which *Trichoderma* acts on immunocompromised cells have not been studied, there is some evidence that substances emitted by *Trichoderma* may be the cause [14]. Volatile organic compounds from a strain of *Trichoderma viride* were also reported to cause histamine release from human bronchoalveolar cells [15]. In addition, the genus *Trichoderma* is maintained within the group of trichothecene producers (see Section 31.5.3), although this could only be confirmed for one small phylogenetic group, the *T. brevicompactum* clade [8]. However, the identity of *Trichoderma* species reported as mycotoxin producers or associated with clinical problems, particularly in the *pre-PCR* era, should be treated with caution. This will be explained as follows:

31.2 TAXONOMY AND IDENTIFICATION OF *TRICHODERMA*

The name *Trichoderma* was introduced more than 200 years ago by Persoon [16] on the basis of material collected in Germany. He included four species in the genus, but only *T. viride* proved to be *Trichoderma* based on subsequent investigations. Bisby [17] proposed that it consisted of a single species, *T. viride*. After 1969, Rifai [18] recognized nine *aggregate species* based on morphological characters. Importantly, he proposed that they contain more than one species and surmised that these would become distinguishable with more refined methods. Bissett [19–23] refined and extended this to sections with several biological species. However, his key was based solely on morphological characters.

Since PCR tools and DNA sequence analysis became available and have been applied to *Trichoderma* and *Hypocrea*, more than 200 *Trichoderma* species have been recognized [24–36]. The results from molecular taxonomy and phylogeny have made clear that the classical phenotypic approach for the identification of *Trichoderma* is severely impaired by the homoplasy of morphological characters, making morphological species identification problematic to even *Trichoderma* specialists. Since the genus is now exceptionally well characterized by gene sequences, molecular identification at the species level has become the standard. Virtually, every species is documented with diagnostic sequences from at least one to two genes in the database of the National Center for Biotechnology Information (NCBI). Researchers are theoretically able to identify all known species using nucleotide blast [37]. However, high similarity of sequences does not necessarily confirm species identity because the intraspecific variability of this sequence can be different in different species. Hence, Druzhinina et al. [38,39] established the oligonucleotide barcode program *TrichO*Key and a BLAST search tool *Tricho*BLAST using only vouchered sequences as the database, which are available online [40]. A problem with this program is that it is exclusively based on the internal transcribed spacers of the ribosomal RNA gene cluster (ITS1 and ITS2), which are now known to be insufficiently polymorphic in all species. To eliminate these problems, *Tricho*BLAST further incorporated sequences of four independent loci from all genetically characterized *Trichoderma* and *Hypocrea* species: (1) ITS1 and ITS2, (2) introns *tef1*_int4[large] and *tef1*_int5[short], (3) exon [*tef1*_exon6[large]] of the gene encoding translation elongation factor 1-alpha [*tef1*], and (4) a portion of the exon between the fifth and seventh eukaryotic conserved amino acids of subunit 2 of the RNA polymerase gene [*rpb2*_exon]. The respective primers and PCR conditions are similar to those used for many other fungi and can be retrieved from [26,32,34]. Also, a summary of the primers and PCR conditions for the amplification of ITS1 and ITS2, *tef1*_int4, *cal1*, and *chi18-5* from human pathogenic *Trichoderma* spp. has been published [41]. Unfortunately, sequences for the latter two genes are not available for all *Trichoderma* spp., but by using the others listed, every known species of *Trichoderma* can be identified. Attempts to develop methods for the specific amplification of genes from individual *Trichoderma* spp. by species-specific primers from any of the aforementioned genes have so far failed, due to the fact that there are no nucleotide regions that vary in all species and that could be used for the design species-specific primers (I.S. Druzhinina, unpublished data). Unfortunately, some journals still accept papers where *Trichoderma* strains have been identified by phenotypic methods only (particularly when, e.g., metabolite production is the subject of the paper), thus adding to the confusion about *Trichoderma* taxonomy.

31.3 CLINICAL ASPECTS OF *TRICHODERMA*

Species of *Trichoderma* can cause invasive mycoses in mammals and humans under conditions where the immune system has been severely compromised. Typically, *Trichoderma* has been isolated from the peritoneal effluent of dialysis patients, infections of transplant recipients receiving immunosuppressive treatment, and patients suffering from leukemia, brain abscesses, and HIV [42–46]. The infections frequently colonize body cavities such as the peritonea or paranasal sinuses, the pulmonary system, and the lung [46–51]. However, skin lesions have also been reported [52]. While infections by *Trichoderma* are not a major threat and occur infrequently, they nevertheless pose difficult diagnostic and therapeutic challenges because without rapid diagnosis and treatment, their clinical manifestations can be fatal [53–57]. Unfortunately, both requirements are difficult to fulfill: *Trichoderma* spp. can only be safely identified by sequence analysis (see earlier). Unless the infection is superficial or localized, such as sinusitis or skin infections, the diagnosis of *Trichoderma* spp. can only be made by isolation of the fungal material from the affected organs (e.g., lungs, heart, pretracheal abscesses). Therefore, even with the use of molecular identification tools, diagnosis is currently a time-consuming task in the treatment of the patient.

Furthermore, *Trichoderma* recovered from a clinic environment are resistant to many antifungal agents [58], for example, fluconazole, 5-fluorocytosine, and amphotericin B. However, they are susceptible, at least partially, to itraconazole, ketoconazole, miconazole, and voriconazole [41]. As mentioned, voriconazole—a registered agent for the treatment of systemic *Aspergillus*, *Candida*, and *Fusarium* spp. infections—was reported active against *Trichoderma*. Due to lower MIC values, fewer side effects, and higher penetration than amphotericin B, it may be an important drug in the treatment of invasive *Trichoderma* infections [58]. More effective drugs are required as the prognosis for recovering from a *Trichoderma* infection remains only 50% [41].

Importantly, the number of *Trichoderma* spp. that cause infections is very limited. While the literature is full of reports naming only *T. viride* or *T. harzianum* as the causal agent, this is due to the use of poor identification methods. Druzhinina et al. [10] have shown that the majority of infections are due to the closely related species—*T. longibrachiatum* and *H. orientalis*. Interestingly, whereas *T. longibrachiatum* is essentially clonal, *H. orientalis* forms a worldwide recombining population, and this difference may be relevant for antifungal therapy, as the genes encoding factors for antibiotic resistance and virulence could be exchanged during sexual reproduction. Clinical isolates of both species shared identical haplotypes with environmental strains, indicating a threat for nosocomial infections as virtually any strain of these species may cause invasive mycoses [10].

Apart of these two species, there are two reliable case reports for *Trichoderma citrinoviride* [59,60], which is a close relative of *T. longibrachiatum*. In addition, *Hypocrea peltata* has been identified from lung tissue of a patient, but the case study did not reveal whether the fungus was the cause of the disease [61,62]. A common feature of *T. longibrachiatum*, *H. orientalis*, *T. citrinoviride*, and *H. peltata* is that they can grow at 37°C, a property absent from most other *Trichoderma* spp.

Interestingly, there are also two case reports for *T. harzianum* [41,49] that actually consist of nine sibling species [32], and it is not known which of them was found in these case studies. There is also one report for *Trichoderma atroviride* [63]. While these species, which are phylogenetically far from *T. longibrachiatum*, *H. orientalis*, and *H. peltata*, are a minority in the number of cases reported, their detection nevertheless indicates that the general opportunistic nature of *Trichoderma* may also enable other species to establish themselves in a clinic environment.

The source of infection by *T. longibrachiatum* and *H. orientalis* appears to be mainly from the indoor mycoflora in hospitals [9,41]. However, in one case, the infection could be traced back to the water in a humidifier in the patient's household [64].

There has been little attempt toward an understanding of the mechanisms by which particular members of the *Hypocrea/Trichoderma* infect human cells. When *T. longibrachiatum* is confronted with lung cell cultures in vitro, the human cells rapidly start to sediment and lose their adhesive properties [65], suggesting the action of proteases and/or secondary metabolites. No such effect was observed for *T. reesei*, which was used as a nonpathogenic control in this study. The genomes of *T. longibrachiatum* and *T. citrinoviride* have recently been sequenced and annotated [66,67], and their analysis toward understanding the ability to attack immunocompromised humans is in progress (I.S. Druzhinina and C.P. Kubicek, unpublished data).

31.4 *TRICHODERMA* IN FOOD AND WATER

Being an opportunistic organism, one can expect that *Trichoderma* will be found in other materials with which humans are in direct contact. In fact, they were among the most frequently isolated fungi in a survey of fungi in drinking water systems in Norway, albeit only in surface water systems [68]. They were absent from hot-water taps, which is consistent with *Trichoderma* not being thermotolerant. Hageskal et al. [69] isolated 123 *Trichoderma* strains from raw water, treated water, and water from private homes and hospital installations. They identified 11 *Trichoderma/Hypocrea* and one unknown species using molecular identification methods. Approximately 22% of the surface-derived water samples were positive for species from the *T. viride* clade, and the species were frequently isolated throughout the surface-sourced drinking water distribution system. Water treatment had only a minor effect in removing *Trichoderma* from raw water, suggesting that the fungus may actually grow in the water distribution system. One must note that this single study was performed in a geographically small area, and studies in other regions are required. Gashgari et al. [70] recently identified *T. viride* as the dominant fungal species in treated water and household water of Jeddah City, Saudi Arabia. The species identity as *T. viride*, however, must be considered as uncertain because the authors used ITS1 and ITS2 sequence analysis for species determination only, which does not distinguish between *T. viride* and closely related species [71]. *Trichoderma* has been found in bottled water in Brazil, but no species identification was performed [72]. As noted earlier, there is one reported case of occurrence of *Trichoderma* in a humidifier [64].

Food is also a common source for the introduction of fungi. Unfortunately, there has been no systematic study on the occurrence of *Trichoderma* as a contaminant of food. However, some case studies on spices, berries, grapes and citrus fruits, herb tea, and nuts have shown that *Trichoderma* was isolated [73–77]. It should be noted that all these studies were using cultivation-based identifications, and isolates were not identified to species.

31.5 TOXIC SECONDARY METABOLITES PRODUCED BY *TRICHODERMA*

Trichoderma spp. produce a copious array of secondary metabolites [78]. However, these lists are seriously flawed by the high number of incorrect species names for the producing species, which makes an accurate alignment of secondary metabolites to *Trichoderma* spp. impossible. In addition, although toxicity tests (mostly against bacteria and other fungi) have been published for many of them, only a few have been investigated for a possible toxicity to humans and mammals. Toxicological data are available for only three classes of compounds that are produced by *Trichoderma* spp.: (1) peptaibols, (2) epidithiodioxopiperazine (ETP) metabolites, and (3) trichothecenes.

31.5.1 PEPTAIBOLS

Peptaibols are small linear peptides (characterized by molecular weights of 500–2200 Da), which bear some notable differences to other peptides: they contain a high number of nonproteinogenic, α,α'-dialkylated α-amino acids such as isovaline and α-aminoisobutyric acid (Aib); their N-terminal amino acid is acetylated, and the C-terminus is an amino alcohol, mostly phenylalaninol. These properties have given rise to the name peptaibol (*pept*ide, *Aib*, and amino alcoho*l*). They are typically produced by taxa of the Hypocreaceae, Clavicipitaceae, and Bionectriaceae of the order Hypocreales and proliferate most strongly in *Trichoderma* spp. Publications regarding the isolation of peptaibiotics from fungi not belonging to the *Hypocreales* (notably basidiomycetes) should be considered very critically as outlined in [79]. Since *Trichoderma* spp. are mycoparasites, they infect other asco- or basidiomycete's fruiting body but may not have been visible to the naked eye, which may have led to the detection of peptaibols in extracts from these fruiting bodies. It is therefore possible that peptaibols may be present in edible mushrooms that have become infected by *Trichoderma* [80], but this topic has so far not been investigated.

Peptaibols present unique physicochemical and biological activities depending on particular structural properties: they form a helical structure with the hydrophobic side chains exposed to the surface that allows them to interact with natural and artificial bilayers and form pores or voltage-dependent ion channels increasing membrane permeability [81–85].

The structure and properties of more than 700 peptaibols are collected in the Peptaibol Database [86]. Details of their biochemistry and structure are reviewed by Duclohier [85]. Since peptaibols are peptides, their biosynthesis occurs by nonribosomal peptide synthases (NRPSs). These large multifunctional enzymes are encoded by a single gene that encodes coordinated groups of modules, each containing active sites for single-synthesis steps. Each module is required for the catalysis of one single cycle of product length elongation. The number and order of modules and the type of domains that are present within a module determine the structural variation of the resulting peptide product. A minimum set of domains required for one elongation cycle consist of a module with adenylation (A), thiolation (T, using a peptidyl-carrier protein), and condensation (C) domain. The A-domain is responsible for substrate selection and its covalent fixation on the phosphopantetheine (PCP) arm of the T-domain through an AMP-derivative intermediate. The C-domain catalyzes the formation of the peptide bond between the aminoacyl- or peptidyl-S-PCP from the upstream module and the aminoacyl moiety attached to the PCP in the corresponding downstream module. Most peptide synthases also have a C-terminal thioesterase domain in the last module that is responsible for the release/cyclization of the final product. However, this module is absent in peptaibol synthases, and peptaibols are thus linear peptides. Instead, they possess a dehydrogenase domain at the C-terminus that reduces the C-terminal amino acid to the amino acid alcohol. It is also believed that this module affects the release of the product by reductive cleavage.

The peptaibols formed by *Trichoderma* usually comprise between 11 and 23 amino acid residues, with a single central proline moiety [79,83,85]. Genes encoding peptaibol synthases have been cloned and functionally characterized from *Trichoderma virens*, *T. reesei*, *T. asperellum*, and *T. harzianum* [87–92]. Based on these data, the general structure of the *Trichoderma* peptaibol synthases can be grouped into two types: large proteins that consist of 18 or more modules and smaller ones that consist of 14 modules. While the large synthases produce only the high-molecular-weight peptaibols, the small synthases were shown to be responsible for formation of both the 11- and 14-residue peptaibol classes [89,92]. Degenkolb et al. [90] proposed that the formation of two differently sized peptaibols by the 14-module synthases is due to module skipping: the switch follows the completion of the third peptide bond at C-domain 3, with the acylated tripeptide remaining attached to the carrier domain of module 3 and then being transferred to either module 4 (for the 14-mer products) or module 7 (for the 11-mer products), leading to 11- or 14-residue peptaibols, respectively. When compared to other C-domains, the condensation domain 3 has a unique structure, and its 3D structure suggests that it is capable of long-range domain interactions.

The resulting three differently sized peptaibols, however, occur in numerous heterogenous mixtures. Part of this is due to a relaxed specificity of some of the A-domains for their corresponding amino acid substrates, and the one actually incorporated into the peptaibol depends on their relative concentrations in the cellular pool [79]. Also, peptaibols can undergo significant posttranslational processing, particularly by amidation of proline or by cyclization of Aib and proline into a hydroxyketopiperazine ring.

The nature of peptaibols as toxins is currently a matter of debate and has been heated by the claim that peptaibols are the causal agents for health problems in fungus-contaminated buildings when *T. longibrachiatum* is present [93–95]. Experimental evidence for this is not conclusive. The 20-residue peptaibols alamethicin and paracelsin and other 11-residue trichobrachins were reported to be highly toxic in three in vitro invertebrate models, namely, *Crassostrea gigas*, *Artemia salina*, and *Daphnia magna* [93,95]. However, Degenkolb et al. [96] found that the alamethicin standard used was contaminated with the trichothecene mycotoxin harzianum A (see later [8]) that is indeed highly toxic to *Artemia salina* [93]. Oral administration of various peptaibols to rodents and ruminants revealed a very low toxicity (155–165 mg/kg body weight/animal), probably due to the high molecular weight that prevented passing through the intestinal cell [96]. Mikkola et al. [94] reported the mycelia of *T. longibrachiatum* contained 0.5%–2.6% (w/w) of peptaibols, and while this is a high concentration for a secondary metabolite in physiological terms, it would require the uptake of about 11 g peptaibol (corresponding to 430 g dry weight of the fungus) by a human of 70 kg body weight.

Several studies have also reported that peptaibols cause hemolysis of erythrocytes and uncoupling of oxidative phosphorylation in mitochondria [97]. However, these actions require the direct interaction between the peptaibols and the respective membranes and are also observed with many amphiphilic detergent-like molecules (e.g., soaps). Consequently, the so far reported toxic effects of peptaibols are rather theoretical and well below the threshold of human consequence [96].

31.5.2 Epidithiodioxopiperazines

ETPs are produced by some *Trichoderma* species and *Aspergillus fumigatus*, *Eurotium chevalieri*, *Myrothecium verrucaria*, and certain *Penicillium* species. They can be classified as cyclic peptide-derived metabolites that bear a characteristic internal disulfide bridge (Figure 31.1) responsible for all known toxic effects of these molecules such as induction of apoptosis, inhibition of the proteasome preventing NF-κB activation, and inhibition of angiogenesis [98].

The only ETP metabolites that are known from *Trichoderma* spp. are gliotoxin and gliovirin. Gliotoxin was the first ETP reported from any fungus and is the best characterized. However, this was mainly because of its action as the toxic principle of *Aspergillus fumigatus* (see Chapter 13 of this volume). Its name is derived from *Gliocladium fimbriatum* (an earlier synonym of the plant pathogen *Myrothecium verrucaria*) from which it was identified first [99,100]. In addition, gliotoxin was the first metabolite described from *Trichoderma* (initially *Trichoderma lignorum*, an earlier invalid name of *Gliocladium virens* [101,102]). Gliotoxin has a fungistatic effect and was implicated in antagonism of *Rhizoctonia* in the soil in the 1930s [99,100]. Interestingly, the mycotoxin is produced only by *T. virens Q* strains and is not produced by *P* strains of *T. virens* and all other *Trichoderma* spp. that are used as biocontrol agents [103]. Gliotoxin can be detected within 16 h of growth in liquid culture [104] and has also been demonstrated to accumulate in the rhizosphere [105]. There are, however, contradictory reports on its importance in biocontrol under controlled cultural and environmental conditions [106–109].

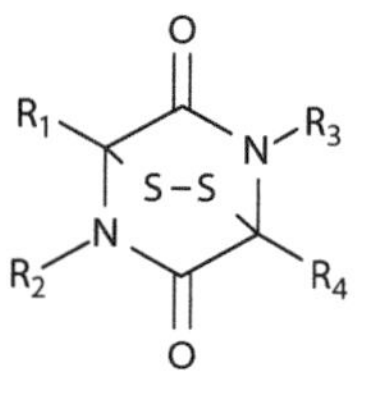

FIGURE 31.1 Basic chemical structure of ETPs. R_1–R_4 represent compound-specific side chains.

As for most secondary metabolites that are produced by fungi, the genes for the biosynthesis of ETP metabolites are clustered in the genome. Patron et al. [110] have shown that putative ETP gene clusters are present in 14 ascomycete taxa, but absent in numerous other ascomycetes examined, so the distribution of the ability to synthesize ETP metabolites appears to be limited. Furthermore, these clusters are discontinuously distributed throughout different ascomycete lineages. Patron et al. [110] found that the gene content and order in these clusters are not absolutely fixed: they identified common genes and performed a phylogenetic analysis of six of them. Thereby, almost all cluster genes form monophyletic clades with noncluster fungal paralogues being the nearest out-groups, which suggests that a progenitor ETP gene cluster is assembled within an ancestral taxon. Within each of the cluster clades, the cluster genes grouped together in consistent subclades, but these relationships were not always concordant with the phylogenetic relationship of the fungi.

Production of ETPs is generally discontinuous among fungal species, for example, among the Aspergillus, only *A. fumigatus* produces gliotoxin. On the other hand, distantly related fungi produce the same ETP, for example, gliotoxin is produced by unrelated fungi such as *A. fumigatus*, *Penicillium* spp., *Candida* spp., and *T. virens* [99,111–113].

The clusters of *Leptosphaeria maculans* and *A. fumigatus* have been investigated in more detail and thus provide a model for its analysis in *Trichoderma* (for a review, see [113]): they are 55 and 28 kb in length, respectively, and ten genes are common to both (Figure 31.2). Eight of these encoded enzymes can be aligned with a function: a two-module nonribosomal peptide synthetase (*gliP*), thioredoxin reductase (*gliT*), O-methyl transferase (*gliM*), a methyl transferase with unknown specificity (*gliN*), glutathione S-transferase (*gliG*), cytochrome P450 monooxygenases (*gliC*), amino cyclopropane carboxylate synthase (*gliI*), and a dipeptidase (*gliJ*). The specific role in gliotoxin biosynthesis can be deduced of a few of these genes (e.g., GliP, which catalyses the condensation of two amino acids in the initial biosynthetic step), but the reaction catalyzed by the others is so far unknown.

Two further genes occur in both clusters: a zinc binuclear ($Zn[II]_2Cys_6$) transcriptional regulator (*gliZ*) that controls expression of the biosynthetic enzymes and a transporter (*gliA*) putatively involved in toxin efflux. Additional genes are probably encoding enzymes involved in the modification of the core ETP moiety, but a functional analyses of them has not been performed as yet.

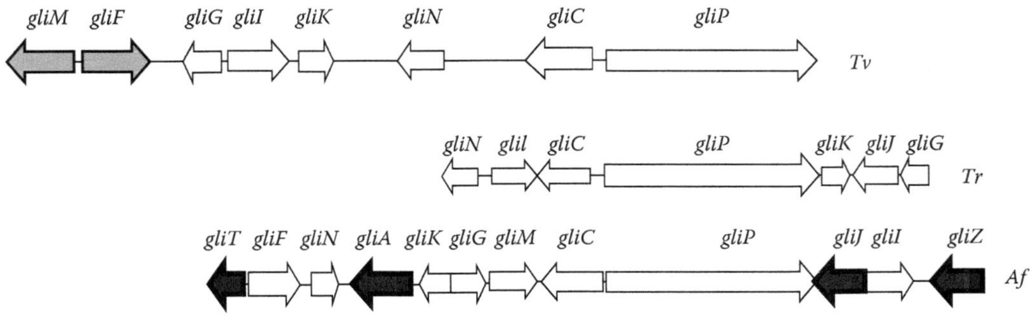

FIGURE 31.2 Structures of the putative gliotoxin biosynthesis clusters of *T. virens* Gv29-8, *T. reesei*, and *A. fumigatus*. Genes present in *A. fumigatus* but not found in *T. virens* Gv29-8 are highlighted in black. Those present in *T. virens* Gv29-8 but absent in *T. reesei* are indicated in gray. Genes, whose functions have been identified, are *gliP*, nonribosomal peptide synthetase; *gliM*, O-methyl transferase; *gliN*, a methyl transferase with unknown specificity; *gliG*, glutathione S-transferase; *gliC*, cytochrome P450 monooxygenases; *gliI*, amino cyclopropane carboxylate synthase; *gliJ*, dipeptidase; *gliT*, thioredoxin reductase; *gliZ*, a zinc binuclear ($Zn[II]_2Cys_6$) transcriptional regulator; and *gliA*, a transporter putatively involved in toxin efflux. Functions have not yet been assessed for the other genes shown. (Data taken and modified from Howell, C.R. and Stipanovic, R.D., *Can. J. Microbiol.*, 29, 321, 1983.)

The GliP cluster of *T. virens* has been compared with the *A. fumigatus* GliP cluster: it contains only eight genes that are closely related to the *A. fumigatus* genes (Figure 31.2). Since scaffold 47 that harbors this cluster is small (about 30 kb [114]), the possibility remains that the sequence is incomplete and poor coverage of this region in the genome might account for the missing genes in this cluster—such as *gliZ*, *gliJ*, *gliA*, and *gliT* (putatively encoding a transcription factor, dipeptidase, transporter, and thioredoxin reductase, respectively).

T. reesei does not produce gliotoxin, but also contains a gliotoxin biosynthesis cluster of however even smaller size than that of *T. virens*. It is possible that the absence of some of the genes present in *T. virens* is the reason why this species is unable to produce gliotoxin. No GliP cluster was detected in the genome of other *Trichoderma* spp. such as *T. atroviride*, *T. asperellum*, or *T. harzianum* (unpublished data).

Atanasova et al. [115] showed that almost all genes required for gliotoxin biosynthesis, that is, a two-module nonribosomal peptide synthetase (*gliP*), thioredoxin reductase (*gliT*), *O*-methyl transferase (*gliM*), a methyl transferase with unknown specificity (*gliN*), glutathione S-transferase (*gliG*), cytochrome P450 monooxygenases (*gliC*), amino cyclopropane carboxylate synthase (*gliI*), and a dipeptidase (*gliJ*) were upregulated during confrontation with the potential host *Rhizoctonia solani*. Interestingly, also genes involved in the provision of the precursor of gliotoxin, L-phenylalanine, and of the glutathione required for the formation of the central disulfide bond were induced at least some time points during this interaction. Provision of sulfur for cysteine and subsequently glutathione biosynthesis thereby appeared to be an essential requirement already before contact, as indicated by the upregulation of two sulfate permeases, one sulfatase, the cysteine biosynthesis genes ATP-sulfurylase and PAPS reductase, and SCON2, an ubiquitin ligase involved in regulating sulfur metabolism under conditions of low sulfate supply. Most of these genes were upregulated only before contact, suggesting that *T. virens* could later fulfill the sulfur requirement by feeding on *R. solani*. In contrast, a shortage in L-phenylalanine appears to take place only during overgrowth as illustrated by the increased expression of 3-deoxy-D-arabino-heptulosonate-7-phosphate synthase, the first enzyme involved in the biosynthesis of aromatic amino acids, and also the regulatory target for this pathway.

Unlike the *T. virens* cluster, in which genes are induced during confrontation with its host, *Rhizoctonia solani*, the genes of the gliotoxin biosynthesis cluster in *T. reesei* are not expressed during confrontation [115].

Gliotoxin is also regulated by the developmental stage such as conidiation: in *T. virens*, deletion of the central fungal developmental regulator (*vel1*) impaired production of conidia on solid medium and chlamydospores in rich medium and impaired expression of the gliotoxin biosynthesis gene *gliP* [116].

In *A. fumigatus*, gliotoxin is thought to contribute to invasive aspergillosis of mammals [117]. Its role as a virulence factor is deduced from a range of studies [118], and its production at zones of cellular detachment of A549 human cells grown in agar with the fungus indicates a role in tissue colonization [119]. Also, the levels of gliotoxin secreted by different *A. fumigatus* strains correlate with virulence on larvae of the insect *Galleria mellonella* (an insect that serves as a model for in vivo testing of virulence of *Aspergillus* spp. [120]). No such data are known for *T. virens*, and the fungus has never been reported as a facultative pathogen of humans or animals.

P strains of *T. virens* do not produce gliotoxin, but do produce gliovirin (Figure 31.3) that has potent antimicrobial properties especially against oomycetes [108,109]. However, the genome of *T. virens P* strains has not been sequenced, and no gliotoxin gene cluster has been identified.

31.5.3 Trichothecenes

Trichothecenes are a family of over 200 secondary metabolites with a common tricyclic 12,13-epoxytrichothec-9-ene (EPT) core structure (Figure 31.4) [120]. These metabolites have been reported from at least six genera of the fungal order Hypocreales (class Sordariomycetes): *Fusarium*, *Myrothecium*, *Spicellum*, *Stachybotrys*, *Trichoderma*, and *Trichothecium*. Based on the

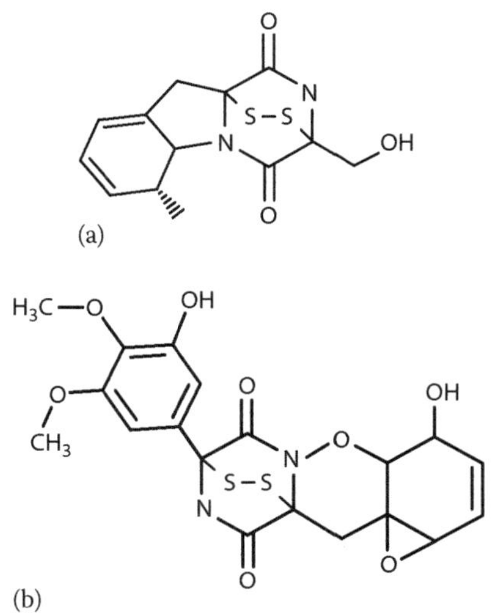

FIGURE 31.3 Chemical structure of gliotoxin (a) and gliovirin (b).

substitution pattern of this core structure, they can be classified into four groups (types A, B, C, and D), which can be fungus specific: type A, to which all *Trichoderma* trichothecenes belong [121–123], encompasses those variants that have either a hydroxyl group or an ester function or no oxygen substitution at C-8. Types B–D are reviewed in [124]. The core chemical structure of trichothecenes (Figure 31.4) identifies them as small, amphipathic molecules that can pass cell membranes by passive diffusion. They are therefore easily absorbed via the skin and gastrointestinal systems, which results in a rapid effect on fast growing cells and tissues [125,126]. Exposure to trichothecenes has been reported cause a number of symptoms, from feed refusal, immunological problems, vomiting, skin dermatitis, to immunosuppressive effects and neurotoxicity [127]. Thereby, the main mechanism of toxic action involves an inhibition of eukaryotic protein synthesis by preventing peptide bond formation at the peptidyl transferase center of the 60S ribosomal subunit [127]. In addition, there is evidence for an inhibition of mitochondrial protein synthesis, interaction with protein sulfhydryl groups, and generation of free oxygen radicals, thus eliciting oxidative stress [126,128].

Trichothecene formation is a major threat to the use of agricultural biomass as a feedstock because of infection by *Fusarium* spp. (see Chapter 21 of this volume), but it has also been shown to be the major reason for damp building–related illnesses caused by *Stachybotrys*, a significant indoor environmental contaminant, which has been found [129]. Yet the actual trichothecene-producing *Trichoderma* spp. have so far never been found as indoor contaminants.

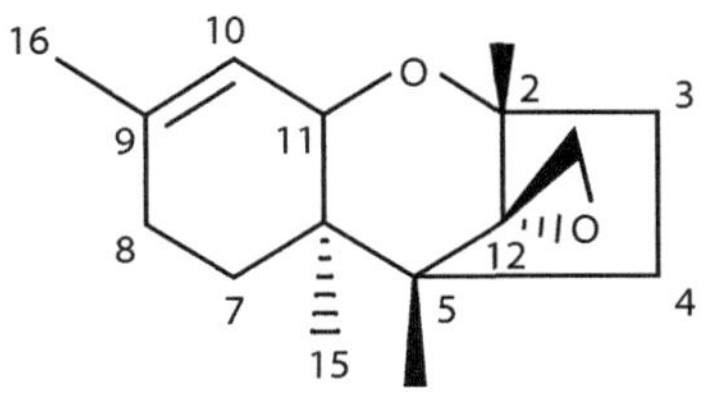

FIGURE 31.4 General chemical structure of trichothecenes. Type A trichothecenes are substituted at position 8.

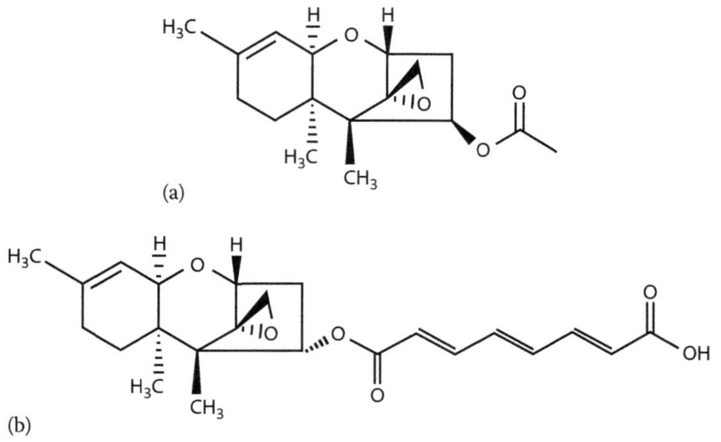

FIGURE 31.5 Chemical structure of trichodermin (a) and harzianum A (b).

Trichoderma is known to produce two type A trichothecenes, trichodermin and harzianum A (Figure 31.5). Trichodermin was originally isolated from *T. viride* by Godtfredsen and Vangedal [122]. Unfortunately, the strain used by Godtfredsen and Vangedal is not available, and its true species identity thus remains unknown. Watts et al. [123] reported the production of trichodermin by a mutant of *T. reesei*, based on the identity of the R_f profiles by TLC and retention time by HPLC. Corley at al. [124] isolated the trichothecene harzianum A from *T. harzianum*. These three papers led to the frequently found statement that *Trichoderma* produces trichothecenes. However, progress in strain identification and genome sequencing led to a revision: *T. reesei* lacks a *tri5* orthologue (the gene encoding the key enzyme of trichothecene biosynthesis; see as follows) and is thus unable to initiate trichothecene biosynthesis, a fact that is also underscored by the absence of trichothecene metabolites in culture filtrates of *T. reesei* QM 9414 grown on various different conditions (M. Sulyok and C.P. Kubicek, unpublished data). The species identity of the *T. reesei* mutant P12 is thus unclear. Nielsen et al. [130] reidentified the *T. harzianum* that produced harzianum A as *T. brevicompactum* by employing metabolite profiles, micromorphology, macromorphology on yeast extract sucrose agar and potato dextrose agar, and DNA sequences of the ITS1/ITS2 regions of the nuclear ribosomal DNA. They determined that only *T. brevicompactum* produced trichodermin and/or harzianum A on all media investigated. Subsequently, Degenkolb et al. [8] split *T. brevicompactum* into four separate species (*T. brevicompactum*, *T. arundinaceum*, *T. turrialbense*, and *T. protrudens*), all of which were found to produce trichothecenes. These four species form a separate phylogenetic clade within *Trichoderma*, and this taxon is considered currently the source of trichothecene production within *Trichoderma*.

The precursors of trichothecene biosynthesis are formed in the isoprenoid biosynthetic pathway, from which the specific trichothecene biosynthetic pathway branches off at farnesyl pyrophosphate by the action of the enzyme trichodiene synthase (termed *Tri5* in *Fusarium* spp.) that catalyzes the cyclization of farnesyl pyrophosphate to trichodiene [121]. Trichodiene then undergoes a series of oxygenations catalyzed by the cytochrome P450 monooxygenase TRI4 to finally form the intermediate isotrichotriol. Detailed reviews on the biosynthesis of trichothecenes have been published, and the reader is referred to these sources for details [118]. Cardoza et al. [131] have recently established the biosynthetic pathway of trichodermin and harzianum A formation in *T. arundinaceum*. It requires only five biosynthetic enzymes: TRI5 for the formation of trichodiene; TRI4 for the oxygenation of trichodiene, which leads to the formation of EPT; TRI11 for the oxygenation of EPT at C-4; and TRI3 for the acylation of the C-4 oxygen (Figure 31.6; [129]). It should be noted that harzianum A contains an additional octatriendioyl

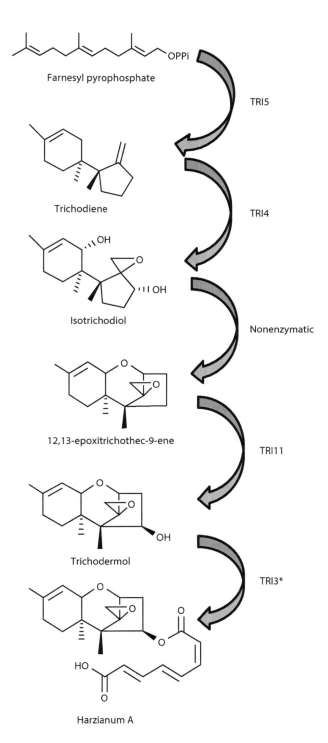

FIGURE 31.6 Hypothetical biosynthetic pathway for harzianum A formation in *T. arundinaceum* [130]. Gene products involved in individual steps are given by their TRI number. Note that proof for the reaction of TRI3* is still missing and this step is therefore still hypothetical.

moiety that is esterified to the C-4 oxygen (Figure 31.5). The biosynthesis of the octadecatrienoic acid has not been studied yet but is probably performed by a polyketide synthase which yet remains to be identified [131].

The knowledge of the trichothecene biosynthetic pathway also offers a quick PCR test for the presence of this pathway (and thus the potential production of trichothecenes) in *Trichoderma* ssp.: the trichodiene synthase gene—due to its location at the beginning of the pathway—is most well suited for this purpose and has already been established for this purpose in *Fusarium* [132,133] and *Stachybotrys* [134] isolates.

31.6 CONCLUDING REMARKS

Species of the genus *Trichoderma* have a number of remarkable properties that made them valuable actors in biotechnology and agriculture. Safety is of course an important issue for any such organism. Now that reliable species identification tools for it are available, it is possible to pinpoint those *Trichoderma* species that actually produce compounds that present a hazard to human and animals. A drawback, however, is that PCR- and sequence analysis–based tools are still slow; recent progress in chemotaxonomy using mass spectrometry–based methods [12,135,136] may become a faster alternative once a metabolite database of all species has been established.

So far, the list of species that indeed form severe toxins (e.g., the trichothecenes) is small and does not justify classifying the whole genus as toxigenic. However, attention has certainly to be applied, particularly when *new* species are introduced into applications that increase the contact with humans. Also, it will be advisable to monitor the fate of antifungal and plant-stimulating compounds like gliotoxin and gliovirin from *T. virens* in the respective plants to be able to maintain the claim of safety of these applications. This also implies the identification and testing of the antifungal principles in other biocontrol strains (e.g., members of the *T. harzianum* species clade, *T. asperellum*, or *T. atroviride*), for which the participation of secondary metabolites and their eventual toxicity is as yet not known with certainty.

However, as we have outlined in this chapter, the opportunistic nature of *Trichoderma* makes species from this genus capable of occupying virtually every niche. There are yet only few data available on the occurrence of *Trichoderma* spp. in food, feed, water, and other materials that are consumed by humans, but what is available shows that *Trichoderma* belongs to the major fungal populations detected. A more thorough examination of the pathogenicity seems to be justified.

ACKNOWLEDGMENTS

The authors' own work on this topic was supported by Austrian Science Foundation grant to CPK [FWF P21266].

REFERENCES

1. Klein, D. and Eveleigh, D.E. 1998. Ecology of *Trichoderma*. In: Kubicek, C.P. and Harman, G.E., eds. *Trichoderma and Gliocladium*, Vol. 1. Taylor & Francis, London, U.K. pp. 57–69.
2. Druzhinina, I.S. et al. 2011. *Trichoderma*: The genomics of opportunistic success. *Nature Rev Microbiol* **16**:749–759.
3. Kubicek, C.P. 2012. *Fungi and Lignocellulosic Biomass*. John Wiley & Sons, Ames, IA. 290pp.
4. Kubicek, C.P. 2012. Systems biological approaches towards understanding cellulase production by Trichoderma reesei. *J Biotechnol*. **163**:133–142.
5. Harman, G.E., Howell, C.R., Viterbo, A., Chet, I., and Lorito, M. 2004. *Trichoderma* species—Opportunistic, avirulent plant symbionts. *Nat Rev Microbiol* **2**:43–56.
6. Goswami, J., Pandey, R.K., Tewari, J.P., and Goswami, B.K. 2008. Management of root knot nematode on tomato through application of fungal antagonists, *Acremonium strictum* and *Trichoderma harzianum*. *J Environ Sci Health Ser B* **43**:237–240.

7. Kyalo, G., Affokpon, A., Coosemans, J., and Coynes, D.L. 2007. Biological control effects of *Pochonia chlamydosporia* and *Trichoderma* isolates from Benin [West-Africa] on root-knot nematodes. *Comm Agric Appl Biol Sci* **72**:219–223.
8. Degenkolb, T. et al. 2008. The *Trichoderma brevicompactum* clade: A new lineage with new species, new peptaibiotics, and mycotoxins. *Mycol Prog* **7**:177–209.
9. Kredics, L. et al. 2003. Clinical importance of the genus *Trichoderma*. A review. *Acta Microbiol Immun Hung* **50**:105–117.
10. Druzhinina, I.S. et al. 2008. Different reproductive strategies of *Hypocrea orientalis* and genetically close but clonal *Trichoderma longibrachiatum*, both capable to cause invasive mycoses of humans. *Microbiology* **154**:3447–3459.
11. Andersen, B., Frisvad, J.C., Søndergaard, I., Rasmussen, I.S., and Larsen, L.S. 2011. Associations between fungal species and water-damaged building materials. *Appl Environ Microbiol* **77**:4180–4188.
12. Thrane, U., Poulsen, S.B., Nirenberg, H.I., and Lieckfeldt, E. 2001. Identification of *Trichoderma* strains by image analysis of HPLC chromatograms. *FEMS Microbiol Lett* **203**:249–255.
13. Portnoy, J.M., Kwak, K., Dowling, P., Van Osdol, T., and Barnes, C. 2005. Health effects of indoor fungi. *Ann Allergy Asthma Immunol* **94**:313–319.
14. Peltola, J. et al. 2001. Toxic-metabolite-producing bacteria and fungus in an indoor environment. *Appl Environ Microbiol* **67**:3269–3274.
15. Larsen, F.O. et al. 1998. Volatile organic compounds from the indoor mould *Trichoderma viride* cause histamine release from human bronchoalveolar cells. *Inflamm Res* **47**(Suppl 1):S5–S6.
16. Persoon, C.H. 1794. Disposita methodical fungorum. *Römers Neues Mag Bot* **1**:81–128.
17. Bisby, G.R. 1939. *Trichoderma viride* Pers. ex Fries, and notes on *Hypocrea*. *Trans Br Mycol Soc* **23**:149–168.
18. Rifai, M.A. 1969. A revision of the genus *Trichoderma*. *Mycol Pap* **116**:1–56.
19. Bissett, J. 1984. A revision of the genus *Trichoderma*. I. Sect. *Longibrachiatum* sect. nov. *Can J Bot* **62**:924–931.
20. Bissett, J. 1991a. A revision of the genus *Trichoderma*. II. Infrageneric classification. *Can J Bot* **69**:2357–2372.
21. Bissett, J. 1991b. A revision of the genus *Trichoderma*. III. Sect. *Pachybasium*. *Can J Bot* **69**:2373–2417.
22. Bissett, J. 1991c. A revision of the genus *Trichoderma*. IV. Additional notes on section *Longibrachiatum*. *Can J Bot* **69**:2418–2420.
23. Bissett, J. 1992. *Trichoderma atroviride*. *Can J Bot* **70**:639–641.
24. Lu, B. et al. 2004. *Hypocrea/Trichoderma* species with pachybasium-like conidiophores: Teleomorphs for *T. minutisporum* and *T. polysporum* and their newly discovered relatives. *Mycologia* **96**:310–342.
25. Chaverri, P., Castlebury, L.A., Overton, B.E., and Samuels, G.J. 2003. *Hypocrea/Trichoderma*: Species with conidiophore elongations and green conidia. *Mycologia* **95**:1100–1140.
26. Chaverri, P., Castlebury, L.A., Samuels, G.J., and Geiser, D.M. 2003. Multilocus phylogenetic structure within the *Trichoderma harzianum/Hypocrea lixii* complex. *Mol Phylogenet Evol* **27**:302–213.
27. Bissett, J. et al. 2003. New species of *Trichoderma* from Asia. *Can J Bot* **81**:570–586.
28. Kraus, G. et al. 2004. *Trichoderma brevicompactum* sp. nov. *Mycologia* **96**:1057–1071.
29. Druzhinina, I., Chaverri, P., Fallah, P., Kubicek, C.P., and Samuels, G.J. 2004. *Hypocrea flavoconidia*, a new species with yellow conidia from Costa Rica. *Stud Mycol* **51**:401–407.
30. Jaklitsch, W.M., Komon, M., Kubicek, C.P., and Druzhinina, I. 2005. *Hypocrea voglmayri*, a new *Hypocrea* species from the Austrian Alps, defines a new phylogenetic clade in *Trichoderma*. *Mycologia* **97**:1391–1404.
31. Druzhinina, I.S., Komon-Zelazowska, M., Atanasova, L., Seidl, V., and Kubicek, C.P. 2010. Evolution and ecophysiology of an industrial producer *Hypocrea jecorina* [Anamorph *Trichoderma reesei*] and a new sympatric agamospecies related to it. *PLoS One* **5**:e9191.
32. Druzhinina, I.S., Kubicek, C.P., Komoń-Zelazowska, M., Mulaw, T.B., and Bissett, J. 2010. The *Trichoderma harzianum* demon: Complex speciation history resulting in coexistence of hypothetical biological species, recent agamospecies and numerous relict lineages. *BMC Evol Biol* **10**:94.
33. Atanasova, L., Jaklitsch, W.M., Komon-Zelazowska, M., Kubicek, C.P., and Druzhinina, I.S. 2010. The clonal species *Trichoderma parareesei* sp. nov., likely resembles the ancestor of the cellulase producer *Hypocrea jecorina/T. reesei*. *Appl Environ Microbiol* **76**:7259–7267.
34. Druzhinina, I.S. et al. 2012. Molecular phylogeny and species delimitation in the *Longibrachiatum* clade of *Trichoderma*. *Fungal Genet Biol* **49**:358–368.
35. Jaklitsch, W.M. 2009. European species of *Hypocrea* Part I. The green-spored species. *Stud Mycol* **63**:1–91.

36. Jaklitsch, W.M. 2011. European species of *Hypocrea* part II: Species with hyaline ascospores. *Fungal Divers* **48**:1–250.
37. Anonymous. BLAST: Basic local alignment search tool. http://blast.ncbi.nlm.nih.gov/Blast.cgi.
38. Druzhinina, I. et al. 2005. An oligonucleotide barcode for species identification in *Trichoderma* and *Hypocrea*. *Fungal Genet Biol* **42**:813–828.
39. Koptchinski, A., Komon, M., Kubicek, C.P., and Druzhinina, I.S. 2005. *Tricho*BLAST: A multiloci database of phylogenetic markers for *Trichoderma* and *Hypocrea* powered by sequence diagnosis and similarity search tools. *Mycol Res* **109**:658–660.
40. Druzhinina, I. and Kopchinskiy, A. International subcommission on *Trichoderma* and *Hypocrea* taxonomy. www.isth.info.
41. Kredics, L., Hatvani, L., Manczinger, L., Vágvölgyi, C., and Antal, Z. 2011. *Trichoderma*. In: Liu, D., ed. *Molecular Detection of Human Fungal Pathogens*. CRC Press, Boca Raton, FL. pp. 509–526.
42. Furukawa, H. et al. 1998. Acute invasive sinusitis due to *Trichoderma longibrachiatum* in a liver and small bowel transplant recipient. *Clin Infect Dis* **26**:487–489.
43. Hennequin, C., Chouaki, T., Pichon, J.C., Strunski, V., and Raccurt, C. 2000. Otitis externa due to *Trichoderma longibrachiatum*. *Eur J Clin Microbiol Infect Dis* **19**:641–642.
44. Munoz, F.M. et al. 1997. *Trichoderma longibrachiatum* infection in a pediatric patient with aplastic anemia. *J Clin Microbiol* **35**:499–503.
45. Myoken, Y. et al. 2002. Fatal necrotizing stomatitis due to *Trichoderma longibrachiatum* in a neutropenic patient with malignant lymphoma: A case report. *Int J Oral Maxillofac Surg* **31**:688–691.
46. Lagrange-Xélot, M., Schlemmer, F., Gallien, S., Lacroix, C., and Molina, J.M. 2008. *Trichoderma* fungaemia in a neutropenic patient with pulmonary cancer and human immunodeficiency virus infection. *Clin Microbiol Infect* **14**:1190–1192.
47. Alanio, A. et al. 2008. Invasive pulmonary infection due to *Trichoderma longibrachiatum* mimicking invasive Aspergillosis in a neutropenic patient successfully treated with voriconazole combined with caspofungin. *Clin Infect Dis* **46**:e116–e118.
48. Santillan Salas, C.F. et al. 2011. Fatal post-operative *Trichoderma longibrachiatum* mediastinitis and peritonitis in a paediatric patient with complex congenital cardiac disease on peritoneal dialysis. *J Med Microbiol* **60**:1869–1871.
49. Kantarcioğlu, A.S. et al. 2009. Fatal *Trichoderma harzianum* infection in a leukemic pediatric patient. *Med Mycol* **47**:207–215.
50. De Miguel, D. et al. 2005. Nonfatal pulmonary *Trichoderma viride* infection in an adult patient with acute myeloid leukemia: Report of one case and review of the literature. *Diagn Microbiol Infect Dis* **53**:33–37.
51. Walsh, T.J. et al. 2004. Infections due to emerging and uncommon medically important fungal pathogens. *Clin Microbiol Infect* **10**(Suppl 1):48–66.
52. Trabelsi, S., Hariga, D., and Khaled, S. 2010. First case of *Trichoderma longibrachiatum* infection in a renal transplant recipient in Tunisia and review of the literature. *Tunis Med* **88**:52–57.
53. Seguin, P. et al. 1995. Successful treatment of a brain abscess due to *Trichoderma longibrachiatum* after surgical resection. *Eur J Clin Microbiol Infect Dis* **14**:445–448.
54. Tanis, B.C., van der Pijl, H., van Ogtrop, M.L., Kibbelaar, R.E., and Chang, P.C. 1995. Fatal fungal peritonitis by *Trichoderma longibrachiatum* complicating peritoneal dialysis. *Nephrol Dial Transplant* **10**:114–116.
55. Richter, S. et al. 1999. Fatal disseminated *Trichoderma longibrachiatum* infection in an adult bone marrow transplant patient: Species identification and review of the literature. *J Clin Microbiol* **37**:1154–1160.
56. Chouaki, T., Lavarde, V., Lachaud, L., Raccurt, C.P., and Hennequin, C. 2002. Invasive infections due to *Trichoderma* species: Report of 2 cases, findings of in vitro susceptibility testing, and review of the literature. *Clin Infect Dis* **35**:1360–1367.
57. Tang, P. et al. 2003. Allergic fungal sinusitis associated with *Trichoderma longibrachiatum*. *J Clin Microbiol* **41**:5333–5336.
58. Kratzer, C., Tobudic, S., Schmoll, M., Graninger, W., and Georgopoulos, A. 2006. In vitro activity and synergism of amphotericin B, azoles and cationic antimicrobials against the emerging pathogen *Trichoderma* spp. *J Antimicrob Chemother* **58**:1058–1061.
59. Kviliute, R., Paskevicius, A., Gulbinovic, J., Stulpinas, R., and Griskevicius, L. 2008. Nonfatal *Trichoderma citrinoviride* pneumonia in an acute myeloid leukemia patient. *Ann Hematol* **87**:501–502.
60. Kuhls, K., Lieckfeldt, E., Börner, T., and Guého, E. 1999. Molecular reidentification of human pathogenic *Trichoderma* isolates as *Trichoderma longibrachiatum* and *Trichoderma citrinoviride*. *Med Mycol* **37**:25–33.

61. Samuels, G.J. and Ismaiel, A. 2011. *Hypocrea peltata*: A mycological Dr Jekyll and Mr Hyde? *Mycologia* **103**:616–630.
62. Druzhinina, I.S. et al. 2008. An unknown species from *Hypocreaceae* isolated from human lung tissue of a patient with non-fatal pulmonary fibrosis. *Clin Microbiol Lett* **29**:180–184.
63. Ranque, S. et al. 2008. Isolation of *Trichoderma atroviride* from a liver transplant. *J Mycol Med* **18**:234.
64. Enríquez-Matas, A., Quirce, S., Cubero, N., Sastre, J., and Melchor, R. 2009. Hypersensitivity pneumonitis caused by *Trichoderma viride*. *Arch Bronconeumol* **43**:304–305.
65. Seibel, C. et al. 2008. Pathogenesis related gene expression in the opportunistic fungal pathogen *Trichoderma longibrachiatum*. Poster at the Ninth European Conference on Fungal Genetics, April 5–8, Edinburgh, U.K.
66. JGI: Joint Genome Institute. US Department of Energy. *Trichoderma citrinoviride*. http://genome.jgi.doe.gov/Trici4/Trici4.home.html.
67. JGI: Joint Genome Institute. US Department of Energy. *Trichoderma longibrachiatum*. http://genome.jgi.doe.gov/Trilo3/Trilo3.home.html.
68. Hageskal, G., Knutsen, A.K., Gaustad, P., de Hoog, G.S., and Skaar, I. 2006. Diversity and significance of mold species in Norwegian drinking water. *Appl Environ Microbiol* **72**:7586–7593.
69. Hageskal, G., Vrålstad, T., Knutsen, A.K., and Skaar, I. 2008. Exploring the species diversity of *Trichoderma* in Norwegian drinking water systems by DNA barcoding. *Mol Ecol Resour* **8**:1178–1188.
70. Gashgari, R.M., Elhariry, H.M., and Gherbawy, Y.A. 2013. Molecular detection of mycobiota in drinking water at four different sampling points of water distribution system of Jeddah City (Saudi Arabia). *Geomicrobiol J* **30**:29–35.
71. Jaklitsch, W.M., Samuels, G.J., Dodd, S.L., Lu, B.S., and Druzhinina, I.S. 2006. *Hypocrea rufa/Trichoderma viride*: A reassessment, and description of five closely related species with and without warted conidia. *Stud Mycol* **56**:135–177.
72. Ribeiro, A. et al. 2006. Fungi in bottled water: A case study of a production plant. *Rev Iberoam Micol* **23**:139–144.
73. Mandeel, Q.A. 2005. Fungal contamination of some imported spices. *Mycopathologia* **159**:291–298.
74. Tournas, V.H. and Katsoudas, E. 2005. Mould and yeast flora in fresh berries, grapes and citrus fruits. *Int J Food Microbiol* **105**:11–17.
75. Halt, M. 1998. Moulds and mycotoxins in herb tea and medicinal plants. *Eur J Epidemiol* **14**:269–274.
76. Kacániová, M. and Fikselová, M. 2007. Mycological flora on tree fruits, crust, leaves and pollen *Sorbus domestica* L. *Ann Agric Environ Med* **14**:229–232.
77. Freire, F.C., Kozakiewicz, Z., and Paterson, R.R. 2000. Mycoflora and mycotoxins in Brazilian black pepper, white pepper and Brazil nuts. *Mycopathologia* **149**:13–19.
78. Sivasithamparam, K. and Ghisalberti, E.L. 1998. Secondary metabolism in *Trichoderma* and *Gliocladium*. In: Kubicek, C.P. and Harman, G.E., eds. *Trichoderma and Gliocladium*, Vol. 1. *Basic Biology, Taxonomy and Genetics*. Taylor & Francis, London, U.K., pp. 139–191.
79. Kubicek, C.P., Zelazowska-Komon, M., Sandor, E., and Druzhinina, I.S. 2007. Facts and challenges in the understanding of the biosynthesis of peptaibols by *Trichoderma*. *Chem Biodiv* **4**:1068–1082.
80. Castle, A., Speranzini, D., Rghei, N., Alm, G., Rinker, D., and Bissett, J. 1998. Morphological and molecular identification of Trichoderma isolates on North American mushroom farms. *Appl Environ Microbiol* **64**:133–137.
81. Rebuffat, S., Goulard, C., Bodo, B., and Roquebert, M.F. 1999. The peptaibol antibiotics from *Trichoderma* soil fungi; structural diversity and membrane properties. *Recent Res Dev Org Bioorg Chem* **3**:65–91.
82. Peltola, J. et al. 2004. Biological effects of *Trichoderma harzianum* peptaibols on mammalian cells. *Appl Environ Microbiol* **70**:4996–5004.
83. Whitmore, L. and Wallace. 2004. Analysis of peptaibol sequence composition: Implications for in vivo synthesis and channel formation. *Eur Biophys J* **33**:233–237.
84. Degenkolb, T. and Brückner, H. 2008. Peptaibiomics: Towards a myriad of bioactive peptides containing C[alpha]-dialkylamino acids? *Chem Biodiv* **5**:1817–1843.
85. Duclohier, H. 2007. Peptaibiotics and peptaibols: An alternative to classical antibiotics? *Chem Biodiv* **4**:1023–1026.
86. Whitmore, L. and Chugh, J. Peptaibol database. http://peptaibol.cryst.bbk.ac.uk/home.shtml.
87. Neuhof, T., Dieckmann, R., Druzhinina, I.S., Kubicek, C.P., and von Döhren, H. 2007. Intact-cell MALDI-TOF mass spectrometry analysis of peptaibol formation by the genus *Trichoderma/Hypocrea*: Can molecular phylogeny of species predict peptaibol structures? *Microbiology* **153**:3417–3437.

88. Wiest, A. et al. 2002. Identification of peptaibols from *Trichoderma virens* and cloning of a peptaibol synthetase. *J Biol Chem* **277**:20862–20868.
89. Mukherjee P.K. et al. 2011. Two classes of new peptaibols are synthesized by a single non-ribosomal peptide synthetase of *Trichoderma virens*. *J Biol Chem* **286**:4544–4554.
90. Chutrakul, C. and Peberdy, J.F. 2005. Isolation and characterisation of a partial peptide synthetase gene from *Trichoderma asperellum*. *FEMS Microbiol Lett* **252**:257–265.
91. Vizcaíno, J.A. et al. 2006. Detection of peptaibols and partial cloning of a putative peptaibol synthetase gene from *T. harzianum* CECT 2413. *Folia Microbiol* **51**:114–120.
92. Degenkolb, T. et al. 2012. The production of multiple small peptaibol families by single 14-module Peptide synthetases in *Trichoderma/Hypocrea*. *Chem Biodiv* **9**:499–535.
93. Favilla, M., Macchia, L., Gallo, A., and Altomare, C. 2006. Toxicity assessment of metabolites of fungal biocontrol agents using two different [*Artemia salina* and *Daphnia magna*] invertebrate bioassays. *Food Chem Toxicol* **44**:1922–1931.
94. Mikkola, R. et al. 2012. 20-Residue and 11-residue peptaibols from the fungus *Trichoderma longibrachiatum* are synergistic in forming Na[+]/K[+]–permeable channels and adverse action towards mammalian cells. *FEBS J* **279**(22):4172–4190.
95. Poirier, L. et al. 2007. Toxicity assessment of peptaibols and contaminated sediments on *Crassostrea gigas* embryos. *Aquat Toxicol* **83**:254–262.
96. Degenkolb, T., von Döhren, H., Nielsen, K.F., Samuels, G.J., and Brückner, H. 2008. Recent advances and future prospects in peptaibiotics, hydrophobin, and mycotoxin research, and their importance for chemotaxonomy of *Trichoderma* and *Hypocrea*. *Chem Biodiv* **5**:671–680.
97. Irmscher, G. and Jung, G. 1977. The hemolytic properties of the membrane modifying peptide antibiotics alamethicin, suzukacillin and trichotoxin. *Eur J Biochem* **80**:165–174.
98. Mullbacher, A., Waring, P., Tiwari-Palni, U., and Eichner, R.D. 1986. Structural relationship of epipolythiodioxopiperazines and their immunomodulating activity. *Mol Immunol* **23**:231–236.
99. Weindling, R. and Emerson, O. 1936. The isolation of a toxic substance from the culture filtrate of *Trichoderma*. *Phytopathology* **26**:1068–1070.
100. Weindling, R. 1934. Studies on a lethal principle effective in the parasitic action of *Trichoderma lignorum* on *Rhizoctonia solani* and other soil fungi. *Phytopathology* **34**:1153.
101. Brian, P.W. 1944. Production of gliotoxin by *Trichoderma viride*. *Nature* **154**:667–668.
102. Brian, P.W. and Hemming, H.G. 1945. Gliotoxin, a fungistatic metabolic product of *Trichoderma viride*. *Ann Appl Biol* **32**:214–220.
103. Howell, C.R. 1996. Simple selective media for isolation of *Trichoderma virens* from soil. *Phytopathology* **86**(11 suppl.):S19.
104. Wilhite, S.E. and Straney, D.C. 1996. Timing of gliotoxin biosynthesis in the fungal biological control agent *Gliocladium virens* (*Trichoderma virens*). *Appl Microbiol Biotechnol* **45**:513–518.
105. Lumsden, R.D., Locke, J.C., Adkins, S.T., Walter, J.F., and Ridout, C.J. 1992. Isolation and localization of the antibiotic gliotoxin produced by *Gliocladium virens* from alginate prill in soil and soilless media. *Phytopathology* **82**:230–235.
106. Howell, C.R. 2006. Understanding the mechanisms employed by *Trichoderma virens* to effect biological control of cotton diseases. *Phytopathology* **96**:178–180.
107. Howell, C.R., Stipanovic, R., and Lumsden, R. 1993. Antibiotic production by strains of *Gliocladium virens* and its relation to biocontrol of cotton seedling diseases. *Biocontrol Sci Technol* **3**:435–441.
108. Howell, C.R. and Stipanovic, R.D. 1983. Gliovirin, a new antibiotic from *Gliocladium virens* and its role in the biological control of *Pythium ultimum*. *Can J Microbiol* **29**:321–324.
109. Wilhite, S.E., Lumsden, R.D., and Straney, D.C. 1994. Mutational analysis of gliotoxin production by the biocontrol fungus *Gliocladium virens* in relation to suppression of *Pythium* damping-off. *Phytopathology* **84**:816–821.
110. Patron, N.J. et al. 2007. Origin and distribution of epipolythiodioxopiperazine [ETP] gene clusters in filamentous ascomycetes. *BMC Evol Biol* **7**:174.
111. Mac Donald, J.C. and Slater, G.P. 1975. Biosynthesis of gliotoxin and mycelianamide. *Can J Biochem* **53**:475–478.
112. Shah, D.T. and Larsen, B. 1991. Clinical isolates of yeast produce a gliotoxin-like substance. *Mycopathology* **116**:203–208.
113. Gardiner, D.M., Waring, P., and Howlett, B.J. 2005. The epipolythiodioxopiperazine [ETP] class of fungal toxins: Distribution, mode of action, functions and biosynthesis. *Microbiology* **151**:1021–1032.
114. JGI: Joint Genome Institute. US Department of Energy. *Trichoderma virens*. http://genome.jgi.doe.gov/TriviGv29_8_2/TriviGv29_8_2.info.html.

115. Atanasova, L. et al. 2013. Comparative transcriptomics reveals different strategies of *Trichoderma* mycoparasitism. *BMC Genomics* **14**:121.
116. Mukherjee, P.K. and Kenerley, C.M. 2010. Regulation of morphogenesis and biocontrol properties in *Trichoderma virens* by a VELVET protein, Vel1. *Appl Environ Microbiol* **76**:2345–2352.
117. Sutton, P., Newcombe, N.R., Waring, P., and Mullbacher, A. 1994. In vivo immunosuppressive activity of gliotoxin, a metabolite produced by human pathogenic fungi. *Infect Immun* **62**:1192–1198.
118. Hope, W.W. and Denning, D.W. 2004. Invasive aspergillosis: Current and future challenges in diagnosis and therapy. *Clin Microbiol Infect* **10**:2–4.
119. Daly, P. and Kavanagh, K. 2002. Immobilization of *Aspergillus fumigatus* colonies in a soft agar matrix allows visualization of A549 cell detachment and death. *Med Mycol* **40**:27–33.
120. Reeves, E.P., Messina, C.G.M., Doyle, S., and Kavanagh, K. 2004. Correlation between gliotoxin production and virulence of *Aspergillus fumigatus* in *Galleria mellonella*. *Mycopathology* **158**:73–79.
121. Godtfredsen, W.O. and Vangedal, S. 1965. *Trichodermin*, a new sesquiterpene antibiotic. *Acta Chem Scand* **19**:1088–1102.
122. Watts, R., Dahiya, J., Chaudhary, K., and Tauro, P. 1988. Isolation and characterization of a new antifungal metabolite of *Trichoderma reesei*. *Plant Soil* **107**:81–84.
123. Corley, D.G., Miller-Wideman, M., and Durley, R.C. 1994. Isolation and structure of harzianum A: A new trichothecene from *Trichoderma harzianum*. *J Nat Prod* **57**:422–425.
124. McCormick, S.P., Stanley, A.M., Stover, N.A., and Alexander, N.J. 2011. Trichothecenes: From simple to complex mycotoxins. *Toxins* **3**:802–814.
125. Wannemacher, R.W. and Winer S.L. 1977. Trichothecene mycotoxins. In: Sidell, R.R., Takafuji, E.T., and Franz, D.R., eds. *Medical Aspects of Chemical and Biological Warfare.* Office of the Surgeon General at TMM Publications, Washington, DC. pp. 655–676.
126. Ueno, Y. 1985. The toxicology of mycotoxins. *Crit Rev Toxicol* **14**:99–132.
127. Cundliffe, E. and Davies J.E. 1977. Inhibition of initiation, elongation, and termination of eukaryotic protein synthesis by trichothecene fungal toxins. *Antimicrob Agents Chemother* **11**:491–499.
128. Ueno, Y. and Matsumoto, H. 1975. Inactivation of some thiol-enzymes by trichothecene mycotoxins from *Fusarium* species. *Chem Pharm Bull* **23**:2439–2442.
129. Straus, D.C. 2011. The possible role of fungal contamination in sick building syndrome. *Front Biosci* **3**:562–580.
130. Nielsen, K.F., Gräfenhan, T., Zafari, D., and Thrane, U. 2005. Trichothecene production by *Trichoderma brevicompactum*. *J Agric Food Chem* **53**:8190–8196.
131. Cardoza, R.E. et al. 2011. Identification of loci and functional characterization of trichothecene biosynthesis genes in filamentous fungi of the genus *Trichoderma*. *Appl Environ Microbiol* **77**:4867–4877.
132. Niessen, L., Schmidt, H., and Vogel, R.F. 2004.The use of *tri5* gene sequences for PCR detection and taxonomy of trichothecene-producing species in the *Fusarium* section *Sporotrichiella*. *Int J Food Microbiol* **95**:305–319.
133. Quarta, A., Mita, G., Haidukowski, M., Logrieco, A., Mulè, G., and Visconti, A. 2006. Multiplex PCR assay for the identification of nivalenol, 3- and 15-acetyl-deoxynivalenol chemotypes in *Fusarium*. *FEMS Microbiol Lett* **259**:7–13.
134. Black, J.A, Dean, T.R., Foarde, K., and Menetrez, M. 2008. Detection of *Stachybotrys chartarum* using rRNA, *tri5*, and beta-tubulin primers and determining their relative copy number by real-time PCR. *Mycol Res* **112**:845–881.
135. Respinis, S.D. et al. 2010. MALDI-TOF MS of *Trichoderma*: Model system for the identification of microfungi. *Mycol Prog* **9**:79–100.
136. Kang, D., Kim, J., Choi, J.N., Liu, K.H., and Lee, C.H. 2011. Chemotaxonomy of *Trichoderma* spp. using mass spectrometry-based metabolite profiling. *J Microbiol Biotechnol* **21**:5–13.

32 Trichosporon, Magnusiomyces, and Geotrichum

Guillermo Quindós, Cristina Marcos-Arias, Elena Eraso, and Josep Guarro

CONTENTS

32.1 Mycology 539
32.2 Human Disease 546
32.3 Diagnosis of Trichosporonosis and Geotrichosis 551
32.4 Therapy for Trichosporonosis and Geotrichosis 553
32.5 Conclusions 556
References 557

Trichosporon, *Magnusiomyces*, and *Geotrichum* are ubiquitous fungi in nature, closely associated with plants and animals, and can be part of the human microbiota. Moreover, these fungi cause human disease when they cause deep infections with a high mortality (40%–80%), although most isolates show a low virulence and infections are superficial. The terms (1) *trichosporonosis* and (2) *geotrichosis* refer to those mycoses caused by (1) *Trichosporon* and (2) *Geotrichum* and *Magnusiomyces*, respectively. In this chapter, to avoid confusion, we will specify the agent causing the geotrichosis. Invasive mycoses caused by these fungi are infrequent in the general population; however, the increase in the number of immunodeficient patients and use of antifungal prophylaxis have converted them into emergent opportunistic agents causing severe nosocomial infections worldwide. Moreover, these infections have also been observed in patients suffering from cancer and in critically ill patients exposed to multiple invasive medical procedures. The ability of *Trichosporon*, *Magnusiomyces*, and *Geotrichum* to adhere to and develop biofilms on implanted devices may account for persistence and dissemination, since this ability promotes (1) escape from host immune responses and (2) resistance to antifungal drugs.[1,2] In addition, the presence of glucuronoxylomannan in the cell wall of *Trichosporon* and the ability of these fungi to produce hydrolytic enzymes are additional virulence factors related to the progression of infections.

32.1 MYCOLOGY

Trichosporon (from Greek, *trichos* meaning hair and *sporon* meaning spores) is a basidiomycetous yeastlike fungus phenotypically characterized by its ability to develop blastoconidia, true hyphae, pseudohyphae, and arthroconidia. Its cell wall is multilamellar, and the presence of dolipores, with or without parenthesomes, is another important distinctive characteristics. Other morphological characteristics such as appresoria, sarcinae, giant cells, and endoconidia can be present. *Trichosporon* grows readily on routine culture media as smooth, shiny gray to cream, cerebriform, yeastlike colonies that become dry and membranous with age (Figure 32.1). *Trichosporon* also hydrolyzes xylose and assimilates different carbohydrates, although glucose fermentation is absent. Moreover, the ability of *Trichosporon* to degrade urea differentiates it from *Geotrichum* and *Magnusiomyces*.[3–8]

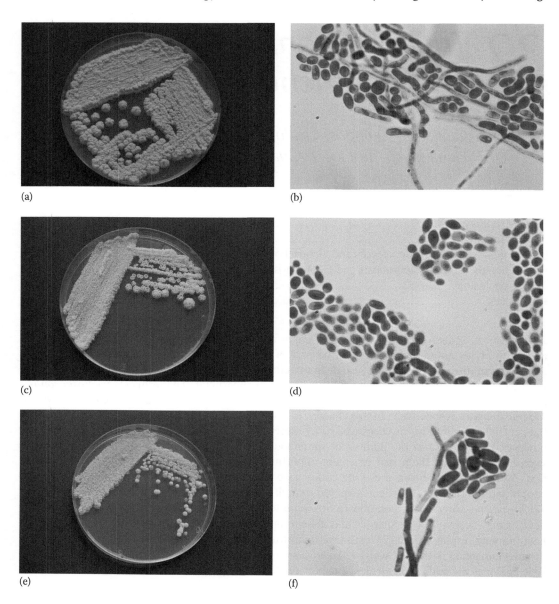

FIGURE 32.1 Macroscopic (a, c, e) and microscopic morphology (b, d, f) of *T. asahii, T. dermatis, and T. inkin*, respectively.

Trichosporon is widely distributed in a variety of habitats in nature, predominantly in tropical and temperate areas. These microorganisms can be isolated from soil, decomposing wood and other organic matter, air, water (rivers, lakes, and seas), scarab beetles, bird droppings, guano, bats, pigeons, cheese, and cattle, among others.[9–11] In humans, *Trichosporon* can be part of oral, gastrointestinal, and genital microbiotas and colonizes transiently the respiratory tract and skin.[7,12–15] *Trichosporon* has been isolated from 4%–10% of specimens from scrotal, perianal, and inguinal skin of healthy males.[16,17] Some species cause superficial and invasive infections in humans and animals.[6,7,12,15,18–25]

There are 38 *Trichosporon* species divided in five phylogenetic clades: Brassicae, Cutaneum, Gracile, Ovoides, and Porosum (Table 32.1).[3,26,27] This division correlates with taxonomic

TABLE 32.1
Trichosporon Species, Habitats and Infections

Clade	Name of Species	Habitat/Source	Diseases	References
Brassicae	*Trichosporon brassicae*	Food (salami, dairy products, cabbage)	—	[3]
	Trichosporon domesticum	Worldwide distribution	Summer-type hypersensitivity pneumonia (H)	[151]
			Chronic cystitis (A)	[223]
	Trichosporon montevideense	Worldwide distribution	Summer-type hypersensitivity pneumonia (H)	[151]
				[224]
				[121]
			Invasive trichosporonosis (H)	[151,224]
			Onychomycosis (A)	
Cutaneum	*Trichosporon scarabaeorum*	Gut of scarab beetle	—	[49]
	Trichosporon cutaneum	Worldwide distribution	White piedra	[193]
	Trichosporon debeurmannianum	Clinical specimen	—	[225]
	Trichosporon dermatis	Clinical specimen	Cutaneous trichosporonosis (H)	[225]
			Summer-type hypersensitivity pneumonia (H)	[151]
	Trichosporon jirovecii	Unknown	Unknown (urine and skin clinical specimens)	[193]
	Trichosporon moniliforme	Soil and water	—	[30]
	Trichosporon mucoides	Wide distribution	Summer-type hypersensitivity pneumonia (H)	[151]
			White piedra (H)	[13]
			Onychomycosis (H)	[22]
			Invasive trichosporonosis (H)	[226]
	Trichosporon smithiae	Brazilian soil	—	[227]
	Trichosporon terricola	Soil and home environment	Summer-type hypersensitivity pneumonia (H)	[151]
Gracile	*Trichosporon mycotoxinivorans*	Cultivated termites/unknown	Invasive trichosporonosis (H)	[128,228]
	Trichosporon dulcitum	Soil	—	[229]
	Trichosporon gracile	Animal sources and sour milk	—	[230]
	Trichosporon laibachii	Soil and animal feces	—	[27]
	Trichosporon vadense	Soil	—	[26]
	Trichosporon veenhuisii	Buffalo dung	—	[227]

(*Continued*)

TABLE 32.1 (Continued)
Trichosporon Species, Habitats and Infections

Clade	Name of Species	Habitat/Source	Diseases	References
Ovoides	Trichosporon aquatile	Soil	Summer-type hypersensitivity pneumonia (H)	[151]
	Trichosporon asahii	Worldwide distribution	Summer-type hypersensitivity pneumonia (H)	[151]
	Trichosporon asteroides	Soil and human skin	Invasive trichosporonosis (H)	[24]
	Trichosporon caseorum	Cheese	Invasive trichosporonosis (H)	[121,231]
	Trichosporon coremiiforme	Soil and home environment	—	[232]
	Trichosporon faecale	Soil and home environment	Summer-type hypersensitivity pneumonia (H)	[151]
			Invasive trichosporonosis (H)	[233]
	Trichosporon inkin	Clinical specimens (unknown ecological niche)	Summer-type hypersensitivity pneumonia (H)	[151]
			Invasive trichosporonosis (H)	[20]
			Endocarditis (H)	[134]
			Invasive trichosporonosis (H)	[19,234]
			White piedra (H)	[100]
	Trichosporon japonicum	Home environment	Summer-type hypersensitivity pneumonia (H)	[151]
	Trichosporon lactis	Cheese	Invasive trichosporonosis (H)	[233]
			—	[232]
	Trichosporon ovoides	Hair, soil, and home environment	Summer-type hypersensitivity pneumonia (H)	[151]
			White piedra (H)	[101]
	Trichosporon insectorum	Brazilian cheese and guts of beetles	—	[12]
Porosum	Trichosporon dehoogii	Unknown	—	[26]
	Trichosporon gamsii	Colombian soil	—	[26]
	Trichosporon guehoae	Soil	—	[26]
	Trichosporon lignicola	Wood pulp (Sweden)	—	[37]
	Trichosporon loubieri	Unknown	Invasive trichosporonosis (H)	[102,205,235]
	Trichosporon porosum	Soil, trees, and bat guano	Nasal granuloma (A)	[23]
	Trichosporon sporotrichoides	Soil (Suriname)	—	[27]
	Trichosporon wieringae	Forest soil (Belgium)	—	[26]

Note: A, animal disease; H, human disease; P, phytopathogen.

characteristics, such as the major ubiquinones (Q-9 or Q-10).[28] However, the classification of *Trichosporon* has been controversial. Beigel described *Trichosporon* in 1865 from a hair infection, but the fungus was misclassified as an alga, *Pleurococcus beigelii*, by Küchenmeister and Rabenhorst. In 1890, Behrend observed the fungus causing irregular nodules along the hair in a beard and named it *Trichosporon ovoides*, which is called white piedra (*piedra* means stone in Spanish). In 1902, Vuillemin renamed all *Trichosporon* species as *Trichosporon beigelii* because he transferred *Pleurococcus beigelii* to the genus *Trichosporon* due to its similarities. In 1909, Beurmann isolated this fungus from a cutaneous lesion and denominated it *Oidium cutaneum*, renamed in 1926 by Ota as *Trichosporon cutaneum*. The genus was represented by only two species, *T. beigelii* and *T. cutaneum*, until 1942 when Diddens and Lodder considered that both fungi were the same species: *T. beigelii* was adopted by physicians and *T. cutaneum* by environmental mycologists.[3,15,29,30]

Gueho et al. revised and reclassified the genus on the basis of the results generated by more recent methods such as guanine–cytosine content, DNA reassociation studies, nutritional profiles, and sequencing of the 26S region of the ribosomal DNA in which 19 taxa and the new species *Trichosporon mucoides* were accepted.[30] The name *T. beigelii* was used predominantly in the medical literature until recently, which has created confusion and limited the value of some clinical cases. Also, there are problems to discriminate between closely related species, and *T. cutaneum* shows a high heterogeneity[24] where *T. cutaneum sensu lato* has been divided into more than 10 species.[3] Middelhoven et al. proposed an identification scheme based on unconventional carbon sources, such as quinic acid, L-4-hydroxyproline, and orcinol as unique carbon sources.[26,27] However, intergenic spacer 1 (IGS1) sequence analysis is currently accepted as the most common and reliable method for the identification of *Trichosporon* species.[3,26,31]

The number of novel species increases continually, and six more new species have been described: *Trichosporon dohaense* (Ovoides clade) isolated from *tinea pedis*, onychomycosis, and catheter-associated fungemia,[32] *Trichosporon chiarelli* (Cutaneum clade) isolated from ant nests,[33] *Trichosporon xylopini* (Porosum clade) isolated from the gut of beetles (Coleoptera),[34] *Trichosporon cacaoliposimilis* (Gracile clade),[35] *Trichosporon oleaginosus* (Cutaneum clade),[35] and *Trichosporon vanderwaltii* isolated from the gut of beetles and forest soil.[36] However, the previously described taxon *Trichosporon pullulans* has been reassigned to the new genus *Guehomyces*, as *Guehomyces pullulans*, based on phylogenetic analysis of rRNA gene sequences.[37]

Trichosporon asahii is the most common species causing invasive trichosporonosis and presents nine different genotypes based on *IGS1* sequences that exhibit substantial variability in their geographic distribution.[15] Sun et al. reported that genotype IV was the most prevalent among *T. asahii* isolates from intensive care units (ICU) in China.[38] *Trichosporon* isolates have different virulence factors, related to colony morphology switching, glucuronoxylomannan production, enzyme expression, and slime/biofilm development. However, these virulence factors are not produced homogenously, and there is a great variability among clinical and environmental isolates. Colonies of environmental isolates were rugose, whereas the majority of the clinical isolates were powdery on Sabouraud dextrose agar. The clinical isolates consisted mainly of blastoconidia and arthroconidia, whereas environmental isolates were predominantly hyphae. Interestingly, all environmental isolates recovered from a mouse infection switched their macromorphologies to powdery colonies, with conidia as the predominant cell type, and the changes appear to be an adaptative response within the host to facilitate dissemination.[39]

Trichosporon expresses glucuronoxylomannan in the cell wall similar to *Cryptococcus neoformans*. Glucuronoxylomannan is a 1,3-α-D-linked mannan backbone with a β-D-glucopyranosyluronic acid residue, as well as β-D-xylopyranosyl residues.[40]

The production of glucuronoxylomannan may be related to the avoidance of phagocytosis and the dissemination by the blood stream of *Trichosporon* cells, as the polysaccharide may attenuate the phagocytic capability of neutrophils and monocytes in vivo. *T. asahii* isolates demonstrated a phenotypic switching and increased the amount of secreted glucuronoxylomannan in mice. All clinical isolates released significantly higher levels of glucuronoxylomannan than environmental isolates.[39] Glucuronoxylomannan production by *T. asahii* is lower than that by *C. neoformans*.[41] The major components of *T. asahii* glucuronoxylomannan are mannose (60%), xylose (24%), and glucose (8%). However, in *C. neoformans*, the third most commonly component is glucuronic acid (10%). *Cryptococcus* cells with undetectable glucuronoxylomannan are rapidly phagocyted in comparison to the polysaccharide-coated cells. Finally, glucuronoxylomannan has a protective role in *T. asahii* similar to that in *C. neoformans*.[41,42]

Clinical isolates of *T. asahii* were not able to produce secreted aspartic proteinases or phospholipases,[38,42] while they can secrete active β-*N*-acetylhexosaminidase.[42] Proteases, lipases, and other enzymes increase fungal pathogenicity by degrading proteins and disrupting cell membranes, and they may be important in *Trichosporon* dissemination.[43] Clinical isolates of *Trichosporon* can produce biofilms that have been associated with recalcitrant intravenous and urinary catheter infections and increased resistance to antifungal drugs.[2,38] Finally, *Trichosporon* is used in biotechnology and aroma enhancement of wine and liquor fermentation with relevance to food-related properties.[44–49]

Geotrichum (from Greek, *geo* as relating to earth and *trichos* meaning hair) (teleomorphs: *Galactomyces* and *Dipodascus*) are ascomycetous yeastlike fungi found in many habitats, such as air, water, plants,[50] and dairy products.[51–55] Most of them have restricted ecological niches often associated with insects and plants (Table 32.2).[56] Moreover, *Geotrichum candidum* and its

TABLE 32.2
Geotrichum/*Magnusiomyces* Species, Habitats and Infections

Name of Species	Habitat/Source	Diseases	References
Geotrichum bryndzae	Bryndza cheese	—	[236]
Geotrichum candidum	Worldwide distribution	Invasive geotrichosis (H)	[58,60,61]
		Oral geotrichosis (H)	[153]
		Dermatomycosis (A)	[63]
		Oral ulcers (A)	[65]
		Tonsillitis (A)	[64]
		Sour rot (P)	[76]
Geotrichum carabidarum	Guts of Coleoptera and Lepidoptera species	—	[237]
Geotrichum candidum var. *citri-aurantii*	Citrus fruit and soil	Citrus sour rot (P)	[238]
Geotrichum cucujoidarum	Guts of Coleoptera and Lepidoptera species	—	[237]
Geotrichum decipiens	Basidiocarps of *Armillaria* species	—	[79]
Geotrichum europaeum	Wheat field soil	—	[83]
Geotrichum fermentans	Grapes, cherries, and oranges	—	[239]
Geotrichum ghanense	Cocoa beans	—	[240]
Geotrichum histeridarum	Guts of Coleoptera and Lepidoptera species	—	[237]
Geotrichum klebahnii	Slime fluxes of trees	—	[241]
Geotrichum phurueaensis	Soil forest	—	[242]
Geotrichum siamensis	Mangrove forest	—	[242]
Geotrichum pseudocandidum	Wood and soil	—	[80]
Geotrichum restrictum	Endophyte of *Picea abies*	—	[83]
Magnusiomyces capitatus	Worldwide distribution	Invasive geotrichosis (H)	[135,154,184,211,222,243]
		Oral geotrichosis (H)	[153]

Note: A, animal disease; H, human disease; P, phytopathogen.

teleomorph *Galactomyces candidus* have been described as human colonizers causing superficial and deep infections[57–65]: Most cases of invasive geotrichosis are related to immunosuppression.[57–61,66] *G. candidum* is used in maturation and flavoring of cheese and in brewing as they inhibit mycotoxins produced by *Fusarium* in contaminated barley.[67,68] The fungus is employed in single cell protein[69] and aroma[70,71] production due to its proteolytic and aromatic properties. Furthermore, some species are useful for bioremediation of wastewater from distilleries, biodegradation of textile dyes, and removal of the phenolic components from olive mills.[72–75] Finally, some species are phytopathogens responsible of sour rot of citrus, tomatoes, and vegetables and the biodeterioration of fruit and vegetable juices.[76,77]

Geotrichum develops white, usually dry and powdery colonies and is characterized by the presence of true hyphae that disarticulate into arthroconidia (Figure 32.2). Septal walls are perforated by micropores. In some species, such as *Geotrichum fermentans*, blastospores can be observed. *Geotrichum* is able to assimilate different carbon sources, although glucose fermentation is often absent, and it does not degrade urea, all of which are useful for the identification of *Geotrichum* species.[78]

Geotrichum was described initially by Link in 1809 as a monotypic genus, *G. candidum*, isolated from soil,[79] and now 15 species of *Geotrichum* have been accepted with teleomorphs in *Galactomyces* Redhead and Malloch and one in *Dipodascus* de Lagerheim.[78,80,81]

Galactomyces geotrichum was considered the teleomorph of *G. candidum*. However, Smith et al. considered *G. geotrichum* as a complex divided in four groups: *G. geotrichum sensu stricto* and *G. geotrichum* groups A, B, and C[82] by DNA melting curves, DNA reassociation studies, and

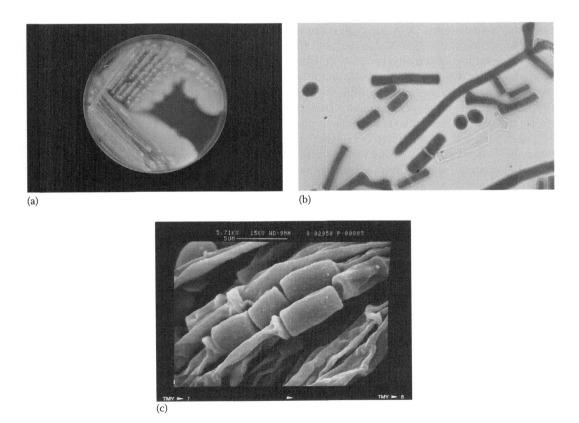

FIGURE 32.2 Macroscopic morphology (a) and micromorphology (b, light microscopy; c, scanning electron microscopy) of *G. candidum*.

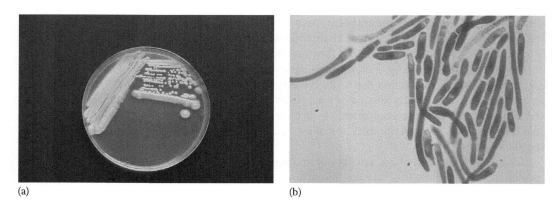

FIGURE 32.3 Macroscopic morphology (a) and micromorphology (b) of *M. capitatus*.

nutritional profiles. In 2004, de Hoog et al. revised *Geotrichum* and its teleomorphs and concluded that *G. candidus* is the teleomorph of *G. candidum*, whereas *G. geotrichum* remains as the teleomorph of an unnamed *Geotrichum* species.[83]

Salkin et al.[84] observed that *G. capitatum* produced a branched conidial apparatus and classified it in the genus *Blastoschizomyces* Salkin in 1985. However, de Hoog et al. did not accept this classification because some species of *Dipodascus* produced a similar structure and transferred this species to *Dipodascus* as *Dipodascus capitatus*.[4] Moreover, in 1995, Kurtzman and Robnett analyzed the 26S ribosomal DNA gene sequences and observed that *Dipodascus* included two genetically distinct species[85]: Ueda-Nishimura and Mikata observed a similar separation in 18S ribosomal DNA sequence data.[86] Finally, in 2004, de Hoog and Smith transferred *D. capitatus* and its anamorph to the genus *Magnusiomyces* (Latin *magnus* meaning abundant or great and Greek *myces* meaning fungi) from phylogenetic analyses of rRNA gene sequences.[83] *Magnusiomyces capitatus* (formerly *Blastoschizomyces capitatus*, *G. capitatum*, *Trichosporon capitatum*, or *D. capitatus*) is urease negative and thermotolerant and can grow in the presence of cycloheximide (Figure 32.3). *M. capitatus* is a ubiquitous microorganism found in soil, water, air, plants, and poultry and dairy products. This fungus was isolated from dishwashers, a man-made ecological niche also accommodating other human opportunistic fungal pathogens, such as *Aspergillus*, *Candida*, *Fusarium*, *Penicillium*, and *Rhodotorula*.[87] Along with colonizing the skin, bronchial, and intestinal tract of healthy people, it can cause disseminated opportunistic infections in patients with cancer, especially acute leukemia.[88–93] *M. capitatus* has been identified in hematology wards as a contaminant of food and drinks as have other fungi.[94,95] There is an apparent differential geographic distribution between *Magnusiomyces* and *Trichosporon*, with *M. capitatus* and *T. asahii* infections more common in Europe and North America, respectively.[7]

32.2 HUMAN DISEASE

Human diseases associated to these three fungal genera differ in their clinical presentations depending upon the immune status of the patients. Invasive mycoses occur mostly in onco-hematological and other patients with immunosuppression, whereas superficial infections and allergic diseases are found predominantly in immunocompetent hosts.[7,8] The most common clinical presentation of trichosporonosis is a benign superficial lesion of hair, called white piedra, characterized by the presence of irregular nodules. These grayish-white or light-brown nodules have a soft texture, are composed of fungal cells, and are loosely adhered to the hair shaft. The nodules are located on the distal half of the hair shaft but can usually be detached. The uncovered part of the hair shafts and the underlying skin appear unaffected, and no broken hairs are observed. White piedra is an uncommon and asymptomatic cosmopolitan disease affecting children and adults from areas with tropical and temperate climates and may affect a variety of hairy body areas, including the scalp, face (beard, moustache, and eyebrows), axilla, and pubic hairs. Interestingly, white piedra has been reported in young females who frequently use hats, headbands, or veils that can lead to higher humidity and limited sunlight exposure.[96,97] Concomitant hair infection by *Trichosporon* and corynebacteria has been observed, and it is still unclear whether the proteolytic capabilities of the corynebacteria may be relevant for the establishment of fungal infection.[98,99] White piedra is caused predominantly by *T. asteroides*, *T cutaneum*, *T. inkin*, *T. loubieri*, and *T. ovoides*.[100–103] Moreover, *T. cutaneum* can also cause onychomycosis and otomycosis, and Mexican authors have documented the isolation of *Trichosporon* from *tinea pedis* and onychomycosis from a significant number of patients (3%–43%) suffering from these diseases.[103–105] The mode of transmission of superficial trichosporonosis remains unclear, but poor hygiene, bathing in contaminated water, and sexual transmission can play roles.[96,106] Close contact, hair humidity, the length of the scalp hair, shared cosmetics, and dermal lotions may serve also for acquiring white piedra.[96,99,107,108]

The clinical manifestations of invasive trichosporonosis are similar to those of invasive candidiasis: Acute and chronic clinical infections have been reported. Acute trichosporonosis frequently has a sudden onset and progresses rapidly. It is observed primarily in neutropenic patients with persistent fever that does not respond to broad-spectrum antibiotic therapies. Blood cultures were demonstrated as positive in ca. 75% of patients. Cutaneous lesions (from erythematous rash, maculopapules, and necrotic ulcers on the trunk and extremities), pulmonary infiltrates (lobar consolidations, bronchopneumonia, or reticulonodular patterns), and hypotension, with renal and ocular involvement, can be common.[109–111] In most cases, the major risk factors included underlying neoplasia (acute and chronic leukemia, multiple myeloma, and solid tumors) and neutropenia.[112–117] In patients without cancer or neutropenia, the predisposing factors include treatment with high doses of corticoids, prosthetic valves, solid organ transplantation, chronic active hepatitis, antimicrobial therapy, cystic fibrosis, chronic granulomatous disease, severe burns, secondary hemochromatosis, hyperimmunoglobulin E syndrome (Job syndrome), and intravenous drug abuse (Tables 32.3 and 32.4).[118–123] *Trichosporon* species are inhabitants of the human microbiota, and many trichosporonosis have an endogenous origin. Gastrointestinal colonization followed by translocation throughout the gut may be considered the source of infection in many invasive trichosporonosis, mainly in cancer patients.[124,125] *Trichosporon* is able to disseminate from the gut to the blood in immunosuppressed rabbits but not in healthy animals.[126] Some trichosporonosis may have an exogenous origin as *Trichosporon* may enter the blood through a contaminated intravascular catheter. Kontoyiannis et al. observed that central venous catheter–related fungemia was the cause of 70% of all episodes in 17 cancer patients with trichosporonosis.[127] In most cases, the portal of entry appears to be the gastrointestinal tract, central venous catheters or percutaneous vascular devices, and the respiratory tract.[119,121,122,128,129] Antifungal prophylaxis with fluconazole or other drugs, as in patients with onco-hematological diseases, may prevent gut translocation of *Trichosporon*, favoring its exogenous acquisition. In very low birth weight neonates suffering from invasive trichosporonosis, the acquisition of the infection can occur by *Trichosporon* gastrointestinal translocation or through the indwelling catheter. Neonate skin colonization may be acquired either from health-care workers or after vaginal delivery, as women may harbor *Trichosporon* in their genital region.[14,130,131]

T. asahii is the most common species of the genus causing disseminated infections.[7,119,121,132,133] *T. asahii* cells rapidly adhere to polystyrene after 30 min of incubation and after 72 h showed different cellular morphologies in the mature biofilm (budding yeasts and hyphae) embedded within an extracellular polysaccharide matrix. *T. asahii* cells in biofilms are more resistant to the antifungal agents tested than planktonic cells.[2,38] Other species associated with invasive infection include *T. inkin*, which has caused peritonitis, osteomyelitis, endocarditis, and urinary tract, cutaneous, and subcutaneous infections[134]; *T. loubieri*, responsible of invasive trichosporonosis and cutaneous ulcers[135,136]; and *T. mycotoxinivorans*, which has been associated with pneumonia in patients with cystic fibrosis.[128]

TABLE 32.3
Risk Factors of Acquiring Trichosporonosis or Geotrichosis for Hospitalized Patients

	Risk Factors	
Clinical Setting	**Trichosporonosis**	**Geotrichosis (Including Infections by *Magnusiomyces capitatus*)**
Immunodeficiency	Neutropenia (intensity, duration, and dynamics)	Neutropenia (intensity, duration, and dynamics)
	Lymphopenia (intensity, duration, and dynamics)	Lymphopenia (intensity, duration, and dynamics)
	Cellular and humoral immunodeficiencies	Cellular and humoral immunodeficiencies
	Malnutrition	High doses of chemotherapy and/or corticosteroids
	Senescence	
	High doses of chemotherapy and/or corticosteroids	Immune suppressors (alemtuzumab, infliximab, antithymocytic globulin)
	Immune suppressors (alemtuzumab, infliximab, antithymocytic globulin)	
Organic dysfunctions	Skin and mucosal disruption and mucositis by surgery, radiotherapy, chemotherapy, GVHD, intravascular catheters	Renal and/or hepatic dysfunction
		Hyposplenia or asplenia
	Renal and/or hepatic dysfunction	
	Hyposplenia or asplenia	
Microbial colonization and reactivation of latent infections	Antimicrobial agents	Antimicrobial agents
	Prolonged hospitalization or ICU stay	Prolonged hospitalization or ICU stay
	Previous invasive mycoses	Previous invasive mycoses
		Contaminated food and drinks

Sources: Refs. [7,15,25,58,62,92,93,95,110,111,132,140,144,244,245].
Note: GVHD, graft-versus-host disease.

Incidence rates of 1.5% and 0.5%, for trichosporonosis and geotrichosis, respectively, by *M. capitatus* have been reported in patients with leukemia.[137] In these patients with hematological malignancies, *Trichosporon* is after *Candida* as the most common yeast causing invasive infections leading to 50%–80% mortality rates despite antifungal therapy.[25,138–140] Indeed, breakthrough trichosporonosis in immunocompromised patients after administration of amphotericin B and/or echinocandins, and more rarely after the use of triazoles, has been reported.[137,141–143] In a large Italian retrospective multicenter study including 287 cases of invasive trichosporonosis and 99 cases of invasive geotrichosis, over 20 years, the most common underlying conditions were hematological diseases, peritoneal dialysis, and solid tumors.[135] Fungemia occurred in 75% of patients and disseminated infection in 50% of cases. Despite the large number of clinical isolates included in the study, only two isolates were accurately identified as *T. loubieri* and other eight as *T. pullulans* (currently *G. pullulans*), and the other 287 isolates were identified only as *Trichosporon* sp. This fact is a clear example of the difficulty to identify *Trichosporon* isolates correctly.

Kontoyiannis et al.[127] described the clinical spectrum and outcome of invasive trichosporonosis in 17 patients with cancer. The overall incidence was 8 cases per 100,000 admitted patients; most infected patients had acute leukemia and/or neutropenia. Most patients (59%) had fungemia as the sole fungal manifestation. Central venous catheter–related infection was present in 70%, and 60% of episodes were breakthrough infections in patients that had received antifungal therapy for more than 60 days. The crude mortality rate after 30 days of admission was 53%.

TABLE 32.4
Risk Factors and Patient Populations at High Risk for Suffering from Invasive Trichosporonosis and Geotrichosis

Predisposing Factors	Trichosporonosis	Geotrichosis (Including Infections by *Magnusiomyces capitatus*)
General factors	Severity of acute disease	Severity of acute disease
	Age (<1 year and >65 years old)	Age (>65 years old)
	Major surgery (mainly gastrointestinal)	Hematological malignancies
	Hematological malignancies	
	ICU stay	
	Indwelling catheters	
	Multiple blood transfusions	
	Parenteral nutrition	
	Mechanical ventilation	
Patients at higher risk	Low weight neonate (<1500 g)	Prolonged neutropenia
	Prolonged neutropenia (<500 cells per mm^3)	Chemotherapy
	Central venous catheter	Treatment with corticosteroids and/or immunosuppressors
	Chemotherapy	
	Treatment with corticosteroids and/or immunosuppressors	Treatment with broad-spectrum antibacterial agents
	Renal failure, hemodialysis	Treatment with polyenes or echinocandins
	Polytraumatisms	
	Treatment with broad-spectrum antibacterial agents	
	Treatment with polyenes or echinocandins	

Sources: Refs. [7,15,17,25,92,93,110,119,132,140,144,196,245–248].

Ruan et al.[121] described 19 patients with invasive trichosporonosis. Cancer was the underlying disease in 58% of the patients, but only four patients were neutropenic at the time of diagnosis. Central venous catheter and antibiotic use were the most common conditions, being present in ≥90% of all patients. Crude mortality rate at 30 days after infection was 42%.

Suzuki et al.[133] evaluated 33 patients with trichosporonemia and hematological malignancies. The majority of them (80%–85%) suffered from acute leukemia and neutropenia, and most of them developed breakthrough trichosporonosis after 5 days of systemic antifungal therapy. Skin lesions were reported in 12 patients (57%) and pneumonia in 19 patients (36%). The mortality that could be attributed was 76%, with 67% of deaths within 10 days of admission.

Additionally, critically ill patients admitted to ICU, and subjected to invasive medical procedures and antibiotic treatment, were at risk for suffering from invasive trichosporonosis.[7,119,132] Chagas-Neto et al.[132] observed that most of 22 Brazilian patients with trichosporonemia were critically ill but without neoplasia. Most infected patients had degenerative diseases with organ failures and were subjected to treatment with broad-spectrum antibiotics and multiple invasive medical procedures before developing fungemia. Crude mortality rates were 30% for children and 87.5% for adults. Underlying disease, persistent neutropenia, and concurrent infections were significant contributing factors to overall mortality.[7,132]

There are reports of organ-specific trichosporonosis and patients with disseminated trichosporonosis presenting with pneumonia, soft tissue lesions, eventually, endophthalmitis, endocarditis (in natural and prosthetic valves), brain abscess, meningitis (including chronic fungal meningitis), arthritis, esophagitis, lymphadenopathy, liver and spleen abscess, peritonitis, and uterine infections.[15,118,132,134,144] Renal involvement in disseminated infection is quite common and occurs in approximately 75% of cases. However, it is difficult to interpret the isolation of *Trichosporon* from urine cultures, as it can be a sign of invasive infection or a simple colonization without clinical importance.[144,145] Ocular alterations (chorioretinitis) are frequent in disseminated trichosporonosis and may cause decreased or complete loss of vision due to retinal vein occlusion and retinal detachment. Unlike candidal endophthalmitis, *Trichosporon* usually infects uveal tissues, including the iris, but spares the vitreous. *Trichosporon* peritonitis is an uncommon complication of peritoneal dialysis, usually in patients with previous bacterial infections treated with broad-spectrum antibiotics.[144] Endocarditis involving native or prosthetic valves of nonneutropenic patients caused by *Trichosporon* has been reported. Patients usually develop large vegetations that may lead to embolism. Valve replacement is mandatory, but recurrences are common, and the prognosis is poor regardless of the antifungal therapy.[134] Chronic hepatosplenic trichosporonosis is very similar to chronic disseminated candidiasis, and it has been described in leukemic patients recovering from neutropenia, showing fever, enlarged liver and spleen, and multiple abscesses. In chronic disseminated trichosporonosis, symptoms may be present for several weeks to months and include persistent fever despite broad-spectrum antimicrobial therapy. Abdominal computerized tomography or magnetic resonance can reveal hepatic or splenic lesions compatible with abscesses, but a biopsy is needed to confirm the diagnosis.[122,146]

Finally, *Trichosporon* is the major etiological agent of summer-type hypersensitivity pneumonitis, and various reports have described that most patients had anti-*Trichosporon* antibodies in their sera.[147–152] *Trichosporon* isolates were found extensively in the patients' houses, and the elimination of these organisms prevented disease. *Trichosporon* species are responsible for the pneumonitis, leading to type III and IV allergies via repeated inhalation of organic dust and arthroconidia that contaminate home environments during the hot, humid, and rainy summer season in western and southern Japan. Mizobe et al. have characterized the antigenic components involved in summer-type hypersensitivity pneumonitis as the *T. cutaneum* glucuronoxylomannan.[40]

There are fewer reports of infections caused by *Geotrichum* and *M. capitatus*, and these infections are less frequent than trichosporonosis. However, *G. candidum* causes superficial and deep infections in immunocompromised patients.[57–66] Bonifaz et al.[153] described 12 patients suffering from oral geotrichosis in a 20-year retrospective study. Nine were females and the associated conditions were diabetes mellitus (67%), leukemia, Hodgkin's lymphoma, and HIV/AIDS infection. Oral geotrichoses presented in three clinical varieties, the most frequent being pseudomembranous (75%), but a hyperplastic presentation and the presence of ulcers were also observed. *G. candidum* and *M. capitatus* were isolated in 11 and 1 cases, respectively. Henrich et al.[60] reported the case of a woman with relapsed acute myelogenous leukemia after allogenic stem cell transplantation that developed disseminated *G. candidum* infection during chemotherapy-induced neutropenia. The isolate was susceptible to voriconazole, amphotericin B, and micafungin in vitro.

Invasive geotrichosis due to *M. capitatus* is not so common as trichosporonosis or candidiasis in the immunocompromised host, but the clinical spectrum is very similar. Fever, unresponsiveness to antibiotic treatment during the period of deep neutropenia, and the presence of fungemia are common. Moreover, metastatic skin lesions are usually observed as generalized maculopapular rashes. Abscesses in the liver and spleen at the time of the recovery from neutropenia are indistinguishable from those seen in chronic disseminated candidiasis or trichosporonosis.[88,92,93,154–157] The involvement of the lung is common in invasive mycoses by *M. capitatus*, and cough, expectoration, chest pain, spontaneous pneumothorax, and pulmonary infiltrates have been described.[93] However, the participation of endovascular catheters as the source of *M. capitatus* has been infrequently reported, and the ability to develop biofilms within the endovascular devices has been described.[1] The mortality attributed to invasive infections by *M. capitatus* is between 50% and 80%.[88,93] García-Ruiz et al.[93] described five cases of proven invasive infections by *M. capitatus* (2.8%) among 176 adult patients diagnosed with acute leukemia or related hematological diseases. A rate of 0.5% was reported for the Italian multicenter retrospective study conducted by Girmenia et al.[135] on 3000 patients with acute leukemia.

32.3 DIAGNOSIS OF TRICHOSPORONOSIS AND GEOTRICHOSIS

The diagnosis of superficial trichosporonosis or geotrichosis is uncomplicated. However, white piedra can be confused with pediculosis (*Phthirus pubis*) and *trichomycosis axillaris* (*Corynebacterium tenuis*), and skin and mucosal infections caused by these fungi could be misdiagnosed as superficial candidiasis or dermatophytoses. Conversely, the clinical diagnosis of invasive infections is particularly difficult because of the nonspecific clinical findings associated with these mycoses. The microbiology (or mycology) laboratory is of great help in supplying specific diagnoses in the cases of superficial or invasive infections.

Microbiological diagnosis of invasive infections is based on the isolation of fungi from sterile clinical specimens, such as blood, cerebrospinal fluid, or tissue biopsies. Phenotypic methods for the identification of clinical isolates are based on the characterization of the micromorphological aspects of colonies and biochemical profiles. The presence of arthroconidia and testing for urease production are very useful tools. However, morphological aspects and biochemical tests do not determine the species, and molecular methods, including sequencing of the IGS region, are necessary to obtain the accurate identification of isolates.[3–7] The presence of these fungi in nonsterile clinical specimens may be due to colonization. However, their presence in (1) sputum, (2) bronchoalveolar lavages, and/or (3) urine from febrile high-risk patients should be taken into account because of the possibility of a disseminated mycosis. Moreover, the histopathologic study of biopsies allows identification of some fungal structures, such as yeasts, arthroconidia, hyphae, and pseudohyphae. Immunohistochemistry can complement the differential histological diagnosis, although specific antibodies to *Trichosporon*, *Magnusiomyces*, or *Geotrichum* are not available commercially.[158]

These fungi grow quickly in conventional microbiological media, such as blood agar, and on fungal-specific media, such as Sabouraud dextrose agar. Chromogenic media, for example, chromID Candida (bioMérieux, France) or CHROMagar Candida (CHROMagar, France), can be helpful for distinguishing polymicrobial mycoses. However, the colonies of *Trichosporon*, *Magnusiomyces*, and *Geotrichum* are very similar in appearance to those of some species of *Candida*.[159] Microscopical identification is straightforward and is based on the observation of septate hyphae, pseudohyphae, arthroconidia, blastoconidia, appresoria, etc. There are several commercial systems based on carbohydrate assimilation: some of them are automated or semiautomated, such as API ID 32C system (bioMérieux), Auxacolor 2 (Bio-Rad, United States), Vitek Systems (bioMérieux), or Baxter Microscan (Baxter Microscan, United States), that can facilitate the identification in 24–48 h of most pathogenic yeasts.[160,161] Moreover, *Trichosporon* hydrolyzes urea, while *Geotrichum* and *Magnusiomyces* do not (Figure 32.4).[4,78] The identification of the different species of *Trichosporon*, *Magnusiomyces*, and *Geotrichum* can be facilitated by using carbohydrate assimilation tests and differential growth at different temperatures in media with cycloheximide.[4,162] Nevertheless, the high variability of phenotypic characters among species makes these methods inaccurate.[163] Most commercial tests are able to identify *T. asahii*, but they do not include methods for the more recently described species.

Genomic studies can be very useful for the identification of the species grouped in these genera[7] although they remain unavailable in most clinical laboratories. Most of these techniques are based on the amplification, detection, and, in some cases, sequencing of highly conserved regions of DNA. They include techniques based on single-step PCR, RAPD-PCR, nested PCR, PCR-RFLP, Luminex technology, and quantitative or real-time PCR.[164] The most frequently used target is the ribosomal multicopy gene (rDNA gene). The ribosomal rDNA region consists of genes encoding for small (18S), 5.8S, and large (28S) subunits of ribosomal DNA separated by the internal transcribed spacers, ITS1 and ITS2. These spacers are more variable domains and contain species-specific sequences. The ITS regions, the large subunit of rDNA (D1/D2 domain), and the IGS1 region localized between the 28S and 5S genes of rDNA remain the most commonly sequenced fungal loci. Recently, a technique based on line blot hybridization and rolling circle amplification has been developed that accurately differentiated *T. asahii*, *T. cutaneum*, *T. dermatis*, *T. domesticum*,

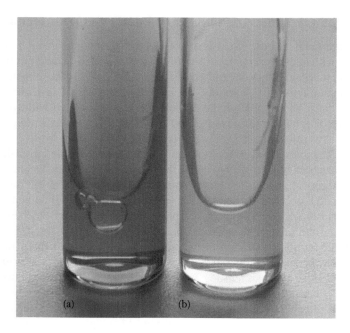

FIGURE 32.4 Urease test: positive (a) and negative (b).

T. inkin, *T. japonicum*, *T. jirovecii*, and *T. laibachii*.[165] Most molecular techniques used for the specific identification of *Geotrichum* are based on ITS rDNA sequences.[83] Finally, Luminex technology has proven to be suitable also for the identification of different species of *Trichosporon*. Specific capture probes designed in the D1/D2 region of ribosomal DNA, ITS regions, and IGS region have been used for this purpose.[166]

Matrix-assisted laser desorption/ionization time-of-flight mass spectrometry (MALDI-TOF MS) is suitable for the fast and secure identification of many pathogens in the clinical and applied laboratory. For example, several studies have reported good performances for *Candida* identification.[167,168] However, the identification performance is lower for other fungi, such as *Cryptococcus*, *Trichosporon*, and *Geotrichum*.[169,170] This may be due to the (1) small number of isolates analyzed, (2) low representation of these genera in the databases, and/or (3) particular composition of the cell walls that are very rich in external sugars, making protein extraction more difficult.[169,170] Therefore, it will be necessary to expand the available databases to improve that aspect of the identification of species of *Trichosporon*, *Magnusiomyces*, and *Geotrichum* at least.

There are no standardized serologic assays for the diagnosis of invasive trichosporonosis. Nevertheless, *Trichosporon* and *C. neoformans* capsular glucuronoxylomannans cross-react, and the serum latex agglutination test for *C. neoformans* may be positive in patients with disseminated trichosporonosis.[41,171,172] Latex agglutination tests for the detection of cryptococcal antigen are commercially available such as Cryptococcal Antigen Latex Agglutination System (CALAS; Meridian Bioscience, Inc., United States), Murex Cryptococcus Test (Remel, United States), and Crypto-LA test (Wampole Laboratories, United States).[173]

Moreover, detection of the panfungal 1,3-β-D-glucan (BG) can be useful for diagnosing of invasive trichosporonosis, as this molecule is detectable in the blood during most invasive mycoses.[174,175] The polysaccharide is part of the cell wall of many pathogenic fungi, including *Trichosporon*, *Magnusiomyces*, and *Geotrichum*. However, *Mucor*, *Rhizopus*, and *Cryptococcus* cell walls contain little or no BG.[176–178] Overall, the utility for the detection of serum BG for fungemia diagnosis remains unclear. In a recent study involving patients with hematologic disorders, the concentrations of BG increased only in half of patients who developed trichosporonemia.[179] In addition, overall survival was significantly worse in patients with low levels of BG although there were no significant differences between patient groups in terms of clinical background. *T. asahii* released BG concentrations comparable with those produced by *Candida* but showed a lower reactivity (62%) with the BG test.[180] There are several commercial techniques for serum quantification of BG: the most frequently used method in Europe and the Americas is Fungitell (Associates of Cape Cod, United States). The detection of glucuronoxylomannan may be a better marker than BG in the early diagnosis of invasive trichosporonosis, although testing of both biomarkers may be a more suitable means of diagnosing and monitoring the progression of invasive trichosporonosis.[181]

M. capitatus produces a soluble antigen that cross-reacts with the *Aspergillus* galactomannan, when using an enzyme immunosorbent assay (Platelia® *Aspergillus*, Bio-Rad, United States).[182–184] This cross-reactivity suggests that galactomannan detection may have a potential role in the diagnosis and management of invasive infection caused by this fungus. However, García-Ruiz et al.[93] reported five invasive *M. capitatus* infections with negative galactomannan test during the follow-up testing of patients. Conversely, the BG determinations were positive throughout the monitoring, and these authors suggest that BG could be a useful biomarker for invasive infections caused by *M. capitatus*.[93,178]

Standardized molecular tests are not yet available for the routine diagnosis of trichosporonosis or geotrichosis.[162,185] There are few reports on the use of PCR for direct detection of fungal DNA in clinical specimens from patients with invasive infections by these fungi. One of them is a nested PCR assay with specific primers that detected *T. asahii* DNA in 9 of 11 serum samples, from 7 patients with invasive infection diagnosed at autopsy.[186] An in-house PCR assay for the detection of DNA of clinically important *Trichosporon* species in serum was developed, and the sensitivity for *T. asahii* infection diagnosis was 64% (7 of 11 patients).[187] The results of a quantitative PCR assay for *T. asahii* DNA detection in the serum of a patient with disseminated trichosporonosis correlated with the clinical course of the disease.[185] A nested PCR for the detection to *Trichosporon* DNA in formalin-fixed and paraffin-embedded tissues that could be useful for supportive diagnosis of disseminated infection was developed by Sano et al.[108] The novel Luminex xMAP hybridization technology has been used successfully for the detection of fungi, including *Trichosporon*, in clinical specimens, such as peripheral blood, lung biopsies, and bronchotracheal secretions from patients with documented invasive mycoses. This nucleotide hybridization assay allows the analysis of different target sequences in a single reaction and species-specific capture probes that are covalently linked to fluorescent microbeads. After hybridization, the microbeads carrying the target amplicons are classified using a flow cytometer.[188]

32.4 THERAPY FOR TRICHOSPORONOSIS AND GEOTRICHOSIS

The management of these infections shows two different faces. The superficial infections do not offer many problems, and shaving, topical disinfection, and other local measures, combined with antifungal treatment, are useful. However, invasive mycoses caused by these fungi have a bad prognosis, and treatment is controversial. The fungal burden is usually very high when the infection is diagnosed, and therapeutic failure is frequent—in many cases related to antifungal resistance.

Treatment of white piedra and other superficial trichosporonosis or geotrichosis is based on shaving or clipping the hairs of the affected regions. However, relapses are common, and topical or oral antifungal therapies using itraconazole, or even voriconazole, are recommended.[96,110]

Treating patients with invasive infections of these fungi remains a medical challenge, and only limited data on the in vitro and in vivo antifungal activities are available. The initial and predominant steps should be directed toward minimizing known risk factors for invasive mycoses, such as neutropenia or immunosuppression, if feasible. All invasive procedures that disrupt the integrity of skin or the gastrointestinal tract should be used with caution. Moreover, the administration of broad-spectrum antibiotics and antifungal agents without activity against these fungi should not be used. Intravenous or parenteral alimentation should be changed as soon as possible to oral feeding. Careful hand hygiene between hospital staff and visitors should be encouraged, and other infection control practices can be important prevention strategies, for example, use of personal protective equipment, cleaning of environment, laundry, disinfection and sterilization of equipment, single-use equipment, waste management, and safe sharp handling.

When the invasive mycosis has been established, antifungal therapy is a key tool for recovery of the patient by eliminating these fungi from blood and tissues. However, optimal antifungal therapy has not been established currently. Amphotericin B, the historical gold standard for many invasive mycoses, is not very efficacious, since resistant isolates of *T. asahii* and other species of *Trichosporon* have been described. Unfortunately, no relevant clinical trials comparing different antifungal therapeutic options have been carried out. The general patterns of in vitro antifungal susceptibility for the most relevant species of *Trichosporon*, *Magnusiomyces*, and *Geotrichum* are shown in Table 32.5. In vitro susceptibility studies and animal models suggest that azoles, such as voriconazole and posaconazole, are more effective than amphotericin B or echinocandins for the eradication of infections caused by *Trichosporon*.[7,121,132,133,152,189–191]

Amphotericin B MIC values (range 1–8 μg/mL) against *Trichosporon*, particularly *T. asahii*, are higher than serum concentrations achievable in most patients receiving amphotericin B deoxycholate.[15,25,192] All *T. asahii* and most *T. coremiiforme* and *T. faecale* isolates tested showed resistance to amphotericin B (MICs ≥ 2 μg/mL), whereas the other species of *Trichosporon* tested had MICs of <1 μg/mL.[193] Fungicidal activity has not been observed in animal models or in neutropenic patients

TABLE 32.5
General Pattern of Antifungal Susceptibility of Relevant Species of *Trichosporon*, *Magnusiomyces*, and *Geotrichum* Compared to the Three Most Frequent Species of *Candida*

Species	Antifungal Agent						
	AMB	FLU	VOR	POS	AND	CAS	MIC
Candida albicans	S	S	S	S	S	S	S
C. parapsilosis	S	S	S	S	S–R	S–R	S–R
C. glabrata	S–I	SDD–R	S–R	S–R	S	S	S
Trichosporon asahii	I–R	S–R	S	S	R	R	R
T. asteroides	I–R	S	S	S	R	R	R
T. coremiiforme	R	S	S	S	R	R	R
T. faecale	R	S	S	S	R	R	R
T. inkin	S–R	S	S	S	R	R	R
T. japonicum	S–R	S	S	S	R	R	R
T. loubieri	S–R	S	S	S	R	R	R
T. montevideense	I–R	S	S	S	R	R	R
T. mucoides	S–R	S	S	S	R	R	R
T. mycotoxinivorans	S–R	S	S	S	R	R	R
M. capitatus	S–R	S	S	S	R	R	R
G. candidum	S–R	S	S	S	I–R	I–R	I–R

Sources: Refs. [7,32,105,132,141,146,152,189–191,194,197,204,209,219–221,226,249–254].

Notes: S, susceptible; SDD, susceptible dose-dependent; I, intermediate susceptible; R, resistant; AMB, amphotericin B; AND, anidulafungin; CAS, caspofungin; FLU, fluconazole; MIC, micafungin; POS, posaconazole; VOR, voriconazole.

although some isolates can be inhibited by low concentrations of amphotericin B. Girmenia et al. reported a clinical response to amphotericin B in only 13 of 55 patients (23.6%) with hematological diseases and disseminated trichosporonosis.[135]

Echinocandins have little activity against *Trichosporon* and are not recommended for the treatment of invasive infections.[25,121,194,195] Moreover, breakthrough trichosporonosis has been reported in immunocompromised patients treated with caspofungin or micafungin.[137,141,194,196]

Triazoles, in general, have demonstrated excellent in vitro activities against most clinical isolates of *Trichosporon* tested. The inhibitory concentrations of fluconazole (MIC range 4–64 μg/mL), isavuconazole (range 0.03–0.25 μg/mL), itraconazole (range 0.12–16 μg/mL), posaconazole (range 0.25–1 μg/mL), and voriconazole (range 0.03–0.25 μg/mL) are within the range of the drug concentrations therapeutically available.[197–201] Voriconazole and posaconazole were the antifungal drugs with the best in vitro activity against most *Trichosporon* isolates.[103] Chagas-Neto et al., using the Clinical and Laboratory Standards Institute broth microdilution test, found that the MICs of triazoles against 22 bloodstream *Trichosporon* isolates were generally lower than those of amphotericin B. Voriconazole showed the lowest MIC values against all the clinical isolates tested.[132] Similar data have been demonstrated in another study that evaluated the in vitro activities of nine antifungal drugs against 43 clinical isolates of *Trichosporon*. Voriconazole exhibited excellent in vitro activity, including against those isolates resistant to fluconazole.[121] Fluconazole was used as the initial therapy in 16 out of 19 patients with invasive trichosporonosis in a single medical center in Taiwan, and a mortality rate of 42% was observed.[121] Suzuki et al. reported a longer survival for patients with hematological malignancies suffering from trichosporonemia treated with azoles than for those treated with other antifungal drugs.[133] Although the available data are scarce, voriconazole (200–300 mg twice daily) may be useful for treating patients with disseminated trichosporonosis (Table 32.6).[202–206] The use of posaconazole is limited to the treatment of invasive trichosporonosis caused by isolates resistant to other antifungal drugs because it is available only as an oral suspension and it has erratic absorption.

The optimal therapeutic strategy for invasive geotrichosis is unknown. Clinical reports are scarce, and in vitro antifungal susceptibility studies show contradictory results. Most *M. capitatus* isolates are susceptible to amphotericin B, itraconazole, posaconazole, and voriconazole. However, they are not usually susceptible to caspofungin or micafungin and show dose-dependent susceptibility for fluconazole (Table 32.5). Based on their excellent in vitro activity,

TABLE 32.6
Tentative Approaches for the Treatment of Invasive Trichosporonosis and Geotrichosis

Clinical Presentation	Treatment	
	First Line	Alternative
Invasive trichosporonosis	VOR ± ABLC or LAMB	VOR + AND or CAS or MIC POS + ABLC or LAMB
Invasive geotrichosis caused by *Magnusiomyces capitatus*	VOR ± AND or CAS or MIC	ABLC or LAMB ± VOR ABLC or LAMB ± AND or CAS or MIC
Invasive geotrichosis caused by *Geotrichum candidum*	VOR ± AND or CAS or MIC	ABLC or LAMB ± VOR ABLC or LAMB ± AND or CAS or MIC
Biofilm and catheter-associated mycoses	Early catheter removal + VOR	Catheter retention + VOR + AND or CAS or MIC

Sources: Refs. [7,15,93,110,132,144,192,202–206].

Notes: ABLC, amphotericin B lipid complex; AMB, amphotericin B deoxycholate; AND, anidulafungin; CAS, caspofungin; FLC, fluconazole; ITR, itraconazole; LAMB, liposomal amphotericin B; MIC, micafungin; POS, posaconazole; VOR, voriconazole.

amphotericin B, voriconazole, and posaconazole are suitable for therapy.[93,207–210] Nevertheless, some strains show decreased susceptibility or poor clinical response to itraconazole or amphotericin B. An isolate of *M. capitatus* in a hematology unit was resistant to amphotericin B, 5-fluorocytosine, caspofungin, and itraconazole.[211]

Breakthrough geotrichosis caused by *M. capitatus* has been reported in patients receiving prophylactic fluconazole treatment or in patients receiving empiric caspofungin or micafungin as empiric treatment.[212–214] Previous exposure of patients with invasive candidiasis to fluconazole or caspofungin can lead to the isolation of species that are less susceptible to these antifungal drugs, such as *C. parapsilosis*, *C. glabrata*, or *C. krusei*, and a similar situation may occur with *M. capitatus*.[88,89,93,184,215]

Current expert consensus recommends the removal of intravenous catheters in patients suffering from invasive mycoses, whenever feasible, due to the high affinity of yeasts for prosthetic material. Catheter preservation has been associated with prolonged candidemia and worst outcomes especially in critically ill unstable patients.[216,217] The same recommendations could perhaps be appropriate for invasive trichosporonosis. Inappropriate antifungal therapy and lack of removal of central lines have been associated with poorer outcomes in nonneutropenic patients with trichosporonosis. Daily blood cultures after initiating therapy are recommended, and a search for metastatic foci if they remain positive, such as deep abscesses or endocarditis, is recommended. Antifungal therapy should be continued for at least 2 weeks after the last positive blood culture and after resolution of all clinical signs and symptoms of infection.

The limited range of available antifungal drugs and the in vitro resistance of many clinical isolates requires that the option of combination therapy to decrease the high mortality rates is considered and to reduce dose and toxicity (Table 32.6). The combination of fluconazole and amphotericin B has been used successfully in a bone marrow transplant recipient who developed breakthrough trichosporonosis while receiving prophylaxis against invasive fungal infection with caspofungin. Interestingly, a combination of echinocandin with amphotericin B or azoles appears to have some in vitro and in vivo synergistic antifungal effects against *Trichosporon*, leading to prolonged survival time and decreased renal fungal burden when compared with either agent alone.[218–221] Voriconazole in combination with liposomal amphotericin B has been the recommended treatment for invasive infections caused by *M. capitatus*.[93] Two clinical cases caused by *M. capitatus* had a good clinical response to combined caspofungin plus voriconazole treatment, despite the clinical isolate from one of the cases showed high echinocandin MICs.[57,222] However, combination therapy should not generally be used outside the context of a clinical trial, until better data on efficacy and safety from prospective, randomized, clinical trials that demonstrate superiority over monotherapy are obtained. However, the prognosis is poor, and antifungal regimens containing triazoles appear to be the best therapeutic approach. Recovery from (1) neutropenia and (2) the poor clinical condition of patients is essential for improving the prognosis of these invasive mycoses.

32.5 CONCLUSIONS

Trichosporon, *Magnusiomyces*, and *Geotrichum* spp. are widely distributed in nature and can colonize and be part of human microbiota. In the immunocompetent host, infections are limited to various superficial clinical conditions, such as white piedra and oral or genital infections. However, these fungi are also an emergent cause of invasive nosocomial infections in hematologic patients suffering from malignancies and in immunocompromised or intensive care patients exposed to invasive medical procedures. These organisms constitute between 1% and 5% of blood fungal isolates. *Trichosporon* adheres to biomedical devices and forms biofilms on catheters and prosthesis, resisting antifungal drug action and escaping from immune responses. *Trichosporon* glucuronoxylomannan and the ability to produce proteases and lipases are other virulence factors involved in the pathogenesis of trichosporonosis. *T. asahii* is the most common species causing invasive infections. Disseminated infections by other species, such as *T. asteroides*, *T. dermatis*, *T. inkin*, *T. jirovecii*,

T. loubieri, *T. mucoides*, and *T. mycotoxinivorans*, have been reported. However, species identification based on morphological aspects and biochemical properties is difficult. The presence of arthroconidia and urease production are useful for identifying the genus, but molecular methods, such as sequencing of the IGS region, are necessary for the identification of the species involved. The most frequent species from *Magnusiomyces* and *Geotrichum* involved in human disease are *M. capitatus* and *G. candidum*. A similar diagnostic approach to *Trichosporon* is needed for the identification of clinical isolates belonging to *Magnusiomyces* and *Geotrichum*. These fungi show a variable resistance to antifungal drugs including echinocandins and polyenes (*Trichosporon* spp.) or azoles (some *M. capitatus* isolates) that could explain the reported breakthrough infections during empiric antifungal therapy in neutropenic patients. Disseminated trichosporonosis and geotrichosis represent a great challenge for diagnosis and treatment. There are no specific diagnostic biomarkers, and blood culture is essential for the detection of invasive infections. Prognosis is poor, and antifungal therapies containing triazoles, such as voriconazole, appear to be the best options. Removal of central venous catheters and control of underlying diseases and risk factors can improve the clinical outcome.

REFERENCES

1. D'Antonio, D. et al. Slime production by clinical isolates of *Blastoschizomyces capitatus* from patients with hematological malignancies and catheter-related fungemia. *Eur. J. Clin. Microbiol. Infect. Dis.* **23**, 787–789 (2004).
2. Di Bonaventura, G. et al. Biofilm formation by the emerging fungal pathogen *Trichosporon asahii*: Development, architecture, and antifungal resistance. *Antimicrob. Agents Chemother.* **50**, 3269–3276 (2006).
3. Sugita, T. Chapter 161—*Trichosporon*. In *The Yeasts* (5th edn.) (eds. Kurtzman, C.P., Fell, J.W., Boekhout, T.) pp. 2015–2061 (Elsevier, London, U.K., 2011).
4. de Hoog, G.S., Guarro, J., Figueras, M.J., and Gené, J. *Atlas of Clinical Fungi* (Centraalbureau voor Schimmelcultures/Universitat Rovira i Virgili, Utrecht/Reus, the Netherlands, 2000).
5. Larone, D.H. *Medically Important Fungi; A Guide to Identification* (ASM Press, Washington, DC, 2002).
6. Boekhout, T. and Guého, E. Basidiomycetous yeasts. In *Pathogenic Fungi in Humans and Animals* (ed. Howard, D.H.) (Marcel Dekker Inc., New York, 2003).
7. Chagas-Neto, T.C., Chaves, G.M., and Colombo, A.L. Update on the genus *Trichosporon*. *Mycopathologia* **166**, 121–132 (2008).
8. Bonifaz, A. *Micología Médica Básica* (McGraw-Hill, México, DF, 2010).
9. Stursova, M., Zifcakova, L., Leigh, M.B., Burgess, R., and Baldrian, P. Cellulose utilization in forest litter and soil: Identification of bacterial and fungal decomposers. *FEMS Microbiol. Ecol.* **80**, 735–746 (2012).
10. Wu, Y., Du, P.C., Li, W.G., and Lu, J.X. Identification and molecular analysis of pathogenic yeasts in droppings of domestic pigeons in Beijing, China. *Mycopathologia* **174**, 203–214 (2012).
11. Zhang, E., Sugita, T., Tsuboi, R., Yamazaki, T., and Makimura, K. The opportunistic yeast pathogen *Trichosporon asahii* colonizes the skin of healthy individuals: Analysis of 380 healthy individuals by age and gender using a nested polymerase chain reaction assay. *Microbiol. Immunol.* **55**, 483–488 (2011).
12. Fuentefria, A.M. et al. *Trichosporon insectorum* sp. nov., a new anamorphic basidiomycetous killer yeast. *Mycol. Res.* **112**, 93–99 (2008).
13. Gueho, E., Improvisi, L., de Hoog, G.S., and Dupont, B. *Trichosporon* on humans: A practical account. *Mycoses* **37**, 3–10 (1994).
14. Quindos, G., Schneider, J., Alvarez, M., Ponton, J., and Cisterna, R. Antibodies against *Trichosporon beigelii* in vaginal washings from asymptomatic women. *J. Med. Microbiol.* **28**, 223–225 (1989).
15. Colombo, A.L., Padovan, A.C., and Chaves, G.M. Current knowledge of *Trichosporon* spp. and trichosporonosis. *Clin. Microbiol. Rev.* **24**, 682–700 (2011).
16. Rose, H.D. and Kurup, V.P. Colonization of hospitalized patients with yeast-like organisms. *Sabouraudia* **15**, 251–256 (1977).
17. Haupt, H.M., Merz, W.G., Beschorner, W.E., Vaughan, W.P., and Saral, R. Colonization and infection with *Trichosporon* species in the immunosuppressed host. *J. Infect. Dis.* **147**, 199–203 (1983).
18. Fadhil, R.A., Al-Thani, H., Al-Maslamani, Y., and Ali, O. *Trichosporon* fungal arteritis causing rupture of vascular anastomosis after commercial kidney transplantation: A case report and review of literature. *Transplant. Proc.* **43**, 657–659 (2011).

19. Macedo, D.P. et al. *Trichosporon inkin* esophagitis: An uncommon disease in a patient with pulmonary cancer. *Mycopathologia* **171**, 279–283 (2011).
20. Moreno-Coutino, G., Aquino, M.A., Vega-Memije, M., and Arenas, R. Necrotic ulcer caused by *Trichosporon asahii* in an immunocompetent adolescent. *Mycoses* **55**, 93–94 (2012).
21. Rota, A. et al. Presence and distribution of fungi and bacteria in the reproductive tract of healthy stallions. *Theriogenology* **76**, 464–470 (2011).
22. Sageerabanoo, Malini, A., Oudeacoumar, P., and Udayashankar, C. Onychomycosis due to *Trichosporon mucoides*. *Indian J. Dermatol. Venereol. Leprol.* **77**, 76–77 (2011).
23. Sharman, M.J., Stayt, J., McGill, S.E., and Mansfield, C.S. Clinical resolution of a nasal granuloma caused by *Trichosporon loubieri*. *J. Feline Med. Surg.* **12**, 345–350 (2010).
24. Sugita, T., Nishikawa, A., Shinoda, T., and Kume, H. Taxonomic position of deep-seated, mucosa-associated, and superficial isolates of *Trichosporon cutaneum* from trichosporonosis patients. *J. Clin. Microbiol.* **33**, 1368–1370 (1995).
25. Walsh, T.J. et al. Infections due to emerging and uncommon medically important fungal pathogens. *Clin. Microbiol. Infect.* **10(Suppl 1)**, 48–66 (2004).
26. Middelhoven, W.J., Scorzetti, G., and Fell, J.W. Systematics of the anamorphic basidiomycetous yeast genus *Trichosporon* Behrend with the description of five novel species: *Trichosporon vadense, T. smithiae, T. dehoogii, T. scarabaeorum* and *T. gamsii*. *Int. J. Syst. Evol. Microbiol.* **54**, 975–986 (2004).
27. Sugita, T. et al. *Trichosporon* species isolated from guano samples obtained from bat-inhabited caves in Japan. *Appl. Environ. Microbiol.* **71**, 7626–7629 (2005).
28. Sugita, T. and Nakase, T. Molecular phylogenetic study of the basidiomycetous anamorphic yeast genus *Trichosporon* and related taxa based on small subunit ribosomal DNA sequences. *Mycoscience* **39**, 7–13 (1998).
29. Gueho, E., de Hoog, G.S., and Smith, M.T. Neotypification of the genus *Trichosporon*. *Antonie Van Leeuwenhoek* **61**, 285–288 (1992).
30. Gueho, E. et al. Contributions to a revision of the genus *Trichosporon*. *Antonie Van Leeuwenhoek* **61**, 289–316 (1992).
31. Sugita, T., Nakajima, M., Ikeda, R., Matsushima, T., and Shinoda, T. Sequence analysis of the ribosomal DNA intergenic spacer 1 regions of *Trichosporon* species. *J. Clin. Microbiol.* **40**, 1826–1830 (2002).
32. Taj-Aldeen, S.J. et al. Molecular identification and susceptibility of *Trichosporon* species isolated from clinical specimens in Qatar: Isolation of *Trichosporon dohaense* Taj-Aldeen, Meis and Boekhout sp. nov. *J. Clin. Microbiol.* **47**, 1791–1799 (2009).
33. Pagnocca, F.C. et al. Yeasts isolated from a fungus-growing ant nest, including the description of *Trichosporon chiarellii* sp. nov., an anamorphic basidiomycetous yeast. *Int. J. Syst. Evol. Microbiol.* **60**, 1454–1459 (2010).
34. Gujjari, P., Suh, S.O., Lee, C.F., and Zhou, J.J. *Trichosporon xylopini* sp. nov., a hemicellulose-degrading yeast isolated from the wood-inhabiting beetle *Xylopinus saperdioides*. *Int. J. Syst. Evol. Microbiol.* **61**, 2538–2542 (2011).
35. Gujjari, P., Suh, S.O., Coumes, K., and Zhou, J.J. Characterization of oleaginous yeasts revealed two novel species: *Trichosporon cacaoliposimilis* sp. nov. and *Trichosporon oleaginosus* sp. nov. *Mycologia* **103**, 1110–1118 (2011).
36. Motaung, T.E. et al. *Trichosporon vanderwaltii* sp. nov., an asexual basidiomycetous yeast isolated from soil and beetles. *Antonie Van Leeuwenhoek* **103**, 313–319 (2013).
37. Fell, J.W. and Scorzetti, G. Reassignment of the basidiomycetous yeasts *Trichosporon pullulans* to *Guehomyces pullulans* gen. nov., comb. nov. and *Hyalodendron lignicola* to *Trichosporon lignicola* comb. nov. *Int. J. Syst. Evol. Microbiol.* **54**, 995–998 (2004).
38. Sun, W., Su, J., Xu, S., and Yan, D. *Trichosporon asahii* causing nosocomial urinary tract infections in intensive care unit patients: Genotypes, virulence factors and antifungal susceptibility testing. *J. Med. Microbiol.* **61**, 1750–1757 (2012).
39. Karashima, R. et al. Increased release of glucuronoxylomannan antigen and induced phenotypic changes in *Trichosporon asahii* by repeated passage in mice. *J. Med. Microbiol.* **51**, 423–432 (2002).
40. Mizobe, T., Ando, M., Yamasaki, H., Onoue, K., and Misaki, A. Purification and characterization of the serotype-specific polysaccharide antigen of *Trichosporon cutaneum* serotype II: A disease-related antigen of Japanese summer-type hypersensitivity pneumonitis. *Clin. Exp. Allergy* **25**, 265–272 (1995).
41. Fonseca, F.L. et al. Structural and functional properties of the *Trichosporon asahii* glucuronoxylomannan. *Fungal Genet. Biol.* **46**, 496–505 (2009).

42. Ichikawa, T. et al. Phenotypic switching and beta-N-acetylhexosaminidase activity of the pathogenic yeast *Trichosporon asahii*. *Microbiol. Immunol.* **48**, 237–242 (2004).
43. Ghannoum, M.A. Potential role of phospholipases in virulence and fungal pathogenesis. *Clin. Microbiol. Rev.* **13**, 122–43, table of contents (2000).
44. Ageitos, J.M., Vallejo, J.A., Veiga-Crespo, P., and Villa, T.G. Oily yeasts as oleaginous cell factories. *Appl. Microbiol. Biotechnol.* **90**, 1219–1227 (2011).
45. Chen, X.F. et al. Oil production on wastewaters after butanol fermentation by oleaginous yeast *Trichosporon coremiiforme*. *Bioresour. Technol.* **118**, 594–597 (2012).
46. Luo, Z.H., Wu, Y.R., Pang, K.L., Gu, J.D., and Vrijmoed, L.L. Comparison of initial hydrolysis of the three dimethyl phthalate esters (DMPEs) by a basidiomycetous yeast, *Trichosporon*DMI-5-1, from coastal sediment. *Environ. Sci. Pollut. Res. Int.* **18**, 1653–1660 (2011).
47. Sollner Dragicevic, T.L. et al. Biodegradation of olive mill wastewater by *Trichosporon cutaneum* and *Geotrichum candidum*. *Arh. Hig. Rada. Toksikol.* **61**, 399–405 (2010).
48. Wang, Y., Xu, Y., and Li, J. A novel extracellular beta-glucosidase from *Trichosporon asahii*: Yield prediction, evaluation and application for aroma enhancement of Cabernet Sauvignon. *J. Food Sci.* **77**, M505–15 (2012).
49. Xiang, W. et al. Microbial succession in the traditional Chinese Luzhou-flavor liquor fermentation process as evaluated by SSU rRNA profiles. *World J. Microbiol. Biotechnol.* **29**, 559–567 (2013).
50. Li, H.Y., Zhao, C.A., Liu, C.J., and Xu, X.F. Endophytic fungi diversity of aquatic/riparian plants and their antifungal activity in vitro. *J. Microbiol.* **48**, 1–6 (2010).
51. Alper, I., Frenette, M., and Labrie, S. Genetic diversity of dairy *Geotrichum candidum* strains revealed by multilocus sequence typing. *Appl. Microbiol. Biotechnol.* **97**, 5907–5920 (2013).
52. Delavenne, E. et al. Fungal diversity in cow, goat and ewe milk. *Int. J. Food Microbiol.* **151**, 247–251 (2011).
53. Desmasures, N., Bazin, F., and Gueguen, M. Microbiological composition of raw milk from selected farms in the Camembert region of Normandy. *J. Appl. Microbiol.* **83**, 53–58 (1997).
54. Mu, Z., Yang, X., and Yuan, H. Detection and identification of wild yeast in Koumiss. *Food Microbiol.* **31**, 301–308 (2012).
55. Pottier, I., Gente, S., Vernoux, J.P., and Gueguen, M. Safety assessment of dairy microorganisms: *Geotrichum candidum*. *Int. J. Food Microbiol.* **126**, 327–332 (2008).
56. Groenewald, M., Coutinho, T., Smith, M.T., and van der Walt, J.P. Species reassignment of *Geotrichum bryndzae*, *Geotrichum phurueaensis*, *Geotrichum silvicola* and *Geotrichum vulgare* based on phylogenetic analyses and mating compatibility. *Int. J. Syst. Evol. Microbiol.* **62**, 3072–3080 (2012).
57. Fianchi, L. et al. Combined voriconazole plus caspofungin therapy for the treatment of probable *Geotrichum* pneumonia in a leukemia patient. *Infection* **36**, 65–67 (2008).
58. Garcia-Lozano, T., Sanchez Yepes, M., Aznar Oroval, E., Ortiz Munoz, B.A., and Guillen Bernardo, I. Intra-abdominal seroma and lymphopenia without leucopenia in a cancer patient. *Geotrichum candidum* infection. *Rev. Iberoam. Micol.* **31**, 152–153 (2012).
59. Heinic, G.S., Greenspan, D., MacPhail, L.A., and Greenspan, J.S. Oral *Geotrichum candidum* infection associated with HIV infection. A case report. *Oral Surg. Oral Med. Oral Pathol.* **73**, 726–728 (1992).
60. Henrich, T.J., Marty, F.M., Milner, D.A., Jr., and Thorner, A.R. Disseminated *Geotrichum candidum* infection in a patient with relapsed acute myelogenous leukemia following allogeneic stem cell transplantation and review of the literature. *Transpl. Infect. Dis.* **11**, 458–462 (2009).
61. Sfakianakis, A. et al. Invasive cutaneous infection with *Geotrichum candidum*: Sequential treatment with amphotericin B and voriconazole. *Med. Mycol.* **45**, 81–84 (2007).
62. Hazen, K.C. and Howell, S.A. Yeasts. *Blastomycetes* and *Endomycetes*. In *Pathogenic Fungi in Humans and Animals* (ed. Howard, D.H.) (Marcel Dekker Inc., New York, 2003).
63. Figueredo, L.A., Cafarchia, C., and Otranto, D. *Geotrichum candidum* as etiological agent of horse dermatomycosis. *Vet. Microbiol.* **148**, 368–371 (2011).
64. Lee, E.J., Gabor, M., Turner, M., Ball, M., and Gabor, L. Tonsillitis in a weaner pig associated with *Geotrichum candidum*. *J. Vet. Diagn. Invest.* **23**, 175–177 (2011).
65. Pal, M. Role of *Geotrichum candidum* in canine oral ulcers. *Rev. Iberoam. Micol.* **22**, 183 (2005).
66. Kantardjiev, T., Kuzmanova, A., Baikushev, R., Zisova, L., and Velinov, T. Isolation and identification of *Geotrichum candidum* as an etiologic agent of geotrichosis in Bulgaria. *Folia. Med. (Plovdiv)* **40**, 42–44 (1998).
67. Boutrou, R. and Gueguen, M. Interests in *Geotrichum candidum* for cheese technology. *Int. J. Food Microbiol.* **102**, 1–20 (2005).

68. Wolf-Hall, C.E. Mold and mycotoxin problems encountered during malting and brewing. *Int. J. Food Microbiol.* **119**, 89–94 (2007).
69. Ziino, M. et al. Lipid composition of *Geotrichum candidum* single cell protein grown in continuous submerged culture. *Bioresour. Technol.* **67**, 7–11 (1999).
70. Celinska, E., Kubiak, P., Bialas, W., Dziadas, M., and Grajek, W. *Yarrowia lipolytica*: The novel and promising 2-phenylethanol producer. *J. Ind. Microbiol. Biotechnol.* **40**, 389–392 (2013).
71. Mdaini, N., Gargouri, M., Hammami, M., Monser, L., and Hamdi, M. Production of natural fruity aroma by *Geotrichum candidum*. *Appl. Biochem. Biotechnol.* **128**, 227–235 (2006).
72. Asses, N., Ayed, L., Bouallagui, H., Sayadi, S., and Hamdi, M. Biodegradation of different molecular-mass polyphenols derived from olive mill wastewaters by *Geotrichum candidum*. *Int. Biodeterior. Biodegrad.* **63**, 407–413 (2009).
73. Bleve, G. et al. Selection of non-conventional yeasts and their use in immobilized form for the bioremediation of olive oil mill wastewaters. *Bioresour. Technol.* **102**, 982–989 (2011).
74. Fitzgibbon, F.J., Nigam, P., Singh, D., and Marchant, R. Biological treatment of distillery waste for pollution-remediation. *J. Basic Microbiol.* **35**, 293–301 (1995).
75. Waghmode, T.R., Kurade, M.B., Kabra, A.N., and Govindwar, S.P. Biodegradation of Rubine GFL by *Galactomyces geotrichum* MTCC 1360 and subsequent toxicological analysis by using cytotoxicity, genotoxicity and oxidative stress studies. *Microbiology* **158**, 2344–2352 (2012).
76. Thornton, C.R., Slaughter, D.C., and Davis, R.M. Detection of the sour-rot pathogen *Geotrichum candidum* in tomato fruit and juice by using a highly specific monoclonal antibody-based ELISA. *Int. J. Food Microbiol.* **143**, 166–172 (2010).
77. Xu, S. et al. In vitro and in vivo antifungal activity of a water-dilutable cassia oil microemulsion against *Geotrichum citri-aurantii*. *J. Sci. Food Agric.* **92**, 2668–2671 (2012).
78. de Hoog, G.S. and Smith, M.T. Chapter 91—Geotrichum link. In *The Yeasts* (eds. Kurtzman, C.P., Fell, J.W., Boekhout, T.) pp. 1279–1286 (Elsevier, London, U.K., 2011).
79. de Hoog, G.S., Smith, M.T., and Guého, E. A revision of the genus *Geotrichum* and its teleomorphs. *Studies Mycol.* **29**, 1–131 (1986).
80. de Hoog, G.S. and Smith, M.T. Chapter 31—Galactomyces Redhead & Malloch (1977). In *The Yeasts* (eds. Kurtzman, C.P., Fell, J.W., Boekhout, T.) pp. 413–420 (Elsevier, London, U.K., 2011).
81. de Hoog, G.S. and Smith, M.T. Chapter 27—Dipodascus. In *The Yeasts* (eds. Kurtzman, C.P., Fell, J.W., Boekhout, T.) pp. 385–392 (Elsevier, London, U.K., 2011).
82. Smith, M.T., Poot, G.A., and de Cock, A.W. Re-examination of some species of the genus *Geotrichum* Link: Fr. *Antonie Van Leeuwenhoek* **77**, 71–81 (2000).
83. de Hoog, G.S. and Smith, M.T. Ribosomal gene phylogeny and species delimitation in *Geotrichum* and its teleomorphs. *Studies Mycol.* **50**, 489–515 (2004).
84. Salkin, I.F., Gordon, M.A., Samsonoff, W.A., and Rieder, C.L. *Blastoschizomyces capitatus*, a new combination. *Mycotaxon* **22**, 375–380 (1985).
85. Kurtzman, C.P. and Robnett, C.J. Molecular relationships among hyphal ascomycetous yeasts and yeast-like taxa. *Can. J. Bot.* **73**, S824–S830 (1995).
86. Ueda-Nishimura, K. and Mikata, K. Two distinct 18S rRNA secondary structures in *Dipodascus* (Hemiascomycetes). *Microbiology* **146 (Pt 5)**, 1045–1051 (2000).
87. Zalar, P., Novak, M., de Hoog, G.S., and Gunde-Cimerman, N. Dishwashers—A man-made ecological niche accommodating human opportunistic fungal pathogens. *Fungal Biol.* **115**, 997–1007 (2011).
88. Martino, P. et al. *Blastoschizomyces capitatus*: An emerging cause of invasive fungal disease in leukemia patients. *Rev. Infect. Dis.* **12**, 570–582 (1990).
89. Martino, R. et al. *Blastoschizomyces capitatus* infection in patients with leukemia: Report of 26 cases. *Clin. Infect. Dis.* **38**, 335–341 (2004).
90. Buchta, V., Zak, P., Kohout, A., and Otcenasek, M. Case report. Disseminated infection of *Blastoschizomyces capitatus* in a patient with acute myelocytic leukaemia. *Mycoses* **44**, 505–512 (2001).
91. Bouza, E. and Munoz, P. Invasive infections caused by *Blastoschizomyces capitatus* and *Scedosporium* spp. *Clin. Microbiol. Infect.* **10(Suppl 1)**, 76–85 (2004).
92. Quindós, G., García-Ruiz, J.C., and Sanz, M.A. Otras infecciones fúngicas invasoras. In *Micosis invasoras en pacientes oncohematológicos*, pp. 79–95 (Revista Iberoamericana de Micología, Bilbao, Spain, 2009).
93. García-Ruiz, J.C. et al. Invasive infections caused by *Saprochaete capitata* in patients with haematological malignancies: Report of five cases and review of antifungal therapy. *Rev. Iberoam. Micol.* **30**, 248–255 (2013).
94. Bouakline, A., Lacroix, C., Roux, N., Gangneux, J.P., and Derouin, F. Fungal contamination of food in hematology units. *J. Clin. Microbiol.* **38**, 4272–4273 (2000).

95. Gurgui, M. et al. Nosocomial outbreak of *Blastoschizomyces capitatus* associated with contaminated milk in a haematological unit. *J. Hosp. Infect.* **78**, 274–278 (2011).
96. Kiken, D.A. et al. White piedra in children. *J. Am. Acad. Dermatol.* **55**, 956–961 (2006).
97. Viswanath, V., Kriplani, D., Miskeen, A.K., Patel, B., and Torsekar, R.G. White piedra of scalp hair by *Trichosporon inkin*. *Indian J. Dermatol. Venereol. Leprol.* **77**, 591–593 (2011).
98. Ellner, K.M., McBride, M.E., Kalter, D.C., Tschen, J.A., and Wolf, J.E., Jr. White piedra: Evidence for a synergistic infection. *Br. J. Dermatol.* **123**, 355–363 (1990).
99. Youker, S.R., Andreozzi, R.J., Appelbaum, P.C., Credito, K., and Miller, J.J. White piedra: Further evidence of a synergistic infection. *J. Am. Acad. Dermatol.* **49**, 746–749 (2003).
100. Richini-Pereira, V.B., Camargo, R.M., Bagagli, E., and Marques, S.A. White piedra: Molecular identification of *Trichosporon inkin* in members of the same family. *Rev. Soc. Bras. Med. Trop.* **45**, 402–404 (2012).
101. Tambe, S.A. et al. Two cases of scalp white piedra caused by *Trichosporon ovoides*. *Indian J. Dermatol. Venereol. Leprol.* **75**, 293–295 (2009).
102. Padhye, A.A. et al. *Trichosporon loubieri* infection in a patient with adult polycystic kidney disease. *J. Clin. Microbiol.* **41**, 479–482 (2003).
103. Araujo Ribeiro, M., Alastruey-Izquierdo, A., Gomez-Lopez, A., Rodriguez-Tudela, J.L., and Cuenca-Estrella, M. Molecular identification and susceptibility testing of *Trichosporon* isolates from a Brazilian hospital. *Rev. Iberoam. Micol.* **25**, 221–225 (2008).
104. Ruiz-Esmenjaud, J., Arenas, R., Rodriguez-Alvarez, M., Monroy, E., and Felipe Fernandez, R. *Tinea pedis* and onychomycosis in children of the Mazahua Indian Community in Mexico. *Gac. Med. Mex.* **139**, 215–220 (2003).
105. Mendez-Tovar, L.J. et al. Micosis among five highly underprivileged Mexican communities. *Gac. Med. Mex.* **142**, 381–386 (2006).
106. Benson, P.M., Lapins, N.A., and Odom, R.B. White piedra. *Arch. Dermatol.* **119**, 602–604 (1983).
107. Roshan, A.S., Janaki, C., and Parveen, B. White piedra in a mother and daughter. *Int. J. Trichol.* **1**, 140–141 (2009).
108. Sano, M. et al. Supplemental utility of nested PCR for the pathological diagnosis of disseminated trichosporonosis. *Virchows Arch.* **451**, 929–935 (2007).
109. Moretti-Branchini, M.L. et al. *Trichosporon* species infection in bone marrow transplanted patients. *Diagn. Microbiol. Infect. Dis.* **39**, 161–164 (2001).
110. Richardson, M.D. and Warnock, D.W. *Fungal Infection. Diagnosis and Management* (Blackwell Publishing, Oxford, U.K., 2003).
111. Binder, U. and Lass-Florl, C. Epidemiology of invasive fungal infections in the Mediterranean area. *Mediterr. J. Hematol. Infect. Dis.* **3**, e20110016 (2011).
112. Keay, S., Denning, D.W., and Stevens, D.A. Endocarditis due to *Trichosporon beigelii*: In vitro susceptibility of isolates and review. *Rev. Infect. Dis.* **13**, 383–386 (1991).
113. Fisher, D.J. et al. Neonatal *Trichosporon beigelii* infection: Report of a cluster of cases in a neonatal intensive care unit. *Pediatr. Infect. Dis. J.* **12**, 149–155 (1993).
114. Mirza, S.H. Disseminated *Trichosporon beigelii* infection causing skin lesions in a renal transplant patient. *J. Infect.* **27**, 67–70 (1993).
115. Hajjeh, R.A. and Blumberg, H.M. Bloodstream infection due to *Trichosporon beigelii* in a burn patient: Case report and review of therapy. *Clin. Infect. Dis.* **20**, 913–916 (1995).
116. Lussier, N., Laverdiere, M., Delorme, J., Weiss, K., and Dandavino, R. *Trichosporon beigelii* funguria in renal transplant recipients. *Clin. Infect. Dis.* **31**, 1299–1301 (2000).
117. Chan, R.M., Lee, P., and Wroblewski, J. Deep-seated trichosporonosis in an immunocompetent patient: A case report of uterine trichosporonosis. *Clin. Infect. Dis.* **31**, 621 (2000).
118. Cawley, M.J. et al. *Trichosporon beigelii* infection: Experience in a regional burn center. *Burns* **26**, 483–486 (2000).
119. Ebright, J.R., Fairfax, M.R., and Vazquez, J.A. *Trichosporon asahii*, a non-*Candida* yeast that caused fatal septic shock in a patient without cancer or neutropenia. *Clin. Infect. Dis.* **33**, E28–E30 (2001).
120. Netsvyetayeva, I. et al. *Trichosporon asahii* as a prospective pathogen in solid organ transplant recipients. *Mycoses* **52**, 263–265 (2009).
121. Ruan, S.Y., Chien, J.Y., and Hsueh, P.R. Invasive trichosporonosis caused by *Tichosporon asahii* and other unusual *Trichosporon* species at a medical center in Taiwan. *Clin. Infect. Dis.* **49**, e11–e17 (2009).
122. Rodrigues Gda, S., de Faria, R.R., Guazzelli, L.S., Oliveira Fde, M., and Severo, L.C. Nosocomial infection due to *Trichosporon asahii*: Clinical revision of 22 cases. *Rev. Iberoam. Micol.* **23**, 85–89 (2006).

123. Dotis, J., Pana, Z.D. and Roilides, E. Non-*Aspergillus* fungal infections in cronic granulomatous disease. *Mycoses* **56**, 449–462 (2013).
124. Colombo, A.L. et al. Gastrointestinal translocation as a possible source of candidemia in an AIDS patient. *Rev. Inst. Med. Trop. Sao Paulo* **38**, 197–200 (1996).
125. Nucci, M. and Anaissie, E. Revisiting the source of candidemia: Skin or gut? *Clin. Infect. Dis.* **33**, 1959–1967 (2001).
126. Walsh, T.J., Orth, D.H., Shapiro, C.M., Levine, R.A., and Keller, J.L. Metastatic fungal chorioretinitis developing during *Trichosporon* sepsis. *Ophthalmology* **89**, 152–156 (1982).
127. Kontoyiannis, D.P. et al. Trichosporonosis in a tertiary care cancer center: Risk factors, changing spectrum and determinants of outcome. *Scand. J. Infect. Dis.* **36**, 564–569 (2004).
128. Hickey, P.W. et al. *Trichosporon mycotoxinivorans*, a novel respiratory pathogen in patients with cystic fibrosis. *J. Clin. Microbiol.* **47**, 3091–3097 (2009).
129. Walsh, T.J., Melcher, G.P., Lee, J.W., and Pizzo, P.A. Infections due to *Trichosporon* species: New concepts in mycology, pathogenesis, diagnosis and treatment. *Curr. Top. Med. Mycol.* **5**, 79–113 (1993).
130. Ellner, K., McBride, M.E., Rosen, T., and Berman, D. Prevalence of *Trichosporon beigelii*. Colonization of normal perigenital skin. *J. Med. Vet. Mycol.* **29**, 99–103 (1991).
131. Salazar, G.E. and Campbell, J.R. Trichosporonosis, an unusual fungal infection in neonates. *Pediatr. Infect. Dis. J.* **21**, 161–165 (2002).
132. Chagas-Neto, T.C., Chaves, G.M., Melo, A.S., and Colombo, A.L. Bloodstream infections due to *Trichosporon* spp.: Species distribution, *Trichosporon asahii* genotypes determined on the basis of ribosomal DNA intergenic spacer 1 sequencing, and antifungal susceptibility testing. *J. Clin. Microbiol.* **47**, 1074–1081 (2009).
133. Suzuki, K. et al. Fatal *Trichosporon* fungemia in patients with hematologic malignancies. *Eur. J. Haematol.* **84**, 441–447 (2010).
134. Ramos, J.M., Cuenca-Estrella, M., Gutierrez, F., Elia, M., and Rodriguez-Tudela, J.L. Clinical case of endocarditis due to *Trichosporon inkin* and antifungal susceptibility profile of the organism. *J. Clin. Microbiol.* **42**, 2341–2344 (2004).
135. Girmenia, C. et al. Invasive infections caused by *Trichosporon* species and *Geotrichum capitatum* in patients with hematological malignancies: A retrospective multicenter study from Italy and review of the literature. *J. Clin. Microbiol.* **43**, 1818–1828 (2005).
136. Rastogi, V. et al. Non-healing ulcer due to *Trichosporon loubieri* in an immunocompetent host and review of published reports. *Mycopathologia* **176**, 107–111 (2013).
137. Goodman, D., Pamer, E., Jakubowski, A., Morris, C., and Sepkowitz, K. Breakthrough trichosporonosis in a bone marrow transplant recipient receiving caspofungin acetate. *Clin. Infect. Dis.* **35**, E35–E36 (2002).
138. Krcmery, V., Krupova, I., and Denning, D.W. Invasive yeast infections other than *Candida* spp. in acute leukaemia. *J. Hosp. Infect.* **41**, 181–194 (1999).
139. Fleming, R.V., Walsh, T.J., and Anaissie, E.J. Emerging and less common fungal pathogens. *Infect. Dis. Clin. North Am.* **16**, 915–33, vi–vii (2002).
140. Pagano, L. et al. The epidemiology of fungal infections in patients with hematologic malignancies: The SEIFEM-2004 study. *Haematologica* **91**, 1068–1075 (2006).
141. Matsue, K., Uryu, H., Koseki, M., Asada, N., and Takeuchi, M. Breakthrough trichosporonosis in patients with hematologic malignancies receiving micafungin. *Clin. Infect. Dis.* **42**, 753–757 (2006).
142. Bayramoglu, G., Sonmez, M., Tosun, I., Aydin, K., and Aydin, F. Breakthrough *Trichosporon asahii* fungemia in neutropenic patient with acute leukemia while receiving caspofungin. *Infection* **36**, 68–70 (2008).
143. Hosoki, K., Iwamoto, S., Kumamoto, T., Azuma, E., and Komada, Y. Early detection of breakthrough trichosporonosis by serum PCR in a cord blood transplant recipient being prophylactically treated with voriconazole. *J. Pediatr. Hematol. Oncol.* **30**, 917–919 (2008).
144. Vazquez, J.A. *Rhodotorula, Saccharomyces, Malassezia, Trichosporon, Blastoschizomyces*, and *Sporobolomyces*. In *Essentials of Clinical Mycology* (eds. Kauffman, C.A., Pappas, P.G., Sobel, J.D., and Dismukes, W.E.) (Springer Science + Business Media, New York, 2011).
145. Silva, V., Zepeda, G., and Alvareda, D. Nosocomial urinary infection due to *Trichosporon asahii*. First two cases in Chile. *Rev. Iberoam. Micol.* **20**, 21–23 (2003).
146. Meyer, M.H. et al. Chronic disseminated *Trichosporon asahii* infection in a leukemic child. *Clin. Infect. Dis.* **35**, e22–e25 (2002).

147. Ando, M. et al. Serotype-related antigen of *Trichosporon cutaneum* in the induction of summer-type hypersensitivity pneumonitis: Correlation between serotype of inhalation challenge-positive antigen and that of the isolates from patients' homes. *J. Allergy Clin. Immunol.* **85**, 36–44 (1990).
148. Nishiura, Y. et al. Assignment and serotyping of *Trichosporon* species: The causative agents of summer-type hypersensitivity pneumonitis. *J. Med. Vet. Mycol.* **35**, 45–52 (1997).
149. Kataoka-Nishimura, S. et al. Invasive infection due to *Trichosporon cutaneum* in patients with hematologic malignancies. *Cancer* **82**, 484–487 (1998).
150. Groll, A.H. and Walsh, T.J. Uncommon opportunistic fungi: New nosocomial threats. *Clin. Microbiol. Infect.* **7(Suppl 2)**, 8–24 (2001).
151. Sugita, T., Ikeda, R. and Nishikawa, A. Analysis of *Trichosporon* isolates obtained from the houses of patients with summer-type hypersensitivity pneumonitis. *J. Clin. Microbiol.* **42**, 5467–5471 (2004).
152. Arikawa, T. et al. Galectin-9 expands immunosuppressive macrophages to ameliorate T-cell-mediated lung inflammation. *Eur. J. Immunol.* **40**, 548–558 (2010).
153. Bonifaz, A. et al. Oral geotrichosis: Report of 12 cases. *J. Oral Sci.* **52**, 477–483 (2010).
154. Villa Lopez, I., Doblas Claros, A., Saavedra, J.M., and Herrera Carranza, M. Multi-organ failure in patient with fungaemia due to *Saprochaete capitata*. *Rev. Iberoam. Micol.* **30**, 261–263 (2013).
155. Chang, W.W. and Buerger, L. Disseminated geotrichosis; Case report. *Arch. Intern. Med.* **113**, 356–360 (1964).
156. Polacheck, I., Salkin, I.F., Kitzes-Cohen, R., and Raz, R. Endocarditis caused by *Blastoschizomyces capitatus* and taxonomic review of the genus. *J. Clin. Microbiol.* **30**, 2318–2322 (1992).
157. Sanz, M.A. et al. Disseminated *Blastoschizomyces capitatus* infection in acute myeloblastic leukaemia. Report of three cases. *Support. Care Cancer* **4**, 291–293 (1996).
158. Guarner, J. and Brandt, M.E. Histopathologic diagnosis of fungal infections in the 21st century. *Clin. Microbiol. Rev.* **24**, 247–280 (2011).
159. Eraso, E. et al. Evaluation of the new chromogenic medium Candida ID 2 for isolation and identification of *Candida albicans* and other medically important *Candida* species. *J. Clin. Microbiol.* **44**, 3340–3345 (2006).
160. Ayats, J. et al. Spanish Society of Clinical Microbiology and Infectious Diseases (SEIMC) guidelines for the diagnosis of invasive fungal infections. 2010 update. *Enferm. Infecc. Microbiol. Clin.* **29**, 39.e1–e39.15 (2011).
161. Quindos, G., Eraso, E., Lopez-Soria, L.M., and Ezpeleta, G. Invasive fungal disease: Conventional or molecular mycological diagnosis? *Enferm. Infecc. Microbiol. Clin.* **30**, 560–571 (2012).
162. Pincus, D.H., Orenga, S., and Chatellier, S. Yeast identification—Past, present, and future methods. *Med. Mycol.* **45**, 97–121 (2007).
163. Gente, S., Sohier, D., Coton, E., Duhamel, C., and Gueguen, M. Identification of *Geotrichum candidum* at the species and strain level: Proposal for a standardized protocol. *J. Ind. Microbiol. Biotechnol.* **33**, 1019–1031 (2006).
164. Balajee, S.A., Sigler, L., and Brandt, M.E. DNA and the classical way: Identification of medically important molds in the 21st century. *Med. Mycol.* **45**, 475–490 (2007).
165. Xiao, M. et al. Practical identification of eight medically important *Trichosporon* species by reverse line blot hybridization (RLB) assay and rolling circle amplification (RCA). *Med. Mycol.* **51**, 300–308 (2013).
166. Diaz, M.R. and Fell, J.W. High-throughput detection of pathogenic yeasts of the genus *Trichosporon*. *J. Clin. Microbiol.* **42**, 3696–3706 (2004).
167. Marklein, G. et al. Matrix-assisted laser desorption ionization-time of flight mass spectrometry for fast and reliable identification of clinical yeast isolates. *J. Clin. Microbiol.* **47**, 2912–2917 (2009).
168. Stevenson, L.G., Drake, S.K., Shea, Y.R., Zelazny, A.M., and Murray, P.R. Evaluation of matrix-assisted laser desorption ionization-time of flight mass spectrometry for identification of clinically important yeast species. *J. Clin. Microbiol.* **48**, 3482–3486 (2010).
169. Lohmann, C. et al. Comparison between the Biflex III-Biotyper and the Axima-SARAMIS Systems for yeast identification by matrix-assisted laser desorption ionization-time of flight mass spectrometry. *J. Clin. Microbiol.* **51**, 1231–1236 (2013).
170. Sendid, B. et al. Evaluation of MALDI-TOF mass spectrometry for the identification of medically-important yeasts in the clinical laboratories of Dijon and Lille hospitals. *Med. Mycol.* **51**, 25–32 (2013).
171. Campbell, C.K., Payne, A.L., Teall, A.J., Brownell, A., and Mackenzie, D.W. Cryptococcal latex antigen test positive in patient with *Trichosporon beigelii* infection. *Lancet* **2**, 43–44 (1985).

172. Lyman, C.A. et al. Detection and quantitation of the glucuronoxylomannan-like polysaccharide antigen from clinical and nonclinical isolates of *Trichosporon beigelii* and implications for pathogenicity. *J. Clin. Microbiol.* **33**, 126–130 (1995).
173. Babady, N.E. et al. Evaluation of three commercial latex agglutination kits and a commercial enzyme immunoassay for the detection of cryptococcal antigen. *Med. Mycol.* **47**, 336–338 (2009).
174. Karageorgopoulos, D.E. et al. Beta-D-glucan assay for the diagnosis of invasive fungal infections: A meta-analysis. *Clin. Infect. Dis.* **52**, 750–770 (2011).
175. De Pauw, B. et al. Revised definitions of invasive fungal disease from the European Organization for Research and Treatment of Cancer/Invasive Fungal Infections Cooperative Group and the National Institute of Allergy and Infectious Diseases Mycoses Study Group (EORTC/MSG) Consensus Group. *Clin. Infect. Dis.* **46**, 1813–1821 (2008).
176. Koo, S., Bryar, J.M., Page, J.H., Baden, L.R., and Marty, F.M. Diagnostic performance of the (1 → 3)-beta-D-glucan assay for invasive fungal disease. *Clin. Infect. Dis.* **49**, 1650–1659 (2009).
177. Obayashi, T. et al. Plasma (1 → 3)-beta-D-glucan measurement in diagnosis of invasive deep mycosis and fungal febrile episodes. *Lancet* **345**, 17–20 (1995).
178. Cuetara, M.S. et al. Detection of (1 → 3)-beta-D-glucan as an adjunct to diagnosis in a mixed population with uncommon proven invasive fungal diseases or with an unusual clinical presentation. *Clin. Vaccine Immunol.* **16**, 423–426 (2009).
179. Nakase, K. et al. Is elevation of the serum beta-D-glucan level a paradoxical sign for *Trichosporon* fungemia in patients with hematologic disorders? *Int. J. Infect. Dis.* **16**, e2–e4 (2012).
180. Odabasi, Z. et al. Differences in beta-glucan levels in culture supernatants of a variety of fungi. *Med. Mycol.* **44**, 267–272 (2006).
181. Liao, Y., Hartmann, T., Ao, J.H., and Yang, R.Y. Serum glucuronoxylomannan may be more appropriate for the diagnosis and therapeutic monitoring of *Trichosporon* fungemia than serum beta-D-glucan. *Int. J. Infect. Dis.* **16**, e638 (2012).
182. Bonini, A. et al. Galactomannan detection in *Geotrichum capitatum* invasive infections: Report of 2 new cases and review of diagnostic options. *Diagn. Microbiol. Infect. Dis.* **62**, 450–452 (2008).
183. Giacchino, M. et al. *Aspergillus* galactomannan enzyme-linked immunosorbent assay cross-reactivity caused by invasive *Geotrichum capitatum*. *J. Clin. Microbiol.* **44**, 3432–3434 (2006).
184. Schuermans, C. et al. Breakthrough *Saprochaete capitata* infections in patients receiving echinocandins: Case report and review of the literature. *Med. Mycol.* **49**, 414–418 (2011).
185. Tsuji, Y. et al. Quantitative PCR assay used to monitor serum *Trichosporon asahii* DNA concentrations in disseminated trichosporonosis. *Pediatr. Infect. Dis. J.* **27**, 1035–1037 (2008).
186. Sugita, T. et al. A nested PCR assay to detect DNA in sera for the diagnosis of deep-seated trichosporonosis. *Microbiol. Immunol.* **45**, 143–148 (2001).
187. Nagai, H., Yamakami, Y., Hashimoto, A., Tokimatsu, I., and Nasu, M. PCR detection of DNA specific for *Trichosporon* species in serum of patients with disseminated trichosporonosis. *J. Clin. Microbiol.* **37**, 694–699 (1999).
188. Landlinger, C. et al. Species-specific identification of a wide range of clinically relevant fungal pathogens by use of Luminex xMAP technology. *J. Clin. Microbiol.* **47**, 1063–1073 (2009).
189. Tokimatsu, I. et al. The prophylactic effectiveness of various antifungal agents against the progression of trichosporonosis fungemia to disseminated disease in a neutropenic mouse model. *Int. J. Antimicrob. Agents* **29**, 84–88 (2007).
190. Carrillo-Munoz, A.J. et al. Activity of caspofungin and voriconazole against clinical isolates of *Candida* and other medically important yeasts by the CLSI M-44A disk diffusion method with Neo-Sensitabs tablets. *Chemotherapy* **54**, 38–42 (2008).
191. Mekha, N. et al. Genotyping and antifungal drug susceptibility of the pathogenic yeast *Trichosporon asahii* isolated from Thai patients. *Mycopathologia* **169**, 67–70 (2010).
192. Adami, F. et al. Successful control of *Blastoschizomyces capitatus* infection in three consecutive acute leukaemia patients despite initial unresponsiveness to liposomal amphotericin B. *Mycoses* **54**, 365–369 (2011).
193. Rodriguez-Tudela, J.L. et al. Susceptibility patterns and molecular identification of *Trichosporon* species. *Antimicrob. Agents Chemother.* **49**, 4026–4034 (2005).
194. Espinel-Ingroff, A. In vitro antifungal activities of anidulafungin and micafungin, licensed agents and the investigational triazole posaconazole as determined by NCCLS methods for 12,052 fungal isolates: Review of the literature. *Rev. Iberoam. Micol.* **20**, 121–136 (2003).
195. Cornely, O.A., Schmitz, K., and Aisenbrey, S. The first echinocandin: Caspofungin. *Mycoses* **45(Suppl 3)**, 56–60 (2002).

196. Akagi, T. et al. Breakthrough trichosporonosis in patients with acute myeloid leukemia receiving micafungin. *Leuk. Lymphoma* **47**, 1182–1183 (2006).
197. Arikan, S. and Hascelik, G. Comparison of NCCLS microdilution method and Etest in antifungal susceptibility testing of clinical *Trichosporon asahii* isolates. *Diagn. Microbiol. Infect. Dis.* **43**, 107–111 (2002).
198. CLSI. *Clinical and Laboratory Standards Institute: Reference Method for Broth Dilution Antifungal Susceptibility Testing of Yeasts: Approved Standard.* (CLSI documents M27-A3 and M27-S, Wayne, 2008).
199. EUCAST. Subcommittee on Antifungal Susceptibility Testing (AFST) of the ESCMID European Committee for Antimicrobial Susceptibility Testing (EUCAST): EUCAST definitive document EDef 7.1: Method for the determination of broth dilution MICs of antifungal agents for fermentative yeasts. *Clin. Microbiol. Infect.* **14**, 398–405 (2008).
200. Arendrup, M.C., Cuenca-Estrella, M., Lass-Florl, C., Hope, W., and EUCAST-AFST. EUCAST technical note on the EUCAST definitive document EDef 7.2: Method for the determination of broth dilution minimum inhibitory concentrations of antifungal agents for yeasts EDef 7.2 (EUCAST-AFST). *Clin. Microbiol. Infect.* **18**, E246-7 (2012).
201. Guinea, J. et al. In vitro antifungal activities of isavuconazole and comparators against rare yeast pathogens. *Antimicrob. Agents Chemother.* **54**, 4012–4015 (2010).
202. Fournier, S. et al. Use of voriconazole to successfully treat disseminated *Trichosporon asahii* infection in a patient with acute myeloid leukaemia. *Eur. J. Clin. Microbiol. Infect. Dis.* **21**, 892–896 (2002).
203. Antachopoulos, C. et al. Fungemia due to *Trichosporon asahii* in a neutropenic child refractory to amphotericin B: Clearance with voriconazole. *J. Pediatr. Hematol. Oncol.* **27**, 283–285 (2005).
204. Asada, N. et al. Successful treatment of breakthrough *Trichosporon asahii* fungemia with voriconazole in a patient with acute myeloid leukemia. *Clin. Infect. Dis.* **43**, e39–e41 (2006).
205. Gabriel, F., Noel, T., and Accoceberry, I. Fatal invasive trichosporonosis due to *Trichosporon loubieri* in a patient with T-lymphoblastic lymphoma. *Med. Mycol.* **49**, 306–310 (2011).
206. Hosokawa, K. et al. Successful treatment of *Trichosporon* fungemia in a patient with refractory acute myeloid leukemia using voriconazole combined with liposomal amphotericin B. *Transpl. Infect. Dis.* **14**, 184–187 (2012).
207. Espinel-Ingroff, A. Comparison of in vitro activities of the new triazole SCH56592 and the echinocandins MK-0991 (L-743,872) and LY303366 against opportunistic filamentous and dimorphic fungi and yeasts. *J. Clin. Microbiol.* **36**, 2950–2956 (1998).
208. Espinel-Ingroff, A. In vitro activity of the new triazole voriconazole (UK-109,496) against opportunistic filamentous and dimorphic fungi and common and emerging yeast pathogens. *J. Clin. Microbiol.* **36**, 198–202 (1998).
209. Pfaller, M.A. and Diekema, D.J. Epidemiology of invasive mycoses in North America. *Crit. Rev. Microbiol.* **36**, 1–53 (2010).
210. Pfaller, M.A. et al. Clinical breakpoints for the echinocandins and *Candida* revisited: Integration of molecular, clinical, and microbiological data to arrive at species-specific interpretive criteria. *Drug Resist. Update* **14**, 164–176 (2011).
211. Savini, V. et al. Multidrug-resistant *Geotrichum capitatum* from a haematology ward. *Mycoses* **54**, 542–543 (2011).
212. Chittick, P., Palavecino, E.L., Delashmitt, B., Evans, J., and Peacock, J.E., Jr. Case of fatal *Blastoschizomyces capitatus* infection occurring in a patient receiving empiric micafungin therapy. *Antimicrob. Agents Chemother.* **53**, 5306–5307 (2009).
213. D'Antonio, D. et al. Osteomyelitis and intervertebral discitis caused by *Blastoschizomyces capitatus* in a patient with acute leukemia. *J. Clin. Microbiol.* **32**, 224–227 (1994).
214. D'Antonio, D. et al. Emergence of fluconazole-resistant strains of *Blastoschizomyces capitatus* causing nosocomial infections in cancer patients. *J. Clin. Microbiol.* **34**, 753–755 (1996).
215. Lortholary, O. et al. Recent exposure to caspofungin or fluconazole influences the epidemiology of candidemia: A prospective multicenter study involving 2,441 patients. *Antimicrob. Agents Chemother.* **55**, 532–538 (2011).
216. Garnacho-Montero, J., Diaz-Martin, A., Ruiz-Perez De Piappon, M., and Garcia-Cabrera, E. Invasive fungal infection in critically ill patients. *Enferm. Infecc. Microbiol. Clin.* **30**, 338–343 (2012).
217. Garnacho-Montero, J. et al. Impact on hospital mortality of catheter removal and adequate antifungal therapy in *Candida* spp. bloodstream infections. *J. Antimicrob. Chemother.* **68**, 206–213 (2013).
218. Bassetti, M. et al. *Trichosporon asahii* infection treated with caspofungin combined with liposomal amphotericin B. *J. Antimicrob. Chemother.* **54**, 575–577 (2004).

219. Serena, C., Marine, M., Pastor, F.J., Nolard, N., and Guarro, J. in vitro interaction of micafungin with conventional and new antifungals against clinical isolates of *Trichosporon, Sporobolomyces* and *Rhodotorula. J. Antimicrob. Chemother.* **55**, 1020–1023 (2005).
220. Serena, C., Pastor, F.J., Gilgado, F., Mayayo, E., and Guarro, J. Efficacy of micafungin in combination with other drugs in a murine model of disseminated trichosporonosis. *Antimicrob. Agents Chemother.* **49**, 497–502 (2005).
221. Serena, C., Gilgado, F., Marine, M., Pastor, F.J., and Guarro, J. Efficacy of voriconazole in a guinea pig model of invasive trichosporonosis. *Antimicrob. Agents Chemother.* **50**, 2240–2243 (2006).
222. Etienne, A. et al. Successful treatment of disseminated *Geotrichum capitatum* infection with a combination of caspofungin and voriconazole in an immunocompromised patient. *Mycoses* **51**, 270–272 (2008).
223. Sakamoto, Y. et al. Case report. First isolation of *Trichosporon domesticum* from a cat. *Mycoses* **44**, 518–520 (2001).
224. Alshahni, M.M. et al. A suggested pathogenic role for *Trichosporon montevideense* in a case of onychomycosis in a Japanese monkey. *J. Vet. Med. Sci.* **71**, 983–986 (2009).
225. Sugita, T. et al. Two new yeasts, *Trichosporon debeurmannianum* sp. nov. and *Trichosporon dermatis* sp. nov., transferred from the *Cryptococcus humicola* complex. *Int. J. Syst. Evol. Microbiol.* **51**, 1221–1228 (2001).
226. Lacasse, A. and Cleveland, K.O. *Trichosporon mucoides* fungemia in a liver transplant recipient: Case report and review. *Transpl. Infect. Dis.* **11**, 155–159 (2009).
227. Middelhoven, W.J., Scorzetti, G., and Fell, J.W. *Trichosporon veenhuisii* sp. nov., an alkane-assimilating anamorphic basidiomycetous yeast. *Int. J. Syst. Evol. Microbiol.* **50(Pt 1)**, 381–387 (2000).
228. Hirschi, S. et al. Disseminated *Trichosporon mycotoxinivorans, Aspergillus fumigatus*, and *Scedosporium apiospermum* coinfection after lung and liver transplantation in a cystic fibrosis patient. *J. Clin. Microbiol.* **50**, 4168–4170 (2012).
229. Heneberg, P. and Rezac, M. Two *Trichosporon* species isolated from Central-European mygalomorph spiders (Araneae: Mygalomorphae). *Antonie Van Leeuwenhoek* **103**, 713–721 (2013).
230. Bai, M. et al. Occurrence and dominance of yeast species in naturally fermented milk from the Tibetan Plateau of China. *Can. J. Microbiol.* **56**, 707–714 (2010).
231. Kustimur, S. et al. Nosocomial fungemia due to *Trichosporon asteroides*: Firstly described bloodstream infection. *Diagn. Microbiol. Infect. Dis.* **43**, 167–170 (2002).
232. Lopandic, K. et al. *Trichosporon caseorum* sp.nov. and *Trichosporon lactis* sp.nov., two basidiomycetous yeasts isolated from cheeses. In *Frontiers in Basidiomycote Mycology* (eds. Agerer, R., Piepenbring, M., and Blanz, P.) (IHW-Verlag und Verlagsbuchhandlung, Eching, 2004).
233. Kalkanci, A. et al. Molecular identification, genotyping, and drug susceptibility of the basidiomycetous yeast pathogen *Trichosporon* isolated from Turkish patients. *Med. Mycol.* **48**, 141–146 (2010).
234. Janagond, A., Krishnan, K.M., Kindo, A.J., and Sumathi, G. *Trichosporon inkin*, an unusual agent of fungal sinusitis: A report from south India. *Indian. J. Med. Microbiol.* **30**, 229–232 (2012).
235. Marty, F.M., Barouch, D.H., Coakley, E.P., and Baden, L.R. Disseminated trichosporonosis caused by *Trichosporon loubieri. J. Clin. Microbiol.* **41**, 5317–5320 (2003).
236. Sulo, P., Laurencik, M., Polakova, S., Minarik, G., and Slavikova, E. *Geotrichum bryndzae* sp. nov., a novel asexual arthroconidial yeast species related to the genus *Galactomyces. Int. J. Syst. Evol. Microbiol.* **59**, 2370–2374 (2009).
237. Suh, S.O. and Blackwell, M. Three new asexual arthroconidial yeasts, *Geotrichum carabidarum* sp. nov., *Geotrichum histeridarum* sp. nov., and *Geotrichum cucujoidarum* sp. nov. isolated from the gut of insects. *Mycol. Res.* **110**, 220–228 (2006).
238. McKay, A.H., Forster, H., and Adaskaveg, J.E. Distinguishing *Galactomyces citri-aurantii* from *G. geotrichum* and characterizing population structure of the two postharvest sour rot pathogens of fruit crops in California. *Phytopathology* **102**, 528–538 (2012).
239. Omemu, A.M., Oyewole, O.B., and Bankole, M.O. Significance of yeasts in the fermentation of maize for ogi production. *Food Microbiol.* **24**, 571–576 (2007).
240. Nielsen, D.S., Jakobsen, M., and Jespersen, L. *Candida halmiae* sp. nov., *Geotrichum ghanense* sp. nov. and *Candida awuaii* sp. nov., isolated from Ghanaian cocoa fermentations. *Int. J. Syst. Evol. Microbiol.* **60**, 1460–1465 (2010).
241. Randhawa, H.S., Mussa, A.Y., and Khan, Z.U. Decaying wood in tree trunk hollows as a natural substrate for *Cryptococcus neoformans* and other yeast-like fungi of clinical interest. *Mycopathologia* **151**, 63–69 (2001).

242. Kaewwichian, R., Yongmanitchai, W., Srisuk, N., Fujiyama, K., and Limtong, S. *Geotrichum siamensis* sp. nov. and *Geotrichum phurueaensis* sp. nov., two asexual arthroconidial yeast species isolated in Thailand. *FEMS Yeast Res*. **10**, 214–220 (2010).
243. Ikuta, K. et al. Successful treatment of systemic *Geotrichum capitatum* infection by liposomal amphotericin-B, itraconazole, and voriconazole in a Japanese man. *Intern. Med*. **49**, 2499–2503 (2010).
244. Ersoz, G. et al. An outbreak of *Dipodascus capitatus* infection in the ICU: Three case reports and review of the literature. *Jpn. J. Infect. Dis*. **57**, 248–252 (2004).
245. Pagano, L. et al. Fungal infections in recipients of hematopoietic stem cell transplants: Results of the SEIFEM B-2004 study—Sorveglianza Epidemiologica Infezioni Fungine Nelle Emopatie Maligne. *Clin. Infect. Dis*. **45**, 1161–1170 (2007).
246. Amft, N., Miadonna, A., Viviani, M.A., and Tedeschi, A. Disseminated *Geotrichum capitatum* infection with predominant liver involvement in a patient with non Hodgkin's lymphoma. *Haematologica* **81**, 352–355 (1996).
247. Chamilos, G. et al. Invasive fungal infections in patients with hematologic malignancies in a tertiary care cancer center: An autopsy study over a 15-year period (1989–2003). *Haematologica* **91**, 986–989 (2006).
248. Batlle, M., Quesada, M.D., Moreno, M., and Ribera, J.M. Outbreak of blastoschysomics spp. infection in a haematology unit: Study of 6 cases and finding of the source of infection. *Med. Clin. (Barc)* **135**, 672–674 (2010).
249. Girmenia, C., Pizzarelli, G., D'Antonio, D., Cristini, F., and Martino, P. in vitro susceptibility testing of *Geotrichum capitatum*: Comparison of the E-test, disk diffusion, and Sensititre colorimetric methods with the NCCLS M27-A2 broth microdilution reference method. *Antimicrob. Agents Chemother*. **47**, 3985–3988 (2003).
250. Serena, C., Rodriguez, M.M., Marine, M., Pastor, F.J., and Guarro, J. Combined therapies in a murine model of blastoschizomycosis. *Antimicrob. Agents Chemother*. **51**, 2608–2610 (2007).
251. Serena, C., Marine, M., Marimon, R., Pastor, F.J., and Guarro, J. Effect of antifungal treatment in a murine model of blastoschizomycosis. *Int. J. Antimicrob. Agents* **29**, 79–83 (2007).
252. Gadea, I. et al. Genotyping and antifungal susceptibility profile of *Dipodascus capitatus* isolates causing disseminated infection in seven hematological patients of a tertiary hospital. *J. Clin. Microbiol*. **42**, 1832–1836 (2004).
253. Dannaoui, E. et al. Comparative in vitro activities of caspofungin and micafungin, determined using the method of the European Committee on Antimicrobial Susceptibility Testing, against yeast isolates obtained in France in 2005–2006. *Antimicrob. Agents Chemother*. **52**, 778–781 (2008).
254. Cejudo, M.A. et al. Evaluation of the VITEK 2 system to test the susceptibility of *Candida* spp., *Trichosporon asahii* and *Cryptococcus neoformans* to amphotericin B, flucytosine, fluconazole and voriconazole: A comparison with the M27-A3 reference method. *Med. Mycol*. **48**, 710–719 (2010).

33 Wallemia

Janja Zajc, Sašo Jančič, Polona Zalar, and Nina Gunde-Cimerman

CONTENTS

33.1 Introduction 569
33.2 Phylogeny and Identification of the Genus *Wallemia* 570
33.3 Ecology of the Genus *Wallemia* 570
      33.3.1 Xerophily of the Genus *Wallemia* 571
33.4 Molecular and Physiological Adaptations of the *Wallemia* spp. to Life at
Low Water Activity 572
33.5 Bioactive Metabolites Produced by the *Wallemia* spp. 574
33.6 Pathogenesis, Clinical Features, and Diagnosis of *WALLEMIA sebi* 575
33.7 Methods 575
      33.7.1 Cultivation 575
      33.7.2 Identification of *Wallemia* spp. 576
      33.7.3 Detection of *W. sebi* by Conventional and Real-Time PCR 577
Acknowledgments 577
Abbreviations 577
References 578

33.1 INTRODUCTION

Food is an ecosystem that includes higher levels of nutrients for growth of potential spoilage microorganisms than their natural environments such as soil and water.[1] The spoilage of fresh, living food, particularly fresh fruits, vegetables, and also grains and nuts before harvest, is limited to those microorganisms that can overcome defense mechanisms, while the spoilage of processed, dormant, and nonliving food depends on physical and chemical factors.[1] Prevention of food spoilage traditionally depends on a reduction of the biologically available water, using drying or freezing or by adding different solutes.

As the central molecule for biological processes, low amounts of biologically available water (i.e., a low water activity [a_w]) represent one of the most pervasive stresses for biological systems, and only specially adapted organisms can thrive under such conditions. Food with a a_w below 0.9 will only support the growth of a few highly resistant bacterial pathogens,[2] whereas many xerotolerant and xerophilic fungal species can grow. Indeed, tolerance to low a_w is apparent in only 10 of the 140 known orders of fungi, with most of them belonging to the Ascomycota,[3] and xerotolerance is extremely rare in the division Basidiomycota. Therefore, the genus *Wallemia* represents a rare genus of cosmopolitan, xerotolerant, and xerophilic basidiomycetous fungi that are frequently involved in food spoilage of, in particular, sweet, salty, and dried food and that have mycotoxigenic and mycotic potential.

33.2 PHYLOGENY AND IDENTIFICATION OF THE GENUS *WALLEMIA*

Based on the analysis of the nuclear small subunit ribosomal RNA gene (SSU rRNA) and ultrastructural characteristics (e.g., dolipore septum), *Wallemia* is included in the phylum Basidiomycota. The unique phylogenetic position of *Wallemia*, accompanied by the rarely encountered conidiogenesis and extreme xerotolerance, resulted in the description of the new class of Wallemiomycetes and the new order of Wallemiales.[4] This taxonomic placement was later supported by molecular analyses of three ribosomal RNA genes (18S, 25S, and 5.8S) and three nuclear protein-coding genes (rpb1, rpb2, and tef1). The class Wallemiomycetes was described as an early diverging lineage of Basidiomycota. It has a basal position near the Entorrhizomycetidae and might be a sister group of the Agaricomycotina and Ustilaginomycotina.[5]

The hypothesis that *Wallemia* is closely related to Agaricomycotina[6] was confirmed in a detailed genome study of *Wallemia sebi*.[7] This phylogenetic analysis of a 71-protein dataset supported the position of *Wallemia* as the earliest diverging lineage of Agaricomycotina. This was confirmed also by the septal pore ultrastructure, which showed the septal pore apparatus as a variant of the *Tremella* type. The relationships between the three subphyla of Basidiomycota have been difficult to resolve. Recent data support Ustilaginomycotina as the sister group to the branch that unites Wallemiomycetes and the remaining Agaricomycotina, and Pucciniomycotina as the sister group to the rest of Basidiomycota.[7,8]

Based on differences in conidial size, on their xerotolerance, and on sequence data of the internal transcribed spacer region (ITS) rDNA, three *Wallemia* species have been identified: *W. ichthyophaga*, *W. sebi*, and *W. muriae*. *W. ichthyophaga* differs from the other two species in many of the nucleotides of the SSU rDNA and the ITS rDNA.[4] In addition, based on molecular analyses of mcm7, tsr1, rpb1, rpb2 and hal2, four new species are indicated. Three of these are closely related to *W. sebi*, while the fourth (with five known isolates) represents a putative link between *W. ichthyophaga* and the other species of this genus (Jančič, Gunde-Cimerman et al., unpublished data).

On standard mycological media, the *Wallemia* spp. can be recognized by small, walnut-brown colonies. Despite the considerable molecular distances between the individual *Wallemia* spp., they all show a unique form of conidiogenesis, which is seen as the basauxic development of fertile hyphae, and the basipetal segregation of the conidial units and their disarticulation into mostly four arthrospore-like conidia (Figure 33.1e and f). Nevertheless, the different species within the genus *Wallemia* can be distinguished by morphological and physiological characteristics, such as the size range of the conidia (Figure 33.1e), the presence of sarcina-like structures (Figure 33.1g) in *W. ichthyophaga*, and the differences in their degrees of xerophily.[4] *W. sebi* is currently the only recognized *Wallemia* spp. that can grow on MEA without the addition of NaCl or sugar, and it has smaller conidia than the other two species.

33.3 ECOLOGY OF THE GENUS *WALLEMIA*

The *Wallemia* spp. are frequently involved in food spoilage and particularly of dried, salty, and sweet foods like chocolate (Figure 33.1c).[9] They have often been isolated from indoor[10] and outdoor[11] air in urban and agricultural environments[12,13] and from cereal grains, like rice,[14] wheat, barley, and corn. The *Wallemia* spp. have also been detected in hay, pollen, and soil, bound to sea-salt crystals (Jančič, Gunde-Cimerman et al., unpublished data), and in sea sediments,[15] seawater organisms,[16] and hypersaline water of man-made salterns (Figure 33.1a) on different continents. Salters are therefore proposed as a natural habitat for *Wallemia* spp.[4] The *Wallemia* spp. are also agents of degradation of cultural heritage objects.[17]

To date, only the species *W. sebi* has been described as the common cause of allergological problems, which are better known as farmer's lung disease (FLD),[18,19] and although rarely, it has been shown to be the causative agent of cutaneous and subcutaneous infections in humans.[20] However, it is worth noting that none of the isolates involved in the aforementioned studies were identified using molecular methods, and therefore, their designation as *W. sebi* must remain questionable.

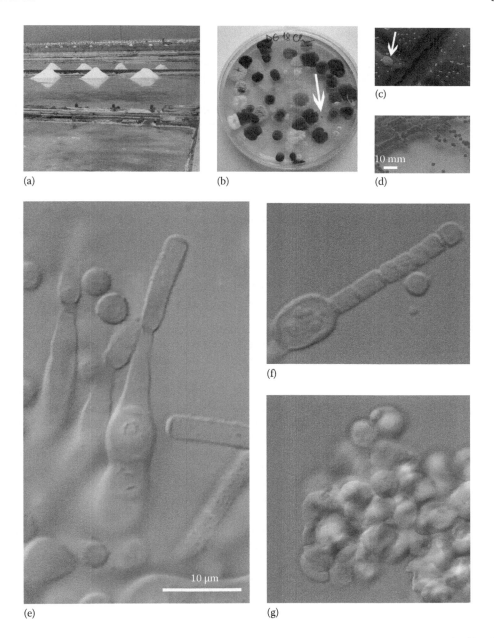

FIGURE 33.1 (a) Salterns as a habitat for *Wallemia* spp. (b) DG18 medium after outdoor air filtration, with small *Wallemia* colony indicated by an arrow. (c) *Wallemia* colony growing on a piece of chocolate (see arrow). (d) Culture of *W. sebi* growing on MY30G culture medium. (e and f) *W. ichthyophaga* conidiogenous apparatus. (g) Meristematic clumps of *W. ichthyophaga*. Scale bar in panel (e) also applies to (f) and (g).

33.3.1 Xerophily of the Genus *Wallemia*

The genus *Wallemia* is one of the best known low-a_w-tolerant groups of fungi, and it is distributed in well-defined habitats.[21] Out of 140 known orders of fungi, only 10 include species that tolerate low a_w, with most of these belonging to the Ascomycota.[3] Xerotolerance, and even more xerophily, is relatively rare in Basidiomycota, and therefore, it is surprising that the *Wallemia* spp. represent one of the most xerophilic fungal taxa.

Two out of the three described *Wallemia* spp., *W. muriae* and *W. ichthyophaga*, require media with low a_w, and thus, these can be considered as obligate xerophiles. *W. sebi* can also thrive on media without additional solutes,[4] and it can grow over a wider range of a_w (0.997–0.690) in glucose/fructose media.[22] However, in media with added NaCl as the major solute, the lowest a_w for the growth of *W. sebi* was reported to be 0.80,[4,22] which corresponds to 4.5 M NaCl. The a_w growth ranges of *W. muriae* and *W. ichthyophaga* are 0.984–0.805 and 0.959–0.771, respectively.[4] *W. muriae* can tolerate 0.7–4.3 M NaCl, while *W. ichthyophaga* can thrive in media with NaCl above 1.7 M and up to NaCl saturation (5.3 M). Due to its obligate requirement of at least 10% NaCl in the medium, *W. ichthyophaga* is the most halophilic fungus known to date. The halophilic versus xerophilic nature of *W. ichthyophaga* was further demonstrated by its considerably more rapid and more abundant growth on media with added NaCl as the solute, thus lowering the a_w, in comparison with media with high concentrations of either glucose or honey.[4,23,24] Interestingly, all of the *Wallemia* spp. show optimal growth in media with low a_w, with all three showing their greatest colony diameters at lowered a_w: *W. sebi* and *W. muriae* at 0.96 and *W. ichthyophaga* at 0.88.[4]

33.4 MOLECULAR AND PHYSIOLOGICAL ADAPTATIONS OF THE *WALLEMIA* SPP. TO LIFE AT LOW WATER ACTIVITY

Microbial survival in different environments depends on the ability of an organism to sense and to respond to environmental factors. Such responses involve complex alterations in gene expression, which can lead to metabolic changes and the subsequent adaptation to the new conditions.[25–27]

In environments with low a_w, organisms are exposed to turgor-related stress, due to high concentrations of solutes and to the toxicities of certain ions. The adaptations of fungi to high concentrations of NaCl have been well studied, while osmotic stress induced by high concentrations of sugar has received little attention. The mechanisms of salt tolerance in fungi have been mostly studied in the salt-sensitive *Saccharomyces cerevisiae*,[28–30] the halotolerant yeast *Debaryomyces hansenii*,[31–36] and the extremely halotolerant black yeast *Hortaea werneckii*.[37–41] Only more recently have the physiological and molecular adaptations of the genus *Wallemia* become the focus of studies and especially of the most halophilic representative, *W. ichthyophaga*.

The pathway for the sensing of osmolarity changes in fungi (e.g., *S. cerevisiae* and *H. werneckii*) and for the facilitation of adaptation of cells to increased osmolarity of the environment is known as the high-osmolarity glycerol (HOG) signaling pathway. This is one of the best understood mitogen-activated protein kinase (MAPK) cascades.[30] Homology searches in the genome of *W. sebi* have shown that the HOG pathway is mostly conserved[7] as it encodes: (1) two Hog1 (MAPK) homologues that appear to function in osmotolerance, (2) putative homologues of genes that encode various proteins that are involved in the activation of the upstream HOG pathway (e.g., Ste11p, Cla4p),[7] and (3) the downstream target orthologues Rck2p[42] and Sgd1p.[43] Genes involved in the ability to live under osmotic stress have also been investigated using *in silico* analyses, which identified 93 putative osmotic-stress proteins, including the two Hog1-like genes.[7] Finally, the functional analysis of HOG pathway of the halophilic *W. ichthyophaga* is under investigation. So far it is known that it possesses two Hog1 homologues that have the lowest transcription under optimal salinity, whereas at limiting salinities, the transcription is highly induced. Interestingly, the pattern of Hog1 phosphorylation to both hypo- and hyperosmotic shocks is opposite to the pattern observed in model yeast *S. cerevisiae*.[44,45]

As with other halophilic and halotolerant microorganisms, most of the halotolerant and halophilic fungi prevent the loss of internal water and achieve osmotic balance in hypersaline environments by the synthesis and/or accumulation of small organic molecules that are known as compatible solutes. At the same time, they maintain low concentration of salt in their cell cytoplasm.[46]

Investigations of the osmotic strategies of the *Wallemia* spp. have confirmed these adaptations. Atomic absorption spectroscopy of cell extracts has revealed very low intracellular concentrations of Na^+ and K^+, while nuclear magnetic resonance and high-performance liquid chromatography/mass spectrometry analyses have shown that all three *Wallemia* spp. accumulate a mixture of three polyols: glycerol, arabitol, and mannitol. Glycerol has been shown to be the main compatible solute of *Wallemia* spp., not only because its levels are the highest of all of these solutes, but also because its intracellular levels increase with increasing salinity.[47] Increased production and accumulation of glycerol in environments with high concentrations of NaCl is a common strategy of osmoadaptation in different halotolerant fungi and also in the model organism, the extremely halotolerant black yeast *H. werneckii*.[27,37] This appears to be the preferred strategy among eukaryotes as the production of glycerol is energetically the most reasonable.[46] NAD-dependent glycerol-3-phosphate dehydrogenase (GPD) is the key enzyme in the synthesis of glycerol from the glycolytic intermediate dihydroxyacetone phosphate.[48,49] Recently, GPD1 was identified and characterized for the first time in the halophilic fungus *W. ichthyophaga* (WiGPD1). This GPD1 was shown to be salt inducible, and therefore to contribute to osmoadaptation, as the transcript numbers of the mRNA from Wi*GPD1* in salt-adapted cells showed a gradual increase at increasing salinities, as a response to saline stress.[50] Furthermore, although comparison of GPD1 from the salt-sensitive *S. cerevisiae* to WiGPD1 showed high overall amino-acid similarity, a significant difference was also revealed. The N-terminal PTS2 sequence that is important for peroxisome localization [51] is lacking in the WiGPD1 *W. ichthyophaga* homologue, which appears to be because of its function in osmotic stress: the constant cytosolic localization of GPD1 is beneficial to organisms that live in extremely saline environments.[50]

Morphological adaptations of the *Wallemia* spp. to moderate and high NaCl concentrations have been studied using combinations of light microscopy and focused ion beam/scanning and transmission electron microscopy.[23] *W. sebi* and *W. muriae* have thicker and shorter hyphal compartments and larger mycelial pellets at high salinity, while *W. ichthyophaga* forms sarcina-like multicellular clumps that are composed of compactly packed spherical cells, with no hyphae. The sizes of these cells did not correlate with the increased salinity, whereas their multicellular clumps become significantly larger. Meristematic growth in the form of multicellular clumps and additional cover with extracellular polysaccharides appear to greatly enhance the survival of the *Wallemia* spp. in stress environments.[52–55] The cells of all three of the *Wallemia* spp. are covered with extracellular polysaccharides, which, while protecting these cells against desiccation, might also protect at high salinity and high sugar concentrations.[23,24]

To prevent damage to these cells, additional adaptations at the levels of the plasma-membrane composition[56–59] and the cell-wall structure[23] are also required. An interesting and unique phenomenon has been seen at the level of the ultrastructure of the cell wall of *Wallemia* spp. at higher salinity, with the thickness of multilayered cell walls increasing. The cell-wall thickness of *W. sebi* and *W. muriae* at low salinity is approximately 0.2 µm, and of *W. ichthyophaga*, 0.6 µm. At high salinity, the cell-wall thicknesses of both *W. sebi* and *W. muriae* increase only slightly, while in *W. ichthyophaga*, the thickness increases up to 1.6 µm, thus resulting in a 1.67-fold increase in thickness.[60]

Also the genome and transcriptome analysis suggested the important role of the cell wall in the adaptation of *W. ichthyophaga* to life at high salinity. There were 26 genes coding for the so-called halophilic hydrophobins found in the genome, and they were among the relatively few differentially expressed genes. Hydrophobins are cell-wall proteins with various roles in fungi, namely, they affect the permeability of the cell wall for solutes, give the cell-wall strength and rigidity, and are sometimes responsible for connecting the cells. The transcriptional response of hydrophobin-coding genes shows their important role in the adaptation to salinity, either by modulation of the cell walls or by aggregating the cells into multicellular clumps.[8]

Such significant cell-wall thickening[23] and extensive production of cell-wall protein hydrophobins[8] described for *W. ichthyophaga* as a response to extremely saline conditions are unique and previously undescribed fungal responses. It appears that the different morphological phenomena have important roles for the successful growth of *Wallemia* spp. at extreme salinity.

33.5 BIOACTIVE METABOLITES PRODUCED BY THE *WALLEMIA* SPP.

The search for the production of biologically active compounds, including mycotoxins, has been mostly focused on cosmopolitan mesophilic fungi, with the more rare halotolerant and halophilic fungi being overlooked. As fungi adapted to grow at low a_w can contaminate food preserved with high concentrations of salt or sugar, or by desiccation, as shown for *Wallemia* spp., these fungi might represent a serious threat to human and animal safety.

Our knowledge of the mycotoxigenic potential of *Wallemia* spp. is at present limited, as the presence of mycotoxins has been studied exclusively in *W. sebi*, which until 2005[4] represented the only known species of this genus. The toxicity of *W. sebi* culture filtrates has been shown on the HeLa cell lineage, in mice,[61] and in other biological assays.[62] In 1990, two related tricyclic dihydroxysesquiterpenes that are known as walleminol A and walleminol B or walleminone were isolated from *W. sebi* from a contaminated cake. Both of these are toxic to certain cell lines, protozoa, and brine shrimps. The minimum inhibitory dose of walleminol A in the bioassays is comparable to a number of mycotoxins, such as citrinin and penicillic acid (approximately 50 μg/mL).[63] As well as mycotoxins, two other bioactive components have been identified in *W. sebi*: the azasteroids UCA 1064-B[64] and UCA 1064-B.[65] They both exhibit antibacterial and antimycotic activities, while only UCA 1064-B also shows antitumor activity.

Another new bioactive compound, cyclopentanopyridine alkaloid (3-hydroxy-5-methyl-5,6-dihydro-7H-cyclopenta[b]pyridin-7-one), was isolated when *W. sebi* were grown in a medium with 10% NaCl. This showed antimicrobial activity against *Enterobacter aerogenes*, with a minimum inhibition concentration of 76.7 μM. Also, 11 known aromatic secondary metabolites were detected under saline conditions.[66]

All three of the *Wallemia* spp. have been investigated recently for the production of hemolytic and antibacterial compounds, with the screening of fungi isolated from different extreme environments. Organic extracts of the three species have shown high hemolytic and moderate antibacterial activities against Gram-positive *Bacillus subtilis*, particularly under stress conditions.[67] In *W. ichthyophaga*, there was a higher hemolytic activity in extracts from cultures grown with high concentrations of glucose, while the hemolytic potential of the organic extracts of *W. muriae* and *W. sebi* considerably increased when the cultures were exposed to low temperatures (10°C).[67]

Gas chromatography–mass spectrometry analysis of a *W. sebi* ethanol extract revealed a complex mixture of 21 sterols and fatty acids. The most intense chromatographic peaks corresponded to palmitic acid (C16:0), a mixture of linoleic (C18:2) and oleic (C18:1) acids, and ergosterol. Unsaturated fatty acids were responsible for hemolytic activities toward red blood cells and artificial small lipid vesicles with various lipid compositions. The study showed concentration-dependent hemolysis and preference for lipid membranes with higher fluidity.[68]

In summary, low a_w conditions have been shown to induce the production of bioactive metabolites in the xerophilic *Wallemia* spp. These might have protective roles in adaptation to these environments.[67] As has been shown for free fatty acids, some bioactive compounds might contribute to territorial competition in aquatic ecosystems, by affecting the growth of phytoplankton, algae, and cyanobacteria.[69]

It is of note that the *Wallemia* spp. that are commonly involved in food spoilage of low a_w foods can synthesize mycotoxins, such as walleminol A, which can be excreted into food contaminated by *W. sebi*.[63] Thus, the bioactive and mycotoxigenic potential of all three of the *Wallemia* spp. should be considered in food quality control, as food and feed contaminated with *W. sebi* might represent a health risk.[68]

33.6 PATHOGENESIS, CLINICAL FEATURES, AND DIAGNOSIS OF *WALLEMIA SEBI*

Similar to the production of bioactive compounds, our knowledge of the involvement of *Wallemia* spp. in human infections and pathogenesis has so far been limited to only *W. sebi*, which until 2005[4] was the only known species of this genus.

Wallemia sebi is a ubiquitous mold that can thrive in low a_w foods and feed, as well as in other dry environments, such as dust. It is commonly reported as the cause of respiratory allergies, such as bronchial asthma[18,70] and FLD,[18,71] and albeit rarely, cutaneous and subcutaneous infections in humans.[20]

It was also reported to be a causative agent in atopic diseases as some asthmatic individuals showed immediate-type hypersensitivity to *W. sebi*. The skin prick tests of its extract elicited positive reactions in 5.4% asthmatic patients, and radioallergosorbent test showed positive results in 18.9%.[72,73]

FLD is a form of occupational hypersensitivity pneumonitis (extrinsic allergic alveolitis) caused by chronic inhalation of microorganisms (antigens) from moldy hay, straw, or grain.[74] Its clinical expression is characterized by symptoms of dyspnea, cough, tiredness, headaches, occasional fever/night sweats, and general feeling of sickness. Any one or all of the symptoms may be apparent depending on the severity of FLD: acute, subacute, or chronic.[19,71,75–77] *W. sebi*, together with *Eurotium amstelodami* and *Absidia corymbifera*, are likely to be the main cause of FLD[19,71] in eastern France. Fungi involved in FLD reached a peak in January and February, which corresponded to the period when the number of FLD cases in the region was the highest. The main factor of their proliferation in hay is bad harvest conditions (rain during harvest, soil in the hay).[71]

Two cases of cutaneous and subcutaneous infections described in 1909 and another one in 1950 were named as *hemisporiosis*, after the synonymous species *Hemispora stellata*. No clinical features were given. Other infections caused by *W. sebi* were described more than 50 years later, in only 2008.[20] The reason for these rare reports might be the extremely slow growth of the fungus, which on mesophilic media can be quickly overgrown by contaminants, and/or the misidentification of cultures. A rapid and precise alternative to conventional morphological and biochemical detection methods of *W. sebi* is PCR amplification and sequence analysis of the ITS rDNA.[4]

Infections by *Wallemia* spp. are either infrequent or underdiagnosed, and therefore, there is little information on their clinical features and treatments. In a study by Guarro et al.,[19] a case of subcutaneous phaeohyphomycosis, of a 43-year-old woman in northern India, was described as a non-healing ulcer on the dorsum of the foot. The erythematous lesion, which started as an itchy papule, lasted 8 months and gradually increased in size. The patient could not recall any prior injury. The diagnosis was based on the histological demonstration of septate hyphae and the recovery of the fungus in culture. The patient was treated with itraconazole, but did not return for evaluation. It should be highlighted that the infected patient was immunocompetent and diabetes and HIV free. This case report added the genus *Wallemia* to the relatively short list of basidiomycetous fungi that have been reported to be the causative agents of infections in humans.

33.7 METHODS

33.7.1 Cultivation

For the isolation of all three *Wallemia* species, selective media with sugar (10% glucose–12% NaCl; MY10–12, $a_w = 0.916$) are used.[1] Cultures of *W. ichthyophaga* grow well on solid malt extract medium

(2% malt extract) with at least 10% NaCl and those of *W. muriae* and *W. sebi* on malt extract, yeast extract, and 50% glucose agar (MY50G; 2% malt extract, 0.5% yeast extract).[1] *Wallemia* species can also be cultivated in liquid and on solid yeast nitrogen base (YNB) medium (1.7 g of YNB, 5 g of $(NH_4)_2SO_4$ per liter, 0.8 g of complete supplement mixture per liter, 20 g of glucose per liter, pH 7.0) supplemented with NaCl.[23] For the cultivation of *W. sebi*, also dichloran–18% glycerol (DG18) agar (Figure 33.1b) is used.[13] On these media, the *Wallemia* spp. can be recognized by small, walnut-brown colonies (Figure 33.1b through d). A biopsy sample of subcutaneous infection can be cultured on media supplemented with antibiotic such as Sabouraud dextrose agar (SDA; Difco) containing chloramphenicol (0.05 mg/mL) and SDA with chloramphenicol and cycloheximide (0.5 mg/mL).[20] The cultivation should last for 14 days at room temperature (22°C–24°C).[4] Cultures in liquid media are incubated in the dark at 28°C with constant shaking at 180 rpm.[23]

33.7.2 Identification of *Wallemia* spp.

For the identification of the *Wallemia* spp., the following dichotomous key that is based on phenotypic characteristics has been proposed[4]:

1. Colonies grow well at 24°C on MEA and reach 3–6 mm in diameter in 14 days; conidia are short and cylindrical, 1.5–2.5 μm in diameter, with no sarcina-like structures   *W. sebi*
2. Colonies only grow on MEA with additional solutes (NaCl, glucose); conidia are larger than 2.5 μm in diameter, with or without sarcina-like structures   2
3. Colonies grown on malt extract yeast extract with 50% glucose (MY50G) agar are dark brown, with a cerebriform surface; conidia 3.5–5.0 μm in diameter, with sarcina-like structures   *W. ichthyophaga*
4. Colonies grown on MY50G agar are walnut brown, with a powdery surface; conidia 2.5–3.0 μm in diameter, without sarcina-like structures   *W. muriae*

However, for the precise identification of the level of species, the aforementioned morphological and physiological characteristics do not suffice, and they need to be complemented by the ITS** rDNA and SSU* rDNA sequence analysis and compared to the type strains or other reference strains: *W. sebi* (AY328915**, AY741379*), *W. muriae* (AY302534**, AY741381*), and *W. ichthyophaga* (AY302523**, AY741382*).[4] DNA is extracted from ca. 1 cm² of 14-day-old cultures by mechanical lysis.[78] Amplification of ITS rDNA is performed by using primers V9G (TTAAGTCCCTGCCCTTTGTA)[79] and LS266 (GCATTCCCAAACAACTCGACTC).[80] Amplification of SSU rDNA is performed by using primers NS1 (GTAGTCATATGCTTGTCT)[81] and NS24 (AAACCTTGTTACGACTTTTA).[82] PCR amplification is performed in a 50 μL reaction volume containing 10–100 ng of nuclear DNA, 50 pmol of each primer, 0.5–2 U of *Taq* DNA polymerase, and 200 μM each deoxynucleoside triphosphate; as follows: 2–5 min of initial denaturation at 94°C, followed by 30 cycles of 30 s at 94°C, 1 min at 45°C–60°C, and 1–2 min at 72°C. Finally, 2–5 min at 72°C of extension is performed. The annealing temperature depends on the primer combination used.[83] PCR fragments are purified using the commercial kit, and sequence reactions are analyzed on an ABI Prism 3700 (Applied Biosystems). Sequences are assembled and edited using SeqMan 3.61 (DNASTAR, Inc., Madison, United States).[4]

None of the ITS rDNA sequence of any *Wallemia* strains could be compared with any fungal sequence published so far. ITS rDNA sequences of *W. sebi* and *W. muriae* strains are well alignable, while sequences of *W. ichthyophaga* are 63–78 bp longer with multiple gaps at various positions and consequently hardly alignable to those of *W. sebi* and *W. muriae* strains. ITS rRNA sequences of *W. ichthyophaga* appear unrelated to those of other *Wallemia* taxa; therefore, it is possible that the genus *Wallemia* comprises a complex of phylogenetically remote genera and that taxa in between the extant *Wallemia* species have become extinct or have not yet been isolated and identified.[4]

33.7.3 Detection of *W. sebi* by Conventional and Real-Time PCR

For the rapid detection and quantification of *W. sebi* in environmental samples, two sets of PCR primers specific to *W. sebi* were designed: Wall-SYB4 (5′-GTAGTGAACTATATTGAAGAA-3′) and Wall-SYB6 (5′-ATGAGTCAATAATATAACGTC-3′) (Wall-SYB4/6) and Wall-SYB7 (5′-GATTGGATGACGTTATATTAT-3′) and Wall-SYB8 (5′-ACAACAAAATGTCGTACCG-3′) (Wall-SYB7/8). Primer pair Wall-SYB4/6 covers nucleotide positions 621–991 in the *W. sebi* 18S rDNA sequence (GenBank accession number AF548107), and the pair Wall-SYB7/8 covers nucleotides 963–1290. These pairs can be applied in either conventional PCR or real-time PCR, both PCR systems are proved to be highly specific and sensitive for the detection of *W. sebi* even in high background of other fungal DNAs.

Conventional PCR is performed in a 25 μL reaction volume containing 1–5 ng of template DNA, 10 pmol of each primer, 0.75 U of *Taq* DNA polymerase (Invitrogen Life Technologies, Carlsbad, CA), 200 μM each deoxynucleoside triphosphate (Amersham Pharmacia Biotech, Uppsala, Sweden), and 1.5 mM $MgCl_2$. The thermal profile of the reaction is as follows: 3 min of initial denaturation at 94°C, followed by 30 cycles of 30 s at 94°C, 30 s at 54°C, and 30 s at 72°C. At the end, 3 min at 72°C of final extension step is performed.

Real-time PCR is performed in 25 μL of a reaction volume consisting of 1–5 ng of template DNA, 10 pmol of each primer, and 12.5 μL of iQ SYBR Green Supermix (Bio-Rad) with an iCycler iQ real-time PCR detection system (Bio-Rad). The thermal profile of the reaction is as follows: 3 min at 95°C and 40 cycles consisting of 10 s at 95°C and 60 s at 60°C, followed by a dissociation curve. After amplification, the melting curve analysis is run for 60 s at 95°C, 60 s at 60°C, and a slow rise in temperature to 95°C at a rate of 0.5°C/10 s with continuous acquisition of fluorescence decline. The real-time PCR conditions for the two primer pairs are the same except that the annealing and extension times at 60°C were 60 and 30 s for Wall-SYB4/6 and Wall-SYB7/8, respectively. Each DNA sample, including the negative control, is analyzed by three replicate assays.

Assuming one fungal genome is ca. 4.0×10^{-5} ng of DNA, the conventional PCR could potentially detect 5–10 fungal spores in a reaction, while the real-time PCR system described in this study can potentially detect one spore in a PCR. Similar to the conventional PCR, the detection limit of the real-time PCR system is not affected by the presence of nontarget DNAs. In conclusion, these analytical methods facilitate the rapid detection and quantification of *W. sebi* in environmental samples, thus providing information about its distribution and ecology.[13]

ACKNOWLEDGMENTS

This work was supported in part by research grants J4–1019 and in part by Young Researcher Fellowships for J. Zajc and S. Jančič, from the Slovenian Research Agency (ARRS). It was also partly financed by the *Centre of Excellence for Integrated Approaches in Chemistry and Biology of Proteins*, number OP13.1.1.2.02.0005, financed by European Regional Development Fund (85% share of financing) and by the Slovenian Ministry of Higher Education, Science and Technology (15% share of financing).

ABBREVIATIONS

a_w	Water activity
GPD1	Glycerol-3-phosphate dehydrogenase
ITS rDNA	Internal transcribed spacer region ribosomal deoxyribonucleic acid
MEA	Malt extract agar
MY50G	Malt extract yeast extract 50% glucose agar
PTS2	Peroxisomal targeting sequence
SSU rDNA	Small subunit ribosomal deoxyribonucleic acid
UV	Ultraviolet

REFERENCES

1. Pitt, J.I. and Hocking, A.D. *Fungi and Food Spoilage* (Blackie Academic & Professional, London, U.K., 2009).
2. Brewer, M.S. Traditional preservatives—Sodium chloride. In *Encylopaedia of Food Microbiology*, Vol. 3 (eds. Robinson, R.K., Blatt, C.A., and Patel, P.D.) pp. 1723–1728 (Academic, London, U.K., 1999).
3. de Hoog, G.S., Zalar, P., van den Ende, B.G., and Gunde-Cimerman, N. Relation of halotolerance to human pathogenicity in the fungal tree of life: An overview of ecology and evolution under stress. In *Adaptation to Life at High Salt-Concentration in Archaea, Bacteria and Eukarya* (eds. Gunde-Cimerman, N., Oren, A., and Plemenitaš, A.) pp. 373–395 (Springer, Dordrecht, the Netherlands, 2005).
4. Zalar, P., de Hoog, G.S., Schroers, H.J., Frank, J.M., and Gunde-Cimerman, N. Taxonomy and phylogeny of the xerophilic genus *Wallemia* (Wallemiomycetes and Wallemiales, cl. et ord. nov.). *Antonie Van Leeuwenhoek International Journal of General and Molecular Microbiology* **87**, 311–328 (2005).
5. Matheny, P.B., Gossmann, J.A., Zalar, P., Kumar, T.K.A., and Hibbett, D.S. Resolving the phylogenetic position of the Wallemiomycetes: An enigmatic major lineage of Basidiomycota. *Canadian Journal of Botany-Revue Canadienne De Botanique* **84**, 1794–1805 (2006).
6. Hibbett, D.S. A phylogenetic overview of the Agaricomycotina. *Mycologia* **98**, 917–925 (2006).
7. Padamsee, M. et al. The genome of the xerotolerant mold *Wallemia sebi* reveals adaptations to osmotic stress and suggests cryptic sexual reproduction. *Fungal Genetics and Biology* **49**, 217–226 (2012).
8. Zajc, J. et al. Genome and transcriptome sequencing of the halophilic fungus *Wallemia ichthyophaga*: Haloadaptations present and absent. *BMC Genomics* **14**, 617 (2013).
9. Samson, R.A., Hoekstra, E.S., Frisvad, J.C., and Filtenborg, O. *Introduction to Food- and Airborne Fungi* (Centraalbureau voor Schimmelcultures, Baarn, the Netherlands, 2002).
10. Amend, A.S., Seifert, K.A., and Bruns, T.D. Quantifying microbial communities with 454 pyrosequencing: Does read abundance count? *Molecular Ecology* **19**, 5555–5565 (2010).
11. Fröhlich-Nowoisky, J., Pickersgill, D.A., Despres, V.R., and Poschl, U. High diversity of fungi in air particulate matter. *Proceedings of the National Academy of Sciences of the United States of America* **106**, 12814–12819 (2009).
12. Kristiansen, A., Saunders, A.M., Hansen, A.A., Nielsen, P.H., and Nielsen, J.L. Community structure of bacteria and fungi in aerosols of a pig confinement building. *FEMS Microbiology Ecology* **80**, 390–401 (2012).
13. Zeng, Q.Y., Westermark, S.O., Rasmuson-Lestander, A., and Wang, X.R. Detection and quantification of *Wallemia sebi* in aerosols by real-time PCR, conventional PCR, and cultivation. *Applied and Environmental Microbiology* **70**, 7295–7302 (2004).
14. Fredlund, E. et al. Moulds and mycotoxins in rice from the Swedish retail market. *Food Additives & Contaminants. Part A, Chemistry, Analysis, Control, Exposure & Risk Assessment* **26**, 527–533 (2009).
15. Singh, P., Raghukumar, C., Verma, P., and Shouche, Y. Fungal community analysis in the deep-sea sediments of the Central Indian Basin by culture-independent approach. *Microbial Ecology* **61**, 507–517 (2011).
16. Liu, W.C., Li, C.Q., Zhu, P., Yang, J.L., and Cheng, K.D. Phylogenetic diversity of culturable fungi associated with two marine sponges: *Haliclona simulans* and *Gelliodes carnosa*, collected from the Hainan Island coastal waters of the South China Sea. *Fungal Diversity* **42**, 1–15 (2010).
17. Michaelsen, A., Pinar, G., and Pinzari, F. Molecular and microscopical investigation of the microflora inhabiting a deteriorated Italian manuscript dated from the thirteenth century. *Microbial Ecology* **60**, 69–80 (2010).
18. Lappalainen, S., Pasanen, A.L., Reiman, M., and Kalliokoski, P. Serum IgG antibodies against *Wallemia sebi* and Fusarium species in Finnish farmers. *Annals of Allergy Asthma & Immunology* **81**, 585–592 (1998).
19. Reboux, G. et al. Role of molds in farmer's lung disease in Eastern France. *American Journal of Respiratory and Critical Care Medicine* **163**, 1534–1549 (2001).
20. Guarro, J. et al. Subcutaneous phaeohyphomycosis caused by *Wallemia sebi* in an immunocompetent host. *Journal of Clinical Microbiology* **46**, 1129–1131 (2008).
21. Cannon, P.F. and Sutton, B.C. Microfungi on wood and plant debris. In *Biodiversity of Fungi Inventory and Monitoring Methods* (eds. Mueller, G.M., Bills, G.F., and Foster, M.S.) pp. 217–240 (Elsevier Academic Press, San Diego, CA, 2004).
22. Pitt, J.I. and Hocking, A.D. Influence of solute and hydrogen-ion concentration on water relations of some xerophilic fungi. *Journal of General Microbiology* **101**, 35–40 (1977).

23. Kralj Kunčič, M., Kogej, T., Drobne, D., and Gunde-Cimerman, N. Morphological response of the halophilic fungal genus *Wallemia* to high salinity. *Applied and Environmental Microbiology* **76**, 329–337 (2010).
24. Kralj Kunčič, M., Zajc, J., Drobne, D., Pipan Tkalec, Ž., and Gunde-Cimerman, N. Morphological responses to high sugar concentrations differ from adaptations to high salt concentrations in xerophilic fungi *Wallemia* spp. *Fungal Biology* **117**, 466–478 (2013).
25. Yale, J. and Bohnert, H.J. Transcript expression in *Saccharomyces cerevisiae* at high salinity. *Journal of Biological Chemistry* **276**, 15996–16007 (2001).
26. Vaupotič, T. and Plemenitaš, A. Differential gene expression and Hog1 interaction with osmoresponsive genes in the extremely halotolerant black yeast *Hortaea werneckii*. *BMC Genomics* **8**, 280 (2007).
27. Petrovič, U., Gunde-Cimerman, N., and Plemenitaš, A. Cellular responses to environmental salinity in the halophilic black yeast *Hortaea werneckii*. *Molecular Microbiology* **45**, 665–672 (2002).
28. Blomberg, A. and Adler, L. Physiology of osmotolerance in fungi. *Advances in Microbial Physiology* **33**, 145–212 (1992).
29. Blomberg, A. Metabolic surprises in *Saccharomyces cerevisiae* during adaptation to saline conditions: Questions, some answers and a model. *FEMS Microbiology Letters* **182**, 1–8 (2000).
30. Hohmann, S. Osmotic stress signaling and osmoadaptation in yeasts. *Microbiology and Molecular Biology Reviews* **66**, 300–372 (2002).
31. Andre, L., Nilsson, A., and Adler, L. The role of glycerol in osmotolerance of the yeast *Debaryomyces hansenii*. *Journal of General Microbiology* **134**, 669–677 (1988).
32. Larsson, C. and Gustafsson, L. Glycerol production in relation to the ATP pool and heat-production rate of the yeasts *Debaryomyces hansenii* and *Saccharomyces cerevisiae* during salt stress. *Archives of Microbiology* **147**, 358–363 (1987).
33. Larsson, C., Morales, C., Gustafsson, L., and Adler, L. Osmoregulation of the salt-tolerant yeast *Debaryomyces hansenii* grown in a chemostat at different salinities. *Journal of Bacteriology* **172**, 1769–1774 (1990).
34. Larsson, C. and Gustafsson, L. The role of physiological-state in osmotolerance of the salt-tolerant yeast *Debaryomyces hansenii*. *Canadian Journal of Microbiology* **39**, 603–609 (1993).
35. Prista, C., Almagro, A., Loureiro-Dias, M.C., and Ramos, J. Physiological basis for the high salt tolerance of *Debaryomyces hansenii*. *Applied and Environmental Microbiology* **63**, 4005–4009 (1997).
36. Almagro, A. et al. Effects of salts on *Debaryomyces hansenii* and *Saccharomyces cerevisiae* under stress conditions. *International Journal of Food Microbiology* **56**, 191–197 (2000).
37. Kogej, T. et al. Osmotic adaptation of the halophilic fungus *Hortaea werneckii*: Role of osmolytes and melanization. *Microbiology* **153**, 4261–4273 (2007).
38. Kogej, T., Gorbushina, A.A., and Gunde-Cimerman, N. Hypersaline conditions induce changes in cell-wall melanization and colony structure in a halophilic and a xerophilic black yeast species of the genus Trimmatostroma. *Mycological Research* **110**, 713–724 (2006).
39. Kogej, T., Gostinčar, C., Volkmann, M., Gorbushina, A.A., and Gunde-Cimerman, N. Mycosporines in extremophilic fungi—Novel complementary osmolytes? *Environmental Chemistry* **3**, 105–110 (2006).
40. Kogej, T., Ramos, J., Plemenitaš, A., and Gunde-Cimerman, N. The halophilic fungus *Hortaea werneckii* and the halotolerant fungus *Aureobasidium pullulans* maintain low intracellular cation concentrations in hypersaline environments. *Applied and Environmental Microbiology* **71**, 6600–6605 (2005).
41. Plemenitaš, A., Vaupotič, T., Lenassi, M., Kogej, T., and Gunde-Cimerman, N. Adaptation of extremely halotolerant black yeast *Hortaea werneckii* to increased osmolarity: A molecular perspective at a glance. *Studies in Mycology* **61**, 67–75 (2008).
42. Teige, M., Scheikl, E., Reiser, V., Ruis, H., and Ammerer, G. Rck2, a member of the calmodulin-protein kinase family, links protein synthesis to high osmolarity MAP kinase signaling in budding yeast. *Proceedings of the National Academy of Sciences of the United States of America* **98**, 5625–5630 (2001).
43. Lin, H.L., Nguyen, P.N., and Vancura, A.V. Phospholipase C interacts with Sgd1p and is required for expression of GPD1 and osmoresistance in *Saccharomyces cerevisiae*. *Molecular Genetics and Genomics* **267**, 313–320 (2002).
44. Konte, T. and Plemenitaš, A. The HOG signal transduction pathway in the halophilic fungus *Wallemia ichthyophaga*: Identification and characterisation of MAP kinases WiHog1A and WiHog1B. *Extremophiles* **17**, 623–636 (2013).
45. Plemenitaš, A. et al. Adaptation to high salt concentrations in halotolerant/halophilic fungi: A molecular perspective. *Frontiers in Microbiology* **5**, 199 (2014).

46. Oren, A. Bioenergetic aspects of halophilism. *Microbiology and Molecular Biology Reviews* **63**, 334–348 (1999).
47. Zajc, J., Kogej, T., Ramos, J., Galinski, E.A., and Gunde-Cimerman, N. The osmoadaptation strategy of the most halophilic fungus *Wallemia ichthyophaga*, growing optimally at salinities above 15% NaCl. *Applied and Environmental Microbiology* **80**, 247–256 (2014).
48. Albertyn, J., Hohmann, S., Thevelein, J.M., and Prior, B.A. *GPD1*, which encodes glycerol-3-phosphate dehydrogenase, is essential for growth under osmotic stress in *Saccharomyces cerevisiae*, and its expression is regulated by the high-osmolarity glycerol response pathway. *Molecular and Cellular Biology* **14**, 4135–4144 (1994).
49. Norbeck, J., Pahlman, A.K., Akhtar, N., Blomberg, A., and Adler, L. Purification and characterization of two isoenzymes of DL-glycerol-3-phosphatase from Saccharomyces cerevisiae. Identification of the corresponding GPP1 and GPP2 genes and evidence for osmotic regulation of Gpp2p expression by the osmosensing mitogen-activated protein kinase signal transduction pathway. *Journal of Biological Chemistry* **271**, 13875–13881 (1996).
50. Lenassi, M. et al. Adaptation of the glycerol-3-phosphate dehydrogenase Gpd1 to high salinities in the extremely halotolerant *Hortaea werneckii* and halophilic *Wallemia ichthyophaga*. *Fungal Biology* **115**, 959–970 (2011).
51. Jung, S., Marelli, M., Rachubinski, R.A., Goodlett, D.R., and Aitchison, J.D. Dynamic changes in the subcellular distribution of Gpd1p in response to cell stress. *The Journal of Biological Chemistry* **285**, 6739–6749 (2010).
52. Palkova, Z. and Vachova, L. Life within a community: Benefit to yeast long-term survival. *FEMS Microbiology Reviews* **30**, 806–824 (2006).
53. Wollenzien, U., de Hoog, G.S., Krumbein, W.E., and Urzí, C. On the isolation of microcolonial fungi occurring on and in marble and other calcareous rocks. *Science of the Total Environment* **167**, 287–294 (1995).
54. Selbmann, L., de Hoog, G.S., Mazzaglia, A., Friedmann, E.I., and Onofri, S. Fungi at the edge of life: Cryptoendolithic black fungi from Antarctic desert. *Studies in Mycology* **51**, 1–32 (2005).
55. Mishra, A. and Jha, B. Isolation and characterization of extracellular polymeric substances from microalgae *Dunaliella salina* under salt stress. *Bioresource Technology* **100**, 3382–3386 (2009).
56. Petrovič, U., Gunde-Cimerman, N., and Plemenitaš, A. Salt stress affects sterol biosynthesis in the halophilic black yeast *Hortaea werneckii*. *FEMS Microbiology Letters* **180**, 325–330 (1999).
57. Turk, M. et al. Salt-induced changes in lipid composition and membrane fluidity of halophilic yeast-like melanized fungi. *Extremophiles* **8**, 53–61 (2004).
58. Gostinčar, C. et al. Expression of fatty-acid-modifying enzymes in the halotolerant black yeast *Aureobasidium pullulans* (de Bary) G. Arnaud under salt stress. *Studies in Mycology* **61**, 51–59 (2008).
59. Gostinčar, C., Turk, M., Plemenitaš, A., and Gunde-Cimerman, N. The expressions of Δ^9-, Δ^{12}-desaturases and an elongase by the extremely halotolerant black yeast *Hortaea werneckii* are salt dependent. *FEMS Yeast Research* **9**, 247–256 (2009).
60. Kralj Kunčič, M. Adaptations of fungal genus *Wallemia* spp. on growth in extreme saline and extreme sweet environment at the level of cell wall and metabolite productions. Doctoral dissertation, Univerza v Ljubljani, Medicinska fakulteta, Ljubljana, Slovenia (2010).
61. Saito, M. et al. Screening tests using HeLa cells and mice for detection of mycotoxin-producing fungi isolated from foodstuffs. *Japanese Journal of Experimental Medicine* **41**, 1–20 (1971).
62. Wood, G.M. Assessment of toxigenic moulds in foods by means of a biological screening method. *Proceedings of the 5th Meeting on Mycotoxins in Animal and Human Health* (ed. Moss, M.O.), Edinburgh (1985).
63. Wood, G.M., Mann, P.J., Lewis, D.F., Reid, W.J., and Moss, M.O. Studies on a toxic metabolite from the mold *Wallemia*. *Food Additives and Contaminants* **7**, 69–77 (1990).
64. Chamberlin, J.W. et al. Structure of antibiotic A 25822 B, a novel nitrogen-containing C28-sterol with antifungal properties. *The Journal of Antibiotics* **27**, 992–993 (1974).
65. Takahashi, I. et al. UCA1064-B, a new antitumor antibiotic isolated from Wallemia sebi: Production, isolation and structural determination. *Journal of Antibiotics* **46**, 1312–1314 (1993).
66. Peng, X.P. et al. Aromatic compounds from the halotolerant fungal strain of *Wallemia sebi* PXP-89 in a hypersaline medium. *Archives of Pharmacal Research* **34**, 907–912 (2011).
67. Sepčić, K., Zalar, P., and Gunde-Cimerman, N. Low water activity induces the production of bioactive metabolites in halophilic and halotolerant fungi. *Marine Drugs* **9**, 59–70 (2011).

68. Botić, T., Kunčič, M.K., Sepčić, K., Knez, Z., and Gunde-Cimerman, N. Salt induces biosynthesis of hemolytically active compounds in the xerotolerant food-borne fungus *Wallemia sebi*. *FEMS Microbiology Letters* **326**, 40–46 (2012).
69. Wu, J.T., Chiang, Y.R., Huang, W.Y., and Jane, W.N. Cytotoxic effects of free fatty acids on phytoplankton algae and cyanobacteria. *Aquatic Toxicology (Amsterdam, Netherlands)* **80**, 338–345 (2006).
70. Hanhela, R., Louhelainen, K., and Pasanen, A.L. Prevalence of microfungi in finnish cow barns and some aspects of the occurrence of wallemia-sebi and fusaria. *Scandinavian Journal of Work Environment & Health* **21**, 223–228 (1995).
71. Roussel, S. et al. Microbiological evolution of hay and relapse in patients with farmer's lung. *Occupational and Environmental Medicine* **61**, e3 (2004).
72. Sakamoto, T. et al. Allergenic and antigenic activities of the osmophilic fungus *Wallemia sebi* asthmatic patients. *Arerugi = [Allergy]* **38**, 352–359 (1989).
73. Sakamoto, T. et al. Studies on the osmophilic fungus *Wallemia sebi* as an allergen evaluated by skin prick test and radioallergosorbent test. *International Archives of Allergy and Applied Immunology* **90**, 368–372 (1989).
74. American Thoracic Society. Respiratory health hazards in agriculture. *American Journal of Respiratory and Critical Care Medicine* **158**, S1–S76 (1998).
75. Roussel, S., Reboux, G., Dalphin, J.C., Laplante, J.J., and Piarroux, R. Evaluation of salting as a hay preservative against farmer's lung disease agents. *Ann Agric Environ Med* **12**, 217–221 (2005).
76. Roussel, S. et al. Farmer's lung disease and microbiological composition of hay: A case-control study. *Mycopathologia* **160**, 273–9 (2005).
77. Gbaguidi-Haore, H., Roussel, S., Reboux, G., Dalphin, J.C., and Piarroux, R. Multilevel analysis of the impact of environmental factors and agricultural practices on the concentration in hay of microorganisms responsible for farmer's lung disease. *Annals of Agricultural and Environmental Medicine* **16**, 219–225 (2009).
78. Gerrits van den Ende, A.H.G. and de Hoog, G.S. Variability and molecular diagnostics of the neurotropic species *Cladophialophora bantiana*. *Studies in Mycology* **43**, 151–162 (1999).
79. de Hoog, G.S. and Gerrits van den Ende, A.H. Molecular diagnostics of clinical strains of filamentous Basidiomycetes. *Mycoses* **41**, 183–189 (1998).
80. Masclaux, F., Gueho, E., de Hoog, G.S., and Christen, R. Phylogenetic relationships of human-pathogenic Cladosporium (Xylohypha) species inferred from partial LS rRNA sequences. *Journal of Medical and Veterinary Mycology* **33**, 327–338 (1995).
81. White, T.J., Burns, T., Lee, S., and Taylor, J. Amplification and direct sequencing of fungal ribosomal RNA genes for phylogenies. In *PCR Protocols. A Guide to Methods and Amplifications* (eds. Innis, M.A., Gelfand, D.H., Sninsky, J.J., and White, T.J.) pp. 315–322 (Academic, San Diego, CA, 1990).
82. Gargas, A. and Taylor, J.W. Polymerase chain reaction (PCR) for amplifying and sequencing nuclear 18S rDNA from lychenized fungi. *Mycologia* 589–592 (1992).
83. de Hoog, G.S., Guarro, J., Gené, J., and Figueras, M.J. *Atlas of Clinical Fungi*, (Centraalbureau voor Schimmelcultures/Universitat Rovira i Virgili, Utrecht/Reus., the Netherlands, 2000).

34 Vaccine Development against Fungi

James I. Ito

CONTENTS

34.1 Introduction 583
34.2 Pathophysiology of Invasive Fungal Infections 584
34.3 Immunology of Invasive Fungal Infections 584
    34.3.1 Invasive Candidiasis 584
    34.3.2 Invasive Aspergillosis 585
34.4 Vaccines 586
    34.4.1 *Candida* Vaccines 586
        34.4.1.1 Antibody-Mediated Passive Protection 586
        34.4.1.2 Antigens Used for Active Vaccination 587
    34.4.2 *Aspergillus* Vaccines 589
        34.4.2.1 Vaccination of the Immunosuppressed Host: Proof of Principal 589
        34.4.2.2 Identifying Immunogenic/Protective *Aspergillus* Antigens 589
        34.4.2.3 Other Approaches 590
        34.4.2.4 Cross-Protective Antigens 590
        34.4.2.5 Requirement for a Safe and Effective Adjuvant 591
        34.4.2.6 Problems and Obstacles on the Road to an *Aspergillus* Vaccine ... 591
34.5 Summary 592
References 592

34.1 INTRODUCTION

Invasive fungal infections (IFIs) occur in most immunocompromised patients such as the neutropenic patient undergoing chemotherapy for hematologic malignancy (HM) and the recipients of hematopoietic cell transplantation (HCT) or solid organ transplantation (SOT). While invasive mold infections (IMIs), predominantly invasive aspergillosis (IA), occur only in severely immunocompromised hosts, invasive candidiasis (IC) occurs in severely ill, non-immunocompromised patients.

The mortality rates remain high for IFI despite the availability of potent antifungal agents and the routine use of antifungal prophylaxis in settings of high risk, for example, in HCT. For example, the mortality rate for IA was 57.8% for HCT recipients during a recent (2001–2006) survey period.[1] And there are very few antifungal agents in the pharmaceutical *pipeline*.

Thus, there is an urgent need for new approaches in the treatment and prevention of IFI. Areas that have received the most attention recently have been immunotherapy and immunoprophylaxis, that is, vaccination. Probably the reason for the delay in attempting vaccination against IFI, especially IA, was the knowledge that patients who were severely immunocompromised did not respond to vaccination. In fact, HCT recipients are not immunized with bacterial or viral vaccines for months or even years until they are immune reconstituted and no longer immunosuppressed.[2]

However, vaccination against IA was deemed feasible 15 years ago in neutropenic[3] and corticosteroid (CS) immunosuppressed[4] mice.

There is no U.S. Food and Drug Administration–approved antifungal vaccine available for clinical use despite the research focussed on vaccines for endemic fungal infections (e.g., coccidioidomycosis, blastomycosis, histoplasmosis), cryptococcosis, and IC. The most developed vaccine is an anti-IC vaccine, rAls3p-N, that has progressed through two phase I studies.[5]

There are many recent excellent reviews on fungal vaccines against many opportunistic and endemic fungi.[6–11] However, we will focus on the efforts to develop vaccines against IC and IA in this chapter and discuss (1) the pathophysiology of IFI, (2) the immunology of IFI, (3) *Candida* vaccines, (4) *Aspergillus* vaccines, (5) cross-reactive vaccines, and (6) the obstacles to the development of safe and effective human antifungal vaccines.

34.2 PATHOPHYSIOLOGY OF INVASIVE FUNGAL INFECTIONS

Candida species are yeasts whereas *Aspergillus* species are molds. *Candida* is a commensal or normal colonizer, whereas *Aspergillus* is an environmentally acquired organism. *Candida* has to penetrate the protective mucous membrane layer and invade the bloodstream where it disseminates, whereas *Aspergillus* is inhaled into the sinopulmonary tract and invades by direct extension into the lungs, sinuses, and brain. The fungus rarely, if ever, disseminates via the bloodstream. Thus, they have very different modes of pathogenesis.

34.3 IMMUNOLOGY OF INVASIVE FUNGAL INFECTIONS

The immunology of IFI has been extensively reviewed recently.[12–15] However, here, we will focus on the immunology of IC and IA.

34.3.1 INVASIVE CANDIDIASIS

Candida organisms have been found in soil, food, animals, inanimate objects, and hospital environments. *Candida* species can be normal colonizers or commensals of the mucous membranes of the gastrointestinal or female genital tract. Disease can take the form of mucocutaneous candidiasis (MC) where the yeast proliferates at the surface of the mucous membranes or IC where it invades the bloodstream. Thus, the first lines of defense are the skin and mucous membranes. If these barriers are breached, the second line of defense is the phagocytes: polymorphonuclear neutrophils (PMNs), monocytes, macrophages, and dendritic cells (DCs). IC requires colonization of the mucous membranes, damaged mucous membranes whereby *Candida* translocates across damaged mucous membranes into the vascular space, and defective phagocytes.[16] The major risk factors for IC include neutropenia, CS therapy, cytotoxic chemotherapy, broad-spectrum antibiotics, central venous catheters, and gastrointestinal surgery. Each one of these risk factors affects one or more requirements for IC, that is, colonization, damaged barrier, and/or phagocyte function. T cell immunity plays the major role in defense against IC primarily by enhancing the fungicidal effect of the phagocyte effector cells.

The mechanism(s) whereby adaptive T cell immunity comes into play in IC has been thoroughly studied over the past 20 years and has recently been reviewed.[15,17,18] The classic Th1/Th2 paradigm of Th1 responses mediating protection and Th2 responses causing detrimental inflammation is much more complex and is still being worked out.

Briefly, DCs are primarily responsible for acquiring antigens, decoding this information, and then, via specific cytokine signals, stimulating various T cell pathways including Th1, Th2, Th17, and Treg T cells. The T cell subsets in turn secrete cytokines that mediate protective or detrimental/pathogenic effects on phagocytes and the inflammatory process. The primary protective response against IC is the Th1 response. Th1 lymphocytes produce IFN-γ that stimulates the antifungal activity of PMN and macrophages.[19] Conversely, the Th2 response mediated by the cytokines IL-4 and IL-10 exacerbates disease by deactivating fungicidal effector cells (e.g., by inhibiting the IFN-γ-dependent NO production by macrophages). The importance of the balance between Th1 and Th2 responses has been recently emphasized.[20] More recently, the role of Th17 cells has been studied, but the contribution of Th17 responses to IC immunity is unclear. Animal studies have shown that mice deficient in IL-17RA have an increased susceptibility to disseminated candidiasis[21] and IL-17A induced by dectin-2 is necessary for defense against IC.[22] However, other studies have shown a detrimental effect of IL-17.[23,24]

The most telling evidence for a more important role of Th17 in MC and a lesser role in IC is the report of patients with hereditary diseases affecting the Th17 signaling pathway (e.g., the hyper-IgE syndrome and the Q295X mutation in CARD9) who present with MC but not systemic candidiasis.[25] Tregs suppress immunity against IC and result in higher susceptibility to IC.[26,27] At mucosal sites, however, Tregs promote tolerance and allow long-term memory that enhances resistance to reinfection. Finally, CD8+ cytotoxic T cells and γδ T cells may play a role in candidiasis, but it is probably restricted to protecting against MC and not IC. In summary, the major protective T cell response against IC is Th1-mediated phagocyte activation via Th1 cytokine production.

34.3.2 Invasive Aspergillosis

IA is the most common IMI in the most severely immunocompromised hosts. These include patients with HM and neutropenia and SOT and HCT transplant recipients. The HCT recipient is especially at high risk during the two phases of HCT: (1) the neutropenic, preengraftment period and (2) the subsequent postengraftment period where the neutrophil count has recovered but where graft-versus-host disease (GVHD) and, more importantly, its therapy with CSs begin. Thus, the two major risk factors for IA are neutropenia and CS therapy in the setting of GVHD. More recently, IA has become the major complication late in HCT as it has moved from the neutropenic to the late postengraftment period.[28] Thus, currently, the primary risk factor for IA in HCT is CS therapy/immunosuppression.

Aspergillus species are ubiquitous organisms found in the air, soil, food, and water. The size of the conidia (2–4 μm) allows these airborne spores to be inhaled through the sinopulmonary tract and settle in the pulmonary alveoli. However, hundreds to thousands of conidia are inhaled daily without consequence in the immunocompetent host.[29] From earlier animal studies, it was concluded that the first line of defense against inhaled conidia was the alveolar macrophage[30] and that these cells could phagocytize and kill resting conidia. It was also concluded that neutrophils were active against germinating conidia, thus forming a second line of defense against invading hyphae. These observations were made in animals treated with either cytotoxic agents (inducing neutropenia) or CSs (impairing macrophage function). However, recent studies have demonstrated that there is an early (less than 3 h) influx of neutrophils into the respiratory tree capable of preventing conidial germination and hyphal germination in the absence of alveolar macrophages.[31] These observations do not preclude an important role for alveolar macrophages, but rather reveal the important role of the neutrophil in anticonidial and antihyphal activity. The neutrophil and macrophage constitute the first line of defense in the innate immune system, and pathogen recognition is based upon pattern recognition receptors (PRRs) that recognize pathogen-associated molecular patterns.

But the crucial initial encounter between pathogen and host that will determine the subsequent T cell response is between conidia and the DC.[32] Soon after inhalation of conidia, uptake of conidia or hyphae by pulmonary DC takes place within 3 h.[33] Within 6 h, these specific DCs migrate to the lung-draining lymph nodes and orchestrate the T cell response. Like the macrophage, the DC depends upon PRR to decode the antigens of the invading pathogen. These PRRs include membrane-bound PRR toll-like receptors (TLRs) and soluble receptors such as pentraxin-3.[29] Triggering of TLRs on DC by live conidia leads to the production of inflammatory cytokines TNF and IL-12, which leads to the priming and differentiation of *Aspergillus*-specific Th1 CD4+ T cell clones.[18] These primed and differentiated CD4+ T cells then traffic back to the site of infection and release their Th1 cytokines TNF and IFN-γ that in turn stimulate the killing of conidia and hyphae by macrophages and neutrophils.

The type of specific T cell response to *Aspergillus* elicited depends on many factors. One of these is the viability of the conidia to which the host is exposed. Exposure of DCs to live *Aspergillus* conidia results in the production of IL-12, IL-18, and IFN-γ, in turn promoting Th1 differentiation.[34,35] Intratracheal inoculation of live conidia results in a Th1 response,[3] while exposure to killed spores resulted in a T2 response.[36] The Th1/Th2 paradigm[37] promulgated over 25 years ago is still the basis for our understanding of the differential T cell responses against microbial pathogens. Th1 responses lead to protection, while Th2 responses are detrimental to the host and are associated with allergy. Each lineage tends to inhibit the other. But there are other T cell lineages that affect the immune response to *Aspergillus*. The roles of Th17 T cells and IL-17/IL-23 are not clear. One study demonstrated a detrimental effect of IL-17 (decreased neutrophil killing and clearance of *Aspergillus*),[23] while another suggested a protective role.[38] A lack of stimulation of IL-17 by *Aspergillus* was reported and suggested that protection does not rely on Th17 responses.[39] Finally, the role of CD4+ CD25+ T cells (Tregs) is still being studied. Tregs suppressed neutrophil function during early intranasal inoculation of spores, and subsequently, Tregs inhibited Th2 allergic responses. Finally, swollen, but not resting, conidia counteracted Treg activity.[40]

Thus, a robust Th1 response is the major protective T cell response against pulmonary *Aspergillus* infection. This is mediated by the production of Th1 cytokines IFN-γ, IL-2, IL-12, and IL-18 that stimulate macrophage activation, generation of cytotoxic CD4+ T cells, production of opsonizing antibody, and delayed-type hypersensitivity.[41] One of the major immunosuppressive effects of CSs is considered to be its suppression of macrophage function through suppression of (1) Th1 proliferation and Th1 cytokines (IFN-γ)[42] and (2) the release of oxidative and nonoxidative metabolites.[43] More importantly, it was demonstrated that these CS immunosuppressed macrophages still responded to IFN-γ, restoring its anti-*Aspergillus* activity.[44] CSs have also been shown to impair neutrophil function by inducing defects in trafficking, adherence, and release of antimicrobial effector molecules.[45]

34.4 VACCINES

IFI remains the most frequent infectious complication in the most immunocompromised hosts with continuing high mortality rates despite the development of new and more effective antifungal agents over the past two decades.[1] Thus, much research has focussed on new approaches such as immunotherapy and vaccines. But the paradox is that there is evidence that vaccination is ineffective in this setting.[46] Although T cell–mediated immunity in the form of a robust Th1 response is felt to be the major component of protective immunity against IFI, this does not preclude other types of immunity, such as antibody, from mediating protection.

34.4.1 Candida Vaccines

As *Candida* species are saprophytes of the gastrointestinal tract, antibodies are acquired early in life. But this antibody response is heterogeneous and polyclonal and does not confer protection against IC. But while the polyclonal response does not protect, it has been demonstrated that protective monoclonal antibodies can be generated.[47,48] Thus, while T cell–mediated immunity is considered paramount in the defense against IC, much effort has been devoted toward passive vaccination with monoclonal antibodies.

34.4.1.1 Antibody-Mediated Passive Protection

Mycograb is one of the most developed monoclonal antibodies as an anti-Hsp90 directed at a common heat shock protein of fungi.[49,50] It has been clinically tested in patients with candidemia in combination with lipid-associated amphotericin B.[51] Other monoclonal antibodies that were protective in preclinical studies are (1) anti-β-1,3-glucan mAb 2G8,[52] (2) antimannoprotein mAb C7,[53] (3) anti-idiotypic antibodies,[54] (4) antimannan mAb,[48,55] and (5) antiglycosyl mAb.[56] The mechanisms of action appear to be candidacidal (2,3,5), growth inhibitory (1), or neutralization of Hsp90.

34.4.1.2 Antigens Used for Active Vaccination

The antigens from which some of these monoclonal antibodies have been generated have also been used as vaccines including the (1) 65 kDa mannoproteins,[57] (2) β-1,3-glucan,[52] and (3) β-1,2-mannosides.[48,55] More recently, vaccination with a recombinant Hyr1p, a virulence factor for *Candida albicans* mediating resistance to phagocyte killing, protected mice against candidemia.[58] Also, using an immunoproteomic approach, Bar et al. have discovered a Th cell epitope that protected mice against candidemia after vaccination.[59]

But the vaccine candidates that have been studied most intensely and have come closest to clinical trials are a (1) synthetic glycopeptide vaccine, β-(Man)3-Fba-TT, developed by the J. E. Cutler laboratory,[60] and (2) recombinant N-terminus of Als3p, which is an adhesin of *Candida* called rAls1p-N developed by the J. E. Edwards group.[6] The latter vaccine, rAls1p-N, has actually just completed two phase I trials.[5]

34.4.1.2.1 β-(Man)3-Fba-TT Vaccine

In 1995, Han and Cutler first demonstrated that a mannan adhesin fraction of *C. albicans* encapsulated into liposomes and used to vaccinate mice resulted in protection against disseminated candidiasis.[61] They also showed that antiserum from vaccinated animals when administered intraperitoneally to mice prior to intravenous challenge with *Candida* resulted in protection. Two monoclonal antibodies were identified in the protective sera, and both were IgM and specific for cell wall mannan, but only one, MAb B6.1, protected animals against IC.[61]

Biochemical characterization of the B6.1 epitope revealed that it was a mannotriose.[62] Nuclear magnetic resonance spectroscopic analysis of the epitope yielded data consistent with a β-(1 → 2)-linked mannotriose (β-(Man)3).[62] Because the protective epitope was identified as an oligosaccharide, it was considered that it would be poorly immunogenic. Thus, peptides in *C. albicans* cell wall proteins were investigated utilizing algorithm peptide epitope searches,[63] and 6 T cell peptides were selected and conjugated to β-(Man)3. All six glycopeptide vaccines were tested in animals, but only three demonstrated strong protection against hematogenous challenge with *C. albicans*.[63] One of these glycoconjugates, β-(Man)3-Fba, was selected as an ideal vaccine because (1) the 14-mer Fba peptide sequence was unique to *C. albicans* and would not be expected to cross-react with any human enzyme sequences and (2) the parent protein (Fba1p) is an enzyme essential for the viability of *C. albicans*.[63] However, this *C. albicans*–specific epitope may not be protective against other *Candida* species.

Recent work has demonstrated that the peptide Fba, without conjugation to mannotriose, can confer protection in animals by using the human-approved adjuvant, alum, or by a DC-based immunization method.[60] Finally, the β-(Man)3-Fba glycopeptide conjugate was coupled to tetanus toxoid (TT) and resulted in protection almost equal to that of vaccination without adjuvants.[60] Thus, it appears that the glycopeptide conjugate β-(Man)3-Fba coupled to TT (β-(Man)3-Fba-TT) is ready for human trials without the need of adjuvants.

34.4.1.2.2 rAls3p-N Vaccine

The Edwards group at Harbor/UCLA Medical Center and the Los Angeles Biomedical Research Institute have made the most progress toward realizing a protective vaccine against IC. This effort has been ongoing since the late 1990s and has been nicely reviewed.[6]

Hoyer et al.[64] first cloned the *C. albicans* ALS1 gene that coded for an adhesin protein, Als1p. Fu et al. demonstrated that this cell surface protein was an effector of filamentation and adherence.[65] Also, Als1p was required for virulence in a mouse model of IC. Also, Loza et al. indicated that the adherence characteristic and endothelial cell binding region is localized within its N-terminus.[66] This suggested that a recombinant N-terminal domain of Als1p (rAls1p-N) would be a good candidate for a vaccine against IC.

Ibrahim et al.[67] vaccinated BALB/c mice with rAls1p-N and subsequently challenged the animals with intravenous *C. albicans* blastospores. Vaccinated animals were significantly protected compared to nonvaccinated animals. Vaccination resulted in increased *Candida* stimulation of Th1 splenocytes and increased in vivo delayed-type hypersensitivity, although antibody titers did not correlate with protection. The vaccine was not protective in T cell–deficient mice but was in B cell deficient, suggesting that protection was T cell (Th1) mediated and not antibody mediated.

Spellberg et al.[68] demonstrated that the rAls1p-N vaccine was effective in neutropenic and CS-treated immunocompromised mice against IC. Furthermore, Ibrahim et al. indicated that this vaccine was effective against other *Candida* species including *C. glabrata*, *C. krusei*, *C. parapsilosis*, and *C. tropicalis*.[69] These are the next most common causes of IC and, together with *C. albicans*, account for 99% of *Candida* isolates from patients with IC.

Vaccination with rAls3p-N resulted in comparable protection to rAls1p-N against IC but also significantly more protection against oropharyngeal and vaginal candidiasis.[70] Because of the possible advantage that a vaccine could protect against mucosal and invasive disseminated candidiasis, the development of the rAls3p-N vaccine was chosen.

An even more significant observation was that the candidal adhesins, Als1p and Als3p, were predicted to have 3D structural similarity to clumping factor (ClfA), a member of a family of surface adhesions expressed by *Staphylococcus aureus* known as microbial surface components that recognized adhesive matrix molecules.[71] In a series of experiments, vaccination with rAls3p-N protected mice against intravenous (i.e., tail vein) challenge with multiple strains of *S. aureus*, including methicillin-resistant *S. aureus*.[72] This dramatically expanded the potential for the rAls3p-N vaccine as it protected against two of the most common microbiological pathogens.

Further research by the Edwards group delineated the role of Als3p in the pathophysiology of infection and the mechanism of protection when it was used as a vaccine. Als3 was required for *Candida* to be endocytosed by human endothelial and epithelial cells.[73] Als3p was also required for binding to multiple host cell surface proteins, including N-cadherin on endothelial cells and E-cadherin on epithelial cells. Als3p mimics host cadherins and functions as an invasin by binding to host cadherins and inducing endocytosis by host cells.

To define the protective mechanisms of immunity induced by rAls3p-N, further studies were carried out in murine models of *C. albicans* and *S. aureus* sepsis.[74] rAls3p-N vaccination induced a Th1/Th17 response, resulting in recruitment and activation of phagocytes at sites of infection and more effective clearance of *C. albicans* and *S. aureus* from tissues. Thus, vaccine-induced immunity protects by stimulating enhanced recruitment and killing of pathogens by effector cells. There was one surprising and contradictory finding in this study. Whereas in prior studies,[68] rAls1p-N induced protection against IC in neutropenic mice, the current study[74] demonstrated no protection in rAls3p-N-vaccinated mice. The latter study demonstrates the important role of neutrophils in protection but also raises an important dilemma. This vaccine may not work in the setting of neutropenia. The authors attribute this discrepancy to the use of a different vaccine (rAls1p-N) and adjuvant (complete Freund's adjuvant) for the disparate results. But it does raise a very important issue for this vaccine.

The aforementioned studies have led to a phase I clinical trial of rAls3p-N that has recently been completed.[75] Forty healthy, adult subjects were randomized to receive one dose of NDV-3 containing either 30 or 300 μg of Als3p, or placebo. NDV-3 at both dose levels was safe and generally well-tolerated. Anti-Als3p total IgG and IgA1 levels for both doses reached peak levels by day 14 postvaccination, with 100% conversion of all vaccinated subjects. On average, NDV-3 stimulated peripheral blood mononuclear cell production of both IFN-γ and IL-17A, which peaked at day 7 for subjects receiving the 300 μg dose and at day 28 for those receiving the 30 μg dose. Six months after receiving the first dose of NDV-3, 19 subjects received a second dose of NDV-3 identical to their first dose to evaluate memory B- and T-cell immune responses. The second dose resulted in a significant boost of IgG and IgA1 titers in >70% of subjects, with the biggest impact in those receiving the 30 μg dose. A memory T-cell response was also noted for IFN-γ in almost all subjects and for IL-17A in the majority of subjects. Thus, it appears that this vaccine stimulates in humans a robust T-cell and B-cell response. Further clinical studies are anticipated.

34.4.1.2.3 Obstacles to the Development of a Candida *Vaccine*

The major problems associated with the development of a *Candida* vaccine have been reviewed.[6] There is a lack of monetary support from government and industry, that is, pharmaceutical companies. It would cost approximately US$3–4 million to support the development of a vaccine from the laboratory to a phase I clinical trial that tends to discourage the pharmaceutical industry as the market is perceived as being too small.

34.4.2 ASPERGILLUS VACCINES

34.4.2.1 Vaccination of the Immunosuppressed Host: Proof of Principal

IA occurs with greatest frequency in the most severely immunocompromised hosts. Most severely affected are patients with HM and profound and prolonged neutropenia. Also, the allogeneic HCT recipient suffering from GVHD and being treated with high and prolonged doses of CSs is greatly at risk. Recent data[1] demonstrate that the vast majority of IMI in HCT occurs after neutropenia has resolved, that is, postengraftment, where the major risk factors are GVHD and CS therapy. Vaccination prior to immune reconstitution does not result in immunity when vaccinating against bacteria and viruses.[2] And it has been assumed that vaccination prior to immunosuppression (e.g., HCT) would fail to confer protection after immunosuppression. These suppositions might be the reason for the delay in attempting vaccination against IMI in the setting of severe immunosuppression.

Cenci et al.[3] were the first to demonstrate that vaccination of animals with a crude culture filtrate of *Aspergillus fumigatus* prior to administration of chemotherapy, subsequently resulting in neutropenia, protected against subsequent lethal challenge with inhaled *Aspergillus* conidia. They also demonstrated that adoptive transfer of *Aspergillus* antigen-specific CD4+ T cells resulted in some protection, albeit in a model of infection whereby animals are challenged with conidia injected intravenously. Also, local CD4+ cells proliferated in response to the antigen, and local IFN-γ and IL-2 production increased in vaccinated animals. A recruitment of lymphocytes and macrophages to the site (lungs) of infection in vaccinated animals was observed despite a profound systemic neutropenia suggesting the availability of the effector cells necessary for protection.

Ito and Lyons[4] were able to demonstrate protection against IA by vaccinating mice subcutaneously with *A. fumigatus* sonicated hyphae (HS) or culture filtrate prior to cortisone acetate therapy and subsequent intranasal challenge with *Aspergillus* conidia. They also demonstrated that prior intranasal challenge with viable conidia (VC) also resulted in protection and that systemic (subcutaneous) vaccination was superior to local (intranasal) vaccination. Thus, proof of principal was demonstrated that vaccination prior to immunosuppression (neutropenia and CS) can confer protective immunity.

34.4.2.2 Identifying Immunogenic/Protective *Aspergillus* Antigens

Much effort has been directed toward identifying the immunogenic and protective antigens of *Aspergillus*.

Bozza et al. demonstrated that the allergen Asp f16 (now designated as Asp f9/16), but not Asp f3, when administered with CpG oligodeoxynucleotides (ODNs) as adjuvants intranasally, was protective against IA in neutropenic mice.[33] Ito et al.[76] analyzed the sera from mice vaccinated with crude vaccine (HS) and sera from mice exposed to VC. Immune sera were tested for reactivity against hyphal protein extracts by Western blots. The Western blots of HS-vaccinated mice showed a multitude of bands at many molecular weights. In contrast, the Western blot of VC-immunized mice showed bands primarily at 19 kDa. Through a combined immunochemical and mass spectrometric approach, a 19 kDa antigen was identified as the allergen Asp f3. As VC-challenged mice reacted strongly and almost exclusively to Asp f3, it suggested that Asp f3 would be an ideal candidate for a protective *Aspergillus* subunit vaccine. Recombinant Asp f3 (rAsp f3) was synthesized and compared to HS as vaccines administered prior to CS treatment in subsequently VC-challenged mice. Protection was equally significant for both the rAsp f3 and HS vaccines. In order to avoid the possibility of inducing an allergic reaction (or state such as allergic bronchopulmonary aspergillosis) to a known *Aspergillus* allergen, truncated forms of Asp f3 were synthesized that eliminated one or both of the two IgE binding sites. Both the C-truncated and N-truncated forms of Asp f3 were demonstrated to be protective vaccines in CS animals, but not the bitruncated version. Also, two specific T cell epitopes of Asp f3, an 11-mer and a 13-mer polypeptides, were identified that might be candidates for an *Aspergillus* vaccine.[77]

In an ambitious study to survey a large number of antigen candidates for their abilities to activate adaptive Th cell responses and Th cytokine production and protect animals against IA, Bozza et al. assessed the immunogenicity and protection induced by various antigens including secreted proteins, membrane-anchored proteins, glycolipids, and polysaccharides.[78] They demonstrated that, in general, secreted proteins induced Th2 cell activation, membrane proteins induced Th1/Treg cell activation, glycolipids induced Th17 activation, and polysaccharides induced predominantly IL-10 production. More importantly they demonstrated that one secreted protein (protease), Pep 1p, and the two membrane-anchored proteins, Gel 1p and Crf 1p, were immunogenic, activated Th1/Treg, and protected animals against IA. Furthermore, these three proteins retained their Th1/Treg activating potential in human Ag-specific T cell clones. It is interesting to note that Asp f16 has recently been found to be identical to Asp f9[79] and that Asp f9 and Crf 1 are splicing variants of the same gene, crf1.[80] Also, Chaudhary et al. demonstrated that Asp f3 and Asp f9 elicit strong Th1-directed CD4+ responses.[81]

Thus, there are at least five proteins, two of which are probably splicing variants of the same gene, that appear to be excellent candidates for a protective *Aspergillus* vaccine in humans: Asp f3, Asp f9, Pep 1p, Gel 1p, and Crf 1p.

34.4.2.3 Other Approaches

Of course the most comprehensive approach to discovering the protective antigen(s) of *Aspergillus*, now that the entire genome has been sequenced, is to synthesize each protein and test it in an animal model of IA. However, this would be unrealistically time-consuming and expensive. Some have suggested focussing this proteomic approach by selecting proteins that would be more likely to be immunogenic and protective. Chaudhuri et al.[82] have suggested targeting only those proteins that are likely to be adhesins, a putative virulence factor of microbes. However, this assumes that protection is mediated by neutralization of this virulence factor but, as noted earlier, innate and acquired immunity is mediated by Th1 cell immunity.

In another approach,[83] the sera from intravenously infected rabbits were run against water-soluble *A. fumigatus* proteins that were separated electrophoretically,[83] and 59 proteins were identified. These have yet to be tested in an inhalational model of IA under immunosuppressive conditions. But challenging animals with conidia administered intravenously is not the portal of entry of infection nor is the response to this intravenous challenge likely to be similar to that elicited through an intranasal or inhalational challenge. Again, protection is mediated through a Th1 response whereby the protective *Aspergillus* antigen induces the *Aspergillus*-specific Th1 cell to produce cytokines, which, in turn, activate the effector cells (macrophages).

34.4.2.4 Cross-Protective Antigens

It would be ideal if any future *Aspergillus* vaccine would also be protective against other fungal pathogens. The recent rise in the incidence of mucormycosis has been well documented[84] and candidiasis is the most common of IFIs. There are two possible approaches to this: (1) create a polyvalent vaccine consisting of protective antigens from all fungal species to which protection is desired and (2) select antigens that are cross protective, that is, homologues, if they exist, of the protective *Aspergillus* antigen. This latter approach is already being used.

Asp f3 has homologues in other fungal species: (1) PMP20 from *Aspergillus nidulans*, (2) cDNA from *Aspergillus oryzae*, (3) Pmp1 from *Coccidioides posadasii*, (4) PMPA and PMPB (PMP20) from *Candida boidinii*, (5) alkyl hydroperoxide reductase from *C. albicans*, (6) Pen c3 from *Penicillium citrinum*, and (7) alkyl hydroperoxide peroxidase (Ahp1) from *Saccharomyces cerevisiae*.[76] However, there does not appear to be an Asp f3 homologue in the Mucorales (i.e., *Rhizopus*, *Mucor*, *Cunninghamella*).

However, the *Aspergillus* protective secreted aspartic protease Pep 1p has homologues in *Candida* species (Bozza et al.[78]), and these have been shown to be protective against candidiasis in mice.[85] The GPI-anchored protein Gel 1p has a homologue in *C. posadasii* and is protective against coccidioidomycosis in mice.[86] The GPI-anchored protein Crf 1p is homologous to the Crh 1 protein in *Candida* species and has been shown to be protective against IC.[87] While the glycolipids did not confer protection, the polysaccharides were effective in this regard.[78]

Interestingly, laminarin (a glucan from the alga *Laminaria digitata*) conjugated with the diphtheria toxoid CRM197 was reported to be protective against disseminated candidiasis and an intravenous challenge of *Aspergillus*.[52,88] A heat-killed *Saccharomyces* vaccine has been shown to protect against intravenous challenge with *Coccidioides*, *Aspergillus*, and *Candida*.[89] But the protective antigen has not been defined. Most recently, a proteomic approach to surveying all of the *A. fumigatus* cell wall proteins for homology (percentage identity and length) to other fungal pathogens has been undertaken.[90] The survey included comparison of known proteomes in *C. albicans*, *Mucor circinelloides*, *Rhizopus oryzae*, *Fusarium oxysporum*, *Acremonium alcalophilum*, *Cryptococcus neoformans*, *C. posadasii*, *Penicillium marneffei*, and *Mus musculus*. There were a number of proteins that shared significant homology to most of these fungi and lacked homology to human proteins and therefore were the best candidates for a cross-protective fungal vaccine. These included Crf1, Ecm33, Alp2, Gel4, and Crf2. But these homologues have yet to be tested for their immunogenicity and protective capability.

34.4.2.5 Requirement for a Safe and Effective Adjuvant

In order to induce a Th1-directed immune response via vaccination, an adjuvant will probably be required. Vaccination with rAsp f3 and truncated versions of rAsp f3 in mice required the use of TiterMax, an adjuvant not approved for use in humans.[76] However, a particulate form of full-length rAsp f3 that was adjuvant free conferred protection against IA in mice.[76] Thus, the form of the protein vaccine may obviate the need of any adjuvant.

Unmethylated CpG ODNs were shown to act as potent adjuvants when used with Asp f16 in a neutropenic mouse model of IA.[33] Again, ODNs have not been approved for use in humans. Bozza et al. have also demonstrated that DC pulsed with conidia or transfected with conidial RNA and adoptively transferred into mice that were recipients of HCT resulted in significant protection against subsequent IA. This protection was superior to that conferred by adoptive transfer of *Aspergillus*-specific T cells.[91]

34.4.2.6 Problems and Obstacles on the Road to an *Aspergillus* Vaccine

Assuming that an immunogenic and protective antigen has been identified and a safe and effective adjuvant has been selected, there remain major obstacles before such a vaccine proceeds to phase I clinical trials. The major obstacle is cost.[9] It will be expensive to develop good manufacturing practice–compliant manufacturing and preclinical toxicity studies. Further costs are incurred with Investigational New Drug filing, protocol development, and, finally, the phase I, II, III, and IV clinical trials. For such a small anticipated market, no pharmaceutical company would likely undertake such an expensive endeavor.

A more practical problem, however, is how we would administer the vaccine in the setting of HCT. Vaccination has not yet been shown to be effective in the HCT setting although different

Aspergillus vaccines have been shown to be effective when administered prior to immunosuppression (neutropenia and CS) in animals. Since the immune system of the HCT recipient is effectively ablated by day 0 of HCT and T cell immune reconstitution (of donor origin) does not take place for many months (and longer if there is GVHD and continuing CS therapy), it is not expected that recipient vaccination prior to HCT will confer protection after HCT. One possible way to circumvent this problem is to vaccinate the donor with the hope that transfer of *Aspergillus*-specific T cells will take place and confer immediate protection against IA. But if this does not occur, then protection may not be conferred until full immune reconstitution takes place many months later. What may have to fill this *gap* before immune reconstitution may be adoptively transferred *Aspergillus*-specific T cells or, better yet, DCs (of donor origin) pulsed with *Aspergillus* antigen or transfected with RNA.[91] But this latter option is labor intensive and time consuming and, again, very expensive. In any case, it is anticipated that an effective antigen(s) will have to be administered to the recipient, pre- and post-HCT, and donor on multiple occasions. Only clinical trials will tell us to whom (donor and/or recipient), when (prior and/or post HCT), and how many vaccinations will have to be administered to confer protection in the recipient.

34.5 SUMMARY

In the past decade, there has been an increased interest and effort in developing vaccines against IFIs, especially IC and IA. This resurgence came about because of the continuing dismal outcomes of IFI despite new and more powerful antifungal agents and with the realization that vaccination can confer protection against IFI even in those severely immunocompromised who are not expected to mount such an immune response. Despite the many obstacles to developing such vaccines, one anticandidal vaccine, rAls3p-N, has made it to the stage of phase I–II clinical trials and, if successful in protecting against IC, will be approved and clinically available, hopefully within the next 5 years. The development of a protective anti-*Aspergillus* vaccine is not yet in sight. However, there are a number of good candidates being studied. The hope is that there is an antigen or antigens that are not only protective against IA but are common antigens that could protect against a broad spectrum of other molds and even *Candida* species. There are many obstacles that remain including whether an adjuvant will be required for these vaccines and the question of who will be candidates for this vaccine and how and when it be administered. For the anticandidal vaccine, the ideal candidates would be the (1) high-risk ICU patient and (2) immunocompromised neutropenic patient with HM or those undergoing HCT. These patients would receive the vaccine prior to their at-risk period. For the anti-*Aspergillus* vaccine, the candidates would be those patients with hematologic malignancies undergoing chemotherapy resulting in prolonged neutropenia and those undergoing HCT or SOT. The question of whom (donor and/or recipient) and when to vaccinate in the setting of HCT remains to be determined. Whether a *bridging* period of passive immunity via T cell transfer will be required has yet to be determined. Despite all of these obstacles, the future remains bright for the eventual development of a protective antifungal vaccine(s) against IC, IA, and, hopefully, all IFIs.

REFERENCES

1. Baddley, J.W. et al. Factors associated with mortality in transplant patients with invasive aspergillosis. *Clin Infect Dis* **50**, 1559–1567 (2010).
2. Tomblyn, M. et al. Guidelines for preventing infectious complications among hematopoietic cell transplantation recipients: A global perspective. *Biol Blood Marrow Transplant* **15**, 1143–1238 (2009).
3. Cenci, E. et al. T cell vaccination in mice with invasive pulmonary aspergillosis. *J Immunol* **165**, 381–388 (2000).
4. Ito, J.I. and Lyons, J.M. Vaccination of corticosteroid immunosuppressed mice against invasive pulmonary aspergillosis. *J Infect Dis* **186**, 869–871 (2002).

5. Hennessey, J.P. et al. A phase 1 clinical evaluation of NDV3, a vaccine to prevent disease caused by *Candida* spp. and *Staphylococcus aureus*. In *51st Interscience Conference on Antimicrobial Agents and Chemotherapy* (American Society of Microbiology, Chicago, IL, 2011).
6. Edwards, J.E., Jr. Fungal cell wall vaccines: An update. *J Med Microbiol* **61**, 895–903 (2012).
7. Cassone, A. and Casadevall, A. Recent progress in vaccines against fungal diseases. *Curr Opin Microbiol* **15**, 427–433 (2012).
8. Fidel, P.L., Jr. and Cutler, J.E. Prospects for development of a vaccine to prevent and control vaginal candidiasis. *Curr Infect Dis Rep* **13**, 102–107 (2011).
9. Spellberg, B.J. Vaccines for invasive fungal infections. *F1000 Med Rep* **3**, 13 (2011).
10. Cassone, A. Fungal vaccines: Real progress from real challenges. *Lancet Infect Dis* **8**, 114–124 (2008).
11. Cutler, J.E., Deepe, G.S., Jr., and Klein, B.S. Advances in combating fungal diseases: Vaccines on the threshold. *Nat Rev Microbiol* **5**, 13–28 (2007).
12. Wuthrich, M., Deepe, G.S., Jr., and Klein, B. Adaptive immunity to fungi. *Annu Rev Immunol* **30**, 115–148 (2012).
13. Brown, G.D. and Netea, M.G. Exciting developments in the immunology of fungal infections. *Cell Host Microbe* **11**, 422–424 (2012).
14. Romani, L. Immunity to fungal infections. *Nat Rev Immunol* **11**, 275–288 (2011).
15. van de Veerdonk, F.L. and Netea, M.G. T-cell subsets and antifungal host defenses. *Curr Fungal Infect Rep* **4**, 238–243 (2010).
16. Koh, A.Y., Kohler, J.R., Coggshall, K.T., Van Rooijen, N., and Pier, G.B. Mucosal damage and neutropenia are required for *Candida albicans* dissemination. *PLoS Pathog* **4**, e35 (2008).
17. Romani, L. Cell mediated immunity to fungi: A reassessment. *Med Mycol* **46**, 515–529 (2008).
18. Rivera, A. and Pamer, E.G. CD4+ T-cell responses to *Aspergillus fumigatus*. In *Aspergillus fumigatus and Aspergillosis* (eds. Latgé, J.-P. and Steinbach, W.J.), pp. 263–277 (ASM Press, Washington, DC, 2009).
19. Mencacci, A. et al. Innate and adaptive immunity to *Candida albicans*: A new view of an old paradigm. *Rev Iberoam Micol* **16**, 4–7 (1999).
20. Haraguchi, N. et al. Impairment of host defense against disseminated candidiasis in mice overexpressing GATA-3. *Infect Immun* **78**, 2302–2311 (2010).
21. Huang, W., Na, L., Fidel, P.L., and Schwarzenberger, P. Requirement of interleukin-17A for systemic anti-*Candida albicans* host defense in mice. *J Infect Dis* **190**, 624–631 (2004).
22. Saijo, S. et al. Dectin-2 recognition of alpha-mannans and induction of Th17 cell differentiation is essential for host defense against *Candida albicans*. *Immunity* **32**, 681–691 (2010).
23. Zelante, T. et al. IL-23 and the Th17 pathway promote inflammation and impair antifungal immune resistance. *Eur J Immunol* **37**, 2695–2706 (2007).
24. Bozza, S. et al. Lack of Toll IL-1R8 exacerbates Th17 cell responses in fungal infection. *J Immunol* **180**, 4022–4031 (2008).
25. Glocker, E.O. et al. A homozygous CARD9 mutation in a family with susceptibility to fungal infections. *N Engl J Med* **361**, 1727–1735 (2009).
26. Netea, M.G. et al. Toll-like receptor 2 suppresses immunity against *Candida albicans* through induction of IL-10 and regulatory T cells. *J Immunol* **172**, 3712–3718 (2004).
27. Sutmuller, R.P. et al. Toll-like receptor 2 controls expansion and function of regulatory T cells. *J Clin Invest* **116**, 485–494 (2006).
28. Kontoyiannis, D.P. et al. Prospective surveillance for invasive fungal infections in hematopoietic stem cell transplant recipients, 2001–2006: Overview of the Transplant-Associated Infection Surveillance Network (TRANSNET) Database. *Clin Infect Dis* **50**, 1091–1100 (2010).
29. Hohl, T.M. and Feldmesser, M. *Aspergillus fumigatus*: Principles of pathogenesis and host defense. *Eukaryot Cell* **6**, 1953–1963 (2007).
30. Schaffner, A., Douglas, H., and Braude, A. Selective protection against conidia by mononuclear and against mycelia by polymorphonuclear phagocytes in resistance to *Aspergillus*. Observations on these two lines of defense in vivo and in vitro with human and mouse phagocytes. *J Clin Invest* **69**, 617–631 (1982).
31. Mircescu, M.M., Lipuma, L., van Rooijen, N., Pamer, E.G., and Hohl, T.M. Essential role for neutrophils but not alveolar macrophages at early time points following *Aspergillus fumigatus* infection. *J Infect Dis* **200**, 647–656 (2009).
32. Romani, L. Dendritic cells in *Aspergillus* infection and allergy. In *Aspergillus fumigatus and Aspergillosis* (eds. Latgé, J.-P. and Steinbach, W.J.), pp. 247–261 (ASM Press, Washington, DC, 2009).

33. Bozza, S. et al. Vaccination of mice against invasive aspergillosis with recombinant *Aspergillus* proteins and CpG oligodeoxynucleotides as adjuvants. *Microbes Infect* **4**, 1281–1290 (2002).
34. Brieland, J.K. et al. Cytokine networking in lungs of immunocompetent mice in response to inhaled *Aspergillus fumigatus*. *Infect Immun* **69**, 1554–1560 (2001).
35. Gafa, V. et al. Human dendritic cells following *Aspergillus fumigatus* infection express the CCR7 receptor and a differential pattern of interleukin-12 (IL-12), IL-23, and IL-27 cytokines, which lead to a Th1 response. *Infect Immun* **74**, 1480–1489 (2006).
36. Rivera, A., Van Epps, H.L., Hohl, T.M., Rizzuto, G., and Pamer, E.G. Distinct CD4+-T-cell responses to live and heat-inactivated *Aspergillus fumigatus* conidia. *Infect Immun* **73**, 7170–7179 (2005).
37. Mosmann, T.R. and Coffman, R.L. TH1 and TH2 cells: Different patterns of lymphokine secretion lead to different functional properties. *Annu Rev Immunol* **7**, 145–173 (1989).
38. Werner, J.L. et al. Requisite role for the dectin-1 beta-glucan receptor in pulmonary defense against *Aspergillus fumigatus*. *J Immunol* **182**, 4938–4946 (2009).
39. Chai, L.Y. et al. Anti-*Aspergillus* human host defence relies on type 1 T helper (Th1), rather than type 17 T helper (Th17), cellular immunity. *Immunology* **130**, 46–54 (2010).
40. Montagnoli, C. et al. Immunity and tolerance to *Aspergillus* involve functionally distinct regulatory T cells and tryptophan catabolism. *J Immunol* **176**, 1712–1723 (2006).
41. Antachopoulos, C. and Roilides, E. Cytokines and fungal infections. *Br J Haematol* **129**, 583–596 (2005).
42. Hebart, H. et al. Analysis of T-cell responses to *Aspergillus fumigatus* antigens in healthy individuals and patients with hematologic malignancies. *Blood* **100**, 4521–4528 (2002).
43. Roilides, E., Blake, C., Holmes, A., Pizzo, P.A., and Walsh, T.J. Granulocyte-macrophage colony-stimulating factor and interferon-gamma prevent dexamethasone-induced immunosuppression of antifungal monocyte activity against *Aspergillus fumigatus* hyphae. *J Med Vet Mycol* **34**, 63–69 (1996).
44. Schaffner, A. Therapeutic concentrations of glucocorticoids suppress the antimicrobial activity of human macrophages without impairing their responsiveness to gamma interferon. *J Clin Invest* **76**, 1755–1764 (1985).
45. Barnes, P.J. Anti-inflammatory actions of glucocorticoids: Molecular mechanisms. *Clin Sci (Lond)* **94**, 557–572 (1998).
46. Tomblyn, M. et al. Guidelines for preventing infectious complications among hematopoietic cell transplant recipients: A global perspective. Preface. *Bone Marrow Transplant* **44**, 453–455 (2009).
47. Casadevall, A. Antibody immunity and invasive fungal infections. *Infect Immun* **63**, 4211–4218 (1995).
48. Cutler, J.E. Defining criteria for anti-mannan antibodies to protect against candidiasis. *Curr Mol Med* **5**, 383–392 (2005).
49. Matthews, R.C. et al. Preclinical assessment of the efficacy of mycograb, a human recombinant antibody against fungal HSP90. *Antimicrob Agents Chemother* **47**, 2208–2216 (2003).
50. Matthews, R.C. and Burnie, J.P. Recombinant antibodies: A natural partner in combinatorial antifungal therapy. *Vaccine* **22**, 865–871 (2004).
51. Pachl, J. et al. A randomized, blinded, multicenter trial of lipid-associated amphotericin B alone versus in combination with an antibody-based inhibitor of heat shock protein 90 in patients with invasive candidiasis. *Clin Infect Dis* **42**, 1404–1413 (2006).
52. Torosantucci, A. et al. A novel glyco-conjugate vaccine against fungal pathogens. *J Exp Med* **202**, 597–606 (2005).
53. Moragues, M.D. et al. A monoclonal antibody directed against a *Candida albicans* cell wall mannoprotein exerts three anti-*C. albicans* activities. *Infect Immun* **71**, 5273–5279 (2003).
54. Magliani, W. et al. Engineered killer mimotopes: New synthetic peptides for antimicrobial therapy. *Curr Med Chem* **11**, 1793–1800 (2004).
55. Han, Y., Ulrich, M.A., and Cutler, J.E. *Candida albicans* mannan extract-protein conjugates induce a protective immune response against experimental candidiasis. *J Infect Dis* **179**, 1477–1484 (1999).
56. Kavishwar, A. and Shukla, P.K. Candidacidal activity of a monoclonal antibody that binds with glycosyl moieties of proteins of *Candida albicans*. *Med Mycol* **44**, 159–167 (2006).
57. Sandini, S., La Valle, R., De Bernardis, F., Macri, C., and Cassone, A. The 65 kDa mannoprotein gene of *Candida albicans* encodes a putative beta-glucanase adhesin required for hyphal morphogenesis and experimental pathogenicity. *Cell Microbiol* **9**, 1223–1238 (2007).
58. Luo, G. et al. Candida albicans Hyr1p confers resistance to neutrophil killing and is a potential vaccine target. *J Infect Dis* **201**, 1718–1728 (2010).

59. Bar, E. et al. A novel Th cell epitope of *Candida albicans* mediates protection from fungal infection. *J Immunol* **188**, 5636–5643 (2012).
60. Xin, H. et al. Self-adjuvanting glycopeptide conjugate vaccine against disseminated candidiasis. *PLoS One* **7**, e35106 (2012).
61. Han, Y. and Cutler, J.E. Antibody response that protects against disseminated candidiasis. *Infect Immun* **63**, 2714–2719 (1995).
62. Han, Y., Kanbe, T., Cherniak, R., and Cutler, J.E. Biochemical characterization of *Candida albicans* epitopes that can elicit protective and nonprotective antibodies. *Infect Immun* **65**, 4100–4107 (1997).
63. Xin, H., Dziadek, S., Bundle, D.R., and Cutler, J.E. Synthetic glycopeptide vaccines combining beta-mannan and peptide epitopes induce protection against candidiasis. *Proc Natl Acad Sci USA* **105**, 13526–13531 (2008).
64. Hoyer, L.L., Scherer, S., Shatzman, A.R., and Livi, G.P. Candida albicans ALS1: Domains related to a Saccharomyces cerevisiae sexual agglutinin separated by a repeating motif. *Mol Microbiol* **15**(1), 39–54 (1995).
65. Fu, Y. et al. *Candida albicans* Als1p: An adhesin that is a downstream effector of the EFG1 filamentation pathway. *Mol Microbiol* **44**, 61–72 (2002).
66. Loza, L. et al. Functional analysis of the *Candida albicans* ALS1 gene product. *Yeast* **21**, 473–482 (2004).
67. Ibrahim, A.S. et al. Vaccination with recombinant N-terminal domain of Als1p improves survival during murine disseminated candidiasis by enhancing cell-mediated, not humoral, immunity. *Infect Immun* **73**, 999–1005 (2005).
68. Spellberg, B.J. et al. The anti-*Candida albicans* vaccine composed of the recombinant N terminus of Als1p reduces fungal burden and improves survival in both immunocompetent and immunocompromised mice. *Infect Immun* **73**, 6191–6193 (2005).
69. Ibrahim, A.S., Spellberg, B.J., Avanesian, V., Fu, Y., and Edwards, J.E., Jr. The anti-*Candida* vaccine based on the recombinant N-terminal domain of Als1p is broadly active against disseminated candidiasis. *Infect Immun* **74**, 3039–3041 (2006).
70. Spellberg, B.J. et al. Efficacy of the anti-*Candida* rAls3p-N or rAls1p-N vaccines against disseminated and mucosal candidiasis. *J Infect Dis* **194**, 256–260 (2006).
71. Sheppard, D.C. et al. Functional and structural diversity in the Als protein family of *Candida albicans*. *J Biol Chem* **279**, 30480–30489 (2004).
72. Spellberg, B.J. et al. The antifungal vaccine derived from the recombinant N terminus of Als3p protects mice against the bacterium *Staphylococcus aureus*. *Infect Immun* **76**, 4574–4580 (2008).
73. Phan, Q.T. et al. Als3 is a *Candida albicans* invasin that binds to cadherins and induces endocytosis by host cells. *PLoS Biol* **5**, e64 (2007).
74. Lin, L. et al. Immunological surrogate marker of rAls3p-N vaccine-induced protection against *Staphylococcus aureus*. *FEMS Immunol Med Microbiol* **55**, 293–295 (2009).
75. Schmidt C. S. et al. NDV-3, a recombinant alum-adjuvanted vaccine for Candida and Staphlococcus aureus is safe and immunogeneic in healthy adults. *Vaccine* **30**, 7594–7600 (2012).
76. Ito, J.I. et al. Vaccinations with recombinant variants of *Aspergillus fumigatus* allergen Asp f 3 protect mice against invasive aspergillosis. *Infect Immun* **74**, 5075–5084 (2006).
77. Ito, J.I., Lyons, J.M., Diaz-Arevalo, D., Hong, T.B., and Kalkum, M. Vaccine progress. *Med Mycol* **47**(Suppl. 1), S394–S400 (2009).
78. Bozza, S. et al. Immune sensing of *Aspergillus fumigatus* proteins, glycolipids, and polysaccharides and the impact on Th immunity and vaccination. *J Immunol* **183**, 2407–2414 (2009).
79. Bowyer, P. and Denning, D.W. Genomic analysis of allergen genes in *Aspergillus* spp.: The relevance of genomics to everyday research. *Med Mycol* **45**, 17–26 (2007).
80. Schutte, M. et al. Identification of a putative Crf splice variant and generation of recombinant antibodies for the specific detection of *Aspergillus fumigatus*. *PLoS One* **4**, e6625 (2009).
81. Chaudhary, N., Staab, J.F., and Marr, K.A. Healthy human T-cell responses to *Aspergillus fumigatus* antigens. *PLoS One* **5**, e9036 (2010).
82. Chaudhuri, R., Ansari, F.A., Raghunandanan, M.V., and Ramachandran, S. FungalRV: Adhesin prediction and immunoinformatics portal for human fungal pathogens. *BMC Genomics* **12**, 192 (2011).
83. Asif, A.R., Oellerich, M., Amstrong, V.W., Gross, U., and Reichard, U. Analysis of the cellular *Aspergillus fumigatus* proteome that reacts with sera from rabbits developing an acquired immunity after experimental aspergillosis. *Electrophoresis* **31**, 1947–1958 (2010).
84. Park, B.J. et al. Invasive non-*Aspergillus* mold infections in transplant recipients, United States, 2001–2006. *Emerg Infect Dis* **17**, 1855–1864 (2011).

85. Vilanova, M. et al. Protection against systemic candidiasis in mice immunized with secreted aspartic proteinase 2. *Immunology* **111**, 334–342 (2004).
86. Delgado, N., Xue, J., Yu, J.J., Hung, C.Y., and Cole, G.T. A recombinant beta-1,3-glucanosyltransferase homolog of *Coccidioides posadasii* protects mice against coccidioidomycosis. *Infect Immun* **71**, 3010–3019 (2003).
87. Stuehler, C. et al. Cross-protective TH1 immunity against *Aspergillus fumigatus* and *Candida albicans*. *Blood* **117**, 5881–5891 (2011).
88. Torosantucci, A. et al. Protection by anti-beta-glucan antibodies is associated with restricted beta-1,3 glucan binding specificity and inhibition of fungal growth and adherence. *PLoS One* **4**, e5392 (2009).
89. Stevens, D.A., Clemons, K.V., and Liu, M. Developing a vaccine against aspergillosis. *Med Mycol* **49**(Suppl. 1), S170–S176 (2011).
90. Champer, J. et al. Protein targets for broad-spectrum mycosis vaccines: Quantitative proteomic analysis of *Aspergillus* and *Coccidioides* and comparisons with other fungal pathogens. *Ann N Y Acad Sci* **1273**(1), 44–51 (2012).
91. Bozza, S. et al. A dendritic cell vaccine against invasive aspergillosis in allogeneic hematopoietic transplantation. *Blood* **102**, 3807–3814 (2003).

35 Fungi in Drinking Water

Ida Skaar and Gunhild Hageskal

CONTENTS

35.1 Diversity of Mold Species in Drinking Water 597
35.2 Significance of Mold Species for Water Quality and Human Health 599
35.3 Drinking Water Fungi Associated with Allergy 599
35.4 Mycotoxin-Producing Fungi in Drinking Water 600
35.5 Drinking Water Fungi Asssociated with Mycoses 600
35.6 Source and Distribution of Mold Species in Drinking Water 601
35.7 Methods for Fungal Analyses of Drinking Water 601
35.8 Guidelines and Microbiological Analyses of Drinking Water 603
35.9 Conclusions 603
References 604

Molds are receiving increased attention as agents of human health problems. Although several studies have shown that distribution systems regularly contain molds, the significance of molds in drinking water has been paid only modest attention. However, it has been established that water may be a possible dissemination route for potentially pathogenic, toxigenic, and allergenic mold species. The link between waterborne molds and human health is still not fully understood, but it is important to be aware that species of clinical concern in immunosuppressed individuals are also present in drinking water. Molds may be aerosolized in indoor air when water passes installations such as taps or showers. Elevated levels of certain mold species may therefore constitute a potential health risk in susceptible groups. For these reasons, the mycobiota of water should be considered when microbiological safety and quality of drinking water are assessed.

35.1 DIVERSITY OF MOLD SPECIES IN DRINKING WATER

Various investigations have established the presence of molds in water distribution systems and have indicated that water may disseminate a huge diversity of potentially allergenic, toxigenic, and opportunistic fungal species to hospitals and private homes.[1–18] Table 35.1 summarizes mold genera demonstrated in drinking water in some studies. Nagy[2] found that the four most frequently occurring genera in Canadian chlorinated and unchlorinated drinking water were *Penicillium*, *Periconia*, *Acremonium*, and *Paecilomyces*. The mean number of filamentous fungal colony-forming units (CFUs) per 100 mL of drinking water was almost twice as high in the chlorinated system compared to the unchlorinated system. In drinking water from the East Coast of the United States, *Aspergillus* was found to be the genus most frequently isolated, but also *Penicillium*, *Cladosporium*, and *Alternaria* were commonly demonstrated.[6] In drinking water in Nevada, *Cladosporium*, *Phoma*, and *Alternaria* were the top three genera found.[7] Anaissie and Costa claimed that nosocomial aspergillosis is waterborne and reported the rank-order distribution of aspergilli concentration gradient by species to be *Aspergillus niger*, *A. fumigatus*, and *A. flavus*, in descending order.[12] In another study by Anaissie and colleagues, *Fusarium* species was recovered from 57% of hospital water system samples from Arkansas, with *Fusarium solani* and *F. oxysporum* as the most common.[13] Hinzelin and Block[19] found that the genera *Penicillium*,

TABLE 35.1
Results from Mycological Analyses of Drinking Water Given as Percent of Samples Where the Genera Were Demonstrated

	Chlorinated Drinking Water				Unchlorinated Drinking Water	Chlorinated Drinking Water
	Hageskal et al. (Norway)[18]	Arvanitidou et al. (Greece)[11]	Frankova and Horecka (Slovakia)[10]	Hinzelin and Block (France)[5]	Åkerstrand (Sweden)[4]	Nagy and Olson (California)[2]
Number of Samples	273	126	140	38	42	32
Acremonium	4.4	8.0	3,8	2,3	33	15
Alternaria	ND	1.6	9.4	ND	ND	1.7
Aspergillus	16.8	42.1	9.4	16.3	5	1.1
Beauveria	6.6	ND	ND	ND	ND	ND
Chrysosporium	0.4	7.2	ND	ND	ND	ND
Cladosporium	8.8	1.8	39.6	4.6	31	ND
Exophiala	ND	0.8	ND	ND	ND	ND
Fusarium	2.2	0.8	17.0	ND	17	1.8
Microsporum	ND	ND	D	ND	ND	ND
Mucor	6.2	0.8	ND	7	14	ND
Paecilomyces	6.2	ND	15.1	ND	17	10
Penicillium	66.3	50.8	35.8	23.3	65	27.5
Phialophora	11.7	0.8	ND	ND	69	ND
Phoma	14.3	ND	ND	ND	29	ND
Rhizopus	ND	3.2	D	13.9	ND	ND
Scopulariopsis	0.7	4.0	ND	ND	2	ND
Stachybotrys	ND	0.8	ND	ND	ND	ND
Trichoderma	26.7	5.6	D	11.6	21	ND
Trichothecium	ND	4.8	ND	ND	ND	ND
Trichophyton	ND	ND	3.8	ND	ND	ND
Verticillium	0.7	4.0	0.7	ND	19	ND
Number of genera demonstrated	30	27	39	7	17	19
Number of species demonstrated	94	Not reported	64	Not reported	Not reported	Not reported
Mean CFU/100 mL	9	36.6	Not reported	2.1	40	18 (unchlorinated) 34 (chlorinated)

CFU, colony-forming units; ND, not detected; D, detected.

Aspergillus, and *Rhizopus* represented more than 50% of the isolated fungal strains from French drinking water. In drinking water in Bratislava were *Cladosporium*, *Penicillium*, *Fusarium*, and *Paecilomyces* reported as the most common genera.[8,10] In a Greek study, *Penicillium* and *Aspergillus* were isolated from 64% and 53% of the samples, respectively.[11] From groundwater-derived public water in Germany, *Phialophora* was reported as the by far most demonstrated fungal genus, followed by *Acremonium* and *Exophiala*.[14] In drinking water in Portugal were *Penicillium* and *Acremonium* reported as the most frequently isolated fungi. There was an interesting difference in the pattern of isolation of the key taxa with season: penicillia predominated in early summer and *Acremonium* in winter.[17] Kelley and colleagues[15] reported melanized, thick-walled fungal species to predominate after water treatment. In a Turkish study, sixteen species of fungi were isolated, and *Penicillium*, *Aspergillus*, and *Acremonium* were reported to be the genera most frequently isolated.[16] Niemi et al.[3] reported that mesophilic fungi, which were not further identified, were common in Finnish drinking water, whereas thermophilic fungi were present only in low concentrations. This was later confirmed by Zacheus and Martikainen.[9] From Sweden, Åkerstrand[4] reported *Phialophora*, *Penicillium*, *Acremonium*, and *Cladosporium* as the most frequently found fungal genera. Ormerod[1] reported high numbers of fungi in Norwegian drinking water, but no identification of the fungi was performed. Later, Hageskal and colleagues identified 94 mold species,[18] later increased to 106 species,[20] belonging to 30 genera from Norwegian drinking water. The mycobiota was dominated by species of *Penicillium*, *Trichoderma*, and *Aspergillus*, with some of them occurring throughout the drinking water system.[18]

35.2 SIGNIFICANCE OF MOLD SPECIES FOR WATER QUALITY AND HUMAN HEALTH

Several of the species identified in drinking water may have the potential to cause allergies or infections in humans if susceptible individuals are exposed, and several of the molds are potentially toxin producers. The genera *Aspergillus* and *Penicillium* include species of most importance in terms of pathogenicity, allergic potential, and mycotoxin production. Since these genera were among the most frequently isolated from drinking water,[2,4,5,10,11,18] the likelihood of water as a dissemination route for problem molds is significant. Additionally, other demonstrated genera in drinking water, such as *Absidia*, *Acremonium*, *Mucor*, *Paecilomyces*, and *Trichoderma*,[18,21] also include potentially pathogenic species.

35.3 DRINKING WATER FUNGI ASSOCIATED WITH ALLERGY

The implication of mold species in allergy, asthma, or other respiratory problems has been a subject for several studies worldwide. Various mold genera are suspected of being important causative factors in the growing epidemic of allergy and asthma observed among humans.[22–26] The evidence linking molds such as *Aspergillus*, *Cladosporium*, and *Penicillium* species with severe asthma has been found to be strong.[27] The role of *Penicillium* and *Aspergillus* species in sick building syndromes was reviewed by Schwab and Straus.[28] They concluded that these genera play a major role in causing many of the allergic and respiratory symptoms seen in humans. Several of the same genera and species as recovered from water have been associated with similar symptoms in water-damaged buildings and mold-contaminated indoor environments.[27–29] In addition, allergic reactions and respiratory effects caused by mold-contaminated water have been indicated.[30–32]

Outbreaks of skin irritations experienced in association with taking showers and baths were reported from Råbacka, Sweden.[32] Water samples recovered from 77 up to 3100 CFU/100 mL of *Phialophora richardsiae* and the mold-contaminated waters were suggested to be the causative factor for these skin problems. A case report from Finland indicated that elevated levels of *Aureobasidium pullulans* in water used in a home sauna were causing hypersensitivity pneumonitis, and the symptoms were described as sauna taker's disease.[30] Another case report from Finland was reported by Muittari et al.[31] More than 100 people showed symptoms similar to hypersensitivity pneumonitis after taking saunas, hot bath, or showers. Symptoms were even reported after laundering and dishwashing. The following investigation indicated that the epidemic was caused by water contaminated by *A. fumigatus*, *Mucor*, *Absidia*, and *Candida*.

Finally, elevated levels of potentially allergenic mold species may become a problem. The possibility that fungi in water may cause allergic reactions or respiratory problems in susceptible individuals is valid, although the knowledge about such correlations is limited and requires epidemiological studies.

35.4 MYCOTOXIN-PRODUCING FUNGI IN DRINKING WATER

Several of the mold species demonstrated in drinking water[18,20] are known as potent mycotoxin producers. Potentially aflatoxin-producing *A. flavus* strains were isolated from a cold water storage tank, and aflatoxins were detected in the water samples in a study from England.[33] It was demonstrated that *Fusarium graminearum* is able to produce zearalenone in water.[15,34,35] Interestingly, Gromadzka[36] and colleagues reported that the highest zearalenone concentration in water samples from Poland was detected at a time of reduced biological activity of fungi responsible for the biosynthesis of this toxin. Another study indicated *Penicillium citrinum* strains producing the toxin citrinin in bottled mineral water.[37] Mycotoxins can induce toxic responses in vertebrates or other animals when fed at low concentrations. Mycotoxins produced in water will of course be extremely diluted and would perhaps be of minor concern. However, in some occasions, water is stored in cisterns or reservoirs for prolonged periods, or water may be stagnated in piping systems, so that concentrations of mycotoxins may increase. Water is consumed in large amounts every day, and small amounts of mycotoxins over several years could possibly have negative effect on the immune system, in addition to all the other factors affecting human health. The mycotoxin-producing potential of the species isolated from drinking water should be further investigated.

35.5 DRINKING WATER FUNGI ASSSOCIATED WITH MYCOSES

It is unlikely that the occurrence of molds in water, at the concentrations observed in these studies, would cause mycoses in healthy individuals. On the other hand, if the appropriate conditions are present and regrowth of molds occurs in the water systems, exposure of humans to large amounts of potentially harmful mold species could occur.

The focus on molds as agents of infections in immunocompromised individuals, such as AIDS, cancer, or transplant recipient patients that undergo hospital treatment, is increasing.[38-42] Costs in use of antifungal medicine have exploded in many hospitals recently. The treatment is extensive and often long term, and some fungal species are resistant to standard antifungal agents. The mortality rates of fungal infections are high, in some cases exceeding 90%.[43]

Several attempts to determine the source of these mold infections have been conducted.[44-47] Samples were collected from hospital environments, but clear links between the environmental source and the patients are still lacking, although it is generally assumed that the primary route of mold infections, such as aspergillosis, is inhalation of conidia from contaminated indoor air. However, installation of air filters has unaffected infections, and so the search for other sources is required.

Water has also been investigated as a source of infection, and two studies have reported genetic relatedness between water-related and clinical mold strains.[48,49] Anaissie et al.[48] showed that an isolate of *A. fumigatus* recovered from a patient with aspergillosis was genotypically identical to an isolate recovered from the shower wall in the patient's room. Warris et al.[49] proved that genotypic relatedness between clinical and environmental isolates suggests that patients with invasive aspergillosis can be infected by strains originating from water. Clinical isolates recovered from patients matched those recovered from water sources in two clusters.

Conidia or small hyphal fragments in water are also able to aerosolize when the water passes through installations such as taps and showers.[12] However, results on a limited number of strains do not allow definite identification of the source of contamination. A single source of infection may not be possible to identify; thus, multiple sources, including water, should be considered in each case. Because epidemiological evidence is lacking, water cannot be excluded as a possible source of mold infections.

Species identification in Norwegian studies has demonstrated that clinically concerning mold species, such as *Aspergillus ustus* and *A. fumigatus*, are present and able to multiply in the water systems.[18,21,50] If such molds become aerosolized into the indoor air when water passes through water installations such as taps and showers, water may be considered as a possible reservoir for potentially harmful mold species. An outbreak of invasive infections caused by *A. ustus* was reported, where a common source of infections was suggested.[51] Clinical isolates from six patients were genetically similar, but an exact source of infection was not determined. Analyses of the water system were merited in this outbreak, since it may have provided a common source of infection of all patients. Since *Aspergillus* species in *section Usti* seems to be able to establish in heated water installations,[18,50] the hospital warming tank could have been a possible source for the fungal infections, and water samples should be investigated.

Air levels of *Fusarium* and *Aspergillus* were observed to increase in hospital environments after running showers multiple times.[48,52] Several studies have suggested that respiratory exposure to aerosolized molds from water may occur,[12,52,53] and genetic relatedness between clinical mold strains and water-related mold strains has been indicated.[48,49] It has also been recommended that hospitalized patients at high risk for infections should avoid exposure to hospital water and rather use sterile water, not only because of fungal contamination of the water but also due to the risk of other waterborne pathogens such as *Pseudomonas aeruginosa*, *Aeromonas hydrophila*, and *Mycobacterium* spp.[54]

35.6 SOURCE AND DISTRIBUTION OF MOLD SPECIES IN DRINKING WATER

Hageskal et al.[18,21] found that the risk to recover molds is higher in surface-sourced than groundsourced water. Water sampled from cold water taps more often had molds than the showers and hot water tap samples,[21] which is congruent with the results of a study conducted on hot and cold water in Finland.[9] Surprisingly, high mold numbers were observed in the groundwater-derived hot tap samples compared to the surface-derived hot taps, indicating that thermotolerant species, especially *Aspergillus calidoustus*,[18,50] established in these facilities.[21] Distribution trends of molds at the different sampling points in the water systems suggested that a larger part of the samples from hospital installations than from private home installations were positive for molds.[21] Although no statistically significant results were obtained, this trend leads to possible explanations. Hospitals often have long and complex water pipe systems, with blind ends where water may stagnate, which provide excellent conditions for molds.

The water source may therefore not be the only contributor to mold contamination of the water reaching taps and showers. The levels of molds seemed to increase during transportation in the groundwater-sourced distribution system.[21] This may be explained by leakages of soil and water into the water pipes combined with biofilm formation. Molds established in biofilms could serve as reservoirs of conidia. Additionally, fragments of biofilm may be released resulting in increased contamination. Biofilm formation could also be the case in the surface water systems, where the mold levels had a tendency to be sustained in the distribution system, although water treatment and the heating of water had some effect in reducing the mold levels.[18,21]

35.7 METHODS FOR FUNGAL ANALYSES OF DRINKING WATER

A critical point with respect to the study of fungi in water is how the analyses are performed. Unfortunately, there is no international standard method described for analyses of fungi in drinking water; hence, several isolation procedures are used in various studies. The choice of method is naturally dependent of the purpose of the analyses. If the aim is to investigate water as a source for a clinical fungal strain, the medium should be as selective as possible and the incubation temperature should be 37°C. On the other hand, if the objective is to survey the fungi in water, methods that

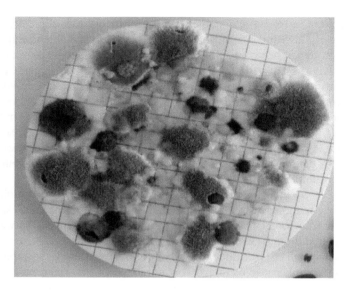

FIGURE 35.1 Membrane filter with mold recovery on DG18 medium after filtration of 100 mL water and incubation for one week at 20°C. The water sample was collected from surface-derived raw water.

allow a broad diversity of fungi to grow should be employed. However, we have to keep in mind that regardless of the method used, there will always be fungi that will go undetected.

Different isolation procedures may result in different detection limits of fungi in the water. The most commonly used methods are membrane filter techniques (Figure 35.1), applying volumes of 10–1000 mL water.[1–3,5,9,11,15–17,21,48,52,55–59] Another technique performed is direct plate spread with volumes of 0.1–1 mL water.[8,10,14] Centrifugation of water to obtain the fungal propagules or direct microscope observations are also applicable methods.[60]

The isolation medium also varies between the investigations performed. High nutritional media are often used, such as Sabouraud dextrose agar,[16] Sabouraud glucose agar,[58] and malt extract agar.[3] Others use media that inhibit overgrowth of fast-growing fungi, such as dichloran 18% glycerol agar (DG18).[21] DG18 is now recommended as a general medium for the isolation and enumeration of fungi in foods with high water activity (aw > 0.90)[61] and may, therefore, also be suitable for water analyses. Half-strength corn meal agar (CMA/2) is an example of the use of a low nutritional medium.[17] The use of different media may result in selectivity toward some fungal species and loss of others. Kinsey and colleagues[62] recommended a combination of both high and low nutritional media, to obtain as broad cross section of fungi present in water as possible.

The incubation temperature is also important for what fungal species are obtained. The incubation temperatures range from 20°C to 37°C in the studies performed and will consequently result in the recovery of different genera and species.

In order to assess the contribution of fungi to problems in drinking water, the quantification is essential. Quantification of fungi in water is mostly assessed by counting total fungal colonies on agar plates, referred to as CFUs per volume of water sample investigated.[60] Due to the nature of filamentous fungi, the data generated in this way may be difficult to interpret. Fungi have a tendency to be unevenly distributed in water, which may cause difficulty in ensuring representative samples. A frequent problem in quantifying fungi is overgrowth of the filter, which can occur very quickly. Therefore, it is very important to monitor the plates and remove new colonies as soon as they appear. Another problem is that not all fungi are able to grow under laboratory conditions. In consequence, the currently used quantification methods will never provide the precise number of fungi but may give a fair indication of the level of fungi present in water. Chemical methods, such as analyses of ergosterol for quantification of total fungal biomass, as suggested by Kelley and Paterson,[63] could be useful as supplemental methods.

Identification to the species level should preferably be done, at least for the most dominant species. The identification of fungi has mainly been based on morphological identification keys. This identification is subjective, relying on the knowledge and experience of the individual researcher, and also on good and updated identification keys. A frequent problem is that some fungi never sporulate, and thus cannot be identified morphologically. DNA sequencing is not dependent on sporulating fungi or even viable fungi and is a good supplement to morphological identification. Hence, this method relies on the accuracy of public databases, such as GenBank. Analyses of fungal metabolites could also contribute to ensure correct identification. In many of the studies performed, the fungal colonies have only been identified to genus rather to the species level. This makes the consideration of concerning species difficult, as species of the same genera may have very different characters with respect to, for example, toxin production, pathogenicity, and allergenic potential.

In consequence, these limits in the methodology make comparisons between the different studies difficult and may explain much of the variation in the results obtained. The methods for analyzing fungi in drinking water should thus be standardized.

35.8 GUIDELINES AND MICROBIOLOGICAL ANALYSES OF DRINKING WATER

The purpose of monitoring the microbiological quality of drinking water is to obtain a basis for judging related health risk and implementing, when necessary, adequate protective measures. However, there are few guidelines regarding what is considered as normal or acceptable levels of fungi in water. Limited information is given in the literature regarding this issue. From the mold problem case in Råbacka, Sweden,[32] one of the concluding comments was that mold levels of 1000 CFUs/L in water can be compared to those levels found in air in certain working environments (10^6 CFU/m^3), which are known to give immune response by exposure. Pursuant to the Swedish drinking water regulations,[64] the limit of microfungi in water before esthetical and technical impact is 100 CFU/100 mL sample. Hageskal et al.[18] argue that this limit is too high and claim that CFU counts alone may not be sufficient for the mold analyses of water; the species identity and prevalence should also be established, because levels far less than 100 CFU/100 mL may reduce the water quality. If the mycobiota in a water sample constitute one single or a few species, 100 CFU/100 mL water could indicate regrowth in the distribution system and exposure of humans to large amounts of potentially harmful species. Since the levels of molds vary considerably from one sampling to the next, the mean CFU of molds may perhaps not give completely satisfactory information alone. The range (minimum and maximum) levels of molds provided good additional indication of the mold levels in the water and should be considered in addition to the mean. Species identification is also of most importance in this regard, because elevated levels of potentially harmful species could constitute a greater potential health risk.

35.9 CONCLUSIONS

Water distribution systems may serve as a dissemination route of a wide diversity of mold species to private homes and hospitals. The chance to recover molds is higher in surface-sourced water than in ground-sourced water. It is more likely to recover molds from cold water and showers than from hot water, although some thermotolerant species may establish in hot water facilities.[18,21,50]

The health risk related to mold species in drinking water is still not fully understood. Nevertheless, several studies have demonstrated that the same species that are of clinical concern are also present in water.[4,11–13,18,48,50,52–54,65] Drinking water may be a possible source of fungi in indoor environments. Water should be investigated in the same way as air, dust, and indoor surfaces, in the case of mold problems. Since molds are able to aerosolize when water passes through water installations such as taps and showers, the mold content of water should possibly be kept under surveillance in hospital environments where high-risk patients are undergoing treatment.

Several of the genera recovered from water are reported to cause allergic reactions or asthma when humans are exposed to high levels of these genera in mold-contaminated indoor air.[22–27,30,31] Problem species may establish and multiply in biofilms in the water system and result in elevated levels of molds. Aerosolized molds may be inhaled from indoor air and enhance asthmatic or allergic problems in susceptible individuals.

Further perspectives regarding fungi in drinking water should aim to perform epidemiological studies to investigate the impact of waterborne fungi on human health, both with respect to fungal infections and allergies. Management of molds in distribution systems should also be studied, since the current water treatment obviously is not sufficient against fungi. Based on further research, the analyses of fungi should possibly be included in drinking water regulations. The role of molds in biofilms is poorly understood and should be further studied.

REFERENCES

1. Ormerod, K.S. Heterotrophic microorganisms in distribution systems for drinking water. *Vatten* **43**, 262–268 (1987).
2. Nagy, L.A. and Olson, B.H. The occurrence of filamentous fungi in drinking water distribution systems. *Can J Microbiol* **28**, 667–671 (1982).
3. Niemi, R.M., Knuth, S., and Lundstrom, K. Actinomycetes and fungi in surface waters and in potable water. *Appl Environ Microbiol* **43**, 378–388 (1982).
4. Åkerstrand, K. Förekomst av mögelsvampar i dricksvatten. *Vår Föda* **36**, 320–326 (1984).
5. Hinzelin, F. and Block, J.C. Yeasts and filamentous fungi in drinking water. *Environ Technol Lett* **6**, 101–103 (1985).
6. Rosenzweig, W.D., Minnigh, H., and Pipes, W.O. Fungi in potable water distribution systems. *Am Water Works Assoc* **78**, 53–55 (1986).
7. West, P.R. Isolation rates and characterization of fungi in drinking water distribution systems. In *Proceedings of the Water Quality Technology Conference*, pp. 457–473 (Portland, OR, 1986).
8. Franková, E. Isolation and identification of filamentous soil Deuteromycetes from the water environment. *Biológica* **48**, 287–290 (1993).
9. Zacheus, O.M. and Martikainen, P.J. Occurrence of heterotrophic bacteria and fungi in cold and hot water distribution systems using water of different quality. *Can J Microbiol* **41**, 1088–1094 (1995).
10. Frankova, E. and Horecka, M. Filamentous soil fungi and unidentified bacteria in drinking water from wells and water mains near Bratislava. *Microbiol Res* **150**, 311–313 (1995).
11. Arvanitidou, M., Kanellou, K., Constantinides, T.C., and Katsouyannopoulos, V. The occurrence of fungi in hospital and community potable waters. *Lett Appl Microbiol* **29**, 81–84 (1999).
12. Anaissie, E.J. and Costa, S.F. Nosocomial aspergillosis is waterborne. *Clin Infect Dis* **33**, 1546–1548 (2001).
13. Anaissie, E.J. et al. Fusariosis associated with pathogenic *Fusarium* species colonization of a hospital water system: A new paradigm for the epidemiology of opportunistic mold infections. *Clin Infect Dis* **33**, 1871–1878 (2001).
14. Gottlich, E. et al. Fungal flora in groundwater-derived public drinking water. *Int J Hyg Environ Health* **205**, 269–279 (2002).
15. Kelley, J., Kinsey, G., Paterson, R., Brayford, D., Pitchers, R., and Rossmoore, H. *Identification and Control of Fungi in Distribution Systems*. AWWA Research Foundation and American Water Works Association (Denver, CO, 2003).
16. Hapcioglu, B., Yegenoglu, Y., Erturan, Z., Nakipoglu, Y., and Issever, H. Heterotrophic bacteria and filamentous fungi isolated from a hospital water distribution system. *Indoor and Built Environment* **14**, 487–493 (2005).
17. Goncalves, A.B., Paterson, R.R., and Lima, N. Survey and significance of filamentous fungi from tap water. *Int J Hyg Environ Health* **209**, 257–264 (2006).
18. Hageskal, G., Knutsen, A.K., Gaustad, P., de Hoog, G.S., and Skaar, I. Diversity and significance of mold species in Norwegian drinking water. *Appl Environ Microbiol* **72**, 7586–7593 (2006).

19. Hinzelin, F. and Block, J.C. Yeasts and filamentous fungi in drinking water. *Environ Technol Lett* **6**, 101–103 (1985).
20. Hageskal, G., Vralstad, T., Knutsen, A.K., and Skaar, I. Exploring the species diversity of Trichoderma in Norwegian drinking water systems by DNA barcoding. *Mol Ecol Resour* **8**, 1178–1188 (2008).
21. Hageskal, G., Gaustad, P., Heier, B.T., and Skaar, I. Occurrence of moulds in drinking water. *J Appl Microbiol* **102**, 774–780 (2007).
22. Jaakkola, M.S. et al. Indoor dampness and molds and development of adult-onset asthma: A population-based incident case-control study. *Environ Health Perspect* **110**, 543–547 (2002).
23. Green, B.J., Mitakakis, T.Z., and Tovey, E.R. Allergen detection from 11 fungal species before and after germination. *J Allergy Clin Immunol* **111**, 285–289 (2003).
24. Kauffman, H.F. and van der Heide, S. Exposure, sensitization, and mechanisms of fungus-induced asthma. *Curr Allergy Asthma Rep* **3**, 430–437 (2003).
25. Lugauskas, A., Krikstaponis, A., and Sveistyte, L. Airborne fungi in industrial environments—Potential agents of respiratory diseases. *Ann Agric Environ Med* **11**, 19–25 (2004).
26. Hogaboam, C.M., Carpenter, K.J., Schuh, J.M., and Buckland, K.F. *Aspergillus* and asthma—Any link? *Med Mycol* **43**(Suppl. 1), S197–S202 (2005).
27. Denning, D.W., O'Driscoll, B.R., Hogaboam, C.M., Bowyer, P., and Niven, R.M. The link between fungi and severe asthma: A summary of the evidence. *Eur Respir J* **27**, 615–626 (2006).
28. Schwab, C.J. and Straus, D.C. The roles of *Penicillium* and *Aspergillus* in sick building syndrome. *Adv Appl Microbiol* **55**, 215–238 (2004).
29. Cooley, J.D., Wong, W.C., Jumper, C.A., and Straus, D.C. Fungi and the indoor environment: their impact on human health. *Adv Appl Microbiol* **55**, 1–30 (2004).
30. Metzger, W.J., Patterson, R., Fink, J., Semerdjian, R., and Roberts, M. Sauna-takers disease. Hypersensitivity due to contaminated water in a home sauna. *J Am Med Assoc* **236**, 2209–2211 (1976).
31. Muittari, A. et al. An epidemic of extrinsic allergic alveolitis caused by tap water. *Clin Allergy* **10**, 77–90 (1980).
32. Åslund, P. Hudirritasjoner förorsakade av mikrosvampar. *Vår Föda* **36**, 327–336 (1984).
33. Paterson, R.R.M., Kelley, J., and Gallagher, M. Natural occurrence of aflatoxins and Aspergillus flavus (Link) in water. *Lett Appl Microbiol* **25**, 435–436 (1997).
34. Russell, R. and Paterson, M. Zearalenone production and growth in drinking water inoculated with *Fusarium graminearum*. *Mycol Prog* **6**, 109–113 (2007).
35. Paterson, R.R.M. Erratum to Zearalenone production and growth in drinking water inoculated with *Fusarium graminearum*. *Mycol Prog* **6**, 115 (2007).
36. Gromadzka, K., Waskiewicz, A., Golinski, P., and Swietlik, J. Occurrence of estrogenic mycotoxin—Zearalenone in aqueous environmental samples with various NOM content. *Water Res* **43**, 1051–1059 (2009).
37. Criado, M.V., Fernandez Pinto, V.E., Badessari, A., and Cabral, D. Conditions that regulate the growth of moulds inoculated into bottled mineral water. *Int J Food Microbiol* **99**, 343–349 (2005).
38. Gene, J. et al. Cutaneous infection caused by *Aspergillus ustus*, an emerging opportunistic fungus in immunosuppressed patients. *J Clin Microbiol* **39**, 1134–1136 (2001).
39. White, D.A. Aspergillus pulmonary infections in transplant recipients. *Clin Chest Med* **26**, 661–674, vii (2005).
40. Kauffman, C.A. Fungal infections. *Proc Am Thorac Soc* **3**, 35–40 (2006).
41. Nucci, M. and Anaissie, E. Emerging fungi. *Infect Dis Clin North Am* **20**, 563–579 (2006).
42. Marr, K.A., Carter, R.A., Crippa, F., Wald, A., and Corey, L. Epidemiology and outcome of mould infections in hematopoietic stem cell transplant recipients. *Clin Infect Dis* **34**, 909–917 (2002).
43. Pfaller, M.A., Pappas, P.G., and Wingard, J.R. Invasive fungal pathogens: Current epidemiological trends. *Clin Infect Dis* **43**, S3–S14 (2006).
44. Panagopoulou, P., Filioti, J., Farmaki, E., Maloukou, A., and Roilides, E. Filamentous fungi in a tertiary care hospital: Environmental surveillance and susceptibility to antifungal drugs. *Infect Control Hosp Epidemiol* **28**, 60–67 (2007).
45. Panagopoulou, P. et al. Environmental surveillance of filamentous fungi in three tertiary care hospitals in Greece. *J Hosp Infect* **52**, 185–191 (2002).
46. Guarro, J. et al. Use of random amplified microsatellites to type isolates from an outbreak of nosocomial aspergillosis in a general medical ward. *Med Mycol* **43**, 365–371 (2005).
47. Menotti, J. et al. Epidemiological study of invasive pulmonary aspergillosis in a haematology unit by molecular typing of environmental and patient isolates of *Aspergillus fumigatus*. *J Hosp Infect* **60**, 61–68 (2005).

48. Anaissie, E.J. et al. Pathogenic *Aspergillus* species recovered from a hospital water system: A 3-year prospective study. *Clin Infect Dis* **34**, 780–789 (2002).
49. Warris, A. et al. Molecular epidemiology of *Aspergillus fumigatus* isolates recovered from water, air, and patients shows two clusters of genetically distinct strains. *J Clin Microbiol* **41**, 4101–4106 (2003).
50. Hageskal, G., Kristensen, R., Fristad, R.F., and Skaar, I. Emerging pathogen *Aspergillus calidoustus* colonizes water distribution systems. *Med Mycol* **49**, 588–593 (2011).
51. Panackal, A.A., Imhof, A., Hanley, E.W., and Marr, K.A. *Aspergillus ustus* infections among transplant recipients. *Emerg Infect Dis* **12**, 403–408 (2006).
52. Warris, A. et al. Recovery of filamentous fungi from water in a paediatric bone marrow transplantation unit. *J Hosp Infect* **47**, 143–148 (2001).
53. Anaissie, E.J. et al. Pathogenic molds (including *Aspergillus* species) in hospital water distribution systems: A 3-year prospective study and clinical implications for patients with hematologic malignancies. *Blood* **101**, 2542–2546 (2003).
54. Anaissie, E.J., Penzak, S.R., and Dignani, M.C. The hospital water supply as a source of nosocomial infections: A plea for action. *Arch Intern Med* **162**, 1483–1492 (2002).
55. Doggett, M.S. Characterization of fungal biofilms within a municipal water distribution system. *Appl Environ Microbiol* **66**, 1249–1251 (2000).
56. Goncalves, A.B., Santos, I.M., Paterson, R.R., and Lima, N. FISH and Calcofluor staining techniques to detect in situ filamentous fungal biofilms in water. *Rev Iberoam Micol* **23**, 194–198 (2006).
57. Hendrickx, T.L., Meskus, E., and Keiski, R.L. Influence of the nutrient balance on biofilm composition in a fixed film process. *Water Sci Technol* **46**, 7–12 (2002).
58. Kanzler, D. et al. Occurrence and hygienic relevance of fungi in drinking water. *Mycoses* **51**, 165–169 (2008).
59. Varo, S.D. et al. Isolation of filamentous fungi from water used in a hemodialysis unit. *Rev Soc Bras Med Trop* **40**, 326–331 (2007).
60. Mara, D. and Horan, N. *The Handbook of Water and Wastewater Microbiology* (Elsevier Academic Press, London, U.K., 2006).
61. Samson, R.A., Hoekstra, E., and Frisvad, J.C. (eds.). *Introduction to Food- and Airborne Fungi*, p. 389 (Centraalbureau voor Schimmelcultures, Utrecht, the Netherlands, 2004).
62. Kinsey, G.C., Paterson, R.R., and Kelley, J. Methods for the determination of filamentous fungi in treated and untreated waters. *J Appl Microbiol* **85**(Suppl. 1), 214S–224S (1998).
63. Kelley, J. and Paterson, R.R.M. Comparisons of ergosterol to other methods for determination of *Fusarium graminearum* biomass in water as a model system. In *The 22nd European Culture Collections' Organization Meeting* (eds. Lia, N. and Smith, D.), pp. 235–239 (Micoteca da Universidade do Minho, Braga, Portugal, 2003).
64. Anonymous. Drinking water regulations. (National Food Administration, Uppsala, Sweden, 2001).
65. Warris, A., Voss, A., Abrahamsen, T.G., and Verweij, P.E. Contamination of hospital water with *Aspergillus fumigatus* and other molds. *Clin Infect Dis* **34**, 1159–1160 (2002).

Index

A

Absidia-like taxa classification, 355
Acetoxyscirpenediol, 404
Achaetomium, 212
    A. strumarium, 212, 216
    clinically significant, 217
    infections in humans, 213–215
Acremonium
    A. alabamense, 125
    A. alternatum, 115, 117, 122
    A. charticola, 118, 125
    A. collariferum, 125
    A. falciforme, 120
    A. pteridii, 124
    A. recifei, 121–122
    A. sclerotigenum, 120, 122–123
    A. spinosum, 124
    A. strictum, 117–118, 120
        genogroup *IV* sequence, 122
        onychomycosis, causal agent, 123
    in food, 118–120
    history, 115–117
    isolates
        food-related sources, 119
        sequencing of, 117
    in medicine, 120–125
    molecular analysis, 118
    molecular phylogenetic study, 116
    ribosomal internal transcribed spacer
        sequences, 118
Aflatoxicosis
    acute, 168–169
    chronic, 169
Aflatoxin B1 (AFB1), 64
Aflatoxins, 427, 600
    Aspergillus mycotoxins, 177–178
        acute aflatoxicosis, 168–169
        biosynthesis, 169–170
        chronic aflatoxicosis, 169
        molecular detection, 170
        prevalence, 167–168
        producers, 167
AFTOL database, 48
Agglutinin-like sequence (Als) family, 200
Airborne transmission, 303
Alimentary toxic aleukia (ATA), 325
Allergic bronchopulmonary disease, 257
Allergic fungal rhinosinusitis, 254, 256–257
Altenuene (ALT), 140
    chemical structures, 141, 143
    occurrence in food and feed, 145–146
    production, 141
Alternaria alternata f. sp. *lycopersici* toxins
    (AAL toxins), 140, 144
    chemical structures, 141, 143
    occurrence in food and feed, 146
    toxicity, 144
Alternaria spp.
    Basic Local Alignment Search Tool, 135
    clinical presentation, 130–131
    conventional diagnosis, 132–133
    cutaneous and subcutaneous infections, 130
    description, 129
    DNA extraction, 135
    epidemiology, 132
    growth temperature, 129
    microscopic examination, 130
    MicroSeq D2 LSU rDNA sequencing kit, 134
    molecular detection tests, 135
    morphology and biology, 129–130
    mycotoxins (*see* Mycotoxins, *Alternaria* spp.)
    nuclear rRNA, 134
    ocular infections, 131
    onychomycosis, 131
    pathogenesis, 131–132
    phytotoxins, 131–132
    potato carrot agar plates, 132
    real-time RT-PCR, 133
    reverse transcriptase PCR, 133
    rhinosinusitis, 131
    ribosomal rDNA genes, 134
    ribosomal RNA, 134
    18S and 28S rDNA, 134
    spores, 131
    target DNA selection, 134
    V-8 juice agar, 132
Alternariol (AOH)
    biosynthetic pathways and metabolism, 141
    chemical structures, 141, 143
    genotoxicity studies, 142
    human exposure, 146
    occurrence in food and feed, 145–146
    synthesis, 140
Alternariol monomethyl ether (AME), 140
    biosynthetic pathways and metabolism, 141
    chemical structures, 141, 143
    human exposure, 146
    occurrence in food and feed, 145
    production, 141
    toxicity, 142
Altertoxins, 144
Amino acid substitution models, 24–25
Amplified fragment length polymorphism (AFLP), 37
AOH, *see* Alternariol (AOH)
Approximate likelihood ratio, 26

Ascomycota, 7
Ashbya gossypii, 87
Aspergillosis, 167; *see also* Invasive aspergillosis (IA)
   description, 151
   etiological agents, 152
   spectrum, 152–154
Aspergillus
   acute graft-versus-host disease, 153
   aerosolization, 158
   A. flavus, 441–442
   A. fumigates, 61
   A. fumigatus
      amphotericin B, 108
      mitochondrial DNA/ribosomal DNA ratio, 109
      primer sets, 106
   A. nidulans
      metabolic model, 87
      resequencing, 98
   in biofilms, 159–160
   DNA origin, 104–105
   in drinking water
      domestic water supplies, 156–157
      hospital water supplies, 157–159
   ELISA, 160
   in food, 154–156
   immunocompromised patients, 152
   inhalation of *Aspergillus* conidia, 152
   lateral flow devices, 160
   nonimmunocompromised individuals, 152
   real-time PCR, 160–161
   species, 151–152
   vaccines
      cross-protective antigens, 590–591
      immunogenic/protective antigens, 589–590
      immunosuppressed host, 589
      problems and obstacles, 591–592
      proteomic approach, 590
      safe and effective adjuvant, 591
Aspergillus mycotoxins
   aflatoxins
      acute aflatoxicosis, 168–169
      biosynthesis, 169–170
      chronic aflatoxicosis, 169
      molecular detection, 170
      prevalence, 167–168
      producers, 167
   A. fumigatus, 167
   A. nidulans, 166
   A. rugulosus, 166
   aspergilloses, 167
   A. terreus, 166
   chemical structures, 166
   citrinin, 176–177
   climate change, 177–178
   fumonisins, 173–174
   indole-tetramic acid, 176
   koji molds, 166
   microarray-based method, 178
   ochratoxins
      biological effects, 171–172
      biosynthesis, 172
      definition, 170–171
      detection, 171
      producers, 171
      protein-coding genes, 172
      real-time PCR method, 173
   patulin, 175–176
Aureobasidium
   chemotaxonomy, 189
   clinical manifestation, 192–193
   epidemiology, 192
   food contamination, 191
   identification, 189
   molecular biology, 189–191
   morphology, 187–188
   pathogenesis, 193
   prevention, 193
   taxonomy, 189
   treatment, 193
Automated DNA extraction, 107

B

Barcode of Life Database (BOLD), 46, 48
Basic Local Alignment Search Tool (BLAST), 69, 135
Basidiomycota, 7, 18
Bayesian method, 25–26
Beauvericin, 336, 404
Best pair of primers (BPP) method, 44–45
(1,3)-*beta*-D-glucan (BG), 507–508
Bimolecular fluorescence complementation
      (BiFC) assays, 232
Biofilms
   Aspergillus in, 159–160
   Candida, 200–201
   Rhodotorula, 490–491
Biological resource centers, role of, 98–99
Black yeast, 189
Bootstrap, 26
Botrytis cincerea, 23
BRaunschweig ENzyme DAtabase (BRENDA), 77
Brazil nuts, 168
Bronchoalveolar lavage (BAL) fluid, 123–124

C

Caco-2 cells, 327, 332
Candida
   adhesion ability, 198–200
   biofilm formation, 200–201
   C. glabrata, 86, 199–206
   C. norvegensis, 197
   C. orthopsilosis, 192, 201
   host interaction mechanisms, 198
   invasion and damage of host tissues
      filamentous form development, 202–203
      lipases, 204
      phospholipases, 203
      secreted aspartyl proteinases, 203
   laboratory identification
      non-PCR-based molecular methods, 205–206
      PCR-based molecular methods, 204–205
   scanning electron microscopy images, 199, 201
   urinary epithelial cells, 199
   vaccines
      antibody-mediated passive protection,
         586–587
      antigens, 587–589

virulence factors, 197–198
yeast frequency, 198
CBS-KNAW culture collection, 51
Cell-free fungal DNA, 104–105
Central venous catheter (CVC), 502
Chaetomium
    clinical significant, 212–217
    dermatomycotic infections, 216
    diagnosis, 220–221
    granulomatous reaction, 212
    infections in humans, 213–215
    invasive infections, 216–217
    Masson-Fontana stain, 212
    mycotoxins, 217–218
    nail infections, 216
    prevalence in food, 218–219
    prevalence in water, 218
    pulmonary infections, 216
    soil saprophytes, 211
    subcutaneous phaeohyphomycosis, 212
    systemic infection, 217
Chemotaxonomy, 189
Chlamydospores, 188–189
Chokepoint enzymes, 86
Citrinin
    description, 176–177
    production, 431–437
Claviceps purpurea
    Botrytis cincerea, 232
    Clavicipitaceae, 232, 235
    diagnostic tools, 235
    ectopic integration rate, 232
    epifluorescence image, 234
    ergot sclerotium, 239
    flower infection
        molecular mechanisms, 242–244
        phytopathogenic fungi, 240–241
        pollen mimicry, 241–242
    genome-wide bioinformatic analysis, 235
    guanine–cytosine content, 230–231
    integrate exogenous, 230
    knockout mutants, 233–234
    life cell imaging, 234
    molecular characterization, 235
    pathogenic lifestyle, 239–240
    sexual cycle, 239
Claviceps spp.
    convulsive ergotism, 229
    ergot alkaloids, 230
        biosynthesis, 236–239
        bromocriptine, 236
        dihydroergotoxin, 236
        ergotoxines, 236
        natural ergopeptines, 236
        prenylated tryptophan, 235
    ergot sclerotia, 230
    gangrenous ergotism, 229
Cocoa fermentation, 474
Coffee fermentation, 474
Continuous ambulatory peritoneal dialysis (CAPD)-related peritonitis, 488
Conventional PCR methods, 283, 311, 428, 577
Cross-protective antigens, 590–591
Crystal violet (CV) staining, *R. mucilaginosa*, 491
Culture-independent molecular methods, 465
Curvularia
    allergic rhinitis, 260
    clinical presentations
        allergic bronchopulmonary disease, 257
        allergic fungal rhinosinusitis, 256–257
        eye infections, 255
        invasive disease, 258
        skin infections, 256
        Tenckhoff catheters, 257–258, 260
    disseminated infections, 251
    environment, 251–253
    histopathologic examination, 260
    immunohistochemical assays, 260
    keratitis, 259
    lactophenol cotton blue preparation, 252, 259
    molecular methods, 261
    pathogenesis
        allergens, 253–254
        allergic fungal rhinosinusitis, 254
        C. verruculosa, 254–255
        hybrid peptide–polyketides, 255
        melanin, 253
    treatment, 261–262
Cyclopiazonic acid–producing *Penicillium*, 440–442

D

de novo genome assembly, 71
Deoxynivalenol (DON), 326–328
Dermatophytosis, 454
Desferrioxamine (DFO), 391
Deuteromycota, 16
Dibenzo-α-pyrones, 142
Dichloran rose Bengal yeast extract sucrose agar (DRYES), 132
Digital polymerase chain reaction (dPCR), 110–111
Dilution plate technique, *Paecilomyces*, 403
DNA–DNA reassociation technique, 38, 465
DNA reassociation kinetics, 96
Domestic water supplies, *Aspergillus*, 156–157
DON, *see* Deoxynivalenol (DON)
Double-strand breaks (DSB), 230
dPCR, *see* Digital polymerase chain reaction (dPCR)
Drinking water
    Aspergillus
        domestic water supplies, 156–157
        hospital water supplies, 157–159
    diversity of mold species, 597–599
    fungal analyses methods, 601–603
    fungi
        associated with allergy, 599–600
        associated with mycoses, 600–601
    guidelines and microbiological analyses, 603
    mycotoxin-producing fungi, 600
    Paecilomyces, 403
    significance of mold species, 599
    source and distribution of mold species, 601
Droplet digital PCR (ddPCR), 110

E

Electrophoresis, 37–38
Electrophoretic karyotypes, 96

Emerging/minor toxins
  beauvericin, 336
  enniatins, 334–335
  fusaproliferin, 337–338
  moniliformin, 336–337
Encephalitozoon
  biliary microsporidiosis, 280
  cell-mediated immunity, 281
  chronic diarrhea, 280
  clinical signs and symptoms, 279–280
  diagnosis
    cell culture, 283
    electron microscopy, 282
    fluorescent staining techniques, 282
    immunofluorescence test, 282
    microsporidia from water, 283–284
    PCR methods, 283
    serology, 283
    staining methods, 282
  disseminated infections, 280
  genome structure, 272
  hepatitis, 280
  HGT, 273
  human infection, 273–275
  humoral immunity, 282
  intraspecific variability
    E. cuniculi, 277–278
    E. hellem, 278–279
    E. intestinalis, 279
    ITS, 277
  mitosome, 268
  morphology and biology
    electron microscopy, 270
    endospore, 269
    exospore, 269
    extrusion apparatus, 269
    life cycles, 270–271
    microsporidia spore representation, 269
    polyribosomes, 270
    sexual transmission, 271–272
    sporoplasm, 269
  ocular infections, 281
  pathology, 281
  peritonitis, 280
  PFGE, 272
  prevalence, 274–275
  prevention, 285
  recombinant detection, 268
  ribosomal DNA-based phylogenies, 268
  source of infection
    waterborne transmission, 276
    zoonotic transmission, 275–276
  treatment, 284–285
Enniatins, 334–335
Enterocytozoon
  E. bieneusi infection in humans
    airborne transmission, 303
    cell culture, 310
    direct person-to-person transmission, 299–300
    distribution, 298–299
    epidemiology, 303–306
    FISH techniques, 311
    fluorochrome stains, 309
    foodborne transmission, 303
    histological examination, 308–309
    indirect immunofluorescence technique, 309–310
    modified trichrome technique, 309
    pathogenesis, 307
    PCR-based techniques, 311
    prevalence, 299
    prevention, 306–307
    quick-hot Gram-chromotrope technique, 309
    serologic tests, 310
    signs and symptoms, 308
    specimen collection, 308
    transmission electron microscopy, 308
    treatment, 311–312
    waterborne transmission, 300–303
    zoonotic transmission, 300
  endospore, 297
  exospore, 297
  extrusion apparatus, 297
  life cycle, 297–298
  pathogenic species, 294
  pébrine, 293
  taxonomy, 294–296
Entomophthoromycoses, 361
Environmental fungal DNA, 107
Enzyme Commission (EC) numbers, 76
Epidithiodioxopiperazines, 526–528
Epithelial adhesin (Epa) gene family, 200
Ergosterol peroxide, 404
Ergot fungus, *see Claviceps*
Ergotismus convulsivus, 229
Ergotismus gangraenosus, 229
Estrogenic toxins, *Fusarium* spp., 331–334
Eumycetoma agent, 122
Eumycotic mycetoma, 256
Eupenicillium, 402, 423

F

Farmer's lung disease (FLD), 363–364, 570
Fluorescent in situ hybridization (FISH) technique
  E. bieneusi infection in humans, 311
  non-PCR-based molecular methods, 205–206
Flux balance analysis (FBA), 73, 83
Foodborne transmission, *E. bieneusi*, 303
Food mycology, 2–3
Fumonisins
  Aspergillus mycotoxins
    biological effects, 174
    biosynthesis, 174
    molecular detection, 174
    nonaketide-derived mycotoxins, 173
    prevalence, 174
    producers, 173
  Fusarium spp. mycotoxins, 329–331
Fumonisins B_1 (FB_1)
  cytokine production, 330
  cytotoxicity studies, 331
  dosing, 330
  human carcinogen, 329
  nixtamalization, 331
  oral administration, 331
  sphingolipid metabolism, 329–330
Functional genomics, 71–73
Fungal barcoding database, 48

Fungal cell walls, biogenesis, 451
Fungal chromosomes, 93–94
Fungal contamination of food, metabolomics, 62–64
Fungal DNA barcoding
　applications, 39–40
　description, 39
　euKaryote Orthologous Groups, 43–44
　Ideal Locus method, 43
　internal transcribed spacer region, 38
　　and *CO1*, 40
　　intragenomic variation, 42
　　pairwise distances within yeasts species, 42–43
　　RPB1, 40
　primer pairs location/function, 45
　reference sequence availability, 45–46
　second-and third-generation sequencing, 51
　true phylogeny, 43
Fungal Genetics Stock Center (FGSC), 94–96, 98
Fungal Genome Initiative, 96
Fungal genomes, 93–94, 97
Fungal organisms
　abandonment of separate naming, 11–12
　biodiversity of, 451
　classification, 5–7
　conserved and rejected names, 10
　effective publication, 8–9
　prospects, 12
　protected and suppressed names, 11
　sanctioned names, 10–11
　scientific name allocation, 8
　valid publication, 9–10
Fungal taxonomy, 2, 38
Fusaproliferin, 337–338
Fusarium spp.
　F. graminearum, 326, 428, 600
　mycotoxins (*see* Mycotoxins, *Fusarium* spp.)

G

Galactomannans, 154–155, 160–161, 553
Gastrointestinal tract infections, 503–505
GenBank, 47–48, 118, 122, 489, 603
Genitourinary tract infections, 507
Genome-scale metabolic models (GSMMs)
　BRENDA, 77
　conversion to stoichiometric model
　　biomass formation reaction, 80
　　mathematical representation, 83
　　null space of S, 81
　　pseudo-metabolic network system, 81–82
　　steady-state approximation, 81
　fungal reconstructions, 84–85
　genes, proteins, and reactions association, 78–79
　genome annotation, 76–77
　health applications, 86–87
　industrial applications, 87
　localization of enzymes, 79
　manual curation, 79–80
　metabolic genes, 76–77
　metabolic network reconstruction iterative process, 75–76
　reconstruction, 73–75
　spontaneous reactions, 79
　stoichiometry, 79
　validation, 83–84

Genome-scale metabolic networks (GSMNs), 73
Genome structural annotation, 71
Genomic libraries, 95–96
Geotrichum; *see also* Trichosporon
　antifungal susceptibility, 554
　characterization, 545
　G. candidum, 545–546
　geotrichosis, 539
　　diagnosis, 551–553
　　incidence rates, 548
　　invasive, 549–550, 555
　　risk factors, 548–549
　　therapy, 553–556
　habitats and infections, 544
　ITS rDNA sequences, 552
GFP, *see* Green fluorescent protein (GFP)
Giemsa stain, 123
Gliomastix roseogriseum, 123
Gliotoxin, 61, 526–527
　Chaetomium, 218
　chemical structure, 528–529
Gliovirin, 526, 528–529
Global metabolic profiling, *see* Metabolomics
Glucose-regulated protein 78 (GRP78), 362, 392
Glycerol-3-phosphate dehydrogenase (GPD), 573
Green fluorescent protein (GFP), 428
GSMMs, *see* Genome-scale metabolic models (GSMMs)

H

Harzianum A, 530–531
Hazard analysis and critical control point (HACCP) system, 146
Herpes simplex viral keratitis, 454
Highly active antiretroviral therapy (HAART), 285, 312
　Encephalitozoon, 284
　Enterocytozoon, 299
High-osmolarity glycerol (HOG) signal cascading pathway, 430, 572
High-throughput global metabolic profiling, 58–59
Homology, 16
Homoplasy, 16
Horizontal gene transfer (HGT), 27, 32, 273
Hospital water supplies, *Aspergillus* species, 157–159
Host-specific toxins, 139–140
HT-2, 326
Human pathogenic fungi, 2
Hyalohyphomycosis
　description, 404
　P. javanicus, 405
　P. lilacinus
　　animal model, 408
　　human model, 405, 407–408
　　treatment, 411
　P. marquandii
　　human model, 405
　　treatment, 411
　P. variotii
　　animal model, 408
　　human model, 405–406
　　treatment, 411
4-Hydroxy-4*H*-furo[3,2-*c*]pyran-2(6*H*)-one, 175

I

IA, *see* Invasive aspergillosis (IA)
Ideal Locus method, 43
IFIs, *see* Invasive fungal infections (IFIs)
Immunocompromisation, 452
Indirect immunofluorescence (IIF) technique, 282, 309–310
Infective endocarditis (IE), 503
Inferring phylogeny, 22
    distance methods, 23
    evolutionary models
        amino acid substitution models, 24–25
        approximate likelihood ratio, 26
        Bayesian method, 25–26
        bootstrap, 26
        Jack-knife, 26
        maximum likelihood methods, 26
        nucleic acid substitution models, 23–24
        parsimony, 25
        super trees and supermatrices, 27
        tree formats, 27–28
        tree visualization, 28–29
    neighbor joining methods, 23
INSD reference databases, advantages, 47
Intergenic spacers (IGS), 279
Internal transcribed spacer (ITS), 30, 38, 277, 279, 303–304, 551
International Code of Nomenclature for algae, fungi, and plants (ICN), 8, 10
International Commission on the Taxonomy of Fungi (ICTF), 11–12, 46
International Mycological Congresses (IMCs), 8
International Nucleotide Sequence Database Collaboration (INSDC), 38
Interstitial keratitis, 454
Invasive aspergillosis (IA), 103, 151, 586
    diagnosis, 161
    extrinsic risk factor, 154, 162
    HCT transplant recipients, 585
    incidence, 153
    mortality rate, 153
    occurrence rate in neutropenic patients, 152
    prevalence, 153
    prognosis, 110
    qPCR (*see* Quantitative PCR (qPCR))
    risk factors, 110
    sporadic cases, 154
Invasive candidiasis, 198, 547, 584–585
Invasive fungal infections (IFIs), 105
    Aspergillus vaccines
        cross-protective antigens, 590–591
        immunogenic/protective antigens, 589–590
        immunosuppressed host, 589
        problems and obstacles, 591–592
        proteomic approach, 590
        safe and effective adjuvant, 591
    Candida vaccines
        antibody-mediated passive protection, 586–587
        antigens, 587–589
    immunology
        invasive aspergillosis, 585–586
        invasive candidiasis, 584–585
    mortality rates, 583
    pathophysiology, 584
    Phoma species, 454–455
    Saccharomyces, 501
Itraconazole, 261, 395, 411–412
    A. pullulans, 193
    E. hellem infection, 284
ITS, *see* Internal transcribed spacer (ITS)

J

Jack-knife, 26

K

Kashin–Beck disease, 325, 337
Keratitis
    Curvularia, 253, 255, 259, 261
    and endophthalmitis, 507
    Phoma, 454
    P. marquandii, 405
    and *Rhodotorula* scler006itis, 488
Kluyveromyces
    classification, 510
    industrial uses, 510
    K. lactis, 509–510
    microbial food enzymes, 511
    microbiology, 510
    opportunistic infections, 510–511
Kupffer cells, 330

L

Leucinostatin, 404
Lichtheimia
    in agriculture, 360–361
    antifungal susceptibility, 366–367
    biology, 359–360
    carbon source assimilation profiles, 364
    chronological overview, 355–356
    clinical features and pathogenesis, 362–363
    clinical therapy, 367
    dissemination, 363
    epidemiology and risk factors, 361–362
    Farmer's lung disease, 363–364
    in food production, 361
    hybridization-based identification, 366
    life cycle, 359–360
    MALDI-TOF-MS-based methods, 366
    morphological features, 356–359, 364
    oligonucleotide primers, 365
    PCR-based identification, 364–365
    probe-bound PCR fragments, 366
    sequence-based methods, 365
Linkage maps, 94–95
Lymphadenitis, 361, 488

M

Magnusiomyces; *see also Trichosporon*
    antifungal susceptibility, 554
    habitats and infections, 544
    M. capitatus, 546

Index

β-(Man)3-Fba-TT vaccine, 587
Manual curation, GSMN, 79–80
MAP kinase cascade, 429–430
Masked mycotoxins, 338–339
Matrix-assisted laser desorption/ionization time-of-flight mass spectrometry (MALDI-TOF MS), 552
    Lichtheimia identification, 366
    nonculture-based detection, IFIs, 508
Maximum likelihood methods, 26
Melanins
    Alternaria spp., 131
    Curvularia spp., 253
Meningitis, *Rhodotorula* species, 488
merlin 2.0 Java application, 77
Metabolic systems biology, 70
Metabolomics, 72
    description, 57–58
    flexibility, 59
    foods and foodstuffs, biochemical changes, 64
    in fungal biology, 59
    fungal contamination of food, 62–64
    fungal pathogen research, 57
    goal, 57
    human and animal health, biochemical aspects, 65
    mycotoxic fungi-mycotoxin interaction, 64–65
    nuclear magnetic resonance
        advantages, 58
        F. graminearum wheat pathogen *vs. Tri-5* gene deletion, 62
        limitations, 58
        vs. mass spectrometry, 58
    phytopathological systems, 62
    of plant–fungi interactions, 61–62
    rhizoxin toxin source, 64
    S. cerevisiae, 59–61
Metagenomics studies, using NGS methods, 52
6-Methylsalicylic acid (6-MSA), 176, 438
Microarray technology, 429
Microascus species
    antifungal resistance, 381
    culture collections, 379–380
    deep-tissue infection, 381
    detection and identification, 379–380
    diagnosis and treatment, 381–382
    direct/dilution-plating methods, 379
    incidence
        in cheese, 378–379
        in food, 377–379
        in water, 376–377
    keratinous tissue infection, 380
    M. brevicaulis
        in air and soil samples, 377
        arsenic compounds, 380
        in cheese, 378–379
        conidiophore and conidia, 375, 377
    M. cirrosus, 375–376
    M. longirostris, 375–376
    mycotoxins, 380
    onychomycosis, 380
    prevalence, 375
    systemic disease, 381
Microbiological tools, 103
MicroSeq D2 LSU rDNA sequencing kit, 134

Microsporidia
    description, 31, 267–268
    Encephalitozoon (see *Encephalitozoon*)
    Enterocytozoon (see *Enterocytozoon*)
    spore representation, 269
Minimal inhibitory concentrations (MICs), 261–262, 411–412
    amphotericin B, 554
    Saccharomyces, 509
    triazoles, 555
Minimization of metabolic adjustments (MOMA), 84
Mitogen-activated protein kinases (MAPK)
    Claviceps purpurea, 243
    Wallemia spp., 572
Molecular biology, 3
Molecular detection methods
    aflatoxins, 170
    Alternaria, 135
    fumonisins, 174
    mycotoxin-producing Penicillia, 427–429
    ochratoxins, 172–173
    patulin-producing organisms, 176
Molecular markers, 190
    Phoma infection, in humans, 456–457
    selection for species trees, 30
Moniliformin, 336–337
Monophyletic clade, 18
MS/MS ion fragment spectra, 58
Mucor
    combined therapy, 395
    epidemiology, 392–393
    in foods and industry, 395
    infection sites and patterns, 393
    itraconazole, 395
    M. circinelloides, 389–390
    M. ellipsoideus, 390
    M. hiemalis, 390
    M. indicus, 390
    M. irregularis, 390
    molecular identification, 394
    M. racemosus, 390
    M. ramosissimus, 390
    Mucoromycotina identification, 388–389
    M. velutinosus, 390–391
    mycological diagnosis, 393–394
    pathogenicity, 391–392
    posaconazole, 394–395
    risk factors, 392
    taxonomic position and characteristics, 388–389
    Zygomycota, 387–388
Mucoralean fungi, life cycle, 359–360
Mucormycosis; *see also Lichtheimia*
    amphotericin B, 395–396
    angioinvasion and neurotropism, 392
    clinical presentation, 392
    combined therapy, 395
    cutaneous presentation, 392
    description, 362
    desferrioxamine, 391
    diabetes mellitus, 362
    histology/direct microscopic examination, 393
    invasive mycosis, 391
    iron concentration in plasma, 391

iron overload, 367
mortality rates, 363
nonseptate hyphae, 394
treatment, 367
voriconazole, 392
MycAssay *Aspergillus* PCR, 161
MycoBank, 9–10, 46, 48, 51, 356
MycoChip, 429
Mycotoxin-producing fungi, 62
    in drinking water, 600
    Penicillia
        molecular detection and monitoring methods of, 427–429
        and spoilage, 424–425
Mycotoxins; *see also Aspergillus* mycotoxins
    Alternaria spp.
        AAL toxins, 144
        altertoxins, 144
        chemical names, molecular weights, and formulas, 141–142
        control and prevention, 146
        dibenzo-α-pyrones, 142
        food and feed, occurrence in, 144–146
        host-specific toxins, 139–140
        human exposure, 146
        non-host-specific toxins, 139
        physical and chemical properties, 141
        production, physiology of, 141
        synthesis, 140–141
        tenuazonic acid, 142, 144
        toxicity, 144
    biosynthesis regulation, by food-relevant external factors, 429–431
    Chaetomium, 217–218
    fungal organisms, 7
    Fusarium spp., 323
        emerging/minor toxins, 334–338
        estrogenic toxins, 331–334
        masked mycotoxins, 338–339
        nonestrogenic toxins, 324–331
    masked, 338–339
    Microascus spp., 380
    Penicillium spp.
        biological aspects, 424–425
        cyclopiazonic acid–producing *Penicillium*, 440–442
        mycotoxin biosynthesis regulation, 429–431
        mycotoxin-producing Penicillia, 427–429
        ochratoxin A and citrinin production, 431–437
        pathogenicity aspects, 425–426
        patulin production, 438–440
        PR-toxin production, 442
        taxonomic aspects, 423–424
        toxin biosynthesis aspects, 426
    Scopulariopsis spp., 380
    trichothecene mycotoxins, 521

N

Neurospora
    chromosomes, 93–94
    cosmids, 95–96
    genetic map, 94–95
    N. crassa, 98

Next-generation fungal genomics, 97–98
Next-generation sequencing (NGS), 38–39, 42, 51
    de novo sequencing, 97–98
    digital polymerase chain reaction, 110–111
    metagenomics studies, 52
    resequencing fungal genomes, 98
Nivalenol (NIV), 328–329
Nixtamalization, 331
NMR, *see* Nuclear magnetic resonance (NMR)
Nonconventional yeasts, 463–464; *see also Pichia* spp.
Nonestrogenic toxins, *Fusarium* spp.
    fumonisins, 329–331
    trichothecenes
        DON, 326–328
        general structure, 324
        HT-2, 326
        nivalenol, 328–329
        T-2, 324–326
Nonhomologous DNA end joining (NHEJ), *C. purpurea*, 230–231
Non-host-specific toxins, 139
Nonribosomal peptide synthetase (NRPS), 237–238, 431, 525
Nuclear magnetic resonance (NMR)
    advantages, 58
    F. graminearum wheat pathogen *vs. Tri-5* gene deletion, 62
    limitations, 58
    vs. mass spectrometry, 58
Nuclear rRNA, 38, 134
Nucleic acid substitution models, 23–24

O

Ochratoxins
    Aspergillus mycotoxins
        biological effects, 171–172
        biosynthesis, 172
        definition, 170–171
        detection, 171
        producers, 171
        protein-coding genes, 172
        real-time PCR method, 173
    ochratoxin A, 427–429, 442
        chemical structure, 166–167
        production, 431–437
    ochratoxin B, 171–172, 432–433
Olive fermentation, 474–475
Omics technologies, 72–73
Onychomycosis, 454
    Alternaria spp., 131
    A. strictum, 123
    Curvularia, 256
    Microascus spp., 380
    Scopulariopsis spp., 380
Open reading frames (ORFs)
    hidden Markov models, 72
    prokaryotes and eukaryotes, 71
Opisthokonta, 5, 268
Opportunistic fungal infections (OFIs), 451; *see also Phoma* species
Opportunistic invasive fungal infections (OIFIs), 452
OptGene, 86
Organic metabolites, 58

Orthologous sequences, 16
*otapks*PN gene, 431–432, 435–436

P

Paecilomyces, 423
  amphotericin B
    P. lilacinus, 412
    P. marquandii, 411
    P. variotii, 411
  culturing, 409
  dilution plate technique, 403
  in drinking water, 403
  in food-and feedstuffs, 402–403
  heat resistance, 402
  hyalohyphomycosis (*see* Hyalohyphomycosis)
  infections, laboratory diagnosis for, 408–411
  in vitro antifungal susceptibility, 411–412
  on malt extract agar
    P. lilacinus, 409
    P. marquandii, 409
    P. variotii, 409
  micromorphology
    P. lilacinus, 409–410
    P. variotii, 409–410
  molecular identification, 409
  mycotoxins and biological compounds
    P. carneus, 404
    P. lilacinus, 404
    P. marquandii, 404
    P. tenuipes, 404
    P. variotii, 403
  pathogenicity
    P. lilacinus, 412
    P. variotii, 412
  P. javanicus, 405, 409, 411–412
  P. lilacinus, 413
  polymerase chain reaction, 403
  sequencing of amplified fragments, 408–409
  taxonomy and morphological characteristics, 401–402
Paralogs, 16
Paraphyletic clade, 18
Parsimony, 25
Patulin
  Aspergillus mycotoxins, 175–176
  biological effects, 175
  biosynthesis, 176
  chemical structure, 166
  molecular detection, 176
  occurrence, 175
  producers, 175
  production of, 438–440
PD, *see* Peritoneal dialysis (PD)
Pébrine, 293
Penicillium
  conidiophores of, 423
  mycotoxins
    biological aspects, 424–425
    cyclopiazonic acid–producing
      Penicillium, 440–442
    mycotoxin biosynthesis regulation, 429–431
    mycotoxin-producing Penicillia, 427–429
    ochratoxin A and citrinin production, 431–437
    pathogenicity aspects, 425–426
    patulin production, 438–440
    PR-toxin production, 442
    taxonomic aspects, 423–424
    toxin biosynthesis aspects, 426
  P. brevicompactum, 438
  P. camemberti, 426, 441
  P. chrysogenum, 426
  P. citrinum, 600
  P. cyclopium, 440
  P. digitatum, 424–425
  P. expansum, 426, 438, 440
  P. italicum, 424–425
  P. nordicum, 435, 437
  P. oxalicum, 424
  P. roqueforti, 424, 426, 442
  P. verrucosum, 426, 433–437
Peptaibols, 524–526
Peptide nucleic acid (PNA), 205–206
Peritoneal dialysis (PD)
  Curvularia, 258, 260–261
  Saccharomyces, 503, 505
  Trichosporon peritonitis, 550
PFGE, *see* Pulsed-field gel electrophoresis (PFGE)
Phaeohyphomycosis, 212, 216, 256, 453–454
Phoma species
  infection in humans
    cutaneous infections, 454–455
    keratitis, 454
    molecular markers, 456–457
    morphological characters, 456
    phaeohyphomycosis, 453–454
    pulmonary infection, 455–456
    types, 452–453
  P. exigua, 455
  P. eypyrena, 454
  P. hibernica, 454
  P. minutella, 454
  P. oculohominis, 454
Phylogenetic trees, 15
  dataset selection, 20
  description, 17–19
  fungal tree of life, 31–32
  gene clusters, 32
  gene *vs.* species phylogeny, 30
  homology and homoplasy, 16
  horizontal gene transfer, 32
  input format issues, 19–20
  molecular marker selection for species trees, 30
  orthologous sequences, 16
  paralogs, 16
  phylogeography, 30–31
  positive and purifying selection, 21–22
  protein *vs.* DNA sequences, 20
  sequence alignment, 20–21
  species classification and identification, 30
  supernumerary chromosomes, 32
  xenologs, 16
Pichia spp., 463
  cocoa fermentation, 474
  coffee fermentation, 474
  culture-independent molecular methods, 465
  D1D2 sequences, 464
  DNA–DNA reassociation technique, 465
  molecular manipulation tools, 473–474

multigene sequencing, 464
olive fermentation, 474–475
polyphyletic character, 466
taxonomy, 466–468
    Ambrosiozyma, 468
    Babjeviella, 471
    Barnettozyma, 470
    Cyberlindnera, 470
    Hyphopichia, 471
    Kodamaea, 472
    Komagataella, 472–473
    Kregervanrija forms, 466, 468
    Meyerozyma, 472
    Millerozyma, 471–472
    Nakazewaea, 466
    Ogataea, 468–469
    Peterozyma, 469
    P. fermentans, 466
    P. kluyveri, 466
    P. kudriavzevii, 466
    P. membranifaciens, 464, 466, 474
    Pricomyces, 471
    Saturnispora, 468
    Scheffersomyces, 472
    Starmera, 470
    Wickerhamomyces, 469–470
    Yamadazyma, 471
    Zygoascus, 473
Plasmodium falciparum, 86
Platelia Aspergillus ELISA, 154–155
p-450 monooxygenase gene, 440
Polar tube protein (PTP), 277–279
Polymerase chain reaction (PCR), 103–104; *see also* Quantitative PCR (qPCR)
    aflatoxin-producing isolate, 170
    Alternaria spp. in food, 133
    Candida spp., 204–205
    E. bieneusi infection in humans, 311
    Encephalitozoon, 283
    Lichtheimia spp., 364–365
    patulin-producing organisms, 176
Polyphyletic clade, 18–19
Posaconazole, 366–367, 394–395, 555–556
Potato carrot agar (PCA) plates, 132
Prosthetic valve endocarditis (PVE), 503
Protein microarrays, 72
Proteome, 58, 72
PR-toxin production, 442
p450-2 transcript, 440
Public culture collection, 126
Pulmonary infection, 363
    Chaetomium, 216
    Phoma spp., in humans, 455–456
    Saccharomyces spp., 505–506
Pulsed-field gel electrophoresis (PFGE), 272, 465
Pyrenochaeta romeroi, 455

Q

Quantitative PCR (qPCR), 158
    advantages, 105
    clinical performance, 110
    false positives, prevention of, 105–107
    internal controls, 107–108
    panfungal primers, 107, 109
    PCR yield optimization, 108–109
    quality control design, 109–110

R

rAls3p-N vaccine, 584, 587–589
Random amplified polymorphic DNA (RAPD), 37, 221, 508
Reactive oxygen species (ROS), 244, 333, 336
Reference databases, 45, 47–50
RefSeq database, 48
Respiratory tract infections, *Saccharomyces*, 505–507
Restriction enzyme fragment length polymorphism, 465
Restriction fragment length polymorphism (RFLP) mapping, 95
Reverse transcriptase PCR (RT-PCR), 133, 161, 173, 428, 434
Rhizopus microsporus, 64, 359, 363
Rhodotorula species
    animal experimentation, 484
    CAPD-related peritonitis, 488
    central venous catheters, 484, 486–488
    conventional identification techniques, 489
    crystal violet staining, 491
    DNA extraction, 489
    environmental microbiology, 484–486
    epidemiology, 483–486
    fungemia, 484
        after transplantation, 488
        central venous catheters, 487
        incidence, 483
        in preterm neonates, 488
        risk factors, 487
    gene sequencing, 489
    lymphadenitis, 488
    meningitis, 488
    microscopic morphology, 489
    pathogenicity, 484
    peritonitis in liver transplant recipient, 488
    postoperative infections, 488
    risk factors, 486
    scleritis and keratitis, 488
    species-specific oligonucleotide primer, 490
    susceptibility studies, 490–491
    treatment, 490–491
    universal primers, 490
Ribosomal Database Project (RDP), 38
Ribosomal RNA (rRNA), 134, 204, 283, 295

S

Sabouraud dextrose agar (SDA), 187, 204, 455, 498, 507, 543, 551, 576, 602
Saccharomyces
    biotechnology, 496
    clinical syndrome
        endocarditis, 503
        endophthalmitis, 507
        fungal keratitis, 507
        fungemia, 502–503
        gastrointestinal tract infections, 503–505

genitourinary tract infections, 507
invasive fungal infections, 501
respiratory tract infections, 505–507
vaginitis, 505
virulent isolates, 502
ecology and etiology, 500
human population, 500–501
microbiology
baking, 499–500
clade, 497
diploid cells, 498
fermentation, 499
genetic variation, 498
haploid cells, 498
hemiascomycetes, 497
morphology and features, 498–499
nutritional supplements, 500
probiotic supplement, 500
sensu stricto, 497
yeasts, 498
nonculture-based molecular methods, 507–508
numerous genetic fingerprinting methods, 508
ribosomal DNA, 508
S. boulardii, 496
S. cerevisiae, 197, 495–496
GSMM reconstruction, 79
metabolomics, 59–60
vanillin productivity, 84
treatment, 508–509
Sanger sequencing, 38, 96–97
Saprophytic fungi, 451
Sarocladium kiliense, 120–123
Sarocladium strictum, 117, 120, 122–124
Scopulariopsis species
antifungal resistance, 381
culture collections, 379–380
deep-tissue infection, 381
detection and identification, 379–380
diagnosis and treatment, 381–382
direct/dilution-plating methods, 379
incidence
cheese, 378–379
food, 377–379
water, 376–377
keratinous tissue infection, 380
mycotoxins, 380
onychomycosis, 380
prevalence, 375
systemic disease, 381
SDA, *see* Sabouraud dextrose agar (SDA)
Secreted aspartyl proteinases (Saps), 203
Serologic tests
E. bieneusi infection in humans, 310
Encephalitozoon, 283
Spore wall protein (SWP), 277, 279
Sterigmatocystin (ST), 87
aflatoxins, 167
production, 218
Superficial punctate keratitis, 454
Super trees and supermatrices, 27
SYBR Green dye, 107
Systems biology
challenge, 70
description, 69

metabolic, 70
metabolite profiling, 72–73

T

T-2, 324–326
  vs. DON, 327
  nivalenol, 328
Talaromyces, 402, 423
Tandem mass spectrometry, 58
Taxonomic ranks, 6
Teleomorph, 11, 375, 377, 379, 402–403, 545–546
Tentoxin (TEN), 139
  chemical structures, 141, 143
  human exposure, 146
  occurrence in food and feed, 145–146
Tenuazonic acid (TeA), 139
  chemical structures, 141, 143
  mean chronic dietary exposure, 146
  occurrence in food and feed, 145–146
  toxicity, 142, 144
Toxic secondary metabolites, *Trichoderma*
  epidithiodioxopiperazines, 526–528
  peptaibols, 524–526
  trichothecenes, 528–532
Toxigenic fungi identification, 427
Transcriptome, 58, 72, 573
Transmission electron microscopy (TEM), *Enterocytozoon*, 308
Transporter Classification (TC) numbers, 76
Traumatic keratitis, 454
Tree formats, 27–28
Tree visualization, 28–29
Trichoderma
  clinical aspects, 523–524
  food and water, 524
  harmless fungi, 521
  taxonomy and identification, 522
  toxic secondary metabolites
    epidithiodioxopiperazines, 526–528
    peptaibols, 524–526
    trichothecenes, 528–532
  trichothecene mycotoxins, 521
Trichodermin, 530
Trichosporon
  antifungal susceptibility, 554
  characterization, 539
  glucuronoxylomannan, 543–544
  habitats and infections, 540–542
  hair infection, 547
  morphological characteristics, 539–540
  trichosporonosis, 539
    acute, 547
    chronic hepatosplenic, 550
    clinical presentation, 546
    diagnosis, 551–553
    incidence rates, 548
    invasive, 547–549, 552–553, 555–556
    risk factors, 547–549
    therapy, 553–556
Trichothecenes
  DON, 326–328
  fumonisins, 329–331
  HT-2, 326

nivalenol, 328–329
T-2, 324–326
Trichoderma, 528–532
Trihydroxynaphthalene reductase, 253

U

UNITE database, 38, 48
Unrooted/rooted phylogenetic trees, 18

V

Vaginitis, 505
Viable but nonculturable (VBNC) bacterial cells, 133
Vomitoxin, *see* Deoxynivalenol (DON)

W

Wallemia
    bioactive metabolites, 574–575
    ecology, 571–572
    methods
        cultivation, 575–576
        detection, 577
        identification, 576
    molecular and physiological adaptations, 572–574
    pathogenesis, clinical features, and diagnosis, 575
    phylogeny and identification, 570
Wardomyces anamorph, 375
Waterborne transmission
    Encephalitozoon, 276
    Enterocytozoon, 300–303
Water quality and human health, mold species, 599
Web Accessible Sequence Analysis for Biological Inference (WASABI), 38

X

Xenologs, 16
Xerophily, 571–572

Y

Yeast Metabolome Database, 60

Z

Zearalenone (ZON)
    α-ZOL and β-ZOL, 332
    chemical structure, 331
    conjugation, 332
    cytokine production, 333
    daily intake, 332
    dosages, 333–334
    estrogenic properties, 332
    reproductive process in animals, 333
Zinniol, 139–140
Zoonotic transmission
    Encephalitozoon, 275–276
    Enterocytozoon, 300–301
Zygomycosis, 361–362, 393
Zygomycota, 32, 387–388

An environmentally friendly book printed and bound in England by www.printondemand-worldwide.com

PEFC Certified

This product is
from sustainably
managed forests
and controlled
sources

www.pefc.org

This book is made of chain-of-custody materials; FSC materials for the cover and PEFC materials for the text pages.